HANDBOOK OF TOPOLOGICAL FIXED POINT THEORY

Handbook of Topological Fixed Point Theory

Edited by

R. F. Brown
University of California,
Los Angeles, U.S.A.

M. Furi
Department of Applied Mathematics 'G. Sansone',
Florence, Italy

L. Górniewicz
Juliusz Schauder Center of the Nicolaus Copernicus University,
Poland

and

B. Jiang
Department of Mathematics,
Peking University, Beijing, China

Springer

A C.I.P. Catalogue record for this book is available from the Library of Congress.

ISBN-10 1-4020-3221-8 (HB)
ISBN-10 1-4020-3222-6 (e-book)
ISBN-13 978-1-4020-3221-9 (HB)
ISBN-13 978-1-4020-3222-6 (e-book)

Published by Springer,
P.O. Box 17, 3300 AA Dordrecht, The Netherlands.

www.springeronline.com

Printed on acid-free paper

Printed in the Netherlands.

TABLE OF CONTENTS

PREFACE

Fixed point theory concerns itself with a very simple, and basic, mathematical setting. For a function f that has a set X as both domain and range, a *fixed point* of f is a point x of X for which $f(x) = x$. Two fundamental theorems concerning fixed points are those of Banach and of Brouwer. In Banach's theorem, X is a complete metric space with metric d and $f\colon X \to X$ is required to be a contraction, that is, there must exist $L < 1$ such that $d(f(x), f(y)) \leq Ld(x, y)$ for all $x, y \in X$. The conclusion is that f has a fixed point, in fact exactly one of them. Brouwer's theorem requires X to be the closed unit ball in a Euclidean space and $f\colon X \to X$ to be a map, that is, a continuous function. Again we can conclude that f has a fixed point. But in this case the set of fixed points need not be a single point, in fact every closed nonempty subset of the unit ball is the fixed point set for some map. The metric on X in Banach's theorem is used in the crucial hypothesis about the function, that it is a contraction. The unit ball in Euclidean space is also metric, and the metric topology determines the continuity of the function, but the focus of Brouwer's theorem is on topological characteristics of the unit ball, in particular that it is a contractible finite polyhedron. The theorems of Banach and Brouwer illustrate the difference between the two principal branches of fixed point theory: metric fixed point theory and topological fixed point theory. The *Handbook of Metric Fixed Point Theory*, edited by Art Kirk and Brailey Sims and published by Kluwer in 2001, presented that portion of the subject and, in this companion volume, we take up the other part of the fixed point story.

The classification of mathematical content is seldom easy. For instance, the distinction between the metric and topological fixed point theories is far from precise and it can be difficult to determine to which a specific topic belongs. In the same way, although fixed point theory is generally considered a branch of topology, the influence of nonlinear analysis, and the related subject of dynamics, is so profound that much of fixed point theory could just as well be considered a part of analysis. The papers in this Handbook reflect the varied, and not easily classified, nature of the mathematics that makes up topological fixed point theory. To impose some structure on its contents, the papers have been divided into four "chapters", each of which consists of papers that have something important in common with each other.

The title of Chapter I, homological methods in fixed point theory, points out the importance of algebraic topology, specifically homology theory, as the source of many of the mathematical tools used in fixed point theory. This chapter also illustrates the fact that topological fixed point theory is not just about the equation $f(x) = x$. For instance, if the function f is multi-valued, taking points of X to subsets of the same space, a *fixed point* is a point such that $x \in f(x)$. The pioneering work of Lefschetz was in the context of coincidence theory. For two maps $f, g: X \to Y$ between closed orientable manifolds of the same dimension, a nonzero value of the homotopy invariant that Lefschetz introduced implies the existence of a *coincidence*, that is, a point $x \in X$ such that $f(x) = g(x)$. Another modification of the fixed point equation is $f^n(x) = x$ where f^n denotes the n-times iteration of a map $f: X \to X$. A point x such that $f^n(x) = x$ is called a *periodic point.* The iterates of f constitute a discrete dynamical system on X and the periodic points can furnish important dynamical information. The influence of dynamics can also be observed in the fact that such homogenous spaces as nilmanifolds and solvmanifolds are the setting for several of the papers in this chapter.

Since homogeneous spaces are spaces of cosets, algebra plays an important role in the study of maps on such spaces. Another way in which algebra impacts fixed point theory is through the study of equivariant maps. If a space is acted on by a group, then an equivariant map is one that respects the action. Chapter II is devoted to the fixed point theory of equivariant maps and its application to analysis.

Topological fixed point theory is often referred to as "Nielsen theory". This terminology reflects the importance of the concepts introduced by Jacob Nielsen that furnish a homotopy invariant lower bound for the number of solutions to an equation. All the papers in Chapter III contain the name of Nielsen, or of Wecken who expanded Nielsen's ideas, in their title. Again the objects of study are not just fixed points. Coincidences, periodic points and fixed points of multivalued maps all make their appearance in this chapter, as do *roots*, the solutions to the equation $f(x) = a$ where $f: X \to Y$ is a map and $a \in Y$. Nielsen theory is particularly interesting, and challenging, when the spaces are compact surfaces, as some of the papers in Chapter III demonstrate.

The substantial size, and content, of Chapter IV indicates the importance of the applications of topological fixed point theory to nonlinear analysis and dynamics. Problems are formulated in terms or fixed or periodic points, coincidences and roots. The tools of fixed point theory are those of the previous chapters: the Lefschetz number, fixed point index, Nielsen number and, for root problems, the topological degree. Again the fuctions considered may be multivalued as well as single valued. However, a notable difference between the papers in this chapter and many of those earlier in the Handbook is the extention of the tools to more general settings than those of the purely topological investigations. These powerful tools are then employed to obtain results about periodic solutions and solutions satisfying boundary conditions and other constraints, for differential equations and differential inclusions.

We have not attempted a definition of the "topological fixed point theory" that is the subject of this Handbook; neither will we try to define precisely what a "handbook" is. A handbook contains information that will furnish the mathematician reader, whether an established researcher or a graduate student, with a foundation in its subject and a guide to further study. An up-to-date handbook also gives its readers a sense of the present state of the art and, ideally, offers some clues as to where the subject will go in the future. But a handbook is not a textbook in which the reader starts on the first page expecting to find a complete and detailed exposition following a clearly indicated line of development that extends to the very last page. Instead, the reader of a handbook is invited to view its table of contents as a buffet from which to taste some items while perhaps consuming others that seem particularly attractive. Each of the 28 authors who contributed to this handbook was asked to do so because the editors consider that person an expert on the topic that he or she was invited to write about. The style of presentation and level of mathematical detail was determined by the authors, based on their own mathematical taste and their judgment of the best way to present their specialty.

This handbook is the sum of the contributions of its authors. It exists because these busy people were willing to expend a considerable amount of time and effort and we are grateful to them for doing it. We very much appreciate the help of the Juliusz Schauder Center of the Nicolaus Copernicus University in Toruń. Mariusz Czerniak managed the collection of the papers and Jolanta Szelatyńska converted them into a uniform style. We are grateful for the support of our editors at Kluwer: Liesbeth Mol who initiated the project and Lynn Brandon who saw if through to completion and Marlies Vlot who supervised the producion of the handbook.

The Editors

CHAPTER I

HOMOLOGICAL METHODS IN FIXED POINT THEORY

It is well known that homology theory plays a fundamental role in fixed point theory or, more precisely, in topological fixed point theory. With the application of homology theory one can obtain:

(a) global fixed point theorems,

(b) local fixed point theorems.

In 1923 S. Lefschetz proved a global fixed point theorem, now called Lefschetz fixed point theorem. This theorem is still studied by a number of mathematicians and it is an important part of topological fixed point theory. The second important part of topological fixed point theory is mainly connected with local fixed point problems and it is called fixed point index theory or, in particular, topological degree theory. Fixed point index theory was introduced by H. Hopf in the late 1920s for maps on finite polyhedra and essentially developed by B. O'Neill in 1953. Note that a modern definition of this notion was presented by A. Dold in 1965.

Topological degree theory was initited by E. L. Brouwer in 1912 in the finite dimensional case and extended by J. Leray and J. P. Schauder in 1934 to the infinite dimensional case.

The purpose of this chapter is to present, in the most general form, both the global and local cases studied in topological fixed point theory.

1. COINCIDENCE THEORY

Daciberg L. Gonçalves

1. Introduction

This survey article analyzes the problem of minimizing the number of coincidence points of two mappings by deforming them through homotopies. We present the main invariants which have been used in the study of this problem, trying to illustrate with examples which give a better understanding of how to develop the theory. We will stress the particular features of coincidence theory and very seldom will we refer to fixed point and root theory, which are of course related. The reader will find enough material about these two topics elsewhere, including in this handbook. We also include a very brief exposition of some topics that have been treated only very recently.

The paper is divided into six sections. In Section 2, the main problems are stated, a few generalities are discussed, and several examples are presented to illustrate which kind of result one might expect. In Section 3, we present the Lefschetz–Hopf trace formula. We consider this formula in several contexts such as, for maps between closed orientable manifolds, for maps between compact manifolds with boundary and for maps between nonorientable manifolds. Section 4 is devoted to the Nielsen coincidence classes and the Reidemeister coincidence classes. They can be defined in a quite general context without any conceptual difficulty. In Section 5, we deal with the index of an isolated set of coincidence points. Two types of index are analyzed, in particular when this set is a Nielsen coincidence class. This can be done in certain categories. We will consider the case where the domain is a complex of dimension n and the target is a manifold also of dimension n. In Section 6, we look at the case where the two spaces involved are complexes of dimension ≤ 2, in particular, the case of surfaces. There is a close relationship between the coincidence problem on surfaces and the existence of solutions of certain equations in the braid groups. We will discuss these points in detail. In Section 7 we describe the recent work which has been done for maps between two manifolds of different dimension.

The author wish to express his sincere thanks to Prof. R. Brown for his careful reading and his valuable suggestions to make the exposition much clearer. Also to J. Guaschi for helping with the presentation.

2. The coincidence problem

The development of coincidence theory has been greatly influenced by fixed point theory, although not only by it. It is fair to start this work by pointing out the great influence of Brouwer's Fixed Theorem, dated 1905, on the development of both fixed point theory and coincidence theory. Whereas the main emphasis of Brouwer's result was on the existence of a fixed point, a little after, in the 1920's, Nielsen in his studies of surfaces (see [Ni]) was interested in estimating the minimal number of fixed points among all maps in a given homotopy class of maps. For this purpose an invariant was defined which we call nowadays the Nielsen number. From 1923 to 1927, in a series of papers [L1]–[L4], Lefschetz greatly generalized Brouwer's result for coincidence under the hypothesis that the spaces involved were compact orientable manifolds of the same dimension. This is perhaps the beginning of coincidence theory. In 1929, Hopf in [H1] extended Lefschetz's result for fixed points where the spaces in question were no longer manifolds but finite complexes. This is the famous Lefschetz–Hopf Theorem. In 1936 Reidemeister described (in [R]) in an algebraic way (in terms of traces) not only the information about the Lefschetz–Hopf number but the Nielsen classes and their indices, at least for fixed points. Wecken (see [We]) in the early 40's showed that for self maps of a complex satisfying mild conditions, the map can always be deformed to have exactly the Nielsen number of fixed points. Then in 1955, Schirmer (see [Sc]), published her thesis which showed a Wecken type result for coincidences of maps between two closed orientable manifolds of the same dimension greater or equal to three. This was the first advance of the theory after Lefschetz. For more details about the above and also the early history of the subject, see the excellent article [Bw2] by Brown. Also for related topics of coincidence theory not covered in this article see [BGZ1]. I hope that the facts described above give us enough motivation and information to understand the more recent development of coincidence theory.

We will consider the study of coincidence theory basically in two contexts. The first one is for a pair of continuous maps $(f,g)\colon X \to Y$, and the second is for a pair of maps in the category of pairs of spaces. In some cases we will have to assume certain conditions, dividing the study into several subcases.

For the first case, we will assume in general that X is at least a finite CW-complex, and Y a (not necessarily orientable) manifold (without boundary). This is not the most general situation that one can imagine, and perhaps not even the one which covers all important applications. Nevertheless these cases reflect the setting where a relevant and substantial amount of work has been done up to now using algebraic and geometric topology. For other cases see [BGZ1]. Several subcases are of interest and they have their own features. Namely:

(2.1.1) X, Y are closed manifolds (compact and without boundary) of the same dimension greater than or equal to three.

(2.1.2) X, Y are closed manifolds of the same dimension equal to two.

(2.1.3) X is either a finite complex or a finite CW-complex, and Y is a closed manifold with $\dim X = \dim Y \geq 3$.

(2.1.4) X is either a finite complex or a finite CW-complex, and Y is a closed manifold with $\dim X = \dim Y = 2$.

(2.1.5) X, Y are compact manifolds where $\dim X \geq \dim Y$.

(2.1.6) X is either a finite complex or a finite CW-complex, and Y is a closed manifold with $\dim X \geq \dim Y$.

For the second situation, we consider the category of pairs of spaces (M_1, M_2) where M_1 is a manifold and M_2 is a submanifold. Consider a pair of maps $f, g\colon M_1 \to N_1$ where the manifolds M_1, N_1 have the same dimension. Coincidence theory may be studied under several different hypotheses on the maps f, g and the kind of submanifold. Few cases have been considered in the literature so far. The one we will be most concerned with is the following: consider closed manifolds with nonempty boundary and a pair of maps $f, g\colon M \to N$ where M, N are manifolds of the same dimension. Assume that one of the maps, the map g for example, is a map of pairs namely $g\colon (M, \partial M) \to (N, \partial N)$. In this category we define a homotopy to be a pair of homotopies f_t, g_t, where f_t is a homotopy of f, and g_t is a boundary preserving homotopy of g i.e. $g_t(\partial(M)) \subset \partial N$. Coincidence theory in this category was studied in [BSc1], [BSc2].

We never consider the situation where $\dim X < \dim Y$. The reason is because under this hypothesis there is no coincidence theory, see Proposition 2.9.

Let X, Y be two topological spaces and $f\colon X \to Y$ a continuous map. For a pair of maps $f, g\colon X \to Y$, let $\mathrm{Coin}(f,g) = \{x \in X : f(x) = g(x)\}$. This set $\mathrm{Coin}(f,g)$ changes when we replace f, g by maps f', g', respectively, where f', g' are homotopic to f, g, respectively. Coincidence theory, generally speaking, studies properties relative to this family of sets. Here we try to present how algebraic and geometric topology has been used as an effective tool to analyze such problems. Based on the development of the theory so far, we can consider the following *basic questions* where each of them is interesting in its own right.

(2.2) QUESTION. For a given pair of maps (f, g) we would like to know if it can be deformed to a pair which is coincidence free, i.e. $\mathrm{Coin}(f', g') = \emptyset$ for some pair (f', g') homotopic to (f, g), or equivalently if the empty set belongs to the family of all sets $\mathrm{Coin}(f', g')$ as (f', g') runs over the set of all maps homotopic to (f, g), respectively.

The next two questions also have to do with the idea of *minimizing the set of coincidence points*,but in a more subtle situation then that of Question (2.2). In contrast to the fixed point case, since the pair of spaces X, Y may be distinct, in principle this will make the study of coincidence more subtle, in particular with respect to the understanding and formulation of the minimizing question. Let us first give a definition. For a set A we denote its cardinality by $|A|$.

(2.3) DEFINITION. Let $\mathrm{MC}[f, g]$ be defined as

$$\begin{aligned}\mathrm{MC}[f,g] &= \min_{f'\simeq f,\ g'\simeq g} |\{x \in X : f'(x) = g'(x)\}| \\ &= \min_{f'\simeq f,\ g'\simeq g} \{|\mathrm{Coin}(f', g')|\} = \mathrm{MC}[g, f],\end{aligned}$$

where $\simeq$ means homotopic.

(2.4) QUESTION. Compute $\mathrm{MC}[f, g]$ by means of homotopy invariants associated to the maps and the spaces involved.

The weaker question, which consists of deciding whether $\mathrm{MC}[f, g]$ is finite or infinite, is already quite interesting, and constitutes a relevant step to the full understanding of the coincidence problem. Of course Question (2.2) is equivalent to knowing if $\mathrm{MC}[f, g] = 0$. We will see that $\mathrm{MC}[f, g]$ is quite often infinite. When this happens, the notion of "minimizing the set of coincidence points" is not quite straightforward, and there are alternatives formulations. Three relevant measures of a set are:

(2.5.1) The number of its connected components.
(2.5.2) The Hausdorff measure, see [HW].
(2.5.3) The cohomological dimension defined below.

Let us consider case (2.5.3) and denote the Čech cohomology with coefficients in a ring R by $\check{H}(\cdot\,, R)$. Because of the use of Poincaré duality in coincidence theory see [V], Čech cohomology is a suitable cohomology theory to measure the set $\mathrm{Coin}(f, g)$.

(2.6) DEFINITION. Given a set A let the *cohomological dimension* of A with respect the coefficient R, denoted by $\mathrm{cohd}(A, R)$, be defined by

$$\mathrm{cohd}(A, R) = \max\{n \in \mathbb{Z} : \check{H}^n(A, R) \neq 0\}.$$

(2.7) QUESTION. Compute the minimum of $\mathrm{cohd}(\mathrm{Coin}(f', g'), R)$ as f', g' runs over the family of all pairs of maps homotopic to f, g, respectively.

I will leave the reader to explore other possibilities of questions concerning the idea of minimization. We hope that the examples given below will provide a better feeling for the above questions. Of course a specific problem may provide the reader with a very natural kind of minimizing problem.

(2.8) REMARK. As a result of the development of the theory, in Section 4 we will raise questions similar to (2.2), (2.4) and (2.7), stated in terms of the notion of Nielsen classes.

For the rest of this section we give some general results which show when $\mathrm{MC}[f,g]$ is finite or infinite, and some examples of the infinite case. These results indicate which are the relevant questions of coincidence theory that should be analyzed in each case.

The first result explains why we consider the case $\dim X \geq \dim Y$.

(2.9) PROPOSITION. *Let X be a* CW*-complex and Y a manifold where* $\dim X < \dim Y$. *Then any pair of maps $f, g\colon X \to Y$ can be deformed to be coincidence free.*

PROOF. Consider the map $f \times g\colon X \to Y \times Y$. The pair can be deformed to be coincidence free if and only if the map $f \times g$ can be compressed to the subspace $Y \times Y - \Delta Y$, where ΔY is the diagonal on $Y \times Y$. By obstruction theory (see [Wh] or [St]) the obstructions for such a deformation lie in the cohomology groups $H^i(X, \pi_i(Y \times Y, Y \times Y - \Delta Y)$. By either [Fad1] or [FH1], the groups $\pi_i(Y \times Y, Y \times Y - \Delta Y))$ vanish for $0 \leq i \leq n-1$, where $\dim Y = n$. Since $\dim X < \dim Y$ it follows that all the groups vanish and the result follows. □

Now suppose that X is a finite CW-complex, Y a manifold and $\dim X = \dim Y$. We show that $\mathrm{MC}[f,g]$ is finite in several cases.

(2.10) PROPOSITION. *Let X be a finite* CW*-complex and Y a manifold where* $\dim X = \dim Y$. *Then any pair of maps $f, g\colon X \to Y$ can be deformed to a pair (f', g') such that* $\mathrm{Coin}(f', g')$ *is finite.*

PROOF. Following [FH1] let us consider a cocycle which represents the obstruction to deforming the pair to be coincidence free. This cocycle may be written as a sum of elementary cocycles. The local coefficient system is given by the group ring $\mathbb{Z}[\pi] \approx \pi_n(Y \times Y, Y \times Y - \Delta Y)$ where $\pi = \pi_1(Y)$. So for each n dimensional cell, we have a map $\partial e^n \to Y \times Y - \Delta Y$ which is the restriction of a map of pairs $(e^n, \partial e^n) \to (Y \times Y, Y \times Y - \Delta Y)$. Now we claim that there is an extension of the map defined on the boundary to the interior, such that the number of coincidences is finite. This follows from the identification and construction of the generators of $\pi_n(Y \times Y, Y \times Y - \Delta Y) \approx \mathbb{Z}[\pi_1(Y)]$. □

The following result also holds.

(2.11) PROPOSITION. *If X is a finite simplicial complex of dimension $\leq n$ and Y is a finite complex whose maximal simplices of all of dimension $\geq n$, then* $\mathrm{MC}[f,g]$ *is finite.*

PROOF. This result is not explicitly stated in [Sc1] but it follows immediately from [Sc, Hilfssatz 1 and Satz IIa]. □

It is natural to ask what happens to the minimal number of coincidence points if we deform only one of the maps. It turns out that in many cases, this minimal number is the same as the case where one deforms both maps. Therefore one may take advantage of this equivalence to make some specific calculations. More precisely, when the target space Y is a manifold, the number $\mathrm{MC}[f,g]$ may also be achieved by deforming only one of the maps. This is useful when studying the minimality of $\mathrm{Coin}(f',g')$.

(2.12) PROPOSITION. *If Y is a manifold (without boundary) then the number* $\mathrm{MC}[f,g]$ *is equal to the minimum of the cardinality of* $\mathrm{Coin}(f',g)$ *for all maps f' homotopic to f:*

$$\mathrm{MC}(g)[f] = \min_{f' \simeq f} |\{x \in X : f'(x) = g(x)\}|.$$

PROOF. Following [G2, Proposition 1.5], consider the fibred pair $(Y\times Y, Y\times Y - \Delta) \xrightarrow{p_2} Y$ (see [Fad1]), where p_2 is the projection on the second coordinate. Certainly we have that $\mathrm{MC}[f,g] \leq \mathrm{MC}(g)[f]$. So it suffices to show that $\mathrm{MC}(g)[f] \leq \mathrm{MC}[f,g]$. For this, let (f',g') be an arbitrary pair. We have that $\overline{g}, \overline{g}' \colon X \to Y \times Y$ are homotopic, where $\overline{g}(x) = (f(x), g(x))$ and $\overline{g}'(x) = (f'(x), g'(x))$. Let H be such a homotopy. The map g' has a lift, namely (f',g') such that $(f',g')(X - \mathrm{Coin}(f',g')) \subset Y \times Y - \Delta$. By the lifting property of fibered pairs, it follows that there is a lift $\widetilde{H}$ of H such that $\widetilde{H}(X - \mathrm{Coin}(f',g')) \subset Y \times Y - \Delta$. So $\widetilde{H}(\,\cdot\,,1) = (f'',g)$ and $\mathrm{Coin}(f'',g) \subset \mathrm{Coin}(f',g')$. Therefore $\#\mathrm{Coin}(f,g) \leq \#\mathrm{Coin}(f',g')$ and the result follows. □

The above result was first proved in [Br1] in a stronger form.

Let M, N be two compact manifolds of the same dimension with boundary, and consider a pair of maps $f, g \colon M \to N$. Coincidence theory can be studied under several different hypotheses on the maps f, g. The case where we assume that one of the maps, the map g for example, is a map of pairs, namely $g\colon (M,\partial M) \to (N,\partial N)$, has been studied in [BSc1] and [BSc2]. In this category we define a homotopy as a pair of homotopies f_t, g_t, where f_t is a homotopy of f and g_t is a boundary-preserving homotopy of g i.e. $g_t(\partial(M)) \subset \partial N$. Then we can define the minimal number of coincidence, also denoted by $\mathrm{MC}[f,g]$, in a given homotopy class as described above. As an immediate consequence of [BSc1, Theorem 6.1] we have:

(2.13) COROLLARY. *The minimal number of coincidence points* $\mathrm{MC}[f,g]$ *in the category of pairs defined above is finite.*

Let us present a simple example which shows that it is not possible to minimize the number of points and the number of connected components of Coin at the same time.

(2.14) EXAMPLE. Let K be the union of the circle S^1 of radius one with an arc that is one of its diameters, and let $N = S^1$. Consider the pair of maps $f, c\colon K \to S^1$ where c is the constant map at the point $1 \in S^1 \subset \mathbb{C}$, and $f(z) = z^2$ if $z \in S^1$ and $f(z) = 1$ for z belonging to the diameter. Observe that $\mathrm{Coin}(f,c)$ is the diameter, so it has one connected component. But by a small deformation of f (for example by composing the map f with a small rotation of the target S^1), we obtain a pair (f',c) such that $\mathrm{Coin}(f',c)$ consists of two points, so it is not connected. We cannot have $\mathrm{Coin}(f_1,c) = \emptyset$ for f_1 homotopic to f, since f is not null homotopic. Also the map f_1 restricted to S^1 is a map of degree 2 which has as preimage of $1 \in S^1$ at least 2 connected components. This implies that we cannot expect to be able to minimize the number of points and the number of components at the same time.

The above example can be modified to the case where the codimension is positive. It will have different features, and will be examined in Example (2.16).

We cannot expect $\mathrm{MC}[f,g]$ to be finite, even when X and Y are finite CW-complexes for which $\dim(X) > \dim(Y)$. When this happens, the two properties that we must have in mind for the purpose of minimizing are the cohomological dimension and the number of components of Coin. In order to illustrate the non-finiteness of $\mathrm{MC}[f,g]$, consider the following example:

(2.15) EXAMPLE. Let $p\colon E \to B$ be a locally-trivial fibration with connected fibre between compact orientable manifolds E, B with $m = \dim(E) > \dim(B) = n$, and let $c\colon E \to B$ be the constant map at a point $b_0 \in B$. By using a Serre spectral sequence argument, we have that $H^{m-n}(\mathrm{Coin}(p,c),\mathbb{Z}) \neq 0$. Hence $\mathrm{Coin}(p,c)$ is not finite. Now let p' be an arbitrary map homotopic to p (not necessarily a fibre map). Let $\check{H}^i(\,\cdot\,,\,\cdot\,)$ denote the i-th Čech cohomology group. From the proof of [G5, Proposition 3.2], using the transfer as defined in [Dol, Chapter VIII, Section 10], we have that

$$\check{H}^{m-n}(p'^{-1}(b_0),\mathbb{Z}) \cong \check{H}^{m-n}(p^{-1}(b_o),\mathbb{Z}) \neq 0.$$

So $\mathrm{MC}[p,c]$ is not finite.

Now we consider an example where we can minimize the cohomological dimension and the number of components at the same time. However, the set is not minimal in the sense that it contains a proper subset which can be realized as the Coin of some pair.

(2.16) EXAMPLE. Let X be the union of the torus $T^{n+2} = S^1 \times S^1 \times T^n$ with a cylinder $S^1 \times [0,1]$, where we glue the boundary of the cylinder $S^1 \times \{0,1\}$ to the circles $S^1 \times \{1,-1\} \times \bar{1} \subset T^{n+2} = S^1 \times S^1 \times T^n$. Define $f: X \to T^{n+1}$ as follows: when restricted to $T^{n+1} = 1 \times T^{n+1}$, f is given by $f(1, z_2, \dots, z_{n+2}) = (z_2^2, \dots, z_{n+2})$ and $f(z) = 1$ for z in the cylinder. Observe that f is well defined and continuous. Let $c: X \to T^{n+1}$ be the constant map. As in Example (2.14), we have that $\mathrm{Coin}(f,c)$ is the cylinder, so it is a connected 2-dimensional manifold with boundary. After a small deformation (just compose f with a small rotation of the circle), we obtain that $\mathrm{Coin}(f',c)$ is the union of two disjoint circles. Now we will show that $\mathrm{Coin}(f'',c)$ can be a connected complex of dimension 1 for some map f'' homotopic to f. More precisely, we can make $\mathrm{Coin}(f'',c)$ to be the union of two circles with a segment from one circle to the other. The cylinder is the quotient of the square under certain identifications on the boundary. The constant map from the square can easily be deformed, relative to the boundary, so that the preimage of the base point of the target is only the boundary. This defines the new map f'' such that $\mathrm{Coin}(f_1'',c)$ is the complex in question. Finally we will show that for any (f'',g'') homotopic to (f,c), $\mathrm{Coin}(f'',g'')$ contains two closed disjoint subsets whose cohomological dimension is at least one. Consider the restriction of f to $T^{n+2} = S^1 \times S^1 \times T^n$, which by abuse of notation we also denote by f. This is a fibration. From the proof of Proposition (2.12), it follows that it suffices just to deform f. The image of the fundamental group $f_\#: \pi_1(T^{n+2}) \to \pi_1(T^{n+1})$ has index 2; we consider the lift to the double covering. Let $\widetilde{f}$ be the lifting of f. From Example (2.15) above it follows that for any deformation of the map $\widetilde{f}$, the preimage of a point has cohomological dimension at least 1. Since the preimage of the base point by f is the same as the preimage of two points by $\widetilde{f}$, the result follows.

(2.17) EXAMPLE. Consider a pair of maps $f, g: X \to Y$ between manifolds of dimension m, n, respectively. We claim that if there is a homology class $\alpha \in H_n(X,\mathbb{Z})$ such that the image of α by the composition

$$j_* \circ (f \times g)_*: H_n(X,\mathbb{Z}) \to H_n(Y \times Y, \mathbb{Z}) \to H_n(Y \times Y, Y \times Y - \Delta; \mathbb{Z})$$

is nontrivial, then the cohomological dimension of $\mathrm{Coin}(f',g')$ is at least $m-n$ for any pair f', g' homotopic to f, g. From the commutative diagram below

$$\begin{array}{ccc} H_n(X;\mathbb{Z}) & \longrightarrow & H_n(Y \times Y;\mathbb{Z}) \\ \downarrow{\scriptstyle j_1} & & \downarrow{\scriptstyle j_2} \\ H_i(X, X - \mathrm{Coin}(f',g'), \mathbb{Z}) & \longrightarrow & H_i(Y \times Y, Y \times Y - \Delta; \mathbb{Z}) \end{array} \qquad (2.18)$$

the map $j_* \circ (f \times g)$ factors through $H_n(X, X - \mathrm{Coin}(f', g'); \mathbb{Z})$ which implies that $H_n(X, X - \mathrm{Coin}(f', g'); \mathbb{Z}) \neq 0$. Using duality we obtain $\check{H}^{m-n}(\mathrm{Coin}(f', g'), \mathbb{Z}) \neq 0$ and the result follows.

Given homeomorphisms $\phi: X \to X$ and $\psi: Y \to Y$, for each map $f: X \to Y$ we may define a new map $\psi \circ f \circ (\phi)^{-1}$. Let Homeo($X$), Homeo($Y$) denote the group of the homeomorphisms of X and Y respectively, under the operation $h * k = h \circ k$. The map $\mathrm{Homeo}(X) \times \mathrm{Homeo}(Y) \times Y^X \to Y^X$ defined above is an action of the group $\mathrm{Homeo}(X) \times \mathrm{Homeo}(Y)$, with the product group structure on the function space Y^X. This action factors through an action on the set of homotopy classes of maps $[X, Y]$. It is straightforward to see that the coincidence set $\mathrm{Coin}(f, g)$ is homeomorphic to the $\mathrm{Coin}(\psi \circ f \circ (\phi)^{-1}, \psi \circ g \circ (\phi)^{-1}))$ of the two new maps. An immediate consequence of this fact is that for all the coincidence questions that have been treated so far, the answer for a pair (f, g) is the same as that for the pair $(\psi \circ f \circ (\phi)^{-1}, \psi \circ g \circ (\psi)^{-1})$. Therefore for practical purposes if we want to verify if a property is true for all pairs, it suffices to do so for a set of homotopy classes $[f, g]$ which contains one representative in each orbit with respect to the action of $\mathrm{Hom}(X) \times \mathrm{Hom}(Y)$ on the set of homotopy classes of pair of maps. This fact can be useful depending on how much knowledge one has about the orbits. In the case of maps between surfaces the naive idea above has been used with success in at least two situations. The first was in [GZ], [BGZ2] to study Coin of pairs of maps between surfaces where the second map is the constant map (or the root case). The second was in [GJi] in order to study bounds of the index on coincidence Nielsen classes.

(2.19) Remark. It is common to say that fixed point theory is a particular case of coincidence theory, it being enough to consider the particular case where the second map $f_2 = \mathrm{id}$. This is in general correct but some concepts defined in coincidence theory when applied to the situation where the second map is the identity, do not always correspond to the concept defined in fixed point theory. See for example [BGZ1] for a discussion about the number MC (see also [BSc1]) defined above and [GKe] for the notion of Wecken homotopies.

3. The Lefschetz–Hopf trace formula

This section is devoted to finding an efficient and computable way to decide whether a pair of maps cannot be deformed to be coincidence free. This is Question (2.2). The great achievement for this problem was a formula given by Lefschetz in the period 1923–1927, see the series of papers [L1]–[L4]. For a given pair of maps $f, g: M \to N$ where M, N are closed orientable manifolds of the same dimension, the Lefschetz formula associates an integer to such a pair. Despite the fact that it would have been more natural in those days to find such a formula to detect fixed points, the Lefschetz formula detects coincidence. This is a formula which clearly depends only on the homotopy classes of the maps f and g. Immediately after Lefschetz found the formula, Hopf extended it to pairs $f, \mathrm{id}: K \to K$ where id is the identity (the fixed point case) and K is a finite complex, not necessarily a manifold. So the formula nowadays is called the Lefschetz–Hopf trace formula. In more modern language, we start by describing the results obtained relative to this formula for closed orientable manifolds. Then we move on to discuss further results for other categories of spaces. The most recent result that we will treat here is the case of pairs of maps $f, g: M \to N$, where both manifolds are compact with nonempty boundary, $g(\partial M) \subset \partial N$ and one of the maps is orientation true (see below for the definition of orientation true).

Let M, N be orientable closed manifolds and $f, g: M \to N$ a pair of maps. (For more details see [V, Chapter VI].)

(3.1) Definition. Let the homomorphism $\vartheta^i: H^i(N, \mathbb{Q}) \to H^i(N, \mathbb{Q})$ be the composition $D_N^{-1} \circ (f)_* \circ D_M \circ (g)^*$, where D is the Poincaré duality homomorphism. The *Lefschetz number* $\widehat{L}(f, g)$ is defined by the Lefschetz–Hopf trace formula

$$\widehat{L}(f, g) = \sum_{i=1}^{n} (-1)^i \operatorname{trace}(\vartheta^i).$$

The Lefschetz number can also be defined in cohomological terms, allowing the use of the multiplicative structure of cohomology and of cohomology operations in the computation of the Lefschetz number, see [Fad2], [GO].

(3.2) Definition (See [V, Chapter VI]). Let the homomorphism

$$\widehat{\vartheta}_i: H_i(M, \mathbb{Q}) \to H_i(M, \mathbb{Q})$$

be the composite $D_M \circ (g)^* \circ D_N^{-1} \circ (f)_*$, where $D_M: H^{n-i}(M, \mathbb{Q}) \to H_i(M, \mathbb{Q})$ is the Poincaré duality homomorphism. The *Lefschetz number* $L(f, g)$ is defined by the Lefschetz–Hopf trace formula

$$L(f, g) = \sum_{i=1}^{n} (-1)^i \operatorname{trace}(\widehat{\vartheta}_i).$$

The equalities $\operatorname{trace}(\vartheta_i) = \operatorname{trace}(\widehat{\vartheta}^{n-i})$, $1 \leq i \leq n$, imply $L(f, g) = (-1)^n \widehat{L}(f, g)$ for the two Lefschetz numbers defined above. The Lefschetz number is anti-commutative: $L(f, g) = (-1)^n L(g, f)$.

The importance of this number is due to the fact that its non-vanishing guarantees the existence of a coincidence point. The number $L(f, g)$ is homotopy invariant; hence the statements about the existence of a coincidence can be reformulated in the following stronger form. More precisely:

(3.3) THEOREM. *Let $f, g: M \to N$ be two continuous maps between closed orientable manifolds of the same dimension n. If $L(f,g) \neq 0$ then* $\mathrm{Coin}(f', g') \neq \emptyset$ *for any f', g' homotopic to f, g, respectively.*

We will see that the development described above, will be repeated in several other categories. They have some similarities. At the end we show how a theorem of the type of Theorem (3.3) above is proved.

In 1980, Nakaoka (see [Na]), studied coincidences of fibre-preserving maps, a subject that we do not treat here. In this work he developed a type of Lefschetz–Hopf formula for fibre-preserving maps, where the fibres are manifolds. However, the results there are also stated for the case where the fibres are compact orientable manifolds with boundary (see [Na, Section 8]). So a Lefschetz–Hopf trace formula for manifolds with boundary appears for the first time in [Na]. Later in 1992, Mukherjea ([Mu]) independently studied the Lefschetz–Hopf trace formula for manifolds with boundary in more detail. We now describe this case.

Consider the category of compact orientable manifolds with boundary, and let M, N be two such manifolds of the same dimension. Let $f, g: M \to N$ be a pair of maps such that the second map satisfies the property $g(\partial M) \subset \partial N$. By a homotopy of the pair we mean a pair of homotopies (H, G), where H is an arbitrary homotopy of f and G is a boundary-preserving homotopy of g i.e. G_t is a boundary-preserving map for all $t \in [0,1]$.

Let $D_M: H^*(M, \partial M; \mathbb{Q}) \to H_{n-*}(M, \mathbb{Q})$ and $D_N: H^*(N, \partial N; \mathbb{Q}) \to H_{n-*}(N, \mathbb{Q})$ be the Poincaré–Lefschetz duality homomorphisms.

(3.4) DEFINITION. Given a pair of maps $f, g: M \to N$, where $g(\partial M) \subset \partial N$, let the homomorphism $\vartheta^i: H^i(N, \partial N; \mathbb{Q}) \to H^i(N, \partial N; \mathbb{Q})$ be the composition $D_N^{-1} \circ (f)_* \circ D_M \circ (g)^*$. The *Lefschetz–Hopf trace formula* is given by:

$$L(f,g) = \sum_{i=1}^{n} (-1)^i \mathrm{trace}(\vartheta^i).$$

Then we have the following result.

(3.5) THEOREM. *Let $f, g: M \to N$ be two continuous maps between compact orientable manifolds of the same dimension n and with boundary. If the Lefschetz–Hopf number above satisfies $L(f,g) \neq 0$ then* $\mathrm{Coin}(f', g') \neq \emptyset$ *for any f', g' homotopic to f, g, respectively, where the homotopy between g and g' is boundary preserving.*

Let us exploit this result a little. The Brouwer Fixed Point Theorem says that every continuous map $f: D^n \to D^n$ (D^n is the n-dimensional disk) has a fixed point. Nevertheless a pair of maps $f, \mathrm{id}: D^n \to D^n$, where the second map is the identity, can be deformed as a pair such that $\mathrm{Coin}(f', \mathrm{id}') = \emptyset$. In fact, this is the case for any pair of maps $f, g\colon D^n \to D^n$, because any map is homotopic to an arbitrary constant map. But the suitable generalization of the fixed point question would be: given any deformation (f', id') of the pair (f, id) where the homotopy connecting id to id′ preserves the boundary S^{n-1} of D^n, is $\mathrm{Coin}(f', \mathrm{id}')$ non empty? From [BrSc1] we find that not only is the answer yes in this case, but there are many other maps $g\colon (D^n, \partial D^n) \to (D^n, \partial D^n)$ which have a similar property to the identity in the above example. In more detail:

(3.6) DEFINITION. A map $g\colon (M, \partial M) \to (N, \partial N)$ is said to be *coincidence producing* if for every map $f\colon M \to N$ we have $\mathrm{Coin}(f, g) \neq \emptyset$.

(3.7) PROPOSITION. *The identity* $\mathrm{id}\colon (D^n, \partial D^n) \to (D^n, \partial D^n)$ *is coincidence producing. Further, any map* id′ *homotopic to* id *as a map of pairs is coincidence producing.*

PROOF. Let us compute $L(f, \mathrm{id})$. Since the pair $(D^n, \partial D^n)$ only has cohomology in dimension n, it follows that the induced homomorphism $D_N^{-1} \circ (f)_* \circ D_M \circ (g)^*\colon H^n(D^n, \partial D^n) \to (D^n, \partial D^n)$ is the identity because $f_*\colon H_0(D^n, \mathbb{Q}) \to H_0(D^n, \mathbb{Q})$ is the identity independent of the map f. So the result follows. □

It would be nice to know if the property of being a coincidence producing map is a property of the homotopy class. See [BwSc1, Section 7].

So far, all the results concern orientable manifolds. In 1997, see [GJ], the Lefschetz–Hopf formula for nonorientable manifolds was treated. The development of the theory in general in this case is similar. But there are differences in the details; we shall make this explicit and give more information. Following [GJ], for an arbitrary manifold M we can define a local orientation along any path, and along a closed path, the orientation is either preserved or reversed. Homotopic closed paths have the same behavior and so this yields a homomorphism from the fundamental group to $\mathbb{Z}_2$. Denote by Γ_W be the local system over a manifold W, where the coefficients are the rationals and the action is given by the orientation behavior of the loop. This local system is called *the orientation system* with rational coefficients of the manifold. It follows that Γ_W is trivial if and only if W is orientable.

Let us recall a concept which was formally defined by P. Olum in [O]. A map $f\colon M \to N$ is called *orientation true* if for every element $\alpha \in \pi_1(M)$, both α and $f_\#(\alpha)$ have the same sign, i.e. either both loops are orientation preserving or both are orientation reversing. If $f\colon M \to N$ is orientation true then the induced local system $f^*(\Gamma_N)$ is given by

$$\alpha \cdot 1 = f_\#(\alpha) \cdot 1 = \mathrm{sign}(f_\#(\alpha)) \cdot 1 = \mathrm{sign}(\alpha) \cdot 1,$$

where 1 is the unit of $\mathbb{Q}$. Hence $f^*(\Gamma_N) = \Gamma_M$, and f induces a map

$$f^*: H^*(N, \Gamma_N) \to H^*(M, \Gamma_M)$$

on the cohomology with local coefficients.

We start with the case of compact manifolds without boundary. Let

$$D_M: H^*(M, \Gamma_M) \to H_{n-*}(M, \mathbb{Q}) \quad \text{and} \quad D_N: H^*(N, \Gamma_N) \to H_{n-*}(N, \mathbb{Q})$$

be the Poincaré duality homomorphisms with local coefficients.

We will consider the situation where the second map g is orientation true. Then we define:

(3.8) DEFINITION. For maps $f, g: M \to N$ where g is orientation true, let the homomorphism $\vartheta^i: H^i(N, \mathbb{Q}) \to H^i(N, \mathbb{Q})$ be the composite $D_N^{-1} \circ (g)_* \circ D_M \circ (f)^*$. The *Lefschetz–Hopf trace formula* is given by

$$L(f, g) = \sum_{i=1}^{n} (-1)^i \mathrm{trace}(\vartheta^i).$$

Let us observe that in the case where both maps are orientation true we can compute $L(f, g)$ and $L(g, f)$. Recall that in the orientable case we showed after Definition (3.2) that $L(f, g) = (-1)^n L(g, f)$ where n is the dimension of the manifold. This is not the same in the nonorientable case and the examples below illustrate this fact.

(3.9) EXAMPLE. Let $\widetilde{f}_k: S^2 \to S^2$ be an odd map ($\widetilde{f}_k(-x) = -\widetilde{f}_k(x)$) of degree k (k must then be an odd number). This map induces a map $f_k: RP^2 \to RP^2$ and we have $\deg(f_k) = \deg(\widetilde{f}_k) = k$. On the other hand, f_k induces an isomorphism of fundamental groups, hence is orientation true. Let us fix two odd integers k, l and compute the Lefschetz number $L(f_k, f_l)$ (with rational coefficients $\mathbb{Q}$). Since RP^2 is $\mathbb{Q}$-acyclic, only the sequence

$$H^0(RP^2;\mathbb{Q}) \xrightarrow{f_k^*} H^0(RP^2;\mathbb{Q}) \longrightarrow H_2(RP^2;\Gamma) \xrightarrow{f_l^*} H_2(RP^2;\Gamma) \longrightarrow H^0(RP^2;\mathbb{Q})$$

may make a non-zero contribution to $L(f_k, f_l)$. Thus $L(f_k, f_l) = \deg(f_l) = l$. But on the other hand, $L(f_l, f_k) = \deg(f_k) = k$.

(3.10) EXAMPLE. The following example shows that $L(f, g) \neq 0$ does not imply $L(g, f) \neq 0$. Let $p, c: S^2 \to RP^2$, where p is the projection and c is the constant map. It is easy to see that $L(p, c) = 0$ but $L(c, p) = 2$.

This type of example can even be given for selfmaps.

(3.11) EXAMPLE. Let K be the Klein bottle. Take $f, g\colon K \to K$ where the induced maps on the fundamental group are given by $f_{\#}(\alpha) = 1$, $f_{\#}(\beta) = \beta$, $g_{\#}(\alpha) = \alpha$ and $g_{\#}(\beta) = \beta^3$. We have that $L(f, g) = 0$ and $L(g, f) = 2$.

Now we move to the case with boundary. Let $f, g\colon M \to N$, where g is orientation true. We have two essentially distinct geometric situations, and for each case a formula may be defined and be used to detect coincidences. The first one is if $g\colon (M, \partial M) \to (N, \partial N)$ is a map of pairs. The second one is if $f\colon (M, \partial M) \to (N, \partial N)$ is a map of pairs.

Let us consider the first case.

(3.12) DEFINITION. For maps $f, g\colon M \to N$ and $g\colon (M, \partial M) \to (N, \partial N)$ a map of pairs, where g is orientation true, let the homomorphism $\vartheta^i\colon H^i(N; \mathbb{Q}) \to H^i(N; \mathbb{Q})$ be the composite $D_N^{-1} \circ (g)_* \circ D_M \circ (f)^*$. The *Lefschetz–Hopf trace formula* is given by

$$L'(f, g) = \sum_{i=1}^{n} (-1)^i \operatorname{trace}(\vartheta^i).$$

Then we have:

(3.13) THEOREM. *Let $f, g\colon M \to N$ be two continuous maps between closed manifolds of the same dimension n. If the Lefschetz–Hopf number above satisfies $L'(f, g) \neq 0$ then $\operatorname{Coin}(f', g') \neq \emptyset$ for any f', g' homotopic to f, g, respectively, where the homotopy between g and g' is boundary preserving.*

Now let us consider the second case: g is orientation true and $f(\partial M) \subset \partial N$.

(3.14) DEFINITION. For maps $f, g\colon M \to N$ and $f\colon (M, \partial M) \to (N, \partial N)$ a map of pairs, where g is orientation true, let the homomorphism $\vartheta^i\colon H^i(N, \partial N; \mathbb{Q}) \to H^i(N, \partial N; \mathbb{Q})$ be the composite $D_N^{-1} \circ (g)_* \circ D_M \circ (f)^*$. The *Lefschetz–Hopf trace formula* is given by

$$L''(f, g) = \sum_{i=1}^{n} (-1)^i \operatorname{trace}(\vartheta^i).$$

Then we have:

(3.15) THEOREM. *Let $f, g\colon M \to N$ be two continuous maps between closed manifolds of the same dimension n. If the Lefschetz–Hopf number above satisfies $L''(f, g) \neq 0$ then $\operatorname{Coin}(f', g') \neq \emptyset$ for any f', g' homotopic to f, g, respectively, where the homotopy between g and g' is boundary preserving.*

It will immediately occur to the reader that if the two maps f, g are orientation true, we have two Lefschetz numbers, and both may be used to detect coincidence points in the homotopy class of the pair (f, g). What is the relation between these two numbers? From [GJ, Theorem 4.9] let $f|_{\partial M}, g|_{\partial M}$ be the restrictions of f, g, respectively, to the boundary of M. By hypothesis it turns out that they are maps $f|_{\partial M}, g|_{\partial M} \colon \partial M \to \partial N$ where $\partial M, \partial N$ are closed manifolds of dimension $n-1$. Since ∂g is also orientation true, the Lefschetz number is defined and we have:

(3.16) PROPOSITION. $L'(f,g) - L''(f,g) = L(f|_{\partial M}, g|_{\partial M})$.

We have seen above that under several different hypotheses, there is a theorem on the existence of coincidence for a given pair of maps. In all cases the proof of such theorems follows the same pattern. So we will illustrate here the case where M, N are manifolds with boundary and $f, g \colon M \to N$, where g is orientation true and boundary-preserving.

From a Thom class in $H^n(\operatorname{int} N \times \operatorname{int} N, \operatorname{int} N \times \operatorname{int} N - \Delta; \mathbb{Q} \times \Gamma_N)$, one can define (see [GJ, Section 3]) a cohomology class

$$\widetilde{U}'_N \in H^n(\operatorname{int} N \times (N, \partial N); \mathbb{Q} \times \Gamma_M).$$

We define the index of the pair f,g as the image of the fundamental class $z_M \in H_n(M, \partial M; \Gamma_M)$ under the sequence of homomorphisms:

$$H_n(M, \partial M; \Gamma_M) \xrightarrow{d'_*} H_n(M \times (M, \partial M); \mathbb{Q} \times \Gamma_M)$$
$$\xrightarrow{f \times g)_*} H_n(\operatorname{int} N \times (N, \partial N); \mathbb{Q} \times \Gamma_N) \xrightarrow{\langle \widetilde{U}'_N, \cdot \rangle} \mathbb{Q}$$

and denote $\operatorname{ind}'(f,g) = \langle \widetilde{U}'_N, (f \times g)_* d'_*(z_M) \rangle$ (see more about indices in the next section).

Here is the main result which is also called the normalization property.

(3.17) THEOREM (Normalization). *Let $f, g \colon M \to N$ be a pair of maps between closed n-manifolds with g orientation true and $g(\partial M) \subset \partial N$. Then*

$$L'(f,g) = \operatorname{ind}'(f,g).$$

The proof of the above result, after all the preparation concerning the Thom class, is similar to the proof of the corresponding result in [V, Chapter 6]. An immediate consequence of the above theorem is the existence of a coincidence in the suitable homotopy class of the pair whenever the number given by the Lefschetz–Hopf trace formula is not zero.

(3.18) COROLLARY. *If the pair (f,g) can be deformed to be coincidence free then $L(f,g)$ is zero.*

PROOF. By properties of homology and cohomology we have the equality

$$\langle \widetilde{U}'_N, (f \times g)_* d'_*(z_M) \rangle = \langle d'^* \circ (f \times g)^* \widetilde{U}'_N, (z_M) \rangle.$$

If the pair (f, g) can be deformed to be coincidence free, then we have a factorization $h: M \to N \times N - \Delta$ of (f, g), and $(f \times g)^* \widetilde{U}'_N$ can be computed as $(h)^* \circ i^* \widetilde{U}'_N$ which is zero since $\widetilde{U}'_N$ is the restriction of a cohomology class of the pair. The result follows. □

The results of this section lead one to pose the question about the necessity of the condition on the Lefschetz–Hopf number, that is, about the validity of the converse to the above assertion. A great part of contemporary coincidence theory deals with results which are, in some sense, converses to this homotopical form of the Lefschetz–Hopf theorems [Fu], [Fad2], [Fad3]. One also seeks a sharper estimate of the minimal number of coincidence points. This is the purpose of the next two sections.

4. Nielsen and Reidemeister coincidence classes

In this section we define Nielsen and Reidemeister coincidence classes. There is a natural way to associate a Reidemeister class to each non-empty Nielsen class of a pair of maps (f, g). The correspondence is injective and suggests that the Reidemeister classes can be regarded as a coordinate system to index the Nielsen classes. We define equivalence relations on the set of Nielsen classes and on the set of Reidemeister class. Two such classes are said to be *(H, G)-related* where (H, G) is a self-homotopy of the pair (f, g).

4.1. Nielsen coincidence classes.

(4.1) DEFINITION. Let $f, g: X \to Y$ be a pair of continuous mappings. Two coincidence points $x_0, x_1 \in \operatorname{Coin}(f, g)$ are *Nielsen equivalent* if there exists a path λ in X such that $\lambda(0) = x_0, \lambda(1) = x_1$ and $f_1(\lambda)$ is homotopic to $f_2(\lambda)$ relative to the end points. An equivalence class is called a *Nielsen class.*

Under very mild conditions, one can show that a Nielsen class is a closed and open subset of $\operatorname{Coin}(f, g)$. For instance suppose that X is locally path connected, and Y is semi-locally simply connected.

(4.2) PROPOSITION. *Each Nielsen class is open in* $\operatorname{Coin}(f, g)$.

PROOF. Let $x_1 \in \operatorname{Coin}(f, g)$. It follows from the hypotheses that there exists a neighbourhood W of $f(x_1) = g(x_1) = y_1$ in Y such that any loop in W with base point y_1 is homotopically trivial. Consider a path-connected neighbourhood $U \subset f^{-1}(W) \cap g^{-1}(W)$ of x_1. Now take $x_2 \in U \cap \operatorname{Coin}(f, g)$ and λ a path in U connecting x_1 to x_2. Then $f \circ \lambda$ and $g \circ \lambda$ are two paths in W with the same end points. Applying the property of W, it follows that the two paths are homotopic relative to the end points. Therefore the two points belong to the same Nielsen class and the result follows. □

If we further assume that Y is Hausdorff we can show that a Nielsen class is closed.

(4.3) PROPOSITION. *Under the above hypothesis, a Nielsen class of* (f, g) *is closed in* $\mathrm{Coin}(f, g)$.

PROOF. Since Y is Hausdorff it follows that the diagonal $\Delta_Y \subset Y \times Y$ is closed. But $\mathrm{Coin}(f, g) = (f \times g)^{-1}(\Delta_Y)$, where $f \times g: X \to Y \times Y$. It follows that $\mathrm{Coin}(f, g)$ is closed. So it suffices to show that one Nielsen class is closed in X. Let x_0 be an accumulation point of a Nielsen class F. By the previous proof there is an open set such that all the coincidence points in the neighbourhood belong to the same Nielsen class of x_0, and the result follows. □

If X and Y satisfy all of the above hypotheses then the previous two results show:

(4.4) COROLLARY. *The set* $\mathrm{Coin}(f, g)$ *is a closed subset of* X, *and each Nielsen class is open and closed in* $\mathrm{Coin}(f, g)$.

The challenge is to define or associate to a Nielsen class an algebraic object which at the same time is computable and which has the property that if this algebraic object is not zero, than we must have a coincidence point. We explore this aspect in Section 5.

4.2. Reidemeister coincidence classes. Reidemeister classes are one of the basic tools in the study of coincidence theory. They can be regarded as a coordinate system to index the Nielsen classes. Motivated by the study of maps on nonorientable manifolds, we give a definition of Reidemeister classes which in this special case will distinguish some of them. Otherwise it will coincide with the usual one.

Let $f, g: X \to Y$ be a pair of maps between arbitrary spaces and let $x_0 \in X$ satisfy $f(x_0) = g(x_0)$. Then we define:

(4.5) DEFINITION. *The Reidemeister classes of the pair* (f, g) are the set of equivalence classes of elements of $\pi_1(Y, y_0)$ given by the relation $\alpha \equiv \beta$ if and only if there exists $\theta \in \pi_1(X, x_0)$ such that $\beta = g_\#(\theta)\alpha(f_\#(\theta))^{-1}$. For maps between manifolds, a class $\alpha \in \pi_1(Y)$ is called a *defective class* if there exist θ such that $\alpha = g_\#(\theta)\alpha(f_\#(\theta))^{-1}$ and $\mathrm{sign}(\theta)\mathrm{sign}((f_\#(\theta)) = -1$.

Suppose that $\mathrm{Coin}(f, g)$ is not empty, x_0 is a coincidence point and $y_0 = f(x_0) = g(x_0)$ is its image. We will define an injection from the set of Nielsen classes into the set $R(f, g)$ of Reidemeister classes. For $x_1 \in \mathrm{Coin}(f, g)$, let λ be any path from x_0 to x_1. The path $g(\lambda) * f(\lambda)^{-1}$ is a loop and it defines an element of $\pi_1(Y, y_0)$. It is straightforward to see that different paths from x_0 to x_1 define elements of the same Reidemeister class, and Nielsen equivalent coincidence points are associated with elements of the same Reidemeister class.

4.3. Related classes. Let (f, g) be a pair of maps. Given a Nielsen class, one would like to know if this class can disappear under homotopy. In order to formalize this idea, we define a notion of related classes. Let (H, G) be a homotopy of the pairs (f, g) and (f_1, g_1).

(4.6) DEFINITION. Two Nielsen classes $C_1 \in \mathrm{Coin}(f, g)$ and $C_2 \in \mathrm{Coin}(f_1, g_1)$ are said to be *(H, G)-related* if there is a path λ starting at a point of C_1 and ending at a point of C_2 such that $H(\lambda(t), t)$ is end point homotopic to $G(\lambda(t), t)$.

For the case of two classes of the same pair of maps we define:

(4.7) DEFINITION. Two classes $C_1, C_2 \in \mathrm{Coin}(f, g)$ are said to be *homotopy related* if there is a pair of self-homotopies (H, G) of (f, g) such that C_1 is (H, G)-related to C_2.

A similar notion can be defined for Reidemeister classes.

(4.8) DEFINITION. Given a self-homotopy (H, G) of the pair (f, g), consider the pair of loops $(w_1, w_2) \in \pi_1(Y) \times \pi_1(Y)$, where $w_1(t) = G(x_0, t)$ and $w_2(t) = H(x_0, t)$. We say that $[\alpha]$ is *(H, G)-related* to $[\beta]$ if $[w_2 \alpha w_1^{-1}] = [\beta]$. They are *homotopy related* if they are (H, G)-related for some pair of self homotopies.

We have a relation on Nielsen classes and another on Reidemeister classes. The map defined above from the set of Nielsen classes into the set of Reidemeister classes preserves these relations.

(4.9) PROPOSITION. *The correspondence which associates a Nielsen coincidence class to a Reidemeister class preserves related classes.*

PROOF. Let $\mathrm{Coin}(f, g)$ be nonempty and let x_0 be a base point. Consider two coincidence points $x_1, x_2 \in \mathrm{Coin}(f, g)$, and suppose that they are related. So there exists a self homotopy (F, G) and a path λ from x_1 to x_2 such that $H(\lambda(t), t)$ is end point homotopic to $G(\lambda(t), t)$. Take α to be any path connecting x_0 to x_1. In the space $X \times I$, the path $\lambda_2 = (x_0, (1-t)) * (\alpha, 0) * (\lambda(t), t)$ connects $(x_0, 1)$ to $(x_2, 1)$, and is end homotopic to the path $\lambda_2' = (\alpha, 1) * (\beta, 1)$. The loop $G(\,\cdot\,, 1)(\lambda_2') * F(\,\cdot\,, 1)(\lambda_2')^{-1})$ defines an element of the Reidemeister class associated to the element x_2 and this element is homotopic to

$$
\begin{aligned}
&G(\,\cdot\,, 1)(\lambda_2) * F(\,\cdot\,, 1)(\lambda_2)^{-1}) \\
&\quad = G((x_0, (1-t)) * (\alpha, 0) * (\lambda(t), t)) F((\lambda(t), t)^{-1} * (\alpha, 0)^{-1} * (x_0, (1-t))^{-1}.
\end{aligned}
$$

Therefore it follows that $\beta = w_2 \alpha w_1^{-1}$ and the result follows. □

Now we may define à la Brooks [Br2] a “geometric essential Nielsen coincidence class”.

(4.10) DEFINITION. A Nielsen coincidence class is said to be *geometric essential* if it is not (H, G)-related to the empty class. Otherwise it is called *geometric inessential.* Therefore we can define the *geometric Nielsen coincidence number* to be the number of geometric essential coincidence classes.

4.4. Jiang subgroup for coincidences. In 1964, Jiang [Ji1] defined a subgroup $J(f) \subset \pi_1(X)$ for every map $f: X \to X$, which was later called the *Jiang subgroup of the map* f. This group was very useful for fixed point theory. In 1965, Gottlieb (see [Gt]) defined and studied the subgroup of the fundamental group corresponding to the above group, for the case where f is the identity map. This group was defined in terms of function spaces. This subgroup, denoted by $G_1(X)$, was later called the *first Gottlieb group* of X. The first Gottlieb group not only has a close relationship with fixed point theory, but also with other branches of mathematics, like the theory of fibrations, rational homotopy theory and group actions (see [Op]).

We adapt the above notion of Jiang subgroup to the coincidence case. Let Z^X be the space of functions from X to Z with the compact-open topology, where X is a space with base point x_0. Then we have the evaluation map $e: Z^X \to Y$ given by $e(g) = g(x_0)$.

(4.11) DEFINITION. The *Jiang subgroup* $J(f)$, corresponding to a map $f: X \to Z$ is the image

$$e_{\#}(\pi_1(Y^X, f)) \subset \pi_1(Z, f(x_0)).$$

We apply the above definition to the coincidence case. Namely, given a pair of maps $f, g: X \to Y$, consider the map $f \times g: X \to Y \times Y$. So we define two groups related to the pair (f, g).

(4.12) DEFINITION. The subgroup $J(f \times g)$ is called the *Jiang subgroup of the pair* (f, g). Also we denote by $J_c(f, g) \subset \pi_1(Y)$ the smallest subgroup which contains the image of $J(f \times g)$ under the map $\pi_1(Y) \times \pi_1(Y) \to \pi_1(Y)$ given by $(\alpha, \beta) \to \beta(\alpha)^{-1}$.

In the next section we show some applications of the Jiang subgroup. We define:

(4.13) DEFINITION. A pair (X, Y) is called a *Jiang pair for coincidences* if $J_c(f, g) = \pi_1(Y)$ for all pairs of maps.

Whenever $J_c(f, g) = \pi_1(Y)$ we have nice properties for the coincidence theory of the pair (f, g). Assuming that X, Y satisfies some mild conditions such as X is compact and that Nielsen classes are closed and open, as a direct consequence of the definitions above we have:

(4.14) THEOREM. *If $J_c(f,g) = \pi_1(Y)$ then*

(4.14.1) *All Reidemeister classes are related. Hence all Nielsen classes are related.*

(4.14.2) *Either $N(f,g) = 0$ or all Nielsen classes are essential.*

(4.14.3) *If the geometric Nielsen number is not zero, then it is finite and is equal to the number of Reidemeister classes (which thus has to be finite).*

PROOF. Part (4.14.1). It suffices to show that the class of the trivial element $1 \in \pi_1(Y)$ is related to the class of an arbitrary element $\alpha \in \pi_1(Y)$. By hypothesis there exist $\theta_1 \times \theta_2 \in J(f) \times J(g)$ such that $\theta_1\theta_2^{-1} = \alpha$. This implies that the classes $[1]$, $[\alpha]$ are (H,G)-related for H, G self-homotopies corresponding to θ_1, θ_2, respectively.

Part (4.14.2) follows directly from part (4.14.1).

Part (4.14.3). Since X is compact and the Nielsen classes are open and closed it follows that the number of Nielsen classes is finite. Since by hypothesis there exists one Nielsen class which is essential, then there is a Reidemeister which is essential. It follows from (4.14.1) that all the other Nielsen classes corresponds to essential Reidemeister classes and conversely. So the result follows. □

We leave to the reader to explore the relation between Jiang spaces and Jiang pairs of spaces. Of course if Y is a Jiang space (for example a Lie group or an H-space) then (X,Y) is a Jiang pair for any space X.

To conclude this section we will show how Reidemeister classes arise naturally in the context of obstruction theory to deform a pair $f,g\colon K \to N$, where N is a n-dimensional manifold and K is a CW-complex. For a set J, denote the sum of $\mathbb{Z}$ indexed by the set J by $\mathbb{Z}[J]$.

(4.15) PROPOSITION. *The group in which the primary obstruction to making f,g coincidence free lies is the direct sum $H^n(K;\mathbb{Z}[\mathbb{R}_i])$ where $\mathbb{R}_i$ runs over all Reidemeister classes. In particular, if K is also an n-dimensional manifold then this group is isomorphic to the direct sum of A's where A is either $\mathbb{Z}$ or $\mathbb{Z}_2$ (not necessarily the same) indexed by the set of Reidemeister classes.*

The main idea behind this Proposition comes from [FH1]. But for further details and in a more explicit form see [G4].

5. Index of an isolated subset of the coincidence set

We start this section by making a few comments about index theory for fixed points. Such an index has a long history dating back to Hopf [H1]. It has been further developed and defined for a quite general family of spaces, including the categories of finite polyhedra and compact metric ANR's. For more details see [BGZ1].

In coincidence theory the situation is more subtle, and we will make more restrictions on the spaces involved when compared with the fixed point case. Consider a pair of maps $f, g: X \to Y$ and an isolated subset $F \subset \mathrm{Coin}(f, g)$. In some situations, we will be able to define *the coincidence index of the subset* F. This will be equivalent to the so-called local coincidence index under certain conditions.

5.1. Maps between orientable manifolds. In coincidence theory for closed orientable manifolds M_1 and M_2 of the same dimension, a local index of a subset F was defined. Consider homology with integral coefficient $\mathbb{Z}$. Following [V, Chapter VI] we define the local coincidence index as follows.

(5.1) DEFINITION. Let M_1, M_2 be n-dimensional closed orientable manifolds, V an open subset of M_1, and $f, g: V \to M_2$ mappings such that $C = \mathrm{Coin}(f, g)$ is compact. The *coincidence index* $i(V; f, g)$ of the pair (f, g) defined on C is the integer given by the image of the fundamental class of M_1 under the composition

$$H_n(M_1) \to H_n(M_1, M_1 \setminus W) \to H_n(V, V \setminus W) \\ \to H_n(M_2 \times M_2, M_2 \times M_2 - \Delta(M_2)) \cong \mathbb{Z},$$

where the second map is the excision, the third is induced by $(f,g)(x)=(f(x),g(x))$, and W satisfies $\mathrm{Coin}(f, g) \subset W \subset \overline{W} \subset V$.

Because of the following property (5.2.1) we can drop the reference to V and write $i(\mathrm{Coin}(f, g))$ instead of $i(V; f, g)$.

(5.2) PROPERTIES. The index satisfies the following properties.

(5.2.1) (Localization) $i(V; f, g) = i(V'; f|_{V'}, g|_{V'})$ for every open set $V' \subset V$ containing C. This allows us to write $i(\mathrm{Coin}(f, g)) = i(V; f, g)$.

(5.2.2) (Additivity) If V is a finite union of open sets V_i, $i = 1, \dots, r$, and C is disjoint union of compact sets C_i with $C_i \subset V_i$, then

$$i(V; f, g) = i(V_1, f_1, g_1) + \dots + i(V_r; f_r, g_r) \quad \text{where } (f_j, g_j) = (f|_{V_j}, g|_{V_j}).$$

(5.2.3) (Homotopy invariance) If $f_t, g_t: V \to M_2$, $0 \le t \le 1$, are homotopies such that $K = \{x \in V : \text{there exists } t \in [0, 1] \text{ with } f_t(x) = g_t(x)\}$ is compact, then $i(\mathrm{Coin}(f_0, g_0)) = i(\mathrm{Coin}(f_1, g_1))$.

(5.2.4) (Multiplicativity) Let $f, g: V \to M_2$, $f', g': V' \to M_2'$ be maps. Then $\mathrm{Coin}(f \times f', g \times g') = \mathrm{Coin}(f, g) \times \mathrm{Coin}(f', g')$ and $i(\mathrm{Coin}(f \times f', g \times g')) = i(\mathrm{Coin}(f, g)) \cdot i(\mathrm{Coin}(f', g'))$.

(5.2.5) (Normalization) For a pair $f, g: M_1 \to M_2$ of globally defined maps the index $i(\mathrm{Coin}(f, g))$ is equal to the Lefschetz number $L(f, g)$ of the pair (f, g).

Let us point out that the commutativity property, which holds for fixed point theory, does not hold for coincidence theory.

(5.3) Example. Consider four maps $f, g, h, k\colon T^2 \to T^2$ from the torus into the torus which in cohomology at dimension one, induce the homomorphisms given by the matrices

$$A = \begin{pmatrix} 1 & -2 \\ 1 & 0 \end{pmatrix}, \quad B = \begin{pmatrix} 3 & -1 \\ 0 & -1 \end{pmatrix}, \quad C = \begin{pmatrix} 0 & -1 \\ 1 & 1 \end{pmatrix}, \quad D = \begin{pmatrix} 1 & 0 \\ 0 & 1 \end{pmatrix}$$

respectively. By the normalization axiom, the index of Coin is given by the Lefschetz number. In the case of maps into the torus, see [BBTP], it follows that $I(f \circ h, g \circ k; T^2) = \det(B - A \circ C) = 0$ and $I(h \circ f, k \circ g; T^2) = \det(B - C \circ A) = 2$.

We can ask about the situation for maps between manifolds that are not necessarily orientable. In 1993, Dobreńko and Jezierski in [DJ] defined an index for the case where the manifolds are not necessarily orientable. Their definition applies to a closed subset F which is a subset of a Nielsen class. The index of such a set F is an element of $\mathbb{Z}$ or $\mathbb{Z}_2$, in particular for the case of a Nielsen class. They adopted the procedure used by Hopf for the fixed point case. More precisely, working in the differentiable category, they first deform the maps so that every coincidence point becomes regular. They then define the index of an isolated coincidence point as an element of either $\mathbb{Z}$ or $\mathbb{Z}_2$ by taking into account the orientation system on the manifold given by the fundamental group. To be more explicit, a coincidence point x has index in $\mathbb{Z}_2$ if there is a loop α based in x such that $f(\alpha)$ is homotopic to $g(\alpha)$ and the loops α, $f(\alpha)$ have different signs. Later in 1997, Gonçalves and Jezierski [GJ] defined the index of an isolated set of coincidence points without any differentiability assumption. In particular, it turns out that for a finite Nielsen coincidence class F, their index coincides with that of [DJ]. More recently, an index has been defined for Nielsen coincidence classes of maps from a finite complex K into a manifold (see [G4]).

We will start by describing a notion of index of an isolated set of coincidence points. Here we basically follows the ideas from [FH2] and [G4]. We will not use the terminology "local index" to highlight the fact that there is a difference from the local index as defined in [FH2].

Let U be an open set of K and $(f, g)\colon U \to N^n$ be a pair of maps where the set of coincidence points is compact. As in [FH2], we consider the diagonal $\triangle$ in $N^n \times N^n$ and replace the inclusion $N^n \times N^n - \triangle \hookrightarrow N^n \times N^n$ by a fiber map $p\colon E \to N^n \times N^n$, where $E = \{(\alpha, \beta) : \alpha(0) \neq \beta(0)\}$, and $p(\alpha, \beta) = (\alpha(1), \beta(1))$. For $b = (x, y)$ in $N^n \times N^n$ and $F_b = p^{-1}(b)$, $\pi_{m-1}(F_b)$ is a local system of coefficients on $N^n \times N^n$. There is an isomorphism of local systems on $N^n \times N^n$

$$\zeta\colon \pi_{m-1}(F_b, b) \to \mathcal{Z}[\pi],$$

where $\pi = \pi_1(N^n, x)$ and the action of $\pi \times \pi$ on $\mathcal{Z}[\pi]$ is given by

$$(\tau, \sigma) \cdot \alpha = \operatorname{sign}(\tau)\sigma \cdot \alpha \cdot \tau^{-1}.$$

We will refer to this system as $\mathcal{B}$. Let the local system on U be that induced from by $f \times g: U \to N^n \times N^n$ from $\mathcal{B}$, and denote it by $\mathcal{B}(f \times g)$. Consider the fiber space $E(f,g)$ obtained by pulling back $p: E \to N^n \times N^n$ over U by $f \times g$.

The obstruction to deforming the pair (f,g) to a coincidence-free pair is related to the obstruction to extending sections of the fiber map $E(f,g) \to U$. Following the steps in [FH2] and making the usual adaptations to the coincidence case, we end up with:

(5.4) DEFINITION. The *coincidence index* of $(f,g): U \to N^n$ is the cohomology class *i(f,g)* in $H^n_c(U; \mathcal{B}(f \times g))$ with the property that (f,g) can be deformed by a compact homotopy to be a coincidence-free pair if and only if $\subset (f,g)$ vanishes.

Now consider an isolated set F of coincidences of (f,g), and let V be an open set of U such that $F = V \cap \operatorname{Coin}(f,g)$. Consider the composition

$$H^n(V, V-F; \mathcal{B}(f \times g)) \xrightarrow{j^{*-1}} H^n(U, U-F; \mathcal{B}(f \times g))$$
$$\xrightarrow{k^*} H^n(U, U - \operatorname{Coin}(f,g); \mathcal{B}(f \times g))$$

where the first arrow is the inverse of the excision isomorphism and the second is induced by the inclusion. Recall that $H^n_c(U; \mathcal{B}(f \times g))$ is the inverse limit of $H^n(U, U-C; \mathcal{B}(f \times g))$, where the limit is taken over all compact subsets C of U. We now define *the index of F with respect to f*, where by f we mean the function globally defined in U. We point out that this is different from the definition of local index given in [FH2].

(5.5) DEFINITION. The *coincidence index* of F with respect to f, g, denoted by *i(f,g;F)*, is defined to be the element in $H^n_c(U; \mathcal{B}(f \times g))$ given by $k^*(j^*)^{-1}(\alpha)$, where $\alpha \in H^n(V, V-F; \mathcal{B}(f \times g))$ corresponds to the coincidence index of (f,g): $V \to N^n$.

The index satisfies the properties (5.2.1)–(5.2.5).

The proofs of these properties are straightforward. The first three were basically proved as Definition–Proposition 2.7, Propositions 2.10 and 2.8, respectively, in [FH2].

The following example from [G4], explains the reason for defining the index of an isolated set F in a slightly different way to that of local index. However, for orientable manifolds, the two definitions lead to the same result. More precisely, the local index is an integer, and the index of F is that integer times the fundamental top cohomology class.

(5.6) EXAMPLE. Let us consider the projective plane $\mathbb{R}P^2$ as the disk $D \subset \mathbb{R}^2$ of radius 1 under the relation $x \sim -x$ for $x \in S^1 \subset D$. Call $X \subset \mathbb{R}P^2$ the open disc of radius 1/2. Let $f\colon \mathbb{R}P^2 \to \mathbb{R}P^2$ be the map given by:

$$f(x) = \begin{cases} x^2 & \text{if } ||x|| \leq 1/2, \\ x^2/(4||x||^2) & \text{if } 1 \geq ||x|| \geq 1/2. \end{cases}$$

This is a well-defined map on $\mathbb{R}P^2$. Now let f_1 be the restriction of f to X. Let us consider the constant map with domain RP^2 or X. In both cases we denote the function by c, where $c(x) = \overrightarrow{0}$. Now one can consider $\mathrm{Coin}(f, c)$ and $\mathrm{Coin}(f_1, c)$. It is easy to see that $\mathrm{Coin}(f, c) = \mathrm{Coin}(f_1, c) = \overrightarrow{0}$. But $I(\overrightarrow{0}, (f, c))$ is not equal to $I(\overrightarrow{0}, (f_1, c))$: the first is an element of $\mathbb{Z}_2$, zero in fact, while the second is the integer 2.

Once we have defined a notion of index of a Nielsen class, one may define the notion of an essential Nielsen coincidence class with respect to this index. Another type of essential class can be defined in the same fashion as in [Br1], independent of index.

(5.7) DEFINITION. A Nielsen coincidence class is called *essential* if its index is different from zero. The number of essential Nielsen classes is called the *Nielsen coincidence number* and is denoted by $N(f_1, f_2)$.

In the rest of this section, we summarize some results about the Nielsen coincidence number. By this number we mean the number defined with respect to the index which is the integral index if we have maps between orientable manifolds or is the index defined in [DJ], which is either $\mathbb{Z}$ or $\mathbb{Z}_2$, for maps between not necessarily orientable manifolds. So we will assume that f, g are a pair of continuous maps between two closed manifolds (not necessarily orientable) of the same dimension.

(5.8) THEOREM. *The Nielsen number is a homotopy invariant, and*

$$N(f, g) \leq \mathrm{MC}[f, g].$$

The given inequality shows that the converse to the homotopical Lefschetz–Hopf Theorem 2.3 can be separated into two independent steps, each of which has an independent interest and allows a quantitative statement.

Step 1. To find the conditions, allowing a connection between the numbers $L(f, g)$ and $N(f, g)$. (To find the conditions when $L(f, g) = 0$ implies $N(f, g) = 0$.)

Step 2. To find the conditions when the equalities $\mathrm{MC}[f, g] = N(f, g)$ hold. (To find the conditions when $N(f, g) = 0$ implies $\mathrm{MC}[f, g] = 0$.)

Step 2 remains as a main open question for maps between surfaces, as well in the fixed point case. For dimension greater or equal to 3, the question was solved in 1955 by Schirmer (see [Sc]) for orientable manifolds, and in 1993 by Dobreńko and Jezierski (see [DJ]) for nonorientable manifolds.

(5.9) THEOREM. *Let $f, g: M \to N$ be maps between closed orientable manifolds of the same dimension ≥ 3. Then* $\mathrm{MC}[f,g] = N(f,g)$.

These results suggest the following definition:

(5.10) DEFINITION. A pair $(f,g): X \to Y$ is said to have the *Wecken property* if $\mathrm{MC}[f,g] = N(f,g)$.

An important algebraic tool was defined in 1936 by Reidemeister in [R]. For a map $f: X \to X$ of a finite simplicial complex X, he considered a lifting $\widetilde{f}: \widetilde{X} \to \widetilde{X}$ of f to the universal cover $\widetilde{X}$ and the induced homomorphism $(\widetilde{f})_*$ on the chain complex $C_*(\widetilde{X})$ of $\widetilde{X}$ considered as a $\mathbb{Z}[\pi]$-module. Here π is the fundamental group of the space X, isomorphic to the group of covering transformations on $\widetilde{X}$. Using traces, Reidemeister defined an element, the so-called *Reidemeister trace*, in the quotient of $\mathbb{Z}[\pi]$ by a certain relation. The quotient by this relation is the sum of $\mathbb{Z}$'s indexed by the Reidemeister classes, and the Reidemeister trace completely determines the Nielsen classes as follows: every non-vanishing coefficient in the Reidemeister trace corresponds to exactly one essential Nielsen class whose index is the value of the coefficient and vice-versa. This provides an alternative way to look at the Nielsen number. A good introduction to the theory of Nielsen and Reidemeister numbers is given in the books of B. Jiang and T. H. Kiang (see [Ji4] and [Ki]).

It would be nice to have the analog of the Reidemeister trace for coincidences.

In 1993 in [DJ] and 1997 in [G4], Theorem (5.9) was extended to non-orientable manifolds.

(5.11) THEOREM. *If $f, g: M \to N$ are maps between closed manifolds of the same dimension n, $n \geq 3$. Then* $\mathrm{MC}[f,g] = N(f,g)$.

The above results show the relevance of the Nielsen coincidence number. We will now try to calculate $N(f,g)$. Conditions on the space are found, allowing one to calculate (estimate) the index of a Nielsen class, Nielsen number, Reidemeister number and Lefschetz–Hopf number. Based on Theorem (4.14) we have:

(5.12) THEOREM. *Let M and N be closed manifolds of the same dimension, and in addition let (M, N) be a Jiang pair for coincidences. If $(f,g): M \to N$ is a pair of maps then either $N(f,g) = 0$, or $N(f,g)$ is the Reidemeister number and all the classes have the same index. Hence the classes are all defective or else all not defective. Further if N is a Jiang space ($J(\mathrm{id}) = \pi_1(Y)$) and $N(f,g) \neq 0$, then $N(f,g) = [\pi_1(N) : (f_\# - g_\#)(\pi_1(M))]$; in this case the index $[\pi_1(N) : (f_\# - g_\#)$ $(\pi_1(M))]$ is finite.*

For the special case of self-maps of the n-dimensional torus T^n, Theorem (5.12) can be considerably strengthened, thanks to the following theorem from [BBPT]:

(5.13) THEOREM. *For a map $f: T^n \to T^n$, all Nielsen classes of f have the same index $+1$, -1 or 0 and the absolute value of the Lefschetz number is equal to the Nielsen number. Furthermore, if the Lefschetz number is different from zero then its absolute value is equal to the absolute value of the determinant of $M - I$ which coincides with the number $|R(f)|$ of Reidemeister classes. Here M and I denote the matrix induced by f on $H_1(T^n, \mathbb{Z})$ and the identity matrix, respectively.*

The above result has been generalized in several directions. One is that of coincidences of maps between homogeneous spaces of the same dimension. This includes tori and nilmanifolds. The reader may find more about this elsewhere including [An], [FH4], [G3], [G5] and [Wo]. The second direction is that of the possible values of the indices of Nielsen classes. In the next section, we shall describe results when the two manifolds are surfaces. Note that the previous result only covers the case where the surface is the two dimensional tori.

To conclude this section we may consider our three main questions stated in Section 2, where we replace $\mathrm{Coin}(f, g)$ by a Nielsen coincidence class. With this formulation, these questions seem very natural to us, and suitable for the study of coincidences in positive codimension, a topic which shall be treated in Section 7.

6. Coincidences on surfaces and equations in braid groups

Coincidence point problems in dimension 2 are in some sense more complicated than those in dimensions ≥ 3 (see [Ji2], [I], [Ji3], [Ji6]). Intuitively this happens because in small dimensions we do not have enough freedom to remove coincidences. So we cannot expect to have the Wecken property. In this section we develop results for coincidence on surfaces, taking into account the fact that we are in dimension 2. We start by describing formulae which enable us to compute in an effective way the Lefschetz–Hopf trace formula. This will be done for maps between closed surfaces that are not necessarily orientable. For this purpose, as in Section 3, we will assume that at least one of the maps is orientation true. These formulas were obtained in [GO]. As usual, this will give a sufficient condition for the existence of a coincidence. Then we consider the problem of estimating $\mathrm{MC}[f_1, f_2]$. As in the fixed point case this is difficult and very little is known. We explore a relation between the coincidence problem and a certain algebraic problem about equations in braid groups. This is the suitable replacement for obstruction theory in the case of coincidences in dimension greater than or equal to three. Despite the complexity of the algebraic problem in the braid groups, we may give some applications. We show that if we compose a pair of maps between surfaces with a pinching map, we obtain a pair which has the Wecken property. We will not explore examples of pairs which do not satisfy the Wecken property, but we describe one situation where the property holds. This corresponds to self-coincidence i.e. (f, f) for f a map between surfaces. More precisely, we can deform the pair to (f_1, f_2) such that $\mathrm{Coin}(f_1, f_2)$ is either empty or has a single point. At the end of the section we study the problem of the possible values of the indices of the coincidence Nielsen classes among all pairs (f_1, f_2) such that the degrees of f_1 and f_2 are two fixed values. To avoid confusion with the genus g of a surface, in this section we denote a pair of maps by f_1, f_2, instead of f, g as in the other sections.

6.1. Lefschetz–Hopf trace formula. We start by describing formulae which enable us to compute in an effective way the Lefschetz-Hopf trace formula. This will be done for maps between closed surfaces not necessarily orientable. These formulae were obtained in [GO]. We first describe the formulae in the orientable case, then the case where the domain is a connected sum of Klein bottles and the target is the Klein bottle. For the other cases, the reader will find more details in [GO].

Following [GO], we start by considering the orientable case. Let T_g be the orientable surface of genus g, where $g \geq 1$. If $\mathbb{Q}$ is the field of rational, let $\alpha_1, \ldots, \alpha_{2g-1}, \alpha_{2g}$ be the generators of $H^1(T_g; Q)$ so that:

$$\alpha_i \cup \alpha_j = \begin{cases} \alpha_1 \cup \alpha_2 & \text{for } i = 2r-1 \text{ and } j = 2r,\ r = 2, \ldots, g, \\ 0 & \text{otherwise.} \end{cases}$$

Let $f_1 \colon T_h \to T_g$ be a map, where $h \geq 1$. The map f_1 induces $f_1^* \colon H^1(T_g; Q) \to H^1(T_h; Q)$. For $u = 1, \ldots, 2g$ let $f_1^*(\alpha_u) = \sum_{v=1}^{2h} a_{vu}\beta_v$, where β_j, $1 \leq j \leq 2h$ are the generators of $H^1(T_h; Q)$. Let $A = (a_{ij})$ be the matrix associated with f_1^*. Observe that it can be written as:

$$A = \begin{pmatrix} A_{1,2}^{1,2} & A_{1,2}^{3,4} & \cdots & A_{1,2}^{2g-1,2g} \\ A_{3,4}^{1,2} & A_{3,4}^{3,4} & \cdots & A_{3,4}^{2g-1} \\ \vdots & \vdots & \ddots & \vdots \\ A_{2h-1,2h}^{1,2} & A_{2h-1,2h}^{3,4} & \cdots & A_{2h-1,2h}^{2g-1,2g} \end{pmatrix},$$

where

$$A_{2m-1,2m}^{2l-1,2l} = \begin{pmatrix} a_{2m-1,2l-1} & a_{2m-1,2l} \\ a_{2m,2l-1} & a_{2m,2l} \end{pmatrix}.$$

If $f_2 \colon T_h \to T_g$ is another map and $B = (b_{ij})$ is the matrix associated with $f_2^* \colon H^1(T_g; Q) \to H^1(T_h; Q)$, from [GO] we have:

(6.1) THEOREM. *The Lefschetz coincidence number of the pair (f_1, f_2) is given by:*

$$\Lambda(f_1, f_2) = \sum_{i=1}^{g} \sum_{r=1}^{h} \det(B_{2r-1,2r}^{2i-1,2i} - A_{2r-1,2r}^{2i-1,2i}).$$

Let $f_1, f_2\colon S_1 \to S_2$ be two maps between compact surfaces (not necessarily orientable) where f_1 is orientation true i.e. $\mathrm{sign}(\alpha)\cdot\mathrm{sign} f_{1\#}(\alpha) = 1$ for $\alpha \in \pi_1(S_1)$. We will compute $L(f_1, f_2)$. In order to do this, we will need to know $f_1^*\colon H^1(S_2;\Gamma_2) \to H^1(S_1;\Gamma_1)$ for Γ the trivial local coefficient system $\mathbb{Q}$, and the coefficient system given by the local orientation with rational coefficients, denoted by Γ_{S_i}. We will present the details for the case where S_2 is the Klein bottle and state the results for the other cases. Since f_1 is orientation true, it follows that f_1 induces a homomorphism in cohomology with the local coefficient system given by local orientation (compare with Section 4). The reader may find more details about cohomology with local coefficients in [Wh].

Now let K_g be the surface which is the connected sum of T_g and one copy of the Klein bottle, i.e. $K_g = T_g \# K$. Let $\overline{\alpha}$ and $\overline{\beta}$ be generators of $H_1(K,\mathbb{Z})$ which arises from the description of K as the square with the identifications given by the relation $\alpha\beta\alpha\beta^{-1}$.

We describe a basis for $H^*(K_g;\mathbb{Q})$ and the dual pairing basis in $H^*(K_g;\Gamma_{K_g})$. Perhaps it is worthwhile to mention that the cocycle dual to $\overline{\beta}$ (in the sense of $\mathrm{Hom}(\,\cdot\,,\mathbb{Q})$) survives as a non-trivial class in $H^1(K_g;\mathbb{Q})$ but the cocycle dual to $\bar{\alpha}$ does not. On the other hand, the dual of $\overline{\beta}$ is not even a cocycle in $H^1(K_g;\Gamma_{K_g})$ but the dual of $\overline{\alpha}$ is a cocycle which survives as a non-trivial element of $H^1(K_g;\Gamma_{K_g})$. So let $\{\widetilde{a}_1, \widetilde{b}_1, \dots, \widetilde{a}_g, \widetilde{b}_g, \widetilde{\alpha}\}$ be a basis for $H^1(K_g;\Gamma_{K_g})$, where $\widetilde{\alpha}$ is the dual of $\overline{\alpha}$ and the others have already been defined. The dual pairing basis of this given basis is $\{b_1, -a_1, \dots, b_g - a_g, \beta\}$.

Now, we will consider the case where the first surface is K_g. Then:

$$\begin{aligned} f_1^*(\widetilde{\alpha}) &= \sum \lambda_i \widetilde{a}_i + \xi_i \widetilde{b}_i + \lambda \widetilde{\alpha}, \\ f_1^*(\beta) &= \sum m_i a_i + n_i b_i + m\beta, \\ f_2^*(\beta) &= \sum r_i a_i + s_i b_i + r\beta. \end{aligned}$$

(6.2) THEOREM. *The Lefschetz coincidence number of the pair (f_1, f_2) is given by:*

$$L(f_1, f_2) = \sum_{i=1}^{g} \det \begin{pmatrix} \lambda_i & \xi_i \\ m_i & n_i \end{pmatrix} + \lambda(m - r) - \sum_{i=1}^{g} \det \begin{pmatrix} \lambda_i & \xi_i \\ r_i & s_i \end{pmatrix}.$$

6.2. Minimal number of coincidence points and equations in the braid groups. We now describe an algebraically equivalent way to obtain $\mathrm{MC}[f_1, f_2] = 0$, or to be able to deform the pair to be coincidence free. This will be done in terms of a solution of certain equations in the braid group. It is more subtle to interpret algebraically the minimal number of coincidence points in the homotopy class of the pair when $\mathrm{MC}[f_1, f_2] \neq 0$. We remark that we do not expect to have the Wecken property for coincidences between surfaces. A similar approach for fixed points has been shown to be useful. For example, see the papers [FH3], [Ji5]–[Ji7], [G1], [GF].

This generalization of the algebraic approach for fixed points to coincidences was carried out in [Je2], [G2] and [BGZ2]. Only the case of maps between orientable closed surfaces will be considered here, the remaining cases being similar. In order to state the result let us fix some notation. Let S_h, S_g be orientable compact surfaces of genus h, g respectively, and let $\overline{e}_1, \ldots, \overline{e}_{2h}$ be a fixed basis of $\pi_1(S_h - \operatorname{int}(D), x_0)$, where D is a closed disk having the base point of S_h in its boundary. Let us assume that the chosen basis satisfies the relation $\prod_{i=1}^{h}[e_{2i-1}, e_{2i}] = [T]$, where $[T]$ is the element given by the boundary of the disk (see [Ji7, p. 125]). Denote by $e_1, \ldots, e_{2h}$ the image of $\overline{e}_1, \ldots, \overline{e}_{2h}$ in $\pi_1(S_h)$. Suppose we are given two maps $f_1, f_2 \colon S_h \to S_g$, where we have $f_{1\#}(e_i) = w_i$, $f_{2\#}(e_i) = v_i$, $i = 1, \ldots, 2h$, and w_i, v_i belong to $\pi_1(S_g, y_1), \pi_1(S_g, y_2)$ respectively, and $f_1(x_0) = y_1 \neq y_2 = f_2(x_0)$. Following [FH3, Section 4], we have two subgroups of $\pi_1(S_g \times S_g - \Delta, (y_1, y_2))$ generated by $(\rho_{1,1}, \rho_{2,1}, \ldots, \rho_{2g-1,1}, \rho_{2g,1})$, $(\rho_{1,2}, \rho_{2,2}, \ldots, \rho_{2g-1,2}, \rho_{2g,2})$, respectively, which we denote by F_1, F_2, respectively. Let W_i, V_i be elements of F_1, F_2 which project onto w_i, v_i, respectively under the map $\pi_1(S_g \times S_g - \Delta, (y_1, y_2)) \to \pi_1(S_g \times S_g, (y_1, y_2))$ induced by inclusion. Let $N(f_1, f_2) = r$, and let $k_1, \ldots, k_r$ be the indices of the essential Nielsen classes. Also, let $\{1, \alpha_2, \ldots, \alpha_r\}$ be a set of representatives of the Reidemeister classes of (f_1, f_2) corresponding to the essential Nielsen classes. Let the class defined by α_i have index k_i and the class defined by 1 have index k_1. (For the definition of the local coincidence index, see Section 4). We choose a base point in the class defined by 1.

From [Je2] we obtain:

(6.3) THEOREM. *The pair $(f_1, f_2)\colon S_h \to S_g$ can be deformed to a pair (f_1', f_2') with k coincidence points with non-vanishing indices $i_1, \ldots, i_k \in \mathbb{Z}$ if, and only if, the equation*

$$\prod_{i=1}^{h}[\vartheta_{2i-1}W_{2i-1}V_{2i-1}, \vartheta_{2i}W_{2i}V_{2i}] = (B)^{i_1}(B^{\alpha_2})^{i_2}\ldots(B^{\alpha_k})^{i_k}$$

has a solution $\vartheta_j \in N$, $j = 1, \ldots, 2h$, for some set $\{1, \alpha_2, \ldots, \alpha_k\}$ of words in $\pi_1(S_g \times S_g \setminus \Delta)$.

In [Ji7] and [Je2], the calculation of $MF[f]$ and $\mathrm{MC}[f_1, f_2]$, respectively, is reformulated as an algebraic optimization problem using a suitable quadratic equation in the special braid group on two strings. A criterion to decide whether a pair (f, g) satisfies the Wecken condition is given in [G2] or [BGZ2]:

(6.4) THEOREM. *The pair* $(f_1, f_2)\colon S_h \to S_g$ *has the Wecken property if and only if the following conditions are fulfilled.*

(6.4.1) *If* $N(f_1, f_2) = 0$, *there is a solution* $\beta_j \in N$, $j = 1, \dots, 2h$, *of the equation*

$$\prod_{j=1}^{h} [\beta_{2j-1} W_{2j-1} V_{2j-1}, \beta_{2j} W_{2j} V_{2j}] = 1.$$

(6.4.2) *If* $N(f_1, f_2) = k \neq 0$ *and* $i_1, \dots, i_k$ *are the indices of the Nielsen classes, there is a solution* $\beta_j \in N$, $j = 1, \dots, 2h$, *of the equation*

$$\prod_{j=1}^{h} [\beta_{2j-1} W_{2j-1} V_{2j-1}, \beta_{2j} W_{2j} V_{2j}] = (B)^{i_1} (B^{\alpha_2})^{i_2} \dots (B^{\alpha_k})^{i_k}$$

for some set $\{1, \alpha_2, \dots, \alpha_k\} \subset F_1$ *where* $\{p_{1\#}(\alpha_j) : j = 1, \dots, k\}$ *are representatives of the Reidemeister classes corresponding to the essential Nielsen classes. (The free group* F_1 *is as defined above.)*

The following application of this result is from [G2]:

(6.5) THEOREM. *For a pair of maps* $(f_1, f_2)\colon S_h \to S_g$, *there exists an integer* n *such that* $\mathrm{MC}[f_1 \circ p, f_2 \circ p] = N(f_1 \circ p, f_2 \circ p)$, *where* $p\colon S_{n+h} \to S_h$ *is a pinching map. The number* $N(f_1 \circ p, f_2 \circ p) = N(f_1, f_2)$ *does not depend on* n.

(6.6) CONJECTURE. $\mathrm{MC}[f_1 \circ p, f_2 \circ p] = \max\{N(f_1, f_2), \mathrm{MC}[f_1, f_2] - 2\}$ *for a given pair of maps* $f_1, f_2\colon S_h \to S_g$ *and a pinching map* $p\colon S_{h+1} \to S_h$.

This conjecture is true for $g = 0, 1$ and, for general g, for a constant map f_2.

For mappings into the sphere $S_0 = S^2$ this statement can be formulated in a stronger form [BGZ2]:

(6.7) PROPOSITION. *For mappings* $f_1, f_2\colon S_h \to S_0 = S^2$, *the following conditions are equivalent:*

(6.7.1) $\mathrm{MC}[f_1, f_2] = 0$;
(6.7.2) $\deg(f_1) + \deg(f_2) = 0$.

If $\mathrm{MC}[f_1, f_2] \neq 0$ *then* $\mathrm{MC}[f_1, f_2] = 1$.

We conclude this subsection with some questions concerning the Wecken property for coincidences.

(6.8) QUESTION. Let us consider a pair of maps $f, g\colon W_1 \to W_2$ between closed surfaces where both maps are coverings. Does the Wecken property hold? The answer is "yes" if the maps are homotopy equivalences. In the latter case this question was raised in 1927 by J. Nielsen (see [N]), i.e. whether $MF[f] = N(f)$ for maps on surfaces in the fixed point case. If f is a homotopy equivalence Ivanov (see [I]) and Jiang (see [Ji3]) gave independently a positive answer. It is easy to see that the fixed point problem is the same as the coincidence problem if the identity is replaced by a homeomorphism.

(6.9) QUESTION. A similar question when only one of the maps is a covering.

(6.10) QUESTION. The similar question when neither map is a covering. One might even consider the particular case where each of the two functions has a curve as its image.

6.3. The Wecken property for self-coincidences. Now we present one situation where the Wecken property holds. The proof will illustrate a method which has also been used successfully for coincidences in positive codimension, at least for self-coincidences. Consider a map $f\colon W_1 \to W_2$ between closed surfaces and the pair (f, f). Let $A(f)$ denote the absolute degree of f (see [Ep], [BSc3]).

(6.11) THEOREM. *If* $A(f) \cdot \chi(W_2) \neq 0$ *the map* f *can be deformed to a map* f' *such that* $\mathrm{Coin}(f, f')$ *has only one point. In the case of equality there exists* f' *such that* $\mathrm{Coin}(f, f') = \emptyset$.

PROOF. If $A(f) \cdot \chi(W_2) = 0$ then either $A(f) = 0$ or $\chi(W_2) = 0$. If $A(f) = 0$ then by [Kn] it follows that we can deform f to a map f' which is not surjective. Suppose that $y_0 \in W_2$ is not in the image, and consider a vector field on W_2 which vanishes at most at y_0. Making a small deformation of f' along this vector field, we obtain f'' such that $\mathrm{Coin}(f', f'') = \emptyset$. The result follows from Proposition (2.12). If $\chi(W_2) = 0$ then the argument is similar and simpler. We consider a vector field in W_2 which is nowhere zero.

Suppose that $A(f) \cdot \chi(W_2) \neq 0$. Consider the pullback of the tangent bundle which is a 2-line bundle over W_1. Let x_0 be a point in the preimage of y_0 and consider the bundle restricted to $W_1 - \{x_0\}$. We claim that this bundle admits a vector field which vanishes at most at x_0. For each fixed first Stiefel–Whitney characteristic class, we know from [Hi] that the number of equivalent vector bundles are in one-to-one correspondence with $H^2(W_1 - \{x_0\}; \mathbb{Z})$ with twisted coefficients. Since the space $W_1 - \{x_0\}$ has the homotopy type of a bouquet of circles, this latter group is zero and so we have only one such bundle. Over each circle there are only two possibilities for the bundle. The first is if the first Stiefel–Whitney class is trivial, so the bundle is trivial and the result follows immediately. The second possibility is when the first Stiefel–Whitney class is non-trivial. In this case, the bundle is obtained from the product $[0, 1] \times \mathbb{R}^2$ by identifying $(0, x, y) \sim (1, x, -y)$. But this bundle admits a nowhere-zero vector field, and as before we can construct a nowhere-zero vector field over $W_1 - \{x_0\}$. We can make a deformation of the vector field keeping it nowhere zero, such that it extends continuously to x_0 by defining it to be the zero vector at x_0. Now define the map g from the vector field $V(x)$ constructed above. For $x \in W_1$ define $g(x)$ to be $\exp(f_*(V(x))$ where exp is the exponential map from the tangent bundle of W_2 to W_2, and f_* is the bundle map. It is clear that the number of coincidence points is at most one and the result follows. □

6.4. Index bounds for Nielsen coincidence classes. Now we study the values of the index of a coincidence Nielsen class for a pair of maps between surfaces. We briefly recall some facts about the fixed point case. First, from [BBPT] observe that for a pair of self maps in the torus the index of a Nielsen class is either 1 or -1. In particular it is bounded. On the closed orientable surface S_g of genus $g > 1$, by Thurston's classification of diffeomorphisms of surfaces, it was proved in [GuJi] that for any diffeomorphism $f\colon S_g \to S_g$, the index of any Nielsen fixed point class $\mathbb{C}$ is bounded, and the bounds depend only on the genus of the surface. More precisely we have that $3 - 4g \leq \operatorname{ind}(\mathbb{C}) \leq 1$. This result has been extended in [Ji8] to arbitrary maps $f\colon S_g \to S_g$ not necessarily homotopic to a diffeomorphism. For the surfaces S^2 and $\mathbb{R}P^2$, the indices are unbounded. Now we consider the coincidence case and we follow [GJi] to describe the results of maps between orientable surfaces.

Given integers g, h, d_1, d_2, consider any pair of surface maps $(f_1, f_2)\colon S_h \to S_g$ with $|\deg(f_1)| = d_1$ and $|\deg(f_2)| = d_2$, and consider the index of any Nielsen coincidence class $\mathbb{C}$. Are these indices bounded? In the case that the answer is "yes", one would like to find such bounds.

More precisely, we define

$$B(g, h, d_1, d_2) := \sup\{|\operatorname{ind}(\mathbb{C}) : |\deg(f_1)| = d_1,\ |\deg(f_2)| = d_2\},$$

where the supremum is taken over all coincidence classes $\mathbb{C}$ of all pairs of maps $(f_1, f_2)\colon S_h \to S_g$ with the given degrees. It is an integer or infinite.

Clearly $B(g, h, d_2, d_1) = B(g, h, d_1, d_2)$. So we may assume $d_1 \leq d_2$.

By Kneser's formula (see [Kn] or [ZVC]) $h - 1 \geq d_i(g-1)$ when $d_i \neq 0$.

By [JiGu], if $g = h > 1$ (hence $d_i \leq 1$), then $B(g, g, 1, 1) \leq 4g - 3$.

By [Ji8], if $g = h > 1$ (hence $d_i \leq 1$), then $B(g, g, 0, 1) \leq 4g - 3$.

By [BBPT] $B(1, 1, d_1, d_2) = 1$.

We start by describing the results where the index is bounded. The first result is about $B(g, g, 1, 1)$. This means that the second map $f_2\colon S_g \to S_g$ has degree ± 1, hence, up to homotopy, can be assumed to be a homeomorphism. By routine arguments, we know that the inclusion $\operatorname{Coin}(f_1, f_2) \to \operatorname{Fix}(f_2^{-1} \circ f_1)$ is a homeomorphism which preserves the indices of the Nielsen classes, and $|\deg(f_1)| =$

$|\deg(f_2^{-1} \circ f_1)|$. So the coincidence question is equivalent to the fixed point question.

(6.12) THEOREM. $B(g, g, 1, 1) = 1 - \chi(S_g) = 4g - 3$.

The proof of the above result contains two parts. The first is to show that $B(g, g, 1, 1) \leq 1 - \chi(S_g) = 4g - 3$. This is done in [Ji8]. The second part is obtained by constructing an example which contains a Nielsen class whose index is $4g - 3$. This is Example 2 in [GJi].

For $B(g, g, 1, 0)$ we have:

(6.13) THEOREM. $B(g, g, 1, 0) = 1 - \chi(S_g) = 2g - 1$.

The inequality $B(g, g, 1, 0) \leq 1 - \chi(S_g) = 2g - 1$ follows from [GJi, Proposition 5] and the fact that the inequality is sharp follows from [GJi, Example 4].

Suppose $h > g > 1$ and $p: S_h \to S_g$ is a covering map. Then $h - 1$ is a multiple of $g - 1$. Set $m = (h-1)/(g-1)$. Let $f: S_h \to S_g$ be a map, and let $\mathbb{C}$ be an arbitrary Nielsen class of the pair (f, p). Consider its index $\operatorname{ind}(\mathbb{C}) = \operatorname{ind}(\mathbb{C}; f, p)$. Are such indices bounded?

We seek to derive boundedness results from those in fixed point theory.

Take a point $x_0 \in \mathbb{C}$ to be the base point in S_h, and take $\overline{x}_0 := f(x_0) = p(x_0)$ to be that in S_g. For the sake of brevity, we shall write $G = \pi_1(S_g)$ and $H = \pi_1(S_h)$. Thus $p_\pi: H \to G$ is injective.

(6.14) PROPOSITION. *Assume that there exists a subgroup $K \subset H$ such that*

$$[H : K] = n < \infty \quad \text{and} \quad f_\pi(K) \subset p_\pi(K). \tag{6.14.1}$$

Then $|\operatorname{ind}(\mathbb{C})| \leq |2n\chi(S_h) - 1|$.

In many situations we do not have bounds on the indices.

(6.15) THEOREM. $B(g, h, d_1, d_2) = \infty$ *in the following two cases:*

(6.15.1) *When $g = 1$ and for $h > 1$.*

(6.15.2) *When $2 \leq g \leq h$ and $0 \leq d_1, d_2 < (h-1)/(g-1)$, hence neither f_1 nor f_2 is homotopic to a covering map.*

The proof of the above result is obtained by constructing maps which have sufficiently large index. Case (6.15.1) is simpler than case (6.15.2). The above results lead to the interesting question which is to know what happens when one of the functions is a covering. An even more general question can be stated thus: Let $p: Y \to X$ be a finite covering, and assume that X and Y have the property that the index of the Nielsen fixed point classes is bounded. What can we say about the index of the coincidence classes when one of the maps is p? For the case where the target is the sphere S^2, we see that $B(0, h, d_1, d_2) = d_2 - d_1$ because S^2 is simply connected and there is only one homotopy class of maps of each degree.

7. Positive codimension

Coincidence theory in positive codimension is not just a generalization of coincidence theory in codimension zero. Several natural situations occur which are related to coincidences in positive codimension: for example the intersection of two submanifolds the sum of whose dimensions is greater than the dimension of the manifold, see [HQ] and [Mc]. It is also related to the problem of removing fixed points of a given homotopy class of a map f, see [DiG] and [Je1]. Another example is that a coincidence problem of maps into nilmanifolds (regarding the spaces as CW-complexes of the same dimension) is closely related to a coincidence between nilmanifolds of different dimension, see [GW]. Now we describe in more detail the development of this subarea. With a few exceptions, the spaces will be closed manifolds.

Let $f, g\colon M^m \to N^n$ be two maps between closed manifolds. To make (f, g) coincidence free amounts to being able to deform f and g to f' and g', respectively, such that the intersection of the diagonal Δ_N of N and the graph of $f' \times g'$ is empty. Equivalently, it is the same as deforming $f \times g$ into the subspace $N \times N - \Delta_N$. This approach to the converse Lefschetz theorem was first studied by Fuller (see [Fu]) via obstruction theory in the case where N is simply connected. Subsequently, obstruction theory has been used to study fixed points and coincidences, even in codimension zero, see [FH1]–[FH3], [G4], [BG1], [BG2]. It turns out that obstruction theory is a suitable tool, not only to detect coincidence points but also to study the number $\mathrm{MC}[f, g]$. Relevant references in the positive codimension case include those by Wyler ([Wy]), Hatcher and Quinn ([HQ]), Dimovski and Geoghegan ([DiG]). There has been more recent progress in this area.

Coincidence problems studied via obstruction theory can be divided in two types. The first is when the primary obstruction does not vanish. Then one may use classical algebraic topology to study Coin. The second type is when the primary obstruction vanishes, and one has to look for higher obstructions. It is a very subtle question to detect a higher obstruction. Classical obstruction theory does not provide an effective method to make such computations. The level of difficulty of the general problem, and the existence of some machinery to deal with obstructions in a certain range, has suggested that one should consider the problem in two subcases: the first is called "in the stable range", namely the situation where the $\dim M^m < 2 \dim N^n - 2$; the second is called "the unstable case". Other branches of homotopy theory including bundle theory have treated such questions. One particular case of the general problem is the "self-coincidence problem" namely the case where the two maps are equal, and results have been obtained. This is called the deformation case.

We now describe some recent progress of coincidence questions in positive codimension. We start with the case where the primary obstruction is non-zero, and then go on to consider the complementary case.

In [DoG] this question was basically analyzed for (p, p), where $p: E \to B$, is a Hopf fibration. A general homotopy approach is used to study this question, and using bundle theory, one shows for which values of n the Hopf fibrations $p_n: S^{4n+3} \to HP^n$ can be made self-coincidence free.

Also in [GR] the same problem has been analyzed for maps $p: E \to S^n$ where E is the total space of a sphere bundle over S^n.

The deformation problem has a natural variation which should have interest for analisys. Namely when it is possible to make the pair (f, f) coincidence free by a small deformation (see the definition of small deformation before Proposition (7.3)). In [GR] a family of examples is given where such a small deformation is not possible. In [Ko], Koschorke studied this problem in the stable case in a very efficient way by means of singular normal bordism. As in the classical case, for the coincidence set, one associates an index which is an element of a certain singular normal bordism group. As long as we are in the stable range, this index is very effective.

The systematic study of coincidence theory in positive codimension, is little by little becoming more of a reality. The many connections of this problem with several branches of geometric topology, homotopy theory and classical algebraic topology, has made the subject quite fascinating.

7.1. Positive codimension and primary obstruction. More recently, the primary obstruction for deforming a pair of maps $f, g: M^m \to N^n$ to be coincidence free was studied in [GJZ] and [GW2]. See also [GW3] for related work. We will assume that the manifolds are orientable, although it should not be necessary. The primary obstruction to deforming f and g on the n-th skeleton of M off the diagonal ΔN is given by a class $o_n(f, g) \in H^n(M, Z[\pi])$ which is the pullback of the twisted Thom class of N. Here $H^n(M, Z[\pi])$ is the cohomology of the manifold M with local coefficients $Z[\pi]$ for $\pi = \pi_1(N)$. If we use a diagram similar to that given in Example (2.17), but with cohomology with local coefficients, we can easily obtain:

(7.1) Proposition. *If the primary obstruction does not vanish then*

$$H^n(M, M - \mathrm{Coin}(f', g'), Z[\pi]) \neq 0$$

for any pair of maps f', g' homotopic to f, g, respectively.

This will give some restrictions on the size of $\mathrm{Coin}(f', g')$.

Following [GJZ], let us assume that $f \times g: M \to N \times N$ is transversal to the diagonal. Then $\mathrm{Coin}(f, g)$ is a submanifold. The main result of [GJW] says:

(7.2) THEOREM. *The coincidence set* Coin(f, g) *determines the homology element dual to the primary obstruction such that*

$$D_M^{-1}(o_n(f, g)) = [z_{C(f,g)}^{\pi^*}] \in H_{m-n}(M; \pi^*).$$

The above result together with some other ideas have been successfully used to study coincidence between nilmanifolds in [GW2].

7.2. Positive codimension and higher obstructions. Obstruction theory is a suitable language to formulate Question (2.2), namely when a pair of maps may be deformed to be coincidence free. But it is not an effective tool fo making explicit calculations. In [DG], the coincidence problem was studied under the hypothesis that the two maps are the same. We call this the *self-coincidence* case. A homotopy-theoretic approach was developed to analyze the self-coincidence problem and some concrete calculations were made. Another question considered for this special case of self-coincidence was to know when a small deformation is possible. More precisely, we say that (f, f) can be deformed to be *coincidence free by small deformation*, if given $\epsilon > 0$ there is an ϵ-homotopy of f to a map f' such that (f, f') is coincidence free. The following three results are from [DG]. Let $f: M^m \to N^n$ be a continuous map (M^m arbitrary), and N^n a compact differentiable manifold. Let τ_N be the tangent bundle of the differentiable manifold N, $S(\tau_N)$ the sphere bundle and $q: S(\tau_N) \to N$ the projection map. We have:

(7.3) PROPOSITION. *The map $f: M^m \to N^n$ admits a lift to $S(\tau_N^n)$ if and only if (f, f) can be deformed to be coincidence free by a small deformation.*

Suppose that $f: M^m \to N^n$ is a map (N^n is a smooth manifolds). The horizontal bundle with respect to the map f, denoted by $f^*(\tau_N^n)$, is the pullback of the tangent bundle of N^n by f. Then we have:

(7.4) PROPOSITION. *If $f: M^m \to N^n$ is a map where N^n is a smooth manifold then the map $f: M^m \to N^n$ admits a lift to $S(\tau_N^n)$ if and only if the horizontal tangent bundle $f^*(\tau_N^n)$ over M^m has a nowhere-zero cross section. Hence (f, f) can be deformed to be coincidence free by a small deformation if and only if the horizontal tangent bundle $f^*(\tau_N^n)$ over M^m has a nowhere-zero cross section.*

For the classical Hopf bundles $p_n\mathbb{H}: S^{4n+3} \to \mathbb{H}P_n$, we have:

(7.5) THEOREM. *The pair $p_n\mathbb{H}$, $p_n\mathbb{H}$ can be deformed to be coincidence free if and only if $24|(n+1)$.*

In [Ko], Koschorke considered the self-coincidence problem for maps $E \to B$ under the hypothesis that $\dim E < 2\dim B - 2$. He defines an element of the bordism group, denoted by $\omega(f) \in \Omega(M^m; f^*(TN) - TM)$, which works as a kind of index. From [Ko] we obtain:

(7.6) THEOREM. *Assume that $m < 2n-2$. Then (f,f) can be deformed to be coincidence free if and only if $\omega(f)=0$.*

About the problem concerning the difference between deformation and small deformation, from either [DG] or [Ko] we obtain:

(7.7) THEOREM. *In the stable range, $\dim M^m < 2\dim N^n - 2$, a pair (f,f) can be deformed to be coincidence free if and only if it can be done by a small deformation.*

Now we consider to the unstable range, namely when $m \geq 2n-2$. In [GR] coincidences of maps between two spheres $S^m \to S^n$ for $m \leq 2n-1$ as well as maps from sphere bundles over spheres into the sphere, are studied. As a consequence of the results about coincidences of maps between spheres, a family of examples is given for which the statement of Theorem (7.6) does not hold in the case $m \geq 2n-2$. It also follows that Theorem (7.6) is sharp. Let $[\iota_{2n}, \iota_{2n}]$ denote the Whitehead product.

(7.8) THEOREM. *Let $f_{2n}\colon S^{4n-1} \to S^{2n}$ be any representative for $[\iota_{2n}, \iota_{2n}]$ with $n>1$. Then (f_{2n}, f_{2n}) can be made coincidence free for all $n>1$. Moreover, (f_{2n}, f_{2n}) can be made coincidence free by a small deformation if and only if n is even.*

The final result affirms that homotopy disjointness does not imply homotopy disjointness by a small deformation even for $m = 4n-2$ in general. Let σ be the element of the stable 7-stem of the homotopy groups of the sphere.

(7.9) PROPOSITION. *Let $f\colon S^{30} \to S^{16}$ represent σ_{16}^2. Then (f,f) can be made coincidence free, but not coincidence free by a small deformation.*

References

[An] D. V. Anosov, *The Nielsen number of maps of nilmanifolds*, Russian Math. Surveys **40** (1985), 149–150.

[BGZ1] S. Bogatyi, D. L. Gonçalves and H. Zieschang, *Coincidence theory: the minimizing problem*, collected papers dedicated to the 60th Birthday of Academician Sergei Petrovich Novikov, Proc. Stekolov Inst. Math. **225**, 45–76 (English version); 52–86 (Russian version).

[BGZ2] ———, *The minimal number of roots of surface mappings and quadratic equations in free products*, Math. Z. **236** (2001), 419–452.

[BG1] L. D. Borsari and D. L. Gonçalves, *A Van Kampen type theorem for coincidence*, Topology Appl. **101** (2000), 149–160.

[BG2] ———, *Obstruction theory and minimal number of coincidences for maps from a complex into a manifold*, Topol. Methods Nonlinear Anal. **21** (2003), 115–130.

[Br1] R. B. S. Brooks, *On removing coincidences of two maps when only one, rather than both of them, may be deformed by a homotopy*, Pacific J. Math. **40** (1972), 45–52.

[Br2] ———, *On the sharpness of the Δ_2 and Δ_1 Nielsen numbers*, J. Reine Angew. Math. **259** (1973), 101–108.

[BBPT] R. B. S. Brooks, R. F. Brown, J. Pak and D. H. Taylor, *Nielsen numbers of maps of tori*, Proc. Amer. Math. Soc. **52** (1975), 398–400.

[Bw1] R. Brown, *The Lefschetz Fixed Point Theorem*, Scott, Foresman Co., Glenview, Ill., London, 1971.

[Bw2] ______, *Fixed point theory*, History of Topology, North-Holland, 1999, pp. 271–299.

[BSc1] R. Brown and H. Schirmer, *Nielsen coincidence theory and coincidence-producing maps for manifolds with boundary*, Topology Appl. **46** (1992), 65–79.

[BSc2] ______, *Correction to: Nielsen coincidence theory and coincidence-producing maps for manifolds with boundary*, Topology Appl. **46** (1992), 65–79; Topology Appl. **67** (1995), 233–234.

[BSc3] ______, *Nielsen root theory and Hopf degree theory*, Pacific J. Math. **198** (2001), 49–80.

[DJ] R. Dobreńko and J. Jezierski, *The coincidence Nielsen theory on non-orientable manifolds*, Rocky Mountain J. Math. **23** (1993), 67–85.

[Dol] A. Dold,, *Lectures on Algebraic Topology*, vol. 200, Grundl. Math. Wiss., Springer, Berlin–Heidelberg–New York, 1972.

[DoG] A. Dold and D. L. Gonçalves, *Self-coincidences of fibre maps*, Osaka J. Math. (to appear).

[DiG] D. Dimovski and R. Geoghegan, *One-parameter fixed point theory*, Forum Math. **2** (1990), 125–154.

[Ep] D. B. A. Epstein, *The degree of a map*, Proc. London Math. Soc. **3** (1966), 369–383.

[Fad1] E. Fadell, *Generalized normal bundles for locally-flat imbeddings*, Trans. Amer. Math. Soc. **114** (1965), 488–513.

[Fad2] ______, *On a coincidence theorem of F. B. Fuller*, Pacific J. Math. **15** (1965), 825–834.

[Fad3] ______, *Recent results in the fixed point theory of continuous maps*, Bull. Amer. Math. Soc. **76** (1970), 10–29.

[FH1] E. Fadell and S. Husseini, *Fixed point theory for non simply connected manifold*, Topology **20** (1981), 53–91.

[FH2] ______, *Local fixed point index theory for non simply connected manifolds*, Illinois J. Math. **25** (1981), 673–699.

[FH3] ______, *The Nielsen number on surfaces*, Topological Methods in Nonlinear Functional Analysis Contemp. Math., vol. 21, 1983, pp. 59–98.

[FH4] same, *On a theorem of Anosov on Nielsen numbers for nilmanifolds*, Nonlinear Functional Analysis and its Applications (Maratea, 1985), Nato Advanced Science Institutes Series C: Mathematical and Physical Science, vol. 173, Kluwer Acad. Publ., Dordrecht, 1986, pp. 47–53.

[Fu] F. B. Fuller, *The homotopy theory of coincidences*, Ann. of Math. **59** (1954), 219–226.

[G1] D. L. Gonçalves, *Braids groups and Wecken pairs*, Contemp. Math. **72** (1988), 89–97.

[G2] ______, *Coincidence of maps between surfaces*, J. Korean Math. Soc. **36** (1999), 243–256.

[G3] ______, *Coincidence Reidemeister classes on nilmanifolds and nilpotent fibrations*, Topology Appl. **83** (1998), 169–183.

[G4] ______, *Coincidence theory for maps from a complex into a manifold*, Topology Appl. **92** (1999), 63–77.

[G5] ______, *The coincidence Reidemeister classes of maps on nilmanifolds*, Topol. Methods in Nonlinear Anal. **12** (1988), 375–386.

[GF] D. L. Gonçalves and D. L. Ferrario, *Homeomorphisms of surfaces locally may not have the Wecken property*, Proceedings of the XI Brazilian Topology Meeting (S. Firmo, D. L. Gonçalves and O. Saeki, eds.), WS Publication publaddr Singapore, 2000, pp. 1–9.

[GJ] D. L. Gonçalves and J. Jezierski, *Lefschetz coincidence formula on non-orientable manifolds*, Fund. Math. **153** (1997), 1–23.

[GJW] D. L. Gonçalves, J. Jezierski and P. Wong, *Obstruction theory and coincidences in positive codimension*, preprint (2002).

[GJi] D. L. Gonçalves and B. Jiang, *The index of coincidence Nielsen classes of maps between surfaces*, Topology Appl. **116** (2001), 73–89.

[GO] D. L. Gonçalves and E. Oliveira, *The Lefschetz coincidence number for maps among compact surfaces*, Far east journal of Mathematical Sciences, special volume, Part II (1997), 147–166.

[GR] D. L. Gonçalves and D. Randall, *Self-coincidence of maps from S^q-bundles over S^n to S^n*, Boletin de la Sociedad Matematica Mexicana (to appear).

[GW1] D. L. Gonçalves and P. Wong, *Nilmanifolds are Jiang type spaces for coincidences*, Forum Math. **13** (2001), 133–141.

[GW2] ———, *Obstructions, coincidences on nilmanifolds, and fibrations*, preprint (2003).

[GW3] ———, *Duality and a general Gottlieb's formula*, preprint (2003).

[GZ] D. L. Gonçalves and H. Zieschang, *Equations in free groups and coincidence of mappings on surfaces*, Math. Z. **237** (2001), 1–29.

[Gt] D. Gottlieb, *A certain subgroup of the fundamental group*, Amer. J. Math. **87** (1965), 840–856.

[HQ] A. Hatcher and F. Quinn, *Bordism invariants of intersections of submanifolds*, Trans. Amer. Math. Soc. **200** (1974), 327–344.

[Hi] J. A. Hilmann, *The Algebraic Characterization of Geometric 4-manifolds*, London Math. Soc. Lecture Note Ser. **198** (1994), Cambridge University Press, Cambridge–New York–Melbourne.

[H1] H. Hopf, *Über die algebraische Anzahl von Fixpunkten*, Math. Z. **29** (1929), 493–524.

[HW] W. Hurewicz and H. Wallman, *Dimension Theory*, Princeton Mathematical Series, vol. 4, Princeton University Press, Princeton N. J., 1941.

[Je1] J. Jezierski, *One codimensional Wecken type theorems*, Forum Math. **5** (1993), 421–439.

[I] N. V. Ivanov, *Nielsen numbers of self-maps of Surfaces*, Zap. Nauch. Sem. Leningrad. Otdel Mat. Ins. Steklov (LOMI) **122** (1982), 56–65. (in Russian)

[Je2] J. Jezierski, *The least number of coincidence points on surfaces*, J. Austral. Math. Soc. Ser. A **58** (1995), 27–38.

[Ji1] B. Jiang, *Estimation of the Nielsen numbers Chinese Math.* **5** (1964), 330–339.

[Ji2] ———, *On the least number of fixed points*, Amer. J. Math. **102** (1980), 749–763.

[Ji3] ———, *The fixed point classes from the differential view point*, Fixed Point Theory. Lectures Notes in Math., vol. 886, 1981, pp. 163–170.

[Ji4] ———, *Lectures on Nielsen fixed point theory*, Contemporary Math. **14** (1983).

[Ji5] ———, *Fixed points and braids*, Inv. Math. **75** (1984), 69–74.

[Ji6] ———, *Fixed points and braids* II, Math. Ann. **272** (1985), 249–256.

[Ji7] ———, *Surface maps and braid equations* I, Lecture Notes in Math. **1369** (1989), 125–141.

[Ji8] ———, *Bounds for fixed points on surfaces*, Math. Ann. **311** (1998), 467–479.

[GuJi] G. H. Guo and B. Jiang, *Fixed points of surface diffeomorphisms*, Pacific J. Math. **160** (1993), 67–89.

[Ki] T. H. Kiang, *The theory of fixed point classes* (1989), Springer–Verlag, Berlin–Heidelberg–New York.

[Kn] H. Kneser, *Die kleinste Bedeckungszahl innerhalb einer Klasse von Flächen abbildungen*, Math. Ann. **103** (1930), 347–358.

[Ko] U. Koschorke, *Self-coincidences in higher codimensions*, preprint (2002).

[L1] S. Lefschetz, *Continuous transformations of manifolds*, Proc. Nat. Acad. Sci. U.S.A. **9** (1923), 90–93.

[L2] ———, *Intersections of complexes on manifolds*, Proc. Nat. Acad. Sci. U.S.A. **11** (1925), 290–292.

[L3] ———, *Intersections and transformations of complexes and manifolds*, Trans. Amer. Math. Soc. **29** (1926), 1–49.

[L4] ———, *Manifolds with boundary and their transformations*, Trans. Amer. Math. Soc. **29** (1927), 429–462.

[Mc] C. K. McCord, *A Nielsen theory for intersection numbers*, Fund. Math. **152** (1997), 117–150.

[Mu] K. Mukherjea, *Coincidence theory for manifolds with boundary*, Topology Appl. **46** (1992), 23–39.

[Na] M. Nakaoka, *Coincidence Lefschetz numbers for fibre-preserving maps*, J. Math. Soc. Japan **32** (1980), 751–779.

[Ni] J. Nielsen, *Untersuchungen zur Topologie der geschlossenen zweiseitigen Flächen*, Acta Math. **50** (1927), 189–358; English transl.: Jakob Nielsen Collected Mathematical Papers, Birkhäuser, Basel, 1986, pp. 223–341.

[O] P. Olum, *Mappings of manifolds and the notion of degree*, Ann. of Math. **58** (1953), 458–480.

[Op] J. Oprea, *Gottlieb Groups, Groups actions and Rational Homotopy*, Lecture Notes Series-Seoul National University, vol. 29, 1995.

[R] K. Reidemeister, *Automorphismen von Homotopiekettenringen*, Math. Ann. **112** (1936), 586–593.

[Sc] H. Schirmer, *Mindestzahlen von Koinzidenzpunkten*, J. Reine Angew. Math. **194** (1955), 21–39.

[St] N. E. Steenrod, *Topology of Fibre Bundles*, Princeton, 1950.

[V] J. Vick, *Homology Theory*, Graduate Text Math., vol. 145, Springer–Verlag, Berlin–Heidelberg–New York, 1994.

[We] F. Wecken, *Fixpunktklassen* I–III, Math Ann. **117** (1941), 659–671; **118** (1942), 216–234; **118** (1942), 544–577.

[Wh] G. Whitehead, *Elements of Homotopy Theory*, Springer–Verlag, New York, 1978.

[Wo] P. Wong, *Reidemeister number, Hirsch rank and coincidences on solvmanifolds*, preprint (1997), Bates College, USA.

[Wy] A. Wyler, *Sur certaines singularités d'applications de variétés topologiques*, Comment. Math. Helv. **42** (1967), 28–48.

[ZVC] H. Zieschang, E. Vogt and H. D. Coldewey, *Surfaces and planar discontinuous groups*, Lecture Notes Math. **835** (1980), Springer–Verlag, Berlin–Heidelberg–New York.

2. ON THE LEFSCHETZ FIXED POINT THEOREM

LECH GÓRNIEWICZ

Introduction

It is well known that one of the most deep applications of the homology theory is connected with the fixed point theory.

In 1923 S. Lefschetz formulated the famous fixed point theorem which is now known as the Lefschetz fixed point theorem. Later, in 1928, H. Hopf gave a new proof of the Lefschetz fixed point theorem for self-mappings of polyhedra. Let us remark that Lefschetz formulated his theorem for compact manifolds. In 1967, A. Granas extended the Lefschetz fixed point theorem to the case of absolute neighbourhood retracts. The proof of the theorem was based on the fact that all compact absolute neighbourhood retracts are homotopically equivalent with polyhedra. Then the case of noncompact absolute neighbourhood retracts was reduced to the compact case by using the generalized trace theory introduced by J. Leray.

In the present paper we would like to present current results concerning this theorem for metric spaces.

We shall formulate this theorem for mappings of metric absolute neighbourhood retracts. The respective classes of mappings are the following:

- compact mappings,
- compact absorbing contractions,
- condensing mappings.

In the last part of our survey we remark some further possibilities. These are:

- the problem of inverse Lefschetz fixed point theorem,
- the Lefschetz fixed point theorem for approximative absolute neighbourhood retracts and $\mathbb{Q}$-simplicial spaces,
- the Lefschetz fixed point theorem for nearly extendable mappings,
- non-metric case; topological vector spaces admissible in the sense of Klee,
- multivalued mappings,

We believe that our article will be useful for analysts applying topological fixed point theory in nonlinear analysis.

ALGEBRAIC PRELIMINARIES

In this section we intend to give a survey of the following notions:

- the trace,
- the generalized trace,
- Leray endomorphisms,
- Lefschetz numbers,
- Euler–Poincaré characteristics,
- Lefschetz power series,
- Lefschetz periodic numbers.

The material of this section will be used in Section 3. For details concerning the above material, we recommend: [Bo1]–[Bo3], [Go1], [Go2], [Gr4] and [Ler].

1. The ordinary and generalized trace

In what follows by V_f we shall denote the category of finite dimensional vector spaces over the fixed field K (usually as K, we shall take the field of rational numbers $\mathbb{Q}$) and linear mappings.

Let $\mathcal{E}_f$ denote the class of all endomorphisms in V_f.

(1.1) Definition. By the *ordinary trace* "tr" we mean a map $\operatorname{tr}: \mathcal{E}_f \to K$, which satisfies the following two conditions:

(1.1.1) (Additivity) $\operatorname{tr}(T + S) = \operatorname{tr}(T) + \operatorname{tr}(S)$, for every two endomorphisms of a vector space E in V_f into itself.

(1.1.2) (Commutativity) For any two linear mappings $T: E \to E_1$ and $S: E_1 \to E$ in V_f, we have $\operatorname{tr}(S \circ T) = \operatorname{tr}(T \circ S)$.

Observe that (1.1.2) can be expressed in terms of commutative diagrams as follows. If the diagram

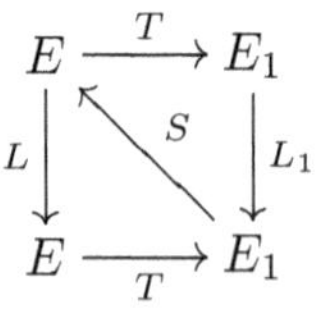

is commutative in V_f, i.e. $L = S \circ T$ and $L_1 = T \circ S$, then $\operatorname{tr}(L) = \operatorname{tr}(L_1)$.

(1.2) LEMMA. *If the following diagram in* V_f

$$\begin{array}{ccc} E & \xrightarrow{T} & E_1 \\ {\scriptstyle L}\downarrow & & \downarrow{\scriptstyle L_1} \\ E & \xrightarrow[T]{} & E_1 \end{array}$$

is commutative and T *is an isomorphism, then* $\mathrm{tr}(L) = \mathrm{tr}(L_1)$.

PROOF. Let $S\colon E_1 \to E$ be the inverse of T. Then from the commutativity property (1.1.2) we get $\mathrm{tr}(L) = \mathrm{tr}(S \circ L_1 \circ T) = \mathrm{tr}(L_1 \circ T \circ S) = \mathrm{tr}(L_1)$ and the proof is completed. □

For given trace $\mathrm{tr}\colon \mathcal{E}_f \to K$ we define a function $t\colon K \to K$ by letting: $t(a) = \mathrm{tr}(L)$, where $L\colon K \to K$ is the endomorphism defined by the formula $L(x) = ax$.

It follows from Lemma (1.1) that if E is an arbitrary one dimensional vector space with a basis $\{e\}$ and $L_1\colon E \to E$, $L_1(e) = ae$, then $\mathrm{tr}(L) = \mathrm{tr}(L_1)$, where L is defined above.

Now, from (1.1), we get:

(1.3) LEMMA. *For every* $a, b \in K$

(1.3.1) $t(a + b) = t(a) + t(b)$,
(1.3.2) $t(ab) = t(ba)$.

Now, assume that $E = E_1 \oplus E_2$ is a direct sum in V_f and let $L\colon E \to E$ be an endomorphism. Let $x \in E$, then x has a unique representation of the type:

$$x = x_1 + x_2 \quad \text{where } x_1 \in E_1 \text{ and } x_2 \in E_2.$$

Consequently $L(x) = L(x_1 + x_2) = (L_{11}(x_1) + L_{12}(x_2), L_{21}(x_1) + L_{22}(x_2))$, where $L_{11}\colon E_1 \to E_1$, $L_{12}\colon E_2 \to E_1$, $L_{21}\colon E_1 \to E_2$ and $L_{22}\colon E_2 \to E_2$. Observe, that we have one to one correspondence between L and $(L_{11}, L_{12}, L_{21}, L_{22})$. In what follows, we shall keep the above notations.

(1.4) PROPERTY. *Assume that* $L\colon E_1 \oplus E_2 \to E_1 \oplus E_2$ *satisfies one of the following two conditions:*

(1.4.1) $L(x_1, x_2) = (L_{11}(x_1), 0)$,
(1.4.2) $L(x_1, x_2) = (0, L_{22}(x_2))$,

for every $(x_1, x_2) \in E_1 \oplus E_2$, *then* $\mathrm{tr}(L) = \mathrm{tr}(L_{11})$ *and* $\mathrm{tr}(L) = \mathrm{tr}(L_{22})$ *according to* (1.4.1) *or* (1.4.2).

PROOF. Assume that L satisfies (1.4.1). For the proof we define:

$$T\colon E_1 \oplus E_2 \to E_1, \quad T(x_1, x_2) = L_{11}(x_1),$$
$$S\colon E_1 \to E_1 \oplus E_2, \qquad S(x_1) = (x_1, 0).$$

Then $L = S \circ T$ and $T \circ S = L_{11}$. So our claim follows from the commutativity property of the trace. The second case is strictly analogous. The proof is completed. □

(1.5) PROPERTY. *If $L: E_1 \oplus E_2 \to E_1 \oplus E_2$ satisfies one of the following two conditions:*

(1.5.1) $L(x_1, x_2) = (L_{12}(x_2), 0)$,
(1.5.2) $L(x_1, x_2) = (0, L_{21}(x_1))$,

for every $(x_1, x_2) \in E_1 \oplus E_2$, then $\mathrm{tr}(L) = 0$ (in both cases).

PROOF. For the simplicity we shall consider only the case (1.5.1) (the second case is strictly analogous). Now, we define:

$$T: E_1 \oplus E_2 \to E_1, \quad T(x_1, x_2) = L_{12}(x_2),$$
$$S: E_1 \to E_1 \oplus E_2, \quad S(x_1) = (x_1, 0).$$

Then we have $S \circ T = L$ and $T \circ S = 0$ and again our claim follows from the commutativity property of the trace. The proof is completed. □

Now, (1.4), (1.5) and the additivity property of the trace implies:

(1.6) PROPERTY. *For any endomorphism $L: E_1 \oplus E_2 \to E_1 \oplus E_2$ in $\mathcal{E}_f$, we have $\mathrm{tr}(L) = \mathrm{tr}(L_{11}) + \mathrm{tr}(L_{22})$.*

Finally, by induction with respect to $n = \dim E$ ($\dim E$ denotes the dimension of E), we get:

(1.7) PROPERTY. *Let $e_1, \ldots, e_n$ be a basis of E and let $L: E \to E$ be defined on the basis as follows:*

$$L(e_i) = \sum_{j=1}^{n} a_i^j e_j, \quad a_i^j \in K.$$

Then

(1.7.1)
$$\mathrm{tr}(L) = \sum_{i=1}^{n} t(a_i^i).$$

Above, having the function $\mathrm{tr}: \mathcal{E}_f \to K$ satisfying (1.1.1) and (1.1.2), we were able to define the function $t: K \to K$ satisfying (1.3.1) and (1.3.2).

Now, from property (1.7) it follows that having $t: K \to K$ with properties (1.3.1) and (1.3.2) we are able to define (by formula (1.7.1)) the function $\mathrm{tr}: \mathcal{E}_f \to K$ with properties (1.1.1) and (1.1.2).

So, we have one to one correspondence between $\mathrm{tr}\colon \mathcal{E}_f \to K$ and $t\colon K \to K$. Recall that a diagram in V_f of the type:

$$\cdots \to E_n \xrightarrow{T_n} E_{n-1} \xrightarrow{T_{n-1}} E_{n-2} \to \cdots, \quad n \in \mathbb{Z},$$

is called an *exact sequence* provided for every $n \in \mathbb{Z}$ we have: $\operatorname{Im} T_n = \operatorname{Ker} T_{n-1}$, where $\mathbb{Z}$ denotes the ring of integers, $\operatorname{Im} T_n = \{T_n(x) : x \in E_n\}$ and $\operatorname{Ker} T_{n-1} = \{y \in E_{n-1} : T_{n-1}(y) = 0\}$. In particular, an exact sequence of the type

$$0 \longrightarrow E_1 \xrightarrow{T} E \xrightarrow{S} E_2 \longrightarrow 0$$

is called a short exact sequence.

Note that if we have a short exact sequence then the space E_1 is isomorphic to the subspace $T(E_1)$ of E and E_2 is isomorphic to the factor space $E|_{\operatorname{Im} T}$.

We prove:

(1.8) PROPOSITION. *If the following diagram with exact rows in* V_f

$$\begin{array}{ccccccccc} 0 & \longrightarrow & E_1 & \xrightarrow{T} & E & \xrightarrow{S} & E_2 & \longrightarrow & 0 \\ & & \downarrow{\scriptstyle L_1} & & \downarrow{\scriptstyle L} & & \downarrow{\scriptstyle L_2} & & \\ 0 & \longrightarrow & E_1 & \longrightarrow & E & \longrightarrow & E_2 & \longrightarrow & 0 \end{array}$$

is commutative, then $\mathrm{tr}(L) = \mathrm{tr}(L_1) + \mathrm{tr}(L_2)$.

PROOF. Since $0 \to E_1 \xrightarrow{T} E \xrightarrow{S} E_2 \to 0$ is an exact sequence, in view of (1.2), we can assume without loss of generality that E_1 is a subspace of E and $E_2 = E/E_1$. Consequently, we can assume that $E = E_1 \oplus E_2$, then $L_1 = (L_{11}, L_{22})$ (comp. (1.6)) and our claim follows from (1.6). The proof is completed. □

Now, we are going to define the generalized trace. Let us denote by V the category of vector spaces over the field K (usually $\mathbb{Q}$) and linear mappings. Of course, V_f is a subcategory of V. Let $L\colon E \to E$ be an endomorphism in V. We let $N(L) = \{x \in E : L^n(x) = 0 \text{ for some natural } n \in \mathbb{N}\}$, here $L^n\colon E \to E$ is nth iteration of L, i.e.

$$L^n = \underbrace{L \circ \ldots \circ L}_{n\text{-times}}.$$

Then $N(L)$ is called the generalized kernel of L. Evidently, $N(L)$ is a subspace of E which is invariant under L, i.e. $L(N(L)) \subset N(L)$.

Let $\widetilde{E} = E|N(L)$. We define $\widetilde{L}\colon \widetilde{E} \to \widetilde{E}$ by putting $\widetilde{L}([x]) = [L(x)]$ for every $x \in E$. Then $\widetilde{L}$ is a well defined endomorphism on $\widetilde{E}$.

(1.9) DEFINITION. An endomorphism $L: E \to E$ is called a *Leray endomorphism* provided $\widetilde{E}$ is an object in V_f. Then we put $\mathrm{Tr}(L) = \mathrm{tr}(\widetilde{L})$ and $\mathrm{Tr}(L)$ is called the *generalized trace* of L.

(1.10) DEFINITION. An endomorphism $L: E \to E$ is called *weakly nilpotent* provided $N(L) = E$.

(1.11) REMARK. Since for a weakly nilpotent $L: E \to E$, we have $\widetilde{E} = 0$ therefore every weakly nilpotent endomorphism $L: E \to E$ is a Leray endomorphism and $\mathrm{Tr}(L) = 0$.

Observe that for any endomorphism $L: E \to E$ the induced endomorphism $\overline{L}: N(L) \to N(L)$, $\overline{L}(x) = L(x)$, for every $x \in N(L)$, is weakly nilpotent.

Assume that $\dim E < \infty$ and $L: E \to E$ is an endomorphism. Of course, L is a Leray endomorphism. So we have two traces for L namely the ordinary trace $\mathrm{tr}(L)$, and the generalized trace $\mathrm{Tr}(L)$. We claim that

$$\mathrm{Tr}(L) = \mathrm{tr}(L). \tag{1.12}$$

In fact, we have the following commutative diagram with exact rows:

$$\begin{array}{ccccccccc} 0 & \longrightarrow & N(L) & \longrightarrow & E & \longrightarrow & \widetilde{E} & \longrightarrow & 0 \\ & & \Big\downarrow{\scriptstyle \overline{L}} & & \Big\downarrow{\scriptstyle L} & & \Big\downarrow{\scriptstyle \widetilde{L}} & & \\ 0 & \longrightarrow & N(L) & \longrightarrow & E & \longrightarrow & \widetilde{E} & \longrightarrow & 0. \end{array}$$

Consequently from (1.8), we get: $\mathrm{tr}(L) = \mathrm{tr}(\overline{L}) + \mathrm{tr}(\widetilde{L})$. It follows from (1.11) that $\mathrm{tr}(\overline{L}) = 0$. Consequently $\mathrm{tr}(L) = \mathrm{tr}(\widetilde{L}) = \mathrm{Tr}(L)$ and the proof is completed.

From (1.12) we deduce that the definition of the generalized trace presented in (1.9) is correct.

Below we shall list some important properties of the generalized trace. Note that the respective proofs are straightforward by reduction to the ordinary trace.

(1.13) PROPERTIES.

(1.13.1) *If (in V) the following diagram is commutative:*

$$\begin{array}{ccc} E & \xrightarrow{\ T\ } & E_1 \\ {\scriptstyle L}\Big\downarrow & \overset{S}{\swarrow} & \Big\downarrow{\scriptstyle L_1} \\ E & \xrightarrow[\ T\ ]{} & E_1 \end{array}$$

then L is a Leray endomorphism if and only if L_1 is a Leray endomorphism and in that case we get $\mathrm{Tr}(L) = \mathrm{Tr}(L_1)$.

(1.13.2) *Assume that (in V) the following diagram*

$$\begin{array}{ccccccccc} 0 & \longrightarrow & E_1 & \longrightarrow & E & \longrightarrow & E_2 & \longrightarrow & 0 \\ & & \downarrow{\scriptstyle L_1} & & \downarrow{\scriptstyle L} & & \downarrow{\scriptstyle L_2} & & \\ 0 & \longrightarrow & E_1 & \longrightarrow & E & \longrightarrow & E_2 & \longrightarrow & 0 \end{array}$$

with exact rows is commutative. Then, if two of the three endomorphisms L_1, L, L_2 are Leray endomorphisms, then so is the third one and in such a case we have: $\mathrm{Tr}(L) = \mathrm{Tr}(L_1) + \mathrm{Tr}(L_2)$.

(1.13.3) *Assume that $L_i: E_i \to E_i$, $1 \leq i \leq n$ are given endomorphisms and $L: \bigoplus_{i=1}^n E_i \to \bigoplus_{i=1}^n E_i$ is defined as follows:*

$$L(e_1, \dots, e_n) = (L(e_1), \dots, L(e_n)).$$

Then L is a Leray endomorphism if and only if L_i is a Leray endomorphism for every $i = 1, \dots, n$ and in such a case we have:

$$\mathrm{Tr}(L) = \sum_{i=1}^{n} \mathrm{Tr}(L_i).$$

2. The Lefschetz number

In this section we shall consider the category GV of graded vector spaces and linear mappings of degree zero. Thus $E = \{E_n\}_{n\in\mathbb{N}}$ is a graded vector space and $L: E \to E_1$ is a linear map of degree zero provided $L = \{L_n\}_{n\in\mathbb{N}}$ and $L_n: E_n \to E_{1n}$ is a linear map for every $n \in \mathbb{N}$. In particular $L = \{L_n\}: E \to E$ is an endomorphism (of degree zero).

(2.1) DEFINITION. A graded vector space $E = \{E_n\}$ is of a *finite type* provided $E_n = 0$, for almost all $n \in \mathbb{N}$ and $\dim E_n < \infty$, for all $n \in \mathbb{N}$.

(2.2) DEFINITION. Let $L = \{L_n\}: E \to E$ be an endomorphism and let E be of a finite type. Then we define the *Lefschetz number* $\lambda(L)$ of L by putting:

$$\lambda(L) = \sum_{n\in\mathbb{N}} (-1)^n \mathrm{tr}(L_n).$$

Now, we are going to define the generalized Lefschetz number $\Lambda(L)$ of L. Assume that $L: E \to E$ is an endomorphism of a graded vector space E. So we have an endomorphism $L_n: E_n \to E_n$, for every $n \in \mathbb{N}$, where $E = \{E_n\}_{n\in\mathbb{N}}$ and $L = \{L_n\}_{n\in\mathbb{N}}$.

For every $n \in \mathbb{N}$, we can consider the vector space $\widetilde{E}_n = E_n/N(L_n)$. Consequently, we get the following graded vector space: $\widetilde{E} = \{\widetilde{E}_n\}$.

(2.3) DEFINITION. We shall say that an endomorphism $L\colon E \to E$ of a graded vector space E is a *Leray endomorphism* provided the space $\widetilde{E} = \{\widetilde{E}_n\}$ is of finite type, for a Leray endomorphism L, we define the generalized Lefschetz number $\Lambda(L)$ by putting

$$\Lambda(L) = \sum_{n\in\mathbb{N}} Tr(L_n).$$

It follows from (1.12) that if E is of a finite type, then $\Lambda(L) = \lambda(L)$.

Below, we shall collect properties of the Lefschetz number. Note that they follow from the respective properties of the trace.

(2.4) PROPERTIES.

(2.4.1) *If in the category* GV *the following diagram is commutative:*

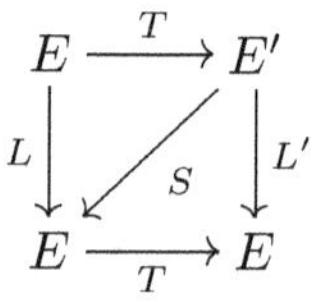

then L is a Leray endomorphism if and only if L' is and in such a case we have $\Lambda(L) = \Lambda(L')$.

(2.4.2) *If in the category* GV *the g diagram with exact rows*

$$\begin{array}{ccccccccc} 0 & \longrightarrow & E' & \longrightarrow & E & \longrightarrow & E'' & \longrightarrow & 0 \\ & & \downarrow{\scriptstyle L'} & & \downarrow{\scriptstyle L} & & \downarrow{\scriptstyle L''} & & \\ 0 & \longrightarrow & E' & \longrightarrow & E & \longrightarrow & E'' & \longrightarrow & 0 \end{array}$$

is commutative then, if two from the following three endomorphisms L', L'', L are Leray endomorphisms, then so is the third one and in such a case we have: $\Lambda(L) = \Lambda(L') + \Lambda(L'')$.

(2.4.3) *If $L\colon E \to E$ is a weakly nilpotent endomorphism, i.e. L_n is weakly nilpotent, for every $n \in \mathbb{N}$, then L is a Leray endomorphism and $\Lambda(L) = 0$.*

3. Periodic invariants, the Euler–Poincaré charactersitic

Let $L\colon E \to E$ be a Leray endomorphism in GV. Using notations of the preceding section we define the Euler–Poincaré characteristic $\chi(L)$ of L by putting:

$$\chi(L) = \sum_n (-1)^n \dim(\widetilde{E}_n). \tag{3.1}$$

We start with the following lemma:

(3.2) LEMMA. *Let $L\colon E \to E$ be an endomorphism and let*

$$L^m = \underbrace{L \circ \ldots \circ L}_{m\text{-times}}$$

be its mth iterate, i.e. $L^m = \{L_n^m\}_{n\in\mathbb{N}}$, *for every* $n \in \mathbb{N}$. *Then L is a Leray endomorphism if and only if* L^m *is a Leray endomorphism, for* $m = 1, 2, \ldots$ *and*

$$\chi(L) = \chi(L^m), \quad m \geq 1.$$

PROOF. It is sufficient to observe that for every $m = 1, 2, \ldots$ and for every $n \in \mathbb{N}$ we have $N(L_n) = N(L_n^m)$. □

Starting from now until the end of this section we shall assume that K is the field of complex numbers.

(3.3) LEMMA. *Let $L\colon E \to E$ be an endomorphism in V_f and let λ_j, $1 \leq j \leq \dim E$ be roots of the charactersitic polynomial of L. Then*

$$\mathrm{tr}(L^m) = \sum_j (-\lambda_j)^m, \quad m = 1, 2, \ldots$$

PROOF. We can represent L as a triangular matrix with roots $\lambda_1, \ldots, \lambda_s$, $s = \dim E$, on the main diagonal. Then for the mth iteration L^m of L the matrix is also traingular (keeping the same basis) and on the main diagonal, we get mth powers $(\lambda_1)^m, \ldots, (\lambda_s)^m$ of $\lambda_1, \ldots, \lambda_s$, respectively. It proves our lemma. □

Now, let $L = \{L_n\}\colon E \to E$ be a Leray endomorphism in GV. For such a L, we are able to assign the power series $S(L)$ in K by letting:

$$\text{(3.4)} \quad S(L)(z) = \Lambda(L) + \sum_{m=1}^{\infty} \Lambda(L^m)\cdot z^m = \sum_{m=0}^{\infty} \Lambda(L^m)\cdot z^m, \quad \text{where } L^0 = \mathrm{id}_E.$$

(3.5) THEOREM. *Let $L = \{L_n\}\colon E \to E$ be a Leray endomrphism. Let W_n be the charactersitic polynomial of $\widetilde{L}_n\colon \widetilde{E}_n \to \widetilde{E}_n$ with roots $\lambda_{n,j} \neq 0$, $1 \leq j \leq \dim \widetilde{E}_n$, let $W(\lambda) = \Pi_n(-1)^n W_n$ and $T(\lambda) = (\log W(\lambda))' = W'(\lambda)/W(\lambda)$. Then, for $0 < |z| < \min|\lambda_{n,j}|^{-1} = r$, we get*

$$S(L)(z) = \sum_{n,j} \frac{(-1)^n}{1-\lambda_{n,j}} = \frac{1}{z}T\Big(\frac{1}{z}\Big).$$

PROOF. By using (3.3), for $|z| < r$, we get

$$S(L)(z) = \sum_{m=0}^{\infty} \lambda(\widetilde{L}^m)\cdot z^m = \sum_{m=0}^{\infty}\sum_i (-1)^i \mathrm{tr}(\widetilde{L}_i^m) = \sum_{m=0}^{\infty}\sum_{i,j}(-1)^i \lambda_{i,j}^m z^m.$$

So taking the logaritmic derivative, we get

$$T(\lambda) = \sum_{i,j} (-1)^i \frac{(-1)^i}{z - \lambda_{i,j}}$$

and consequently, we have

$$\sum_{i,j} (-1)^i \frac{(-1)^i}{1 - \lambda_{i,j} z} = \sum_{i,j} \frac{1}{z} \frac{(-1)^i}{z^{-1} - \lambda_{i,j}} = \frac{1}{z} \cdot T\left(\frac{1}{z}\right),$$

for $0 < |z| < r$. The proof is completed. □

It implies that $S(L)$ can be represented in a unique form as a factor of two polynomials W_1, W_2, i.e., $S(L) = W_1/W_2$ such that $(W_1, W_2) = 1$ and $\deg W_1 < \deg W_2$. This allows us to define the natural number $P(L)$ of L by putting

$$P(L) = \begin{cases} \deg W_2 & \text{if } S(f) \neq 0, \\ 0 & \text{if } S(f) = 0. \end{cases} \tag{3.6}$$

Below we shall summarize the properties of the above considered invariants Λ, χ, P and S.

(3.7) PROPERTIES.

(3.7.1) *There exists a natural number m such that $\Lambda(L^m) \neq 0$ if and only if $P(L) \neq 0$.*

(3.7.2) *If $\chi(L) \neq 0$ then $P(L) \neq 0$.*

(3.7.3) *If $P(L) \neq 0$, then for every natural number l there is a natural number m such that $l \leq m \leq l + P(L)$ and $\Lambda(L^m) \neq 0$.*

(3.7.4) *Let $L\colon E \to E$ and $L'\colon E' \to E'$ be two Leray endomorphisms. If there exists a natural number l such that $\Lambda(L^m) = \Lambda((L')^m)$, for every natural number $l \leq m < l + P(L) + P(L')$, then $\Lambda(L^k) = \Lambda((L')^k)$, for all natural k and $\chi(L) = \chi(L')$.*

The proof of (3.7) is straightforward. For details concerning Sections 1–3 we recommend [Bow1]–[Bow3] and [Gr1], [Gr2].

4. Homological invariants of continuous mappings

In what follows by Top_2 we shall denote the category of pairs of topological spaces and continuous maps. For simplicity, we let $(X, \emptyset) = X$.

By $H = \{H_n\}\colon \mathrm{Top}_2 \to \mathrm{GV}$, we shall denote the homology functor with coefficients in K (either the singular homology functor or the Čech homology functor with compact carriers).

We shall assume that the above functors are well known. We recomend [Do] and [SP], for the singular homology theory and [Go1]–[Go3], for the Čech homology with compact carriers. Thus for a pair (X, A) by $H(X, A) = \{H_n(X, A)\}$ we denote the graded vector space, $H_n(X, A)$ being the n-dimensional relative homology. For a continuous map $f\colon (X, A) \to (Y, B)$, $H(f) = f_* = \{f_n\}$ is the linear induced map, where $f_n\colon H_n(X, A) \to H_n(Y, B)$, for every $n \in \mathbb{N}$.

(4.1) DEFINITION. A continuous map $f\colon (X, A) \to (X, A)$ is called a *Lefschetz map* provided $f_*\colon H(X, A) \to H(X, A)$ is a Leray endomorphism, for such f we define:

(4.1.1) the *Lefschetz number* $\Lambda(f)$ of f by $\Lambda(f) = \Lambda(f_*)$,
(4.1.2) the *Euler–Poincaré characteristic* $\chi(f)$ of f by $\chi(f) = \chi(f_*)$,
(4.1.3) the *Lefschetz power series* $S(f)$ of f by $S(f) = S(f_*)$,
(4.1.4) the *Lefschetz number of periodicity* $P(f)$ of f by $P(f) = P(f_*)$.

Since for homotopic mappings f and g (written $f \sim g$) the induced linear maps are equal, we obtain:

(4.2) PROPERTY. *If $f \sim g$, then f is a Lefschetz map if and only if g is a Lefschetz map and in such a case we get $\Lambda(f) = \Lambda(g)$.*

From (2.4.1) immediately follows:

(4.3) PROPERTY. *Assume that in* Top_2 *the following diagram is commutative*

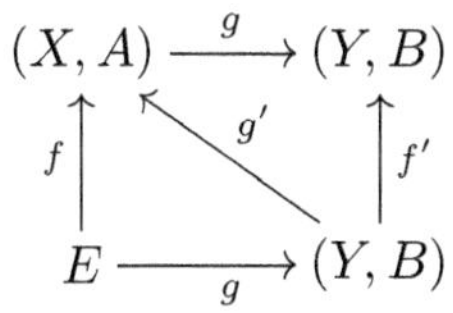

Then f is a Lefschetz map if and only if f' is a Lefschetz map and in such a case we have $\Lambda(f) = \Lambda(f')$.

Let $f\colon (X', A') \to (X, A)$ be a continuous map of pairs and $(X', A') \subset (X, A)$, i.e. $X' \subset X$ and $A' \subset A$. Then we let $\mathrm{Fix}(f) = \{x \in X' : f(x) = x\}$, in what follows $\mathrm{Fix}(f)$ is called the *set of fixed points* of f.

(4.3.1) REMARK. Observe that under assumptions of (4.3), we get

$$\mathrm{Fix}(f) \neq \emptyset \Leftrightarrow \mathrm{Fix}(f') \neq \emptyset.$$

Observe that (4.4) can be applied to the following two particular cases:

(4.3.2) Remark. Let $f: X \to X$ be a continuous map such that $f(X) \subset B \subset X$. Let $f_B: B \to B$ be defined by $f_B(x) = f(x)$, for every $x \in B$. Then the following diagram is commutative:

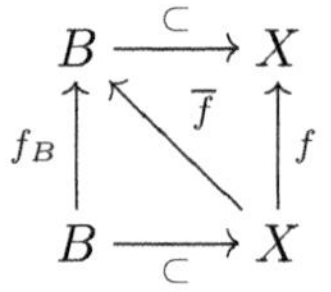

(4.3.3) Remark. Assume that $r: Y \to X$ and $s: X \to Y$ are continuous mappings such that $r \circ s = \mathrm{Id}_X$, i.e. $(r \circ s)(x) = x$, for every $x \in X$. Then X is said to be r-*dominated* by Y and r is said to be an r-map. If $f: X \to X$ is a continuous map, then we get the following commutative diagram:

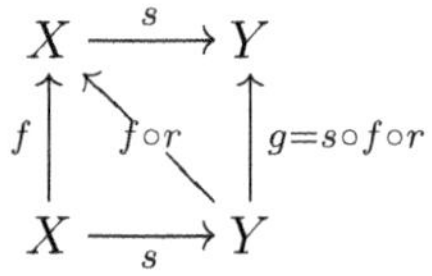

Having a continuous map $f: (X, A) \to (X, A)$, we denote by $f_X: X \to X$ and $f_A: A \to A$ the respective contractions of f.

In view of (2.4.2) and the exact sequence of homologies we infer:

(4.4) Property. *Let $f: (X, A) \to (X, A)$ be a continuous map. If two of the maps f, f_X, f_A are Lefschetz maps, then so is the third one and in such a case we have $\Lambda(f) = \Lambda(f_X) - \Lambda(f_A)$.*

Assume $f: (X, A) \to (X, A)$ is a continuous map. We shall say that A *absorbs compact sets* (with respect to f) provided for every compact $K \subset X$ there is a natural number $n = n(K)$ such $f^n(K) \subset A$.

Since H is a homology funtor (in both cases) with *compact carriers* (comp. [Sp] or [Go2]) we deduce that, if for a map $f: (X, A) \to (X, A)$ the set A absorbs compact sets, then f_* is weakly nilpotent. Finally, in view of (2.4.3) we get:

(4.5) Property. *If $f: (X, A) \to (X, A)$ is a map in* Top_2 *such that A absorbs compact sets, then f is a Lefschetz map and $\Lambda(f) = 0$.*

A nonempty space X is called *acyclic* (with respect to H) provided

$$H_n(X) = \begin{cases} 0 & \text{for } n \geq 1, \\ K & \text{for } n = 0. \end{cases}$$

A continuous map $f: X \to X$ is called *homologically trivial* provided

$$f_n = \begin{cases} 0 & \text{for } n \geq 1, \\ \mathrm{id}_{H_0(X)} & \text{for } n = 0. \end{cases}$$

As an easy consequence of the homology functor and properties of the Lefschetz number we get:

(4.6) PROPERTY. *Let $f: X \to X$ be a continuous map and assume that one of the following three conditions is satisfied for some $n \geq 0$:*

(4.6.1) *$f^n(X) \subset A$ and A is an acyclic subset of X,*
(4.6.2) *$f^n: X \to X$ is homologically trivial and $H_0(X) = K$,*
(4.6.3) *$f^n \sim \text{const}$ and $H_0(X) = K$.*

Then f is a Lefschetz map and $\Lambda(f) = 1$.

Now, from (3.2) immediately follows:

(4.7) PROPERTY. *A map $f: (X, A) \to (X, A)$ is a Lefschetz map if and only if $f^n: (X, A) \to (X, A)$, $n \geq 1$ is a Lefschetz map and in such a case we have $\chi(f) = \chi(f^n)$.*

(4.8) PROPERTY. *If $f: (X, A) \to (X, A)$ is a Lefschetz map, then we have:*

(4.8.1) *$\chi(f) \neq 0$ if and only if $P(f) \neq 0$,*
(4.8.2) *$P(f) = 0$ if and only if $\Lambda(f^n) = 0$, for some n,*
(4.8.3) *If $P(f) \neq 0$ then for every $m \geq 0$ at least one of the numbers*

$$\Lambda(f^{m+1}), \dots, \Lambda(f^{m+P(f)})$$

has to be different from zero.

APPRIOPRIATE CLASSES OF MAPPINGS

The aim of this part is to define possibly large classes of mappings for which the Lefschetz fixed point theorem is true; these are:

- compact mappings,
- eventually compact mappings,
- asymptotically compact mappings,
- mappings with compact attractors,
- compact absorbings contractions,
- k-set contractions,
- condensing mappings.

A classification of the above classes of mappings will be presented. For more precise information, we recommend: [AG], [FG], [FGo], [Go2], [Go3], [Gr4], [Kr] and [Nu].

5. Compact type mappings

In this section we shall assume that all topological spaces are Hausdorff and all mappings between topological spaces are continuous.

(5.1) DEFINITION ([1]). A map $f: X \to Y$ is called *compact* provided the set $\mathrm{cl}_Y f(X)$ is compact, where $\mathrm{cl}_Y f(X)$ is the closure of $f(X)$ in Y, i.e. the set $f(X)$ is relatively compact.

We let $\mathrm{C}(X) = \{f: X \to X : f \text{ is compact}\}$.

(5.2) DEFINITION. A map $f: X \to Y$ is called *locally compact* provided for every $x \in X$ there exists an open neighbourhood U_x of x in X such that the set $\mathrm{cl}_Y f(U_x)$ is compact.

We let $\mathrm{LC}(X) = \{f: X \to X : f \text{ is locally compact}\}$. Evidently, we have

$$\mathrm{C}(X) \subset \mathrm{LC}(X). \tag{5.3}$$

(5.4) DEFINITION. A map $f: X \to X$ is called *eventually compact* provided there exists a natural number n such that the nth iteration $f^n: X \to X$ of f is compact.

We let $\mathrm{EC}(X) = \{f: X \to X : f \text{ is eventually compact}\}$ and

$$\mathrm{EC}_0(X) = \mathrm{EC}(X) \cap \mathrm{LC}(X).$$

Note that the class of eventually compact mappings is a natural generalization of the class of compact mappings. For example, let $f: \mathbb{R} \to \mathbb{R}$ ($\mathbb{R}$ is the Euclidean real line) be defined

$$f(x) = \begin{cases} -x & \text{for } x \geq 0, \\ 0 & \text{for } x < 0. \end{cases}$$

Then $f^2: \mathbb{R} \to \mathbb{R}$ is a compact map and f is not compact.

(5.5) DEFINITION. A map $f: X \to X$ is called *asymptotically compact* provided the following two conditions are satisfied:

(5.5.1) for every $x \in X$ the orbit $\{x, f(x), f^2(x), \dots\}$ is relatively compact,

(5.5.2) the core $C_f = \bigcap_{n \geq 1} \overline{f^n(X)}$ of f is nonempty and compact.

We let $\mathrm{ASC}(X) = \{f: X \to X : f \text{ is asymptotically compact}\}$ and

$$\mathrm{ASC}_0(X) = \mathrm{ASC}(X) \cap \mathrm{LC}(X).$$

([1]) In what follows for a subset $A \subset X$ the following two notations are equivalent $\mathrm{cl}_X A$ and $\overline{A}$.

(5.6) DEFINITION. We shall say that the map $f: X \to X$ has a *compact attractor* provided there exists a compact nonempty set $K \subset X$ such that for every open neighbourhood U of K in X and for every $x \in X$ there exists a natural number n_x such that $f^n(x) \in U$, for every $n \geq n_x$.

We let $\mathrm{CA}(X) = \{f: X \to X : f \text{ has a compact attractor}\}$ and

$$\mathrm{CA}_0(X) = \mathrm{CA}(X) \cap \mathrm{LC}(X).$$

Note, that the notion of compact attractor comes from physics and is important in chaos theory. Observe also that if X is a complete metric space and $f: X \to X$ is a contraction map, then f has a compact attractor. Namely, then the set $\mathrm{Fix}(f)$ is a singleton and of course it is a compact attractor for f. Finally, we shall introduce the following class of mappings.

(5.7) DEFINITION. A map $f: X \to X$ is called a *compact absorbing contraction* (written CAC-map), if there exists an open subset U of X such that the following conditions are satisfied:

(5.7.1) $f(U) \subset U$ and the map $\widetilde{f}: U \to U$, $\widetilde{f}(x) = f(x)$, is compact,

(5.7.2) for every $x \in X$ there exists a natural number $n = n_x$ such that $f^n(x) \in U$.

We let $\mathrm{CAC}(X) = \{f: X \to X : f \text{ is CAC-map}\}$.

(5.8) REMARK.

(5.8.1) It is easy to see that if $f \in \mathrm{CAC}(X)$ and U is chosen according to the definition (5.7), then U absorbs compact sets.

(5.8.2) Observe that if $f \in \mathrm{CAC}(X)$ and U is chosen as in (5.7), then $\mathrm{cl}_U f(U)$ is compact attractor for f, so $\mathrm{CAC}(X) \subset \mathrm{CA}(X)$.

(5.8.3) Observe, that if $f \in \mathrm{CA}(X)$ and A is an attractor of f, then $\mathrm{Fix}(f) \subset A$.

(5.9) PROPOSITION. $\mathrm{EC}(X) \subset \mathrm{ASC}(X)$.

PROOF. Let $f \in \mathrm{EC}(X)$ and assume that $K = f^{n_0}(X)$ is a compact set. Then, for every $x \in X$, we have:

$$\bigcup_{i=1}^{\infty} f^i(x) \subset \{x\} \cup \{f(x)\} \cup \ldots \cup \{f^{n_0-1}(x)\} \cup f^{n_0}(X).$$

Thus, the set $\{x\} \cup \{f(x)\} \cup \ldots \cup \{f^{n_0-1}(x)\} \cup \overline{f^{n_0}(X)}$ is compact, i.e. every orbit is a relatively compact set. Moreover, we have:

$$f^{n_0+i}(K) = f^{n_0}(f^i(K)) \subset f^{n_0}(X) \subset K \quad \text{for all } i \geq 0.$$

Consequently,

$$\emptyset \neq \bigcap_{i=0}^{\infty} f^{n_0+i}(K) \subset \bigcap_{i=0}^{\infty} f^{n_0+i}(X) = \bigcap_{i=0}^{\infty} f^{i}(X)$$

which implies that the core C_f of f is nonempty and compact. So the proof of (5.9) is completed. □

(5.10) PROPOSITION. $\mathrm{ASC}(X) \subset \mathrm{CA}(X)$.

PROOF. Let $f \in \mathrm{ASC}(X)$. Then the set C_f is nonempty compact. It is enough to show that C_f is an attractor of f. Let V be an open neighbourhood of C_f in X and let $x \in X$. We put

$$L = \overline{\bigcup_{i=1}^{\infty} f^{i}(x)}.$$

Then, for $0 \leq j \leq n$, and for arbitrary n we get:

$$f^{2n}(x) \subset \bigcup_{i=0}^{\infty} f^{i+j}(x) = f^{j}\left(\bigcup_{i=0}^{\infty} f^{i}(x)\right) \subset f^{j}(L),$$

and so

$$f^{2n}(x) \subset \bigcap_{j=0}^{n} f^{j}(L).$$

Therefore, it is enough to show that there is a natural number n_x such that

$$\bigcap_{j=0}^{n_x} f^{j}(L) \subset V.$$

In fact, we obtain:

$$\bigcap_{n=0}^{\infty} A_n \subset \bigcap_{n=0}^{\infty} f^{n}(X) = C_f \subset V, \quad \text{where } A_n = \bigcap_{i=0}^{n} f^{i}(L).$$

Since A_n is a decreasing sequence of compact sets, this implies that there are natural numbers $n_1 < \ldots < n_k$ such that $A_{n_1} \cap \ldots \cap A_{n_k} \subset V$, but $A_{n_1} \cap \ldots \cap A_{n_k} = A_{n_k}$, and so $n_x = n_k$ is the required natural number. The proof of (5.10) is completed. □

Summing up the above, we get:

$$\begin{array}{ccccc} \mathrm{EC}(X) & \subset & \mathrm{ASC}(X) & \subset & \mathrm{CA}(X) \\ \cup & & & & \cup \\ K(X) & & \subset & & \mathrm{CAC}(X) \end{array} \tag{5.11}$$

We start by the following:

(5.12) LEMMA. $\mathrm{EC}_0(X) \subset \mathrm{CAC}(X)$.

PROOF. Let $f \in \mathrm{EC}_0(X)$, $M = f^{n_0}(X)$ be compact and let $K = \bigcup_{i=0}^{n_0-1} f^i(M)$. Then K is compact and

$$f(K) = \bigcup_{i=1}^{n_0} f^i(M) \subset K \cup M = K.$$

Since f is locally compact, there exists an open neighbourhood W of K in X such that $L = \overline{f(W)}$ is compact. Now, we define open sets $V_0, \dots, V_{n_0}$ such that $L \cap \overline{f(V_i)} \subset V_{i-1}$ and $K \cup f^{n_0-i}(L) \subset V_i$, $i = 1, \dots n_0$. Namely, we put $V_0 = W$. If $V_0, \dots, V_i$ are the required sets then $(K \cup f^{n_0-i}(L)) \cap (L \setminus V_i) = \emptyset$. Therefore, there is an open set V such that $K \cup f^{n-i}(L) \subset V \subset \overline{V} \subset V_i \cup (X \setminus L)$. We let $V_{i+1} = f^{-1}(V)$. We obtain

$$f(K \cup f^{n_0-(i+1)}(L)) = f(K) \cup f(f^{n_0-(i+1)}(L)) \subset K \cup f^{n_0-i}(L) \subset V,$$

and so $K \cup f^{n_0-(i+1)}(L) \subset V_{i+1}$. Moreover, we have $f(V_{i+1}) \subset \overline{V} \subset V_i \cup (X \setminus L)$, so $L \cap \overline{f(V_{i+1})} \subset V_i$. Letting $U = V_0 \cap \dots \cap V_{n_0}$, we have $M \subset K \subset U$ and

$$f(U) \subset f(V_0) \cap \dots \cap f(V_{n_0}) \subset L \cap \overline{f(V_1)} \cap \dots \cap \overline{f(V_{n_0})}.$$

Consequently, we obtain

$$\overline{f(U)} \subset (L \cap \overline{f(V_1)}) \cap \dots \cap (L \cap \overline{f(V_{n_0})} \cap L \subset V_0 \cap \dots \cap V_{n_0} = U,$$

and $\overline{f(U)}$ is compact. Since $M \subset U$, for every $x \in X$, we get $f^{n_0}(x) \subset U$, and the proof is completed. □

The following example shows that $\mathrm{EC}(X) \not\subset \mathrm{CAC}(X)$.

(5.13) EXAMPLE. Let $C = \{\{x_n\} \subset \mathbb{R} : \{x_n\}$ is a bounded sequence$\}$ be the space of bounded sequences with the usual supremum norm. We define $F\colon C \to C$ as $F(\{x_n\}) = \{0, x_1, 0, x_3, 0, x_5, 0, \dots\}$. Then $F^2 = 0$, so $F \in \mathrm{EC}(C)$, but $F \notin \mathrm{CAC}(C)$.

Now, we prove

(5.14) PROPOSITION. $\mathrm{CA}_0(X) \subset \mathrm{CAC}(X)$.

PROOF. Let $f \in \mathrm{CA}_0(X)$ and let K be a compact attractor for f. Since f is locally compact, there exists an open neighbourhood W of K in X such that $L = \overline{f(W)}$ is a compact set. We have

$$L \subset X \subset \bigcup_{i=0}^{\infty} f^{-i}(W).$$

It implies that $\{f^{-i}(W)\}_{i=1,2,\dots}$ is an open covering of L in X. Therefore, there is a finite subcovering $f^{-i_1}(W),\dots,f^{-i_j}(W)$.

Let $n=\max\{i_1,\dots,i_j\}$ and $V=\bigcup_{i=0}^{n} f^{-i}(W)$. Then we have $L\subset V$, $W\subset V$ and $f^{-i}(W)\subset f^{-i}(V)$. Consequently,

$$X\subset\bigcup_{i=0}^{\infty} f^{-i}(W)\subset\bigcup_{i=0}^{\infty} f^{-i}(V).$$

We get

$$f(V)=\bigcup_{i=0}^{n} f^{-i+1}(W)=f(W)\cup\bigcup_{i=1}^{n-1} f^{-i}(W)\subset f(W)\cup V\subset L\cup V=V$$

and, moreover, we have

$$f^{n+1}(V)\subset\bigcup_{i=0}^{n} f^{n-i+1}(W)=\bigcup_{i=0}^{n} f^{i}(f(W))\subset\bigcup_{i=0}^{n} f^{i}(L).$$

Consequently, the set $\bigcup_{i=0}^{n} f^{i}(L)$ is a compact subset of V, and so we have shown that $\widetilde{f}\in \mathrm{EC}(X)$, where $\widetilde{f}=f|_V$. In view of (5.12), we infer $\widetilde{f}\in\mathrm{CAC}(V)$, but it immediately implies that $f\in\mathrm{CAC}(X)$. It fact, if U is an open set of V such that

$$V\subset\bigcup_{i=0}^{\infty}\widetilde{f}^{-i}(U)\subset\bigcup_{i=0}^{\infty} f^{-i}(U,)$$

then

$$X\subset\bigcup_{i=0}^{\infty} f^{-i}(V)\subset\bigcup_{i=0}^{\infty} f^{-i}(U),$$

and $\overline{\widetilde{f}(U)}=\overline{f(U)}$ is a compact subset of V. The proof is completed. □

Summing up the above, we get

(5.15) $$K(X)\subset\mathrm{EC}_0(X)\subset\mathrm{ASC}_0(X)\subset\mathrm{CA}_0(X)\subset\mathrm{CAC}(X).$$

Finally, we show that all of the above inclusions are proper.

(5.16) Example. Let $F\colon C\to C$ be defined as $F(\{x_n\})=\{x_2,x_3,x_4,\dots\}$. Then $F\in\mathrm{CA}(C)$ with the compact attractor $K=\{0\}$, but $F\notin\mathrm{CAC}(C)$.

(5.17) Example. Let $f\colon\mathbb{R}\to\mathbb{R}$ be defined as follows

$$f(x)=\begin{cases} x-1 & \text{for } x>1,\\ 0 & \text{for } -1\le x\le 1,\\ x+1 & \text{for } x<-1,\end{cases}$$

and let $F\colon C\to C$ be given by the formula $F(\{x_n\})=\{f(x_1),f(x_2),\dots\}$. Then $F\in\mathrm{CAC}(C)\cap\mathrm{CA}(C)$ with the compact attractor $K=\{0\}$, but $F\notin\mathrm{CA}_0(C)$.

(5.18) EXAMPLE. Let $f\colon [0,\infty) \to [0,\infty)$ be defined

$$f(x) = \begin{cases} \dfrac{1}{x} & \text{for } x < 1, \\ 1 & \text{for } x \geq 1. \end{cases}$$

Then $f^2(x) = \{1\}$, for every $x \in [0,\infty)$. Therefore, $f \in \mathrm{EC}_0([0,\infty))$, and $f \notin K([0,\infty))$.

(5.19) EXAMPLE. Let

$$\begin{aligned} A = \{(x,y) \in \mathbb{R}^2 : (x < 0 \vee y < 0) \wedge x^2 + y^2 \leq 1\} \\ \cup \{(x,y) \in \mathbb{R}^2 : (x \geq 0 \wedge y \geq 0) \wedge (y \leq 1 \wedge x \leq 1) \vee y > 1 \wedge 0 < x \leq 1)\}. \end{aligned}$$

We define $f\colon A \to A$ by putting $f(x,y) = (x/2, y/y)$. Then $f \in \mathrm{EC}(A)$ and $f \in \mathrm{ASC}_0(X)$.

(5.20) EXAMPLE. Let $f\colon \mathbb{R} \to \mathbb{R}$ be defined $f(x) = x/2$. Evidently, $f \in \mathrm{CA}_0(\mathbb{R})$ with the compact attractor $K = \{0\}$, but $f \notin \mathrm{ASC}(\mathbb{R})$, because $C_f = \mathbb{R}$.

6. Condensing mappings

Throughout this subsection let (X,d) be a complete metric space. If A is a bounded subset of X the *diameter* $\delta(A)$ of A is defined by

$$\delta(A) = \sup\{d(x,y) : x,y \in A\}.$$

For each bounded $A \subset X$ the *Kuratowski measure of noncompactness* $\alpha(A)$ of A is defined by

$$(6.1)\quad \alpha(A) = \inf\{\varepsilon > 0 : A \text{ is covered by a finite number of sets each with diametr no longer than } \varepsilon\}.$$

In particular, for each $\varepsilon > \alpha(A)$ there is a finite number of subsets $B_1,\dots,B_n \subset A$ such that $\delta_i(B_i) < \varepsilon$ and $\bigcup_{i=1}^n B_i = A$. Note also that $0 \leq \alpha(A) \leq \delta(A)$, for each bounded set $A \subset X$. Some fundamental properties of α are given by the following proposition.

(6.2) PROPOSITION. *Let A, B, K_n, $n = 1,2,\dots$ be bounded subsets of X then we have:*

(6.2.1) $\alpha(A \cup B) = \max\{\alpha(A), \alpha(B)\}$,

(6.2.2) $\alpha(N_\varepsilon(A)) \leq \alpha(A) + 2\varepsilon$, *where* $N_\varepsilon(A) = \{x \in X : \mathrm{dist}(x,A) = \inf\{d(x,y) : y \in A\} < \varepsilon\}$,

(6.2.3) $\alpha(A) = 0$ *if and only if A is relatively compact,*

(6.2.4) *if* $K_1 \supset \ldots \supset K_n \ldots$ *and K_i is closed nonempty for every $i = 1,2,\dots$, then* $\lim_n \alpha(K_n) = 0$ *implies that* $K_\infty = \bigcap_{n=1}^\infty K_n$ *is compact and nonempty,*

(6.2.5) $\alpha(\mathrm{cl}_X A) = \alpha(A)$,

(6.2.6) *if* $A \subset B$ *then* $\alpha(A) \leq \alpha(B)$.

Moreover, if we assume that X is a Banach space and $r \in \mathbb{R}$ is a real number, then we have:

(6.2.7) $\alpha(rA) = |r|\alpha(A)$,
(6.2.8) $\alpha(A + B) = \alpha(A) + \alpha(B)$, *where* $A + B = \{x + y : x \in A \text{ and } y \in B\}$,
(6.2.9) $\alpha(\operatorname{conv}(A)) = \alpha(A)$, *where* $\operatorname{conv}(A) = \bigcap\{B : B \text{ is convex and } A \subset B\}$.

As a simple corollary from (6.2.3) and (6.2.5), we get:

(6.3) COROLLARY. *If A is a relatively compact subset of a Banach space X, then* $\operatorname{conv}(A)$ *is relatively compact, too.*

(6.4) DEFINITION. A continuous map $f: X \to X$ is called *condensing* (γ-set contraction) provided:

(6.4.1) if $A \subset X$ is bounded and $\alpha(A) > 0$, then $\alpha(f(A)) < \alpha(A)$ (there exists $\gamma \in [0, 1)$ such that $\alpha(f(A)) \le \gamma \cdot \alpha(A)$).

Of course any compact map is γ-set contraction and any γ-set contraction map is condensing. We let

$$\begin{aligned} \mathrm{CD}(X) &= \{f: X \to X : f \text{ is condensing}\}, \\ \mathrm{CD}_0(X) &= \{f: X \to X : f \text{ is } \gamma\text{-set contraction for some } \gamma \in [0, 1)\}, \\ \mathrm{CT}(X) &= \{f: X \to X : f \text{ is contraction map}\}. \end{aligned}$$

Then, we get:

$$\begin{array}{ccccc} \mathrm{CT}(X) & \subset & \mathrm{CD}_0(X) & \subset & \mathrm{CD}(X) \\ & & \cup & & \\ & & \mathrm{C}(X) & & \end{array} \tag{6.5}$$

(6.6) LEMMA. *If $f \in \mathrm{CD}(X)$, then* $\operatorname{Fix}(f)$ *is compact provided it is a bounded set.*

PROOF. Of course $f(\operatorname{Fix}(f)) = \operatorname{Fix}(f)$. So $\alpha(f(\operatorname{Fix}(f))) = \alpha(\operatorname{Fix}(f))$ but $\alpha(f(\operatorname{Fix}(f))) < \alpha(\operatorname{Fix}(f))$. So $\alpha(\operatorname{Fix}(f)) = 0$ and $\operatorname{Fix}(f)$ is compact because it is closed. □

(6.7) LEMMA. *If (X, d) is a complete bounded space and $f: X \to X$ is a condensing map, then we have*

$$\lim_{n \to \infty} \alpha(\overline{f^n(X)}) = 0.$$

The above lemma is classical; for the proof see, for example, [AG1].

(6.8) LEMMA. *Let (X, d) be a complete bounded space. If $f \in \mathrm{CD}(X)$, then $f \in \mathrm{ASC}(X)$, in particular $f \in \mathrm{CA}(X)$.*

PROOF. First observe that (6.7) implies that the core

$$C_f = \bigcap_{n=1}^{\infty} \overline{f^n(X)}$$

is compact and nonempty. Moreover, let $0(x) = \{x, f(x), f^2(x), \dots\}$ be the orbit of $x \in X$ with respect to f. Then, we have $0(x) = \{x\} \cup f(0(x))$. So if we assume that $\alpha(0(x)) > 0$ then we get $\alpha(0(x)) = \alpha(f(0(x))) < \gamma(0(x))$, a contradiction. So $\alpha(0(x)) = 0$ and our claim follows from (6.2.3). The proof is completed. □

THE LEFSCHETZ FIXED POINT THEOREM

This section is fundamental in our considerations. We would like to show that the Lefschetz fixed point theorem is true for CAC-maps of arbitrary ANRs (absolute neighbourhood retracts) and condensing maps of some special ANRs (which we shall define later). Instead of this we shall discuss relative versions and periodic versions of the Lefschetz fixed point theorem. Some further problems will be treated in the last chapter.

For further information see: [Bo2], [Bow1]–[Bow5], [Br1], [Do], [FG], [FGo], [Gr1]–[Gr4], [Le], [Kr], [Nu], [Th].

7. The case of open subsets in Euclidean spaces

Let $\mathbb{R}^n$ denote the n-dimensional Euclidean space. Let $U \subset \mathbb{R}^n$ be an open subset. We shall consider the following class of mappings

$$M_c(U, \mathbb{R}^n) = \{f: U \to \mathbb{R}^n : f \text{ is continuous and } \mathrm{Fix}(f) \text{ is compact}\}.$$

Following A. Dold (see [Do]) we shall define a map ind: $M_c(U, \mathbb{R}^n) \to \mathbb{Z}$ called the *fixed point index function.*

In this order we proceed as follows. For given $f \in M_c(U, \mathbb{R}^n)$, we have

$$\begin{array}{ccccc} S^n & \xrightarrow{\ l\ } & (S^n, S^n \setminus \mathrm{Fix}(f)) & \xleftarrow{\ j\ } & (U, U \setminus \mathrm{Fix}(f)) \\ & & & & \Big\downarrow {\scriptstyle i-f} \\ & & & & (\mathbb{R}^n, \mathbb{R}^n \setminus \{0\}), \end{array}$$

where l, j are the respective inclusions, $S^n = \mathbb{R}^n \cup \{\infty\}$ is the one point compactyfiction of $\mathbb{R}^n$ and $(i - f)(x) = x - f(x)$, for every $x \in U$.

Now, we apply to the above diagram the n-dimensional singular homology functor with integer coefficients and we get

$$\begin{array}{ccccc} \mathbb{Z} \approx H_n(S^n) & \xrightarrow{l_n} & H_n(S^n, S^n \setminus \mathrm{Fix}(f)) & \xleftarrow[j_n]{\sim} & H_n(U, U \setminus \mathrm{Fix}(f)) \\ & & & & \Big\downarrow {\scriptstyle (i-f)_n} \\ & & & & \mathbb{Z} = H_n(\mathbb{R}^n, \mathbb{R}^n \setminus \{0\}). \end{array}$$

Observe that using the excision axiom (for $B = S^n \setminus U$) we conclude that j_n is an isomorphism. We let

$$\text{(7.1)} \qquad \mathrm{ind}(f) = [(i-f)_n \circ (j_n)^{-1} \circ l_n](1).$$

Consequently, our fixed point index $\mathrm{ind}(f)$ of f is an integer.

Again, following A. Dold, we shall list some properites of the fixed point index defined above.

(7.2) PROPERTIES.

(7.2.1) (Existence) *If* $\mathrm{ind}(f) \neq 0$, *then* $\mathrm{Fix}(f) \neq \emptyset$.

(7.2.2) (Localization) *If* V *is an open subset of* $\mathbb{R}^n$ *such that* $\mathrm{Fix}(f) \subset V \subset U$, *for some* $f \in M_c(U, \mathbb{R}^n)$, *then* $\widetilde{f} \in M_c(V, \mathbb{R}^n)$, $\widetilde{f}(x) = f(x)$, *for every* $x \in V$ *and* $\mathrm{ind}(f) = \mathrm{ind}(\widetilde{f})$.

(7.2.3) (Additivity) *Assume* $f \in M_c(U, \mathbb{R}^n)$, U_1, U_2 *are two open subsets of* $\mathbb{R}^n$ *such that* $U = U_1 \cup U_2$ *and* $U_1 \cap U_2 = \emptyset$. *If* $f_1 \in M_c(U_1, \mathbb{R}^n$ *and* $f_2 \in M_c(U_2, \mathbb{R}^n)$, $f_1(x) = f(x)$ *and* $f_2(x) = f(x)$, *for every* $x \in U_1$ *or* $x \in U_2$, *respectively, then* $\mathrm{ind}(f) = \mathrm{ind}(f_1) + \mathrm{ind}(f_2)$.

(7.2.4) (Homotopy) *If* $f, g \in M_c(U, \mathbb{R}^n)$ *are homotopic (in the class of* $M_c(U, \mathbb{R}^n)$*), i.e. there exists homotopy* $h: U \times [0,1] \to \mathbb{R}^n$ *such that*

$$h(x,0) = f(x), \quad h(x,1) = g(x), \quad \textit{for every } x \in U$$

and the set $\mathrm{Fix}(h) = \{x \in U : h(x,t) = x, \textit{ for some } t \in [0,1]\}$ *is compact, then* $\mathrm{ind}(f) = \mathrm{ind}(g)$.

(7.2.5) (Unity) *If* $f: U \to \mathbb{R}^n$ *is a constant map, i.e.* $f(x) = x_0$, *for every* $x \in U$, *then*

$$\mathrm{ind} = \begin{cases} 1 & \textit{if } x_0 \in U, \\ 0 & \textit{if } x_0 \notin U. \end{cases}$$

(7.2.6) (Multiplicativity) *Assume that* $f \in M_c(U, \mathbb{R}^n)$, $g \in M_c(W, \mathbb{R}^m)$. *Then* $f \times g \in M_c(U \times W, \mathbb{R}^{n+m})$, $(f \times g)(x,y) = (f(x), g(y))$, *for every* $x \in U$ *and* $y \in W$, *and* $\mathrm{ind}(f \times g) = \mathrm{ind}(f) \cdot \mathrm{ind}(g)$.

(7.2.7) (Commutativity) *Assume that* $U \subset \mathbb{R}^n$, $V \subset \mathbb{R}^m$ *are open and* $f: U \to \mathbb{R}^m$, $g: V \to \mathbb{R}^n$ *are two continuous maps. Consider* $g \circ f: f^{-1}(U) \to \mathbb{R}^n$ *and* $f \circ g: g^{-1}(U) \to \mathbb{R}^m$. *Then*

$$g \circ f \in M_c(f^{-1}(V), \mathbb{R}^n) \Leftrightarrow f \circ g \in M_c(g^{-1}(U), \mathbb{R}^m)$$

and in such a case we have $\operatorname{ind}(g \circ f) = \operatorname{ind}(f \circ g)$.

(7.2.8) (Normalization) *Assume* $f: U \to U$ *is a compact map and consider the map* $\overline{f}: U \to \mathbb{R}^n$, $\overline{f}(x) = f(x)$, *for every* $x \in U$. *Then*

(a) $\overline{f} \in M_c(U, \mathbb{R}^n)$,

(b) f *is a Lefschetz map,*

(c) $\operatorname{ind}(\overline{f}) = \Lambda(f)$.

At first, we would like to remark that propeties (7.2.1)–(7.2.5) are simple consequences of (7.1) and some another properties of the homology functor.

The proof (7.2.7) is geometrical and can be obtained by using (7.2.2), (7.2.4)–(7.2.6).

For the proof (7.2.6) and (7.2.8) we need more homology theory, namely the universal coefficients theorem and the Künneth theorem are used. All details are contained in Dold's book [Do].

(7.3) Remark. Let us observe that the Lefschetz number was defined as an element of the given field K but the fixed point index as an integer. If we shall take $K = \mathbb{R}$, then, in view of the universal coefficients theorem, (7.2.8)(c) says that the Lefschetz number $\Lambda(f)$ of f is an integer and is equal to $\operatorname{ind}(\overline{f})$.

Finally, from (7.2.8)(c) and (7.2.1), we obtain the following special case of the Lefschetz fixed point theorem.

(7.4) Theorem. *Let* U *be an open subset of* $\mathbb{R}^n$ *and let* $f: U \to U$ *be a compact map. Then:*

(7.4.1) f *is a Lefschetz map, and*

(7.4.2) $\Lambda(f) \neq 0$ *implies that* $\operatorname{Fix}(f) \neq \emptyset$.

8. Compact maps of arbitrary ANRs

In this section we shall present the Lefschetz fixed point theorem for compact mappings of arbitrary ANRs as it was done by A. Granas (see [Gr1], comp. also [Gr4]).

We shall make use of the following theorem.

(8.1) THEOREM (Schauder Approximation Theorem). *Let U be an open subset of a normed space E and let $f: X \to U$ be a compact mapping. Then for every $\varepsilon > 0$ there exists a natural number $n = n(\varepsilon)$, an n-dimensional subspace $E^n \subset E$ and a compact map $f_\varepsilon: X \to U$ such that:*

(8.1.1) $\|f_\varepsilon(x) - f(x) < \varepsilon$, *for every* $x \in X$,
(8.1.2) $f_\varepsilon(X) \subset E^n$,
(8.1.3) f_ε *is homotopic to* f.

SKETCH OF THE PROOF. We can assume that $\varepsilon > 0$ is small enough, i.e. $f(X)$ is contained in the finite union of open balls $B(y_i, \varepsilon) \subset U$, $i = 1, \ldots, k$.

We let $\mu_i: U \to [0, \infty)$, $\mu_i(x) = \max\{0, \varepsilon - \|f(x) - y_i\|\}$. Then we are able to define $f_\varepsilon: X \to U$ by putting:

$$f_\varepsilon(x) = \sum_{j=1}^{k} \lambda_j(x) \cdot y_j \quad \text{where } \lambda_j(x) = \frac{\mu_j(x)}{\sum_{i=1}^{k} \mu_i(x)},$$

for every $x \in X$. The verification that f_ε satisfies (8.1.1)–(8.1.3) is immediate. □

(8.2) THEOREM. *Let U be an open subset of a normed space and let $f: U \to U$ be a compact map. Then:*

(8.2.1) f *is a Lefschetz map, and*
(8.2.2) $\Lambda(f) \neq 0$ *implies that* $\mathrm{Fix}(f) \neq \emptyset$.

PROOF. Take $\varepsilon_m = 1/m$ and according to (8.1) let f_m be ε_m-approximation of f. Then for every $m = 1, 2, \ldots$ we have $n = n(m)$ such that $f_m(U) \subset E^{n(m)}$. Let $U_{n(m)} = U \cap E^{n(m)}$ and $\widetilde{f}_m(x) = f_m(x)$, for every $x \in U_{n(m)}$ be defined as follows:

$$\widetilde{f}_m(x) = f_m(x), \quad \text{for every } x \in U_{n(m)}.$$

We have a diagram (for every $m = 1, 2, \ldots$):

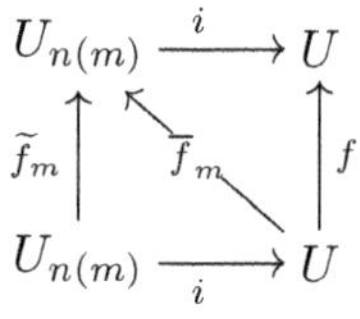

in which i is the inclusion map and $\overline{f}_m(x) = f_m(x)$, for every $x \in U$. It follows from (8.1.3) that the following diagram is commutative

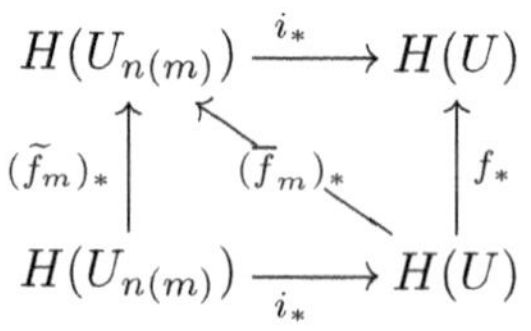

But, in view of (7.4), $(\widetilde{f}_m)_*$ is a Leray endomorphism. So from (2.4) we deduce that f_* is a Leray endomorphism. Consequently, f is a Lefschetz map.

Now, assume that $\Lambda(f) \neq 0$. So the commutativity of the above diagram implies that $\Lambda(\widetilde{f}_m) \neq 0$, for every $m = 1, 2, \ldots$ Using Theorem (7.4), for every m, we get a point $x_m \in U_{n(m)} \subset U$ such that $f_m(x_m) = x_m$.

In view of (8.1.1), we have:

$$\|f(x_m) - f_m(x_m)\| = \|f(x_m) - x_m\| < \frac{1}{m}. \tag{8.3}$$

Since f is compact, we can assume without loss of generality that $\lim_m f(x_m) = x_0$, consequently from (8.3), we get that $\lim_m x_m = x_0$.

Finally, from the continuity of f we deduce that $f(x_0) = x_0$ and the proof is completed. □

Following K. Borsuk (see [Bo1]), we define:

(8.4) Definition. A metric space X is called an *absolute neighbourhood retract* (written $X \in \mathrm{ANR}$) provided there exists a normed space E and an open subset $U \subset E$ such that X is r-dominated by U.

(8.5) Definition. A metric space X is called an *absolute retract* (written $X \in \mathrm{AR}$) provided there exists a normed space E such that X is r-dominated by E.

We have:

(8.6) If $X \in \mathrm{AR}$ then $X \in \mathrm{ANR}$.

It can be proved that if a metric space X is r-dominated by a convex set C in a normed space E, then $X \in \mathrm{AR}$ (this can be proved by using the Dugundji extension theorem).

Since every normed space is contractible, so every $X \in \mathrm{AR}$ is contractible. Therefore, we have:

(8.7) If $X \in \mathrm{AR}$, then

$$H_m(X) = \begin{cases} \mathbb{Q} & \text{for } m = 0, \\ 0 & \text{for } m > 0, \end{cases}$$

where we consider the homology functor with coeficients in $\mathbb{Q}$.

Consequently, in view of (4.6), we obtain:

(8.8) If $X \in \mathrm{AR}$ and $f\colon X \to X$ is a continuous map, then f is a Lefschetz map and $\Lambda(f) = 1$.

Now, we are able to formulate the main result of this section (see [Gr1]).

(8.9) THEOREM (The Lefschetz Fixed Point Theorem for Compact Mappings of ANRs). *Let* $X \in \text{ANR}$ *and let* $f: X \to X$ *be a compact map. Then we have:*

(8.9.1) f *is a Lefschetz map, and*
(8.9.2) $\Lambda(f) \neq 0$ *implies that* $\text{Fix}(f) \neq \emptyset$.

PROOF. By definition there exists an open subset U in a normed space E and two continuous mappings: $r: U \to X$ and $s: X \to U$ such that $r \circ s = \text{id}_X$. We have the following commutative diagram:

$$\begin{array}{ccc} X & \xrightarrow{s} & U \\ {\scriptstyle f}\uparrow & \nwarrow{\scriptstyle f\circ r} & \uparrow{\scriptstyle s\circ f\circ r} \\ X & \xrightarrow[s]{} & U \end{array}$$

in which $s \circ f \circ r$ is a Lefschetz map (comp. (8.2)). Therefore our claim follows from (4.3) and (4.4.1). The proof is completed. □

(8.10) COROLLARY. *If* X *is an acyclic* ANR *or* $X \in \text{AR}$, *then any compact map* $f: X \to X$ *has a fixed point.*

From (8.9) and (4.6) we get:

(8.11) COROLLARY. *Let* $f: X \to X$ *be a compact map of an* ANR*-space* X *into itself. If there exists an acyclic subset* $A \subset X$ *and a natural number* n *such that* $f^n(X) \subset A$, *then* f *has a fixed point.*

(8.12) REMARK. We would like to remark that the fixed point index considered in Section 7 can be taken up for arbitrary ANRs (see [Gr3]). Namely, assume that $X \in \text{ANR}$, U is an open subset of X and $f: U \to X$ is a compact map with $\text{Fix}(f)$ to be compact. Then using the Schauder approximation theorem the fixed point index $\text{ind}(f)$ of f can be defined and it satisfies all properties contained in (7.2).

9. CAC-mappings of arbitrary ANRs

In this section we shall formulate the Lefschetz fixed point theorem in the most general form which is known up to date.

(9.1) THEOREM (The Lefschetz Fixed Point Theorem). *Let* $X \in \text{ANR}$ *and let* $f \in \text{CAC}(X)$. *Then* f *is a Lefschetz map and* $\Lambda(f) \neq 0$ *implies that* $\text{Fix}(f) \neq \emptyset$.

PROOF. We choose an open subset $U \subset X$ according to the definition (5.7). Then $\widetilde{f}: U \to U$, $\widetilde{f}(x) = f(x)$ is a compact map. We consider the map $\overline{f}: (X, U) \to (X, U)$, $\overline{f}(x) = f(x)$, for every $x \in X$.

Since U is an open set which absorbs points, it is easy to proove that U absorbs compact sets (see (5.8.1)). So, in view of (4.5), we obtain that $\overline{f}$ is a Lefschetz map and $\Lambda(\overline{f}) = 0$ (observe that $\overline{f}_*$ is weakly nilpotent). Now U as an open subset of ANR-space X is an ANR, too, and $\widetilde{f}: U \to U$ is compact from (8.9) we obtain that $\widetilde{f}$ is a Lefschetz map.

Consequently, by applying (4.4), we conclude that f is a Lefschetz map and

$$0 = \Lambda(\overline{f}) = \Lambda(f) - \Lambda(\widetilde{f}).$$

Now, if we assume that $\Lambda(\widetilde{f}) \neq 0$ then by using (8.9) $\mathrm{Fix}(\widetilde{f}) \neq \emptyset$. But $\mathrm{Fix}(f) = \mathrm{Fix}(\widetilde{f})$, so the proof is completed. □

From (9.1) in particular we get:

(9.2) COROLLARY. *Let* $X \in \mathrm{ANR}$. *If one of the following conditions is satisfied:*

(9.2.1) $$f \in \mathrm{C}(X),$$

(9.2.2) $$f \in \mathrm{EC}_0(X),$$

(9.2.3) $$f \in \mathrm{ASP}_0(X),$$

(9.2.4) $$f \in \mathrm{CA}_0(X),$$

then f is a Lefschetz map and $\Lambda(f) \neq 0$ implies that $\mathrm{Fix}(f) \neq \emptyset$. *In particular, if* $X \in \mathrm{AR}$ *and f satisfies one of the conditions* (9.2.1)–(9.2.4), *then* $\mathrm{Fix}(f) \neq \emptyset$.

(9.3) OPEN PROBLEM. Is it possible to prove (3.3) for $f \in \mathrm{EC}(X)$, or $f \in \mathrm{ASP}(X)$ or $f \in \mathrm{CA}(X)$?

The above problem is quite old and in our opinion it seems to be very difficult. The following open problem seems to be less difficult:

(9.4) OPEN PROBLEM. Assume that $X \in \mathrm{AR}$ and $f \in \mathrm{EC}_0(X)$ or $f \in \mathrm{ASP}_0(X)$, or $f \in \mathrm{CA}_0(X)$, or $f \in \mathrm{CAC}(X)$. Prove that $\mathrm{Fix}(f) \neq \emptyset$ without using the homology functor.

10. Condensing mappings

The problem to prove the Lefschetz fixed point theorem for condensing mappings of arbitrary complete ANRs is quite complicated.

The first partial result was proved by R. Nussbaum in 1975 (see [Nu]) and can be formulated as follows.

(10.1) THEOREM. *Let U be an open subset of a Banach space and assume that*

$$f \in \mathrm{CD}(U) \cap \mathrm{CA}(U).$$

Then f is a Lefschetz map and $\Lambda(f) \neq 0$ implies that $\mathrm{Fix}(f) \neq \emptyset$.

Below we shall try to omit the assumption that $f \in \mathrm{CA}(U)$.

(10.2) DEFINITION. A complete bounded space (X, d) is called a *special* ANR (written $X \in \mathrm{ANR}_s$) provided there exists an open subset U of a Banach space E and two continuous maps $r: U \to X$ and $s: X \to U$ such that

(10.2.1) $r \circ s = \mathrm{id}_X$,

(10.2.2) r and s are non-expansive, i.e. $\alpha(r(B)) \leq \alpha(B)$ and $\alpha(s(A)) \leq \alpha(A)$, for every bounded sets A and B.

We prove the following Lefschetz type fixed point theorem.

(10.3) THEOREM. *Let $X \in \mathrm{ANR}_s$ and $f \in \mathrm{CD}(X)$. Then f is a Lefschetz map and $\Lambda(f) \neq 0$ implies that* $\mathrm{Fix}(f) \neq \emptyset$.

PROOF. From (6.7) we deduce that f has a compact attractor $A \subset X$. Let U, $r: U \to X$ and $s: X \to U$ be chosen according to (10.2). We define a map $\widetilde{f}: U \to U$ by putting $\widetilde{f} = s \circ f \circ r$.

Let B be a bounded subset of U, then we get:

$$\alpha(\widetilde{f}(B)) = \alpha(s(f(r(B)))) \leq \alpha(f(r(B))) < \alpha(r(B)) \leq \alpha(B)$$

and therefore $\widetilde{f} \in \mathrm{CD}(U, U)$.

Now, observe that $s(A)$ is a compact attractor for $\widetilde{f}$. Consequently, $\widetilde{f}$ satisfies all assumptions of (10.1), in particular $\widetilde{f}$ is a Lefschetz map.

Moreover, we have the following commutative diagram:

$$\begin{array}{ccc} X & \xrightarrow{\;s\;} & U \\ {\scriptstyle f}\Big\uparrow & \nwarrow^{f \circ r} & \Big\uparrow{\scriptstyle \widetilde{f} = s \circ f \circ r} \\ X & \xrightarrow[\;s\;]{} & U \end{array}$$

and hence our claim follows from (10.1) and (4.3) (see also Remark (4.3.1)). □

We shall make use of the following:

(10.4) LEMMA. *Let $f: X \to X$ be a continuous map. Assume further that A is a compact attractor of f and V is an open neighbourhood of A in X. Then there exists an open $U \subset X$ such that:*

(10.4.1) $$f(U) \subset U,$$

(10.4.2) $$A \subset U \subset V.$$

PROOF. Let $U = \bigcap_{n=0}^{\infty} f^{-n}(V)$. Then $f(U) \subset U$ and $A \subset U$. We only need to show that U is an open subset of X. On the contrary, suppose that there exists a sequence $\{x_n\} \subset X \backslash U$ such that $\lim_{n\to\infty} x_n = x$ and $x \in U$. Let $K = \{x_n\} \cup \{x\}$. Then K is compact and consequently there exists m such that $f^i(K) \subset V$, for all $i \geq m$. Hence $x_n \in \bigcap_{i=m}^{\infty} f^{-i}(V)$. But $x_n \notin U$ so $x_n \notin \bigcap_{i=m}^{\infty} f^{-i}(V)$ and from the continuity of f it follows $x \notin \bigcap_{i=0}^{m} f^{-i}(V)$ contradicts the fact that $x \in U$. □

Now, we shall prove the following result which is additional to (10.1) and (10.3).

(10.5) THEOREM. *Assume that X is a nonexpansive retract of some open subset W of a Banach space E. Assume further that $f \in \mathrm{CA}(X)$ and A is a compact attractor of f. If there exists an open nieghbourhood V of A in X such that the retraction $f|_V \colon V \to X$ of f to V is a condensing map; then f is a Lefschetz map and $\Lambda(f) \neq 0$ implies that* $\mathrm{Fix}(f) \neq \emptyset$.

PROOF. For the proof consider the following commutative diagram:

$$\begin{array}{ccc} X & \xrightarrow{\ i\ } & W \\ {\scriptstyle f}\uparrow & & \uparrow{\scriptstyle g=i\circ f\circ r} \\ X & \longrightarrow & W \end{array}$$

in which $r\colon W \to W$ is a nonexpansive retraction and $i\colon X \to W$ is the inclusion map. So f is a Letschetz map iff g is a Lefschetz map. Observe that $A = i(A)$ is an compact attractor of g and moreover, the retraction

$$g|_{r^{-1}(V)}\colon r^{-1}(V) \to W$$

of g to $r^{-1}(V)$ is a condensing map. So by using (10.4) we get an open U of W such that $\widetilde{g}\colon U \to U$, $\widetilde{g}(u) = g(u)$ is a condensing map with compact attractor. Consequently (10.1) holds true for $\widetilde{g}$. On the other hand $g\colon W \to W$ has the following properties:

(a) $g \in \mathrm{CA}(W)$,
(b) for some open subset $U \subset W$ containing an attractor for g we have $f(U) \subset U$ and $\widetilde{g}\colon U \to U$, $\widetilde{g}(x) = g(x)$ is a Lefschetz map,
(c) U absorbs points and consequently compact sets.

Therefore, we can consider the map $\overline{g}\colon (W, U) \to (W, U)$, $\overline{g}(x) = g(x)$, for every $x \in X$ and we can proceed as in the proof of (9.1). We get that $(\overline{g})_*$ is weakly nilpotent and hence $\Lambda(g) = \Lambda(\widetilde{g})$.

Finally, we have $\Lambda(f) = \Lambda(g) = \Lambda(\widetilde{g})$ and our theorem follows from the commutativity of the above diagram and (10.1). The proof is completed. □

(10.6) REMARK. Observe that any γ-set contraction map is condensing. Therefore all results of this section remain true for γ-set contraction mappings.

The following problem is open.

(10.7) OPEN PROBLEM. Assume that $X \in \mathrm{ANR}$ and $f \in \mathrm{CD}(X)$. Does the Lefschetz fixed point theorem hold for such an f?

11. Relative versions of the Lefschetz fixed point theorem

From the point of view of applications in dynamical systems (for example) the relative version of the Lefschetz fixed point theorem is important. In the relative version we get not only the existence of fixed points but also some information about their localization. For the respective proofs, instead of the Lefschetz number (the normalization property), we need the fixed point index for the appropriate class of mappings.

At first, we would like to remark the following two facts:

(11.1) the fixed point index is well defined for CAC-mappings on ANRs (see [AG] or [Go2]),

(11.2) the fixed point index is well defined for condensing CA-mappings on open subset of Banach spaces (see [AG]).

We have the following three versions of relative Lefschetz fixed point theorem:

(11.3) THEOREM. *Let* $X_0 \subset X$ *and* $X, X_0 \in \mathrm{ANR}$. *Assume that* $f\colon (X, X_0) \to (X, X_0)$ *is a map such that* f_X *and* f_{X_0} *are* CAC*-mappings. Then the Lefschetz number* $\Lambda(f)$ *of* f *is well defined and* $\Lambda(f) \neq 0$ *implies that*

$$\mathrm{Fix} \cap (\overline{X \setminus X_0}) \neq \emptyset.$$

(11.4) THEOREM. *Let* W *be an open subset of a Banach space* E *and* W_0 *be an open subset of* W, *and let* $f\colon (W, W_0) \to (W, W_0)$ *be a mapping such that:*

(11.4.1) f_W *and* f_{W_0} *are condensing mappings with compact attractors.*

Then the Lefschetz number $\Lambda(f)$ *of* f *is well defined and* $\Lambda(f) \neq 0$ *implies that*

$$\mathrm{Fix} \cap (\overline{W \setminus W_0}) \neq \emptyset.$$

Similarly, for k-set contraction mappings we get:

(11.5) THEOREM. *Let* W *and* W_0 *be the same as in* (11.4) *and* $f\colon (W, W_0) \to (W, W_0)$ *be a mapping such that:*

(11.5.1) f_W *and* f_{W_0} *are* k*-set contractions with relatively compact orbits.*

Then the Lefschetz number $\Lambda(f)$ *of* f *is well defined and* $\Lambda(f) \neq 0$ *implies that*

$$\mathrm{Fix} \cap (\overline{W \setminus W_0}) \neq \emptyset.$$

As we mentioned above more details concerning the relative version of the Lefschetz fixed point theorem can be found in [AG], [GD] and [Go2].

12. Existence of periodic points

Let $f: X \to X$ be a continuous map. A point $x \in X$ is called *periodic for f with period n* provided $f^n(x) = x$. Observe that any fixed point of f is periodic with the period n, for arbitrary $n \geq 1$.

In what follows we shall assume that $X \in$ ANR, Now, we are able to prove:

(12.1) Theorem (Periodic Point Theorem). *Let $f \in$ CAC(X). If $\Lambda(f) \neq 0$ or $P(f) \neq 0$, then f has a periodic point with period n, where $m+1 \leq n \leq P(f)$ and m is an arbitrary natural number ($m \geq 0$).*

Sketch of Proof. It follows from Theorem (9.1) that f is a Lefschetz morphism. In view of Theorem (4.8) it is sufficient to assume that $P(f) \neq 0$. Applying Theorem (4.8), for any $m \geq 0$, we get n such that $\Lambda(f^n) \neq 0$, where $m+1 \leq n \leq m + P(f)$.

Since the composition of CAC-morphisms is CAC-morphism again, we deduce from Theorem (9.1) that the set Fix(f^n) of fixed points of f^n is nonempty. Of course, any $x \in$ Fix(f^n) is n-periodic point of f. Hence the proof is completed. □

Using Theorem (11.3) (instead of (9.1)) we get:

(12.2) Theorem. *Assume that $A \subset X$ is an* ANR*-space. Assume further that $f \in M((X,A),(X,A))$ is a* CAC*-morphism of pairs. If $\Lambda(f) \neq 0$ or $P(f) \neq 0$, then f has an n-periodic point in $\overline{X \setminus A}$, where $m+1 \leq n \leq m + P(f)$ and $m \geq 0$ is an arbitrary natural number.*

The proof of Theorem (12.2) is strictly analogous to the proof of Theorem (12.1).

(12.3) Remark. It can be easily checked that $\Lambda(f)$ and $P(f)$ are homotopic invariants.

(12.4) Remark. For more details we recommend [Ag], [GD] and [Go2].

FINAL REMARKS AND COMMENTS

In this part we are going to signalize some other possibilities of generalizations of the Lefschetz fixed point theorem.

13. Other generalizations of the Lefschetz fixed point theorem

There are several possibilities of generalizing the Lefschetz fixed point theorem to compact spaces. Some of them will be presented in this section. For further details we recommend: [Bo2], [EM], [Gr2], [Gr4], [Th].

(13.1) DEFINITION. We shall say that the *metric compact space* X *is* ε-*dominated by a metric space* Y provided there are two continuous mappings $s_\varepsilon: X \to Y$, $r_\varepsilon: Y \to X$ such that, for every $x \in X$, we have $d(x, r_\varepsilon(s_\varepsilon(x))) < \varepsilon$, in other words, the mappings $r_\varepsilon \circ s_\varepsilon$ and id_X are ε-close (written $r_\varepsilon \circ s_\varepsilon \sim_\varepsilon \mathrm{id}_X$).

(13.2) DEFINITION. We shall say that a compact metric space (X, d) is an *approximative* ANR (written $X \in$ AANR) provided there exists an open subset U in a Banach space E such that X is ε-dominated by U for every $\varepsilon > 0$.

(13.3) REMARK. Since every compact metric space can be embedded into the Hilbert cube J^∞, we can assume in (13.2) that U is an open subset of J^∞.

The following two pieces of information are taken from [Go1].

(13.4) PROPERTY. *If* $X \in$ AANR, *then* $H(X)$ *is of finite type (here it is important to consider the Čech homology functor).*

(13.5) PROPERTY. *If* (X, d) *is a compact metric space such that* $H(X)$ *is a finite type. Then there exists* $\varepsilon_0 > 0$ *such that for every metric space* Y *and for any two continuous mappings* $f, g: Y \to X$, *if* $f \sim_\varepsilon g$, $0 < \varepsilon \le \varepsilon_0$, *then* $f_* = g_*$.

(13.6) THEOREM (The Lefschetz Fixed Point Theorem for AANRs). *Let* $X \in$ AANR *and let* $f: X \to X$ *be a continuous map. If* $\lambda(f) \neq 0$, *then* f *has a fixed point.*

PROOF. It follows from (13.4) that $H(X)$ is a space of finite type so $\lambda(f) = \lambda(f_*)$ is well defined. Now, choose U, r_ε, s_ε according to (13.2) and assume that $\varepsilon \le \varepsilon_0$, where ε_0 is taken as in (13.5). We let $e_n = 1/n$ and assume that $\varepsilon_n \le \varepsilon_0$, for every $n = 1, 2, \ldots$ Then for every n we have the following diagram:

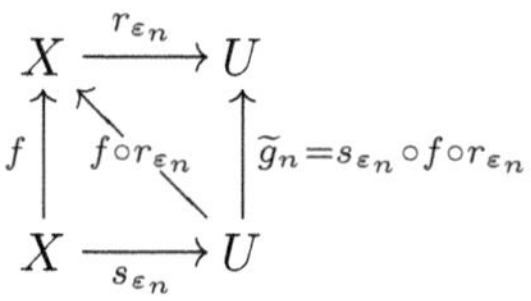

which, in view of (13.5), is homologically commutative.

So $\lambda(f) = \lambda(g_n)$ and g_n is a compact map for every n. If we assume that $\lambda(f) \neq 0$ then $\Lambda(g_n) \neq 0$ and from (8.2) we deduce that there is $u_n \in U$ such that

$$g_n(x_n) = (s_{\varepsilon_n} \circ f \circ r_{\varepsilon_n})(u_n) = u_n.$$

Consequently, we get $r_{\varepsilon_n} \circ s_{\varepsilon_n} \circ f \circ r_{\varepsilon_n}(u_n) = r_{\varepsilon_n}(u_n)$ and hence

$$d(f(r_{\varepsilon_n}(u_n)), r_{\varepsilon_n}(u_n)) < 1/n.$$

So if we put $r_{\varepsilon_n}(u_n) = x_n$ and $\lim_n x_n = x$, then from the above inequality and continuity of f we obtain $f(x) = x$ and the proof is completed. □

(13.7) REMARK. The above notion of an AANR-space can be generalized as follows. A compact metric space (X, d) is a *weak approximative* ANR (written $X \in$ w-AANR) provided for every $\varepsilon > 0$ there exists an open subset U_ε in a Banach space which ε-dominates X.

Of course, we have AANR $\subset$ w-AANR.

Now, the proof of (13.6) can be adopted to the following situation.

(13.8) THEOREM (The Lefschetz Fixed Point Theorem for w-AANNRs). *Assume that $X \in$ w-AANR and that $H(X)$ is of a finite type. If $f: X \to X$ is a continuous map with $\lambda(f) \neq 0$, then* $\mathrm{Fix}(f) \neq \emptyset$.

A compactum (i.e. a compact metric space) is called *appropriate* if $H(X)$ is of finite type, where H denotes as above the Čech homology functor with coefficients in $\mathbb{Q}$.

For a compactum X by $\mathrm{Cov}_f(X)$ we shall denote the family of all finite open covers of X. For $\alpha \in \mathrm{Cov}_F(X)$ by X_α we shall denote the nerve of α and by $C_\#(X_\alpha)$ its chain complex.

A map f from a compactum X into another compactum Y is said to be an *NE-map* if there exist AR-spaces M, N containing X and Y, respectively, and there is a map $\widetilde{f}: M \to N$ with $\widetilde{f}(x) = f(x)$, for every $x \in X$, such that the following condition is satisfied:

(13.9) for every $\varepsilon > 0$ there is a neighbourhood U of X in M such that for every neighbourhood V of Y in N there is a map $g: U \to V$ with $\rho(\widetilde{f}(x))$, $g(x) < \varepsilon$, for every point $x \in U$.

One shows (see [Bo2]) that the choice of spaces $M, N \in$ AR containing X and Y, respectively, and also the choice of a map $\widetilde{f}: M \to M$ satisfying the condition $\widetilde{f}(x) = f(x)$, for every $x \in X$ are immaterial (are of little significance).

The class of NE-maps is quite general. In particular, if X or Y is an ANR, then every map $f: X \to Y$ is nearly extendable. Moreover, if X, Y are compacta and if all values of a map $f: X \to Y$ belong to an ANR-set $A \subset Y$, then f is an NE-map. One also shows that for all compacta X, Y the subset of the functional space Y^X consisting of all NE-maps is closed in Y^X. Finally, if the shape of a compactum X is trivial, then every NE-map $f: X \to X$ has a fixed point (see: [Bo1] and [Bo2]).

(13.10) THEOREM. *Let X be an appropriate compactum. Then every* NE-*map $f: X \to X$ with $\Lambda(f) \neq 0$ has a fixed point.*

A continuous function $f: X \to X$ is said to be a $\mathbb{Q}$-simplicial map if for each $\alpha \in \mathrm{Cov}_f(X)$, there exists an $\alpha \in \mathrm{Cov}_f(X)$ with $\alpha' > \alpha$ such that for each

$\beta \in \mathrm{Cov}_f(X)$ there exists a chain map α': $\varphi: C_\#(X_{\alpha'}) \to C_\#(X_\beta)$ such that $\sup(\varphi(c)) \subseteq ST_\beta St_\alpha f(\sup(c))$, for $c \in C_\#(X_{\alpha'})$.

(13.11) THEOREM. *Let X be an appropriate compactum and let $f: X \to X$ be a $\mathbb{Q}$-simplicial map, then $\lambda(f) \neq 0$ implies that* $\mathrm{Fix}(f) \neq \emptyset$.

One can prove that if $\mathrm{id}_X: X \to X$ is a $\mathbb{Q}$-simplicial map, then any continuous $f: X \to X$ is $\mathbb{Q}$-simplicial.

For further information we recommend [Th].

14. The converse problem

The converse of the Lefschetz fixed point theorem is false in general, even for polyhedra. In this section we shall see what additional assumptions are needed to get the converse to the Lefschetz fixed point theorem. For more details we recommend [Br1].

We say that a connected geometric complex $|K|$ *is of type* S if the dimension of K is at least three and if, for each vertex $v \in K$ $|St(v)|$ is connected, where, we recall, $St(v)$ denotes the collection of all simplices of K which contain v.

A polyhedron X *is of type* S if it is homeomorphic to a geometric complex of type S.

A polyhedron X is an H-space if there exists a point $e \in X$ and a continuous map $\mu: X \to X$ such that $\mu(x, e) = \mu(e, x) = x$, for all $x \in X$.

Note that if X is an H-space then $T(X)$ [the so called Jiang subgroup (see [Br1] for details)] is equal to $\pi_1(X)$, where $\pi_1(X)$ is the fundamental group of X.

We have the following result:

(14.1) THEOREM. *If X is a simply connected H-space and $f: X \to X$ is a continuous map such that* $\mathrm{Fix}(f) = \emptyset$, *then* $\lambda(f) = 0$.

(14.2) REMARK. The assumption that X is an H-space can be replaced by the condition $T(X) = \pi_1(X)$.

15. Non-metric case

In this section we shall survey current results concerning the Lefschetz fixed point theorem in non-metric spaces. For more details we recommend [AG], [BEM], [FG], [FGO] and [Gr4].

(15.1) DEFINITION. A Hausdorff space X is called a *Lefschetz space* (denoted by $X \in \mathcal{L}$) provided for every $f \in \mathrm{CAC}(X)$ we have:

(15.1.1) f is a Lefschetz map, and
(15.1.2) if $\Lambda(f) \neq 0$, then $\mathrm{Fix}(f) \neq \emptyset$.

By $\mathrm{Cov}(X)$ we shall denote the family of all open coverings of X.

(15.2) DEFINITION. Let $\alpha \in \mathrm{Cov}(X)$. We say that X is *α-dominated by Y* provided there exist two continuous maps $s_\alpha : X \to Y$ and $r_\alpha : Y \to X$ such that $r_\alpha \circ s_\alpha \sim_\alpha \mathrm{id}_X$, i.e. there exists a homotopy $h: X \times [0,1] \to X$ such that:

(15.2.1) $h(x,0) = x$, $h(x,1) = r_\alpha(s_\alpha(x))$, for every $x \in X$, and

(15.2.2) for every $x \in X$ there is an $U_x \in \alpha$ such that $h(x,t) \in U_x$, for every $t \in [0,1]$.

Evidently, if $f \sim_\alpha g$, then f and g are α-close, i.e. for every $x \in X$, there is $U_x \in \alpha$ such that $f(x), g(x) \in U$.

By $\mathcal{R}(\mathcal{Z})$ we denote the class of all Hausdorff spaces which are r-dominated by a space in $\mathcal{Z}$. We say that $X \in \mathcal{D}(\mathcal{Z})$, if for every $\alpha \in \mathrm{Cov}(X)$ there exists a space $Y_\alpha \in \mathcal{Z}$ such that X is α-dominated by Y_α.

Clearly, $\mathcal{Z} \subset \mathcal{R}(\mathcal{Z}) \subset \mathcal{D}(\mathcal{Z})$. In fact, we are able to prove the following:

(15.3) THEOREM. $\mathcal{D}(\mathcal{Z}) = \mathcal{R}(\mathcal{Z}) = \mathcal{Z}$.

Let U be a neighbourhood of the origin in a linear topological space E. Then E is *shrinkable* provided for any $x \in \overline{U}$ and $0 < \lambda < 1$ the point $\lambda x \in U$.

It is known that shrinkable neighbourhoods form the base of E at 0. It follows that given an arbitrary neighbourhood W of 0 there is a shrinkable neighbourhood V of 0 such that $V + V \subset W$ and any interval $tx + (1-t)y$, $0 \le t \le 1$, with x and y in V is entirely contained in W. This allows us to prove the following lemma:

(15.4) LEMMA. *Let U be an open subset of a linear topological space E. Then for each $\alpha \in \mathrm{Cov}\, U$ there exists a refinement $\beta \in \mathrm{Cov}\, U$ such that any two β-close maps of any space X into U are stationary α-homotopic, i.e. there exists an α-homotopy h, such that $h(x,t) \in V_\alpha$ for some $V_\alpha \in \mathcal{L}$ and every $0 \le t \le 1$ and moreover, it is constant whenever $f(x) = g(x)$.*

(15.5) DEFINITION. Let E be a linear topological space. We say that E is *Klee admissible* provided for any compact $K \subset E$ and any $\alpha \in \mathrm{Cov}_E(K)$ there is a map $\pi_\alpha : K \to E$ such that:

(15.5.1) $\pi_\alpha(K)$ is contained in a finite dimensional subspace E^n of E,

(15.5.2) the inclusion $i: K \to E$ and π_α are α-close.

(15.6) REMARK. Evidently, every locally convex spaces and in particular every normed space is Klee admissible.

The author doesn't know an example of a topological vector space which is not Klee admissible!

Having (15.4) and (15.5) the proof of the following theorem is strictly analogous to the proof of (8.2) and (9.1) if we are doing it in two steps:

(a) for compact maps (comp. (8.2)),
(b) for arbitrary CAC-maps (comp. (9.1)).

(15.7) THEOREM. *If E is a Klee admissible space and U is an open subset of E, then $U \in \mathcal{Z}$.*

A Hausdorff space X is a *neighbourhood extension space for compact spaces* (resp. *for compact metric spaces*) provided for any pair (Y, A), where Y is a compact Hausdorff space and A a closed subset of Y (resp. Y is a compact metric space and A a closed subset of Y), and any map $f_0: A \to X$ there is an extension $f: U \to X$ of f_0 to an open neighbourhood of A in Y. The class of the neighbourhood extension spaces for compact spaces (resp. for compact metric spaces) will be denoted by NES (compact) (resp. NES (compact metric)). Clearly, NES (compact) $\subset$ NES (compact metric).

(15.8) LEMMA. *Every Tichonoff cube is a retract of a locally convex topological space.*

(15.9) THEOREM. *Every open subset of a Tichonoff cube is a Lefschetz space.*

Theorem (15.9) clearly follows from (15.7) and (15.8). Finally, we get the following two formulations of the Lefschetz fixed point theorem.

(15.10) THEOREM. NES $\subset \mathcal{L}$.

(15.11) THEOREM. *Let $X \in$ NES (compact, metric) and let $f: X \to X$ be an admissible map such that $f(X)$ is contained in a compact metrizable subset of X. Then f is a Lefschetz map and $\Lambda(f) \neq \{0\}$ implies that f has a fixed point.*

16. Multivalued mappings

A special part of the topological fixed point theory is connected with multivalued mappings. In fact, to such problems these are devoted 3 monographs: [AG], [Go1], [Go2]. In this section we shall briefly sketch some aspects of this subject which are concerned with the Lefschetz fixed point theorem. In this section as the homology functor $H = \{H_n\}$ we shall understand the Čech homology functor with compact carriers and coefficients in the field of rationals $\mathbb{Q}$.

A continuous map $p: Y \to X$ is called a *Vietoris map* provided the following conditions are satisfied:

(16.1.1) p is onto, i.e. $p(Y) = X$,
(16.1.2) p is perfect, i.e. p is closed and for every $x \in X$ the set $p^{-1}(x)$ is compact and acyclic.

Note that if X, Y are metric spaces then usually the assumption that p is perfect is replaced by the following two assumptions: p is proper and $p^{-1}(x)$ is acyclic, for every $x \in X$. In what follows the symbol $p: Y \Rightarrow X$ is used for Vietoris mappings.

(16.2) THEOREM (Vietoris Mapping Theorem). *If $p: Y \Rightarrow X$, then $p_*: H(Y) \xrightarrow{\sim} H(X)$ is a linear isomorphism.*

The symbol $\varphi: X \multimap Y$ we shall keep for multivalued mappings. We shall assume that $\varphi(x)$ is nonempty and compact subset of Y, for every $x \in X$.

We let $\Gamma_\varphi = \{(x,y) \in X \times Y : y \in \varphi(x)\}$ and

$$\begin{aligned} p_\varphi : \Gamma_\varphi \to X, \quad p_\varphi(x,y) &= x, \\ q_\varphi : \Gamma_\varphi \to Y, \quad q_\varphi(x,y) &= y. \end{aligned}$$

A multivalued map $\varphi: X \multimap Y$ is called *acyclic* provided φ is upper semicontinuous (u.s.c.) and $\varphi(x)$ is acyclic, for every $x \in X$.

Observe, that if $\varphi: X \multimap Y$ is acyclic then the natural projection $p_\varphi: \Gamma_\varphi \Rightarrow X$ is a Vietoris map and, in view of (16.2), $(p_\varphi)_*$ is an isomorphism.

In 1946, S. Eilenberg and D. Montgomery (see [EM]) proved the following famous theorem:

(16.3) THEOREM. *Assume that $X \in$ ANR and X is compact. If $\varphi: X \multimap X$ is an acyclic map, then:*

(16.3.1) *the Lefschetz number of $\lambda(\varphi)$ is φ is well defined, and*
(16.3.2) $\lambda(\varphi) \neq 0$ *implies that* $\mathrm{Fix}(\varphi) = \{x \in X : x \in X\} \neq \emptyset$.

Note that for any acyclic map it is possible to define the induced linear map $\varphi_*: H(X) \to H(X)$ by putting $\varphi_* = (q_\varphi)_* \circ (p_\varphi)_*^{-1}$ and consequently the Lefschetz number $\lambda(\varphi)$ of φ by putting $\lambda(\varphi) = \lambda(\varphi_*)$.

The original proof of (16.3) presented in [EG] is quite complicated. The modern proof was presented, for the first time, in [Go1], where the Dold method was taken up for multivalued mappings.

In [Go1] there was introduced a larger, than acyclic, class of multivalued mappings the so called *admissible mappings.*

Below we shall recall some notions and properties connected with admissible mappings.

(16.4) DEFINITION. A multivalued map $\varphi: X \multimap Y$ is called *admissible* provided there exists a diagram in Top_2:

$$X \overset{p}{\Longleftarrow} \Gamma \overset{q}{\longrightarrow} Y$$

such that $\varphi(x) = q(p^{-1}(x))$, for every $x \in X$. Such a pair (p,q) is called a *selected pair* for φ and we write $(p,q) \subset \varphi$.

Evidently, any acyclic map φ is admissible and we have $(p_\varphi, q_\varphi) \subset \varphi$.

Note, that for given admissible map $\varphi: X \multimap Y$, we can get many, even infinitely many, selected pairs. It is not difficult to see that the map $\varphi: S^1 \multimap S^1$, $\varphi(x) = S^1$, for every $x \in S^1$, posses infinitely many selected pairs (see [Go1]), where S^1 is the unit sphere in two dimensional Euclidean space $\mathbb{R}^2$.

(16.5) PROPERTY. *If $\varphi: X \multimap Y$ and $\psi: Y \multimap Z$ are two admissible mappings, then the composition $(\psi \circ \varphi): X \multimap Z$, $\psi(\varphi(x)) = \bigcup\{\psi(y) : y \in \varphi(x)\}$ is admissible too.*

SKETCH OF THE PROOF. Let $(p,q) \subset \varphi$ and $(p_1, q_1) \subset \psi$. We have the followin diagram

$$X \stackrel{p}{\Longleftarrow} \Gamma \stackrel{q}{\longrightarrow} Y \stackrel{p_1}{\Longleftarrow} \Gamma_1 \stackrel{q_1}{\longrightarrow} Z.$$

Let us define $\Gamma \otimes \Gamma_1 = \{(u,v) \in \Gamma \times \Gamma_1 : p_1(v) = q(u)\}$ and

$$\overline{p}: \Gamma \otimes \Gamma_1 \Rightarrow \Gamma, \quad \overline{p}(u,v) = u, \qquad \overline{q}: \Gamma \otimes \Gamma_1 \to \Gamma_1, \quad \overline{q}(u,v) = v.$$

Since $\overline{p}$ is a Vietoris map and the composition of two Vietoris map is Vietoris, too, we get $(p \circ \overline{p}, q_1 \circ \overline{q}) \subset (\psi \circ \varphi)$ and the proof is completed. □

(16.6) EXAMPLE. Let $\psi: S^1 \multimap S^1$ be defined as follows:

$$\psi(x) = \{y \in S^1 : \|x - y\| \leq 19/20\}$$

and let $\varphi: S^1 \to S^1$ be defined by $\varphi(x) = S^1$, for every $x \in S^1$.

Evidently, ψ is acyclic but $(\psi \circ \psi)(x) = \varphi(x) = S^1$, for every $x \in S^1$, is no longer acyclic (only admissible!). Observe, that $((-\psi) \circ \psi) = \varphi$, too. It is easy to see that φ has many acyclic decompositions.

Assume that $\varphi: X \multimap X$ is an admissible map. We can define the induced set $\{\varphi_*\}$ of φ by putting

$$\{\varphi_*\} = \{q_* \circ p_*^{-1} : (p,q) \subset \varphi\},$$

We shall say that the admissible map $\varphi: X \multimap X$ is a Lefschetz map provided for every $(p,q) \subset \varphi$ the linear endomorphism $q_* \circ p_*^1$ is a Leray endomorphism.

If $\varphi: X \multimap X$ is a Lefschetz map then we define the Lefschetz set $\{\Lambda(\varphi)\}$ of φ by

$$\{\Lambda(\varphi)\} = \{\Lambda(q_* \circ p_*^{-1} : (p,q) \subset \varphi\}.$$

Note that, if additionaly φ is acyclic, then $\{\Lambda(\varphi)\}$ is a singleton (as defined by S. Eilenberg and D. Montgomery).

Therefore the Lefschetz fixed point theorem for multivalued mappings consists of two claims, namely:

(16.7) φ is a Lefschetz map, and
(16.8) if $\{\Lambda(\varphi)\} \neq \{0\}$, then $\mathrm{Fix}(\varphi) \neq \emptyset$.

Let $\varphi: X \multimap X$ be an admissible map. It is evident how to define the following classes of multivalued maps:

(16.9) $C_A(X)$ – compact, admissible mappings,

(16.10) $EC_A(X)$ – eventually compact and admissible mappings,

(16.11) $ASC_A(X)$ – asymptotically compact and admissible mappings,

(16.12) $CA_A(X)$ – admissible mappings with compact attractor,

(16.13) $CAC_A(X)$ – compact absorbing admissible contractions,

(16.14) $CD_A(X)$ – condensing and admissible mappings.

Finally, we would like to describe that all the results presented in Sections 8–12 and 15 can be taken up to the above classes of admissible mappings (see [Go2] for metric case and [AG] for non-metric case).

The respective formulations are left to the reader.

References

[AG1] J. Andres and L. Górniewicz, *Topological Principles for Boundary Value Problems*, Kluwer, Dordrecht, 2003.

[AB] M. Arkowitz and R. Brown, *The Lefschetz–Hopf theorem and axioms for the Lefschetz number* (to appear).

[BEM] H. Ben-El-Mechaiekh, *Spaces and maps approximation and fixed points*, J. Comp. Appl. Math. **113** (2000), 283–308.

[Bo1] K. Borsuk, *Theory of Retracts*, vol. 44, Monografie Matematyczne, PWN, Warsaw, 1967.

[Bo2] ______, *On the Lefschetz fixed point theorem for nearly extendable maps*, Bull. Acad. Polon. Sci. **12** (1975), 1273–1279.

[Bow1] C. Bowszyc, *Fixed points theorem for the pairs of spaces*, Bull. Polish Acad. Sci. Math. **16** (1968), 845–851.

[Bow2] ______, *On the Euler–Poincaré characteristic of a map and the existence of periodic points*, Bull. Acad. Polon. Sci. **17** (1969), 367–372.

[Bow3] ______, Some Theorems in the Theory of Fixed Points, (Thesis), Warszawa, 1969. (in Polish)

[Br1] R. F. Brown, *The Lefschetz Fixed Point Theorem*, Scott, Foresman and Co., Glenview, Ill., 1971.

[Br2] ______, *Fixed Point Theory*, History of Topology, Elsevier, 1999, pp. 271–299.

[Do] A. Dold, *Lectures on Algebraic Topology*, Springer–Verlag, Berlin, 1972.

[DG] J. Dugundji and A. Granas, *Fixed Point Theory*, I, Monograf. Mat, vol. 61, PWN, Warszawa, 1982.

[EM] S. Eilenberg and D. Montgomery, *Fixed point theorems for multivalued transformations* **58** (1946), Amer. Jour. Math, 214–222.

[FGo] G. Fournier and L. Górniewicz, *Survey of some applications of the fixed point index*, Sem. Math. Superiore **96** (1985), Montreal, 95–136.

[Go1] L. Górniewicz, *Homological Methods in Fixed Point Theory of Multivalued Maps*, Dissertationes Math., vol. 129, Warsaw, 1976.

[Go2] ______, *Topological Fixed Point Theory of Multivalued Mappings*, Kluwer, Dordrecht, 1999.

[Go3] ______, *On the Lefschetz fixed point theorem*, Math. Slovaca **52** (2002), no. 2, 221–233.

[Gr1] A. Granas, *Generalizing the Hopf–Lefschetz fixed point theorem for non-compact* ANRs, Symp. Inf. Dim. Topol., Báton-Rouge, 1967; Ann. Math. Studies **69** (1972), 119–130.

[Gr2] ______, *Fixed point theorems for approximative* ANRs, Bull. Polish Acad. Sci. Math. **16** (1968), 15–19.

[Gr3] ———, *The Leray–Schauder index and the fixed point theory for arbitrary* ANR*s*, Bull. Soc. Math. France **100** (1972), 209–228.

[Gr4] ———, *Topics in the Fixed Point Theory*, Sem. J. Leray, Paris, 1969/70.

[Kr] W. Kryszewski, *The Lefschetz type theorem for a class of noncompact mappings*, Rendiconti del Circolo Matematico di Palermo, Serie II **19** (1987), 365–384.

[Le] S. Lefschetz, *Algebraic Topology*, Amer. Math. Soc., Providence, R. I., 1942.

[Ler] J. Leray, *Théorie des pointes fixés: indice total et nombre de Lefschetz*, Bull. Soc. Math. France **87** (1959), 221–233.

[Nu] R. D. Nussbaum, *Generalizing the fixed point index*, Math. Ann. **228** (1979), 259–278.

[Sp] E. H. Spanier, *Algebraic Topology*, McGraw-Hill, New York, 1966.

[Th] R. B. Thompson, *A unified approach to local and global fixed point indices*, Adv. Math. **3** (1969), 1–71.

3. LINEARIZATIONS FOR MAPS OF NILMANIFOLDS AND SOLVMANIFOLDS

EDWARD C. KEPPELMANN

1. Circle maps

Let S^1 be the unit circle and suppose that $f: S^1 \to S^1$ is a map. It is clear from the definitions that the Nielsen theory (both ordinary and periodic) is described completely by the homotopy class of f. We wish to delve deeper into this statement by picking a specific representative of this homotopy class as follows. The fundamental group $\pi_1(S^1) \cong \mathbb{Z}$. The set of homomorphisms $\varphi: \mathbb{Z} \to \mathbb{Z}$ is in one-to-one correspondence with the integers under the bijection $\varphi \leftrightarrow \varphi(1)$. In addition, since the higher homotopy groups of S^1 are trivial we know that the homotopy classes of self-maps of S^1 are in one-to-one correspondence with these homomorphisms so we see that the homotopy class of f is completely described by a single integer called the degree $m = \deg(f)$ of f.

We can view S^1 as the coset space $p: \mathbb{R} \to \mathbb{R}/\mathbb{Z}$ so that the linear function $\lambda_m: \mathbb{R} \to \mathbb{R}$ given by $\lambda_m(x) = mx$ induces a well defined function on S^1 of degree m. This is our special representative (called the *model circle map* or *linearization*) in the homotopy class of all self-maps on S^1 of degree m.

(1.1) LEMMA. *If $f: S^1 \to S^1$ is the model circle map of degree $m \neq 1$, then the $|m-1|$ fixed points of f all belong to distinct essential Nielsen classes. Therefore, $N(f) = |\text{fix}(f)| = |m-1|$.*

PROOF. The set of lifts of the model degree m map f all have the form $\lambda_{m,k}(x) = mx + k$ as k varies over all possible integers. This means that $x, y \in \text{fix}(f)$ are Nielsen equivalent (see [Jb]) if and only if there is an integer k so that $mx + k = x$ and $my + k = y$. Subtracting the two equations gives that $y - x = m(y - x)$. Since $m \neq 1$ this means that $y = x$. This proves that all Nielsen classes consist of singletons. To see that all these are essential we note that they must all have the same index since the local *linear* behavior of λ_m near each element of $p^{-1}(\text{fix}(f))$ is constant. Since the Lefschetz number $L(f) = |1 - m| \neq 0$, these classes must all be essential. $\square$

Suppose that $g\colon S^1 \to S^1$ is any map of degree d. By the term *linearization of* g we can mean either of the following:

(1.2.1) The model circle map of degree m.
(1.2.2) The integer m (perhaps thought of as a 1×1 matrix).

(1.3) REMARK. In the case where $m \neq 1$ we note that the linearization of g (in sense (1.2.1) above) is a Wecken map. For $m = 1$ the linearization of g is the identity and this can easily (by an irrational rotation) be perturbed to be periodic point free.

2. Chapter outline

The study of linearizations involves a hierarchy of complexity of both spaces and maps as we move up the chain from S^1 to tori to nilmanifolds and finally to solvmanifolds. At each level, we will describe how the linear algebra theoretic data involved in linearizations provides all the necessary ingredients in the formulas for self-maps needed to calculate both the ordinary and periodic Nielsen numbers as well as the coincidence and root theory on these spaces. We will also discuss the simultaneous minimization of periodic points of all periods for self-maps of these spaces and how linearization data describes when this is possible.

From the simple description of a single integer for maps on S^1, the linearization of self-maps on tori consists of square matrices with integer entries. As we move up this hierarchy of spaces, this complexity continues to increase. For linearizations on nilmanifolds and solvmanifolds we must change not only the maps but the spaces as well. These model solvmanifolds and their associated model maps provide a rich source of examples. Because they carry all the Nielsen theory complexity that can ever occur between self-maps on any solvmanifold or nilmanifold, they give an excellent machine by which general theorems can be proved. As each new element of our space hierarchy is considered, we hope that the reader will take advantage of the insight offered by considering the similarities and differences with what has been discussed previously.

The reader should recall the following standard notation and definitions that we will use throughout the chapter. If $f\colon X \to X$ then the set of *periodic points of* f is $\operatorname{per}(f) = \bigcup_{n=1}^{\infty} \operatorname{fix}(f^n)$. A point $x \in \operatorname{per}(f)$ has *minimal period* m (and is said to be irreducible at level m) provided that m is the smallest natural number with $x \in \operatorname{fix}(f^m)$. When X is suitable (e.g. any of the spaces in this chapter), the periodic Nielsen numbers $NP_n(f)$ and $N\Phi_n(f)$ provide homotopy invariant lower bounds for the number of periodic points of f which have minimal period n and minimal period which divides n, respectively.

The following theorem is useful to give the reader the appropriate intuition about these periodic numbers on the spaces considered here. This is significantly less complicated than the general situation on arbitrary spaces. Just like points, Nielsen classes of the iterates of f can have a minimal period n at which level they are referred to as irreducible (See [BJ] and [HPY], [HY], [HK1] for details on the Nielsen periodic numbers.)

(2.1) THEOREM. *Suppose that $f: X \to X$ is a map on a torus, nilmanifold, or solvmanifold. Then $NP_n(f)$ consists of those essential Nielsen classes of f^n which are irreducible and $N\Phi_n(f) = \sum_{m|n} NP_m(f)$.*

3. The p dimensional torus T^p

Recall that the p dimensional torus $S^1 \times \ldots \times S^1 = (S^1)^p$ can be viewed as the coset space $T^p = \mathbb{R}^p/\mathbb{Z}^p$ where $\mathbb{R}^p$ is given its usual abelian group structure. The fundamental group $\pi_1(T^p)$ is isomorphic to $\mathbb{Z}^p$ and because T^p is aspherical we know that the set of homomorphisms from $\mathbb{Z}^p$ to $\mathbb{Z}^p$ is in one-to-one correspondence with the collection of homotopy classes of self-maps on T^p. In analogy with Section 1, if $f: T^p \to T^p$ is a self map then we think of it's *linearization* as the $p \times p$ matrix F over $\mathbb{Z}$ which describes the homotopy class of f with respect to some standard basis for $\pi_1(T^p)$. In analogy with Section 1 we can also think of the *linearization* of f as the map induced on T^p by the linear transformation $\lambda_F: \mathbb{R}^p \to \mathbb{R}^p$ given by $\lambda_F(\overrightarrow{x}) = F\overrightarrow{x}$ which (because F has integer entries) is invariant on $\mathbb{Z}^p$. As before, we have the following theorem:

(3.1) THEOREM ([BBPT], [Ha]). *Suppose that f is a self map on the torus with linearization matrix F. Then the Lefschetz number and Nielsen numbers are related to each other and F by $N(f) = |L(f)| = |\det(F - I)|$. The linearization map of f (i.e. the map induced by F) has isolated singleton essential Nielsen classes if and only if $|\det(F - I)| \neq 0$. In this case if F induces f, then $N(f) = |\mathrm{fix}(f)|$.*

In analogy with (1.3) we have the following:

(3.2) THEOREM. *Suppose that $f: T^p \to T^p$ has linearization matrix F with $\det(F - I) = 0$. Then there is a map $g \sim f$ with no periodic points. The map g is induced by a correspondence of the form $G: \mathbb{R}^p \to \mathbb{R}^p$ given by $G(\overrightarrow{v}) = F\overrightarrow{v} + \overrightarrow{w}$ where $\overrightarrow{w} \notin (F - I)(\mathbb{R}^p)$ has certain "irrational" properties* ([1]).

PROOF SKETCH. By a change of basis over $\mathbb{Z}$ it is possible, using generalized eigenspaces, to view the linearization F of f as a cross product of the form $\mathrm{id} \times \widetilde{f}$. Applying the homotopy of (1.3) to the identity factor then produces a map with no periodic points. □

([1]) This is the so called "Anosov trick" – see [K1].

In terms of the ordinary Nielsen numbers we have the following important computational formula arising from the principle of inclusion/exclusion.

(3.3) THEOREM ([HY], [HPY]). *Suppose that* $f\colon T^p \to T^p$. *If* $N(f^k) \neq 0$, *then* $N(f^k) = N\Phi_k(f)$ *and*

$$NP_n(f) = \sum_{\tau \subset \mathrm{p}(n)} (-1)^{|\tau|} N\Phi_{n:\tau}(f)$$

where $\mathrm{p}(n)$ *is the set of prime divisors of* n *and* $n : \tau = n \prod_{p\in\tau} p^{-1}$.

(3.4) EXAMPLE. The following three calculations help to illustrate the basic ideas behind the formula in Theorem (3.3). Theorem (3.1) helps us intuitively by allowing us to associate the essential fixed point classes of any f^n with the number of singleton fixed point classes of the linearization of f^n. For the purposes of pedagogical insight, in the discussion which follows assume that none of $N(f)$, $N(f^2)$, $N(f^3)$, $N(f^4)$, $N(f^6)$ or $N(f^{12})$ is zero.

(3.4.1) $NP_3(f) = N(f^3) - N(f)$.

(3.4.2) $NP_6(f) = N(f^6) - N(f^3) - N(f^2) + N(f)$.

(3.4.3) $NP_{12}(f) = N(f^{12}) - N(f^6) - N(f^4) + N(f^2)$.

In calculation (3.4.1) we see how the formula basically captures all the points of minimal period 3 by removing those fixed points of f^3 which are also fixed points of f. To capture the points of minimal period 6 in (3.4.2) we must remove both the points of minimal period 3 and those of minimal period 2 but by subtracting both $N(f^3)$ and $N(f^2)$ we have taken away the fixed points of f once too many. This all increases in complexity when we consider the periodic points of minimal period 12 in (3.4.3). This is determined by removing from $\mathrm{fix}(f^{12})$ those points of minimal period < 12. While such points could have a minimal period of $1, \dots, 4$ or 6 they each must be fixed points of either f^6 or f^4. Those which are fixed points of both f^6 and f^4 have to have minimal period either 2 or 1 and hence must also belong to $\mathrm{fix}(f^2)$.

The next result explains how to calculate $N\Phi_n(f)$ on tori when $N(f^n) = 0$.

(3.5) THEOREM ([HY]). *Suppose that* $f\colon T^p \to T^p$. *If* $N(f^n) = 0$, *then*

$$N\Phi_n(f) = \sum_{\emptyset \neq \mu \subseteq M(f,n)} (-1)^{\#\mu - 1} N(f^{\xi(\mu)}),$$

where $M(f,n)$ *consists of those maximal divisors* m *of* n *with the property that* $N(f^m) \neq 0$ *and* $\xi(\mu)$ *is the greatest common divisor of all elements of* μ.

(3.6) EXAMPLE. Suppose $f: T^4 \to T^4$ has the following linearization matrix

$$F = \begin{bmatrix} 0 & 1 & 0 & 0 \\ 0 & 0 & 1 & 0 \\ 0 & 0 & 0 & 1 \\ -1 & 0 & 1 & 0 \end{bmatrix}.$$

We note that $N(f^{12}) = 0$ so to calculate $N\Phi_{12}(f)$ we must use the formula given above. We can check that F is periodic with minimal period 12. In fact, the matrix F has 4 distinct eigenvalues $\lambda_1 = e^{\pi i/6}$, $\lambda_2 = e^{5\pi i/6}$, $\lambda_3 = e^{7\pi i/6}$, $\lambda_4 = e^{11\pi i/6}$ which are exactly the 4 primitive 12th roots of unity. Thus $N(f^k) = |\prod_{i=1}^{4}(1-\lambda_i^k)|$ will be nonzero for all $k < 12$. This means that $M(f,n) = \{6,4\}$ consists of the maximal proper divisors of 12. Hence the formula for $N\Phi_{12}(f)$ has summands for $\{6\}, \{4\}$, and $\{4,6\}$. This gives (respectively) $N\phi_{12}(f) = N(f^6) + N(f^4) - N(f^2) = 16+9-1 = 24$. In other words, while f^{12} has no essential Nielsen classes, we cannot expect to deform f so that f^{12} is fixed point free. We would be restricted in such an attempt by the essential classes of f^6 and f^4 (i.e. $\text{fix}(f^4), \text{fix}(f^6) \subseteq \text{fix}(f^{12})$). The formula considers those classes and subtracts those common to both these iterates (i.e. those of f^2) which would otherwise be counted twice.

If $N(f^n) \neq 0$, Theorems (3.1) and (3.3) make it clear that the linearization of f realizes the lower bound $N\Phi_n(f)$ (i.e. by performing the homotopy which amounts to linearizing f we obtain a map which has exactly $N\Phi_n(f)$ periodic points of period n). That this can also be accomplished when $N(f^n) = 0$ is an important result that took some serious ingenuity to prove.

(3.7) THEOREM ([Y1], [Y2]). *Suppose $f: T^p \to T^p$ and $n > 0$ are fixed. Then there is a $g \sim f$ so that for all $k|n$ we have that*

(3.7.1) *the number of periodic points of minimal period k for g equals $NP_k(f)$,*
(3.7.2) $|\text{fix}(g^k)| = N\Phi_k(f)$.

Work by Jezierski (see [Je1], [Je2]) to extend a version of Theorem (3.7) to a broad class of manifolds (that certainly includes all manifolds discussed in this chapter) is one of the profound accomplishments in this field.

4. Nilmanifolds

Suppose that G is a connected, simply connected Lie group. This means that G is diffeomorphic to some Euclidean space and is equipped with some group operation $*$. Furthermore, the functions $*: G \times G \to G$ and $I: G \to G$ given by $*(g,h) = g * h$ and $I(g) = g^{-1}$ are smooth. We say that G is *nilpotent* provided that the descending central series $G = G_0 \supseteq G_1 \supseteq \ldots$ terminates at the identity in finitely many steps. (Recall that $G_0 = G$ and $G_{i+1} = [G, G_i] = \{xyx^{-1}y^{-1} : x \in G,\ y \in G_i\}$.) We say that a subgroup $\Gamma \subset G$ is *uniform* provided that the space of cosets G/Γ is compact. Although not originally defined this way ([2]), it is equivalent to say that a *nilmanifold* is a coset space of the form G/Γ where G is a connected simply connected nilpotent Lie group and Γ is a uniform discrete subgroup.

(4.1) EXAMPLE. If $G = \mathbb{R}^p$ is equipped with the usual group structure of componentwise addition, then we can see that the p-torus $T^p = G/\mathbb{Z}^p$ is indeed a nilmanifold. The descending central series terminates immediately.

(4.2) EXAMPLE. Let G be the collection of all upper triangular $n \times n$ matrices with 1s along the diagonal. It is not hard to check that such a collection of invertible matrices forms a connected simply connected nilpotent Lie group under matrix multiplication. Furthermore, since integer matrices of determinant $+1$ are invertible over $\mathbb{Z}$, it is also clear that the subset Δ of matrices of G which have integer entries is a discrete uniform subgroup of G and hence that G/Δ is a nilmanifold.

(4.3) REMARK. When $p = 3$ we get the so-called *baby nil* nilmanifold (see [HK1]). The name comes from the fact that there are no nilmanifolds of dimension 1 or 2 which are not homeomorphic to S^1 or T^2. Thus baby nil is in some sense the simplest non-torus nilmanifold there is ([3]).

(4.4) REMARK. The second example above is in some sense as general as possible. As shown in [Ma] it turns out that for any nilmanifold H/J there is an integer p so that H and J are subgroups of the full Lie group G of $p \times p$ upper triangular matrices with ones on the diagonal.

The analogy of nilmanifolds to tori begins with (but is by no means limited to) the following:

(4.5) THEOREM. *Suppose that $f : M \to M$ is a map on a nilmanifold $M = G/\Delta$. Then, up to homotopy, we can assume that f is induced by a homomorphism $\mathcal{F} : G \to G$ which is invariant on Δ.*

(4.6) EXAMPLE. Let $g = (a_1, \dots, a_5)$ and $h = (b_1, \dots, b_5)$ in $G = \mathbb{R}^5$ and define a binary operation $*$ by

$$g * h = (a_1 + b_1, a_2 + b_2 + a_1 b_4, a_3 + b_3 + a_1 b_5, a_4 + b_4, a_5 + b_5).$$

([2]) Traditionally a nilmanifold is defined to be smooth transitive G-space in which G is a nilpotent Lie group. The equivalence is proved in [Ma].

([3]) There are other 3-dimensional nilmanifolds besides baby nil that can be formed by taking uniform discrete subgroups other than the integer lattice (see [JM]).

It is possible to check that this operation gives G the structure of a Lie group. In fact, the operation arises from the multiplication of 4×4 matrices of the form

$$g = \begin{bmatrix} 1 & a_1 & a_2 & a_3 \\ 0 & 1 & a_4 & a_5 \\ 0 & 0 & 1 & 0 \\ 0 & 0 & 0 & 1 \end{bmatrix}.$$

Thus $g^{-1} = (-a_1, a_1a_4 - a_2, a_1a_5 - a_3, -a_4, -a_5)$. It is also clear that G is nilpotent since $G_1 = [G, G] = \{(0, x, y, 0, 0) \in G\}$ and $[G, G_1] = \{(0, 0, 0, 0, 0)\}$. The collection of elements whose entries belong to $\mathbb{Z}$ is clearly a discrete uniform subgroup Δ of G and the quotient $M = G/\Delta$ is therefore a nilmanifold of dimension 5. Software like Maple can help you check that the following correspondence $\mathcal{F}\colon G \to G$ given by $\mathcal{F}(g) = (\mathcal{F}_1(g), \dots, \mathcal{F}_5(g))$ where $g = (a_1, \dots, a_5)$ is a homomorphism which is invariant on Δ:

$$\begin{aligned}
\mathcal{F}_1(g) &= a_1 - 42a_5 + 6a_4, \\
\mathcal{F}_2(g) &= 306a_5 - 294a_5^2 + 13a_4 - 6a_4^2 + 84a_5a_4 + 2a_1^2 \\
&\quad - 168a_1a_5 + 24a_1a_4 - 26a_2 + 182a_3, \\
\mathcal{F}_3(g) &= -1457a_5 + 1470a_5^2 - 22a_4 + 30a_4^2 - 420a_5a_4 \\
&\quad + \frac{1}{2}a_1 + \frac{5}{2}a_1^2 - 210a_1a_5 + 30a_1a_4 - 20a_2 + 140a_3 \\
\mathcal{F}_4(g) &= 14a_5 - 2a_4 + 4a_1, \\
\mathcal{F}_5(g) &= -70a_5 + 10a_4 + 5a_1.
\end{aligned}$$

This means that $\mathcal{F}\colon G \to G$ induces a self map $f\colon M \to M$.

In analogy with what happens on tori, the self map f of Example (4.6) is in some sense the best representative in its homotopy class. However, since the underlying group G is not abelian, such a representative is far from linear as the formulae show. In order to properly define the linearization of such an f we need first to consider the following construction of Fadell–Husseini [FH] (and in some sense also Anosov [A]).

Since G is nilpotent it has $G_k \subseteq \text{center}(G)$ for some k (thus $G_{k+1} = [G, G_k] = \{0\}$). It can be shown that if Δ is uniform in G, then Δ_i is uniform in G_i for each i. This means that G_k/Δ_k is a compact manifold and in fact since G_k is actually abelian this must be a torus T. Furthermore, the quotient G/G_k is connected, simply connected and nilpotent with a descending series with one fewer nonzero term. It is also the case that Δ/Δ_k is uniform in G/G_k so that $(G/G_k)/(\Delta/\Delta_k)$ is a nilmanifold B of strictly smaller dimension than $M = G/\Delta$. Of course since $\mathcal{F}$ preserves commutators it is also defined on all these subgroups and quotient groups. This all gives us the following:

(4.7) THEOREM ([FH]). *Suppose that* $f\colon M \to M$ *is a self map of a compact nilmanifold. Then there is a torus* T *and nilmanifold* B *with* $\dim B < \dim M$ *so that, up to homotopy, we can assume that* f *is fibre preserving on the following fibration of Fadell–Husseini*

$$\begin{array}{ccc} T & \xrightarrow{f_0} & T \\ \downarrow & & \downarrow \\ M & \xrightarrow{f} & M \\ \downarrow & & \downarrow \\ B & \xrightarrow[\overline{f}]{} & B \end{array}$$

Because the Fadell–Husseini fibration of Theorem (4.7) satisfies the so called *naïve addition conditions* (see [HKW]), we have the following Nielsen theory for these fibrations:

(4.8) THEOREM ([FH], [HKW]). *Suppose that* $(f_0, f, \overline{f})$ *is a fibre preserving map of a Fadell–Husseini fibration as in Theorem* (4.7). *Then* $N(f) = N(f_0)N(\overline{f})$.

PROOF. The naïve addition conditions of [HKW] mean that

$$N(f) = \sum_{x \in \xi} N(f_x)$$

where $\xi \subseteq \mathrm{fix}(\overline{f})$ consists of one fixed point from each essential Nielsen class of $\overline{f}$ and f_x denotes f restricted to the fibre over $x \in B$. Since $|\xi| = N(\overline{f})$ the specified formula then follows by showing that, for any $x, y \in \xi$, $N(f_x) = N(f_y)$ (4). This follows by making the following observations:

(4.8.1) If $M = G/\Delta$ and we denote T (the fibre over the identity coset of B) by G_k/Δ_k as discussed above, then the fibre over some other $x \in \mathrm{fix}(\overline{f})$ is in diffeomorphic correspondence to T by the rule $\alpha\Delta_k \Leftrightarrow g\alpha\Delta_k$ where $\alpha \in G_k$ and $g \in G$ fixed is any value such that the coset of g in $(G/G_k)/(\Delta/\Delta_k)$ is x.

(4.8.2) Since f is induced by the homomorphism F we know that f_0 sends $\alpha\Delta_k$ to $F(\alpha)F(\Delta_k) = F(\alpha)\Delta_k$ and f_x sends $g\alpha\Delta_k$ to $F(g)F(\alpha)F(\Delta_k) = F(g)F(\alpha)\Delta_k$. Furthermore, $\alpha \in G_k$ implies that $F(\alpha) \in G_k$.

(4.8.3) Therefore up to the indicated correspondence between fibres the two fibre maps are related by F and $F(g)F$.

(4) Such a fibration is said to exhibit Nielsen number uniformity. Since f_x and f_y actually have the same homotopy type this happens in a very strong way on these fibrations. We will see in our discussion of solvmanifold fibrations that there are situations of Nielsen number fibre uniformity where the homotopy type of the map on different setwise fixed fibres does vary.

(4.8.4) Since G'/Δ is connected $F(g)F$ is homotopic to F and hence f_x has the same homotopy type as f_0. □

Having produced the Fadell–Husseini fibration in which f is fibre preserving on M, we can now repeat the process by forming the Fadell–Husseini fibration in which $\overline{f}$ is fibre preserving on B. The process continues and will, since the base manifolds in each such fibration are nilmanifolds of strictly decreasing dimensions, eventually get us to a fibration where the base is a torus. In the process, we will obtain a finite list of maps $f_0, \ldots, f_{k-1}$ on tori with the property that $N(f) = \prod_{i=0}^{k-1} N(f_i)$ where k is the nilpotency class of G.

(4.9) DEFINITION. Suppose that $f: M \to M$ is a self map on a nilmanifold with tori decomposition maps $f_0, \ldots f_{k-1}$ on tori of dimensions $d_0, \ldots, d_{k-1}$ with respective linearization matrices $F_0, \ldots F_{k-1}$. Then the *linearization* of f is the $m \times m$ matrix F in block diagonal form with blocks $F_0, \ldots F_{k-1}$. (Here $m = \dim M = \sum_{i=0}^{k-1} d_i$). In alignment with what we will do for solvmanifolds we can think of the *linearization* or *linear model* for M and f as the m dimensional torus T^m with the linear self map induced by the matrix F.

Since in the above description we have that $\det(F - I) = \prod_{i=0}^{k-1} \det(F_i - I)$ and F^n is block diagonal with blocks $F_0^n, \ldots, F_{k-1}^n$ we have the following.

(4.10) THEOREM ([FH], [HKW], [HK1]). *Suppose that $f: M \to M$ is a self map on a nilmanifold with linearization matrix F. Then for any n we have that $N(f^n) = |L(f^n)| = |\det(F^n - I)|$ and the conclusions and formulas of Theorems* (2.1), (3.1), (3.2) (5), (3.3) *and* (3.5) *are all valid.*

(4.11) REMARK. As Anosov showed, the linearization F of f can also be produced as the derivative $\mathcal{F}': \mathcal{G} \to \mathcal{G}$ on the Lie algebra $\mathcal{G}$ of G of the homomorphism $\mathcal{F}: G \to G$ which induces the model map in the homotopy class of f (6). Suppose that $\mathcal{F}'(\text{center}(\mathcal{G}))$ is not all of center$(\mathcal{G})$. In this case the "Anosov trick" extension of Theorem (3.2) to nilmanifolds can be viewed as using the exponential map $\exp: \mathcal{G} \to G$ to produce a flow and subsequent isotopy of the identity which when composed with F gives a deformation of F by an "irrational" amount in the direction of any vector not in $(\mathcal{F}' - I)(\text{center}(\mathcal{G})) \cap \text{center}(\mathcal{G})$.

(5) Strictly speaking the "Anosov trick" for nilmanifolds must be applied to a torus. In general then this must be done either on the center of the nilpotent Lie group or in one of the centers of the pieces of the Fadell–Husseini decomposition and the homotopy lifted to the whole space (see [K1]).

(6) This matrix may in fact be slightly different and not of the block diagonal form in the construction described above. However, it will be upper block triangular and the Nielsen calculations involving $\det(F - I)$ will not be changed.

(4.12) EXAMPLE. We would like to calculate the linearization of the homomorphism induced map of Example (4.6). Since G_1 is abelian of dimension 2 and $G_2 = \{0\}$ it follows that the Fadell–Husseini fibration for the nilmanifold M of dimension 5 in (4.6) fibres M over T^3 with fibres homeomorphic to T^2. Since G_1 comprises coordinates 2 and 3 of the Euclidean space $\mathbb{R}^5$ for G, we can compute the linearization of f_0 by simply setting $a_1 = a_4 = a_5 = 0$ and looking at $\mathcal{F}_2$ and $\mathcal{F}_3$. To examine the linearization of $\overline{f}$ we notice that in this case $\mathcal{F}_1$, $\mathcal{F}_4$ and $\mathcal{F}_5$ are already linear in only a_1, a_4 and a_5. Therefore the linearization of $\mathcal{F}$ is (⁷)

$$F = \begin{bmatrix} -26 & 182 & 0 & 0 & 0 \\ -20 & 140 & 0 & 0 & 0 \\ 0 & 0 & 1 & 6 & -42 \\ 0 & 0 & 4 & -2 & 14 \\ 0 & 0 & 5 & 10 & -70 \end{bmatrix}.$$

Thus $N(f^6) = N\Phi_6(f) = |\det(F^6 - I)| = 4,573,592,631,247,055,962,586,040$ and $NP_6(f) = 4,573,592,631,244,366,597,922,496$. Hence less than 6×10^{-11} percent of the fixed points of f^6 have minimal period less than 6.

5. Solvmanifolds

5.1. Background. In many ways solvmanifolds, both in definition and form, share some common characteristics with nilmanifolds and tori. However, as we shall see, there are some important complicating differences as well. Rather than simply sketch the traditional complete and definitive approach to solvmanifolds which is readily available in the literature [Mo], [Mc], [K1], [K2], [HK1]–[HK4], [HKW], we will illustrate these differences in a concrete way with examples and intuition that is more easily accessible. We will begin this section in a leisurely way with some important background material. Our overall goal in this first part is to gain some insight into why the Klein bottle is a solvmanifold. We do this, not because it is an essential fact for the remainder of this exposition, but because the path to this goal will reveal many hidden treasures along the way.

A connected, simply connected Lie group G (like those discussed in Section 4) is *solvable* provided that the *derived series* $G_{(0)} = G, G_{(1)} = [G, G], \ldots, G_{(i+1)} = [G_{(i)}, G_{(i)}]$ terminates with $\{0\}$ in finitely many steps. Comparing with the descending central series G_i which defines nilpotency as in Section 4, we note that $G_1 = G_{(1)}$ and in general if you assume that $G_{(i)} \subseteq G_i$ then $G_{(i+1)} = [G_{(i)}, G_{(i)}] \subseteq [G, G_{(i)}] \subseteq [G, G_i] = G_{i+1}$. This implies that if G is nilpotent (so some $G_k = \{0\}$ and hence $G_{(k)} = \{0\}$ as well) then it is also solvable. The converse is false.

(⁷) Since the ordering of the bases for producing F must be done according to the tori decomposition for F the columns from left to right and the rows from top to bottom are in order a_2, a_3, a_1, a_4, a_5.

As we did with nilmanifolds, there is a practical definition of a solvmanifold ([8]) as a coset space of the form G/Γ where G is a connected, simply connected solvable Lie group and Γ is a closed uniform subgroup ([9]). Notice that unlike the case of a nilmanifold we cannot assume that Γ is discrete in the formulation of a solvmanifold.

(5.1) DEFINITION. Suppose that G and H are groups and let $\varphi: H \to \mathrm{Aut}(G)$ be a homomorphism. For $h \in H$ and $g \in G$ we shall denote the image of g under $\varphi(h)$ by $\varphi^h(g)$. The *semidirect product* of G and H (denoted $G \times_\varphi H$) is the group formed from the set $G \times H$ with the group operation

$$(g_1, h_1) * (g_2, h_2) = (g_1\varphi^{h_1}(g_2), h_1h_2).$$

(5.2) EXAMPLE. Give $G = \mathbb{R}^3 = \mathbb{R}^2 \times_\phi \mathbb{R}$ the semidirect product group structure where each factor is abelian and ϕ operates as a circle group of rotations by the rule that (where we express linear transformations via matrix multiplication)

$$\phi(t) = \begin{bmatrix} \cos(\pi t) & -\sin(\pi t) \\ \sin(\pi t) & \cos(\pi t) \end{bmatrix}.$$

We note that $G_{(1)} = [G, G] = \mathbb{R}^2 \times_\phi \{0\}$ and $G_{(2)} = [G_{(1)}, G_{(1)}] = \{(\overrightarrow{0}, 0)\}$. Thus G is indeed solvable but it is not nilpotent since for all $i \geq 1$ we have that $G_i = G_{(1)} = \mathbb{R}^2 \times_\varphi \{0\}$.

Let Γ denote the subgroup $(\mathbb{Z} \times \mathbb{R}) \times_\phi \mathbb{Z}$. Then in fact $K = G/\Gamma$ is the Klein bottle with fundamental group $\mathbb{Z} \times_\psi \mathbb{Z}$ where $\psi: \mathbb{Z} \to \mathrm{Aut}(\mathbb{Z})$ is multiplication by $\psi(m) = (-1)^m$. In fact, K is exactly our model solvmanifold construction (see Example (5.23)) of the twisted product of two circles.

This demonstrates conclusively that the Klein bottle is indeed a solvmanifold. However, even though it is a surface, we claim that it cannot be represented in the form G/Γ where $G \cong \mathbb{R}^2$ and Γ is discrete. As a vehicle for describing some of the basic theory behind Lie algebras and Lie groups (see [Ja], [W]), we will now take a little space to prove this fact.

(5.3) DEFINITION. A *Lie algebra* $\mathcal{G}$ is a vector space equipped with a bilinear bracket operation $[\cdot, \cdot]: \mathcal{G} \times \mathcal{G} \to \mathcal{G}$ such that for all $X, Y, Z \in \mathcal{G}$ we have that

(5.3.1) (anti-commutativity) $[X, Y] = -[Y, X]$,
(5.3.2) (Jacobi identity) $[[X, Y], Z] + [[Y, Z], X] + [[Z, X], Y] = 0$.

---

(8) In this chapter we always assume that our manifolds are compact.

(9) This is equivalent to the traditional definition that a solvmanifold is a smooth transitive G-space where G is a solvable Lie group.

(5.4) DEFINITION. If $\mathcal{G}$ is a Lie algebra then $[\mathcal{G},\mathcal{G}]$ denotes $\{[X,Y]: X,Y \in \mathcal{G}\}$. As in the case with Lie groups we can form the derived and lower central series from $\mathcal{G}$ using these brackets in the place of commutators. Thus $\mathcal{G}_0 = \mathcal{G}_{(0)} = \mathcal{G}$, $\mathcal{G}_{i+1} = [\mathcal{G},\mathcal{G}_i]$ and $\mathcal{G}_{(i+1)} = [\mathcal{G}_{(i)},\mathcal{G}_{(i)}]$. If there is a k so that $\mathcal{G}_k = \{0\}$ (respectively $\mathcal{G}_{(k)} = \{0\}$) then $\mathcal{G}$ is *nilpotent* (respectively *solvable*). We say that $\mathcal{G}$ is *abelian* when $[\mathcal{G},\mathcal{G}] = \{0\}$.

We will use insight from, but will not attempt to prove, the following foundational result (see [Ja, Chapter 6, Theorem of Ado] and [W] for details). Hopefully, the examples which follow will give the reader some important insight into the correspondence.

(5.5) THEOREM. *There is a bijective structure preserving correspondence between finite dimensional Lie algebras and simply connected Lie groups. Suppose that under this association that the Lie algebras $\mathcal{G}$, $\mathcal{H}$ are associated to the simply connected Lie groups G, H, respectively. A function $\varphi: G \to H$ is a homomorphism of Lie groups if and only if its derivative $\varphi': \mathcal{G} \to \mathcal{H}$ is a homomorphism of Lie algebras. In this case the following diagram commutes:*

$$\begin{array}{ccc} \mathcal{G} & \xrightarrow{\varphi'} & \mathcal{H} \\ {\scriptstyle \exp}\downarrow & & \downarrow{\scriptstyle \exp} \\ G & \xrightarrow[\varphi]{} & H \end{array}$$

This implies that under this correspondence $\mathcal{G}$ is nilpotent, solvable, or abelian if and only if (respectively) G is nilpotent, solvable, or abelian. In fact, the correspondence matches $\mathcal{G}_i$ and $\mathcal{G}_{(i)}$ with G_i and $G_{(i)}$, respectively.

The nature of the exp map is explained by the following cannonical example and theorem.

(5.6) EXAMPLE. Let $\mathrm{GL}(n,\mathbb{R})$ denote the Lie group of invertible $n \times n$ matrices with real entries under the operation of matrix multiplication. Let $\mathcal{G}(n,\mathbb{R})$ denote the collection of all $n \times n$ matrices. Then $\mathcal{G}(n,\mathbb{R})$ is a Lie algebra with bracket defined by $[A,B] = AB - BA$. The function $\exp: \mathcal{G}(n,\mathbb{R}) \to \mathrm{GL}(n,\mathbb{R})$ is defined by $\exp(A) = I + \sum_{j=1}^{\infty} A^j/j!$ and sends sub algebras of $GL(n,\mathbb{R})$ to subgroups of $\mathcal{G}(n,\mathbb{R})$ and commutes with homomorphisms and their derivatives on these subobjects in the way described for Theorem (5.5).

(5.7) THEOREM. *Example (5.6) is the most cannonical example in the sense that every finite dimensional Lie algebra $\mathcal{G}$ imbeds faithfully (i.e. 1-to-1) into $\mathcal{G}(n,\mathbb{R})$ for some n. This implies that the correspondence described by Theorem (5.5) can be realized with sub-Lie groups and sub-Lie algebras of* $\mathrm{GL}(n,\mathbb{R})$ *and $\mathcal{G}(n,\mathbb{R})$, respectively.*

(5.8) EXAMPLE. Suppose that $\mathcal{G}$ is the abelian Lie algebra of dimension n. Then its associated simply connected Lie group is $\mathbb{R}^n$ under the usual operation of coordinate wise addition.

(5.9) EXAMPLE. Suppose that $\mathcal{G}$ is spanned by X, Y, Z with bracket operation defined by $[X, Z] = [Y, Z] = 0$ and $[X, Y] = Z$. Then $\mathcal{G}$ is a nilpotent Lie algebra which imbeds in $\mathcal{G}(3, \mathbb{R})$ as the set of upper triangular matrices with zeros on the diagonal. The associated simply connected Lie subgroup of $\mathrm{GL}(3, \mathbb{R})$ is the baby nil group arising from the product of upper triangular 3×3 matrices with ones on the diagonal.

(5.10) LEMMA. *Up to isomorphism, there is only one non-abelian Lie algebra $\mathcal{J}$ of dimension 2. $\mathcal{J}$ is solvable but not nilpotent. It's associated simply connected Lie group is not connected.*

PROOF. Suppose that $\mathcal{G}_2$ has basis $\{X, Y\}$. From anti-commutativity we must have that $[X, X] = [Y, Y] = 0$. Since $\mathcal{G}_2$ is nonabelian we must have $[X, Y] \neq 0$. By rescaling and permuting basis vectors we can therefore assume without loss in generality that $[X, Y] = Y$. Thus $\mathcal{J}_{(1)} = [\mathcal{J}, \mathcal{J}]$ has dimension one and is spanned by Y. Since $\mathcal{J}_{(2)} = \{[Y, Y]\} = \{0\}$ but $\mathcal{J}_2 = [\mathcal{J}, \mathcal{J}_1] = \mathrm{span}\{Y\} = \mathcal{J}_1$ we readily see that $\mathcal{J}$ is solvable but not nilpotent. Example (5.11) below shows the simply connected Lie group J corresponding to $\mathcal{J}$. The uniqueness of the correspondence and the fact that J is not connected guarantees that there can be no other associated simply connected and connected Lie group. □

(5.11) EXAMPLE. Suppose that X and Y are the following matrices:

$$X = \begin{bmatrix} \frac{1}{2} & 0 \\ 0 & -\frac{1}{2} \end{bmatrix}, \qquad Y = \begin{bmatrix} 0 & 1 \\ 0 & 0 \end{bmatrix}$$

Then $[X, Y] = XY - YX = Y$ so this is a representation of $\mathcal{J}$ as a sub-Lie algebra of $\mathcal{G}(2, \mathbb{R})$. We claim that the corresponding Lie group J equals $(\mathbb{R} - \{0\}) \times \mathbb{R}$ under the operation of $(a_1, a_2) * (b_1, b_2) = (a_1 b_1, a_1 b_2 + a_2/b_1)$. Since this J is simply connected but not connected (it consists of the cartesian plane minus the x-axis) this will complete the proof of Lemma (5.10).

To see the claim we note that using matrix exponentiation as described in (5.6) above we have that

$$\exp(2\ln(a)X)\exp\left(\frac{b}{a}Y\right) = \begin{bmatrix} a & 0 \\ 0 & \frac{1}{a} \end{bmatrix} \begin{bmatrix} 1 & \frac{b}{a} \\ 0 & 1 \end{bmatrix} = \begin{bmatrix} a & b \\ 0 & \frac{1}{a} \end{bmatrix}$$

which corresponds precisely to the tuple (a, b) in the above specified J. The fact that this is indeed a Lie group follows from the observations that matrix multiplication is associative, the identity is $(1, 0)$, and $(a, b)^{-1} = (1/a, -b)$.

(5.12) COROLLARY. *The Klein bottle cannot be expressed in the form* G/Γ *where* G *is a simply connected, connected solvable Lie group of dimension* 2 *and* Γ *is discrete.*

PROOF. If this were possible then the simply connected connected solvable G would have to be different than the Lie group J shown in Example (5.11) above. However, this would mean by Theorem (5.5) that the Lie algebra $\mathcal{G}$ is not isomorphic to $\mathcal{J}$ of (5.10). Since the only other dimension 2 Lie algebra is the abelian Lie algebra this would force $\mathcal{G}$ to be abelian and hence G to be abelian so the quotient G/Γ would be a torus and not the Klein bottle. □

5.2. Maps and fibrations. Like the Fadell–Husseini fibration for a nilmanifold, solvmanifolds can be decomposed as the following theorem of Mostow ([Mo]) explains:

(5.13) THEOREM. *Suppose that* S *is a compact solvmanifold. Then there is a fibration* $M \hookrightarrow S \to T$ *in which* M *is a nilmanifold and* T *is a torus. Furthermore, if* $f: S \to S$ *is a self map, then up to homotopy we can assume that* f *is fibre preserving on this fibration.*

(5.14) DEFINITION. Now of course a given solvmanifold may have many associated Mostow fibrations. The one described by Theorem (5.13) that works for all homotopy classes of self-maps, is referred to as the *minimal Mostow fibration* of the solvmanifold S.

As in the previous section, rather than providing a general proof, we will attempt to illustrate the reasons behind Theorem (5.13) with thre enlightening example of the Klein bottle.

(5.15) EXAMPLE. Suppose that we view the Klein bottle

$$K = \frac{\mathbb{R}^3}{\Delta} = \frac{\mathbb{R}^2 \times_\phi \mathbb{R}}{\Delta}$$

as in Example (5.2). As we discussed in that example $\pi_1(K) = \mathbb{Z} \times_\psi \mathbb{Z}$ where $\psi(m)$ is multiplication by $(-1)^m$. Thus $\pi_1(K)_{(1)} = [\pi_1(K), \pi_1(K)] \cong \mathbb{Z} = \pi_1(S^1)$. Now if $f: K \to K$ is a map then $f_\#(\pi_1(K)_{(1)}) \subseteq \pi_1(K)_{(1)}$. This means that $f_\#$ will also induce a self homomorphism on the quotient $\pi_1(K)/[\pi_1(K), \pi_1(K)] \cong \mathbb{Z} = \pi_1(S^1)$. Therefore, up to homotopy, f is fibre preserving on the fibration of (5.13) for the Klein bottle which is $S^1 \hookrightarrow K \to S^1$ ([10]).

([10]) In a general solvmanifold $S = G/\Gamma$ it may be the case that $\pi_1(S)/[\pi_1(S), \pi_1(S)]$ is not torsion free and hence does not induce a fibration with torus as base as described in (5.13). However, there is always a minimal universally invariant nilpotent subgroup of $\pi_1(S)$ containing $[\pi_1(S), \pi_1(S)]$ which will do the job. The details of how this corresponds to a nilmanifold formed from a nilpotent subgroup of G with a discrete subgroup of Γ is described in [Mc, Theorem 1.1].

This formulation of the Klein bottle also gives us insight into why on a solvmanifold $S = G/\Gamma$ we cannot assume that every self map comes (even up to homotopy) from a homomorphism on G. For example, suppose that on $\pi_1(K) = \mathbb{Z} \times_\psi \mathbb{Z}$ we define

$$f_\#(m_1, m_2) = (km_1, pm_2).$$

In order for this to be a homomorphism we must have that

$$f_\#(m_1, m_2)f_\#(n_1, n_2) = f_\#(m_1 + (-1)^{m_2}n_1, m_2 + n_2).$$

Therefore,

$$\begin{aligned}(km_1, pm_2)(kn_1, pn_2) &= (km_1 + (-1)^{pm_2}kn_1, pm_2 + pn_2)\\ &= (km_1 + (-1)^{m_2}kn_1, pm_2 + pn_2)\end{aligned}$$

so that in order for this correspondence $f_\#$ to be a homomorphism we must have that for all m_1, m_2, n_1, n_2 that $(-1)^{pm_2}kn_1 = (-1)^{m_2}kn_1$. In other words, our only constraint is that p be odd. However, on $G = \mathbb{R}^2 \times_\phi \mathbb{R}$ it is impossible to define in an interesting way a correspondence $F: G \to G$ given by $(\overrightarrow{v}, t) \to (A\overrightarrow{v}, pt)$ (where A is a 2×2 matrix and $p \in \mathbb{R}$) which is a homomorphism. Since

$$(\overrightarrow{v}, t)(\overrightarrow{w}, s) = (\overrightarrow{v} + \phi(t)\overrightarrow{w}, t + s),$$

with $\phi(t)$ as in (5.2), we must have that

$$(A\overrightarrow{v}, pt)(A\overrightarrow{w}, ps) = (A\overrightarrow{v} + \phi(pt)A\overrightarrow{w}, pt + ps) = (A\overrightarrow{v} + A\phi(t)\overrightarrow{w}, pt + ps)$$

and hence we need that for all t that $A\phi(t) = \phi(pt)A$. A check of the individual matrix entries in this equation shows that unless $p = 1$, $A = 0$. Thus we certainly cannot produce an F on G which will restrict to the $f_\#$ on Γ shown above.

The ordinary Nielsen theory of these Mostow fibrations is captured by the following theorem. Along with (5.13) and our ability to calculate ordinary Nielsen numbers of self-maps on nilmanifolds and tori, this shows that the calculation of ordinary Nielsen numbers on solvmanifolds is essentially straightforward. However, we will defer the presentation of examples until after Section 5.4 when we have indicated the recipe which is, in a real sense, the motivation behind all examples.

(5.16) THEOREM ([Mc], [HK1], [HKW]). *Mostow fibrations satisfy the naïve addition conditions. This means that if $(f, \overline{f})$ is a fibre preserving map on a Mostow fibration $M \hookrightarrow S \to T$ of a solvmanifold S, then $N(f) = \sum_{x\in\xi} N(f_x)$ where $\xi \subseteq \mathrm{fix}(\overline{f})$ consists of one point from each essential class of $\overline{f}$.*

5.3. The Nielsen periodic numbers for weakly Jiang maps. As this subsection and Section 8 will illustrate, Theorem (5.16) forms the basis of the calculation of the Nielsen periodic numbers for maps on solvmanifolds. However, depending on the nature of the fibration and the map f (especially with regard to how much $N(f_x)$ can vary for different $x \in \operatorname{fix}(\overline{f})$), the calculation in specific cases can be anything between relatively painless to barely algorithmic. We start with something (see [H] for more details) that is true for all self-maps. It is the same result we had for nilmanifolds and tori and it shows that once we can find either of $NP_m(f)$ or $N\Phi_m(f)$ for all $m|n$ then we can get the other at level n.

(5.17) THEOREM. *Suppose that $f\colon S \to S$ is a map on a solvmanifold S. Then $N\Phi_n(f) = \sum_{m|n} NP_m(f)$ and hence*

$$NP_n(f) = \sum_{\tau \subseteq \mathrm{P}(n)} (-1)^{|\tau|} N\Phi_{n:\tau}(f)$$

where $\mathrm{P}(n)$ is the set of prime divisors of n and $n\colon\tau = n/\prod_{p\in\tau} p$.

The simpler cases of finding these periodic numbers occurs when the appropriate iterates of the self map are weakly Jiang so we define this term next. Recall that $\mathcal{R}(f)$ denotes the number of Reidemeister classes of the map $f\colon X \to X$ (i.e. the number of twisted conjugacy classes of $f_\#$ on $\pi_1(X)$).

(5.18) DEFINITION. Suppose that X is a space where Nielsen theory can be defined (e.g. a compact manifold). Then we say that a self map $f\colon X \to X$ is *weakly Jiang* provided that either $N(f) = 0$ or $N(f) = \mathcal{R}(f)$. In other words, all algebraic classes share the same status of essentiality or inessentiality.

(5.19) REMARK. One of the pioneering works in the theory of calculations of Nielsen numbers was the discovery by Boju Jiang (see [J]) that all tori are Jiang spaces (i.e. all self-maps are weakly Jiang). A study of [H] reveals that on both the Fadell–Husseini fibration of a nilmanifold $f\colon M \to M$ or the Mostow fibration of a solvmanifold $f\colon S \to S$, we have that any Reidemeister class α of f can be projected to a Reidemeister class $\overline{\alpha}$ for $\overline{f}$. If in addition we know that there is a geometric realization of $\overline{\alpha}$ which contains some $b \in \operatorname{fix}(\overline{f})$, then this projection is also part of a decomposition of α to both a Reidemeister class of $\overline{f}$, and a Reidemeister class α_b of f_b. Using this decomposition idea, there are naïve type addition formulas for $\mathcal{R}(f)$ when it is finite. Because of the strong fibre uniformity of Fadell–Husseini fibrations (i.e. for all $x, y \in \operatorname{fix}(\overline{f})$, $\mathcal{R}(f_x) = \mathcal{R}(f_y)$ – see Theorem (4.8)), it is also true that all maps on nilmanifolds are weakly Jiang. This reasoning extends to the following theorem which provides an easy way to tell if arbitrary maps on Mostow fibrations are weakly Jiang or not.

(5.20) THEOREM ([HK1], [HK2]). *Suppose that f is a fibre preserving self map on a Mostow fibration $M \hookrightarrow S \to T$. Then f is weakly Jiang if and only if one of the following holds:*

(5.20.1) $N(\overline{f}) = 0$ *(so* $N(f) = 0$*).*
(5.20.2) *For all* $x \in \text{fix}(\overline{f})$, $N(f_x) \neq 0$ *(so* $N(f) \geq N(\overline{f})$*).*
(5.20.3) *For all* $x \in \text{fix}(\overline{f})$, $N(f_x) = 0$ *(so* $N(f) = 0$*).*

As we indicated before, a huge class of examples of solvmanifolds and their maps are coming in this chapter. However, the classic example of non-weakly Jiang maps is on the Klein bottle (see [H]) where one can arrange for situations where the restriction to fibres (which are S^1) all have degrees which are equally distributed among the set $\{+1, -1\}$. On S^1, we have $N(\deg(-1)) = 2$ and $N(\deg 1) = 0$ so this is precisely an example where the conditions of Theorem (5.20) fail. The following result shows why weakly Jiang maps are so easy to work with.

(5.21) THEOREM ([HK1], [HK2]). *Let S be a solvmanifold and $f: S \to S$. Define $M(f, n)$ to be the set of maximal divisors m of n with the property that $N(f^m) \neq 0$ (e.g. if $N(f^n) \neq 0$ then $M(f, n) = \{n\}$ and if $N(f^m) = 0$ for all $m|n$ then $M(f, n) = \emptyset$). Then if f^m is weakly Jiang for each $m \in M(f, n)$ then*

$$N\Phi_n(f) = \sum_{\emptyset \neq \mu \subseteq M(f,n)} (-1)^{|\mu|-1} N(f^{\text{GCD}(\mu)}).$$

(5.22) COROLLARY ([HK1]). *If $f: S \to S$ is such that f^n is weakly Jiang and $N(f^n) \neq 0$, then $N\Phi_n(f) = N(f^n)$. If $N(f^m) = 0$ for all $m|n$, then $N\Phi_n(f) = N(f^n) = 0$.*

5.4. Model solvmanifolds. The following construction due to Heath and Keppelmann in [HK4] provides one with an infinite variety of solvmanifolds and their associated (in the sense of (5.14)) minimal Mostow fibrations. Although we will use these as our main source of examples, it will also be important to explain how these models in a very significant way describe all the Nielsen theory (both ordinary, periodic, coincidence and root theory) that can occur for self-maps of and maps between any solvmanifolds where base and fibre of domain and range have matching dimensions.

(5.23) EXAMPLE. Fix natural numbers p and u and let $A: \mathbb{Z}^p \to \text{Aut}(\mathbb{Z}^u)$ be a homomorphism of groups. For $\overrightarrow{l} \in \mathbb{Z}^p$ we shall let $A^{\overrightarrow{l}}$ denote the resulting invertible linear map from $\mathbb{Z}^u$ to itself. Define $\sim_A$ to be the equivalence relation on $\mathbb{R}^u \times \mathbb{R}^p$ generated by the following:

(5.23.1) $(\overrightarrow{x}, \overrightarrow{y}) \sim_A (\overrightarrow{x} + \overrightarrow{k}, \overrightarrow{y})$ for any $\overrightarrow{k} \in \mathbb{Z}^u$ (i.e. on the first coordinate the relation is exactly like that which forms the torus T^u).

(5.23.2) $(\vec{x}, \vec{y}) \sim_A (A^{\vec{l}}(\vec{x}), \vec{y} + \vec{l})$ for any $\vec{l} \in \mathbb{Z}^p$ (i.e. on the second coordinate the relation is also like that which forms T^p but such equivalences act to "twist" the T^u torus of the first coordinate).

It can be shown (for all the details see [HK4]) that the resulting quotient space S is indeed a solvmanifold with minimal Mostow fibration $T^u \hookrightarrow S \to T^p$ where the projection to the base is induced by projection onto the second coordinate. While we won't prove this completely, it is indeed clear that $\pi_1(S) \cong \mathbb{Z}^u \times_A \mathbb{Z}^p$ and hence that $\mathbb{Z}^u \cong \pi_1(T^u) = \pi_1(S)_{(1)}$. This helps explain why the indicated fibration is minimal (see Example (5.15)) and also shows that $\pi_1(S)$ is indeed solvable but not generally nilpotent since $\pi_1(S)_{(2)} = [\pi_1(T^u), \pi_1(T^u)] = \{0\}$ but in general $\pi_1(S)_2 = [\pi_1(T^u), \pi_1(S)]$ could be all of $\pi_1(T^u)$.

(5.24) DEFINITION. The *gluing data* of a model solvmanifold S as constructed in Example (5.23) is the homomorphism $A: \mathbb{Z}^p \to \mathrm{Aut}(\mathbb{Z}^u)$ which defines how the base and fibre are "twisted together" to form the total space S. In an arbitrary Mostow fibration $M \hookrightarrow S \xrightarrow{p} T$, the gluing data is the linearization of the homomorphism from $\pi_1(T)$ to $\mathrm{Aut}(\pi_1(M))$ obtained from path lifting. (I.e. we linearize each automorphism. Since linearization is a functorial process we do still get a homomorphism in a way which is consistent with what happens on the models.)

We now consider a construction to produce model maps on model solvmanifolds.

(5.25) EXAMPLE. Suppose that S is the solvmanifold produced by the construction of (5.23). Let $X: \mathbb{Z}^u \to \mathbb{Z}^u$ and $Y: \mathbb{Z}^p \to \mathbb{Z}^p$ be linear maps with the property that for all $\vec{l} \in \mathbb{Z}^p$ we have that $XA^{\vec{l}} = A^{Y\vec{l}}X$. (We shall call this the *commuting constraint*). Then it is not hard to see that the correspondence $(\vec{x}, \vec{y}) \to (X\vec{x}, Y\vec{y})$ respects the equivalence relation $\sim_A$ and induces a self map f on S.

Since the constructed model solvmanifold fibration of (5.23) uses ordinary projection, we can certainly see that $\overline{f}$ is already linearized as the map Y. Now if $\vec{0}$ denotes the identity coset of $\mathbb{Z}^p$ then $f[(\vec{x}, \vec{0})] = [(X\vec{x}, Y\vec{0})] = [(X\vec{x}, \vec{0})]$ so that X is the linearization of f_0. To consider f on the fibres of other fixed points of $\overline{f}$ suppose that $\vec{b} \in \mathbb{R}^p$ belongs to the T^p coset of a fixed point of $\overline{f}$ (i.e. $\vec{b} - Y\vec{b} \in \mathbb{Z}^p$). Then we observe that

$$(\vec{x}, \vec{b}) \to (X\vec{x}, Y\vec{b}) \sim_A (A^{\vec{b}-Y\vec{b}}X\vec{x}, Y\vec{b} + \vec{b} - Y\vec{b}) = (A^{\vec{b}-Y\vec{b}}X\vec{x}, \vec{b})$$

so that the linearization of $f_{[\vec{b}]}$ is $A^{\vec{b}-Y\vec{b}}X$. The Nielsen class of $\vec{b}$ corresponds to the Reidemeister class of $\vec{b} - Y\vec{b}$. The fact that T^p is a Jiang space means that every coset of $\mathbb{Z}^p/((Y - I)\mathbb{Z}^p)$ (i.e. every Reidemeister class of $\overline{f}$) is represented by some such $\overrightarrow{b}$ and hence from Theorems (5.16) and (3.1) we have that

$$N(f) = \sum_{\overrightarrow{k} \in \mathbb{Z}^p/((Y-I)\mathbb{Z}^p)} |\det(A^{\overrightarrow{k}} X - I)|.$$

We shall now proceed to give a fairly large collection of examples which use the construction of (5.23) in a variety of ways. For a full analysis of these the reader is invited to consult [HK1]–[HK5].

(5.26) Example. With the construction of (5.23) we can take $A\colon \mathbb{Z}^p \to \mathrm{Aut}(\mathbb{Z}^u)$ to be trivial (i.e. for all $\overrightarrow{l} \in \mathbb{Z}^p$ we have that $A^{\overrightarrow{l}} = I$). In this case then the resulting space S is just T^{p+u} and the commuting constraint will hold for any X and Y. Every summand in the formula of (5.25) is the same and we get that $N(f) = |\det(X - I)||\det(Y - I)|$ in complete agreement with the theory for a torus (3.1).

(5.27) Example. The Klein bottle discussed in (5.2) can be formulated in terms of (5.23) using $p = u = 1$ and defining $A\colon \mathbb{Z} \to \mathrm{Aut}(\mathbb{Z})$ so that A^l is multiplication by $(-1)^l$. For a complete analysis of the resulting periodic and ordinary Nielsen theory see [H] and [HK1]–[HK3], [HKW].

(5.28) Example. With the construction of (5.23), using $p = 1$ and $u = 4$, define $A\colon \mathbb{Z} \to \mathrm{Aut}(\mathbb{Z}^4)$ to be generated by

$$A^1 = A = \begin{bmatrix} 0 & 1 & 0 & 0 \\ 0 & 0 & 1 & 0 \\ 1 & 0 & 1 & -1 \\ 1 & 0 & 1 & 0 \end{bmatrix}, \qquad X = \begin{bmatrix} 1 & -2 & 3 & -4 \\ -1 & 1 & -3 & -3 \\ -6 & -1 & -5 & 3 \\ -3 & 3 & 1 & -6 \end{bmatrix}.$$

For the matrix X given it is easy to check that $XA = A^{11}X$ so that the correspondence $(\overrightarrow{x}, t) \to (X\overrightarrow{x}, 11t)$ for $\overrightarrow{x} \in \mathbb{R}^4$ and $t \in \mathbb{R}$ gives a self map f on the resulting solvmanifold S.

Since $\overline{f}$ is a map of degree 11 on S^1, the Reidemeister classes of $\overline{f}$ in $\mathbb{Z}$ can be denoted using their most popular representatives by $0, \ldots, 9$. Routine calculations yield the following:

j	0	1	2	3	4
$N(f_j) = \lvert\det(A^j X - I)\rvert$	751	605	1181	775	311

j	5	6	7	8	9
$N(f_j) = \lvert\det(A^j X - I)\rvert$	605	691	355	661	1255

Since all of these are nonzero, Theorem (5.20) tells us that f is indeed weakly Jiang. Furthermore, we can calculate $N(f) = \sum_{j=0}^{9} N(f_j) = 7190$. Likewise, f^2 is also weakly Jiang but in this case the correspondence $(\overrightarrow{x}, t) \to (X^2\overrightarrow{x}, 121t)$ which induces f^2 requires us to check that

$$N((f^2)_j) = |\det(A^j X^2 - I)| \neq 0 \quad \text{for } j = 0, \dots, 119.$$

We can thus get that

$$N(f^2) = \sum_{j=0}^{119} N((f^2)_j) = 52661880.$$

Since this is nonzero, Corollary (5.22) tells us that $N\Phi_2(f) = 52661880$ as well. Finally, Theorem (5.17) and the fact that $N(f) = N\Phi_1(f)$ tells us that

$$NP_2(f) = N(f^2) - N(f) = 52654690$$

and hence that less than 0.02% of the fixed points of f^2 are fixed points of f.

(5.29) EXAMPLE. Using the construction of (5.23) let $p = 4$ and $u = 1$. For $\overrightarrow{q} = (q_1, q_2, q_3, q_4) \in \mathbb{Z}^4$ define $A\colon \mathbb{Z}^4 \to \mathrm{Aut}(\mathbb{Z})$ by $A(\overrightarrow{x}) =$ multiplication by $(-1)^{q_1+q_2+q_3+q_4}$. Let

$$Y = \begin{bmatrix} 1 & 0 & 1 & 0 \\ 0 & -1 & 0 & -1 \\ 2 & 0 & 2 & 0 \\ 0 & -2 & 0 & -2 \end{bmatrix}.$$

We claim that for $t \in \mathbb{R}$ and $\overrightarrow{y} \in \mathbb{R}^4$ that the correspondence $(t, \overrightarrow{y}) \to (3t, Y\overrightarrow{y})$ induces a well defined self map f on the resulting solvmanifold S. In this case, the commuting constraint is that for all $\overrightarrow{q} \in \mathbb{Z}^4$ we have that $3A^{\overrightarrow{q}} = A^{Y\overrightarrow{q}}3$. Since this is just an equation of integers we really need that $A^{\overrightarrow{q}} = A^{Y\overrightarrow{q}}$ for all $\overrightarrow{q} \in \mathbb{Z}^4$. This will follow once we see that for the standard basis vectors ε_i in $\mathbb{Z}^4$ we have that $-1 = A^{\varepsilon_i} = A^{Y\varepsilon_i} = A^{\text{column } i \text{ of } Y}$ for $i = 1, \dots, 4$. This follows by noting that each column of Y has an odd sum.

We notice that there are $N(\overline{f}) = |\det(Y - I)| = 8$ cosets of $(Y - I)\mathbb{Z}^4$ in $\mathbb{Z}^4$ for which we need to calculate the linearization of f restricted to the fibre above that coset. However, it is clear that these linearizations will all be integers with two possible values: $N(f_0) = |3 - 1| = 2$ or $|(-1)3 - 1| = 4$. This confirms from (5.20) that f is weakly Jiang. Furthermore, which value occurs above a given coset or Reidemeister class of $\overline{f}$ is determined by the parity of that coset ([11]). We claim

([11]) Notice that since $Y - I$ has even column sums, every $\mathbb{Z}^4$ vector in a given coset of $Y - I$ will have the same parity.

that the collection of even cosets (denoted $\mathcal{K}_e$) and the collection of odd cosets (denoted $\mathcal{K}_o$) are equal. To see this just consider the correspondence $\mathcal{K}_o \to \mathcal{K}_e$ given by the rule $\overrightarrow{z} \to \overrightarrow{z} + \varepsilon_1$. This is a bijection with an inverse given by $\overrightarrow{w} \to \overrightarrow{w} - \varepsilon_1$. Therefore, $8/2 = 4$ of the fixed points of $\overline{f}$ lie below fibres where f has degree 3 and 4 of the fixed points of $\overline{f}$ lie below fibres where $\overline{f}$ has degree -3. Therefore, $N(f) = N(\overline{f})\,\text{average}\{N(f_b) : b \in \text{fix}(\overline{f})\} = 8 * 3 = 24$.

In a similar fashion, we see that all iterates of f are weakly Jiang, $\deg((f_0)^n) = 3^n$, and the average Nielsen fibre number is $(3^n - 1 + 3^n + 1)/2 = 3^n$. Thus the following calculations result from using (5.22) and (5.17).

n	$N(\overline{f})$	$N(f^n) = N\Phi_n(f)$	$NP_n(f)$
1	8	24	24
2	64	576	552
3	728	19656	19632
4	6400	518400	517824
6	529984	386358336	386338128
12	282428473600	1500940704384576000	150094070051581440

(5.30) EXAMPLE. Using the construction of (5.23) let $p = 2$ and $u = 3$. Define $A\colon \mathbb{Z}^2 \to \text{Aut}(\mathbb{Z}^3)$ to be generated by the following two matrices:

$$A^{(1,0)} = \begin{bmatrix} 3 & 1 & 5 \\ 3 & 1 & 6 \\ -2 & -1 & -4 \end{bmatrix}, \qquad A^{(0,1)} = \begin{bmatrix} 2 & -1 & 1 \\ 0 & -2 & -3 \\ -1 & 1 & 0 \end{bmatrix}.$$

This makes sense since

$$\det(A^{(1,0)}) = \det(A^{(0,1)}) = 1 \quad \text{and} \quad A^{(1,0)}A^{(0,1)} = A^{(0,1)}A^{(1,0)} = A^{(1,1)}.$$

Therefore, we have that for any $(j,k) \in \mathbb{Z}^2$ that $A^{(j,k)} = (A^{(1,0)})^j(A^{(0,1)})^k$.

We claim that the correspondence from $\mathbb{R}^3 \times \mathbb{R}^2$ to $\mathbb{R}^3 \times \mathbb{R}^2$ given by $(\overrightarrow{v}, \overrightarrow{w}) \to (X\overrightarrow{v}, Y\overrightarrow{w})$ gives a well defined map f on the corresponding solvmanifold:

$$X = \begin{bmatrix} 1 & -2 & 0 \\ -1 & -5 & -7 \\ -1 & 2 & 0 \end{bmatrix}, \qquad Y = \begin{bmatrix} 2 & 3 \\ 1 & 1 \end{bmatrix}.$$

The claim is established by observing that $XA^{(1,0)} = A^{(2,1)}X$ and $XA^{(0,1)} = A^{(3,1)}X$.

We notice that Y is the linearization of $\overline{f}$ and so $N(\overline{f}) = |\det(Y - I)| = 3$. We observe that $a = [0,0]^T, b = [1,0]^T$, and $c = [0,1]^T$ represent the 3 different cosets of $(Y - I)(\mathbb{Z}^2)$. This follows since none of $(Y - I)^{-1}(a - b)$, $(Y - I)^{-1}(a - c)$ nor $(Y - I)^{-1}(b - c)$ belong to $\mathbb{Z}^2$. These are, respectively, $(Y - I)^{-1}[-1,0]^T = [0,-1/3]^T$, $(Y - I)^{-1}[0,-1]^T = [-1,1/3]^T$ and $(Y - I)^{-1}[1,-1]^T = [-1,2/3]^T$. Thus we can calculate $N(f)$ using the fibre linearizations X, $A^{(1,0)}X$, $A^{(0,1)}X$ which give Nielsen numbers of 12, 3 and 9, respectively. Therefore, $N(f) = 12 + 3 + 9 = 24$.

For f^2, we first note that $N(\overline{f}^2) = |\det(Y^2 - I)| = 9$. After calculations like the above we determine a complete set of Reidemeister representatives for $\overline{f}$ and we compute the following (we shall use $[a,b]^T$ to denote the coset $[a,b]^T + (Y^2 - I)(\mathbb{Z}^2)$):

$b \in \text{fix}(\overline{f}^2)$	$[0,0]^T$	$[1,0]^T$	$[2,0]^T$	$[0,1]^T$	$[0,2]^T$
$N((f^2)_b)$	48	63	39	39	63

$b \in \text{fix}(\overline{f}^2)$	$[1,1]^T$	$[1,2]^T$	$[2,1]^T$	$[2,2]^T$	total
$N((f^2)_b)$	48	39	63	48	$N(f^2) = 450$

Since all the entries for $N((f^2)_b)$ and $N(f_b)$ are nonzero, it is clear that f^2 and f are weakly Jiang. Therefore, $NP_2(f) = N\Phi_2(f) - N\Phi_1(f) = N(f^2) - N(f) = 426$.

6. Linearizations on solvmanifolds

The following theorem explains why the model construction of (5.23) is so universally important.

(6.1) THEOREM ([HK4]). *Suppose that $(g, \overline{g})$ is a fibre preserving self map of a Mostow fibration $M \hookrightarrow S \to T$ for a solvmanifold S. If $N(\overline{g}) = 0$ then $N(g) = 0$. Otherwise, without loss of generality, we may assume that the identity element 0 in the compact abelian group T is fixed by $\overline{g}$. Let g_0 denote the restriction of g to the fibre over 0. Define a homomorphism $B: \pi_1(T) \to \text{Aut}(\pi_1(M))$ by the path lifting on the (Hurewicz) Mostow fibration. Let $u = \dim M$ and $p = \dim T$. Define $\widehat{S}$ to be the solvmanifold formed using the construction of (5.23) where the gluing data A is the linearization of B. Let f_0, $\overline{f}$ denote the respective linearizations of g_0 and $\overline{g}$. Then the diagonal map $f = (f_0, \overline{f})$ will be a fibre (as in Example (5.25)) preserving self map of $\widehat{S}$. Furthermore, for all n, $N(f^n) = N(g^n)$, $L(f^n) = L(g^n)$ $NP_n(f) = NP_n(g)$, and $N\Phi_n(f) = N\Phi_n(g)$. Coincidence and root theories for self maps between S and $\widehat{S}$ are also the same.*

(6.2) DEFINITION. We say that $(\widehat{S}, f)$ is the *linearization* of (S, g).

(6.3) REMARK. Theorem (6.1) demonstrates the somewhat complex but yet also straightforward way in which the linearizations of maps on nilmanifolds and tori combine together to yield the linearization of a map on a solvmanifold. We note that this all works because there is a Mostow fibration in which every homotopy class of self maps has a fibre preserving representative. However, it is true that some solvmanifolds will have more than one Mostow fibration and some maps of these fibrations may be fibre preserving on several of these fibrations. In such cases it is important to realize that strictly speaking the "linearization of f on S" may be ambiguous since we really are defining linearizations with respect to a given fibering.

(6.4) REMARK. As we noted in our discussion of nilmanifolds, when we linearize, the nilmanifold is replaced by the much simpler space, the torus of the same dimension. In general a torus will have many more homotopy classes of self-maps than a nilmanifold because the maps on the nilmanifolds must arise from homomorphisms of the underlying Lie group and Lie algebra. If the underlying Lie group is nonabelian then certainly not just any integer entry matrix will give a self map like it would on the torus. Similarly, Theorem (6.1) reminds us that while every self map of an arbitrary solvmanifold has a corresponding map on a model solvmanifold, the model solvmanifold in general will have many more self-maps than the original solvmanifold. As the following shows, this would especially occur, for example, if the fibre of the original Mostow fibration were not a torus.

(6.5) EXAMPLE. In this example we will construct a model nilmanifold Mostow fibration, with a self map that we can linearize. Let H be the connected simply connected nilpotent Lie group of matrices of the form

$$H = \left\{ \begin{bmatrix} 1 & x & y \\ 0 & 1 & z \\ 0 & 0 & 1 \end{bmatrix} : x, y, z \in \mathbb{R} \right\}.$$

Then the subcollection Γ of H consisting of matrices with integer entries is a uniform subgroup of H and the coset space H/Γ is the standard nilmanifold M called baby nil (see [HK1]). Now it is possible to use a software package like Maple to check that the following functions are homomorphisms (φ is a periodic period 6 isomorphism) of H which are invariant on Γ:

$$\varphi\left(\begin{bmatrix} 1 & x & y \\ 0 & 1 & z \\ 0 & 0 & 1 \end{bmatrix}\right) = \begin{bmatrix} 1 & 73z + 9x, & 302z - 292z^2 + \frac{209}{2}x - \frac{9}{2}x^2 - 73xz + y \\ 0 & 1 & -8z - x \\ 0 & 0 & 1 \end{bmatrix},$$

$$\Theta\left(\begin{bmatrix} 1 & x & y \\ 0 & 1 & z \\ 0 & 0 & 1 \end{bmatrix}\right)$$
$$= \begin{bmatrix} 1 & -56z + x, & -256144z + 28z^2 - \frac{63641}{2}x - \frac{1}{2}x^2 + 56xz - 57y \\ 0 & 1 & -z - x \\ 0 & 0 & 1 \end{bmatrix}.$$

Furthermore, it is also possible to check that $\Theta\varphi = \varphi^5\Theta$. Now on $\mathbb{R}^3$ define the group operation $*$ by the rule

$$(x_1, y_1, z_1) * (x_2, y_2, z_2) = (x_1 + x_2, y_1 + y_2 + x_1 z_2, z_1 + z_2).$$

In other words, this is just a different notation for the operation on H discussed above. Define an equivalence relation on $\mathbb{R}^4 = \mathbb{R}^3 \times \mathbb{R}$ to be generated by the rules

(6.5.1) $(\overrightarrow{v}, t) \sim (\overrightarrow{k} * \overrightarrow{v}, t)$ when $\overrightarrow{k} \in \mathbb{Z}^3$.
(6.5.2) $(\overrightarrow{v}, t) \sim (\varphi^k(\overrightarrow{v}), t + k)$ when $k \in \mathbb{Z}$.

The resulting space is a solvmanifold S with minimal Mostow fibration $M \hookrightarrow S \to S^1$. Furthermore, the above mentioned relationship between φ and Θ means that the correspondence $F(\overrightarrow{v}, t) = (\Theta(\overrightarrow{v}), 5t)$ induces a well defined fibre preserving map $f\colon S \to S$.

We wish to compute the linearization of (S, f). We begin this process by first considering the linearizations of φ and Θ. Since the Fadell-Husseini fibration for M involves S^1 (for the y factor) fibres over T^2 (for the x and z coordinates), these linearizations are block diagonal matrices with a 1×1 and 2×2 block. However, if we maintain the x, y, z ordering then these are no longer quite block diagonal and we get the following linearizations (which we denote with $\widehat{\ }$):

$$\widehat{\varphi} = \begin{bmatrix} 9 & 0 & 73 \\ 0 & 1 & 0 \\ -1 & 0 & -8 \end{bmatrix}, \qquad \widehat{\Theta} = \begin{bmatrix} 1 & 0 & -56 \\ 0 & -57 & 0 \\ -1 & 0 & -1 \end{bmatrix}.$$

This means that the linearization of S (on which the linearization $\widehat{f}$ of f operates) is the model solvmanifold $\widehat{S}$ produced from the construction of (5.23) with $u = 3$, $p = 1$ and gluing data $A\colon \mathbb{Z} \to \mathrm{Aut}(\mathbb{Z}^3)$ generated by $A^1 = \widehat{\varphi}$. (Notice that $\det(\widehat{\varphi}) = 1$ so $\widehat{\varphi}$ is indeed invertible over the integers). Then $\widehat{f}\colon \mathbb{R}^3 \times \mathbb{R} \to \mathbb{R}^3 \times \mathbb{R}$ is the model diagonal map $\widehat{f}(\overrightarrow{v}, t) = (\widehat{\Theta}\overrightarrow{v}, 5t)$. Thus $\widehat{f}_0 = \widehat{\Theta}$ and $\overline{\widehat{f}}$ is the 1×1 matrix $[5]$.

Just as in the original more complicated S we have a special commuting constraint between φ and Θ, we also have here that $\widehat{\Theta}\widehat{\varphi} = \widehat{\varphi}^5\widehat{\Theta}$. In other words, $\widehat{f}$ satisfies the requirements for a model map on $\widehat{S}$.

It is clear from this construction that there will be many model maps on $\widehat{S}$ which do not arise from the linearization of maps on S. In fact, any 3×3 matrix B with the property $BA^1 = A^r B$ will yield a map h on $\widehat{S}$ of the form $h(\overrightarrow{v}, t) = (B\overrightarrow{v}, rt)$. For example, any B of the following form (with $r = 5$) will give such a map

$$B = \begin{bmatrix} a & 0 & 17a + 73b \\ 0 & e & 0 \\ b & 0 & -a \end{bmatrix}.$$

However, unless $e = -a^2 - 17ab - 73b^2$ such a B does not arise as the linearization of a map on M ([12]).

This discussion should be compared with that of [HK4] where a solvmanifold (similar to what is done here) called grumpy sol is considered. In that solvmanifold, the possible self-maps are so restrictive that the only possibilities involve situations where (in the notation of this setting) Θ is a power of φ and $r = 1$. However, just as we see here, it turns out that there are many self-maps on the linearization of this so called grumpy (i.e. uncooperative) solvmanifold.

7. $\mathcal{N}R$ solvmanifolds

As we commented in Section 4, any self map $f\colon M \to M$ on a nilmanifold M satisfies the Anosov theorem which says that the Lefschetz and Nielsen numbers are related by $N(f) = |L(f)|$. Recalling that any self map $f\colon M \to M$ is, up to homotopy and hence without loss of generality to the Nielsen Lefschetz theory, fibre preserving on the Fadell–Husseini fibration for M, this result follows by induction on the dimension of M, the naïve addition conditions of the Fadell–Husseini fibration, and the following two important facts (see [FH]):

(7.1.1) The Anosov theorem holds for all maps of tori.
(7.1.2) The homotopy type of f_x is independent of $x \in \mathrm{fix}(\overline{f})$.

Together these details mean that when $N(\overline{f}) \neq 0$ and $\overline{f}$ has been linearized, that

$$N(f) = \sum_{x \in \mathrm{fix}(\overline{f})} N(f_x) = N(\overline{f})N(f_0) = |L(\overline{f})||L(f_0)| = |L(\overline{f})L(f_0)| = |L(f)|.$$

The purpose of this section is to describe an important class of solvmanifolds where, even though the ingredients of the proof are slightly different, the same conclusions hold. Although not originally formulated in [KMc] in this way, the following definition is equivalent:

([12]) It is known that on the Fadell–Husseini fibration of baby nil, where every self map f will be fibre preserving, we always have that the determinant of the linearization of f_0 equals the degree of $\overline{f}$ (see [JM]). It is this reasoning that is generating the condition on e.

(7.2) DEFINITION. Suppose that S is a solvmanifold. Let $A\colon \mathbb{Z}^p \to \operatorname{Aut}(\mathbb{Z}^u)$ denote the gluing data of the linearization $\widehat{S}$ of the minimal Mostow fibration of S. We say that S is an $\mathcal{N}R$ *solvmanifold* provided that no matrix $A^{\overrightarrow{v}}$ for $\overrightarrow{v} \in \mathbb{Z}^p$ has eigenvalues of magnitude 1 which are not equal to 1. ([13])

The following important result holds for $\mathcal{N}R$ solvmanifolds.

(7.3) LEMMA. *Suppose that S is an $\mathcal{N}R$ solvmanifold and that $f\colon S \to S$ is a fibre preserving map on a Mostow fibration of S. Assume that $N(\overline{f}) \neq 0$. Then although the homotopy type of f_x may vary, $N(f_x)$ will be independent of $x \in \operatorname{fix}(\overline{f})$.*

The proof of Lemma (7.3) proceeds by reducing all solvmanifold maps to their linearized models and using the following purely algebraic result to make the whole machinery tick:

(7.4) LEMMA ([KMc]). *Let $\mathcal{A}\colon \mathbb{Z}^p \to \operatorname{Aut}(\mathbb{Z}^u)$ be a homomorphism with the property:*

(7.4.1) *if $\overrightarrow{v} \in \mathbb{Z}^p$ and $\lambda \in \mathbb{C}$ are such that λ is an eigenvalue of $\mathcal{A}(\overrightarrow{v})$ with $|\lambda| = 1$, then $\lambda = 1$.*

Suppose that matrices X $(u \times u)$ and Y $(p \times p)$ have entries in $\mathbb{Z}$ with the property that, for any $1 \leq i \leq p$, $X\mathcal{A}(\varepsilon_i) = \mathcal{A}(Y_i)X$ where ε_i denotes the ith standard basis vector of $\mathbb{R}^p$ and $Y_i = Y\varepsilon_i$ is the ith column of Y. Then $\det(\mathcal{A}(\overrightarrow{v})X - I)$ is independent of $\overrightarrow{v} \in \mathbb{Z}^p$.

(7.5) COROLLARY. *The Anosov theorem $N(f) = |L(f)|$ holds for all maps of $\mathcal{N}R$ solvmanifolds.*

While we will not attempt to prove this lemma here, it is our hope that Examples (7.9) and (7.10) (when considered in the appropriate bases) reveal the guts of what is going on with this result. An awareness of the following lemma is critical to having the big picture:

(7.6) LEMMA ([KMc], [HK3], [HK4]). *Suppose that X, A and B are $n \times n$ matrices with $XA = BX$. If $\overrightarrow{v}$ belongs to the generalized eigenspace of A with eigenvalue λ (i.e. there is a minimal k so that $(A - \lambda I)^k \overrightarrow{v} = 0$), then $(B - \lambda I)^k X \overrightarrow{v} = 0$. In other words, the commuting constraint for X guarantees that generalized eigenvectors are sent to generalized eigenvectors of the same eigenvalue with equal or lower level exponent k. Thus, for example, if λ is not an eigenvalue of B then the generalized eigenspace of A with eigenvalue λ must belong to the kernel of X. Thus, when A and B share no eigenvalues, $X = 0$.*

([13]) An exponential solvmanifold is defined as one where the expotential map from the associated Lie algebra to the associated Lie group is onto. All expotential maps are $\mathcal{N}R$ solvmanifolds but not vice versa.

(7.7) EXAMPLE. The simplest example of a non-$\mathcal{N}R$ solvmanifold is without a doubt the Klein bottle of Example (5.27). Here the gluing data produces 1×1 matrices with the most famous root of unity (i.e. -1) different than 1. As a result, if d is the degree of f_0, then we get maps of the fibres with possible degrees of $\pm d$. These indeed exhibit non-Nielsen number fibre uniformity with Nielsen numbers of either $|d-1|$ or $|d+1|$. Example (5.29) where the eigenvalues are also ± 1 is another example of a non-$\mathcal{N}R$ solvmanifold.

(7.8) EXAMPLE. The solvmanifold S and self map of Example (5.30) do not demonstrate fibrewise Nielsen number uniformity. Therefore S cannot possibly be $\mathcal{N}R$. This is confirmed when we examine the eigenvalues of the gluing data matrices. The set of eigenvalues of both $A^{(1,0)}$ and $A^{(1,0)}$ are $\{1, -1/2 \pm i\sqrt{3}/2\}$ which indeed includes the two primitive cube roots of unity.

The following example is yet a more interesting version of a highly non-$\mathcal{N}R$ solvmanifold. It illustrates the somewhat special conditions needed for fibre non-uniformity.

(7.9) EXAMPLE. We construct a model solvmanifold S using $u = 8$ and $p = 1$ as in Example (5.23). We define the gluing data $A\colon \mathbb{Z} \to \mathrm{Aut}(\mathbb{Z}^8)$ to be generated by

$$A = \begin{bmatrix} 0 & 1 & 0 & 0 & 0 & 0 & 0 & 0 \\ 0 & 0 & 1 & 0 & 0 & 0 & 0 & 0 \\ 0 & 0 & 0 & 1 & 0 & 0 & 0 & 0 \\ 0 & 0 & 0 & 0 & 1 & 0 & 0 & 0 \\ 0 & 0 & 0 & 0 & 0 & 1 & 0 & 0 \\ 0 & 0 & 0 & 0 & 0 & 0 & 1 & 0 \\ 0 & 0 & 0 & 0 & 0 & 0 & 0 & 1 \\ 1 & 0 & 0 & 0 & 0 & 0 & 0 & 0 \end{bmatrix}.$$

We can then define $f\colon S \to S$ with $\overline{f}$ of degree 3 and f_0 with linearization

$$X = \begin{bmatrix} a & b & c & d & w & x & y & z \\ x & y & z & a & b & c & d & w \\ c & d & w & x & y & z & a & b \\ z & a & b & c & d & w & x & y \\ w & x & y & z & a & b & c & d \\ b & c & d & w & x & y & z & a \\ y & z & a & b & c & d & w & x \\ d & w & x & y & z & a & b & c \end{bmatrix}.$$

In other words, $XA = A^3X$. In fact, one can check that the characteristic polynomial of A is $\lambda^8 - 1$ so that the eigenvalues of A are $\lambda_0, \ldots, \lambda_7$ with $\lambda_j = e^{j2\pi i/8}$ for each j. This confirms that S is not $\mathcal{N}R$. The operation of $\lambda_j \to \lambda_j^3$ forms a permutation on this set of eigenvalues $\{\lambda_0, \lambda_1, \ldots, \lambda_7\}$ given by $(\lambda_1, \lambda_3)(\lambda_2, \lambda_6)(\lambda_5, \lambda_7)$. The existence of this permutation means X can exist with nonzero entries. The fact that $\lambda_4 = \lambda_1\lambda_3 = \lambda_5\lambda_7 = -1$ rather than $+1 = \lambda_0 = \lambda_2\lambda_6$ gives our Nielsen number fibre non-uniformity. It is all much clearer when we work in a diagonal basis for A:

$$A = \begin{bmatrix} \lambda_0 & 0 & 0 & 0 & 0 & 0 & 0 & 0 \\ 0 & \lambda_1 & 0 & 0 & 0 & 0 & 0 & 0 \\ 0 & 0 & \lambda_2 & 0 & 0 & 0 & 0 & 0 \\ 0 & 0 & 0 & \lambda_3 & 0 & 0 & 0 & 0 \\ 0 & 0 & 0 & 0 & \lambda_4 & 0 & 0 & 0 \\ 0 & 0 & 0 & 0 & 0 & \lambda_5 & 0 & 0 \\ 0 & 0 & 0 & 0 & 0 & 0 & \lambda_6 & 0 \\ 0 & 0 & 0 & 0 & 0 & 0 & 0 & \lambda_7 \end{bmatrix},$$

$$X = \begin{bmatrix} m & 0 & 0 & 0 & 0 & 0 & 0 & 0 \\ 0 & 0 & 0 & p & 0 & 0 & 0 & 0 \\ 0 & 0 & 0 & 0 & 0 & 0 & q & 0 \\ 0 & r & 0 & 0 & 0 & 0 & 0 & 0 \\ 0 & 0 & 0 & 0 & s & 0 & 0 & 0 \\ 0 & 0 & 0 & 0 & 0 & 0 & 0 & t \\ 0 & 0 & u & 0 & 0 & 0 & 0 & 0 \\ 0 & 0 & 0 & 0 & 0 & v & 0 & 0 \end{bmatrix}.$$

The nonzero entries of X correspond to the cycle mentioned above: Only entries (shift all j subscripts by 1) $X_{1,1}$, $X_{5,5}$, $X_{2,4}$, $X_{4,2}$, $X_{3,7}$, $X_{7,3}$, $X_{6,8}$, $X_{8,6}$ can be nonzero for the equation $XA = A^3X$ to hold.

Since $\overline{f}$ has degree 3 we see that $N(\overline{f}) = |3-1| = 2$ and thus the fixed point set of $\overline{f}$ consists of two points $\{x_0, x_1\}$. This means that $N(f) = N(f_{x_0}) + N(f_{x_1})$. The linearization of these two maps on fibres are X and AX. Now

$$N(f_{x_0}) = |\det(X - I)| = |(m-1)(s-1)(-1+pr)(-1+vt)(-1+qu)|$$

whereas in this factored form the other fibre has a similar looking but different formulation given by

$$\begin{aligned} N(f_{x_1}) &= |\det(AX - I)| \\ &= |(m\lambda_0 - 1)(s\lambda_4 - 1)(-1 + \lambda_1\lambda_3 pr)(-1 + \lambda_7\lambda_5 vt)(-1 + \lambda_2\lambda_6 qu)| \\ &= |(m-1)(-s-1)(-1-pr)(-1-vt)(-1+qu)|. \end{aligned}$$

The next example contrasts the above one with the case of an $\mathcal{NR}$ solvmanifold. This comes from [KMc] although the exact nature of all self-maps was not studied there.

(7.10) EXAMPLE. We construct a solvmanifold S with precisely the same setup as (7.9) but this time we take the gluing data to be generated by the matrix

$$A = \begin{bmatrix} 0 & 1 & 0 & 0 & 0 & 0 & 0 & 0 \\ 0 & 0 & 1 & 0 & 0 & 0 & 0 & 0 \\ 0 & 0 & 0 & 1 & 0 & 0 & 0 & 0 \\ 0 & 0 & 0 & 0 & 1 & 0 & 0 & 0 \\ 0 & 0 & 0 & 0 & 0 & 1 & 0 & 0 \\ 0 & 0 & 0 & 0 & 0 & 0 & 1 & 0 \\ 0 & 0 & 0 & 0 & 0 & 0 & 0 & 1 \\ -1 & 0 & 12 & 0 & -6 & 0 & 12 & 0 \end{bmatrix}.$$

The eigenvalues of A are $\pm\alpha, \pm 1/\alpha, \pm\beta$, and $\pm 1/\beta$ where $\alpha = \sqrt{2-\sqrt{2}} + i\sqrt{\sqrt{2}-1}$ and $\beta = \sqrt{2+\sqrt{2}} + \sqrt{\sqrt{2}+1}$. Even though $|\alpha| = 1$, α is not a root of unity because its polar angle is an irrational fraction of 2π. The eigenvalue β is real with $|\beta| > 1$. Therefore, the gluing data has no roots of unity as eigenvalues so S is indeed $\mathcal{NR}$.

The only integers r for which eigenvalues $(A) \cap$ eigenvalues $(A^r) \neq \emptyset$ are $r = \pm 1$ ([14]). Of course if $f\colon S \to S$ has $\deg(\overline{f}) = 1$ then $N(\overline{f}) = 0$ so $N(f) = 0$. Therefore, let us consider the production of such maps f with $\overline{f}$ of degree -1 and f_0 with linearization X. Again, working in a diagonal basis for A we have the following where the nonzero entries of X correspond to the permutation of the eigenvalues of A which take place when each eigenvalue is replaced by its reciprocal.

$$A = \begin{bmatrix} \alpha & 0 & 0 & 0 & 0 & 0 & 0 & 0 \\ 0 & \frac{1}{\alpha} & 0 & 0 & 0 & 0 & 0 & 0 \\ 0 & 0 & -\alpha & 0 & 0 & 0 & 0 & 0 \\ 0 & 0 & 0 & -\frac{1}{\alpha} & 0 & 0 & 0 & 0 \\ 0 & 0 & 0 & 0 & \beta & 0 & 0 & 0 \\ 0 & 0 & 0 & 0 & 0 & \frac{1}{\beta} & 0 & 0 \\ 0 & 0 & 0 & 0 & 0 & 0 & -\beta & 0 \\ 0 & 0 & 0 & 0 & 0 & 0 & 0 & -\frac{1}{\beta} \end{bmatrix},$$

([14]) From Lemma (7.6) this means that if $f\colon S \to S$ is such that $\deg(\overline{f}) \neq \pm 1$ then the linearization of any f_x for $x \in \mathrm{fix}(\overline{f})$ is the zero matrix. This has Nielsen number 1 so in this case $N(f) = N(\overline{f})$.

$$X = \begin{bmatrix} 0 & a & 0 & 0 & 0 & 0 & 0 & 0 \\ b & 0 & 0 & 0 & 0 & 0 & 0 & 0 \\ 0 & 0 & 0 & c & 0 & 0 & 0 & 0 \\ 0 & 0 & d & 0 & 0 & 0 & 0 & 0 \\ 0 & 0 & 0 & 0 & 0 & w & 0 & 0 \\ 0 & 0 & 0 & 0 & x & 0 & 0 & 0 \\ 0 & 0 & 0 & 0 & 0 & 0 & 0 & y \\ 0 & 0 & 0 & 0 & 0 & 0 & z & 0 \end{bmatrix}.$$

As in Example (7.9) we again have that $N(\overline{f}) = 2$ and so with $\mathrm{fix}(\overline{f}) = \{x_0, x_1\}$ we have as before that f_{x_0} and f_{x_1} have linearizations of X and AX. Here we have fibre uniformity since

$$N(f_{x_0}) = |\det(X - I)| = |(1 - ab)(1 - cd)(1 - xw)(1 - yz)|$$

whereas

$$\begin{aligned} N(f_{x_1}) &= |\det(AX - I)| \\ &= \left|\left(1 - \left(\alpha\frac{1}{\alpha}\right)ab\right)\left(1 - (-\alpha)\left(-\frac{1}{\alpha}\right)cd\right)\right. \\ &\quad \cdot \left.\left(1 - \left(\beta\frac{1}{\beta}\right)xw\right)\left(1 - (-\beta)\left(-\frac{1}{\beta}\right)yz\right)\right| \\ &= |(1 - ab)(1 - cd)(1 - xw)(1 - yz)|. \end{aligned}$$

It is important to point out that unlike the strong uniformity which occurs in self-maps of Fadell–Husseini fibrations for nilmanifolds where every self map on fibres over fixed points of $\overline{f}$ have the same homotopy types and hence linearizations, self-maps of $\mathcal{NR}$ solvmanifolds exhibit a weaker kind of uniformity. Namely, while for $x, y \in \mathrm{fix}(\overline{f})$ the linearizations of f_x and f_y can differ, we will still get that $N(f_x) = N(f_y)$.

Suppose that $f\colon S \to S$ is a self map of an $\mathcal{NR}$ solvmanifold. Let X denote the linearization of f_0 and Y the linearization of $\overline{f}$. Then Nielsen number fibre uniformity means that, when $\overline{f}$ is linear,

$$\begin{aligned} N(f) &= \sum_{x \in \mathrm{fix}(\overline{f})} N(f_x) \\ &= \sum_{x \in \mathrm{fix}(\overline{f})} N(f_0) = N(f_0)N(\overline{f}) = |\det(X - I)\det(Y - I)|. \end{aligned}$$

This readily gives us the following which shows that linearizations of maps on $\mathcal{NR}$ solvmanifolds can be given, depending on the application, an extremely simple form.

(7.11) THEOREM. *Suppose that $f\colon S \to S$ is a self map of an $\mathcal{N}R$ solvmanifold of dimension n. Let X and Y denote the linearizations of f_0 and $\overline{f}$, respectively. Then the linearized self map $g\colon T^n \to T^n$ with the block matrix given below has exactly the same Nielsen periodic theory as f:* ([15])

$$g = \begin{bmatrix} X & 0 \\ 0 & Y \end{bmatrix}.$$

We now close this section with some different examples on roots and coincidences. Recall that if T is a torus and $f, g\colon T \to T$ are self-maps with linearizations X and Y, respectively, then the Nielsen coincidence number $N(f,g)$ of f and g is given by $|\det(X-Y)|$ (see [B]). This easily generalizes for self maps to nilmanifolds and in a naive sum way to Mostow fibrations (see [Mc1], [Mc2]). In a similar way, the Nielsen root number of f is given by $\mathrm{root}(f) = |\det(X)|$ (see [B]). This is also computed on nilmanifold and solvmanifold fibrations in the usual naive sum way (see [BBS], [BO]) although since the gluing data matrices A always satisfy $\det(A) = \pm 1$, we will automatically get fibre wise uniformity on any fibration since $|\det(AX)| = |\det(A)\det(X)| = |\det(A)||\det(X)| = |\det(X)|$. ([16])

The following type of calculations are justified by the semi-index product formula of [Je3] whether or not the manifolds involved are orientable.

(7.12) EXAMPLE. Let S be the solvmanifold constructed using (5.23) with $p = 1$ and $u = 4$ and gluing data $\mathcal{A}\colon \mathbb{Z} \to \mathrm{Aut}(\mathbb{Z}^4)$ generated by

$$A^1 = A = \begin{bmatrix} 10 & 1 & -23 & 0 \\ 0 & 0 & 1 & 14 \\ 1 & 0 & 0 & 1 \\ -1 & 0 & 0 & 0 \end{bmatrix}.$$

The eigenvalues of A are the following

$$\beta_1 = \frac{3+\sqrt{5}}{2}, \quad \beta_2 = \frac{3-\sqrt{5}}{2}, \quad \beta_3 = \frac{7+3\sqrt{5}}{2} \quad \text{and} \quad \beta_4 = \frac{7-3\sqrt{5}}{2}.$$

We note that $\beta_1^2 = \beta_3$ and $\beta_2^2 = \beta_4$. Also, $1/\beta_1 = \beta_2$ and $1/\beta_3 = \beta_4$. This means

([15]) Since on an $\mathcal{N}R$ solvmanifold the homotopy type of the various f_x can differ, it is conceivable that such simplified models may not work for certain coincidence theories, especially in the case of self maps.

([16]) [B], [BS] show that on a compact manifold, the Nielsen root number is independent of the location of the target point.

that we can define two maps on S: f of type $(2, X)$ and g of type $(-1, Y)$ where

$$X = \begin{bmatrix} 669 & 96 & -2205 & -102 \\ -1281 & -15 & 1521 & -2337 \\ 159 & 30 & -639 & -132 \\ -255 & -33 & 783 & -15 \end{bmatrix},$$

$$Y = \begin{bmatrix} -2 & -396 & 46 & 11 \\ -121885 & -5493 & 126350 & 5051 \\ -5283 & -17 & -3 & -5120 \\ -15 & 2 & 350 & 5498 \end{bmatrix}.$$

The Nielsen coincidence number is then defined by using $N(\overline{f}, \overline{g}) = |2 - (-1)| = 3$ and so

$$N(f, g) = |X - Y| + |AX - Y| + |A^2X - Y|.$$

In this case we do have fibrewise coincidence uniformity (i.e. all these terms are the same) so we get

$$N(f, g) = N(\overline{f}, \overline{g})N(X, Y) = 3|X - Y| = 4,082,913,519,431,883.$$

(7.13) EXAMPLE. Using the solvmanifold S of Example (7.12) and the remark made before this example about fibre uniformity, we can easily compute that

$$\text{root}(g) = \text{root}(\overline{g})\text{root}(g_0) = |-1||\det(Y)| = 1,360,971,173,143,961.$$

Similarly, but because the operation $\alpha \to \alpha^2$ does not give a permutation on the eigenvalues of A, we have that $\text{root}(f) = |2||\det(X)| = 0$.

8. Nielsen periodic theory for non-weakly Jiang maps

Suppose that $f: S \to S$ is a map on a solvmanifold and n is a fixed positive integer with the property that f^n is not weakly Jiang. In other words, if $\{b_1, \ldots, b_k\} = \text{fix}(\overline{f}^n)$, then it means that $N(f^k)$ may not equal $N\Phi_k(f)$ and that $N((f^n)_{b_i})$ varies with i in such a way that sometimes this is zero and sometimes it isn't. Even if this doesn't happen for f^n but does occur for some f^k with $k|n$, then this means we cannot use Möbius inversion of the $\{N(f_k) : k|n\}$ to find $NP_n(f)$. Instead we must rely on Theorem (8.1) below from [HK2]. This result is a consequence of the naive addition formula for Mostow fibrations of solvmanifolds and it basically says that if $b \in \text{fix}(\overline{f}^n)$ has minimal period m then $(f^m)_b$ is a self map on the fibre over b and the level n/m irreducible periodic point classes of $(f^m)_b$ inject into the level n irreducible periodic point classes of f. Furthermore, all level n irreducible periodic point classes of f arise in this way (from various b and m).

(8.1) THEOREM ([HK2]). *Suppose that $f: S \to S$ is a map on a solvmanifold and n is such that $\overline{f}^n$ has $N(\overline{f}^n) = k \neq 0$ fixed points $b_1, \dots, b_k$. Let p_i denote the minimum period of b_i under $\overline{f}$. Then*

$$NP_n(f) = \sum_{i=1}^{k} NP_{n/p_i}((f^{p_i})_{b_i}).$$

Of course Theorem (8.1) will work on any self map f not just the non-weakly Jiang maps. However, when f is weakly Jiang we do not have to work so hard to find $NP_n(f)$ since we can first use the $\{N(f^k) : k|n\}$ to get the $\{N\Phi_k(f) : k|n\}$. Möbius inversion then gives us $\{NP_k(f) : k|n\}$. This process is reversed for non-weakly Jiang maps since we must first get the $\{NP_k(f) : k|n\}$ using (8.1). The $N\Phi_n(f)$ then follow using that $\sum_{k|n} NP_k(f) = N\Phi_n(f)$.

Example (8.3) below concerns a self map f on a model solvmanifold which factors over a circle. The following ideas will be important considerations in this example for deciding what the minimal periods p_i of the various b_i are and how these relate to the linearizations of $(f^{p_i})_{b_i}$.

(8.2) LEMMA ([HY], [HPY], [HK3]). *Suppose that g is the linear map on the circle S^1 of degree $d \neq 1, -1$. Then* $\text{fix}(g^n)$ *consists of $|d^n - 1|$ points equally spaced on the circle $[0,1]/\{0 \sim 1\}$. If we label these points by the values $\{0, 1, \dots, |d^n - 1| - 1\}$ consecutively in a counter clockwise fashion starting at the coset of zero, and if $x_j \in [0,1]$ represents the point j, then $dx_j - x_j = j$. Furthermore, if $m|n$, then the indices of* $\text{fix}(g^m) = \Delta_m = \{0, 1, \dots, |d^m - 1| - 1\} \subset \text{fix}(g^n) = \Delta_n = \{0, \dots, |d^n - 1| - 1\}$ *correspond to one another by the rule $\beta_{m,n}: \Delta_m \to \Delta_n$ consisting of multiplication by the number $\iota_{m,n} = 1 + d^m + d^{2m} + \dots + d^{n-m}$.*

The techniques in the following computation are typical of the methods presented in [HK3].

(8.3) EXAMPLE. Consider the model solvmanifold S of Example (5.28) but define a different self map f than the one given by the matrix X specified there. Instead we shall let $X = I$ be the 4×4 identity matrix. Since the matrix A generating the gluing data has period 10 we do indeed get a model map f in which $\overline{f}$ has degree 11. ($XA = IA = AI = AII = AA^{10}I = A^{11}I = A^{11}X$). In computing $N(f)$ we denote $\text{fix}(\overline{f})$ by $\{0, \dots, 9\}$. The linearization of f_i is A^i. These have the following Nielsen numbers:

$N(I)$	$N(A)$	$N(A^2)$	$N(A^3)$	$N(A^4)$
0	1	5	1	5

$N(A^5)$	$N(A^6)$	$N(A^7)$	$N(A^8)$	$N(A^9)$
16	5	1	5	1

The fact that some of these values are zero and some are nonzero confirms that f is non-weakly Jiang. Furthermore, all of the A^i (including $i = 0$) do occur as the linearizations of f^n restricted to various fibres over points of $\text{fix}(\overline{f}^n)$ so that every f^n is also non-weakly Jiang.

While a general formula for $NP_n(f)$ is too convoluted to be worth the trouble, the following procedure shows that it's computation is algorithmic (many of our comments will apply more generally):

Function $NP_n(f)$
INPUT: positive integer n
OUTPUT: $NP_n(f)$

$NP_n(f) = 0$

START FOR LOOP. For each $b \in \text{fix}(\overline{f}^n) = \{0, \ldots, 11^n - 2\}$ compute the contribution of $(f^n)_b$ to $NP_n(f)$ as follows:

(8.3.1) find the smallest $k|n$ so that $\iota_{k,n} = 1 + 11^k + 11^{2k} + \cdots + 11^{n-k}$ divides b. Thus b for $\overline{f}^n$ represents a periodic point for $\overline{f}$ of minimal period k.

(8.3.2) b is represented by the index $s = b/\iota_{k,n} \in \text{fix}(\overline{f}^k) = \{0, 1, \ldots, 11^k - 2\}$ for $\overline{f}^k$.

(8.3.3) The linearization of f^k on the principal fiber is the 4×4 identity matrix I. The gluing data matrix for S is generated by A, and $\overline{f}^k$ is a map on S^1 of degree 11^k. The commuting constraint for f^k is that $IA = A^{11^k} I$ (which follows from the fact that $11 \equiv 1 (\text{mod } 10)$ implies $11^k \equiv 1^k (\text{mod } 10) = 1 (\text{mod } 10)$). Thus $(f^k)_s$ has linearization A^s. ([17])

(8.3.4) Increment $NP_n(f)$ by the contribution of b which is $NP_{n/k}(A^s)$. ([18])

END FOR LOOP

Return $NP_n(f)$

END OF ALGORITHM

An examination of the various $NP_r(A^s)$ reveals that these are often zero and hence the above algorithm is actually less complicated than it seems. The zeros occur because of the following facts:

(8.3.5) $A^s = A^p$ when $s \equiv p (\text{mod } 10)$.

([17]) In the general case where f_0 has linearization X we note that f^k on the principal fibre has linearization X^k and $(f^k)_s$ has linearization $A^s X^k$.

([18]) In general we increment by $NP_{n/k}(A^s X^k)$.

(8.3.6) For $s > 0$, A^s has a period given by $10/\mathrm{GCD}(s \bmod 10, 10)$. In fact, the possible periods of A^s are 1 (for $s = 0$ only), 2 (for $s = 5$ only), 5 (for $s = 2, 4, 6, 8$), and 10 (for $s = 1, 3, 7, 9$). We note that $NP_r(A^s)$ will be zero for $s \geq$ period of A^s. ([19])

(8.3.7) Even when $r <$ period of A^s, $NP_r(A^s) = 0$ can often be zero. In particular, when A^s has period 10, $NP_r(A^s)$ is nonzero only for $r \in 1, 2, 5$ and when A^s has period 2 or 5 then $NP_r(A^s)$ is nonzero only when $r = 1$.

(8.3.8) This is all summarized in the following table which computes $NP_r(A^s)$ for the congruence class of s modulo 10:

$NP_r(A^s)$	$r = 1$	$r = 2$	$r = 3$	$r = 4$	$r = 5$	$r \geq 6$
$s \equiv 0$	0	0	0	0	0	0
$s \equiv 1, 3, 7$ or 9	1	4	0	0	15	0
$s \equiv 2, 4, 6$ or 8	5	0	0	0	0	0
$s \equiv 5$	16	0	0	0	0	0

We will now consider the calculation of $NP_n(f)$ for some special values of n.

Case 1. n is a prime p.

Since p is prime, all points of $\mathrm{fix}(\overline{f}^p)$ have a minimal period of either p or 1. Those of period 1 contribute terms of the form $NP_p(A^s)$ for $s = 0, \ldots, 9$ corresponding to the points of $\mathrm{fix}(\overline{f})$.

Those elements of $\mathrm{fix}(\overline{f}^p)$ of minimal period p contribute terms of the form

$$NP_1(A^s) = N(A^s).$$

There is such a term for each $0 \leq s \leq 11^p - 2$ whenever s is not divisible by $\iota_{1,p} = 1 + 11^1 + \ldots + 11^{p-1}$. We will make this calculation by first removing from

$$N(f^p) = \sum_{s=0}^{11^p-2} N(A^s)$$

the reducible terms. Since $|\mathrm{fix}(\overline{f}^p)| = 11^p - 1 \equiv 0 (\mathrm{mod}\ 10)$ all possible linearizations on fibres over fixed points of $\overline{f}^p$ occur with equal frequency. Thus

$$N(f^p) = \frac{11^p - 1}{10} \sum_{s=0}^{9} N(A^s) = \frac{11^p - 1}{10} 40 = 4(11^p - 1).$$

([19]) A periodic map of period k on any space will never have periodic points of a minimal period which does not divide k because every point in the space is a periodic point of period (perhaps not minimally) k. This means that any map homotopic to a irreducible map of period k will never have periodic Nielsen classes of period greater than k. On a solvmnifold where the Euler characteristic is zero, there won't even be any essential classes of minimal period greater than half the period of the map.

Since every summand of $\iota_{1,p}$ is congruent to 1 modulo 10 we have that

$$\iota_{1,p} = 1 + \sum_{i=1}^{p-1} 11^i \equiv 1 * p = p(\text{mod } 10).$$

Thus ,the congruence class of $\text{fix}(\overline{f}) = \{0, \ldots, 9\}$ gets, modulo 10, multiplied by p when the index is boosted to $\text{fix}(\overline{f}^p)$ and multiplied by $\iota_{1,p}$. Therefore, the required correction to $N(f^p)$ from the points in the base of period p equals

$$N(f^p) - \sum_{s=0}^{9} N(A^{ps}).$$

This last summation is the same as $\sum_{s=0}^{9} N(A^s) = 40$ for all p which do not divide 10 and it happens to also be 40 for $p = 2$ as well. These calculations are all tabulated below ([20]):

$$NP_p(f) = N(f^p) - \sum_{s=0}^{9} N(A^{ps}) + \sum_{s=0}^{9} NP_p(A^s)$$

p	2	3	5	≥ 7
$\sum_{s=0}^{9} NP_p(A^s)$	16	0	60	0
$\sum_{s=0}^{9} N(A^{ps})$	40	40	80	40
$NP_p(f)$	$4(11^p) - 28$	$4(11^p) - 44$	$4(11^p) - 24$	$4(11^p) - 44$

Case 2. $n = p^k$ for p a prime and $k > 1$.

In this case, we first note that any fixed point of $\overline{f}^{p^k}$ reduces to be a fixed point of $\overline{f}^{p^{k-1}}$ if and only if it does not have minimal period $n = p^k$. This follows by observing that if the point reduces to some level $t|p^k$ then $t = p^q$ for some $q < k$ and hence the point can be boosted from level t to level p^{k-1} to provide a reduction to level p^{k-1}. Furthermore, each such reduction is uniquely represented as an index in $\text{fix}(\overline{f}^{p^{k-1}}) = \{0, 1, \ldots, 11^{p^{k-1}} - 2\}$ (see [HY], [HPY]).

Now $\iota_{p^{k-1},p^k} = 1 + 11^{p^{k-1}} + 11^{2p^{k-1}} + \ldots + 11^{p^k - p^{k-1}}$ consists of $p^k/p^{k-1} = p$ terms congruent to 1 modulo 10 and hence $\iota_{p^{k-1},p^k} \equiv p(\text{mod } 10)$. Therefore, the contributions of these reductions to $NP_n(f)$ are of the form $NP_m(A^{ps})$ with $m \geq 1$ and $0 \leq s \leq 11^{p^{k-1}} - 1$.

([20]) Weakly Jiang theory would give that $NP_p(f) = N(f^p) - N(f)$. While this does in fact agree with our formula for most p, it is clearly incorrect for $p = 2, 5$.

We claim that, for $m > 1$, $NP_m(A^{ps})$ is always zero. To see this first note that $NP_p(A^{ps}) = 0$. This is certainly true for $p > 5$ by the calculations made above.

For $p = 5$ we check that $NP_5(I) = NP_5(A^5) = 0$.

For $p = 3$ we have that $NP_3(\text{anything}) = 0$.

For $p = 2$ we see that $NP_2(A^2) = NP_2(A^4) = NP_2(A^6) = NP_2(A^8) = 0$.

Now if $p = m = 2$ then since $NP_4(\text{anything}) = 0$ the result follows. Otherwise if $m \geq 2$ then $p^m \geq 6$ so $NP_{p^m}(\text{anything}) = 0$).

Therefore, we can conclude that the only contribution from $\overline{f}^n$ comes from those points of minimal period n. These will contribute all terms of the form

$$NP_1(A^s) = N(A^s) \quad \text{for } 0 \leq s \leq 11^{p^k} - 2$$

and s not reducible (i.e. s is not divisible by ι_{p^{k-1},p^k}). Now $\sum_{s=0}^{p^k-2} N(A^s)$ is just $N(f^n)$ and the total sum of the terms which we don't want to count is just equal to

$$\sum_{s=0}^{11^{p^{k-1}}-2} N(A^{\iota_{p^{k-1},p^k}(s)}) = \frac{11^{p^{k-1}}}{10} \sum_{s=0}^{9} N(A^{ps}).$$

Hence we may conclude that ([21])

$$p \text{ prime and } k > 1, \quad NP_{p^k}(f) = N(f^{p^k}) - \frac{11^{p^{k-1}} - 1}{10} \sum_{s=0}^{9} N(A^{ps}).$$

Case 3. $n = pq$ for p, q distinct primes.

As the reader will see, the complexity involved in this calculation is significantly greater than what we have done so far! ([22]) Although some of these numbers were computed above, we will reorganize ourselves and make note of the following quantities:

$\alpha_{r,t} = \sum_{s=0}^{9} NP_r(A^{ts})$

$\alpha_{r,t}$	$t=1$	$t=2$	$t=3$	$t=5$	$t \geq 6$
$r=1$	40	40	40	80	40
$r=2$	16	0	16	0	16
$r=3$	0	0	0	0	0
$r=5$	60	0	60	0	60
$r \geq 7$	0	0	0	0	0

([21]) $\sum_{s=0}^{9} N(A^{ps})$ was calculated in Case 1.

([22]) The case of general n and arbitrary f is left as an exercise for the insane reader.

$\alpha_{rt,1} = \sum_{s=0}^{9} NP_{rt}(A^s)$

$\alpha_{rt,1}$	$t=1$	$t=2$	$t=3$	$t=4$	$t=5$	$t\geq 7$
$r=1$	40	16	0	60	0	0
$r=2$	16	0	0	0	0	0
$r=3$	0	0	0	0	0	0
$r=5$	60	0	0	0	0	0
$r\geq 7$	0	0	0	0	0	0

In this case the possible minimal periods of a reducible point $b \in \text{fix}(\overline{f}^n)$ are p, q, or 1. Furthermore, b reduces to level 1 if and only if it reduces to both levels p and q (see [HPY], [HY]).

If b has minimal period p or q then it contributes a term to $NP_n(f)$ of the form $NP_q((f^p)_b)$ or $NP_p((f^q)_b)$ respectively. These are equal to $NP_q(A^s)$ for $0 \leq s \leq 11^p - 2$ with $\iota_{1,p} \not| s$ and $NP_p(A^s)$ for $0 \leq s \leq 11^q - 2$ and $\iota_{1,q} \not| s$, respectively. Now because $11^q - 1, 11^p - 1 \equiv 0 (\text{mod } 10)$ and $\iota_{1,p} \equiv p (\text{mod } 10)$ and $\iota_{1,q} \equiv q (\text{mod } 10)$ we can make the following simplifications

$$\text{For period } p \text{ points of } \overline{f} = \sum_{s=0,\,\text{irreducible}}^{11^p-2} NP_q(A^s) = \frac{11^p - 1}{10}\alpha_{q,1} - \alpha_{q,p}.$$

Similarly, the period q points of $\overline{f}$ contribute $(11^q - 1)\alpha_{p,1}/10 - \alpha_{p,q}$.

If b belongs to $\text{fix}(\overline{f})$ and thus has minimal period 1, then its total contribution to $NP_n(f)$ is of the form $\sum_{s=0}^{9} NP_{pq}(A^s) = \alpha_{pq,1}$.

Finally, if b has minimal period $n = pq$ under $\overline{f}$ then its contribution to $NP_n(f)$ is $NP_1(A^s) = N(A^s)$ for $0 \leq s \leq 11^{pq} - 2$ and $\iota_{1,pq}, \iota_{p,pq}, \iota_{q,pq} \not| s$. Now the indices which are divisible by $\iota_{p,pq} = 1 + 11^p + 11^{2p} + \ldots + 11^{p(q-1)} \equiv q \text{ (mod } 10)$ will have (modulo 10) the form qs for $0 \leq s \leq 11^p - 2$. Similarly, indices divisible by $\iota_{q,pq}$ will have the form ps for $0 \leq s \leq 11^q - 2$. Finally, indices divisible by both $\iota_{p,pq}$ and $\iota_{q,pq}$ will also be divisible by $\iota_{1,pq} = 1 + 11 + 11^2 + \ldots + 11^{pq-1} \equiv pq (\text{mod } 10)$ and will have (modulo 10) the form pqs for $0 \leq s \leq 9$. Thus the total contribution comes from (all s) minus (s reducible to level p only) minus (s reducible to level q only) minus (s reducible to level 1). Writing this out gives:

$$N(f^n) - \left[\frac{11^p - 1}{10}\alpha_{q,1} - \alpha_{q,p}\right] - \left[\frac{11^q - 1}{10}\alpha_{p,1} - \alpha_{p,q}\right] - \alpha_{pq,1}$$

Using that $N(f^n) = (11^{pq} - 1)\alpha_{1,1}/10$ we get

$$\frac{11^{pq} - 1}{10}\alpha_{1,1} - \frac{11^p - 1}{10}\alpha_{q,1} - \frac{11^q - 1}{10}\alpha_{p,1} + \alpha_{p,q} + \alpha_{q,p} - \alpha_{pq,1}.$$

9. Periodic points on solvmanifolds

Suppose that $f: T \to T$ is a map on a torus. While the periodic Nielsen numbers $NP_n(f)$ and $N\Phi_n(f)$ of f focus on what is potentially possible (through a homotopy of f) for eliminating the periodic points related to the specific iterate f^n, in this section we will look at results which consider what kinds of minimums can occur simultaneously over all the iterates of f. The following results from [Ha] will explain what we have in mind. The results for nil- and solvmanifolds are available in [K1], [K2].

For a self map $f: M \to M$ recall that $\text{per}(f)$ denotes the set $\bigcup_{n=1}^{\infty} \text{fix}(f^n)$ of all periodic points of f.

(9.1) THEOREM ([Ha]). *Suppose that $f: T \to T$ is a self map on a torus. Then there exists a $g \sim f$ with $\text{per}(g)$ finite (denoted $\text{per}(f) \sim$ finite) if and only if the sequence $\{N(f^n)\}_{n=1}^{\infty}$ of Nielsen numbers for the iterates of f is bounded.*

Since for any $g \sim f$ we have that $f^n \sim g^n$ we know that $N(f^n) \leq \text{fix}(g^n)$. Thus clearly when $\{N(f^n)\}_{n=1}^{\infty}$ is unbounded there can be no such g. This reasoning would hold on any space where the Nielsen number is defined.

To get a sense for the converse result suppose that $f: T^k \to T^k$ has $k \times k$ linearization matrix A. Thus

$$N(f^n) = \det(A^n - I) = \prod_{j=1}^{k} (\lambda_j^n - 1)$$

where $\{\lambda_1, \ldots, \lambda_k\}$ denotes the eigenvalues (with repetitions as necessary) of A. [Ha] showed that because these eigenvalues come from a matrix with integer entries, unless such a product is identically zero for all n (which will happen if and only if one of the $\lambda_j = 1$), such a sum will be bounded if and only if all the nonzero eigenvalues are roots of unity. In this case, it is then possible to do an integer entries change of basis to an upper triangular matrix so that the result can be proved (with lots of effort!) inductively on the dimension of the torus.

The induction is started by noting that if $f: S^1 \to S^1$ is such that $\{N(f^n)\}_{n=1}^{\infty}$ is bounded, then $\deg(f) \in \{-1, 0, 1\}$. Each of these three cases provides some food for an insightful discussion as to how this result works in general:

(0) When $\deg(f) = 0$ then f is homotopic to a constant map g which has one fixed point and no other periodic points. In fact, the existence of one or more zero eigenvalues for a general $f: T^k \to T^k$ implies that we can homotope f to map T^k onto a proper subtorus of itself. Subsequent iterations then essentially only need work on this smaller dimensional sub torus where induction results can already be assumed to hold.

(−1) When $\deg(f) = -1$ one can describe the linearization of f by $f(e^{i\theta}) = e^{-i\theta}$. Consider the map $\delta\colon S^1 \to S^1$ which is homotopic to the identity and "stretches" S^1 in the sense that it leaves ± 1 fixed but moves all other points away from -1 and towards $+1$. By doing this in an identical fashion in the upper and lower hemispheres of S^1 and maintaining the ordering of points in the sense that if two points $a = e^{i\theta_1}, b = e^{i\theta_2}$ are such that $0 < \theta_1 < \theta_2 < \pi$ then $\delta(a) = e^{i\beta_1}, \delta(b) = e^{i\beta_2}$ will be such that $\theta_i > \beta_i$ for $i = 1, 2$ and $\beta_1 < \beta_2$. Because $\delta \sim 1_{S^1}$ we see that $g = \delta \circ f \sim f$. Furthermore, for a point $x \neq \pm 1$ on S^1, the orbit $\{g^n(x)\}$ will consist of a sequence of points which alternate between hemispheres and monotonically approach but never reach 1. Therefore, the only periodic points of g are the fixed points $\{\pm 1\}$ so $\mathrm{per}(g)$ is indeed finite. It turns out that virtually all of Halpern's homotopies as well as those used to prove extensions of his results to nil and solvmanifolds, rely on this notion of post- and pre-composing the linearization of f by certain carefully prescribed isotopies of the identity.

(+1) When $\deg(f) = +1$ we know that $N(f^n) = 0$ for all n. The linearization of f is the identity which is homotopic to a small rotation g by an irrational multiple of 2π. Such a map g will have $\mathrm{per}(g) = \emptyset$. This technique can in fact be generalized to prove a similar result for nilmanifolds.

The following theorem generalizes the above considerations to nilmanifolds (see also Theorems (3.2), (4.10) and Remark (4.11)).

(9.2) THEOREM ([K1]). *Suppose that $f\colon G/\Gamma = M \to M$ is a self map induced by a homomorphism on a nilmanifold. Suppose that $\{N(f^n)\}_{n=1}^{\infty}$ is bounded. Then there exists $u, v \sim 1_M$ so that $f \sim g = ufv$ has* $\mathrm{per}(g)$ *finite. Furthermore, for all n in which $N(f^n) \neq 0$ we will have that $|\mathrm{fix}(g^n)| = N(f^n)$. If $N(f) = 0$ (which implies that $N(f^n) = 0$ for all n) then such a construction will produce a map g with per$(g) = \emptyset$ (denoted* $\mathrm{per}(f) \sim \emptyset$*).*

PROOF SKETCH. The proof uses induction based on [Ha] and the Fadell–Husseini fibration of a nilmanifold. If $f_0\colon T \to T$ is the map on the principal fibre and $\overline{f}\colon B \to B$ is the map on the nilmanifold base of the Fadell–Husseini fibration for M, then $N(f^n) = N(f_0^n)N(\overline{f}^n)$. Thus when $\{N(f^n)\}_{n=1}^{\infty}$ is bounded it means that either both $\{N(\overline{f}^n)\}_{n=1}^{\infty}$ and $\{N(f_0^n)\}_{n=1}^{\infty}$ are bounded or else one of $N(f_0^n)$ or $N(\overline{f}^n)$ is zero for all n. In the first and last cases we can use the induction hypothesis to deform $\overline{f}$ and then lift this to a homotopy of f which will produce a map with finitely many (perhaps none) periodic fibres. We then use the form of the homotopy to adjust the identity in a neighborhood of a few of the fibres from each orbit of $\overline{f}$ to produce the desired result for f.

In the second case the proof is made possible by the "Anosov trick" from [K1] which we have discussed previously ([23]). The basic idea behind the "Anosov trick" is the following. Recall that on a nilmanifold the linearization of f can be thought of as being based on a homomorphism induced map $F: G \to G$. The linearization matrix for F is then the derivative $F': \mathcal{G} \to \mathcal{G}$ on the Lie algebra of G (see (4.11)). When $N(f) = \det(F' - I) = 0$ it means that $F' - \mathrm{id}: \mathcal{G} \to \mathcal{G}$ is not onto. As a result, we can choose a vector $\vec{V} \notin (F' - I)\mathcal{G}$. The exponential map $\exp: \mathcal{G} \to G$ can then be used to produce a set of elements in G of the form $\exp(t\vec{V})$ for $t \in \mathbb{R}$. We can then create a map $\delta_t \sim 1_M$ which comes from multiplication by $\exp(t\vec{V})$. When $\vec{V}$ belongs to the center of $\mathcal{G}$ (as it can when $N(f_0^n) = 0$ for all n) it turns out that for all but at most countably many choices of t, the composition $\delta_t \circ f$ will be periodic point free. This is exactly the generalization of what we did up above on S^1 using an irrational rotation. (Only the countably many rational rotations of the identity on S^1 produce maps with periodic points). □

Completely generalizing these results to solvmanifolds is highly nontrivial for the following reasons:

(9.3.1) Suppose that $f: S \to S$ is a fibre preserving map on a Mostow fibration $N \hookrightarrow S \to T$. Mostow fibrations of solvmanifolds are non-orientable in the following sense. If $x \in \mathrm{fix}(\overline{f}^k)$ and $y \in \mathrm{fix}(\overline{f}^l)$ then there is no immediate relationship between $\{N(((f^k)_x)^n)\}_{n=1}^{\infty}$ and $\{N(((f^l)_y)^n)\}_{n=1}^{\infty}$. One could potentially be bounded or zero while the other is unbounded.

(9.3.2) The deformation $\overline{g} \sim \overline{f}$ using the results for nilmanifolds described above may not be enough by itself to carry an inductive proof along. Although we will have that $|\mathrm{fix}(\overline{g})| = N(\overline{f}^n)$ when $N(\overline{f}^n) \neq 0$, we cannot say much about what happens when p is such that $N(\overline{f}^p) = 0$. For such a value of p there could potentially be a periodic point b of $\overline{g}$ which has minimal period p. In this case, the values $\{N(((f^p)_b)^n)\}_{n=1}^{\infty}$ have no effect on the sequence $\{N(f^n)\}_{n=1}^{\infty}$. Thus it is conceivable that the first sequence could be unbounded even though the second sequence is not.

(9.3.3) It may indeed be the case that $\{N(\overline{f}^n)\}_{n=1}^{\infty}$ is unbounded while $\{N(f^n)\}_{n=1}^{\infty}$ remains bounded. If this happens then we need to know what is possible among the various sequences $\{N(((f^k)_b)^n)\}_{n=1}^{\infty}$ for $b \in \mathrm{fix}(\overline{f}^k)$. When $\{N(f^n)_{n=0}^{\infty} = \{0\}$ in theis setting knowing whether we can find $g \sim f$ with $\mathrm{per}(f) = \emptyset$ is problematic.

These issues were extensively studied in the unpublished paper [K2] ([24]).

([23]) This trick was first presented in [A] to eliminate fixed points.

([24]) This paper was originally written in 1996 when the constructs of model solvmanifolds, the

Recall that the matrix theory of all self maps on solvmanifolds can be completely understood by examining that of their associated linearized models. [K2] studied these relationships and revealed the following (among other things):

(9.4) THEOREM. *Suppose that $f: S \to S$ is a fibre preserving map on a Mostow fibration $N \hookrightarrow S \to T$. Then we have $\{N(f^n)\}_{n=1}^{\infty}$, $\{N(\overline{f}^n)\}_{n=1}^{\infty}$ and for each $b \in \mathrm{fix}(\overline{f}^k)$, $\{N(((f^k)_b)^n)\}_{n=1}^{\infty}$. The following relationships hold among these sequences:*

(9.4.1) *If $N(\overline{f}) = 0$ then $\{N(\overline{f}^n)\}_{n=1}^{\infty} = \{0\}$ so $\{N(f^n)\}_{n=1}^{\infty} = \{0\}$. In this case the result for tori applied to $\overline{f}$ and lifted to a homotopy of f will give a $g \sim f$ with $\mathrm{per}(g) = \emptyset$.*

(9.4.2) *If $\{N(\overline{f}^n)\}_{n=1}^{\infty}$ is bounded then the results for tori can be applied to give a map $\overline{g} \sim \overline{f}$ which can then be lifted to a homotopy $g \sim f$. The map g will have finitely many periodic fibres. What is further possible with g depends on the nature of the various $\{N(((f^k)_b)^n)\}_{n=1}^{\infty}$.*

(9.4.3) *If there exists a point $b \in \mathrm{fix}(\overline{f}^k)$ with $\{N(((f^k)_b)^n)\}_{n=1}^{\infty} \neq \{0\}$ and bounded, then every $\{N(((f^k)_b)^n)\}_{n=1}^{\infty}$ is bounded.*

(9.4.4) *Suppose that $\{N(f^n)\}_{n=1}^{\infty}$ is bounded. Then each $\{N(((f^k)_b)^n)\}_{n=1}^{\infty}$ is also bounded. All $\{N(((f^k)_b)^n)\}_{n=1}^{\infty}$ are identically zero if in addition $\{N(\overline{f}^n)\}_{n=1}^{\infty}$ is unbounded.*

Point (9.3.2) above is well handled in [Y1], [Y2] where it was shown that tori are periodic Wecken. In other words, given $h: T \to T$ and n, there is a map $w \sim h$ so that w has exactly $NP_n(h)$ periodic points of minimal period n. We know that when $\{N(h^n)\}_{n=1}^{\infty}$ is bounded, then there are clearly only finitely many n for which $NP_n(h) \neq 0$. In an obvious way [K2] extends [Y2] to show that it is possible to deform h so that $NP_n(f)$ cn be realzied for all n (especially when this is identically zero). Because the base of any Mostow fibration is always a torus, this is all we need to apply the basic constructions to Mostow fibrations to eliminate the possibility of points in the base giving the concerns of (9.3.2). However, a command of Jezierski's results for nilmanifolds in this regard would go a long way in determining exactly what kinds of minimizations are possible on the fibres and for f in total. In particular, we would need to know to what extent the periodic Wecken realizations on the nilmanifold fibres can be realized by post- and pre-composing f with deformations of the identity. (This would lend

homotopic periodicity findings of Marzantowicz and Jezierski, and the periodic Wecken results of Jezierski were yet to be discovered. As a result, the paper had a highly technical clumsiness that referees found (rightly so) to be impalatable. Advances in the last 8–10 years with these results and further insights into the matrix theory of model solvmanifolds have made a much more modern treatment with even better results now possible. Such an improved paper is waiting to be written.

understanding to what extent such a result can simultaneously realize $NP_n(f)$ for different values of n.)

The case where $\{N(f^n)\}_{n=1}^{\infty}$ is bounded but $\{N(\overline{f}^n)\}_{n=1}^{\infty}$ is unbounded and all $\{N(((f^k)_b)^n)\}_{n=1}^{\infty} = \{0\}$ still remains, to a large extent, the most mysterious. Because the collection of periodic points of $\overline{f}$ can have many limit points, traditional homotopies on the various fibres cannot be done individually with a finite homotopy and in such a way that they do not interfere with each other. Since the question being dealt with here is one of realizing a homotopy and not merely one of calculation or the extraction of structure, it is not enough to merely produce a result that works for model solvmanifold fibrations. We need to do this in greater generality and such constructions would seem to require a more intimate knowledge of the Lie algebra of the general solvmanifold. The general philosophy of one possible approach is part of our closing example.

(9.5) EXAMPLE. Consider the 9 dimensional solvmanifold S discussed in Example (7.9) and the diagonalized form of A and X for the map $f: S \to S$ presented there. Since $\deg(\overline{f}) = 3$, there are exactly two types of linearizations for the maps f_0 and f_1 on the fibres over the two fixed points of $\overline{f}$. These are X and AX. We note that the eigenvalues of X are $\{n, s, \pm\sqrt{pr}, \pm\sqrt{vt}, \pm\sqrt{qu}\}$ while the eigenvalues of AX are $\{m, -s, \pm i\sqrt{pr}, \pm i\sqrt{vt}, \pm\sqrt{qu}\}$. The absolute values of the eigenvalues in the two sets are equal and so this confirms the above mentioned general result that of both sequences are nonzero then either both of $\{N(f_0^n)\}_{n=1}^{\infty}$ and $\{N(f_1^n)\}_{n=1}^{\infty}$ are bounded or both are unbounded. A similar analysis will apply to f^k although the number of possible linearizations on the fibres will be much greater. (The possibilities are A^jX^k for $0 \le j < 3^k - 1 = |\mathrm{fix}(\overline{f}^k)|$). From this example one might naively imagine that if we had taken a different solvmanifold where the eigenvalues of the gluing data matrix A were not of absolute value one then we might see some significant differences in the unboundedness/boundedness of these Nielsen number sequences for the maps on fibres. However, this in fact will not happen. For example, we know that if the eigenvalues of A are not roots of unity other than 1 then the solvmanifold is $\mathcal{NR}$ and we will have that $N(A^jX^k)$ is independent of j ([25]).

Now suppose that X is such that $\{N(f^n)\}_{n=1}^{\infty}$ is bounded. Then obviously this implies that for all but finitely many choices of j, k with $0 \le j < 3^k - 1$ that $N(A^jX^k) = 0$. Thus for all but finitely many choices of j, k one of the eigenvalues of A^jX^k (which are m^k, $(-1)^js^k$, $e^{\pi ij/4}(pr)^{k/2}$, $e^{3\pi ij/4}(pr)^{k/2}$, $e^{5\pi ij/4}(vt)^{k/2}$, $e^{7\pi ij/4}(vt)^{k/2}$, $e^{\pi ij/2}(qu)^{k/2}$, $e^{3\pi ij/2}(qu)^{k/2}$) equals 1. Because these

([25]) Recall also that $\det(A) = \pm 1$ so eigenvalues with magnitude greater than 1 must be compensated by the existence of eigenvalues with magnitude less than 1. Thus what the McCord–Keppelmann result is really saying here is that the two effects nullify each other.

modulus 1 multipliers $e^{lj\pi i/4}$ have only finitely many possibilities which are all equally represented, we can see that the only possibility is for $m = 1$ and hence that $\{N(f^n)\}_{n=1}^{\infty} = \{0\}$ thus confirming (9.4.4).

Thus suppose $m = 1$ so that $N(f^n) = 0$ for all n. We will demonstrate how to deform f to a map g with $\mathrm{per}(g) = \emptyset$. Recall the construction of (5.23) which produces S as a quotient space from $\mathbb{R}^8 \times \mathbb{R}$ by the rules that $(\overrightarrow{w}, t) \sim (\overrightarrow{w} + \overrightarrow{k}, t)$ and $(\overrightarrow{w}, t) \sim (A^l \overrightarrow{w}, t + l)$ where $l \in \mathbb{Z}$ and $\overrightarrow{k} \in \mathbb{Z}^8$. Because we know that $XA = A^3X$ we have that the correspondence $(\overrightarrow{w}, t) \to (X\overrightarrow{w}, 3t)$ on $\mathbb{R}^8 \times \mathbb{R}$ is well defined and induces a map homotopic to f. The nth iterate of f equals model map $(\overrightarrow{w}, t) \to (X^n \overrightarrow{w}, 3^n t)$. Thus if $(\overrightarrow{w}, t)$ is a period n periodic point then $(X^n \overrightarrow{w}, 3^n t) \sim (\overrightarrow{w}, t)$. Hence $t - 3^n t = p \in \mathbb{Z}$ and $(X^n \overrightarrow{w}, 3^n t) \sim (A^p X^n \overrightarrow{w}, 3^n t + t - 3^n) = (A^p X^n \overrightarrow{w}, t)$. Therefore, as a map on T^8, $A^p X^n \overrightarrow{w} = \overrightarrow{w}$.

Since $m = 1$ and the $(1, 1)$ entry of A is also 1 we see that A and X share a common eigenvalue 1 eigenspace. Furthermore, since A is diagonizable, this eigenspace is not in the image of any $(A^p X^n - I)\mathbb{R}^8$ for any $p \in \mathbb{Z}$ and $n > 0$. Hence essentially by the "Anosov trick" we can translate by a carefully chosen multiple $\eta \overrightarrow{v}$ of $\overrightarrow{v}$ in this direction (i.e. replace multliplication by X with multiplication by X plus $\eta \overrightarrow{v}$ to make every $A^p X^n$ fixed point free). The correspondence $g(\overrightarrow{w}, t) = (X\overrightarrow{w} + \eta \overrightarrow{v}, 3t)$ then induces a periodic point free map homotopic to f.

References

[A] D. V. Anosov, *The Nielsen number of maps of nil-manifolds*, Russian Math. Surveys **40** (1985), 149–150.

[B] R. Brooks, *Coincidences, Roots and Fixed Points*, Ph. D. dissertation, UCLA, 1967.

[BBPT] R. Brooks, R. Brown, J. Pak and D. Taylor, *Nielsen numbers of maps on tori*, Proc. Amer. Math Soc. **52** (1975), 398–400.

[BBS] R. Brooks, R. Brown and H. Schirmer, *The absolute degree and Nielsen root number of a fibre-preserving map*, Topology Appl. **125** (2002), 1–46.

[BS] R. Brown and H. Schirmer, *Nielsen root theory and Hopf degree theory*, Pacific J. Math. **198** (2001), 49–80.

[BO] R. Brooks and C. Odenthal, *Nielsen numbers for roots of maps of aspherical manifolds*, Pacific J. Math. **170** (1995), 405–420.

[FH] E. Fadell and S. Husseini, *On a theorem of Anosov on Nielsen numbers for nilmanifolds*, Funct. Anal. Appl. (Maratea) (1985), 47–53; NATO Adv. Sci. Inst. Ser. C Math. Phys. Sci. **173** (1986), Reidel, Dordrecht–Boston, Mass..

[Ha] B. Halpern, *Periodic points on tori*, Pacific J. Math. **83** (1979), 117–133.

[H] P. Heath, *Fibre techniques in Nielsen theory calculations*, this handbook.

[HK1] P. Heath and E. Keppelmann, *Fibre techniques in Nielsen periodic point theory on nil- and solvmanifolds* I, Topology Appl. **76** (1997), 217–247.

[HK2] ———, *Fibre techniques in Nielsen periodic point theory on nil- and solvmanifolds* II, Topology Appl. **106** (2000), 149–167.

[HK3] ———, *Fibre techniques in Nielsen periodic point theory on solvmanifolds* III: *Calculations*, Quaestiones Math. **25** (2002), 177–208.

[HK4] ———, *Model solvmanifolds for Lefschetz and Nielsen theories*, Quaestiones Math. **25** (2002), 483–501.

[HKW] P. Heath, E. Keppelmann and P. Wong, *Addition formulae for Nielsen numbers and for Nielsen type numbers of fibre preserving maps*, Topology Appl. **67** (1995), 133–157.

[HPY] P. Heath, R. Piccinini and C. You, *Nielsen type numbers for periodic points* I, Topological Fixed Point Theory and Applications, Lecture Notes in Math. (B. Jiang, ed.), vol. 1411, Springer, Berlin, 1989.

[HY] P. Heath and C. You, *Nielsen type numbers for periodic points* II, Topology Appl. **43** (1992), 219–236.

[Ja] N. Jacobson, *Lie Algebras*, Interscience Publishers, New York, 1962.

[J] B. Jiang, *Lectures on Nielsen Fixed Point Theory*, Contemporary Mathematics, vol. 14, Amer. Math. Society, Providence, Rhode Island,, 1983.

[Jb] B. Jiang, *Introduction to Nielsen fixed point theory*, this handbook.

[Je1] J. Jezierski, *Cancelling periodic points*, Math. Ann. **321** (2001), 107–130.

[Je2] ———, *Wecken's theorem for periodic points*, Topology **42** (2003), 1101–1124.

[Je3] ———, *The semi-index product formula*, Fund. Math. **140** (1992), 99–120.

[JM] J. Jezierski and M. Marzantowicz, *Homotopy minimal periods for maps of three-dimensional nilmanifolds*, Pacific J. Math. **209** (2003), 85–101.

[K1] E. Keppelmann, *Periodic points on nilmanifolds and solvmanifolds*, Pacific J. Math. **164** (1994), 105–128.

[K2] ———, *Periodic points on solvmanifolds*, never published.

[KMc] E. Keppelmann and C. McCord, *The Anosov theorem for exponential solvmanifolds*, Pacific J. Math. **170** (1995), 143–159.

[Ma] A. I. Mal'cev, *On a class of homogeneous spaces*, Amer. Math. Soc. Transl. **39** (1951), 276–307.

[Mc] C. McCord, *Nielsen numbers and Lefschetz numbers on solvmanifolds*, Pacific J. Math. **147** (1991), 153–164.

[Mc1] ———, *Lefschetz and Nielsen coincidence numbers on nilmanifolds and solvmanifolds*, Topology Appl. **43** (1992), 249–261.

[Mc2] ———, *Lefschetz and Nielsen coincidence numbers on nilmanifolds and solvmanifolds* II, Topology Appl. **75** (1997), 81–92.

[Mo] G. D. Mostow, *Factor spaces of solvable groups*, Ann. of Math. **60** (1954), 1–27.

[R] A. H. Rarivoson, *Calculation of the Nielsen numbers on nilmanifolds and solvmanifolds*, Ph.D. dissertation (1993), University of Cincinnati.

[W] F. Warner, *Foundations of differentiable manifolds and Lie groups*, Graduate Texts in Mathematics, corrected reprint of the 1971 edition, vol. 94, Springer–Verlag, New York–Berlin, 1983.

[Y1] C. You, *The least number of periodic points on tori*, Adv. Math. **24** (1995), 155–160.

[Y2] ———, *A note on periodic points on tori*, preprint.

4. HOMOTOPY MINIMAL PERIODS

Wacław Marzantowicz

1. Introduction

Let $f\colon X \to X$ be a self-map of a topological space, in particular connected polyhedron e.g. a compact connected smooth manifold. We shall use the following notation. For a given $m \in \mathbb{N}$, by $P^m(f)$ we denote the fixed point set $\operatorname{Fix}(f^m)$. A point $x \in P^m(f)$ is called point *of period* m. The set

(1.1) $$P(f) := \bigcup_{m\in\mathbb{N}} P^m(f) = \bigcup_{m\in\mathbb{N}} \operatorname{Fix}(f^m)$$

we call *the set of periodic points of* f.

Next, for a given $m \in \mathbb{N}$, by $P_m(f)$ we denote the set

(1.2) $$P_m(f) = P^m(f) \setminus \bigcup_{k<m} P^k(f).$$

A point $x \in P_m(f)$ is called *m-periodic point of f* and m is called *a minimal period of f*.

(1.3) Remark. It is easy to check that for $k < m$ we have $P^k(f) \subset P^m(f)$ if and only if $k|m$. Consequently

$$P_m(f) = P^m(f) \setminus \bigcup_{k|m, k\neq m} P^k(f).$$

Note also that the set $P(f)$ of all periodic points can be represented as a disjoint union $P(f) = \bigcup_{m\in\mathbb{N}} P_m(f)$. By the definition the set $P(f) \subset X$ describes a dynamics of f, i.e. depends on all the iterations of f but could be very complicated set in general. It is convenient to study the dynamics of f by another simpler invariant.

Research supported by KBN grant No. 2/PO3A/04/522.

(1.4) DEFINITION. For a given self-map $f: X \to X$ we define

$$\mathrm{Per}\,(f) := \{m \in \mathbb{N} : P_m(f) \neq \emptyset\},$$

called the set of minimal periods of f.

It is also of interest to study, as an invariant of the dynamics of f two arithmetic functions associated with the map. For a given set A by $\#A$ we mean the cardinality of A with the notation $\#A = \infty$ if A is infinite. With this convention, we can consider two functions from $\mathbb{N}$ to $\mathbb{N} \cup \infty$ given as

$$m \mapsto \#P^m(f) \quad \text{and} \quad m \mapsto \#P_m(f). \tag{1.5}$$

The information that $P(f)$, correspondingly $\mathrm{Per}\,(f)$, is infinite, or that $\limsup \#P_m(f) = \infty$, respectively $\limsup \#P^m(f) = \infty$, states that the dynamics of f is rich. Unfortunately all the above invariants of dynamics of f are not stable in the following sense:

We say that an invariant of self-map f of a compact metric space X is stable if for every map g sufficiently close to a has the same value as for f.

A sufficiently close means here that the distance $\mathrm{dist}(f,g) := \sup \rho(f(x), g(x))$ is small. To illustrate it we give a simple example.

(1.6) EXAMPLE. Let $f = \mathrm{id}\,|_{S^1}$ be the map of the circle $S^1 = \{\mathbb{R}/\mathbb{Z}\}$. Take $0 < \theta < 1$ an irrational number and consider $g_\theta: S^1 \to S^1$ a map of the circle defined as $g_\theta(s) := s + \theta \mod 1$.

It is easy to check that $P^1(f) = P(f) = S^1$, $\mathrm{Per}\,(f) = \{1\}$ and $P(g_\theta) = \emptyset$. On the other hand $\mathrm{dist}(f, g_\theta) = \theta$.

We would like to remind that if self-maps of a good space are sufficiently close then they are homotopic. We present a statement of it formulated not in the most general version, leaving a proof to the reader (it could be found in many positions of the literature as well).

(1.7) PROPOSITION. *Let $f: X \to X$ be a map of a compact smooth manifold. Then there exists $\varepsilon > 0$ such that for a map $g: X \to X$*

$$\mathrm{dist}(f,g) < \varepsilon \Rightarrow g \sim f.$$

In the next definition by an invariant we mean a general notion not specifying its values which can be elements of any object.

(1.8) DEFINITION. We say that an invariant characterizes the homotopy dynamics of $f \in \mathrm{Map}(X, X)$ if it depends on the all iterations of f has the same value for every g which is homotopic to f.

As a direct consequence of the definition, Proposition (1.7) and Example (1.6) is the following.

(1.9) COROLLARY. *The functions* $m \mapsto \#P^m(f)$, $m \mapsto \#P_m(f)$, *and the sets* $P(f)$, $\mathrm{Per}\,(f)$ *are not homotopy dynamics invariants.*

Since $f \sim g$ implies $f^m \sim g^m$ for every m, the sequences $\{L(f^m)\}_1^\infty$ of Lefschetz number, and $\{N(f^m)\}_1^\infty$ of the Nielsen number correspondingly, of all iterations of f are homotopy dynamics invariants (cf. [Br2], [Ji4] for definitions). To use the sequence $\{L(f^m)\}$ to study any of the invariants mentioned in Corollary (1.9) one have to pose additional geometric assumption on the map f as the smoothness, complex analycity, or transversality. In particular the below Example (1.12) shows that without any such assumption we have $L(f^m) \xrightarrow{m\to\infty} \infty$ but $P(f)$ consists of two fixed points. We shall discuss it later.

Also a use of the sequence $\{N(f^m)\}_1^\infty$ of the Nielsen numbers of iterations of $f\colon X \to X$ to study the dynamics of f requires a geometric assumption on the space X.

Anyway there is a direct consequence of definition of Nielsen number.

(1.10) PROPOSITION. *Let* $f\colon X \to X$ *be a continuous map of a compact CW-complex (e.g. a finite polyhedron). If* $\{N(f^m)\}$ *is unbounded then* $P(f)$ *is infinite.*

PROOF. By the main property of Nielsen number, $\#P(f^m) \geq N(f^m)$, and statement follows. □

Now we give an example which illustrate more persuasively than Example (1.6) that the functions $m \mapsto \#P^m(f)$, $m \mapsto \#P_m(f)$, set $P(f)$ and the set $\mathrm{Per}\,(f)$, are not stable invariants.

First consider the map $z \mapsto z^r$, $r \geq 2$, of the circle $S^1 = \{z \in \mathbb{C} : |z| = 1\}$. By the definition of f $\#P^m(f) = \#\{z : z^{r^m} = z\} = r^m - 1$. Furthermore, $P_m(f)$ is equal to the set of primitive roots of unity of degree $r^m - 1$. Consequently $\#P_m(f) = \phi(r^m - 1)$ where $\phi(m)\colon N \to N$ is the Euler function

$$\phi(n) := \#\{k \leq n : (k,n) = 1\}, \tag{1.11}$$

and (n,k) is the greater common divisor of n and k. It is known that for $n = p_1^{\alpha_1} \dots p_s^{\alpha_s}$ we have $\phi(n) = \prod_{i=1}^s p_i^{\alpha_i}(1 - 1/p_i)$.

(1.12) EXAMPLE. Let $h_r\colon S^1 \to S^1$ be a map of the circle of degree $r \geq 2$, e.g. $h(z) := z^r$. Let next $\zeta\colon [0,1] \to [0,1]$ be a map given as $\zeta(t) := \sqrt{t}$. Representing S^2 as the suspension of S^1, i.e. $S^2 = S^1 \times [0,1]/\sim$ where $S^1 \times \{0\} \sim *$, $S^1 \times \{1\} \sim *$ and $(x,t) \sim (x,t)$ if $t \neq 0, 1$, we define a map

$$f([z,t]) := [(h_d(z), \zeta(t))].$$

Then $\deg f = \deg h_r = r$. The set of non-wandering points of f (thus also periodic points) consists of two (fixed) points $[S^1 \times \{0\}]$ and $[S^1 \times \{1\}]$. The same construction works in any dimension n.

Note that this map is locally near the South Pole equivalent to $2z^r|z|^{-1}$, thus not differentiable. Moreover, observe that every map $f:\mathcal{U} \to \mathbb{C}$ of a neighbourhood $\mathcal{U}$ of 0 which in the polar coordinates is of the form $f(\theta,\rho) = (r\cdot\rho\theta, \zeta(\rho)), \zeta(\rho) > \rho$, for $\rho > 0$, can not be smooth at 0. Indeed, then $|\det Df(0)| \geq 1$, because $D_\rho \subset f(D_\rho)$ for every disc D_ρ. But the later yields that f is a local diffeomorphism at 0 contrary to its form along the angle coordinate.

If we take $h_r = z^r$ and $\zeta(t) = t - \varepsilon(t - \sqrt{t})$ then the defined map $f: S^2 \to S^2$ is ε-perturbation of $\Sigma(z^r)$ of the suspension of z^r. Consequently, for every m $P_m(\Sigma(z^r)) \neq \emptyset$ (here $P_m(z^r)$ is a subcomplex of dimension 1, thus infinite). This yields $\mathrm{Per}\,(\Sigma(z^r)) = \mathbb{N}$.

All the above examples legitimate the introduction of the following notion which is stable and homotopy invariant by its definition.

(1.13) DEFINITION. Define the *set of homotopy minimal periods* to be the set

$$\mathrm{HPer}(f) := \bigcap_{g\simeq f} \mathrm{Per}\,(g),$$

i.e. $m \in \mathbb{N}$ is a *homotopy minimal period* of f if it is minimal period for every $g \sim f$ homotopic to f.

By definition $\mathrm{HPer}(f) \subset \mathrm{Per}\,(f)$ and the inclusion is proper in general.

(1.14) EXAMPLE. For the map g_θ of (1.6) we have $\mathrm{Per}\,(g_\theta) = \emptyset$, thus

$$\mathrm{HPer}(\mathrm{id}\,|_{S^1}) \subsetneq \mathrm{Per}\,(\mathrm{id}\,|_{S^1}) = \{1\}.$$

Similarly for the map $\Sigma(z^r), |r| > 1$, we have

$$\{1\} = \mathrm{HPer}(\Sigma z^r) \subsetneq \mathrm{Per}\,(\Sigma z^r) = \mathbb{N}$$

as follows from Example (1.12).

For a smooth manifold map any homotopy dynamics invariant reflects an information about the rigid part of dynamics, because a small perturbation of a map f is homotopic to it by Proposition (1.7). In particular

(1.15) REMARK. If $f: X \to X$ is a map of smooth compact manifold then $\mathrm{HPer}(f) = \mathrm{HPer}(h)$ for any small perturbation h of f.

Let us try to describe the set of homotopy minimal periods for simplest nontrivial manifold the circle. Since every map $f: S^1 \to S^1$ is homotopic to z^r it is enough to study $\mathrm{HPer}(z^r)$.

We have:

(1.16.1) for $r = 1$ $\mathrm{HPer}(f) = \emptyset$, by Example (1.14),

(1.16.2) for $r = 0$ $\mathrm{HPer}(f) = \{1\}$, because $\mathrm{Per}\,(*) = \{1\}$ for the constant map,

(1.16.3) for $r = -1$ $\mathrm{HPer}(f) = \{1\}$, because for $f(z) = \overline{z}$, $\mathrm{Per}\,(f) = \{1, 2\}$ and for the map:

$$f(\theta) = \begin{cases} 2\theta^2 & \text{if} \quad 0 \le \theta \le 1/2, \\ 1/2 + 2(\theta - 1/2)^2 & \text{if} \quad 1/2 \le \theta \le 1, \end{cases}$$

$\theta \in \mathbb{R}/\mathbb{Z} = S^1$, we have $\mathrm{Per} = \{1\}$.

On the other hand it is well known (see [Ji4]) that for a map $f\colon S^1 \to S^1$ of degree r we have $N(f) = r$, thus $N(f^m) = r^m$. This yields $\#P^m(f) \ge r^m$, and together with the obvious $\mathrm{Per}\,(z^r) = \mathbb{N}$ for $|r| > 1$ makes reasonable the question whether $\mathrm{HPer}(z^r) = \mathbb{N}$ for $|r| > 1$.

First the set of homotopy minimal periods was studied (under another name) for a selfmap of the circle $M = S^1$ by L. S. Efremova in [Ef] and L. Block, J. Guckenheimer, M. Misiurewicz, L. S. Young in [BGMY]. They wanted to prove an analog of the Šarkovskii theorem of [Sz] for the circle maps. As the result they got the following theorem which states that the answer to our question is almost affirmative – with only one exception.

(1.17) THEOREM. *Let $f\colon S^1 \to S^1$ be a map of the circle and $A_f = r \in \mathbb{Z} = \mathcal{M}_{1\times 1}(\mathbb{Z})$ the degree of f. There are three types for the minimal homotopy periods of f:*

(E) $\mathrm{HPer}(f) = \emptyset$ *if and only if* $r = 1$.

(F) $\mathrm{HPer}(f) \ne \emptyset$ *and is finite if and only if* $r = -1$ *or* $r = 0$. $\mathrm{HPer}(f) = \{1\}$ *then.*

(G) $\mathrm{HPer}(f) = \mathbb{N}$ *for the remaining* r, *i.e.* $|r| > 1$, *with except one special case:* $r = -2$ *when* $\mathrm{HPer}(f) = \mathbb{N} \setminus \{2\}$.

The original method of proof of Theorem (1.17) is elementary but complicated. We get it, as well as the next theorem stated below as a consequence of a general statement. As the next, L. Alsedá, S. Baldwin, J. Llibre, R. Swanson and W. Szlenk examined the case $M = \mathbb{T}^2$ in [ABLSS1]. To give a description of the set of the homotopy minimal periods (called them "the minimal set of periods") they first used the Nielsen theory. We present their main theorem, after a reformulation in our terms, in the Section 5.

(1.18) EXAMPLE. Let us consider the map $f(z) = \overline{z}^2$ of the circle S^1. Then we have $\deg(f) = -2$, $\mathrm{Fix}\,(f) = \{z : z^3 = 1\}$, but the equation $\overline{\overline{z}^2}^2 = z$ is equivalent to $z^4 = z$, thus $\mathrm{Fix}\,(f) = \mathrm{Fix}\,(f^2)$ and $2 \notin \mathrm{Per}\,(f)$.

A qualitative progress of methods had been made by B. Jiang and J. Llibre who gave a description of the set of homotopy minimal periods for the torus $M = \mathbb{T}^d$, with any $d \in \mathbb{N}$ ([JiLb]). To prove a general theorem (cf. (4.4)) they made use of a fine combinatorics argument and a deep algebraic number theory theorem they proved (close to the A. Schinzel theorem on prime divisors, cf. [Sch]). To complete the demonstration they had to employ a topological result of You ([Yu1], [Yu2]) on the periodic points on tori.

It was a natural question to extend this theorem onto larger classes of compact manifolds which have a similar structure as the tori namely: compact nilmanifolds, compact completely solvmanifolds, exponential solvmanifolds. We will give a survey of recent results in this direction general description of the set of homotopy minimal periods for the maps of compact nilmanifold (Theorem (4.4)), and compact completely solvable solvmanifold, their exemplification for the dimension three (Theorems (5.2), (5.7)) with a detailed list, specification for homeomorphisms (Theorems (5.3), (5.8)).

As applications of the main theorem and its specifications for 3-dimensional manifolds we present theorems which ensure that $\mathrm{HPer}(f) = \mathbb{N}$ provided given small homotopy period exists (Theorems (6.2), (6.4)). We call them the Šarkovskii type theorems. They are consequences of a fine geometrical structure of these manifolds which precipitate special properties of the Nielsen number that are not true in general. In the case of classes of solvmanifolds the set $\mathrm{HPer}(f)$ is known the best, but it was also studied in other cases, e.g. it was derived for the self-maps of the real projective space (Theorem (6.6)) in [Je3]. It is possible to use a similar approach to derive the existence of infinitely many periodic points for a continuous map of the sphere provided the map commutes with a free homeomorphism of a finite order (Theorems (6.10), (6.12)). The result gives also an asymptoptic estimate of the functions defined in (1.5) (cf. [JeMr3]).

As the last group of applications we present relations between the topological entropy and the stuff discussed here. The linearization matrix appears in a formula for the lower estimate of the entropy of a map of compact nilmanifold (Theorems (6.15)–(6.17)). In particular the infiniteness of $\mathrm{HPer}(f)$ implies that entropy is nonzero (Corollary (6.18)).

2. Nil- and solvmanifolds

In this subsection we give an overlook on basic notions and definitions which concerning nil- and solvmanifolds. The part of our presentation follows mainly [GOV], [Ra] and [VGS]. The most of references are taken from these books.

A group G is called *nilpotent* if its central tower is finite i.e.

$$G_0 = G \supset G_1 \supset \ldots \supset G_{k-1} \supset G_k = \mathbf{e}, \tag{2.1}$$

where $G_i := [G, G_{i-1}]$.

A group G is called *solvable* if its normal tower is finite i.e.

$$G_0 = G \supset G_1 \supset \ldots \supset G_{k-1} \supset G_k = \mathbf{e}, \tag{2.2}$$

where $G_i := [G_{i-1}, G_{i-1}]$. Obviously every nilpotent group is solvable.

(2.3) DEFINITION. A homogeneous space $X = G/H$ of a nilpotent or solvable group G is to be said a *nilmanifold* or a *solvmanifold*, respectively.

Now we give examples of compact nilmanifolds.

(2.4) EXAMPLE. The simplest compact nilmanifold is the torus $X = \mathbb{T}^d := \mathbb{R}^d/\mathbb{Z}^d \equiv (S^1)^d$, which corresponds to the abelian group. If $\dim X \leq 2$ then it is the only example of a nilmanifold.

For a ring $\mathcal{R}$ with unity (e.g. $\mathcal{R} = \mathbb{R}, \mathcal{R} = \mathbb{C}$) let $\mathbf{N}_n(\mathcal{R})$ denote the group of all unipotent upper triangular matrices whose entries are elements of the ring $\mathcal{R}$, i.e.

$$\begin{bmatrix} 1 & r_{12} & \cdot & \cdots & \cdot & r_{1n} \\ 0 & 1 & r_{23} & \cdot & \cdots & r_{2n} \\ 0 & 0 & 1 & r_{34} & \cdots & r_{3n} \\ \vdots & \vdots & \vdots & \ddots & \ddots & \vdots \\ \cdot & \cdot & \cdot & \cdots & 1 & r_{n-1n} \\ 0 & 0 & \cdot & \cdots & 0 & 1 \end{bmatrix}.$$

This matrix group let us define *Iwasawa manifolds*:

$$\mathbf{N}_n(\mathbb{R})/\mathbf{N}_n(\mathbb{Z}) \quad \text{and} \quad \mathbf{N}_n(\mathbb{C})/\mathbf{N}_n(\mathbb{Z}[\imath]),$$

where $\mathbb{Z}[\imath]$ is the ring of Gaussian integers. For $n = 3$ they are examples of nilmanifolds of dimension 3 not diffeomorphic to the torus. They are called Heisenberg manifolds. The Iwasawa 3-manifold $\mathbf{N}_3(\mathbb{R})/\mathbf{N}_3(\mathbb{Z})$, is called "Baby Nil".

It is known ([Au] that for $d = 3$ every nilmanifold is, up diffeomorphism, one of: $\mathbf{N}_3(\mathbb{R})/\Gamma_{p,q,r}$, where the subgroup $\Gamma_{p,q,r}$, with fixed $p, q, r \in \mathbb{N}$ consists of all matrices of the form

$$\begin{bmatrix} 1 & k/p & m/(p \cdot q \cdot r) \\ 0 & 1 & l/q \\ 0 & 0 & 1 \end{bmatrix}, \quad \text{where } k, l, m \in \mathbb{Z}.$$

Moreover, it is diffeomorphic to one of $\mathbf{N}_3(\mathbb{R})/\Gamma_{1,1,r}$.

As we noted above in all our examples of nilmanifolds the subgroup $H \subset G$ is discrete. It is consequence of a general fact. Recall that a discrete subgroup $\Gamma \subset G$ of a Lie group G is called a *lattice* if there exists a finite G-invariant measure on G/Γ (see [Ra]). A closed subgroup $H \subset G$ is called a uniform subgroup if G/H is compact. Due to the Mostow theorem, a closed subgroup $H \subset G$ of a solvable group G is uniform if and only if there exists a finite G-invariant measure on G/H (cf. [Ra, Theorem 3.1]). Consequently, a discrete subgroup $\Gamma \subset G$ of a solvable group G is a lattice if it is uniform.

It is known (see [Ml], [GOV]) that every compact nilmanifold is diffeomorphic to a nilmanifold of the form G/Γ, where G is a simple-connected nilpotent Lie group and $\Gamma \subset G$ is a lattice. This statement leads to an equivalent definition of the nilmanifold.

(2.5) DEFINITION. We say that a compact manifold X of dimension d is a *nilmanifold* if it is the quotient space G/Γ of a simple-connected nilpotent group G of dimension d by a lattice $\Gamma \subset G$ of rank d ([Au], [Ml], [Ra]).

By the definition a compact nilmanfold $X = G/\Gamma$ is a $K(\pi, \Gamma)$ space i.e. $\pi_1(X) = \Gamma$ and $\pi_k(x) = 0$ for $k > 1$, because a nilpotent, connected, simple-connected group G is contractible (cf. [Ra], [GOV]).

The same is not true for solvmanifolds. For example the Klein bottle is a compact solvmanifold defined as the homogeneous space $SO(2) \ltimes \mathbb{R}^2 / SO(1) \ltimes (\mathbb{Z} \times \mathbb{R})$ (cf. [GOV, p. 165]). It could be also represented as the $K(\pi, 1)$ space with fundamental group $\pi_1 = \pi = \mathbb{Z} \times_\xi \mathbb{Z}$, where $\xi \colon \mathbb{Z} \to \{1, -1\} = O(1) \subset \mathbb{Z}$ is given by the canonical epimorphism $\mathbb{Z} \to \mathbb{Z}_2$ (cf. [Mo]), but it is not of the form G/Γ, for simple-connected solvable Lie group G and a lattice $\Gamma \subset G$. If it were of the form G/Γ it would be parallelizable which is obviously not true.

(2.6) DEFINITION. A solvmanifold of the form G/Γ, G simple-connected and $\Gamma \subset G$ a lattice, is called the *special* solvmanifold.

Note that every solvmanifold is finitely covered by a special solvmanifold (see [Au], [GOV]). This implies that solvmanifolds are aspherical, that is $\pi_i(X) = 0$ for $i \geq 2$ for any solvmanifold X.

Let $\mathcal{G} \simeq T_e G$ be the Lie algebra of a given Lie group G. Now we designate a few classes of solvmanifolds that distinguish with respect to the spectrum of adjoint operator $\mathrm{ad}_\mathcal{X} \colon \mathcal{G} \to \mathcal{G}$ (cf. [GOV]). Since some authors define these classes in terms of the operator $\mathrm{Ad}_x \colon \mathcal{G} \to \mathcal{G}$, $x \in G$ (cf. [KMC]), we recall the following commutative diagram which allows to show that these two definitions are equivalent

$$
\begin{array}{ccc}
\mathcal{G} & \xrightarrow{\mathrm{ad}} & \mathrm{End}\,(T_e) \\
{\scriptstyle \exp}\downarrow & & \downarrow{\scriptstyle \exp} \\
G & \xrightarrow[\mathrm{Ad}]{} & \mathrm{Aut}\,(T_e)
\end{array}
$$

2.1. Completely solvable groups and manifolds. If for any $X \in \mathcal{G}$ all eigenvalues of the operator ad_X are real then the group G and the Lie algebra $\mathcal{G}$ as well as every solvmanifold G/Γ is called completely solvable. Note that every nilmanifold belongs to this class, since ad_X is a nilpotent matrix, and so it hasonly 0 as the eigenvalue. The above remark says that every nilmanifold is completely solvable. The main property of completely solvable groups is the following *rigidity of lattices* property (see [VGS], [Sa]):

(2.7) THEOREM. *Let G and G' be simple-connected completely solvable Lie groups and $\Gamma \subset G$ be a lattice. Then every homomorphism $\Phi: \Gamma \to G'$ can be extended to a unique homomorphism $\Phi: G \to G'$.*

For our purposes, there will be relevant the following immediate consequence of Theorem (2.7) and the fact that G/Γ is a $K(\pi, 1)$-space with $\pi = \Gamma$.

(2.8) COROLLARY. *Every continuous map $G_1/\Gamma_1 \to G_2/\Gamma_2$ between compact special completely solvable solvmanifolds is homotopic to the map induced by a homomorphism $G_1 \to G_2$.*

An example of completely solvable Lie group G of dimension 3 which is not nilpotent is given in (5.6). There are infinitely many not isomorphic completely solvable compact solvmanifold of dimension 3 which are not diffeomorphic neither to a torus nor to a non-abelian nilmanifold (Example (5.9)).

2.2. Exponential groups and manifolds. If for any $X \in \mathcal{G}$ there is no pure imaginary eigenvalue of the operator ad_X then the group G and the Lie algebra $\mathcal{G}$ as well as every solvmanifold G/Γ is called *exponential*. This is equivalent to the fact that the exponential map $\exp: \mathcal{G} \to G$ is injective or is a diffeomorphism, provided that G is simple-connected. Notice that every completely solvable group is exponential. In fact, if a Lie group is exponential then it is solvable (cf. [[VGS]]).

2.3. NR-manifolds. A class of solvamanifolds called NR-solvmanifolds was introduced by Keppelmann and McCord, see [KMC] for the definition which we omit.

They contain the class of special exponential solvmanifolds. The main property of NR-solvmanifolds is that the Anosov Theorem (3.24) can be generalized to them.

At the end we formulate a property of solvmanifolds we shall use in next.

3. Properties of Nielsen numbers for solvmanifolds

In this subsection we first define the notion of linearization of a map which we need for the formulation of our theorem. Then we present Anosov theorem that comparing the Lefschetz and Nielsen numbers which implies nice arithmetic properties of the sequence $\{N(f^m)\}$. Finally we present the summantion formula for expressing the full Nielsen–Jiang number $NF_m(f)$ as a sum of the prime Nielsen–Jiang numbers $NP_k(k)$, $k|m$. This formula guarantees nice combinatorial properties of the sequence $\{N(f^m)\}$.

3.1. The linearization matrix. We give now different definitions of a notion of linearization of a map of nilmanifold. Since this concept it less known, we do it in detail. For another, more geometrical and subtle, exposition and examples we refer to the chapter "Linearization for maps of nil- and solvmanifolds" of this book (Chapter I.3).

By the definition the central towers $\{G_i\}$, $G_i := [G, G_{i-1}]$, of a nilpotent group G is finite. It is easy to check that G_i are normal subgroups, moreover they are preserved by every endomorphism of G (cf. [Ml], [Ra]). Note also that $G_{k-1} \subset Z(G)$, where $Z(G)$ is the center of G. The number k is called the *length of nilpotency* of G.

For a given lattice $\Gamma \subset G$ we define a decreasing sequence of subgroups of of Γ putting

$$\Gamma_i := \Gamma \cap G_i. \tag{3.1}$$

It is easy to see that $\Gamma_i \supset \widetilde{\Gamma}_i$, where $\widetilde{\Gamma}_i := [\Gamma, \widetilde{\Gamma}_{i-1}]$. By the corresponding property of G_i, every Γ_i is preserved by any homomorphism of G preserving Γ, called *fully invariant* subgroup. In particular it is a normal subgroup of Γ.

Moreover, it is known that Γ_i is a lattice in G_i (cf. [Ml] [Ra, II, Corollary 1]). Furthermore, the rank of every lattice $\Gamma \subset G$ of a nilpotent simple-connected subgroup is equal to $d = \dim G$ ([Ra, II, Theorem 2.21]).

Put $s_i := \dim G_i = \operatorname{rank} \Gamma_i$, and $d_i := s_{i-1} - s_i$. We have $s_k = d_k = 0$, $s_{k-1} = d_{k-1}$, $s_0 = d$, $d_0 = d - d_1$ and $\sum_{i=1}^k d_i = d$.

Observe, that by the definition of central tower, the quotient groups G_{i-1}/G_i, and consequently $\Lambda_i := \Gamma_{i-1}/\Gamma_i$, are abelian. Here and in next we shall use the following well-known fact.

(3.2) PROPOSITION. *Let H be a connected subgroup of connected simply connected solvable group G. Then H is also simply connected.*

PROOF. The space G/H is aspherical as a solvmanifold (see a remark after Definition (2.6)). Then the statement follows from the long exact sequence of homotopy groups of the fibration $H \to G \to G/H$. Indeed for $i \geq 3$ all groups in this sequence are zero, and the exactness of $0 \to \pi_2(H) \to \pi_2(G) \to \pi_2(G/H) \to \pi_1(H) \to \pi_1(G) \to \pi_1(G/H) \to \pi_0(H) \to \pi_0(G) \to \pi_0(G/H) \to 0$ gives $\pi_2(G/H) = \pi_1(H) = 0$. □

In respect of Proposition (3.2) all G_i are simple connected, thus also the quotient groups G_{i-1}/G_i. Consequently $G_{i-1}/G_i \equiv \mathbb{R}^{m_i}$ as an abelian simple-connected Lie group. Moreover, $\Lambda_i = \Gamma_{i-1}/\Gamma_i$ is a lattice in it, thus $\Lambda_i \equiv \mathbb{Z}^{m_i}$. Furthermore the action of Γ on each Λ_i is trivial ([KMC, Theorem 2.1]).

Dividing each term G_i of the central tower (2.1) by Γ i.e. each G_i by Γ_i, we get a decreasing filtration of $X = G/\Gamma$ by sub-nilmanifolds:

$$(3.3) \qquad X_0 = X \supset X_1 \supset \ldots \supset X_{k-1} \supset X_k = *, \quad X_i := G_i/\Gamma_i,\ 0 \le i \le k,$$

Furthermore, dividing G_{i-1} by G_i and next by Γ we see that X_{i-1} forms a fibration $F_i \subset X_{i-1} \xrightarrow{p_i} B_{i-1}$ with the fiber $F_i = X_i$ over the base

$$(3.4) \qquad B_i = (G_{i-1}/G_i)/(\Gamma_{i-1}/\Gamma_i) \equiv \mathbb{R}^{d_i}/\mathbb{Z}^{d_i} \equiv \mathbb{T}^{d_i},$$

in particular $X = G/\Gamma$ is a fibration over the torus. Moreover, every map $f\colon X \to X$ induced by a preserving Γ homomorphism of G preserves also this filtration, consequently it is a fiber map $f_{i-1} = (f_i, \widetilde{f_i})$ of each fibration $X_i \subset X_{i-1} \xrightarrow{p_i} \mathbb{T}^{m_i}$.

$$(3.5) \qquad \text{Here } f_i := f|_{X_{i-1}}, \text{ and } \widetilde{f_i} \text{ is induced by } f_{i-1} \text{ on } B_i \equiv \mathbb{T}^{d_i}.$$

Furthermore, every preserving Γ homomorphisms of G, i.e. homomorphism $\Phi\colon G \to G$ such that $\Phi(\Gamma) \subset \Gamma$, preserves each G_i and Γ_i, and consequently induces a factor homomorphism

$$(3.6) \qquad \psi_i := \psi_{i-1}/\psi_i\colon \Lambda_i \equiv \mathbb{Z}^{d_i} \to \Lambda_i \equiv \mathbb{Z}^{d_i}.$$

We are in position to define the notion of linearization of a map of a nilmanifold (cf. [KMC]).

(3.7) DEFINITION. Let $f\colon X \to X$ be a map of a compact nilmanifold of dimension d. Not changing the homotopy class we can assume that f is induced by a preserving Γ homomorphism $\Phi\colon G \to G$ (Corollary (2.8)). We define the linearization of f as an integral matrix $A_f \in \mathcal{M}_{d\times d}(\mathbb{Z})$ given as the direct sum $A_f := \bigoplus_{i=1}^k A_i$ where $A_i \in \mathcal{M}_{d_i\times d_i}(\mathbb{Z})$ is the matrix of $\Phi_i\colon \mathbb{Z}^{d_i} \to \mathbb{Z}^{d_i}$ defined in (3.6).

The following is a direct consequence of the Definition (3.7).

(3.8) LEMMA. *The integral matrix A_i is equal to the matrix of homomorphism $H_1(\widetilde{f_i})\colon H_1(\mathbb{T}^{d_i};\mathbb{R}) \to H_1(\mathbb{T}^{d_i};\mathbb{R})$ induced by the map $\widetilde{f_i}$ of base $B_i \equiv \mathbb{T}^{d_i}$ of the corresponding fibration (see (3.4)) on the first homology group with real coefficients. The same is true if we take the real cohomology of B_i.*

PROOF. The statement follows from the following well-known facts. $H_1(\mathbb{T}^{d_i};\mathbb{Z}) \equiv \pi_1(\mathbb{T}^{d_i}) \equiv \mathbb{Z}^{d_i} \equiv \Lambda_i$ and the induced maps are the same. The natural map

$H_*(\mathbb{T}^{d_i};\mathbb{Z}) \subset H_*(\mathbb{T}^{d_i};\mathbb{R})$ is a monomorphism of abelian groups and the map induced by $\widetilde{f}_i$ on $H_*(\mathbb{T}^{d_i};\mathbb{R})$ is of the form

$$H_1(\widetilde{f}_i) = H_1(\widetilde{f}_i) \otimes_{\mathbb{Z}} \mathrm{id}_{\mathbb{R}},$$

where on the right hand side is the homomorphism induced on $H_*(\mathbb{T}^{d_i};\mathbb{R})$. Since the real cohomology spaces are conjugated to the real homology spaces, the last part follows by duality. □

(3.9) REMARK. For $X = \mathbb{T}^d$ then the integral matrix of linearization A_f gives preserving the integral lattice linear map of $\mathbb{R}^d$, such that induces a map $[A_f]\colon \mathbb{T}^d \to \mathbb{T}^d$, is homotopic to f, which was well known and used by several authors.

We have already represented a nilmanifold X as the total space of the fibration over a torus, with the fiber being a nilmanifold of a lower dimension (cf. (3.3) and (3.4)). For some reason it is also convenient to represent it in the dual way i.e. as the total space of a fibration over a nilmanifold of lower dimension with the fiber being a torus. To do it we divide G by $G_{k-1} \subset Z(G)$ and get a simple-connected nilpotent group $\widetilde{G}; = G/G_{k-1}$ (cf. Proposition (3.2)). Next dividing $\widetilde{G}$ by $\widetilde{\Gamma} = \Gamma/\Gamma_{k-1} = \Gamma/\Gamma \cap G_{k-1}$, we get a nilmanifold $\widetilde{X} := \widetilde{G}/\widetilde{\Gamma}$ of dimension $d - d_k < d$. Observe that X, $\widetilde{X}$ and the map $p\colon X \to \widetilde{X}$ induced by the canonical projection $G \to G/G_{k-1}$ form a fibration with the fiber $F = G_{k-1}/\Gamma_{k-1} \equiv \mathbb{T}^{d_k}$ an d the base $B := \widetilde{X}$. We call it the *Fadell–Husseini fibration.*

Moreover, a map $f\colon X \to X$ induced by a homomorphism $\Phi\colon G \to G$ is a fiber map of this fibration, i.e is of the form $f = (f', \widetilde{f})$, where $\widetilde{f}$ is a self-map of the base and $f'\colon \mathbb{T}^{d_k} \to \mathbb{T}^{d_k}$ is a map of the fiber.

We are now able to give an alternative definition of the linearization of a map o nilmanifold (cf. [FaHu], [JeMr1]).

(3.10) DEFINITION. Let $f\colon X \to X$ be a map of a compact nilmanifold $M = G/\Gamma$ of dimension d. We define an integral matrix $A_f \in \mathcal{M}_{d\times d}$ by an induction with respect to the length of nilpotency of G.

If X is a torus $\mathbb{T}^d$ then A_f is, by definition, the matrix $A_f\colon \mathbb{Z}^d \to \mathbb{Z}^d$ induced by f on $\pi_1(X) \equiv \mathbb{Z}^d$, or equivalently the matrix of homomorphism induced by f on $H_1(\mathbb{T}^d;\mathbb{Z}) \equiv \mathbb{Z}^d$.

Assume that A_f is defined for self-maps of all nilmanifolds of the length of nilpotency $\leq k-1$ and X is a nilmanifold of the length equal to k. Taking the fibration $(F = \mathbb{T}^{d_k}) \subset X \xrightarrow{p} \widetilde{X}$ and the fiber map $f = (f', \widetilde{f})$ homotopic to f induced by a homomorphism of G, we put

$$A_f \colon= A_{f'} \oplus A_{\widetilde{f}} \in \mathcal{M}_{d\times d}(\mathbb{Z}).$$

It is easy show the following.

(3.11) PROPOSITION. *The matrices A_f defined in Definitions* (3.7) *and* (3.10) *are the same.*

(3.12) REMARK. By the above construction a homotopy class of $f: X \to X$ determines uniquely (up to conjugacy by a modular matrix caused by a change of basis if the free abelian groups) an integral matrix of linearization, The converse is not true, i.e. the linearization does not determine a homotopy class.

(3.13) REMARK. In [FaHu] Fadell and Husseini showed by the following inductive construction of a class of manifolds: starting form the tori and next consecutively adding all total spaces of all compact toral fibrations over what had been already constructed, we get no more than nilmanifolds.

To define the linearization in the general case we need the notion of Mostow fibration (cf. [KMC]). It is that a solvable group π is a fundamental group of a compact solvmanifold iff there exists a exact sequence $\Delta \subset \pi \to \mathbb{Z}^s$ where Δ is finitely generated torsion free nilpotent group ([GOV]). Such a group is called *Wang group* ([GOV]), or $\mathcal{S}$-group ([KMC]). Suppose π is a finitely generated torsion free Wang group and $\Gamma \lhd \pi$ any nilpotent normal subgroup. Let $\Lambda = \pi/\Gamma$. The factorization $\Gamma \to \pi \to \Lambda$ can be realized as the fundamental group sequence of a fibration $N \to X \to \mathbb{T}$, with N a nilmanifold and $\mathbb{T}$ a torus (cf. [Mo], [GOV]), called a *Mostow fibration.* For the definition of linearization we need one such Mostow fibration. Let $\rho: \pi \to \pi_{ab}$ be the abelianization of π. Decompose the abelianization π_{ab} as $\mathcal{F} \oplus \mathcal{T}$ with $\mathcal{F}$ free abelian and $\mathcal{T}$ finite, and let $\Gamma := \rho^{-1}(\mathcal{T})$. Since π is a strongly torsion free Wang group, Γ is nilpotent, and $\Lambda_0 = \pi/\Gamma$ is clearly torsion free abelian. Since $[\pi, \pi]$ is fully invariant, as is $\mathcal{T}$ in the product $\mathcal{F} \oplus \mathcal{T}$, we see that Γ is a fully invariant subgroup of π. This led to the following theorem (cf. [MC1]).

(3.14) THEOREM. *Let X be a compact solvmanifold with $\pi = \pi_1(X)$. Suppose that Γ is the (unique) nilpotent subgroup of π such that $[\pi, \pi]$ is a subgroup of finite index in Γ and $\Lambda_0 = \pi/\Gamma$ is torsion free. Then there is a Mostow fibration $N \to X \to \mathbb{T}_0$ in which N is a nilmanifold with $\pi_1(N) \cong \Gamma$ and $\mathbb{T}_0$ is a torus with $\pi_1(\mathbb{T}_0) = \Lambda_0$. Furthermore, in every homotopy class of self maps on X there is a fiber preserving map of this Mostow fibration.*

Using Theorem (3.14) we can assume that f is a fiber preserving map i.e. the following diagram commutes

$$
\begin{array}{ccccc}
N_b & \longrightarrow & X & \longrightarrow & \mathbb{T}_0 \\
\downarrow{\scriptstyle f_b} & & \downarrow{\scriptstyle f} & & \downarrow{\scriptstyle f_0} \\
N_b & \longrightarrow & X & \longrightarrow & \mathbb{T}_0
\end{array}
\tag{3.15}
$$

If $N(f_0) = L(f_0) = 0$ then f_0, thus f is homotopic to a fixed point free map. Consequently $N(f^m) = 0$ for every m. If $N(f_0) \neq 0$ then we can assume that $[0] = \mathbf{e} \in \mathrm{Fix}\,(f_0)$ ([KMC]).

(3.16) DEFINITION. For a given map $f\colon X \to X$ of a compact NR-solvmanifold, of dim d $N \subset X \xrightarrow{p} \mathbb{T}_0$ be the Mostow fibration given by Theorem (3.14). Assuming that $f = (f_N, f_0)$ is a fiber map, suppose that $L(f_0) \neq 0$ and $f_0([0]) = [0]$. We define the *linearization matrix* of f putting

$$A_f = A_{f_N} \oplus A_{f_0} \in \mathcal{M}_{d\times d}(\mathbb{Z}),$$

where $A_{f_N} = \bigoplus_{i=1}^k A_i$ given by Definition (3.7), and $A_0 := A_{f_0}$ is the linearization of f_0.

(3.17) REMARK. Note that the linearization matrix can be defined in this way for a map f of any solvmanifold X with respect to Theorem (3.14). However if X is not an NR-solvmanifold, then such a matrix does not express the Lefschetz and Nielsen number by the formula of Theorem (3.14) in general (cf. [KMC]).

For a map of a compact completely solvable manifold the matrix A_f of linearization of f can be defined in an analytic way described below (cf. [JeKdMr]).

We start with the following result of Hattori ([Ht], [Sa]) for completely solvable solvmanifold that generalized a previous result of Nomizu [No] for nilmanifolds. We recall that for a given Lie algebra $\mathcal{G}$ the Chevalley–Eilenberg complex $(\Lambda^*\mathcal{G}^*, \delta)$ associated with $\mathcal{G}$ consists of the exterior algebra $\Lambda^*\mathcal{G}^*$ of the dual space $\mathcal{G}^*$ considered as a complex of vector spaces with the jth, $0 \le j \le m$, gradation equal to $\wedge^j\mathcal{G}^*$ and the differential $\delta\colon \wedge^j\mathcal{G}^* \to \wedge^{j+1}\mathcal{G}^*$ defined as

$$\delta(\mathcal{X}^*)(\mathcal{X}_1, \dots, \mathcal{X}_{j+1}) \\ := \sum (-1)^{s+t-1}\mathcal{X}^*([\mathcal{X}_s, \mathcal{X}_t], \mathcal{X}_1, \dots, \widehat{\mathcal{X}_s}, \dots, \widehat{\mathcal{X}_s}, \dots, \mathcal{X}_{j+1}),$$

where the sum is taken over all $1 \le s \le t \le j+1$.

(3.18) THEOREM. *Let $(\Lambda^*\mathcal{G}^*, \delta)$ denotes the Chevalley–Eilenberg complex associated to the Lie algebra $\mathcal{G}$ of a simply connected completely solvable Lie group G. If $\Gamma \subset G$ is a discrete uniform subgroup, then $H^*(G/\Gamma; \mathbb{R}) \cong H^*(\Lambda^*\mathcal{G}^*, \delta)$.*

Note that the Chevalley–Eilenberg complex can by identified with a subcomplex of the de Rham complex consisting of invariant forms. This result together with the Hopf lemma (see [Sp]) leads to the following.

(3.19) PROPOSITION. *Let $f\colon G/\Gamma \to G/\Gamma$ be a self map of a compact special completely solvable solvmanifold of dimension d. Let next $\Phi\colon G \to G$ be a homomorphism such that its factor map $\psi\colon G/\Gamma \to G/\Gamma$ is homotopic to f. Then for the Lefschetz number we have $L(f^m) = \det(\mathrm{I} - (D\Phi(e))^m)$ for every $m \in \mathbb{N}$.*

PROOF. By the Nomizu–Hattori theorem the spectral radius and Lefschetz number of f can be derived by use of the map $D\Phi(e)^*$ of the Chevalley–Eilenberg complex. Since $D\Phi(e)$ is a homomorphism of the Lie algebra, the linear subspaces of co-boundaries, co-cycles are preserved by $\wedge D\Phi(e)^*$. Consequently the cohomology spaces can be identified with the factors of subspaces preserved by $\wedge D\Phi(e)^*$. The inclusion, and consequently inequality, follows from the fact that the spectrum of an operator restricted to an invariant subspace is a subset of the spectrum of entire operator and the same is for the factor operator induced on the factor of an invariant subspace. The second equality

$$L(f^n) = \sum_{k=0}^{k=m} (-1)^k \mathrm{tr}\, H^k(f^n) = \sum_{k=0}^{k=m} (-1)^k \mathrm{tr}\, \Lambda^k (D\Phi(e)^*)^n = \det(\mathbb{I} - (D\Phi(e))^n),$$

is a direct consequence of the Hopf lemma and linear algebra. □

Now we show that the matrix $D\Phi(e)$ has the same spectrum, denoted by σ, as an integral $m \times m$-matrix A_f.

(3.20) PROPOSITION. *Let $M, f\colon M \to M$ be a map of a compact completely solvable solvable manifold, e.g. a nilmanifold, which is induced by an endomorphism $\Phi\colon G \to G$. We have $\sigma(D\Phi(e) = \sigma(A_f)$. Consequently* $\mathrm{sp}\,(D\Phi(e)) = \mathrm{sp}\,(A_f)$ *and* $\mathrm{sp}\,(\wedge D\Phi(e)) = \mathrm{sp}\,(\wedge A_f)$.

PROOF. To shorten notation put $D := D\Phi(e)$. Assuming that M is a nilmanifold, we show that these two matrices have the same characteristic polynomial using an induction over the length on nilpotency of G. If $M = \mathbb{T}^d$, thus $\mathcal{G} = \mathbb{R}^d$, then Φ is a linear map and $D = \Phi = A_f$.

In the general case note that the central tower of G gives a descending tower of ideals of $\mathcal{G}$:

$$\mathcal{G}_k = \mathbf{e} \lhd \mathcal{G}_{k-1} \lhd \mathcal{G}_{k-2} \lhd \cdots \lhd (\mathcal{G}_1 = [\mathcal{G}, \mathcal{G}]) \lhd \mathcal{G}_0 = \mathcal{G},$$

which preserved by any homomorphism of $\mathcal{G}$, where $\mathcal{G}_i$ is the Lie algebra of G_i. Since D is a homomorphism, the matrix of D in a basis $\mathcal{G}$ formed by $\exp^{-1}$ of generators Γ has the form

$$D = \begin{bmatrix} \widetilde{D}_1 & 0 \\ \breve{D}_1 & D_1 \end{bmatrix},$$

where $\widetilde{D}_1 = D\Phi(e)|_{\mathcal{G}_1}$ and D_1 is the matrix induced by D on the quotient (abelian) algebra $\mathcal{G}_0/\mathcal{G}_1$. Consequently, for the characteristic polynomial $\chi_D(t)$ we have $\chi_D(t) = \chi_{D_1}(t)\chi_{\widetilde{D}_1}(t)$, and the statement follows by the induction argument. □

As a consequence of the above construction we get the following.

(3.21) COROLLARY. *Let $f: X \to X$ be a map of a compact nilmanifold $X = G/\Gamma$ and $\Phi: G \to G$ an endomorphism of G such that its factor map is homotopic to f. Let next $D = D\Phi(e)$ be corresponding endomorphism of the Lie algebra $\mathcal{G} = T_eX$ of G and D_i, $1 \le i \le k$ the corresponding endomorphisms of the factor algebras $\mathcal{G}_i/\mathcal{G}_{i-1}$. Then $D_i = A_i$ and consequently $A_f = \bigoplus_{i=1}^k D_i$. Furthermore, if $f = (f_N, f_0)$ is a map of the Mostow fibration $N \subset X \xrightarrow{p} \mathbb{T}_0$ of an NR-solvmanifold X, as in Theorem (3.14), then $A_f = \bigoplus_{i=0}^k D_i$, where D_i, $1 \le i \le k$ are matrices defined above for f_N, and D_0 is the linearization of f_0.*

3.2. Anosov theorem. The basic property of the linearization matrix is the fact that it gives a simple formula for the Lefschetz number of a selfmap of an NR-solvmanifold and all its iteration (cf. [KMC]).

(3.22) THEOREM. *Let $f: X \to X$ be a map of an NR-solvmanifold and $A = A_f$ the linearization matrix defined in Definition (3.16). Then $L(f^m) = \det(\mathrm{I} - A^m)$ for every $m \in \mathbb{N}$.*

PROOF. Let us remind that for the case of torus $X = \mathbb{T}^d$, $A = H^1(f)$ (see Lemma (3.8)) and the statement follows by an elementary linear algebra, because $H^*(\mathbb{T}^d) = \wedge H^1(\mathbb{T}^d)$ is the exterior algebra of $H^1(\mathbb{T}^d)$. Next if X is a compact nilmanfold then the statement follows from Proposition (3.19) (see also [JeMr1]). Suppose that $f = (f_N, f_0)$ is a fiber map of the corresponding Mostow fibration. First observe that if $L(f_0) = 0$ implies $L(f^m) = 0$ for every $m \in \mathbb{N}$. Indeed, then $N(f_0) = 0$ (see the proof of Theorem (3.24) below) and f_0, thus f_0^m, is homotopic to fixed points free map. Consequently, f, and f^m is homotopic to a fixed points free map. Notice also, that any map preserving the Mostow fibration is homotopic to the map which is linear on the base i.e. induced by a linear map of A_{f_0} of $\mathbb{R}^{d_0}$, $d_0 = \dim \mathbb{T}$, which preserves $\mathbb{Z}^{d_0}$ (Remark (3.9)). The fixed points of f

$$\begin{array}{ccccc} N_b & \longrightarrow & X & \longrightarrow & \mathbb{T} \\ \downarrow{\scriptstyle f_b} & & \downarrow{\scriptstyle f} & & \downarrow{\scriptstyle f_0} \\ N_b & \longrightarrow & X & \longrightarrow & \mathbb{T} \end{array}$$

are in the fibres N_b over the fixed points of the map f_0 induced on the base. Let us compute the value of the Lefschetz number of f.

$$\text{(3.23)} \quad \begin{aligned} L(f) &= \sum_{b \in \mathrm{Fix}\,(f_0)} \mathrm{Ind}\,(f_0, b) L(f_b) = \pm 1 \Big(\sum_{b \in \mathrm{Fix}\,(f_0)} \det(\mathrm{I} - A_{f_b}) \Big) \\ &= L(f_0) \det(\mathrm{I} - A_{f_e}) = \det(\mathrm{I} - A_0) \det(\mathrm{I} - A_{f_e}) = \det(\mathrm{I} - A_f). \end{aligned}$$

Here A_{f_b} denotes the matrix for the map of the fiber $f_b: N_b \to N_b$ over the point $b \in \mathrm{Fix}\,(f_0)$. The first equality follows from the fact that the indices of the fixed points of the base map f_0 (induced by a linear one) are all either $+1$ or all are -1 and from the Fixed Point Index Product Formula. The second one follows from Proposition (3.19). The third is the result of Keppelmann and McCord (see [KMC]), which states that for every fixed point $b \in \mathbb{T}$ there exists an integral matrix A_{f_e} such that $\det(\mathrm{I} - A_{f_b}) = \det(\mathrm{I} - A_{f_e})$, thus equal to $L(f_N)$, provided that the solvmanifold X is an NR-solvmanifold. The equality for f^m follows from the functoriality of the linearization i.e. $A_{f^m} = A_f^m$. The last follows from the definition of A_f (cf. Definition (3.16)), because as well the matrix A_0 as the matrices A_i, $1 \le i \le k$ are functorial. □

A generalization of above used property of a map of torus (here linear) which says that local fixed point indices at all fixed points are all either $+1$ or all are -1, or all are 0 led to the Anosov theorem (cf. [An], [FaHu]) which says that that Nielsen number is equal to the absolute value of the Lefschetz number of a map of compact nilmanifold. We give this theorem in the most general version of [KMC].

(3.24) THEOREM. *Let $f\colon X \to X$ be a map of a compact NR-solvmanifold. Then $N(f) = |L(f)|$.*

PROOF. A proof follows the proof of Theorem (3.22). For a self map f of the torus the equality $N(f) = |L(f)|$ was shown in [BBPT]. Suppose that $f = (f_N, f_0)$ is a fiber map of the corresponding Mostow fibration. First observe that if $N(f_0) = 0 = |L(f_0)|$ implies $N(f^m) = L(f^m) = 0$ for every $m \in \mathbb{N}$. Indeed, then $N(f_0) = 0$ and f_0, thus f_0^m, is homotopic to fixed points free map. Consequently, f, and f^m is homotopic to a fixed points free map.

Next, if X is a nilpotent manifold then $N(f) = |L(f)|$ was proved by Anosov in [An] and Fadell and Husseini in [FaHu]. The simple but important observation is that the Fadell–Husseini fibration we have $N(f) = N(f')N(\widetilde{f})$ for a fiber preserving map $f = (f', \widetilde{f})$, and the statement follows by the induction.

To complete the proof for the general case we have to take the absolute value of the first and the third to last term of the equality (3.23). Indeed $|\det(\mathrm{I} - A_{f_e})| = |L(f_e)|$ which is equal to $N(f_e)$ by the Anosov theorem for nilmanifolds. We got $|L(f)| = N(f_0)N(f_e)$. On the other hand in [MC1] it was shown that $N(f) = \sum_{b\in \mathrm{Fix}\,(f_0)} N(f_b)$. But $N(f_b) = |\det(\mathrm{I} - A_b)|$ does not depend on $b \in \mathrm{Fix}\, f_0$ ([KMC]), thus it is equal to $N(f_e)$, and $\#|\mathrm{Fix}\,(f_0) = N(f_0)$, because for the fixed point index $\mathrm{ind}(f_0, b) = \pm 1$ here. This gives $N(f) = N(f_0)N(f_e)$ and the proof is complete. □

3.3. Summation formula. In next we need also properties of Nielsen type invariants of a map which estimate from below the amount of points of period m and the amount of m-periodic points. Their definitions and properties are described in detail in Chapter III.15. We shall recall them briefly.

Let $f: X \to X$ be a self-map of a compact space. There exists an invariant $NF_m(f) \in \mathbb{N} \cup \{0\}$, which we call the *full Nielsen–Jiang periodic number*, and which has the following properties (cf. [Ji4]):

$$0 \le N(f^m) \le NF_m(f) \le \#P^m(f), \quad NF_m(f) \text{ is a homotopy invariant.} \tag{3.25}$$

In [Ji4] Boju Jiang introduced also another invariant of $NP_m(f) \in \mathbb{N} \cup \{0\}$, which we call the *prime Nielsen–Jiang periodic number*, and which has the following properties (cf. [Ji4]):

$$0 \le NP_m(f) \le \#P_m(f) \quad \text{and} \quad NP_m(f) \le NF_m(f), \quad NP_m(f) \text{ is a homotopy invariant.} \tag{3.26}$$

A nontrivial property of the first invariant says that $NF_m(f)$ is the best homotopy invariant estimating the number of points of period m provided X is the PL-manifold. This fact called the Wecken theorem for periodic points, and known also as the Halpern conjecture, was shown by Jezierski.

(3.27) THEOREM. *Let $f: X \to X$ be a map of a compact* PL-*manifold of* $\dim X \ge 3$. *Then there exists $g: X \to X$ homotopic to f and such that $\#P^m(f) = NF_m(f)$.*

(cf. [Je2] for the case $\dim X \ge 4$, [Je4], [Je5] and Chapter III.15 for the case $\dim X \ge 3$).

A nontrivial property of the second invariant says that $NP_m(f)$ is the best homotopy invariant estimating the number of m-periodic points provided X is the PL-manifold. It was also proved by Jezierski.

(3.28) THEOREM. *Let $f: X \to X$ be a map of a compact* PL-*manifold of* $\dim X \ge 3$. *Then f is homotopic to a map $g: X \to X$ satisfying $P_m(g) = \emptyset$ if and only if $NP_m(f) = 0$.*

(cf. [Je2] for the case $\dim X \ge 4$ and Chapter III.15 for the case $\dim X \ge 3$).

(3.29) REMARK. The statement of these two theorems is not true in dimension 2 (cf. Chapter III.15), but counterexamples are non-trivial as for the classical Wecken theorem.

For a map $f: X \to X$ of any space X we have a summation $\#P^m(f) = \sum_{k|m} \#P_k(f)$ over divisors of m, by an obvious combinatorial argument, with the convention that if ∞ appears at the one side it appears also at he second. Suppose for a while that X is a finite set. Then in this formula we can replace $\#P^m(f)$ by $NP_m(f)$ and $\#P_k(f)$ by $NP_k(f)$ respectively. Thus one could expect that the similar summation formula with the full Nielsen–Jiang periodic number and primes Nielsen–Jiang periodic numbers as the summands. Unfortunately, we have only inequality (cf. Chapter III.15).

$$\sum_{k|m} NP_k(f) \leq NF_m(f). \tag{3.30}$$

It is not difficult to show that the inequality is sharp in general. Indeed, it is enough to take the antipodal map $f(x) = -x$ of the sphere S^{2d} (cf. Chapter III.15).

Anyway, if X is a compact solvmanifold then the equality, called the summation formula holds due to the toral essentiality of any map (cf. [HeKa1]–[HeKa3]).

(3.31) THEOREM. *For a map $f: X \to X$ of compact solvmanifold we have*

$$NF_m(f) = \sum_{k|m} NP_k(f) \quad \textit{and consequently} \quad NP_m(f) = \sum_{k|m} \mu(m/k) NF_k(f),$$

where $\mu: \mathbb{N} \to \{-1, 0, 1\}$ is the Möbius function.

We recall that

$$\mu(n) = \begin{cases} 1 & \text{if } n = 1, \\ (-1)^r & \text{if } n = p_1 \cdot \ldots \cdot p_r,\ p_i \text{ distinct primes}, \\ 0 & \text{if } p^2 | n,\ p \quad \text{a prime}. \end{cases}$$

The equivalence of the two equalities of Theorem (3.31) holds for any arithmetic functions and is called the Möbius inversion formula.

For maps of NR-solvmanifolds the inequality $N(f^m) \leq NF_m(f)$ of (3.25) becomes the equality, due the the toral essentiality and the fact that every iteration f^k of f is weakly Jaing map (cf. [HeKa1]–[HeKa3]).

(3.32) THEOREM. *For a map $X \to X$ of an NR-solvmanifold such that $N(f^m)$ $\neq 0$ we have $NF_m(f) = N(f^m)$.*

Summing up, for a map f of an NR-solvmanifold we have $N(f^m) = |\det(\mathrm{I} - A^m)|$, where A the integral matrix of linearization by Theorems (3.22) and (3.24). Consequently the sequence $\{N(f^m)\}$ has nice arithmetical properties. Additionally it expresses the sequence $\{NP_m(f)\}$ by the the Möbius inversion as follows from Theorems (3.31) and (3.32).

4. Main theorem for NR-solvmanifolds

In this section we present a theorem which describes the set of homotopy minimal periods for a map of an NR-solvmanifold. It is based on the proof of the corresponding theorem for tori maps given by Jiang and Llibre in [JiLb] in 1998.

The proof used all the properties of Nielsen numbers of iteration presented in the Subsections 3.2 and 3.3, which were know for maps of torus earlier. A brilliancy of their approach was, for a given map $f\colon \mathbb{T}^d \to \mathbb{T}^d$, to pick up all m for which $NP_m(f) = 0$. It is given in terms of a condition on $N(f^m)$, by a fine combinatorial argument. Then they could use the You theorem (cf. [Yu1], [Yu2]) which stated (among others) that for a torus map $NP_m(f) = 0$ implies that f is homotopic to a map g such that $P_m(f) = \emptyset$ excluding such m from $\mathrm{HPer}(f)$. Remark that the mentioned You theorem is a correspondent of the general Theorem (3.28) in this special torus case. Another distinction of the paper [JiLb] was the observation that there exists $Q(d)$, (i.e. depending on d only) such that for all $m > Q(d)$ the mentioned above condition can not be satisfied if $N(f^m) \neq 0$, i.e. $\mathrm{HPer}(f)$ is equal to the set $\{m : N(f^m) \neq 0\}$ out of the interval $[1, Q(d)]$. To show it Jiang and Llibre used an estimate of the algebraic number theory proved by them. Meantime their theorem was extended onto the case of maps of a nilmanifold [JeMr1] and consecutively maps of an NR-solvmanifold of dimension > 3 or a completely solvable solvmanifold of dimension ≥ 3 in [JeKdMr]. These generalization were possible due extensions of the You theorem from tori onto the nilmanifold, and correspondingly completely solvable solvmanifold of dimension ≥ 3 shown in [JeMr1] and [JeKdMr] respectively. Nowadays all these facts are consequences of the general Theorem (3.28) (cf. [Je2], [Je4], [Je5] or Chapter III.15) which does not need any special structure of a manifold. At this point it would be enough to say that to prove the below theorem one should use all the quoted properties the sequence of Nielsen numbers $\{N(f^m)\}$, Theorem (3.28), and repeat almost verbatim the arguments of paper [JiLb]. Anyway we include them, in a little bit extracted form, to expose the main line of the proof of theorem.

Before stating the theorem we need next notion.

(4.1) Definition. Let $f\colon X \to X$ be a map of compact an NR-solvmanifold of dimension $d \geq 3$ and A its linearization $d \times d$ matrix (Definitions (3.7), (3.16)). Put

$$\mathbb{N} \supset T_A := \{m : \det(\mathrm{I} - A^m) \neq 0\}.$$

We call T_A set of *algebraic homotopy miniml periods.*

By an obvious reason if $m \notin T_A$ then $m \notin \mathrm{HPer}(f)$, i.e.

$$\mathrm{HPer}(f) \subset T_A \tag{4.2}$$

Indeed, then $L(f^m) = N(f^m) = 0$ ((3.22), (3.24)), thus $NP_k(f) = 0$ for every $k|m$ by the property of $NP_m(f)$ and Theorem (3.32). This implies $m \notin \mathrm{HPer}(f)$ in view of Theorem (3.28). Observe also that

$$N(f) = 0 \Leftrightarrow L(f) = 0 \Leftrightarrow 1 \in \sigma(A_f). \tag{4.3}$$

We are in position to formulate our main theorem, which characterizes $\mathrm{HPer}(f)$ by T_{A_f}.

(4.4) THEOREM. *Let $f: X \to X$ be a map of a compact NR-solvmanifold of dimension $d \geq 3$, $A = A_f$ its linearization, and $T_A \subset \mathbb{N}$ as above. Then* $\mathrm{HPer}(f) \subset T_A$ *and it is in one of the following three (mutually exclusive) types:*

(E) $\mathrm{HPer}(f) = \emptyset$ *if and only if* $N(f) = 0$,

(F) $\mathrm{HPer}(f) \neq \emptyset$ *but finite set if and only if all eigenvalues of A are either zero or roots of unity,*

(G) $\mathrm{HPer}(f)$ *is infinite and* $T_A \setminus \mathrm{HPer}(f)$ *is finite if and only if there exists* $\lambda \in \sigma(A)$ *such that* $|\lambda| > 1$.

Moreover, for every d there exist constants $P(d), Q(d) \in \mathbb{N}$ depending on d only such that $\mathrm{HPer}(f) \subset [1, P(d)]$ *in type* (F) *and* $T_A \setminus \mathrm{HPer}(f) \subset [1, Q(d)]$ *in type* (G).

It is worth of pointing out that the statement of this theorem is the same as that of it correspondent in the case when X is the torus $\mathbb{T}^d$ (cf. [JiLb]).

As we already mentioned, the description of $\mathrm{HPer}(f)$ is attainable due an observation established by Boju Jiang and Llibre (cf. [JiLb]).

(4.5) THEOREM. *Let $f: X \to X$ be a map of a compact NR-solvmanifold. Then $NP_m(f) = 0$ if and only if either $N(f) = 0$ or $N(f^m) = N(f^{m/p})$ for some prime factor p of m.*

(4.6) REMARK. In fact the statement of Theorem (4.5) remains true for every self-map of a compact manifold for which hold the summantion formula (Theorem (3.31)), the statement of Theorem (3.32), and the equality $N(f^m) = |\det(\mathrm{I} - A^m)|$, with an integral matrix A.

Finally, a nontrivial estimate from below of the rate of convergence of an algebraic number of module 1 is necessary to show that there exists $Q(d)$ such that $N(f^m) > N(f^{m/p})$ for all $m \in T_A$ satisfying $m > Q(d)$. To write it down we need a new notion.

(4.7) DEFINITION. Let α be an algebraic number of degree d and $a_0 x^d + a_1 x^{d-1} + \ldots + a_d$ its minimal polynomial with roots $\alpha_1, \ldots, \alpha_d$. The measure of α is defined as

$$M(\alpha) := a_0 \prod_{i=1}^{d} \max\{1, |a_i|\}.$$

The crucial is a characterization of an algebraic number given by Jiang and Llibre, also Mignotte ([JiLb]):

(4.8) THEOREM. *For every algebraic number α of degree d and every $m \in \mathbb{N}$ such that $\alpha^m \neq 1$, we have*

$$|1 - \alpha^m| > \frac{1}{2} e^{-9\alpha H^2},$$

where

$$a = \max\left\{20, 12.85|\log\alpha| + \frac{1}{2}\log M(\alpha)\right\}, \quad H = \max\left\{17, \frac{d}{2}\log m + 0.66d + 3.25\right\}.$$

As a consequence the following was derive in [JiLb] that allows to prove the last part of Theorem (4.4).

(4.9) COROLLARY. *Let A be integral $d \times d$ matrix and $\rho := \mathrm{sp}\,(A)$ its spectral radius. Suppose that $\det(\mathrm{I} - A^m) \neq 0$, $\rho > 1$ and $m \geq 5000$. Then for every $n|m$ we have*

$$\frac{|\det(\mathrm{I} - A^m)|}{|\det(\mathrm{I} - A^n)|} > \frac{\rho^{m/2} - 1}{e^{9d(41.4 + (d/2)\log\rho)(d\log m)^2}}.$$

We postpone a discussion of Theorems (4.5), (4.8), and Corollary (4.9) to next subsections.

PROOF OF THEOREM (4.4). The proof is almost literally taken form [JiLb]. With respect to Theorem (3.28) we have $\mathrm{HPer}(f) = \{m : NP_m(f) \neq 0\}$. By inclusion (4.2) the set $\mathrm{HPer}(f)$ is equal to the set of all $m \in T_A$ for which the condition of statement of Theorem (4.5) is violated.

Let $\psi_k(z)$ be the kth cyclotomic polynomial, and let $\Psi_k := |\det\psi_k(A)|$. By the identity $z^m - 1 = \prod_{k|m}\psi_k(z)$ we have

$$N(f^m) = |\det(A^m - \mathrm{I})| = \prod_{k|m} |\det\psi_k(A)| = \prod_{k|m}\Psi_k. \tag{4.10}$$

Note the coefficients of ϕ_k are integers and A is an integer matrix, all these Ψ_k are nonnegative integers. As a direct consequence of (4.10) we get. □

(4.11) PROPOSITION. *If $n|m$ and $N(f^n) = 0$ then $N(f^m) = 0$, if $n|m$ and $N(f^n) \neq 0$ then $N(f^n)|N(f^m)$.*

(4.12) PROPOSITION. *Let χ_A be the characteristic polynomial of an integral matrix A. Put $D_A := \{k \in \mathbb{N} : \psi_k|\chi_A\}$. Then*

(4.12.1) *D_A is a finite set, and $T_A = \mathbb{N} \setminus \bigcup_{k\in D_A} k\mathbb{N}$.*
(4.12.2) *T_A is empty when $\det(\mathrm{I} - A) = 0$, or is infinite when $|\det(\mathrm{I} - A)| > 0$.*

PROOF. (4.12.1) If $m \notin T_A$ then there exists $\lambda \in \sigma(A)$ such that $\lambda^m = 1$. Then λ is a primitive root of unity of degree $k|m$ i.e. a root of the cyclotomic polynomial ψ_k. Since ψ_k is irreducible, we have $\psi_k | \chi_A$, i.e. $k \in D_A$. It is know that ψ_k is of degree $\phi(k)$, where ϕ is the Euler function defined by the formula (1.11). It is also known that for a given d there is only finite number of k such that $\phi(k) \leq d$, which shows that D_A is finite.

(4.12.2) When $\det(\mathrm{I} - A) = 0$ then $T_A = \emptyset$ by the definition and (4.12.1). If $\det(\mathrm{I} - A) \neq 0$ let m be the least common multiple of the set D_A. By (4.12.1) we see that $km + 1 \in T_A$ for every $k \in \mathbb{N}$.

Suppose that the eigenvalues $\lambda_1, \dots, \lambda_d$ of A are indexed such that $\mathrm{sp}\,(A) = \rho = |\lambda_1| \geq |\lambda_2| \geq \cdots \geq |\lambda_d|$. Then each λ_i is an algebraic integer of degree $k_i \leq d$ with the measure (cf. Definition (4.7)) $M(\lambda_i) \leq \rho^{k_i}$. The later follows from the fact that for a monic polynomial (i.e with the leading coefficient equal to 1) with integral coefficients we have this inequality.

Let us consider the cases of Theorem (4.4) depending the form of spectrum of the linearization A.

Type (E): $\det(\mathrm{I} - A) = N(f) = 0$. Then, by Theorem (4.12) $T_A = \emptyset$ and from (4.2) it follows that $\mathrm{HPer}(f) = \emptyset$. Note that it is the first case of Theorem (4.5).

When f is not of Type (E), then there are two possibilities

Type (F): $N(f) = \det(\mathrm{I} - A) \neq 0$ and $\rho \leq 1$. Then, by the Kronecker theorem all the nonzero eigenvalues $\lambda_1, \dots, \lambda_q$ are roots of unity. The Kronecker theorem states that if the all roots of a monic polynomial with integral coefficients are in the unit disc then they must be roots of unity or zero.

Each λ_i, $1 \leq i \leq q$, is a root of some cyclotomic polynomial $\phi_{k_i}(\lambda)$ of degree $k_i = \phi(d_i) \leq d$. Let $h = h(d)$ be the least common multiple of the set $\{k \in \mathbb{N} : \phi(k) \leq d\}$. Then $\lambda_i^h = 1$ for all $1 \leq i \leq q$. Note that $h \notin \mathrm{HPer}(f)$ by its definition ($N(f^h) = 0$) and the sequence $|\det(\mathrm{I} - A^m)| = \prod_{i=1}^q |1 - \lambda_i^m|$ is periodic of period h, i.e. $N(f^{m+h}) = N(f^m)$ for all m. We show that if $m \in \mathrm{HPer}(f)$ then $m|h$, hence $\mathrm{HPer}(f)$ is contained in the set of all proper divisors of $h = h(d)$, thus contained in the interval $[1, P(d)]$, where $P(d)$ is the largest proper divisor of h.

Now take $\delta = (m, h)$ the greatest common divisor of $m \in T_A$ and h. Then there exists natural numbers a, b such that $am - bh = \delta$. Since am is a multiple of n, by Proposition (4.11) $N(f^m)$ divides $N(f^{am}) = N(f^{bh+\delta}) = N(f^\delta)$. But $N(f^\delta) | N(f^m)$, because $\delta | m$, which leads to $N(f^m) = N(f^\delta)$. If $m \neq \delta$ then $NP_m(f) = 0$, as follows from the summation formula (Theorem (3.31)), since all the summands are nonnegative. Consequently $m \notin \mathrm{HPer}(f)$, which completes the proof of this case.

Type (G): Suppose $N(f) \neq 0$ and $\rho > 0$. Let $m \in T_A$ be such that $N(f^m) \neq 0$ thus also $N(f^n) \neq 0$ for every $n|m$. Applying the inequality of Corollary (4.9) we have

$$\frac{N(f^m)}{N(f^n)} > \frac{\rho^{m/2} - 1}{e^{9d(41.4+(d/2)\log\rho)(d\log m)^2}} \quad \text{provided} \quad m \geq 5000. \tag{4.13}$$

Consequently if

$$\frac{\rho^{m/2} - 1}{e^{9d(41.4+(d/2)\log\rho)(d\log m)^2}} > 1 \tag{4.14}$$

then $N(f^m) > N(f^n)$ for every proper divisor n of m, so $m \in \mathrm{HPer}(f)$ by Theorem (4.5). Since the rate of growth of the numerator of the expression of (4.14) as a function of m is smaller than of its denominator, for any fixed $\rho > 1$ the inequality (4.14) is valid for sufficiently large m. Therefore the set $T_A \setminus \mathrm{HPer}(f)$ is finite for every matrix A with $\rho > 1$. We are left with the task to find a common bound valid for all such $d \times d$ matrices. Consider first the case $\rho \geq e^{82.8/d}$. Then $41.4 + \log(d/2)\rho \leq d\log\rho$, hence (4.14) is valid if

$$\frac{m}{2} > 9d^4(\log m)^2. \tag{4.15}$$

Let $m_1 \geq 5000$ be an integer that satisfies (4.15). Then (4.15) and consequently (4.14) holds for all $m \geq m_1$.

Finally consider the remaining case of $\rho < e^{82.8/d}$. Since the coefficients in the characteristic polynomial $\chi_A(\lambda)$ are the elementary symmetric polynomials of the eigenvalues we have

$$|a_i| \leq \binom{m}{i}\rho^i \quad \text{for all } 1 \leq i \leq d. \tag{4.16}$$

So there is only a finite number of possibilities for the integral sequences $(a_1, \ldots, a_d)$ with $\rho < e^{82.8/d}$. Let $\rho > 1$ be the smallest of corresponding $\rho' s$ and let m_2 be the smallest m that validates (4.14) for $\rho = \rho_1$. Thus, when $m \geq m_2$, the condition (4.14) is true whenever $\rho \geq e^{82.8/d}$.

Finally, let us put $m_0(d) := \max\{m_1, m_2\}$. By the above inequality (4.14) is satisfied if $m \geq m_0$, which shows that $T_A \setminus \mathrm{HPer}(f) \subset [1, \ldots, m_0]$. Now take $Q(d) = m_0$. The proof of Theorem (4.4) is complete. □

4.1. Combinatorics and number theory. As we already said the proof of Theorem (4.5) is the a combinatorics argument proved in ([JiLb, Proposition 3.3]) which we only quote here referring the reader to [JiLb] for details.

Let ω be a nonempty finite set. By $C := \{\omega, C_\tau\}$ we denote a function $C\colon 2^\omega \to [1, \infty) \subset \mathbb{R}$.

(4.17) PROPOSITION. *For a given $\tau \subset \omega$ and a function $C\colon 2^{\omega} \to [1,\infty)$ define*

$$D_\tau := \prod_{\xi\subset\tau} C_\xi, \quad \text{for any } \tau\subset\omega; \quad E_\omega := \sum_{\tau\subset\omega}(-1)^{|\omega|-|\tau|} D_\tau,$$

where $|\tau|$ denotes here the cardinality of τ. Then

(4.17.1) $E_\omega \geq 0$,
(4.17.2) $E_\omega = 0$ *if and only if* $D_\omega = D_{\omega\setminus\{\alpha\}}$ *for some element* $\alpha \in \omega$.

PROOF OF THEOREM (4.5). We may assume that $N(f^m) > 0$. Let the prime factorization of m is

$$m = p_1^{\alpha_1}\dots p_s^{\alpha_s}, \quad s\geq 1,\ p_i\text{-prime},\ \alpha_i \geq 1. \tag{4.18}$$

Let $\omega := \{1,\dots,s\}$ and $\tau\subset\omega$ its subset. Put $p_\tau := \prod_{i\in\tau} p_i$ and $m_\tau := m/p_{\omega\setminus\tau} = mp_\tau/p_\omega$. The Möbius inversion formula for $NP_m(f)$ can be written in the following equivalent form (cf. [HeKa1] or Chapter I.3).

$$NP_m(f) = \sum_{\tau\subset\omega}(-1)^{s-|\tau|} N(f^{m_\tau}).$$

In view of Theorem (3.28) $m \notin \mathrm{HPer}(f)$ if and only if $NP_m(f) = 0$. Define

$$C_\tau := \prod_j \Psi_j \tag{4.20}$$

where j is such that $j = p_1^{\beta_1}\dots p_s^{\beta_s}$, with

$$\beta_i = \begin{cases} \alpha_i & \text{if } i\in\tau,\\ 0\leq\beta_i\leq\alpha_i & \text{if } i\notin\tau.\end{cases}$$

Note that $\det(\mathrm{I}-A^m)\neq 0$ implies $\Psi_k\geq 1$ for all $k|m$, hence all $C_k \geq 1$. It is easy to check that

$$\begin{aligned} D_\tau &= \prod_{\xi\subset\tau} C_\xi = \prod_{j|m_\tau}\Psi_j = N(f^{m_\tau}),\\ E_\omega &= \sum_{\tau\subset\omega}(-1)^{|\omega|-|\tau|}D_\tau = NP_m(f).\end{aligned} \tag{4.21}$$

Now applying Proposition (4.17), we conclude that $N(f^m) = E_\omega = 0$ if and only if $N(f^m) = D_\omega = D_{\omega\setminus\{i\}} = N(f^{m/p_i})$ for some $1\leq i\leq s$. □

Let us turn back to the number theory part of the proof of Theorem (4.4). Since it is taken from [JiLb], we refer there the reader for details. In particular we are not going to show how Theorem (4.8) implies the inequality (4.9). Note that Theorem (4.8) says that the powers of an algebraic number of absolute value equal to 1, but not being a root of unity, must not tend to 1 to fast. However remind that the aim was to prove that $N(f^m)/N(f^n) > 1$ for $m > m_0(d)$, d the dimension of the matrix A. It was pointed out to the author by J. Browkin and A. Schinzel that the inequality $N(f^m)/N(f^n) > 1$ follows also from a result of Schinzel ([Sch]). To show it we need new notions and definitions. Let α, β be non-zero integers of an algebraic number field $\mathcal{K}$ of degree d. A prime ideal $\mathfrak{B}$ of $\mathcal{K}$ is called a primitive divisor of $\alpha^m - \beta^m$ if $\mathfrak{B}|\alpha^m - \beta^m$, but $\mathfrak{B}$ does not divide $\alpha^n - \beta^n$ if $n < m$. In 1974 A. Schinzel proved the following theorem (cf. [Sch, Theorem I]).

(4.22) THEOREM (Schinzel Theorem). *If $(\alpha, \beta) = 1$ and α/β is not a root of unity then $\alpha^m - \beta^m$ has a primitive divisor for all $m > m_0(d)$, where d is the degree of α/β and $m_0(d)$ is effectively computable.*

Suppose that $N(f^m) = |\det(A^m - 1)|$, $A \in M_{d\times d}(\mathbb{Z})$ is an integer. For every eigenvalue $\lambda_j \in \sigma(A, 1 \leq j \leq d$, take $\alpha_j := \lambda_j$ and $\beta_j := 1$. By the definitions α_j and β_j are integers of the algebraic field given by the characteristic polynomial of A. If $m \in T_A$ i.e. $N(f^m) \neq 0$ then the hypothesis of the Schinzel theorem is satisfied. Note that if $n|m$, $k = m/n$, then

$$(\lambda^m - 1) = (\lambda^n - 1)(1 + \lambda^n + \lambda^{2n} + \ldots + \lambda^{(k-1)n}). \tag{4.23}$$

Consequently, for any $1 \leq j \leq d$ such that $|\lambda_j| > 1$ and $m \in T_A$, $m > m_0(d_j)$ there exists a primitive ideal $\mathfrak{B}_j \subset \mathcal{K}$ such that $\mathfrak{B}_j|1 + \lambda_j^n + \lambda_j^{2n} + \ldots + \lambda_j^{(k-1)n}$ as follows from the Schinzel theorem. Observe also that $d(\alpha_j) = d(\lambda_j)$ is a divisor of $d :=$ degree $\mathcal{K}$ and $m_0(d_j) \leq m_0(d)$, by an argument of proof of Theorem 3.4 of [Sch]. From this, for every $1 \leq j \leq d$, it follows that

$$\mathfrak{B}_j \quad \text{divides} \quad \frac{N(f^m)}{N(f^n)} = \prod_1^j (1 + \lambda_j^n + \lambda_j^{2n} + \ldots + \lambda_j^{(k-1)n}).$$

The above implies that then there exists a prime $q \in \mathcal{P} \subset \mathbb{N}$ such that q divides $N(f^m)/N(f^n)$ provided $m \in T_A, m > m_0(d)$. Indeed it is enough to take $q \in \mathcal{P}$, where q^{f} is the norm $|\mathfrak{B}_j|$ of the ideal $\mathfrak{B}_j$, since $\mathfrak{B}_j \cap \mathbb{Z}$ is a prime ideal of $\mathbb{Z}$. This shows that $N(f^m)/N(f^n) > 1$ then.

It is worth of pointing out that either the proof of referred Schinzel Theorem (4.22) or the Jiang–Llibre Theorem (4.8) are based on the Baker logarithm inequality which is a deep theorem of the number theory. This leads to a natural question

(4.24) PROBLEM. Compare the constant $m_0(d)$ of the proof of Theorem (4.8) of [JiLb] with $m_0(d)$ given by the proof of Theorem (4.22) of [Sch].

Let us only remark that an estimate of $m_0(d)$ of (4.22) is connected with the theory of Lehmer–Lucas number theory. Although the problem of effective estimate of $m_0(d)$ is of interest, it is a number theory problem and for a derivation of the set $\mathrm{HPer}([A])$, for a given A it is enough to use weaker estimate based on the inequality (4.9). We will discuss it in the next section.

4.2. Computation of the set $\mathrm{HPer}(f)$**.** In this subsection we would like discuss the problems: How for a given compact nilmanifold $X = G/\Gamma$ of dimension d (or a completely solvable solvmanifold X) verify whether a $d \times d$ integral matrix is the linearization of a map $f\colon X \to X$. Next how to derive the set $\mathrm{HPer}(f) = \mathrm{HPer}([A])$ such a matrix A.

We begin with well-known fact

(4.25) PROPOSITION. *For any monic polynomial of degree d with integral coefficients $\lambda^d + a_{d_1}\lambda^{d-1} + \ldots + a_1\lambda^1 + a_0$, $a_i \in \mathbb{Z}$, there exists integral $d \times d$ matrix A such that*

$$\lambda^d + a_{d-1}\lambda^{d-1} + \ldots + a_1\lambda^1 + a_0 = \chi_A(\lambda).$$

PROOF. It is enough to check that for the $d \times d$ matrix

$$A := \begin{bmatrix} 0 & 1 & 0 & \ldots & 0 & 0 \\ 0 & 0 & 1 & \ldots & 0 & 0 \\ \vdots & \vdots & \vdots & \ddots & \vdots & \vdots \\ 0 & 0 & 0 & \ldots & 0 & 1 \\ -a_0 & -a_2 & -a_2 & \ldots & -a_{d-2} & -a_{d-1} \end{bmatrix}$$

we have $\chi_A(\lambda) = \lambda^n + a_{d-1}\lambda^{d-1} + \ldots + a_1\lambda + a_0$. □

The Proposition (4.13) shows that the dimension of characteristic polynomial is the only condition on possible $\mathrm{HPer}(f)$ if f is a self-map of torus.

If $f\colon X \to X$ is a self-map of a compact nilmanifold $X = G/\Gamma$ then there is an algebraic condition on the form of A (cf. Definition (3,7)). Indeed for the decreasing filtration of $X = G/\Gamma$ by sub-nilmanifolds:

$$X_0 = X \supset X_1 \supset \cdots \supset X_{k-1} \supset X_k = *,$$
$$X_i := G_i/\Gamma_i, \quad 0 \le i \le k,$$

of the length k, with $\dim X_i = s_i$. Let $d_i = s_{i-1} - s_i$. If X is an NR-solvmanifold then A_0 is the linearization of the base map $f_0\colon \mathbb{T}^{d_0} \to \mathbb{T}^{d_0}$ (cf. Definition (3.16)). By definition we have

(4.26) $$A = A_f = \bigoplus_{i=0}^{k} A_i, \quad \text{where } A_i \text{ is } d_i \times d_i \text{ matrix.}$$

A direct consequence of (4.26) and the proof of Theorem (4.4) is the following

(4.27) COROLLARY. *The constants* $P(d)$, $Q(d)$ *of the statement of Theorem* (4.4) *depend only on* $\widetilde{d}$, *where* $\widetilde{d} := \max_{0 \le i \le k} d_i$.

We have also another consequence of (4.26) (cf. [JeMr2] for the nilmanifold case).

(4.28) PROPOSITION. *Let* $f: X \to X$ *be a map of a compact* NR*-solvmanifold* X *of dimension* d *and* $A = \bigoplus_{i=0}^{k} A_i$ *the linearization matrix of* f, *where for* $0 \le i \le k$, A_i *is the linearization matrix of* $f_i: \mathbb{T}^{d_i} \to \mathbb{T}^{d_i}$, *induced by* f *as in Definition* (3.16). *Then*

$$T_A = \bigcap_{i=0}^{k} T_{A_i}$$

and

$$T_A \cap \left(\bigcup_{i=0}^{k} \mathrm{HPer}(f_i) \right) = T_A \cap \left(\bigcup_{i=0}^{k} \mathrm{HPer}(A_i) \right) \subset \mathrm{HPer}(f).$$

PROOF. By the definition of $m \in T_A$ if and only if $\det(\mathrm{I} - A^m) = \chi_{A^m}(1) \neq 0$. But

$$\chi_A(1) = \prod_{i=0}^{k} \chi_{A_i}(1)$$

which proves the first equality.

To prove the second formula, first note that $|\chi_{A^n}(1)|$ divides $|\chi_{A^m}(1)|$ if $n|m$ (provided $\chi_{A^n}(1) \neq 0$) for every integral matrix A (cf. Proposition (4.11) or equality (4.23)). Consequently, by Theorem (4.5) it follows that $m \in \mathrm{HPer}(f)$ if $m \in T_A$ and there exists $0 \le i_0 \le k$ such that

$$|\chi_{A_{i_0}^m}(1)| > |\chi_{A_{i_0}^{m/p}}(1)|$$

for every prime $p|m$, since

$$|\chi_{A_j^m}(1)| \ge |\chi_{A_j^{m/p}}(1)|$$

for the remaining i. This shows the statement. □

A property of $\mathrm{HPer}(f)$, derived in a similar way, says that

(4.29) PROPOSITION. *Suppose* $m_1, m_2 \in \mathrm{HPer}(f)$ *and their least common multiple* $m \in T_A$. *Then* $m \in \mathrm{HPer}(f)$.

PROOF. See [JiLb, Corollary 3.6] for a proof of it. □

Note that the formula (4.29) gives a necessary condition on form the linearization matrix A of a map of a given NR-solvmanifold X. Opposite to the torus case this condition is not sufficient in general i.e. not every direct sum $A = \bigoplus_i A_i$ of integral $d_i \times d_i$ matrices is the linearization of a self-map of a nilmanifold $X = G/\Gamma$ (see the proof of Theorem (5.2), Lemma (5.5)) for examples). The additional restriction is a consequence of the fact that A is derived from a homomorphism of the Lie algebra $\mathcal{G}$ of G (cf. Corollary (3.21)).

(4.30) REMARK. Since the set $\mathrm{HPer}(f)$ is determined by a purely algebraic procedure in terms of the linearization matrix, it was natural to make an effort implementing this scheme. In [KoMr] is presented a computer program written as a notebook of "Mathematica", *deriving* the set $\mathrm{HPer}([A])$ for a given integral matrix A. It is based on the Theorem (4.5) and inequality (4.9), since the dependence of $m_0(d) = Q(d)$ "only on" d is not necessary then. It uses also the all mentioned above information e.g. Corollary (4.27), Proposition (4.28). As the first step it checks whether $\sigma(A) \cap \{z = 1\}$ consists of roots from unity only. Under this assumption one can use a modification (simplification) of th inequality (4.9), which drastically cuts the interval of searching $m \in T_A \setminus \mathrm{HPer}([A])$ ([KoMr]).

Summing up our discussion we can formulate a "metatheorem" which says that the homotopy dynamics of every map of any NR-solvmanifold is described by the homotopy dynamics of a map of the torus.

(4.31) THEOREM. *Let $f: X \to X$ be a map of a compact NR-solvmanifold of dimension d. Then there exists a self-map $[A]: \mathbb{T}^d \to \mathbb{T}^d$ of the torus induced by the linearization of f such that that* $\mathrm{HPer}([A]) = \mathrm{HPer}(f)$. *The set* $\mathrm{HPer}([A])$ *can be derived by the procedure described above.*

5. Lower dimensions – a complete description

It is natural to give a complete list of all sets of homotopy minimal periods of maps of a given NR-solvmanifold X in the case if the dimension d of X is small.

Case 1. A list for the case $d = 1$ is given in Theorem (1.17) and can be derived from Theorem (4.4) by an easy computation.

Case 2. For the case $d = 2$ we have only one, up to isomorphism, compact NR-solvmanifold, namely the torus $\mathbb{T}^2$.

As we already said the main theorem of [ABLSS1], [ABLSS2] contains such a description. Here we present it in the terms used by us.

(5.1) THEOREM. *Let $f: \mathbb{T}^2 \to \mathbb{T}^2$ be a map of the torus, $A \in \mathcal{M}_{2\times 2}(\mathbb{Z})$ the linearization of f, and $\chi_A(\lambda) = \lambda^2 - a\lambda + b$ be its characteristic polynomial. There are three types for the minimal homotopy periods of f:*

(E) $\mathrm{HPer}(f) = \emptyset$ *if and only if* $-a + b + 1 = 0$.

(F) $\mathrm{HPer}(f)$ *is nonempty and finite for* 6 *cases corresponding to one of the six pairs* (a, b): $(0, 0)$, $(-1, 0)$, $(-2, 1)$, $(0, 1)$, $(-1, 1)$, $(1, 1)$.

We have $\mathrm{HPer}(f) \subset \{1,2,3\}$ *then. Moreover, the sets* T_A *and* $\mathrm{HPer}(f)$ *are the following:*

(a,b)	T_A	$\mathrm{HPer}(f)$
$(0,0)$	$\mathbb{N}$	$\{1\}$
$(0,1)$	$\mathbb{N}\setminus 4\mathbb{N}$	$\{1,2\}$
$(-1,0)$	$\mathbb{N}\setminus 2\mathbb{N}$	$\{1\}$
$(-1,1)$	$\mathbb{N}\setminus 3\mathbb{N}$	$\{1\}$
$(-2,1)$	$\mathbb{N}\setminus 2\mathbb{N}$	$\{1\}$
$(1,1)$	$\mathbb{N}\setminus 6\mathbb{N}$	$\{1,2,3\}$

Cases of Type (F)

(G) $\mathrm{HPer}(f)$ *is infinite for the remaining* a, *and* b. *Furthermore,* $\mathrm{HPer}(f)$ *is equal to* $\mathbb{N}$ *for all pairs* $(a,b) \in \mathbb{Z}^2$ *with except the following special cases listed below. We say that a pair* $(a,b) \in \mathbb{Z}^2$ *satisfies condition*
(G1) *if* $a \neq 0$ *and* $a+b+1=0$,
(G2) *if* $a+b=0$,
(G3) *if* $a+b+2=0$ *respectively,*
and (a,b) *is not one of the pairs of case* (E) *and* (F).

We have the following table of special cases:

(a,b)	T_A	$\mathrm{HPer}(f)$
$(-2,2)$	$\mathbb{N}$	$\mathbb{N}\setminus\{2,3\}$
$(1,2)$	$\mathbb{N}$	$\mathbb{N}\setminus\{3\}$
$(0,2)$	$\mathbb{N}$	$\mathbb{N}\setminus\{4\}$
(a,b) : (a,b) *satisfies* (G1)	$\mathbb{N}\setminus 2\mathbb{N}$	$\mathbb{N}\setminus 2\mathbb{N}$
(a,b) : (a,b) *satisfies* (G2)	$\mathbb{N}$	$\mathbb{N}\setminus\{2\}$
(a,b) : (a,b) *satisfies* (G3)	$\mathbb{N}$	$\mathbb{N}\setminus\{2\}$

Case 3. In the paper [JiLb] Jiang and Llibre gave such a list for maps of $M = \mathbb{T}^3$ including a separate table for homeomorphisms. As we already emphasized (Theorem (4.31)) a discussion of this case is the most complicated from the algebraic point of view, because there is not a condition on the linearization matrix A then. Consequently for the torus family of the sets of homotopy minimal periods is the largest possible (see Theorem (4.31)).

The aim of the work [JeMr2] was to give such a list for a three dimensional nilmanifold not homeomorphic to the torus. The corresponding theorem says the following.

(5.2) THEOREM. *Let $f: X \to X$ be a map of three-dimensional compact nilmanifold X not diffeomorphic to $\mathbb{T}^3$. Let $A = A_1 \oplus \bar{A} \in \mathcal{M}_{3\times 3}(\mathbb{Z})$ be the matrix induced by the fibre map $f = (f_1, \bar{f})$ and $\chi_A(\lambda) = \chi_{A_1}(\lambda) \cdot \chi_{\bar{A}}(\lambda) = (\lambda - c)(\lambda^2 - a\lambda + b)$ be its characteristic polynomial. Then $c = b$ and there are three types for the minimal homotopy periods of f:*

(E) $\mathrm{HPer}(f) = \emptyset$ *if and only if or* $c = 1$ *or* $-a + c + 1 = 0$.

(F) $\mathrm{HPer}(f)$ *is nonempty and finite only for 2 cases corresponding to* $c = 0$ *combined with one of the two pairs* (a, b): $(0,0)$ *and* $(-1,0)$. *We have* $\mathrm{HPer}(f) = \{1\}$ *then. Moreover, the sets* T_A *and* $\mathrm{HPer}(f)$ *are the following:*

(c, a, ab)	T_A	$\mathrm{HPer}(f)$
$(0,0,0)$	$\mathbb{N}$	$\{1\}$
$(0,-1,0)$	$\mathbb{N} \setminus \mathbb{N}$	$\{1\}$

(G) $\mathrm{HPer}(f)$ *is infinite for the remaining* $(c, a, b = d)$.

Furthermore, $\mathrm{HPer}(f)$ *is equal to* $\mathbb{N}$ *for all triples* $(c, a, b = c) \in \mathbb{Z}^3$ *with except the following special cases listed below.*

(c, a, b)	T_A	$\mathrm{HPer}(f)$
$a + c + 1 = 0$ *with* $a \neq 0$ *and* $d \notin \{-1, -1, 0, 1\}$	$\mathbb{N} \setminus 2\mathbb{N}$	$\mathbb{N} \setminus 2\mathbb{N}$
$(0, -2, 0)$	$\mathbb{N}$	$\mathbb{N} \setminus \{2\}$
$(-1, 1, -1)$	$\mathbb{N} \setminus 2\mathbb{N}$	$\mathbb{N} \setminus 2\mathbb{N}$
$(-1, -1, -1)$	$\mathbb{N} \setminus 2\mathbb{N}$	$\mathbb{N} \setminus 2\mathbb{N}$
$(-2, 1, -2)$	$\mathbb{N} \setminus 2\mathbb{N}$	$\mathbb{N} \setminus 2\mathbb{N}$
$(-2, 0, -2)$	$\mathbb{N}$	$\mathbb{N} \setminus \{2\}$
$(-2, 2, -2)$	$\mathbb{N}$	$\mathbb{N} \setminus \{2\}$

Special cases of type (G)

Moreover, for every pair subset $\mathcal{S}_1 \subset \mathcal{S}_2 \subset \mathbb{N}$, *appearing as* $\mathrm{HPer}(f)$ *and* T_A *listed above there exists a map* $f: X \to X$ *such that* $\mathrm{HPer}(f) = \mathcal{S}_1$ *and* $T_A = \mathcal{S}_2$.

PROOF. A proof is based of a classification of all homomorphisms of the nilpontent group $\Gamma_{1,1,r}$ (cf. Example (2.4)). It is remarkable that a condition on an integer 3×3 matrix A to be a linearization of a map does not depend on r, and consequently relies upon a condition on a matrix to be homomorphism of the Heisenberg algebra (cf. Example (2.4)). Note also that the algebraic condition on linearization for $M \not\cong \mathbb{T}^3$ mentioned after Proposition (4.28) says $\chi_A(\lambda) = \chi_{A_1}(\lambda)\chi_{\overline{A}}(\lambda) = (\lambda - c)(\lambda^2 - a\lambda + b)$ here. But additionally the topology (i.e. that A is a homomorphism of the corresponding Lie algebra) yields that now $c = \deg f_1 = \deg \overline{f} = \det \overline{A} = b$, where f_1 is the fiber map, and $\overline{f}$ is the base map for a fiber map f of corresponding Mostow, Fadell–Husseini fibration associated to $X = G/\Gamma_{1,1,r}$ (cf. [JeMr2]). □

As for the torus case (cf. [JiLb]) we specify the classification for homeomorphisms.

(5.3) THEOREM. *Let $f: X \to X$ be a homeomorphism of three-dimensional compact nilmanifold X not diffeomorphic to $\mathbb{T}^3$. Let $A = A_1 \oplus \overline{A} \in \mathcal{M}_{3\times 3}(\mathbb{Z})$ be the linearization matrix and $\chi_A(\lambda) = (\lambda - c)(\lambda^2 - a\lambda + b)$ its characteristic polynomial. Then $c = b = \pm 1$ and consequently* $\mathrm{HPer}(f) = \emptyset$ *if and only if $c = 1$ (i.e. if f preserves the orientation), or $c = -1$ and $a = 0$. In particular* $\mathrm{HPer}(f) = \emptyset$ *for every preserving orientation homeomorphism. For $c = -1$ (i.e. if f reverse the orientation) and the remaining a we have* $\mathrm{HPer}(f) = \mathbb{N}$ *with the only two exceptions being when $a = 1$ or $a = -1$. For these special cases* $T_A = \mathrm{HPer}(f) = \mathbb{N} \setminus 2\mathbb{N}$.

PROOF. The statement follows from Theorem (5.2) and the fact that $c = \pm 1$ (cf. [JeMr2]). □

We are left with a discussion for these 3-dimensional NR-solvmanifold which are not the nilmanifolds. As a first step one can show that there is only one, up to isomorphism, three dimensional solvable, but not nilpotent, Lie algebra $\mathcal{G}$ for which the corresponding unique simple-connected Lie group G admits a lattice and the quotient is an NR-solvmanifold. It is the following completely solvable Lie algebra $\mathcal{G}$ given by the generators $\mathcal{X}$, $\mathcal{Y}$, $\mathcal{Z}$ and commutators

$$[\mathcal{X}, \mathcal{Y}] = \mathcal{Y}, \quad [\mathcal{X}, \mathcal{Z}] = -\mathcal{Z}, \quad [\mathcal{Y}, \mathcal{Z}] = 0. \tag{5.4}$$

A direct computation shows (cf. [JeKdMr]).

(5.5) LEMMA. *Let $A: \mathcal{G} \to \mathcal{G}$ be any endomorphism of completely solvable Lie algebra (5.4). Then it has the following form with respect to the basis $\{\mathcal{X}, \mathcal{Y}, \mathcal{Z}\}$*

$$A = \begin{bmatrix} a & 0 & 0 \\ b & r & s \\ c & u & v \end{bmatrix},$$

where the coefficients satisfy the following conditions:

(5.5.1) *either* $r = v = s = u = 0$ *and* $a \in \mathbb{Z}$ *is an arbitrary integer,*
(5.5.2) *or there exists a coefficient* r, u, s, v *different from* 0 *and* $a \in \{-1, 1\}$.

Moreover, we have:

(5.5.3) *if* $a = -1$ *then* $r = v = 0$,
(5.5.4) *if* $a = 1$ *then* $s = u = 0$.

This algebra is the Lie algebra of connected, simple-connected, completely solvable three-dimensional Lie group:

$$G := \mathbb{R} \times_{\kappa(t)} \mathbb{R}^2, \quad \text{where } \kappa(t) := \begin{bmatrix} e^t & 0 \\ 0 & e^{-t} \end{bmatrix}. \tag{5.6}$$

This leads to the correspondent specification of the Theorem (4.4) for maps of three dimensional special NR-solvmanifolds (cf. [JeKdMr]).

(5.7) THEOREM. *Let* $f \colon X \to X$ *be a map of a compact three dimensional special* NR*-solvmanifold, thus a completely solvable solvmanifold, which is not diffeomorphic to a nilmanifold. Let next* $A \in \mathcal{M}_{3\times 3}(\mathbb{Z})$ *be its linearization as in Lemma* (5.5). *Then we have three mutually disjoint cases:*

(E) $\mathrm{HPer}(f) = \emptyset$ *if and only if* $N(f) = 0$ *if and only if* ($a = 1$ *or* ($a = -1$ *andsu* $= 1$)).
(G) $\mathrm{HPer}(f) = \mathbb{N}$ *if and only if* $a \neq \{-2, -1, 0 + 1\}$ *and* $r = s = u = v = 0$,
$\mathrm{HPer}(f) = \mathbb{N} \setminus \{2\}$ *if and only if* $a = -2$, $r = s = u = v = 0$,
$\mathrm{HPer}(f) = \mathbb{N} \setminus 2\mathbb{N}$ *if and only if* $a = -1$, $|su| \geq 2$ *and* $r = v = 0$.
(F) $\mathrm{HPer}(f) = \{1\}$ *in the remaining cases.*

(5.8) COROLLARY. *For any homeomorphism* $f \colon X \to X$ *of a compact special three dimensional completely solvable solvmanifold which is not diffeomorphic to a nilmanifold* $\mathrm{HPer}(f)$ *is either empty or consists of the single number* 1.

At the end of this section we give an example of a countable family of not isomorphic compact completely solvable solvmanifolds, each of them is the quotient of the group G defined in (5.6).

(5.9) EXAMPLE. First for $a + a^{-1} = n$ and $2 < n \in \mathbb{N}$ we define a family of simply connected solvable Lie group by $G_n := \mathbb{R} \times_{\kappa_n} \mathbb{R}^2$, where

$$\kappa_n(t) = \begin{bmatrix} \dfrac{a^{t+1} - a^{-t-1}}{a - a^{-1}} & \dfrac{a^{-t} - a^t}{a - a^{-1}} \\[2ex] \dfrac{a^t - a^{-t}}{a - a^{-1}} & \dfrac{a^{1-t} - a^{t-1}}{a - a^{-1}} \end{bmatrix}$$

is a family of 1-parameter subgroups of $SL_2(\mathbb{R})$. Each group G_n is isomorphic to G_3, or equivalently to the group defined in (5.6) (cf. [VGS]). It is easy to see that each of them contains a lattice of the form $\Gamma_n := \mathbb{Z} \times_{\kappa_n(1)} \mathbb{Z}^2$, where

$$\kappa_n(1) = \begin{bmatrix} n & -1 \\ 1 & 0 \end{bmatrix}.$$

Thus we obtain a family of solvmanifolds $X_n := G_n/\Gamma_n$. They are not homeomorphic, since the groups $\Gamma_n = \pi_1(X_n)$ are pairwise non-isomorphic. Indeed, representation of Γ_n on its commutator subgroup factors through $\mathbb{Z}$. If we consider this representation tensored by $\mathbb{Q}$, then it is generated by the matrix $\kappa_n(1)$. For the indices $n \neq m$ these representations are not equivalent because the traces of its generators are different.

One should remark that case of compact solvmanifold which are not NR-solvmanifolds has been also discussed. First of all we should mention the work [Ll] of Llibre where he gave a description of the possible sets $\mathrm{HPer}(f)$ for maps f of the Klein bottle K, i.e. the nonabelian solvmanifold of the dimension 2. (Note that the Klein bottle is not an NR-solvmanifold, it is also not orientable and not special). It is based on a never published work of B. Halpern "Periodic points on the Klein bottle" (preprint 1979), however the computation of the formula for the Nielsen number for a map of the Klein bottle had been in detail shown in the paper [DHT].

(5.10) THEOREM. *Let $f: K \to K$ be a continuous map of the Klein bottle. There are three types for the minimal homotopy periods of f:*

(E) $\mathrm{HPer}(f) = \emptyset$,

(F) $\mathrm{HPer}(f) \neq \emptyset$ *and is finite and* $\mathrm{HPer}(f) = \{1\}$ *then,*

(G) $\mathrm{HPer}(f) = \mathbb{N}$ *or* $\mathrm{HPer}(f) = \mathbb{N} \setminus 2\mathbb{N}$ *for the remaining maps, with except one special case when* $\mathrm{HPer}(f) = \mathbb{N} \setminus \{2\}$.

6. Applications and final discussion

6.1. Šarkovskii type theorems. We first remark that Theorem (4.4) gives us to state the existence of infinitely many homotopy minimal period, thus infinitely many periodic provided that for a map of NR-solvmanifold we know that there exists a sufficiently large homotopy minimal period.

For a given d let $\ell(d)$ is equal to the least common multiple of all $\{k \in \mathbb{N} : \phi(k) \leq d\}$, where ϕ is the Euler function.

(6.1) THEOREM. *Let $f: X \to X$ be a map of compact NR-solvmanifold of dimension $d \geq 3$ and $\widetilde{d} = \max_i d_i$ the constant defined in Corollary* (4.27). *If* *there exists a homotopy minimal period* $m \nmid \ell(\widetilde{d})$, $m < \ell(\widetilde{d})$, *then the set* $\mathrm{HPer}(f)$ *is infinite.*

PROOF. Suppose that $\mathrm{HPer}(f) \neq \emptyset$. By Theorem (4.4) if $\mathrm{HPer}(f)$ is finite then $\chi_A(\lambda)$ decomposes as the product of a power of λ and cyclotomic polynomials each of them is of degree $l_i = \phi(k_i) \mid d_i \mid \widetilde{d}$ for $0 \leq i \leq k$. Then for every $\lambda \in \sigma(A) = \bigcup_i \sigma(A_i)$ we have $\lambda^{\ell(\widetilde{d})} = 1$. From it follows that the sequence $\{N(f^m)\}$ is ℓ-periodic. Then $m \notin \mathrm{HPer}(f)$ if and only if $m \nmid \ell(\widetilde{d})$, as in the proof of Theorem (4.4), Type (F). Consequently $\mathrm{HPer}(f)$ is a subset of the set of all proper divisors of $\ell(\widetilde{d})$. □

Note that a map of $\mathbb{T}^3$ (thus also any three dimensional special solvmanifold X) Theorem (6.1) guarantees that $\mathrm{HPer}(f)$ is infinite provided $m \geq 4$, belongs to $\mathrm{HPer}(f)$, since $\ell(3) = 6$. In this case this condition is sharp. Indeed reading the table of all possible sets of homotopy minimal periods of maps of the type (F) of $\mathbb{T}^3$ given in [JiLb] we see that if $4 \leq m \in \mathrm{HPer}(f)$ then $\mathrm{HPer}(f)$ is infinite. On the other hand for a three dimensional nilmanifold and completely solvable special solvmanifold the existence of $m \geq 2$ belonging to $\mathrm{HPer}(f)$ implies that $\mathrm{HPer}(f)$ is infinite. It is a consequence of the fact that for the torus any polynomial is the characteristic polynomial of linearization of a map (Proposition (4.25)) which is not true for maps of the other classes of discussed manifolds.

In the case of non-abelian 3-dimensional NR-solvmanifolds we can say much more. A detailed description of all the possible sets of homotopy minimal periods appearing for maps of a given class of 3-dimensional NR-solvmanifolds led to the following theorems ([JeMr2], [JeKdMr]) which are proved by an analysis of the table of sets of homotopy minimal periods. (cf. Theorems (5.2), (5.3) for non-abelian nilmanifolds, (5.7), Corollary (5.8) for special not nilpotent NR-solvmanifolds).

(6.2) PROPOSITION. *If a self map of a 3-nilmanifold different than 3-torus is such that* $3 \in \mathrm{HPer}(f)$ *then* $\mathbb{N} \setminus 2\mathbb{N} \subset \mathrm{HPer}(f) \subset \mathrm{Per}\,(f)$. *If* $2 \in \mathrm{HPer}(f)$ *then* $\mathbb{N} = \mathrm{HPer}(f) = \mathrm{Per}\,(f)$. *In particular, the first assumption is satisfied if* $L(f^3) \neq L(f)$ *and the second if* $L(f^2) \neq L(f)$.

(6.3) COROLLARY. *Let* $f\colon X \to X$ *be a homeomorphism, or more general a homotopy equivalence, of a compact three dimensional nilmanifold* X *not diffeomorphic to the torus. If* $\mathrm{HPer}(f) \neq \emptyset$ *then* $\mathbb{N} \setminus 2\mathbb{N} \subset \mathrm{HPer}(f)$. *Moreover, if* $2 \in \mathrm{HPer}(f)$, *e.g. if* $L(f^2) \neq L(f)$, *(or if any* $2k \in \mathrm{HPer}(f)$*) then* $\mathrm{HPer}(f) = \mathbb{N}$.□

(6.4) PROPOSITION. *For a map* $f\colon X \to X$ *of a special* NR*-solvmanifold of dimension* 3 *not diffeomorphic to a nilmanifold we have Šarkovskii type implications:*

(6.4.1) $2 \in \mathrm{HPer}(f)$ *implies* $\mathrm{HPer}(f) = \mathbb{N}$,

(6.4.2) *If* $\mathrm{HPer}(f)$ *contains an even number then* $\mathbb{N}\setminus\{2\}\subset\mathrm{HPer}(f)$,
(6.4.3) *If* $\mathrm{HPer}(f)$ *contains at least two numbers then* $\mathbb{N}\setminus 2\mathbb{N}\subset\mathrm{HPer}(f)$.

We called all the above facts "the Šarkovskii type theorems", because they state that a given small homotopy period implies all, or all with except one, homotopy periods. Anyway we must emphasize that the nature of this phenomenon is completely different than that of the Šarkovskii theorem (see [Sz]). It is worth of pointing out that to check the supposition of the above theorem we have to find out a periodic point which is not removable along deformation. But as a consequence we get a statement about the set $\mathrm{HPer}(f)$, i.e. a stronger affirmation then in the Šarkovskii theorem.

6.2. Projective spaces and periodic points originated by a symmetry. The theory of homotopy minimal periods is best developed for maps of the solvmanifolds. It is a consequence of the properties of the Nielsen numbers presented in the Section 3. The additional point is the fact that the fundamental group is infinite and the rate of growth of the sequence $\{N(f^m)\}$ can be exponential then. Relatively less is known for maps of manifolds with a finite fundamental group. Anyway for the case of $X=\mathbb{RP}^d$ the real projective space there is a complete description of all sets of homotopy minimal periods given by J. Jezierski in [Je3]. It is based on the following computation of the Nielsen numbers.

(6.5) THEOREM. *Let* $f\colon\mathbb{RP}^d\to\mathbb{RP}^d$, $d\geq 2$, *be a map and* $f_\#\colon\mathbb{Z}_2\to\mathbb{Z}_2$ *the map induced by* f *on the fundamental group. Then*

$$N(f)=\begin{cases} 0 & \textit{if } f\sim\mathrm{id} \textit{ and } d \textit{ is odd},\\ 1 & \textit{if } f_\#=0 \textit{ or } f\sim\mathrm{id} \textit{ and } d \textit{ is even},\\ 2 & \textit{if } f_\#=\mathrm{id} \textit{ and } f \textit{ is not homotopic to } \mathrm{id}.\end{cases}$$

A consequence of the above theorem we have the following one (see [Je3]).

(6.6) THEOREM. *Let* $f\colon\mathbb{RP}^d\to\mathbb{RP}^d$ *be a self-map of a real projective space,* $d\geq 2$. *Then the following formula holds (with one exception)*

$$\mathrm{HPer}(f)=\begin{cases} \emptyset & \textit{if } N(f)=0,\\ \{1\} & \textit{if } N(f)=1,\\ \{1,2,2^2,2^3,\dots\} & \textit{if } N(f)=2.\end{cases}$$

The special case is for d *odd and* $\deg f=-1$ *when* $N(f)=2$ *but* $\mathrm{HPer}(f)=\{1\}$.

We must add that a similar description of the set of homotopy minimal periods seems be true for self-maps of the lens spaces and is just being studied now. But, even so, an analyse in this direction led to an estimate of the number of periodic points of a map of sphere which commutes with a free action of a group.

From Theorem (1.17) it follows that for a circle map if $|\deg f| > 1$ then $\mathrm{HPer}(f) = \mathbb{N}$ (or exceptionally $\mathbb{N} \setminus \{2\}$). Moreover,

$$\#P^m(f) \geq N(f^m) = |1 - \deg f^m|,$$

because S^1 is the torus see [Ji4] (cf. Theorems (3.22), (3.24) and (3.32)). Furthermore, for a prime p we have $\#P_{p^\alpha}(f) \geq NP_{p^\alpha}(f) = p^\alpha - p^{\alpha-1}$ as follows from the Möbius formula. The same is not true for maps of S^d, $d \geq 2$, as follows from the Shub example (cf. Example (1.12)).

One can ask what condition on $f\colon S^d \to S^d$, together with the necessary $|\deg f| > 1$ implies infinitely many periodic points. In [JeMr3] it is shown that every continuous map $f\colon S^d \to S^d$, $n \geq 1$, of $\deg f = r$, where $|r| \geq 2$, has similar to the above properties provided it commutes with a free homeomorphism $g\colon S^d \to S^d$ of a finite order. The sequence $\{\#P^m(f)\}$ is unbounded and Per (f) is infinite then.

(6.7) DEFINITION. Let X be a smooth manifold and $g\colon X \to X$ a homeomorphism of the finite order m. We say that g is free if for every $x \in X$ $g^k(x) = x$ and $1 \leq i \leq k$, $g^i(x) = x$ implies $k = m$. Equivalently we say that an action of the cyclic group $\{g\} \simeq \mathbb{Z}_k$ on X is given by $(i, x) \mapsto g^i(x)$. If g is free then this action is called a *free action*.

(6.8) DEFINITION. Let X be a smooth manifold with an action of a cyclic group $\mathbb{Z}_k$ defined by homeomorphisms $g\colon X \to X$. A map $f\colon X \to X$ is $\mathbb{Z}_k$-equivariant if $f\alpha = \alpha f$, for all $\alpha \in \mathbb{Z}_k$. Note that f is $\mathbb{Z}_k$-equivariant if it commutes with the generator of action i.e. $f(gx) = gf(x)$. A homotopy $H\colon X \times [0,1] \to X$ is equivariant i.e. $z \in X$, $t \in [0,1]$, $\alpha \in \mathbb{Z}_k$ implies $H(\alpha x, t) = \alpha H(x,t)$.

Suppose that we have free action of $\mathbb{Z}_k$ on the sphere S^n, $n \geq 2$, i.e. given a free homeomorphism $g\colon S^n \to S^n$ of order k. To formulate our result we need new a notation.

(6.9) DEFINITION. Let $k = p_1^{a_1} \dots p_s^{a_s}$, $a_i > 0$, be the decomposition of m into prime powers. Let m be a natural number. We represent m as $m = p_1^{b_1} \dots p_s^{b_s} p_{s+1}^{a_{s+1}} \dots p_r^{a_r}$, where $p_1, \dots, p_r$ are all (distinct) primes satisfying $p_i | k$ if and only if $i \leq s$, $b_i \geq 0$. Finally we put $m' := p_1^{b_1} \dots p_s^{b_s}$.

We are in position to formulate the main result of this section.

(6.10) THEOREM. *Let $g\colon S^d \to S^d$ $d \geq 1$ be a free homeomorphism of order $k > 1$, and $f\colon S^d \to S^d$ a map commuting with g. Suppose that* $\deg f \neq -1, 0, 1$. *Then for every $m \in \mathbb{N}$ we have $\#\mathrm{Fix}\, f^{mk} \geq k^2 m'$, where m' is defined above. In particular, for $m = k^s$ we have $\#\mathrm{Fix}\, f^{k^{s+1}} \geq k^{s+2}$.*

To show this theorem we employed (see [JeMr3]) a fine modification of the full Nielsen–Jiang periodic number $NF_m(f)$ which estimates $\#\mathrm{Fix}\,(f^m)$ (cf. (3.25)).

It can be applied to the map f/G induced by f on the quotient space X/G, which is a generalized lens space then. In this way we got an estimate of $\#\mathrm{Fix}\,((f/G)^m)$ (this estimate is not true if a map of X/G is not of the form f/G). Finally, by a geometrical reason these fixed points of $(f/G)^m$ give fixed points of f^{mk}.

As a consequence we get:

(6.11) COROLLARY. *Under the above assumptions* $\limsup_{l\to\infty} \#\mathrm{Fix}(f^l)/l \geq k$.

For a cyclic group of prime order the method let us also to estimate the number of m-periodic points of f, with m being the minimal period. Fix a prime number $p|k$ and restrict the action to $\mathbb{Z}_p \subset \mathbb{Z}_k$.

(6.12) THEOREM. *Let* $f\colon S^d \to S^d$ *be a continuous map which commutes with a free homeomorphism* g *of* S^d *of prime order* p. *If* $\deg(f) \neq \pm 1$ *then for each* $s \in \mathbb{N}$ *there exist at least* $p-1$ *mutually disjoint orbits of* f *of periodic points each of length* p^s. *Thus*

$$\#P_{p^s}(f) \geq (p-1)p^s.$$

The same is true for any map homotopic to f *by an equivariant homotopy.*

6.3. Relations to the topological entropy. At the end we would like to announce a relation between the of linearization of the map of an NR-solvmanifold (Definition (3.16)) and the topological entropy of this map. Recall, that for a continuous self-map $f\colon X \to X$ of a metric space X we can assign a real number $\mathbf{h}(f) \geq 0$, or ∞, which measure the dynamics of f and is called *topological entropy*. Roughly speaking, if $\mathbf{h}(f) > 0$ then the dynamics of f is rich. For the definition and more details on the entropy see [HaKa].

Let $H^*(f)\colon H^*(X;\mathbb{R}) \to H^*(X;\mathbb{R})$ be the linear map induced by f on the cohomology space

$$H^*(X;\mathbb{R}) := \bigoplus_{i=0}^{d} H^i(X;\mathbb{R}).$$

Recall that if X is a compact smooth manifold then the singular, Čech, simplicial, cellular, or the de Rham cohomology theory are equivalent (cf. [Sp]).

Denote by $\sigma(f)$ the spectrum of the linear map $H^*(f)\colon H^*(X;\mathbb{R}) \to H^*(X;\mathbb{R})$ induced by $f\colon X \to X$. Next, by $\mathrm{sp}\,(f)$ we denote the spectral radius of map $H^*(f)$. The both are homotopy invariants.

In [Sh] Michael Shub conjectured that the topological entropy is estimated by the spectral radius of $H^*(f)$ provided $f\colon X \to X$ is a C^1-map.

$$\log \mathrm{sp}\,(f) \leq \mathbf{h}(f). \tag{6.13}$$

Estimate (6.13) was proved by Yomdin (see [Yo]) if f is a C^∞-map and for a few special cases under the general C^1 assumption (cf. [HaKa], [Man]). For example

Misiurewicz and Przytycki showed that $\mathbf{h}(f) \geq \log|\deg(f)|$ for a C^1-map (see [HaKa]). Note that (6.13) is not true for a C^0-map as follows from Example (1.12), because the entropy of that map is equal to 0.

In 1977 Misiurewicz and Przytycki showed that the estimate (6.13) holds for any continuous map of the torus $\mathbb{T}^d$ (see [MiPr]). A conjecture that same estimate holds for every continuous self-maps if we assume a special topological form of the manifold was posed by A. Katok in [Ka]:

Let X be a manifold with the universal cover homeomorphic to the Euclidean space $\mathbb{R}^d$.

$$(6.14) \qquad \log \operatorname{sp}(f) \leq \mathbf{h}(f) \quad \text{for all } C^0\text{-maps } f\colon X \to X.$$

In [MrP] the estimate (6.14) was confirmed for the following class of manifolds.

(6.15) THEOREM. *Let $f\colon X \to X$ be a continuous self-map of a compact nil-manifold. Then*

$$\log(\operatorname{sp}(f)) \leq \mathbf{h}(f).$$

Let G be a connected Lie group Γ its subgroup. For a homomorphism $\Phi\colon G \to G$ such that $\Phi(\Gamma) \subset \Gamma$ by the factor map we mean the map $\Phi/\Gamma\colon G/\Gamma \to G/\Gamma$, denoted by ϕ, induced on the quotient space. For a linear operator by given $d \times d$ matrix we put

$$\wedge A := \bigoplus_{l=0}^{d} \wedge^l A,$$

where $\wedge^l A$ is the lth exterior power of A. A step in the proof of Theorem (6.15) is the following statement. (cf. [MrP]).

(6.16) THEOREM. *Let $X = G/\Gamma$ be a compact homogeneous space of a connected Lie group G by a discrete uniform subgroup Γ, and $\phi\colon X \to X$ the factor map of preserving Γ endomorphism $\Phi\colon G \to G$. Then*

$$\log \operatorname{sp}(\phi) \leq \log \operatorname{sp}(\wedge D\Phi(e)).$$

Remind that for a factor map $\phi = \Phi/G$ of a compact completely solvable solv-manifold X the matrix $D\Phi(e)$ has the same spectrum as the linearization A_f of f (cf. Proposition (3.20)). Furthermore $\sigma(D\Phi(e)) = \sigma(A_\phi)$ implies $\sigma(\wedge D\Phi(e)) = \sigma(\wedge A_\phi)$. On the other hand every map f of a compact completely solvable solv-manifold X is homotopic to a factor map $\phi_f = \Phi_f/\Gamma$ by the rigidity property (cf. Theorem (2.7)), and from Theorem (6.16) it follows that $\operatorname{sp}(f) \leq \operatorname{sp}(\wedge D\Phi_f(e)) = \sigma(\wedge A_f)$.

Next modifying the argument of an early and unpublished version of the proof of main theorem of [MiPr] we get the following (cf. [MrP])

(6.17) THEOREM. *For every map f of a compact nilmanifold $X = G/\Gamma$ and and associated homomorphism $\Phi_f\colon G \to G$, $\Phi(\Gamma) \subset \Gamma$ such that the factor map ϕ_f is homotopic to f we have*

$$\log \mathrm{sp}\,(\wedge D\Phi_f(e)) \leq \mathbf{h}(f).$$

Theorem (6.15) follows from Theorems (6.16) and (6.17). (cf. [MrP]).

Note that in general Theorems (6.16) and (6.17) give a sharper estimate than Theorem (6.15).

Observe next that $\mathrm{sp}\,(\wedge A) > 1$ if and only if $\mathrm{sp}\,(A) > 1$. Now the equality $\sigma(A) = \sigma(D\Phi(e))$, Theorems (6.16), (6.17) and Theorem (4.4) (case (G)) imply the following relation between the set of homotopy minimal periods and topological entropy.

(6.18) COROLLARY. *Let $f\colon X \to X$ be a map of compact nilmanifold. If* $\mathrm{HPer}(f)$ *is infinite then* $\mathbf{h}(f) > 0$.

Observe that the claim that $\mathbf{h}(f) > 0$ implies a nonempty set of periodic points, or the set of minimal periods $\mathrm{Per}\,(f)$ of f is false in view of the following example. Take $M = \mathbb{T}^2 = \mathbb{R}^2/\mathbb{Z}^2$ and the product map $f = (f_1, f_2)$, $f\colon \mathbb{T}^2 \to \mathbb{T}^2$ where f_1 is of degree > 1 and f_2 has not periodic points. Also conversely, for the map of the torus $f(x,y) = (x, x+y) \mod 1$ we have $\mathrm{Per}\,(f) = \mathbb{N}$ but $\mathbf{h}(f) = 0$.

A direct consequence of Theorem (6.17) and estimates of the topological entropy from above given earlier by S. Katok [KaS], and Krzyżewski (see also D. Ruelle in [Ru], and F. Przytycki in [Prz] for more refined estimates on this subject) is the following fact.

(6.19) THEOREM. *Let $M = G/\Gamma$ be a quotient of a connected Lie group by a uniform lattice Γ and $\phi\colon M \to M$ be the factor map of a preserving Γ endomorphism $\Phi\colon G \to G$. Then $\mathbf{h}(\phi) = \log \mathrm{sp}\,(\wedge D\Phi(e))$. If M is a nilmanifold, then ϕ minimizes the entropy in its homotopy class.*

References

[ABLSS1] L. Alsedà, S. Baldwin, J. Llibre, R. Swanson and W. Szlenk, *Torus maps and Nielsen numbers*, Nielsen Theory and dynamical Systems (Ch. McCord, ed.), Contemp. Math. Series, vol. 152, Amer. Math. Sosc, Providence, 1993, pp. 1–7.

[ABLSS2] ———, *Minimal sets of periods for torus maps via Nielsen numbers*, Pacific J. Math. **169** (1995), 1–32.

[An] D. K. Anosov, *Nielsen numbers of mappings of nilmanifolds*, Uspekhi Mat. Nauk **40** (244) (1985), 133–134. (Russian)

[Au] L. Auslander, *An exposition of the structure of solvmanifolds*, Bull. Amer. Math. Soc. **79** (1973), 227–285.

[BGMY] L. Block, J. Guckenheimer, M. Misiurewicz and L. S. Young, *Periodic points and topological entropy of one-dimensional maps*, Lectures Notes in Math. **819** (1983), 18–24.

[BBPT] R. Brooks, R. Brown, J. Pak and D. Taylor, *The Nielsen number of maps of tori*, Proc. Amer. Math. Soc. **52** (1975), 346–400.

[Br1] R. F. Brown, *The Nielsen number of a fibre map*, Ann. of Math. **85** (1967), 483–493.

[Br2] ______, *The Lefschetz Fixed Point Theorem*, Glenview, New York, 1971.

[DHT] O. Davey, E. Hart and K. Trapp, *Computation of Nielsen numbers for maps of closed surfaces*, Trans. Amer. Math. Soc. **48** (1996), 3245–3266.

[Ef] L. S. Efremova, *Periodic orbits and the degree of a continuous map of a circle*, Differential Integral Equations (Gorkiĭ) **2** (1978), 109–115. (Russian)

[Fa] E. Fadell, *Natural fiber splitting and Nielsen numbers*, Houston J. Math. **2** (1976), 71–84.

[FaHu] E. Fadell and S. Husseini, *On a theorem of Anosov on Nielsen numbers for nilmanifolds*, Nonlinear Functional Analysis and its Applications (S. P. Singh, ed.), Reidel, 1986, pp. 47–53.

[GOV] V. V. Gorbatsevich, A. L. Onishchik and E. B. Vinberg, *Foundations of Lie theory and Lie transformation groups*, reprint of the 1993 translation *Lie groups and Lie algebras* I, Encyclopaedia Math. Sci., vol. 20, Springer, Berlin, 1993; Springer–Verlag, Berlin, 1997.

[Hl] B. Halpern, *Periodic points on tori*, Pacific J. Math. **83** (1979), 117–133.

[Ht] A. Hattori, *Spectral sequence in the de Rham cohomology of fibre bundles*, J. Fac. Sci. Univ. Tokyo Sect. I **8** (1960), 289–331.

[HaKa] B. Hasselblatt and A. Katok, *Introduction to the Modern Theory of Dynamical Systems*, Cambridge University Press, 1995.

[He] Ph. Heath, *Product formulae for Nielsen numbers of fibre maps*, Pacific J. Math. **117** (1985), 267–289.

[HeYu] Ph. Heath and C. Y. You, *Nielsen type numbers for periodic points*, Topology Appl. **43** (1992), 219–236.

[HeKa1] Ph. Heath and E. Keppelmann, *Fibre techniques in Nielsen periodic point theory on nil and solvmanifolds* I, Topology Appl. **76** (1997), 217–247.

[HeKa2] ______, *Fibre techniques in Nielsen periodic point theory on nil and solvmanifolds* II, Topology Appl. **106** (2000), 149–167.

[HeKaIII] ______, *Fibre techniques in Nielsen periodic point theory on solvmanifolds* III, Calculations, Quaest. Math. **25** (2002), 177–208.

[Je1] J. Jezierski, *Cancelling periodic points,*, Math. Ann. **321** (2001), 107–130.

[Je2] ______, *Wecken theorem for periodic points*, Topology **42** (2003), 1101–1124.

[Je3] ______, *Homotopy periodic sets for selmaps of real projective spaces*, notes (2002).

[Je4] ______, *Weak Wecken Theorem for periodic points in dimension* 3, submitted, Fund. Math. (2004).

[Je5] ______, *Wecken's Theorem for periodic points in dimension* 3, submitted (2003).

[JeMr1] J. Jezierski and W. Marzantowicz, *Homotopy minimal periods for nilmanifolds maps*, Math. Z. **239** (2002), 381–414.

[JeMr2] ______, *Homotopy minimal periods for maps of three dimensional nilmanifolds*, Pacific J. Math. **209** (2003), 85–101.

[JeMr3] ______, *A symmetry of sphere map implies its chaos*, submitted (2004).

[JeKdMr] J. Jezierski, J. Kędra and W. Marzantowicz, *Homotopy minimal periods for solvmanifolds maps*, Topology Appl. (to appear).

[Ji1] B. J. Jiang, *Estimation of the Nielsen numbers*, Acta Math. Sinica **14** (1964), 304–312; Chineese Math. Acta **5** (1964), 330–339.

[Ji2] ______, *On the least number of fixed points*, Amer. J. Math. **102** (1980), 749–763.

[Ji3] ______, *Fixed point classes from a differential viewpoint*, Lecture Notes in Math., vol. 886, Springer, 1981, pp. 163–170.

[Ji4] B. Jiang, Contemp. Math., vol. 14, Providence, 1983.

[Ji5] ______, *Estimation of the number of periodic orbits*, Pacific J. Math. **172** (1997), 151–185.

[JiLb] B. Jiang and J. Llibre, *Minimal sets of periods for torus maps*, Discrete Contin. Dynam. Systems **4** (1998), 301–320.

[Ka] A. W. Katok, *Entropy Conjecture, Smooth Dynamical Systems*, Mir Publishing, Moskow, 1977, pp. 182–203. (Russian)

[KaS] ———, *The estimation from above for the topological entropy of a diffeomorphism*, Global Theory of Dynamical Systems, Proc. Internat. Conf., Northwestern Univ., Evanston, Ill., 1979, Lecture Notes in Math., vol. 819, Springer, Berlin, 1980, pp. 258–264.

[KMC] E. C. Keppelmann and C. K. McCord, *The Anosov theorem for exponential solvmanifolds*, Pacific J. Math. **170** (1995), 143–159.

[KoMr] R. Komendarczyk and W. Marzantowicz, *Algorithm for deriving homotopy minimal periods of nilmanifold and solvmanifold map*, Proceeding 3rd National Conference on Nonlinear Analysis, Łódź, Lecture Notes of the Juliusz Schauder Center for Nonlinear Studies, vol. 3, 2002, pp. 109–130.

[Ll] J. Llibre, *A note on the set of periods for Klein bottle maps*, Pacific J. Math. **157** (1993), 87–93.

[MrP] W. Marzantowicz and F. Przytycki, *Entropy conjecture for continuous maps of nilmanifolds*, notes (2004).

[Ml] A. Malcev A class of homogenous spaces, Izv. Akad. Nauk SSSR Ser. Mat. **13** (1949), 9–32. (Russian)

[Man] A. Manning, *Toral automorphisms, topological entropy and the fundamental group. Dynamical systems*, Asterisque **50** (1977), 273–281.

[MC1] C. K. McCord, *Nielsen numbers and Lefschetz numbers on nilmanifolds and solvmanifolds*, Pacific J. Math. **147** (1991), 153–164.

[MC2] C. K. McCord, *Estimating Nielsen numbers on infrasolvmanifolds*, Pacific J. Math. **154** (1992), 345–368.

[MC3] ———, *Lefschetz and Nielsen coincidences numbers on nilmanifolds and solvmanifolds*, Topology Appl. **43** (1992), 249–261.

[MC4] ———, *Lefschetz and Nielsen coincidences numbers on nilmanifolds and solvmanifolds* II, Topology Appl. **75** (1997), 81–92.

[MiPr] M. Misiurewicz and F. Przytycki, *Entropy conjecture for tori*, Bull. Acad. Polon. Sci. Ser. Sci. Math. Astronom. Phys. **25** (1977), 575–578.

[Mo] G. Mostow, *Factor spaces of solvable groups*, Ann. Math. **60** (1954), 1–27.

[No] K. Nomizu, *On the cohomology of compact homogeneous spaces of nilpotent Lie groups*, Ann. of Math. (2) **59** (1954), 531–538.

[Prz] F. Przytycki, *An upper estimation for topological entropy of diffeomorphisms*, Invent. Math. **59** (1980), 205–213.

[Ra] M. S. Raghunathan, *Discrete Subgroups of Lie Groups*, Springer–Verlag, 1972.

[Ru] D. Ruelle, *An inequality for the entropy of differentiable dynamical systems*, Bol. Soc. Brasil. Mat. **9** (1978), 83–88.

[Sa] M. Saito, *Sur certains groupes de Lie resolubles* I, II, Sci. Pap. Coll. Gen. Ed. Univ. Tokyo **7** (1957), 157–168.

[Sch] A. Schinzel, *Primitive divisors of the expression $A^n - B^n$ in algebraic number fields*, Crelle Journal für Mathematik **268/269** (1974), 27–33.

[Sh] M. Shub, *Dynamical systems, filtrations and entropy*, Bull. Amer. Math. Soc. **80** (1974), 27–41.

[Sp] E. H. Spanier, *Algebraic Topology*, McGraw–Hill, 1968.

[Sz] A. N. Šarkovskii, *Coexistence of cycles of a continuous map of the line into itself*, Ukraïn. Mat. Zh. **16** (1968), 61–71.

[VGS] E. B. Vinberg, V. V. Gorbatsevich and O. V. Shvartsman, *Discrete subgroups of Lie groups*, Current Problems in Mathematics. Fundamental Directions, vol. 21, Itogi Nauki i Tekhniki, 1988, pp. 5–120, 215. (Russian)

[Yo] Y. Yomdin, *Volume growth and entropy*, Israel J. Math. **57** (1987), 285–300.

[Yu1] C. Y. You, *The least number of periodic points on tori*, Adv. in Math. (China) **24** (1995), 155–160.

[Yu2] ———, *A note on periodic points on tori*, Beijing Math. **1** (1995), 224–230.

5. PERIODIC POINTS AND BRAID THEORY

TAKASHI MATSUOKA

1. Introduction

This article surveys applications of the braid theory to the study of the periodic orbit structure of iterated homeomorphisms on surfaces. In dynamical systems, it is often the case that topological invariants are used to study qualitative and quantitative properties of the system. The braid is one of such invariants characterizing topological behavior of periodic orbits in the case of surface homeomorphisms.

Let M be a compact surface, and $f: M \to M$ a homeomorphism. Let S be a finite invariant set of f, i.e. a union of finitely many periodic orbits of f. Assume that f is isotopic to the identity map id, and choose and fix an isotopy $\{f_t\}_{0 \le t \le 1}$ deforming id to f. Then, the image of S under f_t move on M and return to the initial position while t varies from 0 to 1. This motion defines a braid in M. The conjugacy class of this braid is called the braid type of S with respect to the homeomorphism f. The assumption that f is injective and isotopic to id, which is necessary to define the braid type, is rather natural for dynamics as, for example, the time one map of the solution to a periodically forced differential equation satisfies this assumption.

The study of braid types in surface dynamics was started in the early 1980's by several people, and has been developed to an extensive area in the theory of low dimensional dynamical systems. There are two main streams in this study.

One is the application of matrix representations of braid groups which are closely related to the theory of generalized Lefschetz number established by Fadell, Husseini, and Fried in the early 1980's. In fact, the trace of the matrix of the braid of S coincides with the abelianization of the generalized Lefschetz number of the restriction of f to $M - S$. The generalized Lefschetz number describes how periodic orbits not belonging to S link with S. Hence, by computing the matrix one can obtain information of the existence and linking behavior of periodic orbits. This provides a remarkable application of the theory of generalized Lefschetz number.

We shall explain the detail in Section 3, and give a proof of the main result in Section 4.

The other is the application of the Nielsen–Thurston classification theory for surface homeomorphisms up to isotopy started by Nielsen in the 40's and finished by Thurston in the 70's. This theory provides a canonical representative φ for any isotopy class $[f]$ of homeomorphisms on M relative to a finite subset S. A usefulness of considering the braid type of S is based on the fact that it determines an isotopy class of homeomorphisms relative to S, and hence a canonical homeomorphism φ which is unique up to topological conjugacy. The canonical representative φ has the simplest dynamical property among the homeomorphisms in the isotopy class. For instance, if φ is "pseudo-Anosov", then every periodic orbit of φ is "unremovable" in the sense that the orbit persists in an arbitrarily given homeomorphism $f\colon M \to M$ isotopic to φ relative to S. (The unremovability of orbits is explained in detail in an excellent survey article [Boy4] by Boyland on the topological methods in surface dynamics.) Thus, if the canonical homeomorphism has a pseudo-Anosov component and hence has a complicated periodic orbit structure, then we can conclude that the given map f must have at least the same amount of complexity on the periodic orbit structure. In Section 6, the results concerned will be collected.

In Section 7, the implications of the existence of a periodic orbit of certain type is examined more closely using forcing relations on braid types. One defines a forcing relation on the braid types by saying that one braid type is larger than a second if whenever a homeomorphism has a periodic orbit of the first type it also has one of the second. These forcing relations are generalizations of the order relation that occurs in the famous Sharkovskii's theorem ([S]) about the periods of periodic orbits for maps of an interval.

The canonical homeomorphism φ determined by the braid type of S may imply other kinds of dynamical complexity of the map f such as

(1.1) positiveness of the topological entropy,

(1.2) the existence of infinitely many periodic orbits with different rotation vectors.

The results on the dynamical complexity from these viewpoints will be explained in Sections 8 and 9. The topological entropy $h(f)$ is a well known measure of the complexity of a dynamical system. The rotation vector of a periodic orbit measures the rate of rotation of an orbit in the surface in terms of a homology class. This is derived from the braid type by disregarding the mutual linking information among the strings of the braid type, and leaving only the information on the global behavior of the strings.

In the final section, we introduce the results on the topological structure of the periodic point set.

We begin by precisely specifying the primary objects of study in this article.

Standing assumption. Unless otherwise noted, M denotes a compact, connected, orientable surface possibly with boundary. Any self-maps $f\colon M \to M$ will be homeomorphisms isotopic to the identity map id.

Although almost papers contributed to this subject made the assumption that f is a homeomorphism, the assumption of surjectivity of f is not necessary, and one can extend the results easily for topological embeddings $f\colon M \to M$ isotopic to id. There are some important related topics we shall not mention. These include applications of knot theory and the study of area-preserving surface homeomorphisms.

2. Braids and braid types

We shall recall the definition of a braid on a surface M, and introduce the notion of braid type for a collection of periodic orbits of f. For general references of the braid theory, see [Bi], [Hans], [Mo].

A braid can be defined in different, but essentially equivalent, ways. We introduce two of them which are appropriate to the application for dynamical systems. Let n be a positive integer. Let $S = \{x_1, \ldots, x_n\}$ be a set of n different points in M.

2.1. Braids as unions of strings. The original definition of a braid was given by E. Artin in 1925 as a "weaving pattern" of n strings between two parallel planes in euclidean 3-space. The definition has been generalized to an arbitrary surface M. We call a subset G of the product $M \times [0, 1]$ a *geometric n-braid* in M if the following conditions hold:

(2.1.1) G is a union of mutually disjoint n embedded arcs $s_1, \ldots, s_n$.

(2.1.2) Each arc s_i joins a point $(x_i, 0) \in S \times \{0\}$ to $(x_{\mu(i)}, 1) \in S \times \{1\}$, where μ is a permutation defined on $\{1, \ldots, n\}$.

(2.1.3) Each arc intersects every t-slice $M \times \{t\}$, $0 \leq t \leq 1$, exactly once.

An arc s_i is called a *string* in G. If it is necessary to designate the set S, we say G is based at S. See Figure 1 as an example of geometric 3-braids in the closed disk D.

We give an equivalence relation on geometric braids: Two geometric braids based at S are said to be *isotopic* IF one can be continuously deformed to the other through geometric braids all of which are based at S. An equivalence class under this isotopy relation is called an *n-braid* in M based at S. The set of n-braids

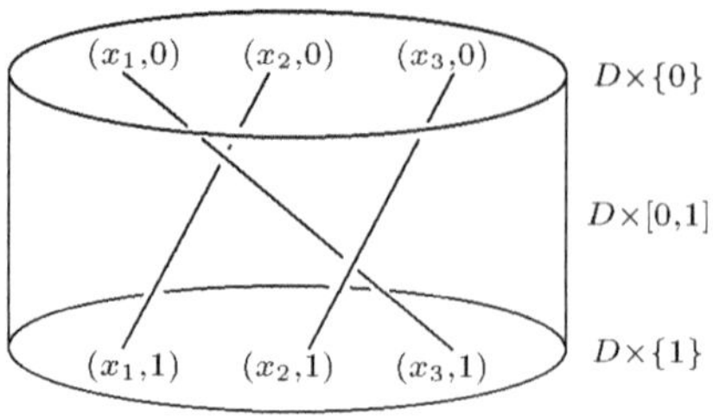

FIGURE 1

based at S forms a group called the braid group, and is denoted by $B_n(M,S)$ or simply by $B_n(M)$.

A geometric braid G is said to be pure (or colored) if the permutation μ corresponding to G is the identity permutation. A braid is said to be *pure* if it is represented by a pure geometric braid. The set $P_n(M,S)$ of pure n-braids becomes a normal subgroup of $B_n(M,S)$ with index $n!$, which is called the pure braid group with n strings.

2.2. Braids as motions of points in M. A geometric braid may be defined as a motion of points on M, more precisely, as a loop in the space of configurations of a set of n points in M. Thus, a braid is considered as an element of the fundamental group of the configuration space.

Let $C_n(M)$ be the set of all subsets S of M with cardinality n, that is,

$$C_n(M) = \{S : S \subset M,\ \sharp S = n\}.$$

This is the set of all possible configurations for an unordered set consisting of n different points in M. This set has a natural topology defined in the following way: Let $F_n(M)$ denote the space $M^n - \Delta$, where Δ is the generalized diagonal of M^n, namely,

$$F_n(M) = \{(x_1, \dots, x_n) \in M^n : x_i \neq x_j \text{ if } i \neq j\}$$

(see [Bi, p. 5], [Hans, p. 91]). $F_n(M)$ is the set of possible configurations for a set of n *ordered* points in M. It has the topology induced from the product topology on M^n. Let Σ_n denote the group of all permutations of the set $\{1, \dots, n\}$. There is a free action of Σ_n on $F_n(M)$ given by $\mu \cdot (x_1, \dots, x_n) = (x_{\mu(1)}, \dots, x_{\mu(n)})$. Let $F_n(M)/\Sigma_n$ be the set of all orbits under this action. This set is given the quotient topology. Note that two ordered configurations are on the same orbit of this action if they coincide after neglecting the order of points. Hence one can think of each orbit as an element of $C_n(M)$. Thus, $C_n(M)$ can be identified with the orbit space $F_n(M)/\Sigma_n$. The projection $F_n(M) \to F_n(M)/\Sigma_n = C_n(M)$ defines an $n!$-fold covering map. The topology on $C_n(M)$ has the property that a map $S\colon [0,1] \to C_n(M)$ is continuous if and only if there exist n paths $x_i(t)$ $(0 \le t \le 1)$ in M with $S(t) = \{x_1(t), \dots, x_n(t)\}$.

The configuration spaces are used to define a notion of braid in M. Let $S(t)$ be a loop in $C_n(M)$. Then, since $S(0) = S(1)$ and $S(t)$ consists of exactly n points for each t, this loop determines a geometric braid G_S in M as the union

$$\bigcup_{0 \le t \le 1} (S(t) \times \{t\}) \subset M \times [0,1].$$

Then two loops $S(t)$, $S'(t)$ in $C_n(M)$ with the same base point are homotopic if and only if the geometric braids $G_S, G_{S'}$ are isotopic. Hence, one can regard an n-braid in M based at S as an element of the fundamental group $\pi_1(C_n(M), S)$, and thus we have

$$B_n(M, S) = \pi_1(C_n(M), S).$$

Let $\xi_0 = (x_1^0, \dots, x_n^0)$ be a point in $F_n(M)$, and let $S = \{x_1^0, \dots, x_n^0\}$. A loop $\xi(t) = (x_1(t), \dots, x_n(t))$ in $F_n(M)$ with base point ξ_0 determines a pure geometric braid which is defined as the union of the n strings $\{(x_i(t), t) : 0 \le t \le 1\}$. This correspondence induces an identification

$$P_n(M, S) = \pi_1(F_n(M), \xi_0).$$

Note that in the case of $n = 1$, we have $C_1(M) = F_1(M) = M$, and hence $B_1(M, \{x_0\}) = P_1(M, \{x_0\}) = \pi_1(M, x_0)$ for any $x_0 \in M$.

2.3. Braid types. In order to apply the braid theory to the study of periodic points, there is a more appropriate notion than a braid. This notion, called a braid type, is defined in the following three equivalent ways.

(2.2) DEFINITION.

(2.2.1) Two geometric n-braids which may be based at different sets are said to be *freely isotopic* if one can be continuously deformed to the other through geometric n-braids, allowing base points vary during the deformation. A free isotopy class of geometric n-braids are called an *n-braid type*. We write $\mathrm{BT}_n(M)$ for the set of all n-braid types on M.

(2.2.2) It is easy to see that two geometric n-braids are freely isotopic if and only if their corresponding loops in $C_n(M)$ are freely homotopic, i.e. there is a homotopy between the two loops which may not keep the base point fixed. Therefore, a braid type may be defined as a free homotopy class of loops in $C_n(M)$.

(2.2.3) Note that there exists a bijective correspondence between the set of free homotopy classes of loops in a connected topological space X and the conjugacy classes of the fundamental group $\pi_1(X, x_0)$ for any base point x_0. Hence, a braid type, which has been defined as a free homotopy class of loops in $C_n(M)$, is regarded as a conjugacy class in the group $\pi_1(C_n(M)) = B_n(M)$.

Given a group G, we denote by G/Conj the set of all conjugacy classes in G. Choose an arbitrary base point S of $C_n(M)$. Then the equivalence of the definitions (2.2.1) and (2.2.3) above leads to the identification

$$\mathrm{BT}_n(M) = B_n(M, S)/\mathrm{Conj}.$$

The braid type which is identified with the conjugacy class of a braid b will be denoted by $[b]$.

2.4. Rotation vectors of braid types. Here, we shall introduce the notion of rotation vector for braid types. We say a braid type is *cyclic* if it is represented by a braid with cyclic permutation μ. In this article, let $\mathrm{CBT}_n(M)$ denote the set of cyclic n-braid types on M. Notice that a number of authors used the symbol $\mathrm{BT}_n(M)$ to denote $\mathrm{CBT}_n(M)$. We shall define a map

$$\kappa\colon \mathrm{CBT}_n(M) \to \pi_1(M)/\mathrm{Conj}$$

as follows: Let $\beta \in \mathrm{CBT}_n(M)$ be represented by a cyclic n-braid b. Then b^n is a pure braid and hence is represented by a loop $(x_1(t), \dots, x_n(t))$ in $F_n(M)$. Choose i, and define $\kappa(\beta)$ as the conjugacy class of the element of $\pi_1(M)$ represented by the loop $x_i(t)$ $(0 \le t \le 1)$ in M. This is well defined, since it does not depend on the choices of b and i. $\kappa(\beta)$ describes the movement of the strings of β projected to M.

(2.3) DEFINITION. Define a map $\rho\colon \mathrm{CBT}_n(M) \to H_1(M; \mathbb{Q})$ by

$$\rho(\beta) = \mathrm{pr}(\kappa(\beta))/n,$$

where $H_1(M; \mathbb{Q})$ is the 1-dimensional homology group with coefficient group $\mathbb{Q}$ and pr is the natural projection from $\pi_1(M)/\mathrm{Conj}$ to $H_1(M)$. Let us call $\rho(\beta)$ the (homological) *rotation vector* of a cyclic braid type β.

2.5. Braids and surface homeomorphisms. We say two homeomorphisms $f, g\colon M \to M$ are isotopic if they are homotopic via a homotopy of homeomorphisms. Braids are closely related to isotopy classes of surface homeomorphisms. Let $\mathrm{Homeo}(M, S; \mathrm{id})$ be the group of homeomorphisms $f\colon M \to M$ isotopic to id with $f(S) = S$, where the group structure is given by composition of maps. We say that $f, g \in \mathrm{Homeo}\,(M, S; \mathrm{id})$ are *isotopic relative to* S if there exists an isotopy $h_t \in \mathrm{Homeo}\,(M, S; \mathrm{id})$ from f to g. The relative isotopy class of $f \in \mathrm{Homeo}\,(M, S; \mathrm{id})$ is denoted by $[f]$. Let

$$\mathrm{Iso}(M, S; \mathrm{id}) = \{[f] : f \in \mathrm{Homeo}\,(M, S; \mathrm{id})\}.$$

This has a group structure induced by that on $\mathrm{Homeo}\,(M, S; \mathrm{id})$.

We shall define a surjective homomorphism

$$\Theta\colon B_n(M, S) \to \mathrm{Iso}(M, S; \mathrm{id}).$$

Let $b \in B_n(M, S)$. Then, there exits an isotopy $f_t\colon M \to M$ with $f_0 = \mathrm{id}$ such that the loop $f_t(S)$ in $C_n(M)$ represents b. Such an isotopy can be constructed in the following way. Let $G = s_1 \cup \ldots \cup s_n$ be a geometric braid representing b. One can extend the velocity vector field of the strings $s_1, \ldots, s_n$ to the whole of $M \times [0, 1]$ so that it is projected to d/dt under the projection $M \times [0, 1] \to [0, 1]$. A desired isotopy $f_t\colon M \to M$ is generated by this vector field. Intuitively, one can think of sliding a rubber sheet down the braid to obtain an isotopy. Let $f = f_1$. Then the relative isotopy class $[f] \in \mathrm{Iso}(M, S; \mathrm{id})$ depends only b. Thus the map Θ is defined by $\Theta(b) = [f]$. This map is a surjective homomorphism, because given $[f] \in \mathrm{Iso}(M, S; \mathrm{id})$, there exists an isotopy f_t from id to f, and it is trivial that the braid b represented by the loop $f_t(S)$ in $C_n(M)$ is sent to $[f]$ under Θ. In fact, it is known that Θ is an isomorphism if the genus of M is greater than one and M is without boundary. Also, the kernel of Θ coincides with the center of $B_n(M)$ if $M = T, n \geq 2$ or $M = S^2, n \geq 3$ (see Theorem 4.3 of [Bi]). For any $S \in C_n(M)$, the map Θ induces a surjective map

$$\overline{\Theta}\colon \mathrm{BT}_n(M) = B_n(M, S)/\mathrm{Conj} \to \mathrm{Iso}(M, S; \mathrm{id})/\mathrm{Conj}.$$

Some authors defined a braid type as an element of the set $\mathrm{Iso}(M, S; \mathrm{id})/\mathrm{Conj}$ as follows:

(2.4) DEFINITION. For $i = 1, 2$, let $S_i \in C_n(M)$ and $f_i \in \mathrm{Homeo}\,(M, S_i; \mathrm{id})$. We say (S_1, f_1) and (S_2, f_2) have the same *braid type* if there exists a homeomorphism $h\colon M \to M$ isotopic to id such that $h(S_1) = S_2$, $[f_1] = [h^{-1} \circ f_2 \circ h] \in \mathrm{Iso}(M, S_1; \mathrm{id})$. The equivalence class of (S, f) will be denoted by $[(S, f)]$.

If we choose $S_0 \in C_n(M)$, then the braid type $[(S, f)]$ can be identified with an element of the set $\mathrm{Iso}(M, S_0; \mathrm{id})/\mathrm{Conj}$. For, if we choose a homeomorphism $h\colon M \to M$ with $h(S_0) = S$, then the conjugacy class of $[h^{-1} \circ f \circ h] \in \mathrm{Iso}(M, S_0; \mathrm{id})$ is uniquely determined by $[(S, f)]$. Hence, we have

$$\{[(S, f)] : S \in C_n(M),\ f \in \mathrm{Homeo}\,(M, S; \mathrm{id})\} = \mathrm{Iso}(M, S_0; \mathrm{id})/\mathrm{Conj}.$$

Moreover, $[(S, f)]$ regarded as an element of $\mathrm{Iso}(M, S; \mathrm{id})/\mathrm{Conj}$ coincides with $\overline{\Theta}(\mathrm{bt}(S, f))$. Thus the braid type defined here is equal to the image of the braid type defined previously.

2.6. Braids on the disk. Now, consider the case where M is a closed disk D. In this case, denote $B_n(D)$, $P_n(D)$, $\mathrm{BT}_n(D)$, $\mathrm{CBT}_n(D)$ simply by B_n, P_n, BT_n, CBT_n, respectively. It follows from the Alexander's trick (see [Bi, Lemma 4.4.1]) that f is isotopic to id if and only if it is orientation-preserving. For $i = 1, \dots, n-1$, we denote by σ_i the ith *elementary braid*, in which the ith string just overcrosses the $(i+1)$th string once and all other strings go straight from the top to the bottom. The braid group B_n admits a presentation with generators $\sigma_1, \dots, \sigma_{n-1}$ and defining relations (see Theorem 1.8 of [Bi]):

$$\begin{aligned} \sigma_i\sigma_j &= \sigma_j\sigma_i && \text{if } |i-j| \geq 2,\ 1 \leq i, j \leq n-1, \\ \sigma_i\sigma_{i+1}\sigma_i &= \sigma_{i+1}\sigma_i\sigma_{i+1} && 1 \leq i \leq n-2. \end{aligned}$$

Define $\rho_n \in B_n$ by $\rho_n = \sigma_1 \dots \sigma_{n-1}$. Let $\theta_n = (\rho_n)^n \in P_n$ and call it the *full-twist* n-braid. Let $\alpha_n \in \mathrm{BT}_n$ be the braid type represented by the braid ρ_n. The kernel Z_n of the homomorphism $\Theta\colon B_n = B_n(D, S) \to \mathrm{Iso}(D, S; \mathrm{id})$ is an infinite cyclic group generated by the full-twist braid θ_n for $n \geq 3$ ([Bi, Corollary 1.8.4]). The action of Z_n on B_n given by multiplication induces its action on BT_n by $\theta_n[b] = [\theta_n b]$. We have the following identifications

$$B_n/Z_n = \mathrm{Iso}(D, S; \mathrm{id}), \quad \mathrm{BT}_n/Z_n = \mathrm{Iso}(D, S; \mathrm{id})/\mathrm{Conj}.$$

(2.5) DEFINITION.

(2.5.1) Let β_1 be a cyclic n braid type and β_2 an m-braid type. We define $[\beta_1, \beta_2] \in \mathrm{BT}_{nm}$ as the braid type obtained by "putting" β_2 inside the braid type β_1 (see Example 2.6 below). This is called the *β_2-extension* of β_1.

(2.5.2) For cyclic braid types $\beta_1, \dots, \beta_{r-1}$ and a braid type β_r, define a braid type $[\beta_1, \dots, \beta_r]$ by $[\beta_1, \dots, \beta_r] = [[\beta_1, \dots, \beta_{r-1}], \beta_r]$ inductively.

(2.6) EXAMPLE. Let $\beta = [\sigma_1\sigma_2^{-1}] \in \mathrm{BT}_3$. Then, $[\alpha_2, \beta] \in \mathrm{BT}_6$ is represented by the braid $(\sigma_1\sigma_2^{-1})(\sigma_3\sigma_4\sigma_5)(\sigma_2\sigma_3\sigma_4)(\sigma_1\sigma_2\sigma_3)$ drawn in Figure 2.

Let $b \in B_n(D, S)$. We introduce the notion of linking number of points in S. There exists a positive integer k with b^k being a pure braid. We shall take the smallest such k. b^k is represented by a loop $(x_1(t), \dots, x_n(t))$ in $F_n(D)$. Let $x_i = x_i(0) \in D$, and $S = \{x_1, \dots, x_n\} \subset D$. For any pair of different points in S, their linking number is defined as follows:

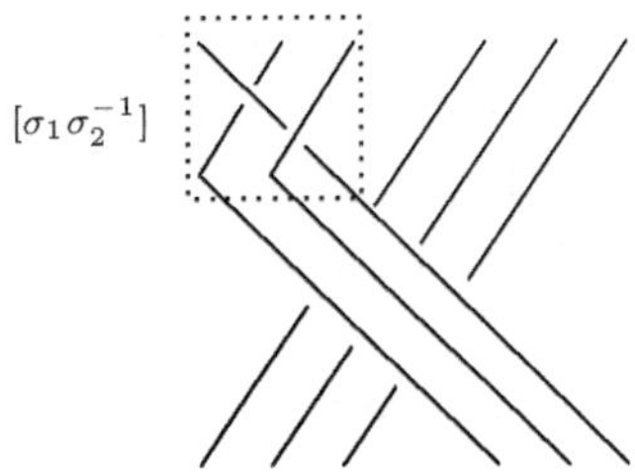

FIGURE 2

(2.7) DEFINITION. Define the *linking number* $\ell k(x_i, x_j; b)$ of x_i and x_j $(i \neq j)$ with respect to b as

$$\ell k(x_i, x_j; b) = \deg(v)/k,$$

where $\deg(v)$ denotes the topological degree of a loop v in $\mathbb{R}^2 - \{0\}$ defined by $v(t) = x_i(t) - x_j(t)$ for $0 \leq t \leq 1$.

For a braid b, define its *exponent sum* $e(b)$ as $e(b) = k_1 + \ldots + k_r$, where $b = \sigma_{i_1}^{k_1} \ldots \sigma_{i_r}^{k_r}$, $r \geq 1$, $1 \leq i_1, \ldots, i_r \leq n-1$, and $k_1, \ldots, k_r$ are integers. Since conjugate braids have the same exponent sum, we can define the exponent sum $e(\beta)$ for a braid type $\beta = [b]$ by $e(\beta) = e(b)$.

Each 3-braid is known to be conjugate, up to multiplication of the full-twist braid θ_3, to σ_1^m, $(\sigma_1\sigma_2)^{\pm 1}$, $\sigma_1\sigma_2\sigma_1$, or an alternating braid $\sigma_1^{i_1}\sigma_2^{-j_1} \ldots \sigma_1^{i_d}\sigma_2^{-j_d}$, where $d \geq 1$, $i_1, \ldots, i_d, j_1, \ldots, j_d$ are positive integers. This representative is unique up to cyclic permutations (cf. [Mu, Proposition 2.1]).

2.7. Braid types of periodic orbits. The orbit of a point $x \in M$ is the set of points $f^p(x)$, $p \in \mathbb{Z}$. A point x in M is a fixed point of f if $f(x) = x$. Let $\mathrm{Fix}(f)$ denote the fixed point set of f. A *periodic point* x of f is a fixed point of f^p for some positive integer p. The least such p is called the *period* of the point. A periodic point of period p will also be called a *p-periodic point.* A periodic orbit is the orbit of a periodic point. Let $\mathrm{Per}_p(f)$ denote the set of periodic points of period p. Note that, in general, $\mathrm{Fix}(f^p)$ may be larger than $\mathrm{Per}_p(f)$.

Let S be a finite invariant set of $f: M \to M$, i.e. it is a finite subset of M with $f(S) = S$. The notion of its braid type was introduced by Boyland [Boy1]. Choose and fix an isotopy $f_t: M \to M$ $(0 \leq t \leq 1)$ from id to f.

(2.8) DEFINITION. Denote by $b(S, f)$ the braid represented by the loop $f_t(S)$ in $C_n(M)$. Let $\mathrm{bt}(S, f)$ be the braid type determined by this braid $b(S, f)$. We call $b(S, f)$ and $\mathrm{bt}(S, f)$ the *braid* and the *braid type* of S, respectively. They will sometimes be written as $b(S)$ and $\mathrm{bt}(S)$ simply.

Since from the dynamics point of view, the actual location of S is irrelevant, the braid type is more appropriate notion than the braid to the topological characterization of periodic orbits. The braid type of S may be defined by adopting the definition of a braid as a union of strings. For a point x of M, define its string by $\operatorname{str}(x) = \{(f_t(x), t) : 0 \leq t \leq 1\}$. This is an embedded arc in $M \times [0,1]$ connecting two points $(x, 0)$ and $(f(x), 1)$. Then the union $\bigcup_{x \in S} \operatorname{str}(x)$ represents $b(S, f)$ and $\operatorname{bt}(S, f)$.

We have seen that, in general, the braid type depends on the choice of an isotopy f_t. There are important situations where an isotopy is provided naturally. Consider a two-dimensional system of ordinary differential equations $dx/dt = F(x,t)$, where $x \in \mathbb{R}^2, t \in \mathbb{R}$ and F is periodic in t with period $T > 0$. We assume the uniqueness for initial value problems. Then the time t-map $f_t \colon \mathbb{R}^2 \to \mathbb{R}^2$ of this system provides an isotopy from id to the Poincaré map $f = f_T \colon \mathbb{R}^2 \to \mathbb{R}^2$. One problem arises here is that $\mathbb{R}^2$ is not compact, so our theory cannot be directly applied. There are several ways to circumvent this difficulty. For instance, if the system is dissipative, then the Poincaré map f has a closed disk D with $f(D) \subset D$. Then the restriction of f to D gives a topological embedding from D into itself isotopic to id ([Ma1], [Ma2]). Also, if the system is a Hamiltonian system and it has an invariant circle, then the closed domain surrounded by this circle becomes an invariant disk. If the Poincaré map is a homeomorphism on $\mathbb{R}^2$, then we are able to compactify the plane $\mathbb{R}^2$ to obtain a sphere S^2 by adding a point at infinity ∞. Then f extends to a homeomorphism of S^2 (see Huang and Jiang [HJ]). Even in the case where f is not surjective, we can still perturb the map f near the infinity, and obtain a homeomorphism on a closed disk (see [Ma4], [HJ]).

If the system is derived from a second-order ordinary differential equation $\ddot{x} = g(x, \dot{x}, t)$ $(x \in \mathbb{R}, t \in \mathbb{R})$, then the braid is determined by the graphs of periodic orbits in the (x,t)-plane uniquely. This is because at any point where two (x,t)-graphs cross, the trajectory with the greater slope $\dot{x}$ passes above the other. Hence the resulting braid has only positive crossings (see [MT], [Ma2]), and is determined uniquely by how the graphs intersect one another.

Peckham ([Pe]) used the linking number of a periodic point x with $f(x)$ to prove the existence of Hopf bifurcation for parametrized periodically forced oscillators. See [BAS], [BSA], [Boy6] for some applications of braid types to fluid mechanics.

3. Matrix representations of braid groups

Some classical matrix representations of braid groups are known to have applications in the study of periodic orbits for surface homeomorphisms. This provides an example of significant applications of the Nielsen fixed point theory.

3.1. Definition of matrix representations. We shall introduce the matrix representations and state a formula which relates the matrix determined by a braid to the linking behavior of periodic orbits. Given a ring Ψ, let $\mathrm{GL}(k, \Psi)$ be the group of invertible $k \times k$ matrices with entries in Ψ. A homomorphism $\nu\colon \Psi \to \Psi'$ to another ring Ψ' induces a homomorphism $\mathrm{GL}(k,\Psi) \to \mathrm{GL}(k,\Psi')$, denoted by the same letter ν, which sends a matrix $R \in \mathrm{GL}(k,\Psi)$ to the matrix obtained from R by replacing each entry to its image under ν. The image $\nu(R) \in \mathrm{GL}(k,\Psi')$ will also be denoted by R^ν.

Now, consider the case of $M = D$. Let Λ be the ring $\mathbb{Z}[a_1, a_1^{-1}, \dots, a_n, a_n^{-1}]$ of integral polynomials in the variables $a_1, \dots, a_n$ and their inverses $a_1^{-1}, \dots, a_n^{-1}$. We shall define a map $R\colon B_n \to \mathrm{GL}(n-1,\Lambda)$. First, for $i = 1, \dots, n-1$, let R_i be the matrix defined by

$$R_1 = \begin{pmatrix} -a_2 & 1 & \\ 0 & 1 & \\ & & I_{n-3} \end{pmatrix}, \qquad R_{n-1} = \begin{pmatrix} I_{n-3} & & \\ & 1 & 0 \\ & a_n & -a_n \end{pmatrix},$$

$$R_i = \begin{pmatrix} I_{i-2} & & & & \\ & 1 & 0 & 0 & \\ & a_{i+1} & -a_{i+1} & 1 & \\ & 0 & 0 & 1 & \\ & & & & I_{n-i-2} \end{pmatrix} \quad \text{for } i = 2, \dots, n-2, \tag{3.1}$$

where I_j denotes the identity matrix of size j. For $b \in B_n$ with permutation μ, let $\nu(b)\colon \Lambda \to \Lambda$ be the ring automorphism given by $\nu(b)(a_i) = a_{\mu(i)}$. The map $R\colon B_n \to \mathrm{GL}(n-1,\Lambda)$ is defined by the rules

$$R(\sigma_i) = R_i, \quad R(bb') = R(b)^{\nu(b')} R(b'). \tag{3.2}$$

Let us call this map R the twisted representation of B_n.

Given a braid $b \in B_n$, one can construct a generalized horseshoe map such that the braid of the set of sinks is equal to b. For the construction, see e.g. [Fr1], [Ma3], [BF, Lecture 4], [Boy4, Section 1.8]. Then the matrix $R(b)$ coincides with the linking transition matrix associated with this map.

We shall state a theorem which relates the representation R to the linking information of fixed points of f. To state it, we need the notion of blow-up of a surface homeomorphism due to Bowen (see [Bow] and also [Boy4, Section 1.6]). Let S be a finite invariant set of f in $\operatorname{Int} M$. Let $\widehat{M}$ be the compactification of $M - S$, that is, the surface obtained from $M - S$ by attaching a boundary circle γ_z at each point z of S. Let Γ be the union of the attached boundary circles, and $\mathrm{pr}\colon \widehat{M} \to M$ the projection. Then $\widehat{M} = (M - S) \cup \Gamma$ and $\Gamma = \mathrm{pr}^{-1}(S) = \partial\widehat{M} - \partial M$. If the restriction of f to $M - S$ is extendable to a homeomorphism $\hat{f}\colon \widehat{M} \to \widehat{M}$, then we call $\hat{f}$ the *blow-up* of f at S. Any homeomorphism which is smooth and non-singular at each point in S has the blow-up at S. $\mathrm{Fix}(\hat{f})$ may be larger than $\mathrm{Fix}(f)$

in general. However, $\widehat{f}$ may have extra points only on Γ, and furthermore the fixed points on Γ can be determined by the differentials $Df(z)$, $z \in S$, explicitly. In particular, if S contains no fixed points of f, or if $Df(z)$ does not have positive real eigenvalues for each $z \in S$, then $\widehat{f}$ has no fixed points on Γ. Therefore, $\mathrm{Fix}(\widehat{f}) = \mathrm{Fix}(f)$.

Consider the case where M is a closed disk D. We shall define the linking number $\ell k(x, A)$ for a point $x \in \mathrm{Fix}(\widehat{f})$ and $A \subset S$. Notice that the compactification $\widehat{D}$ of $D - S$ is homeomorphic to a punctured disk with n punctures. Therefore, we can identify $\widehat{D}$ with $D - \bigcup_{z \in S} \mathrm{Int}\, D_z$, where D_z is a sufficiently small, closed disk in D with center z. Then the boundary circle γ_z can be identified with the boundary ∂D_z. We shall extend the blow-up $\widehat{f}\colon \widehat{D} \to \widehat{D}$ to a homeomorphism $\overline{f}\colon D \to D$ so that $\overline{f}$ coincides with f on S and that $\overline{f}$ has no fixed points on $\mathrm{Int}\, D_z - \{z\}$ for each $z \in S$. We can take an isotopy $\overline{f}_t\colon D \to D$ from id to $\overline{f}$ so that, with respect to this isotopy, the braid $b(S, \overline{f})$ is equal to $b(S, f)$. Since $x \in \widehat{D} \subset D$ and $S \subset \bigcup_z D_z \subset D$, $S \cup \{x\}$ can be regarded as a subset of D. Hence, we have the braid $b_{S \cup x} = b(S \cup \{x\}, \overline{f}) \in B_{n+1}$, and therefore, for $z \in S$, the linking number $\ell k(x, z\,; b_{S \cup x})$ is defined. Define

$$\ell k(x, A) = \sum_{z \in A} \ell k(x, z; b_{S \cup x}).$$

Notice that in the case where the isotopy f_t from id to f is smooth and nonsingular at $z \in S$, for any fixed point x of $\widehat{f}$ in the circle γ_z, $\ell k(x, z; b_{S \cup x})$ is equal to the self-rotation number of z under the isotopy f_t, i.e. the number of full turns of eigenvectors of $Df(z)$ under the homotopy $Df_t(z)\colon \mathbb{R}^2 \to \mathbb{R}^2$.

Let μ be a permutation of $\{1, \dots, n\}$, which is decomposed as a product of m disjoint cycles $\mu_1, \dots, \mu_m$. We take m variables $t_1, \dots, t_m$ and let $\Lambda_\mu = \mathbb{Z}[t_1^{\pm 1}, \dots, t_m^{\pm 1}]$. Consider the surjective homomorphism

$$\pi_\mu\colon \Lambda \to \Lambda_\mu, \quad \pi_\mu(a_i) = t_j,$$

where j is determined by $i \in \mu_j$.

Let $b = b(S, f) \in B_n$, and μ the permutation corresponding to b. Assume μ is decomposed into m cycles. The matrix $R(b)^{\pi_\mu} \in GL(n-1, \Lambda_\mu)$ is useful to obtain information about fixed points. S is decomposed into m periodic orbits P_j. For $I = (i_1, \dots, i_m) \in \mathbb{Z}^m$, let $\mathrm{Fix}_I(\widehat{f})$ be the set of fixed points x of $\widehat{f}$ with

$$(\ell k(x, P_1), \dots, \ell k(x, P_m)) = I.$$

The following theorem follows from Fadell and Husseini (see [FH2]) as an application of the Nielsen fixed point theory. This theorem first appeared expicitly in [Ma1], and a detailed proof has been provided by Huang and Jiang (see [HJ]). Let ind denote the fixed point index.

(3.3) THEOREM. *Let $b = b(S, f)$. Then*

$$-\mathrm{tr}\, R(b)^{\pi_\mu} = \sum_{I \in \mathbb{Z}^m} \mathrm{ind}(\mathrm{Fix}_I(\widehat{f}))\, t_1^{i_1} \ldots t_m^{i_m} \in \mathbb{Z}[t_1^{\pm 1}, \ldots, t_m^{\pm 1}].$$

We shall give a proof of this theorem in the next section.

(3.4) EXAMPLE. If S is a peiodic orbit P of period 3 with $b(P) = \sigma_1\sigma_2^{-1}$ (Figure 1), then $\mu \in \Sigma_3$ is cyclic. Therefore, $m = 1$, $\Lambda_\mu = \mathbb{Z}[t_1^{\pm 1}]$, and $\pi_\mu(a_i) = t_1$ for each i. We have $\mathrm{tr}\, R(\sigma_1\sigma_2^{-1})^{\pi_\mu} = -t_1 + 1 - t_1^{-1}$, and hence $\sum_{i\in\mathbb{Z}} \mathrm{ind}(\mathrm{Fix}_i(\widehat{f}))\, t_1^i = t_1 - 1 + t_1^{-1}$. Since P contains no fixed points, $\mathrm{Fix}(\widehat{f}) = \mathrm{Fix}(f)$. Therefore, by Theorem 3.3, f has at least three fixed points with linking number $1, 0, -1$ with P.

Let $\mathbb{Z}[t, t^{-1}]$ be the ring of integral polynomials in a variable t and its inverse. A homomorphism $\pi\colon \Lambda \to \mathbb{Z}[t, t^{-1}]$ is defined by $\pi(a_i) = t$ for each i. The image of $R(b)$ under this homomorphism π is called the reduced *Burau matrix*, and the homomorphism $\pi \circ R\colon B_n \to \mathrm{GL}(n-1, \mathbb{Z}[t, t^{-1}])$ is called the reduced Burau representation. The matrix $R(b)^\pi$ coincides with the image of $R(b)^{\pi_\mu}$ under the homomorphism from Λ_μ to $\mathbb{Z}[t, t^{-1}]$ sending each t_j to t. The reduced Burau representation is faithful for $n \le 3$, but is not faithful for any $n \ge 5$. The case $n = 4$ is not known (see [R]). Let

$$\mathrm{Fix}_i(\widehat{f}) = \{x \in \mathrm{Fix}(\widehat{f}) : \ell k(x, S) = i\}.$$

As a corollary of Theorem 3.3, we have for $b = b(S, f)$,

(3.5) $$-\mathrm{tr}\, R(b)^\pi = \sum_{i\in\mathbb{Z}} \mathrm{ind}(\mathrm{Fix}_i(\widehat{f}))t^i \in \mathbb{Z}[t, t^{-1}].$$

In the case of a torus T, Huang and Jiang in [HJ] introduced a similar matrix with entries in the ring $\mathbb{Z}[a_2^{\pm}, \ldots, a_n^{\pm 1}, b_1^{\pm 1}, b_2^{\pm 1}]$, where b_1, b_2 correspond to the generators of $H_1(T) \cong \mathbb{Z}^2$, and proved a theorem analogous to Theorem 3.3. Jiang ([J3]) obtained a similar result for an annulus. Shiraki ([Shi2]) considered the case where $M = T$ and S consists of two fixed points, and showed that, putting $b_1, b_2 = 1$, the trace of the matrix introduced in [HJ] becomes a symmetric polynomial in a_2. This is an analogous result to that for $b \in B_3$, $\mathrm{tr}\,(R(b)^\pi) \in \mathbb{Z}[t^{\pm 1}]$ is symmetric, which follows from Matsuoka ([Ma2, p. 429]).

The representation R of B_n is closely related to a classical link invariant called the Alexander polynomial. For a link L with $m \ge 1$ components, let $\Delta_L(t_1, \ldots, t_m)$ be its Alexander polynomial. Let $\widehat{b}$ be the closed braid of b. It is a link consisting of m knots corresponding to the decomposition of the permutation μ into cycles $\mu_1, \ldots, \mu_m$. Let n_j be the length of the j-th cycle μ_j. Then (see Theorem 16.5 of [Mo], Theorem 3.11 of [Bi])

(3.6) $$\det(R(b)^{\pi_\mu} - I) = \begin{cases} (1 + t + \ldots + t^{n-1})\Delta_{\widehat{b}}(t) & \text{if } b \text{ is cyclic,} \\ (t_1^{n_1} \ldots t_m^{n_m} - 1)\Delta_{\widehat{b}}(t_1, \ldots, t_m) & \text{otherwise.} \end{cases}$$

3.2. Applications of Theorem 3.3 to periodic points. As an easy application of Theorem 3.3, we have the following estimate for the number of periodic points. Let μ be the permutation associated with $b(S)$. Assume it is decomposed into m cycles.

(3.7) THEOREM. *Let p be a positive integer. Then*

$$\sharp(\mathrm{Per}_p(f) - S) \geq p(K(S,p) - \nu_p),$$

where $K(S,p)$ is the number of monomials $t_1^{i_1} \dots t_m^{i_m}$ appearing in $\mathrm{tr}\,(R(b(S))^{\pi_\mu})^p$ *with* $\gcd(p, i_1, \dots, i_m) = 1$, *and ν_p is the number of points in S whose period divides p.*

This theorem has been used to obtain estimates for the number of periodic points. Consider the case where $\sharp S = 3$ and the braid type $\mathrm{bt}(S)$ is represented by an alternating braid up to multiplication of the full-twist braid. By computing the number $K(S,p)$ in the above theorem, Matsuoka in [Ma1] proved that if each point of S is an attractive or repelling fixed point, then $\sharp\mathrm{Per}_p(f) \geq p^2$. Also, Matsuoka in [Ma2] proved that $\sharp\mathrm{Per}_p(f) \geq 2\phi(p) - 3$, where $\phi(p)$ is the Euler's function, i.e. the number of integers k such that $1 \leq k \leq p$ and k is relatively prime to p. Moreover, [Ma4] proved that, including the case of $M = \mathbb{R}^2$, $\sharp\mathrm{Per}_p(f)$ is greater than or equal to the number of integers k which satisfy $-(j_1 + \ldots + j_d)p \leq k \leq (i_1 + \ldots + i_d)p$ and k is relatively prime to p and 3.

Andres shows in [A2] that these estimates for $\sharp\mathrm{Per}_p(f)$ can be generalized to admissible multimaps on a surface M under some restrictions, and gives an application to differential inclusions on $\mathbb{R}^2$. Andres in [A1] proved that for some 2-dimensional systems of time-periodic differential equations, Nielsen fixed point theory always implies the existence of three periodic solutions, and indicated the possibility of applying the above estimates to obtain infinitely many periodic solutions. See also Andres and Górniewicz in [AG].

There are further applications of Theorem 3.3 on dynamical properties of surface homeomorphisms. Fried introduced in [Fri1] the twisted Lefschetz zeta function $\zeta(t)$ for differential maps on a manifold X. This is a formal power series on the group ring $\mathbb{Q}H$, where $H = \mathrm{Coker}(f_* - \mathrm{id}) = H_1(X)/\mathrm{Im}(f_* - \mathrm{id})$. It was shown that if f has only finitely many periodic points, then $\zeta(t)$ is a product of the form $(1 - ht^k)^{\pm 1}$, where $h \in H, k \in \mathbb{N}$. If $f\colon D \to D$ is a diffeomorphism isotopic to id with a finite invariant set S, then the zeta function $\zeta(t)$ for the restriction of f to $D - S$ coincides with $\det(tR(b)^{\pi_\mu} - I) \in \mathbb{Z}[t_1^{\pm 1}, \dots, t_m^{\pm 1}, t]$ (see [HJ]), where $b = b(S)$. Thus the matrix $R(b)$ is applicable to show that f has infinitely many periodic orbits. This result can be proved by applying Theorem 3.3 to all iterates of f and combining the information thus obtained in the form of the zeta function.

Matsuoka [Ma5, Proposition 3] assumed that any fixed point of f^n is transversal, and obtained a relationship between the number of periodic points by applying the formula (3.5). Let $\nu_{E,i}(p)$ (resp. $\nu_{O,i}(p)$) be the number of p-periodic points x such that $\ell k(x,S) = i$ with respect to f^p and the differential $Df^p(x)$ has an even (resp. odd) number of eigenvalues > 1. Then, for example, we have that $\sum_{i\in\mathbb{Z}}(\nu_{E,i}(p) - \nu_{O,i}(p))\, t^i$ is equal to $\operatorname{tr} R(b)^\pi - \operatorname{tr}(R(b)^\pi)^p$, where p is an odd prime number and $b = b(S)$.

In [GHal], Hall mentioned that Theorem 3.3 provides an alternative proof for a claim there on the existence of periodic points for annulus twist maps.

The Burau matrix can also be used to give an estimate of the topological entropy. This will be treated later in Section 8.

3.3. Linking number problem. Franks posed the following question: For each periodic orbit P of disk homeomorphisms, does there exist a fixed point which has non-zero linking number with P? The answer is known to be positive in the case where $f - \mathrm{id}$ satisfies a Lipschitz condition (see [BK]). Guaschi in [G3] obtained another partial answer for this problem:

(3.8) THEOREM. *Let P be a periodic orbit of f with period $n \geq 3$. Let q be the least integer greater than or equal to $n/2 - 1$. Then there exists a periodic orbit Q of f with period $\leq q$ such that the linking number of Q with P is not divisible by n.*

This result implies that the linking number problem is affirmative in the case of $n = 3, 4$. The proof is based on Theorem 3.3, the relation (3.6), and one of the basic properties of the Alexander polynomial.

OUTLINE OF PROOF. Let $b = b(P)$. Let $A(t) = R(b)^\pi \in \mathrm{GL}(n-1, \mathbb{Z}[t, t^{-1}])$ be its Burau matrix. We may regard it as a matrix whose entires lie in the field $\mathbb{Q}(t)$ of all rational functions in the indeterminate t over $\mathbb{Q}$. Then there exists a field extension K of $\mathbb{Q}(t)$ such that the characteristic polynomial of $A(t)$ splits over K, and we have the eigenvalues $\lambda_1(t), \ldots, \lambda_{n-1}(t)$ of $A(t)$. Let c be the exponent sum $e(b)$ of b. Then by (3.1) we have $\det(A(t)) = (-t)^c$. The characteristic polynomials satisfy

$$\text{(3.9)}\quad \begin{aligned}\det(A(t) - xI) &= (\lambda_1(t) - x)\ldots(\lambda_{n-1}(t) - x)\\ &= p_{n-1}(t) - p_{n-2}(t)x + \ldots + (-1)^{n-2}p_1(t)x^{n-2} + (-1)^{n-1}x^{n-1},\end{aligned}$$

where for each $1 \leq i \leq n-1$, $p_i(t)$ is the ith elementary symmetric polynomial of $\lambda_1(t), \ldots, \lambda_{n-1}(t)$. Set $\tau_k(t) = \operatorname{tr} A(t)^k = \sum_{i=1}^{n-1} \lambda_i(t)^k$ for $k \geq 1$.

Assume that all periodic orbits Q of f with period $\leq q$ had linking number $\ell k(Q, P)$ divisible by n. Choose an integer k with $1 \leq k \leq q$. Let x be a point in $\mathrm{Fix}(\widehat{f^k})$, and r the period of x. Then k is a multiple of r. Since $k \le q < n$, $\mathrm{Fix}(\widehat{f^k}) = \mathrm{Fix}(f^k)$ and hence $x \in \mathrm{Fix}(f^k)$. Note that $\ell k(x, P)$ with respect to $\widehat{f^k}$ is equal to $(k/r)\ell k(Q, P)$, where Q is the orbit of x, and so it is divisible by n. Therefore, applying the formula (3.5) to $\widehat{f^k}$, we see that $\tau_k(t)$ is a polynomial in t^n. Thus we have proved that $\tau_1(t), \dots, \tau_q(t)$ are polynomials in t^n. This implies that so are $p_1(t), \dots p_q(t)$. Moreover, since $p_{n-q-1}(t), \dots, p_{n-1}(t)$ can be expressed in terms of $\tau_1(t^{-1}), \dots, \tau_q(t^{-1})$ and $p_{n-1}(t) = (-t)^c$, they are polynomials in t^n and t^c. Therefore, since $q+1 \ge n-q-1$, all of $p_1(t), \dots, p_{n-1}(t)$ are polynomilas in t^n and t^c. Then (3.6), (3.9) imply that $\Delta_{\widehat{b}}(1) = 0$. However, it is known (see Note 18.3 [Mo], Corollary 3.11.2 in [Bi]) that $\Delta_{\widehat{b}}(1) = \pm 1$. Hence we get a contradiction. □

4. Nielsen fixed point theory on surfaces

This section is devoted to the proof of Theorem 3.3. The theorem is obtained by applying the theory of generalized Lefschetz numbers, which is one of the most important branches in the Nielsen fixed point theory. In fact, the trace of the matrix appearing in the theorem will be proved to coincide with the abelianization of the generalized Lefschetz number on a punctured surface, and the proof will be finished by computing the abelianization using the free differential calculus on free groups.

4.1. Generalized Lefschetz number. We recall some basic facts about the Nielsen fixed point theory (see [Br] and [J1]), and introduce the notion of generalized Lefschetz number. Let X be a connected finite cell complex, and $f\colon X \to X$ a continuous map. Fixed points x, y of f are said to be *f-Nielsen equivalent* if there is a path l in X from x to y such that l and its image $f \circ l$ are homotopic fixing end points. The equivalence class of $x \in \mathrm{Fix}(f)$ is called the *Nielsen class* of x. Let $\mathrm{NC}(f)$ be the set of f-Nielsen classes. A Nielsen class is essential if it has nonzero fixed point index. The Nielsen number $N(f)$ is the number of essential Nielsen classes of f.

It is a difficult problem, in general, to decide whether two fixed points belong to the same Nielsen class. Reidemeister showed that this geometric problem can be transformed into an algebraic problem. Choose a base point $x_0 \in X$ and a path w from x_0 to $f(x_0)$. Let π be the fundamental group $\pi_1(X, x_0)$ of X, and $\psi\colon \pi \to \pi$ the composition of $f_*\colon \pi_1(X, x_0) \to \pi_1(X, f(x_0))$ with $w_*\colon \pi_1(X, f(x_0)) \to \pi_1(X, x_0)$. Two elements $\lambda_1, \lambda_2 \in \pi$ are said to be *Reidemeister equivalent* if there is a $\lambda \in \pi$ such that

$$\lambda_2 = \psi(\lambda)\lambda_1\lambda^{-1}. \tag{4.1}$$

An equivalence class of elements of π under this equivalence relation is a *Reidemeister class*. Let $\mathcal{R}(\psi)$ denote the set of Reidemeister classes.

For $x \in \mathrm{Fix}(f)$, take a path l from x_0 to x. Then the Reidemeister class represented by $[w(f \circ l)l^{-1}] \in \pi$ is independent of the choice of l. This is denoted by $R(x)$ and is called the Reidemeister class of x. The points in a Nielsen class F have the common Reidemeister class. This common class is denoted by $R(F)$ and is called the Reidemeister class of F. It is easy to see that two fixed points are in the same Nielsen class if and only if they have the same Reidemeister class. Therefore, one can regard $\mathrm{NC}(f)$ as a subset of $\mathcal{R}(\psi)$. Note that the Reidemeister class $R(F)$ is called the *coordinate* of F in [J5]. Guaschi showed in [G4] that the problem of distinguishing Reidemeister classes on a punctued disk is reduced to the conjugacy problem in the braid group.

Given a set U, let $\mathbb{Z}U$ denote the free abelian group generated by the elements of U. If U has a group structure, then $\mathbb{Z}U$ becomes a ring, which is called the group ring of U over $\mathbb{Z}$. If V is another set, and $v\colon U \to V$ is a map, then we use the same symbol v to denote the homomorphism from $\mathbb{Z}U$ to $\mathbb{Z}V$ induced by v. If v is a group homomorphism, $v\colon \mathbb{Z}U \to \mathbb{Z}V$ becomes a ring homomorphism.

For $\alpha \in \mathcal{R}(\psi)$, let $\mathrm{Fix}_\alpha(f) = \{x \in \mathrm{Fix}(f) : R(x) = \alpha\}$. If this is not empty, then it is an f-Nielsen class with Reidemeister class α.

(4.2) Definition. The generalized Lefschetz number $\mathcal{L}(f)$ (also called as the Reidemeister trace) of f is defined as

$$\mathcal{L}(f) = \sum_{F \in \mathrm{NC}(f)} \mathrm{ind}(F) \cdot R(F) = \sum_{\alpha \in \mathcal{R}(\psi)} \mathrm{ind}(\mathrm{Fix}_\alpha(f))\, \alpha \in \mathbb{Z}\mathcal{R}(\psi). \tag{4.3}$$

The Nielsen number $N(f)$ is equal to the number of terms with non-zero coefficient in $\mathcal{L}(f)$. Kelly ([K]) has given an algorithm to compute $\mathcal{L}(f)$ for orientation-preserving homeomorphisms on surfaces with boundary.

It follows from Theorem 4.1 of Jiang and Guo (see [JG]) that if X is a surface M, then every coefficient appearing in $\mathcal{L}(f)$ is at most 1, and almost all coefficients are 1, -1, 0 (see [J5, Corollary 3.2]). More precisely, the following holds:

(4.4) Theorem. *Suppose X is a surface M with negative Euler characteristic. Then*

(4.4.1) *Every f-Nielsen class F has index* $2\chi(M) - 1 \leq \mathrm{ind}(F) \leq 1$.

(4.4.2) *Every f-Nielsen class F has index* $\mathrm{ind}(F) = 1, -1$ *or* 0 *with at most* $-2\chi(M)$ *exceptions.*

This theorem has been generalized to an arbitrary continuous map by Jiang in [J6].

Let $\widetilde{X}$ be the universal covering space of X. The cell structure on X induces a cell structure on $\widetilde{X}$. Assume $f: X \to X$ is a cellular map. Let $\widetilde{f}: \widetilde{X} \to \widetilde{X}$ be a lift of f. Let $C_q(\widetilde{X})$ denote the group of cellular q-chains over $\mathbb{Z}$. Then $\widetilde{f}$ induces the homomorphism $\widetilde{f}_q: C_q(\widetilde{X}) \to C_q(\widetilde{X})$. The group $C_q(\widetilde{X})$ is a free, finitely generated $\mathbb{Z}\pi$-module, and $\widetilde{f}_q$ is a ψ-twisted homomorphism in the sense that

$$\widetilde{f}_q(\lambda u) = \psi(\lambda)\widetilde{f}_q(u), \quad \lambda \in \pi, \ u \in C_q(\widetilde{X}).$$

We can define the trace $\operatorname{tr} \widetilde{f}_q \in \mathbb{Z}\mathcal{R}(\psi)$ as follows: Let $\widetilde{e}_1, \dots, \widetilde{e}_n$ be 1-cells of $\widetilde{X}$ which form a basis of $C_q(\widetilde{X})$ over $\mathbb{Z}\pi$. Let $A = (a_{ij})$ be the matrix over $\mathbb{Z}\pi$ representing the ψ-homomorphism $\widetilde{f}_q$ with respect to these generators, that is, a_{ij}'s are given by $\widetilde{f}_q(\widetilde{e}_i) = \sum_j a_{ij}\widetilde{e}_j$. Let $\operatorname{tr} \widetilde{f}_q = [\operatorname{tr} A]$, where $[\,\cdot\,]$ means the projection of $\mathbb{Z}\,\pi$ to $\mathbb{Z}\mathcal{R}(\psi)$. If $\widetilde{e}'_1, \dots, \widetilde{e}'_n$ is another basis of $C_q(\widetilde{X})$, then this basis is related to $\widetilde{e}_1, \dots, \widetilde{e}_n$ via a matrix C. It is easy to see that the matrix representing $\widetilde{f}_q$ with respect to $\widetilde{e}'_1, \dots, \widetilde{e}'_n$ is given by $C^{\psi}AC^{-1}$, where C^{ψ} means the matrix with entries replaced to their ψ-images. Then by (4.1), $[\operatorname{tr}(C^{\psi}AC^{-1})] = [\operatorname{tr} A] \in \mathbb{Z}\mathcal{R}(\psi)$, which means that $\operatorname{tr} \widetilde{f}_q$ is well defined.

Husseini proved in [Hu] the following trace formula (for partial results see [FH1] and [Fri1]).

$$\mathcal{L}(f) = \sum_q (-1)^q \operatorname{tr} \widetilde{f}_q \in \mathbb{Z}\mathcal{R}(\psi). \tag{4.5}$$

Consider the case of $M = D$. Let $f \in \operatorname{Homeo}(D, S; \mathrm{id})$. Identify $\pi_1(D - S)$ with F_n. Let $x_1, \dots, x_n$ be the generators of F_n. Let $\operatorname{Coker}(f_* - \mathrm{id}) = H_1(D - S)/\operatorname{Im}(f_* - \mathrm{id})$, where f_* is the automorphism of $H_1(D - S)$ induced by f. Let μ be the permutation corresponding to the braid $b(S)$. Let μ be decomposed into m cycles $\mu_1, \dots, \mu_m$. For $i = 1, \dots, n$, let $a_i \in H_1(D - S)$ be the image of x_i under the abelianization homomorphism $\mathrm{Ab}: F_n = \pi_1(D - S) \to H_1(D - S)$, and let $t_j \in \operatorname{Coker}(f_* - \mathrm{id})$ be defined by $t_j = [a_i]$, where $i \in \mu_j$. Then $a_1, \dots, a_n$ and $t_1, \dots, t_m$ give the sets of generators for the groups $H_1(D - S)$ and $\operatorname{Coker}(f_* - \mathrm{id})$ respectively. Hence, we have the following identifications.

$$\Lambda = \mathbb{Z}H_1(D - S), \quad \Lambda_\mu = \mathbb{Z}\operatorname{Coker}(f_* - \mathrm{id}).$$

The abelianization homomorphism $\mathrm{Ab}: F_n \to H_1(D - S)$ induces

$$\begin{gathered} \overline{\mathrm{Ab}}: \mathcal{R}(\psi) \to \operatorname{Coker}(f_* - \mathrm{id}), \\ \mathrm{Ab}: \mathbb{Z}F_n \to \Lambda, \quad \overline{\mathrm{Ab}}: \mathbb{Z}\mathcal{R}(\psi) \to \Lambda_\mu. \end{gathered} \tag{4.6}$$

Let X be the compactification $\widehat{D}$ of $D - S$, and $\widehat{f}: X \to X$ the blow-up of f at S. X is a finite cell complex. Its fundamental group $\pi = \pi_1(X)$ is identified with F_n.

(4.7) LEMMA.

$$\overline{\mathrm{Ab}}(\mathcal{L}(\widehat{f})) = \sum_I \mathrm{ind}(\mathrm{Fix}_I(\widehat{f}))t_1^{i_1} \dots t_m^{i_m} \in \Lambda_\mu.$$

PROOF. Let F be an $\widehat{f}$-Nielsen class. Then, by the definitions of $R(F)$ and the linking number $\ell k(x, P_j)$ for $x \in \mathrm{Fix}(\widehat{f})$, we have

$$\overline{\mathrm{Ab}}(R(F)) = \sum_{j=1}^{m} \ell_j(F)t_j \in \mathrm{Coker}(f_* - \mathrm{id}), \tag{4.8}$$

where $\ell_j(F) = \ell k(x, P_j)$ for $x \in F$. Note that $\ell_j(F)$ is independent of the choice of x in F. (4.8) implies that

$$\begin{aligned}\overline{\mathrm{Ab}}(\mathcal{L}(\widehat{f})) &= \sum_{F \in \mathrm{NC}(\widehat{f})} \mathrm{ind}(F) \cdot \overline{\mathrm{Ab}}(R(F)) \\ &= \sum_F \mathrm{ind}(F)t_1^{\ell_1(F)} \dots t_m^{\ell_m(F)} = \sum_I \mathrm{ind}(\mathrm{Fix}_I(\widehat{f}))t_1^{i_1} \dots t_m^{i_m}.\end{aligned}$$

Thus the lemma is proved. □

4.2. Free differential calculus. We consider the case where the cell complex X is a surface M with boundary. Fadell and Husseini provided in [FH2] a method for calculating $\mathcal{L}(f)$ in terms of $\psi\colon \pi \to \pi$. We identify M with the usual cell structure on M given by a 0-cell x_0, a single 2-cell, and 1-cells $e_1, \dots, e_n$. Then the equality (4.5) becomes

$$\mathcal{L}(f) = [1] - \mathrm{tr}\,\widetilde{f}_1 \in \mathbb{Z}\mathcal{R}(\psi). \tag{4.9}$$

Their method involves the Fox free differential calculus on $\mathbb{Z}F_n$.

Let $x_1, \dots, x_n$ be the generators of $\pi = \pi_1(M, x_0)$ corresponding to $e_1, \dots, e_n$. The fundamental group π can be identified with the free group F_n. The Fox partial derivative operator $\partial/\partial x_j\colon \mathbb{Z}F_n \to \mathbb{Z}F_n$, $j = 1, \dots, n$, is defined by the following rules (see [Bi, Chapter 3.1] and [Mo, Chapter 8]):

$$\frac{\partial}{\partial x_j}(u+v) = \frac{\partial u}{\partial x_j} + \frac{\partial v}{\partial x_j}, \quad u, v \in \mathbb{Z}F_n, \tag{4.10.1}$$

$$\frac{\partial}{\partial x_j}(uv) = \frac{\partial u}{\partial x_j} + u\frac{\partial v}{\partial x_j}, \quad u, v \in \mathbb{Z}F_n, \tag{4.10.2}$$

$$\frac{\partial x_i}{\partial x_j} = \delta_{i,j}, \quad 1 \le i, j \le n, \tag{4.10.3}$$

where $\delta_{ij} = 1, 0$ according to $i = j$, $i \neq j$, and $1 \cdot e \in \mathbb{Z}F_n$ is identified with 1,

$$\frac{\partial 1}{\partial x_j} = 0. \tag{4.10.4}$$

Note that the Fox derivative satisfies $\partial u^{-1}/\partial x_j = -u^{-1}\partial u/\partial x_j$. Given a homomorphism $\psi\colon F_n \to F_n$, let $J(\psi)$ be the *Jacobian matrix* $(\partial\psi(x_i)/\partial x_j)$. We have

$$J(\psi' \circ \psi) = J(\psi)^{\psi'} J(\psi'), \tag{4.11}$$

where $J(\psi)^{\psi'}$ denotes the matrix obtained by $J(\psi)$ by replacing every entry to its ψ'-image.

The main role of the Fox calculus is to provide a formula for describing the lift of a given loop w in M to a path in the universal cover $\widetilde{M}$ in terms of the element of F_n that corresponds to w. More precisely, let $w = w(x_1, \ldots, x_n)$ denote any word in the generators $x_1, \ldots, x_n$. Choose $\widetilde{x}_0$ which covers x_0. If we lift the corresponding loop w in M to a path $\widetilde{w}$ in $\widetilde{M}$ so as to have initial point $\widetilde{x}_0$, the resulting 1-chain $\widetilde{w}$ in $C_1(\widetilde{M})$ coincides with

$$\frac{\partial w}{\partial x_1}\widetilde{e}_1 + \ldots + \frac{\partial w}{\partial x_n}\widetilde{e}_n, \tag{4.12}$$

where $\widetilde{e}_1, \ldots, \widetilde{e}_n$ correspond to lifts of 1-cells $e_1, \ldots, e_n$ with initial point $\widetilde{x}_0$.

We may assume f is cellular and fixes x_0 without loss of generality. Choose a lift $\widetilde{f}\colon \widetilde{M} \to \widetilde{M}$ with $\widetilde{f}(\widetilde{x}_0) = \widetilde{x}_0$. By (4.12), $\widetilde{f}_1(\widetilde{e}_i)$ is equal to the 1-chain $\sum_j (\partial\psi(x_i)/\partial x_j)\widetilde{e}_j$. Therefore the matrix representing $\widetilde{f}_1$ over $\mathbb{Z}\pi$ with respect to $\widetilde{e}_1, \ldots, \widetilde{e}_n$ is given by $(\partial\psi(x_i)/\partial x_j) = J(\psi)$. Therefore by (4.9), we have (see [FH2, Theorem 2.3]),

$$\mathcal{L}(f) = [1 - \operatorname{tr} J(\psi)] \in \mathbb{Z}\mathcal{R}(\psi). \tag{4.13}$$

4.3. Proof of Theorem 3.3. Let $g_i = x_1 \ldots x_i$, $1 \leq i \leq n$. Then $\{g_1, \ldots, g_n\}$ is a basis of F_n and we have (cf. [Bi, p. 121], [Mo, p. 225]) that σ_i acts on these as

$$\begin{aligned} g_i &\to g_{i+1} g_i^{-1} g_{i-1} && \text{if } i \neq 1, \\ g_1 &\to g_2 g_1^{-1} && \text{if } i = 1, \\ g_j &\to g_j && \text{if } j \neq i. \end{aligned} \tag{4.14}$$

Let $b \in B_n$. The action (4.14) defines an automorphism $\psi_b\colon F_n \to F_n$. Define a map $\overline{R}\colon B_n \to \mathrm{GL}(n, \Lambda)$ by $\overline{R}(b) = J(\psi_b)^{\mathrm{Ab}}$, where $J(\psi_b)$ is taken with respect to $\{g_1, \ldots, g_n\}$. Since by (4.14) g_n is fixed under each σ_i, the last row of the matrix $J(\psi_b)$, and hence of $\overline{R}(b)$, is always $(0, \dots, 0, 1)$. Consider the matrix $R'(b)$ obtained from $\overline{R}(b)$ by deleting the last row and column. Then, one can show that this matrix coincides with the matrix $R(b)$ defined by (3.2). In fact, $R'(\sigma_i) = R_i$. Also, since the automorphism $\nu(b)\colon \Lambda \to \Lambda$ is induced by ψ_b and $\mathrm{Ab} \circ \psi_b = \nu(b) \circ \mathrm{Ab}$, the product formula (4.11) implies

$$\begin{aligned} \overline{R}(bb') &= J(\psi_{b'} \circ \psi_b)^{\mathrm{Ab}} = (J(\psi_b)^{\psi_{b'}})^{\mathrm{Ab}} J(\psi_{b'})^{\mathrm{Ab}} \\ &= (J(\psi_b)^{\mathrm{Ab}})^{\nu(b')} J(\psi_{b'})^{\mathrm{Ab}} = \overline{R}(b)^{\nu(b')} \overline{R}(b'). \end{aligned} \tag{4.15}$$

By (4.15) R' satisfies the same defining relations for R, and hence $R'(b) = R(b)$ for any b.

Let $b = b(S, f)$. For $\eta \in \mathbb{Z}F_n$,

$$\overline{\mathrm{Ab}}([\eta]) = (\pi_\mu \circ \mathrm{Ab})(\eta). \tag{4.16}$$

Therefore, by Lemma 4.7 and (4.13), (4.16),

$$\begin{aligned} \sum_{I \in \mathbb{Z}^m} \mathrm{ind}(\mathrm{Fix}_I(\widehat{f}))\, t_1^{i_1} \dots t_m^{i_m} &= \overline{\mathrm{Ab}}(\mathcal{L}(\widehat{f})) = \overline{\mathrm{Ab}}([1 - \mathrm{tr}\, J(\psi_b)]) \\ &= (\pi_\mu \circ \mathrm{Ab})(1 - \mathrm{tr}\, J(\psi_b)) = \pi_\mu(1 - \mathrm{tr}\, \overline{R}(b)) \\ &= -\mathrm{tr}\, R(b)^{\pi_\mu}. \end{aligned}$$

Thus the proof of Theorem 3.3 is completed. □

5. Nielsen–Thurston classification theory

The Nielsen–Thurston classification theory provides a representative φ, called the canonical homeomorphism, for each isotopy class of surface homeomorphisms. The existence of a canonical homeomorphism has an essential importance in the study of perodic orbits of surface homeomorphisms f, namely, the canonical representative φ has the "simplest" dynamical property among the homeomorphisms in the isotopy class, and hence, we can investigate the periodic orbit structure of f by investigating that of the canonical homeomorphism.

5.1. The classification theorem. We give a brief account of the theory. For general references, see [T], [FLP], [CB], [Boy4]. Let $\varphi\colon M \to M$ be a homeomorphism. φ is said to be *finite order* (or periodic) if $\varphi^m = \mathrm{id}$ for some positive integer m. Let S be a set of n points in $\mathrm{Int}\, M$. φ is said to be *pseudo-Anosov relative to* S, if the following conditions hold:

(5.1.1) There exists a pair of transverse foliations $\mathcal{F}^u$, $\mathcal{F}^s$ on M, carrying measures which are uniformly expanded and contracted by φ respectively.

(5.1.2) Each foliation has a finite number of singularities which coincide in the interior $\operatorname{Int} M$ and alternate on the boundary ∂M. Any singularity is p-pronged for some positive integer p, $p \neq 2$.

(5.1.3) Any singularity on ∂M is 3-pronged. (We consider segments of the boundary to be prongs.)

(5.1.4) 1-prongs are permitted only at points of S.

In the case where S is empty, φ is simply said to be pseudo-Anosov. We say φ is *reducible* relative to S if there exists a disjoint union $\Sigma = \Sigma_1 \cup \ldots \cup \Sigma_k$ of simple closed curves in $\operatorname{Int} M - S$, called *reducing curves*, such that

(5.2.1) Σ is an invariant set of φ, and

(5.2.2) each connected component of $M - (\Sigma \cup S)$ has negative Euler characteristic.

(5.3) THEOREM (Nielsen–Thurston Classification Theorem). *Let $f\colon M \to M$ be a homeomorphism, and S a finite invariant set of f in* $\operatorname{Int} M$. *Then f is isotopic relative to S to a homeomorphism φ which is periodic, pseudo-Anosov relative to S, or reducible relative to S. Moreover, in the reducible case, φ is decomposed into finite order and pseudo-Anosov components. More precisely, such φ can be chosen to have an invariant tubular neighborhood $A(\Sigma)$ of Σ such that on each connected component N of the complement of $A(\Sigma)$, $\varphi^k\colon N \to N$ is either periodic or pseudo-Anosov relative to $S \cap N$, where k is the least positive integer such that $\varphi^k(N) = N$. A pseudo-Anosov homeomorphism φ is unique up to topological conjugacy.*

The map φ in the above theorem is called a *canonical homeomorphism* (or a canonical form) relative to S. A component N is called a *component* of φ. It is called a *finite order component* or a *pseudo-Anosov component* if φ^k is finite order or pseudo-Anosov, respectively.

There is an algorithm due to Bestvina and Handel (see [BH]) which determines whether an isotopy class is of reducible, pseudo-Anosov, or finite order. Similar algorithms for the disk case were given by Franks and Misiurewicz in [FM] (see [Hay2] for a revision of this algorithm) and by Los ([Lo1]). Benardete, Gutierrez and Nitecki ([BGN1], [BGN2]) constructed different algorithms for the disk homeomorphisms.

(5.4) EXAMPLE (Pseudo-Anosov disk homeomorphisms). Any pseudo-Anosov homeomorphism on D relative to three points can be obtained from a hyperbolic toral automorphism (see Exposé 13 in [FLP], or Theorem 2.1 and Proposition 2.3 in [LM3]). Regard the torus T as the quotient space $\mathbb{R}^2/\mathbb{Z}^2$. Let $L \in \mathrm{GL}(2, \mathbb{Z})$. Let $\tau\colon T \to T$ be the map induced by a map on $\mathbb{R}^2$ which sends $x \in \mathbb{R}^2$ to $-x$. Then $\operatorname{Fix}(\tau)$ consists of the four points $[(a, b)]$, $a, b = 0, 1/2$. The quotient space

T/τ is homeomorphic to a two sphere S^2, and the quotient map $T \to T/\tau$ is a 2-fold branched cover which has $\mathrm{Fix}(\tau)$ as branch points. Let $f_L: T \to T$ be the homeomorphism induced by $L: \mathbb{R}^2 \to \mathbb{R}^2$. Assume $\operatorname{tr} L > 2$. Then this map is an Anosov diffeomorphism and is called a hyperbolic toral automorphism. f_L induces a homeomorphism $\varphi_L: S^2 \to S^2$. Then the two foliations on T for f_L descend to S^2, and the only singularities introduced are 1-prongs at $\mathrm{Fix}(\tau)$. Thus φ_L becomes a pseudo-Anosov homeomorphism on S^2 relative to $\mathrm{Fix}(\tau)$. Let $x_0 = [(0,0)] \in S^2$ and $S = \mathrm{Fix}(\tau) - \{x_0\}$. Blow up φ_L to $\widehat{\varphi}_L: \widehat{S^2} \to \widehat{S^2}$ at x_0. Then since the resulting space $\widehat{S^2}$ is a closed disk D, we obtain a pseudo-Anosov homeomorphism $\widehat{\varphi}_L: D \to D$ relative to S. L is decomposed as

$$L = L_1^{i_1} L_2^{j_1} \dots L_1^{i_r} L_2^{j_r},$$

where

$$L_1 = \begin{pmatrix} 1 & 1 \\ 0 & 1 \end{pmatrix} = R(\sigma_1)^\pi \,|_{t=-1}, \quad L_2 = \begin{pmatrix} 1 & 0 \\ 1 & 1 \end{pmatrix} = R(\sigma_2^{-1})^\pi \,|_{t=-1}.$$

One can show that $b(S, \widehat{\varphi}_{L_1}) = \sigma_1$, $b(S, \widehat{\varphi}_{L_2}) = \sigma_2^{-1}$ for appropriate isotopies for $\widehat{\varphi}_{L_1}, \widehat{\varphi}_{L_2}$, and therefore we have $b(S, \widehat{\varphi}_L) = \sigma_1^{i_1} \sigma_2^{-j_1} \dots \sigma_1^{i_r} \sigma_2^{-j_r}$.

5.2. Properties of canonical homeomorphisms. We recall some properties of canonical homeomorphisms which we shall need later. Let $\varphi: M \to M$ be an orientation-preserving pseudo-Anosov homeomorphism relative to S. Let $\mathcal{F}$ be either $\mathcal{F}^u$ or $\mathcal{F}^s$. For $x \in M$, let $p(x)$ denote the number of prongs of $\mathcal{F}$ at x. If x is a regular point of $\mathcal{F}$, we regard the two half leaves emanating from x as the prongs at x, and thus set $p(x) = 2$. Let $\mathrm{Sing}(\mathcal{F})$ denote the set of singularities of $\mathcal{F}$. Then we have the following Euler-Poincaré formula (see e.g. [FLP, p. 75]):

$$\sum_{x \in \mathrm{Sing}(\mathcal{F})} (2 - p(x)) = 2\chi(M). \tag{5.5}$$

Notice that a periodic point on ∂M must be a singularity of $\mathrm{Sing}(\mathcal{F}^s)$ or $\mathrm{Sing}(\mathcal{F}^u)$. For the fixed point index of a fixed point x of an iterate φ^m, we have

$$\operatorname{ind}(x, \varphi^m) = \begin{cases} 1 - p(x) & \text{if } x \in \operatorname{Int} M \text{ and } \varphi^m \text{ preserves each prong,} \\ 1 & \text{if } x \in \operatorname{Int} M \text{ and } \varphi^m \text{ rotates the prongs,} \\ -1 & \text{if } x \in \partial M \cap \mathrm{Sing}(\mathcal{F}^s), \\ 0 & \text{if } x \in \partial M \cap \mathrm{Sing}(\mathcal{F}^u). \end{cases} \tag{5.6}$$

In particular, the index of an interior periodic point is zero if and only if it is a one-pronged singularity.

Suppose $g: M \to M$ is a continuous map homotopic to f and $H = \{h_t\}: M \to M$ is a homotopy from f to g. Then $x \in \operatorname{Fix}(f)$ and $y \in \operatorname{Fix}(g)$ are *H-related* if there is a path c from x to y such that c and the path $h_t(c(t))$ are homotopic fixing end points. It is easy to see that if $x, x' \in \operatorname{Fix}(f)$ are H-related to the same point $y \in \operatorname{Fix}(g)$, then x and x' are f-Nielsen equivalent. A fixed point x and an invarinat set Y of f are *f-related* if there is a path l such that $l(0) = x$, $l(1) \in Y$, and l and $f \circ l$ are homotopic via a homotopy l_t of paths with $l_t(0) = x$, $l_t(1) \in Y$ for any t. The following result is contained in Lemma 3.4 of Jiang and Guo ([JG]) or Proposition 1.5 of Boyland ([Boy5]).

(5.7) LEMMA. *Let $\varphi: M \to M$ be a canonical homeomorphism with $S = \emptyset$. Suppose N is a pseudo-Anosov component of φ with $\varphi(N) = N$. Then any fixed point x of φ in $\operatorname{Int} N$ is not φ-Nielsen equivalent to any other fixed point, and moreover x is not φ-related to any boundary circle of M.*

A pseudo-Anosov homeomorphism φ has a Markov partition, i.e. M has a finite cover by (topological) rectangles $R_1, \dots, R_k$, so that for each $i \neq j$, $R_i \cap \operatorname{Int} R_j = \emptyset$ and $f(R_i) \cap R_j$ is either empty or connected. The transition matrix associated with the Markov partition is a $k \times k$ matrix $Q = (q_{ij})$ defined by $q_{ij} = 1$ if $f(R_i) \cap R_j \neq \emptyset$, and 0 otherwise. We can find the periodic points of φ by computing the matrix Q. Sometimes it is useful to think of a pseudo-Anosov homeomorphism φ as being constructed from its Markov partition by gluing rectangles together to get the surface. Then φ is defined so that it acts linearly on every rectangle.

5.3. Unremovability of periodic orbits. Let $h_t: M \to M$ be an isotopy. An isotopy $h'_t: M \to M$ with $h'_0 = h_0$, $h'_1 = h_1$ is a *deformation* of h_t if the arcs corresponding to h_t and h'_t in the space of homeomorphisms of M are homotopic with fixed endpoints. A p-periodic point x_0 of h_0 is said to be *connected* to a p-periodic point x_1 of h_1 by the isotopy h_t if there exist a deformation h'_t of the isotopy h_t and a path c in M with $c(0) = x_0$, $c(1) = x_1$ such that for all t, $c(t)$ is a p-periodic point of h'_t.

(5.8) DEFINITION. Let $f: M \to M$ be a homeomorphism. A p-periodic point x of f is *unremovable* if for any homeomorphism $g: M \to M$ isotopic to f and for any isotopy h_t from f to g, x is connected to a p-periodic point of g by the isotopy h_t.

Thus, if a periodic point x is unremovable, then any g isotopic to f has a periodic point y which shares with x any dynamical property which is preserved as it is followed continuously through a family of homeomorphisms.

Using bifurcation-theoretic techniques, sufficient conditions for the unremovability of a periodic point of a C^1-diffeomorphism were given by Asimov and

Franks (see [AF]). Their work was subsequently generalized to collections of periodic points of homeomorphisms by Hall (see [THal1]). The sufficinet conditions for the unremovability of periodic points x have two requirements. The first requirement is a non-zero fixed point index for an iterate f^p, which detects the fact that g^p has a fixed point y. The second requirement is an "uncollapsibility" of the point x, which implies that this point y has least period p. In terms of the bifurcation theory of dynamical systems, unremovability means that in a one parameter family of homeomorphisms the periodic orbit of x cannot disappear. The index condition insures that the orbit cannot disappear via saddle node. The uncollapsibility condition insures that the orbit cannot collapse to a periodic orbit with lower period. By means of the Nielsen–Thurston classification theorem, necessary and sufficient conditions have been given for the unremovability of fixed points ([JG]) and periodic points ([Boy5]). (Boyland and Hall in [BHa] considered the unremovability relative to a general compact invariant set.) In [Boy5], Boyland proves that for a canonical homeomorphism φ, any periodic point in the interior of a pseudo-Anosov component satisfies the two requirements. (The requirement for the index follows from (5.6) and Lemma 5.7.) Thus, we have

(5.9) THEOREM (Unremovability). *Let φ be a canonical homeomorphism relative to an empty set. Any periodic point in the interior of a pseudo-Anosov component N of φ is unremovable. Moreover, for any homeomorphism f isotopic to φ and any isotopy h_t from φ to f, different periodic points of φ in* Int N *are connected by h_t to different periodic points of f.*

For a homeomorphism f, let $\mathrm{Per}(f)$ be the set of periodic points of f. By Theorem (5.9) we have

(5.10) THEOREM. *Let $f\colon M \to M$ be an orientation-preserving homeomorphism with a finite invariant set S in* Int M. *Let $\varphi\colon M \to M$ be a canonical homeomorphism isotopic to f relative to S. Assume φ has a pseudo-Anosov component. Let L be the union of the interiors of all pseudo-Anosov components of φ. Then, there exists an injective map $\iota\colon \mathrm{Per}(\varphi) \cap L \to \mathrm{Per}(f)$ which is period and braid type preserving in the sense that for any $x \in \mathrm{Per}(\varphi) \cap L$,*

(5.10.1) *$\iota(x)$ has the same period as x, and*

(5.10.2) *the orbit of $\iota(x)$ under f has the same braid type as that of x under φ.*

PROOF. Note that any canonical homeomorphism φ relative to S has the blow-up $\widehat{\varphi}\colon \widehat{M} \to \widehat{M}$ at S. If φ is pseudo-Anosov relative to S, then $\widehat{\varphi}$ becomes a pseudo-Anosov homeomorphism on $\widehat{M}$.

We shall define the map ι. Let $x \in \mathrm{Per}_p(\varphi) \cap L$. If $x \in S$, let $\iota(x) = x$. Suppose $x \notin S$. Then x is a periodic point of $\widehat{\varphi}$ in $\widehat{M} - \mathrm{pr}^{-1}(S)$. For $z \in S$, choose a sufficiently small open disk V_z centered at z, and let $V = \bigcup_{z \in S} V_z$, $\widehat{V} = \mathrm{pr}^{-1}(V)$. We isotope f, relative to S, to a homeomorphism $f' \colon M \to M$ which is smooth and non-singular at every $z \in S$. f' can be chosen so that its p-th iterate $(f')^p$ coincides with f^p away from V. We write $\widehat{f}$ for the blow-up of f' at S. Since φ^p and $(f')^p$ are isotopic relative to S, there exists an isotopy $h_t \colon \widehat{M} \to \widehat{M}$ from $\widehat{\varphi}^p$ to $\widehat{f}^p$. Since $x \in \mathrm{Per}_p(\widehat{\varphi}) \cap L$, applying Theorem 5.9 to $\widehat{\varphi}$, we have that x is connected by h_t to some p-periodic point y of $\widehat{f}^p$.

We show that $y \notin \widehat{V}$. Suppose to the contrary that $y \in \widehat{V}$. Then $y \in \mathrm{pr}^{-1}(V_z)$ for some $z \in S$. Let γ_z be the boundary circle of $\widehat{M}$ corresponding to z. Since V_z is small enough, y is $\widehat{f}^p$-related to γ_z. Since x and y are H-related, where $H = \{h_t\}$, x is $\widehat{\varphi}^p$-related to γ_z. This contradicts with Lemma 5.7. Since f'^p and f^p coincide on $M - V$, y is a p-periodic of f. Let $\iota(x) = y$. By Theorem 5.9, ι is injective and preserves the braid types of periodic orbits. □

6. Applications of Nielsen–Thurston theory

6.1. Classification for braid types. The Nielsen–Thurston theorem leads to the following classification theory for braid types. Let $\beta = [b] \in \mathrm{BT}_n(M)$. Let φ be the canonical homeomorphism in the isotopy class $\Theta(b) \in \mathrm{Iso}(M, S; \mathrm{id})$ corresponding to b.

(6.1) DEFINITION. A braid type $\beta = [b]$ is called *finite order*, *pseudo-Anosov*, or *reducible* if a canonical homeomorphism $\varphi \colon M \to M$ in the isotopy class $\Theta(b)$ is finite order, pseudo-Anosov relative to S, or reducible relative to S, respectively. We say β contains a pseudo-Anosov component if so does φ.

This definition is reasonable because the canonical homeomorphism φ is uniquely determined by β up to topological conjugacy. Note that if two canonical homeomorphisms are topologically conjugate, their decompositions into finite order and pseudo-Anosov components must coincide.

Several results have been obtained on the classification of braid types for some surfaces. Consider first the case of M being the disk D. A unique element of BT_1 is denoted by e.

(6.2) THEOREM.

(6.2.1) *A braid type β is reducible if and only if β is obtained from another braid type β' by replacing a cyclic sub-braid type γ of β' to $[\gamma, \delta]$, where δ is a braid type with at least two strings.*

(6.2.2) *Any cyclic braid type $\beta \neq e$ has a unique expression $\beta = [\beta_1, \dots, \beta_r]$, where each $\beta_i \neq e$ is a cyclic irreducible braid type. Moreover, β is irreducible if and only if $r = 1$.*

(6.2.3) *An irreducible n-braid type β is pseudo-Anosov if and only if $\beta \neq (\alpha_n)^k$ for any integer k.*

(6.2.4) *A cyclic irreducible n-braid type β with exponent sum $e(\beta)$ not divisible by $n-1$ is pseudo-Anosov.*

(6.2.5) *A cyclic n-braid type β with n prime is irreducible.*

(6.2.1) and (6.2.2) follow from the definition of reducibility. The results (6.2.3) and (6.2.4) are due to Boyalnd [Boy1]. ((6.2.4) is an easy consequence of (6.2.3).) (6.2.3) asserts that the problem of classification of irreducible braid types is reduced to the conjugacy problem on the braid group. (6.2.5) is also included in Boyland [Boy1] (see also Benardete, Gutierrez and Nitecki in [BGN2, Theorem 7.3]). (6.2.5) follows from the fact that any reducing curve for the canonical homeomorphism corresponding to β would have to enclose at least 2 points from the orbit, but not all n points, and the fact that the reducing curves must be permuted, which would show that n has non-trivial factors.

The result (6.2.4) is not applicable in the case of $n = 3$, because a cyclic 3-braid type β always has even exponent sum $e(\beta)$, which is divisible by $2 = n-1$. Matsuoka in [Ma3] proved that $\beta \in \mathrm{BT}_3$ is pseudo-Anosov if and only if β is an alternating braid type up to multiplication by the full-twist braid (see also Example 5.4). Generalizing this result to $n \geq 3$, Matsuoka [Ma6] showed that any n-braid is conjugate to $\rho_n\sigma_1^{k_1}\cdots\rho_n\sigma_1^{k_r}$, $k_1,\ldots,k_r > 0$, up to multiplcation by the full-twist braids. Assume that any $n-2$ cyclically consecutive elements of the sequence $k_1,\ldots,k_r$ have sum $\geq n$. Then if $k_i > 3$ for some i or $k_i = 2$ and $k_j = 3$ for some i, j, then β has a pseudo-Anosov component. Song, Ko, and Los in [SKL] gave an explicit representative for any conjugacy class of pseudo-Anosov 4-braids.

Now consider the case where M is an annulus A. Let $\alpha_q = \mathrm{bt}(S, R_{1/q})$, where $R_{1/q}$ is the rigid rotation of A by $1/q$ and S is a periodic orbit with period q of $R_{1/q}$. This clearly is independent of the choice of S. Boyland in [Boy1] showed that if p, q are relatively prime, then a q-braid type β with rotation vector $\rho(\beta) = p/q$ is irreducible, where $H_1(A;\mathbb{Q})$ is identified with $\mathbb{Q}$. Moreover, β is pseudo-Anosov if and only if $\beta \neq [(\alpha_q)^p]$.

Let M be the torus T with an open disk removed, and S consist of a single point x_0. Let $\beta_0 = \mathrm{bt}(\{x_0\}, f) \in \mathrm{BT}_1(M) = \pi_1(M)/\mathrm{Conj}$. The conjugacy class $\kappa(\beta_0)$ introduced in Section 2.4 is represented by the closed loop $C(t) = f_t(x_0)$. Then, Theorem 2' and its addendum of Kra ([Kr]) assert that the classification of β_0 is given by behavior of C as follows: $\beta_0 \neq [e]$ if and only if C is contractible. β_0 is reducible if and only if C is an "essential" curve. All the components of β_0 is finite order if and only if C is homotopic to a power of a simple closed curve.

6.2. Periods of periodic points. The classification for braid types has a great significance in the theory of dynamical systems on surfaces: Theorem (5.10) and the existence of a Markov partition for a pseudo-Anosov homeomorphism imply that if f has a finite invariant set S whose braid type $\mathrm{bt}(S, f)$ contains a pseudo-Anosov component, then f must have dynamical complexity, in fact, it has an infinite number of periodic points with different periods. Various results have been obtained on the periods which appear in f as application of the theorem.

Let $\mathrm{period}(f) \subset \mathbb{N}$ be the set of periods of periodic orbits of f. Given a sequence $\{a_p\}_{p\in\mathbb{N}}$ of non-negative numbers, its *growth rate* $\mathrm{GR}(a_p)$ is defined by

$$\mathrm{GR}(a_p) = \limsup_{p\to\infty}(a_p)^{1/p}.$$

Roughly speaking, a sequence having growth rate λ will grow like λ^p as $p \to \infty$. We have:

(6.3) THEOREM.

(6.3.1) *Suppose the braid type* $\mathrm{bt}(S, f)$ *has a pseudo-Anosov component. Then the growth rate* $\mathrm{GR}(\sharp\mathrm{Per}_p(f))$ *of the number of periodic points of period p is greater than* 1. *Also, there exist infinitely many period doubling sequences of periodic orbits.*

(6.3.2) *If* $\mathrm{bt}(S, f)$ *is pseudo-Anosov, then* $\mathrm{period}\,(f) = \mathbb{N} - E$ *for some finite subset E of* $\mathbb{N}$.

The first half of (6.3.1) is an easy corollary of Theorem 3.8 in Jiang [J5]. See also [J4]. The second half of (6.3.1) is Theorem 4 of Guaschi in [G1]. (6.3.2) is due to Gambaudo and Llibre (see [GL]), and is obtained by investigating periodic points of a pseudo-Anosov homeomorphism by observing the transition matrix associated with it.

More detailed results have been obtained in a special case where M is a closed disk D or a sphere S^2.

(6.4) THEOREM. *Let $M = D$. Then*

(6.4.1) *If* $\mathrm{bt}(S, f)$ *is pseudo-Anosov and* $\sharp S = 3, 4$, *then* $\mathrm{period}(f) = \mathbb{N}$.

(6.4.2) *If P is a periodic orbit with period n with* $\mathrm{bt}(P, f)$ *pseudo-Anosov, then f has periodic points of all periods greater than $n - 3$ or has periodic points of all periods divisible by n.*

(6.4.3) *If* $\mathrm{bt}(S, f)$ *has a pseudo-Anosov component, then f has at least $2p + 3$ periodic points of period $\leq p$ for any $p \geq m(S)$, where $m(S)$ denotes the maximum of the periods of points in S.*

In the case of $M = S^2$, *if* $\mathrm{bt}(S, f)$ *is pseudo-Anosov and if* $\sharp S = 4, 5$, *then* $\mathrm{period}(f) = \mathbb{N}$.

(6.4.1) was proved by Kolev ([K3]) when S is a periodic orbit of period 3, following an example given by Gambaudo, van Strien, and Tresser in [GVT2]. The general form was obtained by Guaschi (see [G2]). Also, Theorem 2 of [Ma3] on embeddings of the plane immediately implies that if $M = D$ or $\mathbb{R}^2$, $\sharp S = 3$, and if f is differentiable, then f has a periodic point of every period greater than three.

The result for S^2 is due to Guaschi ([G2]) and Llibre and MacKay ([LM3]). This result on S^2 and (6.4.1) are the best possible ones. In fact, Guaschi in [G4] presented counter-examples in the case where $M = D$ and $n \geq 5$, or in the case where $M = S^2$ and $n \geq 6$. Also, Llibre and MacKay in [LM3] obtained counter-examples in the case where $M = S^2$ and $n \geq 7$. (6.4.2) is a combination of Propositions 13.2 and 13.4 in Franks and Misiurewicz [FM]. (6.4.3) is in Matsuoka [Ma9]. In the case of $S \subset \mathrm{Fix}(f)$ and $p = 1$, this estimate was obtained in [Ma8, p. 461], and was improved under some transversality condition in [Ma9].

To show some typical arguments in applying Theorem (5.10), we shall prove (6.4.1) in the case where S is a periodic orbit of period 3. This proof follows Kolev in [K3]. By Theorem (5.10), it is enough to prove the result for a pseudo-Anosov homeomorphism φ relative to S. Let $L(\varphi)$ denotes the ordinary Lefschetz number of φ. Since $M = D$ is the disk, $L(\varphi) = L(\varphi^2) = 1$. By (5.6), every fixed point of an iterate of φ has index ≤ 1. Therefore, there exist fixed points x_1 and x_2 for φ and φ^2 with index 1, respectively. By the Euler–Poincaré formula (5.5), the only singularities of the foliations are the 1-prongs at S. Hence, x_2 is a reglular point of the foliations, and hence by (5.6), it must have period 2. Blow up the point x_1 into a circle γ. Then we have an annulus $A = \widehat{D}$. Since the blow-up $\widehat{\varphi}$ has a periodic orbit of period 2 on γ and a fixed point on ∂D, the rotation numbers of $\widehat{\varphi}$ on γ and on ∂D are $1/2$ and $0 \bmod \mathbb{Z}$. Therefore, by Corollary 2.4 of Franks [Fr2], $\widehat{\varphi}$ has a periodic orbit with period $p \geq 3$ in A. These points must be in $D - \{x_0\}$. Thus we have concluded that φ has a periodic orbit for any positive integer p. (We can give an alternative proof by using Example (5.4) (see [LM3] and [Ma3])).

7. Forcing relation on braid types

By Theorem (5.10), the existence of a single periodic orbit with braid type having a pseudo-Anosov component can imply the coexistence of many other periodic orbits. This fact gives rise to the notion of forcing relations, describing which periodic orbits are forced by which other orbits. Forcing relations are defined using an abstraction of the conclusion of the well known Sharkovskiĭ's theorem on the real line.

Let $\mathcal{C}$ be a set of continuous self-maps of a given topological space X, and then introduce a means of specifying periodic orbits of maps in this class: that is, a set $\mathcal{S}$ of specifications and a means of associating to each periodic orbit P of a map $f \in \mathcal{C}$ an element $s(P, f)$ of $\mathcal{S}$. A *forcing relation* $\preceq$ can then be defined on $\mathcal{S}$:

(7.1) Definition. Given $s_1, s_2 \in \mathcal{S}$, we say that $s_1 \succeq s_2$ if and only if every map $f \in \mathcal{C}$ which has a periodic orbit P with $s(P, f) = s_1$ also has a periodic orbit Q with $s(Q, f) = s_2$. If $s_1 \succeq s_2$, we say that s_1 *forces* s_2.

Notice that a forcing relation is necessarily reflexive and transitive. If $s_1 \succeq s_2$ and f_μ is a parametrized family of maps so that f_0 has only trivial dynamics, say a single fixed point, then the parameter values at which a periodic orbit of type s_1 first appears must be after or simultaneous to the value at which s_2 first appears. In this way, forcing relations give a qualitative universality for the birth of periodic orbits in parametrized families.

If we take $\mathcal{C}$ to be the set of continuous maps f on an interval, $\mathcal{S} = \mathbb{N}$, and $s(P, f)$ the period of P, the forcing relation becomes the well known Sharkovskiĭ order (see [S]). More detailed information can be obtained if we take $\mathcal{S}$ to the set all cyclic permutations on $\{1, \dots, n\}$, $n \in \mathbb{N}$, and define $s(P, f)$ as follows: If we label the points of $P = \{x_1, \dots, x_n\}$ so that $x_1 < \dots < x_n$, then define $s(P, f)$ to be the permutation $\mu \in \Sigma_n$ which is defined by $f(x_i) = x_{\mu(i)}$. This forcing relation is a partial order, and has been studied in great detail and much of the work is described in [ALM].

The analogous specification of a periodic orbit P for a surface homeomorphism f is the braid type. This forcing relation was introduced by Boyland in [Boy1]. Let $\mathcal{C}$ be the set $\operatorname{Homeo}(M; \mathrm{id})$ of homeomorphisms on M isotopic to id, $\mathcal{S}$ the set $\mathrm{CBT}(M)$ of all the cyclic braid types, and $s(P, f)$ the braid type of P. For $f \in \operatorname{Homeo}(M; \mathrm{id})$, we set

$$\mathrm{bt}(f) = \{\mathrm{bt}(P, f) : P \text{ is a periodic orbit of } f\} \subset \mathrm{CBT}(M).$$

Then we have

$$\beta_1 \succeq \beta_2 \Leftrightarrow \text{for any } f \in \operatorname{Homeo}(M; \mathrm{id}),\ \beta_1 \in \mathrm{bt}(f) \text{ implies } \beta_2 \in \mathrm{bt}(f).$$

Boyland [Boy3] proved that following:

(7.2) Proposition. *The forcing relation on* $\mathrm{CBT}(M)$ *is a partial order.*

The main ingredient in the proof of this proposition is a theorem of Brunovsky which allows one to find families of diffeomorphisms for which bifurcations of periodic orbits of less than some fixed period take place at different parameter values. This is applied to find an isotopy from f to a simple map, say with its set of periodic orbits consisting of a finite set of fixed points. Since the periodic orbits of period less than a fixed period disappear at different parameter values, the braid types which are present are diminishing one by one. In particular, for any pair of braid types, one can find a map that has one and not the other. A topological proof of this result in the case of $M = D$ is given by Los [Lo2, Lemma 5.2].

In the case of 1-dimensional maps, for each cyclic permutation μ, there exists a piecewise linear map $f_\mu\colon \mathbb{R} \to \mathbb{R}$ which has periodic orbits of precisely those permutations forced by μ. Thus to determine whether or not μ forces ν, we enumerate the periodic orbits of f_μ of the appropriate period, and match the permutation of each in turn against ν (see [ALM]).

A similar method can be used in the two-dimensional case. Boyland [Boy5] constructed a representative φ, called a "condensed" homeomorphism, for any isotopy class of surface homeomorphisms. For a braid type $\beta = [b]$, let φ_β be the condensed homeomorphism in the isotopy class $\Theta(b) \in \mathrm{Iso}(M, S; \mathrm{id})$ corresponding to b. This is uniquely determined up to topological conjugacy. If β is pseudo-Anosov, φ_β is a pseudo-Anosov homeomorphism. An improvement of Theorem (5.10), given in [Boy5], implies:

(7.3) THEOREM ([Boy5]). *Let β be a braid type. Then*

$$\{\gamma \in \mathrm{CBT}(M) : \gamma \preceq \beta\} = \mathrm{bt}(\varphi_\beta).$$

This result reduces the problem of determining the forcing relation on the braid types to determine the braid types in the condensed homeomorphism φ_β. It is possible to decide whether or not β forces γ by using the Markov partition for φ_β to enumerate all of its periodic orbits of the appropriate period, and matching each of these in turn against γ. We can use the algorithms given in [BH], [FM], [Lo1] to find a Markov partition for a pseudo-Anosov homeomorphism. If φ_β is a finite order map, then β only forces itself and the fixed point braid type. It follows from Theorem (7.3) that $\mathrm{bt}(\varphi_\beta) \subset \mathrm{bt}(f)$ for any f in the isotopy class containing φ_β.

Let $\mathrm{FI}(p/q)$ denotes the *Farey interval* of p/q, that is $\mathrm{FI}(p/q) = [a/b, c/d]$, where $a + c = p$, $b + d = q$, $bc - ad = 1$. The endpoints are the nearest rational numbers with denominators less than q. If we write the continued fraction of p/q so that it ends in a one, the endpoints of the Farey interval are the last two convergents of p/q. Consider the braid types $\beta_{p/q}$ on an annulus A obtained by doing a rigid rotation by p/q followed by a Dehn twist around a pair of adjacent points on the orbit. This braid type is represented by a pseudo-Anosov homeomorphism with a Markov partition that is essentially that of a circle map. Using the structute of this Markov partition, Boyland proved in [Boy2] that

$$\beta_{p_1/q_1} \preceq \beta_{p_2/q_2} \Leftrightarrow \mathrm{FI}(p_1, q_1) \subset \mathrm{FI}(p_2, q_2).$$

Also, if k, l are relatively prime and $k/l \in \mathrm{FI}(p/q)$, then $\beta_{p/q} \succ (\alpha_l)^k$.

7.1. Forcing on a disk. The surface where the forcing relation has been most studied is the disk.

(7.4) THEOREM.

(7.4.1) *For cyclic braid types* β_1, β_2,

$$\gamma \prec [\beta_1, \beta_2] \Leftrightarrow \gamma \prec \beta_1 \ or\ \gamma = [\beta_1, \beta_3] \ for\ some\ \beta_3 \prec \beta_2.$$

(7.4.2) $[\beta_1, \dots, \beta_r]$ *is primary, i.e. it does not force any other braid type of the same period, if and only if* $\beta_1, \dots, \beta_{r-1}$ *are finite order and* β_r *is finite order or is pseudo-Anosov such that the corresponding canonical homeomorphism has a single fixed point.*

(7.4.3) *If* $\beta \succ \gamma$ *and* β *is not an extension of* γ, *then* β *forces an infinite number of extensions of* γ.

(7.4.1) and (7.4.2) are due to Theorems 12.1 and 14.4 of Franks and Misiurewicz ([FM]). (7.4.3) is due to Miled and Gambaudo ([MG]). Los in [Lo2] analyzed the structure of the forcing relation by introducing a topology on CBT(D).

Recall that a 3-braid type is pseudo-Anosov if and only if it is represented by a cyclic word consisting solely of the generators σ_1 and σ_2^{-1}, and not their inverses, and has at least one of each of these generators. Extending a result of Matsuoka in [Ma3], Handel in [Han2] obtained the following result:

(7.5) THEOREM. *For cyclic pseudo-Anosov 3-braid types* β_1 *and* β_2, $\beta_1 \succ \beta_2$ *if and only if the cyclic word in* σ_1 *and* σ_2^{-1} *contained in* β_2 *is obtained from that of* β_1 *by deleting generators.*

For example, $[\sigma_1^i \sigma_2^{-1}] \succ [\sigma_1^{i-2} \sigma_2^{-1}]$ for any odd $i \geq 3$.

T. Hall in [THal2], [THal4] and de Carvalho and Hall in [DH1]–[DH3] studied the forcing relation on the braid types appearing in the Smale's horseshoe maps. Their results describe constraints on the order in which periodic orbits can appear when a horseshoe is created. Because any diffeomorphism of a surface with positive topological entropy has a horseshoe in some iterate, the results can be seen as having implications for the mechanism of two-dimensional transitions to chaos in a general context.

Yamaguchi and Tanikawa (see [YT1]–[YT3]) studied the “standard” map, which is the most familiar example of monotone twist maps on an annulus, and found that an infinite sequence of braid types occur as a parameter is increased. It is natural to conjecture that the order of occurrence of braid types would coincide with the forcing relation. Motivated by this conjecture, Kin in [Ki] studied the disk homeomorphisms and obtained results on the forcing relation on a certain class of braid types closely related to the Yamaguchi–Tanikawa's braid types.

In [Ma3], a forcing relation was generalized in two ways. First, it was generalized to some noncompact setting. It treats the plane $\mathbb{R}^2$, and proves that the forcing relations on the cyclic 3-braid types for the disk and the plane have no difference. Secondly, note that Theorem (7.5) is in fact the ordering on the set CBT_3/Z_3 and not on the braid types CBT_3. Thus one may have a finer ordering on CBT_3. In fact, for some 3-braid types, a finer ordering has been obtained in [Ma3]. For instance, for any odd number $i \geq 3$, [Ma3] found a relation $[\sigma_1^i\sigma_2^{-1}] \succ [\theta_3\sigma_1^{i-2}\sigma_2^{-1}]$, which does not follow from Theorem (7.5). Also, for alternating 3-braid types,

$$[\sigma_1^i] \prec [\sigma_1^{i_1}\sigma_2^{-j_1}\dots\sigma_1^{i_r}\sigma_2^{-j_r}] \Leftrightarrow -(j_1+\dots+j_r) \leq i \leq i_1+\dots+i_r.$$

8. Topological entropy

As has been indicated, the existence of a periodic orbit with braid type having a pseudo-Anosov component implies the existence of infinitely many periodic orbits with different periods. This suggests that the existence of such an orbit would imply that f has positive topological entropy. The topological entropy $h(f)$ for a map f is a well known measure of the complexity of a dynamical system (see e.g. [FLP]).

(8.1) Definition. Given a braid type β, define its entropy $h(\beta)$ as the infimum

$$h(\beta) = \inf\{h(f) : f \text{ has a finite invariant set } S \text{ with } \mathrm{bt}(S) = \beta\}.$$

Identify $\pi_1(M-S)$ with a free group F_n with generators $x_1, \dots, x_n$. For $\omega \in F_n$, let $L(\omega) = \min\{r : \omega = x_{i_1}^{\varepsilon_1}\dots x_{i_r}^{\varepsilon_r}, \varepsilon_i = \pm 1\}$, i.e. the minimal length of the word ω. For a homomorphism $\psi\colon F_n \to F_n$, let

$$\mathrm{GR}(\psi) = \max_i \mathrm{GR}(L(\psi^p(x_i))).$$

Call it the *growth rate of* ψ. By (4.14), a braid b induces an automorphism $\psi_b\colon F_n \to F_n$.

(8.2) Theorem.

(8.2.1) $h(\beta) = h(\varphi_\beta) = \log \mathrm{GR}(\psi_b)$, *where* φ_β *is the canonical homeomorphism correponding to* $\beta = [b]$.

(8.2.2) *$h(\beta) > 0$ if and only if β contains a pseudo-Anosov component.*

(8.2.3) *If β_1, β_2 are cyclic pseudo-Anosov braid types and $\beta_1 \succ \beta_2$, then $h(\beta_1) > h(\beta_2)$.*

(8.2.4) *In the case of $M = D$, $h([\beta_1, \beta_2]) = \max\{h(\beta_1), h(\beta_2)/n\}$, where n is the number of strings of β_1.*

(8.2.1) and (8.2.2) are contained in [FLP, Exposé 10]. (8.2.3) is Theorem 9.3 of [Boy4]. (8.2.4) is Theorem 12.2 of Franks and Misiurewicz (see [FM]).

By using the equality $h(\beta) = h(\varphi_\beta)$ in (8.2.1), some estimates for $h(\beta)$ have been given in the case of $M = D$. Theorem 12.9 of Franks and Misiurewicz ([FM]) states that if β is pseudo-Anosov, then $h(\beta) \geq (1/(n-1)) \log 2$. Fehrenbach and Los in [FL] has shown that if a braid type β contains a pseudo-Anosov component, then $h(\beta) \geq (1/n) \log(1 + \sqrt{2})$.

(8.2.2) implies that if f has a periodic orbit with braid type containing a pseudo-Anosov component, then $h(f) > 0$. Conversely, Franks and Handel in [FH] proved that if $h(f) > 0$ and f is smooth, then there exits a finite invariant set S with $\mathrm{GR}(\psi_{b(S)}) > 1$, which implies that $\mathrm{bt}(S, f)$ has a pseudo-Anosov component. Moreover, Gambaudo, Guaschi, and Hall ([GGH]) showed that there exist infinitely many period multiplying cascades with different braid types. Kwapisz and Swanson ([KS]) showed that in the case where f is a diffeomorphism on an annulus A with Hölder continuous derivative, nonvanishing of the "asymptotic entropy" implies that f has a finite invariant set S with $\mathrm{bt}(S, f)$ pseudo-Anosov.

8.1. Burau matrix and entropy. The entropy is related to the Burau matrix of a braid. The following result is contained implicitly in [Fri2]. A proof was given in Kolev [K1].

(8.3) THEOREM. *Let $\beta = [b]$ be a braid type. Then*

$$h(\beta) \geq \sup\{\log(\mathrm{spec}\ R(b)^\pi) : t \in \mathbb{C},\ |t| = 1\},$$

where spec *denotes the spectral radius.*

(8.4) EXAMPLE. Let $b = \sigma_1\sigma_2^{-1} \in B_3$. Then

$$R(b)^\pi = \begin{pmatrix} 1-t & -t^{-1} \\ 1 & -t^{-1} \end{pmatrix}.$$

Therefore, spec $R(b)^\pi$ is the maximum of the absolute values of roots of the quadratic equation $x^2 + (t - 1 + t^{-1})x + 1 = 0$ for $|t| = 1$. Hence, for $t = -1$, $\mathrm{spec}\, R(b)^\pi = (3 + \sqrt{5})/2$ and by Theorem (8.3), $h([\sigma_1\sigma_2^{-1}]) \geq \log((3 + \sqrt{5})/2)$. In fact, this is an equality, since Song, Ko, and Los in [SKL, Corollary 16] has shown that for all 3-braid types the equality in Theorem (8.3) holds at $t = -1$.

More generally, $h([\sigma_1^i\sigma_2^{-1}]) = \log((i+2+\sqrt{i^2+4i})/2)$. Since these values are greater than $\log 2$ which is the topological entropy of the horseshoe map, it implies that such braid types are not realized as periodic orbits of the horseshoe map. Among 3-braids, $[\sigma_1\sigma_2^{-1}]$ is known to have the minimal non-zero entropy (see [Ma3], [Han2]). Song, Ko, and Los [SKL] shows that $[\sigma_3\sigma_2\sigma_1^{-1}]$ reaches the minimum entropy among all pseudo-Anosov 4-braids.

In [SKL, Theorem 15] a braid b' is constructed for a pseudo-Anosov braid type β such that $h(\beta)$ is equal to $\log \operatorname{spec} R(b')^{\pi}$ evaluated at $t = -1$.

Outline of Proof of Theorem (8.3). Let $\psi = \psi_b$. For $w \in F_n$, let $\langle w; x_j\rangle$ denote the number of $x_j^{\pm 1}$ in the word w. Let C_p be the matrix $(\langle \psi^p(x_i); x_j\rangle)$. For a complex matrix $A = (a_{ij})$, define its norm $\|A\|$ by $\|A\| = \max_i \sum_j |a_{ij}|$. Then $\operatorname{spec} A = \operatorname{GR}(\|A^p\|)$. Since $L(w) = \sum_j \langle w; x_j\rangle$, we have

$$\|C_p\| = \max_i \sum_j \langle \psi^p(x_i); x_j\rangle = \max_i L(\psi^p(x_i)),$$

so we have

$$\operatorname{GR}(\|C_p\|) = \operatorname{GR}(\psi). \tag{8.5}$$

$\partial\psi^p(x_i)/\partial x_j$ is written in the form $\sum_{k=1}^{\nu} \varepsilon_k w_k$, where $\nu = \langle \psi^p(x_i); x_j\rangle$, $\varepsilon_k = \pm 1$ and $w_k \in F_n$. Assume $|t| = 1$. Then, since $|(\pi \circ \mathrm{Ab})(w_k)| = |t^{e(w_k)}| = 1$, where $e(w_k)$ is the exponent sum of w_k, we have

$$|(\pi \circ \mathrm{Ab})(\partial\psi^p(x_i)/\partial x_j)| = \left|\sum_{k=1}^{\nu} \varepsilon_k (\pi \circ \mathrm{Ab})(w_k)\right| \leq \nu = \langle \psi^p(x_i); x_j\rangle.$$

Therefore, for any p,

$$\|R(b^p)^{\pi}\| \leq \|C_p\|. \tag{8.6}$$

By (8.5) and (8.6) $\operatorname{GR}(\psi) = \operatorname{GR}(\|C_p\|) \geq \operatorname{GR}(\|(R(b)^{\pi})^p\|) = \operatorname{spec} R(b)^{\pi}$. Thus, by (8.2.1), the conclusion is proved. □

9. Rotation vectors

The rotation number for a circle homeomorphism was introduced by Poincaré. It is well known that if the rotation number of an orientation-preserving circle homeomorphism is rational, then there is a periodic point whose period is given by the denominator of the rational (see [W] for a reference).

This notion has been generalized to surface homeomorphisms. The notion of rotation vector is defined for an orbit which measures the average rate of its motion around the surface in terms of a homology class. An arbitrary orbit will not give a closed loop, so the computation of this vector requires some kind of averaging, and one needs to include irrational coefficients in the homology group. Some orbits will not be assigned a rotation vector. In this case a set of homology classes is assigned, which will be called the rotation set. The size and shape of the union of rotation sets for all points of M give a valuable measure of the complexity of the dynamics.

To define a rotation vector and a rotation set, we shall follow Franks [Fr4]. (See [Po] for a different, but equivalent, definition of a rotation set.) We choose and fix an isotopy f_t from id to f. Let f_t, $t \in \mathbb{R}$, be the extended isotopy $f_t = f_{t-[t]} \circ f^{[t]}$, where $[t]$ is the Gauss symbol, i.e. the greatest integer less than or equal to t. Note that $f_q = f^q$ for integers q.

(9.1) DEFINITION. Choose a base point x_0 of M. For each point x of M, choose a path l_x from x_0 to x. Let $l_{q,x}$ be the product of three paths l_x, $f_{qt}(x)$, and $l^{-1}_{f^q(x)}$. Let $\rho(x,f)$ be the set of accumulation points of the sequence $\{[l_{q,x}]/q\}_{q\in\mathbb{N}}$ in $H_1(M,\mathbb{R})$. This set is called the *rotation set* of x for f. $\rho(x,f)$ will also be denoted by $\rho(x)$. Let $\rho(f) = \bigcup_{x\in M}\rho(x,f)$ and call it the *rotation set* of f. If $\rho(x,f)$ consists of a single point v_0, then we call v_0 the *rotation vector* of x. A periodic point is a *v-periodic point*, where $v \in H_1(M;\mathbb{R})$, if its rotation vector is v.

For any periodic point x, the rotation vector $\rho(x,f)$ exists and it coincides with that of the braid type of its orbit P (see Section 2.4), i.e.

$$\rho(x,f) = \rho(\mathrm{bt}(P,f)) \in H_1(M;\mathbb{R}).$$

9.1. Annulus homeomorphisms. In the case where M is an annulus A, $H_1(A;\mathbb{R}) \cong \mathbb{R}$ and hence the rotation vector is identified with a real number, called the rotation number. Let $A = I \times S^1$ be an annulus, and $\mathrm{pr}\colon \widetilde{A} \to A$ the universal covering map of A, where $\widetilde{A} = I \times \mathbb{R}$. Let $\widetilde{f}_t\colon \widetilde{A} \to \widetilde{A}$ a lift of f_t with $\widetilde{f}_0 = \mathrm{id}$. Let $\widetilde{f} = \widetilde{f}_1$. Suppose $x \in A$, and choose a point $\widetilde{x} \in \widetilde{A}$ with $\mathrm{pr}(\widetilde{x}) = x$. Let $\theta(q)$ be the $\mathbb{R}$-component of $\widetilde{f}^q(\widetilde{x})$, and

$$\rho_+(x) = \limsup_{q\to\infty}(\theta(q) - \theta(0))/q, \quad \rho_-(x) = \liminf_{q\to\infty}(\theta(q) - \theta(0))/q.$$

Then $\rho(x,f) = [\rho_-(x), \rho_+(x)]$. If we change the lift, $\rho(x,f)$ is translated by an integer. We have

(9.2) THEOREM.

(9.2.1) $\rho(f) = \{\rho(x) : x \in A,\ \rho(x)$ *consists of a single number*$\}$.
(9.2.2) $\rho(f)$ *is a closed set.*

(9.2.3) *If φ is a pseudo-Anosov homeomorphism, then $\rho(\varphi)$ is a closed interval, and for any f isotopic to φ, we have* $\operatorname{Int}\rho(\varphi) \subset \rho(f)$.

(9.2.4) *For every rational $p/q \in \rho(f)$, f has p/q-periodic point.*

The proof of (9.2.1) is given in [Boy3, p. 205]. (9.2.2), (9.2.3) are due to Handel ([Han1]). (9.2.4) is due to Franks ([Fr2]).

By (9.2.3) and (9.2.4), we see that if f has a periodic orbit with braid type pseudo-Anosov, then for all the rational numbers in a certain interval are realized as the rotation numbers of periodic orbits of f. This interval was investigated by Boyland in [Boy3] as follows. A periodic orbit is said to be *monotone* if its braid type is equal to $(\alpha_q)^p$ for some rational number p/q. Boyland in [Boy3] showed:

(9.3) THEOREM. *If there exists a non-monotone p/q-periodic orbit, then*

$$\mathrm{FI}(p/q) \subset \rho(f).$$

For example, since $\mathrm{FI}(2/5) = [1/3, 1/2]$, we have by this theorem that if f has a non-monotone $2/5$-periodic orbit, then every rational $k/l \in [1/3, 1/2]$ is contained in $\rho(f)$, and consequently by (9.2.4), there exists a k/l-periodic point. Boyland in [Boy3] showed that $\mathrm{FI}(p/q) = \rho(\varphi)$, where φ is a canonical homeomorphism corresponding to the braid type $\beta_{p/q}$.

OUTLINE OF PROOF OF THEOREM (9.3). Let P be a non-monotone p/q-periodic orbit. Let $\mathrm{FI}(p/q) = [a/b, c/d]$. Since by (9.2.3) $\rho(\varphi)$ is a closed interval, it is enough to show that $a/b, c/d \in \rho(\varphi)$. Assume that $a/b \notin \rho(\varphi)$. Then the quotient space $B = \widetilde{A}/(T^{-a} \circ \widetilde{\varphi}^b)$ becomes an annulus. Let $\pi_B : \widetilde{A} \to B$ be the projection. Let $X = \pi_B^{-1}(P) = \{\widetilde{x}_i : i \in \mathbb{Z}\}$, where $\widetilde{\varphi}(\widetilde{x}_i) = \widetilde{x}_{i+p}$, $T(\widetilde{x}_i) = \widetilde{x}_{i+q}$. Since $bp - aq = 1$, $(T^{-a} \circ \widetilde{\varphi}^b)(\widetilde{x}_i) = \widetilde{x}_{i+1}$. Therefore, $\pi_B(X)$ consists of a single point, say x_0. $T^{-p} \circ \widetilde{\varphi}^q$ induces a pseudo-Anosov homeomorphism $\varphi' \colon B \to B$. Let $\widetilde{\gamma}$ be a path in $\widetilde{A}$ connecting $\widetilde{x_1}$ to $\widetilde{x}_2$. Since φ' fixes x_0 and φ' is isotopic to id relative to x_0, $\gamma = \pi_B(\widetilde{\gamma})$ is homotopic to its image $\varphi'(\gamma)$. This contradicts with a property of a pseudo-Anosov homeomorphism (see [Boy3, p. 207]). □

In Theorem (9.2.3), a point v in the boundary of $\rho(\varphi)$ may not be contained in $\rho(f)$. Therefore, to prove the existence of a v-periodic orbit for f by using (9.2.3), it is necessary to verify that v is contained in the interior of $\rho(\varphi)$. One such sufficient condition is given by Boyland, Guaschi, and Hall in [BGH]. They show that if P is a periodic orbit with pseudo-Anosov braid type, then $\rho(P)$ lies in $\operatorname{Int}\rho(\varphi)$. (This holds also for a torus homeomorphism.) As a corollary, they show that if $\rho(P) = 0$ and the two boundary circles of A are not f-related, then since ∂A is contained in a pseudo-Anosov component of φ, we have $0 = \rho(P) \in \operatorname{Int}\rho(\varphi)$. Therefore $0 \in \rho(f)$.

9.2. Rotation vectors on closed surfaces. Consider the case where M is a closed surface with genus ≥ 1.

(9.4) THEOREM. *Let M be a closed surface of genus $g \geq 1$. Suppose $x_1, \dots, x_{2g+1}$ are periodic points with rotation vectors $\rho_1, \dots, \rho_{2g+1} \in H_1(M; \mathbb{R}) \cong \mathbb{R}^{2g}$. Let Δ be the convex hull $\mathrm{Conv}(\rho_1, \dots, \rho_{2g+1})$ of these vectors. Let P be the union of the orbits of $x_1, \dots, x_{2g+1}$. Assume $\mathrm{Int}\,\Delta$ is not empty. Then*

(9.4.1) *If $g = 1$, $\mathrm{bt}(P, f)$ is pseudo-Anosov and for any point $v \in \mathrm{Int}\,\Delta$ with rational coordinates, there is a v-periodic point.*

(9.4.2) *If $g \geq 2$, then $\mathrm{bt}(P, f)$ contains a pseudo-Anosov component.*

(9.4.3) *If $g \geq 2$ and the condition:*

(*) *There is a periodic point x_0 with $\rho(x_0) \in \mathrm{Int}\,\Delta$, and it is in an essential Nielsen class. Moreover, $0 \notin \partial\Delta - \{\rho_1, \dots, \rho_{2g+1}\}$.*

is satisfied, then $\mathrm{bt}(P, f)$ is pseudo-Anosov, and for any v in a dense subset of $\Delta \cap \mathbb{Q}^{2g+1}$, there exists a v-periodic point.

(9.4.1) is due to Llibre and MacKay (see [LM2]), (9.4.2) is in Pollicott [Po], and (9.4.3) is due to Hayakawa in [Hay1]. (9.4.1) is proved as follows: If $\mathrm{bt}(P)$ were reducible, then there is a reducing curve γ in T. If γ is rotational then Δ is collinear, and if γ is not rotational then some x_i has period greater than q_i, where $\rho_i = p_i/q_i$ in lowest terms. Thus we have a contradiction in either case, and hence $\mathrm{bt}(P)$ is irreducible. If it were of finite order, then $\rho(f)$ has only one point, which is a contradiction. Then Lemma 2.1 in Franks [Fr2] implies that there is a desired periodic point.

Sharp (see [Sh]) strengthened the above theorem in the case of diffeomorphisms and showed that the number of periodic points with prescribed rational rotation vector has exponential growth. The proof uses a Markov partition for a pseudo-Anosov homeomorphism. The result (9.4.1) can be also obtained by combining the following result of Franks [Fr3] and the result of Misiurewicz and Ziemian in [MZ1] which asserts the convex hull $\mathrm{Conv}(\rho(f))$ coincides with the set of accumulation points of the set

$$\{(\widetilde{f}^q(\widetilde{x}) - \widetilde{x})/q : \widetilde{x} \in \mathbb{R}^2,\ q \in \mathbb{N}\},$$

where $\widetilde{f}\colon \mathbb{R}^2 \to \mathbb{R}^2$ is a lift of f.

(9.5) THEOREM ([Fr3]). *For any vector $v = (p_1/q, p_2/q) \in \mathrm{Int}\,\mathrm{Conv}(\rho(f))$ with rational coordinates, where p_1, p_2, q have no common factors, there is a periodic point with period q and rotation vector v.*

There is an example for which a rational point on the boundary $\partial\mathrm{Conv}(\rho(f))$ is not realized by a periodic point [MZ2].

In [M1], Matsumoto gives an example which shows that the hypothesis $(*)$ is necessary in (9.4.3). Pollicott and Sharp in [PoS] showed that any rational points in $\operatorname{Int}\Delta$ is realizable by a periodic orbit under the additional assumption that f is a diffeomorphism. Also, Matsumoto in [M1] proved that the same conclusion holds if $\operatorname{Conv}(\rho(f))$ contains a point realizable by an asymptotic probability. As a corollary, since there always exists a fixed point with rotation vector 0, if $\operatorname{Int}\operatorname{Conv}(\rho(f))$ contains zero, then any rational points of $\operatorname{Int}\operatorname{Conv}(\rho(f))$ is realizable by a periodic orbit. Also, (9.4.2) is improved in the case of diffeomorphisms: It was shown that $g+1$ linearly independent rotation vectors are enough to have the same conclusion as (9.4.2). In [M2], Matsumoto showed that if genus $g(M)$ is positive and S is a finite invariant set of f, and there are periodic points $x_i \in M$ of period p_i $(1 \leq i \leq k)$ such that the set $\{p_i\rho(x_i) : 1 \leq i \leq k\} \subset H_1(M;\mathbb{R})$ is "filling", then f has a contractible fixed point. If genus is zero and if there is a periodic orbit P with $\rho(P) = 0$, then the same conclusion holds.

10. Structure of the set of periodic points

We shall consider the case of $M = D$. Applying the Nielsen–Thurston classification theory and the Nielsen fixed point theory, a variety of results have been obtained on the structure of the set of periodic orbits.

We first collect the results on the structure of the periodic point set for homeomorphisms with zero entropy. If a continuous map on an interval has zero entropy, then every periodic point has period a power of two. In the two-dimensional case, Gambaudo, van Strien, and Tresser in [GVT1] and Llibre and MacKay in [LM1] independently showed that for diffeomorphisms with zero entropy, every braid type is equal to $[(\alpha_{q_1})^{p_1}, \dots, (\alpha_{q_r})^{p_r}]$ for some rational numbers $p_1/q_1, \dots, p_r/q_r$. This result is valid for homeomorphisms (see [FM, Theorem 12.5]). Moreover, [GVT1], [LM1] investigated the periodic orbit structure in terms of braid types. (Guaschi, Llibre, and MacKay in [GLM] obtained an analogous result in the case of genus 1.) Also, for any homeomorphism with zero-entropy, Matsuoka in [Ma11] gives a complete classification of the braid type of the set of periodic points with period $\leq p$, $p \in \mathbb{N}$. (A classification result for fixed points was obtained earlier in [Ma8] by Matsuoka.) For example, it is shown that the pure 4-braid β in Figure 3 cannot be realized as the braid type $\operatorname{bt}(\operatorname{Fix}(f))$ for a disk homeomorphism.

There is a problem related to the linking number problem introduced in Section 3.3. A fixed point x and a periodic orbit P of $f\colon D \to D$ is called *linked* if there are simple closed curves C_1, C_2 in $D - (\{x\} \cup P)$ bounding the two closed subdisks D_1, D_2 such that D_1, D_2 are disjoint, $x \in D_1$, $P \subset D_2$, and $f(C_i)$ is isotopic to C_i in $D - (\{x\} \cup P)$ for $i = 1, 2$.

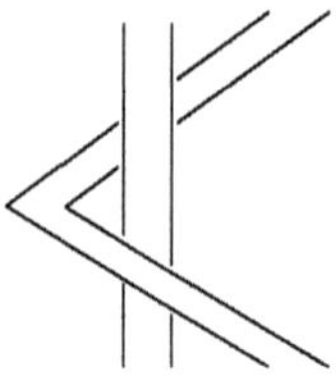

FIGURE 3

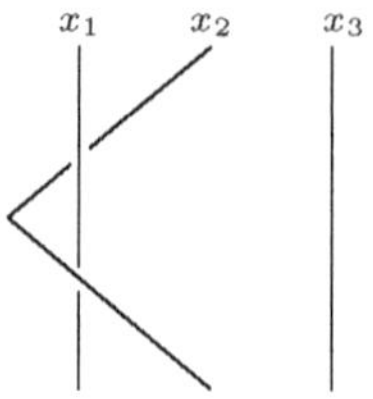

FIGURE 4

(10.1) THEOREM ([G], [K2]). *Let $f: D \to D$ be C^1. Let P be a periodic orbit of f. Then there exists a fixed point of f which is linked with P.*

PROOF. We follow Kolev (see [K2]). We compactify the disk to a sphere S^2. Then f can be extended to a homeomorphism on S^2 having ∞ as a fixed point. Let $\widehat{f}: X \to X$ be the blow-up of the extension $f: S^2 \to S^2$ at P. Let n be the period of P. One can show that $\widehat{f}$ has Nielsen number $N(\widehat{f}) \geq 2$. The result is trivial if $n = 2$. In the case of $n \geq 3$, this is shown as follows: Since f cyclically permutes the points in P, $L(\widehat{f}) = 2$. Since $\chi(X) < 0$, by a result of Nielsen, each Nielsen class of $\widehat{f}$ has index ≤ 1 (see also Theorem (4.4)), and hence $N(\widehat{f}) \geq 2$. If all the fixed points of f were not linked with P, then they were in the same $\widehat{f}$-Nielsen class as ∞. Hence we get $N(\widehat{f}) = 1$, which is a contradiction. (In Gambaudo [G], this result is proved by using Theorem (5.9) and verifying it for canonical homeomorphisms.) □

Matsuoka studied in [Ma7] and [Ma12] the structure of the fixed point set. First, consider the case where all the fixed points are transversal. Then $\mathrm{Fix}(f)$ is decomposed into the two sets $\mathrm{Fix}_+(f)$ and $\mathrm{Fix}_-(f)$ of fixed points with index 1 and -1, respectively. An index is assigned to each fixed point x as follows: If $x \in \mathrm{Fix}_-(f)$, then assign the rotation number of eigenvectors of the differential $Df(x)$ under the isotopy f_t from id to f, and assign a symbol $*$ to any $x \in \mathrm{Fix}_+(f)$. Assume two homeomorphisms f, g have the same number of fixed points, all of which are transversal. Then the following conditions are equivalent ([Ma7]). This result gives a relationship among the fixed point index, braid type, and the self-rotation number of fixed points.

(10.2.1) $\mathrm{Fix}(f)$ and $\mathrm{Fix}(g)$ have the same braid type.
(10.2.2) $\mathrm{Fix}_+(f)$ and $\mathrm{Fix}_+(g)$ have the same braid type.
(10.2.3) $\mathrm{Fix}(f)$ and $\mathrm{Fix}(g)$ have the same indexed braid type.
(10.2.4) $\mathrm{Fix}_-(f)$ and $\mathrm{Fix}_-(g)$ have the same indexed braid type.

[Ma12] introduced an equivalence relation on $\mathrm{Fix}(f)$ defined by using the braid type: Two fixed points are equivalent if the strings corresponding to them are not separated by any other string. Then, any equivalence class E is shown to have index ≤ 1. Hence, the result of Dancer and Ortega ([DO]), which asserts that a Lyapunov stable fixed point has fixed point index 1, implies that any equivalence class E consists of at least two points must contain an unstable fixed point. It is also shown that more than half number of equivalence classes contain a unstable fixed point. Also, the fixed point indices of equivalence classes are proved to be invariant under isotopy.

(10.3) EXAMPLE. If the braid $b(\mathrm{Fix}(f))$ is as in Figure 4, then the equivalence classes are $\{x_1, x_2\}$ and $\{x_3\}$, and for any g with $b(\mathrm{Fix}(g)) = b(\mathrm{Fix}(f))$, the fixed point indices of $\{x_1, x_2\}$ and x_3 must be equal to 1 and 0, respectively. Hence, x_3 and at least one of x_1, x_2 are unstable fixed points for such g.

A similar result on the plane $\mathbb{R}^2$ was obtained in [Ma10]. In this case, the result holds for all equivalence classes except at most one exception.

OUTLINE OF THE PROOF FOR $\mathrm{ind}(E) \leq 1$. To simplify the argument, assume that f has a blow-up $\widehat{f}$ at $\mathrm{Fix}(f)$. It follows from the definition of the equivalence relation that the union γ_E of boundary circles γ_z with $z \in E$ are contained in a component N of φ. This component N must be contained in a φ-Nielsen class F. Then we have

$$\mathrm{ind}(E, f) = \mathrm{ind}(\gamma_E, \widehat{f}) + \sharp E = \mathrm{ind}(F, \varphi) + \sharp E. \tag{10.4}$$

In the case of $\varphi = \mathrm{id}$ on N, Lemma 3.6 in [JG] implies that F is equal to N possibly with finitely many points on boundary circles of pseudo-Anosov components. Therefore, by (5.6) $\mathrm{ind}(F, \varphi) \leq \chi(N) = 1 - \sharp E$, and hence by (10.4), $\mathrm{ind}(E) \leq 1$. On the other hand, in the case where φ is pseudo-Anosov on N, $E = \{x_0\}$ for a point x_0 and $F = \gamma_{x_0} \cap \mathrm{Fix}(\varphi)$, and so by (5.6) $\mathrm{ind}(F, \varphi) \leq 0$. Thus, the result also follows from (10.4) in this case. □

References

[ALM] L. Alsedà, J. Llibre and M. Misiurewicz, *Combinatorial Dynamics and Entropy in Dimension One,*, 2nd ed., Adv. Ser. Nonlinear Dynam., vol. 5, World Scientific, Singapore, 2000.

[A1] J. Andres, *A nontrivial example of application of the Nielsen fixed-point theory to differential systems: Problem of Jean Leray*, Proc. Amer. Math. Soc. **128** (2000), 2921–2931.

[A2] ______, *Nielsen number, Artin braids, Poincaré operators and multiple nonlinear oscillations*, Nonlinear Anal. **47** (2001), 1017–1028.

[AG] J. Andres and L. Górniewicz, *Topological Fixed Point Principles for Boundary Value Problems, "Topological Fixed Point Theory and its Applications"*, vol. 1, Kluwer, Dordrecht, Boston, London, 2003.

[AF] D. Asimov and J. Franks, *Unremovable closed orbits*, preprint (1989), (This is a revised version of a paper which appeared in Lecture Notes in Math. Vol. 1007, Springer–Verlag, 1983).

[BGN1] D. Benardete, M. Gutierrez and Z. Nitecki, *A combinatorial approach to reducibility of mapping classes*, Mapping Class Groups and Moduli Spaces of Riemann Surfaces (C.-F. Bödigheimer and R. Hain, eds.); Contemp. Math. **150** (1993), 1–31.

[BGN2] ———, *Braids and the Nielsen–Thurston classification*, J. Knot Theory Ramifications **4** (1995), 549–618.

[BH] M. Bestvina and M. Handel, *Train-tracks for surface homeomorphisms*, Topology **34** (1995), 109–140.

[Bi] J. S. Birman, *Braids, links and mapping class groups*, Ann. Math. Stud., vol. 82, Princeton Univ. Press, Princeton, 1974.

[BK] C. Bonatti and B. Kolev, *Existence de points fixes enlacés à une orbite périodique d'un homéomorphisme du plan*, Ergodic Theory Dynam. Systems **12** (1992), 677–682.

[Bow] R. Bowen, *Entropy and the fundamental group*, The Structure of Attractors in Dynamical Systems, Lecture Notes in Math. (N. G. Markley et al., eds.), vol. 668, Springer-Verlag, Berlin–Heidelberg–New York, 1978, pp. 21–29.

[Boy1] P. Boyland, *Braid types and a topological method of proving positive entropy* (1984), unpublished.

[Boy2] ———, *An analog of Sharkovski's theorem for twist maps*, Contemp. Math. **81** (1988), 119–133.

[Boy3] ———, *Rotation sets and monotone periodic orbits for annulus homeomorphisms*, Comm. Math. Helv. **67** (1992), 203–213.

[Boy4] ———, *Topological methods in surface dynamics*, Topology Appl. **58** (1994), 223–298.

[Boy5] ———, *Isotopy stability of dynamics on surfaces*, Contemp. Math. **246** (1999), 17–45.

[Boy6] ———, *Dynamics of two-dimensional time-periodic Euler fluid flows*, preprint.

[BAS] P. Boyland, H. Aref and M. Stremler, *Topological fluid mechanics of stirring*, J. Fluid Mech. **403** (2000), 277-304.

[BF] P. Boyland and J. Franks, *Notes on Dynamics of Surface Homeomorphisms*, Informal Lectures Notes, Math. Institute, University of Warwick, 1989.

[BGH] P. Boyland, J. Guaschi and T. Hall, *L'ensemble de rotation des homéomorphismes pseudo-Anosov*, C. R. Acad. Sci. Paris, Sér. I **316** (1993), 1077–1080.

[BHa] P. Boyland and T. Hall, *Isotopy stable dynamics relative to compact invariant sets*, Proc. London Math. Soc. 79 (1999), 673–693.

[BSA] P. Boyland, M. Stremler and H. Aref, *Topological fluid mechanics of point vortex motions*, Phys. D **175** (2003), 69–95.

[Br] R. F. Brown, *The Lefschetz Fixed Point Theorem*, Scott–Foresman, Glenview, 1971.

[CB] A. Casson and S. Bleiler, *Automorphisms of Surfaces After Nielsen and Thurston*, London Math. Soc. Student Texts, No. 9, Cambridge Univ. Press, Cambridge, 1988.

[DO] E. N. Dancer and R. Ortega, *The index of Lyapunov stable fixed points in two dimensions*, J. Dynam. Differential Equations **6** (1994), 631–637.

[DH1] A. de Carvalho and T. Hall, *The forcing relation for horseshoe braid types*, Experiment. Math. **11** (2002), 271–288.

[DH2] ———, *Conjugacies between horseshoe braids*, Nonlinearity **16** (2003), 1329–1338.

[DH3] ———, *Braid forcing and star-shaped train tracks*, Topology **43** (2004), 247–287.

[FH1] E. Fadell and S. Husseini, *Fixed point theory for non-simply connected manifolds*, Topology **20** (1981), 53–92.

[FH2] ———, *The Nielsen number on surfaces*, Topological Methods in Nonlinear Functional Analysis (S. P. Singh et al., eds.), Contemp. Math, vol. 21, Amer. Math. Soc., Providence, 1983, pp. 59–98.

[FLP] A. Fathi, F. Laudenbach and V. Poénaru, *Travaux de Thurston sur les surfaces, Astérisque*, vol. 66–67, Soc. Math. France, Paris, 1979.

[FL] J. Fehrenbach and J. Los, *Une minoration de l'entropie topologique des difféomorphismes du disque*, J. London Math. Soc. **60** (1999), 912–924.

[Fr1] J. Franks, *Knots, links and symbolic dynamics*, Ann. of Math. **113** (1981), 529–552.

[Fr2] ———, *Recurrence and fixed points of surface homeomorphisms*, Ergodic Theory Dynam. Systems **8*** (1988), 99–107.

[Fr3] ______, *Realizing rotation vectors for torus homeomorphisms*, Trans. Amer. Math. Soc. **311** (1989), 107–115.

[Fr4] ______, *Geodesics on S^2 and periodic points of annulus homeomorphisms*, Invent. Math. **108** (1992), 403–418.

[FH] J. Franks and M. Handel, *Entropy and exponential growth of π_1 in dimension two*, Proc. Amer. Math. Soc. **102** (1988), 753–760.

[FM] J. Franks and M. Misiurewicz, *Cycles for disk homeomorphisms and thick trees*, Nielsen Theory and Dynamical Systems (C. McCord, ed.), Contemp. Math., vol. 152, Amer. Math. Soc., Providence, 1993, pp. 69–139.

[Fri1] D. Fried, *Periodic points and twisted coefficients*, Geometric Dynamics (J. Palis ed.), Lecture Notes in Math., vol. 1007, Springer-Verlag, Berlin–Heidelberg–New York, 1983, pp. 261–293.

[Fri2] ______, *Entropy and twisted cohomology*, Topology 25 (1986), 455–470.

[G] J. M. Gambaudo, *Periodic orbits and fixed points of a C^1 orientation-preserving embedding of D^2*, Math. Proc. Cambridge Philos. Soc. **108** (1990), 307–310.

[GGH] J. M. Gambaudo, J. Guaschi and T. Hall, *Period-multiplying cascades for diffeomorphisms of the disc*, Math. Proc. Cambridge Philos. Soc. **116** (1994), 359–374.

[GL] J. M. Gambaudo and J. Llibre, *A note on the periods of surface homeomorphisms*, J. Math. Anal. Appl. **177** (1993), 627–632.

[GVT1] J. M. Gambaudo, S. van Strien and C. Tresser, *The periodic orbit structure of orientation preserving diffeomorphisms on D^2 with topological entropy zero*, Ann. Inst. H. Poincaré Anal. Non Linéaire **49** (1989), 335–356.

[GVT2] ______, *Vers un ordre de Sarkovskii pour les plongements du disque préservant l'orientation*, C. R. Acad. Sci. Paris, Sér. I **310** (1990), 291–294.

[G1] J. Guaschi, *Lefschetz numbers of periodic orbits of pseudo-Anosov homeomorphisms*, Math. Proc. Cambridge Philos. Soc. 115 (1994), 121–132.

[G2] ______, *Pseudo-Anosov braid types of the disc or sphere of low cardinality imply all periods*, J. London Math. Soc. **50** (1994), 594–608.

[G3] ______, *Representations of Artin's braid groups and linking numbers of periodic orbits*, J. Knot Theory Ramifications **4** (1995), 197–212.

[G4] ______, *Nielsen theory, braids and fixed points of surface homeomorphisms*, Topology Appl. **117** (2002), 199–230.

[GLM] J. Guaschi, J. Llibre and R. S. MacKay, *A classification of braid types for periodic orbits of diffeomorphisms of surfaces of genus one with topological entropy zero*, Publ. Mat. **35** (1991), 543–558.

[GHal] G. R. Hall, *A topological version of a theorem of Mather on twist maps*, Ergodic Theory Dynam. Systems **4** (1984), 585–603.

[THal1] T. Hall, *Unremovable periodic orbits of homeomorphisms*, Math. Proc. Cambridge Philos. Soc. **110** (1991), 523–531.

[THal2] ______, *Weak universality in two-dimensional transitions to chaos*, Phys. Rev. Lett. **71** (1993), 58–61.

[THal3] ______, *Fat one-dimensional representatives of pseudo-Anosov isotopy classes with minimal periodic orbit structure*, Nonlinearity **7** (1994), 367–384.

[THal4] ______, *The creation of horseshoes*, Nonlinearity 7 (1994), 861–924.

[Han1] M. Handel, *The rotation set of a homeomorphism of the annulus is closed*, Comm. Math. Phys. **127** (1990), 339–349.

[Han2] ______, *The forcing partial order on the three times punctured disk*, Ergodic Theory Dynam. Systems **17** (1997), 593–610.

[Hans] V. L. Hansen, *Braids and Coverings: Selected Topics*, London Math. Soc. Student Texts, vol. 18, Cambridge Univ. Press, Cambridge, 1989.

[Hay1] E. Hayakawa, *A sufficient condition for the existence of periodic points of homeomorphisms on surfaces*, Tokyo J. Math. **18** (1995), 213–219; Erratum: Tokyo J. Math. **20** (1997), 519.

[Hay2] ______, *Markov maps on trees*, Math. Japonica **51** (2000), 235–240.

[HJ] H. H. Huang and B. J. Jiang, *Braids and periodic solutions*, Topological Fixed Point Theory and Applications (B. J. Jiang, ed.), Lecture Notes in Math., vol. 1411, Springer-Verlag, Berlin–Heidelberg–New York, 1989, pp. 107–123.

[Hu] S. Husseini, *Generalized Lefschetz numbers*, Trans. Amer. Math. Soc. **272** (1982), 247–274.

[J1] B. Jiang, *Lectures on Nielsen Fixed Point Theory*, Contemp. Math., vol. 14, Amer. Math. Soc., Providence, 1983.

[J2] ———, *A characterization of Nielsen classes*, Fixed Point Theory and Its Applications, Contemp. Math., vol. 72, Amer. Math. Soc, Providence, 1988, pp. 157–160.

[J3] ———, *Periodic orbits on surfaces via Nielsen fixed point theory*, Topology-Hawaii (K. H. Dovermann, ed.), World Scientific, Singapore, 1991, pp. 101–118.

[J4] ———, *Nielsen theory for periodic orbits and applications to dynamical systems*, Nielsen Theory and Dynamical Systems (C. McCord, ed., ed.), Contemp. Math., vol. 152, Amer. Math. Soc., Providence, 1993, pp. 183–202.

[J5] ———, *Estimation of the number of periodic orbits*, Pacific J. Math. **172** (1996), 151–185.

[J6] ———, *Bounds for fixed points on surfaces*, Math. Ann. **311** (1998), 467–479.

[JG] B. Jiang and J. Guo, *Fixed points of surface diffeomorphisms*, Pacific J. Math. **160** (1993), 67–89.

[K] M. R. Kelly, *Computing Nielsen numbers of surface homeomorphisms*, Topology **35** (1996), 13–25.

[Ki] E. Kin, *The forcing relation on periodic orbits of pseudo-Anosov braid types for disk automorphisms*, preprint.

[K1] B. Kolev, *Entropie topologique et représentation de Burau*, C. R. Acad. Sci. Paris Sér. I **309** (1989), 835–838.

[K2] ———, *Point fixe lié à une orbite périodique d'un difféomorphisme de* $\mathbb{R}^2$, C. R. Acad. Sci. Paris, Sér. **310** (1990), 831–833.

[K3] ———, *Periodic orbits of period* 3 *in the disc*, Nonlinearity **7** (1994), 1067–1071.

[Kr] I. Kra, *On the Nielsen–Thurston–Bers type of some self-maps of Riemann surfaces*, Acta Math. **146** (1981), 231–270.

[KS] J. Kwapisz and R. Swanson, *Asymptotic entropy, periodic orbits, and pseudo-Anosov maps*, Ergodic Theory Dynam. Systems **18** (1998), 425–439.

[LM1] J. Llibre and R. S. MacKay, *A classification of braid types for diffeomorphisms of surfaces of genus zero with topological entropy zero*, J. London Math. Soc. **42** (1990), 562–576.

[LM2] ———, *Rotation vectors and entropy for homeomorphisms of the torus isotopic to the identity*, Ergodic Theory Dynam. Systems **11** (1991), 115–128.

[LM3] ———, *Pseudo-Anosov homeomorphisms on a sphere with four punctures have all periods*, Math. Proc. Cambridge Philos. Soc. **112** (1992), 539–549.

[Lo1] J. Los, *Pseudo-Anosov maps and invariant train tracks in the disc: A finite algorithm*, Proc. London Math. Soc. **66** (1993), 400–430.

[Lo2] ———, *On the forcing relation for surface homeomorphisms*, Publ. Math. IHES **85** (1997), 5–61.

[M1] S. Matsumoto, *Rotation sets of surface homeomorphisms*, Bol. Soc. Brasil. Mat. **28** (1997), 89–101.

[M2] ———, *Arnold conjecture for surface homeomorphisms*, Topology Appl. **104** (2000), 191–214.

[Ma1] T. Matsuoka, *The number and linking of periodic solutions of periodic systems*, Invent. Math. **70** (1983), 319–340.

[Ma2] ———, *Waveform in the dynamical study of ordinary differential equations*, Japan J. Appl. Math. **1** (1984), 417–434.

[Ma3] ———, *Braids of periodic points and a 2-dimensional analogue of Sharkovskiĭ's ordering*, Dynamical Systems and Nonlinear Oscillations (G. Ikegami, ed.), World Sci. Adv. Ser. Dynam. Systems, vol. 1, World Scientific, Singapore, 1986, pp. 58–72.

[Ma4] ______, *The number and linking of periodic solutions of non-dissipative systems*, J. Differential Equations 76 (1988), 190–201.

[Ma5] ______, *The number of periodic points of smooth maps*, Ergodic Theory Dynam. Systems **9** (1989), 153–163.

[Ma6] ______, *The Burau representation of the braid group and the Nielsen–Thurston classification*, Nielsen Theory and Dynamical Systems (C. McCord, ed.), Contemp Math., vol. 152, Amer. Math. Soc., Providence, 1993, pp. 229–248.

[M7] ______, *Braid type and torsion number for fixed points of orientation-preserving embeddings on the disk*, Math. Japonica **42** (1995), 25–34.

[Ma8] ______, *Braid type of the fixed point set for orientation-preserving embeddings on the disk*, Tokyo J. Math. **18** (1995), 457–472.

[Ma9] ______, *Periodic points of disk homeomorphisms having a pseudo-Anosov component*, Hokkaido Math. J. **27** (1998), 423–455.

[Ma10] ______, *Braid invariants and instability of periodic solutions of time-periodic 2-dimensional ODE's*, Topol. Methods Nonlinear Anal. **14** (1999), 261–274.

[Ma11] ______, *On the linking structure of periodic orbits for embeddings of the disk*, Math. Japonica **51** (2000), 241–254.

[Ma12] ______, *Fixed point index and braid invariant for fixed points of embeddings on the disk*, Topology Appl. **122** (2002), 337–352.

[MT] F. A. McRobie and J. M. T. Thompson, *Braids and knots in driven oscillators*, Internat. J. Bifur. Chaos 3 (1993), 1343–1361.

[MG] S. B. Miled and J. M. Gambaudo, *Cascades of periodic orbits in two dimensions*, Nonlinearity **10** (1997), 1627–1641.

[MZ1] M. Misiurewicz and K. Ziemian, *Rotation sets for maps of tori*, J. London Math. Soc. **40** (1989), 490–506.

[MZ2] ______, *Rotation sets and ergodic measures for torus homeomorphisms*, Fund. Math. 137 (1991), 45–52.

[Mo] S. Moran, *The Mathematical Theory of Knots and Braids: an Introduction*, North-Holland Math. Studies, vol. 82, North-Holland, Amsterdam, 1983.

[Mu] K. Murasugi, *On closed 3-braids*, Mem. Amer. Math. Soc. **151** (1974).

[Pe] B. Peckham, *The necessity of the Hopf bifurcation for periodically forced oscillators*, Nonlinearity **3** (1990), 261–280.

[Po] M. Pollicott, *Rotation sets for homeomorphisms and homology*, Trans. Amer. Math. Soc. **331** (1992), 881–894.

[PoS] M. Pollicott and R. Sharp, *Growth of periodic points and rotation vectors on surfaces*, Topology 36 (1997), 765–774.

[R] D. Rolfsen, *New developments in the theory of Artin's braid groups*, Topology Appl. **127** (2003), 77–90.

[S] A. N. Sharkovskiĭ, *Coexistence of cycles of a continuous map of a line into itself*, Ukrain. Math. Zh. **16** (1964), 61–71. (Russian)

[Sh] R. Sharp, *Periodic points and rotation vectors for torus diffeomorphisms*, Topology **34** (1995), 351–357.

[Shi1] H. Shiraki, *On braid type of fixed points of homeomorphisms defined on the torus*, Mem. Fac. Sci. Kôchi Univ. Ser. A Math. **20** (1999), 113–122.

[Shi2] H. Shiraki, *Fixed point indices of homeomorphisms defined on the torus*, Hokkaido Math. J. **32** (2003), 59–74.

[SKL] W. T. Song, K. H. Ko and J. Los, *Entropies of braids*, J. Knot Theory Ramifications **11** (2002), 647–666.

[T] W. P. Thurston, *On the geometry and dynamics of diffeomorphisms of surfaces*, Bull. Amer. Math. Soc. **19** (1988), 417–431.

[W] J. A. Walsh, *Rotation vectors for toral maps and flows: a tutorial*, Internat. J. Bifur. Chaos **5** (1995), 321–348.

[YT1] Y. Yamaguchi and K. Tanikawa, *Symmetrical non-Birkhoff period-3 orbits in standard-like mappings*, Progr. Theoret. Phys. **104** (2000), 943–954.

[YT2] ———, *Non-symmetric non-Birkhoff period-2 orbits in the standard mapping*, Progr. Theoret. Phys. **106** (2001), 691–696.

[YT3] ———, *Dynamical ordering of non-Birkhoff orbits and topological entropy in the standard mapping*, Progr. Theoret. Phys. **107** (2002), 1117–1145.

6. FIXED POINT THEORY OF MULTIVALUED WEIGHTED MAPS

Jacobo Pejsachowicz — Robert Skiba

1. Introduction

Without doubt the golden age of fixed point theory for multivalued map occurred in the post-war period. Motivated by the recently born disciplines of mathematical economics and game theory, using tools from the flourishing algebraic topology of that time, several well known fixed point theorems were proved. Most of the research was done in the area of fixed points for convex-valued maps, due to their importance in applications. However, the interest of topologists was immediately directed toward more general classes of maps.

The purpose of this paper is to survey the fixed point theory of two very special classes of multivalued maps with weights which were found during the intense research activity of that time. The first, which we will call acyclic weighted carriers, was discovered by Gabriele Darbo in 1950. The second class was introduced by him in 1957 under the name of weighted maps. We will briefly explain how they were found, review the significant results and point out few applications of this theory.

At first glance acyclic weighted carriers look similar to other categories of acyclic maps for which a fixed point theory has been constructed. However, this similarity is only apparent. For example, it is known that several classes of acyclic morphisms are homotopic to a continuous map by a homotopy in this category, see [Kr1] and [Kr2]. This is far from being true for acyclic weighted carriers. In the presence of a nontrivial branching, an acyclic weighted carrier cannot be homotopic to any linear combination of continuous maps.

On the other hand, weighted maps are closely related to maps into the n-th symmetric product of a space (see [Max]). But the concept of weighted map is more flexible and as a consequence the category is considerably larger than that of maps into symmetric products. As we will see, weighted maps and weighted carriers arise quite naturally as solutions of parametrized families of nonlinear equations. This fact makes the corresponding fixed point and degree theories of considerable interest.

In this paper (except when strictly necessary) we will omit any further reference to other classes of multivalued maps considered in literature. Several books and general surveys of fixed point theory for multivalued maps are now available, see [G2], [BGMO], [GrDu].

There are several reasons which led us to write this survey. For the first author it represents a tribute to the memory of a dear friend and teacher, a discrete, quiet person with a special gift: the ability to see mathematics everywhere. It is also an opportunity to integrate old results with new ones, many of them recently obtained by the second author. We also believe that there may be a renewed interest in the algebraic topology of weighted maps because maps of this kind were used in the past years in order to solve some nontrivial and interesting problems in a variety of fields [FO], [FB], [Al1], [Al2], [O].

The presentation is organized respecting the chronological order. Sections 1 to 8 cover old results, mainly those obtained before 1980, while the remaining sections deal with more recent ones.

Not everything here is a survey. In Section 6 the results recently obtained by the second author are used in order to fill a gap in the proof of the main theorem in [CP]. Section 8 contains a reformulation with improvements of some results in [MNP]. The last section is devoted to comments.

2. A finite valued, continuous, fixed point free map from a two-cell into itself

The name of Gabriele Darbo is usually associated to a well known fixed point theorem for α-contractions. However, he never considered this theorem as particularly significant. He wrote only one paper on this subject. As many other mathematicians of his circle, which was deeply influenced by the ideas of Renato Caccioppoli, he developed a wide variety of interests ranging from functional analysis and measure theory to algebraic topology, homological algebra, applications of category theory to networks, elementary number theory. He wrote very few papers, hardly more than four or five in any one area. However, much as in the case of Renato Caccioppoli, his charisma and the influence on his students was, by far, more important than his published work.

One of his first and favorite results is an example of a continuous multivalued fixed point free map sending each point of a two disc to a finite subset of the disc of cardinality between one and three. He found this example during a train trip, when he was working on a question posed by R. Caccioppoli to G. Scorza and which also attracted the interest of E. Magenes among others. The specific question had to do with separation properties of compact subsets of $[0,1]^n$, but counterexamples were found via graphs of fixed-point free maps.

Let us begin the survey with this example presented in the analytic form that was published in [Da1]. An interesting discussion of the geometric ideas behind the construction which were communicated orally by Darbo to Giuseppe Scorza Dragoni can be found in the survey [SD].

Let the two cell be the unit disk E of $\mathbb{R}^2$. We define a map T on the intersection E^+ of E with the first quadrant $\{x : x \geq 0,\ y \geq 0\}$ by making correspond, to a point $P = (x, y)$, the set $T(P)$ of three points $\{Q_1 = (x_1, y_1), Q_2 = (x_2, y_2), Q_3 = (x_3, y_3)\}$ of E defined in coordinates by

$$(2.1)\qquad \begin{cases} x_1 &= 8x^2(1 - x^2 - y^2) - 1,\\ y_1 &= +\sqrt{1 - x_1^2},\\ x_2 &= 8x^2(1 - x^2) - 1,\\ y_2 &= -\sqrt{1 - x_2^2},\\ x_3 &= \begin{cases} +1 & \text{if } 0 \leq x \leq 1/\sqrt{2},\\ 8x^2(1 - x^2) - 1 & \text{if } 1/\sqrt{2} < x \leq 1,\end{cases}\\ y_3 &= -\sqrt{1 - x_3^2}.\end{cases}$$

Since the coordinates of Q_i for $i = 1, 2, 3$ are continuous functions of (x, y), it follows that the points Q_1, Q_2, Q_3 considered separately are continuous maps from E^+ to E.

Let us extend the definition of $T(P)$ to the entire disc E in the following way: if the points P and P' are symmetric with respect to the x-axis (or to the y-axis), then the sets $T(P)$ and $T(P')$ will be in the same relation of symmetry with respect to the same axis. This defines a continuous multivalued map on all of the disc E, because to any point P of E^+ belonging to an axis corresponds a set $T(P)$ which is symmetric with respect to the same axis. This follows easily by inspection.

In this way we obtain a continuous multivalued map of the two cell E into itself taking values in sets of cardinality between one and three (obviously, coincident points have to be considered as the same point). We claim that the map T defined above is without fixed points. To see this we first notice that the image of the map T is completely contained in the unit circle $C \subset E$.

By symmetry, if there is any fixed point of T, then there must be a fixed point $p = (x, y)$ in the first quadrant. But for (x, y) on the arc of C lying on the first quadrant, we have that always $x_1 = -1$, $y_1 = 0$, $x_2 \leq 0$, $y_3 \leq 0$. Thus either $(x, y) = (1, 0)$ or $(x, y) = (0, 1)$. But, both $(1, 0)$ and $(0, 1)$ are sent to $\{(-1, 0)\}$ by T. This proves the claim.

3. Acyclic weighted carriers

At the time that the preceding example was published there was wide interest in the following question: in addition to continuity, what other conditions should be imposed on a multivalued map in order to obtain the fixed point property for maps in the resulting class from an n-ball into itself?

Eilenberg and Montgomery extended the Lefschetz fixed point theorem to acyclic continuous map from an absolute neighbourhood retract into itself. The fixed point property for this class follows from their result. Magenes showed in [M] that continuous maps sending points of a ball into subsets having always the same number of acyclic components also have the fixed point property.

By the results of B. O'Neill in [ON] continuous multivalued maps sending points to sets having either one or k acyclic components have this property too. It still holds for continuous finite-valued maps sending points to sets of cardinality either two or n (see [H]). There are examples (see [ON]) of finite-valued continuous fixed point-free maps from a two-disk into itself with cardinality of $F(x)$ ranging in any subset of $\mathbb{N}$ except $\{n\}$ $\{1, n\}$ and $\{2, n\}$.

In the second part of the paper [Da1], Darbo formulated a possible answer to the question discussed above. He proved that, in the case of a two-disk, fixed point property holds irrespective of the cardinality of the set of acyclic components of the images of the points, if one can assign to each piece of $F(x)$ a weight or multiplicity varying "continuously" with respect to x in the same way as the multiplicities of the roots of a polynomial vary with the coefficients at the polynomial. Let us be more precise about Darbo's result.

A *piece* of a compact space K is any open and closed subset of K. The family of all pieces $\mathcal{P}(K)$ of K is closed under finite unions and intersections. Typically, if K is a compact subspace of a Hausdorff space Y and U is an open subset of Y such that $\partial U \cap K = \emptyset$, then $U \cap K$ is a piece of K. Conversely, any piece C of K is of the form $C = U \cap K$, with U open in X and $\partial U \cap K = \emptyset$.

Let us recall that a multivalued map $F\colon X \to Y$ is called *upper semicontinuous* if for each closed subset $C \subset Y$, its inverse image

$$F^{-1}(C) = \{x \in X\colon F(x) \cap C \neq \emptyset\}$$

is closed. Moreover, F is called *continuous* if the same holds with "closed" interchanged with "open".

(3.1) DEFINITION. Let R be a commutative ring. A multivalued upper semicontinuous map $F\colon X \to Y$ is called a *multivalued map with weights in* R or briefly a *weighted carrier* if for any $x \in X$, the set $F(x)$ is compact and moreover, to any piece C of $F(x)$ is assigned a *weight* or *multiplicity* $m(C, F(x)) \in R$ verifying the following conditions:

(3.1.1) $m(\,\cdot\,, F(x))$ is an additive function on the set $\mathcal{P}(F(x))$, i.e.

$$m(C_1 \cup C_2, F(x)) = m(C_1, F(x)) + m(C_2, F(x))$$

whenever $C_1 \cap C_2 = \emptyset$;

(3.1.2) if U is open in Y, with $\partial U \cap F(x) = \emptyset$, we have that

$$m(F(x) \cap U, F(x)) = m(F(x') \cap U, F(x'))$$

whenever x' is close enough to x.

In order to abbreviate notation, if U is as above, we will write either $m(F(x), U)$ or $m_U(F(x))$ instead of $m(F(x) \cap U, F(x))$ and we will call it the *multiplicity of* $F(x)$ *in* U. If X is connected, then the multiplicity $m(F(x), Y)$ of $F(x)$ with respect to the whole space Y is independent of x. This is an important invariant of the weighted carrier. It will be called the *index* of F and denoted with $I(F)$.

(3.2) EXAMPLE. The map assigning to each n-tuple of complex numbers $(a_0, \ldots, a_n)$ the roots of the polynomial $z^{n+1} + a_n z^n + \ldots + a_0$ is an upper semicontinuous map from $\mathbb{C}^n$ to $\mathbb{C}$ with integral weights given by the multiplicities of the roots. The weight of a piece is the sum of the multiplicities of the points in that piece.

(3.3) EXAMPLE. Let X be a topological space, $O \subset \mathbb{R}^n$ be an open bounded set and let $f\colon X \times \overline{O} \to \mathbb{R}^n$ be a continuous map such that $0 \notin f(X \times \partial O)$. We will consider the topological space X as a parameter space and consider the map f as a family of maps $\{f_x\colon \overline{O} \to \mathbb{R}^n\}_{x \in X}$ where f_x is defined by $f_x(y) = f(x, y)$. Any parametrized family as above defines a multivalued solution map $S\colon X \to O$, sending each $x \in X$ into the set $S(x)$ of solutions y of the equation $f_x(y) = 0$. We claim that it is possible to assign integral weights to pieces of $S(x)$ such that S becomes a weighted carrier.

First we observe that the upper semicontinuity of S is an easy consequence of compactness of $\overline{O}$. For, if $S(x) \subset U$, then $0 \notin f_x(\overline{O} - U)$ which is compact being $\overline{O} - U$ compact. But then, by continuity, $0 \notin f_z(\overline{O} - U)$ for z close enough to z, which implies that $S(z) \subset U$.

To any piece C of $S(x)$ an integral multiplicity can be assigned as follows: take an open set U such that $C = S(x) \cap U$ and such that $0 \notin f_x(\partial U)$, then define $m(C, S(x)) = \deg(f_x, U, 0)$, where deg is the Brouwer topological degree of f_x on U with respect to 0. By the excision property of the degree, the multiplicity $m(C, S(X))$ does not depend on the particular choice of the open set U. Additivity on pieces follows from the additivity of the degree. Moreover, we can use homotopy invariance property in order to verify the last condition.

For this, let U be such that $\partial U \cap S(x) = \emptyset$ and let $\rho = \operatorname{dist}(0, f_x(\partial U)$. The set V of all $z \in X$ such that, $\operatorname{dist}(0, f_z(\partial U) \geq 1/2\rho)$ is an open neighbourhood of x, and, for any $z \in V$, $h(t,y) = tf_x(y) + (1-t)f_z(y)$ is an admissible homotopy between the restrictions of f_x and f_z to $\overline{U}$. Therefore,

$$m(S(x), U) = \deg(f_x, U, 0) = \deg(f_z, U, 0) = m(S(z), U).$$

This proves that S is a weighted carrier.

Notice that, when $S(x) = f_x^{-1}(0)$ is a finite set for every x, then an assignment of multiplicities to pieces of $S(x)$ is equivalent to consider each zero y of f_x with its own multiplicity $m(f_x, y) = \deg(f_x, W, 0)$, where W is any neighbourhood of Y not containing other zeroes of f_x. This is the case of the previous example. As we will see later, this can be better formalized by thinking of this map as a function into the free group generated by O which sends x into the formal combination $\sum_{y \in f_x^{-1}(0)} m(f_x, y)y$.

(3.4) EXAMPLE. In a similar way, let Y be a compact ANR and let $f\colon X \times Y \to Y$ be a family of maps. Then the local fixed point index ([Le1], [GrDu], [Do1]) provides weights for the multivalued map sending $x \in X$ into the set $F(x) \subset Y$ of all fixed points of f_x. The proof that F is a weighted carrier is a word by word restatement of the previous one.

(3.5) EXAMPLE. Let us denote by $c(K)$ the number of connected components of a set K. If $F\colon X \to Y$ is a continuous multivalued map such that for all $x \in X$ $c(F(x))$ is either constantly n or belongs to $\{1, n\}$, then F can be made into a weighted carrier by assigning integral weights in an obvious way. If $c(F(x)) \in \{2, n\}$ and p is a prime, then F can be made a $\mathbb{Z}_p$-carrier provided that an obstruction in $\check{H}^1(X, \mathbb{Z}_{p-1})$ vanishes (see [H]). Darbo's fixed point theorem below and Richard Dunn examples (see [ON]) show that only in this cases a continuous assignment of weights can be made on the grounds of $c(F(x))$ only.

From the above examples it is clear that whenever a function can be constructed having the typical properties of a degree, one is able to assign continuously varying weights of the above type. For example one can use the intersection index of manifold oriented over R as in [CP], the index of a singular point of a vector field and other similar invariants. In the infinite dimensional setting, dealing with families of nonlinear equations on Banach spaces, one can use the Leray–Schauder degree for compact vector-fields (see [MNP]). Also the Caccioppoli's degree provides weights in Z_2 to solutions of equations involving Fredholm maps. It is interesting to notice that Caccioppoli described his degree precisely as an assignment of multiplicities to pieces of $f^{-1}(0)$ (see [CA]).

(3.6) DEFINITION. A weighted carrier $F: X \to Y$ is called *acyclic in positive dimensions* (or simply acyclic) if for each $x \in X$ and all $q > 0$, the q-th Čech homology group of $F(x)$ with coefficients in R, $\check{H}_q(F(x); R) = 0$.

(3.7) REMARK. Notice that for an acyclic weighted carrier F, the set $F(x)$ need not have a finite number of connected components.

(3.8) EXAMPLE.

(3.8.1) If a weighted carrier has a finite number of acyclic components, then it is a acyclic weighted carrier. In particular any finite-valued weighted carrier is acyclic.

(3.8.2) If F is a w-carrier over the ring $\mathbb{Q}$ and $\check{H}_q(F(x), \mathbb{Z})$ are finite groups, for all $q \geq 1$ and for all $x \in X$, then F is an acyclic weighted carrier.

(3.8.3) For any compact subset C of $\mathbb{R}$, $\check{H}_q(C) = 0$ for $q > 0$. Thus any weighted carrier with values in the real line is acyclic.

(3.8.4) A compact subset C of $\mathbb{R}^2$ such that $\mathbb{R}^2 - C$ is connected has $\check{H}_q(C) = 0$ for $q > 0$. Thus, any weighted carrier with values in compact subsets of $\mathbb{R}^2$ with connected complement is an acyclic weighted carrier.

The second part of [Da1], is devoted to weighted carriers defined on a two-disk D in $\mathbb{R}^2$ whose values are compact subsets of $\mathbb{R}^2$ such that for any x the complement of $F(x)$ is connected or, which is the same, acyclic in positive dimensions. Darbo showed that, given a singular one-cycle γ in the complement of $F(x)$, one can define a degree $\omega(F(x), \gamma) \in R$ of $F(x)$ relative to the cycle γ. Using this construction he proved that if $F(\partial D) \subset D$ and $I(F) \neq 0$, then F has a fixed point. Few years later, Darbo's construction was extended by one of his students, Letizia Dal Soglio in [DS2], to $\mathbb{R}^n$. Her results are close in spirit to those in [ONS].

(3.9) REMARK. A completely different approach to this question was taken by R. Connelly in [Co]. He proved the Brouwer fixed point for multivalued upper semicontinuous maps from the ball B^n into itself such that

$$\dim\{x : \check{H}^k(F(x)) \neq 0\} \leq n - k - 2,$$

applying some deep results about Leray spectral sequence of a continuous map to the projection on the domain restricted to the graph of the map.

4. The category of weighted maps

In 1958, Darbo published in [Da2] the first of his three papers devoted to the homology theory of weighted maps. He introduced the category of weighted maps containing the category of topological spaces and continuous map as a proper subcategory, defined a homology functor on this category and related it to singular homology.

Roughly speaking a weighted map is an equivalence class of finite-valued weighted carriers. But Darbo's definition is slightly more elaborate.

Let R be a commutative ring with unit. We denote by $R(X)$ the free R-module generated by a set X. Identifying X with the canonical basis of $R(X)$, any element $\zeta \in R(X)$ can be written in the form $\zeta = \sum \langle \zeta, x \rangle \cdot x$ where $\langle \zeta, x \rangle = \zeta(x)$ is the coefficient of ζ at the point $x \in X$. The set of all x such that the coefficient of ζ at x is different from 0 will be called the support of ζ. Given a subset A of X we shall denote by $m_A(\zeta)$ the sum of coefficients of ζ at the points of A.

(4.1) DEFINITION. Let X, Y be two topological Hausdorff spaces. A *weighted map* $\phi: X \to Y$ with coefficients in R is a homomorphism $\phi: R(X) \to R(Y)$ such that there exists a multivalued upper semicontinuous map $t: X \to Y$ verifying:

(4.1.1) $t(x)$ is a finite subset of Y for any $x \in X$,

(4.1.2) the support of $\phi(x)$ is contained in $t(x)$,

(4.1.3) if U is an open subset of Y and $x \in X$ is such that $t(x) \cap \partial U = \emptyset$ then $m_A(\phi(x)) = m_A(\phi(x'))$ whenever x' is close enough to x.

The multivalued map t in Definition (4.1) is called an upper semicontinuous support of ϕ (u.s.c.-support). Among all of upper semicontinuous supports of ϕ there exists a minimal one defined by $t_\phi(x) = \bigcap_{t \in \mathrm{supp}(\phi)} t(x)$, where we have denoted by $\mathrm{supp}(\phi)$ the set of all usc-supports of ϕ. It is easy to see that the graph of $t_\phi(x)$ is the closure of the set $\{(x,y) : \langle \phi(x), y \rangle \neq 0\}$.

(4.2.1) Any continuous map $f: X \to Y$ can be considered as a weighted map by assigning the coefficient 1 to each $f(x)$.

(4.2.2) Any finite-valued weighted carrier F defines a unique weighted map whose minimal support is contained in F.

(4.2.3) Weighted maps can be added and multiplied by scalars in R. Given two weighted maps $\phi: X \to Y$ and $\psi: X \to Y$ its sum $\phi + \psi: X \to Y$ is a weighted map. Indeed, it is easy to see that the upper semicontinuous map t defined by $t(x) = t_\phi(x) \cup t_\psi(x)$ is a usc-support of the homomorphism $\phi + \psi: R(X) \to R(Y)$. In particular $t_{\phi+\psi} \subset t_\phi \cup t_\psi$. Clearly, if ϕ is a weighted map and $r \in R$, then $r\phi$ is a weighted map. Thus any linear combination of a finite number of continuous maps with coefficients in R is a weighted map.

(4.2.4) The composition two weighted maps is a weighted map. Indeed, if t is a support of $\phi: X \to Y$ and t' is a support of $\psi: Y \to Z$, then it is easy to see that the upper semicontinuous mapping $t't$ is a support of the homomorphism $\psi\phi\colon R(X) \to R(Y)$. The set of all weighted maps from X to Y has a natural structure of R-module that is preserved by the composition.

(4.2.5) The product of two weighted maps $\phi\colon X \to Y$ and $\phi'\colon X' \to Y'$ is the weighted map $\phi \times \phi'\colon X \times X' \to Y \times Y'$ defined by

$$(x, x') \rightsquigarrow \sum_{y \in t_\phi(x); y' \in t_{\phi'}(x')} \langle \phi(x), y \rangle \langle \phi'(x'), y' \rangle (y, y').$$

From (4.1.3) of Definition (4.1) it follows that for any weighted map $\phi\colon X \to Y$ the function $x \rightsquigarrow m_Y(\phi(x))$ is locally constant and hence it is constant on each path component of X. If X is connected, then the element $I(\phi) = m_Y(\phi(x))$ is called the index of ϕ.

Let $\mathcal{T}$ be the category of topological Hausdorff spaces and continuous maps. The weighted category over the ring R is the category $\omega\mathcal{T}$ having the same objects as $\mathcal{T}$ and whose morphisms are weighted maps with coefficients in R. This is an additive category. Clearly $\mathcal{T}$ is a subcategory of $\omega\mathcal{T}$. We shall denote by $[X;Y], (X;Y)$ the morphisms from X to Y in $\mathcal{T}$ and $\omega\mathcal{T}$ respectively.

Let I be the real interval $[0,1]$. Given two weighted maps ϕ and ψ from X to Y we say that ϕ is σ-homotopic to ψ if there exists a weighted map θ from $X \times I$ to Y such that $\theta(x,0) = \phi(x)$ and $\theta(x,1) = \psi(x)$.

It is easy to see that σ-homotopy is an additive equivalence relation in $\omega\mathcal{T}$ which extends the usual homotopy relation on $\mathcal{T}$.

The quotient category of $\omega\mathcal{T}$ defined by this relation will be called the σ-homotopy category $\sigma\mathcal{T}$ of $\omega\mathcal{T}$ and we denote by $\sigma(X;Y)$ the R-module of σ-homotopy classes of weighted maps from X to Y. The inclusion of $\mathcal{T}$ as a subcategory of $\omega\mathcal{T}$ induces a functor J from the homotopy category $\pi\mathcal{T}$ of $\mathcal{T}$ into $\sigma\mathcal{T}$.

If X is a connected space, then the index I is a homomorphism from $(X;Y)$ to R compatible with the relation of σ-homotopy and hence it induces a homomorphism $I\colon \sigma(X;Y) \to R$.

Darbo constructed a homology theory for weighted maps by adapting to this setting the usual construction of the singular homology functor.

Let us denote by Δ_q the canonical q-simplex in $\mathbb{R}^{q+1}$. For any i, with $0 \le i \le q$, consider the map $d_q^i\colon \Delta_{q-1} \to \Delta_q$ given by the inclusion of Δ_{q-1} as the face opposite to the i-th vertex of Δ_q. With them we can define a weighted map $d_q\colon \Delta_{q-1} \to \Delta_q$ $(q = 1, \dots)$ as the linear combination $d_q = \sum_{i=0}^{q} (-1)^i d_q^i$. We extend the definition by setting $\Delta_q = \emptyset$ (empty simplex) for $q < 0$ and then $d_q = 0$ for $q \le 0$. Then, for any $q \in \mathbb{Z}$, we have that $d_q d_{q-1} = 0$.

Given a Hausdorff space X, the R-module *of weighted q-chains on X* is the graded R-module $\mathcal{C}_q(X) = (\Delta_q; X)$. The sequence $d_q\colon \Delta_q \to \Delta_{q+1}$ enables us to define a boundary operator $\partial: \mathcal{C}_q(X) \to \mathcal{C}_{q-1}(X)$ by $\partial_q(\phi) = \phi d_q$. The graded homology module of the complex $(\mathcal{C}(X), \partial)$ will be called the *weighted homology* of the space X. We will denote it by $\mathcal{H}_*(X) = \{\mathcal{H}_q(X)\}$.

Any weighted map $\phi: X \to Y$ induces in a funtorial way a homomorphism $\phi_*: \mathcal{H}_*(X) \to \mathcal{H}_*(Y)$, of degree zero, between the weighted homology of X and that of Y. Two homotopic maps induce the same homomorphism in homology. With this $\mathcal{H}$ becomes additive functor from $\omega\mathcal{T}$ to the category of graded R-modules which is invariant under the σ-homotopy.

The inclusion of $[\Delta_k; X]$ into $(\Delta_k; X)$ extends to a homomorphism from the singular chains with coefficients in R to the weighted ones. It induces a natural transformation $j_X: H_*^{\mathrm{sing}}(X; R) \to \mathcal{H}_*(X)$ from the singular homology with coefficient in R to the weighted homology.

Darbo showed that the functor $\mathcal{H}_*$ satisfies the *Eilenberg–Steenrod axioms* for a homology theory with compact carriers and coefficients in a commutative ring. By the uniqueness of homology theory it follows that, for any CW-complex X, $j_X: H_*^{\mathrm{sing}}(X; R) \to \mathcal{H}_*(X)$ is an isomorphism and in fact a natural equivalence between $H_*^{\mathrm{sing}}(\,\cdot\,; R)$ and $\mathcal{H}_* \circ J(-)$ restricted to the category of CW complexes.

The next paper [Da3] of Darbo was devoted to the coincidence theory for weighted mappings. The construction of the Lefschetz coincidence class is the same as for singlevalued maps.

Let X and Y be Hausdorff spaces and let $\phi, \psi: X \to Y$ be two weighted maps. A point $x \in X$ is a point of *coincidence* of the pair (ϕ, ψ) if $t_\phi(x) \cap t_\psi(x) \neq 0$. By the semi-continuity of the supports the set E of all points of coincidence of the pair (ϕ, ψ) is closed in X.

If A is coincidence free subset of X, denoting by Δ the diagonal in $Y \times Y$, there is an homomorphism

$$\mathfrak{L}^q_{\phi,\psi}: \mathcal{H}_q(X, A) \to \mathcal{H}_q(Y \times Y, Y \times Y - \Delta)$$

induced by the natural inclusions of pairs. It has the usual property: if the pair (ϕ, ψ) is coincidence free than we have $\mathfrak{L}^q_{\phi,\psi} = 0$ for all q.

Taking $X = Y$ and $\psi = \mathrm{id}$ we see that the homomorphism $\mathfrak{L}^q_{\phi,\mathrm{id}}$ detects fixed points.

In order to relate $\mathfrak{L}^q_{\phi,\mathrm{id}}$ to the Lefschetz number Darbo could not use Poincare duality because weighted maps do not behave properly with respect to products. Moreover, there is no natural cohomology theory of singular type in this category. Following an idea of H. Cartan, he restated the Poincare duality in terms of homology only. If W is an orientable manifold over R he defined

$$\mathfrak{D}(W) = \mathcal{H}_q(W \times W, W \times W - \Delta)$$

and proved the following result:

(4.3) THEOREM. *If W is a n-dimensional paracompact orientable manifold which admits a locally finite triangulation, then for any open subset U of W there exists a natural isomorphism $\Psi^q_U: \mathfrak{D}_q(U) = \mathcal{H}_{q-n}(U)$, whenever one of the following conditions holds:*

(4.3.1) *R is a principal ideal domain,*
(4.3.2) *R is a torsion free $\mathbb{Z}$-module.*

Using this isomorphism, he constructed a local fixed point index for weighted maps.

In the last paper of the series [Da4], published in 1961, assuming that R is a field, Darbo computed the homomorphism

$$\mathfrak{L}^n_{\phi,\mathrm{id}}: \mathcal{H}_n(W) \simeq R \to \mathfrak{D}_0(U) \simeq R \quad \text{as} \quad \mathfrak{L}^n_{\phi,\mathrm{id}}(\omega) = (-1)^n \sum (-1)^q \mathrm{Tr}(\phi_{*q})\theta.$$

where θ and ω are generators. This implies the Lefschetz theorem for weighted maps from an compact oriented manifold into itself.

The theorem was then extended to weighted maps from a finite dimensional compact ANR into itself by showing that such spaces are retracts of smooth oriented finite dimensional manifolds.

Darbo proved that any compact ANR is a retract of the product of a finite polyhedron with the Hilbert cube Q^∞ and used this in order to drop the finite-dimensionality hypothesis. The last section is devoted to relax the hypothesis that R is a field.

His final result is:

(4.4) THEOREM. *Let X be a compact absolute neighbourhood retract and let $\phi: X \to X$ be a weighted map with weights in a principal ideal domain R. If the Lefschetz number*

$$L(\phi) = \sum_q (-1)^q \mathrm{Tr}(\phi_{*q}) \neq 0,$$

then ϕ has a fixed point.

Since for contractible spaces the Lefschetz number reduces to $I(\phi)$ we have:

(4.5) COROLLARY. *If X is an absolute retract and $I(\phi) \neq 0$, then ϕ has a fixed point.*

(4.6) EXAMPLE. Let SP^nY be the n-the symmetric product of Y. If $x_1^{k_1} \dots x_s^{k_s}$ denotes an equivalence class of $(x_1 \dots x_n)$ in SP^nY, then the homomorphism $\Pi: SP^nY \to Y$ defined by

$$\Pi(x_1^{k_1} \dots x_s^{k_s}) = \sum_{i=1}^{k_s} k_i x_i.$$

is a weighted map. Thus if $f: X \to SP^n X$ be a singlevalued map, then Πf is a weighted map and fixed point theorems for maps into symmetric products (see [Max]) are consequences of (4.4).

Let us illustrate with some examples taken from [P3] the use of Darbo's fixed point theorem in the study of branch points of maps and correspondences.

(4.7) EXAMPLE. Let M, N be two n-dimensional orientable topological manifolds. Let $f: M \to N$ be a proper continuous map such that the inverse image by f of any point of N is a finite subset of M. Then, denoting with $m(x)$ the multiplicity of the point $x \in f^{-1}(y)$, the map

$$\phi(y) = \sum_{x \in f^{-1}(y)} m(x)x$$

is a weighted map from N to M with coefficients in R.

Clearly $I(\phi) = \deg f$ and $f\phi = \deg f \mathrm{id}_N$. A *branch point of* f is, by definition, any fixed point of the weighted map $\phi f - \mathrm{id}_M$. The number $\delta(f) = L(\phi f - \mathrm{id}_M)$, will be called the *total order of branching* of the map f. Notice that if x is a branch point, then f is cannot be a local homeomorphism at x.

If M, N are surfaces and f is a ramified covering (i.e. f is a covering map in the complement of a finite number of points $\{p_1, \dots, p_r\}$) it can be shown that $\delta(f) = \sum_i^r (m(p_i) - 1)$. On the other hand, by additivity and commutativity of the trace, $L(\phi f - \mathrm{id}_M) = L(f\phi) - L(\mathrm{id}_M) = d\chi(N) - \chi(M)$, where $d = \deg f$ and χ is the Euler–Poincare characteristic. Hence we obtain the Riemann–Hurwitz formula

$$\sum_i^r (m(p_i) - 1) = d\chi(N) - \chi(M).$$

For example, let us consider a two-torus T embedded (canonically) into $\mathbb{R}^3$ and let S be a two-sphere with center p and contained inside T. Let $f: T \to S$ be the map which assigns to each $x \in T$ the intersection point $f(x)$ of the segment $\overline{px}$ with the sphere. The map f is a differentiable map of degree 1 and the inverse image by f of any point of S is finite (it has at most 3 points).

From the previous discussion we obtain a weighted map $\phi: S \to T$ with support f^{-1} and coefficients in $\mathbb{Z}$ such that $f\phi = \mathrm{id}_S$. By the above formula $\delta(f) = 2$ and hence f must have branch points.

Even without a local description of $\delta(f)$ given by the Riemann–Hurwitz formula in the two dimensional case, by Darbo's theorem, if M, N are orientable n-dimensional manifolds and $\deg(f)\chi(N) \neq \chi(M)$, then f must have a branch point. We will see in the next section the effect of this branching on the homotopy theory of weighted maps.

(4.8) EXAMPLE. Let V, W be two n-dimensional nonsingular complex subvarieties of the m-dimensional projective space and let C be a irreducible subvariety of $V \times W$. Then, the correspondence ϕ which sends a point $x \in V$ of the first factor to the projection onto the second factor of the set of intersection points of $\{x\} \times W$ with C, counted with multiplicities, is a weighted map from V to W with graph C. Also $\psi = \phi^{-1}$ is a weighted map since it is defined in the same way.

Non-singular complex varieties are naturally oriented. Hence, denoting with $|V|$ the fundamental class of a manifold, it is possible define the degree $\deg(\phi)$ as the unique integer such that $\phi_*(|V|) = \deg(\phi)|W|$ in $\mathcal{H}_n(W) \simeq H_n^{\text{sing}}(W;\mathbb{Z})$.

If we define the branching order of ϕ and ψ in the same way as before using the commutativity of the trace we obtain

$$\delta(\phi) - \delta(\psi) = \deg(\phi)\chi(W) - \deg(\psi)\chi(V).$$

This is a formula of Zeuthen (see [Se]).

From Darbo's theorem we can conclude that if neither ϕ nor ϕ^{-1} have branch points, then

$$\deg(\phi)\chi(W) = \deg(\psi)\chi(V).$$

Related material can be found in [Se], [L3] and [FB].

(4.9) EXAMPLE. Let G be a finite group acting on a compact Hausdorff space X. Let $\pi\colon X \to X/G$ be the projection to the orbit space. By the factorization lemma (Lemma 2.5 of [P1]) the weighted map $\psi\colon X \to X$ defined by $\psi(x) = \sum_{g\in G} gx$ factorizes through π. It follows that there exist a weighted map $\phi\colon X/G \to X$ such that $\phi\pi(x) = \sum_{g\in G} gx$. This map also verifies $\pi\phi = |G|\mathrm{id}$, where $|G|$ is the order of the group.

If the space X is a finite simplicial complex, the homomorphism $\phi_*\colon \mathcal{H}(X/G) \to \mathcal{H}(X)$ coincides (up to isomorphism) with the Smith transfer. Its very existence implies that π_* is surjective and hence the weighted homology of X/G is finitely generated if X is a finite dimensional ANR. Moreover, $\mathcal{H}_q(X/G) = 0$, for $q \geq 1$, if X is acyclic.

Incidentally, it shows that $\mathcal{H}_*$ is different from the singular theory on general spaces and moreover it behaves better with respect to the existence of transfers. Bredon has given an example of finite dimensional compact G-space for which Smith transfer in singular homology cannot exist (see [S]).

Assuming that X is a finite dimensional ANR, the Lefschetz number is well defined both on X and on X/G. Hence if we define, as before, $\delta(\pi) = L(\phi\pi - \mathrm{id})$ we obtain

$$\delta(\pi) = \sum_{g\neq e} L(g) = |G|\chi(X/G) - \chi(X).$$

In particular if a cyclic group of prime order acts on X, then $L(g) = 0$ for $g \neq e$ and therefore $|G|\chi(X/G) = \chi(X)$.

Weighted maps were rediscovered by Richard Jerrard in 1975 (see [Je2]). He called them m-functions. His definition is very similar to that of Darbo. The basic idea was already present in the paper [Je1], where it was used in order to prove that any closed analytic curve in the plane has a square inscribed in it. The formal definition was suggested to him by P. Young. It is essentially a restatement of Darbo's definition in terms of the graph of the map.

Jerrard defined the homology theory in this category, showed that it verifies all the axioms of a homology theory and proved the Lefschetz fixed point theorem for polyhedra via simplicial approximations. Borrowing an expression from the modern management of scientific research we can say that the results were tested by two independent research groups.

Jerrard applied the above theory to the study of the fixed point property. We will come back to this in Section 8.

Degree theory and linking numbers for weighted maps were studied by Letizia Dal Soglio in [DS3]. Chandan Vora in [Vo1], [Vo2] obtained from Darbo's fixed point theorem various fixed points results of asymptotic type. For example; if $F: X \to X$ is a compact weighted map and there is an integer m such that $F^m(X) \subset A$ where A is acyclic, then F has a fixed point.

5. The σ-homotopy and Mc Cord construction

In this section and the next one we will describe the computation of the σ-homotopy theory of weighted maps obtained in [P1], [P4]. The homotopy theory of weighted maps turns out to be very simple. The σ-homotopy class of a weighted map is essentially determined by the homomorphism induced in homology.

Our approach is through classifying spaces. The σ-homotopy classes of weighted maps into Y correspond to ordinary homotopy classes of maps into a classifying space $B(Y)$. It turns out that $B(Y)$ is a topological R-module, constructed by Mc Cord in [MC], generalizing infinite symmetric products. Since $B(Y)$ is homologically equivalent to a product of Eilenberg–Mac Lane spaces, this completely determines the σ-homotopy category.

The results are better formulated in the homotopy category πW_0 of pointed, CW-complexes and continuous maps preserving base points. We shall indicate with p the base point of any object of W_0 and we denote by the same letter the constant map in this category. The set of all pointed homotopy classes of maps from X to Y will be denoted by $\pi[X;Y]_0$.

Let σW_0 be the category whose morphisms are pointed σ-homotopy classes of weighted maps between CW-complexes having index 0 in any component. Let

$\sigma(X;Y)_0$ be the set of morphisms from X to Y in σW_0. Notice that $\sigma(X;Y)_0$ inherits a natural structure of R-module compatible with composition.

(5.1) DEFINITION. The n-th σ-homotopy module $\sigma_n(Y)$ of a pointed space Y is the module $\sigma(S^n, Y)_0$.

It is easy to see that the group structure of $\sigma_n(Y)$ coincides with the one derived from the co-group structure of the sphere $S^n = S(S^{n-1})$. The relative σ-homotopy groups $\sigma_n(X, A)$ are defined in a similar way.

Clearly the inclusion of W as a subcategory of $\widetilde{W}$ admits a factorization through the respective homotopy categories by means of a functor $J: \pi W_0 \to \sigma W_0$. In particular J induces a natural transformation j_n from the n-th homotopy group $\pi_n(Y)$ into $\sigma_n(Y)$.

The *Hurewicz map* $h_n: \sigma_n(Y) \to \mathcal{H}_n(Y)$ is defined in the usual way. Namely, $h_n(\alpha) = \alpha_*(1_n)$ where 1_n is a generator of $\mathcal{H}_n(S^n) \simeq R$.

The ordinary Hurewicz homomorphism in singular homology induces a homomorphism

$$\widehat{h}_n: \pi_n(Y) \otimes R \to H_n^{\text{sing}}(Y; R).$$

Clearly we have $h_n = \widehat{h}_n(j_n \otimes \text{id})$ up to identification of $H_n^{\text{sing}}(Y; R)$ with the weighted homology $\mathcal{H}_n(Y)$.

The knowledge of $\sigma(X;Y)_0$ completely determines $\sigma(X;Y)$. Indeed, by Lemma 3.8 of [P1], if X, Y are two pointed connected CW-complexes, then any weighted map between them is homotopic to a pointed one. Moreover, two freely homotopic pointed maps are also homotopic in the pointed category. It follows form this that if X, Y are connected, the map sending ϕ to $(\phi - I(\phi)p, I(\phi))$ induces an isomorphism between $\sigma(X;Y)$ and $\sigma(X;Y)_0 \times R$.

Our main tool is the following simplicial approximation theorem (see [P1]).

(5.2) THEOREM. *Let X, Y be polyhedra, X finite. Then, given any weighted mapping $\phi: X \to Y$, there exist a triangulation K of X and weighted map $\psi: X \to Y$, σ-homotopic to ϕ, such that the restriction of ψ to each simplex $|s|$ of $|K|$ has the form $\sum r_i f_i$ where $r_i \in R$ and f_i are singlevalued piecewise-linear maps from $|s|$ to Y.*

Notice that better result should not be expected since it is false that any weighted map $\phi: X \to Y$ is σ-homotopic to a sum $\sum r_i f_i$ with $f_i: X \to Y$ continuous.

(5.3) EXAMPLE (communicated by C. P. Ramanujam). Let $T \subset \mathbb{R}^3$ be a two-torus, S be a two-sphere contained inside T. Let $f: T \to S$ be the map which assign to each $x \in T$ the intersection point $f(x)$ of the segment between x and the center of the sphere. Then, by the Example (4.7) the multivalued map f^{-1} is the support of a weighted map $\phi: S \to T$ such that $f \cdot \phi = \mathrm{id}_S$ (since $\deg f = 1$).

Suppose that ϕ is σ-homotopic to $\sum n_i g_i$, where $g_i: S \to T$ are continuous maps and $n_i \in \mathbb{Z}$. Since $\pi_2(T) = 0$ we have that each g_i is homotopic to a constant map. Moreover, since $\sum n_i = I(\phi) = \deg f = 1$ it follows that ϕ and hence also id_S must be σ-homotopic to a constant map. This is impossible, because $\mathcal{H}_2(S) = \mathbb{Z}$.

Since any finite CW-complex has the homotopy type of a polyhedron via a cellular homotopy equivalence Theorem (5.2) has an interesting corollary.

We say that a weighted map between two CW-complexes is cellular if its minimal support maps the q-skeleton of the domain into the q-skeleton of the range.

(5.4) COROLLARY. *Any weighted map between CW-complexes with finite complex as domain is σ-homotopic to a cellular weighted map. Moreover, if A is a subcomplex and $\varphi: X \to Y$ is such that φ/A is cellular, then φ is σ-homotopic to a cellular map $\psi: X \to Y$ such that $\varphi/A = \psi/A$.*

Let R be a ring. For any pointed space X let us consider the R-module $B(X; R)$ whose elements are functions $\widetilde{n}: X \to R$ such that $\widetilde{n}(p) = 0$ and $\widetilde{n}(x) = 0$, for all but finitely many x. For each $n \geq 0$, let $B_n(X; R)$ be the set of all members of $B(X; R)$ whose value is different from 0 on at most n-elements of X. Thus, $B(X; R)$ is the union of the increasing sequence of subsets $B_0(X; R) \subset B_1(X; R) \subset \ldots$.

The set $X \setminus \{p\}$ can be identified with the canonical basis of the free R-module $B(X; R)$ and $B_n(X; R)$ consist of those elements which belong to the image of the map $\pi_n: (X \times R)^n \to B_n(X; R)$ defined by

$$\pi_n((x_1, \lambda_1), \ldots, (x_n, \lambda_n)) = \lambda_1 x_1 + \ldots + \lambda_n x_n.$$

If we consider R as a discrete space and give to $(X \times R)^n$ the product topology, then $B_n(X; R)$ becomes a topological space by endowing it the quotient topology from the surjective map π_n. Finally, we give $B(X; R)$ the weak topology of the union of the sequence of spaces $B_n(X; R)$. With this topology $B(X; R)$ becomes a functor from the category of pointed spaces to topological R-modules.

In what follows we shall write BX instead of $B(X; R)$ whenever R remains unchanged. The functor B is stable under suspension and exact on co-fibrations. It follows from this that $\pi_* B(Y)$ verifies all the axioms of a homology theory and therefore for any $\pi_n B(Y) \simeq H_n^{\mathrm{sing}}(Y; R)$, where $H_n^{\mathrm{sing}}(Y; R)$ denotes the n-th singular homology group of Y with coefficients in R.

There is a tautological weighted map $\eta: BY \to Y$ defined by

$$\eta(\lambda_1 y_1 + \ldots + \lambda_n y_n) = \sum \lambda_i y_i - \left(\sum \lambda_i\right) p.$$

To see that η is a weighted map, by definition of weak topology, we have only check that η restricted to B_nY is a weighted map. But η restricted to B_nY is a factorization through π_n of the weighted map $\vartheta\colon (Y \times R)^n \to Y$ defined by $\vartheta((y_1, \lambda_1), \ldots, (y_n, \lambda_n)) = \sum \lambda_i y_i$. Since ϑ is a weighted map, the assertion follows from the factorization lemma (see [P1, Lemma 2.5]).

Any pointed weighted map $\varphi\colon X \to Y$ of index 0 at p induces a single-valued map $f_\varphi\colon X \to BY$ in an obvious way. In general, f_φ is not continuous but if X is a simplicial complex with the weak topology and if φ is such that its restriction to any simplex $|s|$ is of the form $\sum \lambda_i f_i$ where f_i are single-valued continuous maps from $|s|$ to Y as in (5.2), then f_φ is continuous since its restriction to any simplex factorizes through $\pi\colon (Y \times R)^n \to B_nY \subset BY$.

This together with Theorem (5.2) gives

(5.5) PROPOSITION. *If $\varphi\colon X \to Y$ is a pointed weighted map of index* 0, *then there is a pointed continuous map $f_\varphi\colon X \to BY$ such that φ is σ-homotopic to $\eta \circ f_\varphi$.*

It follows immediately that $\overline{\eta}\colon \pi[X; BY]_0 \to \sigma(X; Y)_0$ is surjective and, in particular, $\overline{\eta}\colon \pi_n(BY) \to \sigma_n(Y)$ is an epimorphism.

On the other hand the composition

$$\widetilde{H}_n^{\mathrm{sing}}(Y) \simeq \pi_n(BY) \to \sigma_n(Y) \to \widetilde{\mathcal{H}}_n(Y)$$

is the Darbo's isomorphism between the reduced singular homology and the reduced weighted homology. This implies that all the above maps are isomorphisms. In particular:

(5.6) PROPOSITION. *The Hurewicz map $\sigma_n(Y) \to \widetilde{\mathcal{H}}_n(Y)$ is an isomorphism for all $n \geq 0$.*

Moreover, we obtain following theorem by applying the comparison theorem to the homotopy functors $\sigma(JX; Y)_0$ and $\pi[X; B(Y)]_0$:

(5.7) THEOREM. *The map $\overline{\eta}\colon \pi[\,\cdot\,; B(Y)]_0 \to \sigma(J\,\cdot\,; Y)_0$ is a natural equivalence.*

6. Relation between the homotopy and homology theory of weighted maps

In order to compute the homotopy theory of weighted maps let us first define the cellular homology and cohomology functors on the σ-homotopy category of weighted maps. They can be derived from the weighted homology by the usual method.

For $n \geq 0$, let X^n be the n-th skeleton of the CW-complex X. We define the n-th cellular chain module of X by $C_n(X) = \mathcal{H}_n(X^n, X^{n-1})$. Endowing the graded module $C_*(X) = \{C_n(X)\}$ with the boundary operator given by the composition

$$\mathcal{H}_n(X^n, X^{n-1}) \to \mathcal{H}_{n-1}(X^{n-1}) \to \mathcal{H}_{n-1}(X^{n-1}, X^{n-2})$$

we obtain a complex whose homology modules are the *cellular homology modules* $H_*(X)$ of X. The above construction extends to pairs. The homology modules $H_*(X, A)$ obtained in this way are clearly isomorphic to the cellular homology obtained via the singular theory. By the cellular approximation theorem (5.4) any weighted map induces in a functorial way a homomorphism in cellular homology. From (5.6) it follows that

(6.1) PROPOSITION. *For each pair of CW-complexes (X, A), the Hurewicz map $h: \sigma_n(X, A) \to \widetilde{H}_n(X, A)$ is an isomorphism, for all $n \geq 0$.*

(6.2) REMARK. For polyhedral pairs this was proved independently by Jerrard and Meyerson in [JeM] using a different argument. They verified the Eilenberg–Steenrod axioms for $\sigma_n(X, A)$ by localizing weighted maps defined on the simplex in a way that is reminiscent of the proof of the excision property in singular homology.

Given any R-module M *the cellular cohomology of X with coefficients in M* is the homology of the cochain complex $C^*(X) = \mathrm{Hom}(C_*(X); M)$. We will denote it by $H^*(X; M) = \{H^k(X; M)\}_{k \geq 0}$.

Let us consider the homomorphism $\varrho: H^*(X; M) \to \mathrm{Hom}(H_*(X); M)$ induced by the Küneth pairing. Namely, $[\varrho(a)](b) = z'(z)$, where z' is a representative of the cohomology class a and z is the representative of the class b.

An element $a \in H^n(Y; H_n(Y))$ will be called *n-characteristic* if $\varrho(a)$ is the identity map of $H_n(Y)$.

Given a connected CW-complex Y, let $a_n \in H^n(Y; H_n(Y))_{n \geq 1}$ be a sequence of characteristic elements. Let $\varphi: X \to Y$ be a pointed weighted map. We define, the nth *obstruction* of φ, $\varepsilon_n(\varphi) \in H^n(Y; H_n\ (Y))$ by $\varepsilon_n(\varphi) = \varphi^*(a_n)$.

Clearly, for each $n \geq 1$, $\varrho\varepsilon_n(\varphi) = \varphi_*: H_n(X) \to H_n(Y)$.

(6.3) THEOREM. *The natural transformation*

$$\varepsilon = (\varepsilon_n): \sigma(X; Y)_0 \to \prod_{n \geq 1} H^n(X; H_n(Y))$$

is an isomorphism for any connected CW-complexes X, Y.

Indeed, by the comparison theorem for homotopy functors it is enough to check that ε is an isomorphism when $X = S^n$. But this follows immediately from the above discussion and Proposition (6.1).

Using the relation between the pointed and the free homotopy classes of maps the above theorem can be extended to a classification of free σ-homotopy classes of weighted maps from a CW-complexes X into a CW-complex Y such that $\pi_0(Y)$ is finite, see [P4, Theorem (5.8)].

(6.4) THEOREM. *If Y has a finite number of connected components, then there is a natural isomorphism*

$$\varepsilon\colon \sigma(X;Y) \to \prod_{n\geq 0} H^n(X; H_n(Y)).$$

It follows from this that the extension problem for weighted maps with values in such a CW-complex Y has only homological obstructions. Denoting with δ the connecting homomorphism of the exact sequence of a pair, a weighted map $\phi\colon A \to Y$ can be extended to a weighted map from X into Y if and only if the obstruction $\delta\varepsilon_n(\phi) \in H^{n+1}(X, A; H_n(Y))$ vanishes for any $n \geq 0$.

In what follows we always assume that $\pi_0(Y)$ is finite. Using the universal coefficient theorem in cohomology we easily obtain

(6.5) THEOREM. *Let R be a principal integral domain. Then there is an exact sequence*

$$0 \to \prod_{n\geq 0} \mathrm{Ext}(\mathcal{H}_{n-1}(X); \mathcal{H}_n(Y)) \to \sigma(X;Y) \xrightarrow{\mathcal{H}} \prod_{n\geq 0} \mathrm{Hom}(\mathcal{H}_n(X); \mathcal{H}_n(Y)) \to 0$$

where $\mathcal{H}(\varphi) = \varphi_*$. Moreover, the sequence splits.

Thus, if R is a field, then any homomorphism of degree 0 between $\mathcal{H}_*(X)$ and $\mathcal{H}_*(Y)$ is induced in homology by a unique (up to σ-homotopy) weighted map.

(6.6) COROLLARY. *Two CW-complexes are σ-homotopy equivalent if and only if their weighted homology modules are isomorphic. Thus any CW-complex is equivalent to a wedge of Moore spaces $M(\pi; n)$.*

(6.7) COROLLARY. *A subcomplex Y of a complex X is a weighted retract of X (i.e. there is a weighted map $\varphi\colon X \to Y$ such that $\varphi/Y = \mathrm{id}_Y$) if and only if $H_*(Y)$ is a direct summand of $H_*(X)$.*

Next, we use the Theorem (6.3) in order to establish a comparison with stable maps. Let $\{X;Y\} = \lim\operatorname{dir}\pi[S^kX, S^kY]$ be the group of stable homotopy classes of stable maps from X to Y. By Proposition (5.6) and the classical comparison theorem for homotopy functors, the suspension $S\colon \sigma(X,Y) \to \sigma(SX, SY)$ is an isomorphism. It follows from this that J extends to a group homomorphism $j\colon \{X;Y\} \to \sigma(X;Y)$.

(6.8) COROLLARY. *If R is flat over $\mathbb{Z}$ and such that $\otimes R$ kills finite groups (e.g. a $\mathbb{Q}$-algebra), then*

$$j \otimes \mathrm{id}\colon \{X;Y\} \otimes R \to \sigma(X;Y)$$

is an isomorphism.

Let us call a weighted map $\phi\colon X \to Y$ homotopically unbranched if ϕ is σ-homotopic to $\sum_{i=0}^{k} r_i f_i$ where $f_i\colon X \to Y$ are singlevalued continuous maps.

By Freudenthal Suspension Theorem, $S\colon \pi[X,Y] \to \{X;Y\}$ is a bijection whenever Y is n-connected and $\dim X \le 2n$. Thus, from the above corollary we obtain:

(6.9) COROLLARY. *If R is a field, Y is n-connected and* $\dim X \le 2n$, *then any weighted mapping from X to Y is homotopically unbranched.*

7. Approximation of acyclic weighted carriers by weighted maps

In this section we will describe how the results from the preceding section can be used in order to approximate acyclic weighted carriers by weighted maps. This leads to the construction of a well-defined homomorphism induced by a weighted carrier in homology and to a generalization of the Lefschetz fixed point theorem to acyclic weighted carriers and other related classes. Here approximation means not only the graphs of the maps are close enough, which is the usual notion of approximation for multivalued maps, but also a condition regarding to the multiplicities which we are going to explain below.

In the notation in this section we will not make any distinction between weighted maps and finite-valued weighted carriers. Hence we will denote with ϕ both the weighted map ϕ and its minimal support t_ϕ. Let A be a subset of a metric space X. We will denote by $O_\varepsilon(A)$ the ε-neighbourhood of A in X.

(7.1) REMARK. Let us denote by $\widehat{\mathcal{H}}_q(A)$ the inverse limit over the family of ε-neighbourhoods of A of $\mathcal{H}_q(O_\varepsilon(A))$. In [CP], [Sk3] were considered weighted carriers $\Psi : X \to Y$ with acyclic values in the following sense: $\widehat{\mathcal{H}}_q(\Psi(x)) = 0$, for all $x \in X$ and every $q > 0$. However they are the same as weighted carriers with acyclic values in the sense of the *Čech* homology considered here. Indeed, if X is a metric ANR and A is a compact subset of X then, taking into account [Ko] or [Du] and [Ma], [W], we have

$$\begin{aligned}\check{H}_q(A) &= \lim\operatorname{inv} \check{H}_q(O_\varepsilon(A)) = \lim\operatorname{inv} H_q^{\mathrm{sing}}(O_\varepsilon(A))\\ &= \lim\operatorname{inv} \mathcal{H}_q(O_\varepsilon(A)) = \widehat{\mathcal{H}}_q(A).\end{aligned}$$

(7.2) DEFINITION. Let $F: X \to Y$ be a weighted carrier, and let $\varepsilon > 0$. A weighted map $\varphi: X \to Y$ is an ε-approximation of $F: X \to Y$ if:

(7.2.1) $\varphi(x) \subset O_\varepsilon(F(O_\varepsilon(x)))$ for all $x \in X$;
(7.2.2) $m(\varphi(x), U) = m(F(x), U)$ for any piece U of $O_\varepsilon(F(O_\varepsilon(x)))$.

(7.3) THEOREM ([P2], [CP]). *Let $A \subset X$ be a finite polyhedral pair, let Y be a metric ANR and let $F: X \to Y$ be an acyclic w-carrier. Given any $\varepsilon > 0$ there exists a $\delta > 0$ such that any δ-approximation $\varphi: A \to Y$ of F restricted to A can be extended to an ε-approximation $\widetilde{\varphi}: X \to Y$ of F.*

The proof of the theorem is roughly speaking as follows: by the acyclicity of $F(x)$, using an old idea of Lefschetz, one can find a sequence $\delta \leq \varepsilon_1 \leq \ldots \leq \varepsilon_n = \varepsilon$, $n = \dim(X - A)$ such that, for all x, the inclusion of $O_{\varepsilon_i}(F(x))$ in $O_{1/2\varepsilon_{i+1}}(F(x))$ induces a trivial homomorphism in homology. Thanks to the isomorphism between the σ-homotopy and weighted homology groups, applied to the boundary of a simplex, one is able to extend inductively from a q-skeleton to the $(q+1)$-skeleton any δ-approximation of $F|_A$ to an ε-approximation of F. The condition (7.2.2) is fundamental in order to obtain the extension from the 0-skeleton to the next one.

As a consequence we have:

(7.4) COROLLARY. *Any acyclic weighted carrier has an ε-approximation by weighted maps for any $\varepsilon > 0$.*

(7.5) COROLLARY. *Let $F: X \to Y$ be an acyclic w-carrier. Then there exists $\delta > 0$ such that any two δ-approximations φ_1, φ_2 of F are homotopic as weighted maps.*

Now, we are able to define for any acyclic weighted carrier $F: X \to Y$ the homomorphism induced in homology $F_*: \mathcal{H}_*(X) \to \mathcal{H}_*(Y)$.

By the above results any weighted carrier F can be arbitrary approximated by weighted maps. Moreover, any two weighted maps sufficiently close to F are σ-homotopic and hence induce the same homomorphism in the homology theory $\mathcal{H}$. Therefore, we can define F_* as φ_* for φ close enough to F. The same approximation theorem allows us to conclude that F_* is invariant by homotopies of acyclic weighted carriers.

From now on we will assume that R is a principal integral domain. Let $F: X \to X$ be an acyclic weighted carrier from a finite polyhedron X into itself. Then the Lefschetz number of F is defined in the usual way. If a weighted carrier is fixed point free its graph does not intersects the diagonal in $X \times X$. By compactness any close enough approximation in graph is also fixed point free. Therefore from the above approximation and Darbo's fixed point theorem we obtain:

(7.6) THEOREM ([P2]). *Let $F: X \to X$ be a weighted carrier from a finite polyhedron into itself. If $L(F) \neq 0$, then there exists $x \in X$ such that $x \in F(x)$.*

The paper [CP] deals with an extension of the above theorem to a larger class of weighted carriers, called Lefschetz weighted carriers. This class is interesting because it arise quite naturally in applications. Notice that, if $F: X \to Y$ is an acyclic weighted carrier and $f: Z \to X$ (resp. $g: Y \to Z$) are singlevalued continuous maps, then Ff is an acyclic weighted carrier. But gF being a weighted carrier need not be acyclic. However any approximation of F by a weighted map under minimal assumptions will produce also an approximation of $g \circ F$. This motivates the next definition.

(7.7) DEFINITION. A weighted carrier F from a complete metric ANR X into itself will be called a Lefschetz weighted carrier if:

(7.7.1) The closure $\overline{F(X)}$ is compact,

(7.7.2) F can be factorized in the form $F = r \circ G$, where $G: X \to Y$ is an acyclic weighted carrier from X into a complete metric ANR space Y and $r: Y \to X$ is a continuous singlevalued map.

In [CP] the following theorem (Theorem 5.1) was stated.

(7.8) THEOREM. *Let $F: X \to X$ be a Lefschetz weighted carrier of the form $F = r \circ G$. Let $h: \mathcal{H}(X) \to \mathcal{H}(X)$ be defined by $h = r_* G_*$. Then the image of h is finitely generated and hence $L(h))$ is defined. Moreover, if $L(h) \neq 0$ then F has a fixed point.*

However the proof presented in that paper is incomplete. The problem is in the lack of definition of the induced homomorphism $G_*: \mathcal{H}_*(X) \to \mathcal{H}_*(Y)$ for general ANR's. This gap was filled up by Robert Skiba in his Ph. D. thesis. Here is a short survey of his results.

The following theorem is an extension of Theorem (7.3) to the case of a compact ANR.

(7.9) THEOREM ([Sk3]). *Let X be a compact* ANR, *A be a closed* ANR *subset of X, Y be a metric* ANR. *Suppose that $F: X \to Y$ is an acyclic weighted carrier. Given any $\varepsilon > 0$ there exists a $\delta > 0$ such that any δ-approximation $\varphi: A \to Y$ of F restricted to A can be extended to an ε-approximation $\widetilde{\varphi}: X \to Y$ of F.*

The proof of this theorem is based on two lemmas. The first of them is an extension of Dugundji ε-domination theorem to pairs, obtained by Mardesic. The second is a special case of the theorem. Let us underline that a main idea of the proof of the second lemma is taken from [BGK], [GGK2]. Details will appear in [Sk3].

(7.10) REMARK. Many important results concerning approximability of multivalued maps have been obtained by L. Górniewicz, A. Granas, W. Kryszewski and others (see [GGK1], [GGK2], [GL], [Kr4], [Kr5]). For a good survey, we refer the reader to W. Kryszewski [Kr6].

(7.11) LEMMA ([MaS]). *Let (X, A) be a pair of compact* ANR*'s and let $\varepsilon > 0$. Then there is a finite polyhedral pair (P, Q) and maps of pairs $i: (P, Q) \to (X, A)$ and $q: (X, A) \to (P, Q)$ such that $d(iq(x), x) < \varepsilon$ for each $x \in X$.*

(7.12) LEMMA ([Sk3]). *Let (X, A) and Y be as before and let $F: X \to Y$ be an acyclic weighted carrier. Let $B = (X \times \{0\}) \cup (A \times [0, 1])$ and let $H: X \times I \to Y$ be defined by $H(x, t) = F(x)$. Then for every $\varepsilon > 0$ there exists $\delta > 0$ such that any weighted mapping $\varphi: B \to Y$ that is a δ-approximation of H restricted to B can be extended to an ε-approximation of H.*

Now let us complete the proof of Theorem (7.8). Using Theorem (7.9) we can define the homomorphism in homology induced by an acyclic weighted carrier $G: X \to Y$ where X a compact ANR and Y is an ANR in the same way as was used Theorem (7.3) in the case of a polyhedron X.

To define the homomorphism $G_*: \mathcal{H}_*(X) \to \mathcal{H}_*(Y)$ when X is an arbitrary ANR, we will use the fact that $\mathcal{H}_*$ has compact supports. Thus,

$$\mathcal{H}_*(X) = \lim \operatorname{dir}\{\mathcal{H}_*(Z) : Z \text{ compact}\}.$$

Since the family of compact ANR's is cofinal in the family of all compact subset of X, we can compute $\mathcal{H}_*(X)$ as the direct limit of the homology of he directed family $\mathcal{CA}$ of compact ANR subspaces Z of X. If $Z, W \subset X$ are compact ANR's and $Z \subset W$, then by uniqueness $(G|_Z)_* = (G|_W)_* i_*$, where $i: Z \to W$ is the inclusion. Thus

$$\{(G|_Z)_*: \mathcal{H}(Z) \to \mathcal{H}(Y) : Z \in \mathcal{CA}\}$$

is a directed family of homomorphisms. Hence it induces a homomorphism

$$G_*: \mathcal{H}_*(X) = \lim \operatorname{dir} \mathcal{H}_*(Z) \to \mathcal{H}_*(Y).$$

With this definition of G_* one can easily check that the image of $h = r_* G_*$ is finitely generated and the Lefschetz number $L(h)$ can be defined. The rest of the proof of Theorem (7.8) is as in [CP].

(7.13) COROLLARY. *Let C be an acyclic metric* ANR*. Then any Lefschetz weighted carrier F from C into itself with $I(F) \neq 0$ has a fixed point.*

We will close this section with an application of the fixed point theory of Lefschetz weighted carriers to the existence of periodic trajectories of a vector field on a full torus. The problem was firstly investigated by Fuller, in [Fu]. His construction of the homomorphism induced in homology in that paper is very close to what we have discussed in this section and we essentially follow his approach.

Let $C = B^2 \times S^1$ be the full 2-torus, i.e. the product of the 2-ball B^2 with the circle S^1. Let X be a vector field on C pointing inward on the boundary $S^1 \times S^1$ of C. This implies that the trajectory of a point is defined for all time $t \geq 0$. Hence, X generates a semiflow $\Phi(c,t)$ defined on all of $C \times \mathbb{R}^+$. The universal covering of C is a cylinder $D = B^2 \times \mathbb{R}$. The covering map $\pi\colon D \to C$ is defined by $\pi(x,\theta) = (x, e^{i\theta})$. The angular coordinate $e^{i\theta}$ defines a 1-form ω on C such that $\pi^*(\omega) = d\widetilde{\theta}$ where π^* is the induced map on cotangent bundle and $\widetilde{\theta}\colon D \to R$ is given by $\widetilde{\theta}(x,\theta) = \theta$. For $c \in C$ and $t \in \mathbb{R}^+$ let us consider the integral of the form ω over the part of the trajectory going from c to $\Phi(c,t)$, that is,

$$\eta(c,t) = \int_0^t \omega(X_{\Phi(c,t)})\, d\tau.$$

(7.14) THEOREM. *If $\lim_{t\to\infty} \eta(c,t) = \infty$ uniformly in c, then there exists a closed trajectory of the field X.*

PROOF. The vector field X induces a vector field $\widetilde{X}$ whose semiflow $\widetilde{\Phi}(d,t)$ covers the semiflow $\Phi(c,t)$ under the covering projection π. If $d \in D$, $c = \pi(d)$, $t \geq 0$, we have that the integral of the form ω over the trajectory from c to $\Phi(c,t)$ coincides with the integral of the form $d\widetilde{\theta} = \pi^*\omega$ over the path of trajectory from d to $\widetilde{\Phi}(d,t)$. Hence, we get $\eta(c,t) = \eta(d,t) = \widetilde{\theta}(\Phi(d,t)) - \widetilde{\theta}(\Phi(d,0))$. Therefore, the coordinate $\widetilde{\theta}(\widetilde{\Phi}(d,t))$ of the trajectory passing by d must go to infinity with t. Let h_0, h_1 be the embeddings of the ball B^2 in D given by $h_0(x) = (x,0)$, $h_1(x) = (x,1)$. Thus, $S_i = h_i(B^2) = \widetilde{\theta}^{-1}(i)$, $i = 0,1$ are 2-dimensional submanifolds of D. By the above discussion, for each $x \in B^2$, the set

$$\tau(x) = \{t : \widetilde{\theta}\widetilde{\Phi}(h_0(x),t) \in S_1\}$$

is compact. Since $\tau(x)$ is the set of intersection points of a curve with a one-codimensional oriented submanifold S_1 of C, using the intersection index (see Appendix of [CP]), we can assign integral weights to pieces of $\tau(x)$ in a continuous way. With this the map τ becomes an acyclic weighted carrier, since subsets of $\mathbb{R}$ are acyclic in positive dimensions. Moreover, since $\widetilde{\theta}\widetilde{\Phi}(h_0(x),t) \to \infty$ as $t \to \infty$, it is easy to see that the index $I(\tau) = 1$. For each $d \in D$, let us denote by $\Gamma(d)) = \{\widetilde{\Phi}(d,t) : t \geq 0\}$ the trajectory of the vector field $\widetilde{X}$ passing through d. Since the multivalued map $G(x) = \Gamma(h_0(x)) \cap S_1$ can be composed as

$$B^2 \xrightarrow{h_0 \times \tau} D \times R \xrightarrow{\widetilde{\Phi}} D$$

it follows that G is a Lefschetz weighted carrier. But, $G(B^2) \subset S_1$ and hence $F = h_1^{-1} \circ G$ is a Lefschetz weighted carrier from the ball B^2 into itself. Furthermore $I(F) = I(\tau) \neq 0$. Hence, by Corollary (7.13), F has a fixed point. But the fixed points of the map F correspond to the closed trajectories of the vector field X, since $\pi h_1 = \pi h_0$. □

8. Approximation theorems for other classes of weighted carriers

In this section we will consider the class $\mathcal{AW}$ of all weighted carriers that can be approximated in the sense of Definition (7.2) by weighted mappings. In Section 7 we showed that acyclic and Lefschetz weighted carriers belong to $\mathcal{AW}$ under mild assumptions on the domain and range spaces. But weighted carriers arising as zeroes of parameterized families of maps between orientable manifolds of the same dimension and fixed points of maps depending on a parameter (see Examples (3.3), (3.4), (4.7) also belong to the class $\mathcal{AW}$.

The approximation theorems needed here follow easily from a series of results presented by H. Kurland and J. Robbin in their beautiful paper [KuR]. Let us explain what they did there.

By Sard's Theorem, given a point y in the range manifold, any smooth map f between manifolds of the same dimension can be approximated in the C^∞ topology by a map for which y is a regular value. If the domain is compact or, more generally, the map f is proper, then the inverse image of y by the approximating map will be a finite set. This approximation property fails for parameterized families of maps if we expect that y will be a regular value of each map in the family. There are topological obstructions to this. The same happens if we expect to approximate a given single map by a map having only regular values, i.e. a submersion. For example, the map in Example (4.7) cannot be approximated by submersions.

However, if we only want to approximate a given family of maps by families such that, for each map in the family, the inverse image of y is a discrete set, then such approximations exist in great generality. The reason for this is the following: the space $J^k(n)$ of all k-jets of maps from $\mathbb{R}^n$ into itself contains an algebraic variety W^k such that if the k-th jet $j^k f(0)$ of a mapping f at the point 0 is not in W^k, then either $f(0) \neq 0$ or 0 is an isolated zero of f. Moreover, the codimension of W^k goes to infinity with k.

Let P be a finite dimensional manifold. As before we will consider a map $f\colon P \times \mathbb{R}^n \to \mathbb{R}^n$ to be a family of maps $f_p\colon \mathbb{R}^n \to \mathbb{R}^n$ parameterized by P. If we take k large enough so that the codimension of W^k exceeds $\dim P + n$, then any map f such that $j^k f_p(x)$ is transversal to W^k does not intersect W^k and therefore x is at most an isolated zero of f_p. Applying the density of transversal maps (see [AR]) in the way it is done in [KuR], any family $f: P \times \mathbb{R}^n \to \mathbb{R}^n$ can be approximated in the C^∞ topology of $C^\infty(P \times \mathbb{R}^n, \mathbb{R}^n)$ by a map g such that each $g_p^{-1}(0)$ is discrete.

This is a only a prototype of the results in [KuR]. Indeed, we can take instead of $\mathbb{R}^n$ any two manifolds M, N of the same dimension. Moreover, the same holds if we consider fixed points of families of $C^\infty(P \times M, M)$ or singular points of families of vector fields on M parameterized by P.

Let P, X and Y be smooth compact manifolds. Assume that X and Y are orientable over R and that $\dim X = \dim Y$. For simplicity, we will assume that X and Y are without boundary, although all that we are going to say can be extended to manifolds with boundary as well. Let $f: P \times X \to Y$ be a continuous map. Let us consider the solution map

$$F_f: P \times Y \to X \text{ defined by } F_f(p, y) = f_p^{-1}(y).$$

Exactly as explained in Example (3.3), F becomes a weighted carrier with weights in R, assigning weights by use of the degree theory for maps between oriented manifolds.

(8.1) THEOREM. *If P, X, Y are smooth compact manifolds as above and $f: P \times X \to Y$ is a continuous map, then the map $F_f: P \times Y \to X$ defined above is a weighted carrier belonging to $\mathcal{AW}$. More precisely: for any $\varepsilon > 0$ there exist a smooth map $g: P \times X \to Y$ such that the weighted carrier ϕ defined by $\phi(p, y) = g_p^{-1}(y)$ is a weighted map that is an ε-approximation of the weighted carrier F_f.*

PROOF. Let us endow P, X and Y with a metric and their products with the sup metric. Also, the distances between maps are given by the sup metric.

Let $\varepsilon > 0$. We first observe that, since a compact smooth manifold is an ANR, we may choose a δ, with $0 < \delta \leq \varepsilon$, such that any two mappings $f, g: X \to Y$ with $d(f, g) \leq \delta$ are homotopic by a homotopy $\{h_t: X \to Y\}_{t \in [0,1]}$ which verifies $d(f, h_t) \leq \varepsilon$ at each t.

In order to approximate $F = F_f$ by weighted maps we will use the following proposition (see [KuR, Theorem 6.1]).

(8.2) PROPOSITION. *Let P, X and Y be smooth manifolds, with $\dim X = \dim Y$. Then there is an open dense subset $\mathcal{G}$ of $C^\infty(P \times X, Y)$ such that each $g \in \mathcal{G}$ has the property that for each $p \in P$ the map $g_p : X \to Y$ given by $g_p(x) = g(p, x)$ for $x \in X$ is locally finite-to-one.*

It follows from Proposition (8.2) that we may choose a map $g \in \mathcal{G}$ such that $d(f, g) \leq \delta$. Indeed, this can be chosen by first approximating f by a smooth map and then approximating, in the C^0 topology, this smooth map by a map from $\mathcal{G}$. By the compactness of X we see that $g_p^{-1}(y)$ must be a finite set for any $(p,y) \in P \times Y$ and therefore $\phi(p,y) = g_p^{-1}(y)$, considered with multiplicities, is a weighted map (as before, we are denoting by the same symbol both the weighted map and its upper semicontinuous support).

In order to show that ϕ is an ε-approximation of F, we first observe that condition (7.2.1) simply means that the graph of ϕ is a subset of the ε-neighbourhood of the graph of F in the product metric on $P \times X \times Y$. But the isometry $(p,x,y) \to (p,y,x)$ sends the graph of f to the graph of F and the graph of g to the graph of ϕ. Thus (i) follows from the fact that the graph of g belongs to the δ-neighbourhood of the graph of f and $\delta \leq \varepsilon$.

To verify condition (7.2.2), let U be a piece of $O_\varepsilon(F(O_\varepsilon(p,y)))$. Since g and f are δ close, there is a homotopy $h\colon [0,1] \times P \times X \to Y$ between f and g such that $d(f, h_t) \leq \varepsilon$ for each $t \in [0,1]$. We claim that

$$h(t,p,x) \neq y \quad \text{for all } t \in [0,1],\ x \in \partial U,\ p \in P. \tag{8.3}$$

Indeed, suppose that $h(t,p,x) = y$ for some $t \in [0,1]$, $x \in \partial U$, and $p \in P$. Then $d(f(p,x),y) \leq \varepsilon$ and hence $x \in f_p^{-1}(O_\varepsilon(y)) \subset O_\varepsilon(F(O_\varepsilon(p,y)))$. Thus (8.3) holds on and therefore the restriction of h_p to $[0,1] \times \overline{U}$ is an admissible homotopy between f_p and g_p. By the homotopy invariance of degree,

$$m(F(p,y),U) = \deg(f_p,U,y) = \deg(g_p,U,y) = m(\phi(p,y),U),$$

which completes the proof of the theorem. □

Taking P to be a point we get:

(8.4) Corollary. *Let X and Y be smooth compact oriented manifolds of the same dimension and let $f\colon X \to Y$ be a continuous map. Then the inverse map $f^{-1}\colon Y \to X$ is a weighted carrier belonging to $\mathcal{AW}$. Moreover, an approximation is given by the weighted map $\phi(y) = g^{-1}(y)$ where $g\colon X \to Y$ is a single-valued map as close as we wish to f.*

From the above, we can retrieve an old result of H. Hopf in [Ho].

(8.5) Corollary. *Let X and Y be smooth compact oriented manifolds of the same dimension. Let $f\colon X \to Y$ be a map with $\deg f \neq 0$. Then $f_*\colon H_*(X) \to H_*(Y)$ is surjective (and split surjective if $d = 1$).*

Proof. Take g, as above, close enough to f so that $f_* = g_*$ and, in particular, $\deg g = \deg f$. Since $g\phi = [\deg g]\,\mathrm{id}$ we have $g_*\phi_* = [\deg f]\mathrm{id}$ in weighted homology, which, according to Darbo's Theorem, is the same as singular homology. Thus, f_* is surjective. □

It immediately follows from the preceding corollary that there is no map of nonzero degree from S^n into an orientable n-manifold having nontrivial homology in dimension k with $0 < k < n$.

Another consequence of the corollary is that the weighted carriers such as those described in Example (3.3) also belong to $\mathcal{AW}$ provided that the parameter space P is a manifold. To see this, we first observe that the map $f\colon P \times \overline{O} \to \mathbb{R}^n$ can be extended to a map $\overline{f}\colon P \times S^n \to S^n$ such that $0 \notin \overline{f}(P \times (S^n \setminus O))$. From Theorem (8.1), applied to $\overline{f}$, we conclude that the solution map defined by $S(p) = \{x : f_p(,x) = 0\}$ is a weighted carrier in $\mathcal{AW}$.

The foregoing easily generalizes to infinite dimensional spaces using the Leray–Schauder degree. Moreover, one can easily develop variants for noncompact parameter spaces.

Let X be a Banach space, P be a finite dimensional manifold and W be an open subset of $P \times X$ that is locally bounded over P. Let $f\colon \overline{W} \to X$ be a map of the form $f(p,x) = x - c(p,x)$ where $c\colon \overline{W} \to X$ is a compact map. In the terminology of Granas in [Gr1], each map f_p is acompact vector field defined on the set $\overline{W}_p = \{x : (p,x) \in \overline{W}\}$. Thus, we can think of f as a generalized family of compact vector fields parameterized by the space P. Assuming that $0 \notin f(\partial W)$, we can use the Leray–Schauder degree for compact vector fields, as in Example (3.3), in order to assign integral weights to the map defined by $S(p) = \{x : f_p(x) = 0\}$.

(8.6) THEOREM. *If $f\colon \overline{W} \to X$ is a family of compact vector fields parameterized by P, as above, then the restriction of the solution map S to any compact subset of P belongs to $\mathcal{AW}$.*

PROOF. Use steps two and three of the proof of Theorem 1.1 in [MNP] and the proof of (7.2.2) in Theorem (8.1). □

How can all this be used? One way is to use the induced homomorphism in homology by the approximating weighted map as a substitute for the algebraic transfers (see the last section) in order to obtain information about the relation between the homology of the parameter space P and the homology of the total solution set $\mathcal{S} = \{(p,x) : f(p,x) = 0\}$.

Just for simplicity, let us assume that P is compact and connected. Then for any open neighbourhood U of $\mathcal{S}$ we can find a weighted map approximation ϕ of S such that its graph map $\psi(x) = (x, \phi(x))$ has its image contained in U. Let π be the restriction to U of the projection to the parameter space P. Then $\pi\psi = d\,\mathrm{id}$, where $d = \deg_{LS}(f_p W_p, 0)$, for some and hence any, parameter $p \in P$. Thus, if $d \neq 0$, ψ_* injects the weighted homology $\mathcal{H}_*(P)$ of the space P into $\mathcal{H}_*(U)$. On the other hand, π_* is surjective. Taking limits we obtain an injection of $\mathcal{H}_*(P) = \check{H}_*(P)$ into $\check{H}_*(\mathcal{S})$.

The same argument gives also an injection of $\check{H}_*(P, A)$ into $\check{H}_*(\mathcal{S}, \mathcal{S}_{|A})$ for any closed subset A of P such that the closure of the complement of A in P is compact. This has several useful consequences.

First of all, the surjectivity of $\pi_*\colon \check{H}_*(\mathcal{S}, \mathcal{S}_{|A}) \to \check{H}_*(P, A)$ implies the following "continuation" property of the set of solutions $\mathcal{S}$: it contains a connected subset C covering all of the parameter space. Moreover, using the homological characterization of topological dimension it can be shown that C must be of dimension at least $\dim P$ at any of its points.

Results of this type have been obtained in [FMP2], [FMP1], [MP]. However, the above argument was only a heuristic motivation for proofs based on more standard algebraic methods, e.g. using products and umkher homomorphisms.

Another consequence is that one can estimate the cup-length (and hence the category) of the topological space $\mathcal{S}$ in terms of the cup-length of the parameter space P. Working over $\mathbb{Q}$ the surjectivity of π_* implies the injectivity of $\pi^*\colon \check{H}^*(P, \mathbb{Q}) \to \check{H}_*(\mathcal{S}, \mathbb{Q})$ and from this we get that the cup-length of $\mathcal{S}$ is not less than the cup-length of P. This leads to multiplicity results for critical points of various functionals acting on maps with values in topologically nontrivial manifolds, such as the action functional defined of a Hamiltonian vector field (see [F1], [F2], [HZ], [V]).

Weighted mapping approximations can be also used for another purpose. Their mere existence allows us to formulate various geometric conditions under which the weighted carriers obtained as composition of singlevalued maps with the solution map of a parameterized family of equations must vanish at some point. This furnishes a substitution method for solving systems of equations.

Let us illustrate the use of approximations by weighted maps in solving systems of equations with a classical tool in nonlinear analysis, the Liapunov–Schmidt reduction.

Let $h\colon \overline{O} \subset Z \to B$ be a C^1 compact vector field defined on the closure of an open subset O of a Banach space Z. We split the equation $h(z) = 0$ in a neighbourhood of a solution z_0 into a system

$$\begin{cases} f(x, y) = 0, \\ g(x, y) = 0. \end{cases}$$

To do so, we write Z as a product $Z = X \times Y$ where $X = \ker Df(z_0)$ is a finite dimensional subspace of Z and $Y \simeq \operatorname{Im} Df(z_0)$. We take $f = \pi h$ and $g = \mathrm{id} - \pi h$, where π is the projection onto Y along X.

The equation $f(x, y) = 0$ is called *auxiliary equation*. The Liapunov–Schmidt reduction comprises two steps: first, we solve (in a neighbourhood of z_0) the auxiliary equation, using the implicit function theorem. Then we substitute the resulting solution $y = s(x)$ into the second equation obtaining the so called *bifurcation equation* $g(x, s(x)) = 0$. By this substitution, the problem of solving $h(z) = 0$ near z_0 is reduced to finding solutions of the bifurcation equation, which is a finite dimension system of nonlinear equations.

This classical method was generalized, among others, by Lamberto Cesari in order to deal with nonlocal situations. Assuming that the compact part of the equation is Lipschitz, one can still solve the auxiliary equation, not only locally but also in the large. Here, however, the space X is not necessarily the kernel, but is a finite dimensional space chosen so that it permits use of the contraction mapping principle to uniquely solve this equation. Our point is that, by using weighted carriers, there is no need of uniqueness of solutions of the auxiliary equation. Thus, we can drop the Lipschitzianity condition and have a more flexible choice of the finite dimensional space X. Indeed, if O is locally bounded over X and $f(x, y) \neq 0$ for $(x, y) \in \partial O$, then the map $S: X \to Y$, sending x into the set of all solutions $y \in O_x$ of the auxiliary equation, is a weighted carrier which can be approximated by a weighted map on any compact subset of X. Thus, we reduced our original problem $h(z) = 0$ to a finite dimensional multivalued bifurcation equation $0 \in F(x)$ where $F(x) = g(x, S(x))$.

Of course, all this is useful only if we have some good topological principles which lead to a solution of the bifurcation equation. Here we will discuss one such principle of Borsuk—Ulam type.

We will say that two closed subsets A and B of a normed space X are *strictly separated by a hyperplane* if there exists a bounded linear functional on X which is positive on one of them and negative on the other. It is easy to see that this condition is equivalent to $K(\overline{\mathrm{co}}\, A) \cap \overline{\mathrm{co}}\, B = \emptyset$. Here $\overline{\mathrm{co}}\, A$ is the closed convex hull of A and $K(B)$ is the cone from B with vertex at 0.

Let $B \subset \mathbb{R}^n$ be the closed unit ball. We will say that an upper semicontinuous multivalued map $F: B \to \mathbb{R}^n$ verifies the Borsuk–Ulam property on ∂B if, for each $x \in \partial B$, $F(x)$ and $F(-x)$ are strictly separated by a hyperplane.

(8.7) THEOREM. *Let B be the unit ball in $\mathbb{R}^n$ and let $F: B \to \mathbb{R}^n$ be a weighted carrier belonging to $\mathcal{AW}$ such that $I(F) \neq 0$. If F satisfies the Borsuk–Ulam property on ∂B, then there exist a point x in the interior of B such that* $0 \in F(x)$.

PROOF. Observe that the Borsuk–Ulam property, being equivalent to the non-intersection of the graphs of the two upper semicontinuous maps $x \mapsto K(\overline{\mathrm{co}}\, F(x))$ and $x \mapsto \overline{\mathrm{co}}\, F(-x)$, is an open condition. Thus if it holds for F, it must hold for any sufficiently close weighted map. Similarly, if $0 \notin F(B)$, then the same must hold for any weighted map sufficiently close to F. Therefore it is enough to prove the above theorem in the case that F is a weighted map. Suppose F is a weighted map.

We will show that F has a nontrivial degree in the sense of Section 10 by finding an admissible σ-homotopy between F and $I(F)f$ where f is an odd singlevalued map.

This will prove the theorem, since according to the Borsuk–Ulam Theorem odd maps have an odd degree.

We first observe that, by the Borsuk–Ulam condition, for each $x \in \partial B$ there exists $y \in \partial B$ such that

$$\begin{cases} \langle y, z\rangle > 0 & \text{for all } z \in F(x), \\ \langle y, z\rangle < 0 & \text{for all } z \in F(-x). \end{cases} \tag{8.8}$$

For $y \in \partial B$, let $V_y = \{x \in \partial B : (8.8) \text{ holds}\}$. Clearly V_y is an open subset of ∂B for each y. By the Borsuk–Ulam condition we have that $\{V_y\}_{y\in\partial B}$ is an open covering of the compact space ∂B.

Taking a finite sub-cover $\{V_{y_i}\}_{0\le i\le m}$ and a subordinated partition of unity $\{s_i\colon \partial B \to [0,1]\}_{0\le i\le m}$, for $x \in \partial B$, we set

$$f(x) = \sum_{i=0}^{m} (s_i(x) - s_i(-x))y_i.$$

Then $f\colon \partial B \to \mathbb{R}^n$ extends to an odd continuous map defined on B.

Consider $H\colon B \times I \to \mathbb{R}^n$ defined by

$$B \times I \xrightarrow{\Delta\times \mathrm{id}} B \times B \times \mathrm{id} \xrightarrow{F\times f\times \mathrm{id}} \mathbb{R}^n \times \mathbb{R}^n \times I \xrightarrow{\quad g \quad} \mathbb{R}^n$$

where Δ is the diagonal map and $g(x,y,t) = tx + (1-t)y$. It is a matter of calculation [MNP] to check that for all $x \in \partial B$ and all $t \in [0,1]$, $0 \notin H(x,t)$. Thus H is an admissible σ-homotopy between $H(x,0) = I(F)f(x)$ and $H(x,1) = F(x)$. This completes the proof. □

The above result was used in [MNP] to obtain criteria of Landesman–Lazer type for the solvability of semilinear elliptic equations. Papers [CNZ1], [CNZ2], [NOZ], [AC] contain other variants, generalizations to inclusions and applications of the above theory to existence of periodic solutions for differential equations with multivalued right hand side and control theory among others.

Let us discuss the corresponding approximation of fixed points of maps. Let M and P be compact manifolds with P connected, and let $f\colon P \times M \to M$ be a parametrized family of maps from M into itself. By Example (3.4) the map F sending p into the set of fixed points $\mathrm{Fix}(f_p)$ of f_p is an weighted carrier with $I(F) = L(f_p)$. Using Theorem 4.1 of [KuR], for any ε we can find a smooth map g ε-close to f and such that each g_p has a finite number of fixed points. This shows that $F \in \mathcal{AW}$.

The above approximation result can be easily extended to the case of families of maps from a compact finite dimensional ANR X into itself.

For this we will use the following proposition whose proof is taken from [Da4].

(8.9) PROPOSITION. *Compact finite dimensional* ANR *are retracts of oriented and differentiable manifolds.*

PROOF. Let $\mathbb{R}^n$ be the euclidean n-space, embedded in $\mathbb{R}^{n+1}$ as the hyperplane $x_{n+1} = 0$. Let X be a compact subset of $\mathbb{R}^n$ and N a closed and bounded neighbourhood of X in $\mathbb{R}^n$ and let $\varrho\colon N \to X$ be a retraction. The boundary (in $\mathbb{R}^n$) of N, ∂N, is compact and $\partial N \cap X = \emptyset$. Taking a continuous function ψ such that $\psi(x) = 1$ if $x \in X$ and $\psi(x) = 0$ if $x \in \partial N$ and an ε, $0 < \varepsilon < 1/2$, we can approximate ψ by a polynomial P such that

$$|\psi(x_1, \dots, x_n) - P(x_1, \dots, x_n)| < \varepsilon \quad \text{in } \partial N \cup X.$$

This leads to

$$\begin{aligned} P(x_1, \dots, x_n) &< \varepsilon && \text{on } \partial N, \\ P(x_1, \dots, x_n) &> 1 - \varepsilon && \text{on } X. \end{aligned}$$

Since, the set of critical points of a polynomial is finite, almost every c satisfying $\varepsilon < c < 1-\varepsilon$, is a regular value of P. Therefore the set N_c of points of N satisfying $P(x_1, \dots, x_n) \geq c$ is a neighbourhood of X whose boundary in $\mathbb{R}^n$ is a smooth manifold. If c is a regular value of $P(x_1, \dots, x_n) - x_{n+1}^2$, then the submanifold of $\mathbb{R}^{n+1}$, defined by

$$P(x_1, \dots, x_n) - x_{n+1}^2 - c = 0,$$

is orientable. Let W, be the isolated piece of this manifold that projects orthogonally on N_c. We call $\pi\colon W \to N_c$ this orthogonal projection.

For every point $x \in X \subset \operatorname{int} N_c$, $\pi^{-1}(x)$ is made of two points of W. We call $\alpha(x)$ the one, that has positive x_{n+1} coordinate. The map $\alpha\colon X \to W$ is an embedding. Composing the maps

$$W \xrightarrow{\ \pi\ } N_c \xrightarrow{\text{incl}} N \xrightarrow{\ \varrho\ } X$$

we get a map $\beta\colon W \to X$ so that $\beta\alpha = \mathrm{id}_X$. □

(8.10) THEOREM. *If B, X are compact finite dimensional* ANR *and $f\colon B \times X \to X$ is a continuous map, then the map $F\colon P \times X \to X$ defined by $F(b) = \operatorname{Fix}(f_b)$ is a weighted carrier belonging to $\mathcal{AW}$. More precisely: for any $\varepsilon > 0$ there exist a continuous map $g\colon B \times X \to X$ such that the weighted map $\phi(b) = \operatorname{Fix}(g_b)$ is an ε-approximation of the weighted carrier F.*

PROOF. By the previous proposition B and X can be embedded is retracts of finite dimensional compact manifolds P and M respectively. Let i be the embedding of X into M and r be the corresponding retraction. Let j be the embedding of $B \times X$ into $P \times M$ and let $s: P \times M \to B \times X$ be the induced retraction on the product. Take $\delta > 0$ such that r sends δ-close points of M to ε-close points in X. By the preceding discussion we can find a map $g': P \times M \to M$ that is δ-close to ifs such that each g'_p has a finite set of fixed points. Then $g = rg'j$ is ε-close to f and g_b has a finite set of fixed points for each $b \in B$. The proof that $\mathrm{Fix}(g_b)$ is an ε-approximation is as in Theorem (8.1). □

A different approach was used by Jerrard in [Je2]. When X and Y are finite polyhedra, he used piecewise linear approximations in order to approximate the fixed point carrier derived from a family $f: X \times Y \to Y$ with weighted maps. He applied this to the study of the fixed point property of product spaces as follows: if $(f, g): X \times Y \to X \times Y$, then fixed points of (f, g) are in one to one correspondence with fixed points of both $GF: X \to X$ and $FG: Y \to Y$, where F and G are the fixed point carriers associated to f and g respectively. His best result in this direction is the following theorem (see [Je5]).

(8.11) THEOREM. *If X and Y are compact polyhedra such that the Lefschetz number of any self map is nontrivial and if any composition of group homomorphisms $H_n(X) \to H_n(Y) \to H_n(X)$ is zero for $n > 0$, then $X \times Y$ has the fixed point property.*

9. A homological construction of the induced homomorphism

In this section we would like to discuss shortly a different approach to the construction of the induced homomorphism by an acyclic weighted carrier which is due to Haesler and Skordev ([VHS]). They consider a smaller class of acyclic weighted carriers, called m-maps. A m-map is a weighted carrier $\Psi: K \to L$ between finite polyhedra such that each $\Psi(x)$ has only a finite number of acyclic components. They show that any m-map Ψ possesses an associated approximation system (A-system in the terminology of the authors).

It will be out of place to introduce the formal definition of A-system here. It is quite involved. Roughly speaking, taking two decreasing sequences of triangulations of K and L with mesh going to 0 an A-system is a coherent sequence of subsets of the set of all chain maps between the (simplicial) chain complexes of the corresponding triangulation. This concept evolved from the old theory of carriers frequently used before acyclic models theorem appeared. What is important to us is that any A-system defines a homomorphism in simplicial homology.

The following proposition explains to some extent the relation between the homotopical approach to the construction of the induced homomorphism in homology of the previous section and the homological one discussed here.

Let us denote with $C(X)$ the cellular chain complex of a CW-complex X. Given two chain complexes C, C' let us denote by $\pi[C, C']$ the R-module of all chain homotopy equivalence classes of chain homomorphisms (of degree 0) from C to C'. If X, Y are CW-complexes. Then (via approximation by cellular weighted maps) each weighted map $\phi: X \to Y$ induces a chain homomorphism $C(\phi): C(X) \to C(Y)$. By the homotopy invariance $C: (X, Y) \to \mathrm{Hom}(C(X), C(Y))$ factors through $\overline{C}: \sigma(X, Y) \to \pi[C(X), C(Y)]$.

(9.1) PROPOSITION ([P4, Theorem 5.17]). *If the ring R is a principal integral domain, then the homomorphism $\overline{C}: \sigma(X, Y) \to \pi[C(X), C(Y)]$ is an isomorphism.*

A-systems were introduced by Siegberg and Skordev in [SS] as an appropriate setting for the formulation of the multiplicativity and commutativity properties of the local fixed point index for multivalued maps. Using the above theory Haesler and Skordev were able to define a local fixed point index for m-maps which verifies many of the good properties of the fixed point index in the singlevalued case. Here we quote their result:

(9.2) PROPOSITION ([VHS, Proposition 8]). *Let $\Psi: K \to K$ be a m-map with coefficients in a field F ($\Psi \in \mathcal{M}(F)$). Then for any admissible open subset U of K is defined a fixed point index $i(K, \Psi, U) \in F$ such that the following properties hold:*

(9.2.1) (Homotopy) *Let $H \in \mathcal{M}(F)$, $H: \overline{U} \times I \to K$, be a homotopy joining the maps $\Phi_1|_{\overline{U}}$, $\Phi_2|_{\overline{U}} \in \mathcal{M}(F)$ such that for all $t \in I$ and $H_t: U \to K$, $H_t(x) = H(x, t)$, the triple (K, H_t, U) is admissible. Then*

$$i(K, \Phi_1, U) = i(K, \Phi_2, U).$$

(9.2.2) (Additivity) *Let $\Phi \in \mathcal{M}(F)$, $\Phi: K \to K$, and U_1, U_2 be open disjoint subsets of U. If $\mathrm{Fix}(\Phi|_U) \subset U_1 \cup U_2$, then*

$$i(K, \Phi, U) = i(K, \Phi, U_1) + i(K, \Phi, U_2).$$

(9.2.3) (Normalization) *If $\Phi \in \mathcal{M}(F)$, then*

$$i(K, \varphi, K) = \Lambda(\varphi) = \Lambda(A(\Phi)) \quad \text{for } \varphi \in A(\Phi)_k,\ k \geq k_0.$$

(9.2.4) (Commutativity) *Let $\Phi_1: K \to L$, $\Phi_2: L \to K$ be in $\mathcal{M}(F)$, and let $(K, \Phi_2 \circ \Phi_1, U)$ and $(L, \Phi_1 \circ \Phi_2, \Phi_2^1(U))$ be admissible. Assume that for all $y \in \mathrm{Fix}(\Phi_1 \circ \Phi_2) \setminus \Phi_2^{-1}(U)$:*

$$\Phi_2(y) \cap \mathrm{Fix}(\Phi_2 \circ \Phi_1|_{\overline{U}}) = \emptyset.$$

Then

$$i_A(K, \Phi_2 \circ \Phi_1, U) = i_A(L, \Phi_1 \circ \Phi_2, \Phi_2^{-1}(U))$$

(9.2.5) (Mod p-property) *Let $p \in N$ be prime and let $\Phi \in \mathcal{M}(Z_p)$, (K, Φ, U), (K, Φ^p, U) be admissible. Assume that for all $y \in \operatorname{Fix}(\Phi^p) \setminus U$:*

$$\Phi^k(y) \cap \operatorname{Fix}(\Phi^p(\overline{U})) = \emptyset, \quad 1 \leq k < p.$$

Then $i(K, \Phi, U) = i_A(K, \Phi^p, U)$ in Z_p.

(9.2.6) (Multiplicativity) *Let $\Phi_i: K_i \to L_i$, $\Phi_i \in \mathcal{M}(F)$, and (K_i, Φ_i, U_i) be admissible, $i = 1, 2$. Then*

$$i(K_1 \times K_2, \Phi_1 \times \Phi_2, U_1 \times U_2) = i(K_1, \Phi_1, U_1) i(K_2, \Phi_2, U_2).$$

The index i_A in (9.2.5) and (9.2.6) depends upon the construction of A-systems for $\Phi_1 \circ \Phi_2$, $\Phi_2 \circ \Phi_1$ and Φ^p (see [VHS]).

The authors has also established the following version of the Borsuk–Ulam theorem for m-maps.

(9.3) THEOREM. *If $F: S^n \to S^m$ is an m-map with weights in $\mathbb{Z}_2$ that is equivariant under antipodal involution, then $n \leq m$.*

(9.4) REMARK. A recent paper [VHPS] is devoted to the proof of the Lefschetz fixed point theorem for upper semicontinuous maps $\Psi: X \to X$ where X is a compact F-space (ANR or quasicomplex) and each $\Psi(x)$ consists of a finite number of Q-acyclic components. The main tool of this generalization is an extension of the Čech homology functor to the class of m-acyclic maps.

10. Topological degree for weighted maps

Topological degree for weighted maps was studied in various forms in [DS3], [MNP], [JN] Our presentation here follows the lines of the Ph.D. thesis of S. Jodko-Narkiewicz ([JN]). For simplicity we will consider only $R = \mathbb{Z}$. We first recall the definition and properties of the topological degree. It will be defined by means of the Darbo homology functor. Let U be a bounded open subset of R^n. Furthermore, we set

$$A(U, R^n) = \{\varphi: \overline{U} \to R^n \mid \varphi \text{ is a weighted map and } 0 \notin \varphi(\partial U)\}$$

where $\overline{U}$ denotes the closure of U and ∂U is the boundary of U. Now, we shall define a map $\deg: A(U, R^n) \to Z$. Let us recall that we can think of S^n as the one point compactification of R^n, in other words, $S^n = R^n \cup \{\infty\}$. Take any $\varphi \in A(U, R^n)$. Then we have

$$S^n \xrightarrow{k} (S^n, S^n \setminus \varphi_+^{-1}(0)) \xleftarrow{j} (U, U \setminus \varphi_+^{-1}(0)) \xrightarrow{\varphi} (R^n, R^n \setminus \{0\}),$$

where k, j are the respective inclusions Now, we can apply to the above diagram the n-dimensional Darbo homology functor with integer coefficients and we get

$$\begin{array}{ccccc} Z = \mathcal{H}_n(S^n) & \xrightarrow{k_{*n}} & \mathcal{H}_n(S^n, S^n \setminus \varphi_+^{-1}(0)) & \xleftarrow{j_{*n}} & \mathcal{H}_n(U, U \setminus \varphi_+^{-1}(0)) \\ & & & & \downarrow{\scriptstyle \varphi_{*n}} \\ & & & & \mathcal{H}_n(R^n, R^n \setminus \{0\}) = Z, \end{array}$$

where by means of the excision axiom j_{*n} is an isomorphism. Now, we are in a position to define $\deg(\varphi)$ as follows:

$$\deg(\varphi) = [\varphi_{*n} \circ (j_{*n})^{-1} \circ k_{*n})](1).$$

Of course, the definition depends on the choice of the two isomorphisms at the beginning and the end of the above sequence or equivalently on the choice of an orientation on S^n (which induces a unique orientation class $o_0 \in \mathcal{H}_n(R^n, R^n \setminus \{0\})$). Below we shall list some properties of the degree defined above.

(10.1) PROPOSITION ([JN]). *Let U and φ be as above.*

(10.1.1) (Existence) If $\deg(\varphi) \neq 0$, then $0 \in \varphi(U)$.

(10.1.2) (Localization) Let V be an open subset of U containing $\varphi_+^{-1}(0)$, then $\varphi|_{\overline{V}} \in A(V, R^n)$ and $\deg(\varphi) = \deg(\varphi|_{\overline{V}})$.

(10.1.3) (Additivity) Let U_1 and U_2 be disjoint open subsets of U such that $\varphi_+^{-1}(0) \subset U_1 \cup U_2$ and let φ_i denote the restriction of φ to U_i, then $\varphi_i \in A(U_i, R^n)$, for $i = 1, 2$, and $\deg(\varphi) = \deg(\varphi_1) + \deg(\varphi_2)$.

(10.1.4) (Unity) If U contains the origin and $i: U \hookrightarrow R^n$ is the inclusion of U in R^n, then $\deg(i) = 1$.

(10.1.5) (σ-homotopy) Let $\Psi: \overline{U} \times [0, 1] \to R^n$ be a weighted map such that $\Psi_t \in A(U, R^n)$ for every $t \in [0, 1]$, then $\deg(\Psi_0) = \deg(\Psi_1)$.

(10.1.6) (Linearity) If $\varphi_1, \varphi_2 \in A(U, R^n)$, then $\varphi_1 + \varphi_2 \in A(U, R^n)$ and

$$\deg(\varphi_1 + \varphi_2) = \deg(\varphi_1) + \deg(\varphi_2).$$

Moreover, $\deg(k\varphi) = k \deg(\varphi)$ for all $\varphi \in A(U, R^n)$ and $k \in \mathbb{Z}$.

(10.2) DEFINITION ([JN], [MNP]). The *winding number* of a weighted map $\varphi: S^{n-1} \to R^n \setminus \{0\}$ is the unique integer $w(\varphi)$ such that

$$\varphi_{*(n-1)}: \mathcal{H}_{n-1}(S^{n-1}) \to \mathcal{H}_{n-1}(R^n \setminus \{0\})$$

coincides with the multiplication by $w(\varphi)$.

(10.3) PROPOSITION ([JN]). *Assume that* $\varphi \in A(B, R^n)$, *where* B *is the open unit ball* $\in R^n$. *Then* $\deg(\varphi) = w(\varphi|_{S^{n-1}})$.

Now, we give some topological consequences of the topological degree.

(10.4) PROPOSITION ([JN]). *Let* $\varphi\colon K^{n+1} \to R^{n+1}$ *be a weighted map such that* $\varphi(S^n) \subset K^{n+1}$ *and* $I(\varphi) \neq 0$. *Then* $\mathrm{Fix}(\varphi) \neq \emptyset$.

The following is a generalization of the Birkhoff–Kellogg principle in $\mathbb{R}^n$. Let X be a metric space. We will say that $\varphi\colon X \to S^n$ is an algebraically essential map, if $\varphi_{*n}\colon \mathcal{H}_n(X) \to \mathcal{H}_n(S^n)$ is a nontrivial homomorphism.

(10.5) PROPOSITION ([JN]). *Let* p *be an algebraically essential map from* X *into* S^{2k}. *Then for every weighted map* $\varphi\colon X \to S^{2k}$ *with* $I(\varphi) \neq 0$ *there exists a point* $x_0 \in X$ *such that either* $p(x_0) \in \varphi(x_0)$ *or* $-p(x_0) \in \varphi(x_0)$.

11. Topological essentiality

In this section we shall present a concept of topological essentiality which can be interpreted as the topological degree mod 2. We would like to remark that a concept of essentiality for single valued maps was introduced by A. Granas ([Gr1]) and further generalized by L. Górniewicz and M. Ślosarski for some class of multivalued maps ([GS]). Moreover, the above concept was elaborated by M. Furi, M. Martelli and A. Vignoli under the name zero-epi maps ([FMV]).

In what follows E, F are two real normed spaces and U is an open connected bounded subset of E. By $\overline{U}$ we shall denote the closure of U in E. In all of this section we will assume that R is a field and we consider only weighted maps φ with $I(\varphi) \neq 0$. We let:

$$\begin{aligned} W_{\partial U}(U,F) &= \{\varphi\colon \overline{U} \to F \mid \varphi \text{ is a weighted map and } 0 \notin \varphi(\partial U)\};\\ W_C(U,F) &= \{\varphi\colon \overline{U} \to F \mid \varphi \text{ is a weighted map and compact}\};\\ W_0(U,F) &= \{\varphi\colon \overline{U} \to F \mid \varphi \in W_C(U,F) \text{ and } \varphi(x) = \{0\},\ \text{for all } x \in \partial U\}. \end{aligned}$$

(11.1) DEFINITION ([Sk2]). A weighted map $\varphi \in W_{\partial U}(U,F)$ is called essential (with respect to $W_0(U,F)$) provided, for any $\psi \in W_0(U,F)$ there exists a point $x \in U$ such that $\varphi(x) \cap \psi(x) \neq \emptyset$, i.e. φ and ψ have a coincidence in U.

Now, we give some examples of essential weighted maps.

(11.2) EXAMPLE. Let $\varphi \in W_{\partial U}(B,R)$, where B is an open ball at $0 \in E$ with radius $r > 0$. If there exist $x_0, x_1 \in \partial B$ such that

$$u < 0 \quad \text{for every } u \in \varphi(x_0) \quad \text{and} \quad v > 0 \quad \text{for every } v \in \varphi(x_1),$$

then φ is an essential weighted map.

(11.3) EXAMPLE (Essentiality of homeomorphism). Let U be an open and bounded subset of E such that $\overline{U} \in AR$ and let $f: \overline{U} \to f(\overline{U})$ be a homeomorphism such that $f(\overline{U})$ is a closed subset of F. In addition, assume that $f(U)$ is an open subset of F and $0 \in f(U)$. Then f is essential.

(11.4) EXAMPLE (Essentiality of linear isomorphism). Let $L: E \to F$ be a continuous linear isomorphism and let U be an open bounded neighbourhood of the origin in E. Then the restriction $\widetilde{L}: \overline{U} \to F$ of L to $\overline{U}$ is essential.

Let us enumerate several properties of the topological essentiality.

(11.5) PROPOSITION (Existence). *If $\varphi \in W_{\partial U}(U, F)$ is essential, then there exists a point $x \in U$ such that $0 \in \varphi(x)$.*

(11.6) PROPOSITION (Compact perturbation). *If $\varphi \in W_{\partial U}(U, F)$ is essential and $\eta \in W_0(U, F)$, then $\varphi + \eta \in W_{\partial U}(U, F)$ is an essential weighted map.*

(11.7) PROPOSITION (Coincidence). *Assume that $\varphi \in W_{\partial U}(U, F)$ is essential and $\eta \in W_C(U, F)$. Let*

$$B = \{x \in \overline{U} : \varphi(x) \cap (t\eta(x)) \neq \emptyset \text{ for some } t \in [0, 1]\}.$$

If $B \subset U$, then φ and η have a coincidence.

(11.8) PROPOSITION (Localization). *Let $\varphi \in W_{\partial U}(U, F)$ be an essential weighted map. Assume that V is an open subset of U such that $\varphi_+^{-1}(\{0\}) \subset V$ and $\overline{V} \in AR$. Then the restriction $\varphi|_{\overline{V}}$ of φ to $\overline{V}$ is an essential weighted map.*

(11.9) PROPOSITION (σ-homotopy). *Let $\varphi \in W_{\partial U}(U, F)$ be an essential weighted map. If $H: \overline{U} \times [0, 1] \to F$ is a compact weighted map such that:*

(11.9.1) $H(x, 0) = \{0\}$ for every $x \in \partial U$,
(11.9.2) $\{x \in \overline{U} : \varphi(x) \cap H(x, t) \neq \emptyset$ for some $t \in [0, 1]\} \subset U$.

Then the map $\varphi(\cdot) - H(\cdot, 1): \overline{U} \to F$ is an essential weighted map.

Now, we give some topological consequences of the topological essentiality.

(11.10) PROPOSITION (Invariance of domain). *Let $\varphi \in W_{\partial U}(U, F)$ be an essential weighted map and proper. If D is a connected component of the set $F \setminus \varphi(\partial U)$ which contains $0 \in F$. Then $D \subset \varphi(\overline{U})$.*

(11.11) PROPOSITION. *Let $\varphi \in W_{\partial U}(U, F)$ be an essential weighted map and $\psi \in W_C(U, F)$. If $\varphi(x) \cap \psi(x) = \emptyset$ for every $x \in \partial U$, then at least one of the following conditions holds:*

(11.11.1) there exists $x \in U$ such that $\varphi(x) \cap \psi(x) \neq \emptyset$;
(11.11.2) there exists $\lambda \in (0, 1)$ and $x \in \partial U$ such that $\varphi(x) \cap (\lambda\psi(x)) \neq \emptyset$.

(11.12) PROPOSITION (Nonlinear alternative). *Let* $\psi \in W_C(U,F)$ *and* $0 \in U$, *then at least one of the following conditions is satisfied:*

(11.12.1) $\mathrm{Fix}(\psi) \neq \emptyset$,
(11.12.2) there exists $x \in \partial U$ and $\lambda \in (0,1)$ such that $x \in \lambda\psi(x)$.

Let us remark that natural applications of the topological essentiality can be found in control theory (see [GN]).

12. The Lefschetz fixed point theorem for compact absorbing contractions

There are several possibilities of generalizing of the Lefschetz fixed point theorem for weighted maps. Some of them will be presented here (see [VHPS]). In this section we consider the Darbo homology functor with coefficients in the field of rational numbers Q. Moreover, throughout this paragraph all the vector spaces are taken over Q. A graded vector space $E = \{E_n\}_{n\geq 0}$ is said to be *of finite type* if $\dim E_n < \infty$, for every n and $E_n = 0$, for all except a finite number of indices. If $f = \{f_n: E_n \to E_n\}_{n\geq 0}$ is an endomorphism of degree zero, then its Lefschetz number is given by $L(f) = \sum_n (-1)^n \mathrm{tr}(f_n)$. Jean Leray in [Le2] extended the notion of Lefschetz number to a much larger class of graded endomorphisms, using a generalization of the concept of trace. Let $f: E \to E$ be an endomorphism of an arbitrary vector space E. By $f^{(n)}: E \to E$ we will denote the n-th iterate of f. The kernels $\mathrm{Ker} f^{(n)}$, $n \geq 1$ form an increasing sequence and therefore $\mathcal{N}(f) = \bigcup_{n\geq 1} \mathrm{Ker} f^{(n)}$ is a subspace of E. Since f maps $\mathcal{N}(f)$ into itself it induces an endomorphism $\widetilde{f}: \widetilde{E} \to \widetilde{E}$, where $\widetilde{E} = E/\mathcal{N}(f)$ is the factor space. Under assumption that $\dim \widetilde{E} < \infty$ Leray defined the generalized trace $\mathrm{Tr}(f)$ of f by putting $\mathrm{Tr}(f) = \mathrm{tr}(\widetilde{f})$. Let $f = \{f_n\}_{n\geq 0}$ be an endomorphism of degree zero of a graded vector space $E = \{E_n\}_{n\geq 0}$. We say that f is a *Leray endomorphism* if the graded vector space $\widetilde{E} = \{\widetilde{E}_n\}_{n\geq 0}$ is of finite type. For such an f, we can define the *generalized Lefschetz number* $\Lambda(f)$ of f by putting

$$\Lambda(f) = \sum_n (-1)^n \mathrm{Tr}(f_n).$$

It is easy to see that if E is of finite type, then $\Lambda(f) = L(f)$.

A weighted map $\psi: X \to X$ over the field $R = Q$ will be called a *Lefschetz weighted map* if the induced homomorphism $\psi_*: \mathcal{H}_*(X) \to \mathcal{H}_*(X)$ is a Leray endomorphism. For such a weighted map ψ we can define the Lefschetz number $\Lambda(\psi)$ of ψ by putting: $\Lambda(\psi) = \Lambda(\psi_*)$.

Clearly, any weighted map ψ σ-homotopy to a Lefschetz weighted map φ is Lefschetz and $\Lambda(\psi) = \Lambda(\varphi)$. We shall say that a weighted map $\psi: X \to Y$ is a compact weighted map if the closure $\overline{\psi(X)}$ of $\psi(X)$ in Y is a compact set. By straightforward extension of a method due to A. Granas (see [Gr2]) one can prove the following theorem.

(12.1) THEOREM ([Sk1]). *Let $X \in ANR$ and let $\varphi: X \to X$ be a compact weighted map ($\varphi \in K_w(X)$). Then:*

(12.1.1) *φ is a Lefschetz weighted map;*
(12.1.2) *$\Lambda(\varphi) \neq 0$ implies that φ has a fixed point.*

Let us notice that if X is an ANR, then $\mathcal{H}_*(X) = \{\mathcal{H}_n(X)\}_{n\geq 0}$ need not be a graded vector space of finite type. Therefore for a given φ the ordinary Lefschetz number cannot be well defined. So, in the proof of the above theorem we have to consider a concept of the generalized Lefschetz number.

Now, following G. Fournier and L. Górniewicz ([G1], [FG], [G2]) we show how the above theorem can be extended to a class of non-compact mappings. Before doing it, we will recall the necessary notions and facts. Let (X, A) be a pair of spaces. Given a weighted map $\varphi: (X, A) \to (X, A)$ we denote by $\varphi_X: X \to X$ and $\varphi_A: A \to A$ the evident contractions of φ. For a pair (X, A) let us consider the graded vector space $\mathcal{H}_*(X, A) = \{\mathcal{H}_n(X, A)\}_{n\geq 0}$. A weighted map $\varphi: (X, A) \to (X, A)$ is called a Lefschetz weighted map provided $\varphi_*: \mathcal{H}_*(X, A) \to \mathcal{H}_*(X, A)$ is a Leray endomorphism. For a weighted map φ we can define the Lefschetz number $\Lambda(\varphi)$ of φ by putting $\Lambda(\varphi) = \Lambda(\varphi_*)$. The following proposition expresses a basic property of the generalized Lefschetz number:

(12.2) PROPOSITION ([Sk1]). *Let $\varphi: (X, A) \to (X, A)$ be a weighted map. If two of the following weighted maps φ, φ_A, φ_X are the Lefschetz weighted maps, then so is the third one and in such a case we have:*

$$\Lambda(\varphi) = \Lambda(\varphi_X) - \Lambda(\varphi_A).$$

(12.3) DEFINITION. A weighted map $\varphi: X \to X$ is said to be a compact absorbing contraction if there exists an open subset U of X such that $\operatorname{cl}\varphi(U)$ is a compact subset of U and $X \subset \bigcup_{i=0}^{\infty} \varphi^{-i}(U)$.

The set of all compact absorbing contractions will be denoted with: $\mathrm{CAC}_w(X)$. Evidently, any compact weighted map $\varphi: X \to X$ is a compact absorbing contraction (it is enough to take $U = X$). The main property of the compact absorbing contraction is given in the following:

(12.4) PROPOSITION ([Sk1]). *If $\varphi: (X, A) \to (X, A)$ is a weighted map such that A satisfies conditions of Definition* (12.3), *then φ is a Lefschetz weighted map and $\Lambda(\varphi) = 0$.*

After these preliminaries we are able to formulate the following

(12.5) THEOREM ([Sk1]). *Let X be an* ANR *and let $\varphi \in \mathrm{CAC}_w(X)$. Then:*

(12.5.1) *φ is a Lefschetz weighted map;*
(12.5.2) *$\Lambda(\varphi) \neq 0$ implies that φ has a fixed point.*

PROOF. Let $U \subset X$ be chosen according to the Definition (12.3) and let $\overline{\varphi}\colon U \to U$ be a contraction of φ to U. We consider the map $\widetilde{\varphi}\colon (X, A) \to (X, A)$, $\widetilde{\varphi}(x) = \varphi(x)$, for every $x \in X$. From Proposition (12.4) we get that $\widetilde{\varphi}$ is a Lefschetz weighted map and $\Lambda(\widetilde{\varphi}) = 0$. Since $\overline{\varphi}$ is a compact weighted map and $U \in ANR$, we obtain from Theorem (12.1) that $\overline{\varphi}$ is a Lefschetz weighted map. Consequently, by applying Proposition (12.2), we deduce that φ is a Lefschetz weighted map and $\Lambda(\overline{\varphi}) = \Lambda(\varphi)$. Now, if we assume that $\Lambda(\varphi) \neq 0$, then $\Lambda(\overline{\varphi}) \neq 0$ and by using once again Theorem (12.1) we get that $\overline{\varphi}$ has a fixed point but it implies that φ has a fixed point, which completes the proof of our theorem. □

As an immediate consequence of the above theorem, we obtain:

(12.6) COROLLARY. *Let X be an acyclic* ANR *(e.g. a convex subset of a normed space) and let $\varphi\colon X \to X$ be a weighted map with $I(\varphi) \neq 0$. If $\varphi \in \mathrm{CAC}_w(X)$, then φ has a fixed point.*

13. Comments

In this section we will make some comments about possible extensions and applications of the preceding theory and discuss the connection of the approximation theorems in Section 8 with the transfers.

Applications of the fixed point index for weighted maps to branching, only sketched in Section 4, deserve further study. A proof of the commutativity property for the local fixed point index would open the possibility to localize the branching near the set of branch points as in the Riemann–Hurwitz formula.

Fuller's example discussed in Section 7 suggests that Lefschetz carriers may be useful in order to obtain an extension of the well known Wazewski principle to some cases in which not all egress points are strict egress points.

Finally, let us discuss the transfers. A typical transfer is a "wrong way" homomorphism in homology (or cohomology) that is not induced by singlevalued continuous maps. The ones in ordinary homology theory have been studied since the very beginning of algebraic topology. For example, Poincare duality transfer was used by Hopf in [Ho] to obtain results such as Corollary (8.5). Lefschetz (see [L1]–[L3]) and Hopf used them in a more or less explicit way in the fixed point and coincidence formulas. On a more general setting of convexoid spaces a fixed point transfer was introduced by Jean Leray in [Le1] under the name of θ-homomorphism (see also [K1]).

In [S], Schultz generalized the Smith transfer using weighted maps or (better to say) m-functions of Jerrard. We have briefly discussed the relation between weighted maps and Smith transfer in Section 4. To some extent, this relation was used by Oliver in his construction of a transfer for general Lie groups (see [O]). The book [Br] contains a comprehensive discussion of this type of transfers based on sheaf cohomology. In [K2], [K3], [Sm], [Do4], [R], [GoO] can be found many applications of this theory.

We believe that the approach we have been using in Section 8 can be further improved in order to show that all weighted carriers discussed there have a well defined homomorphism in homology, obtained by approximation with special weighted maps, and which depends only on the homotopy class of the family defining the carrier. If this works, one can easily show that the induced homomorphism depends only on the homotopy class of the family f.

Since such a homomorphism is literally (and not only metaphorically, see [Do2]) induced by a multivalued mapping with weights, it could be considered a *geometric transfer*, as opposed to the classical (algebraic) transfer defined in terms of products in various homology and cohomology theories [Do1], [Do2]. The "wrong way" transfer would become the "right way" homomorphism induced by f^{-1}.

We don't know how to compare this type of homomorphisms with the classical ones. There are axiomatic characterizations of transfers that lead to a uniqueness theorem (see [BS2]). This may be helpful.

Another natural question, in this regard, is whether there exists a unifying approach to the approximation of general weighted carriers by weighted maps applicable both to the acyclic weighted carriers and to those in Section 8. This question is closely related to the one raised by Dold in [Do3] about the relation between the coincidence index defined there and the index for acyclic morphisms.

The fixed point theorem in [Co] hints that further improvement in this direction may be possible. Generalizations of the infinite symmetric products to spaces of higher dimensional cycles that arise in [Al1], [Al2], [FB] is also of interest in this regard.

Becker and Gottlieb ([BG1], [Go], [BG2]) showed that a large variety of transfers in generalized cohomology theories are induced by stable maps. The fixed point transfer in [Do2] is also constructed in this way.

A (co)-homology theory defined on the category of CW-complexes extends to the category of weighted maps if and only if it verifies the dimension axiom. This means that our hypothetical geometric theory is limited to transfers in ordinary homology. This is a strong limitation which reduces the interest in this construction, because transfer in extraordinary theories are by far more powerful.

References

[AR] R. Abraham and J. Robbin, *Transversal mappings and flows*, An appendix by Al Kelley (1967), W. A. Benjamin, Inc., New York-Amsterdam.

[Al1] F. J. Almgren, *Applications of multiple-valued functions*, Geometric Modeling, SIAM, Philadelphia, PA, 1987, pp. 43–54.

[Al2] ______, *Multi-functions* mod ν, Geometric Analysis and Computer Graphics (Berkeley, CA, 1988), vol. 17, Math. Sci. Res. Inst. Publ., Springer, New York, 1991, pp. 1–17.

[AC] G. Anichini and G. Conti, *About the existence of solutions of a boundary value problem for a Carathéodory differential system*, Z. Anal. Anwendungen **16** (1997), 621–630.

[BGK] R. Bader, G. Gabor and W. Kryszewski, *On the extension of approximations for set-valued maps and the repulsive fixed points.*, Boll. Un. Mat. Ital. B (7) **10** (1996), 399–416.

[BCG] J. C. Becker, A. Casson and D. H. Gottlieb, *The Lefschetz number and fiber preserving maps*, Bull. Amer. Math. Soc. **81** (1975), 425–427.

[BG1] J. C. Becker and D. H. Gottlieb, *The transfer map and fiber bundles*, Topology **14** (1975), 1–12.

[BG2] ______, *Vector fields and transfers*, Manuscripta Math. **72** (1991), 111–130.

[BS1] J. C. Becker and R. E. Schultz, *Fixed-point indices and left invariant framings*, Geometric Applications of Homotopy Theory, Proc. Conf. Evanston (Ill., 1977), I, vol. 657, Lecture Notes in Math., Springer, Berlin, 1978, pp. 1–31.

[BS2] ______, *Axioms for bundle transfers and traces*, Math. Z. **227** (1998), 583–605.

[BGMO] Yu. G. Borisovich, B. D. Gel'man, A. D. Myshkis and V. V. Obukhovskiĭ, *New results in the theory of multivalued mappings* I *Topological characteristics and solvability of operator relations*, Mathematical Analysis, vol. 25, Itogi Nauki i Tekhniki, pp. 123–197, 200 (Russian); Akad. Nauk SSSR Vsesoyuz. Inst. Nauchn. i Tekhn. Inform. Moscow, 1987; transl. in J. Soviet Math. **49** (1990), 800–855.

[Br] G. E. Bredon, *Sheaf Theory*, Graduate Texts in Mathematics, vol. 170, Springer-Verlag, New York, 1997.

[Ca] R. Caccioppoli, *Funzioni di variabili complesse. Equazioni funzionali*, Opere, vol. II, Edizioni Cremonese, Rome, 1963.

[Co] R. Connelly, *An extension of Brouwer's fixed-point theorem to nonacyclic, set valued functions*, Proc. Amer. Math. Soc. **43** (1974), 214–218.

[CNZ1] G. Conti, P. Nistri and P. Zecca, *Systems of set-valued equations in Banach spaces*, Delay Differential Equations and Dynamical Systems (Claremont, CA, 1990), Lecture Notes in Math., vol. 1475, Springer, Berlin, 1991, pp. 98–109.

[CNZ2] ______, *Systems of multivalued equations for solving controllability problems*, Pure Math. Appl. Ser. B **2** (1992), 215–225.

[CP] G. Conti and J. Pejsachowicz, *Fixed point theorems for multivalued weighted maps*, Ann. Mat. Pura Appl. (4) **126** (1981), 319–341.

[DS1] L. Dal Soglio, *Grado topologico e teoremi di esistenza di punti uniti per trasformazioni plurivalenti di* 3*-celle.*, Rend. Sem. Mat. Univ. Padova **25** (1956), 386–405.

[DS2] ______, *Grado topologico e teoremi di esistenza di punti uniti per trasformazioni plurivalenti di n-celle*, Rend. Sem. Mat. Univ. Padova **27** (1957), 103–121.

[DS3] ______, *Sulla nozione di grado e di coefficiente di allacciamento per mappe ponderate*, Rend. Sem. Mat. Univ. Padova **28** (1958), 280–289.

[Da1] G. Darbo, *Grado topologico e teoremi di esistenza di punti uniti per trasformazioni plurivalenti di bicelle*, Rend. Sem. Mat. Univ. Padova **19** (1950), 371–395.

[Da2] ______, *Teoria dell'omologia in una categoria di mappe plurivalenti ponderate*, Rend. Sem. Mat. Univ. Padova **28** (1958), 188–220.

[Da3] ______, *Sulle coincidenze di mappe ponderate*, Rend. Sem. Mat. Univ. Padova **29**, 256–270, 1959..

[Da4] ______, *Estensione alle mappe ponderate del teorema di Lefschetz sui punti fissi*, Rend. Sem. Mat. Univ. Padova **31** (1961), 46–57.

[Do1] A. Dold, *Lectures on Algebraic Topology*, Springer–Verlag, New York yr 1972; *Die Grundlehren der mathematischen Wissenschaften, Band 200.*

[Do2] ______, *The fixed point transfer of fibre-preserving maps*, Math. Z. **148(3)** (1976), 215–244.

[Do3] ______, *A coincidence-fixed-point index.*, Enseign. Math. (2) **24** (1978), 41–53.

[Do4] ______, *Ramified coverings, orbit projections and symmetric powers*, Math. Proc. Cambridge Philos. Soc. **99** (1986), 65–72.

[Du] J. Dugundji, *Remark on homotopy inverses*, Portugal. Math. **14** (1955), 39–41.

[EM] S. Eilenberg and D. Montgomery, *Fixed point theorems for multi-valued transformations*, Amer. J. Math. **68** (1946), 214–222.

[F1] P. L. Felmer, *Periodic solutions of spatially periodic Hamiltonian systems*, J. Differential Equations **98** (1992), 143–168.

[F2] ______, *Rotation type solutions for spatially periodic Hamiltonian systems*, Nonlinear Anal. **19** (1992), 409–425.

[FMP1] P. M. Fitzpatrick, I. Massabò and J. Pejsachowicz, *Global several-parameter bifurcation and continuation theorems: a unified approach via complementing maps*, Math. Ann. **263** (1983), 61–73.

[FMP2] ______, *On the covering dimension of the set of solutions of some nonlinear equations*, Trans. Amer. Math. Soc. **296** (1986), 777–798.

[FG] G. Fournier and L. Górniewicz, *The Lefschetz fixed point theorem for some non-compact multi-valued maps*, Fund. Math. **94)** (1977), 245–254.

[FB] E. M. Friedlander and H. Blaine Lawson Jr., *A theory of algebraic cocycles*, Ann. of Math. (2) **136** (1992), 361–428.

[FO] K. Fukaya and K. Ono, *Arnold conjecture and Gromov–Witten invariant for general symplectic manifolds*, The Arnoldfest (Toronto, ON, 1997), Fields Inst. Commun. **24** (1999), Amer. Math. Soc., Providence, RI, 173–190.

[Fu] F. B. Fuller, *Fixed points of multiple-valued transformations*, Bull. Amer. Math. Soc. **67** (1961), 165–169.

[FMV] M. Furi, M. Martelli and A. Vignoli, *On the solvability of nonlinear operator equations in normed spaces*, Ann. Mat. Pura Appl. (4) **124** (1980), 321–343.

[G1] L. Górniewicz, *Homological methods in fixed-point theory of multi-valued maps*, Dissertationes Math. (Rozprawy Mat.) **129** (1976), 71.

[G2] ______, *Topological fixed point theory of multivalued mappings*, Mathematics and its Applications, vol. **495**, Kluwer Academic Publishers, Dordrecht, 1999.

[GGK1] L. Górniewicz, A. Granas and W. Kryszewski, *Sur la méthode de l'homotopie dans la théorie des points fixes pour les applications multivoques* II, *L'indice dans les* ANR-*s compacts*, C. R. Acad. Sci. Paris Sér. I Math. **308** (1989), 449–452.

[GGK2] ______, *On the homotopy method in the fixed point index theory of multi-valued mappings of compact absolute neighborhood retracts*, J. Math. Anal. Appl. **161** (1991), 457–473.

[GL] L. Górniewicz and M. Lassonde, *Approximation and fixed points for compositions of R_δ-maps*, Topology Appl. **55** (1994), 239–250.

[GN] L. Górniewicz and P. Nistri, *Topological essentiality and nonlinear boundary value control problems*, Topol. Methods Nonlinear Anal. **13** (1999), 53–72.

[GS] L. Górniewicz and M Ślosarski, *Topological essentiality and differential inclusions*, Bull. Austral. Math. Soc. **45** (1992), 177–193.

[Go] D. Gottlieb, *On realizing Nakaoka's coincidence point transfer as an S-map*, Illinois J. Math. **30** (1986), 689–695.

[GoO] D. H. Gottlieb and M. Özaydin, *Intersection numbers, transfers, and group actions*, Topology Appl. **55** (1994), 87–100.

[Gr1] A. Granas, *The theory of compact vector fields and some of its applications to topology of functional spaces*I, Rozprawy Mat. **30** (1962), 93.

[Gr2] ______, *Generalizing the Hopf–Lefschetz fixed point theorem for non-compact* ANRs, Symposium on Infinite-Dimensional Topology (Louisiana State Univ., Baton Rouge,

La., 1967), pp. 119–130; Ann. of Math. Studies **69** (1972), Princeton Univ. Press, Princeton, N. J.

[GrDu] A. Granas and J. Dugundji, *Fixed Point Theory*, Springer Monographs in Mathematics, Springer–Verlag, New York, 2003.

[GuPo] V. Guillemin and A. Pollack, *Differential Topology*, Prentice-Hall Inc., Englewood Cliffs, N.J., 1974.

[H] B. Halpern, *Algebraic topology and set-valued maps*, Set-Valued Mappings, Selections and Topological Properties of 2^X (Proc. Conf., SUNY, Buffalo, N.Y., 1969), Lecture Notes in Math., vol. 171, Springer, Berlin, 1970, pp. 27–33.

[HZ] H. Hofer and E. Zehnder, *Symplectic invariants and Hamiltonian dynamics*, Birkhäuser Advanced Texts: Basler Lehrbücher (Birkhäuser Advanced Texts: Basel Textbooks), Birkhäuser Verlag, Basel, 1994.

[Ho] H. Hopf, *On some properties of one-valued transformations of manifolds*, Proc. Nat. Acad. Sci. **14** (1928), 206–214.

[HW] W. Hurewicz and H. Wallman, *Dimension Theory*, Princeton Math. Ser., vol. 4, Princeton University Press, Princeton, N. J., 1941.

[Je1] R. P. Jerrard, *On Knaster's conjecture*, Trans. Amer. Math. Soc. **170** (1972), 385–402.

[Je2] ______, *Homology with multiple-valued functions applied to fixed points*, Trans. Amer. Math. Soc. **213** (1975), 407–427.

[Je3] ______, *Erratum to: "Homology with multiple-valued functions applied to fixed points"*, (Trans. Amer. Math. Soc., vol. 213 (1975), 407–427), Trans. Amer. Math. Soc. **218** (1976), 406.

[Je4] ______, *A stronger invariant for homology theory.*, Michigan Math. J. **26** (1979), 33–46.

[Je5] ______, *Fixed points and product spaces*, Houston J. Math. **11** (1985), 191–198.

[JeM] R. Jerrard and M. D. Meyerson, *Homotopy with m-functions*, Pacific J. Math. **84(2)** (1979), 305–318.

[JN] S. Jodko-Narkiewicz, *Topological degree of multivalued weighted mappings*, Ph. D. thesis (1989), Faculty of Mathematics and Computer Science, Nicolaus Copernicus University, Toruń, Poland. (Polish)

[K1] R. J. Knill, *On the homology of a fixed point set*, Bull. Amer. Math. Soc. **77** (1971), 184–190.

[K2] ______, *On the index of isolated closed tori*, Differential Topology, Foliations and Gelfand–Fuks Cohomology, (Proc. Sympos., Pontifícia Univ. Católica, Rio de Janeiro, 1976), Lecture Notes in Math., vol. 652, Springer, Berlin, 1978, pp. 196–211.

[K3] ______, *The Lefschetz fixed point theorem for compact groups*, Proc. Amer. Math. Soc. **66** (1977), 148–152.

[Ko] Y. Kodama, *On* ANR *for metric spaces*, Sci. Rep. Tokyo Kyoiku Daigaku. Sect. A **5** (1955), 96–98.

[Kr1] W. Kryszewski, *Some homotopy classification and extension theorems for the class of compositions of acyclic set-valued maps*, Bull. Sci. Math. **119(1)** (1995), 21–48.

[Kr2] ______, *The fixed-point index for the class of compositions of acyclic set-valued maps on* ANR*s*, Bull. Sci. Math. **120** (1996), 129–151.

[Kr3] ______, *Remarks to the Vietoris theorem*, Topol. Methods Nonlinear Anal. **8** (1997), 383–405.

[Kr4] ______, *Homotopy properties of set-valued mappings*, Ph. D. thesis (1997), Faculty of Mathematics and Computer Science, Nicolaus Copernicus University, Toruń, habilitation dissertation.

[Kr5] ______, *Graph-approximation of set-valued maps on noncompact domains*, Topology Appl. **83** (1998), 1–21.

[Kr6] ______, *Graph-approximation of set valued maps, a survey*, Differential Inclusions and Optimal Control, (J. Andres, L. Góorniewicz, P. Nistri), eds.), vol. 2, Lecture Notes in Nonlinear Analysis, Juliusz Schauder Center for Nonlinear Studies, Nicolaus Copernicus University, Toruń, 1998, pp. 223–235.

[KuR] H. Kurland and J. Robbin, *Infinite codimension and transversality*, Dynamical Systems – Warwick 1974, (Proc. Sympos. Appl. Topology and Dynamical Systems, Univ. Warwick, Coventry, 1973/1974; presented to E. C. Zeeman on his fiftieth birthday), vol. 468, Lecture Notes in Math., Springer, Berlin, 1975, pp. 135–150.

[L1] S. Lefschetz, *Intersections and transformations of complexes and manifolds*, Trans. Amer. Math. Soc. **28** (1926), 1–49.

[L2] ______, *On the fixed point formula*, Ann. of Math. (2) **38** (1937), 819–822.

[L3] ______, *On coincidences of transformations*, Bol. Soc. Mat. Mexicana (2) **2** (1957), 16–25.

[Le1] J. Leray, *Sur les équations et les transformations*, J. Math. Pures Appl. (9) **24** (1946), 201–248.

[Le2] ______, *Théorie des points fixes: indice total et nombre de Lefschetz*, Bull. Soc. Math. France **87** (1959), 221–233.

[M] E. Magenes, *Un'osservazione sui teoremi di esistenza di punti uniti in trasformazioni plurivalenti di una N-cella*, Rend. Sem. Mat. Univ. Padova **19** (1950), 108–113.

[Ma] S. Mardešić, *Equivalence of singular and Čech homology for* ANR*s. Application to unicoherence*, Fund. Math. **46** (1958), 29–45.

[MaS] S. Mardešić and J. Segal, *Shape theory*, North-Holland Mathematical Library, vol. 26, North-Holland Publishing Co., Amsterdam, 1982.

[MNP] I. Massabo, P. Nistri and J. Pejsachowicz, *On the solvability of nonlinear equations in Banach spaces*, Fixed Point Theory (Sherbrooke, Que., 1980), Lecture Notes in Math., vol. 886, Springer, Berlin, 1981, pp. 270–299.

[MP] I. Massabò and J. Pejsachowicz, *On the connectivity properties of the solution set of parametrized families of compact vector fields*, J. Funct. Anal. **59** (1984), 151–166.

[Max] Ch. N. Maxwell, *Fixed points of symmetric product mappings*, Proc. Amer. Math. Soc. **8** (1957), 808–815.

[MC] M. C. McCord, *Classifying spaces and infinite symmetric products*, Trans. Amer. Math. Soc. **146** (1969), 273–298.

[N] M. Nakaoka, *Coincidence Lefschetz number for a pair of fibre preserving maps*, J. Math. Soc. Japan **32(4)** (1980), 751–779.

[NOZ] P. Nistri, V. V. Obukhovskiĭ and P. Zecca, *On the solvability of systems of inclusions involving noncompact operators*, Trans. Amer. Math. Soc. 342(2) (1994), 543–562.

[O] R. Oliver, *A transfer homomorphism for compact Lie group actions*, Math. Ann. **260** (1982), 351–374.

[ON] B. O'Neill, *Induced homology homomorphisms for set-valued maps*, Pacific J. Math. **7** (1957), 1179–1184.

[ONS] B. O'Neill and E. G. Straus, *A fixed point theorem*, Proc. Mmer. Math. Soc. **8** (1957), 1148–1151.

[P1] J. Pejsachowicz, *The homotopy theory of weighted mappings*, Boll. Un. Mat. Ital. B (5) **14** (1977), 702–721.

[P2] ______, *A Lefschetz fixed point theorem for multivalued weighted mappings*, Boll. Un. Mat. Ital. A (5) **14** (1977), 391–397.

[P3] ______, *Teoria di omotopia di mappe ponderate*, Ph. D. thesis (1973), Tesi di Perfezionamento SNS Pisa, Pisa, Italy.

[P4] ______, *Relation between the homotopy and the homology theory of weighted mappings*, Boll. Un. Mat. Ital. B (5) **15** (1978), 285–302.

[R] B. Ralph, *A homotopy transfer for finite group actions*, Pacific J. Math. **160** (1993), 133–141.

[S] R. Schultz, *Homological transfers for orbit space projections*, Manuscripta Math. **24** (1978), 229–238.

[SD] G. Scorza Dragoni, *Il lemma di Sperner: sue implicazioni negli spazi euclidei equestioni connesse*, Contributi del CLISMA **43(2)** (1978), 160–170.

[Se] B. Segre, *Recouvrements de sphères et correspondances entre variétés topologiques*, Colloque sur les Questions de Réalité en Géométrie, Liège, 1955, Georges Thone, Liège, 1956, pp. 149–175.

[SS] H. W. Siegberg and G. Skordev, *Fixed point index and chain approximations*, Pacific J. Math. **102** (1982), 455–486.

[Sk1] R. Skiba, *On the Lefschetz fixed point theorem for multivalued weighted mappings*, Acta Univ. Palack. Olomuc. Fac. Rerum Natur. Math. **40** (2001), 201–214.

[Sk2] ______, *Topological essentiality for multivalued weighted mappings*, Acta Univ. Palack. Olomuc. Fac. Rerum Natur. Math. **41** (2002), 131–145.

[Sk3] ______R. Skiba, *Graph-approximation of multivalued weighted maps* (to appear).

[Sko] G. S. Skordev, *The multiplicativity property of the fixed point index for multivalued maps*, Serdica **15** (1989), 160–170.

[Sm] L. Smith, *Transfer and ramified coverings*, Math. Proc. Cambridge Philos. Soc. **93** (1983), 485–493.

[V] C. Viterbo, *Exact Lagrange submanifolds, periodic orbits and the cohomology of free loop spaces*, J. Differential Geom. **47** (1997), 420–468.

[VHPS] F. von Haeseler, H.-O. Peitgen and G. Skordev, *Lefschetz fixed point theorem for acyclic maps with multiplicity*, Topol. Methods Nonlinear Anal. **19** (2002), 339–374.

[VHS] F. von Haeseler and G. Skordev, *Borsuk–Ulam theorem, fixed point index and chain approximations for maps with multiplicity*, Pacific J. Math. **153** (1992), 369–396.

[Vo1] Ch. S. Vora, *Fixed point theorems for certain compact weighted maps of a manifold*, Math. Student **46** (1979), 81–86.

[Vo2] ______, *Some applications of the fixed point theorems for compact weighted maps*, Math. Student **46** (1979), 87–99.

[W] T. Watanabe, *The continuity axiom and the Čech homology*, Geometric Topology and Shape Theory (Dubrovnik, 1986), Lecture Notes in Math., vol. 1283, Springer, Berlin, 1987, pp. 221–239.

[Wo] Th. J. Worosz, *Multiple-valued functions and their application to the fixed point properties of product spaces*, Bull. Malaysian Math. Soc. (2) **1** (1978), 101–109.

7. FIXED POINT THEORY FOR HOMOGENEOUS SPACES A BRIEF SURVEY

PETER WONG

1. Introduction

In classical Nielsen fixed point theory, the computation of the Nielsen number is very difficult in general and is one of the central issues in the field. W. Franz in [Fr] showed that the fixed point classes of any selfmap f of a lens space have the same fixed point index, in which case, the Nielsen number $N(f)$ is a divisor of the Lefschetz number $L(f)$ and is either zero or equal to the Reidemeister number $R(f)$. In [Ji1], B. Jiang gave conditions on the fundamental group under which all fixed point classes have the same index. If X is a *Jiang space*, i.e. $\pi_1(X)$ satisfies the so-called Jiang condition, then for all $f: X \to X$, (i) if $L(f) = 0$ then $N(f) = 0$ and (ii) if $L(f) \neq 0$ then $N(f) = R(f)$. While the class of Jiang spaces include simply connected spaces, Lie groups, H-spaces, generalized lens spaces, and coset spaces of the form G/G_0 where G_0 denotes a connected closed subgroup of a compact connected Lie group G, such a space necessarily has abelian fundamental group. If $\pi_1(X)$ is finite, it was shown in [Ji1] that (i) and (ii) hold if $\pi_1(X)$ acts trivially on the rational homology of the universal cover of X. A slight generalization of the Jiang condition was also introduced in [FH1] by E. Fadell and S. Husseini. Another computational technique is that for fiber-preserving maps, relating the Nielsen number of a fiber map with the Nielsen numbers of the induced map on the base and of the restriction map on the fiber (see e.g. Chapter 4 of [Ji2] or [HKW]).

In 1984, D. Anosov in [An] showed that for every selfmap $f: N \to N$ of a compact nilmanifold N, $N(f) = |L(f)|$. This result extends the same result for selfmaps of tori by Brooks, Brown, Pak and Taylor (see [BBPT].) By a nilmanifold, we mean a coset space $N = G/\Gamma$ of a nilpotent Lie group G by a closed subgroup Γ. In [NO], B. Norton-Odenthal strengthened Anosov's result by showing that $N(f) > 0 \Rightarrow N(f) = R(f)$, employing techniques of [FH2]. Following the work of A. Mal'cev, we may assume that G is simply-connected and thus

Partially supported by a grant from Bates College.

is homeomorphic to some $\mathbb{R}^n$ and Γ is a cocompact discrete subgroup so that $\Gamma \cong \pi_1(N)$. Anosov's theorem implies that for any selfmap of a nilmanifold, the fixed point classes have the same fixed point index of value 0, +1, or −1. For a non-abelian Γ, N is not a Jiang space while conditions (i) and (ii) hold.

The objective of this paper is to survey some recent results on the Nielsen fixed point and coincidence theory for homogeneous spaces of Lie groups, especially the various new techniques that have been employed in this study.

2. Coset spaces of compact Lie groups

Before D. Anosov (see [An]) proved his theorem about Nielsen numbers of selfmaps on nilmanifolds, most computational results were obtained using Jiang's condition ([Ji1], [Ji2]), product formulas for fiber-preserving maps (see Chapter IV of [Ji2]), or Fox derivatives for surface maps [FH3]. The essence of the Jiang condition is that the Jiang subgroup permutes the fixed point classes. For roots of maps into manifolds, it is known that all root classes are related by the action of the corresponding Jiang subgroup (see [B1]). Given a compact connected Lie group G, let K be a closed subgroup and $M = G/K$ be the space of (left) cosets. If K is connected, then M is known to be a Jiang space. In this section, we investigate the fixed point theory when K is finite by relating a fixed point problem with a root problem.

2.1. Jiang type spaces. The so-called *weak Jiang* condition introduced in [FH1] suggested that one can still compute the Nielsen number (in fact by the Reidemeister number) without having all classes with the same index as long as they have the same sign. E. Hart in [Ha] gave an example, using the Reidemeister trace, of an orbit space of an odd sphere by a finite group and a selfmap such that the fixed point classes do not have uniform index but have the same sign. More generally, when X has finite fundamental group, Jiang type results have already been obtained in [Ji1] and [FH1]. We say that a space X is of *Jiang type* if for every map $f: X \to X$, either $L(f) = 0 \Rightarrow N(f) = 0$ or $L(f) \neq 0 \Rightarrow N(f) = R(f)$.

In [Wo1], we considered a class $\mathcal{H}$ of coset spaces. By definition, a closed orientable manifold M belongs to $\mathcal{H}$ if $M = G/K$ is a coset space of a compact connected Lie group G where K is a closed subgroup and the projection $p: G \to M$ induces a nonzero homomorphism $p_*: H_n(G) \to H_n(M)$ on the nth integral homology with $n = \dim M$. Then, we have

(2.1) THEOREM ([Wo1]). *Let $M = G/K \in \mathcal{H}$. Then M is of Jiang type.*

Lens spaces, which are Jiang spaces, belong to the class $\mathcal{H}$.

(2.2) EXAMPLE ([Wo1]). The classical lens space L_p^{2n-1} is the orbit space of the unit sphere S^{2n-1} by a free action of the cyclic group $\mathbb{Z}_p$ of order p. We may also view L_p^{2n-1} as a homogeneous space of the unitary group $U(n)$ as follows. Let $H = U(n-1)$ be embedded in $U(n)$ as the subgroup of $n \times n$ unitary matrices A with $(A)_{nn} = 1$, $(A)_{in} = 0 = (A)_{nj}$ for $1 \le i, j < n$. Consider $N_1 \cong \mathbb{Z}_p$, the set of $n \times n$ diagonal matrices B_ζ with $\zeta \in \mathbb{Z}_p$ and $(B_\zeta)_{ii} = \zeta$ for $1 \le i \le n$. The set HN_1 is a closed subgroup of $U(n)$ consisting of matrices of the form

$$\begin{pmatrix} \alpha A & 0 \\ 0 & \alpha \end{pmatrix}$$

where $A \in U(n-1)$ and $\alpha \in N_1$. In $U(n)/HN_1$, $\sigma_1 HN_1 = \sigma_2 HN_1$ for $\sigma_i \in U(n)$ if and only if

$$\sigma_2 = \sigma_1 \begin{pmatrix} \alpha A & 0 \\ 0 & \alpha \end{pmatrix}$$

for some A and some α. Since $S^{2n-1} = U(n)/U(n-1)$ under the identification $[\sigma] \leftrightarrow \sigma e_n$ where e_n is the vector $(0, \ldots, 0, 1)$ in $\mathbb{C}^n$ and $\mathcal{M} \in U(n)$, it is straightforward to see that $L_p^{2n-1} = U(n)/HN_1$ as a coset space of $U(n)$.

Similarly, the generalized lens space $L_p^{2n-1}(\ell_1, \ldots, \ell_{n-1})$ can be regarded as a coset space of $U(n)$. The projection $U(n) \to L_p^{2n-1}(\ell_1, \ldots, \ell_{n-1})$ factors into $U(n) \to U(n)/U(n-1) \to L_p^{2n-1}(\ell_1, \ldots, \ell_{n-1})$ where the second map is a p-fold cover. Now we show that $L_p^{2n-1}(\ell_1, \ldots, \ell_{n-1}) \in \mathcal{H}$. It is sufficient to show that the map $p: U(n) \to U(n)/U(n-1) = S^{2n-1}$ induces a nonzero homomorphism in dimension $2n-1$ in homology. In integral cohomology, p induces $p^*: H^{2n-1}(S^{2n-1}) \to H^{2n-1}(U_n)$. An elementary argument in Leray spectral sequence shows that the edge homomorphism p^* is nonzero. Since $H^{2n-1}(U_n)$ is torsion free, we have $H^{2n-1}(U_n) \cong \mathrm{Hom}(H_{2n-1}(U_n), \mathbb{Z})$ and thus the homomorphism $p_*: H_{2n-1}(U_n) \to H_{2n-1}(S^{2n-1})$ is nonzero.

(2.3) Example ([Wo1]). Let $G = S^1 \times SO(3)$, $K = \{1\} \times A_5$. Then $G/K \approx S^1 \times SO(3)/A_5$ whose fundamental group is $\mathbb{Z} \times \mathrm{Icos}$ where Icos denotes the binary icosahedral group of order 120 (the fundamental group of the Poincaré homology 3-sphere). We see that G/K is of Jiang type but not a Jiang space because the fundamental group is not abelian since A_5 is non-abelian. On the other hand, since G/K has infinite fundamental group, neither [FH1] nor [Ji2] applies.

To prove that every $M \in \mathcal{H}$ is of Jiang type, we make use of Fadell's observation [F1]. Given a closed subgroup K of a compact connected Lie group G, K acts freely on G by translation, $k * g \mapsto gk^{-1}$, $k \in K, g \in G$. Moreover, $M = G/K$ admits an action by G via $g_1 \circ g_2 K \mapsto g_1 g_2 K$. By restricting the action to K, M becomes a K-space with fixed point set $M^K = \{eK\}$ where $e \in G$ is the unity.

In [F1], E. Fadell showed that there is a homeomorphism α such that the following diagram is commutative

$$\begin{array}{ccc} (M \times M, M \times M - \Delta_M) & \xrightarrow{\quad\alpha\quad} & G \times_K (M, M - eK) \\ & {}_{p}\searrow \qquad \swarrow_{q} & \\ & M & \end{array}$$

Here, p is the projection on the first factor and q is the fibered product over $M = G/K$. In other words, $(M \times M, M \times M - \Delta_M)$ and $G \times_K (M, M - eK)$ are fiberwise equivalent as bundle pairs over M. This translates into showing that the fixed point proerpty for M, i.e. every $f\colon M \to M$ has a fixed point, is equivalent to the Borsuk-Ulam property, i.e. every K-map $\varphi\colon G \to M$ has nonempty $\varphi^{-1}(eK)$. For every selfmap $f\colon M \to M$, we associate to it the map $\varphi_f\colon G \to M$ given by $\varphi_f(g) = g^{-1}f(gK)$. For any $k \in K$,

$$\varphi_f(k * g) = \varphi_f(gk^{-1}) = kg^{-1}f(gk^{-1}K) = k \circ \varphi_f(g)$$

so φ_f is a K-map. It is easy to see that $f \mapsto \varphi_f$ is a one-to-one correspondence between $\mathrm{Map}(M, M)$ and $\mathrm{Map}_K(G, M)$. Furthermore, $\varphi_f(g) = eK$ if and only if$f(gK) = gK$.

Not only are the fixed points of f in one-to-one correspondence with the K-orbits of *roots* of φ_f, i.e. points in $\varphi_f{}^{-1}(eK)$, but also the fixed point classes of f and the K-equivariant root classes of φ_f are in one-to-one correspondence. By showing that the K-equivariant root index is equal to the fixed point index for the corresponding classes, we then invoke a result of R. Brooks [B1] who showed that if the target space is a closed orientable topological manifold, then all root classes have the same root index (homomorphism). As it turns out, a K-equivariant root class of φ_f consists of disjoint union of (ordinary) root classes (also of φ_f) and its index is essentially the sum of the root indices of these classes. If two K-equivariant root classes of φ_f contain different number of root classes then they would have different K-equivariant root index but of the same sign. (See also [Wo6].)

2.2. $\mathcal{C}$-nilpotent spaces. Besides introducing the so-called Jiang condition, B. Jiang also proved in [Ji1] the following

(2.4) THEOREM ([Ji1]). *Let $\widetilde{X}$ be the universal cover of a compact connected polyhedron X with finite fundamental group π. If π acts trivially on $H_*(\widetilde{X};\mathbb{Q})$ then X is of Jiang type.*

This homological condition of Jiang is similar but weaker than the weak Jiang condition in [FH1]. In an attempt to extend a result of [FH1], D. Gonçalves (see [G2]) studied fixed point theory for selfmaps on $\mathcal{C}$-nilpotent spaces.

The notion of a *class* is a generalization of the definition of a class of abelian groups introduced by J.-P. Serre and that of a Serre class of nilpotent groups given by P. Hilton and J. Roitberg (see [HR]). More precisely, a family $\mathcal{C}$ of groups is called a *class of groups* if it satisfies the following property:

Given a short exact sequence of groups

$$1 \to A \to B \to C \to 1$$

we have $A, C \in \mathcal{C}$ if and only if $B \in \mathcal{C}$.

Let $\mathcal{C}$ be a class of groups. A group π is said to be *$\mathcal{C}$-nilpotent* if $\gamma^n(\pi) \in \mathcal{C}$ for some n where $\gamma^{n-1}(\pi)$ denotes the nth term in the lower central series of π:

$$\pi = \gamma^0(\pi) \supset [\pi, \pi] = \gamma^1(\pi) \supset [\pi, \gamma^1(\pi)] = \gamma^2(\pi) \supset \dots$$

An action $\theta\colon \pi \to \mathrm{Aut}(G)$ of a group π on a group G is said to be *$\mathcal{C}$-nilpotent* if $\gamma^n_\pi(G) \in \mathcal{C}$ for some positive integer n, where $\gamma^n_\pi(G)$ is the smallest normal π-subgroup that contains $[G, \gamma^{n-1}_\pi(G)]$ and the set $\{(\alpha \cdot g)g^{-1} : \alpha \in \pi,\ g \in \gamma^{n-1}_\pi(G)\}$. Moreover, a space X is *$\mathcal{C}$-nilpotent* if (i) $\pi_1(X, x_0)$ is a $\mathcal{C}$-nilpotent group and (ii) the action of $\pi_1(X, x_0)$ on $\pi_n(X, x_0)$ is $\mathcal{C}$-nilpotent, for all $x_0 \in X$ and for all $n \geq 1$. In particular, *nilpotent* spaces are $\mathcal{C}$-nilpotent where $\mathcal{C} = \{1\}$ is the trivial class.

It has been shown in [G1] that many classical theorems in algebraic topology such as Hurewicz and Whitehead theorems are also valid for $\mathcal{C}$-nilpotent spaces. For our purposes, $\mathcal{C}$ will always denote the class of finite groups.

(2.5) EXAMPLE. Jiang spaces are $\mathcal{C}$-nilpotent spaces. The Jiang subgroup $J(X)$ (or the Gottlieb subgroup $G_1(X)$) of $\pi_1(X)$ is the set of elements (when identified with deck transformations) that are $\pi_1(X)$-equivariantly homotopic to the identity $1_{\widetilde{X}}$, where $\widetilde{X}$ denotes the universal cover of X. It follows from Theorem I.4 of [Go] that $G_1(X)$ is a subgroup of the group of elements in $\pi_1(X)$ which act trivially on $\pi_*(X)$. Recall that a space X is Jiang if $G_1(X) = \pi_1(X)$. Since $G_1(X)$ is central in $\pi_1(X)$, it follows that a Jiang space must be $\mathcal{C}$-nilpotent (in fact, nilpotent).

(2.6) EXAMPLE. Let G be a finite group acting freely on an odd sphere S^{2n-1}. The orbit space $X = S^{2n-1}/G$ is $\mathcal{C}$-nilpotent. We should point out that J. Oprea showed in [O] that X is a Jiang space if and only if G is abelian.

The technique of $\mathcal{C}$-nilpotent action was employed in [GW1] to generalize to coincidences a result in [Wo1] for fixed points . The coincidence problem is then reduced to considering coincidences of lifts to finite covers. When the target manifold is $\mathcal{C}$-nilpotent, the action of the group of deck transformations on the rational homology of the cover is shown to be trivial. As a result, using the Lefschetz coincidence numbers of the lifts, the Reidemeister classes are either all essential or all inessential and we showed the following.

(2.7) THEOREM ([GW1]). *Let X and M be closed connected orientable manifolds of the same dimension such that M is $\mathcal{C}$-nilpotent and $[\pi_1(M): Z(\pi_1(M))] < \infty$ where $Z(\pi_1(M))$ denotes the center of $\pi_1(M)$. For any two maps $f, g: X \to M$, the Reidemeister coincidence classes of f, g have coincidence index of the same sign. In particular, M is of Jiang type for coincidences.*

(2.8) REMARK. In the fixed-point case where $X = M$ and g is the identity map, if $\pi_1(X)$ is finite, then Theorem (2.7) follows from [Ji1] since $\pi_1(X)$ acts nilpotently and hence trivially on $H_*(\widetilde{X}; \mathbb{Q})$, the rational homology of the universal cover $\widetilde{X}$.

Another large class of $\mathcal{C}$-nilpotent spaces is that of coset spaces of compact connected Lie groups by finite subgroups. In fact, we have

(2.9) THEOREM ([GW1]). *Let $M = G/K$ be the coset space of a compact connected Lie group where K is a finite subgroup. Then M is $\mathcal{C}$-nilpotent and $[\pi_1(M): Z(\pi_1(M))] < \infty$.*

Thus, for such a coset space M, the hypothesis of Theorem (2.7) is satisfied. We showed in [GW1] that for any two maps $f, g: X \to M$, the Reidemeister coincidence classes of f, g have coincidence index of the same sign. In particular, either $L(f, g) = 0 \Rightarrow N(f, g) = 0$ or $L(f, g) \neq 0 \Rightarrow N(f, g) = R(f, g)$, i.e. M is of Jiang type for coincidences.

2.2. Simple formula for Lefschetz number. For certain $M \in \mathcal{H}$, there is a simple formula for the Lefschetz number of $f: M \to M$ in terms of the topological degree of the associated K-map φ_f. To see this, we begin with a simple example.

(2.10) EXAMPLE. Let $f: S^1 \to S^1$ be a map of degree k on the unit circle $S^1 \subset \mathbb{C}$. It is well known that $L(f) = 1 - \deg f = 1 - k$. We may assume, up to homotopy, that $f(z) = z^k$. Consider the map

$$\mu_f(z) = z \cdot [f(z)]^{-1} = z^{1-k}$$

where $x \cdot y$ denotes the group multiplication in S^1 and the inverse

$$[f(z)]^{-1} = \frac{1}{f(z)} = z^{-k}.$$

It follows that $L(f) = \deg \mu_f$.

In [D], H. Duan showed that if $f: G \to G$ is a selfmap of a compact connected Lie group G, then $L(f) = \deg \mu_f$. While Duan proved this formula using a different method, it is apparent from the last example that the existence of multiplication on G allows one to relate fixed points of f with roots of μ_f. In fact, one of the consequences of relating fixed points of f on $M = G/K$ with the K-orbits of roots of φ_f is the following simple formula for the Lefschetz number, generalizing the result of H. Duan.

(2.11) THEOREM ([Wo1]). *Let $M = G/K \in \mathcal{H}$. Suppose that K is finite. For any $f: M \to M$,*

$$L(f) = (-1)^n \frac{\deg \varphi_f}{|K|}$$

where $|K|$ is the order of K, $n = \dim M$, and $\deg \varphi_f$ is the topological degree of the associated K-map φ_f.

When K is trivial, $\mu_f = (-1)^n \varphi_f$ so this formula reduces to that obtained by H. Duan in [D]. In fact, a similar formula can be obtained for the Lefschetz coincidence number of two maps from a closed orientable n-manifold to an n-manifold $M = G/K \in \mathcal{H}$ with finite K but the proof is different from that in [Wo1] since the domain is no longer a coset space as in the fixed point situation.

(2.12) THEOREM ([Wo2]). *Let X and M be closed connected orientable n-manifolds. Suppose $M = G/K$ is the homogeneous space of left cosets of a compact connected Lie group G and K is a finite subgroup. For any maps $f_1, f_2: X \to M$, let $q: \widetilde{X} \to X$ be the pullback of the projection $G \to G/K$ over f_1 and define $\eta: \widetilde{X} \to G$ by $\eta(\widetilde{x}) = [\widetilde{f}_1(\widetilde{x})]^{-1} \cdot f_2(q(\widetilde{x}))$ where $\widetilde{f}_1: \widetilde{X} \to G$ is a lift of f_1. Then*

$$L(f_1, f_2) = \frac{\deg \eta}{\deg q}.$$

Using similar techniques, one can obtain a similar formula for $L(f, g)$ if the target space is a certain type of H-space. A manifold M is said to be *suitable* if there exists a multiplication together with an element $e \in M$ such that

(2.13.1) $x \cdot e = x$ for all $x \in M$;

(2.13.2) given $a, b \in M$, there exists an $x \in M$ such that $a \cdot x = b$;

(2.13.3) $x \cdot y = x \cdot z \Rightarrow y = z$ for all $x, y, z \in M$.

(2.14) THEOREM ([Wo2]). *Let X and M be closed connected orientable n-manifolds. Suppose M is suitable. Then for any $f, g: X \to M$,*

$$L(f, g) = \deg \varphi \quad \textit{where } \varphi(x) = [f(x)]^{-1} \cdot g(x).$$

3. Classical aspherical manifolds

Of the three classical homotopy invariants (Lefschetz, Nielsen, and Reidemeister numbers) in fixed point or coincidence theory, the Lefschetz trace is the most computatble for it is homological while the Nielsen number is notoriously difficult to compute. The Reidemeister number lies between the Lefschetz and the Nielsen number in terms of level of difficulty in computation. In [BBPT], Brooks, Brown, Pak and Taylor showed that for selfmaps on a torus, the Nielsen number is the absolute value of the Lefschetz number. The fact that a torus is a Jiang space alone only implies that the Nielsen number is a divisor of the Lefschetz number. The main result of [BBPT] asserts more, namely, the fixed point classes have indices 0, +1 or −1. Furthermore, they showed that the torus is the only compact connected Lie group with this property for all selfmaps.

Nilmanifolds are natural generalization of tori especially from the viewpoint of dynamics. By a nilmanifold, we mean a coset space G/K where G is a connected nilpotent Lie group and K is a closed subgroup. By a classical result of A. Mal'cev, G can be chosen to be connected and simply connected so that G is homeomorphic (as topological spaces) to $\mathbb{R}^n$ for some n and K can be chosen as a cocompact discrete subgroup. A nilmanifold is a $K(\pi, 1)$-manifold whose fundamental group π is finitely generated torsion-free nilpotent. In particular, a nilmanifold cannot be a Jiang space unless it is a torus. In [An], D. Anosov proved that $N(f) = |L(f)|$ for any selfmap f on a compact nilmanifold, thereby extending the result of Brooks et al. Shortly after [An], E. Fadell and S. Husseini in [FH2] gave an independent proof of the result using product formulas for Nielsen numbers of fiber preserving maps. Since then, Anosov (e.g. [An]) has prompted many new works and generalizations to the classical aspherical manifolds such as flat, almost flat, and infra-solvmanifolds (see [FJ]) which we shall briefly discuss in this section.

3.1. Nilmanifolds. After D. Anosov proved his result, many results that hold for tori become prime candidates for generalization to nilmanifolds. When N is a torus, it is well known that $L(f) \neq 0$ if and only if$R(f) < \infty$. In [NO], it was shown that for a nilmanifold N, if $L(f) \neq 0$ then $N(f) = R(f)$. The finiteness of $R(f)$ was shown to be equivalent to the non-vanishing of the Lefschetz number, just like the case for tori. More precisely, we have

(3.1) THEOREM ([FeHW]). *Let $f\colon N \to N$ be a selfmap on a compact nilmanifold N. Then $L(f) = 0 \Rightarrow R(f) = \infty$.*

For coincidences of two selfmaps of a compact nilmanifold, D. Gonçalves showed the following

(3.2) THEOREM ([G3]). *Let $f, g\colon N \to N$ be two selfmaps on a compact nilmanifold N. Then $L(f,g) \neq 0 \Leftrightarrow N(f,g) > 0 \Leftrightarrow R(f,g) < \infty$.*

(3.3) REMARK. Theorem (3.2) also holds if the domain and target are different nilmanifolds of the same dimension. However, the proof of that fact requires the study of coincidences of maps between nilmanifolds of different dimensions (see [Wo3]).

In relative Nielsen fixed point theory, the equivalences among the non-vanishing of the Lefschetz number, the non-vanishing of the Nielsen number, and the finiteness of the Reidemeister number have also been generalized to similar equivalences for maps of pairs of nilmanifolds.

(3.4) THEOREM ([CW]). *Let (X, A) be a compact polyhedral pair of the homotopy type of a nilmanifold pair. For any selfmap $f\colon (X, A) \to (X, A)$, the following conditions are equivalent.*

$$R(f; X, A) < \infty, \tag{3.4.1}$$

$$\#\mathrm{Fix}\, f_{\#} \cdot \prod_{k=1}^{l} \#\mathrm{Fix}\, f_{k_{\#}} = 1, \tag{3.4.2}$$

$$N(f) \cdot \prod_{k=1}^{l} N(f_k) \neq 0. \tag{3.4.3}$$

Here, k ranges over the number of components of A and $R(f; X, A)$ denotes the relative Reidemeister number of f.

Other Anosov type results for various relative Nielsen type numbers can be found in [Wo7].

3.2. Infra-nilmanifolds. As already indicated by Anosov in [An], his result does not hold in general for infra-nilmanifolds (these are aspherical manifolds that are finitely covered by nilmanifolds). Shortly after [An] and [FH2], S. Kwasik and K. Lee (see [KwL]) attempted to generalize the theorem for infra-nilmanifolds. From the point of view of crystallographic groups, they proved that $N(f) = L(f)$ if f is a homotopically periodic map on an infra-nilmanifold. Subsequently, this was extended to solvmanifolds by K. Lee ([L1]) and to infra-solvmanifolds by C. McCord ([Mc6]) using different methods.

Let G be a connected, simply connected nilpotent Lie group and $\pi \subset G \rtimes \mathrm{Aut}(G)$ a torsion-free cocompact discrete subgroup such that $\Gamma = \pi \cap G$ is of finite index in π. The group π acts on G as a subgroup of $G \rtimes \mathrm{Aut}(G)$. The orbit space $M = \pi \setminus G$ is called an infra-nilmanifold and the finite group $\Phi = \pi/\Gamma$ is called the holonomy group of M. The group π is an *almost Bieberbach* group. When Φ is trivial, M is a nilmanifold.

If we replace $\mathrm{Aut}(G)$ with $\mathrm{End}(G)$ then $\mathrm{aff}(G) = G \rtimes \mathrm{End}(G)$ is a semi-group such that there is a well defined map $\mathrm{aff}(G) \times G \to G$ as follows. For any $(d, D) \in \mathrm{aff}(G)$ and $x \in G$, $((d, D), x) \mapsto d \cdot Dx$. These maps in $\mathrm{aff}(G)$ are not necessarily homeomorphisms. In [L2], K. Lee proved a Bieberbach's rigidity theorem for infra-nilmanifolds, in particular, he showed that every endomorphism of π is semi-conjugate to an element in $\mathrm{aff}(G)$. In fact, he proved the same semi-conjugation result for homomorphisms between almost Bieberbach (or more generally almost crystallographic) groups of different Hirsch rank (see [L3]). As an application, he proved the following result.

(3.5) THEOREM ([L2]). *Let* $f: M \to M$ *be a map on a compact infra-nilmanifold* $M = \pi \setminus G$. *Let* $g = (d, D) \in \text{aff}(G)$ *be a homotopy lift of* f *such that* g *is an affine endomorphism of* G. *Then* $N(f) = \pm L(f)$ *if and only if* $\det(I - D_* A_*) \geq 0$ *or* ≤ 0 *for all* $A \in \Phi$, *respectively, where* Φ *denotes the holonomy group of* M.

This generalizes the main result in [KwL] for Nielsen numbers of homotopically periodic maps. The proof of this result requires certain averaging formula for the Nielsen number in terms of the fixed point information on the (finite) lifts of the given map. An averaging formula for the Nielsen number of maps of infra-nilmanifolds has been obtained in [KLL]. Applications to periods of expanding maps on infra-nilmanifolds can be found in [LL].

Following the same approach as in [L2], W. Malfait (see [M]) further extended Kwasik–Lee's result by considering a class of maps on infra-nilmanifolds which he called *virtually unipotent* maps. More precisely, let L denote the universal cover of an infra-nilmanifold M. An affine diffeomorphism φ of M is said to be virtually unipotent if and only if there is a lifting $(\ell_f, \alpha_f) \in L \rtimes \text{Aut}(L)$ of φ such that $(\alpha_f)^k$ has eigenvalues equal to 1 for some positive integer k. Then, a map $f: M \to M$ is vitually unipotent if it is homotopic to a virtually unipotent affine diffeomorphism. Clearly, a homotopically periodic map is virtually unipotent. W. Malfait showed the following

(3.6) THEOREM ([M]). *If* $f: M \to M$ *is virtually unipotent on a compact infra-nilmanifold* M, *then* $N(f) = L(f)$.

This geometric approach has been used to study fixed point theory on Seifert manifolds modelled on certain geometries (see [K], [KL1], [KL2]).

3.3. Solvmanifolds and infra-solvmanifolds. One naturally asks to what extent does Anosov's theorem hold for solvmanifolds, that is, for coset spaces of connected solvable Lie groups. C. McCord showed in [Mc] that while the equality between $|L(f)|$ and $N(f)$ does not hold in general, the fixed point classes have index $0, \pm 1$ and thus $|L(f)| \leq N(f)$. The approach was to use the fact that every selfmap $f: M \to M$ of a compact solvmanifold M is homotopic to a fiber preserving map $f': M \to M$ where M fibers over a torus with typical fiber a nilmanifold. Subsequently, E. Keppelmann and C. McCord (see [KeM]) identified a class of solvmanifolds including exponential solvmanifolds (i.e. the exponential map is surjective) for which $N(f) = |L(f)|$.

While, for solvmanifolds, the Lefschetz number is only a lower bound for the Nielsen number, the Reidemeister number can sometimes give more information and be equal to the Nielsen number. In this direction, we have

(3.7) THEOREM ([GW2]). *Let* M *be a compact solvmanifold. For any selfmaps* $f, g: M \to M$, *if* $R(f, g) < \infty$ *then* $R(f, g) = N(f, g)$.

In [Mc3], C. McCord intended to show that the inequality $|L(f,g)| \leq N(f,g)$ would hold for maps between closed orientable solvmanifolds of the same dimension. Although there was a gap in the proof, the validity of the inequality ws verified for certain special cases in its subsequent correction [Mc4]. The use of the Reidemeister number was crucial in [Wo3] in verifying McCord's claim. There, we first consider fiber preserving maps

$$\begin{array}{ccccc} N_1 & \xrightarrow{j_1} & M_1 & \xrightarrow{q_1} & T_1 \\ f' \big\downarrow \big\downarrow g' & & f \big\downarrow \big\downarrow g & & \overline{f} \big\downarrow \big\downarrow \overline{g} \\ N_2 & \xrightarrow[j_2]{} & M_2 & \xrightarrow[q_2]{} & T_2 \end{array}$$

where M_i are compact orientable solvmanifolds of the same dimension, N_i are nilmanifolds, and T_i are tori with $\dim T_1 \geq \dim T_2$.

Note that coincidences of $\overline{f}, \overline{g}$ can be transformed into roots of the map $\sigma = f \cdot g^{-1}$ since the torus is a group. In [Wo3], we showed that if $\sigma\colon T_1 \to T_2$ is a map between two tori with $\dim T_1 \geq \dim T_2$ then the finiteness of the Reidemeister root number $R(\sigma, 0)$ is equivalent to showing that σ is a fibration $T_1 \to T_2$ followed by a finite cover $T_2 \to T_2$, up to homotopy. This allows us to use induction and show the following

(3.8) THEOREM ([Wo3]). *Let X and Y be compact nilmanifolds and* $\dim X \geq \dim Y$. *Given any two maps $f, g\colon X \to Y$, $N(f,g) > 0 \Rightarrow R(f,g) < \infty$.*

(3.9) REMARK. This result has been generalized to coincidences of maps into finite complexes that fiber over a nilmanifold (see [Wo4]).

Furthermore, we were able to give conditions when $N(f,g) = R(f,g)$ and we showed the following

(3.10) THEOREM ([GW2]). *Let $f, g\colon X \to Y$ be two maps between two closed connected orientable n-manifolds, Y be a solvmanifold and $R(f,g) < \infty$. Then there exist finite covers $p_X\colon \widehat{X} \to X$, $p_Y\colon \widehat{Y} \to Y$ and lifts $\widehat{f}, \widehat{g}\colon \widehat{X} \to \widehat{Y}$ of f, g, respectively, such that $L(\widehat{\alpha}\widehat{f}, \widehat{g}) \neq 0$ for all $\widehat{\alpha} \in$* Cov p_Y *if and only if $N(f,g) = R(f,g)$.*

For maps between infra-solvmanifolds, we also have

(3.11) THEOREM ([GW2]). *Suppose $f, g\colon M_1 \to M_2$ are two maps between two closed connected infra-solvmanifolds of the same dimension and $R(f,g) < \infty$. There exist lifts $\widehat{f}, \widehat{g}\colon \widehat{M_1} \to \widehat{M_2}$ of f, g on the finite covers $\widehat{M_i}$ each of which is a solvmanifold such that $L(\widehat{\alpha}\widehat{f}, \widehat{g}) \neq 0$ for all $\widehat{\alpha}$ if and only if $N(f,g) = R(f,g)$.*

Moreover, $L(\widehat{\alpha}\widehat{f}, \widehat{g}) \geq 0$ (≤ 0) *for all deck transformation* $\widehat{\alpha}$ *if and only if* $N(f,g) = L(f,g)$ $(-L(f,g)$*, respectively).*

The two results stated above were proven using techniques of separating Reidemeister classes and factorization of group homomorphisms which will be discussed in the next section.

(3.12) Remark. In the special case when the domain and the target are the same infra-nilmanifold, Theorem (3.11) improves Theorem (3.5) in that $N(f)$ can be computed even when the Lefschetz numbers of the lifts need not have the same sign.

3.4. Reidemeister numbers. The Reidemeister number, which is always an upper bound for the Nielsen number, can be computed at the fundamental group level (see [Fer] for the computation of the Reidemeister number for certain spaces or for certain fundamental groups). Unfortunately, the Reidemeister number need not be finite so the determination of the finiteness of the Reidemeister number is important. We will address in this section the question of the finiteness of the Reidemeister number in the fixed point situation.

Let $\varphi\colon \pi \to \pi$ be a group endomorphism. Recall that the Reidemeister number of φ, denoted by $R(\varphi)$, is the cardinality of the set of equivalence classes under the relation $\alpha \sim \sigma\alpha\varphi(\sigma)^{-1}$ on π. For a selfmap f on a connected topological space, $R(f) := R(f_{\#})$ where $f_{\#}$ is the induced homomorphism on the fundamental group.

The growth of the fundamental group π of a Riemannian manifold M has long been studied. M. Gromov showed that π has polynomial growth if and only if M is an infra-nilmanifold (or almost flat). In establishing a connection between Nielsen fixed point theory and Reidemeister torsion, A. Fel'shtyn and R. Hill conjectured [FeH] that $R(\varphi) = \infty$ if π is a finitely generated torsion-free group with exponential growth and $\varphi\colon \pi \to \pi$ is injective.

If π is a non-elementary Gromov hyperbolic group, $\varphi\colon \pi \to \pi$ an automorphism and Φ the corresponding outer automorphism, G. Levitt and M. Lustig in [LeLu] showed that the set of *isogredience classes* $\mathcal{S}(\Phi)$ is infinite. A. Fel'shtyn observed (see [Fe]) that the infiniteness of $\mathcal{S}(\Phi)$ implicitly implies that automorphisms of π have infinite Reidemeister number. Furthermore, using co-Hopfian property, A. Fel'shtyn showed in [Fe] that if in addition π is torsion-free and freely indecomposable then for every injective φ, $R(\varphi)$ is infinite. This result gives supportive evidence to the conjecture of [FeH] since π has exponential growth.

Based upon techniques used in [GW2] on coincidence theory for solvmanifolds, we showed in [GW5] that the conjecture of Fel'shtyn and Hill does not hold in general (even for automorphisms). The counter-examples in [GW5] are groups that are exponential but *toroidal*, i.e. contain $\mathbb{Z}^2$ and thus cannot be Gromov hyperbolic. In fact, these groups are the fundamental groups of solvmanifolds. The simplest one such is a three dimensional solvmanifold obtained as the mapping torus of an Anosov homeomorphism on the 2-torus. Therefore, the finiteness of the Reidemeister number does not depend solely on the growth rate of the fundamental group. We conjecture in [GW5] that the Fel'shtyn–Hill conjecture would hold if π has exponential growth but is atoroidal. (Note that in dimension three, it has been conjectured that if M is a closed aspherical 3-manifold, then $\pi_1(M)$ is either toroidal or Gromov hyperbolic.)

4. Positive codimension coincidence theory

In Section 2, we have seen that the fixed point problem of a selfmap $f: G/K \to G/K$ on a coset space can be transformed into an equivariant root problem of the associated K-map $\varphi_f: G \to G/K$. When $\dim K > 0$, the root problem then becomes the coincidence problem of φ_f and of $p: G \to G/K$ (projection) between two manifolds of different dimension. The fact that $N(f,g) > 0$ is equivalent to $R(f,g) < \infty$ for maps between nilmanifolds of the same dimension is a consequence of examining maps between nilmanifolds of different dimensions.

As seen in [Wo3], for maps between orientable solvmanifolds of the same dimension, the inequality $|L(f,g)| \leq N(f,g)$ was established by considering fibrations where the base spaces are tori of positive codimension. These are some examples of how positive codimensional coincidence problem naturally arises from zero-codimensional situations. In this final section, we further illustrate this phenomenon when studying coincidence theory of homogeneous spaces.

For manifolds of different dimension, there is obvious difficulty in defining the Lefschetz number, the coincidence index, and thus the Nielsen number. Using a geometric notion of essentiality, R. Brooks extended the classical Nielsen number as follows (see [B2]). Let $f, g: M \to N$ be maps between two spaces and let $\mathrm{Coin}\,(f,g) = \{x \in M : f(x) = g(x)\}$ be the set of coincidences. Suppose $x_1, x_2 \in \mathrm{Coin}\,(f,g)$. Then we say that x_1 and x_2 are *Nielsen equivalent* as coincidences with respect to f and g if there exists a path $\sigma: [0,1] \to M$ such that $\sigma(0) = x_1$, $\sigma(1) = x_2$ and $f \circ \sigma$ is homotopic to $g \circ \sigma$ relative to the endpoints. The equivalence classes of this relation are called coincidence classes. A coincidence class $\mathcal{F}$ is *essential* if for any $x \in \mathcal{F}$ and for any homotopies $\{f_t\}$, $\{g_t\}$ of $f = f_0$ and $g = g_0$, there exist $x' \in \mathrm{Coin}\,(f_1, g_1)$ and a path $\gamma: [0,1] \to X$ with $\gamma(0) = x$, $\gamma(1) = x'$ such that $f_t \circ \gamma$ is homotopic to $g_t \circ \gamma$ relative to the endpoints. We say that $x \in \mathcal{F}$ is $\{f_t\}, \{g_t\}$*-related* to a coincidence of f_1 and g_1. In other words, a class is inessential if it disappears under some homotopy. The Nielsen coincidence number $N(f,g)$ of f and g is defined to be the number of essential coincidence classes. This

Nielsen number coincides with the classical one defined by H. Schirmer [Sch] for maps between closed orientable manifolds of the same dimension.

4.1. Coincidences of maps into nilmanifolds. Let $f, g: M \to N$ be maps between two closed orientable n-manifolds. It is well-known that if N is a torus then all coincidence classes of f, g have the same coincidence index. Given the knowledge about the fixed point theory on nilmanifolds and the fact that nilpotent groups behave in a similar way as abelian groups, it is natural to ask whether the property of obtaining the same coincidence index would hold true if the torus is replaced by a nilmanifold. While the coincidence problem of two maps between two nilmanifolds of the same dimension is well understood ([G3], [G4], [Wo3]), it is not clear how to proceed when M is arbitrary.

Given a pair of maps $f, g: M \to N$, since $\pi_1 N$ is finitely generated torsion-free nilpotent, it was shown in [GW3] that the induced homomorphisms $\varphi, \psi: \pi_1 M \to \pi_1 N$ factor into $\varphi = \overline{\varphi} \circ \varepsilon$, $\psi = \overline{\psi} \circ \varepsilon$ where $\varepsilon: \pi_1 M \to \pi_1 \widehat{N}$ is a surjective homomorphism, and $\widehat{N}$ is a nilmanifold (possibly of different dimension). Since ε is onto, it follows that $R(\varphi, \psi) = R(\overline{\varphi}, \overline{\psi})$. Thus, the coincidence problem can be reduced to the corresponding coincidence problem of $\overline{f}, \overline{g}: \widehat{N} \to N$, particularly the case when $\dim \widehat{N} \geq \dim N$. (If $\dim \widehat{N} < \dim N$, then $\overline{f}$ and $\overline{g}$ are deformable to be coincidence free.)

The factorization of φ and of ψ can be extended to homomorphisms where the target group is virtually polycyclic, i.e. the fundamental group of an infrasolvmanifold. We showed in [GW4] the following result.

(4.1) THEOREM ([GW4]). *Let $1 \to N \to G \to Q \to 1$ be an extension of groups where G is torsion-free, N (strongly) polycyclic and Q finite. Suppose π is a finitely generated group. Then there exists an extension $1 \to \overline{N} \to \overline{\pi} \to \overline{Q} \to 1$ where $\overline{\pi}$ is finitely generated torsion-free, $\overline{N}$ (strongly) polycyclic, $\overline{Q}$ finite and an epimorphism $\varepsilon: \pi \to \overline{\pi}$ with $c_2(\overline{N}) \leq c_2(N)$ such that for every homomorphism $\varphi: \pi \to G$ there is a homomorphism $\overline{\varphi}: \overline{\pi} \to G$ such that $\varphi = \overline{\varphi} \circ \varepsilon$.*

In order to find a finite cover so that the coincidences of lifts project to a single coincidence class, we made use of a very strong separability property (or certain residually finiteness) for virtually polycyclic groups due to P. Stebe [St]. As a result, we obtain the following

(4.2) THEOREM ([GW4]). *Let $f, g: X \to Y$ be two maps between two closed connected orientable n-manifolds. Suppose that Y is an infrasolvmanifold and $R(f, g) < \infty$. Then there exist finite covers $p_X: \widehat{X} \to X$, $p_Y: \widehat{Y} \to Y$ and lifts $\widehat{f}, \widehat{g}: \widehat{X} \to \widehat{Y}$ of f, g, respectively such that $L(\alpha \widehat{f}, \widehat{g}) \neq 0$ for all $\alpha \in \operatorname{Cov} p_Y$ if and only if $N(f, g) = R(f, g)$.*

When Y is a nilmanifold, the finite group $\operatorname{Cov} p_Y$ acts nilpotently and hence trivially on the homology of $\widehat{Y}$. It follows that $L(\alpha\widehat{f}, \widehat{g})$ is independent of α and so we proved the following result.

(4.3) THEOREM ([GW3]). *Nilmanifolds are Jiang type spaces for coincidences. That is, for any maps $f, g\colon X \to N$ from a closed connected orientable n-manifold to a compact k-dimensional nilmanifold, we have $L(f,g) = 0 \Rightarrow N(f,g) = 0$ or $L(f,g) \neq 0 \Rightarrow N(f,g) = R(f,g)$.*

4.3. Obstructions and Reidemeister number. Let X and Y be closed orientable manifolds with $\dim X \geq \dim Y = n$. Given two maps $f, g\colon X \to Y$, the *Lefschetz coincidence index* (see also [B2]) is defined to be the homomorphism

$$\lambda(f,g;X)\colon H^n(Y^\times) \xrightarrow{[(f\times g)d]^*} H^n(X, X - C(f,g)) \to H^n(X)$$

where $Y^\times = (Y \times Y, Y \times Y - \Delta_Y)$ and the cohomology groups are taken to be with integer coefficients. When using local coefficients, the corresponding homomorphism is the primary obstruction of deforming f and g to be coincidence free on the nth skeleton of X, denoted by $o_n(f,g)$. F. B. Fuller (see [Fu]) first made use of obstruction theory to show that if $\dim X = \dim Y$ and Y is simply connected then the converse of the Lefschetz coincidence theorem holds (see also [F2]).

For positive codimension, obstruction theory is used to extend coincidence free maps from one skeleton to the next skeleton. The first is the primary obstruction $o_n(f,g)$ to deforming f and g to be coincidence free on the nth skeleton of X where $n = \dim Y$. This primary obstruction was studied in [GJW]. As an application, we showed the following

(4.4) THEOREM. *Let $f, g\colon T \to N$ be maps from an m-dimensional torus to an n-dimensional nilmanifold where $m \geq n$. If $R(f,g) < \infty$ then $\lambda(f,g) \neq 0$ so that $o_n(f,g) \neq 0$, and $N(f,g) = R(f,g)$.*

One naturally asks whether this theorem would hold if T is replaced by a nilmanifold. The next example shows that the finiteness of the Reidemeister number does not imply the non-vanishing of the Lefschetz coincidence index.

(4.5) EXAMPLE ([GJW]). Consider the following torsion-free nilpotent groups

$$\pi_1 = \langle a, b, c, d, e \,|\, [a,b] = c,\ [a,d] = e,\\ [a,c] = [a,e] = [b,c] = [b,d] = [c,d] = [d,e] = 1\rangle$$

and

$$\pi_2 = \langle \alpha, \beta, \gamma, \delta \,|\, [\alpha,\beta] = \gamma,\ [\alpha,\delta] = [\beta,\delta] = [\alpha,\gamma] = [\beta,\gamma] = 1\rangle.$$

The subgroup $\pi_2' = \langle \gamma, \delta | [\gamma, \delta] = 1 \rangle$ is central in π_2. Consider

$$\pi_1' = \langle c, e | [c, e] = 1 \rangle \subset \pi_1$$

and the homomorphism $\varphi: \pi_1 \to \pi_2$ given by

$$a \mapsto \alpha, \quad b \mapsto \beta, \quad c \mapsto \gamma, \quad d \mapsto \delta, \quad e \mapsto 1.$$

Then $\pi_1' = \varphi^{-1}(\langle c \rangle)$ is normal in π_1 and $\ker \varphi = \langle e \rangle$. We have the following commutative diagram of short exact sequences of groups with torsion-free abelian quotients.

$$\begin{array}{ccccc} \pi_1' & \longrightarrow & \pi_1 & \longrightarrow & \pi_1/\pi_1' \\ \downarrow{\scriptstyle \varphi'} & & \downarrow{\scriptstyle \varphi} & & \downarrow{\scriptstyle \overline{\varphi}} \\ \pi_2' & \longrightarrow & \pi_2 & \longrightarrow & \pi_2/\pi_2' \end{array}$$

Suppose that $f: N_1 \to N_2$ realizes the homomorphism φ so that N_1 is a 5-dimensional nilmanifold with fundamental group π_1 and N_2 is a 4-dimensional nilmanifold with fundamental group π_2. The map f is a fibration whose typical fiber is a circle with fundamental group $\langle e \rangle$. Using Shapiro's Lemma and primary obstructions, we showed in [GJW] that the Lefschetz index homomorphism vanishes in integral cohomology but it (this is the primary obstruction) is non-trivial in cohomology with local coefficients. Thus, f must have roots in the 4th skeleton of N_1.

While $R(f, g) < \infty$ does not imply $\lambda(f, g; M) \neq 0$ when M and N are both nilmanifolds, we proved in [GW6] the following result which answers a question posed in [GW3].

(4.6) THEOREM. *Let $f, g: M \to N$ be maps between two compact nilmanifolds with $\dim M = m \geq n = \dim N$. If $R(f, g) < \infty$ then the primary obstruction $o_n(f, g) \neq 0$ and $N(f, g) = R(f, g) > 0$.*

As a consequence of the study of the primary obstruction, we obtained the following interesting result which states that for a fibration of a manifold over an aspherical manifold, the primary obstruction must be non-trivial and thus a typical fiber must intersect the n-skeleton of the total space where n is the dimension of the base.

(4.7) THEOREM. *Given a map $f: E \to B$ from a closed connected oriented manifold E to a closed connected oriented aspherical manifold B with connected typical fiber, if the primary obstruction $o_n(f, \overline{b}) = 0$ where $\overline{b}$ denotes the constant path at $b \in B$, then f cannot a fibration.*

References

[An] D. Anosov, *The Nielsen number of maps of nilmanifolds*, Russian Math. Surveys **40** (1985), 149–150.

[B1] R. Brooks, *Certain subgroups of the fundamental group and the number of roots of* $f(x) = a$, Amer. J. Math. **95** (1973), 720–728.

[B2] ______, *On the sharpness of the* Δ_2 *and* Δ_1 *Nielsen numbers*, J. Reine Angew. Math. **259** (1973), 101–108.

[BBPT] R. Brooks, R. F. Brown, J. Pak and D. Taylor, *The Nielsen number of maps on tori*, Proc. Amer Math. Soc. **52** (1975), 398–400.

[BO] R. Brooks and C. Odenthal, *Nielsen numbers for roots of maps of aspherical manifolds*, Pacific J. Math. **170** (1995), 405–420.

[BW] R. Brooks and P. Wong, *On changing fixed points and coincidences to roots*, Proc. Amer. Math. Soc. **115** (1992), 527–533.

[Br] K. S. Brown, *Cohomology of Groups*, Springer, New York, 1982.

[Bro] R. F. Brown, *The Lefschetz Fixed Point Theorem*, Scott–Foresman, Glenview, Illinois, 1971.

[CW] F. Cardona and P. Wong, *On the computation of the relative Nielsen number*, Topology Appl. **116** (2001), 29–41.

[Ch] L. Charlap, *Bieberbach Groups and Flat Manifolds*, Springer, New York, 1986.

[CJL] D. Chun, C. Jang and S. Lee, *On the fixed-point theorems on the infrasolvmanifolds*, Comm. Korean Math. Soc. **10** (1995), 681–688.

[D] H. Duan, *The Lefschetz number of selfmaps of Lie groups*, Proc. Amer Math. Soc. **104** (1988), 1284–1286.

[F1] E. Fadell, *Two vignettes in fixed point theory*, Proceedings of the conference on "Topological Fixed Point Theory and Applications" (Tianjin, 1988) (B. Jiang, ed.), vol. 1411, Lecture Notes in Math., 1989, pp. 46–51.

[F2] ______, *On a coincidence theorem of F. B. Fuller*, Pacific J. Math. **15** (1965), 825–834.

[FH1] E. Fadell and S. Husseini, *Fixed point theory for non-simply connected manifolds*, Topology **20** (1981), 53–92.

[FH2] ______, *On a theorem of Anosov on Nielsen numbers for nilmanifolds*, Nonlinear Funct. Anal. Appl. **173** (1986), 47–53.

[FH3] *Nielsen numbers on surfaces*, Contemp. Math. **173** (1986), 47–53.

[FJ] F. T. Farrell and L. E. Jones, *Classical Aspherical Manifolds*, vol. 75, CBMS, Amer. Math. Soc., Providence, 1990.

[Fe] A. L. Fel'shtyn, *The Reidemeister number of any automorphism of a Gromov hyperbolic group is infinite*, Zap. Nauchn. Sem. S.-Petersburg. Otdel. Mat. Inst. Steklov. (POMI) **279** (2001), 229–241.

[FeH] A. L. Fel'shtyn and R. Hill, *The Reidemeister zeta function with applications to Nielsen theory and a connection with Reidemeister torsion*, K-theory **8** (1994), 367–393.

[FeHW] A. L. Fel'shtyn, R. Hill and P. Wong, *Reidemeister numbers of equivariant maps*, Topology Appl. **67** (1995), 119–131.

[Fer] D. Ferrario, *Computing Reidemeister classes*, Fund. Math. **158** (1998), 1–18.

[Fr] W. Franz, *Abbildungsklassen und fixpunktklassen dreidimensionaler Linsenräume*, J. Reine Angew. Math. **185** (1943), 65–77.

[Fu] F. B. Fuller, *The homotopy theory of coincidences*, Ann. Math. **59** (1954), 219–226.

[GN] R. Geoghegan and A. Nicas, *Parametrized Lefschetz–Nielsen fixed point theory and Hochschild homology traces*, Amer. J. Math. **116** (1994), 397–446.

[G1] D. L. Gonçalves, *Generalized classes of groups, $\mathcal{C}$-nilpotent spaces and the Hurewicz Theorem*, Math. Scand. **53** (1983), 39–61.

[G2] ______, *On the Nielsen number of maps on nilpotent spaces*, Atas do Décimo Quarto Colóquio Brasileiro de Matemática (1985), 331–337.

[G3] ______, *Coincidence Reidemeister classes on nilmanifolds and nilpotent fibrations*, Topology Appl. **83** (1998), 169–186.

[G4] ———, *The coincidence Reidemeister classes of maps on nilmanifolds*, Topol. Methods Nonlinear Anal. **12** (1998), 375–386.

[GJW] D. L. Gonçalves, J. Jezierski and P. Wong, *Obstruction theory and coincidences in positive codimension*, preprint (2002).

[GW1] D. L. Gonçalves and P. Wong, *Homogeneous spaces in coincidence theory*, Conference Proceedings of the Tenth Brazilian Topology Meeting (São Carlos, 1996) (P. Schweitzer, ed.); Mat. Contemp. **13** (1997), 143–158.

[GW2] ———, *Homogeneous spaces in coincidence theory* II, Forum Math. (to appear).

[GW3] ———, *Nilmanifolds are Jiang type spaces for coincidences*, Forum Math. **13** (2001), 133–141.

[GW4] *Factorization of group homomorphisms*, J. Pure Appl. Algebra **171** (2002), 197–204.

[GW5] ———, *Twisted conjugacy classes in exponential growth groups*, Bull. London Math. Soc. **35** (2003), 261–268.

[GW6] ———, *Obstruction theory nd coincidences of mps between nilmanifolds*, Arch. Math. (Basel) (to appear).

[Go] D. Gottlieb, *A certain subgroup of the fundamental group*, Amer. J. Math. **87** (1965), 840–856.

[Ha] E. Hart, *Ph. D. Thesis* (1991), University of Wisconsin, Madison.

[HK1] P. Heath and E. Keppelmann, *Fiber techniques in Nielsen periodic point theory on nil and solvmanifolds* I, Topology Appl. **76** (1997), 217–247.

[HK2] ———, *Fiber techniques in Nielsen periodic point theory on nil and solvmanifolds* II, Topology Appl. **106** (2000), 149–167.

[HK3] ———, *Fiber techniques in Nielsen periodic point theory on solvmanifolds* III, Calculations, Quaestiones Math. **25** (2002), 177–208.

[HKW] P. Heath, E. Keppelmann and P. Wong, *Addition formulae for Nielsen numbers and for Nielsen type numbers of fibre-preserving maps*, Topology Appl. **67** (1995), 133–157.

[HR] P. Hilton and J. Roitberg, *Generalized C-theory and torsion phenomena in nilpotent spaces*, Houston J. Math. **2** (1976), 529–559.

[J1] J. Jezierski, *The Nielsen number product formula for coincidence*, Fund. Math. **134** (1989), 183–212.

[J2] ———, *One codimensional Wecken type theorems*, Forum Math. **5** (1993), 421–439.

[Ji1] B. Jiang, *Estimation of the Nielsen numbers*, Acta Math. Sinica **14** (1964), 304–312.

[Ji2] ———, *Lectures on Nielsen Fixed Point Theory*, Contemp. Math. **14** (1983), Amer. Math. Soc., Providence.

[K] E. Kang, *Fixed points theory on closed 2-dimensional manifolds*, Comm. Korean Math. Soc. **15** (2000), 675–681.

[KL1] E. Kang and K. B. Lee, *Fixed point theory on Seifert manifolds modeled on* $S^2 \times \mathbb{R}$, Kyungpook Math. J. **37** (1997), 123–134.

[KL2] ———, *Fixed points theory on geometric Seifert manifold*, Proceedings of the International Conference on "Homotopy Theory and Nielsen Fixed Point Theory" (Seoul, 2000) (E. Kang and M. Woo, eds.), Institute of Science and Technology, Korea University, 2000, pp. 13–20.

[KeM] E. Keppelmann and C. McCord, *The Anosov theorem for exponential solvmanifolds*, Pacific J. Math. **170** (1995), 143–159.

[Ki] T. Kiang, *The Theory of Fixed Point Classes*, Springer, New York, 1987.

[KLL] S. W. Kim, J. B. Lee and K. B. Lee, *Averaging formula for Nielsen numbers*, preprint (2002).

[KwL] S. Kwasik and K. B. Lee, *The Nielsen numbers of homotopically periodic maps of infranilmanifolds*, J. London Math. Soc. **38** (1988), 544–554.

[LL] J. Lee and K. B. Lee, *Lefschetz numbers for continuous maps, and periods for expanding maps on infra-nilmanifolds*, preprint (2001).

[L1] K. B. Lee, *Nielsen numbers of periodic maps on solvmanifolds*, Proc. Amer. Math. Soc. **116** (1992), 575–579.

[L2] ———, *Maps on infra-nilmanifolds* I, Pacific J. Math. **168** (1995), 157–166.

[L3] ———, *Maps on infra-nilmanifolds* II, unpuiblished.

[LeLu] G. Levitt and M. Lustig, *Most automorphisms of a hyperbolic group have very simple dynamics*, Ann. Sci. École Norm. Sup. (4) **33** (2000), 507–517.

[M] W. Malfait, *The Nielsen numbers of virtually unipotent maps on infra-nilmanifolds*, Forum Math. **13** (2001), 227–237.

[Mc] C. K. McCord, *Nielsen numbers and Lefschetz numbers on solvmanifolds*, Pacific J. Math. **147** (1991), 153–164.

[Mc2] ______, *Estimating Nielsen numbers on infrasolvmanifolds*, Pacific J. Math. **154** (1992), 345–368.

[Mc3] ______, *Lefschetz and Nielsen coincidence numbers on nilmanifolds and solvmanifolds*, Topology Appl. **43** (1992), 249–261.

[Mc4] ______, *Lefschetz and Nielsen coincidence numbers on nilmanifolds and solvmanifolds* II, Topology Appl. **75** (1997), 81–92.

[Mc5] ______, *Computing Nielsen numbers*, Proceedings of the Conference on "Nielsen Theory and Dynamical Systems" (Mt. Holyoke, 1992) (C. McCord, ed.), vol. 152, Contemp. Math., 1993, pp. 249–267.

[Mc6] ______, *Nielsen numbers of homotopically periodic maps on infrasolvmanifolds*, Proc. Amer. Math. Soc. **120** (1994), 311–315.

[NO] B. Norton-Odenthal, *Ph. D. Thesis* (1991), University of Wisconsin, Madison.

[O] J. Oprea, *Finite group actions on spheres and the Gottlieb group*, J. Korean Math. Soc. **28** (1991), 65–78.

[Sch] H. Schirmer, *Mindestzahlen von Koinzidenzpunkten*, J. Reine Angew. Math. **194** (1955), 21–39.

[St] P. Stebe, *A residual property of certain linear groups*, Trans. Amer. Math. Soc. **302** (1987), 333–340.

[V] J. Vick, *Homology Theory*, Academic Press, New York, 1973.

[W] F. Wecken, *Fixpunktklassen* II, Math. Ann. **118** (1942), 216–234.

[Wo1] P. Wong, *Fixed point theory for homogeneous spaces*, Amer. J. Math. **120** (1998), 23–42.

[Wo2] ______ Coincidences of maps into homogeneous spaces, Manuscripta Math. **98** (1999), 243–254.

[Wo3] ______, *Reidemeister number, Hirsch rank, coincidences on polycyclic groups and solvmanifolds*, J. Reine Angew. Math. **524** (2000), 185–204.

[Wo4] ______, *Coincidence theory for spaces which fiber over a nilmanifold*, Fixed Point Theory Appl. **2** (2004), 89–95.

[Wo5] ______, *Equivariant Nielsen numbers*, Pacific J. Math. **159** (1993), 153–175.

[Wo6] ______, *Equivariant Nielsen theory*, Nielsen theory and Reidemeister torsion (J. Jezierski, ed.), (Warsaw, 1996); Banach Center Publ. **49** (1999), 253–258; (1997), 143–158.

[Wo7] ______, *Fixed points on pairs of nilmanifolds*, Topology Appl. **62** (1995), 173–179.

CHAPTER II

EQUIVARIANT FIXED POINT THEORY

Topological fixed point theory can be considered in G-topology, i.e. for equivariant mappings of G-topological spaces, where G is a given group acting on a topological space X. The problems discussed in Chapter I can be studied in this setting. In what follows, such fixed point problems are called equivariant fixed point problems.

The purpose of this chapter is to survey current results and problems in the equivariant theory.

8. A NOTE ON EQUIVARIANT FIXED POINT THEORY

DAVIDE L. FERRARIO

1. Introduction

Let G be a finite group acting on a topological space X, which is termed a *G-space*. We recall a few basic facts about equivariant spaces, which can be found in all classical books on the subject (for example in [Br], [TD]). For every $x \in X$ the *isotropy subgroup* (also termed *fixer* or *stabilizer*) of x is $G_x = \{g \in G : gx = x\}$. If $H \subset G$ is a subgroup of G, then the space fixed by H in X is $X^H = \{x \in X : Hx = x\} = \{x \in X : H \subset G_x\}$. If X and Y are G-spaces, then a G-map (i.e. an *equivariant* map) $f: X \to Y$ is a map which commutes with the G-action: for every $g \in G$ and $x \in X$, $f(gx) = gf(x)$. If $x \in X$ and G_x is its isotropy,

$$g \in G_x \Rightarrow gfx = fgx = fx \Rightarrow g \in G_{fx},$$

thus $G_{fx} \supset G_x$. This implies that if f is equivariant, then for every $H \subset G$, $fX^H \subset Y^H$. The restriction of f to the fixed subspace X^H is denoted by $f^H: X^H \to Y^H$. A G-homotopy $H: f_0 \sim f_1$ is a G-map $H: X \times I \to Y$, where the action of G on $X \times I$ is trivial on the I-component. In equivariant fixed point theory one studies the *fixed point set* of the map f, provided that $X \subset Y$:

$$\mathrm{Fix}(f) = \{x \in X : f(x) = x\}.$$

The fixed point set $\mathrm{Fix}(f)$ is a G-subset of X, and hence the disjoint union of G-orbits. The normalizer of a subgroup $H \subset G$ is

$$N_G H = \{g \in G : gHg^{-1} = H\},$$

and the quotient with H yields the *Weyl group* $W_G H = N_G H/H$, which acts on X^H. Also, the restricted map f^H is a $W_G H$-equivariant map.

For a given subgroup $H \subset G$, let (H) denote the class of all the subgroups conjugated to H in G (that is, $H' = gHg^{-1}$ for some $g \in G$). We write $(H) \leq (K)$ when H is subconjugated to K, i.e. when there is $K' = gKg^{-1}$ such that $H \subset K'$.

Let $X_H = \{x \in X : G_x = H\}$ denote the subspace of points with isotropy H, and $X_{(H)} = \{x \in X : (G_x) = (H)\}$ the subspace of points with *isotropy type* (H). The projection $X_{(H)} \to X_{(H)}/G$ is a fibre bundle with fibre homeomorphic to G/H, with associated principal bundle $X_H \to X_H/W_GH$ (the Weyl group W_GH acts freely on X_H). Consider the poset $\mathrm{iso}(X)$ of isotropy types (H) of points in X. The isotropy stratification of X is the stratification given by strata $X_{(H)}$ with isotropy type (H), for $(H) \in \mathrm{iso}(X)$. It is a Thom–Mather stratification (or a CW-stratification), provided that either X is a a smooth G-manifold or G-CW-complex. It is a Siebenmann cone-like stratification if X satisfies only the weaker condition of being a locally smooth G-manifold. Since any finite poset can be endowed with a compatible total order, there is an indexing of $\mathrm{iso}(X) = \{(H_1), \dots, (H_l)\}$ such that if $(H_i) \leq (H_j)$ then $j \leq i$. Thus, the isotropy stratification of X induces a filtration

$$\emptyset = X_0 \subset X_1 \subset \ldots \subset X_l = X,$$

where $X_i = \bigcup_{j=1}^{i} X_{(H_j)}$. For the purpose of equivariant fixed point theory, we require that $X_i \subset X_{i+1}$ is a G-cofibration for every i, since in this case the G-homotopy extension property allows one to build homotopies inductively on isotropy strata (see for example [FW], [Wo1], [Wo4]). This is the case for G-ENRs, G-CW-complexes or locally smooth G-manifolds. To consider the simplest assumption, *throughout the paper we will assume that X is an open G-invariant subset of a smooth G-manifold Y*. Moreover, any G-map $f\colon X \to Y$, where $X \subset Y$ is an open G-subset, will be assumed to be *compactly fixed*: the fixed point set $\mathrm{Fix}(f)$ is compact. Similarly, all G-homotopies will be assumed to be compactly fixed.

Now, there are several approaches to equivariant fixed point theory. The purpose of this note is not to give a (not complete) account or a survey about all the results and theories, since it would go far beyond the limits of this chapter. We will be addressed only to the *naïve* equivariant fixed point theory approach, in the unstable homotopy realm. That is, we will try to sketch answers to the following questions: when, given $f\colon X \subset Y \to Y$, all G-maps f' G-homotopic (compactly fixed) to f will have fixed points? Or, on the contrary, when f is G-homotopic (compactly fixed) to a fixed point free G-map? Is there a structure of $\mathrm{Fix}(f)$ which is invariant under G-homotopies (e.g. the fixed point index or the Reidemeister trace)? And, accordingly, we will need to connect equivariant and non-equivariant fixed point theories, by asking: what is the equivariant corresponding of the Lefschetz number, fixed point index, Reidemeister trace, Nielsen number? As pointed out above, this is a *naïve* approach. There are no generally accepted answers to these questions even in the non-equivariant case.

2. Equivariant Lefschetz classes and the Burnside ring

Consider the fixed point index Ind(f) as defined by Dold in [D1], for a map $f: X \to Y$, with $X \subset Y$ open subspace. The index is an integer with the following properties:

(2.1) if f is a constant map, then Ind$(f) = 1$ if $fX \in X$ and Ind$(f) = 0$ if $fX \notin X$;

(2.2) if $X = \bigcup_j X_j$, such that Fix$(f) \cap X_j$ are mutually disjoint, then

$$\operatorname{Ind}(f) = \sum_j \operatorname{Ind}(f|X_j);$$

(2.3) for every $f: X \to Y$ and $f': X' \to Y'$, $\operatorname{Ind}(f \times f) = \operatorname{Ind}(f) \times \operatorname{Ind}(f')$;

(2.4) moreover, if $X \subset Y$ and $X' \subset Y'$ and given maps $f: X \to Y'$, $g: X' \to Y$, then $\operatorname{Ind}(gf|f^{-1}X') = \operatorname{Ind}(fg|g^{-1}X)$.

These properties (for ENR's) characterize the index and, furthermore, the Lefschetz–Hopf theorem (see Theorem 4.1 of [D1] and also [Br1], [Br2]) can be proved: the Lefschetz number $L(f)$ of a self-map $f: X \to X$ is equal to the fixed point index Ind(f) (which, in turn, is the sum of the indices around its fixed point components). Actually, as pointed out in (5.1) of [D3], assuming additivity and homotopy invariance suffices to characterize Ind(f) for ENRs. Other (axiomatic and not) approaches can be found in the works of M. Arkowitz and R. F. Brown, T. tom Dieck, K. Komiya, E. Laitinen, W. Lück, W. Marzantowicz, C. Prieto, and H. Ulrich (see [AB], [TD], [K3], [L2], [HL], [L1], [LL], [Lu], [MP1], [M2], [PU], [Ul]). We will come back to this topic in Section 5.

Let $A(G)$ denote the Burnside ring of the finite group G, i.e. the Grothendieck ring of finite G-sets. Elements of $A(G)$ are formal differences $S - T$ of (G-isomorphism classes of) finite G-sets S, T, where $S - T = S' - T'$ if and only if $S + T' \cong S' + T$ (the disjoint union are G-isomorphic). Addition and multiplication in $A(G)$ are given by the disjoint sum and the cartesian product. As a consequence of the properties of the stable equivariant cohomotopy theory, one can define the $A(G)$-valued *equivariant Lefschetz number* $\operatorname{Ind}_G(f) = L_G(f) \in A(G)$ (see [TD]). The idea is the following: first, let ω_G^0 denote the stable 0-the dimensional cohomotopy group of S^0 (i.e. the limit under G-suspensions of the homotopy groups $[S^n, S^n]_G$, where S^n are endowed with linear G-actions; in other notation, $\omega_G^0 = \{S^0, S^0\}_G$). To be more precise, consider the poset of all isomorphism classes of orthogonal G-representations, where the inclusion $i_V^W: V \subset W$ holds when $W = V + V'$ for a suitable V'. Let $S(V)$ denote the unit G-sphere in V. Then the sets of G-homotopy classes $[S(V), S(V)]_G$ yield a direct system, with arrows induced by

$$i_V^W: [S(V), S(V)]_G \to [S(W), S(W)]_G,$$

where the arrow is given by the join of a map $S(V) \to S(V)$ with the identity map on $S(V')$ (since $S(V) * S(V') = S(W)$). Such a join is also termed G-suspension. Now, as V ranges over all representations, the limit $\{S^0, S^0\}_G = \lim_V [S(V), S(V)]_G$ exists and is termed the 0th dimensional (co)homotopy group of S^0, and denoted by ω_G^0.

In [TD] it is proved that $\omega_G^0 \cong A(G)$ (Segal's theorem), while full details on the construction of the ω_G^0-cohomotopical index can be found in [D2]. The underlying idea is simply to take the cohomotopy class of the map given by $i - f$, where i means the inclusion $X \subset Y$ and f is the map $f: X \to Y$, after a suitable extension to some sphere. Also, it can be defined as the sum of the equivariant fixed point indices of the (finite number of) components of $\mathrm{Fix}(f)$, or as a collection of traces of Bredon–Illman homology theory (see [L2]).

Roughly speaking, the $A(G)$-valued Lefschetz class $L_G(f)$ turns out to be a function $\mathrm{iso}(X) \to \mathbb{Z}$ which maps the conjugacy class of an isotropy subgroup $H \subset G$ to the integer $\mathrm{Ind}(f^H)$, which is the fixed point index of the restricted map $f^H: X^H \to Y^H$ (the Lefschetz number coincides with Ind_G in a natural way when $X = Y$; otherwise one has to consider the Lefschetz number arising from homology with integer coefficients of the pair $(Y^H, Y^H \setminus X^H)$). Such Lefschetz number (also termed as *equivariant Lefschetz class*) arises also by choosing some universal properties for L_G (Laitinen and Lück, see [LL]). It is the way to *count* fixed points in the Burnside ring, which is the natural ring for equivariant (co)-homotopy theory.

Actually, in spite of the deep algebraic properties of Ind_G and L_G, the questions above are of an *unstable* homotopy nature, hence stable invariants will not yield complete answers. We will see the (two-folded) reason of this in the next sections.

3. Fixed points on pairs

Consider a given G-map $f: X \to Y$, and a subgroup $H \subset G$. Define the *singular set* $X_s^H = \{x \in X^H : G_x \neq H\}$. Since $X^H = \{x \in X : G_x \supset H\}$, one can alternatively write $X_s^H = \{x \in X : G_x \supsetneq H\}$ and $X_s^H = X^H \setminus X_H$. For example, if $H = 1$, these are the points fixed by some element of G; the action is free if and only if $X_s^1 = \emptyset$ (it follows from the definition that if H is not isotropy of some point in X, then $X_s^H = X^H$).

The restricted map $f^H: X^H \to Y^H$ is actually a W_GH-map of pairs

$$(f^H, f_s^H): (X^H, X_s^H) \to (Y^H, Y_s^H),$$

since $f^H X_s^H \subset Y_s^H$. Moreover, every G-homotopy $f_t: X \to Y$ induces a W_GH-homotopy of maps of pairs $((f_t)^H, (f_t)_s^H)$. In particular, if one forgets the W_GH-action, a G-map induces maps of pairs (f^H, f_s^H). Hence, if f can be G-deformed to be fixed point free, then for every H the map of pairs (f^H, f_s^H) can be deformed to a fixed point free map. When one comes to the algebraic invariants that count fixed points (or, equivalently, to obstruction elements in a suitable obstruction theory), it would be reasonable to consider the relative *Schirmer–Nielsen number* for the pair (X^H, X^H_s), or one of its later variants due to J. Jezierski, B. Norton–Odenthal, P. Wong, X. Zhao (see [Sch], [Je], [Zh], [NW]). In fact, the Schirmer–Nielsen number, under suitable conditions (a dimension assumption and a by-passing condition; see below), is the obstruction to deform *a map of pairs* to a fixed point free map. It is easy to find examples where the map f^H can be deformed, while the map of pairs (f^H, f^H_s) might not, while, of course, if f^H can be deformed, so the map of pairs does.

By defining an equivariant Nielsen number of f as the collection of Schirmer–Nielsen numbers $N(f^H; X^H, X^H_s)$ one not only can give prove a converse of the Lefschetz property, but also can compute the minimum number of fixed points in a G-homotopy class of maps (compactly fixed), provided that the *by-passing* condition holds (i.e. $X_H = X^H \setminus X^H_s$ is path-connected and the induced homomorphism $\pi_1(X_H) \to \pi(X^H)$ is onto) and all the fixed manifolds X_H have dimensions at least 3 (see E. Fadell, T. Fomenko, A. Vidal, P. Wong and others [FW], [Fo], [Vi], [Wo2]–[Wo5]). Moreover, P. Wong in [Wo6] generalizes all these constructions by considering the definition of the Nielsen number of a poset of maps (actually, a directed system of subspaces of X and maps there defined). The idea is to consider the inclusion order of all the spaces X^H and to relate the fixed point classes arising at different strata. See also Remark (6.2). See also J. Better's thesis in [Be] for results and detailed references on relative equivariant Nielsen numbers; recent results on equivariant coincidences include J. Guo and P. Heath ([GH]) and P. Fagundes ([F]). For results on fixed points in Jiang spaces the reference is P. Fagundes and D. Gonçalves (see [FG]).

The by-passing condition, since $X \subset Y$ are G-manifolds, immediately implies that the codimension of X^H_s in X^H has to be at least 2. In the equivariant setting this assumption is too restrictive, unless the manifolds are complex, since many important group representations (of Coxeter groups, for example) fix subspaces of codimension 1. Now, the first impression about this problem is that a wall-crossing formula could easily be found: consider a connected G-manifold M with trivial isotropy type. If the action is not free $M^1_s \subset M^1 = M$ is a proper subspace, invariant under $W_G 1 = G$ (the Weyl group of the trivial subgroup $1 \in G$ is G). All the components of M_s of codimension 1 in M, by the smoothness of the action, are fixed by reflections (involutions), hence even if $M_1 = M \setminus M_s$ can be not connected, the quotient M_1/G is connected. Let $D \subset M_1$ be a component. By transitivity of the action on the components $M_1 = GD$, and by equivariance all fixed points in M_1 of a G-map $f\colon M \to M$ are images under G of fixed points contained in D.

Thus all the components of M_1 share the same structure and the same definition of the G-map f.

It is important to know that this phenomenon is not true in general: for $H \subset G$ and X a smooth G-manifold, one cannot prove that X_H/W_GH is connected, provided that X^H is connected. That is, even if X^H is connected, the components of X_H can be not homeomorphic each other by the elements of the Weyl group W_GH. So, one cannot expect in general, even in the connected and simply connected case, the stable fixed point index Ind_G (or, equivalently, the equivariant Lefschetz number L_G) to fulfill a converse of the Lefschetz property, unless G is of special type. The following proposition explains the meaning of this "special type".

(3.1) LEMMA. *Let G be a finite group. Then the following statements are equivalent:*

(3.1.1) *The group G is the direct product of a 2-group and an odd-order group.*
(3.1.2) *There exists a smooth compact G-manifold X and an isotropy $H \subset G$ such that X^H is connected while X_H/W_GH is not connected.*
(3.1.3) *There exists a smooth compact G-manifold X such that for every $H \subset G$ the fixed submanifold X^H can be deformed to a fixed point free map, while X cannot be G-deformed to a fixed point free map.*

PROOF. It follows from Theorem 1.1 of [Fe3] and the construction in the proof of Proposition 4.1. □

4. An equivariant fixed point index via Reidemeister traces

We have seen the first reason for which a stable G-index does not have the converse of the Lefschetz property, even for identity maps (where the fundamental group does not play a role). It is also possible to define counter-examples as in (3.1.3) of the previous proposition, with all the fixed subspaces X^H simply-connected. So, we need to take into account the fact that (f^H, f^H_s) is a W_GH-map of pairs. This was first done by K. Wilczyński (see [Wi]), who defined an equivariant Lefschetz number un terms of relative Lefschetz numbers (collecting the components of the pairs (X^H, X^H_s)).

The main idea, which can be described as well in terms of fixed point indices of map of pairs, is the following. Consider a manifold (possibly with boundary) M and a union of submanifolds $A \subset M$. A map $f\colon (M, A) \to (M, A)$ is termed *taut* if there is a retraction $r\colon W \to A$ from a neighbourhood W of A in M to A such that $f|W = fr$.

(4.1) LEMMA. *If* $f: (M, A) \to (M, A)$ *is taut and* $x \in \mathrm{Fix}(f) \cap A$, *then the index of* x *as a fixed point of* $f_A: A \to A$ *and the index of* x *as a fixed point of* $f: M \to M$ *coincide.*

PROOF. It follows from the commutative property; without loss of generality we can assume that $W \cap \mathrm{Fix}(f) = \{x\}$; let $i: A \to M$ denote the inclusion. We have: $\mathrm{Ind}(x, f|W) = \mathrm{Ind}(x, i f_A r) = \mathrm{Ind}(x, f_A r i) = \mathrm{Ind}(x, f_A)$. □

One can define in a similar way taut maps for G-equivariant maps (in this case the retraction has to be G-equivariant) and taut maps *around* a compact subset (like a component of the fixed point set of a compactly fixed G-map), where the above definition is taken only in a suitable local neighbourhood. This idea, which dates back to K. Wilczyński (see [Wi]), K. Komiya ([K1], [K2]) and W. Marzantowicz ([M1]), has been also recently used by Zhao (in [Zh]) in the definition of a fixed point index for maps of pairs and by the author in [Fe1] for the definition of the index that we are describing in this section. The steps for the definition are the following. We use again the word *taut* with a slightly different meaning: we say that a compactly fixed G-map $f: X \to Y$ is G-taut if for every isotropy $H \in \mathrm{iso}(X)$ the $W_G H$-map of pairs $(f^H, f^H_s): (X^H, X^H_s) \to (Y^H, Y^H_s)$ is taut (around the fixed point set of f^H_s) equivariantly with respect to the Weyl group $W_G H$.

(4.2) LEMMA. *Every* G*-map* $f: X \to Y$ *has an* ε*-approximation* f' *which is compactly fixed* G*-homotopic to* f *and* G*-taut.*

PROOF. See Proposition 4.4 of [Fe1]. □

(4.3) LEMMA. *Any two approximations* f'_0, f'_1 *of* (4.2) *of a* G*-map* $f: X \to Y$ *are* G*-homotopic through a* G*-homotopy* f'_t *which is taut for every* t.

PROOF. See the proof of Lemma 4.8 of [Fe1]. □

So, let $\mathcal{I}_G(f)$ denote the following index. First, it is defined for G-taut maps (which have the property that each restriction f_H is compactly fixed in X_H). The equivariant index $\mathcal{I}_G(f)$ is the collection of Reidemeister traces parameterized as follows. For every $H \in \mathrm{iso}(X)$ let $\mathcal{I}_G(f)(H)$ be the Reidemeister trace $\mathcal{L}(f_H) \in \mathbb{Z}\mathcal{R}(f_H)$ of the restriction $f_H: X_H \to Y^H$ of the G-map $f: X \to Y$ to the subspace $X_H \subset X^H$ (see Remark (6.4) below for a comment on the Reidemeister trace, generalized Lefschetz number for local non-connected maps; see also (6.3) for the differences with the results of [Fe1]). Since a G-homotopy which is G-taut for every $t \in I$ preserves the traces $\mathcal{L}(f_H)$, it is proved that $\mathcal{I}_G(f)$ can be defined for all compactly fixed G-maps, as the index of any G-taut ε-approximation (see (4.3)).

This index has the suitable obstruction property for fixed point free deformations:

(4.4) LEMMA. *If Y^H is a manifold of dimension at least 3 for every H, then $\mathcal{I}_G(f) = 0$ if and only if there exists a G-map f' equivariantly (and compactly fixed) homotopic to f with* $\mathrm{Fix}(f') = \emptyset$.

PROOF. It follows from Corollary 4.23 of [Fe1]. □

Furthermore, by the additivity of the Reidemeister trace one can prove the following formula (which holds for G-taut maps):

(4.5)
$$\mathcal{L}(f^H) = \sum_{K \supset H} i_H^K \mathcal{L}(f_K),$$

where $i_H^K \colon \mathbb{Z}\mathcal{R}(f_K) \to \mathbb{Z}\mathcal{R}(f^K) \to \mathbb{Z}\mathcal{R}(f^H)$ is the homomorphism functorially induces by the inclusion $X_K \subset X^H$ (for $H \subset K$). Formula (4.5) is a generalization of Komiya formula in [K4] to Reidemeister trace. For details and a discussion about its Möbius inversion, see [Fe4]. An immediate consequence of (4.5) is that if $\mathcal{L}(f_H) = 0$ for all H, then $\mathcal{L}(f^H) = 0$ for all H. A Möbius inversion would imply the converse, i.e. that if $\mathcal{L}(f^H) = 0$ for all H, then $\mathcal{L}(f_H) = 0$ for all H. As shown in Example (6.1), this cannot be true in general. Either the singular sets can be by-passed, or some further hypotheses on the map have to be assumed.

A definition of local Reidemeister trace (in terms of Dennis traces of cellular chain maps) can be found also in [FH]. Furthermore, in recent papers of W. Marzantowicz and C. Prieto (see [MP3], [MP2]) a definition equivalent to the definition of taut maps (which are there are named *co-normal* maps) is proposed and used on the problem of classification of G-maps (see also [Fe5] on the subject).

5. Universal invariants and K-theory

We recall the definition of Lefschetz invariant for G-maps, as given in [LL]. Given a finite group G, a *Lefschetz invariant* is an abelian group A (depending on G) and a function L which assigns to a G-map $f \colon X \to X$ the element $L(f)$ such that if f and g are G-homotopic, then $L(f) = L(g)$ (homotopy property), if $f, g \colon X \to X$ are self-maps then $L(fg) = L(gf)$ (commutativity), if $X \subset Y \to Y/X$ is a cofibre sequence and $(f_X, f_Y, f_{Y/X})$ is a self-map of the cofibre sequence, then $L(f_X) - L(f_Y) + L(f_{Y/X}) = 0$ (additivity for cofibre sequences); $f, g \colon \Sigma X = S^1 \wedge X \to \Sigma X$ are maps, then $L(f+g) = L(f) + L(g)$ (linearity: here the sum $f+g$ is defined by the coaction map $S^1 \to S^1 \vee S^1$ on the S^1-component of the suspension $S^1 \wedge X$). These axioms are parallel to the axioms of *trace invariant* for finitely generated projective R-modules (where R is a ring, in this case the group-ring $\mathbb{Z}G$): it is an abelian group A (depending on G) with a function T assigning $T(f) \in A$ to every R-modules endomorphism $f \colon M \to M$ with the properties: if $0 \to M_1 \to M \to M_2 \to 0$ is an exact sequence and (f_1, f, f_2) an endomorphism of the sequence (i.e. vertical morphisms f_1, f, f_2 which commute with the exact sequence), then $T(f) = T(f_1) + T(f_2)$ (additivity); $f, g\colon M \to M \Rightarrow T(f+g) = T(f) + T(g)$ (additivity); $f\colon M \to M'$, $g\colon M' \to M \Rightarrow T(fg) = T(gf)$ (commutativity).

These properties define, in a categorical and standard way, the notions of *universal* trace invariant and *universal* Lefschetz invariant. By considering cellular chain complexes of the strata X_H (which are $W_G H$-free), in [LL] it is shown that there exists a universal Lefschetz invariant, which in turn is the unique extension of the universal trace invariant to chain maps of $\mathbb{Z}G$-modules on chain complexes. It is well-known that the universal trace invariant is the Dennis trace (trace defined K-theoretically as an element in the 0-dimensional Hochschild homology of the ring R), also known as Hattori-Stallings trace. See also R. Geoghegan and A. Nicas on this topic (see [GN]).

This K-theoretical framework, originally motivated by equivariant fixed point theory, has been applied also by Lück in [Lu], where the idea is to consider that Nielsen fixed point theory (and the Reidemeister trace) are of an equivariant nature: the Reidemeister trace of a self-map $f\colon X \to X$ is defined in terms of twisted Hattori-Stallings traces of the chain maps, which are $\pi_1(f)$-twisted projective $\mathbb{Z}\pi_1(X)$-homomorphisms. So, the universal covering space $\widetilde{X}$ is seen as a $\pi_1(X)$-equivariant space, and the arguments above are adapted to deal with the case of non-trivial $\pi_1(f)$-twisting (that is, the fact that the induced homomorphism $\pi_1(f)\colon \pi_1(X) \to \pi_1(X)$ can be different than the identity). The difference is here that the linearity axiom is not considered in [Lu], and therefore the resulting *universal functorial Lefschetz invariant* is, *a priori*, finer than the Reidemeister trace (which, in turn, can be seen as a universal Lefschetz trace with local coefficients in $\pi_1(X)$). The difference is easily understood in the context of linear algebra (over a field) and homomorphisms: the universal functorial Lefschetz invariant of a homomorphism corresponds to its characteristic polynomial, while the universal Lefschetz-trace invariant is the trace of the homomorphism (see Remark (6.5)). In any case, both approaches share the same scheme: use a categorical-universal approach to define the trace or the additive invariant, and then the cellular decompositions of the spaces (in one case, equivariant with respect to G, in the other with respect to $\pi_1(X)$). Now then, it is possible to combine the two approaches for a smooth G-manifold (or a G-CW-complex) X if one considers the isotropy stratification of X and the stratified homology theory defined by Baues and the author in [BF]. In this case it is possible to mimic the axiomatic approach of Laitinen and Lück, and to define universal invariants for $\Pi_G(X)$-cellular chains, yielding naturally a Lefschetz–Hopf type of result (i.e. the invariant as an alternate sum of traces in homology), homotopy invariance, and Lefschetz property (immediately from definition). For another abstract approach (computing the generalized Lefschetz number in generalized equivariant cohomology theories) see also the recent paper of W. Marzantowicz and C. Prieto (see [MP1]).

6. Examples and remarks

(6.1) EXAMPLE. Here we recall briefly the core of example 5.2 of [Fe1]. Let G be the group of order 2 acting on $\mathbb{R}^3$ by $(x, y, z) \to (-x, y, z)$. Let a and b denote the unit circles contained in the plane $z = 0$ and centers in $(-1, 0, 0)$, $(1, 0, 0)$, and c the unit circle contained in the plane $x = 0$ and center in $(0, 0, -1)$. Consider the union X_0 of the three circles a, b and c with basepoint $x_0 = 0$. We denote with the same symbols a, b and c the loops around a, b and c. The induced action of the non-trivial element $g \in G$ is as follows: $ga = b$, $gb = a$, $g|c = 1$. Now, let $f: X_0 \to X_0$ be a map with the following properties: if $x \in c$ then $fx = x$ and $fa = a^{-1}b^{-1}ab$ and $fb = b^{-1}a^{-1}ba$. By a small rotation on c the G-map f can be G-deformed in a way that it has no fixed points in c, and there are two fixed points in a (of index $+1$ and -1) and two corresponding fixed points in b (of index $+1$ and -1). Let x_+ and x_- denote the fixed points in a and gx_+ and gx_- their images in b. The fact is that, in X_0, x_+ and gx_- belong to the same Nielsen class (and the same for x_- and gx_+), which is therefore inessential. Thus $N(f^G) = N(f) = 0$. On the other hand, even considering the regular neighbourhood of X_0 in $\mathbb{R}^3$ (or in any bigger euclidean space), it is not possible to deform f to a G-map which is fixed point free, since the equivariant fixed point index $\mathcal{I}_G(f)$ of [Fe1] is not zero. For a similar example see also Vidal and Izydorek in [IV2]. See also [IV1] and [Vi].

(6.2) REMARK. Dold ([D4]), extending previous results of Zabreĭko-Krasnosel'skiĭ ([ZK]) and Steinlein ([St]), proved the following beautiful theorem. Let $f: X \to Y$ a map (with Y ENR and $X \subset Y$ open) such that all iterates f^n have compact fixed point set $\mathrm{Fix}(f^n)$. The *Lefschetz power series* (also known as *Lefschetz zeta function*) is defined by

$$L(f; t) = \exp\bigg(-\sum_{k=1}^{\infty} \frac{\mathrm{Ind}(f^k)}{k} t^k\bigg).$$

Theorems (1.8) and (1.9) of [D4] state that a sequence of integers coincide with the sequence of fixed point indices of iterates $\mathrm{Ind}(f^k)$ of some compactly fixed map $f: X \to Y$ if and only if the coefficients of the corresponding zeta function are integers. Moreover, the sequence coincide with the sequence of fixed point indices of iterates of a self-map $f: Y \to Y$ of a *compact* Y if and only if $L(f; t)$ is an integral rational function (which can be written as quotient of products of determinants of endomorphisms induced in homology with rational coefficients). The proof is rather simple, and its basic idea comes from the observation that the additive group $\mathbb{Z}$ of integers acts on the set P of periodic points of f (with their fixed point indices), and that hence a Möbius type formula for the poset of subgroups of $G = \mathbb{Z}$ holds:

$$|P^{n\mathbb{Z}}| = \bigsqcup_{k|n} |P_{k\mathbb{Z}}|.$$

Here $P^{n\mathbb{Z}}$ is the set of periodic points with period at most n (i.e. periodic points fixed by $n\mathbb{Z} \subset \mathbb{Z}$), while $P_{k\mathbb{Z}}$ is the set of periodic points with period exactly k (i.e. periodic points with isotropy $k\mathbb{Z}$). Such a formula can be written also as

$$\mathrm{Ind}(1^H) = \sum_{K \supset H} \mathrm{Ind}(1_K),$$

where 1^H and 1_H are the restrictions to P^H or P_H of the identity 1, and can be generalized as Komiya theorem of [K4] for every $f\colon X \to Y$:

$$\mathrm{Ind}(f^H) = \sum_{K \supset H} \mathrm{Ind}(f_K).$$

By using the additive property for the Reidemeister traces, in [Fe4] this result is extended to the case of Reidemeister traces, instead of fixed point indices. Unfortunately I was not able to prove such Möbius-sum formula (nor to invert it in a Möbius fashion) with universal invariants instead of Reidemeister traces.

(6.3) REMARK. One of the side-effects of the functoriality of the Reidemeister trace — generalized Lefschetz number is that $\mathcal{L}(f)$ is an element of an abelian group which depends on the map f. A combinatorial way to encode the information about the number of classes of a given index is to associate to a map (or to its Reidemeister trace) the function which assigns to a fixed point class its fixed point index. As remarked in 3.1 of [Fe1], the set of isomorphism classes of functions with values in $\mathbb{Z} \setminus 0$ and finite domain is a commutative monoid, with respect to the disjoint union and the cartesian product, so that the Grothendieck ring R is well-defined. This concept, mutated from the Burnside ring definition and applied to the context of Nielsen fixed points, allows to define an index which is in an intermediate position between the Nielsen number (which counts the essential fixed point classes) and the Reidemeister trace (which counts the essential fixed point classes, together with their fixed point indices, together with their coordinates in the Reidemeister twisted conjugacy classes). The advantage is that R does not depend upon f and shares some algebraic properties with the Nielsen number. On the other hand, the book-keeping of fixed point indices on the poset structure $\{X_H, X^H\}$ needs the knowledge of coordinates of fixed point classes, and not only their cardinality. Thus an R-valued index can play a significant role only outside equivariant fixed point theory. Actually, in [Fe1] such index was denoted by $\mathcal{I}_G$, while in the present paper this symbol is used for the equivariant index of Section 4.

(6.4) REMARK. The good functorial properties of the set of Reidemeister classes can be encompassed in a straightforward way as follows: consider a map $f: X \to Y$, where $X \subset Y$. Consider the space of all paths $\gamma: I \to Y$ such that $f(\gamma(0)) = \gamma(1)$ and $\gamma(0) \in X$ with the compact-open topology. Now consider its quotient $P(X, Y)$, under the equivalence relation given by homotopies rel. endpoints. The evaluation at 0 and 1, $\varepsilon_0 \times \varepsilon_1: P(X, Y) \to X \times Y$ is a union of coverings (universal coverings) of the graph of $f: X \to Y$ in $X \times Y$. The Reidemeister set $\mathcal{R}(f)$ is just the set of such components. Given a fixed point $fx = x$, the constant path c_x yields the *coordinate* of the fixed point x in $\mathcal{R}(f)$ by $\mathrm{cd}(x) = [c_x]$, where the homotopy class of the constant path c_x belongs to $P(X, Y)$. Adding the index it is easy to define the Reidemeister trace as a formal sum of components of $P(X, Y)$ (i.e. Reidemeister classes). The problem arises only when one wants to relate such a sum with a trace-like quantity, in homology or cellular chains. Then fixed points and base-paths (or choices in the components of the coverings) have to be chosen. An alternative approach, using projective finitely generated modules over a suitable ring, instead of free modules over the group ring $\mathbb{Z}\pi_1(X)$, can be found in [Fe2].

(6.5) REMARK. Consider a self-map $f: X \to X$ such that $\pi_1(X)$ is abelian and $f_\pi: \pi_1(X) \to \pi_1(X)$ is the identity (e.g. if X is simply-connected). Then the universal functorial invariant of f is determined by the product

$$\prod_i \det(1 - tH_i(\widetilde{f}))^{(-1)^i},$$

which is an element of the ring of rational functions with coefficients in $R = \mathbb{Z}\pi_1(X)$ (see Example 1.7 of [Lu]). Here $\widetilde{f}$ is any map induced on the universal covering space $\widetilde{X}$. As we noticed before, the characteristic polynomials $\det(1 - tH_i(\widetilde{f}))$ encompass the trace (in R), while the vice-versa is not true.

References

[AB] M. Arkowitz, and R. F. Brown, *The Lefschetz–Hopf theorem and axioms for the Lefschetz number*, Preprint.

[BF] H.-J. Baues and D. L. Ferrario, *Homotopy and homology of fibred spaces*, Topology Appl. (2003) (to appear).

[Be] J. Better, *Relative equivariant Nielsen fixed point theory*, Ph.D. Thesis (2002), UCLA..

[Br] G. E. Bredon, *Equivariant Cohomology Theories*, Springer–Verlag, Berlin, 1967.

[Br1] R. F. Brown, *On the Lefschetz number and the Euler class*, Trans. Amer. Math. Soc. **118** (1965), 174–179.

[Br2] ______, *The Lefschetz Fixed Point Theorem*, Scott, Foresman and Co., Glenview, Ill.–London, 1971.

[D1] A. Dold, *Fixed point index and fixed point theorem for Euclidean neighbourhood retracts*, Topology **4** (1965), 1–8.

[D2] ______, *K-theory of non-additive functors of finite degree*, Math. Ann. **196** (1972), 177–197.

[D3] ______, *The fixed point index of fibre-preserving maps*, Invent. Math. 25 (1974), 281–297.

[D4] ______, *Fixed point indices of iterated maps.*, Invent. Math. **74** (1983), 419–435.

[FW] E. Fadell and P. Wong, *On deforming G-maps to be fixed point free*, Pacific J. Math. **132** (1988), 277–281.

[F] P. L. Fagundes, *Equivariant Nielsen coincidence theory*, Mat. Contemp. 13 (1997), 117–142, 10th Brazilian Topology Meeting (Sao Carlos, 1996).

[FG] P. L. Fagundes and D. L. Gonçalves, *Fixed point indices of equivariant maps of certain Jiang spaces*, Topol. Methods Nonlinear Anal. **14** (1999), 151–158.

[FH] J. S. Fares and E. L. Hart, *A generalized Lefschetz number for local Nielsen fixed point theory*, Topology Appl. **59** (1994), 1–23.

[Fe1] D. L. Ferrario, *A fixed point index for equivariant maps*, Topol. Methods Nonlinear Anal. **13** (1999), 313–340.

[Fe2] ______, *Generalized Lefschetz numbers of pushout maps defined on non-connected spaces*, Nielsen Theory and Reidemeister Torsion (Warsaw 1996), Polish Acad. Sci., Warsaw, 1999, pp. 117–135.

[Fe3] ______, *Equivariant deformations of manifolds and real representations*, Pacific J. Math. **196** (2000), 353–368.

[Fe4] ______, *A Möbius inversion formula for generalized Lefschetz numbers*, Osaka J. of Math. **4** (2003), 1–27.

[Fe5] ______, *On the equivariant Hopf theorem*, Topology **42** (2003), 447–465.

[Fo] T. N. Fomenko, *On the least number of fixed points of an equivariant mapping*, Mat. Zametki **69** (2001), 100–112.

[GN] R. Geoghegan and A. Nicas, *Trace and torsion in the theory of flows*, Topology **33** (1994), 683–719.

[GH] J .Guo and P. R. Heath, *Equivariant coincidence Nielsen numbers*, Topology Appl. **128** (2003), 277–308.

[HL] H. Honkasalo and E. Laitinen, *Equivariant Lefschetz classes in Alexander–Spanier cohomology*, Osaka J. Math. **33** (1996), 793–804.

[IV1] M. Izydorek and A. Vidal, *A note on the converse of the Lefschetz theorem for G-maps*, Ann. Polon. Math. **58** (1993), 177–183.

[IV2] ______, *An example concerning equivariant deformations*, Topol. Methods Nonlinear Anal. **15** (2000), 187–190, Dedicated to Juliusz Schauder, 1899–1943.

[Je] J. Jezierski, *A modification of the relative Nielsen number of H. Schirmer*, Topology Appl. **62** (1995), 45–63.

[K1] K. Komiya, *A necessary and sufficient condition for the existence of non-singularG-vector fields on G-manifolds*, Osaka J. Math. **13** (1976), 537–546.

[K2] ______, *G-manifolds and G-vector fields with isolated zeros*, Proc. Japan Acad. Ser. A Math. Sci. **54** (1978), 124–127.

[K3] ______, *The Lefschetz number for equivariant maps*, Osaka J. Math. **24** (1987), 299–305.

[K4] ______, *Fixed point indices of equivariant maps and Möbius inversion*, Invent. Math. **91** (1988), 129–135.

[L1] E. Laitinen, *On the Burnside ring and stable cohomotopy of a finite group*, Math. Scand. **44** (1979), 37–72.

[L2] ______, *Unstable homotopy theory of homotopy representations*, Transformation Groups (Poznań, 1985), Lecture Notes in Math., vol. **1217**, Springer, Berlin, 1986, pp. 210–248.

[LL] E. Laitinen and W. Lück, *Equivariant Lefschetz classes*, Osaka J. Math. **26** (1989), 491–525.

[Lu] W. Lück, *The universal functorial Lefschetz invariant*, Algebraic Topology (Kazimierz Dolny, 1997), Fund. Math. **161** (1999), 167–215.

[M1] W. Marzantowicz, *A formula for the G-Euler characteristic*, Bull. Acad. Polon. Sci. Sér. Sci. Math. Astronom. Phys. **23** (1975), 565–570.

[M2] ———, *The Lefschetz number in equivariant K-theory*, Bull. Acad. Polon. Sci. Sér. Sci. Math. Astronom. Phys. **25** (1977), 901–906.

[MP1] W. Marzantowicz and C. Prieto, *Generalized Lefschetz numbers for equivariant maps*, Osaka J. Math. **39** (2002), 821–841.

[MP2] ———, *A decomposition of equivariant stable homotopy classes and a computation of the first equivariant stem*, Preprint.

[MP3] ———, *The unstable equivariant fixed point index and the equivariant degree.*, J. London Math. Soc. (2003) (to appear).

[NW] B. Norton-Odenthal and P. Wong, *A relative generalized Lefschetz number*, Topology Appl. **56** (1994), 141–157.

[PU] C. Prieto and H. Ulrich, *Equivariant fixed point index and fixed point transfer in nonzero dimensions*, Trans. Amer. Math. Soc. **328** (1991), 731–745.

[Sch] H. Schirmer, *A relative Nielsen number*, Pacific J. Math. **122** (1986), 459–473.

[St] H. Steinlein, *Ein Satz über den Leray-Schauderschen Abbildungsgrad*, Math. Z. **126** (1972), 176–208.

[TD] T. tom Dieck, *Transformation Groups*, Walter de Gruyter & Co., Berlin, 1987.

[Ul] H. Ulrich, *Fixed point theory of parametrized equivariant maps*, Lecture Notes in Math., vol. 1343, Springer–Verlag, Berlin, 1988.

[Vi] A. Vidal, *On equivariant deformation of maps*, Publ. Mat. **32** (1988), 115–121.

[Wi] D. Wilczyński, *Fixed point free equivariant homotopy classes*, Fund. Math. **123** (1984), 47–60.

[Wo1] P. Wong, *Equivariant Nielsen fixed point theory for G-maps*, Pacific J. Math. **150** (1991), 179–200.

[Wo2] ———, *Equivariant path fields on G-complexes*, Rocky Mountain J. Math. **22** (1992), 1139–1145.

[Wo3] ———, *Equivariant Nielsen fixed point theory and periodic points*, Nielsen Theory and Dynamical Systems (South Hadley, MA, 1992), Contemp. Math. **152** (1993), 341–350.

[Wo4] ———, *Equivariant Nielsen numbers*, Pacific J. Math. **159** (1993), 153–175.

[Wo5] ———, *Equivariant Nielsen theory*, Nielsen Theory and Reidemeister Torsion (Warsaw, 1996), Banach Center Publ. **49** (1999), Polish Acad. Sci., Warsaw, 253–258.

[Wo6] ———, *Nielsen fixed point theory for partially ordered sets*, Topology Appl. **110** (2001), 185–209.

[ZK] P. P. Zabreĭko and M. A. Krasnosel'skiĭ, *Iterations of operators, and fixed points*, Dokl. Akad. Nauk SSSR **196** (1971), 1006–1009.

[Zh] X. Zhao, *On minimal fixed point numbers of relative maps*, Topology Appl. **112** (2001), 41–70.

9. EQUIVARIANT DEGREE

JORGE IZE

1. Introduction

Topological degree has been one of the main tools in the study of nonlinear problems. It provides useful information on the existence and sometimes multiplicity of solutions, as well as on topological properties of the set of zeros of parametrized problems, such as connectivity, global behavior and dimension.

One of the advantages of the method is that the topological invariants associated to a given problem may be computed by deforming complicated equations to much simpler ones. Thus, the idea of deformation or of homotopies is central to the use of topologival methods.

On the other hand, properties of symmetries have been used in many different ways in order to simplify the study of certain problems, from separation of variables to cancellation of certain terms in series expansions. For equations which commute with the action of a Lie group, that is for equivariant maps, one may look for solutions which are invariant under some subgroup or, for parametrized problems, when one has a symmetry breaking.

There has been a rather large number of papers on the computation of the ordinary degree of equivariant maps, of the Borsuk–Ulam type. Equivariant variational problems have been studied for more than 20 years. On the side of algebraic topology, a good deal of recent research is concerned with equivariant situations, in general in a more abstract context than the one encountered in most applications to differential equations, in particular the linear character of the underlined spaces is ignored.

However, the purpose of this paper is to present a relatively new degree theory, where the allowed maps and deformations are equivariant on sets which are invariant, in a setting which is more appropriate for applications.

Research partially supported by CONACyT, grant G25427-E and KBN-CONACyT agreement.

In the next section, I shall introduce two complementary definitions of the degree and give its basic properties. Section 3 will give topological results and Section 4 is concerned with some applications. Clearly, the main reference is the book [IV], together with results from more recent papers and from the books [KB], [KW], [B], [Fi] and [BD]. Familiarity with some concepts of representation theory, as given in [Br], is assumed.

2. Equivariant degree

A definition of an equivariant degree through a generic construction, as in the case of the classical Brouwer degree, meets several serious difficulties: a "good" definition of genericity, a density result similar to Sard's lemma, a consistent definition of the invariants and of their sum. The construction below avoids most of these difficulties and may also be used in the non-equivariant case.

The setting of this section is the following: Let Γ be a compact Lie group and let B and E be two finite dimensional representations of Γ, with the linear action of Γ denoted by γ on B and by $\widetilde{\gamma}$ on E. It is known that one may renormalize B and E so that the group Γ acts via isometries.

(2.1) Definition. A map $f\colon B \to E$ is said to be *equivariant* if $f(\gamma x) = \widetilde{\gamma} f(x)$, for all γ in Γ and x in the domain of f. If the action of Γ on E is trivial ($\widetilde{\gamma} = \mathrm{id}$) then f is said to be *invariant.*

The purpose of the equivariant degree is to define a topological invariant for equivariant maps and homotopies defined on invariant sets. More precisely, let Ω be a bounded, open, Γ-invariant subset of B and consider a continuous equivariant map $f(x)$, from $\overline{\Omega}$ into E, which is non-zero on $\partial\Omega$. The construction of the equivariant degree is given in the following steps:

(2.2.1) Since Ω is bounded, let B_R be a closed ball of radius R and centered at the origin, containing Ω. Since the action on B is an isometry, B_R is Γ-invariant.

(2.2.2) Let a Dugundji–Gleason Γ-extension $\widetilde{f}(x)$, from B_R into E, of $f(x)$.

(2.2.3) Let then N be a Γ-invariant neighbourhood of $\partial\Omega$, such that N is open, contained in B_R and $\widetilde{f}(x) \neq 0$ on $\overline{N}$.

(2.2.4) Let $\varphi(x)$, from B_R into [0,1], be a Γ-invariant Uryson function with value 0 in $\overline{\Omega}$ and 1 outside $\Omega \cup N$.

(2.2.5) Let $F(t,x)\colon [0,1] \times B_R \to \mathbb{R} \times E$ be the map defined by

$$F(t,x) = (2t + 2\varphi(x) - 1, \widetilde{f}(x)).$$

It is clear that F is Γ-equivariant, where the action on t in $[0,1]$ and on the first component of $\mathbb{R}\times E$ is trivial. Furthermore, $F(t,x) \neq 0$ if x is in $\overline{N}$ (since $\widetilde{f}(x) \neq 0$

there) and if x is outside $\Omega \cup N$ (there $\varphi(x) = 1$ and the first component of $\overline{F}$ reduces to $2t + 1 \geq 1$). Hence, if $F(t, x) = 0$, then x is in Ω, $\widetilde{f}(x) = f(x) = 0$, $\varphi(x) = 0$ and $t = 1/2$. In particular,

$$F(t, x) \colon S^B \equiv \partial([0, 1] \times B_R) \to \mathbb{R} \times E \setminus \{0\},$$

defines an element, $[F]_\Gamma$, of $\Pi^\Gamma_{SB}(S^E)$, where S^E is defined as $\mathbb{R} \times E \setminus \{0\}$.

Furthermore, it is easy to see that if $f(x)$ is deformable to $g(x)$, via an equivariant map which is non-zero on $\partial\Omega$, then $F(t, x)$ will be Γ-homotopic to the map $G(t, x)$ corresponding to $g(x)$. Thus, one obtains an element $[F]_\Gamma$, of the abelian group $\Pi^\Gamma_{SB}(S^E)$, of all Γ-homotopy classes of maps from S^B into S^E.

(2.3) Definition. The *equivariant degree* of f with respect to Ω, is defined as $[F]_\Gamma$ in $\Pi^\Gamma_{SB}(S^E)$, and denoted by $\deg_\Gamma(f; \Omega)$.

(2.4) Remark. It is clear that up to here we have not used the finite dimensionality of B and E. Thus, one may define the Γ-degree either in general or for maps which are compact perturbations of the identity (or k-set-contractions).

It is easy to see that this definition is independent of the elements of the construction and that, if $B = E$ and Γ is trivial, then one recovers the Brouwer degree, since then

$$\deg_{\{e\}}(f; \Omega) = [F] = \deg_B(F; [0, 1] \times B_R).$$

This last degree, by excision, is equal to

$$\deg_B((2t - 1, f(x)); I \times \Omega) = \deg_B(2t - 1; I) \deg_B(f(x); \Omega) = \deg_B(f; \Omega)$$

recalling that $\varphi(x) = 0$ on Ω and using the product formula for the Brouwer degree.

Furthermore, the Γ-degree has the following properties:

(2.4.1) (Existence) If $\deg_\Gamma(f; \Omega)$ is non-trivial, then there exists x in Ω such that $f(x) = 0$.

(2.4.2) (Γ-homotopy invariance) Let $f_\tau \colon \overline{\Omega} \to E$, $0 \leq \tau \leq 1$, be a continuous one-parameter family of Γ-equivariant maps not vanishing on $\partial\Omega$ for all τ in I. Then, the Γ-degree $\deg_\Gamma(f_\tau; \Omega)$ does not depend on τ.

(2.4.3) (Excision) Let $f \colon \overline{\Omega} \to E$ be a continuous Γ-equivariant map such that $f(x) \neq 0$ in $\overline{\Omega} \setminus \Omega_0$, where $\Omega_0 \subset \Omega$ is open and Γ-invariant. Then

$$\deg_\Gamma(f; \Omega) = \deg_\Gamma(f|_{\overline{\Omega}_0}; \Omega_0).$$

(2.4.4) (Suspension) If there is a Γ-extension $\widetilde{f}$ to B_R of f, such that $\widetilde{f}(x) \neq 0$ on $\overline{B}_R \setminus \Omega$ (in particular, if $\Omega = B_R$), then

$$\deg_\Gamma(f; \Omega) = \deg_\Gamma(\widetilde{f}; B_R) = \Sigma_0[\widetilde{f}]_\Gamma,$$

where Σ_0 is the suspension (one-dimensional) homomorphism, by $2t-1$.

(2.4.5) (Hopf property)property!Hopf If Ω is a ball and Σ_0 is one-to-one, then $\deg_\Gamma(f;\Omega) = \deg_\Gamma(g;\Gamma)$ if and only if $f|_{\partial\Omega}$ is Γ-homotopic to $g|_{\partial\Omega}$.

(2.4.6) (Additivity up to one suspension) If $\Omega = \Omega_1 \cup \Omega_2$, Ω_i open with $\overline{\Omega}_1 \cap \overline{\Omega}_2 = \phi$, then

$$\Sigma_0 \deg_\Gamma(f;\Omega) = \Sigma_0 \deg_\Gamma(f;\Omega_1) + \Sigma_0 \deg_\Gamma(f;\Omega_2),$$

where Σ_0 is again the suspension by $2t-1$.

If Ω_2 is a ball (hence, by the invariance, centered at the origin), then the addition formula is true without a suspension. On the other hand, there are examples where the suspension is necessary.

(2.4.7) (Universality) If $\Delta(f;\Omega)$ is any other Γ-degree with the properties (2.4.1)–(2.4.3) and Σ_0 is one to one, then if $\Delta(f;\Omega)$ is non-trivial, this is also the case for $\deg_\Gamma(f;\Omega)$.

Given the simplicity of this construction, it is natural to look for extensions to infinite dimensions or to other classes of maps. The main points to check are the existence of extensions of maps and the group structure of the set $\Pi^\Gamma_{SB}(S^E)$.

If B and E are infinite dimensional it is easy to prove that the above construction is valid, but it is most likely that the Γ-homotopy group of spheres is trivial. Hence, as for the Leray-Schauder degree, some compactness of the maps will be required. More precisely, one will restrict B and E to be of the form

$$B = U \times W, \quad E = V \times W$$

where U and V are finite dimensional Γ-representations and W is an infinite dimensional Γ-space. The maps and homotopies will be of the form:

$$f(x) = f(u,w) = (g(u,w), w - h(u,w))$$

where $g(u,w)$ is in V, h is compact and g and h are Γ-equivariant. Homotopies will affect only g and h.

Then one may show that such maps may be uniformly approximated by equivariant maps of the form

$$f_n(x) = f_n(u,w) = (g(u,w), w - h_n(u,w)),$$

where h_n is in a finite dimensional subrepresentation of W. This approximation allows the definition of the equivariant degree for infinite dimensional spaces provided that the suspension by subrepresentations of W is eventually one-to-one, as it is in the case of the Leray–Schauder degree.

Extensions to other classical situations (k-set contractions, Fredholm maps, maps on manifolds) should not present more than technical difficulties.

Motivated by variational problems, one may look at gradients of invariant functionals. In the simplest setting, let V be a finite dimensional representation of a compact abelian Lie group Γ. For parameters λ in $\mathbb{R}^k$, consider gradients of Γ-invariant functionals. That is, if $\Phi(\lambda, x)$ is Γ-invariant, then it is immediate to prove that $f(\lambda, x) = \nabla_x \Phi(\lambda, x)$ is Γ-equivariant. In this case, one could reduce the class of maps to gradients and define a degree in the following way: Assume that $f(\lambda, x)$ is non-zero on $\partial\Omega$ and let B_R be the ball containing Ω. Let $\widetilde{\Phi}(\lambda, x)$ be an invariant extension of Φ to B_R. By using mollifiers, one may assume that $\widetilde{\Phi}$ is C^1 in x and that $\nabla_x\widetilde{\Phi}(\lambda, x) \equiv \widetilde{f}(\lambda, x)$ is arbitrarily close to $f(\lambda, x)$.

As in the previous construction, consider the gradient, with respect to (t, x), of the functional

$$\widehat{\Phi}(t, \lambda, x) = \varepsilon(t^2 + t(2\varphi(\lambda, x) - 1)) + \widetilde{\Phi}(\lambda, x).$$

For $\varepsilon > 0$, small enough, such that

$$4\varepsilon\|\nabla_x\varphi(\lambda, x)\| \leq \|\widetilde{f}(\lambda, x)\|$$

for all (λ, x) in N, the neigborhood of $\partial\Omega$, the zeros of this gradient are such that $\widetilde{f}(\lambda, x) = 0$ and $t = 1/2$. It is clear that if one has a gradient Γ-homotopy on $\partial\Omega$, the corresponding gradients of $\widehat{\Phi}$ will be Γ-homotopic as maps from $\partial(I \times B_R)$ into $\mathbb{R} \times V \setminus \{0\}$.

(2.5) DEFINITION. Let $\Pi^{\Gamma}_{\nabla S^U}(S^V)$ be the set of Γ-homotopic gradients (with respect to t and x) from $S^U = \partial(I \times B_R)$ into $S^V \equiv \mathbb{R} \times V \setminus \{0\}$. Define the *gradient degree* of $\nabla_x\Phi(\lambda, x)$ with respect to Ω as $\deg^{\Gamma}_{\nabla}(\nabla_x\Phi(\lambda, x); \Omega) \equiv [\nabla_{(t,x)}\widehat{\Phi}(t, \lambda, x)]_{\nabla}$.

However, at this point, we don't know if $\Pi^{\Gamma}_{\nabla S^U}(S^V)$ is a group, since it is not clear that the Borsuk extension theorem holds for gradient maps. On the other hand, the fact that for the non-equivariant case and for a sphere, there is no gain in considering gradient maps and homotopies (see [Pa]), indicates that one should study a more general situation, that of orthogonal maps.

(2.6) DEFINITION. Let $\Gamma = T^n \times A$ be an abelian Lie group, with A a finite group and let $U = \mathbb{R}^k \times V$ be a finite dimensional representation of Γ (with trivial action on $\mathbb{R}^k$), Ω be an open Γ-invariant subset of U, then a map $f(\lambda, x)$, from $\overline{\Omega}$ into V, will be said to be Γ-*orthogonal* if:

$$\begin{aligned} f(\lambda, \gamma x) &= \gamma f(\lambda, x), & \\ f(\lambda, x) \cdot A_j x &= 0, & j = 1, \dots, n, \\ f(\lambda, x) &\neq 0 & \text{if } (\lambda, x) \in \partial\Omega, \end{aligned}$$

for A_j the infinitesimal generators of the action of the torus part, T^n of Γ.

In the case of a gradient, $f(x) = \nabla\phi(x)$, with $\phi(\gamma x) = \phi(x)$, in particular for $\gamma(t)$ a one-parameter subgroup of Γ, and by differentiating with respect to t, one sees that such a gradient is orthogonal.

For an orthogonal map, it is possible to prove that one may extend $f(\lambda, x)$ to a Γ-orthogonal map $\widetilde{f}(\lambda, x)$, for (λ, x) in B_R, a large ball, centered at the origin and containing Ω.

It is then clear that one may repeat the construction for the Γ-degree: take an invariant neighbourhood N of $\partial\Omega$ where $\widetilde{f}(\lambda, x)$ is non-zero, construct a Γ-invariant Uryson function and define

$$F(t, \lambda, x) = (2t + 2\varphi(\lambda, x) - 1, \widetilde{f}(\lambda, x))$$

which will be a Γ-orthogonal map on $I \times B_R$ and non-zero on its boundary, thus, defining an element of the abelian group $\Pi^{\Gamma}_{\perp S^U}(S^V)$.

(2.7) DEFINITION. Define the *orthogonal degree* $\deg^{\Gamma}_{\perp}(f; \Omega)$ of f as $[F(t, \lambda, x)]_{\Gamma}$ in $\Pi^{\Gamma}_{\perp S^U}(S^V)$.

It is easy to see that this orthogonal degree is independent of the construction, since all the deformations can be chosen to be Γ-orthogonal.

(2.8) THEOREM. *The orthogonal degree has all the properties* (2.4.1)–(2.4.7), *i.e. existence, homotopy invariance (for Γ-orthogonal deformations), excision, suspension, the Hopf property, additivity and universality.*

We leave to the reader the task of extending this degree to infinite dimensions for Γ-orthogonal and compact perturbations of the identity. The examples we have looked at so far can be studied by a global reduction to finite dimensions, avoiding in this way some of the technicalities necessary for the infinite dimensional setting.

For gradient maps one may consider the orthogonal degree of $\nabla_x\Phi(\lambda, x)$, which is an easier object to study. Of course, one could also forget the orthogonality and consider only $\deg_{\Gamma}(\nabla_x\Phi(\lambda, x); \Omega)$, obtaining the following maps:

$$\Pi^{\Gamma}_{\nabla S^U}(S^V) \xrightarrow{\perp} \Pi^{\Gamma}_{\perp S^U}(S^V) \xrightarrow{\Pi} \Pi^{\Gamma}_{S^U}(S^V),$$

where $\perp$ consists in forgetting the gradient character but retaining the orthogonality and Π corresponds to maintaining only the equivariance. It is clear that Π is a morphism of abelian groups, and one may show that Π is onto if $k = 0$. On the other hand, one may conjecture, if $k = 0$, that $\perp$ is one to one and onto, extending the non-equivariant result of [Pa].

(2.9) REMARK. Associated to any degree theory there are a certain number of standard applications and operations, such as continuation and bifurcation on one hand and products and composition on the other hand. In the case of problems with symmetries one may also have a change of group, or symmetry breaking, and considerations with suspensions. All these properties will be considered in the next sections.

(2.10) REMARK. The literature on the "classical" degree theory is very extensive. For a reader with interest in analysis, the most accessible texts are the books by Nirenberg, Berger and Krasnosel'skiĭ–Zabreĭko. For a survey of the russian literature, the reader may consult the paper [Z], by Zabreĭko.

On the equivariant side, the situation is scarcer. There are some indices coming from Algebraic Topology, where very often the action has to be free.

For the case of autonomous differential equation, Fuller has introduced in [F], a degree which is a rational. The relation between the Fuller degree and ours has been shown in [IMV3]. Dancer, in [Da], has defined a degree for S^1-gradient maps, which is also a rational, and can be shown to follow from the S^1-degree with a "Lagrange multiplier", see [I1] and [IMV3].

The equivariant degree was first defined, for a general Lie group, in [IMV3]. Gęba and al. have defined a S^1-degree in [DGJM] and then, independently, a degree for a general Lie group in [GKW] and for $B = \mathbb{R}^k \times E$, which corresponds to the "free part" of the degree. Their definition, using the "normal map" approach will be related to the one presented here in the next section. Finally, Rybicki has also defined a degree for S^1-orthogonal maps in [R] and Gęba for Γ-gradient maps, for a general Γ, in [G]. The case of an abelian group was studied in [IV3].

Proofs for the above results may be found in [IV].

3. Classification of equivariant maps

Since the equivariant degree in an element of $\Pi^{\Gamma}_{S^B}(S^E)$, the group of all equivariant homotopy classes of Γ-maps from S^B into S^E, the next step is to compute this group in the largest possible number of cases (as well as the corresponding orthogonal degree), to know its generators and to understand the effect of operations. Two complementary approaches have been used successfully for these computations: one based on extension of maps, what is known as obstruction theory, and the second on approximations by normal maps.

The obstruction approach is based on the study of the topological invariants which prevent an equivariant map from S^B into S^E to be extended to an equivariant map from $I \times B_R$ into S^E. The abstract arguments from Algebraic Topology may be found in [Ko] or [Br]. Concrete computations for S^1-maps were first done in [I1], [IMV2] and [IMV3], while the case of a general abelian group was treated in [IV2].

Using the fact that if a subgroup H of Γ fixes the subspace B^H, the fixed point subspace under H, of B, then the equivariance of any map F implies that F maps B^H into E^H. The step by step extension process begins with $H = \Gamma$ where, in this case, F^Γ sends S^{B^Γ} into S^{E^Γ} and extends to a non-zero map from $I \times B_R^\Gamma$ into S^{E^Γ} if and only if the class of F^Γ in the ordinary homotopy group $\Pi_{S^{B^\Gamma}}(S^{E^\Gamma})$ is trivial. For a given H, one may study the case of a map which sends S^{B^H} into S^{E^H} and $I \times B_R^K$ into S^{E^K} for any isotropy groupsubgroup!isotropy K containing H (hence B^K is a strict subspace of B^H). This extension problem is further simplified by reducing it to a *fundamental cell.* This notion was defined independently by Z. Balanov for a general group, with an inductive construction, and by the author for an abelian group and an explicit construction. In fact, for an abelian group, let $x_1, \ldots, x_n$ be the coordinates in B^H, corresponding to irreducible representations of Γ, with the action of the abelian group Γ on x_j with isotropy H_j (hence Γ/H_j is either trivial or isomorphic to $\mathbb{Z}_2$, with x_j real, or isomorphic to $\mathbb{Z}_m$ or S^1, with x_j complex) then $H = \bigcap H_j$. Let $k = \dim \Gamma/H$, define $\widetilde{H}_j = H_1 \cap \ldots \cap H_j$ and let k_j be the order of $\widetilde{H}_{j-1}/\widetilde{H}_j$. There are exactly k complex coordinates for which $k_j = \infty$, i.e. where $\widetilde{H}_{j-1}/\widetilde{H}_j$ acts as S^1, while on the other coordinates it acts as a cyclic group with k_j elements.

(3.1) DEFINITION. The *fundamental cell* for H is the set

$$\mathcal{C}_H = \{x_j,\ j = 1, \ldots, n : |x_j| \le R, 0 \le \operatorname{Arg} x_j < 2\pi/k_j\}$$

Hence, if $k_j = 1$, there are no limitations on x_j, while on a real x_j with $k_j = 2$, one has $x_j \ge 0$ (there are no other possibilities for the real coordinates) and, if $k_j = \infty$, then x_j is real and positive. Hence the dimension of $\mathcal{C}_H$ is $\dim I \times B^H - \dim W(H)$, where $W(H) = N(H)/H = \Gamma/H$ is the Weyl group of H and $N(H)$ is the normalizer of H in Γ (in the case of an abelian group it is is the whole group). In particular, the dimension of $W(H)$ is the number of x_j with $k_j = \infty$. The fundamental cell has the property that its images under the elements of Γ cover properly $I \times B_R^H$. This implies that any equivariant map is determined completely by its restriction to $\mathcal{C}_H$. Furthermore, if an equivariant map is non-zero on $\partial\mathcal{C}_H$, then it will have a non-zero equivariant extension to $I \times B_R^H$ if and only if its restriction to $\partial\mathcal{C}_H$ has a continuous extension to $\mathcal{C}_H$, i.e. if it is trivial in $\Pi_s(S^{E^H})$, where $s = \dim B^H - \dim W(H)$. Note that the map on the part of $\partial\mathcal{C}_H$ corresponding to some argument equal to $2\pi/k_j$ is given by the map on $\mathcal{C}_H \cap \partial\mathcal{C}_H$.

If an equivariant map F is non-zero on S^{B^H} and on $I \times B_R^K$, for any isotropy subgroup K containing H, then the extension problem may be handled by looking for an extension to edges and faces of $\mathcal{C}_H$. This will be possible provided the dimension of a face is less that $\dim E^H$, hence, if $\dim B^H - \dim W(H) = \dim \mathcal{C}_H <$

$\dim E^H$, one will get a continuous extension to $\mathcal{C}_H$ and, by the action of the group, to $I \times B_R^H$, while, if the two dimensions are equal, we will have as an obstruction the Brouwer degree of the map on $\partial\mathcal{C}_H$. In order to implement this program, one has to be careful with the extensions to the faces of the fundamental cell which have to be done in a consistent sequence, respecting the group action on the boundary of that cell, see [IV, Theorem 1.1, p. 87]. If $\dim \mathcal{C}_H = \dim E^H + 1$, one will get as obstructions some integers for the extensions to the faces of the fundamental cell and another obstruction for the extension to the cell itself (these obstructions are not independent and generate, as we shall state below, a finite group which is isomorphic to $2\Gamma/H$). With these arguments, one can prove that if for all isotropy subgroups H for the action of Γ on B, one has $\dim B^H < \dim E^H + \dim \Gamma/H$, then $\Pi^{\Gamma}_{SB}(S^E) = 0$.

(3.2) Definition. Let $z_1, \ldots, z_k$ be the complex coordinates with $k_j = \infty$, in the decomposition of the fundamental cell $\mathcal{C}_H$, ($z_1, \ldots, z_k$ are not necessarily consecutive). The ball $B_k = \{x \in I \times B^H, z_j$ real and non-negative, $j = 1, \ldots, k\} \cap B_R$ will be called the *global Poincaré section*. Note that B_k has dimension $\dim I \times B^H - \dim \Gamma/H$. Let $H_0 = H_1 \cap \ldots \cap H_k$, with H_j the isotropy of the coordinate $z_j, j = 1, \ldots, k$, then H_0, which leaves B_k globally invariant, will be called the isotropy of the Poincaré section B_k.

H_0 acts as a finite group on B_k and $|H_0/H| = \Pi k_j$, for those x_j with $k_j < \infty$. The fundamental cell for this action of H_0 on B_k is $\mathcal{C}_H$. Furthermore, Γ-equivariant maps on B^H correspond to H_0-equivariant map on B_k. In particular, if the following condition hold:

$(\widetilde{\mathrm{H}})$ for all γ in Γ one has $\det \gamma \cdot \det \widetilde{\gamma} > 0$,

then, if $\dim B^H - \dim W(H) = \dim E^H$ and F_0 is any H_0-equivariant extension to B_k of F, a Γ-equivariant map which is non-zero in S^{B^H} and in B^K, for all K containing H, then the obstruction given by the degree of the extension to $\partial\mathcal{C}_H$ is denoted by $\deg_E(F)$ and is such that:

$$\deg(F_0; B_k) = |H_0/H| \deg_E(F, \widetilde{F}).$$

See [IV, Theorem 1.2, p. 94].

This is the first step in order to prove that the degree of the extension is independent of the previous extensions. In order to continue in this process, the following notion is quite important:

(3.3) Definition. If K contains H, denoted by $K > H$, then one has a *complementing map* in B^H if there is a non-zero equivariant map $F_\perp$ from $B^H \cap (B^K)^\perp \setminus \{0\}$ into $E^H \cap (E^K)^\perp \setminus \{0\}$, with $F_\perp(0) = 0$.

One has then the following result:

(3.4) THEOREM. *Let* $\Pi(H)$ *be the set of all* Γ*-homotopy classes of maps from* S^{B^H} *into* S^{E^H} *which have non-zero* Γ*-extensions to* $I \times B_R^K$*, for all* K *containing* H*. Assume the following hypothesis:*

$$\text{(K)} \qquad \text{Any minimal } K > H \text{ has a complementing map in } B^H$$

where minimal means that adding a variable to B^K*, the isotropy of the new space is* H*. Then* $\Pi(H)$ *is independent of previous extensions and if furthermore, hypothesis* $(\widetilde{\mathrm{H}})$ *holds and* $\dim B^H = \dim E^H + \dim \Gamma/H$*, then the extension degree is independent of* $\widetilde{F}$*, extension of* F *to* $\bigcup B^K$*, and* $\Pi(H) \cong \mathbb{Z}$*.*

See [IV, Theorem 1.4, p. 101].

The same sort of arguments may be used to extend the previous result to the set $\Pi(k)$ of all Γ-homotopy classes of maps

$$F: \bigcup S^{B^H} \to E \setminus \{0\},$$

for isotropy subgroups H with $\dim \Gamma/H = k$, which have Γ-extensions

$$\widetilde{F}: \bigcup I \times B_R^K \to E \setminus \{0\},$$

for all K with $\dim \Gamma/K < k$ and containing H.

(3.5) THEOREM. *Assume that every* H*, with* $\dim \Gamma/H = k$*, satisfies*

$$(\widetilde{\mathrm{K}}) \qquad \begin{cases} \text{(a) Any minimal } K > H \text{ has a complementing map in } V^H, \\ \text{(b) } H \text{ has a complementing map } F_\perp \text{ in } V. \end{cases}$$

Then

$$\Pi(k) \cong \bigoplus_{\dim \Gamma/H = k} \Pi(H).$$

If furthermore, hypothesis $(\widetilde{\mathrm{H}})$ *holds and* $\dim B^H \leq \dim E^H + \dim \Gamma/H$ *for all* H *with* $\dim \Gamma/H = k$*, then* $\Pi(k) \cong \mathbb{Z} \times \ldots \times \mathbb{Z}$*, where there is one* $\mathbb{Z}$ *for each* H_j *such that* $\dim \Gamma/H_j = k$ *and* $\dim B^{H_j} = \dim E^{H_j} + k$*.*

See [IV, Theorem 2.1, p. 103].

Finally, one has the complete picture:

(3.6) THEOREM. *If hypothesis* $(\widetilde{\mathrm{K}})$ *holds for all isotropy subgroups, then*

$$\Pi_{S^B}^\Gamma(S^E) \cong \bigoplus_H \widetilde{\Pi}(H),$$

where $\widetilde{\Pi}(H)$ *stands for the suspension by the corresponding complementing map. Recall that* $\Pi(H) = 0$ *if* $\dim B^H - \dim W(H) < \dim E^H$. *In particular, let* $\widetilde{\Gamma} = \Gamma/T^n$, $\widetilde{B} = B^{T^n}$, $\widetilde{E} = E^{T^n}$, *then*

(3.6.1) *If, for all* H, *one has* $\dim V^H \leq \dim W^H$, *then*

$$\Pi^{\Gamma}_{SB}(S^E) \cong \Pi^{\widetilde{\Gamma}}_{S\widetilde{B}}(S^{\widetilde{E}}) \cong \mathbb{Z} \times \ldots \times \mathbb{Z},$$

with one $\mathbb{Z}$ *for each* H *with* $\dim \Gamma/H = 0$ *and* $\dim B^H = \dim E^H$. *One has* $[F] = \sum d_H[\widetilde{F}_H]$, *where* d_H *is the extension degree and* $\widetilde{F}_H$ *is the generator suspended by its complementing map.*

(3.6.2) *If, for all* H, *one has* $\dim B^H \leq \dim E^H + 1$, *then*

$$\Pi^{\Gamma}_{SB}(S^E) \cong \Pi^{\widetilde{\Gamma}}_{S\widetilde{B}}(S^{\widetilde{E}}) \times \mathbb{Z} \times \ldots \times \mathbb{Z},$$

with one $\mathbb{Z}$ *for each* H *with* $\dim \Gamma/H = 1$ *and* $\dim B^H = \dim E^H + 1$. *One has* $[F] = [\widetilde{F}] + \sum d_H[\widetilde{F}_H]$, *where* $\widetilde{F}$ *is the suspension of* F^{T^n}.

These results are proved by using the group structure of the different sets of Γ-homotopy classes and by working up on dimensions.

A complementary way of computing $\Pi^{\Gamma}_{SB}(S^E)$ is with the introduction of the notion normal maps: assume that hypothesis ($\widetilde{\mathrm{K}}$) holds for all H, isotropy subgroups for B, that is, decomposing B as $B^H \oplus B_{\perp H}$, E as $E^H \oplus E_{\perp H}$, one has a complementing Γ-map $F^{\perp}_H$, from $B_{\perp H}$ into $E_{\perp H}$ with its only zero at 0. We shall assume that these complementing maps are compatible, i.e. that $F^{\perp}_H | B_{\perp K} = F^{\perp}_K | B_{\perp H}$.

This will be the case for $B = \mathbb{R}^k \times E$, since $F^{\perp}_H$ is the identity, or when one has hypothesis (H) for U and E with $B = \mathbb{R}^k \times U$ (see below). Write

$$X = X_H \oplus X_{\perp H} \quad \text{and} \quad F = (F^H, F_{\perp H}).$$

(3.7) DEFINITION. An equivariant map $\widetilde{F}$ in $\Pi^{\Gamma}_{SB}(S^E)$ will be called a *normal map* if for all H, one has $\widetilde{F}_{\perp H}(t, X) = F^{\perp}_H(X_{\perp H})$, if $\|X_{\perp H}\| \leq \varepsilon$.

It is easy to show that any F in $\Pi^{\Gamma}_{SB}(S^E)$ is Γ-homotopic to a normal map $\widetilde{F}$.

A convenient hypothesis, under which one has complementary maps for all isotropy groups and such that hypothesis ($\widetilde{\mathrm{H}}$) holds, is the following: assume that $B = \mathbb{R}^k \times U$ and that U and E satisfy (H), i.e.

(H) for all isotropy subgroup H and K for U, one has

$$\dim U^H \cap U^K = \dim W^H \cap W^K,$$

Or equivalently:

(a) $\dim U^H = \dim E^H$,

(b) There is a Γ-equivariant map: $(x_1, \dots, x_s) \to (x_1^{l_1}, \dots, x_s^{l_s})$, from U into E, where l_j are integers and x^l, for negative l, means $\overline{x}^{|l|}$, ($l_j = 1$ on U^Γ and on the real representations of Γ).

If $U = E$, then $l_j = 1$ for all j.

(3.8) THEOREM. *If $B = \mathbb{R}^k \times U$, where U and E satisfy* (H), *then*

$$\Pi_{SB}^\Gamma(S^E) \cong \Pi_{k-1} \times \mathbb{Z} \times \dots \times \mathbb{Z},$$

with one $\mathbb{Z}$ for each H with $\dim \Gamma/H = k$ and Π_{k-1} corresponds to all H with $\dim W(H) < k$. For each H, with $\dim W(H) = k$ one has an explicit generator $[\widetilde{F}_H]_\Gamma$.

One may compute the components of $[F]_\Gamma$ in $\Pi_{SB}^\Gamma(S^E)$ in the following situation:

(3.9) THEOREM. *Assume $B = \mathbb{R}^k \times U$, where U and E satisfy* (H). *Let $z_1, \dots, z_k$ with isotropy H_0, with $\dim \Gamma/H_0 = k$ and global Poincaré section $B_k = \{(t, X)$ in $I \times B_R$, with $z_j \geq 0$, for $j = 1, \dots, k\}$. Then, if $F: I \times B_R \to \mathbb{R} \times E$, is a Γ-map which is non-zero on $\partial(I \times B_R) \cup \partial B_k$, one has*

$$[F]_\Gamma = \sum_{\underline{H} \leq H \leq H_0} d_H [\widetilde{F}_H]_\Gamma,$$

where $\underline{H}$ is the torus part of H_0 (that is the unique smallest isotropy subgroup with $\dim \Gamma/\underline{H} = k$, contained in H_0). Furthermore, the integers d_H are given by a triangular relation, coming from the equalities:

$$\deg(F^H; B_k^H) = \sum_{H \leq K \leq H_0} \beta_{HK} d_K |H_0/K|,$$

where, if $(x_1^{l_1}, \dots, x_s^{l_s})$ is the complementing map of B^K in B^H (i.e. an equivariant map from $(B^K)^\perp \cap B^H$ into $(E^K)^\perp \cap E^H$), then $\beta_{HK} = \Pi l_j$. In particular, $\beta_{HH} = 1$. In the case when $U = E$, one has that $\beta_{H,K} = 1$ if H is a subgroup of K and 0 otherwise.

See [IV, Theorem 3.4, p. 114].

Note that $B_0 = I \times B_R$ and then $[F]_\Gamma$ is completely determined by the Brouwer degrees of F^H on B^H, recovering well known results in [BD], for instance.

In many examples, $\deg(F^H; B_k^H)$ corresponds to the index of an isolated solution in B_k, that is an isolated orbit in B, and this index is computed via the sign of the determinant of the linearization of some transversal map, hence with values $i_H = \pm 1$, for all H. In this case, one has the result:

(3.10) Proposition. *Assuming* $B = \mathbb{R}^k \times E$ *and* $i_H = \deg(F^H; B_k^H) = \pm 1$ *for all* H, *with* $\underline{H} \leq H \leq H_0$, *for a map satisfying the conditions of Theorem* (3.5), *then*

$$\begin{aligned} d_H &= 0 \quad \text{if } B^H \text{ has a coordinate where } H_0 \text{ acts as } \mathbb{Z}_m,\ m \geq 3, \\ d_{H_0} &= i_{H_0}, \\ d_{H_j} &= (i_{H_j} - i_{H_0})/2 \quad \text{for all maximals } H_j \text{ with } H_0/H_j \cong \mathbb{Z}_2, \end{aligned}$$

d_H *and* i_H *are completely determined by* i_{H_j}, *the above* H_j, *for all* H *not included in the above list.*

This proposition implies that any change of the Γ-degree, at this stage, is detected by changes of d_{H_0} or on d_{H_j}, with $H_0/H_j \cong \mathbb{Z}_2$. This fact will lead to period doubling.

In the case of a non-abelian group and $B = \mathbb{R}^k \times E$, one may define the notion of normal map in the following way:

(3.11) Definition. For an isotropy subgroup H, let B_H be the set of points in B with isotropy H and let $B(H) = \Gamma B_H$ be the set of points in B with isotropy type H, i.e. with isotropy conjugate to H. An equivariant map F is said to be (H)-*normal on* Ω (an open, bounded, invariant subset of B) if there is a $\delta > 0$, such that for all x_0 in $\overline{\Omega}$, with $F(x_0) = 0$ and of type (H), one has that $F(x_0 + v) = v$ for all v orthogonal to $T_{x_0}B_{(H)}$, the tangent space to $B_{(H)}$ at x_0, with $\|v\| \leq \delta$. Such a map will be said to be *normal in* Ω if it is (H)-normal for all orbit types (H) in $\overline{\Omega}$ and it will be called *regular normal in* Ω if it is normal in Ω, of class C^1 and for any orbit type (H) in $\overline{\Omega}$, the map $F_{(H)}$, from $\overline{\Omega} \cap B_{(H)}$ into E^H has 0 as a regular value.

It is easy to see that if Γ is abelian then the two definitions of normality are Γ-homotopic. In the non-abelian case, one may prove:

(3.12) Proposition ([KX]). *If* F, *from* $\mathbb{R}^k \times E$ *into* E, *is an equivariant map, which is non-zero on* $\partial\Omega$, *then, for any small enough* $\varepsilon > 0$, *there is a regular normal map* $\widetilde{F}$, *which is non-zero on* $\partial\Omega$ *and* ε-*close to* F *on* $\overline{\Omega}$.

If the suspension Σ_0 is an isomorphism (true if $\dim E^\Gamma > k+2$), for each isotropy type (H) define $\Pi(H)$ as the subgroup of $\Pi^\Gamma_{SB}(S^E)$ given by the Γ-homotopy classes of regular normal maps in $I \times B_R$ whose zeros are all of type (H). One has then the result:

(3.13) Theorem ([BK]). *Under the above hipothesis, one has*

$$\Pi^\Gamma_{SB}(S^E) \cong \bigoplus_{\dim W(H) \leq k} \Pi(H),$$

If $\dim W(H) = k$, *then* $\Pi(H) \cong \mathbb{Z}$ *if* $W(H)$ *is bi-orientable (i.e. it has an orientation which is invariant under all left and right tranlations) and* $\mathbb{Z}_2$ *if not.*

If $\widetilde{F}$ belongs to $\Pi(H)$, with $\dim W(H) = k$, and 0 is a regular value of $\widetilde{F}_H$ then, for each orbit of zeros of $\widetilde{F}_H$ one has a well defined local index for the Poincaré section: this will define the class of $\widetilde{F}$ in $\Pi(H)$, see [KX]. Furthermore, if $k = 0$ or 1 ($B = E$ or $B = \mathbb{R} \times E$), then [KVW] presents formulae for the components in $\Pi(H)$, with $\dim W(H) = k$, in terms of the fixed point index on $B_{(H)}$ and on $\bigcup B_{(K)}$ for all K containing H, using results by Ulrich.

The above ideas on extensions of equivariant maps and complementing maps may be used in order to compute the ordinary degree of a map from S^B into S^E, if they have the same dimension, of Borsuk–Ulam type. For instance:

(3.14) THEOREM. *If* $\dim B = \dim E$ *and assume that the abelian group* Γ *acts on* B^{T^n} *and on* E^{T^n} *in such a way that hypothesis* (H) *holds on* B^{T^n}, *(in particular these two spaces have the same dimension), then, if* m *is the greatest common divisor of* $\{(\Pi l_j)|\Gamma/H|$, *for* $T^n \le H < \Gamma$ *and* x_j *in* $(B^H)^\perp \cap B^{T^n}\}$, *one has, for any equivariant map* F *from* S^B *into* S^E*:*

$$\begin{aligned}\deg(F; I \times B) &= \beta \deg(F^{T^n}; I \times B^{T^n}),\\ \deg(F^{T^n}; I \times B^{T^n}) &= (\Pi l_j) \deg(F^\Gamma; I \times B^\Gamma) + dm,\end{aligned}$$

where B *stands for the ball* B_R *and the integer* β *is independent of* F *and is given explicitely in terms of the action of* Γ *on* $(B^{T^n})^\perp$ *and* $(E^{T^n})^\perp$ *and any integer* d *is achieved. In particular, if* $B = E$, *then* $\beta = 1$, $l_j = 1$ *and* m *is the GCD of all* $|\Gamma/H|$. *The term* $\deg(F^\Gamma; I \times B^\Gamma)$ *is replaced by* 1 *if* $B^\Gamma = \{0\}$.

For instance, if $\Gamma \cong \mathbb{Z}_m$ acts freely on B and $E^\Gamma = \{0\}$, with $\dim B = \dim E$, then any equivariant map $\partial B_R \to E \setminus \{0\}$ has a degree equal to $\Pi l_j + dm$, where any d is achieved.

More general situations are considered in [IV] for abelian actions, in [KB] for a non-abelian group and [B, Chapter 3] explores the case when the degree may be 0 and $\dim B > \dim E$.

Let now $B = \mathbb{R} \times U$ and assume that U and E satisfy hypothesis (H), i.e. $\dim U^H = \dim E^H$, for all H, isotropy subgroups for U, and there is a Γ-equivariant map $\{x_j^{l_j}\}$ from U into E. From Theorem 3.6, one has

$$\Pi_{S^B}^\Gamma(S^E) = \Pi_{S^{\widetilde{B}}}^{\widetilde{\Gamma}}(S^{\widetilde{E}}) \times \mathbb{Z} \times \ldots \times \mathbb{Z},$$

with one $\mathbb{Z}$ for each isotropy subgroup H with $\dim \Gamma/H = 1$, and $\widetilde{\Gamma} = \Gamma/T^n$, $\widetilde{B} = B^{T^n}$, $\widetilde{E} = E^{T^n}$. The group $A = \Pi_{S^{\widetilde{B}}}^{\widetilde{\Gamma}}(S^{\widetilde{E}})$ is finite and is the product of subgroups $\Pi(H)$ corresponding to H with finite Weyl group. More precisely:

(3.15) THEOREM. *If $B = \mathbb{R} \times E$, then, if $|\Gamma/H|$ is finite, one has that $\Pi(H) \cong Z_2 \times \Gamma/H$, with explicit presentations in terms of the generators of Γ/H. In particular, if $\Gamma/H \cong \mathbb{Z}_n$, then*

$$\Pi(H) \cong \begin{cases} \mathbb{Z}_2 \times \mathbb{Z}_n & \text{if } n \text{ is even,} \\ \mathbb{Z}_{2n} & \text{if } n \text{ is odd.} \end{cases}$$

See [IV, Chapter 3, Theorems 5.5–5.7].

For non-abelian groups, [BKS] give an exact sequence for $\Pi(H)$, which is

$$0 \to \mathbb{Z}_2 \to \Pi(H) \to W(H)/[W(H), W(H)] \to 0$$

if $(\mathbb{R}^n \times B)_H$ is simply connected, with $n > 4$ and $[A, A]$ is the commutator of the group A.

In the case $B = \mathbb{R}^k \times E$, for $k > 1$, the computation of $\Pi(H)$, for $\dim W(H) < k$ is completely open.

In the case of orthogonal maps from V into V, with an abelian group, one has the following result:

(3.16) THEOREM. $\Pi^{\Gamma}_{\perp S^V}(S^V) \cong \mathbb{Z} \times \ldots \times \mathbb{Z}$, *with one $\mathbb{Z}$ for each isotropy subgroup of Γ and explicit generators.*

This theorem says that the topological structure for orthogonal maps is much richer than the one for plain equivariant ones. The idea of the proof is based on the following considerations: one looks at the zeros of the parametrized maps

$$F(x) + \sum \lambda_j A_j x,$$

where A_j are the infinitesimal generators of the action of the group Γ. Since $A_j x$ is orthogonal to $F(x)$, a zero of the above map implies that $F(x)$ is 0 as well as $\sum \lambda_j A_j x$. If x is such that the $A_j x$ in the sum are linearly independent, then $\lambda_j = 0$. Hence, on each isotropy subspace, the problem is equivalent to a parametrized problem, where one may use Theorem 3.8.

The case of a non-abelian group is still open and the case of gradients, with the approximation by normal gradient maps, was treated in [G].

One of the advantages of the definition of the degree as an element of an equivariant homotopy group is that the usual homotopy operations are easily defined. In particular, it is important for infinite dimensional problems to know when the suspension by the identity doesn't affect the homotopy class, that is when it is one-to-one. Besides the abstract result of [N], a useful special case is the following:

If Γ is abelian and $B = \mathbb{R}^k \times U$, where U and E satisfy hypothesis (H), and furthermore if, for all K, H isotropy subgroups for U, with $H < K$, one has

$$\dim E^\Gamma \geq k+2, \quad \dim E^H - \dim E^K \geq k + 2 - \dim \Gamma/H,$$

then any suspension will be one to one. These conditions are achieved if one has repetition of coordinates.

Another important operation is that of symmetry breaking, i.e. when one views a Γ-equivariant map as a Γ_0-equivariant map, where Γ_0 is a subgroup of Γ. This is particularly relevant when one pertubs a Γ-map with a Γ_0-map.

If Γ is abelian and $B = \mathbb{R}^k \times E$, it is easy to guess what will happen to a generator of $\Pi(H)$, for H with $\dim \Gamma/H = k = \dim \Gamma_0/H \cap \Gamma_0$: see [IV, Corollary 7.3, p. 174]. However, the case when $\dim \Gamma_0/H \cap \Gamma_0 < k$ is much more intriguing and happens when one forces an autonomous differential equation with a time-periodic perturbation. For instance, if $\Gamma = S^1$ and $\Gamma_0 \cong \mathbb{Z}_p$, $B = \mathbb{R} \times E$, then one has the result:

(3.17) THEOREM. *Under the above hypothesis,*

$$\textit{if} \quad [F]_{S^1} = \sum_{m\geq 0} d_m[\eta_m]_{S^1}, \quad \textit{then} \quad [F]_{\mathbb{Z}_p} = \sum_{p'|p} d'_{p/p'}[\eta'_{p/p'}]_{\mathbb{Z}_p},$$

where

$$d'_{p/p'} = \sum_j n_j \Big(\sum_{k\geq 0} d_{m_j p/p' + kp} \Big)$$

with $|n_j|$ *odd,* $n_j m_j \equiv 1$*, modulo* p'*, and* $1 \leq m_j < p'$*, with* m_j *and* p' *relatively prime.* p' *divides* p *and the number* $d'_{p/p'}$ *is in* $\mathbb{Z}_{p'}$ *if* p' *is even and in* $\mathbb{Z}_{2p'}$ *if* p' *is odd. The number* d'_p *is in* $\mathbb{Z}_2$*, corresponds to* $H_0 \equiv \mathbb{Z}_p$ *and is*

$$d'_{\Gamma_0} = d'_p = \sum_{k\geq 0} d_{kp}.$$

For instance, if $p = 2$, then one has $d'_{\Gamma_0} = \sum d_{2k}$, mod 2, $d'_{\{e\}} = \sum d_{2k+1}$, mod 2.

For $p = 3$, one has $d'_{\Gamma_0} = \sum d_{3k}$, mod 2, $d'_{\{e\}} = \sum (d_{3k+1} - d_{3k+2})$, mod 6.

For $p = 4$, one has $d'_{\Gamma_0} = \sum d_{4k}$, mod 2, $d'_{\mathbb{Z}_2} = \sum d_{4k+2}$, mod 2, $d'_{\{e\}} = \sum (d_{4k+1} - d_{4k+3})$, mod 4.

For $p = 5$ one has $d'_{\Gamma_0} = \sum d_{5k}$, mod 2, $d'_{\{e\}} = \sum (d_{5k+1} - d_{5k+4}) + 3\sum (d_{5k+2} - d_{5k+3})$, mod 10.

More classical operations concern products and compositions of maps. In the case of products of spaces one has first to identify the isotropy subgroups for the product together with the dimension of the Weyl groups. If Γ is abelian, if (H)

holds for $B_i = \mathbb{R}^{k_i} \times U_i$ and E_i, for $i = 1, 2$, and assuming that $\dim \Gamma/H_i = k_i$ and $\dim \Gamma/H = k_1 + k_2$, then one has explicit formulae relating the generators of $\Pi(H_i)$ and that of $\Pi(H_1 \cap H_2)$. See [IV, Proposition 7.5, p. 182].

If $B_i = E_i$ (hence $k_i = 0$) one may use the algebraic structure of the Burnside ring in the case of a non-abelian group in order to identify the isotropy subgroups of the product and the coefficients of the multiplication formula. See [KW], Section 8.2 and [KVW]. The general case for non-abelian actions is still in process.

The situation for composition is well understood for abelian groups, with explicit relations in top dimensions of the Weyl groups ([IV, Proposition 7.8, p. 191]) but the non-abelian case seems to be untouched. For all these operations, in the case of orthogonal maps one has complete results for abelian groups.

(3.18) REMARK. Besides the several places in this section where open questions were mentionned, the reader may guess easily many more interesting generalizations from the topological viewpoint.

4. Applications

In the case of the classical degree, one of the generic situations is when one has an isolated zero which is a regular point, i.e. when the local index is just the sign of the jacobian of the linearization. For an equivariant map, the linearization of a map F at x_0 has the property that

$$DF(\gamma x_0)\gamma = \widetilde{\gamma} DF(x_0),$$

for all γ in Γ. In particular, DF is H-equivariant, where H is the isotropy subgroup of x_0. Furthermore, if $B = B^H \oplus B_\perp$, $E = E^H \oplus E_\perp$, where $B_\perp$ and $E_\perp$ are $N(H)$-topological complements and the map F may be written as $F = F^H \oplus F_\perp$, then, at any x_H in B_H, the operator $DF(x_H)$ has the diagonal structure

$$DF(x_H) = \operatorname{diag}(D_H F^H, D_\perp F_\perp).$$

See [IV, Property 3.4, p. 11].

Finally, using Schur's lemma and the decomposition of B and E in irreducible representations, one has the decomposition of B into real irreducible representations, typically repeated n_i-times and of real dimension m_i, complex irreducible representations, repeated n_j-times and of complex dimension m_j, and quaternionic irreducible representations, repeated n_l-times and of quaternionic dimension m_l. Any Γ-equivariant matrix has then a block-diagonal form.

More precisely, if B is given by

$$B = \bigoplus_{i=1}^{i=I} (V_i^{\mathbb{R}})^{n_i} \bigoplus_{j=1}^{j=J} (V_j^{\mathbb{C}})^{n_j} \bigoplus_{l=1}^{l=L} (V_l^{\mathbb{H}})^{n_l},$$

then, there are bases of B such that any equivariant matrix, A, has a block diagonal form

$$A = \operatorname{diag}(A_i^{\mathbb{R}}, A_j^{\mathbb{C}}, A_l^{\mathbb{H}}),$$

where $A_i^{\mathbb{R}}$ are real $n_i \times n_i$ matrices repeated m_i times, $A_j^{\mathbb{C}}$ are complex $n_j \times n_j$ matrices, repeated m_j times and $A_l^{\mathbb{H}}$ are $n_l \times n_l$ quaternionic matrices repeated m_l times. See [IV, Appendix A].

If the matrix is invertible, one may deform each piece (as real, complex or quaternionic matrices) by repeating the deformation on the repeated parts and obtain in this way a classification of the homotopy classes of equivariant matrices.

Since the set $GL(n.K)$, of $n \times n$-invertible matrices over the field K, is connected if K is $\mathbb{C}$ or $\mathbb{H}$ (and simply connected in the case of quaternions), if an equivariant matrix A is invertible, then it may be deformed to the suspension of matrices of the form $\operatorname{diag}(\varepsilon_i, \varepsilon_i, \ldots)$, where $\varepsilon_i = \operatorname{sign}\det A_i$, repeated m_i-times and A_i is the matrix on real equivalent representations. This will enable the computation of the Γ-index of Ax at the point $x = 0$.

Note that in the case of an abelian action, there is no quaternionic part. From Proposition 3.9, it is then easy to see that the Γ-index is given by

$$i_\Gamma(Ax; 0) = \varepsilon_0 \Big([F_0]_\Gamma + \sum (\varepsilon_j - 1)/2 [F_j]_\Gamma + \sum d_H [F_H]_\Gamma \Big)$$

where $\varepsilon_0 = \operatorname{sign}\det A^\Gamma, \varepsilon_0\varepsilon_j = \operatorname{sign}\det A^{H_j}$, where $\Gamma/H_j \cong \mathbb{Z}_2$, and d_H are completely determined by ε_0 and $\{\varepsilon_j\}$, for H which are intersections of more than one of the H_j. See [IV, Proposition 2.1, p. 212].

More generally, for the abelian case, if $B = \mathbb{R}^k \times E$ and let 0 be a regular value of a C^1-equivariant map f on Ω with an isolated orbit $(\lambda_0, \Gamma X_0)$ with isotropy H such that $\dim \Gamma/H = k$. Then E is a H-representation, $\operatorname{Ker} Df(\lambda_0, X_0)$ is k-dimensional and is generated by k vectors among $A_1X_0, \ldots, A_nX_0$, with $A_jX = \partial(\gamma X)/\partial\varphi_j|_{\gamma=Id}$, the infinitesimal generators of the action of T^n, the torus part of Γ. Furthermore, $Df(\lambda_0, X_0)|_{B_k}$ is invertible, where B_k is the global Poincaré section, and the Poincaré index i_K of f at (λ_0, X_0) on $B^K \cap B_k$ is

$$i_K = \operatorname{sign}\det Df(\lambda_0, X_0)^K|_{B_k} = i_H \operatorname{sign}\det Df^\perp(\lambda_0, X_0)^K.$$

Then, the Γ-index of the orbit is given by $(d_H, d_{K_1}, \ldots)$ such that $d_H = i_H$, $d_K = (i_K - i_H)/2$, if $H/K \cong \mathbb{Z}_2, d_K$ is completely determined by the above integers if $H/K \cong \mathbb{Z}_2 \times \ldots \times \mathbb{Z}_2$ with more than one factor, and $d_K = 0$ otherwise (see [IV, Theorem 2.3, p. 217]).

As an abstract application, assume that $f(\mu, \lambda, X)$ is a family, parametrized by μ, of Γ-equivariant functions from $\mathbb{R}^k \times E$ into E, with 0 as a regular value for

$\mu \neq \mu_0$. Assume there is a known curve of zeros of $f(\mu, \lambda, X), \lambda_0(\mu), X_0(\mu)$ with common isotropy H, with $\dim \Gamma/H = k$. Then $i_H(\mu)$ and $i_K(\mu)$ are well defined for $\mu \neq \mu_0$ and $K < H$. Then:

(4.1.1) If $i_H(\mu)$ changes sign at μ_0, then one has a global bifurcation at $(\mu_0, \lambda_0(\mu_0), X_0(\mu_0))$ in B^H.

(4.1.2) If $i_H(\mu)$ remains constant but $i_K(\mu)$ changes sign at μ_0 for some K with $H/K \cong \mathbb{Z}_2$, then there is global bifurcation in B^K, i.e. with a period doubling. Topologically all bifurcations are in maximal isotropy subgroups, i.e. with $H/K \cong Z_2$.

Note that for a normal map one has that $i_K = i_H$ and the only component of the Γ-index is d_H.

In order to relate the Γ-index to the more classical Poincaré index for ODE's, one may define "Floquet multipliers" and "hyperbolic orbits" in a general abstract setting. In fact, let W be an infinite dimensional Banach Γ-space and set $V = \mathbb{R}^k \times W$. Suppose that $f(\lambda, X) = X - F(\lambda, X)$, from V into W, is C^1 and $F(\lambda, X)$ is a compact map with $f(\lambda_0, X_0) = 0$ for X_0 with isotropy H such that $\dim \Gamma/H = k$.

(4.2) Definition. (λ_0, X_0) is said to be *hyperbolic* if and only if

(4.2.1) $\dim \operatorname{Ker}(I - F_X(\lambda_0, X_0)) = k$,

(4.2.2) $F_\lambda(\lambda_0, X_0)\colon \mathbb{R}^k \to W$ is one to one,

(4.2.3) $\operatorname{Range} F_\lambda(\lambda_0, X_0) \cap \operatorname{Range}(I - F_X(\lambda_0, X_0)) = \{0\}$.

For instance, in the case of autonomous differential equations, one may choose $W = H^1(S^1)$ and $F(X, \nu) = g(X)/\nu$ corresponding to 2π-periodic solutions to the system

$$\nu \frac{dX}{dt} - g(X) = 0, \quad X \text{ in } \mathbb{R}^N.$$

Let $(\nu_0, X_0(t))$ be a $(2\pi/p)$-periodic solution of $\nu X' - g(X) = 0$. Then, if W is the space $H^1(S^1)$, (ν_0, X_0) is hyperbolic if and only if 0 is a simple eigenvalue of the operator $\nu_0 d/dt - g_X(X_0)$ in W, that is 1 is a simple Floquet multiplier of $\Phi(2\pi)$, where $\Phi(t)$ is the fundamental matrix of the linear system.

Let (λ_0, X_0) be hyperbolic, then one has

(4.3.1) i_H is essentially given by $(-1)^{n_H}$, where n_H is the number of eigenvalues of F_X^H, counted with algebraic multiplicity, which are larger than or equal to 1.

(4.3.2) $i_K = (-1)^{n'_K} i_H$, where n'_K in the number of eigenvalues of $F_X^{\perp K}$, counted with algebraic multiplicity, which are larger than 1, for any isotropy subgroup K contained in H.

Let us return to the system $\nu X' - g(X) = 0$, X in $\mathbb{R}^N$, with a hyperbolic solution (ν_0, X_0), i.e. if $A(t) = g_X(X_0(t))/\nu_0$ and $\Phi(t)$ is the fundamental matrix of the linearization

$$LX = X' - A(t)X$$

then 1 is a simple eigenvalue of $\Phi(2\pi) = \Phi(2\pi/p)^p$, with X_0' as only solution of $LX = 0$, where $2\pi/p$ is the least period of $X_0(t)$.

(4.4) THEOREM. *If 1 is a simple eigenvalue of $\Phi(2\pi)$, let σ_+ be the number of real eigenvalues, counted with algebraic multiplicity, of $\Phi(2\pi/p)$ which are larger than 1 and let σ_- be the number of real eigenvalues of $\Phi(2\pi/p)$ which are less than -1, then, on $W^K = \{X(t)$ in $H^1(S^1)$, which are $2\pi/p'$-periodic, p' dividing $p\}$,*

$$\begin{aligned} i_K &= -(-1)^{\sigma_+} && \text{if } p/p' \text{ is odd,} \\ i_K &= -(-1)^{\sigma_- + \sigma_+} && \text{if } p/p' \text{ is even.} \end{aligned}$$

In particular, the S^1-index has at most two non-zero components: $d_H = i_H$ and, if p is even, $d_K = (i_K - i_H)/2$, for $|H/K| = 2$ or $p' = p/2$, corresponding to period doubling.

See [IV, Proposition 2.6, p. 229].

Similar results may be given if the dependence on ν is more general, [IV, Proposition 2.7, p. 233], or if one has a first integral, that is

$$X' = g(X), \text{ for } X \text{ in } \mathbb{R}^N, \text{ has a first integral } V(X),$$

if $V(X(t))$ remains constant on solutions of the equation, or equivalently that

$$\nabla V(X) \cdot X' = \nabla V(X) \cdot g(X) = 0.$$

Consider the problem of finding 2π-periodic solutions to the equation

$$X' = g(X) + \nu \nabla V(X) = g(X, \nu).$$

If $X_0(t)$ is such a solution, then

$$X_0' \cdot \nabla V(X_0) = \nu \|\nabla V(X_0)\|^2 = \frac{d}{dt} V(X_0(t)).$$

Integrating over a period, one has $\nu \|\nabla V(X_0(t))\| \equiv 0$, thus $\nu = 0$ if, on the orbit $\nabla V(X_0) \not\equiv 0$, or $\nabla V(X_0) \equiv 0$ on the orbit and, in both cases, $X_0(t)$ is a 2π-periodic solution of the original problem.

The components of the S^1-index of $(X_0(t), \nu = 0)$, in case $\nabla V(X_0) \not\equiv 0$, are given in [IV, Proposition 2.8, p. 236] and one may compare to [IMV3], [DT1] and [DT2].

(4.5) REMARK. If one has a time dependent equation

$$\frac{dX}{dt} = f(X, t),$$

where $f(X,t)$ is $2\pi/p_0$-periodic in t, then, one may look for 2π-periodic solutions. If $X_0(t)$ is a $2\pi/p$-periodic solution of the equation, with p dividing p_0, then $X_0(t)$ is hyperbolic if and only if the linearization has no 2π-periodic solutions. Then, i_K has exactly the same form as given in Theorem 4.1, but now the equivariant degree is for $\Gamma = \mathbb{Z}_{p_0}$ and there is no parameter ν.

There is an intermediate case, when an autonomous system is forced by a time-periodic perturbation. That is, one may look at the problem of finding 2π-periodic solutions to the equation

$$\frac{dX}{dt} = g(X, \nu) + \tau h(t, X, \nu),$$

for small τ and where h is $2\pi/p_0$-periodic in t. Hence, the S^1-symmetry is broken to a $\mathbb{Z}_{p_0}$-symmetry, for $\tau \neq 0$. This is an entrainment or phase locking problem and solutions of the perturbed problem are called p_0-subharmonics.

Assume that (ν_0, X_0) is a hyperbolic $2\pi/q$-periodic solution of the autonomous equation $X' = g(X, \nu)$, Let p be the largest common divisor of q and p_0, then the isotropy subgroup of X_0 is $\mathbb{Z}_p$, with $W^{\mathbb{Z}_p}$ corresponding to $2\pi/p$-periodic functions.

Then the non-autonomous equation will have a global continuum of $(2\pi/p)$-periodic solutions (ν, X) going through (ν_0, X_0) and parametrized by τ, where p is the largest common divisor of q and p_0, provided, in the case $p_0 = p$ and q/p_0 even, one has that the sum of the algebraic multiplicities of real eigenvalues of $\Phi(2\pi/q)$ which are less than -1 is even.

In this case, the equivariant degree is in $\Pi^{\mathbb{Z}_{p_0}}_{S^{\mathbb{R}} \times W}(S^W) = \bigoplus \Pi(H)$, where $\Pi(H) \cong \mathbb{Z}_2 \times \Gamma/H$ for each isotropy subgroup of $\mathbb{Z}_{p_0}$, hence if $H \cong \mathbb{Z}_{p_0/p'}$, where p' divides p_0, then $\Gamma/H \cong \mathbb{Z}_{p'}$ and $\Pi(H) \cong \mathbb{Z}_2 \times \mathbb{Z}_{p'}$, if p' is even and $\mathbb{Z}_{2p'}$, if p' is odd. The components of the degree are then given in Theorem 3.15. See [IV, Proposition 2.10, p. 238].

For a non-hyperbolic orbit, one may destroy the solution of the autonomous system with a time-periodic perturbation. For instance, consider the pair of averaged Van der Pol's equations,

$$x'' - x'\left(1 - \frac{1}{2\pi}\int_0^{2\pi} (x^2 + y^2)\, dt\right) + (1+\nu)x = f(x, y),$$

$$y'' - y'\left(1 - \frac{1}{2\pi}\int_0^{2\pi} (x^2 + y^2)\, dt\right) + (1+\nu)y = g(x, y).$$

Thus, using Fourier series, it is easy to prove that, if ν is close to 0, the only non-trivial solutions will be $\nu = 0$, $x(t) = \alpha\cos(t+\varphi), y(t) = \beta\cos(t+\psi)$, with $\alpha^2+\beta^2 = 2$. One has a torus of non-trivial solutions.

The S^1-degree of the non-trivial solution is easily seen to be equal to 2. Hence, any small *autonomous* perturbation of the system will have solutions near the torus. On the other hand, one proves that the $\mathbb{Z}_2$-perturbation

$$\begin{aligned} f(x,y) + \tau(3\cos 2t\, y + \sin 2ty') &= 0, \\ g(x,y) - \tau(3\cos 2t\, x + \sin 2tx') &= 0. \end{aligned}$$

is such that, for τ small and non-zero, the only solution is $x = y = 0$.

More generally, if there is an abelian group Γ_0 such that the map $g(X)$ is Γ_0-equivariant, then one may look for 2π-periodic solutions to the problem

$$\frac{dX}{dt} = g(X,\nu), \quad X \text{ in } \mathbb{R}^N,$$

where ν could be the frequency. Then, one has an equivariant problem for the group $\Gamma = \Gamma_0 \times S^1$. If $(\nu_0, X_0(t))$ is a solution, which is truly periodic in time (i.e. it is not a time-stationary solution, or a rotating wave, which correspond to a Hopf bifurcation and will be considered later), then $X_0(t)$ is non-constant and $(2\pi/p)$-periodic, with isotropy $H = \mathbb{Z}_p \times H_0$, where $\dim \Gamma_0/H_0 = 0$. Assume that $X_0(t)$ is hyperbolic, i.e. $X_0'(t)$ generates the kernel of the linearization $X' - A(t)X$, with $A(t) = Dg(X_0(t),\nu_0)$ and the equation $X' - A(t)X = g_\nu(X_0(t),\nu_0)$ has no 2π-periodic solution. Then, as seen above, the only relevant isotropy subgroups of $\Gamma = S^1 \times \Gamma_0$ are H and K's, with $K < H$ and $H/K \cong \mathbb{Z}_2$. If T^k is the torus part of H, all the important information will be given by orbits which lie in $V_0 \equiv (\mathbb{R}^N)^{T^k}$ for all time. Then V^H is the space of all $(2\pi/p)$-periodic functions with $X(t)$ in $V_0^{H_0}$ for all t and $X(t) = \gamma_0 X(t + 2\pi/q)$, where $\gamma_0^{q_0}$ is in H_0 and $q = pq_0$. The element γ_0 of Γ_0 and the integer q are determined by $X_0(t)$.

Furthermore, for each K_j, with $H/K_j \cong \mathbb{Z}_2$, one has a subgroup K_{0j} of H_0 such that $H_0/K_{0j} \cong \mathbb{Z}_2$ or $H_0 = K_0$, with $V_j = V_0^{K_{0j}} = V_j^+ \oplus V_j^-$ where $\gamma_0^{q_0}$ acts as $\pm Id$ on $V_j^\pm$. The elements of V^{K_j} are those 2π-periodic functions $X(t)$, with $X(t)$ in V_j for all t, and $X(t) = \gamma_0^2 X(t+4\pi/q)$. This behavior, with a combination of geometric and time periodicity, is called a twisted orbit.

Now, the matrix $A(t)$ is H_0-equivariant for each t. Thus, on V_j one has

$$A(t) = \begin{pmatrix} A_0(t) & 0 \\ 0 & A_j(t) \end{pmatrix},$$

where A_0 corresponds to $V_0^{H_0}$ and $A_j(t)$ to V_j^- or to the complement of $V_0^{H_0}$ in V_j^+ (if $K_{0j} = H_0$, the matrix A_j is not present). If $\Phi(t)$ is the fundamental matrix for the problem $X' - A(t)X$, one has $\Phi(t) = \operatorname{diag}(\Phi_0(t), \ldots, \Phi_j(t), \ldots)$ on the decomposition of $\mathbb{R}^N$ on irreducible representations of H_0.

(4.6) PROPOSITION. *Let $(\nu_0, X_0(t))$ be a hyperbolic solution of $X' = g(X, \nu)$ and let k is the algebraic multiplicity of 1 as eigenvalue of $\Phi_0(2\pi/p)$. Let $\sigma_j^{\pm}$ be the number of real eigenvalues, counted with algebraic multiplicity, of $\gamma_0\Phi_j(2\pi/q)$ which are larger than 1, for σ_j^+, or less than -1, for σ_j^-, where $j = 0$ for H and $j \geq 1$ for each K_j with $H/K_j \cong \mathbb{Z}_2$. Then*

$$i_H = (-1)^k(-1)^{\sigma_0^+},$$

$$i_H i_{K_j} = \begin{cases} (-1)^{\sigma_j^+} & \text{if } q \text{ is odd and } V_j^- = \{0\}, \\ (-1)^{\sigma_j^-} & \text{if } q \text{ is odd and } V_j^- \neq \{0\}, \\ (-1)^{\sigma_0^-} & \text{if } q \text{ is even, } p \text{ is odd and } V_j^- \neq \{0\}, \\ (-1)^{\sigma_0^- + \sigma_j^+ + \sigma_j^-} & \text{if } q \text{ is even, } p \text{ is odd and } V_j^- = \{0\} \text{ or } p \text{ is even.} \end{cases}$$

See [IV, Proposition 2.11, p. 242] and compare with [Fi, Definition 6.1].

For non-abelian actions, most applications are for stationary solutions and Hopf bifurcation, where the isotropy subspaces are easier to identify. These results will be reviewed at the end of this paper.

On the other hand, for an abelian action and orthogonal maps, one has the following result:

(4.7) THEOREM. *Let Γx_0 be an isolated orbit of dimension k and isotropy H. Assume that $f(x_0) = 0$, for the orthogonal map f and that* $\dim \operatorname{Ker} Df(x_0) = k$. *Then, the orthogonal index is well defined and is equal to the product*

$$i_\perp(f^{\underline{H}}(x_{\underline{H}}); x_0) i_\perp(Df_\perp(x_0)\overline{X}; 0),$$

where $\underline{H}$ is the torus part of H and $Df_\perp(x_0)\overline{X}$ is the linearization on $(V^{\underline{H}})^\perp$, which is complex self-adjoint and H-orthogonal. The first index has components $d_H = (-1)^{k(k+1/2}(-1)^{n_H}$, where n_H is the number of negative eigenvalues of $Df^H(x_0)$. The integer $d_{H_i} = d_H((-1)^{n_{H_i}} - 1)/2$, where n_{H_i} is the number of negative eigenvalues of $Df_\perp^{H_i}(x_0)$ and $d_{\widetilde{H}_i}$ is completely determined by d_H and d_{H_j}. Furthermore, the second index is essentially given by $n(K_i)$, where K_i are the irreducible representations of H in $(V^{\underline{H}})^\perp$, i.e. $H/K_i \cong S^1$ and $Df_\perp(x_0)$, which is block-diagonal on these representations, has a complex Morse number $n(K_i)$.

This implies that if one has a family $f(\lambda, x)$ of Γ-orthogonal C^1 maps, with $f(\lambda, 0) = 0$, $\lambda \in \mathbb{R}$, $x \in V$ and $B(\lambda) = Df(\lambda, 0)$ invertible for $\lambda \neq 0$ small, then, if $\det Df^\Gamma(\lambda, 0)$ changes sign, as λ goes through 0, one has a bifurcation of non-trivial solutions in V^Γ, while if $\det Df_\perp^{H_i}(\lambda, 0)$, for H_i such that $\Gamma/H_i \cong \mathbb{Z}_2$, changes sign then there is bifurcation in V^{H_i}, or if $n_j(\lambda)$, the complex Morse number of $Df_\perp^{K_j}(\lambda, 0)$ for K_j isotropy of a coordinate such that $\Gamma/K_j \cong S^1$, changes, then one has bifurcation in V^{K_j}.

The quite interesting fact is that, contrary to the usual generic case of the classical degree theory, where the index is the sign of a determinant, here one has a whole integer, the Morse number, giving part of the orthogonal index, a situation which one encounters when treating variational problems.

This result applies readily to periodic solutions of autonomous Hamiltonian systems, of first or second order.

For a first order system, one looks at

$$f(X) = JX' + \nabla H(X) = 0,$$

where $X = (Y, Z)$ is in $\mathbb{R}^{2N}$, J is the standard symplectic matrix and H is C^2. Assume that the abelian group Γ_0 acts symplectically on $\mathbb{R}^{2N}$, i.e. it commutes with J. Suppose that H is Γ_0-invariant and autonomous. Then, $\Gamma = S^1 \times \Gamma_0$ acts on spaces of 2π-periodic functions, and $f(X)$ is Γ-orthogonal. Here the infinitesimal generators for Γ will be $AX \equiv X'$ for the action of S^1 and A_jX, $j = 1, \ldots, n = \operatorname{Rank} \Gamma_0$.

For the second order Hamiltonian equation

$$E(X) \equiv -X'' + \nabla H(X) = 0,$$

for X in $\mathbb{R}^N$ and a C^2 potential H which is Γ_0-invariant, one has that $E(X)$ is Γ-orthogonal, with infinitesimal generators $AX \equiv X'$, A_jX, $j = 1, \ldots, n$.

For instance, for a k-dimensional orbit of stationary solutions, satisfying the hyperbolicity hypothesis of Theorem (4.7), with isotropy $H = S^1 \times H_0$, $B \equiv D^2H(X_0)$ has the form $\operatorname{diag}(B^H, B^{\mathbb{R}}_m, B^{\mathbb{C}}_l, B^{\mathbb{C}}_s)$, where, on each B_m the group H acts as $\mathbb{Z}_2$, on the complex B_l as $\mathbb{Z}_p$, $p \geq 3$, and on the complex B_s as S^1. Each of these matrices is self-adjoint. Furthermore, B_s is complex self-adjoint and H-orthogonal.

The orthogonal index is given by

(4.8.1) $d_H = \eta(-1)^{n_H}$, with n_H the Morse number of B^H and $\eta = (-1)^{k(k+1)/2}$,

(4.8.2) $d_{H_j} = d_H((-1)^{n_j} - 1)/2$, with $(-1)^{n_j} = \operatorname{sign} \det B^{\mathbb{R}}_j$,

(4.8.3) the Morse index of $inJ + \widetilde{B}$ where $\widetilde{B}$ is any of the matrices B^H, $B^{\mathbb{R}}_m$, $B^{\mathbb{C}}_l$ or $B^{\mathbb{C}}_s$, for the mode $n > 0$ and the decomposition of $\mathbb{C}^{2N}$ (induced by that of $\mathbb{R}^{2N}$) in irreducible representations of H. For the second order system, the matrices are $-n^2I + \widetilde{B}$.

(4.8.4) The Morse index of $B^{\mathbb{C}}_s$.

See [IV, Proposition 3.2, p. 258].

On the other hand, for the non-stationary case, assume that X_0 is a $(2\pi/p)$-periodic hyperbolic solution of any of the above Hamiltonian systems, with isotropy H and $\dim \Gamma/H = k$. One has that $H = \mathbb{Z}_p \times H_0$, the torus part of H is

$\underline{H} = \underline{H}_0$, $V^{\underline{H}} = \{X(t) \in V_0 = (\mathbb{R}^N)^{\underline{H}_0}\}$. Furthermore, one has

$$V^H = \{X(t) \in V_0^{H_0} : (2\pi/p)\text{-periodic, } X(t) = \gamma_0 X(t + 2\pi/q))\},$$

where $\gamma_0^{q_0}$ is in H_0 and $q = pq_0$.

Also, for each K_j, with $H/K_j \cong \mathbb{Z}_2$, one has a subgroup K_{0j} of H_0 such that $H_0/K_{0j} \cong \mathbb{Z}_2$ or $H_0 = K_{0j}$ with $V_j = V_0^{K_{0j}} = V_j^+ \oplus V_j^-$ where $\gamma_0^{q_0}$ acts as $\pm Id$ on $V_j^{\pm}$. Then,

$$V^{K_j} = \{X(t) \in V_j : 2\pi\text{-periodic, } X(t) = \gamma_0^2 X(t + 4\pi/q)\}.$$

Finally, for each set of equivalent irreducible representations $\widetilde{V}_l$ of H_0 in $V_0^{\perp}$, with complex coordinates $X^0, \dots, X^r$ and action of γ_0 on X^j as $e^{2\pi i \alpha_j}$, then for each $n_0 = 0, \dots, q-1$, there is a different set of equivalent irreducible representations of H, with isotropy K_{n_0}, in $(V^{\underline{H}})^{\perp}$, and

$$V^{K_{n_0}} = \{X(t) = (X^0(t), \dots, X^r(t)), R_{-2\pi(n_0/q+\alpha_0)}\gamma_0 X^j(t + 2\pi/q) = X^j(t)\},$$

when R_φ is a rotation of angle φ of the coordinates of X^j.

One has then the following result

(4.9) Proposition. *If $X_0(t)$ is an hyperbolic $(2\pi/p)$-truly periodic solution of the system $-X'' + \nabla H(X) = 0$, then the orthogonal index is given by*

(4.9.1) $d_H = \eta(-1)^{n_H}$, *where n_H is the Morse number of $-X'' + B_0(t)X$, with $B_0 = D^2 H(X_0)$ restricted on $V_0^{H_0}$ and X in V^H, i.e. $X(t)$ is in $V_0^{H_0}$, $(2\pi/p)$-periodic and $X(t) = \gamma_0 X(t + 2\pi/q)$, with $\gamma_0^{q_0}$ in H_0 and $q = pq_0$. Here $\eta = (-1)^{k(k+1)/2}$.*

(4.9.2) $d_{K_j} = d_H((-1)^{n_{K_j}} - 1)/2$, *where n_{K_j} is the Morse number of $-X'' + B_j(t)X$, with $B_j = D^2 H(X_0)$ restricted to $V_j \cap (V_0^{H_0})^{\perp}$ and X in V^{K_j}, i.e. $X(t)$ is in this subspace and $X(t) = \gamma_0^2 X(t + 4\pi/q)$.*

(4.9.3) $d_{K_{n_0}}$ *is the complex Morse index of $-X'' + \widetilde{B}_l(t)X$, for each $n_0 = 0, \dots, q-1$ and $X(t)$ in $V^{K_{n_0}}$, i.e. $X(t)$ is in $\widetilde{V}_l$ and $R_{-2\pi(n_0/q+\alpha_0)}\gamma_0 X^j(t + 2\pi/q) = X^j(t)$ for $j = 1, \dots, r$ the complex coordinates of $X(t)$ in $\widetilde{V}_l$, and $\widetilde{B}_l(t)$ is $D^2 H(X_0)$ restricted to $\widetilde{V}_l$.*

For the system $JX' + \nabla H(X) = 0$, the situation is slightly different and requires the use of a global Liapunov–Schmidt truncation. One proves then that the parities of n_H and n_{K_j} do not depend on the level of truncation (hence giving the same indices), while $d_{K_{n_0}}$ is incremented by multiples of the even number, $\dim \widetilde{V}_l$. This comes from the fact that the infinitesimal generators are not compact and, thus, the indices do not stabilize as for the Leray–Schaueder degree. This is a classical difficulty with non-definite hamiltonians (see, for instance [Sz]).

These computations have been applied in [I4] and [IV, Example 3.6, p. 266–286], to two different spring-pendulum settings, with $\Gamma_0 = SO(2)$ and where one bifurcates from a one-dimensional orbit to a two-dimensional one, with a change in $d_{K_{n_0}}$ in the space $V^{K_{n_0}}$ where one has a mixture of spatial and time periodicity in the new patterns. Simpler but also interesting examples are given in the papers by Rybicki.

A final source of classical applications is the case of a Hopf bifurcation or, more generally, the problem of finding non-trivial solutions to the equivariant problem $F(\lambda, X) = 0$, with $F(\lambda, 0) = 0$ $DF(\lambda, 0)$, invertible for λ close to 0 but different from 0 and $DF(0,0) = 0$, as it happens after a Liapunov–Schmidt reduction. One may then look at the Γ-degree of the pair $(\|X\|^2 - \varepsilon^2, F(\lambda, X))$ on the ball

$$\Omega = \{(\lambda, X) : \|\lambda\| \leq 2\rho,\ \|X\| \leq 2\varepsilon\}.$$

Due to the hypothesis, one may deform the pair to the map $(\rho^2 - \|\lambda\|^2, DF(\lambda, 0)X)$. If the parameter space is one-dimensional, one may easily prove that the Γ-degree of the pair is

$$i_\Gamma(DF(-\rho, 0)X) - i_\Gamma(DF(\rho, 0)X)$$

and thus, one has bifurcation if these indices are different. For more parameters, say k of them and for X in the Γ-space V, for instance for a Hopf bifurcation, one may look at the family of invertible Γ-matrices $DF(\lambda, 0)$, for $\|\lambda\| = \rho$ and consider the Whitehead map, of J^Γ-homomorphism, from the set of all Γ-homotopy classes from S^{k-1} into $GL^\Gamma(V)$ into the group $\Pi^\Gamma_{S^{\mathbb{R}^{k-1}\times V}}(S^V)$:

$$\begin{aligned} [S^{k-1} \to GL^\Gamma(V)]_\Gamma &\xrightarrow{J^\Gamma} \Pi^\Gamma_{S^{\mathbb{R}^{k-1}\times V}}(S^V), \\ [A(\lambda)]_\Gamma &\longrightarrow (\|X\| - \varepsilon, A(\lambda)X). \end{aligned}$$

For instance, in the classical Hopf bifurcation, consider the problem of finding 2π-periodic solutions to the autonomous system

$$(\nu_0 + \nu)\frac{dX}{dt} - L(\mu)X - f(X, \mu) = 0, \quad X \text{ in } \mathbb{R}^N,$$

where $f(X, \lambda) = 0(\|X\|^2)$. Thus, on Fourier series, one has

$$\text{in}(\nu_0 + \nu)X_n - L(\mu)X_n - f_n(X, \mu) = 0, \quad n \geq 0.$$

Clearly, the resonant modes are those for which $in\nu_0 I - L(0)$ are not invertible. While the linear parts will be invertible, for (μ, ν) small but not 0, if the eigenvalues of $L(\mu)$ corresponding to the resonant modes go off the imaginary axis, for μ not 0. The classical Hopf case is when there is only one simple resonant mode and such that the eigenvalue crosses the imaginary axis with non-zero speed. In that case, one uses the S^1-equivariance in order to reduce the bifuraction equation to just two components, which are solved, via the implicit function theorem, for the parameters in terms of the norm of the eigenvector.

In the general case, the homomorphism J^Γ will be a group morphism (and sending $[AB]$ into $J^\Gamma[A] + J^\Gamma[B]$), if the product of matrices gives the sum in $\Pi_{k-1}(GL^\Gamma(V))$. This happens if one restricts the set to $GL_+^\Gamma(V)$, consisting of matrices with positive determinants on the real representations (on the complex and quaternionic representations, this is always true). Furthermore, one may obtain a positive determinant by changing the sign of one of the coordinates or of one of the equations, in the real representations. The effect on the Γ-degree may be important, but this operation doesn't affect the kernel of J^Γ, since this operation is an involution. The effect on the degree is given, for an abelian group and $k = 2$, in [IV, Theorem 4.1, p. 294].

The study of the kernel of J^Γ is particularly important since, if $k \leq \dim V^\Gamma$ and Σ_0 is one-to-one, then one may construct an equivariant non-linearity $g(\lambda, X)$ such that $A(\lambda)X + g(\lambda, X)$ has no other solution than $X = 0$, for small λ, if and only if $J^\Gamma[A(\lambda)] = 0$: see [I3, Proposition 6.3, p.475]. In order to study $\Pi_{k-1}(GL_+^\Gamma(V))$ one may use the block-diagonal structure of $A(\lambda)$, [IV, Appendix A], and the Bott periodicity for real, complex and quaternionic matrices.

For a general group, there are few topological results and most papers are concerned with one or two irreducible representations with only one or two parameters: see [CH], [GS], [CL], [V] and so on. Topological results are given in [BKS1], [BKS2], [BKR] for the groups $SO(3)$ and $O(2) \times S^1$. For abelian groups, one may give an explicit characterization in terms of the elements of an exact sequence [I3, Theorem 6.1], and also for $k = 1$ or 2.

In this last case, abelian group and $k = 2$, one has the complete result:

(4.10) THEOREM. *Assume that*

$$\dim V^\Gamma \geq 3, \quad \dim V_j \geq 3, \quad j = 1, \dots, r, \quad \dim_{\mathbb{C}} V_l \geq 2,$$

where V_j are equivalent real irreducible representations and Γ acts on V_l as $\mathbb{Z}_p$. Then, for the family of matrices $A(\lambda) = \operatorname{diag}(A^\Gamma, A_j, A_l, A_k)$, for $\|\lambda\| = \rho$, one has that $J^\Gamma[A]$ is given by:

(4.10.1) *d_0, the class of (A^Γ) in $\Pi_1(GL_+(V^\Gamma)) \cong \mathbb{Z}_2$,*

(4.10.2) *d_j, the class of (A_j) in $\Pi_1(GL_+(V_j)) \cong \mathbb{Z}_2$,*

(4.10.3) *$\sum n_s d_s$, corresponding to a fixed p and subspaces Z_{ls}, with action as $\exp(2\pi i m_s/p)$, with m_s and p relatively prime. The number $|n_s|$ is an odd integer such that $n_s m_s \equiv 1$, mod p, and d_s is the winding number of $\det(A_{ls})$, as a mapping from the loop onto $\mathbb{C} \setminus \{0\}$,*

(4.10.4) *Finally, d_k, the winding number of* $\det(A_k)$ *where A_k corresponds to the space where Γ acts as* $\exp(\pm i m_k \varphi)$.

If these numbers are equal to 0 or congruent to 0, then one may construct a nonlinearity without bifurcation. See [IV, Theorem 4.1, p. 294].

In the case of the differential equation

$$g(\mu, \nu, X) = (\nu + \nu_0)\frac{dX}{dt} - L(\mu)X - f(X, \mu, \nu) = 0, \quad X \text{ in } \mathbb{R}^N,$$

for (μ, ν) close to $(0,0)$ and $f(X, \mu, \nu) = o(\|X\|)$ and with $L(0)$ invertible, we assume that $g(X, \mu, \nu)$ is Γ_0-equivariant. Then, the winding numbers of the different parts of the linearization are given in terms of the number of eigenvalues crossing the imaginary axis in one direction minus the number of those crossing in the other direction: see [IV, Propositions 4.1 and 4.2, p. 303], generalizing the well known criterium for the case where there is no Γ_0.

If one perturbs the above equation by $\varepsilon_0 h(X, \mu, \nu, t)$, where the non-autonomous term has $h(0, \mu, \nu, t) = 0$ and is $(2\pi/p)$-periodic in t, then for $\varepsilon_0 = 0$, one has an S^1-action, while for $\varepsilon_0 \neq 0$, the action is reduced to a $\mathbb{Z}_p$-action.

If d_n denotes the net crossing number of eigenvalues of $in(\nu + \nu_0)I - L(\mu)$, one has new components of J^Γ (now for $\Gamma = \mathbb{Z}_p$), given by

$$d_\Gamma \equiv \sum_{k=1}^{\infty} d_{kp} \qquad \text{modulo } 2,$$

$$d_H \equiv \sum_j n_j \sum_{k=1}^{\infty} d_{m_j p/p' + kp} \qquad \text{modulo } 2p' \text{ if } p' \text{ is odd}$$

$$\text{and modulo } p \text{ if } p \text{ is even.}$$

Here, $\Gamma/H \cong \mathbb{Z}_{p'}$, for any divisor p' of p, the sum is over all m_j, relatively prime to p', with $1 \leq m_j < p'$, and $|n_j|$ is odd such that $n_j m_j \equiv 1$, modulo p'. If d_Γ is odd, one obtains Hopf bifurcation of $(2\pi/p)$-periodic solutions, while if d_H is not congruent to 0, one has Hopf bifurcation of $(2\pi p'/p)$-periodic solutions.

If d_Γ and d_H are all congruent to 0, one may construct a perturbation which destroys the Hopf bifurcation. For instance, take p any integer larger than 1 and consider the following system for 2π-periodic functions:

$$\begin{aligned} x'' - \mu x' + \nu x + 2\varepsilon((p+1)y\cos pt + y'\sin pt) &= 0, \\ y'' - (p-1)\mu y' + (p-1)^2\nu y - 2\varepsilon(p-1)((2p-1)x\cos pt + x'\sin pt) &= 0. \end{aligned}$$

For $\varepsilon = 0$, μ close to 0 and ν close to 1, one has a vertical Hopf bifurcation for $(x, 0)$ with $n = 1$ and for $(0, y)$ with $n = p - 1$. The winding numbers are all 0, except $d_1 = d_{p-1} = 1$. Hence, any autonomous perturbation will undergo this Hopf bifurcation.

But, for $\varepsilon \neq 0$, but small, the system has no non-trivial periodic solution, for (μ, ν) close to $(0, 1)$, [IV, p. 306].

For $p = 1$, one takes out the factors $p - 1$, in the second equation, and one has $d_1 = 2$ but the same result holds.

One may also apply these arguments to problems with first integrals, by looking at the problem of finding 2π-periodic solutions to the problem

$$\frac{dX}{dt} - L(\mu)X - f(X, \mu) - \nu \nabla V(X, \mu) = 0,$$

for which one has a family of first integrals $V(X, \mu)$.

Here, $X = 0$ is assumed to be a solution with $\nabla V(0, \mu) = 0$. Let

$$\nabla V(X, \mu) = H(\mu)X + k(X, \mu), \quad \text{with } k(X, \mu) = o(\|X\|).$$

Assume that $L(0)$ has eigenvalues $\pm i m_1, \ldots, \pm i m_s$, with $0 < m_1 \leq \ldots \leq m_s$, counted with multiplicities. Let $\lambda_j(\mu) = \alpha_j(\mu) + i\beta_j(\mu)$ be the eigenvalues of $L(\mu)$, for μ close to 0, such that $\alpha_j(0) = 0$, $\beta_j(0) = \beta_j$. We shall impose the following hypothesis

$$(\mathrm{H}_j) \qquad \begin{cases} \text{(a) if } \lambda_j(\mu) = i m_j, \text{ for } \mu \text{ close to } 0, \text{ then } \mu = 0, \\ \text{(b) } \operatorname{Ker} H(0) \cap \operatorname{Ker}(i m_j I - L(0)) = \{0\}, \text{ for } j = 1, \ldots, s. \end{cases}$$

It is then possible to prove that hypothesis (H_j) is equivalent to have $i m_j I - L(\mu) - \nu H(\mu)$ invertible for $(\mu, \nu) \neq (0, 0)$, but close to (0,0).

Furthermore, hypothesis (H_j) implies that, if $i\beta_j = \lambda_j(\mu)$ is an eigenvalue of $L(\mu)$, with generalized eigenspace $\operatorname{Ker}(L(\mu) - i\beta_j I)^k$, then $H(\mu)$ is complex self-adjoint and induces a non-degenerate quadratic form on this eigenspace, with a well defined signature.

We shall denote, for $\mu \neq 0$, by $\sigma_j^{\pm}(\mu)$ the sum of the signatures of H on $\bigcup \operatorname{Ker}(L(\mu) - i\beta_j(\mu))^k$, for $\beta_j(\mu) > m_j$ and close to m_j (for $\sigma_j^{+}(\mu)$) and for $\beta_j(\mu) < m_j$ and close to m_j (for $\sigma_j^{-}(\mu)$).

Assume (H_j) hold for $j = 0, \ldots, s$, with $m_0 = 0$. Then, it is possible to show that the bifurcation S^1-index is given by

$$d_j = \operatorname{sign} \det L(\rho_0)(\sigma_j^{+}(-\rho_0) - \sigma_j^{+}(\rho_0)).$$

Hence, a change of signature implies bifurcation. This is the case if $L(\mu) = (\mu + \lambda_0)L$, with $\lambda_0 > 0$ See [IV, Proposition 4.5, p. 315], [DT1], [DT2]. The computation of d_0 is given in [IV, Proposition 4.6, p. 318], in terms of dimensions of different kernels.

In the case of orthogonal maps with one parameter, it is enough to look at the change of the orthogonal index (giving results comparable to the ones achieved by variational methods, see [Sz]). For more parameters, one arrives at an orthogonal J-homomorphism, i.e. for matrices which are Γ-orthogonal. For the blocks corresponding to V^{T^n}, one has then the usual J^Γ-homomorphism which gives necessary and sufficient conditions for the real representations and a result in terms of the exact sequence of [I3, Theorem 6.1, p. 436] already mentionned. For the blocks on the orthogonal complement of the preceding subspace, where Γ acts as S^1 on the irreducible representations, the invertible matrix $A_s(\lambda)$ is complex self-adjoint on $\mathbb{C}^{n_s+m_s}$, with a constant Morse number n_s and defines an element of $\pi_{k-1}(GLS(n_s, m_s))$, of homotopy classes of complex self-adjoint matrices, with n_s negative eigenvalues and m_s positive eigenvalues. This group is isomorphic to $\pi_k(U(n_s + m_s))$ via a map which gives the complex Bott periodicity. It is then possible to prove that $A_s(\lambda)X$ can be perturbed by a Γ-orthogonal map without zeros, but $X = 0$, if and only if $A_s(\lambda)$ is deformable to $\mathrm{diag}(-I_{n_s}, I_{m_s})$ in $\pi_{k-1}(GLS(n_s, m_s))$, if one is in the stable situation, that is if $k-1 \leq 2n_s, 2m_s$. If k is even, the above group is trivial, while it is isomorphic to $\mathbb{Z}$ if k is odd (see [IV3, Theorem 5.2]).

(4.11) Remark. There are many examples on which the theory should be tested, in particular for cases where one has to go beyong a simple Poincaré section approach. One of these non-trivial cases was when one forces an autonomous differential equation with a time-periodic perturbation. A definite characterization of the J^Γ-homomorphism is still lacking for the case of a non-abelian action. Furthermore, the relationship with the different Bott multiplicities is certainly intriguing. Applications of the orthogonal map approach in the case of a non-abelian action and more examples away from the classical equivariant branching lemma should be done in the spirit of [I4] and for many more mechanical systems. The problem of several first integrals is a challenging one, while the extension to retarded differential equations and to more general functional differential equations has alrerady been done in many cases and the extension to PDEs should be straightforward. More complicated extensions should concern quasi-periodic problems, with the possibility of having more parameters and small divisors.

In conclusion, the reader should be able to find a wealth of interesting problems around the equivariant degree.

References

[AMO] R. P. Agarwal, M. Meehan and D. O'Regan, *Fixed point theory and applications*, Cambridge Tracts in Mathematics, vol. 141, Cambridge University Press, 2001.

[AKPR] R. R. Akhmerov, M. I. Kamenskiĭ, A. S. Potapov and A. E. Rodkina, *Measures of Noncompactness and Condensing Operators*, Operator Theory: Advances and Applications, vol. 55, Birkhäuser–Verlag, Basel, 1992.

[A1] J. C. Alexander, *Bifurcation of zeros of parametrized functions*, J. Funct. Anal. **29** (1978), 37–53.

[A2] ______, *Calculating bifurcation invariants as elements of the homotopy of the general linear group*, Illinois J. Pure Appl. Algebra **17** (1980), 117–125.

[A3] ______, *A primer on connectivity*, Lecture Notes in Math. **886** (1981), Springer–Verlag, 445–483.

[AA1] J. C. Alexander and S. S. Antman, *Global and local behavior of bifurcating multidimensional continua of solutions for multiparameter nonlinear eigenvalue problems*, Arch. Rational. Mech. Anal. **76** (1981), 339–354.

[AA2] ______, *Global behavior of solutions of nonlinear equations depending on infinite dimensional parameter*, Indiana Univ. Math. J. **32** (1983), 39–62.

[AF] J. C. Alexander and P. M. Fitzpatrick, *The homotopy of certain spaces of nonlinear operators and its relation to global bifurcation of the fixed points of parametrized condensing operators*, J. Funct. Anal. **34** (1979), 87–106.

[AMP] J. C. Alexander, I. Massabó and J. Pejsachowicz, *On the connectivity properties of the solution set of infinitely parametrized families of vector fields*, Boll. Un. Mat. Ital. A (6) **1** (1982), 309–312.

[AY1] J. C. Alexander and J. A. Yorke, *The implicit function theorem and global methods of cohomology*, J. Funct. Anal. **21** (1976), 330–339.

[AY2] ______, *Parametrized functions, bifurcation and vector fields on spheres*, Anniversary Volume in Honor of Mitropolsky, vol. 275, Naukova Dumka, 1977, pp. 15–17.

[AY3] ______, *Global bifurcation of periodic orbits*, Amer. J. Math. **100** (1978), 263–292.

[AY4] ______, *Calculating bifurcation invariants as elements of the general linear group* I, J. Pure Appl. Algebra **13** (1978), 1–8.

[AMY] K. T. Alligood, J. Mallet-Paret and J. A. Yorke, *Families of periodic orbits: local continuability does not imply global continuability*, J. Differential Geometry **16** (1981), 483–492.

[AlY] K. T. Alligood and J. A. Yorke, *Hopf bifurcation: the appearance of virtual periods in cases of resonance*, J. Differential Equations **64** (1986), 375–394.

[A] H. Amann, *Fixed point equations and Nonlinear eigenvalue problems in ordered Banach spaces*, SIAM Rev. **18** (1976), 620–709.

[Ap] J. Appell (Ed), *Recent trends in nonlinear analysis*, Progress in Nonlinear Differential Equations and their Applications, vol. 40, Birkhäuser, 2000.

[Ar] V. I. Arnold, *Geometrical methods in the theory of differential equations*, Springer–Verlag, New York, 1983.

[Ba] Z. Balanov, *Equivariant Hopf theorem*, Nonlinear Anal. **30** (1997), 3463–3474.

[BK] Z. Balanov and W. Krawcewicz, *Remarks on the equivariant degree theory*, Topol. Methods Nonlinear Anal. **13** (1999), 91–103.

[BKK] Z. Balanov, W. Krawcewicz and A. Kushkuley, *Brouwer degree, equivariant maps and tensor powers*, Abstr. Appl. Anal. **3** (1998), 401–409.

[BKR] Z. Balanov, W. Krawcewicz and B. Rai, *Taylor–Couette problem and related topics*, Nonlinear Analysis: Real World Applications **4** (2003), 541–559.

[BKS1] Z. Balanov, W. Krawcewicz and H. Steinlein, *Reduced $SO(3) \times S^1$-equivariant degree with applications to symmetric bifurcation problems: the case of one free parameter*, Nonlinear Anal. **47** (2001), 1617–1628.

[BKS2] ______, *$SO(3) \times S^1$-equivariant degree with applications to symmetric bifurcation problems: the case of one free parameter*, Topol. Methods Nonlinear Anal. **20** (2002), 335–374.

[BaK] Z. Balanov and A. Kushkuley, *On the problem of equivariant homotopic classification*, Arch. Math. **65** (1995), 546–552.

[B1] T. Bartsch, *A global index for bifurcation of fixed points*, J. Reine Math. **391** (1988), 181–197.

[B2] ______, *The role of the J-homomorphism in multiparameter bifurcation theory*, Bull. Sci. Math. **112** (1988), 177–184.

[B3] ______, *A simple proof of the degree formula for $\mathbb{Z}/p$-equivariant maps*, Comment. Math. Helv. **65** (1990), 85–95.

[B4] ______, *The global structure of the zero set of a family of semilinear Fredholm maps*, Nonlinear Anal. **17** (1991), 313–332.

[B5] ______, *The Conley index over a space*, Math. Z **209** (1992), 167–177.

[B] ______, *Topological Methods for variational problems with symmetries*, Lect. Notes **1560** (1993), Springer–Verlag.

[BC] T. Bartsch and M. Clapp, *Bifurcation theory for symmetric potential operators and the equivariant cup-length*, Math. Z. **204** (1990), 341–356.

[Be1] V. Benci, *A geometrical index for the group S^1 and some applications to the study of periodic solutions of ordinary equations*, Comm. Pure Appl. Math. **34** (1981), 393–432.

[Be2] ______, *Introduction to Morse theory: a new approach*, Topological Nonlinear Analysis (M. Matzeu and A. Vignoli, eds.), Progress in Nonlinear Differential Equations and Their Applications, vol. 15, Birkhaüser, 1994, pp. 37–177.

[Ber] M. S. Berger, *Nonlinearity and Functional Analysis*, Academic Press, 1977.

[BB] M.S. Berger and M. Berger, *Perspectives in nonlinearity. An introduction to nonlinear analysis*, Benjamin Inc., 1968.

[Bo1] Y. G. Borisovich, *Topology and nonlinear functional analysis*, Russian Math. Surveys **37** (1979), 14–23.

[Bo2] ______, *Topological characteristics of infinite dimensional mappings and the solvability of nonlinear boundary value problems*, Proc. Steklov Inst. Math. **3** (1992), 43–50.

[BZS] Y. G. Borisovich, V. G. Zvyagin and Y. I. Sapronov, *Non linear Fredholm maps and the Leray Schauder theory*, Russian Math. Surveys **32** (1977), 1–54.

[Br] G. E. Bredon, *Introduction to Compact Transformation Groups*, Academic Press, New York, 1980.

[Bre] H. Brezis, *Degree theory: old and new*, Topological Nonlinear Analysis II (M. Matzeu and A. Vignoli, eds.), Progress in Nonlinear Differential Equations and Their Applications, vol. 27, Birkhaüser, 1996, pp. 87–108.

[BD] T. Bröcker and T. tom Dieck, *Representations of compact Lie groups*, Grad. Texts in Math. **98** (1985).

[Bro] F. E. Browder, *Nonlinear operators in Banach spaces*, Proc. Sympos. Pure Math **18** (1976).

[Ch] K. C. Chang, *Infinite Dimensional Morse Theory*, Birkhaüser, 1993.

[CI] P. Chossat and G. Iooss, *The Couette–Taylor problem*, Appl. Math. Sci. **102** (1994).

[CL] P. Chossat and R. Lauterbach, *Methods in equivariant bifurcations and dynamical systems*, Adv. Ser. Nonlinear Dynam. **15** (2000).

[CH] S. N. Chow and J. K. Hale, *Methods of Bifurcation Theory*, Grundlehren Math. Wiss., vol. 251, Springer–Verlag, 1982.

[CMP] S. N. Chow and J. Mallet-Paret, *The Fuller index and global Hopf bifurcation*, J. Differential Equations **29** (1978), 66–85.

[CV] M. A. Ciocci and A. Vanderbauwhede, *Bifurcation of periodic orbits for symplectic mappings*, J. Differential Equations Appl. **3** (1998), 485–500.

[CM] M. Clapp and W. Marzantowicz, *Essential equivariant maps and Borsuk–Ulam theorems*, J. London Math. Soc. **61** (2000), 950–960.

[Co] C. Conley, *Isolated invariant sets and the Morse index*, vol. 38, CBMS regional Conf. Series in Math., 1978.

[D] J. Damon, *Applications of singularity theory to the solutions of nonlinear equations*, Topological Nonlinear Analysis (M. Matzeu and A. Vignoli, eds.), Birkhäuser; Progr. Nonlinear Differential Equations Appl. **15** (1994), 178–302.

[Da1] E. N. Dancer, *On the existence of bifurcating solution in the presence of symmetries*, Proc. Roy. Soc. Edinburg A **85** (1980), 321–336.

[Da2] ______, *Symmetries, degree, homotopy indices and asymptotically homogeneous problems*, Nonlinear Anal. **6** (1982), 667–686.

[Da3] ______, *On the indices of fixed points of mappings in cones and applications*, J. Math. Anal. Appl. **91** (1983), 131–151.

[Da4] ______, *Perturbation of zeros in the presence of symmetries*, J. Austral. Math. Soc. **36** (1984), 106–125.

[Da] ______, *A new degree for S^1-invariant gradient mappings and applications*, Ann. Inst. H. Poincaré, Anal. Non. Linéaire **2** (1985), 329–370.

[Da5] ______, *Fixed point index calculations and applications*, Topological Nonlinear Analysis (M. Matzeu and A. Vignoli, eds.), Birkhaüser; Progr. Nonlinear Differential Equations Appl. **15** (1994), 303–340.

[DT1] E. N. Dancer and J. F. Toland, *Degree theory for orbits of prescribed period of flows with a first integral*, Proc. London Math. Soc. **60** (1990), 549–580.

[DT2] ______, *Equilibrium states in the degree theory of periodic orbits with a first integral*, Proc. London Math. Soc. **61** (1991), 564–594.

[De] K. Deimling, *Nonlinear Functional Analysis*, Springer–Verlag, 1985.

[DMM] M. Dellnitz, I. Melbourne and J. E. Marsden, *Generic bifurcation of Hamiltonian vector fields with symmetry*, Nonlinearity **15** (1992), 979–996.

[Di] W. Y. Ding, *Generalizations of the Borsuk theorem*, J. Math. Anal. Appl. **110** (1985), 553–567.

[Do] A. Dold, *Fixpunkttheorie*, Kurseinheiten 1–7. Hagen: Fernuniversität-Gesamthochschule, Fachbereich Mathematik und Informatik, 1983.

[DG] J. Dugundji and A. Granas, *Fixed Point Theory I*, PWN (Polish Scientific Publications), Warszawa, 1982.

[DGJW] G. Dylawerski, K. Gęba, J. Jodel and W. Marzantowicz, *An S^1-equivariant degree and the Fuller index*, Ann. Pol. Math. **52** (1991), 243–280.

[EGKW] L. Erbe, K. Gęba, W. Krawcewicz and J. Wu, *S^1-degree and global Hopf bifurcation theory of functional differential equations*, J. Differential Equations **98** (1992), 277–298.

[FR] E. R. Fadell and P. H. Rabinowitz, *Generalized cohomological index theories for Lie group actions with application to bifurcation questions for Hamiltonian systems*, Invent. Math. **45** (1978), 139–174.

[FHR] E. R. Fadell, S. Husseini and P. Rabinowitz, *Borsuk–Ulam theorems for arbitrary S^1-action and applications*, Trans. Amer. Math. Soc. **274** (1982), 345–360.

[FiM1] M. Field, *Symmetry breaking for compact Lie groups*, vol. 574, Mem. Amer. Math. Soc., 1996.

[FiM2] ______, *Symmetry breaking for equivariant maps*, Algebraic groups and Lie groups (G. I. Lehrer, ed.), Cambridge Univ. Press, 1997, pp. 219–253.

[FiM3] ______, *Symmetry breaking for equivariant maps*, Austral. Math. Soc. Lect. Ser. **9** (1997), 219–253.

[FR] M. Field and R. W. Richardson, *Symmetry breaking and branching problems in equivariant bifurcation theory*, I. Arch. Mat. Mech. Anal. **118** (1992), 297–348.

[Fi] B. Fiedler, *Global Bifurcation of periodic solutions with symmetry*, Lecture Notes in Math., vol. 1309, Springer–Verlag, 1988.

[FH] B. Fiedler and S. Heinze, *Homotopy invariants of time reversible periodic orbits* I. *Theory*, J. Differential Equations **126** (1996), 184–203.

[FPM] P. M. Fitzpatrick, *Homotopy, linearization and bifurcation*, Nonlinear Anal. **12** (1988), 171–184.

[FMP] P. M. Fitzpatrick, I. Massabó and J. Pejsachowicz, *Complementing maps, continuation and global bifurcation*, Bull. Amer. Math. Soc. **9** (1983), 79–81.

[FP1] P. M. Fitzpatrick and J. Pejsachowicz, *The fundamental group of the space of linear Fredholm operators and the global analysis of semilinear equations*, Contemp. Math. **72** (1988), 47–87.

[FP2] ______, *Parity and generalized multiplicity*, Trans. Amer. Math. Soc. **326** (1991), 281–305.

[FP3] ______, *The Leray–Schauder theory and fully non-linear elliptic boundary value problems*, vol. 483, Mem. Amer. Math. Soc. Vol. 101, 1993.

[FPR] P. M. Fitzpatrick, J. Pejsachowicz and P. J. Rabier, *The degree of proper C^2 Fredholm maps: covariant theory*, Topol. Methods Nonlinear Anal. **3** (1993), 325–367.

[FPRe] P. M. Fitzpatrick, J. Pejsachowicz, and L. Recht, *Spectral flow and bifurcation of critical points of strongly indefinite functionals* I. *General theory*, J. Funct. Anal. **162** (1999), 52–95.

[FG] I. Fonseca and W. Gangbo, *Degree Theory in Analysis and Applications*, Oxford Lecture Series in Mathematics and its Applications 2., Clarendon Press, 1995.

[F] F. B. Fuller, *An index of fixed point type for periodic orbits*, Amer. J. Math. **89** (1967), 133–148.

[FMV] M. Furi, M. Martelli and A. Vignoli, *On the Solvability of nonlinear operator equations in normed spaces*, Ann. Mat. Pura Appl. **124** (1980), 321–343.

[G] K. Gęba, *Degree for gradient equivariant maps and equivariant Conley index*, Topological Nonlinear Analysis II, Progress in Nonlinear Differential Equations and their Applications (M. Matzeu and A. Vignoli, eds.), vol. 27, Birkhaüser, 1997, pp. 247–272.

[GKW] K. Gęba, W. Krawcewicz and J. Wu, *An equivariant degree with applications to symmetric bifurcation problems, Part I, Construction of the degree*, Bull. London Math. Soc. **69** (1994), 377–398.

[GW] K. Gęba and W. Marzantowicz, *Global bifurcation of periodic orbits*, Topol. Methods Nonlinear Anal. **1** (1993), 67–93.

[GMV] K. Gęba, I. Massabó and A. Vignoli, *On the Euler characteristic of equivariant vector fields*, Boll. Un. Mat. Ital. A **4** (1990), 243–251.

[GS] M. Golubitsky and D. G. Schaeffer, Appl. Math. Sci. **51** (1986), Springer–Verlag.

[GSS] M. Golubitsky, D. G. Schaeffer and I. N. Stewart, *Singularities and groups in bifurcation theory* II, Springer–Verlag, 1988.

[Gr] A. Granas, *The Leray–Schauder index and the fixed point theory for arbitrary* ANR*s*, Bull. Soc. Math. France **100** (1972), 209–228.

[HO] J. K. Hale and J. C. F. De Oliviera, *Hopf bifurcation for functional equations*, J. Math. Anal. Appl. **74** (1980), 41–59.

[H1] H. Hauschild, *Äquivariante homotopie I.*, Arch. Math. **29** (1977), 158–167.

[H2] ———, *Zerspaltung äquivarianter Homotopiemengen*, Math. Ann. **230** (1977), 279–292.

[HIR] D. Hyers, G. Isac and T. M. Rassias, *Topics in nonlinear analysis and applications*, World Scientific, 1997.

[I1] J. Ize, *Obstruction theory and multiparameter Hopf bifurcation*, Trans. Amer. Math. Soc. **289** (1985), 757–792.

[I2] ———, *Necessary and sufficient conditions for multiparameter bifurcation*, Rocky Mountain J. Math. **18** (1988), 305–337.

[I3] ———, *Topological bifurcation*, Topological Nonlinear Analysis (M. Matzeu and A. Vignoli, eds.), Birkhaüser; Progr. Nonlinear Differential Equations Appl. **15** (1994), 341–463.

[I4] ———, *Two mechanical systems and equivariant degree theory*, Progr. Nonlinear Differential Equations Appl. **40** (2000), Birkhaüser, 177–190.

[IMPV] J. Ize, I. Massabó, J. Pejsachowicz and A. Vignoli, *Structure and dimension of global branches of solutions to multiparameter nonlinear equations*, Trans. Amer. Math. Soc. **291** (1985), 383–435.

[IMV1] J. Ize, I. Massabó and A. Vignoli, *Global results on continuation and bifurcation for equivariant maps*, NATO Adv. Sci. Inst. **173** (1986), 75–111.

[IMV2] ———, *Degree theory for equivariant maps I*, Trans. Amer. Math. Soc. **315** (1989), 433–510.

[IMV3] ———, *Degree theory for equivariant maps: The general S^1-action.*, vol. 481, Mem. Amer. Math. Soc., 1992.

[IV1] J. Ize and A. Vignoli, *Equivariant degree for abelian action, Part* I. *Equivariant homotopy groups*, Topol. Methods Nonlinear Anal. **2** (1993), 367–413.

[IV2] ______, *Equivariant degree for abelian actions, Part* II. *Index computations*, Topol. Methods Nonlinear Anal. **7** (1996), 369–430.

[IV3] ______, *Equivariant degree for abelian actions, Part* III. *Orthogonal maps*, Topol. Methods Nonlinear Anal. **13** (1999), 105–146.

[IV] ______, *Equivariant degree theory*, de Gruyter Ser. Nonlinear Anal. Appl. **8** (2003).

[J] J. Jaworowski, *Extensions of G-maps and Euclidean G-retracts*, Math. Z. **146** (1976), 143–148.

[K1] H. Kielhöfer, *Multiple eigenvalue bifurcation for potential operators*, J. Reine Angew. Math. **358** (1985), 104–124.

[K2] ______, *A bifurcation theorem for potential operators*, J. Funct. Anal. **77** (1988), 1–8.

[K3] ______, *Hopf bifurcation from a differentiable viewpoint*, J. Diff. Eq. **97** (1992), 189–232.

[KK] K. Komiya, *Fixed point indices of equivariant maps and Moebius inversion*, Invent. Math. **91** (1988), 129–135.

[Ko] C. Kosniowski, *Equivariant cohomology and stable cohomotopy*, Math. Ann. **210** (1974), 83–104.

[Kr] M. A. Krasnosel'skiĭ, *Topological Methods in the Theory of Nonlinear Integral Equations*, Pergamon Press, MacMillan, New York, 1964.

[KZ] M. A. Krasnosel'skiĭ and P. P. Zabreĭko, *Geometrical Methods in Nonlinear Analysis*, Grund Math. Wiss, vol. 263, Springer–Verlag, 1984.

[KV] W. Krawcewicz and P. Vivi, *Equivariant degree and normal bifurcations*, B. N. Prasad birth centenary commemoration volume II, Indian J. Math. **42** (2000), 55–67.

[KVW] W. Krawcewicz, P. Vivi and J. Wu, *Hopf bifurcations of functional differential equations with dihedral symmetries*, J. Differential Equations **146** (1998), 157–184.

[KX] W. Krawcewicz and H. Xia, *Analytic definition of an equivariant degree*, Russian Math. **40** (1996), 34–49.

[KW] W. Krawcewicz and J. Wu, *Theory of Degrees with Applications to Bifurcations and Differential Equations*, Canad. Math. Soc. Ser. Monogr. Adv. Texts, John Wiley & Sons, Inc., New York, 1997.

[KS] W. Kryszewski and A. Szulkin, *An infinite dimensional Morse theory with applications*, Trans. Amer. Math. Soc. **349** (1997), 3181–3234.

[KB] A. Kushkuley and Z. Balanov, *Geometric Methods in Degree Theory for Equivariant Maps*, Lecture Notes in Math., vol. 1632, Springer–Verlag, Berlin, 1996.

[LLH] A. Lari-Lavassani, W. F. Langford and K. Huseyin, *Symmetry breaking bifurcations on multidimensional fixed point subspaces*, Dynam. Stability Systems **9** (1994), 345–373.

[LMSM] L. G. Lewis, J. P. May, M. Steinberger and J. E. McClure, *Equivariant stable homotopy theory*, vol. 1213, Springer Lect. Notes in Math., 1986.

[MR] A. Maciejewski and S. Rybicki, *Global bifurcations of periodic solutions of Hénon-Heiles system via degree for S^1-equivariant orthogonal maps*, Rev. Math. Phys. **10** (1998), 1125–1146.

[MPN] J. Mallet-Paret and R. Nussbaum, *Boundary layer phenomena for differential-delay equations with state dependent time delays*, Arch. Rational Mech. Anal. **120** (1992), 99–146.

[MPY] J. Mallet-Paret and J. A. Yorke, *Snakes: oriented families of periodic orbits, their sources, sinks and continuation*, J. Diferential Equations **43** (1982), 419–450.

[MM] J. E. Marsden and M. McCracken, *The Hopf Bifurcation and its Applications*, Springer–Verlag, 1976.

[M1] W. Marzantowicz, *On the nonlinear elliptic equations with symmetries*, J. Math. Anal. Appl. **81** (1981), 156–181.

[M2] ______, *Borsuk–Ulam theorem for any compact Lie group*, J. London Math. Soc. **49** (1994), 195–208.

[M] T. Matsuoka, *Equivariant function spaces and bifurcation points*, J. Math. Soc. Japan **35** (1983), 43–52.

[MV1] M. Matzeu and A. Vignoli (Eds), *Topological Nonlinear Analysis*, Progr. Nonlinear Differential Equations Appl., vol. 15, Birkhaüser, 1994.

[MV2] ———, *Topological Nonlinear Analysis* II, Progr. Nonlinear Differential Equations Appl., vol. 27, Birkhaüser, 1996.

[MW] J. Mawhin and M. Willem, *Critical Point Theory and Hamiltonian Systems*, Springer, New York, 1989.

[Maw] J. Mawhin, *Leray–Schauder degree: a half century of extensions and applications*, Topol. Methods Nonlinear Anal. **14** (1999), 195–228.

[MRS] J. Montaldi, M. Roberts and I. Stewart, *Existence of nonlinear normal modes of symmetric Hamiltonian systems*, Nonlinearity **3** (1990), 695–730.

[N] V. Namboodiri, *Equivariant vector fields on spheres*, Trans. Amer. Math. Soc. **278** (1983), 431–460.

[Ni1] L. Nirenberg, *Topics in nonlinear functional analysis*, Lect. Notes Courant Institute, New York Univ., 1974.

[Ni2] ———, *Variational and topological methods in nonlinear problems*, Bull. Amer. Math. Soc. **4** (1981), 267–302.

[Nu] R. D. Nussbaum, *Differential-delay equations with two time lags*, vol. 205, Memoirs Amer. Math. Soc., 1978.

[Pa] A. Parusinski, *Gradient homotopies of gradient vector fields*, Studia Mathematica **46** (1990), 73–80.

[P] J. Pejsachowicz, *K-theoretic methods in bifurcation theory*, Contem. Math. **72** (1988), 193–205.

[PR] J. Pejsachowicz and P. Rabier, *Degree theory for C^1 Fredholm mappings of index* 0, J. Anal. Math. **76** (1998), 289–319.

[Pe] G. Peschke, *Degree of certain equivariant maps into a representation sphere*, Topology Appl. **59** (1994), 137–156.

[Pe1] W. V. Petryshyn, *Approximation-Solvability of Nonlinear Functional and Differential Equation*, M. Dekker, 1992.

[Pe2] ———, *Generalized Topological Degree and Semilinear Equations*, Cambridge Tracts in Math., vol. 117, Cambridge Univ. Press..

[PU] C. Prieto and H. Ulrich, *Equivariant fixed point index and fixed point transfer in nonzero dimension*, Trans. Amer. Math. Soc. **328** (1991), 731–745.

[Rab1] P. J. Rabier, *Generalized Jordan chains and two bifurcation theorems of Krasnosel'skiĭ*, Nonlinear Anal. **13** (1989), 903–934.

[Rab2] ———, *Topological degree and the theorem of Borsuk for general covariant mappings with applications*, Nonlinear Anal. **16** (1991), 393–420.

[Ra1] P. H. Rabinowitz, *Some global results for nonlinear eigenvalue problems*, J. Funct. Anal. **7** (1971), 487–513.

[Ra2] ———, *Theorie du Degree Topologique et Applications* (1975), (Lectures Notes).

[Ra3] ———, *A bifurcation theorem for potential operators*, J. Funct. Anal. **25** (1977), 412–424.

[Ra4] ———, *Critical point theory*, Topological Nonlinear Analysis (M. Matzeu and A. Vignoli, eds.), Birkhaüser; Progr. Nonlinear Differential Equations Appl. **15** (1994), 464–513.

[Ro] E. H. Rothe, *Introduction to Various Aspects of Degree Theory in Banach Spaces*, Math. Surveys Monogr., vol. 23, Amer. Math. Soc., Providence, 1986.

[Ru] R. L. Rubinstein, *On the equivariant homotopy of spheres*, Dissertationes Math. **134** (1976), 1–48.

[R1] S. Rybicki, *S^1-degree for orthogonal maps and its applications to bifurcation theory*, Nonlinear Anal. **23** (1994), 83–102.

[R2] ———, *Applications of degree for S^1-equivariant gradient maps to variational problems with S^1-symmetries*, Topol. Methods Nonlinear Anal. **9** (1997), 383–417.

[R3] ———, *On periodic solutions of autonomous Hamiltonian systems via degree for S^1-equivariant gradient maps*, Nonlinear Anal. **34** (1998), 537–569.

[R4] ———, *On bifurcation from infinity for S^1-equivariant potential operators*, Nonlinear Anal. **31** (1998), 343–361.

[R5] ______, *Degree for S^1-equivariant strongly indefinite functionals*, Nonlinear Anal. **43** (2001), 1001–1017.

[S1] D. H. Sattinger, *Group Theoretic Methods in Bifurcation Theory*, Lecture Notes in Math., vol. 762, Springer–Verlag, 1979.

[S2] ______, *Branching in the Presence of Symmetry*, Wiley, 1983.

[ScWa] K. Schmitt and Z. Q. Wang, *On bifurcation from infinity for potential operators*, Differential Integral Equations **4** (1991), 933–943.

[SW] J. Smoller and A. G. Wasserman, *Bifurcation and symmetry-breaking*, Inv. Math. **100** (1990), 63–95.

[S] H. Steinlein, *Borsuk's antipodal theorem and its generalizations and applications: a survey*, Topol. Anal. Non Linéaire (1985), Press de l'Univ. de Montreal, 166–235.

[St] C. A. Stuart, *Bifurcation from the essential spectrum*, Topological Nonlinear Analysis II (M. Matzeu and A. Vignoli, eds.), Birkhaüser; Progr. Nonlinear Differential Equations Appl. **27** (1996), 397–444.

[Sz] A. Szulkin, *Index theories for indefinite functionals and applications*, Pitman Res. Notes Math. Ser. **365** (1997), 89–121.

[SZ] A. Szulkin and W. Zou, *Infinite dimensional cohomology groups and periodic of assymptotically linear Hamiltonian systems*, J. Differential Equations **174** (2001), 369–391.

[tD1] T. tom Dieck, *Transformation groups and representation theory*, Lecture Notes in Math., vol. 766, Springer–Verlag, 1979.

[tD2] ______, *Transformation Groups*, de Gruyter, Berlin, 1987.

[T] J. Tornehave, *Equivariant maps of spheres with conjugate orthogonal actions*, Algebraic Topology, Proc. Conf. London Ont. 1981,, vol. 2, Canadian Math. Soc. Conf. Proc., 1982, pp. 275–301.

[U] H. Ulrich, *Fixed point theory of parametrized equivariant maps*, Lecture Notes in Math., vol. 1343, Springer–Verlag, 1988.

[V] A. Vanderbauwhede, *Local bifurcation and symmetry*, vol. 75, Pitman Research Notes in Math., 1982.

[W] Z. Q. Wang, *Symmetries and the calculation of degree*, Chinesw Ann. Math. Ser. B **10** (1989), 520–536.

[W1] G. W. Whitehead, *On the homotopy groups of spheres and rotation groups*, Ann. of Math. **43** (1942), 634–640.

[W2] ______, *Elements of homotopy theory*, Grad. Texts in Math., vol. 61, Springer–Verlag, 1978.

[Z1] E. Zeidler, *Nonlinear Functional Analysis and its Applications*, vol. IV, Springer Verlag, 1985.

[Z2] ______, *Nonlinear functional analysis and its applications*, I. *Fixed-Point Theorems*, Springer–Verlag, 1993.

[Z] P. P. Zabreĭko, *Rotation of vector fields: definition, basic properties and calculations*, Topological Nonlinear Analysis II (M. Matzeu and A. Vignoli, eds.), Birkhaüser; Progr. Nonlinear Differential Equations Appl. **27** (1996), 449–601.

[Zv] V. G. Zvyagin, *On a degree theory for equivariant $\Phi_0 C^I BH$-mappings*, Dokl. Akad. Nauk **364** (1999), 155–157. (Russian)

10. BIFURCATIONS OF SOLUTIONS OF $SO(2)$-SYMMETRIC NONLINEAR PROBLEMS WITH VARIATIONAL STRUCTURE

Sawomir Rybicki

1. Introduction

One of the strongest and the most popular tools in topological nonlinear analysis are the Brouwer topological degree and its infinite-dimensional generalization the Leray–Schauder topological degree, see [De], [Ll].

Usually the Leray–Schauder degree is used to prove the existence, local bifurcations, global bifurcations and continuation of solutions of nonlinear problems. The Krasnosel'skii local bifurcation theorem, see for example [Bro], [De], [Iz1], [Kras], [Ni], ensures the existence of bifurcation points of nontrivial solutions of parameterized nonlinear problems. By the Rabinowitz global bifurcation theorem, see for example [Bro], [De], [Iz1], [Ni], [Ra2], [Ra3], [Ra4], we obtain the existence of branching points of nontrivial solutions of parameterized nonlinear problems. Finally the Leray–Schauder continuation theorem, see [De], ensures the existence of continua of nontrivial solutions of parameterized nonlinear problems.

All the theorems mentioned above play the crucial role in topological nonlinear analysis.

The Brouwer degree is defined for continuous, admissible maps. On the other hand, the Leray–Schauder degree is defined for the class of admissible operators in the form of a compact perturbation of the identity. Therefore the natural question is the following.

Is it possible to define, for a restricted class of operators, degree theory stronger than the Brouwer or the Leray–Schauder degree?

Facing the problem of the existence of solutions of many nonlinear differential equations (Hamiltonian systems, wave equations, elliptic differential equations) one can study these solutions as critical points of functionals defined on suitably

Partially supported by KBN under grant number 2/PO3A/047/12.

chosen functional spaces, see for example [MW], [Ra1], [St]. Therefore the interesting question is if the class of gradient operators is suitable for our purpose. The answer is no. The following result due to Parusinski [Pa] explains the reason. If two continuous, gradient maps $\nabla f_1, \nabla f_2\colon (D^n, S^{n-1}) \to (\mathbb{R}^n, \mathbb{R}^n \setminus \{0\})$ are homotopic in the class of continuous, admissible maps, then they are homotopic in the class of continuous, admissible, gradient maps. In other words, if $\deg_B(\nabla f_1, D^n, 0) = \deg_B(\nabla f_2, D^n, 0)$ ($\deg_B$ is the Brouwer degree), then by the Hopf classification theorem $\nabla f_1, \nabla f_2$ are in the same homotopy class of continuous, admissible maps. Consequently, by the Parusinski theorem, $\nabla f_1, \nabla f_2$ are in the same homotopy class of continuous, admissible, gradient maps. Suppose that degree for continuous, admissible, gradient maps exists. Let us denote it by $\nabla - \deg$. Then from the discussion above we obtain $\deg_B(\nabla f_1, D^n, 0) = \deg_B(\nabla f_2, D^n, 0)$ if and only if $\nabla - \deg(\nabla f_1, D^n, 0) = \nabla - \deg(\nabla f_2, D^n, 0)$. In other words, there is no degree theory (stronger than the Brouwer degree) for the class of continuous, admissible, gradient maps. Moreover, it was shown in [Pa] that in the homotopy class of continuous admissible maps of any continuous map $f\colon (D^n, S^{n-1}) \to (\mathbb{R}^n, \mathbb{R}^n \setminus \{0\})$ there is a gradient map.

It is worth to point out that Rabinowitz has applied in [Ra6] the Leray–Schauder degree to potential operators in the form of a compact perturbation of the identity. The additional gradient structure has allowed to compute the local index at a degenerate minimum of a functional.

On the other hand, there are already developed many theories which allow us to prove the existence and bifurcation of critical points of functionals. Let us list some of them: Morse theory, see [Cha], [KSZ], Conley index theory see [Co], [GIP], [Izy2], [Sm], Lusternik–Schnirelmann theory; linking theory, see [Sch], mountain-pass theorem, see [BB], [Ra1], [Ki1]. Unfortunately, we cannot use these methods to prove the existence of branching points and continuation of critical points of functionals. A change of the Morse index implies only the existence of bifurcation points. It can happen that these points are not branching points, see [Am], [Boh], [Iz3], [Mari], [Ta] for examples and discussion.

If we restrict our considerations to the periodic solutions of autonomous ordinary differential equations or if we study partial differential equations on invariant domains then in a natural way can be introduced an action of a compact Lie group G on a functional space. Borsuk is the mathematician who first observed that the symmetries can lead to the restriction of possible values of the degree. Namely, he established that the degree of an odd map from finite-dimensional sphere into itself is odd. The results concerning the Brouwer and the Leray–Schauder degrees, in the case of G-equivariant maps, can be found in [JS], [R1], [R2], [Wa1], [Wa2]. It is worth nothing that if $\Omega \subset V$ is an open, bounded and $SO(2)$-invariant subset of an orthogonal representation of the group $SO(2)$ such that there are no fixed point of the action of the group $SO(2)$ in Ω, then $\deg_B(\nabla f, \Omega, 0) = 0$.

The proper class of maps for definition of a degree stronger than the Brouwer degree and the Leray–Schauder degree is the class of continuous G-equivariant admissible gradient maps, where G is any compact Lie group.

Therefore from now on we restrict our considerations to G-invariant potentials defined on orthogonal representations of a compact Lie group G. i.e. potentials fixed on the orbits of an action of a compact Lie group G. In this case gradients, whose zeros are in one-to-one correspondence with the solutions of a nonlinear differential equation, are G-equivariant. There are already developed theories which allow us to give the lower estimation of the number of critical orbits of G-invariant functionals. Let us list some of these tools: G-equivariant Morse theory, see [BP], [Bot], [Wa3], [Was], G-equivariant Conley index theory, see [Ba], [FM], [Izy1], [SW], G-equivariant index theory, see [AZ1], [AZ2], [Be1], [Be2], [Fa1], [Fa2], [Fa3], [FH1], [FH2], [FR], G-equivariant Lusternik–Schnirelmann category, see [Ba], [Fa2], [Marz].

We are going to study the existence of branching points, global bifurcations and continuation of critical points of $SO(2)$-invariant functionals. The only, suitable for this purpose, tool is the degree theory for $SO(2)$-equivariant gradient maps. The first degree theory for $SO(2)$-equivariant gradient maps is due to Dancer, see [Da2]. The work was further improved in [Ry1] for general $SO(2)$-actions and in [Ge2] for general compact Lie group actions. Degree theories for the class of $SO(2)$-equivariant transversal and orthogonal maps have been constructed in [HR] and [Ry1], respectively.

Many results concerning equivariant bifurcation theory one can find in [CL].

Finally we would like to point out that there are constructed degree theories for G-equivariant operators without variational structure, where G is any compact Lie group. The best source of references concerning this subject is the book by Ize and Vignoli [IV].

This article is organized in the following way.

In Section 2 we summarize without proofs the relevant material on equivariant topology. In Theorem (2.7) we classify finite-dimensional representations of the group $SO(2)$. Next we define the tom Dieck ring $U(SO(2))$ of the group $SO(2)$. In Lemma (2.11) we study some properties of this ring.

In Section 3 we present properties of the degree for $SO(2)$-equivariant gradient maps. In Theorem (3.1) and Remark (3.2) we collect basic properties of the finite-dimensional version of the degree for $SO(2)$-equivariant gradient maps. To simplify computation of index of an isolated critical point of an $SO(2)$-invariant functional we can combine product formula for the degree for $SO(2)$-equivariant gradient maps presented in Theorem (3.3) with splitting lemma formulated in Lemma (3.4). To apply successfully topological invariants we need computational formulas for them. In Lemma (3.5) we show how to compute the degree of $SO(2)$-equivariant, linear, self-adjoint isomorphisms. In Lemma (3.6) we present some properties of the degree of $SO(2)$-equivariant, self-adjoint isomorphisms. Definition of an infinite-dimensional version of the degree for $SO(2)$-equivariant gradient maps is given by formula (3.11). The basic properties of an infinite-dimensional version of the degree for $SO(2)$-equivariant gradient maps are given in Theorem (3.12). In Theorem (3.13)and Corollary (3.14) we show how to compute the degree of $SO(2)$-equivariant, self-adjoint isomorphism in the form of a compact perturbation of identity.

Two theorems lying in the heart of topological nonlinear analysis are the local Krasnosel'skii and the global Rabinowitz bifurcation theorems, see [Iz1], [Ni]. Bifurcation indices computed in these theorems are computed in terms of the Leray–Schauder degree, which is an element of the ring of integers $\mathbb{Z}$. The algebraic structure of the tom Dieck ring $U(SO(2))$ is much more complicated than the structure in $\mathbb{Z}$. Since the degree of $SO(2)$-equivariant gradient maps is an element of the tom Dieck ring $U(SO(2))$, we expect stronger bifurcation theorems for the class of $SO(2)$-equivariant gradient maps than the classical ones.

Section 4 is devoted to the study of local and global bifurcations of critical points of $SO(2)$-invariant functionals. In Definitions (4.1), (4.2) we introduce the notion of bifurcation point and branching point, respectively. Theorems (4.4), (4.6) are the local and global bifurcation theorems for the general class of $SO(2)$-equivariant, gradient operators in the form of a compact perturbation of the identity, respectively. In Lemma (4.8) we formulate necessary condition for the existence of bifurcation point. The notion of bifurcation index is introduced in Definition (4.9). In Lemma (4.10) we compute bifurcation indices in terms of linearizations of considered operators. Notice that the classical bifurcation index (defined in terms of the Leray–Schauder degree) computed at characteristic value of even multiplicity is trivial. On the other hand it can happen that our bifurcation index is nontrivial. Theorems (4.15), (4.16) are the local and global bifurcation theorems, respectively, where the asumptions are expressed in terms of linearization of considered operators. We finish this section with global bifurcation theorems (4.17), (4.18), (4.20). Notice that in Theorems (4.18), (4.20) we prove unboundedness of continua of critical points of $SO(2)$-invariant potentials.

In Section 5 we present some applications of the degree for $SO(2)$-equivariant gradient maps to elliptic differential equations considered on $SO(2)$-invariant domains. Namely, we consider system of elliptic differential equations (5.1) whose weak solutions can be considered as critical points of $SO(2)$-invariant functional defined by (5.2). In Theorems (5.7), (5.8) we describe continua of weak solutions of systems of elliptic differential equations. It is worth to point out that in Theorem (5.8) we formulate sufficient conditions for the existence of unbounded continua of weak solutions. Since we assume in Theorem (5.8) that the number m of equations in system (5.1) is even, we can not apply the classical Rabinowitz alternative and the Leray–Schauder degree. The reason is that in this case the bifurcation index computed in terms of the Leray–Schauder degree is trivial. In Examples (5.9), (5.10) we illustrate Theorems (5.7), (5.8) assuming that Ω is a two-dimensional and three-dimensional disc, respectively. The last application we consider in this section is the nonlinear eigenvalue problem (5.12) for the Laplace–Beltrami operator $\Delta_{\mathbb{S}^{n-1}}$ defined on the sphere $S^{n-1} \subset \mathbb{R}^n$. In Theorem (5.21) we prove that continua of weak solutions of problem (5.12) are unbounded in $H^1(S^{n-1}) \times \mathbb{R}$.

2. Preliminaries

In this section we summarize without proofs the relevant material on equivariant topology.

We start with a definition of a compact Lie group.

(2.1) DEFINITION. A pair $(G, \cdot)$ is said to be a *Lie group*, if

(2.1.1) $(G, \cdot)$ is a group,
(2.1.2) G is a smooth manifold,
(2.1.3) a map $G \times G \to G$ given by the formula $(g_1, g_2) \to g_1 \cdot g_2^{-1}$ is smooth.

For simplicity of notation, we write G instead of $(G, \cdot)$.

(2.2) REMARK. It is easy to see that if G is a Lie group then maps $g \to g^{-1}$ and $(g_1, g_2) \to g_1 \cdot g_2$ are smooth.

Let us denote by

- $M(n, \mathbb{R})$ the set of real $(n \times n)$-matrices,
- $GL(n, \mathbb{R}) = \{L \in M(n, \mathbb{R}) : \det L \neq 0\}$ the group of real non-degenerate matrices,
- $O(n, \mathbb{R}) = \{L \in GL(n, \mathbb{R}) : M^T = M^{-1}\}$ the group of real orthogonal matrices,
- $SO(n, \mathbb{R}) = \{L \in O(n, \mathbb{R}) : \det M = 1\}$ the group of real special orthogonal matrices.

(2.3) THEOREM ([Bre]). *G is a compact Lie group if and only if it is isomorphic to a closed subgroup of $O(n, \mathbb{R})$ for some n.*

(2.4) DEFINITION. We say that a compact Lie group G *acts on a topological space* X (or X is a G-space) if there is a continuous map $\mu: G \times X \to X$ such that the following conditions are fulfilled

(2.4.1) $\mu(e, x) = x$), for any $x \in X$,
(2.4.2) $\mu(g_1, \mu(g_2, x)) = \mu(g_1 \cdot g_2, x)$, for any $x \in X$ and $g_1, g_2 \in G$.

For abbreviation, we let gx stand for $\mu(g, x)$.

For a given $x \in X$ the subgroup $G_x = \{g \in G : gx = x\}$ is called the isotropy group of x and the set $Gx = \{gx : g \in G\}$ is called the orbit of x. By $\Psi(G)$ we will denote the collection of all closed subgroups of G. A subset $\Omega \subset X$ is said to be G-invariant provided that for any $x \in \Omega$ we have $Gx \subset \Omega$.

Given a closed subgroup $H \subset G$ we let

$$X^H = \{x \in X : hx = x \text{ for all } h \in H\} = \{x \in X : H \subset G_x\},$$
$$X_H = \{x \in X : G_x = H\}.$$

Let X, Y be two G-spaces. W say that a map $f: X \to Y$ is G-equivariant if $f(gx) = gf(x)$ for all $x \in X, g \in G$. Notice that $G_x \subset G_{f(x)}$.

(2.5) DEFINITION. A *representation of a compact Lie group* G is a pair $V = (\mathbb{R}^n, \rho)$, where $\rho: G \to GL(n, \mathbb{R})$ is a continuous homomorphism. If $\rho: G \to O(n, \mathbb{R})$, then the representation V is said to be *orthogonal.* The number n is called the *dimension of the representation* $V = (\mathbb{R}^n, \rho)$.

(2.6) LEMMA ([Bre]). *Each representation of a compact Lie group G is equivalent to an orthogonal representation.*

Notice that if $V = (\mathbb{R}^n, \rho)$ is a representation of G, then letting $gv = \rho(g)(v)$ we obtain a linear G-action on $\mathbb{R}^n$. On the other hand it is evident that any linear G-action on $\mathbb{R}^n$ defines a representation of a compact Lie group G on $\mathbb{R}^n$.

To simplify notation, we continue to write V for $V = (\mathbb{R}^n, \rho)$.

Two G-representations V and W are equivalent if there exists an G-equivariant, linear isomorphism $T: V \to W$.

For $j \in \mathbb{N}$ define a continuous homomorphism $\rho^j: SO(2) \to GL(2, \mathbb{R})$ as follows

$$\rho^j(e^{i \cdot \theta}) = \begin{bmatrix} \cos j \cdot \theta & -\sin j \cdot \theta \\ \sin j \cdot \theta & \cos j \cdot \theta \end{bmatrix} \quad \text{for } 0 \leq \theta < 2\pi.$$

For $k, j \in \mathbb{N}$ we denote by $\mathbb{R}[k, j]$ the direct sum of k copies of $(\mathbb{R}^2, \rho^j)$, we also denote by $\mathbb{R}[k, 0]$ the trivial k-dimensional representation of $SO(2)$. The following classical result gives a complete classification (up to an equivalence) of finite–dimensional representations of the group $SO(2)$.

(2.7) THEOREM ([Ad]). *If V is a representation of $SO(2)$ then there exist finite sequences $\{k_i\}$, $\{j_i\}$ satisfying*

(2.7.1) $j_i \in \{0\} \cup \mathbb{N}$, $k_i \in \mathbb{N}$, $1 \le i \le r$, $j_1 < \ldots < j_r$ *such that V is equivalent to* $\bigoplus_{i=1}^{r} \mathbb{R}[k_i, j_i]$.

Moreover, the equivalence class of V ($V \approx \bigoplus_{i=1}^{r} \mathbb{R}[k_i, j_i]$) is uniquely determined by $\{j_i\}$, $\{k_i\}$ satisfying (2.7.1).

Fix $k \in \mathbb{N}$ and define

$$C^k_{SO(2)}(V, \mathbb{R}) = \{f \in C^k(V, \mathbb{R}) : f(gv) = f(v) \text{ for any } g \in SO(2),\ v \in V\},$$
$$C^k_{SO(2)}(V, V) = \{h \in C^k(V, V) : h(gv) = gh(v) \text{ for any } g \in SO(2),\ v \in V\}.$$

Let us denote by ∇f the gradient of $f \in C^k_{SO(2)}(V, \mathbb{R})$, $k \in \mathbb{N}$.

(2.8) LEMMA. *Let V be an orthogonal representation of the group $SO(2)$ and let $f \in C^k_{SO(2)}(V, \mathbb{R})$, $k \in \mathbb{N}$. Then, $\nabla f \in C^{k-1}_{SO(2)}(V, V)$.*

Degree for $SO(2)$-equivariant gradient maps is an element of tom Dieck ring $U(SO(2))$. Therefore below we define this ring and prove some of its basic properties.

Put $U(SO(2)) = \mathbb{Z} \oplus (\bigoplus_{k=1}^{\infty} \mathbb{Z})$ and define the actions

$$+, *\colon U(SO(2)) \times U(SO(2)) \to U(SO(2)),$$
$$\cdot\colon \mathbb{Z} \times U(SO(2)) \to U(SO(2))$$

as follows:

(2.9.1) $\alpha + \beta = (\alpha_0 + \beta_0, \alpha_1 + \beta_1, \ldots, \alpha_k + \beta_k, \ldots)$,

(2.9.2) $\alpha * \beta = (\alpha_0 \cdot \beta_0, \alpha_0 \cdot \beta_1 + \beta_0 \cdot \alpha_1, \ldots, \alpha_0 \cdot \beta_k + \beta_0 \cdot \alpha_k, \ldots)$,

(2.9.3) $\gamma \cdot \alpha = (\gamma \cdot \alpha_0, \gamma \cdot \alpha_1, \ldots, \gamma \cdot \alpha_k, \ldots)$,

where $\gamma \in \mathbb{Z}$, $\alpha = (\alpha_0, \ldots, \alpha_k, \ldots)$, $\beta = (\beta_0, \ldots, \beta_k, \ldots) \in U(SO(2))$. It is easy to check that $(U(SO(2)), +, *)$ is a commutative ring with unit. Ring $(U(SO(2)), +, *)$ is known as the tom Dieck ring. For definition of a tom Dieck ring $U(G)$ for any compact Lie group G we refer the reader to [Di].

Additionally, define $U_\pm(SO(2)) \subset U(SO(2))$ as follows

$$U_\pm(SO(2)) = \{\alpha \in U(SO(2)) : \pm\alpha_k \ge 0 \text{ for all } k \in \mathbb{N} \cup \{0\}\}.$$

(2.10) REMARK. It is easy to check that

$$*\colon U_\pm(SO(2)) \times U_\pm(SO(2)) \to U_+(SO(2)),$$
$$*\colon U_\pm(SO(2)) \times U_\mp(SO(2)) \to U_-(SO(2)).$$

In the following lemma we collect some basic properties of the tom Dieck ring.

(2.11) LEMMA. *Let $n \in \mathbb{N}$, $\alpha, \beta \in U(SO(2))$. Then:*

(2.11.1) $\mathbb{I} = (1, 0, \dots) \in U(SO(2))$ *is the unit in* $U(SO(2))$,

(2.11.2) $\alpha^n = (\alpha_0^n, n \cdot \alpha_0^{n-1} \cdot \alpha_1, \dots, n \cdot \alpha_0^{n-1} \cdot \alpha_k, \dots) \in U(SO(2))$,

(2.11.2) *if* $\alpha \in U_\pm(SO(2))$, *then* $\alpha^{2n} \in U_+(SO(2))$,

(2.11.4) *if* $\alpha \in U_\pm(SO(2))$, *then* $\alpha^{2n+1} \in U_\pm(SO(2))$,

(2.11.5) α *is invertible in* $U(SO(2))$ *if and only if* $\alpha_0 = \pm 1$; *moreover, if* $\alpha_0 = \pm 1$, *then* $\alpha^{-1} = \pm 2 \cdot \mathbb{I} - \alpha$,

(2.11.6) *if* $\alpha_0 = \beta_0 = 0$, *then* $\alpha * \beta = \Theta \in U(SO(2))$.

PROOF. (2.11.1) Fix any $\alpha \in U(SO(2))$. By formula (2.9.2) we obtain the following

$$\begin{aligned} \alpha * \mathbb{I} =& (\alpha_0, \alpha_1, \dots, \alpha_k, \dots) * (1, 0, \dots, 0, \dots) \\ =& (\alpha_0 \cdot 1, \alpha_0 \cdot 0 + 1 \cdot \alpha_1, \dots, \alpha_0 \cdot 0 + 1 \cdot \alpha_k, \dots) = \alpha. \end{aligned}$$

(2.11.2) The proof is by induction on n. First of all notice that

$$\alpha^1 = (\alpha_0^1, \alpha_0^{1-1} \cdot \alpha_1, \dots, \alpha_0^{1-1} \cdot \alpha_k, \dots) = \alpha.$$

Next fix $n_0 \in \mathbb{N}$ and assume that for any $n < n_0$

$$\alpha^n = (\alpha_0^n, n \cdot \alpha_0^{n-1} \cdot \alpha_1, \dots, n \cdot \alpha_0^{n-1} \cdot \alpha_k, \dots).$$

What is left is to show that the above formula holds true for $n = n_0$. In fact

$$\begin{aligned} \alpha^{n_0} &= \alpha^1 * \alpha^{n_0-1} = (\alpha_0, \alpha_1, \dots, \alpha_k, \dots) \\ &\quad * (\alpha_0^{n_0-1}, (n_0 - 1) \cdot \alpha_0^{n_0-2} \cdot \alpha_1, \dots, (n_0 - 1) \cdot \alpha_0^{n_0-2} \cdot \alpha_k, \dots) \\ &= (\alpha_0^{n_0}, (n_0 - 1) \cdot \alpha_0^{n_0-1} \cdot \alpha_1 + \alpha_0^{n_0-1} \cdot \alpha_1, \dots, \\ &\quad (n_0 - 1) \cdot \alpha_0^{n_0-1} \cdot \alpha_k + \alpha_0^{n_0-1} \cdot \alpha_k, \dots) \\ &= (\alpha_0^{n_0}, n_0 \cdot \alpha_0^{n_0-1} \cdot \alpha_1, \dots, n_0 \cdot \alpha_0^{n_0-1} \cdot \alpha_k, \dots), \end{aligned}$$

which completes the proof.

(2.11.3) Direct consequence of (2.11.2). (2.11.4) Direct consequence of (2.11.2).

(2.11.5) If α is invertible in $U(SO(2))$ then there is $\gamma \in U(SO(2))$ such that $\alpha * \gamma = \mathbb{I}$. Since $\alpha_0 * \gamma_0 = 1$ and $\alpha_0, \gamma_0 \in \mathbb{Z}$, $\alpha_0 = \gamma_0 = \pm 1$. If $\alpha_0 = \pm 1$ then define $\gamma = \pm 2 \cdot \mathbb{I} - \alpha$ and notice that

$$\begin{aligned} \alpha * \gamma &= \alpha * (\pm 2 \cdot \mathbb{I} - \alpha) = \pm 2 \cdot \alpha - \alpha^2 \\ &= (2, \pm 2 \cdot \alpha_1, \dots, \pm 2 \cdot \alpha_k, \dots) - (1, \pm 2 \cdot \alpha_1, \dots, \pm 2 \cdot \alpha_k, \dots) = \mathbb{I}, \end{aligned}$$

which completes the proof.

(2.11.6) Direct consequence of formula (2.9.2). □

If $\beta_1, \dots, \beta_q \in U(SO(2))$, then $\prod_{j=1}^q \beta_j = \beta_1 * \dots * \beta_q$. Moreover, it is understood that $\prod_{j \in \emptyset} \beta_j = \mathbb{I} \in U(SO(2))$.

3. $\nabla_{SO(2)}$-degree and its properties

In this section we present the main properties of the degree for $SO(2)$-equivariant gradient maps. We denote it briefly by $\nabla_{SO(2)} - \deg$. Throughout this section, without loss of the generality, we assume that V, V_1, V_2 are orthogonal representations of the group $SO(2)$.

Let $f \in C^1_{SO(2)}(V, \mathbb{R})$. Since V is an orthogonal representation, we have $\nabla f \in C^0_{SO(2)}(V, V)$. Choose an open, bounded and $SO(2)$-invariant subset $\Omega \subset V$ such that $(\nabla f)^{-1}(0) \cap \partial\Omega = \emptyset$. Under these assumptions we have defined in [Ry1] the degree theory for $SO(2)$-equivariant gradient maps $\nabla_{SO(2)} - \deg(\nabla f, \Omega) \in U(SO(2))$ with coordinates

$$\begin{aligned}\nabla_{SO(2)} - \deg(\nabla f, \Omega) = (&\nabla_{SO(2)} - \deg_{SO(2)}(\nabla f, \Omega),\\ &\nabla_{SO(2)} - \deg_{\mathbb{Z}_1}(\nabla f, \Omega), \dots, \nabla_{SO(2)} - \deg_{\mathbb{Z}_k}(\nabla f, \Omega), \dots).\end{aligned}$$

Set $D_\alpha(V, v_0) = \{v \in V : \|v - v_0\| < \alpha\}$. For simplicity of notation we write $D_\alpha(V)$, instead of $D_\alpha(V, 0)$. The boundary of $\Omega \subset V$ will be denoted by $\partial\Omega$.

In the following theorem we formulate the main properties of $\nabla_{SO(2)} - \deg$.

(3.1) THEOREM ([Ry1]). *The degree for $SO(2)$-equivariant gradient maps has the following properties:*

(3.1.1) *Fix $f \in C^1_{SO(2)}(V, \mathbb{R})$. Let $\Omega \subset V$ be an open, bounded and $SO(2)$-invariant subset such that $(\nabla f)^{-1}(0) \cap \partial\Omega = \emptyset$. Then,*

(a) *if $\nabla_{SO(2)} - \deg(\nabla f, \Omega) \neq \Theta$, then $(\nabla f)^{-1}(0) \cap \Omega \neq \emptyset$,*
(b) *if $H \in \Psi(SO(2))$ and $\nabla_{SO(2)} - \deg_H(\nabla f, \Omega) \neq 0$, then*

$$(\nabla f)^{-1}(0) \cap \Omega^H \neq \emptyset,$$

(c) *if $\Omega = \Omega_0 \cup \Omega_1$ and $\Omega_0 \cap \Omega_1 = \emptyset$, then*

$$\nabla_{SO(2)} - \deg(\nabla f, \Omega) = \nabla_{SO(2)} - \deg(\nabla f, \Omega_0) + \nabla_{SO(2)} - \deg(\nabla f, \Omega_1),$$

(d) *if $\Omega_0 \subset \Omega$ is open, $SO(2)$-invariant and $(\nabla f)^{-1}(0) \cap \Omega \subset \Omega_0$, then*

$$\nabla_{SO(2)} - \deg(\nabla f, \Omega) = \nabla_{SO(2)} - \deg(\nabla f, \Omega_0),$$

(e) *if W is an orthogonal representation of the group $SO(2)$ and $\alpha > 0$, then $\nabla_{SO(2)} - \deg((\nabla f, \mathrm{id}), \Omega \times D_\alpha(W)) = \nabla_{SO(2)} - \deg(\nabla f, \Omega)$,*
(f) *if $f \in C^2_{SO(2)}(V, \mathbb{R})$ is such that $\nabla f(0) = 0$ and $\nabla^2 f(0)$ is an $SO(2)$-equivariant self-adjoint isomorphism then there is $\alpha_0 > 0$ such that for any $\alpha < \alpha_0$ we have*

$$\nabla_{SO(2)} - \deg(\nabla f, D_\alpha(V)) = \nabla_{SO(2)} - \deg(\nabla^2 f(0), D_\alpha(V)).$$

(3.1.2) *Fix* $f \in C^1_{SO(2)}(V \times [0,1], \mathbb{R})$ *such that* $(\nabla_v f)^{-1}(0) \cap (\partial\Omega \times [0,1]) = \emptyset$, *then* $\nabla_{SO(2)} - \deg(\nabla_v f(\cdot, 0), \Omega) = \nabla_{SO(2)} - \deg(\nabla_v f(\cdot, 1), \Omega)$.

(3.2) REMARK. Fix $f \in C^1_{SO(2)}(V, \mathbb{R})$. Let $\Omega \subset V$ be an open, bounded and $SO(2)$-equivariant subset such that $(\nabla f)^{-1}(0) \cap \partial\Omega = \emptyset$. Directly from the construction of $\nabla_{SO(2)} - \deg$ it follows that

(3.2.1) if $H \in \Psi(SO(2))$ and $\Omega_H = \emptyset$, then $\nabla_{SO(2)} - \deg_H(\nabla f, \Omega) = 0$,
(3.2.2) if $0 \in \Omega$ and $V^{SO(2)} = \{0\}$, then $\nabla_{SO(2)} - \deg_{SO(2)}(\nabla f, \Omega) = 1$.

The principal significance of the theorem stated below is that it allows us to compute the $\nabla_{SO(2)} - \deg$ of a product map in terms of $\nabla_{SO(2)} - \deg$ of coordinate maps.

(3.3) THEOREM ([Ry3]). *Let* $\Omega_i \subset V_i, i = 1, 2$, *be open, bounded and* $SO(2)$-*invariant subsets of* $SO(2)$-*representations* V_1, V_2. *Let* $f_i \in C^1_{SO(2)}(V_i, \mathbb{R})$, $i = 1, 2$, *be such that* $(\nabla f_i)^{-1}(0) \cap \partial\Omega_i = \emptyset$, $i = 1, 2$. *Then,*

$$\nabla_{SO(2)} - \deg((\nabla f_1, \nabla f_2), \Omega_1 \times \Omega_2) = \nabla_{SO(2)} - \deg(\nabla f_1, \Omega_1) * \nabla_{SO(2)} - \deg(\nabla f_2, \Omega_2).$$

The point of this lemma is that it allows us to simplify the computation of index of an isolated degenerate critical point.

(3.4) LEMMA ([Ry3]). *Let* $f \in C^2_{SO(2)}(V, \mathbb{R})$ *be such that:*

(3.4.1) $\det(\nabla^2 f(0)) = 0$,
(3.4.2) *there is* $\varepsilon > 0$ *such that* $(\nabla f)^{-1}(0) \cap D_\varepsilon(V) = \{0\}$.

Then there are $\alpha > 0$ *and* $H \in C^2_{SO(2)}(V \times [0,1], \mathbb{R})$ *such that:*

(3.4.3) $\varepsilon > \alpha$,
(3.4.4) $\nabla_v H\colon (D_\alpha(V) \times [0,1], \partial D_\alpha(V) \times [0,1]) \to (V, V \setminus \{0\})$,
(3.4.5) $\nabla_v H(\cdot, 0) = \nabla f(\cdot)$,
(3.4.6) $(\nabla_v H)^{-1}(0) \cap (D_\alpha(V) \times [0,1]) = \{0\} \times [0,1]$,
(3.4.7) *there is* $\varphi_0 \in C^2_{SO(2)}(\ker \nabla^2 f(0), \mathbb{R})$ *such that*

$$\nabla_v H(v_1, v_2, 1) = (\nabla \varphi_0(v_1), \nabla^2 f(0) v_2),$$

where $(v_1, v_2) \in D_\alpha(\ker \nabla^2 f(0)) \times D_\alpha(\operatorname{im} \nabla^2 f(0))$.

Let $GL^{\nabla}_{SO(2)}(V)$ denote the set of self-adjoint, $SO(2)$-equivariant isomorphisms of V. We will denote by $m^-(L)$ the Morse index of a symmetric matrix L.

It is essential to have some computational formulas for $\nabla_{SO(2)} - \deg$. The following theorem yields information about $\nabla_{SO(2)} - \deg$ of isomorphism.

(3.5) LEMMA ([Ry1]). *If* $V \approx \mathbb{R}[k_0, 0] \oplus \mathbb{R}[k_1, m_1] \oplus \ldots \oplus \mathbb{R}[k_r, m_r]$ *and* $L \in GL^{\nabla}_{SO(2)}(V)$ *then:*

(3.5.1) $L = \operatorname{diag}(L_0, \ldots, L_r)$,

(3.5.2) $\nabla_{SO(2)} - \deg_H(L, D_\alpha(V))$

$$= \begin{cases} (-1)^{m^-(L_0)} & \textit{for } H = SO(2), \\ (-1)^{m^-(L_0)} m^-(L_i)/2 & \textit{for } H = \mathbb{Z}_{m_i}, \\ 0 & \textit{for } H \notin \{SO(2), \mathbb{Z}_{m_1}, \ldots, \mathbb{Z}_{m_r}\}; \end{cases}$$

it is understood that if $k_0 = 0$, *then*

$$\nabla_{SO(2)} - \deg_H(L, D_\alpha(V)) = \begin{cases} 1 & \textit{for } H = SO(2), \\ m^-(L_i)/2 & \textit{for } H = \mathbb{Z}_{m_i}, \\ 0 & \textit{for } H \notin \{SO(2), \mathbb{Z}_{m_1}, \ldots, \mathbb{Z}_{m_r}\}; \end{cases}$$

(3.5.3) *in particular, if* $L = -\mathrm{id}$, *then*

$$\nabla_{SO(2)} - \deg_H(-\mathrm{id}, D_\alpha(V)) = \begin{cases} (-1)^{k_0} & \textit{for } H = SO(2), \\ (-1)^{k_0} k_i & \textit{for } H = \mathbb{Z}_{m_i}, \\ 0 & \textit{for } H \notin \{SO(2), \mathbb{Z}_{m_1}, \ldots, \mathbb{Z}_{m_r}\}. \end{cases}$$

The following lemma is a direct consequence of Lemmas (2.11) and (3.5).

(3.6) LEMMA. *If* $V \approx \mathbb{R}[k_0, 0] \oplus \mathbb{R}[k_1, m_1] \oplus \ldots \oplus \mathbb{R}[k_r, m_r]$ *and* $L \in GL^{\nabla}_{SO(2)}(V)$ *then:*

(3.6.1) $\nabla_{SO(2)} - \deg(L, D_\alpha(V))$ *is invertible in* $U(SO(2))$,

(3.6.2) $\nabla_{SO(2)} - \deg(L, D_\alpha(V)) \in U_-(SO(2)) \cup U_+(SO(2)) \subset U(SO(2))$,

(3.6.3) $(-1)^{m^-(L_0)} \cdot \nabla_{SO(2)} - \deg(L, D_\alpha(V)) \in U_+(SO(2))$,

(3.6.4) *if* $n \in \mathbb{N}$, *then*

$$((\nabla_{SO(2)} - \deg(-\mathrm{id}, D_\alpha(V)))^{2n})_H = \begin{cases} 1 & \textit{for } H = SO(2), \\ 2 \cdot n \cdot k_i & \textit{for } H = \mathbb{Z}_{m_i}, \\ 0 & \textit{for } H \notin \{SO(2), \mathbb{Z}_{m_1}, \ldots, \mathbb{Z}_{m_r}\}. \end{cases}$$

(3.7) REMARK. Computing the Brouwer degree $\deg_B(L, \Omega, 0) = \operatorname{sign} \det L$ of a linear isomorphism $L\colon \mathbb{R}^n \to \mathbb{R}^n$ on an open, bounded set $\Omega \subset \mathbb{R}^n$ containing the origin we take into account only the multiplicities of negative eigenvalues of L. In other words, $\deg_B(L,\Omega,0)$ depends only on L. On the other hand, computing the degree for $SO(2)$-equivariant gradient maps $\nabla_{SO(2)}-\deg(L,\Omega)$ of an isomorphism $L \in GL^{\nabla}_{SO(2)}(V)$ on an open, bounded, $SO(2)$-invariant set $\Omega \subset V$ containing the origin, we take into consideration the multiplicities of negative eigenvalues of L and the structure of representation V of the group $SO(2)$.

The natural generalization of the Brouwer degree is the Leray–Schauder degree. Below we define $\nabla_{SO(2)} - \deg$ in an infinite-dimensional case.

Let $(\mathbb{H}, \langle\,\cdot\,,\,\cdot\,\rangle_{\mathbb{H}})$ be an infinite-dimensional, separable Hilbert space which is an orthogonal representation of the group $SO(2)$. Let $\Gamma = \{\pi_n\colon \mathbb{H} \to \mathbb{H} : n \in \mathbb{N}\cup\{0\}\}$ be a sequence of $SO(2)$-equivariant orthogonal projections.

(3.8) Definition. Γ is an $SO(2)$-*equivariant approximation scheme* on $\mathbb{H}$ if the following conditions are fulfilled:

(3.8.1) $\mathbb{H}_n = \operatorname{im}\pi_n$, is an finite-dimensional orthogonal $SO(2)$-subrepresentation of $\mathbb{H}$ for any $n \in \mathbb{N}\cup\{0\}$,

(3.8.2) $\mathbb{H}_n \subset \mathbb{H}_{n+1} = \mathbb{H}_n \oplus (\mathbb{H}_{n+1}\ominus\mathbb{H}_n)$, $\mathbb{H}_n \perp (\mathbb{H}_{n+1}\ominus\mathbb{H}_n)$ for any $n \in \mathbb{N}\cup\{0\}$,

(3.8.3) $\lim_{n\to\infty}\pi_n u = u$, for any $u \in \mathbb{H}$.

Define

$$\begin{aligned} C^k_{SO(2)}(\mathbb{H},\mathbb{R}) &= \{\Phi \in C^k(\mathbb{H},\mathbb{R}) : \Phi(gh) = \Phi(h) \text{ for any } g \in SO(2), h \in \mathbb{H}\}, \\ C^k_{SO(2)}(\mathbb{H}\times\Lambda,\mathbb{R}) &= \{\Phi \in C^k(\mathbb{H}\times\Lambda,\mathbb{R}) : \Phi(\,\cdot\,,\lambda) \in C^k_{SO(2)}(\mathbb{H},\mathbb{R}) \text{ for any } \lambda \in \Lambda\}. \end{aligned}$$

Fix $\Phi \in C^1_{SO(2)}(\mathbb{H},\mathbb{R})$ such that

$$\nabla\Phi(u) = u - \nabla\eta(u) \tag{3.9}$$

where $\nabla\eta\colon \mathbb{H} \to \mathbb{H}$ is an $SO(2)$-equivariant compact operator.

(3.10) Lemma ([Da2]). *Let $\Phi_n = \Phi_{|\mathbb{H}_n}\colon \mathbb{H}_n \to R$ be the restriction of Φ to $\mathbb{H}_n$. Then, $\nabla\Phi_n\colon \mathbb{H}_n \to \mathbb{H}_n$ satisfies the following equality*

$$\nabla\Phi_n(u) = \pi_n(\nabla\Phi(u)) = u - \pi_n\nabla\eta(u).$$

Let $\Omega \subset \mathbb{H}$ be an open bounded and $SO(2)$-invariant set. The boundary of $\Omega \subset \mathbb{H}$ will be denoted by $\partial\Omega$. Assume that $(\nabla\Phi)^{-1}(0)\cap\partial\Omega = \emptyset$. In this situation $\nabla_{SO(2)} - \deg(\nabla\Phi,\Omega) \in U(SO(2))$ is well-defined by the following formula

$$\nabla_{SO(2)}-\deg(\nabla\Phi,\Omega) = \lim_{n\to\infty}\nabla_{SO(2)}-\deg(\mathrm{id}-\pi_n\nabla\eta,\pi_n(\Omega)) \in U(SO(2)), \tag{3.11}$$

see [Ry1] for details.

This theorem yields information about the infinite-dimensional version of the degree for $SO(2)$-equivariant gradient maps.

(3.12) THEOREM ([Ry1]). *The infinite-dimensional generalization of the degree for $SO(2)$-equivariant gradient maps has the following properties:*

(3.12.1) *Fix $\Phi \in C^1_{SO(2)}(\mathbb{H}, \mathbb{R})$ satisfying (3.9). Let $\Omega \subset \mathbb{H}$ be an open, bounded and $SO(2)$-invariant subset such that $(\nabla\Phi)^{-1}(0) \cap \partial\Omega = \emptyset$. Then,*

(a) *if $\nabla_{SO(2)} - \deg(\nabla\Phi, \Omega) \neq \Theta$, then $(\nabla\Phi)^{-1}(0) \cap \Omega \neq \emptyset$; more precisely, if $\nabla_{SO(2)} - \deg_H(\nabla\Phi, \Omega) \neq \Theta$, then $(\nabla\Phi)^{-1}(0) \cap \Omega^H \neq \emptyset$.*

(b) *if $\Omega = \Omega_0 \cup \Omega_1$ and $\Omega_0 \cap \Omega_1 = \emptyset$, then*

$$\nabla_{SO(2)} - \deg(\nabla\Phi, \Omega) = \nabla_{SO(2)} - \deg(\nabla\Phi, \Omega_0) + \nabla_{SO(2)} - \deg(\nabla\Phi, \Omega_1).$$

(c) *if $\Omega_0 \subset \Omega$ is open, $SO(2)$-invariant and such that $(\nabla\Phi)^{-1}(0) \cap \Omega \subset \Omega_0$, then $\nabla_{SO(2)} - \deg(\nabla\Phi, \Omega) = \nabla_{SO(2)} - \deg(\nabla\Phi, \Omega_0)$.*

(d) *assume that $\Phi \in C^2_{SO(2)}(\mathbb{H}, \mathbb{R})$, $\nabla\Phi(0) = 0$ and that $1 \notin \sigma(\nabla^2\eta(0))$. Then, there is $\alpha_0 > 0$ such that for any $\alpha_0 > \alpha$ we have*

$$\nabla_{SO(2)} - \deg(\nabla\Phi, D_\alpha(\mathbb{H})) = \nabla_{SO(2)} - \deg(\mathrm{id} - \nabla^2\eta(0), D_\alpha(\mathbb{H})).$$

(3.12.2) *Let $\Psi \in C^1_{SO(2)}(\mathbb{H} \times [0,1], \mathbb{R})$ be such that $\nabla_u\Psi(u,t) = u - \nabla_u\zeta(u,t)$, where $\nabla_u\zeta \colon \mathbb{H} \times [0,1] \to \mathbb{H}$ is an $SO(2)$-equivariant compact operator. If $\Omega \subset \mathbb{H}$ is an open, bounded and $SO(2)$-invariant subset such that $(\nabla\Psi)^{-1}(0) \cap \partial\Omega = \emptyset$, then $\nabla_{SO(2)} - \deg(\nabla\Psi(\,\cdot\,,0), \Omega) = \nabla_{SO(2)} - \deg(\nabla\Psi(\,\cdot\,,1), \Omega)$.*

(3.12.3) *Fix $\Psi \in C^1_{SO(2)}(\mathbb{H} \times [0,1], \mathbb{R})$ such that $\nabla_u\Psi(u,t) = u - \nabla_u\zeta(u,\lambda)$, where $\nabla_u\zeta \colon \mathbb{H} \times [0,1] \to \mathbb{H}$ is an $SO(2)$-equivariant compact operator. Let $\Omega \subset \mathbb{H} \times [0,1]$ be an open, bounded, $SO(2)$-invariant subset and let $\Omega_\lambda = \{h \in \mathbb{H} : (h,\lambda) \in \Omega\}$ for any $\lambda \in [0,1]$. If $\Omega_0, \Omega_1 \neq \emptyset$ and $(\nabla\Psi(\,\cdot\,,\lambda))^{-1}(0) \cap \partial\Omega_\lambda = \emptyset$ for any $\lambda \in [0,1]$, then $\nabla_{SO(2)} - \deg(\nabla\Psi(\,\cdot\,,0), \Omega_0) = \nabla_{SO(2)} - \deg(\nabla\Psi(\,\cdot\,,1), \Omega_1)$.*

Let $L \colon \mathbb{H} \to \mathbb{H}$ be a linear, bounded, compact, self-adjoint, $SO(2)$-equivariant operator with spectrum $\sigma(L) = \{\lambda_i\}$. By $V_L(\lambda_i)$ we will denote the eigenspace of L corresponding to the eigenvalue λ_i.

In the theorem below we derive a formula for $\nabla_{SO(2)} - \deg$ of an isomorphism in the form of a compact perturbation of the identity. In fact, we reduce the computations to finite-dimensional case.

(3.13) THEOREM ([Ry1]). *Under the above assumptions, if additionally $1 \notin \sigma(L)$, then*

$$\nabla_{SO(2)} - \deg(\mathrm{id} - L, D_\alpha(\mathbb{H})) = \nabla_{SO(2)} - \deg\left(-\mathrm{id}, D_\alpha\left(\bigoplus_{\lambda_1 > 1} V_L(\lambda_i) \oplus \mathbb{R}[2,0]\right)\right).$$

(3.14) COROLLARY. *Combining Theorems* (3.3) *and* (3.13) *we obtain:*

$$\nabla_{SO(2)}-\deg(\mathrm{id}-L, D_\alpha(\mathbb{H})) = \prod_{\lambda_i>1} \nabla_{SO(2)}-\deg(-\mathrm{id}, D_\alpha(V_L(\lambda_i))).$$

4. Krasnosel'skiĭ–Rabinowitz bifurcation theorems

One of the most important theorems in topological nonlinear analysis are the Krasnosel'skii local bifurcation theorem and the Rabinowitz global bifurcation theorem. Proofs and some applications of these theorems one can find, for example, in [Iz1], [Kras], [Ni], [Ra2]–[Ra6]. The classical versions of these theorems have been proved with the use of the Leray–Schauder topological degree.

Roughly speaking, Krasnosel'skiĭ has proved that bifurcation index computed at the eigenvalue of odd multiplicity is nontrivial (in fact it is equal to ± 2) and that the nontriviality of bifurcation index implies the existence of bifurcation point of nontrivial solutions of a nonlinear eigenvalue problem.

On the other hand, Rabinowitz discovered that under certain circumstances a local linearized analysis forces the existence of a branching point of nontrivial solutions of a nonlinear eigenvalue problem. This is a very powerful result that is quoted very often. More precisely, Rabinowitz has proved that nontriviality of the bifurcation index implies the existence of a closed connected set of nontrivial solutions, branching from the set of trivial solutions, which is either unbounded or comes back to the set of trivial solutions.

In this section we restrict our attention to the class of gradient, $SO(2)$-equivariant operators in the form of a compact perturbation of the identity. Such restriction allows us to apply the degree for $SO(2)$-equivariant gradient maps.

The aim of this section is to formulate some versions of the Krasnosel'skiĭ bifurcation theorem and the Rabinowitz alternative for $SO(2)$-equivariant gradient operators.

Let $\mathbb{H}$ be a separable Hilbert space which is an orthogonal representation of the group $SO(2)$. From now on we assume that

(A1) $\Phi \in C^1_{SO(2)}(\mathbb{H}\times\mathbb{R}, \mathbb{R})$,

(A2) $\Phi(u,\lambda) = 1/2\|u\|^2 - \eta(u,\lambda)$,

(A3) $\nabla_u\eta\colon (\mathbb{H}\times\mathbb{R}, \{0\}\times\mathbb{R}) \to (\mathbb{H}, \{0\})$ is a continuous, $SO(2)$-equivariant compact operator.

Put $\mathcal{N}(\Phi) = \{(u,\lambda)\in\mathbb{H}\times\mathbb{R} : \nabla_u\Phi(u,\lambda) = 0 \text{ and } u\neq 0\}$. The closure of A will be denoted by $\overline{A}$.

(4.1) DEFINITION. A point $(0,\lambda_0)\in\mathbb{H}\times\mathbb{R}$ is said to be a *bifurcation point* of solutions of the equation $\nabla_u\Phi(u,\lambda)=0$, if $(0,\lambda_0)\in\overline{\mathcal{N}(\Phi)}$. We will denote by $\mathcal{BIF}(\Phi)$ the set of bifurcation points of solutions of the equation $\nabla_u\Phi(u,\lambda)=0$.

Let $\mathcal{C}(\lambda_0)$ denote a connected component of $\overline{\mathcal{N}(\Phi)}$ such that $(0, \lambda_0) \in \mathcal{C}(\lambda_0)$. A set $\mathcal{C}$ is said to be continuum if it is closed and connected.

(4.2) DEFINITION. A bifurcation point $(0, \lambda_0) \in \mathcal{BIF}(\Phi)$ is said to be a *branching point* of solutions of the equation $\nabla_u \Phi(u, \lambda) = 0$, if $\mathcal{C}(\lambda_0) \setminus \{(0, \lambda_0)\} \neq \emptyset$. We will denote by $\mathcal{BRA}(\Phi)$ the set of branching points of solutions of the equation $\nabla_u \Phi(u, \lambda) = 0$.

(4.3) REMARK. It is clear that $\mathcal{BRA}(\Phi) \subset \mathcal{BIF}(\Phi)$. Moreover, using the implicit function theorem one can show that if $\Phi \in C^2_{SO(2)}(\mathbb{H} \times \mathbb{R}, \mathbb{R})$ and $(0, \lambda_0) \in \mathcal{BIF}(\Phi)$, then $\nabla^2_u \Phi(0, \lambda_0) \colon \mathbb{H} \to \mathbb{H}$ is not an isomorphism.

Set $D_\varepsilon(\mathbb{H}, h_0) = \{h \in \mathbb{H} : \|h - h_0\| < \varepsilon\}$. For simplicity of notation we write $D_\varepsilon(\mathbb{H})$ instead of $D_\varepsilon(\mathbb{H}, 0)$.

Theorems (4.4) and (4.6) are the general local and global bifurcation theorems for the class of $SO(2)$-equivariant gradient operators, respectively. The proofs of these theorems are standard. Since we work in the class of $SO(2)$-equivariant gradient operators we replaced the Leray–Schauder degree by the degree for $SO(2)$-equivariant gradient maps. In fact, analogous theorems can be proved for any class of operators with reasonable degree theory defined for this class.

(4.4) THEOREM (Local Bifurcation Theorem I). *Let Φ satisfies assumptions* (A1)–(A3). *Fix $(0, \lambda_\pm) \in (\{0\} \times \mathbb{R}) \setminus \mathcal{BIF}(\Phi)$ such that $\lambda_- < \lambda_+$. Then there is $\varepsilon > 0$ such that $(\nabla_u \Phi(\,\cdot\,, \lambda_\pm))^{-1}(0) \cap D_\varepsilon(\mathbb{H}) = \{0\}$. Moreover, if*

$$(4.4.1)\quad \nabla_{SO(2)}-\deg(\nabla_u \Phi(\,\cdot\,, \lambda_+), D_\varepsilon(\mathbb{H})) \neq \nabla_{SO(2)}-\deg(\nabla_u \Phi(\,\cdot\,, \lambda_-), D_\varepsilon(\mathbb{H}))$$

then $\mathcal{BIF}(\Phi) \cap (\{0\} \times (\lambda_-, \lambda_+)) \neq \emptyset$.

PROOF. Suppose, contrary to our claim, that (4.4.1) holds true and that

$$\mathcal{BIF}(\Phi) \cap (\{0\} \times (\lambda_-, \lambda_+)) = \emptyset.$$

Since $[\lambda_-, \lambda_+]$ is compact, there is $\varepsilon > 0$ such that

$$(\nabla_u \Phi)^{-1}(0) \cap (D_\varepsilon(\mathbb{H}) \times [\lambda_-, \lambda_+]) = \{0\} \times [\lambda_-, \lambda_+].$$

Consequently, by Theorem (3.12) we obtain

$$\nabla_{SO(2)}-\deg(\nabla_u \Phi(\,\cdot\,, \lambda_+), D_\varepsilon(\mathbb{H})) = \nabla_{SO(2)}-\deg(\nabla_u \Phi(\,\cdot\,, \lambda_-), D_\varepsilon(\mathbb{H})),$$

contrary to (4.4.1). □

The following Whyburn lemma is a topological tool which will be used in the proofs of the global bifurcation theorems.

(4.5) LEMMA ([Wh]). *Let K be a compact space and let $A, B \subset K$ be closed, disjoint subsets such that no component of K intersects both A and B. Then, there are compact subsets $K_A, K_B \subset K$ such that:*

(4.5.1) $A \subset K_A, B \subset K_B$,
(4.5.2) $K_A \cap K_B = \emptyset$,
(4.5.3) $K_A \cup K_B = K$.

(4.6) THEOREM (Global Bifurcation Theorem I). *Let Φ satisfies assumptions (A1)–(A3). Fix $(0, \lambda_\pm) \in (\{0\} \times \mathbb{R}) \setminus \mathcal{BIF}(\Phi)$ such that $\lambda_- < \lambda_+$. Then there is $\alpha > 0$ such that $(\nabla_u \Phi(\cdot, \lambda_\pm))^{-1}(0) \cap D_\alpha(\mathbb{H}) = \{0\}$ and if*

(4.6.1) $\nabla_{SO(2)} - \deg(\nabla_u \Phi(\cdot, \lambda_+), D_\alpha(\mathbb{H})) \neq \nabla_{SO(2)} - \deg(\nabla_u \Phi(\cdot, \lambda_-), D_\alpha(\mathbb{H}))$

then $\mathcal{BRA}(\Phi) \cap (\{0\} \times (\lambda_-, \lambda_+)) \neq \emptyset$. Put $\mathcal{S} = \overline{\mathcal{N}(\Phi)} \cup (\{0\} \times [\lambda_-, \lambda_+])$ and let $\mathcal{C} \subset \mathcal{S}$ be a continuum containing $\{0\} \times [\lambda_-, \lambda_+]$. Then,

(4.6.2) *either the continuum $\mathcal{C} \subset \mathbb{H} \times \mathbb{R}$ is unbounded,*
(4.6.3) *or the continuum $\mathcal{C}$ is bounded and moreover $\mathcal{C} \cap (\{0\} \times (\mathbb{R} \setminus [\lambda_-, \lambda_+])) \neq \emptyset$.*

PROOF. Suppose, contrary to our claim, that the continuum $\mathcal{C}$ is bounded and that $\mathcal{C} \cap (\{0\} \times (\mathbb{R} \setminus [\lambda_-, \lambda_+])) = \emptyset$. Since $\mathcal{BIF}(\Phi)$ is closed and $(0, \lambda_\pm) \in (\{0\} \times \mathbb{R}) \setminus \mathcal{BIF}(\Phi)$, there is $\varepsilon > 0$ such that $\big(\{0\} \times [\lambda_\pm - \varepsilon, \lambda_\pm + \varepsilon]\big) \cap \mathcal{BIF}(\Phi) = \emptyset$. Let $U \subset \mathbb{H} \times \mathbb{R}$ be an open bounded $SO(2)$-invariant subset such that

$$\mathcal{C} \subset U, \quad \partial U \cap (\{0\} \times [\lambda_- - \varepsilon, \lambda_+ + \varepsilon]) = \{(0, \lambda_\pm \pm \varepsilon/2)\}.$$

Notice that $\overline{U} \cap (\nabla_u \Phi)^{-1}(0)$ is compact and put in Lemma (4.5)

$$K := \overline{U} \cap (\nabla_u \Phi)^{-1}(0), \quad A := \mathcal{C}, \quad B := \partial U \cap \overline{\mathcal{N}(\Phi)}.$$

Applying Lemma (4.5) we obtain compact sets K_A, K_B satisfying the statement of Lemma (4.5). Choose sufficiently small $\gamma > 0$ such that γ-neighbourhoods $K_A(\gamma)$, $K_B(\gamma)$ of sets K_A, K_B, respectively, are disjoint. Define an open, bounded, $SO(2)$-invariant set $\mathcal{W}$ as follows

$$\mathcal{W} := SO(2) \cdot (U \setminus \overline{K_B(\gamma)}) \cup (D_\alpha(\mathbb{H}) \times [\lambda_- - \varepsilon, \lambda_+ + \varepsilon]),$$

where α is sufficiently small. Notice that $(\nabla_u \Phi)^{-1}(0) \cap \partial \mathcal{W} = \{(0, \lambda_\pm \pm \varepsilon)\}$. Using the homotopy invariance of $\nabla_{SO(2)} - \deg$ we obtain the following

$$\begin{aligned}
\nabla_{SO(2)} - \deg(\nabla_u \Phi(\cdot, \lambda_-), D_\alpha(\mathbb{H}))& \\
&= \nabla_{SO(2)} - \deg(\nabla_u \Phi(\cdot, \lambda_- - \varepsilon), D_\alpha(\mathbb{H})) \\
&= \nabla_{SO(2)} - \deg(\nabla_u \Phi(\cdot, \lambda_- - \varepsilon), \mathcal{W}_{\lambda_- - \varepsilon}) \\
&= \nabla_{SO(2)} - \deg(\nabla_u \Phi(\cdot, \lambda_+ + \varepsilon), \mathcal{W}_{\lambda_+ + \varepsilon}) \\
&= \nabla_{SO(2)} - \deg(\nabla_u \Phi(\cdot, \lambda_+ + \varepsilon), D_\alpha(\mathbb{H})) \\
&= \nabla_{SO(2)} - \deg(\nabla_u \Phi(\cdot, \lambda_+), D_\alpha(\mathbb{H})),
\end{aligned}$$

contrary to (4.6.1). □

In fact, Theorem (4.6) can be proved in a little bit stronger version. Below we formulate a corollary which is a direct consequence of Lemma (4.5) and Theorem (4.6). The proof of this corollary is standard, therefore we omit it.

(4.7) COROLLARY. *Under the assumptions of Theorem* (4.6) *There is* $(0, \lambda_0) \in \mathcal{BRA}(\Phi) \cap (\{0\} \times (\lambda_-, \lambda_+))$ *such that:*

(4.7.1) *either the continuum* $\mathcal{C}(\lambda_0) \subset \mathbb{H} \times \mathbb{R}$ *is unbounded,*
(4.7.2) *or the continuum* $\mathcal{C}(\lambda_0) \subset \mathbb{H} \times \mathbb{R}$ *is bounded and in addition*

$$\mathcal{C}(\lambda_0) \cap (\{0\} \times (\mathbb{R} \setminus [\lambda_-, \lambda_+])) \neq \emptyset.$$

Let Φ satisfy the following assumptions:

(A4) $\Phi \in C^2_{SO(2)}(\mathbb{H} \times \mathbb{R}, \mathbb{R})$,
(A5) $\Phi(u, \lambda) = 1/2\|u\|^2 - \eta(u, \lambda) = 1/2\|u\|^2 - \lambda/2 < Lu, u > -\psi(u, \lambda)$, where

(a) $L: \mathbb{H} \to \mathbb{H}$ is linear, bounded, self-adjoint, $SO(2)$-equivariant, completely continuous operator,
(b) $\nabla_u \psi: (\mathbb{H} \times \mathbb{R}, \{0\} \times \mathbb{R}) \to (\mathbb{H}, \{0\})$ is a continuous, $SO(2)$-equivariant compact operator,
(c) $\nabla \psi(u, \lambda) = o(\|u\|)$ uniformly for λ in compact subsets of $\mathbb{R}$.

(4.8) LEMMA. *Let* Φ *satisfy assumptions* (A4), (A5). *Then:*

(4.8.1) $\lambda_0^{-1} \notin \sigma(L)$ *if and only if* $\mathrm{id} - \lambda_0 L$ *is an isomorphism,*
(4.8.2) *if* $(0, \lambda_0) \in \mathcal{BIF}(\Phi)$, *then* $\lambda_0^{-1} \in \sigma(L)$.

Fix $\lambda_{k_0} \in \sigma(L)$. Choose $\varepsilon > 0$ such that $\lambda_{k_0}^{-1}$ is the only characteristic value of L in the interval $[\lambda_{k_0}^{-1} - \varepsilon, \lambda_{k_0}^{-1} + \varepsilon]$.

(4.9) DEFINITION. Element $\mathrm{BIF}(\lambda_{k_0}^{-1}) \in U(SO(2))$ defined by

$$\begin{aligned}\mathrm{BIF}(\lambda_{k_0}^{-1}) &= (\mathrm{BIF}_{SO(2)}(\lambda_{k_0}^{-1}), \mathrm{BIF}_{\mathbb{Z}_1}(\lambda_{k_0}^{-1}), \dots, \mathrm{BIF}_{\mathbb{Z}_k}(\lambda_{k_0}^{-1}), \dots) \\ &= \nabla_{SO(2)} - \deg(\mathrm{id} - (\lambda_{k_0}^{-1} + \varepsilon)L, D_\alpha(\mathbb{H})) \\ &\quad - \nabla_{SO(2)} - \deg(\mathrm{id} - (\lambda_{k_0}^{-1} - \varepsilon)L, D_\alpha(\mathbb{H}))\end{aligned}$$

is said to be a *bifurcation index* at $(0, \lambda_{k_0}^{-1}) \in \mathbb{H} \times \mathbb{R}$.

Denote

$$\begin{aligned}\sigma_+(L) &= \sigma(L) \cap (0, +\infty) = \{\lambda_1^+, \lambda_2^+, \dots, \lambda_k^+, \dots\}, \\ \sigma_-(L) &= \sigma(L) \cap (-\infty, 0) = \{\lambda_1^-, \lambda_2^-, \dots, \lambda_k^-, \dots\},\end{aligned}$$

with order $\lambda_1^- < \ldots < \lambda_k^- < \ldots < 0 < \ldots < \lambda_k^+ < \ldots < \lambda_1^+$.

(4.10) Lemma. *Let Φ satisfy assumptions* (a4), (a5). *Then:*

(4.10.1) $\mathrm{BIF}((\lambda_1^+)^{-1}) = \nabla_{SO(2)} - \deg(-\mathrm{id}, D_\alpha(V_L(\lambda_1^+))) - \mathbb{I}$,

(4.10.2) $\mathrm{BIF}((\lambda_1^-)^{-1}) = \mathbb{I} - \nabla_{SO(2)} - \deg(-\mathrm{id}, D_\alpha(V_L(\lambda_1^-)))$,

(4.10.3) *for* $k \geq 2$,

$$\begin{aligned}\mathrm{BIF}((\lambda_k^+)^{-1}) &= \nabla_{SO(2)} - \deg\bigg(-\mathrm{id}, D_\alpha\bigg(\bigoplus_{i=1}^{k-1} V_L(\lambda_i^+)\bigg)\bigg) \\ &\quad * (\nabla_{SO(2)} - \deg(-\mathrm{id}, D_\alpha(V_L(\lambda_k^+))) - \mathbb{I}) \\ &= \bigg(\prod_{i=1}^{k-1} \nabla_{SO(2)} - \deg(-\mathrm{id}, D_\alpha(V_L(\lambda_i^+)))\bigg) \\ &\quad * (\nabla_{SO(2)} - \deg(-\mathrm{id}, D_\alpha(V_L(\lambda_k^+))) - \mathbb{I})\end{aligned}$$

(4.10.4) *for* $k \geq 2$,

$$\begin{aligned}\mathrm{BIF}((\lambda_k^-)^{-1}) &= \nabla_{SO(2)} - \deg\bigg(-\mathrm{id}, D_\alpha\bigg(\bigoplus_{i=1}^{k-1} V_L(\lambda_i^-)\bigg)\bigg) \\ &\quad * (\mathbb{I} - \nabla_{SO(2)} - \deg(-\mathrm{id}, D_\alpha(V_L(\lambda_k^-)))) \\ &= \bigg(\prod_{i=1}^{k-1} \nabla_{SO(2)} - \deg(-\mathrm{id}, D_\alpha(V_L(\lambda_i^-)))\bigg) \\ &\quad * (\mathbb{I} - \nabla_{SO(2)} - \deg(-\mathrm{id}, D_\alpha(V_L(\lambda_k^-)))).\end{aligned}$$

Proof. (4.10.1) First of all we notice that for a sufficiently small $\varepsilon > 0$ we have

$$1-((\lambda_1^+)^{-1} + \varepsilon) \cdot \lambda_i^+ \begin{cases} < 0 & \text{for } i = 1, \\ > 0 & \text{for } i > 1, \end{cases}$$

$$1-((\lambda_1^+)^{-1} - \varepsilon) \cdot \lambda_i^+ > 0 \quad \text{for any } i \in \mathbb{N},$$
$$1-((\lambda_1^+)^{-1} \pm \varepsilon) \cdot \lambda_i^- > 0 \quad \text{for any } i \in \mathbb{N}.$$

Combining (3.1.1)(f), formula (3.11) and the above we obtain the following

$$\begin{aligned}\mathrm{BIF}((\lambda_1^+)^{-1}) &= \nabla_{SO(2)} - \deg(\mathrm{id} - ((\lambda_1^+)^{-1} + \varepsilon)L, D_\alpha(\mathbb{H})) \\ &\quad - \nabla_{SO(2)} - \deg(\mathrm{id} - ((\lambda_1^+)^{-1} - \varepsilon)L, D_\alpha(\mathbb{H})) \\ &= \nabla_{SO(2)} - \deg(-\mathrm{id}, D_\alpha(V_L(\lambda_1^+))) - \mathbb{I},\end{aligned}$$

which completes the proof.

(4.10.2) The same proof as in (4.10.1) works for $\lambda_1^- \in \sigma_(L)$.

(4.10.3) First of all, let us notice that for sufficiently small $\varepsilon > 0$ we have

$$1-((\lambda_k^+)^{-1}+\varepsilon)\cdot\lambda_i^+ \begin{cases} <0 & \text{for } i \leq k, \\ >0 & \text{for } i > k, \end{cases}$$

$$1-((\lambda_k^+)^{-1}-\varepsilon)\cdot\lambda_i^+ \begin{cases} <0 & \text{for } i < k, \\ >0 & \text{for } i \geq k, \end{cases}$$

$$1-((\lambda_k^+)^{-1}\pm\varepsilon)\cdot\lambda_i^- > 0 \quad \text{for any } i \in \mathbb{N}.$$

Combining (3.1.1)(f), formula (3.11) and the above we obtain the following

$$\begin{aligned} \mathrm{BIF}((\lambda_k^+)^{-1}) &= \nabla_{SO(2)}-\deg(\mathrm{id}-((\lambda_k^+)^{-1}+\varepsilon)L, D_\alpha(\mathbb{H})) \\ &\quad -\nabla_{SO(2)}-\deg(\mathrm{id}-((\lambda_k^+)^{-1}-\varepsilon)L, D_\alpha(\mathbb{H})) \\ &= \nabla_{SO(2)}-\deg\Big(-\mathrm{id}, D_\alpha\Big(\bigoplus_{i=1}^{k} V_L(\lambda_i)\Big)\Big) \\ &\quad -\nabla_{SO(2)}-\deg\Big(-\mathrm{id}, D_\alpha\Big(\bigoplus_{i=1}^{k-1} V_L(\lambda_i)\Big)\Big). \end{aligned}$$

Finally, applying Theorem (3.3), we obtain the following

$$\begin{aligned} \mathrm{BIF}((\lambda_k^+)^{-1}) = \nabla_{SO(2)}-\deg\Big(-\mathrm{id}, D_\alpha\Big(\bigoplus_{i=1}^{k-1} V_L(\lambda_i^+)\Big)\Big) \\ * (\nabla_{SO(2)}-\deg(-\mathrm{id}, D_\alpha(V_L(\lambda_k^+))) - \mathbb{I}), \end{aligned}$$

which completes the proof.

(4.10.4) The same proof as in (4.10.3) works for $\lambda_k^- \in \sigma_+(L)$. □

As a direct consequence of the above lemma we obtain the following corollary.

(4.11) Corollary. *Taking into account Lemmas* (3.5) *and* (4.10) *we obtain the following. If* $\lambda_k \in \sigma(L)\setminus\{0\}$, *then* $\mathrm{BIF}(\lambda_k^{-1}) = \Theta \in U(SO(2))$ *if and only if* $\dim V_L(\lambda_k)$ *is even and* $V_L(\lambda_k)^{SO(2)} = V_L(\lambda_k)$. *In other words, bifurcation index vanishes at characteristic value if and only if the corresponding eigenspace is an even-dimensional trivial representation of the group* $SO(2)$.

(4.12) Remark. We can also compute the bifurcation index $\mathrm{BIF}_{LS}(\lambda_k^{-1}) \in \mathbb{Z}$ in terms of the Leray–Schauder degree, see [Iz1], [Ni]. It is known that $\mathrm{BIF}_{LS}(\lambda_k^{-1}) = 0 \in \mathbb{Z}$ if and only if $\dim V_L(\lambda_k^{-1})$ is even. In other words if $V_L(\lambda_k^{-1})$ is a nontrivial even-dimensional representation of the group $SO(2)$ then $\mathrm{BIF}_{LS}(\lambda_k^{-1}) = 0 \in \mathbb{Z}$ and $\mathrm{BIF}(\lambda_k^{-1}) \neq \Theta \in U(SO(2))$.

(4.13) LEMMA. *Let* Φ *satisfy assumptions* (A4), (A5). *Assume additionally that* $\dim V_L(\lambda_k)$ *is even for any* $\lambda_k \in \sigma(L)$. *Then:*

(4.13.1) $\mathrm{BIF}((\lambda_k^+)^{-1}) \in U_+(SO(2))$ *for any* $\lambda_k^+ \in \sigma_+(L)$,
(4.13.2) $\mathrm{BIF}((\lambda_k^-)^{-1}) \in U_-(SO(2))$ *for any* $\lambda_k^- \in \sigma_-(L)$.

PROOF. (4.13.1) $\dim \bigoplus_{i=1}^{k-1} V_L(\lambda_i^+)$ is even since all the eigenspaces of L are even-dimensional. Therefore from Lemma (3.5)it follows that

$$\nabla_{SO(2)} - \deg\Big(-\mathrm{id}, D_\alpha\Big(\bigoplus_{i=1}^{k-1} V_L(\lambda_i^+)\Big)\Big) \in U_+(SO(2)),$$

$$\nabla_{SO(2)} - \deg_{SO(2)}\Big(-\mathrm{id}, D_\alpha\Big(\bigoplus_{i=1}^{k-1} V_L(\lambda_i^+)\Big)\Big) = 1 \in \mathbb{Z},$$

$$\nabla_{SO(2)} - \deg(-\mathrm{id}, D_\alpha(V_L(\lambda_k^+))) \in U_+(SO(2)),$$

$$\nabla_{SO(2)} - \deg_{SO(2)}(-\mathrm{id}, D_\alpha(V_L(\lambda_k^+))) = 1 \in \mathbb{Z},$$

and consequently

$$\nabla_{SO(2)} - \deg(-\mathrm{id}, D_\alpha(V_L(\lambda_k^+))) - \mathbb{I} \in U_+(SO(2)),$$

$$\nabla_{SO(2)} - \deg_{SO(2)}(-\mathrm{id}, D_\alpha(V_L(\lambda_k^+))) - 1 = 0 \in \mathbb{Z}.$$

Finally, combining Remark (2.10) with (4.10.4) we obtain

$$\mathrm{BIF}((\lambda_k^+)^{-1}) \in U_+(SO(2)) \quad \text{for any } \lambda_k^+ \in \sigma_+(L).$$

(4.13.2) The same proof works for $\lambda_k^- \in \sigma_-(L)$. □

We are interested in characteristic values at which the bifurcation index is nontrivial. Therefore below we introduce the notion of an essential eigenvalue.

(4.14) DEFINITION. An eigenvalue $\lambda_k \in \sigma(L)$ is said to be *essential*, if at least one of the following conditions is satisfied

(4.14.1) $\dim V_L(\lambda_k^{-1})$ is odd,
(4.14.2) $V_L(\lambda_k^{-1})$ is a nontrivial representation of the group $SO(2)$.

(4.15) THEOREM (Local Bifurcation Theorem II). *Let* Φ *satisfy assumptions* (A4), (A5) *and let* $\lambda_{k_0} \in \sigma(L)$ *be an essential eigenvalue of* L. *Then,* $(0, \lambda_{k_0}^{-1}) \in \mathcal{BIF}(\Phi)$.

PROOF. From Corollary (4.11) it follows that $\mathrm{BIF}(\lambda_{k_0}^{-1}) \neq \Theta \in U(SO(2))$. The rest of the proof is a direct consequence of Theorem (4.4). □

Below we formulate the global bifurcation theorem. Since $\mathcal{BRA}(\Phi) \subset \mathcal{BIF}(\Phi)$, Theorem (4.15)is a direct consequence of Theorem (4.16).

(4.16) THEOREM (Global Bifurcation Theorem II, [Ry1]). *Let Φ satisfy assumptions* (A4), (A5) *and let $\lambda_{k_0} \in \sigma(L)$ be an essential eigenvalue of L. Then, $(0, \lambda_{k_0}^{-1}) \in \mathcal{BRA}(\Phi)$ and, moreover,*

(4.16.1) *either the continuum $\mathcal{C}(\lambda_{k_0}^{-1}) \subset \mathbb{H} \times \mathbb{R}$ is unbounded,*

(4.16.2) *or the continuum $\mathcal{C}(\lambda_{k_0}^{-1}) \subset \mathbb{H} \times \mathbb{R}$ is bounded and, in addition,*

$$\mathcal{C}(\lambda_{k_0}^{-1}) \cap (\{0\} \times \mathbb{R}) = \{0\} \times \left(\bigcup_{i=0}^{p} \{\lambda_{k_i}^{-1}\}\right) \subset \{0\} \times \{\lambda_k^{-1} : \lambda_k \in \sigma(L) \setminus \{0\}\},$$

(4.16.2.1)
$$\sum_{i=0}^{p} \mathrm{BIF}(\lambda_{k_i}^{-1}) = \Theta \in U(SO(2))$$

(4.17) THEOREM (Global Bifurcation Theorem III). *Let Φ satisfy assumptions* (A4), (A5) *and let*

(4.17.1) $\dim V_L(\lambda_k)$ *be even for any $\lambda_k \in \sigma(L)$,*

(4.17.2) *$\lambda_{k_0} \in \sigma(L)$ be an essential eigenvalue of L.*

If the continuum $\mathcal{C}(\lambda_{k_0}^{-1}) \subset \mathbb{H} \times \mathbb{R}$ is bounded then

(4.17.3) *if $\lambda_{k_0} < 0$ then $\mathcal{C}(\lambda_{k_0}^{-1}) \cap (\mathbb{H} \times (0, +\infty)) \neq \emptyset$,*

(4.17.4) *if $\lambda_{k_0} > 0$ then $\mathcal{C}(\lambda_{k_0}^{-1}) \cap (\mathbb{H} \times (-\infty, 0)) \neq \emptyset$.*

PROOF. Suppose that $\lambda_{k_0} < 0$. From Lemma (4.13) it follows that $\mathrm{BIF}(\lambda_{k_0}^{-1}) \in U_-(SO(2))$. Since λ_{k_0} is an essential eigenvalue of L, $\mathrm{BIF}(\lambda_{k_0}^{-1}) \neq \Theta \in U(SO(2))$. The rest of the proof is a direct consequence of Lemma (4.13) and Theorem (4.17). The same proof works for $\lambda_{k_0} > 0$. □

(4.18) THEOREM (Global Bifurcation Theorem IV, [Ry4]). *Let Φ satisfy assumptions* (A4), (A5) *and let*

(4.18.1) *either $\sigma(L) = \sigma_+(L)$ or $\sigma(L) = \sigma_-(L)$,*

(4.18.2) $\dim V_L(\lambda_k)$ *be even for any $\lambda_k \in \sigma(L)$,*

(4.18.3) *$\lambda_{k_0} \in \sigma(L)$ be an essential eigenvalue of L.*

Then the continuum $\mathcal{C}(\lambda_{k_0}^{-1}) \subset \mathbb{H} \times \mathbb{R}$ is unbounded.

PROOF. Suppose, contrary to our claim, that continuum $\mathcal{C}(\lambda_{k_0}^{-1})$ is bounded and $\sigma(L) = \sigma_+(L)$. From Theorem (4.17) it follows that $\mathcal{C}(\lambda_{k_0}^{-1}) \cap (\mathbb{H} \times (-\infty, 0)) \neq \emptyset$. Therefore from Lemma (4.8) it follows that $\mathcal{BIF}(\Phi) \cap (\{0\} \times (-\infty, 0)) \neq \emptyset$ and $\sigma_-(L) \neq \emptyset$, a contradiction. The same proof works for $\sigma(L) = \sigma_-(L)$. □

(4.19) REMARK. Notice that, since eigenspaces of operator L in Theorems (4.17) and (4.18) are even-dimensional, $\mathcal{BIF}_{LS}(\lambda_k^{-1}) = 0 \in \mathbb{Z}$. In other words, we can not prove these theorems using the Leray–Schauder degree.

(4.20) THEOREM (Global Bifurcation Theorem V, [Ry2]). *Let Φ satisfy assumptions* (A4), (A5) *and let* $\sigma(L) = \sigma_+(L)$. *Assume additionally that for any* $k \in \mathbb{N}$ *there are*

$$p_k, p_1^k, \dots, p_{r_k}^k \in \mathbb{N}, \quad 0 \le m_1^k < \dots < m_{r_k}^k < k$$

such that $V_L(\lambda_k^+) \approx R[p_k, k] \oplus (\bigoplus_{i=1}^{r_k} \mathbb{R}[p_i^k, m_i^k])$. *Then for any* $\lambda_k^+ \in \sigma(L)$ *the continuum* $\mathcal{C}((\lambda_k^+)^{-1}) \subset \mathbb{H} \oplus \mathbb{R}$ *is unbounded.*

PROOF. Fix $\lambda_{k_0}^+ \in \sigma(L) = \sigma_+(L)$ and notice that

(1) since $p_{k_0} > 0$, by Lemma (3.5) we obtain the following

$$\nabla_{SO(2)} - \deg_{\mathbb{Z}_{k_0}}(-\mathrm{id}, D_\alpha(V_L(\lambda_{k_0}^+))) = \pm k_0 \neq 0,$$

(2) since $m_1^{k_0}, \dots, m_{r_{k_0}}^{k_0} < k_0$, by Lemma (3.5) we obtain

$$\nabla_{SO(2)} - \deg_{\mathbb{Z}_k}\bigg(- \mathrm{id}, D_\alpha\bigg(\bigoplus_{i=1}^{k_0-1} V_L(\lambda_i^+)\bigg)\bigg) = 0, \quad \text{for any } k > k_0,$$

(3) from Lemma (3.5) it follows that

$$\nabla_{SO(2)} - \deg_{SO(2)}\bigg(- \mathrm{id}, D_\alpha\bigg(\bigoplus_{i=1}^{k_0-1} V_L(\lambda_i^+)\bigg)\bigg) = \pm 1,$$

(4) from Lemma (3.5) it follows that

$$\nabla_{SO(2)} - \deg_{\mathbb{Z}_k}\bigg(- \mathrm{id}, D_\alpha\bigg(\bigoplus_{i=1}^{k_0-1} V_L(\lambda_i^+)\bigg)\bigg) = 0, \quad \text{for } k \ge k_0.$$

Now combining (4.10.4) and the above we obtain that

$$\mathrm{BIF}_{\mathbb{Z}_{k_0}}(\lambda_{k_0}^{-1}) = \pm k_0 \neq 0 \quad \text{and} \quad \mathrm{BIF}_{\mathbb{Z}_k}(\lambda_{k_0}^{-1}) = 0, \quad \text{for any } k > k_0.$$

Suppose, contrary to our claim, that the continuum $\mathcal{C}((\lambda_{k_0}^+)^{-1}) \subset \mathbb{H} \oplus \mathbb{R}$ is bounded. The rest of the proof is a direct consequence of Theorem (4.16), because formula (4.16.2.1) can never be satisfied. □

5. Applications

Let us consider system of elliptic differential equations of the form

$$\text{(5.1)} \qquad \begin{cases} -\Delta u, = \nabla_u F(u, \lambda) & \text{in } \Omega, \\ u = 0 & \text{on } \partial\Omega \end{cases}$$

where

(F1) $\Omega \subset \mathbb{R}^N$ is an open, bounded, $SO(2)$-invariant subset of an orthogonal $SO(2)$-representation $\mathbb{R}^N$, with a boundary of the class C^{1-},

(F2) $F \in C^2(\mathbb{R}^m \times \mathbb{R}, \mathbb{R})$,

(F3) $\nabla_u F(u, \lambda) = \lambda u + \nabla_u \eta(u, \lambda)$, where

(a) $\eta \in C^2(\mathbb{R}^m \times \mathbb{R}, \mathbb{R})$,
(b) $\nabla_u \eta(0, \lambda) = 0$, for any $\lambda \in \mathbb{R}$,
(c) $\nabla_u^2 \eta(0, \lambda) = 0$, for any $\lambda \in \mathbb{R}$,

(F4) for any $\lambda \in \mathbb{R}$ there are $C_\lambda > 0$ and $1 \leq p_\lambda < (N+2)(N-2)^{-1}$ such that

$$|\nabla_u F(u, \lambda)| \leq C_\lambda (1 + |u|)^{p_\lambda}.$$

Consider the Hilbert space $\mathbb{H} = \bigoplus_{i=1}^m \mathbb{H}_0^1(\Omega)$ with the scalar product

$$\langle u_1, u_2 \rangle_{\mathbb{H}} = \sum_{i=1}^m \langle u_{1,i}, u_{2,i} \rangle_{\mathbb{H}_0^1(\Omega)} = \sum_{i=1}^m \int_\Omega (\nabla u_{1,i}(x), \nabla u_{2,i}(x))\, dx,$$

where $u_i = (u_{i,1}, u_{i,2}, \dots, u_{i,m}) \in \mathbb{H}$, $i = 1, 2$ and $(\,\cdot\,,\,\cdot\,)$ is the usual scalar product in $\mathbb{R}^m$. It is known that $(\mathbb{H}, \langle\,\cdot\,,\,\cdot\,\rangle_{\mathbb{H}})$ is an orthogonal $SO(2)$-representation with $SO(2)$-action given by $g(u(x)) = u(gx)$.

Define a functional $\Phi\colon \mathbb{H} \times \mathbb{R} \to \mathbb{R}$ as follows

$$(5.2)\quad \Phi(u, \lambda) = \frac{1}{2} \int_\Omega \sum_{i=1}^m (|\nabla u_i|^2)\, dx - \int_\Omega F(u, \lambda)\, dx = \frac{1}{2} \langle u, u \rangle_{\mathbb{H}} - \int_\Omega F(u, \lambda)\, dx.$$

Under assumptions (F1)–(F4) the functional Φ satisfies assumptions (A4), (A5). Moreover, the critical points of Φ (with respect to u) are in the one-to-one correspondence with the weak solutions of (5.1). Therefore from now on we will study the critical points of Φ.

Notice that for $u = (u_1, \dots, u_m), \varphi = (\varphi_1, \dots, \varphi_m) \in \mathbb{H}$ we have

$$\langle \nabla_u \Phi(u, \lambda), \varphi \rangle_{\mathbb{H}} = D_u \Phi(u, \lambda)(\varphi) = \langle u - \lambda L(u) - \nabla_u \psi(u, \lambda), \varphi \rangle_{\mathbb{H}}.$$

where

(5.3.1) $L = ((-\Delta)^{-1}, \dots, (-\Delta)^{-1})\colon \mathbb{H} \to \mathbb{H}$ is a linear, bounded, compact, self--adjoint and $SO(2)$-equivariant operator given by

$$\langle Lu, v \rangle_{\mathbb{H}} = \int_\Omega u(x) v(x)\, dx,$$

(5.3.2) $\nabla_u \psi\colon \mathbb{H} \times \mathbb{R} \to \mathbb{H}$ is a compact, $SO(2)$-equivariant operator such that $\nabla_u \psi(0, \lambda) = 0$ and $\nabla_u^2 \psi(0, \lambda) = 0$ for any $\lambda \in \mathbb{R}$.

Let us consider the eigenvalue problem

$$(5.4)\qquad \begin{cases} -\Delta u = \lambda u & \text{in } \Omega, \\ u = 0 & \text{on } \partial\Omega. \end{cases}$$

By $\sigma(-\Delta) = \{\lambda_k\}$ we will denote the set of eigenvalues of $-\Delta$. The eigenspace of $-\Delta$ corresponding to the eigenvalue λ_k will be denoted by $V_{-\Delta}(\lambda_k)$. Finally, put $\mu(\lambda_k) = \dim V_{-\Delta}(\lambda_k)$.

From now on we will study zeroes of operator $\nabla_u \Phi(u, \lambda) = u - \lambda L u - \nabla_u \psi(u, \lambda)$. Directly from the definition of the operator L it follows that

$$\sigma(L) = \sigma_+(L) = \left\{\lambda_k^{-1} : \lambda_k \in \sigma(-\Delta)\right\}, \tag{5.5.1}$$

$$V_L(\lambda_k^{-1}) = \bigoplus_{i=1}^{m} V_{-\Delta}(\lambda_k) \quad \text{for any } \lambda_k^{-1} \in \sigma(L). \tag{5.5.2}$$

From Lemma (4.8) and the above it follows that $\mathcal{BIF}(\Phi) \subset \{0\} \times \sigma(-\Delta)$.

Combining Theorem (3.3), Lemma (4.10) and the above we obtain that

$$\mathrm{BIF}(\lambda_1) = (\nabla_{SO(2)} - \deg(-\mathrm{id}, D_\alpha(V_{-\Delta}(\lambda_1))))^m - \mathbb{I}, \tag{5.6.1}$$

$$\begin{aligned} \mathrm{BIF}(\lambda_k) = {} & \left(\prod_{i=1}^{k-1} (\nabla_{SO(2)} - \deg(-\mathrm{id}, D_\alpha(V_{-\Delta}(\lambda_i))))^m\right) \\ & * ((\nabla_{SO(2)} - \deg(-\mathrm{id}, D_\alpha(V_{-\Delta}(\lambda_k))))^m - \mathbb{I}). \end{aligned} \tag{5.6.2}$$

The following theorem is the Rabinowitz global bifurcation theorem for systems of elliptic differential equations. It yields information about global behaviour of connected sets of weak solutions of system (5.1).

(5.7) THEOREM ([Ry1]). *Assume that* (5.1) *satisfies assumptions* (F1)–(F4). *Fix $\lambda_{k_0} \in \sigma(-\Delta)$ such that $m \cdot \mu(\lambda_{k_0})$ is odd or $V_{-\Delta}(\lambda_{k_0})$ is a nontrivial $SO(2)$-representation. Then $(0, \lambda_{k_0})$ is a branching point of weak solutions of system* (5.1) *and, moreover,*

(5.7.1) *either the continuum $\mathcal{C}(\lambda_{k_0})$ is unbounded in $\mathbb{H} \times \mathbb{R}$,*

(5.7.2) *or the continuum $\mathcal{C}(\lambda_{k_0})$ is bounded in $\mathbb{H} \times \mathbb{R}$, and, in addition, the following conditions are satisfied*

$$\mathcal{C}(\lambda_{k_0}) \cap (\{0\} \times \mathbb{R}) = \{0\} \times \left(\bigcup_{i=0}^{p} \{\lambda_{k_i}\}\right) \subset \{0\} \times \sigma(-\Delta), \tag{5.7.2.1}$$

$$\sum_{i=0}^{p} \mathrm{BIF}(\lambda_{k_i}) = \Theta \in U(SO(2)). \tag{5.7.2.2}$$

PROOF. In order to prove this theorem it is enough to study critical points of the functional $\Phi\colon \mathbb{H} \times \mathbb{R} \to \mathbb{R}$ given by formula (5.2). The functional Φ satisfies assumptions (A4), (A5). Directly from the assumptions it follows that $\lambda_{k_0}^{-1} \in \sigma(L)$ is an essential eigenvalue of L. The rest of the proof is a direct consequence of Theorem (4.16). □

(5.8) THEOREM ([Ry6]). *Assume that* (5.1) *satisfies assumptions* (F1)–(F5) *and that m is even. Fix $\lambda_{k_0} \in \sigma(-\Delta)$ such that $V_{-\Delta}(\lambda_{k_0})$ is a nontrivial $SO(2)$-representation. Then $(0, \lambda_{k_0})$ is a branching point of weak solutions of system* (5.1) *and, moreover, the continuum $\mathcal{C}(\lambda_{k_0})$ is unbounded in $\mathbb{H} \times \mathbb{R}$.*

PROOF. In order to prove this theorem it is enough to study critical points of the functional $\Phi: \mathbb{H} \times \mathbb{R} \to \mathbb{R}$ given by formula (5.2). The functional Φ satisfies assumptions (A4), (A5). Directly from the assumptions it follows that $\lambda_{k_0}^{-1} \in \sigma(L)$ is an essential eigenvalue of L. Since m is even, $\dim V_L(\lambda_k^{-1})$ is even for any $\lambda_k \in \sigma(-\Delta)$. Since $\sigma(-\Delta) = \sigma_+(-\Delta)$, $\sigma(L) = \sigma_+(L)$. The rest of the proof is a direct consequence of Theorem (4.18). □

Denote by $\mu_j^{(k)}$, $j \in \mathbb{N}$, $k \in \mathbb{N} \cup \{0\}$, the positive roots of the Bessel function $J_k(\mu)$.

(5.9) EXAMPLE. Put $\Omega = D^2 \subset \mathbb{R}[1,1]$. Then problem (5.4) possesses the following solutions

$$\sigma(-\Delta) = \bigcup_{k \in \mathbb{N} \cup \{0\}} \bigcup_{j \in \mathbb{N}} \{(\mu_j^{(k)})^2\}, \tag{5.9.1}$$

$$V_{-\Delta}((\mu_j^{(k)})^2) = \operatorname{span}\left\{ \frac{J_k(\mu_j^{(k)} r)}{\sqrt{\pi}|J_k'(\mu_j^{(k)})|} \cos(k\varphi), \frac{J_k(\mu_j^{(k)} r)}{\sqrt{\pi}|J_k'(\mu_j^{(k)})|} \sin(k\varphi) \right\}, \tag{5.9.2}$$

for any $(k, j) \in (\mathbb{N} \cup \{0\}) \times \mathbb{N}$.

Taking into account that $D^2 \subset \mathbb{R}[1,1]$ we obtain that

$$V_{-\Delta}((\mu_j^{(k)})^2) \approx \begin{cases} \mathbb{R}[1,0] & \text{for } k = 0, \\ \mathbb{R}[1,k] & \text{for } k > 0. \end{cases} \tag{5.9.3}$$

If m is odd then from Theorem (5.7) and (5.9.3) it follows that for any $(k, j) \in (\mathbb{N} \cup \{0\}) \times \mathbb{N}$ the point $(0, (\mu_j^{(k)})^2) \in \mathbb{H} \times \mathbb{R}$ is a branching point of weak solutions of system (5.1). Moreover, the continuum $\mathcal{C}((\mu_j^{(k)})^2) \subset \mathbb{H} \times \mathbb{R}$ satisfies the statement of Theorem (5.7).

Fix $j \in \mathbb{N}$. Since $\mu((\mu_j^{(0)})^2) = 1$ and m is odd, $\mathrm{BIF}_{LS}((\mu_j^{(0)})^2) = \pm 2 \in \mathbb{Z}$. Therefore we can apply the classical Rabinowitz global bifurcation theorem (the bifurcation index is computed in terms of the Leray–Schauder degree). From the Rabinowitz alternative we obtain that $(0, (\mu_j^{(0)})^2) \in \mathbb{H} \times \mathbb{R}$ is a branching point of weak solutions of system (5.1) and the continuum $\mathcal{C}((\mu_j^{(0)})^2) \subset \mathbb{H} \times \mathbb{R}$ satisfies the statement of the classical Rabinowitz alternative. Fix $(k, j) \in \mathbb{N} \times \mathbb{N}$. Since $\mu((\mu_j^{(k)})^2) = 2$, $\mathrm{BIF}_{LS}((\mu_j^{(0)})^2) = 0 \in \mathbb{Z}$. That is why the Rabinowitz alternative is not applicable at the eigenvalue $(\mu_j^{(k)})^2$.

If m is even then from Theorem (5.8) and (5.9.3) it follows that for any $(k,j) \in \mathbb{N} \times \mathbb{N}$ the point $(0, (\mu_j^{(k)})^2) \in \mathbb{H} \times \mathbb{R}$ is a branching point of weak solutions of system (5.1). Moreover, continuum $\mathcal{C}((\mu_j^{(k)})^2) \subset \mathbb{H} \times \mathbb{R}$ is unbounded.

Fix $j \in \mathbb{N}$. Since the eigenspace $V_{-\Delta}((\mu_j^{(0)})^2)$ is a trivial representation of the group $SO(2)$ and m is even, we obtain from (5.6.1) and (5.6.2) that $\mathrm{BIF}((\mu_j^{(0)})^2) = \Theta \in U(SO(2))$. In other words, the assumptions of Theorems (5.7), (5.8) are not satisfied at eigenvalues $(\mu_j^{(0)})^2$, $j \in \mathbb{N}$. On the other hand, using the Morse theory, one can prove that for any $j \in \mathbb{N}$ the point $(0, (\mu_j^{(0)})^2) \in \mathbb{H} \times \mathbb{R}$ is a bifurcation point of weak solutions of system (5.1). Moreover, notice that since m is even, $\mathrm{BIF}_{LS}((\mu_j^{(k)})^2) = 0 \in \mathbb{Z}$, for any $(k,j) \in (\mathbb{N} \cup \{0\}) \times \mathbb{N}$. In other words the Rabinowitz alternative is not applicable in this situation.

Let $\mu_j^{(l+1/2)}$, $j \in \mathbb{N}$, $l \in \mathbb{N} \cup \{0\}$ denote the positive roots of the Bessel function $J_{l+1/2}(\mu)$.

(5.10) Example. Put $\Omega = D^3 \subset \mathbb{R}[1,1] \oplus \mathbb{R}[1,0]$. Then problem (5.4) possesses the following solutions

$$\sigma(-\Delta) = \bigcup_{l \in \mathbb{N} \cup \{0\}} \bigcup_{j \in \mathbb{N}} \left\{ (\mu_j^{(l+1/2)})^2 \right\}, \tag{5.10.1}$$

$$V_{-\Delta}((\mu_j^{(l+1/2)})^2) = \mathrm{span}\Bigg(\bigcup_{m=0}^{l} \left\{ \frac{1}{\sqrt{r}} J_{l+1/2}(\mu_j^{(l+1/2)} r) P_l^m(\cos\psi) \cos(m\varphi) \right\} \cup \bigcup_{m=1}^{l} \left\{ \frac{1}{\sqrt{r}} J_{l+1/2}(\mu_j^{(l+1/2)} r) P_l^m(\cos\psi) \sin(m\varphi) \right\} \Bigg), \tag{5.10.2}$$

for any $(l,j) \in (\mathbb{N} \cup \{0\}) \times \mathbb{N}$. Taking into account that $D^3 \subset \mathbb{R}[1,1] \oplus \mathbb{R}[1,0]$ we obtain for any $(l,j) \in (\mathbb{N} \cup \{0\}) \times \mathbb{N}$ that

$$V_{-\Delta}((\mu_j^{(l+1/2)})^2) \approx \mathbb{R}[1,0] \oplus \mathbb{R}[1,1] \oplus \ldots \oplus \mathbb{R}[1,l] \tag{5.11}$$

If m is odd then from Theorem (5.7)and (5.11) it follows that for any $(l,j) \in (\mathbb{N} \cup \{0\}) \times \mathbb{N}$ the point $(0, (\mu_j^{(l+1/2)})^2) \in \mathbb{H} \times \mathbb{R}$ is a branching point of weak solutions of system (5.1). Moreover, the continuum $\mathcal{C}((\mu_j^{(l+1/2)})^2) \subset \mathbb{H} \times \mathbb{R}$ satisfies the statement of Theorem (5.7). On the other hand, since $m \cdot \mu((\mu_j^{(l+1/2)})^2)$ is odd, $\mathrm{BIF}_{LS}((\mu_j^{(l+1/2)})^2) = \pm 2 \in \mathbb{Z}$. From the Rabinowitz alternative it follows that the point $(0, (\mu_j^{(l+1/2)})^2) \in \mathbb{H} \times \mathbb{R}$ is a branching point of weak solutions of system (5.1). It is worth to point out that Theorem (5.7) gives better description of the continuum $\mathcal{C}((\mu_j^{(l+1/2)})^2) \subset \mathbb{H} \times \mathbb{R}$ than the Rabinowitz theorem.

If m is even then from Theorem (5.8) and (5.11) it follows that for any $(l,j) \in \mathbb{N} \times \mathbb{N}$ the point $(0, (\mu_j^{(l+1/2)})^2) \in \mathbb{H} \times \mathbb{R}$ is a branching point of weak solutions of system (5.1). Moreover, the continuum $\mathcal{C}((\mu_j^{(l+1/2)})^2) \subset \mathbb{H} \times \mathbb{R}$ is unbounded. Fix any $j \in \mathbb{N}$. Unfortunately, since the eigenspace $V_{-\Delta}((\mu_j^{(1/2)})^2)$ is a trivial representation of the group $SO(2)$ and m is even, we obtain from (5.6.1) and (5.6.2) that $\mathrm{BIF}((\mu_j^{(1/2)})^2) = \Theta \in U(SO(2))$. In other words, the assumptions of Theorems (5.7), (5.8) are not satisfied at eigenvalues $(\mu_j^{(1/2)})^2, j \in \mathbb{N}$. On the other hand, using the Morse theory, one can prove that for any $j \in \mathbb{N}$ the point $(0, (\mu_j^{(1/2)})^2) \in \mathbb{H} \times \mathbb{R}$ is a bifurcation point of weak solutions of system (5.1). Moreover, notice that since m is even, $\mathrm{BIF}_{LS}((\mu_j^{(l+1/2)})^2) = 0 \in \mathbb{Z}$, for any $(l, j) \in (\mathbb{N} \cup \{0\}) \times \mathbb{N}$. In other words, the Rabinowitz alternative is not applicable in this situation.

From now on we will study the Laplace–Beltrami operator $\Delta_{\mathbb{S}^{n-1}}$ considered on the unit sphere $\mathbb{S}^{n-1} = \{x \in \mathbb{R}^n : \|x\| = 1\}, n \geq 2$. Let us consider an equation

$$(5.12) \qquad -\Delta_{\mathbb{S}^{n-1}} u = f(u, \lambda), \quad (u, \lambda) \in H^1(S^{n-1}) \times \mathbb{R}$$

where $H^1(\mathbb{S}^{n-1})$ is the first Sobolev space with inner product given by the formula

$$(5.13) \qquad \langle u, v \rangle = \int_{\mathbb{S}^{n-1}} (\nabla u \cdot \nabla v + u \cdot v)\, d\sigma$$

and $f: \mathbb{R} \times \mathbb{R} \to \mathbb{R}$ is a C^1-map such that:

(5.14.1) $f(x, \lambda) = \lambda \cdot x + g(x, \lambda)$ for all $(x, \lambda) \in \mathbb{R} \times \mathbb{R}$,
(5.14.2) $g(0, \lambda) = 0$ for all $\lambda \in \mathbb{R}$,
(5.14.3) $D_x g(0, \lambda) = 0$ for all $\lambda \in \mathbb{R}$.

Let H_m^n denote the linear space of homogeneous, harmonic polynomials in n variables and degree m. It is known that

$$\dim H_m^n = \frac{(n + m - 3)! \cdot (n + 2 \cdot m - 2)}{(n - 2)! \cdot m!}.$$

The linear space of restrictions of elements of H_m^n to S^{n-1} will be denoted by $\mathcal{H}_m^n$. It is evident that $\dim H_m^n = \dim \mathcal{H}_m^n$.

(5.15) LEMMA ([Gu], [Shi]). *Under the above notations the following holds true:*

(5.15.1) $L^2(\mathbb{S}^{n-1}) = \bigoplus_{m=0}^{\infty} \mathcal{H}_m^n$,
(5.15.2) *for any* $(n, m) \in (\mathbb{N} \setminus \{1\}) \times (\mathbb{N} \cup \{0\})$ $\mathcal{H}_m^n$ *is an eigenspace of the Laplace–Beltrami operator* $\Delta_{\mathbb{S}^{n-1}}$,
(5.15.3) $(\Delta_{\mathbb{S}^{n-1}})_{|\mathcal{H}_m^n} = -m \cdot (m + n - 2) \cdot \mathrm{id}_{|\mathcal{H}_m^n}$.

(5.16) REMARK. The unit sphere $\mathbb{S}^{n-1}$ will be considered with the following parametrization

$$\begin{aligned} x_1 &= \sin\theta_{n-1}\cdot\ldots\cdot\sin\theta_2\cdot\sin\varphi,\\ x_2 &= \sin\theta_{n-1}\cdot\ldots\cdot\sin\theta_2\cdot\cos\varphi,\\ &\ldots\ldots\ldots\ldots\ldots\ldots\ldots\ldots\ldots\ldots\\ x_{n-1} &= \sin\theta_{n-1}\cdot\cos\theta_{n-2},\\ x_n &= \cos\theta_{n-1}, \end{aligned}$$

where $0\le\varphi<2\pi$ and $0\le\theta_k<\pi$ for $k=2,\ldots,n-1$.

From now on $\mathbb{R}^n$ will be considered as a representation of the group $SO(2)$ of the form $\mathbb{R}[1,1]\oplus\mathbb{R}[n-2,0]$. Of course, $\mathbb{S}^{n-1}$ is an $SO(2)$-invariant subset of $\mathbb{R}[1,1]\oplus\mathbb{R}[n-2,0]$. Moreover, it is obvious that the action of the group $SO(2)$ on $\mathbb{S}^{n-1}$ is given by the formula

$$(5.16.1)\quad e^{\sqrt{-1}\psi}x=(e^{\sqrt{-1}\psi}x_1,e^{\sqrt{-1}\psi}x_2,\ldots,e^{\sqrt{-1}\psi}x_n) = \begin{bmatrix} \sin\theta_{n-1}\cdot\ldots\cdot\sin\theta_2\cdot\sin(\varphi+\psi)\\ \sin\theta_{n-1}\cdot\ldots\cdot\sin\theta_2\cdot\cos(\varphi+\psi)\\ \ldots\\ \sin\theta_{n-1}\cdot\cos\theta_{n-2}\\ \cos\theta_{n-1}\end{bmatrix}$$

$$\text{for } e^{\sqrt{-1}\psi}=\begin{bmatrix}\cos\psi & -\sin\psi\\ \sin\psi & \cos\psi\end{bmatrix}\in SO(2) \text{ and } (x_1,\ldots,x_n)\in\mathbb{S}^{n-1}.$$

(5.17) REMARK. It is well known that $\mathcal{H}^2_m=\operatorname{span}_R\{\cos(m\cdot\varphi),\sin(m\cdot\varphi)\}$ and that

$$\begin{aligned}\mathcal{H}^3_m=\operatorname{span}_R\{&P^0_m(\cos\theta_2),P^1_m(\cos\theta_2)\cdot\cos\varphi,P^1_m(\cos\theta_2)\cdot\sin\varphi,\\ &P^2_m(\cos\theta_2)\cdot\cos(2\cdot\varphi),P^2_m(\cos\theta_2)\cdot\sin(2\cdot\varphi),\ldots,\\ &P^m_m(\cos\theta_2)\cdot\cos(m\cdot\varphi),P^m_m(\cos\theta_2)\cdot\sin(m\cdot\varphi)\},\end{aligned}$$

where P^i_m are the Gegenbauer polynomials.

Remarks (5.16), (5.17) allow us to give the complete description of the spaces $\mathcal{H}^2_m$ and $\mathcal{H}^3_m$ as representations of the group $SO(2)$.

(5.18) REMARK. Using the action of the group $SO(2)$ on $\mathbb{S}^{n-1}$ given in Remark (5.16) one can introduce an action of the group $SO(2)$ on $\mathcal{H}^n_m$ in the following way

$$(5.18.1)\quad e^{\sqrt{-1}\psi}\cdot\Phi(\varphi,\theta_2,\ldots,\theta_{n-1})=\Phi(\varphi+\psi,\theta_2,\ldots,\theta_{n-1})$$

$$\text{for } e^{\sqrt{-1}\psi}=\begin{bmatrix}\cos\psi & -\sin\psi\\ \sin\psi & \cos\psi\end{bmatrix}\in SO(2) \text{ and } \Phi\in\mathcal{H}^n_m.$$

Taking into account Remark (5.17) and the above defined action of the group $SO(2)$ one can easily describe the spaces $\mathcal{H}^2_m$ and $\mathcal{H}^3_m$ as representations of the group $SO(2)$. Namely, it is evident that

(5.18.2) $$\mathcal{H}^2_m \approx \mathbb{R}[1, m] \qquad \text{for all } m \in \mathbb{N} \cup \{0\},$$

(5.18.3) $$\mathcal{H}^3_m \approx \bigoplus_{i=0}^{m} \mathbb{R}[1, i] \quad \text{for all } m \in \mathbb{N} \cup \{0\}.$$

From now on we will consider spaces $\mathcal{H}^n_m$ with the action of the group $SO(2)$ given by (5.18.1). In the following lemma we describe a basis of the space $\mathcal{H}^n_m$.

(5.19) LEMMA ([Wi]). *Any element of the space $\mathcal{H}^n_m$ can be expressed in the unique way as a linear combination of functions of the form*

$$\Phi_K(\varphi, \theta_2, \dots, \theta_{n-1}) = \prod_{j=0}^{n-3} C^{(n-j-2)/2+k_{j+1}}_{k_j - k_{j+1}}(\cos(\theta_{n-j-1})) \cdot \sin^{k_{j+1}}(\theta_{n-j-1}) \cdot e^{\pm i \cdot k_{n-2} \cdot \varphi},$$

where $K = (k_0, \dots, k_{n-2}) \in (\mathbb{N} \cup \{0\})^{n-1}$, $m = k_0 \geq k_1 \geq \dots \geq k_{n-2} \geq 0$ and $C^{(n-j-2)/2+k_{j+1}}_{k_j - k_{j+1}}(\,\cdot\,)$ are the Gegenbauer polynomials.

Notice that for arbitrary $(n, m) \in (\mathbb{N} \setminus \{1\}) \times (\mathbb{N} \cup \{0\})$ it is not obvious how to obtain the complete description of the space $\mathcal{H}^n_m$ as a representation of the group $SO(2)$. Nevertheless, for our purpose it is enough to prove the following lemma.

(5.20) LEMMA ([Ry2]). *For any pair $(n, m) \in (\mathbb{N} - \{1\}) \times (\mathbb{N} \cup \{0\})$ there exist*

(5.20.1) $r_{(n,m)} \in \mathbb{N}$, $p^n_m, p^{(n,m)}_0, \dots, p^{(n,m)}_{r_{(n,m)}} \in \mathbb{N}$,

(5.20.2) $m^{(n,m)}_0, m^{(n,m)}_1, \dots, m^{(n,m)}_{r_{(n,m)}} \in \mathbb{N} \cup \{0\}$,

such that

(5.20.3) $0 \leq m^{(n,m)}_0 < m^{(n,m)}_1 < \dots < m^{(n,m)}_{r_{(n,m)}} < m$,

(5.20.4) $\mathcal{H}^n_m \approx \mathbb{R}[p^n_m, m] \oplus \bigoplus_{i=0}^{r_{(n,m)}} \mathbb{R}[p^{(n,m)}_i, m^{(n,m)}_i]$.

We are in a position to formulate the main theorem describing closed connected components of nontrivial solutions of equation (5.12) bifurcating from the set of trivial solutions.

(5.21) THEOREM ([Ry2]). *Assume that equation* (5.12) *satisfies assumptions* (A1)–(A3). *Then for any $(n, m) \in (\mathbb{N} \setminus \{1\}) \times (\mathbb{N} \cup \{0\})$ $(0, m(m + n - 2)) \in H^1(\mathbb{S}^{n-1}) \times \mathbb{R}$ is a branching point of weak solutions of equation* (5.12) *and moreover continuum $\mathcal{C}(m(m + n - 2)) \subset H^1(\mathbb{S}^{n-1}) \times \mathbb{R}$ is unbounded.*

PROOF. Fix $n \in N - \{1\}$. First of all let us observe that weak solutions of problem (5.12) one can consider as critical points of a C^2-functional $\Phi\colon H^1(\mathbb{S}^{n-1}) \times$

$\mathbb{R} \to \mathbb{R}$ defined as follows:

$$\Phi(u, \lambda) = \frac{1}{2} \int_{S^{n-1}} |\nabla u|^2 d\sigma - \int_{S^{n-1}} F(u, \lambda)\, d\sigma,$$

where $F(u, \lambda) = \int_0^u f(t, \lambda)\, dt$. Since $SO(2)$ has a natural orthogonal action on $\mathbb{S}^{n-1}$ defined by (5.16.1), we then introduce an $SO(2)$-action on $H^1(\mathbb{S}^{n-1}) \times \mathbb{R}$ as follows:

$$\psi \cdot (u(x), \lambda) = (u(e^{\sqrt{-1}\psi} \cdot x), \lambda).$$

The functional Φ is fixed on the orbits of the action of the group $SO(2)$ on the space $H^1(\mathbb{S}^{n-1}) \times \mathbb{R}$. Notice that taking into account (5.13) we obtain the following

$$\Phi(u, \lambda) = \frac{1}{2}\|u\|^2 - (\lambda + 1) \int_{S^{n-1}} \frac{u^2}{2}\, d\sigma - \int_{S^{n-1}} G(u, \lambda)\, d\sigma,$$

where $G(u, \lambda) = \int_0^u g(t, \lambda)\, dt$.

From the above it follows that the operator $\nabla_u \Phi\colon H^1(\mathbb{S}^{n-1}) \times \mathbb{R} \to H^1(\mathbb{S}^{n-1})$ is of the form

$$\nabla_u \Phi(u, \lambda) = u - (\lambda + 1) \cdot L(u) - \nabla\psi(u, \lambda),$$

where

(5.22.1) $L\colon H^1(\mathbb{S}^{n-1}) \to H^1(\mathbb{S}^{n-1})$ is an $SO(2)$-equivariant, gradient, linear, bounded, compact operator,

(5.22.2) $\nabla\psi\colon H^1(\mathbb{S}^{n-1}) \times \mathbb{R} \to H^1(\mathbb{S}^{n-1})$ is an $SO(2)$-equivariant, gradient, compact operator,

(5.22.3) $\nabla\psi(0, \lambda) = 0$ for all $\lambda \in \mathbb{R}$,

(5.22.4) $\nabla^2\psi(0, \lambda) = 0$ for all $\lambda \in \mathbb{R}$,

(5.22.5) $m(m + n - 2) \in \sigma(-\Delta_{\mathbb{S}^{n-1}})$ if and only if $1/m(m + n - 2) + 1 \in \sigma(L)$,

(5.22.6) $V_{-\Delta_{\mathbb{S}^{n-1}}}(m(m + n - 2)) = V_L(1/m(m + n - 2) + 1)$.

In order to complete this proof it suffices to show that the assumptions of Theorem (4.20) are fulfilled. From (1)–(4) it follows that the functional Φ satisfies assumptions (A4), (A5) of Theorem (4.20). Combining (5.22.5), (5.22.6) with Lemma (5.20) we show that all the assumptions of Theorem (4.20) are fulfilled. The rest of the proof is a direct consequence of Theorem (4.20). □

References

[Ad] J. F. Adams, *Lectures on Lie Groups*, W. A. Benjamin Inc., New York–Amsterdam, 1969.

[Am] A. Ambrosetti, *Branching points for a class of variational operators*, J. Anal. Math. **76** (1998), 321–335.

[AZ1] H. Amann and E. Zehnder, *Periodic solutions of asymptotically linear Hamiltonian systems*, Manuscripta Math. **32** (1980), 149–189.

[AZ2] ______, *Multiple periodic solutions for a class of nonlinear autonomous wave equations*, Houston J. Math. **7** (1981), 147–174.

[Ba] T. Bartsch, *Topological Methods for Variational Problems With Symmetries*, Lecture Notes in Math., vol. 1560, Springer–Verlag, Berlin, 1993.

[Be1] V. Benci, *A geometrical index for the group S^1 and some applications to the study of periodic solutions of ordinary differential equations*, Comm. Pure Appl. Math. **34** (1981), 393–432.

[Be2] ______, *On critical point theory for indefinite functionals in the presence of symmetries*, Trans. Amer. Math. Soc. **274** (1982), 533–572.

[BP] V. Benci and F. Pacella, *Morse theory for symmetric functionals on the sphere and an application to a bifurcation problem*, Nonlinear Anal. **9** (1985), 763–773.

[BB] B. Blanchard and E. Brüning, *Variational Methods in Mathematical Physics. An Unified Approach*, Texts and Monographs in Physics, Springer–Verlag, Berlin–Heidelberg–New York, 1992.

[Boh] R. Böhme, *Die Lösung der Versweigungsgleichungen für Nichtlineare Eigenwert-Probleme*, Math. Z. **127** (1972), 105–126.

[Bot] R. Bott, *Lectures on Morse theory, old and new*, Bull. Amer. Math. Soc. **7** (1982), 331–358.

[Bre] G. Bredon, *Introduction to Compact Transformation Groups*, Academic Press, Inc. LTD, 1972.

[Bro] R. F. Brown, *A Topological Introduction to Nonlinear Analysis*, Birkhäuser, Boston, 1993.

[Cha] K. C. Chang, *Infinite-dimensional Morse Theory and Multiple Solution Problems*, Progr. Nonlinear Differential Equations Appl., vol. 6, Birkhäuser Boston, Inc., Boston, MA, 1993.

[CL] P. Chossat and R. Lauterbach, *Methods in Equivariant Bifurcations and Dynamical Systems*, Adv. Ser. Nonlinear Dynam., vol. 15, World Scientific, Singapore, New Jersey–London–Hong Kong, 2000.

[Co] Ch. Conley, *Isolated Invariant Sets and the Morse Index*, CBMS Regional Conf. Ser. in Math., vol. 38, Amer. Math. Soc., Providence, R. I., 1978.

[Da1] E. N. Dancer, private communication.

[Da2] ______, *A new degree for $SO(2)$-invariant mappings and applications*, Ann. Inst. H. Poincaré Anal. Non Linéaire **2** (1985), 473–486.

[DGR] E. N. Dancer, K. Gęba and S. Rybicki, *Classification of Homotopy Classes of G-equivariant Gradient Maps*, Fund. Math. (to appear).

[DR] E. N. Dancer and S. Rybicki, *A note on periodic solutions of autonomous Hamiltonian systems mmanating from degenerate stationary solutions*, Differential Integral Equations **12** (1999), 1–14.

[De] K. Deimling, *Nonlinear Functional Analysis*, Springer–Verlag, Berlin, 1985.

[Di] T. tom Dieck, *Transformation Groups*, Walter de Gruyter, Berlin–New York, 1987.

[Fa1] E. Fadell, *The relationship between Lusternik–Schnirelmann category and the concept of genus*, Pacific. J. Math. **89** (1980), 33–42.

[Fa2] ______, *The equivariant Lusternik–Schnirelmann method for invariant functionals and relative cohomological index theories*, Méth. Topologiques en Analyse non Linéaires (A. Granas, ed.), Sémin. Math. Sup., vol. 95, Montréal, 1985, pp. 41–70.

[Fa3] ______, *Cohomological methods in non-free G-spaces with applications to general Borsuk–Ulam theorems and critical point*, Theorems for Invariant Functionals, Nonlinear Funct. Anal. and Its Appl. (S. P. Singh, ed.), Proc. Maratea 1985, NATO ASI Ser. C, vol. 173, Reidel, Dordrecht, 1986, pp. 1-45.

[FH1] E. Fadell and S. Husseini, *Relative cohomological index*, Adv. Math. **64** (1987), 1–31.

[FH2] ______, *An ideal valued cohomological index theory with applications to Borsuk–Ulam and Bourgain–Yang theorems*, Ergodic Theory Dynam. Systems **8** (1988), 73–85.

[FR] E. Fadell and P. H. Rabinowitz, *Generalized cohomological index for Lie group actions with an application to bifurcations questions for Hamiltonian systems*, Invent. Math. **45** (1978), 139–174.

[FM] B. Fiedler and K. Mischaikow, *Dynamics of bifurcations for variational problems with $O(3)$-equivariance: a Conley index approach*, Arch. Rational Mech. Anal. **119** (1992), 145–190.

[GAR] J. Gawrycka and S. Rybicki, *Solutions of systems of elliptic differential equations on circular domains*, presented for publication.

[Ge1] K. Gęba, private communication.

[Ge2] ______, *Degree for Gradient Equivariant Maps and Equivariant Conley Index*, Topological Nonlinear Analysis, Degree, Singularity and Variations (M. Matzeu and A. Vignoli, eds.), Birkhäuser; Progr. Nonlinear Differential Equations Appl. **27** (1997), 247–272.

[GIP] K. Gęba, M. Izydorek and A. Pruszko, *The Conley index in Hilbert spaces and its applications*, Studia Math. **134** (1999), 217–233.

[GR] K. Gęba and S. Rybicki, *Algorithm of computation of $SO(3)$-equivariant Euler characteristic*, in preparation.

[Gu] D. Gurarie, *Symmetries and Laplacians*, Introduction to Harmonic Analysis, Group Representations and Applications, vol. 174, North-Holland Mathematics Studies, Amsterdam–London–New York–Tokyo, 1992.

[HR] N. Hirano and S. Rybicki, *Remark on degree theory for $SO(2)$-equivariant transversal maps*, Topol. Methods Nonlinear Anal. **22** (2003), 253–272.

[Iz1] J. Ize, *Bifurcation Theory for Fredholm Operators*, Mem. Amer. Math. Soc., vol. 174, 1976.

[Iz2] ______, *Equivariant degree for Abelian actions* I, *Equivariant homotopy groups*, Topol. Methods Nonlinear Anal. **2** (1993), 367–413.

[Iz3] ______, *Topological bifurcation*, Topological Nonlinear Analysis, Degree, Singularity and Variations (M. Matzeu and A. Vignoli, eds.), Progr. Nonlinear Differential Equations Appl., vol. 15, Birkhäuser, 1995, pp. 341–463.

[IMV1] J. Ize, I. Massabo and A. Vignoli, *Degree theory for euivariant maps* I, Trans. Amer. Math. Soc. **315** (1989), 433–510.

[IMV2] ______, *Degree Theory for Equivariant Maps, the General S^1-Action*, Mem. Amer. Math. Soc., vol. 100, 1992.

[IV] J. Ize & A. Vignoli, *Equivariant Degree Theory*, de Gruyter Series in Nonlinear Analysis and Applications, vol. 8, Walter de Gruyter, Berlin–New York, 2003.

[Izy1] M. Izydorek, *A cohomological Conley index in Hilbert spaces and applications to strongly indefinite problems*, J. Differential Equations **170** (2001), 22–50.

[Izy2] ______, *Equivariant Conley index in Hilbert spaces and applications to strongly indefinite problems*, Nonlinear Anal. **51** (2002), 33–66.

[JS] W. Jäger and K. Schmitt, *Symmetry breaking in semilinear elliptic problems*, Analysis, et Cetera, Research Papers (P. H. Rabinowitz and E. Zehnder, eds.), Published in Honor of Jürgen Moser 60th Birthday, Academic Press, Inc., 1990, pp. 451–470.

[Ki1] H. Kielhöfer, *A bifurcation theorem for potential operators*, J. Funct. Anal. **77** (1988), 1–8.

[Ki2] ______, *Bifurcation Theory. An Introduction with Applications to* PDEs, Appl. Math. Sci., vol. 156, Springer, 2003.

[Kras] M. A. Krasnosel'skiĭ, *Topological Methods in the Theory of Nonlinear Integral Equations*, Pergamon Press, Oxford–London–New York, 1964.

[KrWu] W. Krawcewicz and J. Wu, *Theory of Degrees with Applications to Bifurcations and Differential Equations*, Canad. Math. Soc. Ser. Monogr. Adv. Texts, John Wiley and Sons, Inc., 1997.

[KSZ] W. Kryszewski and A. Szulkin, *An infinite-dimensional Morse theory with applications*, Trans. Amer. Math. Soc. **349** (1997), 3181–3234.

[LS] V. K. Le and K. Schmitt, *Global Bifurcation in Variational Inequalities. Applications to Obstacle and Unilateral Problems*, Appl. Math. Sci., vol. 123, Springer-Verlag, New York, 1997.

[Ll] N. G. Lloyd, *Degree Theory*, Cambridge Tracts in Math., vol. 73, Cambridge University Press, Cambridge–New York–Melbourne, 1978.

[MR1] A. Maciejewski and S. Rybicki, *Global bifurcations of periodic solutions of the Hill Lunar problem*, Celestial Mech. Dynam. Astronom. **81** (2001), 279–297.

[MR2] ______, *Global bifurcations of periodic solutions of Hénon–Heiles system via degree for S^1-equivariant orthogonal maps*, Rev. Math. Phys. **10** (1998), 1125–1146.

[Mari] A. Marino, *La biforcazione Nel Caso Variazionale*, Conf. Sem. Mat. Univ. Bari **132** (1977).

[Marz] W. Marzantowicz, *A G-Lusterik–Schnirelmann category of space with an action of a compact Lie group*, Topology **28** (1989), 403–412.

[MW] J. Mawhin and M. Willem, *Critical Point Theory and Hamiltonian Systems*, Springer–Verlag, Berlin–Heidelberg–New York, 1989.

[Mi] S. G. Michlin, *Linear Equations of Mathematical Physics*, Moscow, Nauka, 1964. (Russian)

[Ni] L. Nirenberg, *Topics in Nonlinear Functional Analysis*, Courant Institute of Mathematical Sciences, New York University, 1974.

[Pa] A. Parusiński, *Gradient homotopies of gradient vector fields*, Studia Math. **XCVI** (1990), 73–80.

[Po] A. Pomponio, *Asymptotically linear cooperative elliptic system: existence and multiplicity*, Nonlinear Anal. **52** (2003), 989–1003.

[R1] P. Rabier, *Topological degree and the theorem of Borsuk for general covariant mappings with applications*, Nonlinear Anal. **16** (1991), 399–420.

[R2] ______, *Symmetries, topological degree and a theorem of Z. Q. Wang*, Rocky Mountain J. Math. **24** (1994), 1087–1115.

[Ra1] P. H. Rabinowitz, *Minimax Methods in Critical Point Theory with Applicatons to Differential Equations*, Reg. Conf. Ser. Math., vol. 35, Amer. Math. Soc., Providence, R.I., 1986.

[Ra2] ______, *Nonlinear Sturm–Liouville problems for second order ordinary differential equations*, Comm. Pure Appl. Math. **23** (1970), 939–961.

[Ra3] ______, *Some global results for nonlinear eigenvalue problems*, J. Funct. Anal. **7** (1971), 487–513.

[Ra4] ______, *Global theorems for nonlinear eigenvalue problems and applications*, Contributions to Nonlinear Functional Analysis (E. H. Zarantonello, ed.), Academic Press, New York, 1971, pp. 11–36.

[Ra5] ______, *On bifurcation from infinity*, J. Differential Equations **14** (1973), 462–475.

[Ra6] ______, *A note on topological degree for potential operators*, J. Differential Equations **51** (1975), 483–492.

[Ry1] S. Rybicki, *$SO(2)$-degree for orthogonal maps and its applications to bifurcation theory*, Nonlinear Anal. **23** (1994), 83–102.

[Ry2] ______, *On Rabinowitz alternative for the Laplace–Beltrami operator on S^{n-1}: continua that meet infinity*, Differential Integral Equations **9** (1996), 1267–1277.

[Ry3] ______, *Applications of degree for $SO(2)$-equivariant gradient maps to variational nonlinear problems with $SO(2)$-symmetries*, Topol. Methods Nonlinear Anal. **9** (1997), 383–417.

[Ry4] ______, *On bifurcations from infinity for S^1-equivariant potential operators*, Nonlinear Anal. **31** (1997), 343–361.

[Ry5] ______, *On periodic solutions of autonomous Hamiltonian systems via degree for S^1-equivariant gradient maps*, Nonlinear Anal. **34** (1998), 537–569.

[Ry6] ______, *Global bifurcations of solutions of elliptic differential equations*, J. Math. Anal. Appl. **217** (1998), 115–128.

[Ry7] ______, *Global bifurcations of solutions of Emden–Fowler type equation $-\Delta u(x) = \lambda f(u(x))$ on an annulus in R^n, $n \geq 3$*, J. Differential Equations **183** (2002), 208–223.

[Sch] M. Schechter, *Linking Methods in Critical Point Theory*, Birkhäuser, Boston, 1999.

[Shi] N. Shimakura, *Partial Differential Operators of Elliptic Type*, Translations of Mathematical Monographs, vol. 99, Amer. Math. Soc., Providence, R. I., 1992.

[Sm] J. Smoller, *Shock Waves and Reaction-Diffusion Equations*, Fundamental Principles of Mathematical Science, vol. 258, Springer–Verlag, New York–Berlin, 1983.

[SW] J. Smoller and A. G. Wasserman, *Bifurcation and symmetry-breaking*, Inv. Math. **100** (1990), 63–93.

[St] M. Struwe, *Variational Methods. Applications to Nonlinear Partial Differential Equations and Hamiltonian Systems*, Springer, 1996.

[Ta] F. Takens, *Some remarks on the Böhme–Berger bifurcation theorem*, Math. Z. **125** (1972), 359–364.

[Wa1] Z. Q. Wang, *Degree of Equivariant Gradient Mappings via Morse Theory*, Northeast. Math. J. **6** (1990), 339–345.

[Wa2] ———, *Symmetries and the Calculation of Degree*, Chinese Ann. of Math. Ser. B **4** (1989), 519–536.

[Wa3] ———, *Equivariant Morse theory for isolated critical orbits and its applications to nonlinear problems*, Lecture Notes in Math., vol. 1306, Springer, Berlin, 1988, pp. 202–223.

[Was] A. G. Wasserman, *Equivariant differential topology*, Topology **8** (1969), 127–150.

[Wh] G. Whyburn, *Topological Analysis*, Princeton University Press, 1958.

[Wi] N. Wilenkin, *Special Functions and Representation Theory Nauka, Moscow*, 1965. (Russian)

[Wl] W. S. Wladimirov, *Equations of Mathematical Physics Nauka, Moscow*, 1981. (Russian)

CHAPTER III

NIELSEN THEORY

When emploing homological methods only existence results are obtained.

In 1927 J. Nielsen defined, for a given mapping f, a number $N(f)$ that is now so called the Nielsen number of f. The Nielsen number is a lower bound for the number of fixed points for all maps homotopic to f.

This theory is being intensly developed by numerous researchers. In this chapter we describe the current trends in Nielsen theory.

11. NIELSEN ROOT THEORY

ROBIN BROOKS

1. Introduction

Let $f: X \to Y$ be a map of topological spaces and let $y_0 \in Y$. A *root* of f at y_0 is a point $x \in X$ such that $f(x) = y_0$. In root theory we are interested in finding a lower bound for the number of roots of f at y_0. Ideally, the lower bound should be invariant under homotopies of f; it should be sharp, in the sense that while every map homotopic to f should have at least that number of roots, there should be at least one map homotopic to f having that many roots; and it should be easily computable in terms of algebraic properties of f, e.g. in terms of the homomorphisms induced by f on homology groups and the fundamental group.

There is no difficulty in defining a lower bound satisfying the first two criteria. The following definition is the obvious one:

(1.1) DEFINITION. Let $f: X \to Y$ be a map of topological spaces and let $y_0 \in Y$. We define the *minimum root number for f at y_0* to be the minimum cardinality of $f'^{-1}(y_0)$, where the minimum is taken over all maps f' homotopic to f. It is denoted by $MR(f, y_0)$.

$$MR(f, y_0) = \min_{f' \text{ homotopic to } f} \#(f'^{-1}(y_0)),$$

where $\#(f'^{-1}(y_0))$ denotes the number of elements in $f'^{-1}(y_0)$.

But, in addition to defining this lower bound, we are of course interested in finding techniques to compute it, and Definition (1.1) gives us no means for computing $MR(f, y_0)$.

Motivated by the success of Nielsen fixed point theory, our strategy is to proceed as much as possible by analogy with that theory. We first group the roots of f at y_0 into equivalence classes, called Nielsen root classes. We then define the notion of an essential class and define the Nielsen root number of f at y_0, $NR(f, y_0)$, to be the number of essential root classes. These definitions are formulated in a way that $NR(f, y_0)$ is a homotopy invariant of f. It is then clear that $NR(f, y_0) \leq MR(f, y_0)$, so we are interested in conditions under which this inequality is sharp. This question then breaks into two questions:

(1.2.1) Under what conditions is there map homotopic to f that has exactly $NR(f, y_0)$ Nielsen classes of roots?

(1.2.2) Under what conditions is there a map homotopic to f so that not only does it have $NR(f, y_0)$ Nielsen classes of roots, but also each class contains only one root?

Then when these conditions are satisfied, we clearly have $MR(f, y_0) = NR(f, y_0)$. Even when these conditions are not met, we still know that $NR(f, y_0) \leq MR(f, y_0)$ so techniques for computing $NR(f, y_0)$ are still of interest.

The rest of this chapter is organized as follows: In the next section we go over some of the notation and conventions that will be used in the rest of the chapter.

Section 3 contains the definitions of Nielsen root classes, essentiality of a class, and the Nielsen root number, and includes a theorem that states sufficient conditions for the sharpness of $NR(f, y_0)$ as a lower bound on the number of Nielsen classes. This section also introduces the "Hopf covering" and lift as very useful tools in Nielsen root theory, and the Root Reidemeister number, a useful tool for computing the Nielsen root number.

One of the ways in which Nielsen root theory departs from fixed point theory, is that in fixed point theory there is a Lefschetz number and a related local fixed point index which are defined for a wide class of spaces and maps. The fixed point index can then be used to define essentiality of fixed point classes. In root theory, the natural analog to the Lefschetz number and local index is the Brouwer degree and local degree of a map. However, the Brouwer degree and local degree are defined only for maps of manifolds, and even then only under suitable orientability assumptions. Because of this, we have kept our definition of root class essentiality independent of the existence or choice of a local root index. Nevertheless, it is possible to define the notion of a root index axiomatically, and show, as a theorem rather than definition, that if a root class has a non-zero index, then it is essential. We may also use homology and/or cohomology theory to construct root indexes satisfying the axioms and these specialize for maps of compact orientable manifolds to the well known Brouwer and local degree. Section 4 is devoted to this treatment of root indexes. It concludes with a number of examples applying the theory and also a remark on the role of compactness.

Section 5, is devoted to maps of manifolds and Hopf degree theory. In this section, Hopf's *absolute degree* for a map $f\colon X \to Y$ of n-manifolds is defined. The absolute degree is a generalization to nonorientable manifolds of the Brouwer degree. The two main theorems in Hopf degree theory are stated and proved, in dimensions other than two.

There are, of course, many topics in Nielsen root theory that we are unable to include. The last section of the chapter, Notes and Additional Topics, briefly describes some of these topics and includes references to the literature.

In preparing this chapter, I benefited from early conversations with Daciberg Gonçalves and Peter Wong. I've also benefited from the careful reading, advice and encouragement of Robert Brown. I thank the three of them. Of course I claim complete credit for mistakes and obscurities.

2. Conventions and notation

Throughout this chapter, unless otherwise stated, all topological spaces denoted by X or Y will be assumed to be normal, Hausdorff, connected, locally path connected, and semilocally simply connected. All manifolds, unless mentioned explicitly otherwise, are topological manifolds (locally Euclidean paracompact spaces) without boundary.

The closed unit interval $[0,1]$ will be denoted by I, n-dimensional Euclidean space by $\mathbf{R}^n$, the complex numbers by $\mathbb{C}$, the closed unit n-ball by $\mathbf{B}^n = \{x \in \mathbf{R}^n : \|x\| = 1\}$. A subspace $B \subset X$ will be called an n-ball if it is homeomorphic to $\mathbf{B}^n$.

If $A: I \to X$ is a path in the space X, then $[A]$ denotes its fixed-end-point homotopy class. We use juxtaposition to denote path multiplication, $(AB)(t) = A(2t)$ for $0 \leq 1/2$, and $(AB)(t) = B(2t-1)$ for $1/2 \leq t \leq 1$, assuming $A(1) = B(0)$. The reverse of the path A is denoted by A^{-1}, $A^{-1}(t) = A(1-t)$. We regard a homotopy as a collection $\{h_t: X \to Y : t \in I\}$ of maps indexed by the unit interval such that $(x,t) \mapsto h_t(x)$ is a continuous function from $X \times I$ into Y. We will usually write the homotopy more simply as $\{h_t: X \to Y\}$, or even more simply as $\{h_t\}$.

If $\phi: G \to H$ is a group homomorphism, we will have occasion to refer to the set $H/\phi(G)$ of left cosets $h\phi(G)$ of $\phi(G)$. When $\phi(G)$ is normal in H, this set of cosets is a group commonly called the *cokernel* of ϕ and denoted by $\operatorname{Coker}(\phi)$. We extend this terminology and call $H/\phi(G)$ the cokernel of ϕ and write $\operatorname{Coker}(\phi)$ even when $\phi(G)$ is not normal in H. We use $\#\operatorname{Coker}\phi$ to denote the cardinality of $\operatorname{Coker}\phi$. More generally, $\#S$ denotes the cardinality of S for any set S.

We say that the map $f: (X,A) \to (Y,B)$ *defines* a map $f': (X',A') \to (Y',B')$ if the two maps are the same except for modifications of domain and codomain, more precisely, if $X' \subset X$, $f(X') \subset Y'$, $f(A') \subset B'$, and $f'(x) = f(x)$ for all $x \in X'$.

3. The Nielsen root number and Reidemeister number

(3.1) Definition. Let $f: X \to Y$ be a map and $y_0 \in Y$. A root x_0 of f at y_0 is *Nielsen root f equivalent* to another root x_1 of f at y_0 if there is a path C in X from x_0 to x_1 such that the loop $f \circ C$ in Y at y_0 is fixed-end-point homotopic to the constant path at y_0. This relation is easily seen to be an equivalence relation (Theorem (3.4) below); the equivalence classes are called *Nielsen root classes of f at* y_0. The set of these classes is denoted by $f^{-1}(y_0)/NR$.

The phrase "Nielsen root f equivalent" is terribly awkward; we will omit some of the words "root", "Nielsen", or f, when they are clear from context.

Now suppose $\{f_t: X \to Y\}$ a homotopy. We need a way to relate the roots of the map f_0 at the beginning of the homotopy to the roots of the map f_1 at the end of the homotopy. That is provided by the following definition, which actually generalizes Definition (3.1).

(3.2) Definition. Suppose $\{f_t: X \to Y\}$ a homotopy and $y_0 \in Y$. Then a root x_0 of f_0 at y_0 is Nielsen root $\{f_t\}$ related to a root x_1 of f_1 at y_0 if there is a path C in X from x_0 to x_1 such that the loop $\{f_t \circ C(t)\}$ is fixed-end-point homotopic to the constant path at y_0.

Note that a root x_0 of $f: X \to Y$ is Nielsen equivalent to another root x_1 if and only if x_0 is related to x_1 by the constant homotopy at f. It turns out that if $\{f_t\}$ is a homotopy and one root in a Nielsen class α of f_0 is $\{f_t\}$ related to a root in root class β of f_1, then every root in α is $\{f_t\}$ related to every root in β, so a $\{f_t\}$ relation induces a relation between the root classes of f_0 and f_1. This fact is not hard to prove directly, it is even easier to prove using the *Hopf covering* and *lift* for f defined as follows:

(3.3) Definition. Let $f: X \to Y$ be a map. Then we know from covering space theory that there is a covering $\widehat{q}: \widehat{Y} \to Y$ and a lift $\widehat{f}: X \to \widehat{Y}$ of f through $\widehat{q}$ with the property that for any $x \in X$ the image of the fundamental group $\pi(\widehat{Y}, \widehat{f}(x))$ under the homomorphism induced by $\widehat{q}$ is the same as the image of $\pi(X, x)$ under the homomorphism induced by f. We call $\widehat{f}$ a *Hopf lift of f*, and call $\widehat{q}: \widehat{Y} \to Y$ a *Hopf covering for f*.

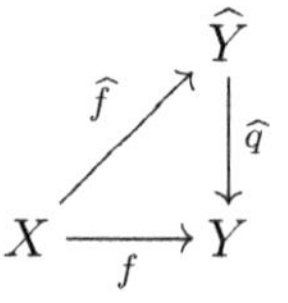

We call this covering a "Hopf" covering because Hopf was the first to exploit it in Nielsen root theory (see [H1], [H2]). Note that although the covering is not unique, it is unique up to covering space isomorphism — a homeomorphism of the covering spaces that commutes with the covering maps. Nor is the lift $\widehat{f}: X \to \widehat{Y}$ unique; however, given any $x \in X$ there is for each $\widehat{y} \in \widehat{q}^{-1}(f(x)) \subset \widehat{Y}$ a unique lift $\widehat{f}$ such that $\widehat{f}(x) = \widehat{y}$. Note also that if $\widehat{q}: \widehat{Y} \to Y$ is a Hopf covering for f, then it is a Hopf covering for any map homotopic to f.

The following two theorems illustrate the extreme usefulness of the Hopf coverings and lifts in Nielsen root theory.

(3.4) THEOREM. *Suppose $f: X \to Y$ a map and $y_0 \in Y$. Let $\widehat{q}: \hat{Y} \to Y$ be a Hopf covering for f, and let $\widehat{f}: X \to \widehat{Y}$ be a Hopf lift of f through $\widehat{q}$. Let $\{f_t: X \to Y\}$ be a homotopy from $f = f_0$ to f_1 and let $\{\widehat{f_t}: X \to \widehat{Y}\}$ be a lift of $\{f_t\}$ through $\widehat{q}$. Then:*

(3.4.1) *A root x_0 of f at y_0 is f equivalent to another root x_1 of f at y_0, if and only if $\widehat{f}(x_0) = \widehat{f}(x_1)$.*

(3.4.2) *Consequently, Nielsen root f equivalence is in fact an equivalence relation, and the equivalence classes (called Nielsen root classes) are precisely those non-empty sets of the form $\widehat{f}^{-1}(\widehat{y})$ where $\widehat{y} \in \widehat{q}^{-1}(y_0)$.*

(3.4.3) *More generally, a root x_0 of $f = f_0$ at y_0 is $\{f_t\}$ related to a root x_1 of f_1 at y_0 if and only if $\widehat{f_0}(x_0) = \widehat{f_1}(x_1)$.*

(3.4.4) *Consequently, the $\{f_t\}$ relation from the roots of f_0 to those of f_1 induces a one-to-one relation from the Nielsen root classes of f_0 at y_0 to the root classes of f_1 at y_0 whereby a root class α_0 of f_0 is $\{f_t\}$ related to a root class α_1 of f_1 if at least one (and therefore every) root $x_0 \in \alpha_0$ is related to at least one (and hence every) root $x_1 \in \alpha_1$, and this is true if and only if $\widehat{f_0}(\alpha_0) = \widehat{f_1}(\alpha_1)$.* [1]

PROOF. We begin by proving (3.4.3). First, suppose $\widehat{f_0}(x_0) = \widehat{f_1}(x_1)$; we will show x_0 is $\{f_t\}$ related to x_1. Let A be a path in X from x_0 to x_1. Then $\{\widehat{f_t} \circ A(t)\}$ is a loop in $\widehat{Y}$ at $\widehat{f}(x_0)$, so its projection $\{\widehat{q} \circ \widehat{f_t} \circ A(t)\} = \{f_t \circ A(t)\}$ is a loop in Y at y_0, and therefore, since $\widehat{q}$ is a Hopf lift for f, it is fixed-end-point homotopic to $f \circ B$ for some loop B in X at x_0. Thus $[(f \circ B)^{-1}\{f_t \circ A(t)\}] = [f \circ B]^{-1}[\{f_t \circ A(t)\}] = [y_0]$. The path $(f \circ B)^{-1}\{f_t \circ A(t)\}$ is easily shown to be fixed-end-point homotopic to the path $\{f_t \circ (B^{-1}A)(t)\}$. Thus $B^{-1}A$ is a path from x_0 to x_1 such that $\{f_t \circ (B^{-1}A)(t)\}$ is fixed-end-point null homotopic. It follows that x_0 is $\{f_t\}$ related to x_1.

---

[1] There are analogous coverings and lifts in coincidence theory: Given maps $f, g: X \to Y$ define $(f,g): X \to Y \times Y$ by $(f,g)(x) = (f(x), g(x))$. There is a covering $\widehat{q}: \widehat{Y} \to Y \times Y$ and lift $\widehat{f}: X \to \widehat{Y}$ such that $\widehat{q}_{\#}(\pi(\widehat{Y}, \widehat{f}(x))) = (f,g)_{\#}(\pi(X,x))$ for any $x \in X$. The Nielsen coincidence classes of f and g are precisely the nonempty sets of the form $\widehat{f}^{-1}(P)$ where P is a path component of $\widehat{q}^{-1}(D)$, and D is the diagonal in $Y \times Y$. See [Br4, Section 2] for details.

To complete the proof of (3.4.3), suppose that x_0 is $\{f_t\}$ related to x_1. There is then a path A in X from x_0 to x_1 such that $\{f_t \circ A(t)\}$ is fixed-end-point null homotopic. The path $\{\widehat{f_t} \circ A(t)\}$ is a lift of $\{f_t \circ A(t)\}$ through $\widehat{q}$, so it must also be fixed-end-point null homotopic, in particular it is a loop. Since this path starts at $\widehat{f_0} \circ A(0) = \widehat{f_0}(x_0)$, ends at $\widehat{f_1} \circ A(1) = \widehat{f_1}(x_1)$, and is a loop we have $\widehat{f_0}(x_0) = \widehat{f_1}(x_1)$. This completes the proof of (3.4.3).

Statement (3.4.1) follows from (3.4.3) by letting $\{f_t\}$ be the constant homotopy at f. Statement (3.4.2) follows immediately from (3.4.1), and (3.4.4) follows immediately from (3.4.1), (3.4.2), and (3.4.3). □

(3.5) THEOREM. *Let $f: X \to Y$ be a map and $y_0 \in Y$. Then there is a family $\{U_\alpha\}$ of mutually disjoint open sets, one for each Nielsen root class α of f at y_0, such that $\alpha \subset U_\alpha$. Consequently, each root class α is both open and closed in $f^{-1}(y_0)$.*

PROOF. Since $\widehat{q}$ is a covering, y_0 has an open neighbourhood V evenly covered by $\widehat{q}$. Then we may write $\widehat{q}^{-1}(V) = \bigsqcup_{\widehat{y} \in \widehat{q}^{-1}(y_0)} V_{\widehat{y}}$, where $\widehat{y} \in V_{\widehat{y}}$ for each $\widehat{y}$. Therefore by (3.4.2) the collection $\{\widehat{f}^{-1}(V_{\widehat{y}}) : \widehat{y} \in \widehat{q}^{-1}(y_0) \text{ and } \widehat{f}^{-1}(\ y) \neq \emptyset\}$ is the desired family. □

(3.6) COROLLARY. *Let $f: X \to Y$ be a map and $y_0 \in Y$. If X is compact, then there are only a finite number of root classes.* [2]

PROOF. Since Y is Hausdorff, then y_0 is closed in Y so $f^{-1}(y_0)$ is closed in X. Thus if X is compact, so is $f^{-1}(y_0)$, and it follows from (3.5) that there are only finitely many root classes. □

According to Theorem (3.4), the $\{f_t\}$ relation is a one-one-to *relation*, so a root class of f_0 is related to *at most* one root class of f_1, and a root class of f_1 has *at most* one root class related to it. It is quite possible, however, for a root of f_0 to not be related to *any* root class of f_1, and it is also possible that a root class of f_1 has no root class of f_0 related to it. That is the foundation for the following definition.

(3.7) DEFINITION. Suppose $f: X \to Y$ a map and $y_0 \in Y$. A root of f at y_0 is *essential* if for any homotopy $\{f_t\}$ beginning at f, it is $\{f_t\}$ related to a root of f_1 at y_0. Similarly, a root class of f at y_0 is *essential* if given any homotopy $\{f_t\}$ from f it is $\{f_t\}$ related to a root class of f_1 at y_0. The number of essential root classes of f at y_0 is the *Nielsen root number of f at y_0*; it is denoted by $NR(f, y_0)$.

From Theorem (3.4), we have an alternative description of essentiality in terms of Hopf lifts:

---

(2) The corollary is true even when X is not compact provided that f is *proper* ($f^{-1}(C)$ is compact whenever C is).

(3.8) THEOREM. *Let $f: X \to Y$ be a map, $y_0 \in Y$, $\widehat{q}: \widehat{Y} \to Y$ and $\widehat{f}: X \to \widehat{Y}$ a Hopf covering and lift for f. Then for any $\widehat{y} \in \widehat{q}^{-1}(y_0)$, $\widehat{f}^{-1}(\widehat{y})$ is an essential root class if and only if $\widehat{f_1}^{-1}(\widehat{y}) \neq \emptyset$ for any homotopy $\{\widehat{f_t}\}$ beginning at $\widehat{f}$.*

Clearly, a $\{f_t\}$ relation restricts to a bijection of the set of essential root classes of f_0 onto the set of essential root classes of f_1. We therefore have

(3.9) THEOREM. *Let $f: X \to Y$ be a map and $y_0 \in Y$. Then $NR(f, y_0)$*

(3.9.1) *is a homotopy invariant of f,*

(3.9.2) *is a lower bound on the number of root classes of f' at y_0 among all those maps f' homotopic to f, and*

(3.9.3) *is therefore a lower bound on the number of roots of f' at y_0 among all those maps f' homotopic to f.*

The Nielsen root number is not only a homotopy invariant, but is also easily seen to be a topological invariant in the following sense:

(3.10) THEOREM. *Suppose*

$$\begin{array}{ccc} X & \xrightarrow{\ f\ } & (Y, Y - y_0) \\ {\scriptstyle g}\downarrow & & \downarrow{\scriptstyle h} \\ X' & \xrightarrow[\ f'\ ]{} & (Y, Y - y_0') \end{array}$$

commutes and both g and h are homeomorphisms. Then g takes the set of Nielsen root classes of f at y_0 bijectively onto the set of those of f' at y_0', and $g(\alpha)$ is an essential class of f if and only if α is an essential class of f'. Hence $NR(f, y_0) = NR(f', y_0')$.

(3.11) REMARK. It is also true that $N(f, y_0)$ is a homotopy type invariant, in the sense that if g and h are not necessarily homeomorphisms, but do have homotopy inverses making the above diagram commute up to homotopy, then one can show that g takes the set of essential Nielsen classes of f at y_0 onto the set of essential classes of f' at y_0' and therefore $NR(f, y_0) = NR(f', y_0')$. It is interesting that $MR(f, y_0)$ additional index entry for minimum root number is a topological invariant, and homotopy invariant but not, in this sense, a homotopy type invariant (see Example (4.22) below).

In Definition (3.7), we have departed from our analogy with Nielsen fixed point theory. In fixed point theory one defines a local fixed point index which may be used to associate with each Nielsen fixed point class of a self-map of a compact ANR an integer called its index. A fixed point class is then defined to be *essential* if its index is non-zero. One then proves fairly easily that if a fixed point class is essential in this sense, then it is essential in a sense analogous to Definition (3.7). One also proves — with much more difficulty — that, under fairly general circumstances, if a class is essential in the sense analogous to Definition (3.7), then its index is non-zero. One of the important ways that root theory differs from fixed point theory is that, except in the case where both X and Y are manifolds, there does not seem to be a well developed index theory. For this reason, we have adopted a definition that does not depend upon a particular choice of index or even on the existence of an index.

Our definition of essentiality says, very roughly speaking, that a Nielsen root class of a map f is essential if it cannot be eliminated by changing f by a homotopy. One might think then that it should be possible to change a map f by a homotopy so as to eliminate all non-essential classes and thereby find a map f' with only $NR(f') = NR(f)$ root classes. This then "proves" that $NR(f)$ is a sharp lower bound on the number of root classes of maps homotopic to f. This argument, however, misses the very important possibility that in eliminating one inessential root class, one might create a new (inessential) root class. The next theorem gives fairly general (but by no means completely general) conditions under which this cannot happen, so the Nielsen root number is a sharp lower bound on the number of root classes. Before stating the theorem we recall a definition.

(3.12) DEFINITION ([Hu, p. 32]). A topological space X is *dominated* by a space D if there are maps $r\colon D \to X$ and $i\colon X \to D$ such that $r \circ i\colon X \to X$ is homotopic to the identity on X.

Note that if we replace the phrase "is homotopic to" by "is equal to", we have the definition of a retract. So "domination" may be viewed as the homotopy version of "retraction".

(3.13) THEOREM ([B4, Theorem 1]). *Let $f\colon X \to Y$ be a map (of path connected and locally path connected spaces) and let $y_0 \in Y$. Suppose that for some integer $n > 2$, X is dominated by a (not necessarily compact) polyhedron of dimension n or less, and that the mth relative homotopy group $\pi_m(Y, Y - y_0) = 0$ for all m with $0 < m < n$. Then there is a map $f'\colon X \to Y$ homotopic to f that has only $NR(f', y_0) = NR(f, y_0)$ root classes.*

We outline a proof of Theorem (3.13) after Example (3.15) below.

The dimensional hypotheses for Theorem (3.13) have the drawback that it is the dominating polyhedron that is assumed to have dimension n or less rather than X — the space of interest. According to [GD, Theorem 17.7.16(c), p. 483], however, if an n-dimensional paracompact space is dominated by a polyhedron, then it is dominated by a polyhedron of dimension n or less. Hence

(3.14) Corollary. *Let $f: X \to Y$ be a map (of path connected and locally path connected spaces) and let $y_0 \in Y$. Suppose that for some integer $n > 2$, X is a paracompact space of (covering) dimension n or less, dominated by a (not necessarily compact) polyhedron, and, $\pi_m(Y, Y - y_0) = 0$ for all m with $0 < m < n$. Then there is a map $g: X \to Y$ homotopic to f that has only $NR(g, y_0) = NR(f, y_0)$ root classes.*

Every ANR X is dominated by a polyhedron, see [Hu, p. 32]. Also $\pi_m(Y, Y - y_0) = 0$ for all m with $0 < m < n$ whenever y_0 has a Euclidean neighbourhood of dimension n or more, so (3.14) applies whenever X is an ANR of dimension n or less and y_0 has a Euclidean neighbourhood of dimension $n > 2$ or more.

Perhaps the most serious limitations of Theorem (3.13) and its corollary are the dimensional constraints. Very little is known when the dimension of X is bigger than n, although some promising results have been obtained for maps of nilmanifolds (see [GW]). Also, it would be nice either to extend the theorem to the $n = 2$ case or find a counter-example. Finally, we note that even when $NR(f, y_0)$ is a sharp lower bound on the number of Nielsen classes, it may often fail to be a sharp lower bound on the number of roots. It may be impossible to deform f so that it has only one root in each class.

(3.15) Example. Let $S^n = \{\mathbf{s} \in \mathbf{R}^n : \|\mathbf{s}\| = 1\}$ be the n sphere. Let X be the space obtained by joining two copies of S^n by an n dimensional cylinder whose ends have been glued to their equators:

$$X = \{(s_1, \ldots, s_{n+1}, t) \in S^n \times [0, 1] : t \in \{0, 1\} \quad \text{or} \quad s_{n+1} = 0\}.$$

Let $Y = S^n$ and $y_0 = (1, 0, \ldots, 0)$. Define $f: X \to Y$ by $f(\mathbf{s}, t) = \mathbf{s}$ for all $(\mathbf{s}, t) \in X$. Since Y is simply connected, then f has only one Nielsen root class at y_0. On the other hand every map homotopic to f has at least *two* roots at y_0. To see this, suppose f' homotopic to f and define $i_0: S^n \to X$ and $i_1: S^n \to X$ by

$$i_0(\mathbf{s}) = (\mathbf{s}, 0) \quad \text{and} \quad i_1(\mathbf{s}) = (\mathbf{s}, 1).$$

Then $f \circ i_0$ is the identity on S^n, so $f' \circ i_0$ is homotopic to the identity on S^n and therefore there is $\mathbf{s}_0 \in S^n$ with $f' \circ i_0(\mathbf{s}_0) = y_0$. Similarly $f' \circ i_1(\mathbf{s}_1) = y_0$ for some $\mathbf{s}_1 \in S^n$. Thus $i_0(\mathbf{s}_0) = (\mathbf{s}_0, 0)$ and $i_1(\mathbf{s}_1) = (\mathbf{s}_1, 1)$ are two distinct roots of f' at y_0.

It appears that to get results analogous to those in fixed point theory, we need to assume much more about the spaces X and Y. We will see in Section 5 that when X and Y are manifolds of dimension not equal to two, then it is possible to deform the map f so that the root classes are all singletons.

We now outline a proof of Theorem (3.13). For a complete proof, see [B4, Section 1].

Outline of Proof of Theorem (3.13). By hypothesis, there is a polyhedron P and maps $r: P \to Y$ and $i: X \to P$ such that $r \circ i$ is homotopic to the identity on X. One first shows that if $g': P \to Y$ is homotopic to $g = f \circ r: P \to Y$ and has only $NR(g, y_0)$ root classes, then $f' = g \circ i: X \to Y$ is homotopic to f and has only $NR(f, y_0)$ root classes. This is done by showing that r takes the set of root classes of g bijectively onto those of f, and also takes the essential root classes of g bijectively onto those of f. This reduces the proof of Theorem 3.13 to the case where X is itself a polyhedron.

The rest of the proof is based on the Hopf covering and the obstruction theory of deformation [Hu, p. 197]. Assume X a polyhedron. Let $\widehat{q}: \widehat{Y} \to Y$ and $\widehat{f}: X \to \widehat{Y}$ be a Hopf covering and lift for f. Let $D \subset \widehat{q}^{-1}(y_0)$ be the set of all $\widehat{y} \in \widehat{q}^{-1}(y_0)$ for which $\widehat{f}^{-1}(\widehat{y})$ is either empty or an inessential Nielsen root class of f at y_0. Then the problem of deforming f so that it has only essential root classes, is equivalent to deforming $\widehat{f}$ from $\widehat{Y}$ into $\widehat{Y} - D$.

Because of the dimensional assumptions the only obstruction to this deformation is the primary obstruction ω_D in dimension n. The obstruction ω_D is an element of the cohomology group $H^n(X, \Gamma_D)$ of X with coefficients in the local system of groups Γ_D obtained by using $\widehat{f}$ to pull back the local system $\{\pi_n(\widehat{Y}, \widehat{Y} - D, \widehat{y}') : \widehat{y}' \in \widehat{Y} - D\}$ from $\widehat{Y} - D$. Similarly, for any $\widehat{y} \in D$ the obstruction $\omega_{\widehat{y}}$ to deforming $\widehat{f}$ into $\widehat{Y} - \widehat{y}$ is an element of $H^n(X, \Gamma_{\widehat{y}})$ where $\Gamma_{\widehat{y}}$ is obtained from the local system $\{\pi_n(\widehat{Y}, \widehat{Y} - \widehat{y}, \widehat{y}') : \widehat{y}' \in \widehat{Y} - \widehat{y}\}$ on $\widehat{Y} - \widehat{y}$. One uses the universal cover of $\widehat{Y}$, the relative Hurewicz isomorphism, and direct sum theorems for homology to show that for every $\widehat{y} \in D$ the inclusions $\widehat{Y} - D \subset \widehat{Y} - \widehat{y}$ induce an isomorphism

$$\pi_n(\widehat{Y}, \widehat{Y} - D, \widehat{y}') \approx \sum_{\widehat{y} \in D} \pi_n(\widehat{Y}, \widehat{Y} - \widehat{y}, \widehat{y}'),$$

which in turn induce an isomorphism

$$\phi: H^n(X, \Gamma_D) \approx \sum_{\widehat{y} \in D} H^n(X, \Gamma_{\widehat{y}}).$$

By naturality, $\phi(\omega_D) = \sum_{\widehat{y} \in D} \omega_{\widehat{y}}$. But each $\widehat{f}^{-1}(\widehat{y})$ is either empty or inessential, so $\omega_{\widehat{y}} = 0$ for each $\widehat{f}^{-1}(\widehat{y})$, therefore $\omega_D = 0$, and therefore $\widehat{f}$ may be deformed into $\widehat{Y} - D$. □

We turn now to the problem of computing the Nielsen root number. The most useful tool in this regard is the root Reidemeister number, which we now examine.

As usual, suppose $f: X \to Y$ a map and $y_0 \in Y$. Let $\widehat{q}: \widehat{Y} \to Y$ be a Hopf covering for f, and let $\widehat{f}: X \to \widehat{Y}$ be a Hopf lift of f through $\widehat{q}$. According to

Theorem (3.4), the set of Nielsen root classes is in one-to-one correspondence with a subset of the fiber $\widehat{q}^{-1}(y_0)$, and therefore in one to one correspondence with a subset of the fiber $\widehat{q}^{-1}(y)$ for any $y \in Y$. But according to covering space theory, for any $x \in X$ the fiber over $f(x)$ is in one-to-one correspondence with the set $\pi(Y, f(x))/\widehat{q}_{\#}(\pi(\widehat{Y}, \widehat{f}(x)))$ of left cosets of the fundamental group $\pi(Y, f(x))$ by the image $\widehat{q}_{\#}(\pi(\widehat{Y}, \widehat{f}(x)))$ of the fundamental group $\pi(\widehat{Y}, \widehat{f}(x))$ of $\widehat{Y}$ at $\widehat{f}(x)$ ([3]). By the choice of the covering $\widehat{q}$, we have $\widehat{q}_{\#}(\pi(\widehat{Y}, \widehat{f}(x))) = f_{\#}(\pi(X, x))$, and therefore the set of Nielsen root classes of f at y_0 is in one-to-one correspondence with a subset of the set $\pi(Y, f(x))/f_{\#}(\pi(X, x))$ of left cosets of $\pi(Y, f(x))$ by $f_{\#}(\pi(X, x))$. According to our conventions, this set is called Coker $f_{\#}$. The number of these cosets is therefore an upper bound on the number of Nielsen root classes of f at y_0, and therefore an upper bound on $N(f, y_0)$. We are thus led to the following definition.

(3.16) DEFINITION. Let $f: X \to Y$ be a map and choose a point $x \in X$. Then f induces a fundamental group homomorphism $f_{\#}: \pi(X, x) \to \pi(Y, f(x))$. The elements of Coker $f_{\#}$ are called *root Reidemeister classes of f (at x)*, the number of these elements is called the *root Reidemeister number* of f and denoted by $RR(f)$, so $RR(f) = \#\,\mathrm{Coker}(f_{\#})$.

The actual set $\mathrm{Coker}(f_{\#}) = \pi(Y, f(x))/f_{\#}(\pi((X, x))$ depends on the choice of the base point $x \in X$; however, its cardinality, $RR(f)$, does not. For this reason we will usually omit mention of the point x. Also we will often omit the word the word "root", when the context makes that clear, and talk of the Reidemeister classes of f, and the Reidemeister number of f.

In his original papers [H1], [H2], Hopf used the symbol j instead of $RR(f)$, and that symbol is also used in subsequent literature referencing Hopf, e.g. [BBS1], [BBS2], [BrS2], [Ep], [Li] and [O]. We have used the phrases "Reidemeister class" and "Reidemeister number" to maintain the connection with fixed point theory and coincidence theory. In particular, what we have called the "root Reidemeister number of f" is the same, in coincidence theory, as the "(coincidence) Reidemeister number of f and [the constant map into] y_0". In symbols: $RR(f) = R(f, y_0)$.

From the above discussion and Definition (3.16) we have,

(3.17) THEOREM. *Let $f: X \to Y$ be a map and $y_0 \in Y$. Let $\widehat{q}: \widehat{Y} \to Y$ and $\widehat{f}: X \to \widehat{Y}$ be a Hopf covering and lift for f. Then:*

(3.17.1) *$RR(f)$ is the cardinality of $\widehat{q}^{-1}(y)$ for any $y \in Y$,*

(3.17.2) *f has at most $RR(f)$ Nielsen root classes, and*

(3.17.3) *$RR(f) \geq NR(f, y_0)$.*

([3]) We could, of course, just as well use right cosets.

In the above discussion, we first made a correspondence from the Nielsen root classes to the fiber $\widehat{q}^{-1}(y_0)$, and then from the fiber to the Reidemeister classes. It is also of interest to exhibit an injection from the Nielsen classes directly into the Reidemeister classes. To do this, suppose that $f^{-1}(y_0) \neq \emptyset$ (otherwise there is not much to do), and let x_0 be a root of f at y_0. We define the injection $\phi: f^{-1}(y_0)/NR \to \pi(Y, y_0)/f_\#(\pi((X, x_0))$ as follows: Let α be a Nielsen root class, choose a root $x \in \alpha$ and a path A in X from x to x_0. Then $f \circ A$ is a loop in Y at y_0 and therefore represents an element $[f \circ A] \in \pi(Y, y_0)$; let $\phi(\alpha) = [f \circ A]f_\#(\pi(X, x_0))$. It must be shown that the definition of $\phi(\alpha)$ does not depend upon the choice of $x \in \alpha$ or the choice of the path A from x to x_0 (so ϕ is well defined), and then that ϕ is injective. We leave both of these tasks to the interested reader.

It will sometimes be more convenient to use the first homology groups rather than the fundamental groups to compute $RR(f)$. We can do this whenever $\pi(Y)$ is abelian.

(3.18) THEOREM. *Let $f: X \to Y$ be a map and suppose that $\pi(Y, y)$ is abelian for some (and therefore every) $y \in Y$. Then $RR(f) = \#\operatorname{Coker} f_1$ where $f_1: H_1(X; \mathbf{Z}) \to H_1(Y; \mathbf{Z})$ is the homomorphism induced by f on the first homology groups with integer coefficients.*

PROOF. Let $h_{(X,x)}: \pi(X, x) \to H_1(X; \mathbf{Z})$ denote the Hurewicz homomorphism. Then $h_{(X,x)}$ is an epimorphism and, since $\pi(Y, y)$ is abelian, $h_{(Y,y)}$ is an isomorphism. Also the following diagram commutes.

$$\begin{array}{ccc} \pi(X, x) & \xrightarrow{f_\#} & \pi(Y, f(x)) \\ {\scriptstyle h_{(X,x)}}\big\downarrow {\scriptstyle \text{epi}} & & {\scriptstyle h_{(Y,f(x))}}\big\downarrow {\scriptstyle \approx} \\ H_1(X; \mathbf{Z}) & \xrightarrow[f_1]{} & H_1(Y, \mathbf{Z}) \end{array}$$

It follows from this and a little algebra that $h_{(Y,(f(x))}$ induces a bijection $\operatorname{Coker} f_\# \approx\!> \operatorname{Coker} f_1$, so $RR(f) = \#\operatorname{Coker} f_\# = \#\operatorname{Coker} f_1$. □

Our main tool for computing the Nielsen root number is the following theorem and its corollary.

(3.19) THEOREM. *Suppose Y a space, $y_0 \in Y$ and that Y and y_0 have the property that for any loop A at y_0 there is a homotopy $\{h_t: Y \to Y\}$ such that*

(3.19.1) *$h_0: Y \to Y$ is the identity on Y,*
(3.19.2) *$h_1: Y \to Y$ is a homeomorphism,*
(3.19.3) *the path $\{h_t(y_0)\}$ is fixed-end-point homotopic to A.*

Then for any map $f: X \to Y$, either $NR(f, y_0) = 0$ or $NR(f, y_0) = RR(f)$.

PROOF. Let $\widehat{q}: \widehat{Y} \to Y$ and $\widehat{f}: X \to \widehat{Y}$ be a Hopf covering and lift for f, and suppose that $NR(f, y_0) \neq 0$. We will show that $\widehat{f}^{-1}(\widehat{y})$ is an essential Nielsen root class for every $\widehat{y} \in \widehat{q}^{-1}(y_0)$. Let $\widehat{y} \in \widehat{q}^{-1}(y_0)$.

Since $NR(f, y_0) \neq 0$, there is a $\widehat{y}_0 \in \widehat{q}^{-1}(y_0)$ such that $\widehat{f}^{-1}(\widehat{y}_0)$ is an essential Nielsen root class of f at y_0. Let $\widehat{A}$ be a path in $\widehat{Y}$ from $\widehat{y}$ to $\widehat{y}_0$. Then $A = \widehat{q} \circ \widehat{A}$ is a loop in Y at y_0, so there is a homotopy $\{h_t: Y \to Y\}$ satisfying (3.19.1), (3.19.2), and (3.19.3) above. Let $\{\widehat{h}_t: \widehat{Y} \to \widehat{Y}\}$ be a lift of $\{h_t\}$ with $\widehat{h}_0$ the identity on $\widehat{Y}$. Since A and $\{h_t(y_0)\}$ are fixed-end-point homotopic, so are $\widehat{A}$ and $\{\widehat{h}_t(\widehat{y})\}$, and therefore $\widehat{h}_1(\widehat{y}) = A(1) = \widehat{y}_0$. Now $\{h_t \circ f\}$ is a homotopy beginning at $h_0 \circ f = f$ and $\{\widehat{h}_t \circ \widehat{f}\}$ is its lift beginning at $\widehat{f}$. Moreover, $\widehat{f}^{-1}(\widehat{y}_0)$ is an essential root class of f, so $(\widehat{h}_1 \circ \widehat{f})^{-1}(\widehat{y}_0) \neq \emptyset$. There is therefore an $x_1 \in X$ such that $\widehat{h}_1 \circ \widehat{f}(x_1) = \widehat{y}_0$. Since h_1 is a homeomorphism, so is its lift $\widehat{h}_1$. Also $\widehat{h}_1(\widehat{y}) = \widehat{A}(1) = \widehat{y}_0$. It follows that $\widehat{f}(x_1) = \widehat{y}$. Thus $\widehat{f}^{-1}(\widehat{y}) \neq \emptyset$ and is therefore a Nielsen root class of f at y_0. To see that it is essential, let $\{f_t\}$ be a homotopy with $f_0 = f$, and $\{\widehat{f}_t: X \to \widehat{Y}\}$ its lift beginning at $\widehat{f}$. Since $\widehat{f}^{-1}(y_0)$ is an essential Nielsen class of f, then $\widehat{f}_1^{-1}(y_0)$ is an essential Nielsen class of f_1 so we may apply exactly the same argument as above to f_1 to show that $\widehat{f}_1^{-1}(\widehat{y}) \neq \emptyset$. Thus $\widehat{f}^{-1}(\widehat{y})$ is essential. □

The requirement that h_1 is a homeomorphism is stronger than necessary. All that is actually required is that h_1 define a homotopy equivalence of the triple $(Y, Y - y_0, y_0)$ into itself. See [B3] for details.

The most important application of (3.19) is to the case where Y is a manifold. For this, and later applications, we need

(3.20) THEOREM. *Suppose Y is manifold, A is any path in Y, and N is a neighbourhood of $A(I)$. Then there is an isotopy $\{h_t: Y \to Y\}$ such that h_0 is the identity on Y, h_t is the identity outside of N, and $h_t(A(0)) = A(t)$ for all $t \in [0, 1]$.*

PROOF. First suppose that the path A is contained entirely in the interior of an n-ball $B \subset N$ and let $\phi: \mathbf{B}^n \to B$ be a homeomorphism. Let $\psi: \operatorname{int} \mathbf{B}^n \to \mathbf{R}^n$ be the homeomorphism given by $\psi(x) = (1/(1 - \|x\|))x$ with inverse $\psi^{-1}(x) = (1/(1 + \|x\|))x$. Then for any $v \in \mathbf{R}^n$ and x_0 on the boundary of $\mathbf{B}^n$, it is straightforward to show that $\lim_{x \to x_0} \psi^{-1}(\psi(x) + v) = x_0$, so we may define $h_t: Y \to Y$ continuously by

$$h_t(y) = \begin{cases} \phi \circ \psi^{-1}(\psi \circ \phi^{-1}(y) + \psi \circ \phi^{-1}(A(t)) - \psi \circ \phi^{-1}(A(0))) & \text{for } y \in \operatorname{int} B, \\ y \quad \text{for } y \in Y - \operatorname{int} B. \end{cases}$$

Then h_0 is the identity, h_t is the identity outside of N, and $h_t(A(0)) = A(t)$ for all $t \in I$, so $\{h_t\}$ is the desired isotopy.

In general, A may be broken into a sequence $A^1, \dots, A^m$ of paths

$$A^k(t) = A\left(\frac{k - 1 + t}{m}\right), \quad \text{for } t \in I,$$

such that each A^k is interior to an n-ball in N. For each A^k construct an isotopy $\{h_t^k\}$ as above, and piece these isotopies together

$$h_t = \begin{cases} h^1_{mt} & \text{for } 0 \le t \le 1/m \\ h^k_{mt-k} \circ h_1^{k-1} \circ \cdots \circ h_1^1 & \text{for } (k-1)/m \le t \le k/m \text{ and } k = 2, \dots, m, \end{cases}$$

to form the desired $\{h_t\}$. □

Theorem (3.20) implies that if Y is a manifold, then it satisfies the hypothesis of Theorem (3.19) for any $y_0 \in Y$.

(3.21) COROLLARY. *Suppose $f: X \to Y$ a map, Y a manifold, and $y_0 \in Y$. Then either $NR(f, y_0) = 0$, or $NR(f, y_0) = RR(f)$.*

Theorem (3.20) also implies, as we will now show, that when Y is a manifold, then $NR(f, y_0)$ is independent of the point $y_0 \in Y$. For this reason, when Y is a manifold, we will often write $NR(f, y_0)$ more simply as $NR(f)$.

(3.22) COROLLARY. *Suppose $f: X \to Y$ a map and Y a manifold. Then $NR(f, y_0)$ is independent of the choice of $y_0 \in Y$, that is, $NR(f, y_0) = NR(f, y_1)$ for all $y_0, y_1 \in Y$.*

PROOF. Let $y_0, y_1 \in Y$. By (3.20) there is an isotopy $\{h_t: Y \to Y\}$ such that h_0 is the identity and $h_1(y_0) = y_1$. Then $\{h_t \circ f\}$ is a homotopy from f to $h_1 \circ f$, so $NR(f, y_1) = NR(h_1 \circ f, y_1)$. But from (3.10), $NR(h_1 \circ f, y_1) = NR(f, h_1^{-1}(y_1)) = NR(f, y_0)$. □

From (3.6) and (3.21) we have

(3.23) COROLLARY. *Suppose $f: X \to Y$ a map, Y a manifold, X compact, and $R(f) = \infty$. Then $NR(f) = 0$.*

Theorem (3.19) and Corollary (3.21) are the main tools for computing the Nielsen root number. In order to apply them, however, we need a method for determining whether or not the Nielsen number is non-zero. This is the goal of index theory, which we now examine.

4. Root indexes

We continue to proceed by analogy with fixed point theory. In fixed point theory, a triple (X, f, U) is *admissible* if X is a suitably nice space, $f: X \to X$ is a self map, and f has no fixed points on the boundary of U. A fixed point index ι is a function assigning to each admissible triple (X, f, U) a rational number $\iota(X, f, U)$ that, very roughly speaking, is an algebraic measure of the number of fixed points of f lying in U, see [Br2, Chapter IV]. In our definition of admissibility, we find it convenient to relax the requirement that the subset be open.

(4.1) DEFINITION. Let X and Y be topological spaces and $y_0 \in Y$. A pair (f, A) is an *admissible pair for* X, Y, y_0 if $f: X \to Y$ is a map, $A \subset X$, and A has a closed neighbourhood N such that $N - A$ contains no roots of f at y_0: $(N - A) \cap f^{-1}(y_0) = \emptyset$.

Notice that if U is open, then (f, U) admissible is equivalent to $\operatorname{Bd} U \cap f^{-1}(y_0) = \emptyset$.

For an admissible pair (f, A), we will sometimes require N to be a neighbourhood not just of A, but also of $\operatorname{cl} A$. That we may do so is an easy consequence of normality:

(4.2) THEOREM. *Suppose (f, A) is an admissible pair for X, Y, y_0. Then $\operatorname{cl} A$ has a closed neighbourhood N such that $N - A$ contains no roots of f at y_0.*

PROOF. Let N' be a closed neighbourhood of A such that $(N' - A) \cap f^{-1}(y_0) = \emptyset$. Then $f^{-1}(y_0) - N' = f^{-1}(y_0) - \operatorname{int} N'$ is a closed set disjoint from the closed set N'. By normality, there are disjoint open neighbourhoods U of $f^{-1}(y_0) - N'$ and V of N'. Let $N = \operatorname{cl} V$. It is straightforward to show N has the desired properties. □

As examples of admissible pairs we have:

(4.3) THEOREM. *Suppose $f: X \to Y$ a map and $y_0 \in Y$. Then (f, X) and $(f, \emptyset)$ are admissible, and so is (f, α) for every Nielsen root class α of f at y_0.*

PROOF. For (f, X) the required neighbourhood N of X is X, for $(f, \emptyset)$ let $N = \emptyset$. For (f, α) let $N = \operatorname{cl} U_\alpha$, where U_α is the open neighbourhood of α guaranteed by Theorem (3.5). □

In fixed point theory, an index is required to satisfy five axioms [Br2, p. 52–53]. However, some of these axioms are included in order to make sure that the axioms are categorical, i.e. there is at most one index. At the present state of root theory, we hesitate to settle on any one index, hence we analogize only the two axioms most important for our purposes: Additivity and Homotopy. We also relax the requirement that the index be rational valued, and require only that it be a function into an abelian group.

(4.4) DEFINITION. Let X and Y be topological spaces and $y_0 \in Y$. A *root index* for X, Y is a function ω from the set of admissible pairs for X, Y, y_0 into an abelian group satisfying:

(4.4.1) (Additivity) Suppose $A \subset X$ and $A_1, \ldots, A_n$ are subsets of A such that
- (a) (f, A) is admissible and (f, A_i) is admissible for each i,
- (b) $f^{-1}(y_0) \cap (A - \bigcup_i A_i) = \emptyset$, and
- (c) $A_i \cap A_j = \emptyset$ for $i \neq j$.

Then $\omega(f, A) = \sum_i \omega(f, A_i)$.

(4.4.2) (Homotopy) Suppose $\{f_t: X \to Y\}$ a homotopy, A is open in X, and (f_t, A) is admissible for every $t \in I$. Then $\omega(f_1, A) = \omega(f_0, A)$.

As an easy consequence of additivity we have

(4.5) THEOREM. *Suppose (f, A) an admissible pair for X, Y, y_0. Then $\omega(f, A) \neq 0$ implies that $f(x) = y_0$ for some $x \in A$.*

PROOF. We prove the contrapositive, so suppose that f has no roots at y_0 in A. Then applying additivity with $A_1 = \emptyset$ we have $\omega(f, A) = \omega(f, \emptyset)$, and applying it with $A_2 = \emptyset$ as well we have, $\omega(f, A) = \omega(f, \emptyset) + \omega(f, \emptyset)$. Therefore $\omega(f, A) = 0$.□

Of particular interest is the index of a root class. In particular we would like to show that the index for homotopy related root classes is the same. The following theorem establishes this, but only in the case where X is compact (4).

(4.6) THEOREM. *Suppose X compact (5), $\{f_t: X \to Y\}$ a homotopy, $y_0 \in Y$, and α a root class of f_0 at y_0. Then:*

(4.6.1) *If α is $\{f_t\}$ related to a root class α_1 of f_1, then $\omega(f_0, \alpha_0) = \omega(f_1, \alpha_1)$.*

(4.6.2) *Otherwise, α is not $\{f_t\}$ related to any root class of f_1, in which case $\omega(f_0, \alpha_0) = 0$.*

PROOF. Let $\widehat{q}: \widehat{Y} \to Y$ and $\{\widehat{f_t}: X \to \widehat{Y}\}$ be a Hopf covering and lift for $\{f_t\}$. Then from Theorem (3.4) $\alpha = \widehat{f_0}^{-1}(\widehat{y}_0)$ for some $\widehat{y}_0 \in \widehat{q}^{-1}(y_0)$, and we need only show that $\omega(f_0, \widehat{f_0}^{-1}(\widehat{y}_0)) = \omega(f_1, \widehat{f_1}^{-1}(\widehat{y}_0))$. Since I is connected, it suffices to show that $\omega(f_t, \widehat{f_t}^{-1}(\widehat{y}_0))$ is a locally constant function of t. So let $t \in I$; we will find a neighbourhood J of t such that $\omega(f_s, \widehat{f_s}^{-1}(\widehat{y}_0))$ is constant for $s \in J$.

The set $\{(x, t) \in X \times I : f_t(x) = y_0\}$ is a closed subset of the compact space $X \times I$ and therefore compact. Let $C = \bigcup_t f_t^{-1}(y_0) \subset X$ be its projection into X. Then C is the continuous image of a compact set and therefore compact.

As in the proof of Theorem (3.5), there is a family $\{U_{\widehat{y}} : \widehat{y} \in \widehat{q}^{-1}(y_0)\}$ of mutually disjoint open subsets of X such that $\widehat{f_t}^{-1}(\widehat{y}) \subset U_{\widehat{y}}$ for every $\widehat{y} \in \widehat{q}^{-1}(y_0)$. Give the set $\mathrm{Map}(X, \widehat{Y})$ of maps from X to $\widehat{Y}$ the compact open topology, then [Hu, Proposition 9.4, p. 75] the assignment $s \mapsto \widehat{f_s}$ is a continuous function from I to $\mathrm{Map}(X, \widehat{Y})$. The set $C - U_{\widehat{y}_0}$ is compact and $\widehat{f_t}$ takes $C - U_{\widehat{y}_0}$ into the open set $\widehat{Y} - \widehat{y}_0$, so t has a neighbourhood $J_0 \subset I$ such that $\widehat{f_s}(C - U_{\widehat{y}_0}) \subset \widehat{Y} - \widehat{y}_0$ for all $s \in J_0$, and therefore.

$$\widehat{f_s}^{-1}(\widehat{y}_0) \subset U_{\widehat{y}_0} \quad \text{for all } s \in J_0. \tag{4.7}$$

(4) But see following notes.

(5) The theorem remains true for noncompact X if we require $f: X \to Y$ to be *proper* and require the homotopies here and in Definition (3.7) to be proper. A map f is proper if $f^{-1}(D)$ is compact whenever D is. A homotopy $\{f_t\}$ is proper if the map $(x, t) \mapsto f_t(x)$ is proper. We have structured the proof so that it is still valid in this more general setting.

Similarly, there is another neighbourhood J_1 of t such that

$$\widehat{f}_s\Big(C - \bigcup_{\widehat{y}\neq\widehat{y}_0} U_{\widehat{y}}\Big) \subset \widehat{Y} - (\widehat{q}^{-1}(y_0) - \widehat{y}_0)$$

for all $s \in J_1$, and therefore

$$f_s^{-1}(y_0) - \widehat{f}_s^{-1}(\widehat{y}_0) \subset \bigcup_{\widehat{y}\neq\widehat{y}_0} U_{\widehat{y}}, \quad \text{for all } s \in J_1. \tag{4.8}$$

Let $J = J_0 \cap J_1$. Then from (4.7) and (4.8), $\operatorname{Bd} U_{\widehat{y}_0} \cap f_s^{-1}(y_0) = \emptyset$ and therefore $(f_s, U_{\widehat{y}_0})$ is admissible for all $s \in J$. Also (4.7) and (4.8) imply that $U_{\widehat{y}_0} \cap f_s^{-1}(y_0) = \widehat{f}_s^{-1}(\widehat{y}_0)$ so by additivity $\omega(f_s, \widehat{f}_s^{-1}(\widehat{y}_0)) = \omega(f_s, U_{\widehat{y}_0})$ for all $s \in J$. By homotopy, $\omega(f_s, U_{\widehat{y}_0})$ is constant for $s \in J$. Therefore $\omega(f_s, \widehat{f}_s^{-1}(\widehat{y}_0))$ is constant for $s \in J$. (6) □

(4.9) Corollary. *Suppose X compact, $f: X \to Y$ a map, $y_0 \in Y$, and ω a root index for X, Y, y_0. Then for any root class α of f at y_0, if $\omega(f, \alpha) \neq 0$ then α is essential.*

Proof. We prove the contrapositive. If α is not essential, then by Definition (3.7), there is a homotopy $\{f_t\}$ beginning at f such that α is not $\{f_t\}$ related to any root class of f_1, so from Theorem (4.6), $\omega(f, \alpha) = 0$. □

Were we to work by complete analogy to fixed-point theory, we would have used Corollary (4.9) as a definition of essentiality. That is, we would define an index that satisfies additivity and homotopy and then define a root class to be essential if its index was non-zero. Hopefully, one would then be able to prove that if a class has index zero, then it is not essential in the sense of Definition (3.7). Unfortunately there does not yet appear to be any such index defined even, e.g. for the case of maps of finite polyhedra, powerful enough to detect essentiality in the sense of Definition (3.7) — with the notable exception, as we will see in Section 5, of maps of manifolds of the same dimension.

We turn now to the construction of a Nielsen root index. We begin with a simple construction that uses ordinary homology theory — any theory H_* satisfying the Eilenberg–Steenrod axioms (see [ES]) — to detect the existence of roots. Let, as usual, $f: X \to Y$ be a map and $y_0 \in Y$, and consider the composition:

$$X \xrightarrow{f} Y \overset{j}{\subset} (Y, Y - y_0).$$

(6) Note that in the above proof it is only C, not X, that needs to be compact, and that C is easily seen to be compact provided $\{f_t\}$ is proper.

This induces homology homomorphisms of the total homology groups

$$H_*(X) \xrightarrow{f_*} H_*(Y) \xrightarrow{j_*} H_*(Y, Y - y_0).$$

We claim that if the composition $j_* \circ f_* \neq 0$, then f — and any map homotopic to f — has at least one root at y_0. To see this, suppose that f has no roots at y_0. Then we may restrict the codomain of f to get a map $f'\colon X \to Y - y_0$. The above composition of maps may then be rewritten as

$$X \xrightarrow{f'} Y - y_0 \overset{j'}{\subset} (Y - y_0, Y - y_0) \overset{k}{\subset} (Y, Y - y_0).$$

But this means that the induced homomorphism $j_* \circ f_* = k_* \circ j'_* \circ f'_*$ may be factored through $H_*(Y - y_0, Y_* - y_0) = 0$, so $j_* \circ f_* = 0$. We are thus led to use $j_* \circ f_*$ as an algebraic measure of whether f can be deformed to be root free. This also motivates us to localize the above construction in order to define a root index.

Here is the localized construction: Suppose that (f, A) is an admissible pair for X, Y, y_0. Let N be a closed neighbourhood of A such that $N - A$ contains no roots of f at y_0, so f defines a pair map

$$f'\colon (N, N - A) \to (Y, Y - y_0).$$

If f has no roots in A, then f' can be factored through the pair $(Y - y_0, Y - y_0)$ whose homology is trivial. Thus, if f has no roots in A, then the induced homomorphism

$$f'_*\colon H_*(N, N - A) \to H_*(Y, Y - y_0)$$

is trivial. We might therefore use f'_* as an algebraic measure of the number of roots in A. The problem is, however, that although f'_* is a member of a group, namely the group of homomorphisms from $H_*(N, N - A)$ into $H_*(Y, Y - y_0)$, this group depends on the set A, whereas for f'_* to be an index, the group should be the same for different sets A and depend only on X, Y, y_0. The solution is to use an enlarged diagram

$$X \overset{i}{\subset} (X, X - A) \overset{e}{\supset} (N, N - A) \xrightarrow{f'} (Y, Y - y_0).$$

By (4.2), we may assume N to be a neighbourhood of $\operatorname{cl} A$ as well as A, so then the inclusion $e\colon (N, N - A) \subset (X, X - A)$ is an excision in the sense of [ES, p. 12] (see [B1, p. 58] for a proof of this relatively easy fact), and therefore induces a homology isomorphism, so we have the following induced diagram.

$$H_*(X) \xrightarrow{i_*} H_*(X, X - A) \overset{e_*}{\xleftarrow{\approx}} H_*(N, N - A) \xrightarrow{f'_*} H_*(Y, Y - y_0).$$

Since e_* is an isomorphism we may form the composition

$$L_*(f, A) = f'_* \circ e_*^{-1} \circ i_*\colon H_*(X) \to H_*(Y, Y - y_0).$$

(4.10) Theorem and Definition. *In the above discussion, the homomorphism $L_*(f, A)$ is independent of the choice of the closed neighbourhood N of $\operatorname{cl} A$, and therefore depends only on X, Y, y_0 and the pair (f, A). Thus, L_* is a function from the set of all admissible pairs for X, Y, y_0 into the abelian group $\hom(H_*(X), H_*(Y, Y - y_0))$ of all homomorphisms from $H_*(X)$ into $H_*(Y, Y - y_0)$. It satisfies the additivity and homotopy conditions of Definition (4.4) and is therefore a root index, called the homology homomorphism root index.*

We will write $L_*(f, A; G)$, when we wish to indicate that G is the coefficient group. We will also write $L_n(f, A)$ to denote the restriction of $L_*(f, A)$ to the nth homology group. Thus $L_n(f, A; G)\colon H_n(X; G) \to H_n(Y, Y - y_0; G)$. The dual cohomology composition

$$L^*(f, A) = i^* \circ e^{*-1} \circ f'^*\colon H^*(Y, Y - y_0) \to H^*(X)$$

is also independent of N and the function L^* from admissible pairs of X, Y, y_0 into $\hom(H^*(Y, Y - y_0), H^*(X))$ is also a root index, called the cohomology homomorphism root index.

Proof. We prove this for homology, the cohomology proof is entirely dual.

We first prove independence from the choice of N. Let N' be another neighbourhood of $\operatorname{cl} A$ such that $N' - A$ is root-free. Let $N'' = N \cup N'$ and consider the following commutative diagram,

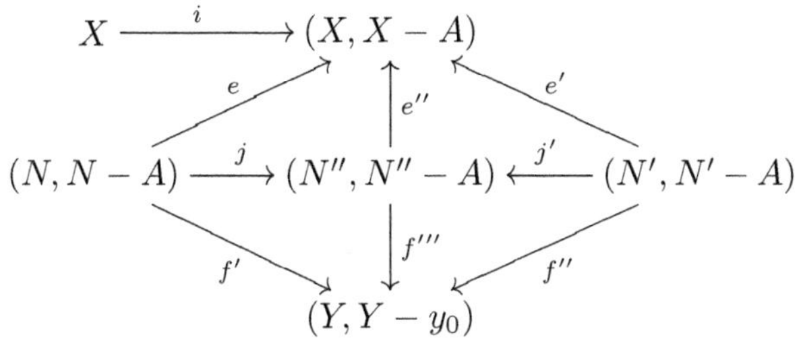

in which f', f'', f''' are defined by f and all other maps are inclusions. A simple diagram chase shows that

$$f'_* \circ e_*^{-1} \circ i_* = f'''_* \circ e''^{-1}_* \circ i_* = f''_* \circ e'^{-1}_* \circ i_*,$$

so $L_*(f, A)$ does not depend upon the choice of N.

For additivity, let A and $A_1, \dots, A_n$ be as in (4.4.1). Let N be a closed neighbourhood of $\operatorname{cl} A$ such that $(N - A) \cap f^{-1}(y_0) = \emptyset$, so we also have

$$\Big(N - \bigcup_i A_i\Big) \cap f^{-1}(y_0) = \emptyset,$$

and consider the diagram

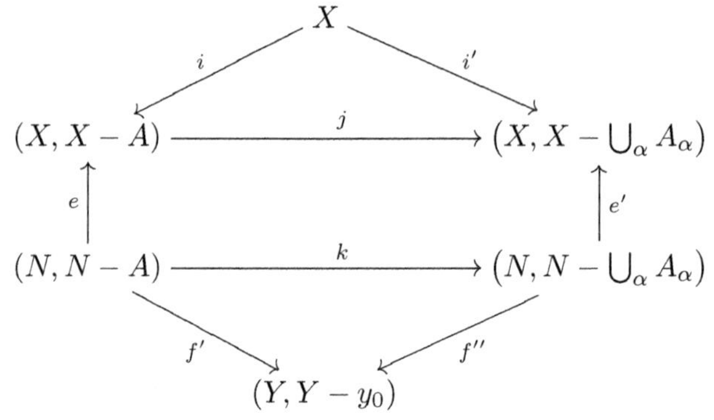

An easy diagram chase shows that

$$L_*(f, A) = f'_* \circ e_*^{-1} \circ i_* = f''_* \circ e'^{-1}_* \circ i_* = L_*(f, \textstyle\bigcup_\alpha A_\alpha).$$

To complete the proof of additivity, we must show that $L_*(f, \bigcup_\alpha A_\alpha) = \sum_\alpha L_*(f, A_\alpha)$. Using normality of X, select for each α a closed neighbourhood N_α of $\operatorname{cl} A_\alpha$ so that the N_α are mutually disjoint and $N_\alpha - A_\alpha$ is root free for each α. Let $N = \bigcup_\alpha N_\alpha$ and $A' = \bigcup_\alpha A_\alpha$ and consider the following diagram:

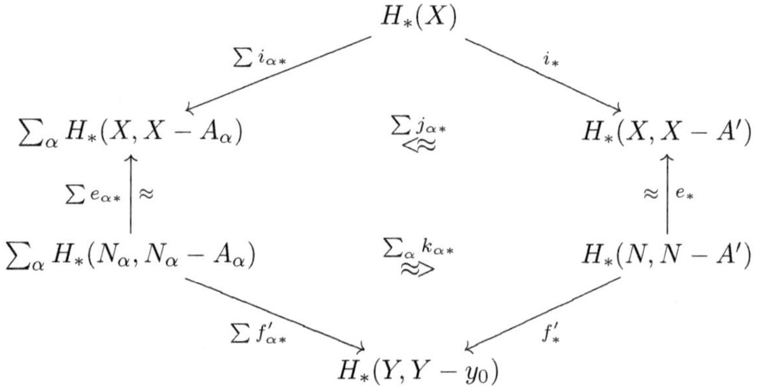

The maps f'_α and f' are defined by f, the maps i_α, i, j_α, e_α, e are all inclusions. By the homology direct sum theorem [ES, p. 33], $\sum_\alpha k_{\alpha *}$ is an isomorphism. The e_α and e are excisions and therefore induce isomorphisms. Thus by commutativity $\sum_\alpha j_{\alpha *}$ is an isomorphism. By commutativity we have $\sum_\alpha e_{\alpha *} = (\sum_\alpha j_{\alpha *}) \circ e_* \circ (\sum_\alpha k_{\alpha *})$. Solving this for e_*^{-1} we get $e_*^{-1} = \sum_\alpha k_{\alpha *} \circ e_{\alpha *}^{-1} \circ j_{\alpha *}$. Thus

$$\begin{aligned} L_*(f, A') &= f'_* \circ e_*^{-1} \circ i_* = f'_* \circ \Big(\sum_\alpha k_{\alpha *} \circ e_{\alpha *}^{-1} \circ j_{\alpha *} \Big) \circ i_* \\ &= \sum_\alpha f'_* \circ k_{\alpha *} \circ e_{\alpha *}^{-1} \circ j_{\alpha *} \circ i_* = \sum_\alpha f'_{\alpha *} \circ e_{\alpha *}^{-1} \circ i_{\alpha *} = \sum_\alpha L_*(f, A_\alpha), \end{aligned}$$

which completes the proof of additivity.

It remains to prove homotopy, let $\{f_t: X \to Y\}$ and A be as in (4.4.2). Then for each $x \in \operatorname{Bd} A$ and $t \in I$, $f_t(x) \in Y - y_0$ so there is a neighbourhood U_{xt} of x and neighbourhood V_{xt} of t such that $f_s(x') \neq y_0$ for all $x' \in U_{xt}$ and $s \in V_{xt}$. For fixed $x \in \operatorname{Bd} A$, the collection $\{V_{xt} : t \in I\}$ is an open cover of I and therefore has a finite subcover $\{V_{xt_1}, \dots, V_{xt_n}\}$. Let $U_x = \bigcap_i U_{xt_i}$ and $U = \bigcup_{x \in \operatorname{Bd} A} U_x \cup A$. Then U is a neighbourhood of $\operatorname{cl} A$ such that $(U - A) \cap f_t^{-1}(y_0) = \emptyset$ for all $t \in I$. By normality of X, U contains a closed neighbourhood N of $\operatorname{cl} A$, and $\{h_t\}$ defines a homotopy $\{f'_t: (N, N - A) \to (Y, Y - y_0)\}$, so $f_{0*} = f_{1*}$. Let $i: X \subset (X, X - A)$ and $e: (N, N - A) \subset (Y, Y - y_0)$ be the indicated inclusions. Then $L_*(f_0, A) = f'_{0*} \circ e_*^{-1} \circ i_* = f'_{1*} \circ e_*^{-1} \circ i_* = L_*(f_1, A)$. □

(4.11) REMARK. Notice that when $A = X$, we may choose $N = X$, and we have

$$L_*(f, X) = j_* \circ f_*: H_*(X) \to H_*(Y, Y - y_0),$$

where $j: Y \subset (Y, Y - y_0)$ is the inclusion. Thus the index L_* is in fact a localization of the global measure we started with in order to detect the existence of roots in X.

We may use L_* and L^* to define other indexes. The most important of these is the *integer root index*, defined below.

(4.12) THEOREM AND DEFINITION. *Let X and Y be spaces and $y_0 \in Y$. Use integer coefficients for homology and assume that for some $n > 0$*

(4.12.1) *$H_n(X; \mathbf{Z})$ is infinite cyclic, and*

(4.12.2) *y_0 has an n-dimensional euclidean neighbourhood.*

Then $H_p(Y, Y - y_0; \mathbf{Z})$ is trivial for $p \neq n$ and is infinite cyclic for $p = n$. Select generators μ of $H_n(X, \mathbf{Z})$ and ν of $H_n(Y, Y - y_0; \mathbf{Z})$ and define an integer valued function λ from the admissible pairs by the formula

$$L_n(f, A; \mathbf{Z})(\mu) = \lambda(f, A)\nu. \tag{4.13}$$

Then λ is a root index for X, Y, y_0 called the integer root index. *The above assumptions are met when X and Y are compact orientable manifolds, and in this case (with proper choice of generators) $\lambda(f, X)$ is the ordinary Brouwer degree of f, $\lambda(f, X) = \deg(f)$.*

Note that like the Brouwer degree, the definition of λ depends upon the choice of the generators. A different choice can result in an index of opposite sign.

PROOF. Since y_0 has an n-dimensional euclidean neighbourhood, it has a neighbourhood B homeomorphic to the closed unit ball in $\mathbf{R}^n$. The inclusion $(B, B - y_0) \subset (Y, Y - y_0)$ is an excision and therefore induces an isomorphism of $H_p(B, B - y_0; \mathbf{Z})$

onto $H^p(Y, Y - y_0; \mathbf{Z})$. Since $H^p(B, B - y_0; \mathbf{Z})$ is trivial for $p \neq n$ and infinite cyclic for $p = n$, the same is true of $H^p(Y, Y - y_0; \mathbf{Z})$. Thus λ is a well defined function from the admissible pairs for X, Y, y_0 into $\mathbf{Z}$. Its additivity and homotopy properties follow directly from those of L_*.

Assume now that X and Y are both compact orientable manifolds. Then both $H_n(X)$ and $H_n(Y)$ are infinite cyclic and the inclusion $j\colon Y \subset (Y, Y - y_0)$ induces an isomorphism in dimension n. Choose $\nu' = j_n^{-1}(\nu)$ as generator of $H_n(Y)$. Recall that with this choice of generators, the Brouwer degree of f, $\deg f$, satisfies the equation $f_n(\mu) = (\deg f)\nu'$. Hence, $j_n \circ f_n(\mu) = j_n((\deg f)\nu') = (\deg f) j_n(\nu') = (\deg f)\nu$. But from (4.11) and (4.13) we have $j_n \circ f_n(\mu) = L_n(f, X)(\mu) = \lambda(f, X)\nu$. Thus $\deg f = \lambda(f, X)$. □

We shall also be interested in the case where X, and possibly Y is a nonorientable manifold. In this case, it will sometimes be more convenient to use integers mod 2 as a coefficient group. With obvious modifications in the proof of (4.12) we have

(4.14) THEOREM AND DEFINITION. *Let X and Y be spaces and $y_0 \in Y$. Use integer* mod 2 *coefficients for homology and assume that for some $n > 0$*

(4.14.1) $H_n(X; \mathbf{Z}/2\mathbf{Z}) \approx \mathbf{Z}/2\mathbf{Z}$, *and*
(4.14.2) y_0 *has an n-dimensional euclidean neighbourhood.*

Then $H_p(Y, Y - y_0; \mathbf{Z}/2\mathbf{Z})$ is trivial for $p \neq n$ and is isomorphic to $\mathbf{Z}/2\mathbf{Z}$ for $p = n$. Let $\mu \in H_n(X, \mathbf{Z}/2\mathbf{Z})$ and $\nu \in H_n(Y, Y - y_0; \mathbf{Z}/2\mathbf{Z})$ be the generators and define an integer mod 2 *valued function λ_2 from the admissible pairs by the formula*

$$L_n(f, A; \mathbf{Z}/2\mathbf{Z})(\mu) = \lambda_2(f, A)\nu. \tag{4.15}$$

Then λ_2 is a root index for X, Y, y_0 called the integer mod 2 root index. *The above assumptions are met when X and Y are compact manifolds, and in this case $\lambda_2(f, X)$ is the ordinary* mod 2 *Brouwer degree of f,* $\deg_2(f)$.

For the integer mod 2 index, there is of course no ambiguity so far as choice of generators.

(4.16) EXAMPLE (The covering of P^2 by S^2). Model the two-sphere S^2 as the unit sphere in $\mathbf{R}^3$ and the projective plane P^2 as S^2 with antipodal points identified. Let $f\colon S^2 \to P^2$ be the identification. Let $n = (0, 0, 1)$, $s = (0, 0, -1) \in S^2$ denote the north and south poles and $y_0 = \{n, s\} = f(\{n, s\}) \in P^2$. Then f has two roots, n and s, at y_0. Let A be any path in S^2 from n to s, then $f \circ A$ is a loop at y_0 whose fixed-end-point homotopy class generates the fundamental group $\pi(Y, y_0) \approx \mathbf{Z}/2\mathbf{Z}$, so n and s must not be Nielsen equivalent. Thus f has exactly two Nielsen classes of f at y_0, namely $\{n\}$ and $\{s\}$. Let N be a closed neighbourhood of n contained in the open northern hemisphere of S^2, and let $f''\colon (N, N-n) \to (f(N), f(N)-y_0)$ be the map defined by f. Consider the maps

$$S^2 \overset{i}{\subset} (S^2, S^2 - n) \overset{e}{\supset} (N, N-s) \overset{f''}{\longrightarrow} (f(N), f(N) - y_0) \overset{e'}{\subset} (P^2, P^2 - y_0).$$

The inclusions e and e' are excisions and f'' is a homeomorphism, so they all induce homology isomorphisms. The inclusion i induces an isomorphism in dimension 2. Note that $f' = e' \circ f''\colon (N, N - \{n\})$ is the map defined by f. Also, for any coefficient group G, $H_2(S^2; G) \approx G$. Thus for any nontrivial G,

$$L_2(f, \{n\}; G) = e_2' \circ f_2' \circ e_2^{-1} \circ i_2\colon H_2(S^2; G) \approx\!\!> H_2(P^2, P^2 - y_0; G)$$

is non-trivial. Thus $\{n\}$ is essential. Similarly (or using (3.21)), $\{s\}$ is essential. therefore $NR(f) = 2$.

Since $L_2(f, \{n\}; \mathbf{Z})$ is an isomorphism, we have $\lambda(f, \{n\}) = \pm 1$, where the sign depends upon the choice of generators for $H_2(S^2; \mathbf{Z})$ and $H_2(P^2, P^2 - y_0; \mathbf{Z})$.

It is instructive to look at $L_*(f, \{s\})$ as well. Let $a\colon S^2 \to S^2$ be the antipodal map, $a(x) = -x$. Then we may use $a(N)$ as a closed neighbourhood of $\{s\}$ in order to compute $L_*(f, \{s\})$. Consider the diagram

$$\begin{array}{ccccccc}
S^2 & \overset{i}{\subset} & (S^2, S^2 - n) & \overset{e}{\supset} & (N, N-n) & \overset{f'}{\longrightarrow} & (P^2, P^2 - y_0) \\
{\scriptstyle a}\updownarrow & & {\scriptstyle a'}\updownarrow & & {\scriptstyle a''}\updownarrow & & \| \\
S^2 & \overset{i'}{\subset} & (S^2, S^2 - n) & \overset{e'}{\supset} & (a(N), a(N) - s) & \underset{f''}{\longrightarrow} & (P^2, P^2 - y_0)
\end{array}$$

where a' and a'' are defined by a and f' and f'' are defined by f. The maps, e, e', a', and a'' all induce isomorphisms so one readily has $e_*'^{-1} = a_*'' \circ e_*^{-1} \circ a_*'$. From this and commutativity

$$L_*(f, \{s\}) = f_*'' \circ e_*'^{-1} \circ i_*' = f_*'' \circ a_*'' \circ e_*^{-1} \circ a_*' \circ i' * = f_*' \circ e_*^{-1} \circ i_* \circ a_* = L_*(f, \{n\}) \circ a_*.$$

In dimension 2 it is well known that $a_2 = -\mathrm{id}_{H_2(S^2)}$, so $L_2(f, \{s\}) = -L_2(f, \{n\})$. In dimensions 0 and 1, $H_0(P^2, P^2 - y_0) = H_1(P^2, P^2 - y_0) = 0$. Thus $L_*(f, \{s\}) = -L_*(f, \{n\})$, and also $\lambda(f, \{s\}) = -\lambda(f, \{n\})$.

Since $\{n\}$ and $\{s\}$ are the only root classes it follows from additivity that $L_*(f, X) = L_*(f, \{s\}) + L_*(f, \{n\}) = 0$. Therefore we cannot always use $L_*(f, X)$ — or any index derived from it — to detect the existence of roots of f. It can happen that the indexes of the root classes cancel with one another so that $L_*(f, X) = 0$ even when some root classes have non-zero indexes. The following theorem shows that this sort of cancelation cannot occur, however, when X is compact and Y is a compact orientable manifold.

(4.17) THEOREM. *Let $f: X \to Y$ be a map, $y_0 \in Y$, and suppose that X is compact and Y is a compact n-manifold. Let α and β be two root classes of f and use either $\mathbf{Z}$ or $\mathbf{Z}/2\mathbf{Z}$ for the coefficient group. Then $L_*(f,\alpha) = \pm L_*(f,\beta)$. Moreover, if Y is orientable, then $L_*(f,\alpha) = L_*(f,\beta)$. Similarly for L^* and, whenever they are defined, for λ and λ_2.*

PROOF. We prove the theorem for L_*, the proof for L^*, with obvious changes, is essentially the same, the case for λ and λ_2 follows easily from the case for L_*.

Let $\widehat{q}: \widehat{Y} \to Y$ and $\widehat{f}: X \to \widehat{Y}$ be a Hopf covering and lift for f, and let $\widehat{y}_0, \widehat{y}_1 \in \widehat{q}^{-1}(y_0)$. It suffices to show that

$$L_*(f, \widehat{f}^{-1}(\widehat{y}_1)) = \begin{cases} L_*(f, \widehat{f}^{-1}(\widehat{y}_0)) & \text{if } Y \text{ is orientable, and} \\ \pm L_*(f, \widehat{f}^{-1}(\widehat{y}_0)) & \text{otherwise.} \end{cases}$$

Let $\widehat{A}$ be a path in $\widehat{Y}$ from $\widehat{y}_0$ to $\widehat{y}_1$, and $A = \widehat{q} \circ \widehat{A}$, so A is a loop at y_0. From (3.20) there is an isotopy $\{h_t: Y \to Y\}$ such that h_0 is the identity and $h_t(y_0) = A(t)$ for all $t \in I$. Lift $\{h_t\}$ to an isotopy $\{\widehat{h}_t: \widehat{Y} \to \widehat{Y}\}$ beginning at the identity on $\widehat{Y}$. Then $\{\widehat{h}_t(\widehat{y}_0)\}$ is a lift of A beginning at $\widehat{y}_0$ and is therefore the same as $\widehat{A}$. In particular $\widehat{h}_1(\widehat{y}_0) = \widehat{A}(1) = \widehat{y}_1$. It follows that $(\widehat{h}_0 \circ \widehat{f})^{-1}(\widehat{y}_1) = \widehat{f}^{-1}(\widehat{y}_1)$ is $\{h_t \circ f\}$ related to $(\widehat{h}_1 \circ \widehat{f})^{-1}(y_1) = \widehat{f}^{-1}(\widehat{h}_1^{-1}(\widehat{y}_1)) = f^{-1}(\widehat{y}_0)$, and therefore by (4.6)

$$L_*(f, \widehat{f}^{-1}(\widehat{y}_1)) = L_*(h_0 \circ \widehat{f}, f^{-1}(\widehat{y}_1) = L_*(h_1 \circ f, \widehat{f}^{-1}(\widehat{y}_0)),$$

so it suffices to show that $L_*(h_1 \circ f, \widehat{f}^{-1}(\widehat{y}_0)) = \pm L_*(f, \widehat{f}^{-1}(\widehat{y}_0)$. Let N be a closed neighbourhood of $\widehat{f}^{-1}(\widehat{y}_0)$ such that $N - \widehat{f}^{-1}(\widehat{y}_0)$ has no roots of f at y_0. Then we have the maps

$$X \overset{i}{\subset} (X, X - \widehat{f}^{-1}(\widehat{y}_0)) \overset{e}{\supset} (N, N - \widehat{f}^{-1}(\widehat{y}_0)) \xrightarrow{f'} (Y, Y - y_0) \xrightarrow{h_1'} (Y, Y - y_0),$$

where f' is the map defined by f, and h_1' is the homeomorphism defined by $h_1: Y \to Y$. Then

$$L_*(f, \widehat{f}^{-1}(\widehat{y}_0)) = f_*' \circ e_*^{-1} \circ i_* \quad \text{and} \quad L_*(h_1 \circ f, \widehat{f}^{-1}(\widehat{y}_0)) = h_{1*}' \circ f_*' \circ e_*^{-1} \circ i_*,$$

so

$$L_*(h_1 \circ f, \widehat{f}^{-1}(\widehat{y}_0)) = h_{1*}' \circ L_*(f, \widehat{f}^{-1}(\widehat{y}_0)).$$

The homomorphism h_{1*}' is induced by a homeomorphism and is therefore an isomorphism. Now for a manifold Y, $H_n(Y, Y - y_0; G) \approx G$ and $H_p(Y, Y - y_0; G) = 0$ for $p \neq n$. Thus for $G = \mathbf{Z}$ or $\mathbf{Z}/2\mathbf{Z}$ the only isomorphism of $H_p(Y, Y - y_0; G)$ onto itself is $\mu \mapsto \mu$ or $\mu \mapsto -\mu$, so $L_*(h_1 \circ f, \widehat{f}^{-1}(\widehat{y}_0)) = \pm L_*(f, \widehat{f}^{-1}(\widehat{y}_0))$ as was to be shown.

Now suppose that Y is orientable, then we will show that h'_{1*} is actually the identity. Since $h_1: Y \to Y$ is homotopic to the identity, the induced map $h_*: H_*(Y) \to H_*(Y)$ is the identity, thus the commutative diagram

$$\begin{array}{ccc} Y & \overset{j}{\subset} & (Y, Y - y_0) \\ {\scriptstyle h_1}\downarrow & & \downarrow{\scriptstyle h'_1} \\ Y & \overset{j}{\subset} & (Y, Y - y_0) \end{array} \qquad \text{induces} \qquad \begin{array}{ccc} H_*(Y) & \xrightarrow{j_*} & H_*(Y, Y - y_0) \\ \| & & \downarrow{\scriptstyle h'_{1*}} \\ H_*(Y) & \xrightarrow[j_*]{} & H_*(Y, Y - y_0) \end{array}$$

Since Y is compact orientable, then $j*$ is an epimorphism which easily implies that h'_{1*} is also the identity, so

$$L_*(h_1 \circ f, \widehat{f}^{-1}(\widehat{y}_0)) = h'_{1*} \circ L_*(f, \widehat{f}^{-1}(\widehat{y}_0)) = L_*(f, \widehat{f}^{-1}(\widehat{y}_0)). \qquad \square$$

We close this section with a few more examples that illustrate the theory.

(4.18) Example (Self maps of S^n, with $n > 1$). Let $f: S^n \to S^n$ be a self map of the n-sphere, with $n > 1$, and let $y_0 \in S^n$. Then S^n is simply connected, so there is at most one Nielsen root class. Since $\lambda(f, S^n) = \deg(f)$, we have $NR(f, y_0)) = 1$ if $\deg f \neq 0$. On the other hand, by a theorem of Hopf, if $\deg(f) = 0$ then f is homotopic to a constant map, and we may choose the constant different from y_0, so $NR(f, y_0) = 0$.

(4.19) Example (Self maps of S^1). Let $f: S^1 \to S^1$ be a self map of the circle, and $y_0 \in S^1$. Since $\pi(S^1, x) \approx \mathbf{Z}$ is abelian, we may use (3.18) to compute $RR(f) = \#(\operatorname{Coker} f_1: H_1(S^1; \mathbf{Z}) \to H_1(S^1; \mathbf{Z}))$. By definition of degree, $f_1(c) = \deg(f)c$ for any $c \in H_1(S^1; \mathbf{Z})$, so

$$RR(f) = \# \operatorname{Coker} f_1 = \#(\mathbf{Z}/\deg(f)\mathbf{Z}) = \begin{cases} \infty & \text{if } \deg(f) = 0, \\ |\deg(f)| & \text{if } \deg(f) \neq 0. \end{cases}$$

According to (4.12), $\lambda(f, X) = \deg(f)$, so if $\deg(f) \neq 0$, then $NR(f, y_0) > 0$ and therefore by (3.21), $NR(f, y_0) = RR(f) = |\deg(f)|$. On the other hand, if $\deg(f) = 0$, then $RR(f) = \infty$ so, by (3.23), $NR(f, y_0) = 0$. Thus, in any case $NR(f, y_0) = |\deg(f)|$.

(4.20) Example (Maps of the figure eight to the circle). Model S^1 as the unit circle in the complex plane, $S^1 = \{z \in \mathbb{C} : \|z\| = 1\}$. Then $X = S^1 \times \{1\} \cup \{1\} \times S^1$ is two copies of S^1 joined at the single point $(1, 1)$ — the figure eight. Let $f: X \to S^1$ be a map and choose $y_0 = 1 \in S^1$. We will find $NR(f, y_0)$.

Let $\ell: S^1 \to X$ and $r: S^1 \to X$ be the injections $\ell(z) = (z, 1)$ and $r(z) = (1, z)$, let $j: S^1 \subset (S^1, S^1 - 1)$ and let μ be a generator of $H_1(S^1)$. Then it is straightforward to show that $H_1(X)$ is the free abelian group generated by $\ell_1(\mu)$ and $r_1(\mu)$, and that

$$f_1(x\ell_1(\mu) + yr_1(\mu)) = (x \deg f \circ \ell + y \deg f \circ r)\mu. \tag{4.21}$$

Since $j_1\colon H_1(S^1) \to H_1(S^1, S^1 - 1)$ is an isomorphism, we have

$$L_*(f, X) = j_* \circ f_* \neq 0,$$

and therefore $NR(f, 1) > 0$, if either $\deg f \circ \ell \neq 0$ or $\deg f \circ r \neq 0$.

Since $\pi(S^1, s)$ is abelian for any s, we have by (3.18) $RR(f) = \#\operatorname{Coker} f_1$. From (4.21) we have $\operatorname{Coker} f_1 \approx \mathbf{Z}/(\deg(f \circ \ell)\mathbf{Z} + \deg(f \circ r)\mathbf{Z})$, and therefore

$$RR(f) = \gcd(|\deg f \circ \ell|, |\deg f \circ r|),$$

if either degree is nonzero, and ∞ if they both are zero. In the former case, $NR(f, 1) > 1$ so by (3.21) $NR(f, 1) = RR(f)$ in the later case, $NR(f, 1) = 0$, by (3.23). Thus

$$NR(f,1) = \begin{cases} 0 & \text{if } \deg f \circ \ell = \deg f \circ r = 0, \\ \gcd(|\deg f \circ \ell|, |\deg f \circ r|) & \text{otherwise.} \end{cases}$$

In particular, if $f(z_1, z_2) = z_1^2 z_2$, then $f \circ \ell(z) = z^2$ and $f \circ r(z) = z$, so $NR(f, 1) = \gcd(2, 1) = 1$. Notice that in this case even though the Nielsen number is one, every map homotopic to f will have at least *two* roots. This is because for any map f' homotopic to f, $f' \circ \ell$ will be of degree 2 and there will be at least two roots z_0 and z_1, say, at 1 and then $(z_0, 1)$ and $(z_1, 1)$ will be roots of f' at 1. Thus, this is another example where, $MR(f, 1) > NR(f, 1)$.

(4.22) Example (Spectacles to S^1). Model a pair of spectacles as $X = \{(z, t) \in \mathbb{C} \times I : z = 1 \text{ or } t \in \{0, 1\}\}$. Then X is a circle joined by the unit interval to another circle. Let $f\colon X \to S^1$ be the left projection. Then f has two roots at 1, namely $(1, 0)$ and $(1, 1)$. Moreover as in Example (3.15) every map homotopic to f has at least two roots, so $MR(f, 1) = 2$. On the other hand the interval joining $(z, 0)$ to $(z, 1)$ is mapped to $1 \in S^1$ so the two roots are Nielsen equivalent. It follows that $NR(f, 1) = 1$, so the Nielsen number is a strict underestimate of $MR(f, 1)$. But now let $X' = S^1 \times \{1\} \cup \{1\} \times S^1$ be the figure eight from (4.20) and define $r\colon X \to X'$ by $r(z, t) = (z, 1)$ if $t = 0$, $r(z, t) = (1, z)$ if $t = 1$, and $r(z, t) = \{1, 1\}$ otherwise. Then it is easy to see (but laborious to construct) a map $j\colon X' \to X$ such that r and j are homotopy inverses, and the diagram

$$\begin{array}{ccc} X' & \xrightarrow{f'} & S^1 \\ {\scriptstyle r}\uparrow\downarrow{\scriptstyle j} & & \| \\ X & \xrightarrow[f]{} & S^1 \end{array}$$

commutes up to homotopy, where $f'(z_1, z_2) = z_1 z_2$. From example (4.20) we have $NR(f') = 1$, so $MR(f') \geq 1$. On the other hand f' has only one root at 1. Therefore $MR(f', 1) = 1 \neq MR(f, 1)$. This shows that MR, unlike NR, is not a homotopy type invariant.

(4.23) EXAMPLE. Let $X = \bigvee_{\alpha \in \mathbf{Z}} S_\alpha$ be a "bouquet" of a countable number of circles S_α, i.e. the union of a countable number of circles joined together at a single point in each. Give X the weak topology — a set is open if and only if its intersection with each S_α is open. Let $f: X \to S^1$ be a map. For each $\alpha \in \mathbf{Z}$ let $i_\alpha: S_\alpha \subset X$ be the inclusion. Then as in Example (4.20), one can show that

$$NR(f, 1) = \begin{cases} 0 & \text{if } \deg f \circ i_\alpha = 0 \text{ for all } \alpha, \\ \gcd_\alpha |\deg f \circ i_\alpha| & \text{otherwise.} \end{cases}$$

This gives us a nontrivial example of a mapping of a noncompact space X. It also provides us with a source of many examples where $NR(f, 1) < \infty$ but $MR(f, 1) = \infty$.

(4.24) EXAMPLE (Maps of the circle to the figure eight). As in example (4.20), let S^1 be the unit circle in $\mathbb{C}$, and model the figure eight as $Y = S^1 \times \{1\} \cup \{1\} \times S^1$. Let $y_0 = (1, 1)$. We analyze the particularly simple map $\ell: S^1 \to Y$, of example (4.20). We have $\ell(1) = y_0$ and 1 is the only root of ℓ at y_0. Thus there is only one Nielsen root class and $N(f, y_0) \leq 1$. The space $Y - y_0$ is the union of two contractible path components, and therefore $H_1(Y - y_0) = 0$. Hence from the portion $0 = H_1(Y - y_0; \mathbf{Z}) \to H_1(Y; \mathbf{Z}) \xrightarrow{j_1} H_1(Y, Y - y_0; \mathbf{Z})$ of the exact sequence of the pair, j_1 is a monomorphism. Thus $L_1(\ell, X) = j_1 \circ \ell_1 = 0$ if and only if $\ell_1 = 0$. Now let $p_\ell: Y \to S^1$ be the left projection, $p_\ell(z_1, z_2) = z_1$. Then $p_\ell \circ \ell$ is the identity on S^1 so, using any nontrivial coefficient group, it follows that $\ell_1 \neq 0$. Thus $L_1(\ell, X) \neq 0$, so $NR(\ell, y_0) > 0$ and therefore $N(\ell, y_0) = 1$.

Notice that in this example $RR(\ell) = \#\operatorname{Coker} \ell_\# = \infty$, so the inequality $NR(\ell, y_0) \leq RR(\ell)$ is sharp despite the fact that $NR(\ell, y_0) > 0$.

(4.25) EXAMPLE (Maps of the annulus to the pinched annulus). Let X be the open annulus $X = \{z \in \mathbb{C} : 1/2 < |z| < 2\}$. Let J be the interval on the x-axis $J = \{x + iy \in X : 1/2 < x < 2 \text{ and } y = 0\} \subset X$. We pinch the set J to a single point in the space X to obtain a space Y by letting $Y = X/J$. Let $f: X \to Y$ be the projection $f(z) = [[z]]$, where $[[z]]$ is the equivalence class containing z, that is, $f(z) = \{z\}$ if $z \notin J$ and $f(z) = J$ if $z \in J$. Let $y_0 = J \in Y$. Then $f^{-1}(y_0) = J \subset X$. But given two points $z_0, z_1 \in J$ we may define the path A in X by $A(t) = (1 - t)z_0 + tz_1$ and then $f \circ A$ is the constant path at y_0. Thus $f^{-1}(y_0) = J \subset X$ is a single Nielsen root class of f. Therefore $NR(f) \leq 1$. The map $r: X \to S^1$ given by $r(z) = z/|z|$ is a deformation retraction and induces deformation retractions $r'\colon Y \to S^1$, and $r''\colon (y, Y - y_0) \to (S^1, S^1 - 1)$ such that the following diagram commutes.

$$\begin{array}{ccccc}
X & \xrightarrow{f} & Y & \overset{j}{\subset} & (Y, Y - y_0) \\
{\scriptstyle r}\big\downarrow & & \big\downarrow{\scriptstyle r'} & & \big\downarrow{\scriptstyle r''} \\
S^1 & \overset{\text{id}}{=\!=\!=} & S^1 & \overset{k}{\subset} & (S^1, S^1 - 1)
\end{array}$$

The deformation retractions r, r', r'' all induce homology isomorphisms, and k induces an isomorphism on the first homology group. It follows that f_1 and j_1 are also isomorphisms so $L_1(f, X) = j_1 \circ f_1\colon H_1(S^1; \mathbf{Z}) \to H_1(Y, Y - y_0; \mathbf{Z})$ is not zero. Thus $N(f, y_0) > 0$ and therefore $N(f, y_0) = 1$. Notice that neither X nor Y are compact ([7]).

Now let y_0 to be any other point in Y, then $y_0 = \{z_0\}$ for a single point $z_0 \in X$. If $|z_0| \neq 1$ then the map $f' = f \circ r$ is homotopic to f but has no roots at y_0. If $|z_0| = 1$, then we may redefine r by $r(z) = 3z/2|z|$, and then $f' = f \circ r$ is homotopic to f but has no roots at y_0. Thus $N(f, y_0) = 0$ unless $y_0 = J$. This shows that, in general, for Y not a manifold, $N(f, y_0)$ depends upon the choice of $y_0 \in Y$.

Most of the above examples have been chosen for one pathology or other. For our last example we choose perhaps the least pathological of examples.

(4.26) Example. Let $X = Y = T$ be an n-torus, the n-fold cartesian product of a circle with itself. Let $f\colon T \to T$ and $y_0 \in T$. Since $\pi(T, x)$ is abelian, from (3.18) we have $RR(f) = \#\operatorname{Coker} f_1$, where f_1 is the induced homomorphism of the first homology groups with integer coefficients. We use the following fact about homomorphisms of free abelian groups: Given a homomorphism $\phi\colon G \to H$ of free abelian groups of the same rank, either $\det A = 0$, in which case $\#\operatorname{Coker}\phi = \infty$, or $\det A \neq 0$, in which case $\#\operatorname{Coker}\phi = |\det A|$, where A is the matrix of ϕ relative to a basis G and a basis for H ([8]). The group $H_1(T; \mathbf{Z})$ is free abelian on n-generators. Choose a basis for $B = \{b_1, \dots, b_n\}$ for $H_1(T; \mathbf{Z})$ and let $A = \{a_{ij}\}$ be the matrix of f_1 relative to this basis. Then

$$RR(f) = \begin{cases} \infty & \text{if } \det A = 0, \\ |\det A| & \text{if } \det A \neq 0. \end{cases}$$

([7]) Nor is f proper, since the singleton set $\{y_0\} = \{J\} \in Y$ is compact, but $f^{-1}(\{J\}) \subset X$ is not compact.

([8]) Sketch of proof: If G and H have rank 1, and $\phi(x) = ax$, then $\operatorname{Coker}\phi \approx \mathbf{Z}/a\mathbf{Z}$ and $A = a$, so the theorem is true for rank $= 1$. If $\phi\colon G \to H$ is a product $\phi_1 \times \phi_2 \to H_1 \times H_2$, then $\#(\operatorname{Coker}\phi_1 \times \phi_2) = (\#\operatorname{Coker}\phi_1)(\#\operatorname{Coker}\phi_2)$. From this it follows that if A is diagonal, then $\#\operatorname{Coker}\phi = |\det A|$. For endomorphisms $\alpha\colon G \to G$ and $\beta\colon H \to H$ one easily has $\#(\operatorname{Coker}\beta \circ \phi \circ \alpha) = \#\operatorname{Coker}\phi$. Now choose endomorphisms α and β so that the matrix of $\beta \circ \phi \circ \alpha$ is diagonal. Then $\#\operatorname{Coker}\phi = \#(\operatorname{Coker}\beta \circ \phi \circ \alpha) = |\det \beta \circ \phi \circ \alpha| = |\pm 1||\det A||\pm 1| = |\det A|$.

But we also have $\deg f = \det A$ ([9]), and since $\lambda(f,T) = \deg f$, $N\!R(f) > 0$ if $\deg f = \det A \neq 0$. On the other hand, as we've just seen, if $\det A = 0$, then $R\!R(f) = \infty$ so by (3.23), $N\!R(f) = 0$. Thus, in any case

$$N\!R(f, y_0) = |\deg f| = |\det A|.$$

(4.27) REMARK. The role of compactness. For most of the theory as developed so far, we have not assumed either X or Y to be compact. In fact, in Example (4.23) the space X — a bouquet of an infinite number of circles — is not compact, yet the map f may be constructed to have any desired (finite) Nielsen number. In example (4.25) neither X nor Y are compact, yet $N\!R(f, y_0) > 1$. On the other hand the important Theorem (4.6) and its corollaries do require X to be compact. Also many important applications, e.g. the fundamental theorem of algebra, require some sort of compactness, and constraints on how drastically we can deform a map by homotopy. (In the usual topological proof of the fundamental theorem, one starts with a polynomial $f\colon \mathbb{C} \to \mathbb{C}$ of positive degree and extends it to self map of the compact space $\mathbb{C} \cup \{\infty\}$.)

One approach that seems particularly fruitful is to require the map $f\colon X \to Y$ to be *proper*, i.e. require $f^{-1}(D)$ to be compact whenever D is, and to require homotopies $\{f_t\}$ to be proper, i.e. the map $(x,t) \mapsto f_t(x)$ should be proper. With these alterations Theorem (4.6) does still remain true (see footnotes ([5]) and ([6]) above). These alterations would probably make it useful to change the homology/cohomology theory used for constructing the indexes to a cohomology theory based on compact supports [D, pp. 288–291] or [Sp, pp. 319–323], since this seems to be a particularly useful theory for proper maps and proper homotopies.

5. Hopf degree theory

In this section we concentrate on maps of manifolds of the same dimension. Our goal is to present two of the most important theorems in topological root theory, Theorems (5.5) and (5.7) below, both of which are originally due to Hopf [H2]. Versions of both theorems are true for maps of arbitrary manifolds of the same dimension, with or without boundary, compact or noncompact, and orientable or nonorientable. We shall restrict our presentation to compact manifolds without boundary. However, the manifolds need not be orientable. Indeed, a large part of this section is devoted to exposing techniques for grappling with nonorientability.

The theorems make use of two as yet undefined concepts associated with a map $f\colon X \to Y$ of two manifolds of the same dimension. The first of these is the property

([9]) Sketch of proof: Let $B^* = \{b_1^*, \dots, b_n^*\}$ be the basis for $H^1(T;\mathbf{Z})$ dual to B. The cup product $\mu = b_1^* \vee \dots \vee b_n^*$ generates $H^n(T;\mathbf{Z})$, and $f^n(\mu) = f^1(b_1^*) \vee \dots \vee f^1(b_n^*) = (\sum_j a_{1j} b_j) \vee \dots \vee (\sum_j a_{nj} b_j^*) = (\det A)(b*_1 \vee \dots \vee b*_n) = (\det A)\mu$.

of f being *transverse to a point* $y_0 \in Y$. The second is the *absolute degree* of f, denoted by $\mathcal{A}(f)$. We give the definition of transversality shortly. The definition of absolute degree, is more involved, and will take some time to develop. Suffice it to say at this point that $\mathcal{A}(f)$ is a homotopy invariant of f, and when f is a map of compact oriented manifolds, then $\mathcal{A}(f) = |\deg(f)|$. Thus $\mathcal{A}(f)$ generalizes the ordinary Brouwer degree to not-necessarily orientable manifolds.

Here is the definition of transverse to a point:

(5.1) DEFINITION. Let $f\colon X \to Y$ be a map of n-dimensional manifolds and $y_0 \in Y$. Then f is *transverse to* y_0 if y_0 has a neighbourhood N *evenly covered* by f, i.e. the preimage $f^{-1}(N)$ of N may be written as the disjoint union

$$f^{-1}(N) = \bigsqcup_{x \in f^{-1}(y_0)} N_x$$

of sets N_x where each N_x is a neighbourhood of $x \in f^{-1}(y_0)$ and is mapped homeomorphically onto N by f.

Thus, being transverse to y_0 is like being a covering at the point y_0. In fact, $f\colon X \to Y$ is a covering if and only if f is transverse to every point $y \in Y$

(5.2) REMARK. In the above definition, we could require the neighbourhoods to be balls if we wished, since every neighbourhood of y_0 contains a ball neighbourhood.

In order for f to be transverse to y_0, it is clearly necessary that f be a local homeomorphism at each point $x \in f^{-1}(y_0)$, i.e. that each $x \in f^{-1}(y_0)$ have a neighbourhood that is mapped homeomorphically onto its image. In general, this condition is not sufficient, but it is sufficient if X is compact (or, more generally, if f is required to be proper), see [BBS2, pp. 37–38].

(5.3) EXAMPLE. Suppose $f\colon X \to Y$ a map of n-manifolds, $y_0 \in Y$, and both X and Y are triangulable. Then we may triangulate Y so that y_0 is in an open simplex s of dimension n. By the simplicial approximation theorem, there is a triangulation of X sufficiently fine so that we may approximate f by a simplicial map g homotopic to f. Then $g^{-1}(s)$ is the disjoint union of open n-simplices each of which is mapped homeomorphically (in fact linearly) onto s, so g is transverse to y_0. This shows that for any map $f\colon X \to Y$ of triangulable n-manifolds and any $y_0 \in Y$, there is a map homotopic to f that is transverse to y.

In fact, we have the following theorem for maps of arbitrary topological manifolds.

(5.4) THEOREM. *Suppose $f: X \to Y$ a map of n-manifolds and $y_0 \in Y$. Then there exists a map $f': X \to Y$ arbitrarily close to f and homotopic to f that is transverse to y_0.*

We will not prove this theorem here. For a proof, see [Ep, Theorem 4.3]

Before moving on to the definition of the absolute degree, we state Hopf's two theorems.

(5.5) THEOREM. *Let f be a map of compact n-manifolds and $y_0 \in Y$. Then:*

(5.5.1) *Every map homotopic to f and transverse to y_0 has at least $\mathcal{A}(f)$ roots at y_0.*

(5.5.2) *There is a map homotopic to f and transverse to y_0 that has exactly $\mathcal{A}(f)$ roots.*

In dimension 1, S^1 is the only compact manifold, and selfmaps of S^1 are homotopic if and only if they have the same degree. We have already seen, Example (4.19), every map $f: S^1 \to S^1$ has at least $|\deg f|$ roots, on the other hand, the map $z \mapsto z^r$ of the unit circle has exactly r roots at y_0, has degree r, and is transverse to y_0, for any $y_0 \in S^1$. Thus the theorem is true in dimension 1. The first proof of (5.5) in dimension 2 was given by Knesser (see [Kn]). For a more modern dimension 2 proof, see [S]. The dimension 2 case requires quite different methods from the $n > 2$ case. We will prove Theorem (5.5) in the case $n > 2$ towards the end of this section.

We will also see (Corollary (5.28)) that $\mathcal{A}(f) > 0$ implies $NR(f) > 0$, so (5.5) has the following important corollary

(5.6) COROLLARY. *Let f be a map of compact n-manifolds, and $y_0 \in Y$. Then the following are equivalent:*

$$\mathcal{A}(f) > 0, \quad NR(f) > 0, \quad MR(f, y_0) > 0.$$

PROOF. By (5.28), $\mathcal{A}(f) > 0$ implies $NR(f) > 0$. We already know $NR(f) > 0$ implies $MR(f, y_0) > 0$. Certainly if $\#g^{-1}(y_0) > 0$ for every map g homotopic to f, then $\#g^{-1}(y_0) > 0$ for every map g homotopic to f and transverse to y_0, thus (5.5) implies that $MR(f, y_0) > 0$ implies $\mathcal{A}(f) > 0$. □

The next Hopf theorem answers one of our important questions in the affirmative for all maps of compact n-manifolds, except for $n = 2$.

(5.7) THEOREM. *Let f be a map of compact n-manifolds with $n \neq 2$ and let $y_0 \in Y$. Then there is a map homotopic to f that has exactly $NR(f)$ roots at y_0. Therefore $NR(f) = MR(f, y_0)$.*

In dimension 1, every map $f: S^1 \to S^1 \subset \mathbb{C}$ is homotopic to a map of the form $f(z) = z^r$, which has exactly r roots at any point $y_0 \in S^1$. From Example (4.19) we have $NR(f) = |\deg f| = r$, so the Theorem is true for $n = 1$. It is one of the remarkable facts in root theory, that the conclusion of the Theorem is false for $n = 2$. For an example of a map from a surface of genus 2 to a surface of genus 1 for which $MR(f, y_0) > NR(f)$, see e.g. [Li, pp. 205–206]. We prove (5.7) for $n > 2$ at the end of this section.

(5.8) EXAMPLE (The fundamental theorem of algebra). Let $p: \mathbb{C} \to \mathbb{C}$ be a polynomial of degree $d > 0$, then we may extend it to a map

$$p: \mathbb{C} \cup \infty \to \mathbb{C} \cup \infty \approx S^2$$

by setting $p(\infty) = \infty$, and the resulting map has Brouwer degree d and therefore $\mathcal{A}(p) = d$.

Since S^2 is simply connected, we have $NR(p) = 1$, and indeed, the map p is homotopic by a straight line homotopy to the map $z \mapsto z^d$ which has only one root at $y_0 = 0$. On the other hand p is also homotopic to the map $z \mapsto \prod_{i=1}^{d}(z - i)$, which has d distinct roots at 0. Moreover, since the derivative is nonzero at each of these roots, it is a local homeomorphism at each root and therefore, by (5.2), it is transverse to 0. In this case the number of roots is equal to the absolute degree, as predicted by (5.5).

In order to proceed further, we recall some terminology and facts we will need about orientation, paths, and covering spaces.

(5.9) REMARK. Let X be a manifold and $p^U: X^U \to X$ its universal covering. Suppose C a loop in X at a point $x \in X$, then from covering space theory, C induces a deck transformation $\tau_C: X^U \to X^U$. We say C is *orientation preserving and reversing loop* or *reversing* according as τ_C is an orientation preserving or reversing homeomorphism. Since τ_C depends only on the fixed-end-point homotopy class of C, so does the property of being orientation preserving (or reversing). If C is orientation preserving (reversing) then so is $A(CA^{-1})$ for any path A in X ending at $C(0) = C(1)$. The fixed-end-point homotopy classes of orientation preserving loops in X at x form a subgroup $\pi^0(X, x)$ of $\pi(X, x)$. If $\pi^0(X, x) = \pi(X, x)$, then X is orientable. Otherwise, $\pi^0(X, x)$ has index two in $\pi(X, x)$ (because the product of two orientation reversing loops is orientation preserving). In this case, there is a double covering $\widetilde{p}: \widetilde{X} \to X$, where $\widetilde{p}(\pi(\widetilde{X}, \widetilde{x})) = \pi^0(X, \widetilde{p}(\widetilde{x}))$. The manifold $\widetilde{X}$ is orientable, the covering $\widetilde{p}: \widetilde{X} \to X$ is unique up to covering space isomorphism and is called the *orientable double covering of* X. In fact, if $\overline{p}: \overline{X} \to X$ is any other two sheeted covering with $\overline{X}$ orientable, then it is covering space isomorphic to $\widetilde{p}: \widetilde{X} \to X$. Notice that $[C] \in \pi^0(X, c)$ if and only if C lifts to a loop in $\widetilde{X}$, so C is orientation preserving if and only if it lifts to a loop in $\widetilde{X}$. This is often a useful criterion for C to be orientation preserving.

The following classification is fundamental to studying maps $f: X \to Y$ of manifolds when either X or Y is nonorientable.

(5.10) DEFINITION. Let $f: X \to Y$ be a map of manifolds. Then:

(5.10.1) We say f is *type* I, if given any loop A in X, $f \circ A$ is orientation preserving if and only if A is orientation preserving. These maps are also called *orientation true.*

(5.10.2) We say f is *type* II if it is not type I, but also there is no orientation reversing loop A in X such that $f \circ A$ is contractible.

(5.10.3) We say f is *type* III if there is an orientation reversing loop A in X such that $f \circ A$ is contractible.

Maps that are type I or type II are called *orientable.* Type III maps are called *nonorientable.*

(5.11) EXAMPLE (Type I maps). Any map $f: X \to Y$ of orientable manifolds is type I. The identity map on any manifold is type I. More generally, any homeomorphism is type I. Still more generally, any covering $q: \overline{Y} \to Y$ of manifolds is type I. To see this, note that if C is a loop in $\overline{Y}$ then both C and $\overline{p} \circ C$ induce the same deck transformation $\tau_C: Y^U \to Y^U$ on their common universal covering space, so $\overline{q} \circ C$ is orientation preserving if and only if C is.

(5.12) EXAMPLE (Type II map). Represent the torus T by the quotient space of the unit square with $(s, 0)$ identified to $(s, 1)$ and $(0, t)$ identified to $(1, t)$. Represent the Klein bottle K by the quotient space of the unit square with $(s, 0)$ identified to $(s, 1)$ and $(0, t)$ identified to $(1, 1-t)$. Define $\widetilde{p}: T \to K$ by

$$\widetilde{p}([(s,t)]) = \begin{cases} [(2s,t)] & \text{for } 0 \le s \le 1/2, \\ [(2s-1,t)] & \text{for } 1/2 \le s \le 1. \end{cases}$$

Then $\widetilde{p}$ is a two fold covering of K by an orientable manifold, and is therefore the orientable covering of K. Let B be the loop in K given by $B(s) = [(s, 0)]$, and let C be the loop $C(t) = [(0, t)]$. Then $\pi(K, [(0, 0)])$ is generated by $[B]$ and $[C]$. The loop B lifts through $\widetilde{p}$ to the path $\widetilde{B}$ given by $\widetilde{B}(s) = [(s/2, 0)]$. Since $\widetilde{B}$ is not a loop, then B must be orientation reversing. On the other hand, C lifts to the path $\widetilde{C}(t) = [(0, t)]$ which is a loop, so C is orientation preserving. Now define $f: K \to T$ by $f([(s, t)]) = [(s, 0)]$. Then $f \circ B(s) = f([(s, 0)]) = [(s, 0)]$, so $[f \circ B]$ is one of the usual generators of $\pi(T, [(0, 0)])$, and therefore not contractible. Also $f \circ B$ is a loop in the orientable space T and must therefore be orientation preserving. Thus f cannot be type I. We also have $f \circ C(t) = f([0, t]) = ([0, 0)]$. Although $\pi(K, [(0, 0)])$ is not abelian the fact that $B(C(B^{-1}C))$ is contractible may be used to write any element in the form $[B]^m[C]^n$. But $f_\#([B]^m[C]^n) = [f \circ B]^m[f \circ C]^n = [f \circ B]^m$

which, since $\pi(T, [(0,0)])$ is free abelian with $[f \circ B]$ as one of its generators, cannot be trivial unless $m = 0$. Thus the only loops in K that are mapped to contractible loops are those fixed-end-point homotopic to C^n for some n, and these are all orientation preserving. Therefore f is not type III, and must therefore be type II.

(5.13) EXAMPLE (Type III map). Represent the 2-sphere S^2 as the closed unit disk D with boundary identified to a point, and let $q: D \to S^2$ be the quotient map. Represent the projective plane P^2 as the disk with opposite points on the boundary identified, and let $p: D \to P^2$ be the quotient map. Then p and q induce a map $f: P^2 \to S^2$ such that $f \circ p = q$.

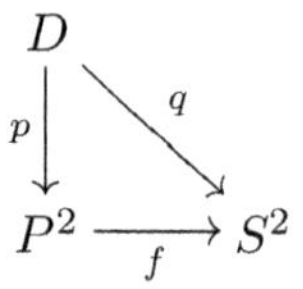

Since P^2 is not orientable, it has an orientation reversing loop B. Since S^2 is simply connected, $f \circ B$ is contractible. Therefore f is type III.

In studying maps $f: X \to Y$ where X is not orientable, we will make frequent use of the orientable double covering $\widetilde{p}: \widetilde{X} \to Y$, and the composed map $f \circ \widetilde{p}: \widetilde{X} \to Y$. The next two theorems bring out the important difference between orientable and nonorientable maps in the way in which roots of f and $f \circ \widetilde{p}$ are related to each other.

(5.14) THEOREM. *Suppose $f: X \to Y$ a map of n-manifolds, X is nonorientable, and $y_0 \in Y$. Let $\widetilde{p}: \widetilde{X} \to X$ be the orientable double covering of X and $\widehat{q}: \widehat{Y} \to Y$ a Hopf covering and lift for f.*

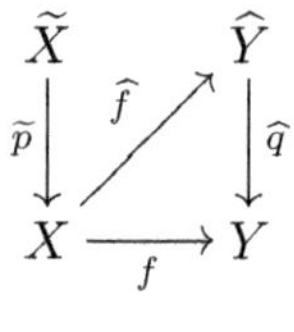

Then:

(5.14.1) *The map f is type* III *if and only if $\widehat{q}$ and $\widehat{f} \circ \widetilde{p}$ are a Hopf covering and lift for $f \circ \widetilde{p}$.*

(5.14.2) *If f is type* III*, then $\widetilde{p}^{-1}(\alpha)$ is a root class of $f \circ \widetilde{p}$ at y_0 if and only if α is a root class of f at y_0. In this case $\widetilde{p}$ sends the set of root classes of $f \circ \widetilde{p}$ bijectively onto the set of Nielsen classes of f.*

PROOF. To prove (5.14.1), let $\widetilde{x}$ be an arbitrary point in $\widetilde{X}$. By definition of Hopf coverings and lifts, $\widehat{q}$ and $\widehat{f} \circ \widetilde{p}$ are a Hopf covering and lift for $f \circ \widetilde{p}$ if and only if $(f \circ \widetilde{p})_{\#}(\pi(\widetilde{X}, \widetilde{x})) = \widehat{q}_{\#}(\pi(\widehat{Y}, \widehat{f} \circ \widetilde{p}(\widetilde{x})))$. Since $\widehat{q}$ is a Hopf covering for f, then $f_{\#}(\pi(X, \widetilde{p}(\widetilde{x})) = \widehat{q}_{\#}(\pi(\widehat{Y}, \widehat{f} \circ \widetilde{p}(\widetilde{x}))$. It follows that $\widehat{q}$ and $\widehat{f} \circ \widetilde{p}$ are a Hopf covering and lift for $f \circ \widetilde{p}$ if and only if

$$(5.15) \qquad (f \circ \widetilde{p})_{\#}(\pi(\widetilde{X}, \widetilde{x})) = f_{\#}(\pi(X, \widetilde{p}(\widetilde{x}))).$$

So we will show that (5.15) holds if and only if f is type III.

Suppose first that f is type III, we will prove (5.15). The left side of (5.15) is clearly included in the right, so we need to show that the right is included in the left. So let $\beta \in f_{\#}(\pi(X, \widetilde{p}(\widetilde{x})))$. Then $\beta = [f \circ A]$ for some loop A in X at $\widetilde{p}(\widetilde{x})$. If A is orientation preserving, then A lifts to a loop $\widetilde{A}$ in $\widetilde{X}$ at $\widetilde{x}$ and then $(f \circ \widetilde{p})_{\#}([\widetilde{A}]) = [f \circ \widetilde{p} \circ \widetilde{A}] = [f \circ A] = \beta$, so $\beta \in (f \circ \widetilde{p})_{\#}((\pi(\widetilde{X}, \widetilde{x}))$. Suppose now that A is orientation reversing. Because f is type III, there is an orientation reversing loop A' at $\widetilde{p}(\widetilde{x})$ such that $[f \circ A']$ is contractible. Then AA' is an orientation preserving loop, so it lifts to a loop $\widetilde{P}$ in $\widetilde{X}$ at $\widetilde{x}$. Then $(f \circ \widetilde{p})_{\#}([\widetilde{P}]) = [f \circ \widetilde{p} \circ \widetilde{P}] = [f \circ (AA')] = [f \circ A][f \circ A'] = [f \circ A] = \beta$, and therefore $\beta \in (f \circ \widetilde{p})_{\#}(\pi(\widetilde{X}, \widetilde{x})$. Thus, if f is type III, then (5.15) is true.

Now suppose that (5.15) is true. To show that f is type III, we need to find an orientation reversing loop A' in X such that $[f \circ A']$ is contractible. Since X is nonorientable, there is an orientation reversing loop A in X at $\widetilde{p}(\widetilde{x})$. From (5.15), there is a loop $\widetilde{P}$ in $\widetilde{X}$ at $\widetilde{x}$ such that $[f \circ \widetilde{p} \circ \widetilde{P}] = [f \circ A]$, and therefore $f \circ ((\widetilde{p} \circ \widetilde{P})A^{-1})$ is contractible. Now $\widetilde{p} \circ \widetilde{P}$ is orientation preserving, since it lifts to the loop $\widetilde{P}$ in $\widetilde{X}$. By choice A and therefore A^{-1} are orientation reversing. Thus $(\widetilde{p} \circ \widetilde{P})A^{-1}$ is the desired A'. This completes the proof of (5.14.1).

The second statement follows easily from the first and the facts that $\widetilde{p}$ is surjective, the Nielsen classes of f at y_0 are the nonempty sets of the form $\widehat{f}^{-1}(\widehat{y}_0)$, and when f is type III the Nielsen classes of $f \circ \widetilde{p}$ are the nonempty sets of the form $(\widehat{f} \circ \widetilde{p})^{-1}(\widehat{y}_0)$ — where $\widehat{y}_0 \in \widehat{q}^{-1}(y_0)$. □

(5.16) THEOREM. *Suppose $f: X \to Y$ an orientable map of n-manifolds, X is not orientable, and $y_0 \in Y$. Let $\widetilde{p}: \widetilde{X} \to X$ be the orientable double covering of X, let $\widehat{q}: \widehat{Y} \to Y$ and $\widehat{f}: X \to \widehat{Y}$ be the Hopf covering and lift for f, and let $\check{q}: \check{Y} \to \widehat{Y}$ and $\breve{f}: \widetilde{X} \to \check{Y}$ be the Hopf covering and lift for $\widehat{f} \circ \widetilde{p}: \widetilde{X} \to \widehat{Y}$, so we have the diagram*

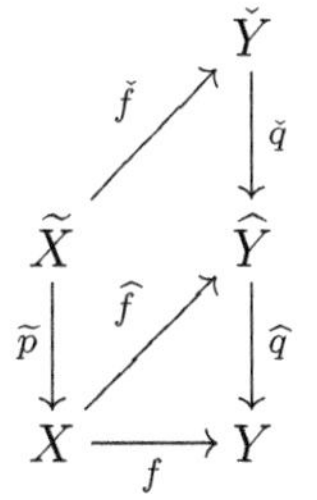

Then:

(5.16.1) *The maps* $\widehat{q} \circ \check{q}$ *and* $\check{f}$ *are a Hopf covering and lift for* $f \circ \widetilde{p}$.

(5.16.2) *The covering* $\check{q}$ *has two sheets.*

(5.16.3) *Let* α *be a Nielsen root class of* f *at* y_0.

(a) *We have* $\widetilde{p}^{-1}(\alpha) = \widetilde{\alpha}_0 \sqcup \widetilde{\alpha}_1$ *for two Nielsen root classes,* $\widetilde{\alpha}_0$ *and* $\widetilde{\alpha}_1$ *of* $f \circ \widetilde{p}$.

(b) *The two classes are related by* $\tau(\widetilde{\alpha}_0) = \widetilde{\alpha}_1$, *where* $\tau: \widetilde{X} \to \widetilde{X}$ *is the period two deck transformation of* $\widetilde{X}$.

(c) *Each of the classes* $\widetilde{\alpha}_0$ *and* $\widetilde{\alpha}_1$ *is mapped bijectively onto* α *by* $\widetilde{p}$.

PROOF. To prove (5.16.1), let $\widetilde{x} \in \widetilde{X}$. Since $\check{q}$ is a Hopf covering for $\widehat{f} \circ \widetilde{p}$, then $(\widehat{f} \circ \widetilde{p})_\#(\pi(X, \widetilde{p}(\widetilde{x}))) = \check{q}_\#(\pi(\check{Y}, \check{f}(\widetilde{x})))$. Hence, using commutativity in the above diagram,

$$\begin{aligned}(f \circ \widetilde{p})_\#(\pi(\widetilde{X}, \widetilde{x})) &= (\widehat{q} \circ \widehat{f} \circ \widetilde{p})_\#(\pi(\widetilde{X}, \widetilde{x})) = \widehat{q}_\#((\widehat{f} \circ \widetilde{p})_\#(\pi(\widetilde{X}, \widetilde{x}))) \\ &= \widehat{q}_\#(\check{q}_\#(\pi(\check{Y}, \check{f}(\widetilde{x}))) = (\widehat{q} \circ \check{q})_\#(\pi(\check{Y}, \check{f}(\widetilde{x})).\end{aligned}$$

Thus $\widehat{q} \circ \check{q}$ is a Hopf covering for $f \circ \widetilde{p}$, and since $\check{f}$ lifts $f \circ \widetilde{p}$ through $\widehat{q} \circ \check{q}$, it is a corresponding Hopf lift.

We now prove (5.16.2). By commutativity, the epimorphism $\widehat{f}_\#: \pi(X, \widetilde{p}(\widetilde{x})) \to \pi(\widehat{Y}, \widehat{f}(\widetilde{p}(\widetilde{x})))$ induces a surjection $\operatorname{Coker} \widetilde{p}_\# \to \operatorname{Coker} \check{q}_\#$. Since $\widetilde{p}$ is two sheeted, then $\# \operatorname{Coker} \widetilde{p}_\# = 2$, so $\# \operatorname{Coker} \check{q}_\# \leq 2$, and therefore $\check{q}$ has either one or two sheets. If $\check{q}$ has just one sheet, then we may assume that $\check{q}$ is the identity. But this would imply that $\widehat{q}$ is a Hopf lift for $f \circ \widetilde{p}$ and therefore, by (5.14), that f is type III. Since this contradicts the hypothesis that f is orientable, it follows that $\check{q}$ has two sheets.

To prove (5.16.3), let α be a root class of f at y_0, so $\alpha = \widehat{f}^{-1}(\widehat{y}_0)$ for some $\widehat{y}_0 \in \widehat{q}^{-1}(y_0)$. For this $\widehat{y}_0$, since $\check{q}$ is two sheeted, there are exactly two points $\check{y}_0$ and $\check{y}_1$, say, in $\check{q}^{-1}(\widehat{y}_0)$. Let $\widetilde{\alpha}_0 = \check{f}^{-1}(\check{y}_0)$ and $\widetilde{\alpha}_1 = \check{f}^{-1}(\check{y}_1)$. Since $\widehat{q} \circ \check{q}$ is a Hopf covering for $f \circ \widetilde{p}$, each $\widetilde{\alpha}_i$ is a root class of $f \circ \widetilde{p}$. Also $\widetilde{p}^{-1}(\alpha) = \widetilde{p}^{-1}(\widehat{f}^{-1}(y_0)) = \check{f}^{-1}(\check{q}^{-1}(\widehat{y}_0)) = \check{f}^{-1}(\{\check{y}_0\} \sqcup \{\check{y}_1\}) = \widetilde{\alpha}_0 \sqcup \widetilde{\alpha}_1$.

We now show that $\widetilde{\alpha}_1 = \tau(\widetilde{\alpha}_0)$. Let $\widetilde{x}_0 \in \widetilde{\alpha}_0$. Then $\widehat{f}(\widetilde{p}(\tau(\widetilde{x}_0)) = \widehat{f}(\widetilde{p}(\widetilde{x}_0)) = \widehat{y}_0$, so either $\tau(\widetilde{x}_0) \in \widetilde{\alpha}_0$ or $\tau(\widetilde{x}_0) \in \widetilde{\alpha}_1$. Let $\widetilde{A}$ be any path in $\widetilde{X}$ from $\widetilde{x}_0$ to $\tau(\widetilde{x}_0)$. Then $\widetilde{p} \circ A$ is a loop in X at $\widetilde{p}(\widetilde{x}_0) = \widetilde{p}(\tau(\widetilde{x}_0))$. Moreover it is an orientation reversing loop, since its lift to $\widetilde{X}$ is the path $\widetilde{A}$. Therefore, since f is orientable, $f \circ \widetilde{p} \circ \widetilde{A}$ cannot be contractible. It follows that $\tau(\widetilde{x}_0)$ cannot be Nielsen related to $\widetilde{x}_0$, and therefore $\tau(\widetilde{x}_0) \in \widetilde{\alpha}_1$. Hence $\tau(\widetilde{\alpha}_0) \subset \widetilde{\alpha}_1$. Similarly $\tau(\widetilde{\alpha}_1) \subset \widetilde{\alpha}_0$, and therefore $\widetilde{\alpha}_1 = \tau(\tau(\widetilde{\alpha}_1)) \subset \tau(\widetilde{\alpha}_0)$, thus $\tau(\widetilde{\alpha}_0) = \widetilde{\alpha}_1$.

Since $\widetilde{p}$ is surjective, the equality $\widetilde{p}^{-1}(\alpha) = \widetilde{\alpha}_0 \cup \widetilde{\alpha}_1$ implies that $\widetilde{p}(\widetilde{\alpha}_0 \cup \widetilde{\alpha}_1) = \alpha$. From this we have $\alpha = \widetilde{p}(\widetilde{\alpha}_0 \cup \widetilde{\alpha}_1) = \widetilde{p}(\widetilde{\alpha}_0 \cup \tau(\widetilde{\alpha}_0)) = \widetilde{p}(\widetilde{\alpha}_0) \cup \widetilde{p}(\tau(\widetilde{\alpha}_0)) =$

$\widetilde{p}(\widetilde{\alpha}_0) \cup \widetilde{p}(\widetilde{\alpha}_0) = \widetilde{p}(\widetilde{\alpha}_0)$. So $\widetilde{p}$ maps $\widetilde{\alpha}_0$ onto α. Similarly $\widetilde{p}$ maps $\widetilde{\alpha}_1$ onto α. Since $\widetilde{\alpha}_0 \cap \widetilde{\alpha}_0 = \emptyset$, and $\widetilde{p}$ is two-to-one, it follows that $\widetilde{p}$ maps each $\widetilde{\alpha}_i$ bijectively onto α. □

We are now in a position to define the *multiplicity* of a map $f: X \to Y$. We will subsequently use the multiplicity to define the absolute degree of f.

(5.17) DEFINITION. Let $f: X \to Y$ be a map of compact n-manifolds. Then we define the *multiplicity of f at y_0*, denoted by $\mathrm{mult}(f, y_0)$, as follows: If f has no roots at y_0, then $\mathrm{mult}(f, y_0) = 0$. Otherwise let α be a root class of f at y_0.

(5.17.1) If X is orientable, then $\mathrm{mult}(f, y_0) = |\lambda(f, \alpha)|$.

(5.17.2) If X is not orientable but f is, then

$$\mathrm{mult}(f, y_0) = |\lambda(f \circ \widetilde{p}, \widetilde{\alpha}_0)| = \mathrm{mult}(f \circ \widetilde{p}, y_0),$$

where $\widetilde{p}$ and $\widetilde{\alpha}_0$ are as in Theorem (5.16).

(5.17.3) If f is not orientable, then $\mathrm{mult}(f) = |\lambda_2(f, \alpha)|$, where $|\lambda_2(f, \alpha)| = 1$ if $\lambda_2(f, \alpha) \equiv 1 \bmod 2$ and is 0 otherwise.

Since we are using absolute values of λ in the first two cases and the mod 2 integer index in the 3rd, the definition is independent of the choice of the generators used to compute λ and λ_2. By (4.17), $|\lambda(f, \alpha)|$ and $|\lambda(f \circ \widetilde{p}, \widetilde{\alpha}_0)|$ are also independent of the choice of α and $\widetilde{\alpha}_0$. Thus $\mathrm{mult}(f, y_0)$ is well defined. We shall see shortly that $\mathrm{mult}(f, y_0)$ is also independent of the point $y_0 \in Y$, so we shall be able to write $\mathrm{mult}(f, y_0)$ more simply as $\mathrm{mult}(f)$.

(5.18) THEOREM. *Suppose $f: X \to Y$ is a map of compact manifolds, $y_0 \in Y$ and* $\mathrm{mult}(f, y_0) \neq 0$. *Then $NR(f) = RR(f)$.*

PROOF. It suffices to show that $NR(f) \neq 0$, for then by (3.21), we have $NR(f) = RR(f)$.

Since $\mathrm{mult}(f, y_0) \neq 0$, there is at least one root class α of f at y_0. If f is nonorientable, then $|\lambda_2(f, \alpha)| = \mathrm{mult}(f, y_0) \neq 0$, so α is essential, and therefore $NR(f) \neq 0$. Similarly, if f is orientable and X is orientable, then $|\lambda(f, \alpha)| = \mathrm{mult}(f, y_0) \neq 0$, so α is essential, and therefore $NR(f) \neq 0$.

It remains to consider the X nonorientable but f orientable case. In this case we use the coverings, lifts and diagram from Theorem (5.16).

Since $\mathrm{mult}(f, y_0) \neq 0$, then f has at least one root x at y_0, and therefore $f \circ \widetilde{p}$ has a root $\widetilde{x} \in \widetilde{p}^{-1}(x)$ at y_0. Let $\widehat{y} = \widehat{f}(x)$, $\breve{y}_0 = \breve{f}(\widetilde{x})$, $\alpha = f^{-1}(\widehat{y}_0)$ and $\widetilde{\alpha} = \breve{f}^{-1}(\breve{y}_0)$. So α is the root class containing x, and $\widetilde{\alpha}$ is the root class containing $\widetilde{x}$. By definition, $|\lambda(f \circ \widetilde{p}, \widetilde{\alpha})| = \mathrm{mult}(f, y_0) \neq 0$, so $\widetilde{\alpha}$ is an essential root class of $f \circ \widetilde{p}$.

Now let g be homotopic to f. Then $g \circ \widetilde{p}$ is homotopic to $f \circ \widetilde{p}$. Let $\widehat{g}$ and $\breve{g}$ be the Hopf lifts of g and $g \circ \widetilde{p}$, so we can use the same diagram but with f, $\widehat{f}$, and $\breve{f}$

replaced by g, $\widehat{g}$, and $\check{g}$. Since $\widetilde{\alpha} = \check{f}^{-1}(\check{y}_0)$ is essential, it follows that $\check{g}^{-1}(\check{y}_0) \neq \emptyset$. Let $\widetilde{x}' \in \check{g}^{-1}(\check{y}_0)$. Then $\widehat{g}(\widetilde{p}(\widetilde{x}')) = \check{q}(\check{g}(\widetilde{x}')) = \check{q}(\check{y}_0) = \widehat{y}_0$. Thus $\widehat{g}^{-1}(\widehat{y}_0) \neq \emptyset$ so α is essential. It follows that $NR(f) \neq 0$. □

The next two theorems show that the multiplicity of f can be computed from the ordinary Brouwer degree of certain lifts.

(5.19) THEOREM. *Let $f\colon X \to Y$ be an orientable map of compact n-manifolds, and $y_0 \in Y$. If f is type I we distinguish two cases:*

(5.19.1) *X is orientable. In this case, let $\widehat{q}\colon \widehat{Y} \to Y$ and $\widehat{f}\colon X \to \widehat{Y}$ be a Hopf covering and lift for f. Then either $RR(f) = \infty$, in which case* $\operatorname{mult}(f, y_0) = 0$, *or $\widehat{Y}$ is compact orientable and* $\operatorname{mult}(f, y_0) = |\deg(\widehat{f})|$.

(5.19.2) *X is nonorientable. As in (5.16), let $\widetilde{p}\colon X \to X$ be the orientable covering of X, and let $\check{q}\colon \check{Y} \to Y$ and $\check{f}\colon \widetilde{X} \to \check{Y}$ be the Hopf covering and lift for $\widehat{f} \circ \widetilde{X} \to hY$. Then either $RR(f) = \infty$, in which case* $\operatorname{mult} f = 0$, *or $\check{Y}$ is compact orientable and* $\operatorname{mult}(f, y_0) = |\deg(\check{f})|$.

If f is type II, *then* $\operatorname{mult}(f) = 0$ *regardless of the orientability of X.*

It is a remarkable fact that type II maps have zero multiplicity. We will see shortly that this implies that they all have zero absolute degree, and this together with (5.5) implies that they are homotopic to root free maps!

PROOF OF (5.19). We first prove the theorem under the assumption that X is orientable. If $RR(f) = \infty$ then by (3.23) we have $NR(f) = 0$ and therefore by (5.18) $\operatorname{mult}(f, y_0) = 0$. So now assume that $RR(f) < \infty$. Then since $q\colon \widehat{Y} \to Y$ has $RR(f) < \infty$ sheets and Y is compact, it follows that $\widehat{Y}$ is compact. To show that $\widehat{Y}$ is orientable and $\operatorname{mult}(f, y_0) = \deg(\widehat{f})$, we begin by showing that $\operatorname{mult}(f, y_0) = |\lambda(\widehat{f}, X)|$.

If f is root free, then $\operatorname{mult}(f) = 0$, but in this case $\widehat{f}$ is also root free and we have $\lambda(\widehat{f}, X) = 0$. So now assume f has at least one root, let α be its root class, so $\widehat{f}(\alpha) = \widehat{y}_0$ for some $\widehat{y}_0 \in \widehat{q}^{-1}(y_0)$. Now construct the diagram

$$\begin{array}{ccccccccc}
 & & & & & & (\widehat{V}, \widehat{V} - \widehat{y}_0) & \overset{\widehat{d}}{\subset} & (\widehat{Y}, \widehat{Y} - \widehat{y}_0) \\
 & & & & & \overset{\widehat{f}'}{\nearrow} & \downarrow{\scriptstyle \widehat{q}'} & & \\
X & \overset{i}{\subset} & (X, X - \alpha) & \overset{e}{\supset} & (N, N - \alpha) & \xrightarrow{f'} & (V, V - y_0) & \overset{d}{\subset} & (Y, Y - y_0)
\end{array}$$

Here V is an evenly covered neighbourhood of y_0, $\widehat{V}$ is mapped homeomorphically onto V by $\widehat{q}$, N is a closed neighbourhood of α that is mapped into V by f, and $\widehat{f}'$, f', and $\widehat{q}'$ are the maps defined by $\widehat{f}$, f, and $\widehat{q}$. By definition,

$L_*(f,\alpha) = d_* \circ f'_* \circ e_*^{-1} \circ i_*$ and $L_*(\widehat{f},\alpha) = \widehat{d}_* \circ \widehat{f}'_* \circ e_*^{-1} \circ i_*$. The inclusions $\widehat{d}$ and d are excisions and therefore induce homology isomorphisms, and $\widehat{q}'$ is a homeomorphism so it also induces an isomorphism. Thus a diagram chase reveals that $L_*(f,\alpha) = d_* \circ \widehat{q}'_* \circ \widehat{d}_*^{-1} \circ L_*(\widehat{f},\alpha)$. Since $d_* \circ \widehat{q}'_* \circ \widehat{d}_*^{-1}$ is an isomorphism, this implies that $\lambda(f,\alpha) = \pm\lambda(\widehat{f},\alpha)$. But α contains all the roots of $\widehat{f}$ at $\widehat{y}_0$, so $\lambda(\widehat{f},\alpha) = \lambda(\widehat{f},X)$. Thus $\text{mult}(f,y_0) = |\lambda(f,\alpha)| = |\lambda(\widehat{f},X)|$.

Now assume that f is type I. Then, since $\widehat{q}$ is orientation true, it follows that $\widehat{f}$ is also type I. Since X is orientable, this means that the loop $\widehat{f} \circ A$ in $\widehat{Y}$ is orientation preserving for every loop A in X, but $\widehat{f}_{\#}\colon \pi(X,x) \to \pi(\widehat{Y}, f(x))$ is an epimorphism for every $x \in X$, so every loop in $\widehat{Y}$ is orientation preserving, and therefore $\widehat{Y}$ is orientable. Since X and $\widehat{Y}$ are both compact orientable, then $\lambda(\widehat{f},X) = \deg(\widehat{f})$, by (4.12), and therefore $\text{mult}(f,y_0) = |\lambda(\widehat{f},X)| = |\deg(\widehat{f})|$. This completes the proof for orientable X and f type I.

Assume now that X is still orientable, but f is type II. Then, because X has no orientation reversing loops there must be an orientation preserving loop A in X such that $f \circ A = \widehat{q} \circ \widehat{f} \circ A$ is orientation reversing. Since $\widehat{q}$ is orientation true, this implies that $\widehat{f} \circ A$ is also orientation reversing, so $\widehat{Y}$ is nonorientable and therefore $H_n(\widehat{Y};\mathbf{Z}) = 0$. But $L_n(\widehat{f},X) = j_n \circ f_n$ where j_n is induced by the inclusion $j\colon \widehat{Y} \subset (\widehat{Y}, \widehat{Y} - \widehat{y}_0)$, so this means that $L_n(\widehat{f},X;\mathbf{Z})$ can be factored through $H_n(\widehat{Y},\mathbf{Z}) = 0$, and therefore $\lambda(f,X) = 0$. It follows that $\text{mult}(f) = |\lambda(f,X)| = 0$. This completes the proof for orientable X.

We now assume that X is nonorientable, so by definition $\text{mult}(f,y_0) = \text{mult}(f \circ \widetilde{p}, y_0)$. If $RR(f) = \infty$, then $\widehat{q}$ has an infinite number of sheets, so $\widehat{q} \circ \check{q}$ also has an infinite number of sheets and therefore $RR(f \circ \widetilde{p}) = \infty$, and therefore, as in the first part of the proof $\text{mult}(f,y_0) = \text{mult}(f \circ \widetilde{p}) = 0$. Henceforth assume that $RR(f) < \infty$, so then $\widehat{Y}$ is compact. Since $\check{Y} \to \widehat{Y}$ is two sheeted, this implies that $\check{Y}$ is also compact.

We first assume that f is type I. Then, since $\widetilde{p}$ is type I, so is $f \circ \widetilde{p}$. So applying the first part of the theorem with $f \circ \widetilde{p}\colon \widetilde{X} \to Y$ in place of f and $\check{f}\colon \widetilde{X} \to \check{Y}$ in place of $\widehat{f}$, we conclude that $\check{Y}$ is orientable and that $\text{mult}(f,y_0) = \text{mult}(f \circ \widetilde{p}) = |\deg(\check{f})|$.

To complete the proof, assume that f is type II. We need to show that $\text{mult}(f,y_0) = 0$, which amounts to showing that $\text{mult}(f \circ \widetilde{p}, y_0) = 0$. By the first part of the theorem, we know that $\text{mult}(f \circ \widetilde{p}, y_0) = 0$ if $f \circ \widetilde{p}$ is type II. So it remains to consider the case where $f \circ \widetilde{p}$ is type I, but f is type II. This can happen only if there is an orientation reversing loop A in X that is mapped to an orientation preserving loop in Y. From covering space theory, the loops A and $f \circ A$ induce deck transformations $\tau_A\colon \widetilde{X} \to \widetilde{X}$ and $\tau_{f\circ A}\colon \check{Y} \to \check{Y}$ such that $\check{f} \circ \tau_A = \tau_{f \circ A} \circ \check{f}$. From the first part of the theorem, $\check{Y}$ is orientable ($f \circ \widetilde{p}$ is type I), so we may compute degrees and find $\deg(\check{f})\deg(\tau_A) = \deg(\tau_{f\circ A})\deg(\check{f})$. Since A is orientation reversing, $\deg(\tau_A) = -1$ and since $f \circ A$ is orientation preserving, $\deg(\tau_{f\circ A}) = 1$. Therefore $\deg \check{f} = 0$, so from the first part of the proof we have $\operatorname{mult}(f \circ \widetilde{p}) = |\deg \check{f}| = 0$. □

Since the mod 2 degree of a map of compact n-manifolds is always defined, the first part of the above proof is easily modified to prove

(5.20) THEOREM. *Let $f: X \to Y$ be a type* III *map of compact n-manifolds, and $y_0 \in Y$. Then either $RR(f) = \infty$, in which case* $\operatorname{mult}(f, y_0) = 0$, *or* $\operatorname{mult}(f) = |\deg_2(\widehat{f})|$, *where as usual $\widehat{f}$ is a Hopf lift for f, $\deg_2(\widehat{f})$ is its* mod 2 *degree, and $|\deg_2(\widehat{f})| = 1$ if $\deg_2(\widehat{f}) \equiv 1$* mod 2 *and is* 0 *otherwise.*

Since $RR(f)$, $\deg(\widehat{f})$, $\deg(\check{f})$, and $\deg_2(\widehat{f})$ are independent of y_0, we have

(5.21) COROLLARY. *Let $f: X \to Y$ be a map of compact n-manifolds. Then* $\operatorname{mult}(f, y_0)$ *is independent of the choice of $y_0 \in Y$.*

Henceforth we will write $\operatorname{mult}(f)$ in place of $\operatorname{mult}(f, y_0)$.

The Hopf coverings in (5.19) depend only on the homotopy type of f, homotopic maps have homotopic lifts, and degree is a homotopy invariant. It follows that $\operatorname{mult} f$ is too:

(5.22) COROLLARY. *If $f, g: X \to Y$ are homotopic maps of compact n-manifolds, then* $\operatorname{mult}(f) = \operatorname{mult}(g)$.

We are finally in a position to define the absolute degree.

(5.23) DEFINITION. Let $f: X \to Y$ be a map of compact n manifolds. The *absolute degree* of f is denoted by $\mathcal{A}(f)$ and is defined to be the product of the multiplicity of f by the number of Nielsen classes of f.

When $\operatorname{mult}(f) > 0$, then every Nielsen class is essential and there are $NR(f)$ of them. Also, in this case, $NR(f) = RR(f)$. Thus,

(5.24) THEOREM. *Let $f: X \to Y$ be a map of compact n manifolds, then*

$$\mathcal{A}(f) = \operatorname{mult}(f) NR(f).$$

When $RR(f) < \infty$, we also have $\mathcal{A}(f) = \operatorname{mult}(f) RR(f)$.

Since both $\operatorname{mult}(f)$ and $N(f)$ are homotopy invariants, so is $\mathcal{A}(f)$.

(5.25) COROLLARY. *Let $f, g: X \to Y$ be homotopic maps of compact n manifolds, then*

$$\mathcal{A}(f) = \mathcal{A}(g).$$

If f is a type III map then by its definition $\operatorname{mult}(f)$ is either 0 or 1, so (5.24) implies

(5.26) COROLLARY. *If f is a type* III *map of n-manifolds and $\mathcal{A}(f) > 0$, then*

$$\mathcal{A}(f) = NR(f).$$

(5.27) REMARK. Once we have proved (5.5), we will know that $NR(f) > 0$ implies $\mathcal{A}(f) > 0$, so for type III maps the equality $\mathcal{A}(f) = NR(f)$ will hold regardless of whether or not $\mathcal{A}(f) > 0$.

Theorem (5.24) also implies

(5.28) COROLLARY. *Let $f: X \to Y$ be a map of compact n manifolds. If $\mathcal{A}(f) > 0$, then $NR(f) > 0$.*

The rest of this section is devoted to proofs of Theorem (5.5) and Theorem (5.7). Most of the proof of (5.5) is contained in the following five lemmas. After these lemmas come the rest of the proof of (5.5) and the proof of (5.7). Our proof of (5.5) is based in large part on the proof in [Ep]. We have, however, filled in a large number of details that [Ep] leaves to the reader.

(5.29) LEMMA. *Suppose X an n manifold with $n > 1$ and $x_0 \in X$. Then:*

(5.29.1) *For any $x \in X - x_0$, the inclusion $k: X - x_0 \subset X$ induces an epimorphism $k_\#: \pi(X - x_0, x) \to \pi(X, x)$, which is an isomorphism if $n > 2$.*

(5.29.2) *If A is any path in X neither of whose endpoints are x_0, then it is fixed-end-point homotopic to a path in $X - x_0$.*

PROOF. Let B be an n-ball with $x_0 \in \operatorname{int} B$, and choose a base point $x \in \operatorname{int} B - x_0$ (the theorem is easily seen to be independent of the choice of basepoint). Because $n > 1$, then $\operatorname{int} B - x_0$, $\operatorname{int} B$ and $X - x_0$ are all connected so we may apply van Kampen's theorem to $\operatorname{int} B$, $X - x_0$, $\operatorname{int} B \cap (X - x_0) = \operatorname{int} B - x_0$ and $\operatorname{int} B \cup (X - x_0) = X$ obtaining the commutative diagram

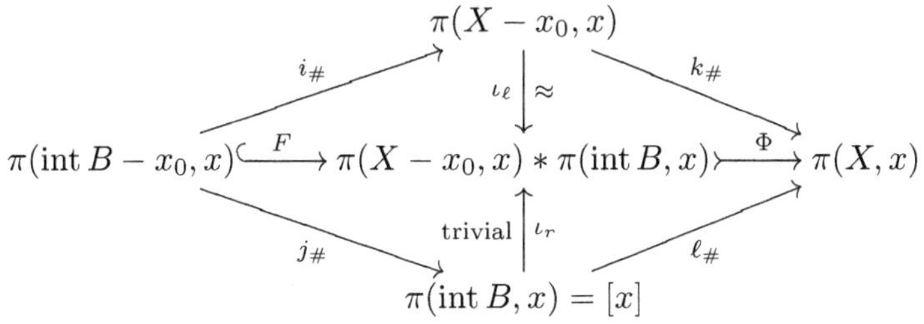

where $i_\#$, $j_\#$, $k_\#$, and $\ell_\#$ are induced by inclusions, ι_ℓ and ι_r are the injections into the free product, F is an injection, and Φ is an epimorphism whose kernel is the normal subgroup generated by the image of F (we are following [L, p. 211]). Since $\operatorname{int} B$ is simply connected, $\pi(\operatorname{int} B, x)$ is trivial so ι_l is an isomorphism, and therefore $k_\#$ is an epimorphism. Now $\operatorname{int} B - x_0$ may be deformation retracted onto an $n-1$-sphere, so when $n > 2$, $\pi(\operatorname{int} B - x_0, x)$ is also trivial and thus $\ker \Phi$ is trivial; consequently Φ and therefore $k_\#$ is an isomorphism. This proves (5.29.1).

For (5.29.2), let A be a path in X neither of whose endpoints are x_0. Let $x = A(0)$ and let A_1 be a path in $X - x_0$ from A(1) back to x. Then AA_1 is a loop in X at x, so since $k_\#$ is an epimorphism, there is a loop C in $X - x_0$ such that $[C] = [AA_1]$. Then CA_1^{-1} is a path in $X - x_0$ with $[CA_1^{-1}] = [A]$. □

(5.30) LEMMA. *Let X be an n-manifold with $n > 1$. Let A be a path in X from x_0 to $x_1 \neq x_0$ Then there is an n-ball $B \subset X$ such that $x_0, x_1 \in \operatorname{int} B$ and $[P] = [A]$ for any path P in B from x_0 to x_1.*

PROOF. Let B_0 be an n-ball with $x_0 \in \int B_0$ and let A_0 be a path in $\int B_0$ from x_0 to a point $x_0' \in \int B_0 - x_0$. Then $A_0^{-1}A$ is a path in X with endpoints in $X - x_0$ so by the previous Lemma there is a path A_1 in $X - x_0$ fixed-end-point homotopic in X to $A_0^{-1}A$, so $[A] = [A_0 A_1]$. Since $X - x_0$ is a neighborhood of A_1, then by Theorem (3.20) there is an isotopy $\{h_t \colon X \to X\}$ such that $A_1 = \{h_t(x_0')\}$, h_0 is the identity on X, and $\{h_t\}$ is constant off $X - x_0$, so $h_t(x_0) = x_0$ for all t. Let $B = h_1(B_0)$. Then both $x_0 = h_1(x_0) \in \int B$ and $x_1 = h_1(x_0') \in B$. Now suppose P is a path in B from x_0 to x_1. It remains to show that $[A] = [P]$.

Define a homotopy $\{H_t \colon I \to X\}$ of paths by

$$H_t(s) = \begin{cases} h_t \circ A_0(s(2-t)) & \text{for } 0 \leq s \leq 1/2, \\ h_{(2s-1)(1-t)+t} \circ A_0(1 + st - t) & \text{for } 1/2 \leq s \leq 1. \end{cases}$$

Both formulas yield $H_t(1/2) = h_t \circ A_0(1 - t/2)$, so $\{H_t\}$ is well defined. It is straightforward to verify that $\{H_t\}$ is a fixed-end-point homotopy from $A_0 A_1$ to $h_1 \circ A_0$, so $[A_0 A_1] = [h_1 \circ A_0]$. Since $h_1 \circ A_0$ and P have the same endpoints and are in the simply connected space B, then $[h_1 \circ A_0] = [P]$. Thus $[A] = [A_0 A_1] = [h_1 \circ A_0] = [P]$. □

The next lemma is solely about the unit n-ball $\mathbf{B}^n$.

(5.31) LEMMA. *Let C_0 and C_1 be two disjoint Euclidean n-balls in $\operatorname{int} \mathbf{B}^n$ and let $z_0 \in \operatorname{Bd} C_0$ and $z_1 \in \operatorname{Bd} C_1$. Then there is an arc a in $\mathbf{B}^n$ from z_0 to z_1 such that $a(I) \cap (C_0 \cup C_1) = \{x_1, x_2\}$ and a deformation retraction $\{r_t\}$ of $(\mathbf{B}^n, \mathbf{B}^n - (\operatorname{int} C_0 \cup \operatorname{int} C_1))$ onto $(C_0 \cup a(I) \cup C_1, \operatorname{Bd} C_0 \cup a(I) \cup \operatorname{Bd} C_1)\}$.*

By "Euclidean ball" we mean an ordinary closed ball in $\mathbf{R}^n$ of the form $B = \{x \in \mathbf{R}^n : \|x - c\| \leq d\}$, so a Euclidean ball is not just a homeomorph of $\mathbf{B}^n$.

PROOF. Let c_0 and c_1 be the centers of C_0 and C_1, let ℓ be the straight line segment joining c_0 to c_1, and let z_0' and z_1' be the points where ℓ intersects $\operatorname{Bd} C_0$

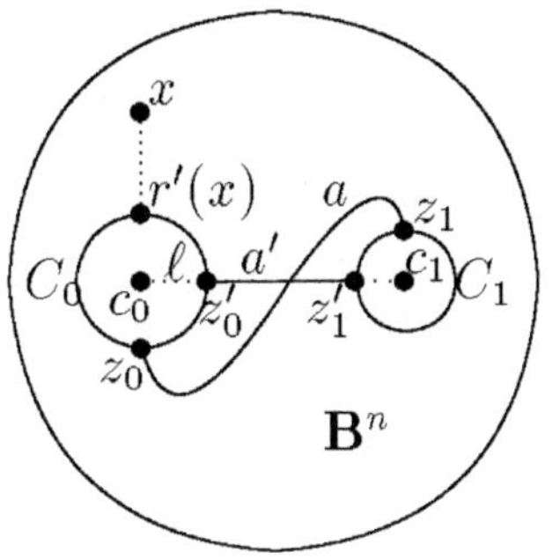

FIGURE 1

and $\operatorname{Bd} C_1$ (see Figure 1). Let a' be the straight line arc from z'_0 to z'_1 (a portion of ℓ), parameterized by $a'(t) = (1-t)z'_0 + tz'_1$.

We first construct a deformation retraction $\{r'_t\}$ of $\mathbf{B}^n$ onto $C_0 \cup a'(I) \cup C_1$. For each $x \in \mathbf{B}^n$ there is an unique point in ℓ that minimizes the distance from x to ℓ. Let $\alpha(x)$ be that point. If $x \in \mathbf{B}^n - (\operatorname{int} C_0 \cup a'(I) \cup \operatorname{int} C_1)$, then the straight line segment from x to $\alpha(x)$ intersects $\operatorname{Bd} C_0 \cup a'(I) \cup \operatorname{Bd} C_1$ in a unique point which we call $r'(x)$. For $x \in C_0 \cup a'(I) \cup C_1$, let $r'(x) = x$. Define r'_t by $r'_t(x) = (1-t)x + tr'(x)$.

Now let z_0 and z_1 be arbitrary points on $\operatorname{Bd} C_0$ and $\operatorname{Bd} C_1$. Construct a homeomorphism $h\colon \mathbf{B}^n \to \mathbf{B}^n$ as follows: Let ρ_i be a rotation of $\mathbf{R}^n$ around c_i such that $\rho_i(z_i) = z'_i$ for $i = 0, 1$. Let $h|C_i = \rho_i|C_i$ for $i = 0, 1$ and then extend h to a homeomorphism on all of $\mathbf{B}^n$, by slowing the rotations down to the identity outside of the C_i ([10]). Define a and $\{r_t\}$ by $a = h^{-1} \circ a'$, and $r_t = h^{-1} \circ r'_t \circ h$. It is easy to check that a and $\{r_t\}$ have the desired properties. □

(5.32) LEMMA. *Suppose $f\colon X \to Y$ is a map of n-manifolds, $n > 2$, $y_0 \in Y$, and $f^{-1}(y_0)$ is finite. Let x_0, x_1 be two roots of f at y_0 that are Nielsen related by a path A in X from x_0 to x_1 and suppose that f is a local homeomorphism at x_0 and x_1. Then there are n-balls B and C in X and D in Y, and a homotopy $\{g_t\colon X \to Y\}$ that have the following properties.*

(5.32.1) $x_0, x_1 \in \operatorname{int} C \subset C \subset \operatorname{int} B$, $B \cap f^{-1}(y_0) = \{x_0, x_1\}$, *and* $y_0 \in \operatorname{int} D$.
(5.32.2) *Any path in B from x_0 to x_1 is fixed-end-point homotopic to A.*
(5.32.3) $g_0 = f$, $\{g_t\}$ *is constant on* $X - \operatorname{int} B$, $g_t^{-1}(y_0) \cap B = \{x_0, x_1\}$ *for all* $t \in I$, *and* $g_1(C) \subset D$.

([10]) Explicitly: For each $i = 0, 1$, construct disjoint Euclidean balls C'_i concentric with C_i such that $C_i \subset \operatorname{int} C'_i$. Let d_i and d'_i be the diameters of C_i and C'_i. Let $\beta_i\colon \mathbf{R} \to I$ be a continuous function such that $\beta_i(t) = 1$ for $t \le d_i$ and $\beta_i(t) = 0$ for $t \ge d'_i$. Let $\{\rho_{it}\}$ be a path from the identity to ρ_i in the group of rotations around the center of C_i. Extend $h|C_i$ to C'_i by $(h|C'_i)(x) = \rho_{i\beta_i(\|x-c_i\|)}(x)$. Finally, let h be the identity on the rest of $\mathbf{B}^n$.

PROOF. The proof requires a large cast of characters; we will describe them as they are introduced. The reader may find it helpful to consult the Figure 2.

- f, X, Y, x_0, x_1, y_0 and A. These are as given to us, except we will also assume that the path A is in $(X - f^{-1}(y_0)) \cup \{x_0\} \cup \{x_1\}$. This assumption is justified by (5.29) and the fact that $f^{-1}(y_0)$ is finite.
- B. The n-ball B is created by applying (5.30) to the manifold $(X - f^{-1}(y_0)) \cup \{x_0\} \cup \{x_1\}$ and the path A. It has x_0 and x_1 in its interior, $B \cap f^{-1}(y_0) = \{x_0, x_1\}$, and B satisfies property (5.32.2) above.
- $\phi\colon \mathbf{B}^n \to B$, is a homeomorphism, $x_0' = \phi^{-1}(x_0)$ and $x_1' = \phi^{-1}(x_1)$.
- C' and C. C' is a Euclidean ball in the interior of and concentric with $\mathbf{B}^n$, and having both x_0' and x_1' in its interior. Let $C = \phi(C')$.
- D, D_0, and D_1. These are n-balls such that $y_0 \in \operatorname{int} D$, $x_i \in \operatorname{int} D_i \subset \operatorname{int} C$, $D_0 \cap D_1 = \emptyset$ and f maps D_i homeomorphically onto D for $i = 0, 1$. They are defined as follows: Since f is a local homeomorphism at x_0 and x_1, there are neighbourhoods U_0 and U_1 of x_0 and x_1 that are mapped homeomorphically onto neighbourhoods V_0 and V_1 of y_0. By choosing the U_i small enough we may assume that they are disjoint and contained in $\operatorname{int} C$. Now choose an n-ball D such that $y_0 \in \operatorname{int} D \subset V_0 \cap V_1$, and let $D_i = (f|U_i)^{-1}(D)$.

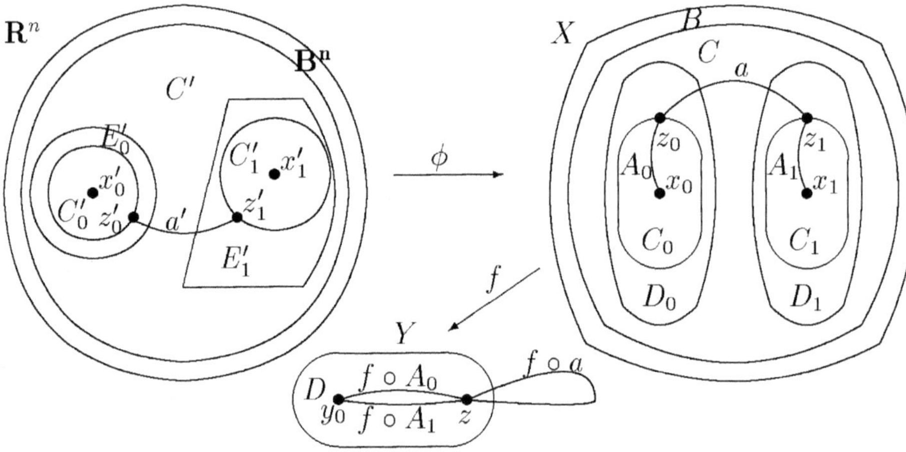

FIGURE 2

Properties (5.32.1), and (5.32.2) are clearly satisfied. It remains to define $\{g_t\}$ satisfying property (5.32.3). To do this we need some more construction that will allow us to apply (5.31).

- E'_0 and E'_1. We construct E'_0 as a Euclidean ball within $\phi^{-1}(D_0)$ and centered on x'_0. Let $E'_1 = \phi^{-1} \circ (f|D_1)^{-1} \circ (f|D_0) \circ \phi(E'_0)$. Then E'_0 and E'_1 are disjoint, since D_0 and D_1 are.
- C'_1, z'_1, z, z'_0, C'_0. Construct C'_1 as a Euclidean ball centered on x'_1 and contained in E'_1. Choose a point z'_1 on its boundary. Let $z = f \circ \phi(z'_1) \in Y$, and $z'_0 = \phi^{-1} \circ (f|D_0)^{-1}(z) \in E'_0$. Then z'_0 is a point in the Euclidean ball E'_0, centered on x'_0, so we may construct a concentric Euclidean ball C'_0 within E'_0 having z'_0 on its boundary.
- a' and $\{r'_t\}$. We can now apply (5.31) to C'_0, C'_1, z'_0, and z'_1 to obtain an arc a' from z'_0 to z'_1 such that $a'(I) \cap C'_i = z'_i$ and a deformation retraction $\{r'_t\}$ of the pair $(\mathbf{B}^n, \mathbf{B}^n - \operatorname{int} C)$ onto the pair $(C'_0 \cup a'(I) \cup C'_1, \operatorname{Bd} C'_0 \cup a'(I) \cup \operatorname{Bd} C'_1)$.
- C_0, C_1, z_0, z_1, a, and $\{r_t\}$. Let $C_i = \phi(C_i)$, $z_i = \phi(z'_i)$, $a = \phi \circ a'$ and $\{r_t\} = \{\phi \circ r'_t \circ \phi^{-1}\}$. Then a is an arc in B from $z_0 \in \operatorname{Bd} C_0$ to $z_1 \in \operatorname{Bd} C_1$ such that $a(I) \cap C_i = z_i$, $\{r_t\}$ is a deformation retraction of the pair $(B, B - \operatorname{int} C)$ onto the pair $(C_0 \cup C_1 \cup a(I), \operatorname{Bd} C_0 \cup a(I) \cup \operatorname{Bd} C_1)$, and $f(z_0) = f(z_1) = z$.

We now construct $\{g_t\}$. We want to show first that $f \circ a$ is contractible in $Y - y_0$ to the point z. Let, for the moment, A_i be a path in C_i from x_i to z_i, for $i = 0, 1$. Then $A_0(aA_1^{-1})$ is a path in B from x_0 to x_1, and therefore by property (5.32.2), $[A_0(aA_1^{-1})] = [A]$. Thus $[f \circ A_0][f \circ a][f \circ A_1]^{-1} = [f \circ A] = [y_0]$, and therefore $[f \circ a] = [(f \circ A_0)^{-1}(f \circ A_1)]$. Now $(f \circ A_0)^{-1}(f \circ A_1)$ is a loop in D at the point z. Since D is contractible, so is $(f \circ A_0)^{-1}(f \circ A_1)$, and therefore $f \circ a$ is contractible to z.

Since $f \circ a$ is a loop in $Y - y_0$ and $n > 2$, by (5.29) we may assume that the contraction takes place in $Y - y_0$ ([11]). So let $\{H_t\}$ be a fixed-end-point homotopy in $Y - y_0$ from $f \circ a$ to the constant path z. Define a homotopy $\{g_t^1 \colon C_0 \cup a(I) \cup C_1 \to Y\}$ by

$$g_t^1(x) = \begin{cases} H_t(a^{-1}(x)) & \text{for } x \in a(I), \\ x & \text{for } x \in C_0 \cup C_1. \end{cases}$$

Here a^{-1} is meant to denote the inverse of a as a function, not as a path. Thus $a^{-1}(x) = s$ means $a(s) = x$. Note that throughout this homotopy $g_t^1(\operatorname{Bd} C_0 \cup a(I) \cup \operatorname{Bd} C_1) \subset Y - y_0$, and at the end g_1^1 maps the pair $(C_0 \cup a(I) \cup C_1, \operatorname{Bd} C_0 \cup a(I) \cup \operatorname{Bd} C_1)$ into $(D, D - y_0)$. Now define a homotopy $\{g_t^2 \colon B \to Y\}$ by

$$g_t^2 = \begin{cases} f \circ r_{2t} & \text{for } 0 \le t \le 1/2, \\ g_{2t-1}^1 \circ r_1 & \text{for } 1/2 \le t \le 1. \end{cases}$$

([11]) This is where the crucial assumption that $n > 2$ is used.

It is straightforward to check that this homotopy is well defined, that at the beginning of the homotopy $g_0^2 = f|B$, that throughout the homotopy $(g_t^2)^{-1}(y_0) \cap B = \{x_0, x_1\}$, and at the end g_1^2 maps the pair $(C, \operatorname{Bd} C)$ into $(D, D - y_0)$.

We now extend $\{g_t^2|C\}$ to all of X by slowing g_t^2 down as one moves from $\operatorname{Bd} C$ to $\operatorname{Bd} B$. To do this, let $\beta\colon B \to I$ be a map that is 1 on C and 0 on $\operatorname{Bd} B$, and define the homotopy $\{g_t\}$ by

$$g_t(x) = \begin{cases} g^2_{\beta(x)t}(x) & \text{for } x \in B, \\ f(x) & \text{for } x \in X - B. \end{cases}$$

Both formulas give $g_t(x) = f(x)$ for $x \in \operatorname{Bd} B$ so the homotopy is well defined. It is straightforward to check that $\{g_t\}$ has the properties listed in (5.32.3). □

(5.33) LEMMA. *Suppose $f\colon X \to Y$ a map of n-manifolds, $n > 2$, $y_0 \in Y$, and X is compact and orientable. Suppose further that there are two roots $x_0, x_1 \in X$ of f at y_0, an n-ball $C \subset X$ and an n-ball $D \subset Y$ such that*

(5.33.1) $x_0, x_1 \in \operatorname{int} C$, $\{x_0, x_1\} = C \cap f^{-1}(y_0)$, *and* $y_0 \in \operatorname{int} D$.
(5.33.2) $\lambda(f, x_1) = -\lambda(f, x_0)$.
(5.33.3) $f(C) \subset D$.

Then there is a homotopy $\{h_t\colon X \to Y\}$ such that $h_0 = f$, $\{h_t\}$ is constant on $X - \operatorname{int} C$, and h_1 has no roots in C.

PROOF. Let $\phi\colon \mathbf{B}^n \to C$ be a homeomorphism. Let $E' \subset \operatorname{int} \mathbf{B}^n$ be a Euclidean ball concentric with $\mathbf{B}^n$ and having both $\phi^{-1}(x_0)$ and $\phi^{-1}(x_1)$ in its interior, and let d denote its radius. Define a deformation retraction $\{r_t\}$ of $(\mathbf{B}^n, \mathbf{B}^n - E')$ onto $(E', \operatorname{Bd} E')\}$ by

$$r'_t(x) = \begin{cases} x & \text{for } x \in E', \\ (1-t)x + t\dfrac{d}{\|x\|}x & \text{for } x \in \mathbf{B}^n - E'. \end{cases}$$

Let $E = \phi(E')$ and define $\{r_t\colon (C, C - E) \to (E, \operatorname{Bd} E)\}$ by $r_t = \phi \circ r'_t \circ \phi^{-1}$. Let $f''\colon (E, \operatorname{Bd} E) \to (D, D - y_0)$ be the map defined by f. Let $s_0 = (1, 0, \ldots, 0) \in \operatorname{Bd} \mathbf{B}^n$. Define a homeomorphism $\phi_E\colon (\mathbf{B}^n, \operatorname{Bd} \mathbf{B}^n, s_0) \to (E, \operatorname{Bd} E, \phi_E(s_0))$ by $\phi_E(x) = \phi(x/d)$. Then $f'' \circ \phi_E$ represents an element $[f'' \circ \phi_E]$ in the nth relative homotopy group $\pi_n(D, D - y_0, f \circ \phi_E(s_0))$. Our first task is to show that this is the trivial element, so that there is a deformation of $f'' \circ \phi_E$ into $D - y_0$. To do this consider the diagram

$$\begin{array}{ccc} \pi_n(E, \operatorname{Bd} E', \phi_E(s_0)) & \xrightarrow{f''_{\pi n}} & \pi_n(D, D - y_0, f'' \circ \phi_E(s_0)) \\ \Big\downarrow h_E & & \Big\downarrow h_D \, \approx \\ H_n(E, \operatorname{Bd} E; \mathbf{Z}) & \xrightarrow[f''_n]{} & H_n(D, D - y_0; \mathbf{Z}) \end{array}$$

The horizontal homomorphisms are induced by f''. The vertical homomorphisms are Hurewicz homomorphisms. Now $H_p(D, D - y_0; \mathbf{Z}) = 0$ for $p < n$, and $D - y_0$ is simply connected $(n > 2)$, therefore h_D is an isomorphism, so it suffices to show that $h_D([f'' \circ \phi_E]) = 0$. Now $h_D([f'' \circ \phi_E]) = h_D(f''_{\pi n}[\phi_E]) = f''_n(h_E([\phi_E]))$, so it suffices to show that $f''_n = 0$. To do this consider another diagram.

$$\begin{array}{ccccccccc} X & \overset{i}{\subset} & (X, X - \operatorname{int} E) & \overset{e}{\supset} & (C, C - \operatorname{int} E) & \overset{f'}{\longrightarrow} & (D, D - y_0) & \overset{j}{\subset} & (Y, Y - y_0) \\ & & & & \cup & \nearrow_{f''} & & & \\ & & & & (E, \operatorname{Bd} E) & & & & \end{array}$$

By additivity we have $\lambda(f, \operatorname{int} E) = \lambda(f, x_0) + \lambda(f, x_1) = 0$, which implies that $L_n(f, \operatorname{int} E; \mathbf{Z}) = 0$. But $L_n(f, \operatorname{int} E) = j_n \circ f'_n \circ e_n^{-1} \circ i_n$, so $j_n \circ f'_n \circ e_n^{-1} \circ i_n = 0$. Now j and e are excisions so both j_n and e_n are isomorphisms. Also $(X, X - \operatorname{int} E)$ is a deformation retract of $(X, X - x_0)$, so the inclusion $(X, X - \operatorname{int} E) \subset (X, X - x_0)$ induces an isomorphism, and therefore since the inclusion $X \subset (X, X - x_0)$ induces an isomorphism in dimension n, so does i. Thus we must have $f'_n = 0$, and therefore $f''_n = 0$. This completes the proof that $[f'' \circ \phi_E] = 0 \in \pi_n(D, D - y_0)$.

Since $[f'' \circ \phi_E] = 0 \in \pi_n(D, D - y_0)$, there is a homotopy $\{h^1_t\colon (\mathbf{B}^n, \operatorname{Bd} \mathbf{B}^n) \to (D, D - y_0)\}$ such that $h^1_0 = f'' \circ \phi_E$ and $h^1_1(\mathbf{B}^n) \subset D - y_0$. Now define a homotopy $\{h^2_t\colon (C, C - \operatorname{int} E) \to (D, D - y_0)\}$ by

$$h^2_t = \begin{cases} f' \circ r_{2t} & \text{for } 0 \le t \le 1/2, \\ h^1_{2t-1} \circ \phi_E^{-1} \circ r_1 & \text{for } 1/2 \le t \le 1. \end{cases}$$

Then $h^2_0 = f'$, $h^2_t(C - \operatorname{int} E) \subset D - y_0$ for all t, and $h^2_1(C) \subset D - y_0$. Now let $\beta\colon C \to I$ be a map that is 0 on $\operatorname{Bd} C$ and 1 on E, and define our final homotopy $\{h_t\colon X \to Y$ by

$$h_t(x) = \begin{cases} h^2_{\beta(x)t}(x) & \text{for } x \in C, \\ f(x) & \text{for } x \in X - C. \end{cases}$$

It is straightforward to verify that $\{h_t\}$ is well defined and has the desired properties. □

We turn now to the proof of the first Hopf theorem.

PROOF OF THEOREM (5.5). We assume that X and Y are compact n-manifolds, $n > 2$, and that f is transverse to y_0. We will also assume that f has at least one root at y_0. Since X is compact, $f^{-1}(y_0)$ is compact also, and since f is transverse to y_0, $f^{-1}(y_0)$ is discrete and therefore finite.

We divide the proof into three cases: X and f orientable, X nonorientable but f orientable, and f (and therefore X) nonorientable.

Case 1. X and f orientable. We will first show that $\lambda(f, x_0) = \pm 1$ for any root x_0. Let U be a neighbourhood of x_0 that is mapped homeomorphically onto a neighbourhood V of y_0 and consider our usual diagram:

$$X \overset{i}{\subset} (X, X - y_0) \overset{e}{\supset} (U, U - x_0) \xrightarrow{f'} (V, V - y - 0) \overset{j}{\subset} (Y, Y - y_0)$$

where f' is defined by f. The inclusion i induces a homology isomorphism in dimension n, e and j are excisions, and f' is a homeomorphism so they all induce isomorphisms. It follows that

$$L_n(f, x_0; \mathbf{Z}) = j_n \circ f'_n \circ e_n^{-1} \circ i_n \colon H_n(X; \mathbf{Z}) \to H_n(Y, Y - y_0; \mathbf{Z})$$

is an isomorphism and therefore $L_n(f, x_0; \mathbf{Z})(\mu) = \pm\nu$ where μ and ν are generators of $H_n(X; \mathbf{Z})$ and $H_n(Y, Y - y_0; \mathbf{Z})$. Therefore, by definition of λ, $\lambda(f, x_0) = \pm 1$.

Let $\alpha = \{x_0, \ldots, x_{d-1}\}$ be a root class of f at y_0. Then

$$\operatorname{mult}(\alpha) = |\lambda(f, \alpha)| = \left| \sum_{i=0}^{d-1} \lambda(f, x_i) \right| \le \sum_{i=0}^{d-1} |\lambda(f, x_i)| = d.$$

Thus $\operatorname{mult}(\alpha)$ is a lower bound on the number of roots in each root class, and therefore $\mathcal{A}(f) = \operatorname{mult}(\alpha)$(number of root classes) is a lower bound on the total number of roots. This proves the first assertion in (5.5).

We now need to show that f can be changed by a homotopy so that it is still transverse to y_0 and has exactly $\mathcal{A}(f)$ roots. So suppose that f has more than $\mathcal{A}(f)$ roots. There is then a class $\alpha = \{x_0, \ldots, x_{d-1}\}$ where $d > \operatorname{mult}(f)$, so there must be two roots, x_0 and x_1 say, such that $\lambda(f, x_1) = -\lambda(f, x_0)$. Apply (5.32) to obtain n-balls B and C in X and an n-ball D in Y, and a homotopy $\{g_t\}$ such that

(5.34.1) $x_0, x_1 \in \operatorname{int} C \subset C \subset \operatorname{int} B$, $B \cap f^{-1}(y_0) = \{x_0, x_1\}$, and $y_0 \in D$.

(5.34.2) $g_0 = f$, $\{g_t\}$ is constant on $X - B$, $g_t^{-1}(y_0) \cap B = \{x_0, x_1,\}$ for all $t \in I$ and $g_1(C) \subset D$.

Then applying the additivity and homotopy properties for λ we have $0 = \lambda(f, x_0) + \lambda(f, x_1) = \lambda(f, \operatorname{int} C) = \lambda(g_1, \operatorname{int} C) = \lambda(g_1, x_0) + \lambda(g_1, x_1)$. So now we may apply (5.33) with g_1 in place of f, to find a new map $h_1 \colon X \to Y$ homotopic to g_1 and therefore f such that $h_1^{-1}(y_0) = f^{-1}(y_0) - \{x_0, x_1\}$. Thus h_1 is homotopic to f, has two fewer roots at y_0 than f, and since the homotopies were constant outside of B, the new map h_1 is still transverse to y_0. We may therefore continue to remove roots until we have a map homotopic to f and transverse to y_0 that has only $\mathcal{A}(f)$ roots.

Case 2. X nonorientable but f orientable. Let $\widetilde{p} \colon \widetilde{X} \to X$ be the orientable double covering. Since f is transverse to y_0, and $\widetilde{p}$ is transverse to each $x \in X$, it is easy to show that $f \circ \widetilde{p}$ is transverse to y_0. So we may apply results from Case 1 to $f \circ \widetilde{p}$.

Let α be a root class of f at y_0. Then from (5.16) there is a root class $\widetilde{\alpha}$ of $f \circ \widetilde{p}$ at y_0 that is mapped bijectively onto α. As in Case 1, $\#\widetilde{\alpha} \geq \text{mult}(f \circ \widetilde{p})$. Since $\widetilde{p}$ sends $\widetilde{\alpha}$ bijectively onto α we have $\#\alpha \geq \text{mult}(f \circ \widetilde{p}) = \text{mult}(f)$, where the equality is by Definition (5.17). Adding this inequality over all Nielsen classes, gives $\#f^{-1}(y_0) \geq \mathcal{A}(f)$, which proves the first assertion in (5.5).

We now need to show that f can be changed by a homotopy to have exactly $\mathcal{A}(f)$ roots and still be transverse to y_0. If $\#f^{-1}(y_0) = \mathcal{A}(f)$, we are done. Otherwise $\#\alpha > \text{mult}(f)$ for some class α. Let $\widetilde{\alpha}$ be one of the two Nielsen classes of $f \circ \widetilde{p}$ that is mapped bijectively onto α by $\widetilde{p}$. Write $\alpha = \{x_0, \dots, x_{d-1}\}$ and $\widetilde{\alpha} = \{\widetilde{x}_0, \dots, \widetilde{x}_{d-1}\}$, where $\widetilde{p}(\widetilde{x}_i) = x_i$ for $i = 1, \dots, d-1$. As in the previous case there are two roots in $\widetilde{\alpha}$, $\widetilde{x}_0$ and $\widetilde{x}_1$ say, such that $\lambda(f \circ \widetilde{p}, \widetilde{x}_1) = -\lambda(f \circ \widetilde{p}, \widetilde{x}_1)$. Let $\widetilde{A}$ be a path in $\widetilde{X}$ from $\widetilde{x}_0$ to $\widetilde{x}_1$ such that $f \circ \widetilde{p} \circ A$ is contractible. Let $A = \widetilde{p} \circ \widetilde{A}$, so A is a path in X from x_0 to x_1 such that $f \circ A$ is contractible. Now apply (5.32) to obtain n-balls B, C, D and a homotopy $\{g_t\}$ having the properties in (5.32). We will now lift the whole picture from X to $\widetilde{X}$.

Since B is simply connected, it is evenly covered by $\widetilde{p}$. Let $\widetilde{B} \subset \widetilde{X}$ be the component of $\widetilde{p}^{-1}(B)$ containing $\widetilde{x}_0$. Then we also have $\widetilde{x}_1 = (\widetilde{p}|\widetilde{B})^{-1}(x_1) \in \widetilde{B}$. To see this, let P be a path in B from x_0 to x_1, so then $[P] = [A]$. Since $(\widetilde{p}|\widetilde{B})^{-1} \circ P$ is a lift of P beginning at $\widetilde{x}_0$, and $\widetilde{A}$ is a lift of A beginning at $\widetilde{x}_0$ it follows that $[(\widetilde{p}|\widetilde{B})^{-1} \circ P] = [\widetilde{A}]$, and in particular, $\widetilde{x}_1 = \widetilde{A}(1) = (\widetilde{p}|\widetilde{B})^{-1} \circ P(1) = (\widetilde{p}|\widetilde{B})^{-1}(x_1) \in \widetilde{B}$. Let $\widetilde{C} = (\widetilde{p}|\widetilde{B})^{-1}(C)$. Since we now know that $\widetilde{x}_i = (\widetilde{p}|\widetilde{B})^{-1}(x_i)$ for $i = 0, 1$, it is straightforward to show that properties (5.32.1)–(5.32.3) hold with $\widetilde{x}_0$, $\widetilde{x}_1$, $\widetilde{B}$, $\widetilde{C}$, and $\{g_t \circ \widetilde{p}\}$ in place of x_0, x_1, B, C, and $\{g_t\}$. Arguing as in Case 1, we still have $\lambda(g_1 \circ \widetilde{p}, \widetilde{x}_1) = -\lambda(f_0 \circ \widetilde{p}, \widetilde{x}_0)$, so we may apply (5.33) to obtain another homotopy, call it $\{\widetilde{h}_t\}$, from $g_1 \circ \widetilde{p}$ to a map $\widetilde{h}_1 \colon \widetilde{X} \to Y$ that is constant outside of $\widetilde{C}$ and has no roots in $\widetilde{C}$. Since $g_1 \circ \widetilde{p}$ had no roots in $\widetilde{B} - \widetilde{C}$, $\widetilde{h}_1$ has no roots in $\widetilde{B}$. Combine the two homotopies $\{g_t \circ \widetilde{p}\}$ and $\{\widetilde{h}_t\}$ by

$$\widetilde{f}_t = \begin{cases} g_{2t} \circ \widetilde{p} & \text{for } 0 \leq t \leq 1/2, \\ \widetilde{h}_{2t-1} & \text{for } 1/2 < t \leq 1. \end{cases}$$

Then $\{\widetilde{f}_t\}$ is a homotopy from $f \circ \widetilde{p}$ that is constant on $\widetilde{X} - \operatorname{int} \widetilde{B}$, and $\widetilde{f}_1$ has no roots in $\widetilde{B}$. Now define a homotopy of f by

$$f_t(x) = \begin{cases} \widetilde{f}_t \circ (\widetilde{p}|\widetilde{B})^{-1}(x) & \text{for } x \in B, \\ f(x) & \text{for } x \in X - B. \end{cases}$$

Then it is easy to check that f_1 has no roots in B and exactly the same roots in $X - B$ as did f. Thus, continuing in this way we may remove roots until we have a map homotopic to f with exactly $\mathcal{A}(f)$ roots.

Case 3. Neither X nor f orientable. In this case, recall that we use the integer mod 2 index λ_2 to define multiplicity, $\operatorname{mult}(f) = |\lambda_2(f,\alpha)| \le 1 \le \#\alpha$, where α is a root class of f at y_0. (Recall also, that $|\lambda(f,\alpha)| = 1$, if $\lambda(f,\alpha) \equiv 1 \bmod 2$ and is 0 otherwise.) Summing this inequality over the root classes of f, we have $\mathcal{A}(f) \le \#f^{-1}(y_0)$, which proves the first assertion in (5.5).

So now we must show that if $\mathcal{A}(f) < \#f^{-1}(y_0)$, then there is a map homotopic to f and transverse to y_0 that has $\mathcal{A}(f)$ roots at y_0. Assume $\mathcal{A}(f) < \#f^{-1}(y_0)$. It suffices to show that there is a map homotopic to f and transverse to y_0 that has fewer roots than f, for then we can remove roots until $\mathcal{A}(f) = \#f^{-1}(y_0)$. We first show that there must be a root class of f that has more than one element. The argument in Case 1 is easily modified to show that because f is transverse to y_0, then $\lambda_2(f,x_0) \equiv 1 \bmod 2$ for every root x_0 of f at x_0. Consequently, if there is a root class of f at y_0 that has only one element, then $\operatorname{mult}(f) = 1$, and if every other root class has only one root, then summing over the root classes we have $\mathcal{A}(f) = \#\text{root classes} = \#\text{roots}$, contrary to hypothesis.

So now suppose a root class $\alpha \subset X$ of f at y_0 has more than one element. Let $\widetilde{p}\colon \widetilde{X} \to X$ be the orientable double covering of X. Let $x_0, x_1 \in \alpha$. Pick $\widetilde{x}_0 \in \widetilde{p}^{-1}(x_0)$ and $\widetilde{x}_1 \in \widetilde{p}^{-1}(x_1)$. Then by (5.14) $\widetilde{x}_0$ and $\widetilde{x}_1$ are Nielsen root related. Since f is transverse to y_0 and $\widetilde{p}$ is a covering, we have as in Case 1 λ, $|\lambda(f\circ\widetilde{p},\widetilde{x}_0)| = |\lambda(f\circ\widetilde{p},\widetilde{x}_1)| = 1$. Notice that we are using the integer index. If $\lambda(f\circ\widetilde{p},\widetilde{x}_0) = -\lambda(f\circ\widetilde{p},\widetilde{x}_1)$, then we may proceed as in Case 2 to find a map homotopic to f, transverse to y_0, but with fewer roots than f, and we are done.

Otherwise, suppose that $\lambda(f\circ\widetilde{p},\widetilde{x}_0) = \lambda(f\circ\widetilde{p},\widetilde{x}_1)$ and consider the following diagram.

$$
\begin{array}{ccccccc}
\widetilde{X} & \overset{i}{\subset} & (\widetilde{X}, \widetilde{X} - \widetilde{x}_1) & \overset{e}{\supset} & (\widetilde{U}, \widetilde{U} - \widetilde{x}_1) & \xrightarrow{\widetilde{f}'} & (Y, Y - y_0) \\
\downarrow{\scriptstyle \tau} & & & & & \nearrow{\scriptstyle \widetilde{f}''} & \\
\widetilde{X} & \overset{i'}{\subset} & (\widetilde{X}, \widetilde{X} - \tau(\widetilde{x}_1)) & \overset{e'}{\supset} & (\tau(\widetilde{U}), \tau(\widetilde{U}) - \tau(\widetilde{x}_1)) & &
\end{array}
$$

Here $\tau\colon \widetilde{X} \to \widetilde{X}$ is the orientation reversing deck transformation of $\widetilde{p}\colon \widetilde{X} \to X$, $\widetilde{U}$ is a neighbourhood of $\widetilde{x}_1$ chosen small enough so that $\widetilde{U}$ has no roots of $f\circ\widetilde{p}$ other than $\widetilde{x}_1$ and $\tau(\widetilde{U})$ has no roots of $f\circ\widetilde{p}$ other than $\tau(\widetilde{x}_1)$, and $\widetilde{f}'$ and $\widetilde{f}''$ are the maps defined by $f\circ\widetilde{p}$. From this diagram, we see that $L_n(f\circ\widetilde{p},\widetilde{x}_1;\mathbf{Z})) = L_n(f\circ\widetilde{p},\tau(\widetilde{x}_1);\mathbf{Z})\circ\tau_n$. Since τ is an orientation reversing homeomorphism, it follows that $L_n(f,\widetilde{x}_0) = -L_n(f,\tau(\widetilde{x}_1))$, and therefore $\lambda(f\circ\widetilde{p},\tau(\widetilde{x}_1)) = -\lambda(f\circ\widetilde{p},\widetilde{x}_1) = -\lambda(f\circ\widetilde{p},\widetilde{x}_0)$. By (5.14) $\tau(\widetilde{x}_1)$ is in the same $f\circ\widetilde{p}$ Nielsen class as $\widetilde{x}_1$ and therefore in the same class as $\widetilde{x}_0$. therefore we may substitute $\tau(\widetilde{x}_1)$ for $\widetilde{x}_1$ and use the same argument as in Case 2 to find a map homotopic to f, transverse to y_0, and with fewer roots at y_0 than f. □

We turn now to the proof of (5.7). The proof is based on the technique we used to prove Theorem (5.5) together with the following Lemma.

(5.35) LEMMA. *Let $f: X \to Y$ be a map of n manifolds, $n > 1$, and let $y_0 \in Y$. Suppose that x_0 is root of f at y_0, that f is a local homeomorphism at x_0, and that $\lambda(x_0, f) = 1$. Let B be an n-ball such that $x_0 \in \operatorname{int} B$ and f maps B homeomorphically onto an n-ball C having y_0 in its interior. Then for any integer $d > 0$ there are points $x'_1, \dots, x'_{d-1} \in \operatorname{int} B$, and a homotopy $\{g_t\}$ such that:*

(5.35.1) $g_0 = f$, $\{g_t\}$ *is constant outside* $\operatorname{int} B$.

(5.35.2) $g_t(B) \subset C$ *and* $g_t(x_0) = y_0$ *for all* $t \in I$.

(5.35.3) $g_1^{-1}(y_0) \cap B = \{x_0, x'_1, \dots, x'_{d-1}\}$, $\lambda(g_1, x_0) = d$, *and* $\lambda(g_1, x'_i) = -1$ *for* $i = 1, \dots, d-1$.

Note that for the ball B we have $\lambda(f, B) = \lambda(g_1, B)$, which of course is a necessary condition for such a homotopy. What we have done is to modify f so that it has $d-1$ new roots each of index -1, and to compensate the index of x_0 has been increased from 1 to d.

PROOF OF THE LEMMA (5.35). This is essentially a lemma about the unit ball. We first prove that there are points $z_1, \dots, z_{d-1} \in \mathbf{B}^n$ and a homotopy $\{h_t: \mathbf{B}^n \to \mathbf{B}^n\}$ such that h_0 is the identity, $\{h_t\}$ is constant on $\operatorname{Bd} B$ and 0, and h_1 is a local orientation reversing homeomorphism at each of the roots $z_1, \dots, z_{d-1}$. Here is the construction: First define $h': \mathbb{C} \to \mathbb{C}$ by $h'(z) = z^d(\overline{z}-1)\dots(\overline{z}-d+1)$. Then h' has $0, \dots, d-1$ for roots at 0 and is a local orientation reversing homeomorphism at each of the roots $1, \dots, d-1$. (The reflection $z \mapsto \overline{z}$ is orientation reversing.) Now define $\psi: \operatorname{int} \mathbf{B}^2 \to \mathbb{C}$ by $\psi(z) = z/(1 - \|z\|)$, so $\psi^{-1}(z) = z/(1 + \|z\|)$, and define $h_1: \mathbf{B}^2 \to \mathbf{B}^2$ by

$$h_1(x) = \begin{cases} \psi^{-1} \circ h' \circ \psi(x) & \text{for } x \in \operatorname{int} \mathbf{B}^2, \\ x & \text{for } x \in \operatorname{Bd} \mathbf{B}^2. \end{cases}$$

One can verify directly that for any $x' \in \operatorname{Bd} \mathbf{B}^2$, $\lim_{x \to x'} \psi^{-1} \circ h' \circ \psi(x) = x'$, so h_1 is well defined. Also, since ψ is a homeomorphism, h_1 is an orientation reversing local homeomorphism at each of its roots $\psi^{-1}(1) = 1/2, \dots, \psi^{-1}(d-1) = (d-1)/d$ at 0. Now define $\{h_t\}$ to be the straight-line homotopy, $h_t(x) = (1-t)x = th_1(x)$. The homotopy $\{h_t\}$ is the desired homotopy when $n = 2$. For $n > 2$, replace $\{h_t\}$ by its $n-2$ fold suspension. Then 0 is a root of h_1 at zero; let $z_1, \dots, z_{d-1}$ be the other roots. Then h_1 is still a local orientation reversing homeomorphism at each of the roots $z_1, \dots, z_{d-1}$. (For an alternative construction see [Li, p. 204].)

Now, to construct $\{g_t\}$, let $\phi: \mathbf{B}^n \to B$ be a homeomorphism such that $\phi(0) = x_0$. Let

$$g_t(x) = \begin{cases} f \circ \phi \circ h_t \circ \phi^{-1}(x) & \text{for } x \in B,\ t \in I, \\ f(x) & \text{for } x \notin B,\ t \in I, \end{cases}$$

and let $x'_i = \phi(z_i)$ for $i = 1, \dots, d-1$. Since $\lambda(f, x_0) = +1$, this means that the generators of $H_n(X; \mathbf{Z})$ and $H_n(Y, Y - y_0; \mathbf{Z})$ have been chosen so that $f|B$ is orientation preserving. Consequently $g_1 = f \circ \phi \circ h_1 \circ \phi^{-1}$ is orientation reversing at each $x'_1, \dots, x'_{d-1}$, and therefore $\lambda(g_1, x'_i) = -1$ for $i = 1, \dots, d-1$. Note that because $\phi(0) = x_0$, x_0 is still a root of g_1 at y_0. By additivity and homotopy we have $\lambda(g_1, x_0) - (d-1) = \lambda(g_1, x_0) + \sum_i \lambda(g_1, x'_i) = \lambda(g_1, B) = \lambda(f, B) = \lambda(f, x_0) = 1$, so $\lambda(g_1, x_0) = d$. □

PROOF OF THEOREM (5.7). We are assuming $f: X \to Y$ is a map of compact n-manifolds, $y_0 \in Y$, and $n > 2$. We must show that f is homotopic to a map having only $N\!R(f)$ roots. We consider three cases: f is nonorientable, X is orientable, and X is nonorientable but f is orientable.

Case 1. f is nonorientable. In this case by (5.26) and the remark following it, we have $N\!R(f) = \mathcal{A}(f)$, so (5.7) follows directly from (5.5).

Case 2. X is orientable. If every Nielsen class has only one root, then there are already only $N\!R(f)$ roots and there is nothing more to prove. So now assume that there is at least one class with more than one root. We will proceed by induction on the number of Nielsen classes that have one root. Our inductive hypothesis is that each class is either a single root or a finite number of roots $x_0, \dots, x_{d-1}$ such that f is a local homeomorphism at each root x_i and $d = \mathrm{mult}(f)$. After each homotopy we will increase the number of classes that have exactly one root.

By (5.5) f is homotopic to a map that is transverse to y_0 and has only $\mathrm{mult}(f)$ roots in each class. Thus the inductive hypothesis can always be satisfied, so suppose that it is. If each Nielsen class contains only one root we are done. So now assume that $\alpha = \{x_0, \dots, x_{d-1}\}$ is a root class with $d = \mathrm{mult}(f) > 1$. Then $\lambda(f, x_i) = \pm 1$ has the same sign for each of the roots x_i. By changing our choice of generator for $H_n(X; \mathbf{Z})$, if necessary, we may assume that $\lambda(f, x_i) = 1$ for each i. Since f is a local homeomorphism at x_0, we may apply Lemma (5.35), to obtain an n-ball neighbourhood B of x_0, an n-ball neighbourhood $C = f(B)$ of y_0, points $x'_1, \dots, x'_{d-1} \in \mathrm{int}\, B$, and a homotopy $\{g_t\}$ as described in the lemma. For any root x of f we have $g_t(x) = y_0$ for all t, from this it's easy to see that two roots of f are Nielsen related if and only they are Nielsen related as roots of g_1. Each of the new roots x'_i of g_1 may be joined by a path in B to x_0 so since $g_1(B) \subset C$ and C is contractible, the new roots x'_i are Nielsen related to x_0. Thus the Nielsen root classes of g_1 are exactly the same as those of f except for α, and this class has been replaced by $\alpha' = \{x_0, \dots, x_{d-1}, x'_1, \dots, x'_{d-1}\}$. For this class we have $\lambda(g_1, x'_i) = -\lambda(g_1, x_i)$ for $i = 1, \dots, d-1$, so we may apply Lemmas (5.32) and (5.33) $d-1$ times as in the proof of (5.5) to eliminate the roots x_i and x'_i and arrive at a new map, homotopic to f that has exactly the same root classes as f

except for α which has been replaced by $\{x_0\}$. Therefore, the new map has one more class consisting of a single root. This completes the proof for Case 2.

Case 3. Let $\widetilde{p}: \widetilde{X} \to X$ be the orientable double covering. We proceed as in Case 2 to reduce the number of roots in each Nielsen class to one. The inductive hypothesis is the same: each class either consists of a single root, or it has mult(f) roots and f is a local homeomorphism at each of these roots. For the inductive step we start with a Nielsen class $\alpha = \{x_0, \dots, x_{d-1}\}$ of the second kind. Select one of the two Nielsen classes of $f \circ \widetilde{p}$ that covers α, and call it $\widetilde{\alpha} = \{\widetilde{x}_0, \dots, \widetilde{x}_{d-1}\}$, where $\widetilde{p}(\widetilde{x}_i) = x_i$ for each i. Let $\widetilde{B}$ be an n-ball neighbourhood of $\widetilde{x}_0$ mapped homeomorphically onto an n-ball neighbourhood C of y_0 by $f \circ \widetilde{p}$, and assume $\widetilde{B}$ small enough so that it has no roots of $f \circ \widetilde{p}$ other than $\widetilde{x}_0$. Apply Lemma (5.35) to find points $\widetilde{x}'_1, \dots, x'_{d-1} \in \widetilde{B}$ and a homotopy $\{\widetilde{g}_t\}$ from f so that the new map $\widetilde{g}_1$ has the same root classes as $f \circ \widetilde{p}$ except for $\widetilde{\alpha}$, which has now been replaced by $\widetilde{\alpha}' = \{\widetilde{x}_0, \widetilde{x}_1, \dots, \widetilde{x}_{d-1}, \widetilde{x}'_1, \dots, \widetilde{x}'_{d-1}\}$, where $\lambda(\widetilde{g}_1, \widetilde{x}'_i) = -\lambda(\widetilde{g}_1, \widetilde{x}_i)$ for each $i = 1, \dots, d-1$. Write $x'_i = \widetilde{p}(\widetilde{x}'_i)$ for each i. Since $f \circ \widetilde{p}|\widetilde{B}$ is a homeomorphism onto its image, so is $\widetilde{p}|\widetilde{B}$, so we may define a homotopy $\{g_t: X \to Y\}$ by

$$g_t(x) = \begin{cases} g_t \circ (\widetilde{p}|\widetilde{B})^{-1}(x) & \text{for } x \in \widetilde{p}(\widetilde{B}), \\ f(x) & \text{for } x \notin \widetilde{p}(\widetilde{B}). \end{cases}$$

Now proceed as in Case 2 of the proof of Theorem (5.5) to eliminate the roots x_i, x'_i one pair at a time (each elimination requires going up to the covering space $\widetilde{X}$ and down again). After $d-1$ such homotopies we are left with a map h homotopic to f that has one more of its Nielsen classes reduced to a single point x_0. The other Nielsen classes will be the same, and each of them will either consist of a single root, or mult(f) roots at each of which h is a local homeomorphism. □

This completes the proof of the two Hopf theorems.

6. Notes and additional topics

The material in Sections 3 and 4 is based largely on Brooks (see [B1], [B3] and [B4]). But see also Kiang in [Ki], and Brown in [Br3] and [Br4] for very readable accounts. The original versions of Theorems (5.5) and (5.7) are due to Hopf [H2] ([12]). The overall discussion in Section 5 is based in broad outline on Brown and Schirmer in [BrS2]. This reference is the best starting point for a more general treatment of Hopf's absolute degree and Theorems (5.5) and (5.7). In their versions of Theorems (5.5) and (5.7), and the definitions of multiplicity and absolute degree the manifolds are not assumed to be compact, and they may

([12]) Much of the material in Sections 3 and 4 was also known to Hopf. At the time the present author wrote [B1]–[B4] he was unfortunately unaware of this earlier work by Hopf.

also have boundary; instead the maps and homotopies are required to be proper. Their proof of Theorem (5.5) is based on microbundles and a version of Whitney's Lemma applicable to topological manifolds. The version of Theorem (5.5) in Lin's paper [Li] requires the manifolds to be compact orientable and differentiable or piecewise linear and makes direct use of Whitney's lemma as stated in [K]. In an effort to keep this chapter self-contained the proof of Theorem (5.5) in Section 5 is a rather extensive rewrite of the proof given by Epstein in [Ep] (whose proof also applies in the noncompact but proper map case). The idea behind our proof of Theorem (5.7) is from Lin [Li].

There are, of course, a large number of topics in root theory not covered here. We close the chapter with a brief description of some of them.

An extremely important topic not treated here is maps of surfaces. Theorem (5.5) is true in dimension $n = 2$, but our proof requires $n > 2$. The original proof in dimension 2 is due to Knesser ([Kn]). A modern proof is given by Skora in [S]. The reader will recall that Theorem (5.7) does not apply in dimension 2, and in fact there are many examples of maps of surfaces where $MR(f, y_0) > NR(f)$. Recent research by Bogatyi, Gonçalves, Kudryavtseva, and Zieschang have produced very nice results on $MR(f, y_0)$ for surface maps that are *not* based on Nielsen theory (see [BGKZ], [BGZ], [GKZ], [GZ]). We have unfortunately not had the space or time to do them justice here.

A topic closely related to root theory is coincidence theory. The most obvious relation is that a *root* of a map $f\colon X \to Y$ at a point $y_0 \in Y$ is also a *coincidence* of the two maps $f, g\colon X \to Y$ where g is the constant map into the point $y_0 \in Y$, so one might expect root theory to be a special case of coincidence theory. However, in coincidence theory one allows both f and g to change by a homotopy, whereas in root theory only f is allowed to change. This can make quite a difference. Any path from y_0 to another point y_1, gives a homotopy of constant maps, so changing the constant map by a homotopy allows us to change the point y_0 to any other point in the path connected space Y. In example (4.25) we saw that changing the point y_0 can change the Nielsen number from one to zero. Remarkably, this problem does not arise when Y is a manifold, even when X is not a manifold. When Y is a manifold, the Nielsen coincidence theory one obtains by allowing both maps to change by a homotopy is essentially the same as when only one is allowed to change [B2]. Another connection arises when Y is a group. In this case a coincidence of $f, g\colon X \to Y$ is a root of $h\colon X \to Y$ at $y_0 \in Y$ where h is the map defined by $h(x) = f(x)g(x)^{-1}$. This connection is explored in [BW].

Brown, Hales, and Stern (see [BrH], [Br1], [BrSt]) have made a series of applications of Nielsen root theory to H-spaces. As an example, a group G is *divisible* if for any $g \in G$ and any $k > 1$ the equation $x^k = g$ has a solution. In 1940, Hopf showed that compact connected Lie groups are divisible. Brown generalizes this result to H-spaces in [Br1] as follows: Let X be an H-space with multiplication $m: X \times X \to X$ and define $m_k: X \to X$ inductively by $m_2(x) = m(x,x)$ and $m_k(x) = m_k(x, m_{k-1}(x))$ for $k > 2$. He shows that if X is a compact manifold without boundary then for any $k \geq 2$ and any $y \in Y$, m_k has at least k^β roots where β is the first Betti number of X (i.e. β is the dimension of the vector space $H_1(X; \mathbf{Q})$). For a very readable account of this research see also [Br3]. Brown, Jiang, and Schirmer conduct a related but different investigation in [BrJS]. There they study roots of the iterates $f^k: X \to X$ of a map $f: X \to X$. They define a homotopy invariant lower bound $NRI_k(f, y_0)$ for the number of *irreducible* roots of f^k at y_0, i.e. roots x_0 such that x_0 is not a root of f^ℓ at y_0 for any $\ell < k$. They show how NRI can be computed in many instances from the homomorphism $f_\#: \pi(X, x) \to \pi(X, f(x))$, and apply the theory to primitive roots of unity in H-manifolds.

In Example (4.26) we saw that for self maps $f: T^n \to T^n$ of the n-torus, we have $NR(f) = |\deg f|$. A natural question is: Under what more general conditions does this happen? Brooks and Odenthal explore this question in [BO], where, for example, it is found that $NR(f) = \deg(f)$ for maps of orientable compact infrasolvmanifolds of the same dimension. The question, and its answers, are extended to nonorientable manifolds in the appendix of [BBS2] where the question becomes: When do we have $NR(f) = \mathcal{A}(f)$? One answer is: whenever f is a map of compact (not necessarily orientable) infrasolvmanifolds of the same dimension.

The class of homotopies we allow is sometimes too large. For example given a map $f: I \to I$ of the unit interval, and a point $y_0 \in I$ it is always possible to deform f so that the new map has no roots at y_0. On the other hand, given a map $f: (I, \mathrm{Bd}\, I) \to (I, \mathrm{Bd}\, I)$ such that $f(\mathrm{Bd}\, I) = \mathrm{Bd}\, I$, there is no way to deform the *pair* map f to be root free (the intermediate value theorem). Brown and Schirmer develop a relative Nielsen theory as follows: Given a map of pairs $f: (X, A) \to (Y, B)$ and a point $y_0 \in Y$, we also have the restriction $f_A: A \to B$, so we have the two Nielsen numbers $NR(f, y_0)$ and $NR(f_A, y_0)$ (the later one is 0 if $y_0 \notin B$). We also have a notion of *relative* essentiality: A Nielsen class of f is relatively inessential if (roughly speaking) it cannot be eliminated by a relative homotopy. Denote the number of relatively essential classes by $NR^+(f, y_0)$. There are some root classes of f at y_0 that contain essential root classes of f_A, denote the number of these by $NR^+(f, f_A, y_0)$. Then Brown and Schirmer define the *relative Nielsen number of roots* of the pair to be

$$NR_{\mathrm{rel}}(f, y_0) = NR(f_A, y_0) + NR^+(f, y_0) - NR^+(f, f_A, y_0).$$

This number is defined and explored by Brown and Schirmer in [BrS1].

The classical Brouwer degree obeys the multiplicative law $\deg(g \circ f) = \deg(f) \cdot \deg(g)$ for the composition of maps $X \xrightarrow{f} Y \xrightarrow{g} Z$ of compact oriented n-manifolds. We also have the multiplicative law $\deg(f_1 \times f_2) = \deg(f_1) \cdot \deg(f_2)$, for the cartesian product of a map $f_1: X_1 \to Y_1$ of n_1-manifolds with a map $f_2: X_2 \to Y_2$ of n_2-manifolds. Are there similarly nice formulas for NR and $\mathcal{A}$? Brooks, Brown and Schirmer address this question in [BBS1], where it is found, for example, that when f, g, f_1, and f_2 are orientable maps of manifolds of the right dimensions, then $\mathcal{A}(g \circ f) = \mathcal{A}(f) \cdot \mathcal{A}(g)$ and $\mathcal{A}(f_1 \times f_2) = \mathcal{A}(f_1) \cdot \mathcal{A}(f_2)$. There are similar multiplicative formulas, some requiring certain correction factors, for nonorientable maps and for NR. The correction factors can all be computed from a knowledge of the fundamental group homomorphisms induced by the maps f, g, f_1, and f_2.

The generalization of a Cartesian product is a locally trivial bundle, $p: M \to X$. Suppose we are given two such bundles $p: M \to X$ and $q: N \to Y$ and a fiber preserving map $f: M \to N$. Then f induces a map $\overline{f}: X \to Y$ such that $\overline{f} \circ p = q \circ f$, and f also restricts to a map $f_x: p^{-1}(x) \to q^{-1}(f(x))$ for each $x \in X$. Suppose that M, X, N, Y are manifolds and $p^{-1}(x)$ and $q^{-1}(y)$ are also manifolds for each $x \in X$ and $y \in Y$, and that $\dim M = \dim Y$, and $\dim X = \dim Y$. What are the relations among the absolute degrees and Nielsen root numbers of f, $\overline{f}$, and f_x? Brooks, Brown, and Schirmer look at this question in [BBS2]. One result (for compact M and N) is that if $NR(f) > 0$ and f is orientable, then $\overline{f}$ is orientable and f_x is orientable for all x, and $\mathcal{A}(f) = \mathcal{A}(\overline{f}) \cdot \mathcal{A}(f_x)$. Again, there are similar multiplicative formulas, some requiring certain correction factors, for nonorientable maps and for NR, and the correction factors can all be computed from a knowledge of the fundamental group homomorphisms induced by the maps f, $\overline{f}$, and f_x.

References

[BGKZ] S. D. Bogatyi, D. Gonçalves, E. A. Kudryavtseva and H. Zieschang, *The minimal number of roots of surface mappings and quadratic equations in free products*, Zametki (to appear).

[BGZ] S. D. Bogatyi and D. Gonçalves and H. Zieschang, *The minimal number of roots of surface mappings and quadratic equations in free products*, Math. Z. **236** (2001), 419–452.

[B1] R. B. S. Brooks, *Coincidences, Roots and Fixed Points*, University of California at Los Angeles, Los Angeles, 1967.

[B2] ———, *On removing coincidences of two maps when only one, rather than both, of them may be deformed by a homotopy*, Pacific J. Math. **39** (1971), 45–52.

[B3] ———, *Certain subgroups of the fundamental group and the number of roots of $f(x)=a$*, Amer. J. Math. **95** (1973), 720–728.

[B4] ———, *On the sharpness of the Δ_1 and Δ_2 Nielsen numbers*, J. Reine Angew. Math. **259** (1973), 101–108.

[BBS1] R. B. S. Brooks, R. F. Brown and H. Schirmer, *The absolute degree and the Nielsen root number of compositions and cartesian products of maps*, Topology Appl. **116** (2001), 5–27.

[BBS2] ______, *The absolute degree and Nielsen root number of a fibre-preserving map*, Topology Appl. **125** (2002), 1–46.

[BO] R. B. S. Brooks and Ch. J. Odenthal, *Nielsen numbers for roots of maps of aspherical manifolds*, Pacific J. Math. **170** (1995), 405–420.

[BW] R. B. S. Brooks and P. N. Wong, *On changing fixed points and concidences to roots*, Proc. Amer. Math. Soc. (1992), 527–533.

[Br1] R. F. Brown, *Divisible H-spaces*, Proc. Amer. Math. Soc. (1970), 185–189.

[Br2] ______, *The Lefschetz Fixed Point Theorem*, Scott, Forseman and Company, Glenview, IL, 1970.

[Br3] ______, *The Topological Theory of Roots*, Lecture Notes for Mathematics, vol. 237, University of California at Los Angeles, Fall 1995, 1995.

[Br4] ______, *A middle-distance look at root theory*, Proceedings of the Workshop on Reidemeister Torsion, Zeta functions, and Nielsen theory, Stefan Banach Research Publications, vol. 45, 1999, pp. 29–41.

[BrH] R. F. Brown and A. W. Hales, *Primitive roots of unity in H-manifolds*, Amer. J. Math. **92** (1970), 612–618.

[BrJS] R. F. Brown, B. Jiang and H. Schirmer, *Roots of iterates of maps*, Topology Appl. **66** (1995), 129–157.

[BrS1] R. F. Brown and H. Schirmer, *Nielsen theory of roots of maps of pairs*, Topology Appl. **92** (1999), 247–274.

[BrS2] ______, *Nielsen root theory and Hopf degree theory*, Pacific J. Math **198** (2001), 49–80.

[BrSt] R. F. Brown and R. J. Stern, *The number of roots in a simply connected H-manifold*, Trans. Amer. Math. Soc. **170** (1972), 499–505.

[D] A. Dold, *Lectures on Algebraic Topology*, 2nd ed., Springer–Verlag, Berlin, 1980.

[ES] S. Eilenberg and N. Steenrod, *Foundations of Algebraic Topology*, Princeton University Press, Princeton, New Jersey, 1952.

[Ep] D. B. A. Epstein, *The degree of a map*, Proc. London Math. Soc. **16** (1966), 369–383.

[GKZ] D. Gonçalves, E. A. Kudryavtseva and H. Zieschang, *Roots of mappings on nonorientable surfaces and equations in free groups*, Manuscr. Math. **107** (2002), 311–341.

[GW] D. Gonçalves and P. Wong, *Obstruction theory, coincidences on nilmanifolds and fibrations over $K(\pi, 1)$-manifolds*, preprint (2003).

[GZ] D. Gonçalves and H. Zieschang, *Equations in free groups and coincidences of mappings on surfaces*, Math. Z. **237** (2001), 1–29.

[GD] A. Granas and Dugundji, *Fixed Point Theory*, Springer–Verlag, New York, 2003.

[H1] H. Hopf, *Zur Topologie der Abbildungen von Mannigfaltigkeiten,* I, Math. Ann. **100** (1928), 579–608.

[H2] ______, *Zur Topologie der Abbildungen von Mannigfaltigkeiten, II*, Math. Ann. **102** (1930), 562–623.

[Hu] Sze-Tsen Hu, *Homotopy Theory*, Academic Press, New York and London, 1959.

[K] M. A. Kervaire, *Geometric and algebraic intersection number*, Comm. Math. Helv. **39** (1965), 271–280.

[Ki] Tsai-han Kiang, *The Theory of Fixed Point Classes*, Springer–Verlag, Berlin, 1989.

[Kn] H. Knesser, *Glättung von Flächenabbildungen*, Math. Ann. **100** (1928), 609–616.

[L] J. M. Lee, *Introduction to Topological Manifolds*, Springer, New York, 2000.

[Li] X. Lin, *On the root classes of mapping*, Acta Math. Sinica (N.S.) **2** (1986), 199–206.

[O] P. Olum, *Mappings of manifolds and the notion of degreee*, Ann. Math. **58** (1953), 458–480.

[S] R. Skora, *The degree of a map between surfaces*, Math. Ann. **276** (1987), 415–423.

[Sp] E. H. Spanier, *Algebraic Topology*, McGraw Hill, New York, 1966.

12. MORE ABOUT NIELSEN THEORIES AND THEIR APPLICATIONS

Robert F. Brown

The topological theory of fixed points has grown into a substantial area of mathematics, a significant portion of which is now called Nielsen theory in honor of a pioneer of the subject whose ideas have proved particularly fruitful. The story of the early development of topological fixed point theory is told in [Br2]. Contributions to this handbook offer detailed descriptions of many of the Nielsen theories. But, when a search of the Mathematical Reviews database turned up almost 300 papers concerned with Niesen theory, I could see that a lot of those papers were about topics that are not included. This chapter will expand, to some extent, the range of Nielsen theories mentioned in this handbook. But, just as practical considerations limited the selection of topics covered elsewhere, the sections that make up this chapter do not by any means fill all the gaps. The selection of the topics, and the amount of detail devoted to each, was determined primarily by the extent of my familiarity with them and certainly not by any judgment as to their present significance or promise for future development. Please note also that results are stated in a form that is convenient for exposition and not necessarily in the greatest generality possible. The rather extensive list of references will permit you to follow up the topics that interest you and there find more complete information.

1. The $\operatorname{mod} H$ Nielsen number

A theorem about multiple fixed points, published by Hirsch in 1940 (see [Hi]), can be viewed as the birth of $\operatorname{mod} H$ Nielsen theory. Let X be a connected finite polyhedron and $f\colon X \to X$ a map. Hirsch proved that if there are distinct lifts $\widetilde{f}_1, \widetilde{f}_2\colon \widetilde{X} \to \widetilde{X}$ to a connected double cover of X such that both have nonzero Lefschetz number, then every map homotopic to f has at least two fixed points. The reason is that, for $j = 1, 2$, the fixed point sets $\operatorname{Fix}(\widetilde{f}_j)$ are nonempty as a consequence of the Lefschetz number hypothesis. Furthermore, their projections

$p(\mathrm{Fix}(\widetilde{f}_j))$ by the covering map $p\colon \widetilde{X} \to X$ are disjoint and are unions of fixed point classes of f. It can be shown that

$$L(\widetilde{f}_j) = 2 \cdot i(f, p(\mathrm{Fix}(\widetilde{f}_j))$$

so each of the sets $p(\mathrm{Fix}(\widetilde{f}_j))$ contains an essential fixed point class and thus $N(f) \geq 2$, from which Hirsch's conclusion follows.

In 1976, McCord ([Mc]) considered more general connected finite regular covering spaces $p\colon \widetilde{X} \to X$ of a finite polyhedron. Letting $H = p_\pi(\pi_1(\widetilde{X}))$, he introduced what is now called the mod H *Nielsen number* $N_H(f)$. McCord defined $N_H(f)$ for a map $f\colon X \to X$ to be the number of conjugacy classes (with respect to covering transformations) of lifts of f to $\widetilde{X}$ that have nonzero Lefschetz numbers and he proved that $N_H(f) \leq N(f)$. The excuse for introducing a homotopy invariant that is a less precise lower bound than the Nielsen number for the number of fixed points of all maps homotopic to f is that it might be easier to compute than $N(f)$, but still give some useful information. McCord illustrated this feature of the mod H Nielsen number by presenting examples of maps of compact, connected manifolds such that $L(f) = 0$ but $N_H(f) = 2$, where $[\pi_1(X) : H] = 2$. The converse Lefschetz theorem, that $L(f) = 0$ implies that f is homotopic to a fixed point free map, was known to hold for a large class of simply-connected polyhedra, including simply-connected triangulated manifolds. It is easy to construct examples on non-simply-connected polyhedra for which the converse Lefschetz theorem fails (see [Br1, p. 111], for instance) but the first manifold examples of the failure of the converse Lefschetz theorem were those of McCord.

A paper of Wang [Wa] published in 1982 presented the mod H theory without the requirement that the covering space be finite. That is, let $f\colon X \to X$ be a map on a compact ANR and suppose H is a normal subgroup of the fundamental group of X that is mapped into itself by the homomorphism induced by f (taking a bit of care about basepoints). It is not required that H be of finite index in the fundamental group. The mod H fixed point classes are then the projections of the conjugacy classes, with respect to covering transformations, of lifts of f to the regular covering space corresponding to H. Equivalently, fixed points x and x' of f are in the same mod H fixed point class if there is a path c from x to x' such that the loop $c(fc)^{-1}$ is in H. The mod H *Nielsen number* $N_H(f)$ is the number of mod H fixed point classes with nonzero fixed point index. An exposition of Wang's theory can be found in the monograph [J5] of Jiang.

A 1982 paper [Y] of You was the first in which the mod H Nielsen number no longer served as merely a lower bound for the classical Nielsen number that might be easier to calculate, but instead it was essential to the development of fixed point theory. Given a space X and a fixed point x of a map $h\colon X \to X$, so that h induces

$h_\pi\colon \pi_1(X,x) \to \pi_1(X,x)$, You defined a subgroup $\mathrm{Fix}(h_\pi)_x$ of $\pi_1(X,x)$ by

$$\mathrm{Fix}(h_\pi)_x = \{\alpha \in \pi_1(X,x) : h_\pi(\alpha) = \alpha\}.$$

Suppose $p\colon E \to B$ is an orientable fiber space where E, B and all the fibers are compact, connected ANRs and the fundamental group of E is abelian. Let $f\colon E \to E$ be a fiber-preserving map which induces $\overline{f}\colon B \to B$ and suppose $N(f) \neq 0$. Choose a fixed point e of f and let $b = p(e)$, then the restriction f_b of f to $p^{-1}(b)$ maps that fiber to itself. Moreover, p induces $p_\pi\colon \pi_1(E,e) \to \pi_1(B,b)$. Let $i\colon p^{-1}(b) \to E$ be the inclusion, inducing $i_\pi\colon \pi_1(p^{-1}(b),e) \to \pi_1(E,e)$, and let H denote the kernel of that homomorphism. Then the mod H Nielsen number $N_H(f_b)$ is well-defined and it appears in the following remarkable formula of You:

$$[\mathrm{Fix}(\overline{f}_\pi)_b\colon p_\pi(\mathrm{Fix}(f_\pi)_e)]N(f) = N(\overline{f}) \cdot N_H(f_b).$$

In the setting of You's result, fixed points x and x' of $f_b\colon p^{-1}(b) \to p^{-1}(b)$ are H-equivalent, for H the kernel of the homomorphism induced by the inclusion $i\colon p^{-1}(b) \to E$, if there is a path c in $p^{-1}(b)$ from x to x' such that ic is homotopic to ifc by a homotopy that fixes the endpoints. In 1992, Woo and Kim introduced in [WK] the following generalization. Let $q\colon X \to Y$ be a map. Fixed points x and x' of a map $f\colon X \to X$ are said to be in the same *q-fixed point class* if there is a path c in X from x to x' such that qc is homotopic to qfc by a homotopy that fixes the endpoints. The q fixed point classes are unions of the usual fixed point classes so a *q-Nielsen number* $N_q(f)$ can be defined as the number of q-fixed point classes with nonzero fixed point index. Then $N_q(f)$ is a homotopy invariant lower bound for $N(f)$. If f is "fiber-preserving" with respect to q in the sense that $x, x' \in X$ with $q(x) = q(x')$ implies $qf(x) = qf(x')$, then $N_q(f) = N_H(f)$ where H is the kernel of the fundamental group homomorphism induced by q. In the case that $q\colon X \to Y$ is a fiber space and the fiber-preserving map $f\colon X \to X$ induces $\overline{f}\colon Y \to Y$, they show that $N_q(f) \leq N(\overline{f})$ and they find conditions for equality.

Also in 1992, Woo and Cho introduced the mod H concept into relative Nielsen theory, see [WC]. Consider (X, A) where X is a finite connected polyhedron and A is a subpolyhedron. In this theory, H is the kernel of the homomorphism of fundamental groups induced by the inclusion of A in X. Let $f\colon (X,A) \to (X,A)$ be a map of pairs, then a fixed point class of $f\colon X \to X$ is called a *common* mod H *fixed point class* if it contains an essential mod H fixed point class of the restriction $f_A\colon A \to A$. By analogy with the relative Nielsen number $N(f; X, A)$ introduced by Schirmer in [S2], and more appropriately called the *Schirmer number*, they defined $N_H(f; X, A)$ which we shall call the mod H *Schirmer number* by

$$N_H(f; X, A) = N(f) + N_H(f_A) - N_H(f; f_A)$$

where $N_H(f_A)$is the mod H Nielsen number and $N_H(f; f_A)$ is the number of common mod H fixed point classes that are essential. They prove that the mod H Schirmer number is a lower bound for the Schirmer number and that it shares the properties of (relative) homotopy invariance, commutativity and homotopy type invariance. They then assume that $p: E \to B$ is a fiber space and that $F_b = p^{-1}(b)$ is a fiber that is mapped to itself by a fiber-preserving map $f: E \to E$. They study conditions under which a product theorem of the form

$$N(f) = N(\overline{f})N_H(f; f_b)$$

holds, where $\overline{f}: B \to B$ is induced by f and $f_b: F_b \to F_b$ denotes the restriction of f.

In their paper [CW] published in 1995, the same authors investigated further the relationship between mod H Nielsen theory and the Schirmer number. Again $f: (X, A) \to (X, A)$ is a map of pairs of finite polyhedra and $f_A: A \to A$ is the restriction of f. The inclusion $i: A \to X$ induces a homomorphism i_π of fundamental groups. Suppose that K and H are normal subgroups of the fundamental groups of A and X respectively such that

$$i_\pi(K) \subseteq H, f_{A\pi}(K) \subseteq K \quad \text{and} \quad f_\pi(H) \subseteq H.$$

Under these conditions, every mod K fixed point class is contained in some mod H fixed point class. A mod H fixed point class is a *common* mod (H, K) *fixed point class* if it contains an essential mod K fixed point class. The Schirmer number $N_K^H(f; X, A)$ in this setting is defined in terms of the appropriate mod H and *modK* Nielsen numbers by

$$N_K^H(f; X, A) = N_H(f) + N_K(f_A) - N_K^H(f; f_A)$$

where $N_K^H(f; f_A)$ denotes the number of essential common mod (H, K) fixed point classes. In the case that K is trivial, this number is written as $N^H(f; X, A)$ and the various theories are related by

$$N_H(f) \leq N_K^H(f; X, A) \leq N^H(f; X, A) \leq N(f; X, A).$$

Additional results on the use of *mod*(H, K) theory to estimate the Schirmer number were published by Chun, Jang and Lee in 1996 [CJL].

A recent (2003) paper of Cardona and Wong (see [CW]) makes use both of a mod H Nielsen number and its relative version, in the setting of fiber-preserving maps. However, a discussion of their results would require a more extensive digression into the Nielsen theory of fiber-preserving maps, a topic which is well covered by P. H. Heath in Chapter III.14.

2. The local Nielsen number

The fixed point index can be thought of as a local version of the fixed Lefschetz number. A non-zero Lefschetz number for a map $f: X \to X$ implies the existence of a fixed point for any map homotopic to f. A nonzero index for f on an open subset U of X implies the existence of a fixed point that lies in U, for any map homotopic to f through homotopies that have no fixed points on the boundary of U and satisfy an appropriate compactness property. In 1981, Fadell and Husseini introduced in [FH1] the local Nielsen number, which produces a homotopy invariant lower bound for the *number* of fixed points on U. The space X is required to be a finite dimensional ANR (but not necessarily compact) and the set of fixed points of a map $f: U \to X$ must be compact, in which case f is called a *compactly fixed map*. Fixed points x and x' of f are in the same (local) fixed point class if there is a path p in U such that p and fp are homotopic in X by a homotopy keeping endpoints fixed. The number of fixed point classes is finite by the compactness of the fixed point set. The classical fixed point index can be applied to define the *local Nielsen number* $n(f, U)$ as the number of fixed point classes with nonzero index. A homotopy $H: U \times I \to X$ is said to be *compactly fixed* if there is a compact subset of U that contains all solutions to $H(x, t) = x$ for all t. Such a homotopy preserves the local Nielsen number, so $n(f, U)$ is a lower bound for the number of fixed points of all maps of U into X that are homotopic to f by a compactly fixed homotopy. Fadell and Husseini introduced a local obstruction index and used it to prove that if X is a connected PL manifold of dimension at least three then, in this setting, the condition $n(f, U) = 0$ is sufficient as well as necessary for the existence of a fixed point free map $g: U \to X$ that is homotopic to f by a compactly fixed homotopy. In a subsequent paper [FH2] of 1983, they showed that this last result does not extend to surfaces by exhibiting a map $f: U \to X$ on a surface such that $n(f, U) = 0$ yet every map homotopic to f by a compactly fixed homotopy has at least two fixed points. By using braid theory arguments, additional examples of surfaces X for which there are maps $f: U \to X$ such that $n(f, U) = 0$ yet every map homotopic to f by a compactly fixed homotopy has fixed points were obtained by Gonçalves in 1986 (see [Gon]).

In 1992, Wong showed in [W] that the local Nielsen number could be related in a useful way to another, apparently quite different, Nielsen theory. Suppose (X, A) is a pair consisting of a connected finite polyhedron and a subpolyhedron, and a map of pairs $f: (X, A) \to (X, A)$ is given. The *extension Nielsen number* $N(f|f_A)$, introduced in [BGS], is a lower bound for the number of fixed points among all maps of pairs $g: (X, A) \to (X, A)$ such that $g(x) = f(x)$ for all $x \in A$, homotopic to f through maps with the same property. Thus, for $f_A: A \to A$ the restriction of f, the map g is, like f, an extension to X of f_A, as is each stage of the homotopy. The number $N(f|f_A)$ is defined to be the number of fixed point classes $\mathbb{F}$ of f that do not intersect the boundary of A and that have the property that their index does not equal the index of the fixed point class $\mathbb{F} \cap A$ of f_A. Even though this definition of the extension Nielsen number does not resemble that of the local Nielsen number, Wong showed that the concepts can be related. For instance, he proved that if $X - A$ is not a 2-manifold and has no local cut points, A can be by-passed and $f\colon (X, A) \to (X, A)$ is compactly fixed on $X - A$, then $N(f|f_A)$ equals the local Nielsen number $n(f|X - A, X - A)$. Results of this kind allow the existing computational theory for the local Nielsen number to be used for the extension Nielsen number as well.

Another connection between the local Nielsen number and a Nielsen theory concerned with extensions of maps was presented by Zhao in 1994 in the paper [Z1]. Consider the universal covering space $p\colon \widetilde{X} \to X$ of a connected finite polyhedron X. For C an open subset of X and $\widetilde{C}$ a component of $p^{-1}(C)$, a *lift* of the restriction $f|C$ of a map $f\colon X \to X$ is a map $\widetilde{f_C}\colon \widetilde{C} \to \widetilde{X}$ such that $p\widetilde{f_C} = (f|C)p$. The local fixed point classes in C, in the sense of [FH1], are the projections of the fixed point sets of lifts. Now suppose $f\colon (X, A) \to (X, A)$ is a map of pairs where A is a subpolyhedron and take C to be a component of $X - A$. There is a unique lift $\widetilde{f}$ of f to the universal covering space whose restriction to $\widetilde{C}$ is $\widetilde{f_C}$. The local fixed point class $\mathbb{F}$ that is the image under p of the fixed point set of $\widetilde{f_C}$ is said to be *special* if the corresponding lift $\widetilde{f}$ has no fixed points on the boundary of $\widetilde{C}$ and *non-special* otherwise. The number of essential special local fixed point classes in $X - A$ is denoted $SN(f|f_A)$. It is an extension Nielsen number because a map $g\colon (X, A) \to (X, A)$ whose restriction to A is f_A and that is homotopic as such to f has at least $SN(f|f_A)$ fixed points in $X - A$; see [Z2]. Moreover, if A can be by-passed, then $SN(f|f_A) = N(f|f_A)$, the extension Nielsen number of [BGS]. Zhao showed in [Z1] that the local Nielsen number $n(f|X - A, X - A)$ is bounded below by $SN(f|f_A)$ and bounded above by the sum of $SN(f|f_A)$ and the number of non-special fixed point classes of f in $X - A$.

A 1994 paper [FHa] of Fares and Hart placed the local Nielsen number in the context of the Reidemeiester trace with the purpose, as in the classical theory, of relating that Nielsen number to a trace-like concept as an aid to calculation. They still consider a compactly fixed map $f\colon U \to X$ with U open in X, where now X is a connected, finite dimensional locally compact polyhedron. A compact subset K of U containing the fixed points of f in its interior is chosen that has the property that the local fixed point classes using paths in K are the same as those, as in [FH1], that use paths in U. Choose lifts $\widetilde{f}$ and $\widetilde{i}$, to the universal covering spaces, of the restriction $f|K$ of f to K and of the inclusion of K in X. A *local setting* for (f, U) consists of those choices: K, $\widetilde{f}$ and $\widetilde{i}$. Let $\widetilde{\pi}_K$ and $\widetilde{\pi}_X$ denote the groups of covering transformations of K and X respectively, then $\widetilde{f}$ and $\widetilde{i}$ induce homomorphisms $\phi, \xi: \widetilde{\pi}_K \to \widetilde{\pi}_X$ by the formulas

$$\widetilde{f}(\tau\widetilde{y}) = \phi(\tau)\widetilde{f}(\widetilde{y}) \quad \text{and} \quad \widetilde{i}(\tau\widetilde{y}) = \xi(\tau)\widetilde{i}(\widetilde{y})$$

for $\tau \in \widetilde{\pi}_K$ and $\widetilde{y} \in \widetilde{K}$. The *Reidemeister action* of $\widetilde{\pi}_K$ on $\widetilde{\pi}_X$ induced by f is defined for $\tau \in \widetilde{\pi}_K$ and $\alpha \in \widetilde{\pi}_X$ by

$$\tau \cdot \alpha = \xi(\tau)\alpha\phi(\tau^{-1}).$$

Representing a lift of f by $\alpha\widetilde{f}$ for some $\alpha \in \widetilde{\pi}_X$, where $\widetilde{f}$ is the lift of the setting, the *coincidence set* of $\alpha\widetilde{f}$ is defined to be

$$\mathrm{Coin}(\alpha\widetilde{f}, \widetilde{i}) = \{\widetilde{x} \in \widetilde{K}: \alpha\widetilde{f}(\widetilde{x}) = \widetilde{i}(\widetilde{x})\}.$$

The projection of $\mathrm{Coin}(\alpha\widetilde{f}, \widetilde{i})$ to K is a local fixed point class N_K^α. Let W denote a subset of $\widetilde{\pi}_X$ consisting of one element from each orbit under the Reidemeister action. The NR (for Nielsen–Reidemeister) *chain* of f with respect to the chosen local setting is defined by

$$N_K(f; \widetilde{f}, \widetilde{i}) = \sum_{\alpha \in W} i(N_K^\alpha)[\alpha]$$

where $i(N_K^\alpha)$ is the fixed point index of the class and $[\alpha]$ denotes the orbit containing α. Thus, $N_K(f; \widetilde{f}, \widetilde{i})$ is an element of $\mathbb{Z}(R)$, the free abelian group generated by the orbits of the Reidemeister action. The local Nielsen number of [FH1] is the number of non-zero terms in $N_K(f; \widetilde{f}, \widetilde{i})$.

The following description of the local Reidemeister trace, also known as the local generalized Lefschetz number, is based on [H1]. We have the universal covering space $p: \widetilde{X} \to X$. Let $C_q(\widetilde{K})$ and $C_q(p^{-1}(K))$ be groups of oriented simplicial integer chains, which are modules over the group ring $\mathbb{Z}(\widetilde{\pi}_X)$. The lift $\widetilde{f}$ induces a $\mathbb{Z}(\widetilde{\pi}_X)$-homomorphism

$$\Phi_q: C_q(\widetilde{K}) \to C_q(p^{-1}(K))).$$

There is a $\mathbb{Z}(\widetilde{\pi}_X)$-module basis B_q for $C_q(\widetilde{K})$ such that $\widetilde{i}(B_q)$ is in one-to-one correspondence with B_q and it is a $\mathbb{Z}(\widetilde{\pi}_X)$-module basis for $C_q(\widetilde{K})$. Identifying an element of B_q with its $\widetilde{i}$ image produces a trace-like element for Φ_q that lies in $\mathbb{Z}(\widetilde{\pi}_X)$. Sending each element of $\mathbb{Z}(\widetilde{\pi}_X)$ to its Reidemeister orbit sends the trace of Φ_q to an element of $\mathbb{Z}(R)$ that is denoted by $T^R(\Phi_q)$. The *local Reidemeister trace* corresponding to the chosen local setting is then defined by

$$L(f; \widetilde{f}, \widetilde{i}) = \sum_q (-1)^q T^R(\Phi_q).$$

The main result is that if X is a manifold of dimension at least three, then

$$N_K(f;\widetilde{f},\widetilde{\imath}) = L(f;\widetilde{f},\widetilde{\imath}) \in \mathbb{Z}(R).$$

That means that, when X is such a manifold, if $L(f;\widetilde{f},\widetilde{\imath})$ can be expressed as an integer linear combination of distinct Reidemeister orbits, then the local Nielsen number is the number of such orbits with non-zero coefficient. A 1995 paper [H2] of Hart presented techniques for using the local Reidemeister trace to calculate the local Nielsen number. These can be applied, for instance, when $\widetilde{\pi}_K$ is finite. The theory presented in [FH1] has been simplified for the purposes of this description. Covering spaces other than the universal covering space are used there in order to include the mod H version of the local Nielsen number, for H a normal subgroup of $\widetilde{\pi}_X$ (that need not be invariant under the fundamental group homomorphism induced by the map).

The examples, given in [FH1] and [Gon], of maps $f: U \to X$ on surfaces where $n(f, U) = 0$ yet f cannot be homotoped to map that is fixed point free on U can be viewed as local versions of the examples of Jiang in [J2] of maps of surfaces with Nielsen number zero that cannot be homotoped to fixed point free maps. None of the examples in [FH1] and [Gon] embed U in X and it follows from [JG] that an embedding of a surface (with boundary) in itself with zero Nielsen number is homotopic to a fixed point free map. Therefore, it was natural to ask whether embeddings of open sets U in surfaces X with $n(f, U) = 0$ would be homotopic to fixed point free maps. However, in 2000 Ferrario and Goncalves provided in [FG] examples to show that, on all but finitely many surfaces X, there exist homeomorphisms $f: X \to X$ and an open subset U of X for which $n(f, U) = 0$, but no compactly fixed homotopy can eliminate all the fixed points on U.

3. Nielsen theories for multivalued functions

The topological fixed point theory of multivalued functions is a very well-developed area that was the subject of the recent (1999) book [G]. What I want to focus on here are the Nielsen theories that exist in the multivalued function setting. Denoting by 2^X the set of subsets of a set X, a *fixed point* of a function $\phi: X \to 2^X$ is a point $x \in X$ such that $x \in \phi(x)$. A Nielsen theory in this setting should obtain a lower bound on the number of points satisfying $x \in \psi(x)$ among $\psi: X \to 2^X$ homotopic to ϕ is some appropriate sense. In order to obtain results about such fixed points it is of course necessary to impose topological conditions on the function ϕ and also on the subsets $\phi(x)$. One such hypothesis is that the sets $\phi(x)$ be closed and we will always make that assumption. A commonly used continuity hypothesis on a multivalued function is that it be *upper semicontinuous* (u.s.c.), that is, if $\phi(x) \subset V$ where V is open in X, then there is an open neighbourhood U of x such that $\phi(U) \subset V$.

The first Nielsen theory for multivalued functions was developed in 1975 by Schirmer (see [S3]) for what she called *small* multivalued functions ϕ on a connected finite polyhedron X. By that she meant that $\phi(x)$ must lie in the star of a vertex of X for each $x \in X$. The previous year in [S4] she had published a simplicial approximation theorem for small u.s.c. multivalued functions. She proved that there is a (single-valued) simplicial map f such that, for v any vertex of an appropriately chosen subdivision, $\phi(v)$ is in the star of $f(v)$ with respect to the original triangulation of X. She defined the Nielsen number $N(\phi)$ to be the usual Nielsen number of f and she proved that it is independent of the choice of simplicial approximation. If the sets $\phi(x)$ are acyclic, that is, have trivial reduced Čech rational cohomology, then $N(\phi)$ is a sharp fixed point lower bound in the sense that there is single-valued map g that is related to ϕ by a homotopy of small acyclic valued maps such that g has exactly $N(\phi)$ fixed points, provided that X is of type S. A finite polyhedron X is type S if X is of dimension at least three and the boundary of the star of every vertex is connected.

Schirmer developed another Nielsen theory, for *n-valued* functions on connected finite polyhedra, in 1984 (see [S5]). In that theory, $\phi: X \to 2^X$ is such that $\phi(x)$ consists of exactly n points for all $x \in X$. The function ϕ is now required to be *continuous*, that is, it is u.s.c. and lower semicontinuous as well, which means that, for each $x \in X$ and open subset V of X that intersects $\phi(s)$, there exists an open neighbourhood U of x such that $\phi(x')$ intersects V for all $x' \in U$. The key to this theory is a "splitting lemma" from a previous paper [S6], that follows from a much earlier (1957) result of O'Neill in [O]. The lemma states that a continuous n-valued function ϕ on a simply-connected space can be split into n single-valued maps. Thus the lemma implies that the n-valued map ϕ on the polyhedron X splits locally. In [S6] it was also proved that ϕ can be homotoped so that its fixed points are finite in number and lie in maximal simplices of X. With the fixed points thus isolated then, splitting ϕ in a neighbourhood of a fixed point x, exactly one of the n distinct maps has x as a fixed point and the index of ϕ at x is defined to be the usual fixed point index of that map at x. Fixed points x and x' of ϕ are defined to be equivalent if there is a path p between them such that some splitting of ϕ restricted to p includes a map for which x and x' are equivalent fixed points in the sense of traditional Nielsen fixed point theory. Thus fixed point classes can be defined for a continuous n-valued function and the Nielsen number $N(\phi)$ defined to be the number of classes of nonzero index. Schirmer proved in [S5] that this procedure produces a well-defined lower bound for the number of fixed points of all continuous n-valued functions homotopic to ϕ by an n-valued continuous homotopy. The following year, she proved in [S7] that Wecken's theorem for manifolds extends to the setting of continuous n-valued functions. That is, if

$\phi\colon X \to 2^X$ is a continuous n-valued function on a compact connected triangulated manifold of dimension at least three, then there is a continuous n-valued function ϕ' homotopic to ϕ such that ϕ' has exactly $N(\phi)$ fixed points.

A 1985 monograph [Dz] by Dzedzej contains an extension of Schirmer's Nielsen theory for n-valued maps to (continouous) maps $\phi\colon X \to 2^X$ on compact connected ANRs such that $\phi(x)$ consists of n acyclic sets rather than n points, provided that each component of $\phi(x)$ has a neighbourhood with the property that every loop in it can be contracted, keeping endpoints fixed, in the entire space. The focus of the monograph is a fixed point index theory for a class of multivalued functions considerably more general than these. Thus, to define a Nielsen number, what is required is an appropriate definition of fixed point class. Let $p\colon \widetilde{X} \to X$ be the universal covering space of the compact ANR on which ϕ is defined and consider $\phi p\colon \widetilde{X} \to X$. The splitting lemma of Schirmer in [S6] applies in this case because it requires only that each $\phi(x)$ consist of exactly n connected sets. Thus ϕp splits into n disjoint acyclic valued maps ψ_j for $j = 1, \dots, n$, each with the property that, for all $x \in X$, there is a neighbourhood of $\psi_j(x)$ such that every loop in it can be contracted in X. With that hypothesis, a result of Jezierski (see below) implies that the ψ_j have lifts to $\widetilde{X}$, that is, there are acyclic-valued functions $\widetilde{\psi}_j$ on $\widetilde{X}$ such that $p(\widetilde{\psi}_j(\widetilde{x})) = \psi_j(\widetilde{x})$ for all $\widetilde{x} \in \widetilde{X}$. This allowed Dzedzej to extend the covering space definition of fixed point classes in this case: Fixed points x and x' of ϕ are in the same class if there is some lift $\widetilde{\psi}_j$ of one of the ψ_j such that both x and x' lie in the image under p of the fixed point set of $\widetilde{\psi}_j$.

The concept of Jezierski that Dzedzej made use of to define his Nielsen theory was published in 1987 (see [Je]). Jezierski defined an *m-map* to be an upper semicontinuous multivalued function ϕ from a space Y to the subsets of a space X such that each subset $\phi(x)$ has a neighbourhood with the property that every loop in it can be contracted in X keeping endpoints fixed. Suppose X admits a universal covering space $p\colon \widetilde{X} \to X$. Jezierski proved that if Y is simply-connected, then for $y_0 \in Y$ and $\widetilde{x}_0 \in \widetilde{X}$ with $p(\widetilde{x}_0) \in \phi(y_0)$, there is a unique lift $\widetilde{\phi}$ of ϕ, that is, $\widetilde{\phi}$ is an m-map from Y to the subsets of $\widetilde{X}$ such that $p\widetilde{\phi}(x) = \phi(x)$ for $x \in X$ and with $\widetilde{x}_0 \in \widetilde{\phi}(y_0)$. Then the definition of fixed point class that Dzedzej used makes sense for any m-map ϕ on a space X with a universal covering space. Since $\widetilde{X}$ is simply-connected, the m-map ϕp has lifts and two fixed points of ϕ are in the same class if they lie in the image under p of the fixed points of a lift of ϕp. If X is a compact connected ANR and ϕ is an m-map on X such that each $\phi(x)$ is acyclic with respect to homology with rational coefficients, then the index theory of Dzedzej can be applied to define a Nielsen number of ϕ that is a homotopy invariant (with respect to homotopies that are m-maps) lower bound for the number of fixed points of ϕ.

A detailed development of this Nielsen theory can be found in the book [G] of Górniewicz.

In 1988, Górniewicz, Granas and Kryszewski introduced in [GGK1] and [GGK2] a class of multivalued functions on compact ANRs that they called J-maps. Denote the set of points in a metric space X that are within ε of a subset A by $U(A, \varepsilon)$. An u.s.c. function $\phi\colon X \to Y$ such that each $\phi(x)$ is compact and connected is a *J-map* if, given $x \in X$ and $\varepsilon > 0$, there exists $0 < \delta \leq \varepsilon$ with the property that, for all $n \geq 1$, the inclusion-induced homomorphism

$$j_\pi\colon \pi_n(U(\phi(x), \delta), y_0) \to \pi_n(U(\phi(x), \varepsilon), y_0)$$

is trivial, for any basepoint y_0 in $U(\phi(x), \delta)$. They proved that, again for any $\varepsilon > 0$, a J-map ϕ can be ε-approximated by a single-valued map $f\colon X \to Y$ in the sense that the graph of f lies in the ε-neighbourhood of the graph of ϕ. Moreover, any two such single-valued approximations are homotopic through these approximations. They defined an *A-map* $\Psi\colon X \to X$ to be a multivalued function that can be expressed as a finite composition of J-maps (not necessarily with X itself as domain or range) that has the property introduced by Jezierski, that is, for each $x \in X$ there is a neighbourhood of $\Psi(x)$ with the property that every loop in it can be contracted in X keeping endpoints fixed. The class of A-maps includes not only those u.s.c. functions such that $\Psi(x)$ is compact and contractible but also those for which

$$\Psi(x) = \bigcap_{j=1}^{\infty} K_j$$

where the K_j are compact and contractible with $K_{j+1} \subset K_j$ (that is, the $\Psi(x)$ are R_δ-sets). The next year, Kryszewski and Miklaszewski proved in [KM] that a valid Nielsen theory for A-maps can be obtained by replacing the J-maps in a composition by their single-valued approximations. That is, suppose we write an A-map in the form $\Psi = \phi_n \dots \phi_1$ where each $\phi_j\colon X_j \to X_{j+1}$ is a J-map (with $X_1 = X_{n+1} = X$). Given $\varepsilon > 0$ there are single-valued ε-approximations $f_j\colon X_j \to X_{j+1}$. The result of Kryszewski and Miklaszewski is that, for ε small enough, the definition

$$N(\Psi) = N(f_n \dots f_1)$$

is independent of all the choices made and it gives a homotopy-invariant, in an appropriate sense, lower bound for the number of fixed points of Ψ.

In 1990, Schirmer (see [S8]) introduced a *bimap*, which she defined to be a continuous (that is both upper and lower semicontinuous) multivalued function $\phi\colon X \to Y$ such that, for each point x in a space X, the image $\phi(x)$ consists of either one or two points of Y and she introduced a fixed point index for such functions. A *bihomotopy* is a continuous family of bimaps $\phi_t: X \to Y$. A bimap $\alpha: I \to X$ is called a *bipath* in X. A bihomotopy $\{\alpha_t\}$ of bipaths is required to fix the endpoints, that is, $\alpha_t(0) = \alpha_0(0)$ and $\alpha_t(1) = \alpha_0(1)$ for all $t \in I$. The next year, she produced a Nielsen number for such bimaps in [S9]. Fixed points x and x' of a bimap $\phi: X \to X$ are in the same *fixed point class* if there is a single-valued path $p: I \to X$ such that the bipath ϕp is bihomotopic to a bipath α such that $\alpha(t) = \{p(t), q(t)\}$ for some single-valued path q. A bimap may be viewed as an n symmetric product map, that is, a map from a space X to the space of unordered sets of n not necessarily distinct points of X in the case $n = 2$. The definition of fixed point classes for a bimap is equivalent to a fixed point class concept for symmetric product maps introduced by Masih in 1979 (see [Ma]). Schirmer defined the Nielsen number $N(\phi)$ in the usual way, as the number of fixed point classes of nonzero index. However a result of Miklaszewski in 1990 (see [Mi]) states that all the fixed points of a symmetric product map must lie in a single fixed point class, so $N(f) \leq 1$. This result would imply that the Nielsen theory of bimaps is uninteresting were it not for the fact that Schirmer proved in [S9] that a Wecken theorem for manifolds holds in her theory. Thus any bimap ϕ on a compact connected triangulated manifold of dimension at least three is bihomotopic to a bimap ψ with a single fixed point and, if the fixed point index of ϕ on X is zero, then ψ can be made fixed point free.

4. Fixed point theory for differentiable functions

Suppose that M is a compact connected differentiable manifold. If we study the fixed points of maps homotopic to a map $f: M \to M$, does it make any difference if we require that f be differentiable and we consider only the differentiable maps homotopic to it? It turns out that the answer can be either yes or no, depending on the circumstances.

It is well known that, in general, the Lefschetz number counts fixed points only in an algebraic sense. That is, if the value of the Lefschetz number is nonzero then every map homotopic to the given one has a fixed point, but usually the value of the Lefschetz number tells us nothing about how many fixed points those maps have. For instance, let $p_2: S^1 \to S^1$ be defined by $p_2(z) = z^2$. As Shub and Sullivan pointed out in 1974 (see [SS]), if the suspension of the p_2 to the 2-sphere is modified to a map f that keeps the poles fixed but moves down the sphere along great circles, then the Lefschetz number of the nth iterate is $L(f^n) = 2^n$ yet the number of fixed points of f^n is two, no matter what the value of n. That map f fails to be differentiable at the poles and they showed that, if only differentiable functions are considered, than this sort of behavior cannot occur. They proved that, for a *differentiable* function $f: M \to M$ on a closed manifold, if the sequence

$\{L(f^n)\}$ is unbounded as n goes to infinity, then the number of fixed points of the f^n is unbounded also. What they actually proved is that, if f is differentiable and x is an isolated fixed point of f^n for all n, then the sequence of values of the fixed point indices $\{i(f^n, x)\}$ is bounded.

The map $p_k\colon S^1 \to S^1$ defined by map $p_k(z) = z^k$, used for an example in the Shub–Sullivan paper when $k = 2$, also played an important role in the discovery of another part of fixed point theory in which differentiability makes a difference. For every k there is a boundary-preserving map $f_k\colon (D^2, S^1) \to (D^2, S^1)$ on the disc which is p_k on S^1, and therefore has $|k-1|$ fixed points there, and f_k has no fixed points on the interior of the disc (see [BG]). However, it was proved by Brown, Greene and Schirmer in 1989 [BGS] that any *differentiable* map of D^2 that restricts to p_k on S^1 for $k \geq 2$ must have at least one fixed point on the interior of the disc. An elementary exposition of this result was published as [BG]. The way differentiability enters is in regard to the difference between the index $i(p_k, x)$ of an isolated fixed point x in S^1 viewed as a fixed point of p_k and the index $i(f, x)$ of x for some map f on D^2 that extends p_k. For continuous functions in general, these integers need not be related but, when f is differentiable, the only possible values of $i(f, x)$ are zero and $i(p_k, x)$. Now $i(p_k, x) = -1$ if $k \geq 2$ but all the $i(f, x)$ must sum to one, the Lefschetz number of f, so there must be at least one fixed point of f that is not in S^1.

This behavior on the 2-disc raised the question of whether what had been observed was a peculiarity of that space or instead was a more widespread phenomenon demonstrating the difference between the fixed point theory of continuous and of differentiable functions. Schirmer explored this question in 1992 [S1]. Let M be a compact connected differentiable manifold with boundary ∂M, let $f\colon (M, \partial M) \to (M, \partial M)$ be a differentiable boundary-preserving map and let $f_\partial\colon \partial M \to \partial M$ be the restriction of f. It had been shown in [BGS] that if, at each fixed point x of f_∂, the transformation of the tangent space $df_{\partial x} - \mathrm{id}\colon T_x(\partial M) \to T_x(\partial M)$ is nonsingular, then either $i(f, x) = 0$ or $i(f, x) = i(f_\partial, x)$. So the setting of boundary-preserving maps that satisfy that hypothesis on the tangent space, called *transversely fixed on the boundary*, was the natural one for Schirmer's investigation. Lower bounds for the number of fixed points in the interior of M for continuous and differentiable extensions of $\overline{f}$ had been introduced in [BGS] and shown to be sharp bounds on manifolds of dimension at least three. The lower bound $N(f|f_\partial)$ in the continuous setting is just the number of essential fixed point classes of f that do not intersect ∂M. In the differentiable case, the bound $N^1(f|f_\partial)$ is the number of essential fixed point classes $\mathbb{F}$ of f that are not representable on ∂M, that is, there is no subset S of $\mathbb{F} \cap \partial M$ such that $i(f, \mathbb{F}) = i(f_\partial, S)$. In the disc example, $N^1(f|f_\partial) = 1$ whereas $N(f|f_\partial) = 0$, so the problem was to learn under what circumstances $N^1(f|f_\partial) - N(f|f_\partial)$ could be nonzero. Schirmer's main result in this regard requires that the transversely fixed map f_∂ satisfy another, very natural, condition which makes it what she called a *sparse map*, namely, that the number of fixed points of f_∂ equal the sum of the $|i(f_\partial, \mathbb{F}_\partial)|$ summed over all the fixed point classes $\mathbb{F}_\partial$ of f_∂. She showed that if f_∂ is sparse then $N^1(f|f_\partial) - N(f|f_\partial)$ is the number of essential common fixed point classes of f that are not transversely common. A common fixed point class of f is one that contains at least one essential fixed point class of f_∂. The definition of transversely common is more complicated. Consider the essential fixed point classes of f_∂ contained in a fixed point class $\mathbb{F}$ of f. Let $u(\mathbb{F})$ (respectively $\ell(\mathbb{F})$) be the sum of the indices of those classes that have positive (respectively negative) index, where both are set to zero if no such fixed point class of f_∂ exists, then $\mathbb{F}$ is transversely common if $\ell(\mathbb{F}) \leq i(f, \mathbb{F}) \leq u(\mathbb{F})$. This characterization allowed Schirmer to demonstrate that $N^1(f|f_\partial) - N(f|f_\partial) \neq 0$ occurs quite frequently. For instance, she proved that that the 2-disc example extends to the n-ball D^n: if $f\colon (D^n, S^{n-1}) \to (D^n, S^{n-1})$ is a map such that f_∂ is a sparse map of degree d where $(-1)^n d \geq 2$, then $N^1(f|f_\partial) - N(f|f_\partial) = 1$. Moreover, for each positive integer n she exhibited a boundary-preserving map f_n of the solid torus $S^1 \times D^2$ for which $N^1(f_n|f_{n\partial}) - N(f_n|f_{n\partial}) = n$.

But not all fixed point behavior is sensitive to differentiability requirements. In an appendix to Schirmer's paper [S1], Greene proved that if M is a C^∞ manifold and $f\colon M \to M$ is continuous, then there is a differentiable function homotopic to f whose fixed point set is identical to that of f. Thus the set of subsets of M that can be the fixed point sets of maps in a given homotopy class is identical to the set of subsets that can be the fixed point sets of the differentiable functions within that class. In 1981, Jiang showed in [J1] that minimization of fixed point sets for the maps of a differentiable manifold in a homotopy class can be carried out smoothly. That is, he proved that in Wecken's theorem that, for a compact connected differentiable manifold M of dimension at least three, any map $f\colon M \to M$ can be homotoped to a map g with exactly $N(f)$ fixed points, the map g can always be constructed to be a differentiable one. Thus the differentiability hypothesis has no more effect on fixed point minimization than it has on fixed point realization. In fact, it is not necessary (or even useful) to assume that the original map f is differentiable. Moreover, this sort of differentiable minimization can be obtained in a relative version, as Greene and Schirmer proved in 1995, see [GS]. Let M be a compact connected manifold and A a submanifold, both of dimension at least three, such that A can be by-passed in M. Schirmer had proved in [S2] that any map of pairs $f\colon (M, A) \to (M, A)$ is homotopic as such to a map g with the number of fixed points equal to the Schirmer number $N(f; M, A)$. It is shown in [GS] that, if M and A are differentiable manifolds and M is connected, then the fixed point minimizing map of pairs g can be constructed to be differentiable under these hypotheses, provided that A is "neatly paired" with M. The neatly paired requirement is that each component of A that does not lie in the interior of M have the properties $\partial A = A \cap \partial M$ and A is not tangent to ∂M at any point of ∂A.

5. Applications of Nielsen theory to differential and integral equations

Leray gave a talk during the 1950 International Congress of Mathematicians about the applications of topology to analysis (see [L1]). He described the work of Nielsen and Wecken that gives a lower bound for the number of fixed points of a map and he suggested that, since solutions to analytic problems can often be formulated as fixed points of functions, that work should furnish a basis for results establishing the existence of multiple solutions to analytic problems. Multiple solution results are often important in analysis, for instance there are problems where a constant function is a trivial solution so fixed point existence theorems are not sufficient to establish the existence of nontrivial solutions. However, he pointed out that the setting of the Nielsen fixed point theory of the time, which was concerned with maps on finite polyhedra or compact ANRs, was not appropriate for analytic problems.

The fixed point setting that describes many types of analytic problems is that of a map $f\colon X \to X$ where X is a connected ANR but X is not compact or even locally compact. However, the type of map f that arises does introduce some compactness into the problem. Specifically, it can be assumed that f is *completely continuous* which means that if S is a bounded subset of X, then there is a compact subset K of X such that $f(S) \subset K$. The first step in obtaining a satisfactory Nielsen fixed point theory for completely continuous maps on ANRs was taken by Leray himself in 1959, in a generalization of the global Lefschetz fixed point theory [L2]. However, the crucial step was taken by Granas in 1972 in his [Gr] when he extended Leray's concept to a fixed point index. Scholz then showed in [Sz] that, in a very general setting, the classical definition of fixed point class could be combined with a suitable index to obtain a Nielsen number that is a lower bound for the number of fixed points of completely continuous maps homotopic, in an appropriate sense, to f.

Thus a Nielsen number $N(f)$ of the type discussed by Scholz would prove the existence of multiple fixed points of a completely continuous map $f\colon X \to X$ of an ANR, provided $N(f) > 1$. However, the usual topological methods for analytic problems lead to self-maps of simply-connected, in fact contractible, ANRs. I will now briefly describe one of these methods because it was the one that suggested a way to overcome the contractible space problem. Suppose X is a normed linear space. If a completely continuous function $f\colon X \to X$ maps some ball C in X into itself, then the Schauder fixed point theorem implies that f has a fixed point, in C. Although such a ball can be found in some circumstances, it greatly restricts the applicability of the theory to require its existence. The Schauder theorem becomes quite a powerful tool of nonlinear analysis when the following less restrictive condition is used. The map f is said to satisfy the *Leray–Schauder boundary condition*, introduced in 1934 (see [LS]), if there exists a positive real number R such that $\|x\| = R$ implies $f(x) \neq \lambda x$ for all real numbers $\lambda > 1$. If a completely continuous map satisfies the Leray–Schauder boundary condition, then it has a fixed point.

To see why, let C be the ball of radius R in X and define a retraction $\rho\colon X \to C$ by setting $\rho(x) = Rx/\|x\|$ for all x with $\|x\| \geq R$. Then $\rho f|C\colon C \to C$ has a fixed point by the Schauder fixed point theorem and it is not difficult to prove that a fixed point of $\rho f|C$ is also a fixed point of f when the Leray–Schauder condition is satisfied. (See [GD] for the details.) The Leray-Schauder boundary condition lends itself well to analytic problems. For instance, if f has "sublinear growth" (roughly, it behaves something like $f(x) = x^q$ for $0 < q < 1$) then for R big enough we have $\|x\| = R$ implies $\|f(x)\| < R$ so the Leray–Schauder boundary condition is satisfied.

On the other hand, since the ball C is contractible, $N(\rho f|C) = 1$ so Nielsen theory cannot tell us anything more about the fixed points of f. But, knowing that it is too restrictive to require that a map $f\colon X \to X$ on a normed linear space send a ball to itself, we can well imagine that the necessity of finding a non-simply-connected subset of X that is mapped to itself by f would demand far too much information for such a theory to be useful in analysis. In 1988, Brown introduced in [Br4] a modification of the Leray–Schauder boundary condition that replaced C by a non-simply connected space and thus allowed for the possibility of Nielsen numbers greater than one. The modification can be motivated by taking X to be the plane so there are rather natural non-simply-connected subsets, namely, annuli. To establish notation, let $0 < r < R$ and define $A_{r,R} = \{x \in \mathbb{R}^2 : r \leq \|x\| \leq R\}$. If we delete the origin $\mathbf{0}$ from the plane, then we can extend the definition of the retraction used in connection with the Leray–Schauder boundary condition to retract onto $A_{r,R}$. Define $\rho\colon \mathbb{R}^2 - \mathbf{0} \to A_{r,R}$ by setting $\rho(x) = rx/\|x\|$ if $0 < \|x\| \leq r$ as well as $\rho(x) = Rx/\|x\|$ for all x with $\|x\| \geq R$. If $f\colon \mathbb{R}^2 \to \mathbb{R}^2$ is a map and there exist r, R such that $f(A) \subset \mathbb{R}^2 - \mathbf{0}$ where $A = A_{r,R}$, then conditions for x with $\|x\| = r$ and $\|x\| = R$ of the Leray–Schauder type, for instance that $\|f(x)\| \geq r$ in the former case and $\|f(x)\| \leq R$ in the latter, will guarantee that fixed points of

$\rho f|A: A \to A$ will be fixed points of f on A, so $N(\rho f|A)$ is a lower bound for the number of such fixed points.

However, it turned out that analytic problems on the plane require a slightly stronger condition, that f is a *μ-annular* map: there exist $0 < \mu < r < R$ such that f satisfies $\|f(x)\| \geq r + \mu$ if $\|x\| = r$, $\|f(x)\| \leq R - \mu$ if $\|x\| = R$ and $\|f(x)\| \geq \mu$ if $r \leq \|x\| \leq R$. The μ-annular condition for maps of the plane is a special case of the *μ-retractibility* of a map $f: X \to X$ of a normed linear space. For a subset Q of X, the μ-neighbourhood is the set of all points $x \in X$ such that $\|x - q\| < \mu$ for some $q \in Q$. Then f is μ-retractible onto a subset S of X if there is a subset W of X containing the union of S and the μ-neighbourhood of $f(S)$ and a retraction $\rho: W \to S$ such that if $y \in W - S$ then $\|y - f(\rho(y))\| > \mu$. The conditions on the map of the plane imply that it is μ-retractible because we can take S to be the annulus A and let $W = \mathbb{R}^2 - \mathbf{0}$.

The main result of [Br4] concerned analytic problems of this form: let E, F be Banach spaces, $L: E \to F$ an isomorphism, $H: E \times \mathbb{R}^n \to F$ a completely continuous map and $B: E \to \mathbb{R}^n$ a continuous linear mapping; find solutions to $Ly = H(y, \lambda)$ that satisfy $By = \mathbf{0}$. Setting $H_\lambda(y) = H(y, \lambda)$, we will need the map $\Pi: \mathbb{R}^n \to \mathbb{R}^n$ defined by

$$\Pi(\lambda) = \lambda + BL^{-1}H_\lambda(\mathbf{0}).$$

Suppose there exists a compact, locally contractible subset S of $\mathbb{R}^n$ and $\mu > 0$ such that (1) Π is μ-retractible onto S with retraction $\rho: W \to S$, (2) for all $\lambda \in S$, the maps $L^{-1}H_\lambda: E \to E$ satisfy the Leray–Schauder boundary condition with respect to the same ball and (3) if λ is in S and y in that ball, then

$$\|H_\lambda(y) - H_\lambda(\mathbf{0})\| < \frac{\mu}{\|BL^{-1}\|}.$$

The conclusion is that there are at least $N(\rho\Pi|S)$ solutions (y, λ) to the problem with y in the ball and λ in S.

Here are some examples from [Br4] of analytic problems of the form $Ly = H(y, \lambda)$, $By = \mathbf{0}$ described above. Let $h: [0.1] \times \mathbb{R}^{2n} \to \mathbb{R}^n$ be continuous. We will seek a map $y = y(t): [0, 1] \to \mathbb{R}^n$ and a vector $\lambda \in \mathbb{R}^n$. One type of problem is a "parametrized Dirichlet problem for an ordinary differential equation":

$$y' = h(t, y, \lambda), \quad y(0) = y(1) = 0.$$

Another type is a "controllability problem" from [FNPZ] where a vector $A \in \mathbb{R}^n$ is given and y and λ must be chosen to satisfy

$$y'' = h(t, y, \lambda), \quad y(0) = y(1) = 0 \quad \text{and} \quad \int_0^1 y(t)\, dt = A.$$

A third type, based on a chemical engineering problem from [LW], requires the vector A and also a positive constant β and seeks y and λ satisfying

$$\beta y'' - y' = h(t, y, \lambda), \quad \beta y'(0) = y(0), \quad y'(1) = 0, \quad y(1) = A.$$

A "three-point boundary value problem" is of this form as well, where now $h\colon [0,1]\times \mathbb{R}^{3n} \to \mathbb{R}^n$, and the problem is to solve

$$y'' = h(t, y, y', \lambda), \quad y(0) = y(1/2) = y(1) = 0.$$

All of the types of problems mentioned so far are based on ordinary differential equations, but the same form also arises in problems involving partial differential equations, as a final type demonstrates. Let Ω be a region of $\mathbb{R}^m$ whose boundary is a smooth, compact $(m-1)$-manifold. There is a given map $h\colon \overline{\Omega}\times\mathbb{R}^{(m+2)n} \to \mathbb{R}^n$. The problem is to find an L^2 function $y(x) = (y_1(x), y_2(x), \ldots, y_n(x))\colon \overline{\Omega} \to \mathbb{R}^n$ and vector $\lambda \in \mathbb{R}^n$ such that the Laplacian satisfies

$$\Delta y = h\left(x, y, \frac{\partial y_1}{\partial x_1}, \ldots, \frac{\partial y_n}{\partial x_m}, \lambda\right), \quad y|\partial\Omega = 0 \quad \text{and} \quad \int_\Omega y(x)\, dx = 0.$$

Specific examples in which problems of these various types that can be shown to satisfy the hypotheses of the main result of [Br4] can be constructed by basing the map h on functions that exhibit the sublinear growth mentioned previously.

In 1990, Brown and Zezza (see [BZ]) used the μ-retractibilty concept in order to apply Nielsen theory to some problems in control theory. The most common type of control problem is based on a differential equation $y' = h(t, y, u)$ where y is a vector-valued function defined on an interval $[0, T]$, so this single equation represents a system of differential equations. Boundary conditions on the differential equation of the form $y(0) = y_0$ and $y(T) = y_1$ can be thought of as specifying the starting point and destination, respectively, of an object whose motion is determined by the differential equation. A solution is a pair (y, c) where c, called the *control*, is restricted to a specified *control space* and the control must be chosen so that the position vector $y(t)$ is determined by a solution to the differential equation that satisfies the boundary conditions. If such a pair (y, c) exists, then the problem is said to be *controllable.* Once a problem is controllable, a criterion is specified and the "best" solution (y, c) according to that criterion is chosen in the resulting *optimal control problem.* For instance, there may be a cost function that depends on the solution and the solution of lowest cost solves the optimal control problem. The Nielsen theory methods are concerned only with the controllability problem but, by demonstrating that there is more than one solution, they also establish the fact that there is a valid optimal control problem in the given setting since there is nothing to optimize when the problem has a unique solution.

The control theory of [BZ] is based on methods from [FNPZ] that permit a general class of controllability problems to be reduced to fixed point problems for maps on finite-dimensional spaces. To avoid a digression into control theory, we will just describe several, closely related, applications of the method. We will not discuss how Nielsen theory was applied except to note that the applications depend on showing that a self-mapping of the plane is μ-annular and then calculating that the corresponding mapping has Nielsen number equal to two. All the applications have a solution (y, c) where $y = (u, v)$ maps $[0, T]$ to the plane and the control $c = (a, b)$ is a point in the plane. The differential system is of the form

$$\begin{aligned} u'(t) &= v(t)^{1/5} + (1 + 2t\Gamma(|c|))a, \\ v'(t) &= u(t)^{1/5} + (1 + 2t\Gamma(|c|))b, \\ u(0) &= v(0) = 0, \\ u(T) &= u_1, \quad v(T) = v_1, \end{aligned}$$

where the function Γ will vary for the three applications we will describe. Let $y_1 = (u_1, v_1) \in \mathbb{R}^2$. For the first example, let $\Gamma_1(x) = x \cos x$, then there is an integer M, depending on $|y_1|$, such that there are at least two solutions (y, c) with $2m\pi \leq |c| \leq 2m\pi + \pi/2$ for each $m \geq M$. Thus there are an infinite number of solutions no matter where the destination y_1 is located. For the second example, take $T = 1$ and let $\Gamma_2(x) = 3x/10 - 7/2$. If $|y_1|$ is sufficiently small, for instance $|y_1| \leq 3/2$ will do, then there are at least two solutions (y, c) with $5 \leq |c| \leq 10$. Thus close destinations can be reached in one unit of time by at least two different routes. For the final application, set $\Gamma_3(x) = (x + 3)/(x^2 + x)$, then it is shown that there are at least two solutions (y, c) for any destination y_1, provided the time interval $[0, T]$ can be chosen with T large enough.

Between 1993 and 1996, Fečkan published a series of papers extending [Br4]. We will mention just a sampling of his results here. A more extensive survey of the contents of those papers can be found in [Fk1]. Recall that the results of [Br4] concern equations of the form $Ly = H(y, \lambda)$ where $L: X \to Y$ is an isomorphism of Banach spaces. Although the isomorphism hypothesis on L readily permits the solutions to such problems to be viewed as fixed points, it is quite a restrictive hypothesis. A technique due to Mawhin produces a fixed point formulation when there is a more general kind of linear function $L: X \to Y$ between Banach spaces that is called a *Fredholm operator of index zero.* This means that the image $\operatorname{im} L$ of L is closed in Y and the kernel of L and its cokernel $Y/\operatorname{im} L$ are finite dimensional, of the same dimension. For a recently published description of Mahwin's theory, see [Br3]. In [Fk2], Fečkan places the μ-retractibility concept in the setting of equations involving Fredholm operators of index zero. One class of applications in [Fk2] is to differential equations of the form $y'' = \varepsilon h(t, y, \lambda)$, for ε small, subject to $y'(0) = y'(1/2) = y'(1) = 0$. These problems appear to be similar to the three-point boundary value problems discussed in [Br4] and they are, in particular because L is defined by $Ly = y''$ in both cases. However, the only solution to $y'' = 0$ that satisfies $y(0) = y(1) = 0$ is the constant function at zero whereas any constant function has the properties $y'' = 0$ and $y'(0) = y'(1) = 0$. Thus, in the three-point boundary value problem, L is one-to-one whereas, in the problem from [Fk2], the linear operator L has a one-dimensional kernel.

Nielsen theory methods are applied for the first time in [Fk2] to establish the existence of multiple periodic solutions to problems in which the operator L is not an isomorphism. For instance, suppose a map $h: \mathbb{R} \times \mathbb{R}^n \to \mathbb{R}^n$ is bounded and 1-periodic in the first variable and consider the differential system $y' = \varepsilon h(t, y)$ for ε small. If the map $\Pi: \mathbb{R}^n \to \mathbb{R}^n$ defined by

$$\Pi(x) = x + \int_0^1 h(s, x)\, ds$$

is μ-retractible onto a compact, locally contractible subset S by a retraction ρ, then $N(\rho\Pi|S)$ is a lower bound for the number of 1-periodic solutions to the system. The general results of [Fk2] are also applied in [Fk3] to study parametrized singularly perturbed nth order boundary value problems, but we will not attempt to describe those problems here.

A further extension of the results in [Fk2], to differential equation problems that arise from vector fields on manifolds, was carried out in [Fk4]. Assume that there is a manifold $M \subset \mathbb{R}^k$ and a function $u: M \times \mathbb{R} \to \mathbb{R}^k$ that assigns a vector tangent to M at each point of the manifold at each time t and that is 1-periodic in t. The corresponding differential equation is $x = \varepsilon u(x, t)$ where a solution $x = x(t): \mathbb{R} \to M$ is a function whose derivative at each time t is determined by the given vector field in the sense that $x(t) = \varepsilon u(x(t), t)$. For μ-retractibility in this setting, a compact, locally contractible subset S of M is required. Suppose for the map $\Pi: M \to \mathbb{R}^k$ defined as above, but with u in place of h, there is a positive number μ and an open subset W of $\mathbb{R}^k$ containing the union of S and the μ-neighbourhood of $f(S)$ such that W is retractible onto S with a retraction $\rho: W \to S$ with the property that $\|y - f(\rho(y))\| > \mu$ for all $y \in W - S$. Fečkan still considers Π to be μ-retractible onto S even though Π is not defined on all of $\mathbb{R}^k$ and, under appropriate hypotheses, he again establishes that the Nielsen number of $\rho\Pi: S \to S$ is a lower bound, now for the number of solutions to $x = \varepsilon u(x, t)$.

Problems in differential equations often require that the solution satisfy a condition that restricts it to a portion C of a (real) linear space X of functions called a *cone* in X. The definition of a cone is that it is a closed convex subset of X

with the properties that $x \in C$ and $t \geq 0$ implies $tx \in C$ whereas $t < 0$ implies $tx \notin C$. An example of a cone is the positive functions in a space of real-valued functions. Once again exploiting the techniques of [Fk2], this time to obtain an abstract result about multiple solutions that lie in a cone, Fečkan obtained in [Fk5] a result about the existence of $\lambda \in \mathbb{R}^2$ for which there are positive solutions to a singular boundary value problem of the form

$$u''(t) = \varepsilon \cdot f(t) \cdot u^{-\alpha}(t) + g(\lambda, t),$$
$$u(0) = u(1) = 0, \quad u(1/2) = a, \quad u(1/5) = b.$$

Here a, b and $\alpha > 0$ are given constants, $0 \leq t \leq 1$, the functions f and g take on positive values and $\varepsilon > 0$ is small. Conditions are obtained that produce a lower bound for the number of distinct vectors λ for which such positive solutions exist. He also described in [Fk5] several other types of singular boundary value problems to which the same technique can be applied.

Recall that a linear function $L: X \to Y$ between Banach spaces is a *Fredholm operator* if the image $\operatorname{im} L$ of L is closed in Y and the kernel and cokernel, $Y/\operatorname{im} L$, of L are finite dimensional. The *index* of a Fredholm operator is defined by

$$\operatorname{index} L = \dim \ker L - \dim Y/\operatorname{im} L.$$

The application of Nielsen theory methods to problems of the form $Lu = F(u)$, where L is Fredholm of positive index, rather than zero index as previously, was initiated in [Fk6]. The problems in [Fk6] are boundary value problems, that is, the solution u is subject to a boundary condition of the type $B(u) = 0$ for a function $B: X \to \mathbb{R}^k$ that, in contrast to the papers previously mentioned, is not necessarily linear though it must satisfy other restrictions.

One parametrized version of the general problem with nonlinear boundary conditions is of the form $u = \varepsilon F(u, \lambda)$ such that $B(u, \lambda) = 0$ where X is a Banach space and $F: X \times \mathbb{R}^m \to X$, $B: X \times \mathbb{R}^m \to \mathbb{R}^m$ are continuous. Let D be a bounded closed convex subset of X. A function $f: D \to X$ is called *weakly inward* if, for each $x \in D$, the equation $f(x) = x + t(y - x)$ has a solution with $t \geq 0$ and $y \in D$. Note that the weakly inward condition is weaker than the requirement that D be mapped to itself by f and in some cases it can be shown to be satisfied when no such D exists that is invariant under f. The paper [Fk7] considers problems of this type where certain of the functions $f_\lambda: D \to X$ defined by $f_\lambda(x) = F(x, \lambda)$ are assumed to be weakly inward. Nielsen theory places a lower bound on the number of vectors λ for which there are solutions in D to the parametrized problem with nonlinear boundary condition.

The Nielsen number can also be used for higher order problems, as illustrated by the following type of result from [Fk8]. Suppose there are positive constants K_1,

K_2, K_3 and continuous functions $f\colon \mathbb{R}\times[0,1]\times\mathbb{R}\to\mathbb{R}$ with $|f(\,\cdot\,,\,\cdot\,,\,\cdot\,)|\le K_1$ and $G\colon \mathbb{R}^2\to\mathbb{R}$ with $K_2\le G(\,\cdot\,,\,\cdot\,)\le K_3$. Consider the problem of finding $x\colon\mathbb{R}\to\mathbb{R}$ that solves

$$\begin{aligned} x''' &= \varepsilon f(x,t,\lambda),\\ \int_0^1 x''(t)^2\,dt &= \varepsilon^2 G(\|x\|,\lambda),\\ x'(1/3) &= x'(1/2)=x'(1)=0. \end{aligned}$$

The number of solutions, that are all nontrivial because in particular $x'(0)\ne 0$, is bounded below by the appropriate Nielsen number.

There are many more results summarized in [Fk1], so the interested reader should refer to that survey. The techniques developed by Fečkan are applied by him only to problems concerned with ordinary differential equations. However, as he notes in [Fk1], there are also problems in partial differential equations for which those same techniques are appropriate.

At about the same time that Fečkan was developing his Nielsen theory for differential equations, Borisovich, Kucharski and Marzantowicz were using Nielsen theory to study integral equations. The paper [BKM] of 1999 offers a good indication of their approach. A system of two nonlinear integral equations of the following form is said to be of *Urysohn type*:

$$\begin{aligned} u(t) &= \int_0^1 K_1(t,s,u(s),v(s))v^{\beta}(s)\,ds,\\ v(t) &= \int_0^1 K_2(t,s,u(s),v(s))u^{\alpha}(s)\,ds. \end{aligned}$$

Suppose further that α and β are positive rational numbers that have odd numerators in reduced form. Let X be the Banach space of pairs of continuous functions on $[0,1]$ with norm the sum of the sup norms of the two functions then, from the system, one has the operator $G\colon X\to X$ defined by

$$G(u,v)(t)=\left(\int_0^1 K_1(t,s,u(s),v(s))v^{\beta}(s)\,ds,\int_0^1 K_2(t,s,u(s),v(s))u^{\alpha}(s)\,ds\right).$$

Let A be the subset of X consisting of pairs (u,v) of functions, each of which takes only non-negative or only non-positive values and are not both the zero function. Sufficient conditions on the functions K_1 and K_2 are presented to imply that the Nielsen number of G restricted to A is defined and equal to the Nielsen number of the selfmap g of the plane with the origin removed defined by $g(c_1,c_2)=(c_2^{\beta},c_1^{\alpha})$. The degree of g is -1 so $N(g)=2$ and therefore the authors may conclude that the system of integral equations has at least two non-zero solutions. They extend this type of result to $2n$ nonlinear integral equations of Urysohn type and include the following specific system in the case $n = 3$ to illustrate their technique:

$$\begin{aligned}
u_1(t) &= \int_0^1 (1 + \sin^2[tv_1^3(s) + u_3^2(s)])v_1^7(s)\,ds,\\
v_1(t) &= \int_0^1 (3 + \cos^2[tu_2(s)])u_1^5(s)\,ds,\\
u_2(t) &= \int_0^1 (1 + t^2 + s^4)v_2^3(s)\,ds,\\
v_2(t) &= \int_0^1 (3 + t\sin[u_2^4(s)])u_2^5(s)\,ds,\\
u_3(t) &= \int_0^1 \ln\left(\frac{1}{10} + \frac{ts}{2}\right)v_3^9(s)\,ds,\\
v_3(t) &= \int_0^1 \arctan(2 + u_1^2(s) + t^3 + v_3^4(s))u_3^4(s)\,ds.
\end{aligned}$$

They prove that this system has at least four non-zero solutions.

Most of the applications of Nielsen theory to nonlinear analysis since the late 1990's have focused on differential inclusions. Let J be a connected subset of the reals and suppose $F\colon J \times \mathbb{R}^n \to 2^{\mathbb{R}^n}$ is a multivalued function. The *differential inclusion* arising from the function F is

$$X \in F(t, X)$$

whose solution is a single-valued function $X\colon J \to \mathbb{R}^n$. Nielsen theory methods, based initially on the multivalued theory of Jezierski from [Je] discussed earlier, have been developed to produce lower bounds for the number of solutions to differential inclusions. There is now an extensive literature of this subject due to Andres, Górniewicz and Jezierski. The subject of differential inclusions is beyond the scope of this section, as its title indicates, but a forthcoming monograph of Andres and Górniewicz [AG] will include an exposition of Nielsen theory applications to differential inclusions. It is noteworthy that when some of the work on differential inclusions is specialized to the case that F is single-valued and thus the differential inclusion becomes a differential equation, the results may still be new. This fact is illustrated by the paper [A]. Andres points out that much of the earlier literature, specifically in the work of Brown and of Fečkan, is limited to problems with a parameter. Moreover, the work on integral equations of Borisovich, Kucharski and Marzantowicz, though concerned with a setting for which Nielsen theory methods are well-suited, could have been carried out using more traditional techniques, as those authors point out at the end of [BKM]. In [A], he exhibits a problem that does not include a parameter and for which the existence of multiple solutions cannot be established by traditional methods. The problem makes use of bounded functions $e, f, g, h\colon \mathbb{R}^3 \to \mathbb{R}$, positive numbers a and b and odd integers m and n greater than or equal to three. The form of the problem is

$$\begin{aligned} u' + ax_2 &= e(x_1, x_2, x_3)v^{1/m} + g(x_1, x_2, x_3), \\ v' + bx_3 &= f(x_1, x_2, x_3)u^{1/n} + h(x_1, x_2, x_3). \end{aligned}$$

The functions e, f, g, h need not be continuous, nor is a solution continuous; we refer the reader to [A] for the specific conditions required. The computation of a Nielsen number establishes the existence of at least two solutions. Moreover, a close look at the proof, using properties of the fixed point index, demonstrates that a third solution to the system of differential equations must also exist.

6. Applications of Nielsen theory to dynamics

Many of the applications of Nielsen theory to dynamics are the work of Jiang and he has written detailed expositions of this subject in [J3], published in 1993, as well as [J4] of 1999. The present, much briefer, discussion is intended merely to give the reader a sample of what Nielsen theory can contribute to this subject.

Let $f\colon X \to X$ be a map, then its iterates are defined by $f^0(x) = x$, $f^1(x) = f(x)$ and, in general, $f^n(x) = f(f^{n-1}(x))$. Dynamics studies the orbit of a point $x \in X$, that is, the set $\{f^n(x)\}$ for all $n \geq 0$. Points that have finite orbits, that is when $f^n(x) = x$ for some n so x is a *periodic point* of f, are very significant in this subject. The *minimum period* of a periodic point x is the n such that $f^n(x) = x$ but $f^j(x) \neq x$ for all $j < n$ and the set $x, f(x), \ldots, f^{n-1}(x)$ is called the *primary orbit* of x. Periodic points can be detected by the Lefschetz number $L(f^n)$ but Nielsen theory can detect more subtle behavior, in particular, the way the number of distinct primary orbits increases as a function of n.

The *growth* of a sequence $\{a_n\}$ is defined by

$$\mathrm{Growth}_{n\to\infty}\, a_n = \max\{1, \limsup_{n\to\infty} |a_n|^{1/n}\}.$$

If $\mathrm{Growth}_{n\to\infty} a_n > 1$, then it is said that the sequence *grows exponentially.* Applying the growth concept to the sequence $\{N(f^n)\}$ of Nielsen numbers of iterates of the map defines the *asymptotic Nielsen number*

$$N^\infty(f) = \mathrm{Growth}_{n\to\infty} N(f^n).$$

Let α be an open cover of a space X, then the *entropy* $H(\alpha)$ of α is defined to be the logarithm of the number of sets in the finite subcover of α that contains the fewest sets.

Let α and β be open covers of a space X, then the *join* $\alpha \vee \beta$ of the covers is the open cover defined by

$$\alpha \vee \beta = \{A \cap B : A \in \alpha,\ B \in \beta\}.$$

For an open cover α of a space X and a map $g\colon X \to X$, the *topological entropy* $h(g, \alpha)$ *of* g *with respect to the cover* α is defined by

$$h(g, \alpha) = \lim_{n\to\infty} \frac{1}{n} H(\alpha \vee g^{-1}(\alpha) \vee \cdots \vee g^{-(n-1)}(\alpha))$$

and $h(g)$, the *topological entropy* of g, is the supremum of the $h(g, \alpha)$ over all open covers α of X. In 1982, Ivanov proved in [I] the entropy theorem which relates Nielsen theory to the entropy concept. It states that if $f\colon X \to X$ is a map of a finite polyhedron, then

$$h(f) \geq \log N^{\infty}(f).$$

The usefulness of the entropy theorem and similar results depends on being able to make computations, or at least good estimates. The setting in which a great deal of such concrete work in dynamics has been carried out is that in which $f\colon X \to X$ is a homeomorphism of a compact surface. In this case, the Thurston canonical form decomposes f into periodic and pseudo-Anosov pieces where there is a stretching factor λ which describes each pseudo-Anosov piece. It was proved in the Jiang–Guo paper [JG] of 1993 that if the Euler characteristic of X is negative, then $N^{\infty}(f)$ equals the largest such stretching factor λ so $h(f) \geq \log \lambda$. It follows that if there is at least one pseudo-Anosov piece, then the number of primary n-orbits grows exponentially in n.

A form of $N^{\infty}(f)$ for maps of pairs is an important tool for applying Nielsen theory to dynamics. Let X be a finite connected polyhedron and A a subpolyhedron. A fixed point x of a map of pairs $f\colon (X, A) \to (X, A)$ is *related to* A if there is a path c from x to A such that c is homotopic to fc through paths from x to A. The *Nielsen number of the complement* $N(f; X - A)$, introduced by Zhao in 1989 (see [Z3]), is the number of essential fixed point classes of f that are not related to A and thus the *asymptotic Nielsen number of the complement* can be defined as

$$N^{\infty}(f; X - A) = \mathrm{Growth}_{n\to\infty} N(f^n; X - A).$$

Now we return to the setting of $f\colon X \to X$ a homeomorphism of a compact surface, which we will further assume is isotopic to the identity map of X. Moreover, we suppose that one primary orbit $P = \{x_0, f(x_0), \ldots, f^{r-1}(x_0)\}$, lying in the interior of X, is already known, where r is greater than the Euler characteristic of X. Thus f may be viewed as a map of pairs $f\colon (X, P) \to (X, P)$. An isotopy $\{h_t\}$, where h_0 is the identity map and $h_1 = f$, determines a geometric braid

$S = \{(h_t(x), t) \in X \times I : x \in P\}$ and thus an element σ_P of the r-string braid group of X.

Suppose a homeomorphism f of the plane $\mathbb{R}^2$ can be extended to the sphere $S^2 = \mathbb{R}^2 \cup \infty$ by setting $f(\infty) = \infty$ and there is an isotopy from the identity map to f keeping ∞ fixed. Assume that there is an orbit of length three and take P to consist of ∞ together with the orbit. The Artin braid group B_3 on three strings has a presentation with two generators σ_1 and σ_2 and one relation. Jiang proved that if the braid σ_P is not conjugate to σ_1^m, $(\sigma_1^2\sigma_2^2)^m$, $(\sigma_1\sigma_2)^m$ or $(\sigma_1\sigma_2\sigma_1)^m$ in $B_3/\mathbb{Z}$ then the number of primary n-orbits of f grows exponentially in n. This result is obtained by proving that $N^\infty(f; X - P)$ is equal to the largest stretching factor of a pseudo-Anosov piece.

In 1996, Jiang proved in [J6] the following entropy theorem for homeomorphisms of punctured surfaces. Let P be a finite subset of the interior of a compact surface X and let $f\colon (X, P) \to (X, P)$ be a homeomorphism, then $h(f) \geq \log N^\infty(f; X - P)$. To illustrate how this entropy theorem is used, suppose f is an orientation-preserving embedding of the 2-disc isotopic to the identity map, which can then be extended to the plane. If f has a periodic orbit of period three whose corresponding braid $\sigma_P = \sigma_1\sigma_2^{-1} \in B_3$, then f has a primary orbit of period n for every integer n. Moreover, it can be shown that $N^\infty(f; X\ P) > 5/2$ in this case, so it follows from the entropy theorem that $h(f) > \log 5/2$.

7. And so on

Given a map $f\colon X \to X$ on a compact ANR, the Nielsen numbers of its iterates $N(f^n)$ are used to define the *Nielsen zeta function*

$$N_f(z) = \exp\left(\sum_{n=1}^{\infty} \frac{N(f^n)}{n} z^n\right)$$

where z is a complex number. Since he introduced the concept in 1984, Fel'shtyn has explored its properties in great detail. A thorough exposition of the subject can be found in his monograph [F].

For a map $f\colon X \times I^n \to X$, the study of solution to $f(x, t) = x$ is called *n-parameter fixed point theory*. Starting with a paper of 1990 for the case $n = 1$, Geoghegan and his coauthors have studied in [DG] this subject and its relationship to other parts of mathematics. The references in [G] include the relevant literature.

For maps $f\colon X \to Z$ and $g\colon Y \to Z$, the subset of $X \times Y$ consisting of solutions to $f(x) = g(y)$ forms the subject matter of *Nielsen intersection theory* which was introduced by McCord in 1997 (see [Mc1]). The theory is concerned with the minimum number of solutions to the equation $f'(x) = g'(y)$ among all maps f' homotopic to f and g' to g. Solutions (x_0, y_0) and (x_1, y_1) to $f(x) = g(y)$ are *equivalent* if there are paths c in X from x_0 to x_1 and d in Y from y_0 to y_1 such that fc and gd are homotopic in Z, keeping the endpoints fixed. An equivalence class is called *essential* if it cannot be eliminated through homotopies of f and g and the *Nielsen intersection number* is the number of essential equivalence classes. An exposition of the subject is given in [Mc2] and its connections to other Nielsen theories, for roots and coincidences, is explained in [Mc3].

I'm sure that many of those almost 300 papers I mentioned at the beginning have still been overlooked in this handbook but I must come to an end somewhere, and this is it.

References

[A] J. Andres, *A nontrivial example of application of the Nielsen fixed-point theory to differential systems: Problem of Jean Leray*, Proc. Amer. Math. Soc. **128** (2000), 2921–2931.

[AG] J. Andres and L. Górniewicz, *Topological Fixed Point Principles for Boundary Value Problems*, Kluwer.

[BKM] A. Borisovich, Z. Kucharski and W. Marzantowicz, *A multiplicity result for a system of real integral equations by use of the Nielsen number*, Banach Center Publ. **49** (1999), 9–18.

[Br1] R. Brown, *The Lefschetz Fixed Point Theorem* (1971), Scott, Foresman.

[Br2] ______, *Fixed point theory*, History of Topology, Elsevier, 1999, pp. 271–299.

[Br3] ______, *A Topological Introduction to Nonlinear Analysis*, Second Edition, Birkhäuser, 2004.

[Br4] ______, *Topological identification of multiple solutions to parametrized nonlinear equations*, Pacific J. Math. **131** (1988), 51–69.

[BG] R. Brown and R. Greene, *An interior fixed point property of the disc*, Amer. Math. Monthly **101** (1994), 39–48.

[BGS] R. Brown, R. Greene and H. Schirmer, *Fixed points of map extensions*, Springer Lecture Notes in Math. **1411** (1989), 24–45.

[BZ] R. Brown and P. Zezza, *Multiple local solutions to nonlinear control processes*, J. Optim. Theory Appl. **67** (1990), 463–485.

[CW] F. Cardona and P. Wong, *The relative Reidemeister number of fiber map pairs*, Topol. Methods Nonlinear Anal. **21** (2003), 131–145.

[ChW] H. Cho and M. Woo, *A relative* mod (H, K) *Nielsen number*, J. Korean Math. Soc. **32** (1995), 371–387.

[CJL] D. Chun, C. Jang and S. Lee, *Estimation of the number of common essential fixed point classes*, Bull. Honam Math. Soc. **13** (1996), 157–163.

[DG] D. Dimovski and R. Geoghegan, *One-parameter fixed point theory*, Forum Math. **2** (1990), 125–154.

[Dz] Z. Dzedzej, *Fixed point index theory for a class of nonacyclic multivalued mappings*, Diss. Math. **253** (1985).

[FH1] E. Fadell and S. Husseini, *Local fixed point index theory for non simply connected manifolds*, Illinois J. Math. **25** (1981), 673–699.

[FH2] ______, *The Nielsen number on surfaces*, Contemp. Math. **21** (1983), 59–98.

[FHa] J. Fares and E. Hart, *A generalized Lefschetz number for the local Nielsen fixed point theory*, Topology Appl. **59** (1994), 1–23.

[Fk1] M. Fečkan, *Multiple solutions of nonlinear equations via Nielsen fixed-point theory: a survey*, Nonlinear Analysis in Geometry and Topology, Hadronic Press, 2000, pp. 77–97.

[Fk2] ______, *Nielsen fixed point theory and nonlinear equations*, J. Differential Equations **106** (1993), 312–331.
[Fk3] ______, *Parametrized singularly perturbed boundary value problems*, J. Math. Anal. Appl. **188** (1994), 417–425.
[Fk4] ______, *Multiple periodic solutions of small vector fields on differential equations*, J. Differential Equations **113** (1994), 189–200.
[Fk5] ______, *Parametrized singular boundary value problems*, J. Math. Anal. Appl. **188** (1994), 426–435.
[Fk6] ______, *Differential equations with nonlinear boundary conditions*, Proc. Amer. Math. Soc. **121** (1994), 103–111.
[Fk7] ______, *Note on weakly inward mappings*, Ann. Polon. Math. **63** (1996), 1–5.
[Fk8] ______, *The interaction of linear boundary value and nonlinear functional conditions*, Ann. Polon. Math. **58** (1993), 103–111.
[F] A. Fel'shtyn, *Dynamical zeta functions, Nielsen theory and Reidemeister torsion*, Mem. Amer. Math. Soc. **699** (2000).
[FG] D. Ferrario and D. Goncalves, *Homeomorphisms of surfaces locally may not have the Wecken property*, XI Brazilian Topology Meeting, 2000, pp. 1–9.
[FNPZ] M. Furi, P. Nistri, M. Pera and P. Zezza, *Topological methods for the global controllability of non-linear systems*, J. Optim. Theory Appl. **45** (1985), 231–256.
[Ge] R. Geoghegan, *Nielsen fixed point theory*, Handbook of Geometric Topology, 2002, pp. 499–521.
[Gon] D. Gonçalves, *Braid groups and Wecken pairs*, Contemp. Math. **72** (1986), 89–97.
[G] L. Górniewicz, *Topological Fixed Point Theory of Multivalued Mappings*, Kluwer Academic Publishers, 1999.
[GGK1] L. Górniewicz, A. Granas and W. Kryszewski, *Sur la méthode de l'homotopie dans la théorie des points fixes pour les applications multivoques,* 1, C. R. Acad. Sci. Paris Sèr. I **307** (1988), 489–492.
[GGK2] ______, *Sur la méthode de l'homotopie dans la théorie des points fixes pour les applications multivoques,* 2, C. R. Acad. Sci. Paris Sèr. I **308** (1989), 449–452.
[Gr] A. Granas, *The Leray–Shauder index and the fixed point theory for arbitrary* ANR*s*, Bull. Soc. Math. France **100** (1972), 209–228.
[GD] A. Granas and J. Dugundji, *Fixed Point Theory*, Springer, 2003.
[GS] R. Greene and H. Schirmer, *Smooth realization of relative Nielsen numbers*, Topology Appl. **66** (1995), 93–100.
[H1] E. Hart, *Local Nielsen fixed point theory and the local generalized H-Lefschetz number*, Contemp. Math. **152** (1993), 177–182.
[H2] ______, *Computation of the local generalized H-Lefschetz number*, Topology Appl. **61** (1995), 115–135.
[Hi] G. Hirsch, *Detérmination d'un nombre minimum de points fixes pour certaines représentations*, Bull. Sci. Math. **64** (1940), 45–55.
[I] N. Ivanov, *Entropy and the Nielsen numbers*, Soviet Math. Dokl. **26** (1982), 63–66.
[Je] J. Jezierski, *The Nielsen relation for multivalued maps*, Serdica **13** (1987), 174–181.
[J1] B. Jiang, *Fixed point classes from a differentiable viewpoint*, Springer Lecture Notes in Math. **886** (1981), 163–170.
[J2] ______, *Fixed points and braids* II, Math. Ann. **272** (1985), 249–256.
[J3] ______, *Nielsen theory for periodic orbits and applications to dynamical systems*, Contemp. Math. **152** (1993), 183–202.
[J4] ______, *Applications of Nielsen theory to dynamics*, Banach Center Publications **49** (1999), 203–221.
[J5] ______, *Lectures on Nielsen fixed point theory*, Contemp. Math. **14** (1983).
[J6] ______, *Estimation of the number of periodic orbits*, Pacific J. Math. **172** (1996), 151–185.
[JG] B. Jiang and J. Guo, *Fixed points of surface diffeomorphisms*, Pacific J. Math. **160** (1993), 67–89.

[KM] W. Kryszewski and D. Miklaszewski, *The Nielsen number of set-valued maps: An approximation approach*, Serdica **15** (1989), 336–344.

[LW] R. Leggett and L. Williams, *Multiple positive fixed points of non-linear operators on ordered Banach spaces*, Indiana Univ. Math. J. **28** (1979), 673–688.

[L1] J. Leray, *Le theorie des points fixes et ses applications en analyse*, Proc. Inter. Congress of Math., 1950 **2** (1952), 202–208.

[L2] ———, *Théorie des pointes fixés: indice total et nombre de Lefschetz*, Bull. Soc. Math. France **87** (1959), 221–233.

[LS] J. Leray and J. Schauder, *Topologie et equations fonctionnelles*, Ann. Sci. École Norm. Sup. (4) **51** (1934), 45–78.

[Ma] S. Masih, *On the fixed point index and Nielsen fixed point theorem for symmetric product mappings*, Fund. Math. **102** (1979), 143–158.

[Mc1] C. McCord, *A Nielsen theory for intersection numbers*, Fund. Math. **152** (1997), 117–150.

[Mc2] ———, *Wecken theorems for Nielsen intersection theory*, Banach Center Publ. **49** (1999), 235–252.

[Mc3] ———, *The three faces of Nielsen: coincidences, intersections and preimages*, Topology Appl. **103** (2000), 155–177.

[Mc] D. McCord, *An estimate of the Nielsen number and an example concerning the Lefschetz fixed point theorem*, Pacific J. Math. **66** (1976), 195–203.

[Mi] D. Miklaszewski, *A reduction of the Nielsen fixed point theorem for symmetric product maps to the Lefschetz theorem*, Fund. Math. **135** (1990), 175–176.

[O] B. O'Neill, *Induced homology homomorphisms for set-valued maps*, Pacific J. Math. **7** (1957), 1179–1184.

[S1] H. Schirmer, *Nielsen theory of transversal fixed point sets*, Fund. Math. **141** (1992), 31–59.

[S2] ———, *A relative Nielsen number*, Pacific J. Math. **122** (1986), 459–473.

[S3] ———, *A Nielsen number for fixed points and near points of small multifunctions*, Fund. Math. **88** (1975), 145–156.

[S4] ———, *Simplicial approximation of small multifunctions*, Fund. Math. **84** (1974), 121–126.

[S5] ———, *An index and Nielsen number for n-valued multifunctions*, Fund. Math. **124** (1984), 207–219.

[S6] ———, Fix-*finite approximations of n-valued multifunctions*, Fund. Math. **121** (1984), 73–80.

[S7] ———, *A minimum theorem for n-valued multifunctions*, Fund. Math. **126** (1985), 83–92.

[S8] ———, *A fixed point index for bimaps*, Fund. Math. **134** (1990), 93–104.

[S9] ———, *The least number of fixed points of bimaps*, Fund. Math. **137** (1991), 1–8.

[SS] M. Shub and D. Sullivan, *A remark on the Lefschetz fixed point formula for differentiable maps*, Topology **13** (1974), 189–191.

[Sz] K. Scholz, *The Nielsen fixed point theory for non-compact spaces*, Rocky Mountain J. Math. **4** (1974), 81–87.

[Wa] S. Wang, *General H-fixed point classes and their application*, Beijing Daxue Xuebao (1982), 11–18.

[W] P. Wong, *A note on the local and extension Nielsen numbers*, Topology Appl. **48** (1992), 207–213.

[WC] M. Woo and H. Cho, *A relative* mod K *Nielsen number*, J. Korean Math. Soc. **29** (1992), 409–422.

[WK] M. Woo and J. Kim, *Note on a lower bound of Nielsen number*, J. Korean Math. Soc. **29** (1992), 117–125.

[Y] C. You, *Fixed point classes of a fiber map*, Pacific J. Math. **100** (1982), 217–241.

[Z1] X. Zhao, *A new Nielsen type number for map extensions*, Far East J. Math. Sci. **2** (1994), 17–26.

[Z2] ———, *Estimation of the number of fixed points of map extensions*, Acta Math. Sinica **8** (1992), 357–361.

[Z3] ———, *A relative Nielsen number for the complement*, Springer Lecture Notes in Math. **1411** (1989), 189–199.

13. ALGEBRAIC TECHNIQUES FOR CALCULATING THE NIELSEN NUMBER ON HYPERBOLIC SURFACES

EVELYN L. HART

1. Introduction

Let X be a hyperbolic surface. That is, let X be a compact connected surface, with or without boundary, for which the Euler characteristic $\chi(X)$ is negative. We consider here algebraic techniques for finding the Nielsen number $N[f]$. In [M] McCord, in his overview of calculating the Nielsen number in a variety of settings, discussed the difficulty of finding Nielsen numbers for hyperbolic surfaces. Since his article was published, new techniques have been found. But the problem is by no means solved.

Although the emphasis here is on algebraic techniques, there are also geometric results. For certain types of homotopy classes of maps that contain a homeomorphism, Kelly presents a geometric algorithm for finding $N[f]$ in [K5]. His algorithm is based in part on the work of Bestvina and Handel from [BH]. For our Example (3.1), Kelly uses this algorithm to find a Nielsen number when algebraic techniques fail.

Another approach is that of Guaschi in [Gu] using the Nielsen–Thurston classification of surface homeomorphisms along with Artin's theory of braids. Guaschi shows that any braid conjugacy invariant can be used to distinguish Reidemeister classes. See also new techniques for authomorphisms in [MV].

The Nielsen number approximates $\mathrm{Min}[f]$, with

$$\mathrm{Min}[f] = \min\{\#\mathrm{Fix}(g) : g \simeq f\}.$$

But in some cases $\mathrm{Min}[f]$ can be calculated. Kelly, in [K1], provides an algebraic algorithm for finding $\mathrm{Min}[f]$ on the twice punctured disk from the induced endomorphism $f_\#$ on the fundamental group. (See [K2] for an extension.) Llibre and Nunes, in [LN], provide the corresponding algorithm for the figure eight. These two spaces share the same fundamental group, yet the two spaces can end up having different minimum numbers for the same endomorphism on the free group on two generators. Llibre and Nunes state that their algorithm could be extended to a bouquet of circles but would become quite complicated. From Kelly's algorithm it is known that the figure eight and the twice punctured disk are totally non-Wecken. That is, for non-homeomorphisms the difference between $\text{Min}[f]$ and $N[f]$ can be arbitrarily large. This has also been proved for a bouquet of circles by Turaev in [T] and by Ferrario in [F2]. (See Wagner's work in [W2] for classes of self–maps on the twice punctured disk for which $N[f] = \text{Min}[f]$.)

One might then ask why we care about the Nielsen number on the figure eight or the twice punctured disk. After all, the Nielsen number is used to approximate $\text{Min}[f]$. One answer comes from Nielsen periodic point theory. For iterates of a map f, the goal of Nielsen periodic point theory is to calculate bounds for the minimum number of fixed points of f^n and also for the minimum number of fixed points of f^n that have least period n. The crucial part is that the minimum is taken over all maps homotopic to f itself rather than maps homotopic to f^n. The results of Kelly, Llibre, and Nunes do not extend to the calculation of these minimum numbers. In other words, knowing that the n-th iterate f^n has fixed points that can be removed by a homotopy of f^n does not guarantee that these fixed points can be removed by a homotopy of f. There are extensions of the Nielsen number that do provide bounds on these minimum numbers for periodic points. See [J1] and other chapters in this volume for details.

Furthermore, Heath and Keppelmann [HK] are studying periodic point theory on fibrations for which the naive addition conditions fail. Key examples come from fibrations over bouquets of circles. On these spaces it is precisely the Nielsen structure in the base that determines relationships between Nielsen classes in distinct fibers in the total space. For a map on a wedge of circles, there can be inessential Nielsen classes that contain fixed points whose indices are non-zero (and cancel). And it is possible that no homotopy will remove these fixed points. Such classes can have a profound impact on the number of periodic points that occur in the fibers over these fixed points.

Thus those techniques that produce the Nielsen number and also preserve information about the structure of the Nielsen classes of fixed points are most valuable to us. We define the Nielsen structure of a map f and the Reidemeister structure of a homomorphism $f_\#$ in Section 4 and discuss which techniques preserve this structure.

In Section 2 we consider the definition of the Reidemeister trace (originally from [R]) and Wecken's proof from [We] that this alternating sum of traces is equal to a sum of Reidemeister classes so that when each class appears at most once in the sum then the index of each class is the coefficient of the class. This means that the sum of the coefficients is the Lefschetz number of the map and that the number of terms with non-zero coefficient is the Nielsen number of the map. (See also [B2], [H1], [G1], [G2] and [Hu].)

In Section 3 we continue by considering the calculation of the Reidemeister trace for self-maps of closed hyperbolic surfaces using the Fox calculus and techniques developed in [FH] and [DHT]. This process is algorithmic except for the last step of reducing the Reidemeister trace so that each term occurs at most once. We then discuss some ad hoc techniques for this last step including abelianization and bounds on the index of a class.

Next, in Section 4 we define the Nielsen structure of a map and the Reidemeister structure of a homomorphism. We continue with a description of techniques for finding the Reidemeister trace for hyperbolic surfaces with boundary.

In Section 4.1 we examine Wagner's method from [W1], which does preserve the Reidemeister structure of a map while finding $N[f]$. Wagner provides significant insight in two new ways. First, for a given fixed point she keeps track of which loop of a bouquet of circles contains that fixed point. This corresponds to remembering which generator gave rise to a particular term of the Fox trace. Second, Wagner realized that in order to calculate $N[f]$ we do not need to be able to distinguish any two elements of the fundamental group. She concentrates on the terms of the Fox trace and provides a simple algorithm for finding the Reidemeister equivalences that are needed. A map is Wagner-characteristic if Wagner's algorithm finds *all* of the Reidemeister equivalences between terms from the Fox trace. Thus the goal becomes determining which homomorphisms are Wagner-characteristic. Wagner defines what it means for an endomorphism to have remnant. Theorem 3.7 of [W1] (attributed to R. F. Brown) says in essence that most endomorphisms on a finitely generated free group do have remnant. We provide here Wagner's proof that any homomorphism with remnant is Wagner-characteristic. Our notation is simpler than in the original, and these changes make it clear how to extend these ideas as we do in Section 4.5 and [H2]. In [W2] Wagner provides a formula for $N[f]$ that depends only on the induced homomorphism $f_{\#}$, but this formula requires that the homomorphism be known to be Wagner-characteristic as well as what she defines as simple. If the homomorphism is known to be Wagner-characteristic, then Wagner's algorithm from [W1] is quite easy to apply to find $N[f]$.

In Section 4.4 we discuss the work of Yi from [Y]. Yi provides a significant extension of Wagner's work by proving that a certain large collection of homomorphisms on the free group with two generators can be changed to homomorphisms that are Wagner-characteristic without changing the Nielsen number. While Wagner's algorithm gives us both the Nielsen number and the Reidemeister structure for a W-characteristic homomorphism, Yi's process (which applies to many maps on the figure eight) does not preserve the Reidemeister structure of the original homomorphism.

We end the paper in Section 4.5 with an introduction of the results from [H2], in which Wagner's definitions of Wagner-characteristic and directly related are extended to the new concepts of n-characteristic and n–related. Here n is the length of the word z that acts on a term of the Fox trace via the Reidemeister action. That is, $z \circ W_i = f_\#(z)W_i z^{-1}$, and for W_i and W_j Reidemeister equivalent terms of the Fox trace we have

$$z = W_j^{-1} f_\#(z) W_i. \tag{1.1}$$

For a limited class of homomorphisms on the figure eight, it is proven in [H2] that the Reidemeister structure can be determined using only short words z to find any Reidemeister equivalences between terms in the Fox trace. This can be extremely useful for using a computer algebra system such as Magma (see [BCM]) to search for equivalences. Results of the search can be confirmed by hand.

The author is indebted to many people for helpful conversations. Thanks go to Robert F. Brown, John Guaschi, Michael Kelly, Davide Ferrario, Peter Yi, Edward Keppelmann, and Phil Heath. Wei Ren provided useful insights while working on an undergraduate summer project related to Section 4.5 that included programming using Magma. R. F. Brown gave an informal talk, at the conference Dynamical Zeta Functions, Nielsen Theory and Reidemeister Torsion, June 1996, in Warsaw, Poland, in which he explained Wagner's algorithm. The talk was quite useful as a starting point for the last part of this paper.

2. The Reidemeister trace

Given a map $f: X \to X$ with X a hyperbolic surface, we define $f_\#$ to be the homomorphism induced by f on the fundamental group of X. The homomorphism $f_\#$ depends on the choice of base path from the base point x_0 to $f(x_0)$. We have $\pi_1(X)$ finitely generated free (for X a a surface with boundary) or we have $\pi_1(X)$ a one-relator group (for X a hyperbolic closed surface of genus $r \geq 2$). The free group on r generators is denoted by F_r.

The Reidemeister action is an action of $\pi_1(X)$ on itself given by the following: for any $z, \alpha \in \pi_1(X)$ we have $z \circ \alpha = f_\#(z)\alpha z^{-1}$. (In some papers the action is $z \circ \alpha = z^{-1}\alpha f_\#(z)$. For this alternate notation, minor changes must be made to the statements below.) The equivalence class of α is $[\alpha]$, and the set of these Reidemeister classes is $R(f_\#)$.

There is a one-to-one function from the set of Nielsen classes of fixed points of the map f into $R(f_\#)$, and the index of a Nielsen class is assigned to the corresponding Reidemeister class. Thus $N[f]$ is the number of elements of $R(f_\#)$ with non-zero index.

Specifically, with $\widetilde{X}$ the universal covering space of X and with $\widetilde{f}$ a lift of f, for every $\alpha \in \pi_1(X)$ the set of points in X given by $p(\mathrm{Fix}(\alpha^{-1}\widetilde{f}))$ is either a Nielsen class of f or is the empty set. And two such sets that are not the empty set are equal if and only if the two group elements are Reidemeister equivalent. The choice of lift $\widetilde{f}$ corresponds to the choice of base path from x_0 to $f(x_0)$ in X. For x a fixed point of f with $x \in p(\mathrm{Fix}(\alpha^{-1}\widetilde{f}))$, the Nielsen class of x corresponds to the Reidemeister class of α.

For the definition of the Reidemeister trace and for the proof that it provides information about $R(f_\#)$ and $N[f]$, we follow [H1].

The algebraic version of Nielsen fixed point theory incorporates the index by considering a generalization of the Lefschetz number that is an element of the free $\mathbb{Z}$ module $\mathbb{Z}(R(f_\#))$. This element is called the Reidemeister trace $\mathrm{RT}(f, \widetilde{f})$ and was introduced by Reidemeister in [R]. This element has also been called the generalized Lefschetz number. The trace $\mathrm{RT}(f, \widetilde{f})$ is defined for any finite connected CW-complex X.

The theorem stated below was proven by Wecken in [We]. It says that when $\mathrm{RT}(f, \widetilde{f})$ is reduced so that each element of $R(f_\#)$ occurs at most once, then the coefficient of a Reidemeister class is the index of the corresponding Nielsen class of fixed points. Thus $L[f]$ is the sum of the indices in $\mathrm{RT}(f, \widetilde{f})$, and $N[f]$ is the number of terms with non-zero coefficient in the reduced form of $\mathrm{RT}(f, \widetilde{f})$. Note that $\mathrm{RT}(f, \widetilde{f}) = 0$ exactly when $N[f] = 0$. The challenge is to distinguish Reidemeister classes and thus determine which terms of the Reidemeister trace can be combined.

2.1. The definition of the Reidemeister trace. First we choose a base point x_0 for X. Then given a map $f: X \to X$ we choose a base path ω from x_0 to $f(x_0)$. We can replace f with a simplicial map that is homotopic to f because $N[f]$ is a homotopy invariant. In fact, we can do this without changing the base path significantly so that the induced homomorphism on $\pi_1(X)$ remains the same. This means that replacing f with such a simplicial approximation will not change the Reidemeister structure. Thus we assume without loss of generality that f is a simplicial map from X to X. We give the universal cover $\widetilde{X}$ the simplicial structure inherited from X, and we choose a base point $\widetilde{x}_0$ for $\widetilde{X}$ such that $\widetilde{x}_0$ projects onto x_0. The unique lift of ω beginning at $\widetilde{x}_0$ determines a lift $\widetilde{f}: \widetilde{X} \to \widetilde{X}$ of f. For each dimension q we choose a $\mathbb{Z}[\pi]$ basis for $C_q(\widetilde{X}, \mathbb{Z})$, the simplicial q chains of the covering space. This basis consists of one lift of each of the q-cells of X. Then $\widetilde{f}_q: C_q(\widetilde{X}, \mathbb{Z}) \to C_q(\widetilde{X}, \mathbb{Z})$ satisfies the following: For each $\widetilde{u}$ in the basis and for each $\sigma \in \pi_1(\widetilde{X})$ we have $\widetilde{f}_q(\sigma\widetilde{u}) = f_\#(\sigma)\widetilde{f}_q(\widetilde{u})$. Then for M_q the matrix over $\mathbb{Z}[\pi]$ for $\widetilde{f}_q$, we are interested in the alternating sums of traces of the M_q. Let $\rho: \mathbb{Z}[\pi] \to \mathbb{Z}(R(f_\#))$ be the linear extension of $\rho(\alpha) = [\alpha]$ for all $\alpha \in \pi_1(X)$. Then we can define the Reidemeister trace by:

(2.1) DEFINITION. The *Reidemeister trace* of f with lift $\widetilde{f}$ is defined to be

$$\mathrm{RT}(f, \widetilde{f}) = \sum_q (-1)^q \rho(\operatorname{trace} M_q) \in \mathbb{Z}(R(f_\#)).$$

2.2. Properties of the Reidemeister trace. If a different lift of f is chosen in the definition of the Reidemeister trace, then the new lift can be expressed in terms of the original lift as $\gamma\widetilde{f}$ for some $\gamma \in \pi_1(X)$. Then we have that for any $\beta \in \pi_1(X)$ the coefficient of $\rho(\beta)$ in $\mathrm{RT}(f, \widetilde{f})$ equals the coefficient of $\rho(\gamma\beta)$ in $\mathrm{RT}(f, \gamma\widetilde{f})$. The definition of $\mathrm{RT}(f, \widetilde{f})$ is independent of the other choices made in the definition, such as the choice of simplicial structure and the choice of transversal W explained in the theorem below.

The Reidemeister trace is homotopy invariant and satisfies the commutativity property. Specifically, if the map g is homotopic to f via a homotopy H, and if we choose as the lift of H the homotopy $\widetilde{H}$ beginning with $\widetilde{f}$ (so that $\widetilde{H}_0 = \widetilde{f}$), then we have $\mathrm{RT}(f, \widetilde{f}) = \mathrm{RT}(g, \widetilde{g})$ whenever we choose $\widetilde{g} = \widetilde{H}_1$. This choosing of a lift $\widetilde{g}$ is the same as choosing as the base path from x_0 to $g(x_0)$ the base path for f followed by the path $H(x_0, t)$. The commutativity property follows from the fact that for X and Y finite, connected CW-complexes and for maps $f: X \to Y$ and $g: Y \to X$ we have $g(\mathrm{Fix}(f \circ g)) = \mathrm{Fix}(g \circ f)$. There is a bijection between the collections of Nielsen classes for $g \circ f$ and $f \circ g$, and thus $\mathrm{RT}(g \circ f, \widetilde{g} \circ \widetilde{f}) = \mathrm{RT}(f \circ g, \widetilde{f} \circ \widetilde{g})$.

(2.2) THEOREM (Wecken). *For X a compact, connected polyhedron and for f a self-map of X,*

$$\mathrm{RT}(f, \widetilde{f}) = \sum_{\alpha \in W} i(\rho(\alpha))\rho(\alpha),$$

where W is a transversal for $R(f_\#)$ containing exactly one element of each Reidemeister class and where $i(\rho(\alpha))$ is the index of the Reidemeister class $\rho(\alpha)$. Thus $L[f]$ is the sum of the coefficients of $\mathrm{RT}(f, \widetilde{f})$, and $N[f]$ is the number of terms with non-zero index in the reduced form of $\mathrm{RT}(f, \widetilde{f})$.

SKETCH OF PROOF. Given a simplicial structure for X, we use the Hopf approximation technique described in [B1] and in Part 1 of [We]. This allows us to assume without loss of generality that f is a simplicial map (from a subdivision of the original simplicial complex to the original complex) for which each fixed point is in a maximal simplex and for which no two fixed points are in the same simplex. This forces $\widetilde{f}$ to be a simplicial map as well. (Because the Hopf approximation gives us a simplicial map that is in some sense close to our original map, the base path has not changed enough to change the induced homomorphism on the fundamental group.)

For any maximal q-simplex u in the simplicial structure for X, let $\widetilde{u}$ be the lift of u that is in the $\mathbb{Z}[\pi]$-basis for $C_q(\widetilde{X};\mathbb{Z})$. For any fixed point x of f in a maximal simplex u, there is a unique point $\widetilde{x} \in \widetilde{u}$ that projects down to x. The lift $\widetilde{f}$ takes this $\widetilde{x}$ to another point in the preimage of x, and hence $\widetilde{f}(\widetilde{x}) = \alpha\widetilde{x} \in \alpha\widetilde{u}$ for some $\alpha \in \pi_1(X)$. Thus the q-chain $\widetilde{f}_q(\widetilde{u})$ in $C(\widetilde{X};\mathbb{Z})$ contains the term $-\alpha\widetilde{u}$ if f reverses the orientation of u and contains the term $\alpha\widetilde{u}$ if f preserves the orientation of u. This means that $\pm\alpha$ is on the diagonal of M_q. The fixed point index of x is $(-1)^q$ (is $(-1)^{q+1}$) if f preserves (reverses) the orientation of u. Thus the occurrence of $\pm\alpha$ in $(-1)^q\text{trace}(M_q)$ that corresponds to x has as its coefficient the fixed point index of x, and we also have that the Reidemeister class of α represents the Nielsen class of x. When we add all of the occurrences of $\rho(\alpha)$ in the Reidemeister trace, the sum of the individual coefficients is the sum of the indices of the fixed points of one Nielsen class for f. □

Now that the connection between the Reidemeister trace and the Reidemeister structure has been established, we can use any simplicial approximation to our map that we choose.

2.3. Calculating the Reidemeister trace. For a closed hyperbolic surface X, we give X the CW-structure of one 0-cell, n 1-cells, and a single 2-cell. For a hyperbolic surface with boundary X, the surface is homotopy equivalent to a wedge of circles so that we do not have a 2-cell.

For any hyperbolic surface, the calculation of the Reidemeister trace involves a free group and a homomorphism on that group. Given a map f, a base point, and a choice of base path, we have a homomorphism $f_{\#}\colon \pi_1(X) \to \pi_1(X)$. If the fundamental group is not free, there are many ways to write $f_{\#}(a_i)$ for a generator a_i. That is, there are many endomorphisms of F_r that induce $f_{\#}$ on the one relator group. For example, suppose that X is the connected sum of two tori, with

$$\pi_1(X) = \langle a,b,c,d \mid aba^{-1}b^{-1}cdc^{-1}d^{-1}\rangle.$$

Then if $f_{\#}(a) = cdc^{-1}d^{-1}$, we could just as easily have said $f_{\#}(a) = bab^{1-}a^{-1}$.

(2.3) DEFINITION (Definition of $f_{\#F}$). Let X be a closed hyperbolic surface. Given a particular representation of $f_{\#}$, let $f_{\#F}\colon F_r \to F_r$ be the homomorphism for which $f_{\#}(a_i)$ and $f_{\#F}(a_i)$ look identical as strings of letters for each generator a_i.

The Fox calculus, introduced in [CF], was first used to calculate the Reidemeister trace for surfaces in [FH]. The derivatives record information about the lifts of cells to the universal cover. For a_i a generator of F_r and for w a word in F_r, the Fox partial derivative $\partial w/\partial a_i$ is determined by the following:

$$\frac{\partial}{\partial a_i}(uv) = \frac{\partial u}{\partial a_i} + u\frac{\partial v}{\partial a_i} \quad \text{for } u, v \in F_r, \tag{2.4.1}$$

$$\frac{\partial a_j}{\partial a_i} = \delta_{i,j} \quad \text{and} \quad \frac{\partial 1}{\partial a_i} = 0. \tag{2.4.2}$$

And as a consequence we have

$$\frac{\partial w^{-1}}{\partial a_i} = -w^{-1}\frac{\partial w}{\partial a_i} \quad \text{for any } w \in F_r. \tag{2.4.3}$$

(2.5) DEFINITION (The Fox trace). Fadell and Husseini, in [FH], defined what we will call the *Fox trace* to be the element of $\mathbb{Z}[F_r]$ given by

$$\mathrm{FT}(f_{\#F}) = 1 - \sum_{i=1}^{r} \frac{\partial f_{\#F}(a_i)}{\partial a_i} + A_2,$$

where we use the Fox calculus and where $A_2 \in \mathbb{Z}[F_r]$ is the trace of the cellular map f_2 in dimension 2 induced by $\widetilde{f}$.

That is, if M_2 is the $\mathbb{Z}[\pi]$ matrix for $\widetilde{f}_2$ on cellular chains, then A is the trace of M_2 interpreted in $\mathbb{Z}[F_r]$. If the surface has boundary, then $A_2 = 0$.

(2.6) THEOREM (Fadell and Husseini, [FH]). *The Reidemeister trace is the image under ρ of the Fox trace. That is,*

$$\mathrm{RT}(f, \widetilde{f}) = \rho(\mathrm{FT}(f_{\#F})) = \sum_{\alpha \in W} i(\rho(\alpha))\, \rho(\alpha).$$

3. Closed hyperbolic surfaces

We consider here X a closed surface of genus $n \geq 2$, and we have

$$\pi_1(X) = \langle a_1, b_1, \dots, a_n, b_n \mid a_1 b_1 a_1^{-1} b_1^{-1} \dots a_n b_n a_n^{-1} b_n^{-1} \rangle,$$

where the relator is the product of n commutators. In this section, the examples will be for $X = T^2 \# T^2$, the closed surface of genus 2. Thus the fundamental group for the examples is

$$G = \langle a, b, c, d \mid abABcdCD \rangle,$$

where we use upper case letters to denote inverses.

We work in the free group as well as in the one-generator group for the first steps of the calculation of $\mathrm{RT}(f, \widetilde{f})$. By choosing the representation of $f_{\#F}$ to be as simple as possible, we can guarantee that the trace in dimension two is a single term in $\mathbb{Z}[\pi]$. This is merely a convenience, because the algorithm below gives the trace in dimension two no matter what representation of $f_{\#}$ is chosen.

(3.1) Example (Part 1). Let X be the connected sum of two tori, and let f be a self-map on X that induces the homomorphism $f_\#$ on the fundamental group $G = \langle a, b, c, d : abABcdCD \rangle$. The following example is similar to those in [DHT] and [FH]. Let $f_{\#F}$ be given by

$$a \to ba^4, \quad b \to A^3B, \quad c \to dc^3, \quad d \to C^2D.$$

The Fox trace is

$$\mathrm{FT}(f_{\#F}) = 1 - (b + ba + ba^2 + ba^3 - A^3B + d + dc + dc^2 - C^2D) + A_2 \in \mathbb{Z}[F_4].$$

We will return to this example as we develop the theory for calculation $\mathrm{RT}(f, \widetilde{f})$.

The trace in dimension 2. For X a closed hyperbolic surface, the following algorithmic method for determining A_2, the trace in dimension 2, was introduced in [DHT]. We have that $f_{\#F}(R)$ is a product of conjugates of R and R^{-1} because $f_\#(R) = 1$ in $\pi_1(X)$. If we write $f_{\#F}(R) = \prod_{i=1}^{k} y_i R^{\lambda_i} y_i^{-1}$, with $y_i \in F_r$ and $\lambda_i \in \mathbb{Z}$, then $A_2 = \sum_{i=1}^{k} \lambda_i y_i \in \mathbb{Z}[F_r]$.

(3.1) Example (Part 2). Continuing the previous example, we have that

$$f_{\#F}(R) = f_{\#F}(abABcdCD) = baBAdcDC = baBA \cdot R^{-1} \cdot abAB,$$

so that $A_2 = -baBA$. Note that we could also have said that $baBAdcDC = cdCD \cdot R^{-1} \cdot dcDC$ and concluded that $A_2 = -cdCD$. Although these two answers for A_2 are different in the free group F_4, they are equal in $\pi_1(X)$. Thus we have that

$$\begin{aligned} \mathrm{RT}(f, \widetilde{f}) &= \rho(\mathrm{FT}(f_{\#F})) \\ &= \rho(1 - b - ba - ba^2 - ba^3 + A^3B - d - dc - dc^2 + C^2D - baBA). \end{aligned}$$

The work that remains. So far, the work we have done has been algorithmic. It is not difficult to program a computer algebra system, such as Magma [BCM], to do the calculations. But the next, and last, step is to determine which of the group elements that appear in $\mathrm{FT}(f_{\#F})$ are Reidemeister equivalent. We can not find the Nielsen number of f unless we have found the reduced form of $\mathrm{RT}(f, \widetilde{f})$. The techniques available for this last step are ad hoc in nature.

3.1. Reducing the Reidemeister trace. The techniques described here work for a surface with boundary as well as for a closed hyperbolic surface.

Suppose that X is a hyperbolic surface and that $f: X \to X$ is a map. Let $G = \pi_1(X)$, and let $\overline{G}$ be the abelianization of G. Then f induces three homomorphisms: $f_{\#F}: F_r \to F_r$, $f_\#: G \to G$, and $\overline{f_\#}: \overline{G} \to \overline{G}$. Given $\alpha, \beta \in G$, let $[\alpha]$

and $[\beta]$ be the Reidemeister classes for $f_\#$ of α and β, respectively. We would like to know whether $[\alpha]=[\beta]$ in $R(f_\#)$.

Whenever we have $\alpha, \beta \in G$ and we determine that $\overline{\alpha}$ and $\overline{\beta}$ are not Reidemeister equivalent in $R(\overline{f_\#})$, then we can conclude that $[\alpha] \neq [\beta]$ in $R(f_\#)$. In an abelian group, detecting Reidemeister equivalence becomes a simple problem in linear algebra with the set of Reidemeister classes given by $\text{Coker}(1-\overline{f_\#})$. This powerful technique is called *abelianization*. We use abelianization in the example below. The same ideas can be applied using any quotient group of G by a normal subgroup N for which $f_\#(N) \subseteq N$. Ferrario makes use of these ideas and extends them in [F1]

(3.1) Example (Part 3). We continue Example (3.1). In the abelianization $\overline{G}$, any solution $\overline{z}$ to $\overline{\alpha} = \overline{f_\#(z)}\overline{\beta}\overline{z}^{-1}$ must be of the form $\overline{z} = \overline{a^i b^j c^k d^l}$ with

$$\overline{\alpha}\overline{\beta}^{-1} = \overline{f_\#(z)}\overline{z}^{-1} = \overline{a^{3i-3j} b^{i-2j} c^{2k-2l} d^{k-2l}}.$$

Let $\overline{\alpha}\overline{\beta}^{-1} = \overline{a^{\Delta a} b^{\Delta b} c^{\Delta c} d^{\Delta d}}$. Whenever $3 \not| \Delta a$ or $2 \not| \Delta c$, then $\overline{\alpha}$ and $\overline{\beta}$ are not Reidemeister equivalent (and thus $[\alpha] \neq [\beta]$ in $R(f_\#)$).

In this example, we conclude from the technique of abelianization that none of the terms ba, ba^2, and dc is Reidemeister equivalent to any other of the terms appearing in the Fox trace.

Other techniques. We continue with a discussion of a few ad hoc techniques. Fadell and Husseini used in [FH] the fact that

$$[g] = [f_\#(g)] \quad \text{for all } g \in \pi_1(X) \quad \text{because } g \circ g = f_\#(g) g g^{-1} = f_\#(g).$$

In fact, for all $g, h \in \pi_1(X)$ we have

$$[f_\#(g)h] = [hg] \quad \text{because } g^{-1} \circ f_\#(g)h = f_\#(g^{-1}) f_\#(g) hg = hg.$$

These ideas work for Reidemeister equivalence in any space and are essentially the same as those used by Wagner in [W1] to show that her W is Reidemeister equivalent to the corresponding $\overline{W}$. (See Theorem (4.2) below.)

(3.1) Example (Part 4). We have $G = \pi_1(X) = \langle a, b, c, d \mid abABcdCD \rangle$ and

$$R(f, \widetilde{f}) = \rho(1 - b - ba - ba^2 - ba^3 + A^3B - d - dc - dc^2 + C^2D - baBA).$$

Because $f(b) = A^3B$, we have $[b] = [A^3B]$. Similarly, $[d] = [C^2D]$. In each case a group element with index 1 has combined with a group element with index -1.

Note also that if we use $z = a$, we have that $[A^3B] = [baBA]$. (That is, $f_\#(a) \cdot A^3B \cdot A = baBA$.)

index	equivalent elements	comments
-1	ba	not equivalent to any other term in $\mathrm{FT}(f_{\#F})$
-1	ba^2	not equivalent to any other term in $\mathrm{FT}(f_{\#F})$
-1	dc	not equivalent to any other term in $\mathrm{FT}(f_{\#F})$
-1	b, A^3B, $baBA$	possibly equivalent to other terms
-1	1, ba^3, dc^2	possibly equivalent to other terms
0	d, C^2D	possibly equivalent to other terms

We now know that $N[f]$ is either five or four, and to determine which it is we must discover whether $[1] = [b]$.

This author has checked using Magma that there are no words z of length 6 or less for which $f_{\#}(z) \cdot 1 \cdot z^{-1} = b$. It appears that the Nielsen number is five.

Algebraic techniques fail at this point. However, Kelly reports that using his geometric algorithm in [K5] he can prove that the homotopy class of f contains a homeomorphism with exactly 5 fixed points and no Nielsen paths between them. Thus the Nielsen number is actually 5, as we suspected.

Bounds on the fixed point index of a class. Bounds on the index of a Reidemeister class are quite useful when determining whether two group elements are Reidemeister equivalent. In the work of Jiang, Guo, and Kelly (see [JG], [J2], [K3] and [K4]), geometric arguments are used to develop the following powerful results for X any compact connected surface with negative Euler characteristic. For each Reidemeister class $\rho(\alpha) = [\alpha]$, let $i([\alpha])$ be the index of the class. Then for each $[\alpha]$ we have $i([\alpha]) \leq 1$, and there is a limit on the number of classes with index less than -1. Let $S = \{[\alpha] \in R(f_{\#}) : i([\alpha]) \leq -2\}$. There is a limit on the size of S because $\sum_S (i([\alpha]) + 1) \geq 2\chi(X)$. These facts imply that $|L[f] - \chi(X)| \leq N[f] - \chi(X)$. While some of these restrictions are not easily checked by computer, they are often helpful once other techniques have been used to find partial information about the Reidemeister classes. These bounds are not helpful for Example (3.1). However, the bounds are crucial for the following example taken from [DHT] and included in [H1]:

(3.2) EXAMPLE. For any $n \geq 2$ let $f_{\#F}$ be given by

$$a \to c^{-n+1}d^{-1}, \quad b \to dc^n, \quad c \to a, \quad d \to b.$$

The Fox trace $\mathrm{FT}(f_{\#F})$ is $1 + A_2$. We have $f_{\#F}(R) = cdCDabAB = cdCD \cdot R \cdot dcDC$. Thus $A_2 = cdCD$, and we have $\mathrm{RT}(f, \widetilde{f}) = \rho(1) + \rho(cdCD)$. Using the bounds on the index of a Reidemeister class, we know that these terms are distinct and that $N[f] = 2$. (Michael Kelly pointed out the index bounds after [DHT] went to press.) Ferrario, using techniques from [F1], was also able to distinguish these classes.

The use of the index bounds does not satisfy our requirement that our techniques be based in algebra because their proof is based in geometry. However, they are useful, easy to apply, and necessary at this stage in the development of the theory.

4. Hyperbolic surfaces with boundary

For X a hyperbolic surface with boundary, X is homotopy equivalent to a wedge of r circles and has fundamental group isomorphic to F_r. The calculation of the Fox trace for a map inducing $f_\#$ on F_r is the same as for a closed surface except that the trace in dimension two, A_2, is zero. As before, the Reidemeister trace is $\mathrm{RT}(f, \widetilde{f}) = \rho(\mathrm{FT}(f_\#))$, and the challenge is to distinguish classes of group elements that appear in the Fox trace so that the Reidemeister trace can be put into reduced form. The techniques described above apply here as well: abelianization, index bounds, and the fact that for all $g, h \in \pi_1(X)$ we have $[f_\#(g)h] = [hg]$.

In the work of Wagner and Yi (see [W1], [W2], and [Y]), the Reidemeister trace is not mentioned, although Wagner's initial work of finding W's is equivalent to finding the Fox trace. Instead, given an endomorphism $f_\#$ on F_r, Wagner exhibits a self-map f on the wedge of r circles for which $f_\#$ is this endomorphism and for which all fixed points are isolated and have index ± 1. Wagner's map f has exactly one fixed point for each of the terms in the unreduced form of the Fox trace $\mathrm{FT}(f_\#)$. Her calculation of Wagner tails (see below) corresponds exactly to the Fox derivative calculations for the Fox trace.

In the introduction we stated that more than the Nielsen number is needed when the figure eight is the base space of a fibration and when the goal is to study periodic points. For a map f, we say that *the Nielsen structure of f* is the collection of all pairs (N, i) with N a Nielsen class of fixed points of f and i the index of that Nielsen class. Some of the Nielsen classes may be inessential, but even those pairs can contain information needed in the study of periodic points of fibrations.

Our techniques will provide only algebraic information. We begin with the terms of the Fox trace. Each group element in the unreduced trace represents a fixed point as discussed above. Our goal is to keep track of which group elements are Reidemeister equivalent as we reduce the Reidemeister trace. We want to keep information about both essential and inessential Reidemeister classes that represent fixed points. To be precise, we form pairs (R, i) for which each R is a subset of a Reidemeister class appearing in $\mathrm{RT}(f, \widetilde{f})$ (and so R is a set of fundamental group elements) and i is the index of R. *The Reidemeister structure of* $f_\#$ is the collection of pairs (R, i) for which each group element from the Fox trace is in some R and for which in any two pairs the two R's are subsets of distinct Reidemeister classes.

In the Reidemeister structure that we find, not every element of the fundamental group is in one of the subsets R. We can find information only for those elements of the fundamental group that appear in the Fox trace. Some of these group elements may very well represent inessential fixed points that can be removed by a homotopy of f. What is important is that any inessential fixed points that can not be removed by a homotopy are represented in the Reidemeister structure as well.

4.1. Wagner's algorithm. Let $f\colon X \to X$ be a map on a hyperbolic surface with boundary. Because X is homotopy equivalent to a wedge of circles, we think of X as such a space. Using Wagner's algorithm from [W1] one can find Reidemeister equivalences between terms in the Fox trace of $f_\#$ and thus find information about the essential Nielsen classes of f. There are classes of maps for which Wagner's algorithm is conclusive. That is, for these maps the algorithm determines any and all Reidemeister equivalences between terms in the Fox trace of $f_\#$. Thus for these maps the algorithm finds $\mathrm{RT}(f, \widetilde{f})$, $N[f]$, and the Reidemeister structure of $f_\#$.

We have $\pi_1(X) = F_r$. Wagner replaces f with a map g homotopic to f that has only isolated fixed points and for which $g_\# = f_\#$. It is not necessary for us to do this because the Fox trace of $f_\#$ contains all the necessary information for finding the Reidemeister structure of f using Wagner's algorithm. But it is useful to know that there is a one-to-one correspondence between the fixed points of g and the terms of the Fox trace and that the terms of the Fox trace are precisely the group elements in F_r that Wagner calls W_i uses in her algorithm.

(4.1) EXAMPLE (Part 1). Consider the wedge of 4 circles. We will call the generators of the fundamental group a, b, c and d. Let $f_\#$ be the endomorphism defined as follows:

$$a \to aaBa, \quad b \to BB, \quad c \to bccbAA, \quad d \to ddBa.$$

Then the Fox trace is $1 - 1 - a \ - aaB + B + BB - b - bc - 1 - d$.

Labeling the terms of the Fox trace. The terms of the Fox trace are elements in F_r, and we seek to determine whether any of these elements are Reidemeister equivalent. We label each term with its order in the trace beginning with zero (so that a group element that occurs twice is considered with multiplicities) and record the generator a_j for which the term appeared as part of $\partial f_{\#}(a_j)/a_j$. The first 1 in the trace (corresponding to dimension zero) is labeled W_0, because it is first, and it has no generator assigned to it because it is not part of one of the partial derivatives.

Suppose that W_i comes from $\partial f_{\#}(a_j)/a_j$. Then W_i is assigned another group element $\overline{W}_i = f_{\#}(a_j^{-1})W_i a_j$. By definition, W_0 is assigned $\overline{W}_0 = 1$. It is clear that W_i is Reidemeister equivalent to $\overline{W}_i$.

(4.1) EXAMPLE (Part 2). We have the following information.

order $= i$	loop $= a_j$	coefficient	W_i	$\overline{W}_i = f_{\#}(a_j^{-1})W_i a_j$
0	—	+1	1	1 (by definition)
1	a	−1	1	AbA
2	a	−1	a	Ab
3	a	−1	aaB	1
4	b	+1	B	bb
5	b	+1	BB	b
6	c	−1	b	$aaBC$
7	c	−1	bc	aaB
8	d	−1	1	AbD
9	d	−1	d	Ab

Wagner tails. Yi calls the W's and $\overline{W}$'s Wagner tails, perhaps because we have $f_{\#}(a_i) = W_i a_j \overline{W}_i^{-1}$. Now that these Wagner tails have been determined, the algorithm continues. We compare all the W's and $\overline{W}$'s. The W's that have been found to be equivalent by the following theorem are combined into partial Reidemeister classes. This theorem is similar to ideas in earlier papers, such as [FH].

(4.2) THEOREM (Wagner). *We have $[W_i] = [W_j]$ whenever we have $W_i = W_j$, $W_i = \overline{W}_j$, or $\overline{W}_i = \overline{W}_j$.*

PROOF. If the W's are equal, then we are done. If $W_i = \overline{W}_j$, then we use $z = a_j^{-1}$ and check that this z is a solution to $W_i = f_{\#}(z)W_j z^{-1}$. If $\overline{W}_i = \overline{W}_j$, then the solution is $z = a_i a_j^{-1}$. □

(4.1) EXAMPLE (Part 3). We have $W_0 = W_1 = \overline{W}_3 = W_8$, $W_3 = W_7$, $\overline{W}_2 = \overline{W}_9$, and $\overline{W}_5 = W_6$. Thus we know that each of the following sets consists of elements that are Reidemeister equivalent: $\{W_0, W_1, W_3, W_7, W_8\}$, $\{W_2, W_9\}$, and

$\{W_5, W_6\}$. We must determine whether these sets are in distinct Reidemeister classes and whether W_4 is equivalent to any of these group elements.

The index bound arguments are not helpful here, but abelianization does show that Wagner's process has found all equivalences for this examples. The Reidemeister class containing W_4 has index $+1$, and the other classes have indices -3, -2 and 0, respectively. Thus $N[f] = 3$ and $L[f] = -4$. If any inessential fixed point classes of f are preserved by homotopy (and thus of interest to those studying periodic points), those fixed points are represented by the Reidemeister class $\{W_5, W_6\}$, which is inessential. Abelianization tells us that W_5 and W_6 are not Reidemeister equivalent to the group elements in the essential class. This seemingly useless information about the Reidemeister structure can be significant for studying the periodic point theory of fibrations, as discussed in the introduction.

We next address the question: For which homomorphisms is Wagner's algorithm sufficient for determining the Reidemeister classes? In other words, under what conditions can we know that Wagner's algorithm will find all Reidemeister equivalent terms of the Fox trace as it did in the example above?

4.2. Wagner-characteristic homomorphisms.

(4.4) Definition (Wagner). W_i and W_j are *directly related* ($W_i \overset{\text{dir}}{\sim} W_j$) whenever $W_i = W_j$, $W_i = \overline{W}_j$, or $\overline{W}_i = \overline{W}_j$. We say that a homomorphism $f\colon F_r \to F_r$ is *Wagner-characteristic* if whenever two terms from the Fox trace, W_x and W_y, are equivalent then there is a sequence $(W_{k_1}, \dots, W_{k_s})$ such that $W_x \overset{\text{dir}}{\sim} W_{k_1}$, $W_{k_s} \overset{\text{dir}}{\sim} W_y$, and $W_{k_t} \overset{\text{dir}}{\sim} W_{k_{t+1}}$ for $t = 1, \dots, s-1$.

We have proven that the homomorphism in Example (4.1) is actually Wagner-characteristic. But in this case knowing that our homomorphism is Wagner-characteristic does not help us at all because we have already found the Nielsen number and the Reidemeister classes. We would like to be able to determine which collections of homomorphisms are Wagner-characteristic when we are unable to distinguish Reidemeister classes in any other way. Once this has been done, it is simple to apply Wagner's algorithm to a Wagner-characteristic homomorphism and find the Reidemeister structure. This algorithm can be programmed in Magma or another computer algebra system.

4.3. Remnant implies Wagner-characteristic.

(4.5) Definition (Wagner). For $f_\#\colon F_r \to F_r$ a homomorphism, and for a_k a generator of F_r, $f_\#(a_k)$ *has remnant* (with respect to $f_\#$) if we can write $f_\#(a_k) = U_k \overline{X}_k V_k$ such that $\overline{X}_k \neq 1$, the product is reduced except that U_k and V_k might be 1, and $\overline{X}_k$ never cancels at all in any of the products $f_\#(a_k)f_\#(a_i)$, $f_\#(a_i)f_\#(a_k)$, $f_\#(a_j^{-1})f_\#(a_k)$, $f_\#(a_k)f_\#(a_j^{-1})$, with $i, j = 1, \dots, r$ and $j \neq k$.

The longest such subword $\overline{X}_k$ in $f_\#(a_k)$ is *the remnant of* $f_\#(a_k)$ (with respect to $f_\#$).

The homomorphism $f_\#$ *has remnant* if $f_\#(a_i)$ has remnant for $i = 1, \dots, r$.

(4.6) Example (Part 1). The endomorphism $f_\#$ on the free group with 3 generators that is defined by

$$a \to aBcA, \quad b \to abb, \quad c \to BccA$$

has remnant. We have $\overline{X}_a = B$, $\overline{X}_b = b$, and $\overline{X}_c = c$.

Wagner includes in [W1] a theorem attributed to R. F. Brown that states in essence that most maps have remnant. Note that the homomorphism in Example (4.1) does not have remnant because $f_\#(a)$ does not have remnant. (Consider $f_\#(ca) = bccbAA \cdot aaBa$ and $f_\#(aD) = aaBa \cdot AbDD$.) Thus Wagner-characteristic does not imply remnant. The following powerful theorem proves that the opposite implication does hold. We provide an outline of Wagner's proof with simpler notation than that used in the original paper.

(4.7) Theorem (Wagner). *Any homomorphism $f_\#\colon F_r \to F_r$ that has remnant is Wagner-characteristic (because for any $z \in F_r$ we have $|z| \le |f_\#(z)|$ and because any term in the Fox trace is either* 1 *or is an initial segment of $f_\#(a_i)$ for some generator a_i).*

Outline of Proof. Given two Reidemeister equivalent terms W_x and W_y from the Fox trace of $f_\#$, we must prove that there is a sequence of W's each directly related to the next that will connect W_x to W_y. We know that there exists a $z \in F_r$ such that $z = W_x^{-1} f_\#(z) W_y$.

Let $z = a_{i_1}^{\varepsilon_1} \dots a_{i_m}^{\varepsilon_m}$ with each a_{i_j} a generator of F_r, with each $\varepsilon_j = \pm 1$, and with this product reduced. We have

$$f_\#(z) = (U_{i_1}\overline{X}_{i_1}V_{i_1})^{\varepsilon_1} \dots (U_{i_m}\overline{X}_{i_m}V_{i_m})^{\varepsilon_m} = M\overline{X}_{i_1}^{\varepsilon_1}P_1\overline{X}_{i_2}^{\varepsilon_2}P_2 \dots P_{m-1}\overline{X}_{i_m}^{\varepsilon_m}N,$$

where $M = U_{i_1}$ or $V_{i_1}^{-1}$, where $N = U_{i_m}^{-1}$ or V_{i_m}, and where P_j is the reduced form of either V_jU_{j+1}, $U_j^{-1}U_{j+1}$, $U_j^{-1}V_{j+1}^{-1}$, or $V_jV_{j+1}^{-1}$, depending on the values of ε_j and ε_{j+1}. We have

$$|f_\#(z)| \ge \sum_j |\overline{X}_{i_j}| \ge |z|.$$

In the product

$$M\overline{X}_{i_1}^{\varepsilon_1}P_1\overline{X}_{i_2}^{\varepsilon_2}P_2 \dots P_{m-1}\overline{X}_{i_m}^{\varepsilon_m}N,$$

the $\overline{X}_{i_j}$ are like boulders in a stream. They never wash away. The P_j are like pebbles. They might be completely washed away and equal 1 when the product is reduced. It is our goal to show that the P_j do all equal 1. This will allow us to find the sequence of W's that we need.

Case 1. In the product

$$W_x^{-1} f_\#(z) W_y = W_x^{-1} M \overline{X}_{i_1}^{\varepsilon_1} P_1 \overline{X}_{i_2}^{\varepsilon_2} P_2 \dots P_{m-1} \overline{X}_{i_m}^{\varepsilon_m} N W_y,$$

we assume that W_x^{-1} does not cancel with any of $\overline{X}_{i_1}^{\varepsilon_1}$ and that W_y does not cancel with any of $\overline{X}_{i_m}^{\varepsilon_m}$. In this case we have

$$|z| = |W_x^{-1} f_\#(z) W_y| \geq |\overline{X}_{i_1}^{\varepsilon_1} P_1 \overline{X}_{i_2}^{\varepsilon_2} P_2 \dots P_{m-1} \overline{X}_{i_m}^{\varepsilon_m}| = \sum_j |\overline{X}_{i_j}| + \sum_j |P_j|.$$

This forces $P_j = 1$ and $\overline{X}_{i_j} = a_{i_j}$ for all j. We also must have $W_x = M$ and $W_y = N^{-1}$.

Recall that $f(a_{i_j}) = U_{i_j} \overline{X}_{i_j} V_{i_j}$ is a reduced product for each j. Thus $f(a_{i_j}) = U_{i_j} a_{i_j} V_{i_j}$ is reduced, and we must have U_{i_j} appearing as one of the W's in the Fox trace. Also we must have V_{i_j} appearing as a $\overline{W}^{-1}$ for each j. So if we think of U_j as W, then $V_j = \overline{U}_j$. Because each P_j is 1 we have the desired sequence of direct relationships connecting W_x to W_y as follows

$$W_x \overset{\text{dir}}{\sim} U_{i_1} \overset{\text{dir}}{\sim} U_{i_2} \overset{\text{dir}}{\sim} \dots \overset{\text{dir}}{\sim} U_{i_m} \overset{\text{dir}}{\sim} W_y.$$

For example, if $P_j = U_{i_j}^{-1} V_{i_{j+1}}^{-1} = 1$ then we know that

$$U_{i_j} = V_{i_{j+1}}^{-1} = \overline{U}_{i_{j+1}}, \quad \text{so } U_{i_j} \overset{\text{dir}}{\sim} U_{i_{j+1}}.$$

Case 2. The assumption in Case 1 does not hold. In this case Wagner replaces W_x, W_y, or both with their corresponding $\overline{W}$'s in order to find an equation using a shorter z for which the argument used in Case 1 does apply. For example, if W_x^{-1} cancels with some of $\overline{X}_{i_1}^{\varepsilon_1}$, then because W_x is an initial segment of some $f_\#(a_k)$ and $f_\#$ has remnant, we can conclude that W_x is an initial segment of $f_\#(a_{i_1})$, and thus $k = i_1$ and $\varepsilon_1 = 1$. Thus we have

$$f_\#(a_k) = U_k \overline{X}_k V_k = W_x a_k \overline{W}_x^{-1},$$
$$z = a_k a_{i_2}^{\varepsilon_2} \dots a_{i_m}^{\varepsilon_m} = a_k \overline{W}_k^{-1} f_\#(a_{i_2}^{\varepsilon_2} \dots a_{i_m}^{\varepsilon_m}) W_y.$$

Wagner replaces z with $a_k^{-1} z$ and proves that $\overline{W}_k^{-1}$ does not cancel with remnant in $\overline{W}_k^{-1} f_\#(a_{i_2}^{\varepsilon_2})$. □

(4.6) EXAMPLE (Part 2). We return to Example (4.6) to apply the theorem and determine the Reidemeister structure of the homomorphism.

We have $f_\#$ given by $a \to aBcA$, $b \to abb$, and $c \to BccA$, and we know that $f_\#$ has remnant. The Fox trace is $1 - 1 + aBcA - a - ab - B - Bc$. We have

order $= i$	loop $= a_j$	coefficient	W_i	$\overline{W}_i = f_\#(a_j^{-1})W_i a_j$
0	—	$+1$	1	1 (by definition)
1	a	-1	1	aCb
2	a	$+1$	$aBcA$	a
3	b	-1	a	B
4	b	-1	ab	1
5	c	-1	B	aC
6	c	-1	Bc	a

Wagner's algorithm tells us that the Reidemeister structure has two subsets of Reidemeister classes: $\{W_0, W_1, W_4\}$ and $\{W_2, W_3, W_5, W_6\}$. Because $f_\#$ has remnant, we know that the two Reidemeister classes are distinct. Both classes are essential, and the Nielsen number is 2.

4.4. Yi's process for calculating $N[f]$ on the figure eight. For a surface with $\pi_1(X) = F_2 = \langle a, b\rangle$, and for a map f that does not have remnant, Yi describes in [Y] a process that replace f with a homomorphism g for which $N[f] = N[g]$ and for which either g has remnant or g has a simple Fox trace. Thus either Wagner's algorithm can be used to find $N[g]$ or else simpler techniques will work. Yi's process is straightforward to apply, once it is understood, and works for a large class of maps. This is a significant extension of Wagner's results for F_2. However, the Reidemeister structure of g need not be the same as that of f, so this technique is of limited interest to those concerned with periodic point theory.

Yi uses the idea of a mutant, introduced by Jiang in [J2]. He also uses versions of ideas introduced by Kelly in [K1]: a standard form of a homomorphism along with certain moves used to change the homomorphism while preserving the Nielsen number. The version of standard form that Yi uses is the following:

(4.8) DEFINITION (Yi's standard form). Suppose that f is a map on the wedge of two circles (or on the twice punctured disc) and that $f_\#(a) = X$ and $f_\#(b) = Y$. Then $f_\#$ is in standard form if the following conditions hold:

(4.8.1) For any $q \in \langle a, b\rangle$ we have $|X| + |Y| \leq |qXq^{-1}| + |qYq^{-1}|$.
(4.8.2) The word Y is cyclically reduced. (That is, YY is reduced.)
(4.8.3) There is no cancellation in the product XY^{-1} nor in the product XY.

Note that the first condition says that no change of base path can shorten the sum of the lengths of the images of a and b.

As in Kelly's work, Yi defines moves that replace a map by a map homotopic to it.

Move 1. (This is used to replace a map with a map homotopic to it that induces a homomorphism in standard form.) Choose some $Q \in F_2$ and let $i_Q(g) = QgQ^{-1}$. Then replace $f_\#$ with a map $f'_\#$ for which $f'_\# = i_Q \circ f_\#$. Then f and f' are freely homotopic so that their Nielsen numbers are equal, but they may have quite different Reidemeister structures.

Move 2. (This move is used if a homomorphism has $f_\#(a)$ shorter than $f_\#(b)$. Yi's result is stated in terms of the opposite being true.) Let ψ be a homeomorphism that induces $\psi_\#$ given by $\psi_\#(a) = b$ and $\psi_\#(b) = a$. Replace f with $\psi f \psi^{-1}$. By commutativity, the Nielsen number and the Reidemeister structure are preserved.

Move 3. (This move is used to lengthen the image words to try to force the new homomorphism to have remnant.) Let ϕ be a homeomorphism that induces $\phi_\#(a) = ba$ and $\phi_\#(b) = b$. Replace f with $\phi f \phi^{-1}$. Again, by commutativity, the Nielsen number and the Reidemeister structure are preserved.

Move 4. Let θ be a homeomorphism that induces $\theta_\#(a) = a$ and $\theta_\#(b) = b^{-1}$. Replace f with $\theta f \theta^{-1}$. The Nielsen number and the Reidemeister structure are preserved.

(4.9) THEOREM (Yi, [Y, Theorems 2 and 3]). *Let f be a map on the twice punctured disc or the figure eight such that $f_\#(a) = U\overline{X}$ and $f_\#(b) = U$ with U and $\overline{X}$ reduced words and $\overline{X}$ not equal to any integral power of U.*

(4.9.1) *If $f_\#$ is in standard form, $\overline{X}$ is not an initial segment of U, and U is not an initial segment of $\overline{X}$, then there is a mutant g of f that is Wagner-characteristic. This map g can be found using a sequence of the moves listed above, and $N[f] = N[g]$.*

(4.9.2) *If $f_\#$ is in standard form, $f_\#(a) = U^n\overline{V}$ with $n \in \mathbb{Z}$ maximal, and $\overline{V}$ is not an initial segment of U, then either there is a mutant g of f that is Wagner-characteristic, implying $N[f] = N[g]$, or else "we can compute the Nielsen number directly".*

Yi also discusses the cases of homomorphisms that do not satisfy the hypotheses of his theorem. They are either homomorphisms for which known techniques can be used to find the Nielsen number or else they are of a particular (unusual) form for which he does not know what to do. But neither does anyone else.

(4.10) EXAMPLE. Let $f_\#(a) = a^4B^2$ and $f_\#(b) = a^2B$. This homomorphism does not have remnant. It is also not in standard form because XY^{-1} is not reduced. But if we use Move (1) with $Q = B$ then $f_1 = i \circ f$ and we have $f_{1\#}(a) = Ba^4B$ and $f_{1\#}(b) = Ba^2$. The homomorphism $f_{1\#}$ still does not have remnant, but it is in Yi's standard form. We have $U = Ba^2$ and $\overline{X} = a^2B$. We now follow Yi's process using Move 3 and let $f_{2\#} = \phi \circ f_{1\#} \circ \phi^{-1}$ so that $f_{2\#}(a) = babaB$ and $f_{2\#}(b) = aba$. Now we see that $f_{2\#}$ has remnant and $\mathrm{FT}(f_{2\#}) = 1 - b - bab - a$. Using Wagner's algorithm we have

order $= i$	loop $= a_j$	coefficient	W_i	$\overline{W}_i = f_{2\#}(a_j^{-1})W_i a_j$
0	—	$+1$	1	1
1	a	-1	b	bAB
2	a	-1	bab	b
3	b	-1	a	A

Because the homomorphism $f_{2\#}$ is W-characteristic, we know that the partial Reidemeister classes are $\{W_0\}$, $\{W_1, W_2\}$, and $\{W_3\}$, with indices $+1$, -2, -1, respectively. We also know that these three sets represent distinct Nielsen classes. Thus $3 = N[f_2] = N[f_1] = N[f]$. But because $i_B \circ f_\#$ does not necessarily have the same Reidemeister structure as $f_\#$, we can not conclude anything about the Reidemeister structure of f. We continue this example below and use a method that does give us the Reidemeister structure of $f_\#$.

4.5. n-characteristic homomorphisms. We turn now to results from [H2]. For a self–map on a hyperbolic surface with boundary, we consider properties of $f_\#$ that can make calculation of the Reidemeister structure possible. We consider the length of the word z needed in the Reidemeister action to connect two Reidemeister equivalent group elements from the Fox trace.

For a free group on two generators, Yi's method often allows us to find easily the Nielsen number of a homomorphism. In general, Yi's method does not preserve Reidemeister structure, as we can see by comparing Example (4.10) with Example (4.16).

When the Reidemeister structure is needed, then another option is another extension of Wagner's work discussed here: n-characteristic homomorphisms. For a certain class of homomorphisms, if two group elements from the Fox trace are Reidemeister equivalent then there is a the solution z to equation (1.1) with $|z| \leq 3$.

When a computer algebra system such as Magma is available, then an upper bound on the length of a solution z to equation (1.1) can be quite helpful. Given two W's, if Magma finds a z, it can be checked easily by hand. If not, then one can begin with confidence to prove that no z exists using the bound on the length of z.

The following definition is an extension of Definition (4.4).

(4.11) DEFINITION (n-related and n-characteristic). Let f be a self–map on a surface with boundary X that has $\pi_1(X) = F_r$, and let the homomorphism $f_\#: F_r \to F_r$ be induced by f. Then two terms W_1 and W_2 of the Fox trace are *n-related* ($W_1 \overset{n}{\sim} W_2$) whenever there is a $z \in F_r$ of length $|z| \leq n$ for which $z = W_1^{-1} f(z) W_2$.

A homomorphism $f_\#: F_r \to F_r$ is *n-characteristic* if whenever we have two Reidemeister equivalent terms, W_x and W_y, of the Fox trace there is a sequence $(W_{k_1}, \dots, W_{k_s})$ of terms from the Fox trace such that $W_x \overset{n}{\sim} W_{k_1}$, $W_{k_s} \overset{n}{\sim} W_y$, and $W_{k_t} \overset{n}{\sim} W_{k_{t+1}}$ for $t = 1, \dots, s-1$.

We note that this definition is for a homomorphism. Two homomorphisms induced by the same map (say with different base points) can have very different values of n in the definition of n-characteristic.

If a homomorphism is n-characteristic, then to find the Reidemeister structure one need only act on each term of the Fox trace by all group elements of length less than or equal to n. For small n, and for free groups with relatively few generators, this is straightforward using Magma.

(4.12) DEFINITION (S-related elements of F_r). Let f be a self-map on a surface with boundary X that has $\pi_1(X) = F_r$. Let the homomorphism $f_\#: F_r \to F_r$ be induced by f. Then given a subset S of the free group F_r, two terms W_i and W_j of the Reidemeister trace are *S-related* ($W_i \overset{S}{\sim} W_j$) when there is some $z \in S$ for which $W_i = f_\#(z) W_j z^{-1}$ or $W_j = f_\#(z) W_i z^{-1}$.

The set S is said to *determine the Reidemeister structure* of the homomorphism $f_\#$ if for any two Reidemeister equivalent terms W_x and W_y of the Fox trace there is a sequence $\{W_{k_i}\}$ of such terms with $W_x \overset{S}{\sim} W_{k_1}$, $W_{k_s} \overset{S}{\sim} W_y$, and $W_{k_t} \overset{S}{\sim} W_{k_{t+1}}$ for $t = 1, \dots, s-1$.

(4.13) REMARK. Any Wagner-characteristic homomorphism on F_r is actually 2-characteristic, but the converse is not true, as we see in Example (4.16) below. For any Wagner-characteristic homomorphism $f_\#$, the Reidemeister structure of $f_\#$ is determined by the set

$$S = \{a_i, a_i a_j^{-1} : i \neq j \text{ and } i, j = 1, \dots, r\},$$

where a_i is a generator of F_r. See the proof of Theorem (4.2).

We next introduce a class of endomorphisms on F_2 that are 2-characteristic and have very small sets that determine the Reidemeister structure of the homomorphism.

(4.14) Definition (Property (MRN) for a homomorphism). A homomorphism $f_\#\colon F_2 \to F_2$ has Property (MRN) if it has the following form: For m, n, r nontrivial elements of F_2, $f(a) = mrn$ and either

(MRN1) $f(b) = mn$, or
(MRN2) $f(b) = n^{-1}m^{-1}$,

with the words $nm, n^{-1}r, rm^{-1}, mrn$ and mn all reduced. In addition, we require that $r = sts^{-1}$ with $s, t \in F_2$ such that $|t| \geq 2$, t is cyclically reduced (this means that tt is reduced), and s possibly equal to 1.

We say that a homomorphism of this form has Property (MRN). To specify one of the particular forms above we say that $f_\#$ has Property (MRN1) or Property (MRN2). The proof of the next theorem for homomorphisms with Property (MRN1) is in [H2] and too involved to include here. The result for homomorphisms with Property (MRN2) follows by an argument inspired by Yi's work.

(4.15) Theorem. *A homomorphism $f_\#$ with Property* (MRN) *is* 2*-characteristic. In addition, for such a homomorphism, two group elements from* FT($f_\#$) *are* 3*-related whenever they are Nielsen equivalent. Specifically, if $f_\#$ has Property* (MRN1) *then the Reidemeister structure of $f_\#$ is determined by the set*

$$S = \{a, b, aB, A, B, bA\}.$$

In fact we have more. Any two Reidemeister equivalent terms from the Fox trace for $f_\#$ are T-equivalent using the set

$$T = \{a, b, aB, A, B, bA, bbA, aBB, aBA, abA\}.$$

The corresponding sets for homomorphisms with Property (MRN2) *are*

$$S = \{a, b, ab, A, B, BA\} \quad \text{and} \quad T = \{a, b, ab, A, B, BA, BBA, abb, abA, aBA\}.$$

Thus to find all 2-equivalences between terms of the Fox trace for a homomorphism with Property (MRN1) (and hence be able to deduce the Reidemeister structure and the Nielsen number), it is sufficient to act on each term in the Reidemeister trace using only each z in the set $\{a, b, aB\}$. If we then form a graph with the W's as vertices and 2-equivalences as directed edges, the path components will be subsets of distinct Reidemeister classes. Also, if two terms from the Fox calculus are Reidemeister equivalent, then the group element z that makes them equivalent is in the set $\{a, b, aB, A, B, bA, bbA, aBB, aBA, abA\}$.

The restriction in the definition of Property (MRN) that $|t| = 1$ is necessary. An example in [H2] has $|t| = 1$ and is 4-characteristic.

In the next example we use the above theorem to find the Reidemeister structure and the Nielsen number.

(4.16) EXAMPLE (Property (MRN1), not W-characteristic). We repeat here Example 4.10, but we determine the Reidemeister structure as well as the Nielsen number. Let $f_\#(a) = a^4B^2$ and $f_\#(b) = a^2B$. This homomorphism has Property (MRN1) with $m = a^2$, $r = t = a^2B$, $s = 1$, and $n = B$. The Fox trace is $1-1-a-a^2-a^3+a^2B$. By abelianization, we see that a^2B is not Reidemeister equivalent to any of the other terms in the Fox trace. Using Wagner's algorithm we find only the obvious equivalence between x_0 and x_1 using the following information:

order $= i$	loop $= a_j$	coefficient	W_i	$\overline{W}_i = f(a_j^{-1})W_ia_j$
0	—	+1	1	1
1	a	−1	1	b^2A^3
2	a	−1	a	b^2A^2
3	a	−1	a^2	b^2A
4	a	−1	a^3	b^2
5	b	+1	a^2B	b

A search using Magma found that W_3 and W_4 are Nielsen equivalent using $z = aB$. This is easily checked by hand, and thus this homomorphism is 2-characteristic but not W-characteristic.

No other equivalences were found by Magma when using $|z| \leq 3$. So far, we have classes $C_1 = \{W_0, W_1\}$ of index 0, $C_2 = \{W_2\}$ of index -1, $C_3 = \{W_3, W_4\}$ of index -2, and $C_4 = \{W_5\}$ of index $+1$, and we do not yet know whether there are any equivalences between fixed points in C_1, C_2 and C_3. The index bounds do not help here. Thus the Nielsen number is either 2 or 3.

If we trust Magma (acting on each of the W_i using all six of the z's listed in Theorem (4.15)), then by Theorem (4.15) all four classes listed must be distinct and $N[f] = 3$.

(We can use Yi's process to find that $N[f] = 3$ and thus conclude that C_2 and C_3 are not subsets of the same Reidemeister class. But we would also like to know whether the group elements in C_1 are Reidemeister equivalent to elements in any of the other sets because the fixed points represented by C_1 are of interest to those studying periodic point theory.)

To confirm Magma's conclusion, we can try all z's necessary by hand or we can use the following modification of the technique of abelianization.

We seek solutions $\overline{z}$ to $\overline{z\,f(z)}^{-1} = \overline{W}_1^{-1}\overline{W}_2$. Let $\overline{z} = \overline{a^ib^j}$. Because we know that we need only solutions with $|z| \leq 3$, we add the restriction $3 \geq |z| \geq |i| + |j|$. We have that $\overline{a^{-3i-2j}b^{2i+2j}} = \overline{W}_1^{-1}\overline{W}_2$. The fact that the exponent of $\overline{b}$ is even is what allowed us to determine that $\{W_5\}$ is a singleton class. For any pair of the other W's, we need the exponent of $\bar{b}$ to be zero. Thus we have $i = -j$ and $3 \geq |z| \geq 2|i|$. This forces $|i|$ to be 1 or 0. So the only possibilities for z are aB and bA (using the restricted set given in Theorem (4.15)). Thus we have $\overline{a^{\pm 1}b^0} = \overline{W}_1^{-1}\overline{W}_2$, and W_0 and W_3 are not equivalent. Now we have a very limited list of equivalences to test: two different pairs, each with two possible z's. It is easy to check that there are no additional equivalences. We have now confirmed Magma's result.

Extensions of the various techniques described in this section, perhaps even extensions of these techniques to the calculations on closed surfaces, would be welcome.

References

[BH] M. Bestvina and M. Handel, *Train-tracks for surface homeomorphisms*, Topology **34** (1995), 109–140.

[BCM] W. Bosma, J. J. Cannon and G. Mathews, *Programming with algebraic structures: Design of the Magma language*, Proceedings of the 1994 International Symposium on Symbolic and Algebraic Computation, Oxford, July 20–22, 1994 (M. Giesbrecht, ed.), Association for Computing Machinery, 1994, pp. 52–57.

[B1] R. F. Brown, *The Lefschetz Fixed Point Theorem*, Scott, Foresman and Co., 1971.

[B2] ———, *Fixed point theory*, History of Topology, Elsevier, 1999, pp. 271–299.

[CF] R. Crowell and R. Fox, *An Introduction to Knot Theory*, Ginn and Co., 1963.

[DHT] O. Davey, E. Hart and K. Trapp, *Computation of Nielsen numbers for maps of closed surfaces*, Trans. Amer. Math. Soc. **348** (1996), 3245–3266.

[FH] E. Fadell and S. Husseini, *The Nielsen number on surfaces*, Contemp. Math. **21** (1983), 59–98.

[F1] D. Ferrario, *Computing Reidemeister classes*, Fund. Math. **158** (1998), 1–18.

[F2] *Fixed points in bouquets of circles*, Far East J. Math. Sci., Special Volume Part II (1997), 129–136.

[G1] R. Geoghegan, *Fixed points in finitely dominated compacta: the geometric meaning of a conjecture of H. Bass*, Lecture Notes in Math., vol. 870, Springer–Verlag, 1981, pp. 6–22.

[G2] ———, *Nielsen fixed point theory*, Handbook of Geometric Topology (R. J. Daverman and R. B. Sher, eds.), North–Holland, 2002, pp. 499–521.

[Gu] J. Guaschi, *Nielsen theory, braids and fixed points of surface homeomorphisms*, Topology Appl. **117** (2002), 199–230.

[H1] E. Hart, *The Reidemeister trace and the calculation of the Nielsen number*, Nielsen Theory and Reidemeister Torsion, vol. 49, Banach Center Publications, Polish Academy of Sciences, 1999, pp. 151–157.

[H2] ———, *Reidemeister classes and Nielsen numbers on wedges of circles and punctured discs*, submitted.

[HK] P. Heath and E. Keppelmann, Work in progress, personal conversation, Fall 2003.

[Hu] S. Husseini, *Generalized Lefschetz numbers*, Trans. Amer. Math. Soc. **272** (1982), 247–274.

[J1] B. Jiang, *Lectures on Nielsen Fixed Point Theory*, Contemp. Math., vol. 14, Amer. Math Soc., Providence, Rhode Island, 1983.

[J2] ———, *Bounds for fixed points on surfaces*, Math. Ann. **311** (1998), 467–479.

[JG] B. Jiang and J. Guo, *Fixed points of surface diffeomorphisms*, Pacific J. Math. **160** (1993), 67–89.

[K1] M. Kelly, *Minimizing the number of fixed points for self-maps of compact surfaces*, Pacific J. Math. **126** (1987), 81–123.

[K2] ———, *Bounds on the fixed point indices for self-maps of certain simplicial complexes*, Topology Appl. **108** (2000), 179–196.

[K3] ———, *Minimal surface maps, fixed point indices and a Jiang–Guo type inequality*, Nielsen Theory and Reidemeister Torsion (Warsaw, 1996), vol. 49, Banach Center Publ., Polish Acad. Sci., Warsaw, 1999, pp. 227–233.

[K4] ———, *A bound on the fixed–point index for surface mappings*, Ergodic Theory Dynam. Systems **17** (1997), 1393–1408.

[K5] ———, *Computing Nielsen numbers of surface homeomorphisms*, Topology **35** (1996), 13–25.

[LN] J. Llibre and A. Nunes, *Minimum number of fixed points for maps of the figure eight space. Discrete dynamical systems*, Internat. J. Bifur. Chaos Appl. Sci. Engrg. **9** (1999), 1795–1802.

[MV] A. Martino and E. Ventura, *The conjugacy problem for free-by-cyclic groups*, preprint (2004).

[M] C. McCord, *Computing Nielsen Numbers*, Nielsen Theory and Dynamical Systems, Contemp. Math. **152** (1993), 249–267.

[R] K. Reidemeister, *Complexes and homotopy chains*, Bull. Amer. Math. Soc. **56** (1950), 297–307.

[T] V. G. Turaev, *Nielsen numbers and fixed points of mappings of bouquets of circles*, Zap. Nauchn. Sem. Leningr. Odtel. Mat. Inst. **122** (1982), 135–136.

[W1] J. Wagner, *An algorithm for calculating the Nielsen number on surfaces with boundary*, Trans. Amer. Math. Soc. **351** (1999), 41–62.

[W2] ———, *Classes of Wecken maps of surfaces with boundary*, Topology Appl. **76** (1997), 27–46.

[We] F. Wecken, *Fixpunktklassen,* I, II, III, Math. Ann. **117** (1941), 659–671; **118** (1942), 216–234, 544–577.

[Y] P. Yi, *An algorithm for computing the Nielsen number of maps on the pants surface*, Ph. D. Thesis, UCLA (2003).

14. FIBRE TECHNIQUES IN NIELSEN THEORY CALCULATIONS

PHILIP R. HEATH

1. Genesis, a brief history of beginnings

There are two basis questions associated with any Nielsen theory. The first is the so called Wecken question, which asks if the homotopy lower bound defined by the given Nielsen theory, is sharp. The second is the computational question, which of course asks how to calculate the given Nielsen numbers. For many years after Nielsen gave his definition of the ordinary Nielsen number, the only tool was "by hand" calculations using his definition on covering spaces. On the other hand, the resurgence of interest in Nielsen theory in the last part of the last century, introduced three new tools for calculations. Firstly there are the Jiang space techniques, where when the Lefschetz number is non-zero, the Nielsen number is equal to the appropriate Reidemeister number. Secondly there are fibre-space techniques, which attempt to give formulae connecting the Nielsen theories of base and fibre to the Nielsen theory of the total space. Finally there is the Fox calculus on surfaces invented by Fadell and Husseini. This chapter, as the title indicates, gives a brief exposition of the middle of the three methods of computation, together with occasional tangential considerations, where I was unable to resist the temptation (i.e. the use of fibre techniques to show that solvmanifolds are Wecken, see Corollary (6.5)).

By way of a historical introduction, the subject was in initiated by Bob Brown in his 1967 paper [B1]. The idea was to give a Nielsen theory analogue of the the Lefschetz product theorem of a fibre preserving ([1]) map $f: E \to E$, of an orientable (see (2.11)) Hurewicz fibration ([2]) $p: E \to B$. If the induced map on the base is denoted by $\overline{f}: B \to B$, and the restriction of f, to the fibre $F_b := p^{-1}(b)$ over

([1]) Fibre preserving means the lower half of the right hand diagram in (1.2) is commutative.

([2]) A Hurewicz fibration is one that has the absolute covering homotopy property, that is given any maps f and H, which make the left hand diagram in (1.2) commutative, where i is the inclusion, then there exists a third map $F: X \times I \to E$ which preserves the commutativity of the diagram.

a fixed point b in the base is denoted by $f_b: F_b \to F_b$, then the Lefschetz product formula ([3]) is $L(f) = L(\overline{f}) \cdot L(f_b)$. Thus [B1] gave the formula

$$N(f) = N(\overline{f}) \cdot N(f_b). \tag{1.1}$$

This idea is very natural and even obvious. For a given fibre preserving map $f: E \to E$, $\overline{f}: B \to B$ with b a fixed point of $\overline{f}$, the right hand diagram below

$$\begin{array}{ccc} X \times \{0\} & \xrightarrow{f} & E \\ {\scriptstyle i}\downarrow & & \downarrow{\scriptstyle p} \\ X \times I & \xrightarrow[H]{} & B \end{array} \qquad\qquad \begin{array}{ccc} F_b & \xrightarrow{f_b} & F_b \\ {\scriptstyle j}\downarrow & & \downarrow{\scriptstyle j} \\ E & \xrightarrow{f} & E \\ {\scriptstyle p}\downarrow & & \downarrow{\scriptstyle p} \\ B & \xrightarrow[\overline{f}]{} & B \end{array} \tag{1.2}$$

is strictly commutative. Now suppose that $x \in E$ is a fixed point of f, then $\overline{f}(p(x)) = p(f(x)) = p(x)$, so $p(x)$ is a fixed point of $\overline{f}$. Thus fixed points of fibre-preserving maps f of E lie over fixed points in B. For orientable fibrations the Nielsen numbers $N(f_b)$ on the restrictions f_b of f to fibres over the various fixed points b of $\overline{f}$, are all the same (see (2.11) and (2.13)). Since fixed points of f_b are also fixed points of f, there must be at least $N(f_b)$ fixed points of f in E over each fixed point $b \in B$. There are at least $N(\overline{f})$ fixed points in the base, so there must be at least $N(\overline{f}) \cdot N(f_b)$ fixed points of f in E. The problem is that this last number may not be the Nielsen number of f. In fact in [BF] counter examples are given on the Hopf fibration $S^1 \to S^3 \to S^2$. In particular fibre preserving maps $f: S^3 \to S^3$ of degree d, are exhibited for all $d \in \mathbb{Z}$. The homotopy ladder shows that the maps on S^1 are also of degree d, so when $d \neq 1$ we have that $N(f) = N(\overline{f}) = 1$, while $N(f_b) = |1-d|$ (see (2.3)). The point is that when the $|1-d|$ distinct Nielsen classes in the fibre are included into S^3, they coalesce into a single Nielsen class there. In addition to pointing out this mistake, the paper [BF] also gave very strong sufficient conditions under which formula (1.1) does hold (essentially the conditions given in Example (2.20)).

In the decade that followed these papers, various attempts at improvement of these results were made. Perhaps the most significant (in terms of the definitive

([3]) This folk-law theorem generalized the Euler–Poincaré formula $\chi(E) = \chi(B)\chi(F_b)$, given by Serre. The idea of the folk-law proof was that Serre's spectral sequence argument of the Euler–Poincaré formula could be generalized to Lefschetz numbers. In fact a proof was widely circulated by M. McCord in 1966, but never published. A modification of his proof can be found in the appendix of [HMP]. However a simpler proof can be found by specializing Jerzy Jezierski's proof of the coincidence version of this formula in Theorem (9.7).

solution (see [Y1]) that was eventually given in the early eighties) is Pak's result on the so called Pak obstruction $\mathcal{P}(\mathcal{F}, f)$ to Brown's formula (see [P]), and Ed Fadell's contribution to its calculation ([F]). For a locally trivial fibre bundle (4) $\mathcal{F} = \{E, p, B\}$, in addition to the usual orientability and $L(f)$ being non-zero, the definition of $\mathcal{P}(\mathcal{F}, f)$ required that E, B and every fibre be Jiang spaces. Then $\mathcal{P}(\mathcal{F}, f)$ is $\#(\operatorname{Ker} \overline{\jmath}_*)$, the cardinality of the kernel of the induced homomorphism $\overline{\jmath}_*$, on the cokernels in the diagram of abelian groups and homomorphisms

$$\begin{array}{ccccc}
\pi_1(F_b) & \xrightarrow{1-f_{b*}} & \pi_1(F_b) & \longrightarrow & \operatorname{Coker}(1 - f_{b*}) \\
\downarrow j_* & & \downarrow j_* & & \downarrow \overline{\jmath}_* \\
\pi_1(E) & \xrightarrow[1-f_*]{} & \pi_1(E) & \longrightarrow & \operatorname{Coker}(1 - f_*)
\end{array}$$

Under these conditions, the definition is independent of b. Pak's result then was that

$$\mathcal{P}(\mathcal{F}, f) \cdot N(f) = N(\overline{f}) \cdot N(f_b).$$

So formula (1.1) holds when $\mathcal{P}(\mathcal{F}, f) = 1$. In the Hopf fibration example from [BF], we have $\operatorname{Coker}(1 - f_{b*}) \cong \mathbb{Z}_{|1-d|}$, and $\operatorname{Coker}(1 - f_*) \cong 1$, so $\mathcal{P}(\mathcal{F}, f) = |1 - d|$.

For $f: (X, x) \to (X, x)$, we use $\operatorname{Fix} f_*^x$, to denote the kernal of the function $1 \cdot f_*^{x-1}: \pi_1(X, x) \to \pi_1(X, x)$, defined on α by $1 \cdot f_*^{x-1}(\alpha) = \alpha f_*^x(\alpha^{-1})$. The commutativity of the right hand diagram in (1.2) now allows us to deduced the existence of an induced homomorphism $p_*: \operatorname{Fix} f_*^x \to \operatorname{Fix} \overline{f}_*^{p(x)}$. So if $K := \ker(j_*) = 1$, where $j_*: \pi_1(F_b) \to \pi_1(E)$, then $\mathcal{P}(\mathcal{F}, f) = [\operatorname{Fix} \overline{f}_*^{p(x)}; p_*(\operatorname{Fix} f_*^x)]$. Ed Fadell's contribution in [F] is that this number is 1 for Hurewicz fibrations $p: E \to B$, when f admits a fibre-splitting (Definition (2.17)).

The definitive solution on product theorems came in the early 80's when Cheng-ye You in [Y1], produced the following formula for orientable Hurewicz fibrations under certain commutativity conditions

$$[\operatorname{Fix} \overline{f}_*^x; p_*(\operatorname{Fix} f_*^{p(x)})] N(f) = N_K(\overline{f}) \cdot N(f_b).$$

Here $N_K(f_b)$ is the mod K Nielsen number of f_b (see Section 2). This is the point at which I entered the game. I had attended a talk by Bob Brown at the 1980 Sherbrook conference, in which he had sketched the pre-You work. I was fascinated, and since I was chair of the colloquium committee at my school that year, I invited Bob to come and talk at M.U.N. (my school). By the time he came he had discovered Cheng-ye's work, and gave a presentation of it. This providential encounter radically changed the direction of my research, which since that time

(4) For such an $\mathcal{F} = \{E, p, B\}$, if B is paracompact, then $p: E \to B$ is a fibration.

has moved almost exclusively into Nielsen theory. By a happy coincidence this was immediately before my sabbatical year, a large part of which I spent trying to understand both Nielsen theory and Cheng-ye's paper. The paper [H1] resulted from that attempt. My previous research enabled me to reformulate Cheng-ye's work using ideas from groupoids and fibration and cofibration theory, and to formulate some "second generation" results. This reformulation also modified the the fundamental group approach, defining as usual the Reidemiester sets as quotients of the fundamental group, but assigning an index to the Reidemeister classes defined this way (see Section 2). These classes then act as weighted "components" for the geometric classes, enabling us to distinguish the various (different) possibly empty fixed point classes, the supposed advantage of the covering space approach (see [J, p. 2]). The modified approach also seeks wherever possible, to encapsulate key ideas within exact sequences and/or commutative diagrams. So running through this exposition, one finds exact Reidemeister and Fix group sequences and their generalizations (see (2.5), (9.5), and (9.18)). These ideas act as a unifying theme, so that each Nielsen formula is seen to arise from an appropriate exact sequence.

2. Product formulae for fibre preserving maps

We fix notation, and recall certain aspects of the modified fundamental group approach. Let $f\colon X \to X$ be a map. We denote the set of fixed points of f by $\Phi(f) = \{x \in X : f(x) = x\}$. Suppose that H is a normal subgroup ([5]) of $\pi_1(X)$, that $x \in \Phi(f)$ ([6]), and that $f_*^x(H) \subseteq H$, then f induces a homomorphism $f_*^x\colon \pi_1(X)/H \to \pi_1(X)/H$. This homomorphism in turn defines the Reidemeister classes of f, by the relation $\alpha, \beta \in \pi_1(X)/H$ are Reidemeister equivalent, if there is a class $\gamma \in \pi_1(X)/H$ such that $\alpha = \gamma\beta f_*^x(\gamma^{-1})$, with classes denoted $\mathcal{R}(f_*^x)$, and with cardinality $R(f)$. Let $\mathrm{Fix}_H f_*^x = \{\alpha H \in \pi_1(X,x)/H : f_*^x(\alpha H) = \alpha H\}$ denote the algebraic set of fixed points. The following sequence is an *exact sequence of based sets and basepoint preserving functions* (with the obvious basepoints).

$$(2.1)\qquad 1 \to \mathrm{Fix}_H f_*^x \to \pi_1(X,x)/H \xrightarrow{1\cdot f_*^{x-1}} \pi_1(X,x)/H \to \mathcal{R}_H(f_*^x) \to 1,$$

where $1 \cdot f_*^{x-1}(\alpha) = \alpha f_*^x(\alpha^{-1})$, When $\pi_1(X,x)/H$ is abelian ([7]), then $1 \cdot f_*^x$ is a homomorphism which we then write additively as $1 - f_*^x$. In this case $\mathcal{R}_H(f_*^x)$ inherits a canonical abelian group structure as the cokernel of $1 - f_*^x$, and of course $\mathrm{Fix}_H(f_*^x)$ is the kernel of this same function. We summerize this in the next proposition.

([5]) We actually need $H \subseteq \pi(X)$ to be a normal subgroupoid, see [HK1], [Jz1].

([6]) This is a simplification made for the purposes of exposition. If x is not a fixed point, then one chooses a "base path" $\omega\colon x \to f(x)$ and uses the homomorphism f_*^ω defined on α to be $\omega f_*(\alpha)\omega^{-1}$ in place of f_*^x etc.

([7]) In fact wherever we find "abelian" in the fixed point parts of this survey, it can be replaced by the "eventual commutativity" of the maps involved (see Chapter III.16).

(2.2) PROPOSITION ([H1]). *If $\pi_1(X, x_0)/H$ is abelian, then the sequence*

$$0 \to \mathrm{Fix}_H(f_*^x) \to \pi_1(X,x)/H \xrightarrow{1-f_*^x} \pi_1(X,x)/H \to \mathcal{R}_H(f_*^x) \to 0$$

is an exact sequence of abelian groups and homomorphisms.

When $H = 1$, we will omit it from this and subsequent notation. We recall next the relation between the geometry and the algebra via the functions ρ_H defined below. We then recall the notion of index on the H-Reidemeister classes (see [H1]), which we can then regard as "coordinates" for the geometric classes, empty or not. This device gets round the difficulty, said to be the disadvantage of the fundamental group approach, of not being able to discuss possibly empty fixed point classes. We denote the mod H Nielsen classes of f by the set $\Phi_H(f)/\sim$, with elements $\mathbf{F}$ etc. There is an injection

$$\rho_H: \Phi_H(f)/\sim \to \mathcal{R}_H(f_*^x)$$

given by $\rho_H(\mathbf{F}) = [c^{-1}f(c)H]$, where $c: x \to y$ is any path from the basepoint x to any representative y of $\mathbf{F}$. Again when $H = 1$ it is omitted. We then define the index of an H Reidemeister class $[\alpha H] \in \mathcal{R}_H(f_*^x)$ by

$$i([\alpha H]) = \begin{cases} \mathrm{ind}(\mathbf{F}) & \text{if } [\alpha H] = \rho_H(\mathbf{F}), \\ 0 & \text{otherwise.} \end{cases}$$

A class $[\alpha H]$ is said to be *essential*, if $i([\alpha H]) \neq 0$. We denote the set of essential classes of f by $\mathcal{E}(f)$, then $N_H(f)$, the mod H Nielsen number is the cardinality $\#(\mathcal{E}(f))$.

Since the torus T^k is a Jiang space, the equation $N(f) = |\det(I - A)|$ when $L(f) \neq 0$ in the next result is an immediate corollary of (2.2) with $H = 1$. That $N(f) = |L(f)|$ is more subtle, it is the main result of [BBPT].

(2.3) THEOREM ([BBPT]). *Let $f: T^k \to T^k$ be a map of a k torus T^k, and let matrix A be the matrix of the induced map on $\mathbb{Z}^k$, then*

$$N(f) = |L(f| = |\det(I - A)|,$$

and if $L(f) \neq 0$, then $N(f) = R(f)$.

Let $F_b \to E \xrightarrow{p} B$ be a fibration, and $K := \mathrm{Ker}\, j_{b*}$ where $j_{b*}: \pi_1(F_b) \to \pi_1(E))$, is induced by the inclusion $j_b: F_b \to E$. Then the sequence

$$1 \to \pi_1(F_{b*})/K \xrightarrow{j_{b*}} \pi_1(E) \xrightarrow{p_*} \pi_1(B) \to 1$$

is exact. Let f be a fibre preserving map of $p\colon E \to B$, using sequence (2.1) on the three maps f_b, f and $\overline{f}$ with $H = K$, the trivial group and the trivial group respectively, we obtain the following commutative diagram of *exact sequences of sets and based functions.*

$$
\begin{array}{ccccccccc}
1 & \longrightarrow & \mathrm{Fix}_K/f^x_{b*} & \xrightarrow{j_{b*}} & \mathrm{Fix}\, f^x_* & \xrightarrow{p_*} & \mathrm{Fix}\, \overline{f}^b_* & & \\
 & & \downarrow & & \downarrow & & \downarrow & & \\
1 & \longrightarrow & \pi_1(F)/K & \xrightarrow{j_{b*}} & \pi_1(E)p_* & \longrightarrow & \pi_1(B) & \longrightarrow & 1 \\
 & & \downarrow & & \downarrow{\scriptstyle 1\cdot f^{x-1}_*} & & \downarrow & & \\
1 & \longrightarrow & \pi_1(F)/K & \xrightarrow{j_{b*}} & \pi_1(E) & \xrightarrow{p_*} & \pi_1(B) & \longrightarrow & 1 \\
 & & \downarrow & & \downarrow & & \downarrow & & \\
 & & \mathcal{R}_K(f^x_{b*}) & \xrightarrow[j_{b*}]{} & \mathcal{R}(f^x_*) & \xrightarrow[p_*]{} & \mathcal{R}(\overline{f}^b_*) & \longrightarrow & 1
\end{array}
$$

We now give a kind of non-abelian "snake" lemma connecting the top and bottom horizontal sequences. To this end, let $\overline{\beta} \in \mathrm{Fix}\, \overline{f}^b_*$. Since $\mathrm{Fix}\, \overline{f}^b_*$ is a subgroup of $\pi_1(B)$ and $p_*\colon \pi_1(E) \to \pi_1(B)$ is surjective, then there is a $\beta \in \pi_1(E)$ with $p_*(\beta) = \overline{\beta}$. Now $(1 \cdot f^{x-1}_*)(\beta) = \beta f^x_*(\beta^{-1})$ lies in the kernel of $p_*\colon \pi_1(E) \to \pi_1(B)$, so there is an $\alpha K \in \pi_1(F)/K$ with $j_{b*}(\alpha K) = \beta f^x_*(\beta^{-1})$. We define $\delta(\overline{\beta})$ to be the projection of αK into $\mathcal{R}_K(f^x_{b*})$. Even though the diagram is only a mixture of groups and homomorphisms and sets and functions, there is enough structure around to prove the following lemma.

(2.4) LEMMA. *δ is a well defined function $\delta\colon \mathrm{Fix}\, \overline{f}^b_* \to \mathcal{R}_K(f^x_{b*})$.*

In practice the "liftings" that occur in the definition of δ can be thought of as arising from a regular lifting function for the fibration in question (more technical versions of this from [H2] allowed for fibre translation function τ_ω in [H1] to be base point preserving). The following exact sequences behave very much like the bottom end of the exact homotopy sequence of a fibration, as it moves from groups to based sets.

(2.5) THEOREM. *Let $f\colon E \to E$ be a fibre preserving map of a Hurewicz fibration $p\colon E \to B$, and let $x \in \Phi(f)$, and $b := p(x) \in \Phi(\overline{f})$. Then the following sequence of groups and homomorphisms (the "Fix" terms), and based sets and base point preserving functions (where the classes of the zero paths, are the base points for the Reidemeister sets), is exact.*

$$1 \to \mathrm{Fix}_K f^x_{b*} \xrightarrow{j_{K*}} \mathrm{Fix}\, f^x_* \xrightarrow{p_*} \mathrm{Fix}\, \overline{f}^b_* \xrightarrow{\delta} \mathcal{R}_K(f^x_{b*}) \xrightarrow{j_*} \mathcal{R}(f^x_*) \xrightarrow{p_*} \mathcal{R}(\overline{f}^b_*) \to 1.$$

When $\pi_1(E)$ is abelian, then so are $\pi_1(F_b)/K$, and $\pi_1(B)$, the above sequence becomes an exact sequence of abelian groups and homomorphisms, and $\mathrm{im}(\delta) \cong \mathrm{Fix}\,\overline{f}^b_*/p_*(\mathrm{Fix}\,f^x_*)$. The following corollary then, is immediate from cardinality considerations.

(2.6) COROLLARY. *If, in addition to the hypothesis of Theorem* (2.5), *we have that* $\pi_1(E)$ *is abelian, then*

$$[\mathrm{Fix}\,\overline{f}^b_*; p_*(\mathrm{Fix}\,f^x_*)]R(f) = R_K(f_b)\cdot R(\overline{f}).$$

In fact Cheng-ye You's formula (1.3) is now immediate for Jiang spaces, since when $L(f) \neq 0$, both $L(f_b)$ and $L(\overline{f})$ are non-zero, and we can simply replace the Reidemeister numbers with the corresponding Nielsen numbers in the formula in (2.6). However when the spaces are not Jiang spaces, we need much more subtle arguments.

Laying behind the product theorems for the Lefschetz and Nielsen numbers is the following product theorem for index. which can be thought of as a local version of the Lefschetz product theorem. Note that it does not require the fibration to be orientable.

(2.7) LEMMA (i.e. [J]). *Let* $f\colon E \to E$ *be a fibre preserving map of a fibration* $p\colon E \to B$, *with induced map* $\overline{f}\colon B \to B$ *on the base. Suppose that* b *is an isolated fixed point of* $\overline{f}$, *and* $A \subseteq \Phi(f_b)$ *be an isolated set of fixed points of* f_b, (*i.e.* A *is open and compact in* $\Phi(f_b)$). *Then* A *is also an isolated set of fixed points of* f, *and*

$$\mathrm{ind}(f, A) = \mathrm{ind}(f_b, A)\cdot \mathrm{ind}(\overline{f}, b).$$

The most important consequence of Lemma (2.7) is the following (see [Y1] and [H1]).

(2.8) LEMMA ([Y1], [H1]). *Let* $[\alpha K] \in \mathcal{R}_K(f^x_{b*})$, $[\beta] \in \mathcal{R}(f^x_*)$ *and* $[\overline{\beta}] \in \mathcal{R}(\overline{f}^b_*)$ *with* $j_{b*}([\alpha K]) = [\beta]$, *and* $p_*([\beta]) = [\overline{\beta}]$. *Then* $[\beta]$ *is essential, if and only if both* $[\alpha K]$ *and* $[\overline{\beta}]$ *are essential.*

One of the implications of Lemma (2.8), is that if x in Theorem (2.5) lies in an essential class $[\beta]$, then every element of $(j_{b*})^{-1}([\beta]) = \mathrm{im}(\delta)$ is essential, and we have the following corollary which is implicit in [H1].

(2.9) COROLLARY. *If, in addition to the hypothesis of* (2.5), *we have that* x *lies in an essential class of* f, *then the following is an exact sequence of groups and based sets.*

$$1 \to \mathrm{Fix}_K f^x_{b*} \xrightarrow{j_{b*}} \mathrm{Fix}\,f^x_* \xrightarrow{p_*} \mathrm{Fix}\,\overline{f}^b_* \xrightarrow{\delta} \mathcal{E}_K(f_b) \xrightarrow{j_{b*}} \mathcal{E}(f) \xrightarrow{p_*} \mathcal{E}(\overline{f}).$$

It is helpful to think of the sequences in Corollary (2.9) as a series of exact sequences, rather than as a single sequence. The reason for this is that for different choices of x, the sequences can be very different, even over the same b in B (see Example (5.1)). More particularly consider the following example of the Klein bottle K^2.

(2.10) EXAMPLE (Jiang, [J, p. 79]). The Klein bottle K^2 fibres over the circle S^1 as shown in the diagram. Here we regard K^2 as the quotient of the rectangle obtained by identifying opposite sides according to the indicated orientation. A self map f is obtained by reflecting about the vertical dotted line. In particular the restriction of f to the fibre over one of the fixed points b in the base is a map on S^1 of degree -1. The map $\overline{f}$ has two fixed points b and d, each in separate essential classes, and there are three non-empty fixed point classes $\{x_1\}$, $\{x_2\}$ and $\{y\}$ of f. The first two are essential and the last is inessential.

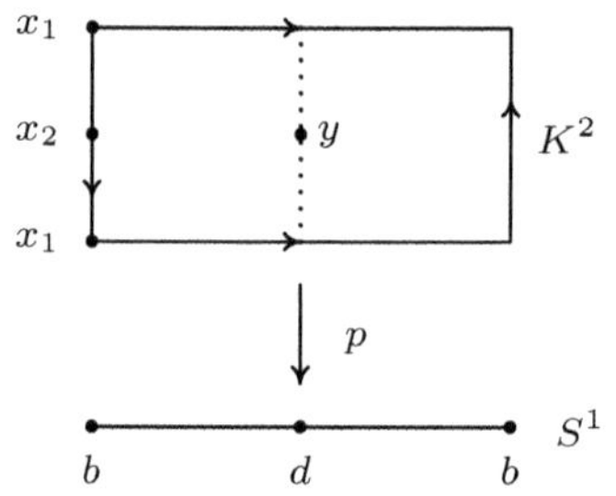

FIGURE 1

If we take x in Corollary (2.9) to be either x_1 or x_2, then the "Fix" terms in the sequence are 1, and the last three terms become

$$\{\{x_1\},\{x_2\}\} \xrightarrow{j_{b*}} \{\{x_1\},\{x_2\}\} \xrightarrow{p_*} \{\{b\},\{d\}\},$$

where j_{b*} is the identity, and $\mathrm{im}(p_*) = \{b\}$. Note that $p_*: \mathcal{E}(f) \to \mathcal{E}(\overline{f})$ need not be surjective. Note also for any $y \in F_d$, that $\mathcal{E}_K(f_d) = \emptyset$.

The purpose of the requirement that p be orientable, in the early results in fibred Nielsen theory, was to ensure the uniformity of the Nielsen number portions of formula (1.3). This condition was present in all early product formula, but of course is absent from Example (2.10).

(2.11) DEFINITION. A fibration $p: E \to B$ is said to be (*homotopically*) *orientable*, if for any two paths λ, μ in B with the same endpoints $\lambda(0) = \mu(0)$ and $\lambda(1) = \mu(1)$, the fibre translations $\tau_\lambda \simeq \tau_\mu: F_{\lambda(0)} \to F_{\lambda(1)}$.

(2.12) LEMMA ([8]). *If $\lambda \sim \mu: b \to d$ as paths in B, then $\tau_\lambda \simeq \tau_\mu: F_b \to F_d$. If $b, d \in \Phi(f)$, then the diagram*

$$\begin{array}{ccc} F_b & \xrightarrow{f_b} & F_b \\ {\scriptstyle \tau_\lambda}\downarrow & & \downarrow{\scriptstyle \tau_{\overline{f}(\lambda)}} \\ F_d & \xrightarrow[f_d]{} & F_d \end{array}$$

is homotopy commutative. Furthermore, $\tau_{\omega\mu} \sim \tau_\omega \tau_\mu$, and $\tau_0 \sim 1$, and so in particular $\tau_{\lambda^{-1}} f_d \tau_\lambda \sim \tau_{\lambda^{-1}\overline{f}(\lambda)} f_b$. That is f_d and $\tau_{\lambda^{-1}\overline{f}(\lambda)} f_b$ are equal up to conjugation.

For $K \neq 1$, the next result is a bit more difficult than it looks. It requires that $f_{b*}(K) \subset J(f_{b*})$ which however is automatic (see [Y1, p. 235] or [H1, 4.7]).

(2.13) PROPOSITION. *Let $f: E \to E$ be a fibre preserving map of a Hurewicz fibration $p: E \to B$, and $b, d \in \Phi(\overline{f})$. If either*

(2.13.1) b and d are in the same Nielsen class, or

(2.13.2) if p is orientable,

then we can replace $\tau_{\overline{f}(\lambda)}$ with τ_λ in Lemma (2.12), so that f_b and f_d have the same homotopy type as maps. Also in either case $L(f_b) = L(f_d)$, $N(f_b) = N(f_d)$, $N_K(f_b) = N_K(f_d)$, $R(f_b) = R(f_d)$ and finally $R_K(f_b) = R_K(f_d)$.

(2.14) THEOREM (You, [Y1]). *If $N_K(f_b)$ is independent of b in any essential class of $\overline{f}$, and if $[\mathrm{Fix}\, \overline{f}_*^{p(x)}; p_*(\mathrm{Fix}\, f_*^x)]$ is independent of x in any essential class of f. Then*

$$[\mathrm{Fix}\, \overline{f}_*^{p(x)}; p_*(\mathrm{Fix}\, f_*^x)] N(f) = N_K(\overline{f}) \cdot N(f_b).$$

In particular the formula holds if $\pi_1(X)$ is abelian, and p is orientable.

We sketch a proof which will accommodate the later "naïve addition theorem".

SKETCH OF PROOF. We make the simplification that $\#(\Phi(\overline{f})) = N(\overline{f})$, i.e. that each fixed point class in the base consists of a single point. This gives the flavour of the proof without getting into the technicalities that are needed to define certain transformations (T) between fibres, and so show the analysis is independent of various choices made ([9]). Now since p_* restricts to a function $\mathcal{E}(f) \to \mathcal{E}(\overline{f})$ we have that

$$N(f) = \#(\mathcal{E}(f)) = \#\Big(\bigcup_{b \in \Phi(\overline{f})} p_*^{-1}(\mathbf{F}_b) \Big) = \sum_{b \in \Phi(\overline{f})} \#(p_*^{-1}(\mathbf{F}_b)),$$

([8]) This simple but fundamental lemma occurs independently all over the place. It was one of the things that made me prick up my ears at Brown's 1980 Sherbrook talk, since I had needed it in my thesis.

([9]) One of the simplifications in [H1] is that such transformations can be taken to be basepoint preserving. The citation for the necessary fibration/cofibration theory for this given in [H1, 4.4], is incorrect. The correct citation should be [H2].

where $\overline{\mathbf{F}}_b$ denotes the fixed point class of b. Now let $x \in \mathbf{F}_x$ in $p_*^{-1}(\overline{\mathbf{F}}_b)$. Since the sequence in (2.9) is exact at $\mathcal{E}(f)$ then $\#(p_*^{-1}(\overline{\mathbf{F}}_b)) = c_b$, where c_b is the number of equivalence classes of $\mathcal{E}_K(f_b)$ under the equivalence relation defined next. Two mod K classes in $\mathcal{E}_K(f_b)$ are said to be equivalent, if they have the same image in $\mathcal{E}(f)$ under j_{b*}. Thus $N(f) = \sum_{b\in\Phi(\overline{f})} c_b$. Let $\mathbf{F}_x$ be a fixed point class of f containing x, and let $b = p(x)$. The the sequence (2.9) is exact with this x and this b. By exactness at $\mathcal{E}_K(f_b)$, the number of points of $\mathcal{E}_K(f_b)$ that are coalesced to $\mathbf{F}_x$ is $[\operatorname{Fix} \overline{f}_*^b; p_*(\operatorname{Fix} f_*^x)]$. But this number is independent of x (and hence b), so that the cardinality c_b of the quotient sets of $\mathcal{E}_K(f_b)$ is $N_K(f_b)/[\operatorname{Fix} \overline{f}_*^b; p_*(\operatorname{Fix} f_*^x)]$ and by hypothesis, this is independent of both x and b. Thus

$$\begin{aligned} N(f) &= \sum_{b\in\Phi(\overline{f})} c_b = \sum_{b\in\Phi(\overline{f})} N_K(f_b)/[\operatorname{Fix} \overline{f}_*^b; p_*(\operatorname{Fix} f_*^x)] \\ &= N(\overline{f}) N_K(f_b)/[\operatorname{Fix} \overline{f}_*^b; p_*(\operatorname{Fix} f_*^x)], \end{aligned}$$

as required. □

We can think of Pak's results as an addendum of our proof of Theorem (2.14). In fact since the function $\mathcal{E}(f_b) \to \mathcal{E}_K(f_b)$, which takes a class to its mod K class is surjective, then the function $\mathcal{E}(f_b) \to \mathcal{E}(f)$ has the same image as $j_{b*}\colon \mathcal{E}_K(f_b) \to \mathcal{E}(f)$. Note also under conditions of commutatativity, for the cokernels given in the introduction in the definition of Pak's number $\mathcal{P}(\mathcal{F})$, we have $\operatorname{Coker}(1 - f_{b*}^x) \cong \mathcal{R}(f_{b*}^x)$, $\operatorname{Coker}(1 - f_*^x) \cong \mathcal{R}(f_*^x)$ and that $\mathcal{P}(\mathcal{F}, f)$ is independent of x and b. Thus under the additional hypothesis that p is orientable, the proof of You's formula can be mimicked to give the following modest generalization (to arbitrary Hurewicz fibrations and to non-Jiang spaces) of Pak's result.

(2.15) ADDENDUM ([P]). *If p is orientable, and $\pi_1(E)$ is abelian, then*

$$\mathcal{P}(\mathcal{F}, f)N(f) = N(\overline{f})N(f_b).$$

(2.16) COROLLARY (You). *If p is orientable, then $N(f) = N(\overline{f})N(f_b)$ if and only if*

(2.16.1) $N_K(f_b) = N(f_b)$ and
(2.16.2) for each $x \in \mathbf{F} \in \mathcal{E}(f)$ we have $[\operatorname{Fix} \overline{f}_*^{p(x)}; p_*(\operatorname{Fix} f_*^x)] = 1$.

The condition that $[\operatorname{Fix} \overline{f}_*^{p(x)}; p_*(\operatorname{Fix} f_*^x)] = 1$, had been studied earlier by Ed Fadell (see [F]). The condition that $N_K(f_b) = N(f_b)$ holds of course, if $K = 1$, but also holds under a number of other conditions (see the concept of nilpotent homomorphisms in [H1, 4.18]). Taken together without orientability these two conditions have been dubbed *the naïve addition conditions.* This originated with

Boju Jiang's description of formula (1.1) as the naïve product formula ([J, p. 87]). The phrase was coined in [HKW], where it gives necessary and sufficient conditions for a "naïve addition" formula ([HKW, p. 87] see also (5.5)). Ed Fadell's definition (omitting his condition that $K = 1$) is the following.

(2.17) DEFINITION ([F]). Let $p: E \to B$ be a fibration, and f be a fibre preserving map. Then p is said to admit *a fibre-splitting with respect to f* if $p_*: \pi_1(E) \to \pi_1(B)$ admits a right inverse (section) σ such that, if $H = \mathrm{im}(\sigma)$, then H is normal, and $f_*(H) \subset H$.

(2.18) PROPOSITION ([F]). *If $p: E \to B$ is a fibration, and f a fibre preserving map which admits a fibre-splitting with respect to f, then*

$$[\mathrm{Fix}\,\overline{f}_*^{p(x)}; p_*(\mathrm{Fix}\, f_*^x)] = 1.$$

PROOF. It is easy to show that a fibre splitting defines a section to $p_*: \mathrm{Fix}\, f_*^x \to \mathrm{Fix}\,\overline{f}_*^b$. □

(2.19) COROLLARY ([F]). *If p is orientable, $K = 1$, and either*

(2.19.1) p admits a fibre-splitting with respect to f, or
(2.19.2) $\mathrm{Fix}\,\overline{f}_*^b = 1$ for all b in $\Phi(\overline{f})$,

then $N(f) = N(\overline{f})N(f_b)$.

(2.20) EXAMPLES ([BF]). Among the conditions then, which imply that $N(f) = N(\overline{f})N(f_b)$ we have

(2.20.1) $\pi_1(B) = \pi_2(B) = 0$;
(2.20.2) $\pi_1(F) = 0$;
(2.20.3) p is a trivial fibration, and $\pi_1(B) = 0$;
(2.20.4) p is a trivial fibration, and $f = f_b \times \overline{f}$.

3. Applications of product formulae to nil and $\mathcal{NR}$ solvmanifolds

In this section we give applications of the naïve product theorem to nilmanifolds and $\mathcal{NR}$ solvmanifolds. We briefly outline the essentials of the fibre theory of nilmanifolds as presented in [FH] (but see also [HK1]). A *nilmanifold* is a homogeneous space of the form $N = G/\Gamma$ where G is a nilpotent Lie group which is homeomorphic (as a space only) to some $\mathbb{R}^n$ and Γ is a uniform (i.e. N is compact) discrete subgroup. The main reference for nilmanifolds is the fundamental work of [M], the main references for Nielsen theory of nilmanifolds are [FH], [A].

The simplest examples of nilmanifolds are the tori where $G = \mathbb{R}^n$ has the usual abelian structure and $\Gamma = \mathbb{Z}^n$. Nilmanifolds are aspherical with $\pi_1(N) \cong \Gamma$ and any self map $f: N \to N$ is, up to homotopy, induced by a homomorphism

$F\colon G \to G$. Furthermore, any homomorphism on Γ extends uniquely to a homomorphism on G. When N is a torus the homomorphism F inducing f is simply a linear map whose matrix with respect to the usual basis has integer entries. For us the most important property of nilmanifolds is the following

(3.1) THEOREM (see [KMc], [HK1], [R]). *Let N be a nilmanifold which is not a torus. Then there is a toral decomposition for N which consists of a sequence of tori $T_1, \dots, T_k$ and a sequence of nilmanifolds $N_2 = T_1, N_3, \dots, N_{k-1}, N_k, N_{k+1} = N$ such that for each $i = 2, \dots, k$ there is an orientable fibration $T_i \to N_{i+1} \to N_i$. That is, there is a tower of orientable fibrations*

$$\begin{array}{ccccccccc}
T_k & T_{k-1} & T_{k-2} & & T_i & & T_3 & T_2 \\
\downarrow & \downarrow & \downarrow & & \downarrow & & \downarrow & \downarrow \\
N & N_k & N_{k-1} & \cdots & N_{i+1} & \cdots & N_4 & N_3 \\
\downarrow & \downarrow & \downarrow & & \downarrow & & \downarrow & \downarrow \\
N_k & N_{k-1} & N_{k-2} & & N_i & & N_3 & T_1
\end{array}$$

decomposing N. Furthermore, each self map f of N is homotopic to a map g which induces fibre preserving maps $g_i\colon N_i \to N_i$ with fibre uniformity on each of the above fibrations.

Since all spaces in the tower are $K(\pi, 1)$'s (Eilenberge–McLane spaces), then (as is a little unusual in algebraic topology) the topology can be thought of as reflecting the algebra exactly. That is whatever can be done algebraically, can also be done topologically. Thus each fibration $T_i \to N_{i+1} \to N_i$ in the tower can be thought of as realizing the exact sequences $[G, G_i] \to [G, G_{i+1}] \to [G, G_{i+1}]/[G, G_i]$ associated with the corresponding term in the lower central series of the covering simply connected nilpotent Lie group (quotiented appropriately). Also, since any algebraic self homomorphisms must take commutators to commutators, it can be realized as a fibre preserving map $f_{i+1}\colon N_{i+1} \to N_{i+1}$ of $T_i \to N_{i+1} \to N_i$. We jokingly refer to taking advantage of this exact reflection of the algebra and the geometry, as chanting the McCord mantra (it was Chris's joke at a talk he gave in Mt. Holyoke). Finally, note that right hand fibration $T_2 \hookrightarrow N_3 \to T_1$ (which we think of as the first in the tower, but coming from the last stage of the lower central series) has a torus in both base and fibre. This is important in inductive arguments based on these finite towers.

(3.2) EXAMPLE (Baby nil). The simplest non-abelian example of a nilmanifold BN occurs in dimension 3, and is called the Heisenberg group (we christened it Baby nil in [HK1]). The nilmanifold BN is a quotient of the group G of all 3×3

upper triangular matrices with 1s

$$G = \begin{pmatrix} 1 & x & y \\ 0 & 1 & z \\ 0 & 0 & 1 \end{pmatrix}$$

along the diagonal, as shown The binary operation is matrix multiplication. With this understood, we can represent these matrices as triples $X = (x, y, z)$, with corresponding group operation $X_1X_2 = (x_1, y_1, z_1)(x_2, y_2, z_2) = (x_1 + x_2, y_1 + y_2 + x_1z_2, z_1 + z_2)$.

Let Γ be the subgroup of G that consists of those matrices with integer entries, then let $BN = G/\Gamma$. Note that $X_1X_2X_1^{-1}X_2^{-1} = (0, -x_2z_1 + x_1z_2, 0)$, so the center of G consists of those matrices of the form $(0, y, 0)$. The fibre-tower follows this decomposition to give the Fadell–Husseini fibration for BN with the form $S^1 \to BN \xrightarrow{p} T^2$, where $T^2 = S^1 \times S^1$ is the two dimensional torus, and p is induced by the projection on the "x and z" factors.

The symbol $\Phi_{\mathcal{E}}(f)$ below, denotes those fixed points of f that are contained in some essential fixed point class of f.

(3.3) Definition ([HKW]). We say that a space X is *essentially* Fix *trivial* provided that for any $f: X \to X$, and any $x \in \Phi_{\mathcal{E}}(f)$, the group $\operatorname{Fix} f_*^x = \{\alpha \in \pi_1(X, x) : f_*^x(\alpha) = \alpha\} = 1$.

It is not hard to see that on any torus T, if $\operatorname{Fix}(f_*^x) \neq 0$ then $\Phi_{\mathcal{E}}(f) = \emptyset$ so the tori are essentially fix trivial (see also [HKW, p. 144]).

(3.4) Lemma ([FH]). *Nilmanifolds are aspherical and essentially fix trivial.*

Proof. Starting with the fact that tori are aspherical and essentially fix trivial, we proceed by induction on the tower in Theorem (3.1). Let f be a map which is fibre preserving with respect to this tower, and let $f_{i+1}: N_{i+1} \to N_{i+1}$ be the induced map on N_{i+1}. The inductive step for essential fix triviality uses the portion (with $K = 1$) of the sequence

$$1 \longrightarrow \operatorname{Fix} f_{ib*}^x \xrightarrow{j_{b*}} \operatorname{Fix} f_{(i+1)*}^x \xrightarrow{p_*} \operatorname{Fix} \overline{f}_{i*}^b$$

from Theorem (2.5). It follows easily that when $\operatorname{Fix} f_{ib*}^x$ and $\operatorname{Fix} \overline{f}_{i*}^b$ are trivial, then so is $\operatorname{Fix} f_{(i+1)*}^x$. □

The next theorem is due to Anasov, but we give Fadell and Husseini's fibre-space proof.

(3.5) THEOREM ([A], [FH]). *For a map f of a nilmanifold N, we have $N(f) = |L(f)|$.*

PROOF. The proof is again by induction, using the result (2.3) for tori. Starting the induction we see that $N(f_3) = N(f_2)N(f_1) = |L(f_2)||L(f_1)| = |L(f_2)L(f_1)| = |L(f)|$. Again the inductive step is identical. □

The reader will have noticed that the Anasov Theorem (3.5), does not have the part about $N(f) = |\det(I - A)|$. In fact this part is also true, but we need to define what we mean by A in the context of nilmanifolds. The appropriate A is the linearization of f, which is defined only up to conjugation. If $f\colon N \to N$ is a fibre preserving map of a nilmanifold N, then *the linearization of f* is the matrix F which results from placing the linearizations of the induced maps on $T_1, \ldots, T_k$ in a big matrix as block diagonals (see Chapter I.3).

(3.6) EXAMPLE (Baby nil, [HK1]). Let $BN = G/\Gamma$ be as in Example (3.2). Consider the function $\widetilde{\phi}$ on G given by

$$\widetilde{\phi}(x, y, z) = \Big(-x + 3z, x^2 + 4x + \frac{1}{2}(21z^2 - 17z) - 6xz - y, -2x + 7z \Big).$$

This map being a homomorphism induces a self map ϕ of BN (see [HK4, 2.2]). The Fadell and Husseini tower of fibrations for BN consists of a single stage, which takes the form $S^1 \to BN \xrightarrow{\pi} T^2$, where π is induced by projection on the first and third coordinates. Note that ϕ is fibre preserving with respect to π. Since S^1 is the fibre over $(0, 0)$, the linearization of the map on the principal fibre is the 1 by 1 matrix (-1). On the other hand the linearization of the induced map on T^2 is the matrix $\overline{F}$ below. The linearization F of ϕ, is as displayed.

$$\overline{F} = \begin{pmatrix} -1 & 3 \\ -2 & 7 \end{pmatrix}, \qquad F = \begin{pmatrix} -1 & 3 & 0 \\ -2 & 7 & 0 \\ 0 & 0 & -1 \end{pmatrix}, \qquad Z = \begin{pmatrix} a & b & 0 \\ d & f & 0 \\ 0 & 0 & af - bd \end{pmatrix}.$$

In fact every matrix Z that is the linearization of a map of BN has the form of Z as shown.

Note that the next results which uses the naïve product formula, gets away from all commutativity assumptions on the spaces involved.

(3.7) THEOREM ([FH]). *Let $f\colon N \to N$ be a map of a nilmanifold N, with linearization F, then $N(f) = |\det(I - F)| = |L(f)|$.*

For the ϕ in Example (3.6) whose linearization is F, we have that $N(f) = |\det(I - F)| = 12$.

PROOF. Let F_i be the linearizations of the restrictions of f to f_i to the tori $T_1, \ldots, T_k$ in the Fadell–Husseini tower for N. Then by Theorems (2.19) and (2.3) we have $N(f) = |\det(I - F_1)| \ldots |\det(I - F_k)| = |\det(I - F)|$. For the second equality the induction step uses equations of the form $N(f) = N(\overline{f})N(f_b) = |L(\overline{f})||L(f_b)| = |L(\overline{f})|$. □

The class of compact nilmanifolds is one of a number of different classes of homogeneous spaces. The, so to speak, "next" class after nilmanifolds, is the class of solvmanifolds (quotient spaces of simply connected solvable Lie groups by uniform subgroups). Chris McCord in [Mc1] set out to see how far the results on nilmanifolds could be generalized to solvmanifolds. The most important property of solvmanifolds (for us) is the following theorem.

(3.8) THEOREM ([Mc1]). *Let S be a compact solvmanifold. Then there is a fibration $N \to S \to T$ (the minimal Mostow fibration), with T a torus, N a nilmanifold and with with the property that any self map of S is, up to homotopy, fibre preserving.*

The algabraic counterpart of Theorem (3.8), over which one chants the McCord mantra, is the existence of a maximal normal analytic nilpotent subgroup of the universal cover of S, whose quotient is free abelian. The appropriate quotient of each of these groups gives N, S and T, respectively. We call the fibration in Theorem (3.8) the minimal Mostow fibration, or simply the Mostow fibration of S for short. The proof of the corollary below uses the Mostow fibration, and follows the inductive step of the proof of Lemma (3.4).

(3.9) COROLLARY ([HK1]). *Solvmanifolds are aspherical and essentially fix trivial.*

The next result is the main theorem of [Mc1].

(3.10) THEOREM ([Mc1]). *Let $f: S \to S$ be a map of a compact solvmanifold S, then*

$$|L(f)| \leq N(f).$$

If f is fibre preserving on the Mostow fibration $N \to S \to T$ of S, then $|L(f)| = N(f)$ if and only if either $L(\overline{f}) = 0$, or $\operatorname{sgn}(L(f_b))$ is independent of $b \in \Phi(\overline{f})$. Furthermore,

$$|L(f)| = |L(\overline{f})L(f_b)| = |N(\overline{f})N(f_b)| = |N(f)|$$

if and only if either $L(\overline{f}) = 0$, or $L(f_b)$ is independent of $b \in \Phi(\overline{f})$.

We sketch the proof of Theorem (3.10), which in [Mc1] is entirely geometric, reducing the fixed point classes to singletons in base fiber and total space. Chris's proof is interesting, in that it contains a prototype addition formula (given in the proof below) for fibre preserving maps of solvmanifolds (see also Theorem (5.5)).

SKETCH OF PROOF. If $L(\overline{f}) = 0$ then $\overline{f}$ (and hence f by the homotopy lifting property) is homotopic to a fixed point free map, and so $|L(f)| = N(f) = 0$. So suppose that $L(\overline{f}) \neq 0$. Without loss of generality, we may assume for the induced map $\overline{f}$ that $\#(\Phi(\overline{f})) = N(\overline{f})$, so that each fixed point is in its own Nielsen class. The proof in [Mc1] gives the equation

$$N(f) = \sum_{b \in \Phi(\overline{f})} N(f_b) = \sum_{b \in \Phi(\overline{f})} |L(f_b)| \geq \left| \sum_{b \in \Phi(\overline{f})} L(f_b) \right| = |L(f)|.$$

The first step is a naïve addition formula for solvmanifolds ([10]) which we will prove formally later. The second equality is from Theorem (3.5), the inequality is obvious. To see the last step we note first that T is a Jiang space. So then all fixed point classes, in this case each with a single element, have the same index. Since $N(\overline{f}) = |L(\overline{f})|$, these indices must all be either 1 or -1. By normalization $L(f_b) = \sum_{j=1,\dots,N(f_b)} i(f_b, \mathbf{F}_{bj})$, where for $j = 1, \dots, N(f_b)$ the $\mathbf{F}_{bj}$ are the fixed point classes of f_b. Again $L(f) = \sum_{b,j} i(f, \mathbf{F}_{bj}) = \sum_{b,j} i(f_b, \mathbf{F}_{bj}) \cdot \mathrm{ind}(\overline{f}, b) = \pm \sum_{b,j} i(f_b, \mathbf{F}_{bj})$ by Lemma (2.7) and above. So $|L(f)| = |\sum_{b,j} i(f_b, \mathbf{F}_{bj})| = |\sum_b L(f_b)|$ as required for part two of the theorem. For the remaining part we merely continue the last equality with "$= |L(\overline{f}) L(f_b)|$". □

Theorem (3.10) raises the question of when $L(f_b)$ is independent of $b \in \Phi(\overline{f})$. A full answer to this question involves some subtle work on how linearizations of the various f_b relate to each other, and the "gluing data" of the Mostow fibration. Such considerations are covered in Chapter I.3. However we state some of the results here.

Let $N \to S \to T^k$ be a Mostow fibration of a solvmanifold S, and let u be the dimension of N. For $j = 1, \dots, k$ let ω_j be a path in $\mathbb{R}^k$ from $\overrightarrow{0}$ to e_j, the jth standard basis element in $\mathbb{R}^k$. Then the projection of ω_j in T^k is a loop at 0 (we abuse notation and also write ω_j as the loop in T^k). Let A_j be the linearization of the induced map τ_{ω_j} on the fibre N, then the A_j induce, in the obvious way, a well defined homomorphism

$$A \colon \mathbb{Z}^k \to \mathrm{Aut}(\mathbb{Z}^u).$$

We refer to A as *the linearization of the gluing data of the Mostow fibration.*

([10]) So here, hidden in Chris McCord's proof as early as 1991, is an implicit addition formula for Nielsen numbers on solvmanifolds. This is explicit in his 1993 paper [Mc2, Algorithm 5.2].

(3.11) LEMMA ([KMc]). *Let $f: S \to S$ be a fibre preserving map of a compact solvmanifold S, with respect to its Mostow fibration $N \to S \to T^k$, and let X be the linearization of f_0, where $0 = \overrightarrow{0}$ is the basepoint of T^k. Let $b \in \Phi(\overline{f})$, and ω_b be any path from 0 to b in T^k, then the linearization of f_b is $A(\omega_b^{-1}\overline{f}(\omega_b))X = A(\rho([b]))X$, where ρ is defined in Section 2.*

We now give the following corollary of Theorem (3.10).

(3.12) COROLLARY. *In the context of Theorem (3.10), let Y be the linearization of $\overline{f}$. Then $|L(f)| = N(f)$, if and only if either*

(3.12.1) $\det(I - Y) = 0$, or

(3.12.2) $\operatorname{sgn}(\det(A(\omega)X - I)) = \operatorname{sgn}(\det(X - I))$ for all $\omega \in \pi_1(T^k)$.

Secondly $N(f) = N(\overline{f})N(f_b) = |L(\overline{f})||L(f_b)| = |L(f)|$ if and only if either

(3.12.3) $\det(I - Y) = 0$, or

(3.12.4) $\det(A(\omega)X - I) = \det(X - I)$ for all $\omega \in \pi_1(T^k)$.

The last condition of Theorem (3.12) is so useful, we give it a name. We say that a map f with linearization X on the principle fibre is *fibre uniform*, if $\det(A(\omega)X - I) = \det(X - I)$ for all $\omega \in \pi_1(T^k)$. We emphasize that this can hold even when maps between the fibres over fixed points are of different same homotopy types (as maps). The next example from [A] shows that the equality $N(f) = |L(f)|$ does not hold for all maps of solvmanifolds.

(3.13) EXAMPLE. For this example we regard the Klein bottle K^2 as a quotient of $\mathbb{R}^2$ under the equivalence relation generated by $(x, y) \sim (x + k, y)$, and $(x, y) \sim ((-1)^\ell x, y + l)$, with Mostow fibration $S^1 \to K^2 \to S^1$ induced by projection on the second factor. Let f be the self map of K^2 induced by the function $\widetilde{f}$ on $\mathbb{R}^2$ given by $\widetilde{f}(x, y) = (2x, -y)$. Thus $\overline{f}$ is the map of degree -1, and f_0 the map of degree 2. So $\overline{f}$ has two fixed points in the base which correspond to 0 and 1 (in $\mathcal{R}(\overline{f}_*^0) = \mathbb{Z}_2$), with corresponding Nielsen numbers $N(f_0) = |(-1)^0 \cdot 2 - 1| = 1$, and $|(-1)^1 \cdot 2 - 1| = 3$ on the fibres. As we shall see later from the naïve addition formula $N(f) = 1 + 3$, and f has 4 fixed points all in essential classes. Three of these classes have index 1, while the last has index -1, so $L(f) = 2$.

We cannot expect the formula $N(f) = |L(f)|$ then, to hold in general for solvmanifolds. Perhaps the most interesting application of Corollary (3.12) is the application to $\mathcal{NR}$ solvmanifolds in [KMc], the main subject of the paper [KMc]. An $\mathcal{NR}$ solvmanifold is a solvmanifold in which none of the A_j have roots of unity other than 1 ($\mathcal{NR}$ stands for no roots). The class of all $\mathcal{NR}$ solvmanifolds includes all exponential solvmanifolds as a subclass.

(3.14) THEOREM ([KMc]). *Let $f: S \to S$ be a map of a compact $\mathcal{NR}$ solvmanifold S. Then $|L(f)| = N(f)$, and $N(f) = N(\overline{f})N(f_b) = |L(\overline{f})||L(f_b)| = |L(f)|$.*

The main ingredients that are new in this theorem from [KMc], are not really fibre techniques, since once the Mostow fibrations are set up, it is not fibre techniques which save the day, but rather the "strong algebraic information available" (which yields that $\det(A(\alpha)X - I) = \det(X - I)$ for all $\alpha \in \pi_1(T^k)$, through martix analysis). We therefore omit the proof. Ed Keppelmann showed me the next example.

(3.15) EXAMPLE. Let k be an integer, A the matrix given below, and $\mathcal{A}: \mathbb{Z} \to \mathrm{Aut}(\mathbb{Z}^2)$ the homomorphism defined by putting $\mathcal{A}(z) = A^z$, for $z \in \mathbb{Z}$. As we shall see later (Theorem (7.1)), the space $S = (\mathbb{R}^2 \times \mathbb{R})/\sim$ is a solvmanifold, where the equivalence relation is definded by requiring that for any $l \in \mathbb{Z}$, and for any $\overrightarrow{m} \in \mathbb{Z}^2$ the pairs $(\overrightarrow{x}, \overrightarrow{y})$ and $(A(l)(\overrightarrow{x}) + \overrightarrow{m}, y + l)$ are equivalent. The induced Mostow fibration is $T^2 \hookrightarrow S \xrightarrow{\pi} S^1$ where π is induced by projection. Let $a, b \in \mathbb{Z}$, in addition to A consider the following matrices over $\mathbb{Z}$

$$A = \begin{bmatrix} k & k-1 \\ 1 & 1 \end{bmatrix} \qquad X = \begin{bmatrix} a & (1-k)(a+b) \\ b & -a \end{bmatrix}$$

$$X_1 = \begin{bmatrix} 2 & -12 \\ 2 & -2 \end{bmatrix} \qquad AX_1 = \begin{bmatrix} 14 & -54 \\ 4 & -14 \end{bmatrix}.$$

As can be seen directly, or from Theorem (7.1), the function on $\mathbb{R}^3$ which takes a pair $(\overrightarrow{x}, y)$ to $(X\overrightarrow{x}, -y)$, induces a well defined fibre preserving map f of S. As an example let $k = 4$ and $a = b = 2$, then X_1 and AX_1 above, are the corresponding linearizations of the fibers over the Reidemeister classes 0 and 1 in the base (see Lemma (3.11)). It is apparent that these restrictions f_0 and f_1 are not of the same homotopy type. However the "strong algebraic information" (no roots of unity other than 1) allows us to deduce that $N(f_0) = N(f_1)$ which, as can be easily checked, is 21. So from Theorem (3.14) we have that $N(f) = 2 \cdot 21 = 42$ for the special case, and $N(f) = 2((k-1)(ab + b^2) - a^2 + 1)$ for unspecified a, b and k. Chapter I.3 gives further insight into what is happening here.

We close this section with a result (conditions under which $N(f) = R(f)$) that comes chronologically later in the literature. This result fits here at this point in the exposition on fixed points, because it is at this point in the parallel theory of coincidences on homogeneous spaces, that the result ($N(f, g) = R(f, g)$) starts to become prominent as an alternative to $N(f, g) = |L(f, g)|$. For nilmanifolds the result here is due originally to Brigitte Norton–Odenthal ([N-O]).

(3.16) THEOREM ([N-O], [HK1]). *Let $f: S \to S$ be a map of a compact nilmanifold or a compact $\mathcal{NR}$ solvmanifold S. If $|L(f)| \neq 0$ then $N(f) = R(f)$.*

This result is not true for arbitrary solvmanifolds as can be seen using Example (2.10). Our proof follows by fibre induction from

(3.17) THEOREM ($N(f) = \mathcal{R}(f)$, [HK1]). *Let f be a fibre preserving map on a fibration $F \overset{j}{\hookrightarrow} E \overset{p}{\longrightarrow} B$ with $K = \ker(j_*) = 0$. If $N(\overline{f}) = \mathcal{R}(\overline{f})$ and $N(f_b) = \mathcal{R}(f_b)$ for every $b \in \Phi_{\mathcal{E}}(\overline{f})$, then $N(f) = \mathcal{R}(f)$.*

The basic idea of the proof is that a Reidemeister class $[\alpha]$ of f is essential if and only if both its "components" are essential (Lemma (2.8)). Of course if $p_*([\alpha])$ is not the basepoint in the Reidemeister classes of $\overline{f}$, then we need to change base points, so we can use the appropriate sequence from (2.5). Our proof will include a modification of sequence (2.5)which we will need later.

PROOF. Let y be a fixed point of some F_d with $d = p(y)$ (so we can form some sequence as in (2.5)), and let $[\alpha] \in \mathcal{R}(f_*^y)$. We need to show that $[\alpha]$ is essential.

If $[\overline{\alpha}] := p_*([\alpha]) = [0_{p(y)}]$ is the basepoint in $\mathcal{R}(\overline{f}_*^{p(y)})$, then by exactness of sequence (2.5) at $\mathcal{R}(f_*^y)$, there is a class $[\beta K] \in \mathcal{R}_K(f_{p(y)*}^y)$ such that $j_*([\beta K]) = [\alpha]$. By hypothesis these components $[\overline{\alpha}]$ and $[\beta K]$ of $[\alpha]$ are essential, so by Lemma (2.8) we have that $[\alpha]$ is essential. If $[\overline{\alpha}]$ is not the basepoint in $\mathcal{R}(\overline{f}_*^{p(y)})$, it is nevertheless essential by hypothesis, so it can be represented by some $b \in \Phi(\overline{f})$.

We build a Lemma (2.5) sequence with fibre F_b. Let $c\colon b \to p(y)$ be a path, then there is an index preserving bijection $c^{\#}\colon \mathcal{R}(\overline{f}_*^b) \to \mathcal{R}(\overline{f}_*^{p(y)})$ defined by $c^{\#}([\gamma]) = [c^{-1}\gamma\overline{f}(c)]$ (see for example [H1, Lemma 2.4] or [HK1, p. 230]) which takes the zero class $[0_b]$ to $[\overline{\alpha}]$ ([H1, Lemma 2.5]).

Next we choose an $x \in F_b$ (we do not assume it is a fixed point). We then choose a path $\mu\colon x \to f_b(x)$ in F_b (so that $p(\mu)$ is a constant path at b), and form the Reidemeister classes $\mathcal{R}(f_{b*}^{\mu})$ and $\mathcal{R}(f_*^{\mu})$ associated with the homomorphisms $f_{b*}^{\mu}\colon \pi_1(F_b)/K \to \pi_1(F_b)/K$ and $f_*^{\mu}\colon \pi_1(E) \to \pi_1(E)$, where for example $f_*^{\mu}(\beta) = \mu^{-1}f(\beta)\mu$.

We come now to the point where we must admit that we were being overly simplistic in using only fixed points in our exposition. In fact the theory so far (including (2.5) and (2.8)) goes through when using paths like μ as opposed to fixed points and constant paths as base points (for details see [HK1]). In particular we have an exact sequence which ends with

$$\mathcal{R}(f_{b*}^{\mu}) \overset{j_*}{\longrightarrow} \mathcal{R}(f_*^{\mu}) \longrightarrow \mathcal{R}(\overline{f}_*^{b}).$$

Let $d\colon x \to y$ be a path in E, we use the index preserving bijection ([11]) $d_{\mu}^{\#}\colon \mathcal{R}(f_*^{\mu}) \to \mathcal{R}(f_*^y)$ given by

$$d_{\mu}^{\#}([\delta]) = [d^{-1}\delta\mu f(d)].$$

([11]) These index preserving bijections are also used to show that the Nielsen numbers, definded via the fundamental group method, are well defined.

We put this data together in the following commutative diagram

$$
\begin{array}{ccc}
\mathcal{R}(f^{\mu}_{b*}) & & \\
\downarrow & & \\
[\alpha'] \in \mathcal{R}(f^{\mu}_{*}) & \xrightarrow{d^{\#}_{\mu}} & \mathcal{R}(f^{y}_{*}) \ni [\alpha] \\
\downarrow & & \downarrow \\
[0_b] \in \mathcal{R}(f^{b}_{*}) & \xrightarrow[c^{\#}]{} & \mathcal{R}(\overline{f}^{p(y)}_{*}) \ni [\overline{\alpha}]
\end{array}
$$

Let $[\alpha']$ be the Reidemeister class in $\mathcal{R}(f^{\mu}_{*})$ that corresponds to $[\alpha] \in \mathcal{R}(f^{y}_{*})$ under the bijection $d^{\#}_{\mu}$. The fact that $c^{\#}$ is a bijection together with the commutativity of the diagram shows that $p_*([\alpha']) = [0_b]$. So we now proceed as in the first case to deduce that $[\alpha']$ and hence $[\alpha]$ are essential. □

Theorem (3.17) provokes the following definition.

(3.18) Definition. We say that a map $f: X \to X$ is *weakly Jiang* provided that either $N(f) = 0$ or else $N(f) = \mathcal{R}(f)$. When, for a given X, all possible f are weakly Jiang, then we say that X *is weakly Jiang.*

The fact that tori are Jiang spaces and $\mathcal{N}R$ solvmanifolds (see Section 4) exhibit fibre uniformity allows us to conclude the first part of the following corollary which also gives a characterization of weakly Jiang maps on solmanifolds.

(3.19) Corollary. *On all nilmanifolds and $\mathcal{N}R$ solvmanifolds $N(f) = \mathcal{R}(f)$ whenever $N(f) = |L(f)| \neq 0$. On a given Mostow fibration for an arbitrary solvmanifold one has $N(f) = \mathcal{R}(f)$ if and only if $N(\overline{f}) \neq 0$ and for all $b \in \Phi_{\mathcal{E}}(\overline{f})$, $N(f_b) \neq 0$, if and only if $\overline{f}$ and restrictions f_b of f to all fibres over a set χ that represents $N(\overline{f})$ are weakly Jiang, and the product $L(\overline{f}) \cdot \Pi_{b\in\chi} L(f_b) \neq 0$.*

4. Other product theorems

In this section, we continue the exposition of product theorems for Nielsen theory, with a product theorem for pullbacks [HMP], with relative product theorems due to Aaron Schusteff (see [Sct]), and with generalizations of this from [CW]. Let $p: E \to B$, and $q: A \to B$ be maps. Then the pullback of f and q, is the subspace $A \sqcap E$ of $A \times E$, which consists of ordered pairs (a, e) with $q(a) = p(e)$. When A is the one point space, we identify q with a point q of B, and then $A \sqcap E$ is (up to homeomorphism) simply F_q. There is a commutative diagram shown in the left hand picture below

$$(4.1)\qquad \begin{array}{ccc} A \sqcap E & \xrightarrow{q_p} & E \\ {\scriptstyle p_q}\downarrow & & \downarrow{\scriptstyle p} \\ A & \xrightarrow[q]{} & B \end{array} \qquad\qquad \begin{array}{ccc} \pi_1(A) \sqcap \pi_1(E) & \xrightarrow{q_{p*}} & \pi_1(E) \\ {\scriptstyle p_{q*}}\downarrow & & \downarrow{\scriptstyle p*} \\ \pi_1(A) & \xrightarrow[q*]{} & \pi_1(B) \end{array}$$

where $q_p(a, e) = a$, and $p_q(a, e) = e$. It is easy to see that p_q is a fibration when p is, and that the fibre of p_q over a point $a \in A$ is homeomorphic to the fibre $F_{q(a)}$ of p. Let f be a fibre preserving map of p with induced map $\overline{f}$, and let $g: A \to A$ be a map with the property that $qg = \overline{f}q$. Then g and f induce a map $g \sqcap f: A \sqcap E \to A \sqcap E$ defined by $(g \sqcap f)(a, y) = (g(a), f(y))$. For a fixed point a of g, $q(a)$ is a fixed point of $\overline{f}$, and with the identification of the fibre of p_q over a with $F_{q(a)}$ over p we have that $(g \sqcap f)_a = f_{q(a)}$.

(4.2) THEOREM ([HMP]). *If p is an orientable fibration, then*

$$L(g \sqcap f)L(\overline{f}) = L(g)L(f). \tag{4.2.1}$$

PROOF. It is not hard to show that p orientable implies that p_q is orientable, and so we have two Lefsechetz product formulae $L(g \sqcap f) = L(g)L(f_{q(a)})$, and $L(\overline{f}) = L(\overline{f})L(f_{q(a)})$. From the first equation, if $L(g) = 0$ then $L(g \sqcap f) = 0$ too. So if $L(g) = 0$ then formula (4.2.1) is identically 0.

Suppose then, that $L(g) \neq 0$. In this case from the first equation $L(f_{q(a)}) = L(g \sqcap f)/L(g)$, substituting this in the second equation gives the result. □

We can obviously try a similar trick on the Reidemeister sequences of the fibrations p and p_q. However we have no common term to eliminate as in the proof of Theorem (4.2). This is because the formula for p is

$$[\operatorname{Fix} \overline{f}_*^{q(a)}; p_*(\operatorname{Fix} f_*^x)]R(f) = R(\overline{f}) \cdot R_K(f_{q(a)}),$$

while the formula for p_q is

$$[\operatorname{Fix} g_*^a; p_{q*}(\operatorname{Fix}(g \sqcap f)_*^{(a,x)})]R(g \sqcap f) = R(g) \cdot R_U(f_{q(a)}),$$

where U is the kernel of $\pi_1(F_{q(a)}) \to \pi_1(A \sqcap E)$. In fact the pullback on the right in diagram (4.1) does have $\pi_1(F_{q(a)})/K$ as the kernel of both vertical homomorphisms, and so this diagram has the right algebraic ingredients needed in order to imitate (4.2). However we need to interpret $\pi_1(A) \sqcap \pi_1(E)$ appropriately. Accordingly let H be the kernel of the homomorphism $\pi_1(A \sqcap E) \to \pi_1(A) \sqcap \pi_1(E)$, which takes α to the pair $(p_{q*}(\alpha), q_{p*}(\alpha))$. Then

$$\pi_1(A) \sqcap \pi_1(E) \cong \pi_1(A \sqcap E)/H.$$

So what is needed now is a "mod H" Reidemeister version of Theorem (4.2).

(4.3) LEMMA. *Let $f\colon E \to E$ be a fibre preserving map of a Hurewicz fibration $p\colon E \to B$, $g\colon A \to A$ commute with $\overline{f}$ via $q\colon A \to B$, and let $\pi_1(A)$ and $\pi_1(E)$ be abelian. Let $(x,a) \in \Phi(g \sqcap f)$, and $b := p(x) \in \Phi(\overline{f})$, then*

$$[\operatorname{Fix} g_*^a; p_{q*}(\operatorname{Fix}(g \sqcap f)_*^{(a,x)})] R_H(g \sqcap f) = R(g) \cdot R_k(f_{q(a)}),$$
$$[\operatorname{Fix} \overline{f}_*^{q(a)}; p_*(\operatorname{Fix} f_*^x)] R(f) R(g) = [\operatorname{Fix} g_*^a; p_{q*}(\operatorname{Fix}(g \sqcap f)_*^{(a,x)})] R_H(g \sqcap f) R(\overline{f}).$$

So $[\operatorname{Fix} \overline{f}_*^{q(a)}; p_*(\operatorname{Fix} f_*^x)]/[\operatorname{Fix} g_*^a; p_{q*}(\operatorname{Fix}(g \sqcap f)_*^{(a,x)})] R(f) R(g) = R_H(g \sqcap f) R(\overline{f})$.

In [HPM] we required that $(g \sqcap f)(H) \subseteq J(g \sqcap f)$ for the next result, where J denotes the Jiang subgroup. However the more careful analysis of these ideas in [Y2] shows that this condition is redundant.

(4.4) THEOREM (see [HMP], [Y2]). *If $\pi_1(E)$ and $\pi_1(A)$ are abelian, and p is orientable, then*

$$[\operatorname{Fix} \overline{f}_*^{q(a)}; p_*(\operatorname{Fix} f_*^x) + p_{q*}(\operatorname{Fix}(g \sqcap f)_*^{(a,x)})] N(f) N(g) = N_H(g \sqcap f) N(\overline{f}),$$

where $[\operatorname{Fix} \overline{f}_*^{q(a)}; p_*(\operatorname{Fix} f_*^x) + p_{q*}(\operatorname{Fix}(g \sqcap f)_*^{(a,x)})]$ is the number of double cosets of $p_*(\operatorname{Fix} f_*^x)$ and $p_{q*}(\operatorname{Fix}(g \sqcap f)_*^{(a,x)})$ in $\operatorname{Fix} \overline{f}_*^{q(a)}$. Furthermore, if

$$[\operatorname{Fix} \overline{f}_*^{q(a)}; p_*(\operatorname{Fix} f_*^x)] = 1,$$

and $H = 1$, then $N(f)N(g) = N(g \sqcap f) N(\overline{f})$.

The proof consists of comparing the final Reidemeister formula of (4.4) with the formula

$$[\operatorname{Fix} \overline{f}_*^{q(a)}; p_*(\operatorname{Fix} f_*^x) + p_{q*}(\operatorname{Fix}(g \sqcap f)_*^{(a,x)})] R(f) R(g) = R_H(g \sqcap f) R(\overline{f})$$

obtained from the Mayor Vietoris type sequence (see [HMP], [H4])

$$\operatorname{Fix}(g_*^a) \times \operatorname{Fix}(f_*^x) \xrightarrow{b} \operatorname{Fix}(\overline{f}_*^b) \xrightarrow{\Delta} \mathcal{R}_H(g_*^a \sqcap f_*^x) \longrightarrow \mathcal{R}(g_*^a) \sqcap \mathcal{R}(f_*^x) \longrightarrow 0.$$

The proof of Theorem (4.4) is now completed by observing that when

$$[\operatorname{Fix} \overline{f}_*^{q(a)}; p_*(\operatorname{Fix} f_*^x)] = 1,$$

then $[\operatorname{Fix} \overline{f}_*^{q(a)}; p_*(\operatorname{Fix} f_*^x) + p_{q*}(\operatorname{Fix}(g \sqcap f)_*^{(a,x)})] = 1$ too.

We come now to the relative product theorems of Aaron Schusteff (see [Sct]). Recall that the relative Nielsen number of a map $f\colon (X, X_0) \to (X, X_0)$ of pairs (see [S2]), is defined to be

$$N(f; X, X_0) = N(f) + N(f_0) - N(f; f_0),$$

where f_0 denotes the restriction of f to X_0, and $N(f; f_0)$ is the number of essential classes of f that contain an essential class of f_0. The aim of [Sct] is to give product theorems in the category of pairs. The starting point is to examine a trivial fibration in this category. Starting with maps $\overline{f}: (B, B_0) \to (B, B_0)$, and $\widehat{f}: (F, F_0) \to (F, F_0)$, we have the

$$f = \widehat{f} \times \overline{f}: (F \times B, F_0 \times B_0) \to (F \times B, F_0 \times B_0),$$

which is the trivial map of the trivial fibration

$$(F, F_0) \xrightarrow{i_F} (F \times B, F_0 \times B_0) \xrightarrow{\pi_B} (B, B_0)$$

of map pairs. Moreover, this fibration should, by all accounts, satisfy the conditions for the naïve product theorem (see Example (2.20)), and the obvious question to ask is, is

$$N(f, F \times B, F_0 \times B_0) = N(\widehat{f}, F, F_0) N(\overline{f}, B, B_0)?$$

Before we answer this question, we give the following lemma which allows a more convenient formulation of $N(f, F \times B, F_0 \times B_0)$.

(4.5) Lemma. *Let f be as described above, and let $(a, x) \in \Phi(f)$. Then the diagram*

$$\begin{array}{ccc} \mathcal{R}(\widehat{f}^a_{0*} \times \overline{f}^x_{0*}) & \xrightarrow{q} & \mathcal{R}(\widehat{f}^a_{0*}) \times \mathcal{R}(\overline{f}^x_{0*}) \\ \downarrow & & \downarrow \\ \mathcal{R}(\widehat{f}^a_{*} \times \overline{f}^x_{*}) & \xrightarrow[q]{} & \mathcal{R}(\widehat{f}^a_{*}) \times \mathcal{R}(\overline{f}^x_{*}) \end{array}$$

is commutative, where q takes a class $[\alpha]$ into the pair $(\pi_F([\alpha]), \pi_B([\alpha]))$ and the vertical functions are induced by inclusion. Furthermore

$$N(f; f_0) = N(\widehat{f}; \widehat{f}_0) N(\overline{f}; \overline{f}_0).$$

Proof. Note that the horizontal functions are bijections, they are index preserving by Lemma (2.8) applied to trivial fibrations. The last part now follows from the definitions. □

Using the definitions and the naïve product theorem for $N(f)$ and $N(f_0)$ we have

$$N(f, F \times B, F_0 \times B_0) = N(\widehat{f}) N(\overline{f}) + N(\widehat{f}_0) N(\overline{f}_0) - N(\widehat{f}; \widehat{f}_0) N(\overline{f}; \overline{f}_0).$$

On the other hand we have that

$$N(\widehat{f}, F, F_0) N(\overline{f}, B, B_0) = (N(\widehat{f}) + N(\widehat{f}_0) - N(\widehat{f}; \widehat{f}_0))(N(\overline{f}) + N(\overline{f}_0) - N(\overline{f}; \overline{f}_0)).$$

The next example gives the answer no in general to the question asked above.

(4.6) EXAMPLE. Let $X = X_1 \cup X_2$, be the wedge of a 2-disc with the circle where $X_1 = \{z \in \mathbb{C} : |z| = 1\}$ and $X_2 = \{w \in \mathbb{C} : |w-2| = 1\}$, and let $X_0 = S^1$ the boundary circle of X_1. For $m, q \in \mathbb{Z}$, we define maps $g_{mq}: X \to X$ by $g_{mq}(z) = z^m$ for $z \in X_1$, and $g_{mq}(w) = (2-w)^q$ for $w \in X_2$.

Now let $(F, F_0) = (X, X_0) = (B, B_0)$ with $\widehat{f} = g_{ab}$ and $\overline{f} = g_{cd}$, be our relative product fibre space example. From above and (2.3) we have respectively

$$N(f, F \times B, F_0 \times B_0) = |b-1||d-1| + |a-1||c-1| - 1 \cdot 1,$$

while $N(\widehat{f}, F, F_0)N(\overline{f}, B, B_0) = (|b-1| + |a-1| - 1)(|d-1| + |c-1| - 1)$. These two expressions are the same if and only if

$$(|a-1|-1)(|d-1|-1) + (|c-1|-1)(|b-1|-1) = 0.$$

In particular whenever $|y-1| \geq 2$ for all $y \in \{a, b, c, d\}$, the two are not the same.

Note that the two expressions for $N(f, F \times B, F_0 \times B_0)$ above coincide whenever either $N(\widehat{f}) = N(\widehat{f}_0) = N(\widehat{f}; \widehat{f}_0)$, or $N(\overline{f}) = N(\overline{f}_0) = N(\overline{f}; \overline{f}_0)$. The most obvious situation in which this occurs is given by the next lemma whose proof is left to the reader. In fact there are other conditions under which the conclusions of this lemma hold, and these make for more subtle versions of Theorem (4.8) below.

(4.7) LEMMA. *Let $f: (X, X_0) \to (X, X_0)$ be a map of pairs in which the inclusion $i_X: X_0 \to X$ is a homotopy equivalence. Then $N(f; X, X_0) = N(f) = N(f_0) = N(f; f_0)$.*

The following result from [Sct] is now immediate.

(4.8) THEOREM. *Let $f = (\widehat{f} \times \overline{f})$ be the trivial map of the trivial fibration $(F, F_0) \to (F \times B, F_0 \times B_0) \to (B, B_0)$. If either of the inclusions $i_B: B_0 \to B$, or $i_F: F_0 \to F$ is a homotopy equivalence, then $N(f, F \times B, F_0 \times B_0) = N(\widehat{f}, F, F_0)N(\overline{f}, B, B_0)$.*

The goal of the rest of [Sct] is to generalize this theorem (4.8) together with more subtle versions of it. Schusteff writes "Since our primary interest lies in studying product formulas *all of whose terms are homotopy type invariants*, there will be no loss of generality — and a considerable gain in convenience — in adopting the 'working hypotheses' $B = B_0$ or $F_0 = F$". Likewise we state the results as homotopy equivalence, but use equality in the proofs.

The context in which we work, reverting to our earlier notation, is that of having a fibre preserving map $f: (E, E_0) \to (E, E_0)$ of the relative fibration

$$(F_b, F_{b0}) \longrightarrow (E, E_0) \xrightarrow{p} (B, B_0)$$

in which b is a fixed point of $\overline{f}_0$, $p\colon E \to B$ and $p_0\colon E_0 \to B_0$ are Hurewicz fibrations, and $F_{b0} = p_0^{-1}(b) \subset E_0$. Let $x \in \Phi(f_{b0})$, in this context we have the following commutative diagram ("ladder") of sequences from Theorem (2.5).

$$\begin{array}{ccccccccccccccc}
1 & \longrightarrow & \mathrm{Fix}_{K_0} f^x_{0b*} & \xrightarrow{j_{K0*}} & \mathrm{Fix}\, f^x_{0*} & \xrightarrow{p_{0*}} & \mathrm{Fix}\, \overline{f}^{\,b}_{0*} & \xrightarrow{\delta_0} & \mathcal{R}_{K_0}(f^x_{0b*}) & \xrightarrow{j_{0*}} & \mathcal{R}(f^x_{0*}) & \xrightarrow{p_{0*}} & \mathcal{R}(\overline{f}^{\,b}_{0*}) & \longrightarrow & 1 \\
 & & \downarrow{\scriptstyle i_{F*}} & & \downarrow{\scriptstyle i_{E*}} & & \downarrow{\scriptstyle i_{B*}} & & \downarrow{\scriptstyle i_{F*}} & & \downarrow{\scriptstyle i_{E*}} & & \downarrow{\scriptstyle i_{B*}} & & \\
1 & \longrightarrow & \mathrm{Fix}_{K} f^x_{b*} & \xrightarrow[j_{K*}]{} & \mathrm{Fix}\, f^x_{*} & \xrightarrow[p_*]{} & \mathrm{Fix}\, \overline{f}^{\,b}_{*} & \xrightarrow[\delta]{} & \mathcal{R}_{K}(f^x_{b*}) & \xrightarrow[j_*]{} & \mathcal{R}(f^x_{*}) & \xrightarrow[p_*]{} & \mathcal{R}(\overline{f}^{\,b}_{*}) & \longrightarrow & 1
\end{array}$$

Here of course, K_0 is the kernel of $\pi_1(F_0) \to \pi_1(E_0)$. The reader will notice that there is something new here (we have ways of making you notice!). For the relative Nielsen number $N(f, X, X_0)$ of a self map f of a pair (X, X_0), the induced function $\mathcal{R}(f^x_{0*}) \to \mathcal{R}(f^x_*)$ leads firstly to the definition of the number $N(f; f_0)$ of essential classes of f that contain essential classes of f_0, and secondly to the relative Nielsen number $N(f, X, X_0)$ itself. In the diagram we have an induced function $i_{F*}\colon \mathcal{R}_{K_0}(f^x_{0b*}) \to \mathcal{R}_K(f^x_{b*})$, and this leads firstly to an obvious generalization of $N(f; f_0)$ to give the number $N_{K_0,K}(f; f_0)$, of essential mod K classes of f that contain essential mod K_0 classes of f_0. Then secondly, it leads to the following definition

$$N_{K_0,K}(f_b, F_b, F_{b0}) := N_K(f_b) + N_{K_0}(f_{b0}) - N_{K_0,K}(f_b; f_{b0}).$$

The number $N_{K_0,K}(f_b, F_b, F_{b0})$ has the usual properties of Nielsen type numbers, it is a homotopy invariant lower bound, homotopy type invariant, satisfies the commutative property, and of course $N_{K_0,K}(f_b, F_b, F_{b0}) \leq N(f_b, F_b, F_{b0})$. Furthermore when $K_0 = K = 1$, then equality holds. Theorem (4.9) below is preceded in [Sct] by a Reidemeister version.

(4.9) THEOREM. *Let $f = (E, E_0) \to (E, E_0)$ be a fibre uniform, fibre preserving map of the relative fibration described above. Suppose further that $L(f) \neq 0$ and $L(f_0) \neq 0$, that F_b, F_{b0}, B and B_0 are Jiang spaces, and that $\pi_1(E)$ and $\pi_1(B)$ are abelian. If $i_B\colon B_0 \to B$ is a homotopy equivalence, and $p_*(\mathrm{Fix}\, f^x_{0*}) = p_*(\mathrm{Fix}\, f^x_*)$, then*

$$[\mathrm{Fix}\, \overline{f}^{b}{}_*; p_*(\mathrm{Fix}\, f^x_*)] N(f, E, E_0) = N_{K_0,K}(f_b, F_b, F_{b0}) N(\overline{f}, B, B_0).$$

If for some $b \in \Phi(\overline{f}_0)$, $i_B\colon B_0 \to B$ is a homotopy equivalence, and

$$[\mathrm{Fix}\, \overline{f}^{b}_*; p_*(\mathrm{Fix}\, f^x_*)] N_{K_0}(f_{b0}) = [\mathrm{Fix}\, \overline{f}^{b}_{0*}; p_*(\mathrm{Fix}\, f^x_0)] N_K(f_b)$$

then

$$[\mathrm{Fix}\, \overline{f}^{b}{}_*; p_*(\mathrm{Fix}\, f^x_*)] N(f, E, E_0) = N_K(f_b) N(\overline{f}, B, B_0).$$

PROOF. For convenience we use Pak type notation, and denote by $P_K(f)$ the order of the kernel of $j_*: \mathcal{R}_K(f^x_{b*}) \to \mathcal{R}(f^x_*)$. For the first part then since $p_*(\operatorname{Fix} f^x_{0*}) = p_*(\operatorname{Fix} f^x_*)$ we have, using the "working hypothesis" $i_B = 1$, that

(4.10.1) $P_K(f) = P_{K_0}(f_0)$, and
(4.10.2) $N(\overline{f}, B, B_0) = N(\overline{f}) = N(\overline{f}_0) = N(\overline{f}; \overline{f}_0)$.

Furthermore from these two equations and the product formulas for $N(f_0)$ and $N(f)$, we have

(4.10.3) $P_K(f)N(f_0) = N_{K_0}(f_{b0})N(\overline{f})$ and
(4.10.4) $P_K(f)N(f) = N_K(f_b)N(\overline{f})$.

Since $\pi_1(E)$ and $\pi_1(B)$ are abelian, then the ladder defined earlier consists of abelian groups. We denote the cokernels of $i_{F*}: \mathcal{R}_{K_0}(f^x_{0b*}) \to \mathcal{R}_K(f^x_{b*})$ and $\mathcal{R}(\overline{f}^b_{0*}) \to \mathcal{R}(\overline{f}^b_*)$ by $\operatorname{Coker}(i_{F*})$ and $\operatorname{Coker}(i_{E*})$, respectively. The surjectivity of $i_{B*}: \operatorname{Fix} \overline{f}^b_{0*} \to \operatorname{Fix} \overline{f}^b_*$ and $p_{0*}\mathcal{R}(f^x_{0*}) \to \mathcal{R}(\overline{f}^b_{0*})$, and the bijectivity of $i_{B*}: \mathcal{R}(\overline{f}^b_{0*}) \to \mathcal{R}(\overline{f}^b_*)$ allows us to show, by a diagram chase, that $\operatorname{Coker}(i_{F*}) \cong \operatorname{Coker}(i_{E*})$. Since all classes are essential, this translates to $N_K(f_b)N(f; f_0) = N(f)N_{K_0,K}(f_b; f_{b0})$. Multiplying by $P_K(f)$ gives

$$\begin{aligned} P_K(f)N_K(f_b)N(f; f_0) &= P_K(f)N(f)N_{K_0,K}(f_b; f_{b0}) \\ &= N_K(f_b)N(\overline{f})N_{K_0,K}(f_b; f_{b0}). \end{aligned}$$

Since $N_K(f_b) \neq 0$, we can cancel it from the two outside expressions to get

$$P_K(f)N(f; f_0) = N(\overline{f})N_{K_0,K}(f_b; f_{b0}). \tag{4.10.5}$$

So

$$\begin{aligned} P_K(f)N(f; E, E_0) &= P_K(f)(N(f) + N(f_0) - N(f; f_0)) \\ &= P_K(f)N(f) + P_K(f)N(f_0) - P_K(f)N(f; f_0) \\ &= N_K(f_b)N(\overline{f}) + N_{K_0}(f_{b0})N(\overline{f})N(\overline{f})N_{K_0,K}(f_b; f_{b0}) \\ &= (N_K(f_b) + N_{K_0}(f_{b0}) - N_{K_0,K}(f_b; f_{b0}))N(\overline{f}). \\ &= N_{K_0,K}(f_b, F_b, F_{b0})N(\overline{f}, B, B_0). \end{aligned}$$

The proof of the second part is similar. □

There are naïve versions of this theorem which we illustrate below, but omit the statements. Before we do so however, we remark that there is an obvious question here. If Jiang space results can be generalized to the highly non-commutative context of nilmanifolds, can these relative results also be generalized to nilmanifolds (and beyond — see Remarks (4.12) and (4.13))? In fact we give as our example, an example from [CW2] which illustrate just such a generalization to nilmanifolds.

(4.11) EXAMPLE (Cardona–Wong, [CW2]). Let $p\colon (BN, T^2) \to (T^2, S^1)$ be the fibration of pairs with BN as in Example (3.6), and with T^2 embedded in BN as $(x, y) \to (0, x, y)$ (not as one might more usually expect as $(x, y) \to (x, 0, y)$). Then $p_0\colon T^2 \to S^1$ takes $(x, y) \to y$. Furthermore let $f\colon (BN, T^2) \to (BN, T^2)$ be determined by $\widetilde{f}(x, y, z) = (-x, -2y, 2z)$, then $N(f; E, E_0) = 6$ and $N(f_b, F_b, F_{b0})N(\overline{f}, B, B_0) = 2 \cdot 3$ giving

$$N(f; E, E_0) = N(f_b, F_b, F_{b0})N(\overline{f}, B, B_0)$$

for nilmanifolds. This example then, illustrates an extension of a naïve version of Theorem (4.9) above, mentioned in [CW2] for "Jiang type" spaces (see the definition below).

(4.12) REMARK. The theorems of Fernanda Cardona and Peter Wong in [CW2] are technical results which extend Schusteff's ideas to the non-fibre uniform Reidemeister setting. So [CW2] does in that context, what the chronically earlier Theorem (5.4) from [HKW] in Section 1.4, does for Nielsen numbers in the non-fibre uniform context. In particular when f is locally fixed group uniform (see Section 5), the authors of [CW2] give addition formulas for $R(f; E, E_0)$ and $R(f; E - E_0)$. They illustrate their addition formulae on nilmanifolds and on a particular map of solvmanifolds where, because it exhibits a "Jiang type condition" (either $L(f) = N(f) = 0$ or $N(f) = R(f)$, and in this case $L(f) \neq 0$) they are able to equate $N(f; E, E_0)$ with $R(f; E, E_0)$, and $N(f; E - E_0)$ with $R(f; E - E_0)$, respectively. An interesting application, which we cannot give here, is also given in [WC2] to the asyompotic Nielsen numbers $NI^\infty(f)$.

(4.13) REMARK. To give an example of Theorem (4.9) as it stands, we could simply replace BN with T^3 in Example (4.11), and as a self map take the linearization of the f given there. Of course calculations are the same since the Nielsen numbers are calculated from linearizations.

5. Addition formulae for fibre preserving maps

Following a remark by Boju Jiang in his book ([J, p. 87]), Bob Brown's formula $N(f) = N(\overline{f}) \cdot N(f_b)$, came to be widely known as the naïve product formula. Perhaps the reason for this was the initial overly optimistic hope for its wide applicability. On the other hand, we seemed initially to be mesmerized by it, not at first perhaps fully realizing its inadequacy, and so concentrating on finding conditions under which it held. On the other hand, several years before McCord's somewhat hidden use of a naïve addition formula for solvmanifolds ([Mc1]), Chengye You had shown me a proof of Halpern's (unpublished) results on the Nilesen numbers of iterates of maps on the Klein bottle using the very same addition formula (see Example (7.3)). It was not, however until Ed Keppelmann, Peter Wong and I got to know each other really for the first time at Mount Holyoke, that it dawned on us that somebody should write a systematic account of addition formulae. It seemed very natural that we should do such a project together. We had a lot of fun. This section records some of the things we discovered along the way.

Speaking intuitively, the product formula holds, when there is a global uniformity in both the indices $[\mathrm{Fix}\,\overline{f}_*^{p(x)}; p_*(\mathrm{Fix}\, f_*^x)]$ and the mod K Nielsen classes $N_K(f_b)$ in the fibre. On the other hand naïve and other nice addition formulae, hold under conditions of "local uniformity". In fact local uniformity of the mod K Nielsen classes $N_K(f_b)$ is automatic. What we mean by this, is that $N_K(f_b)$ in independent of b within its fixed point class $\mathbf{F}$, in B. If, in addition $[\mathrm{Fix}\,\overline{f}_*^{p(x)}; p_*(\mathrm{Fix}\, f_*^x)]$ is independent of b within its class, and of $x \in F_b$, then $N(f)$ would be a sum of the form $\sum N_K(f_b)/[\mathrm{Fix}\,\overline{f}_*^{p(x)}; p_*(\mathrm{Fix}\, f_*^x)]$ (see Theorem (5.4)). In fact, as we will now see, the number $[\mathrm{Fix}\,\overline{f}_*^{p(x)}; p_*(\mathrm{Fix}\, f_*^x)]$ need not even be independent of x in the fiber over a single fixed point in B.

Observe first from the sequence in (2.9), that the number $[\mathrm{Fix}\,\overline{f}_*^{p(x)}; p_*(\mathrm{Fix}\, f_*^x)]$ can be interpreted as the number of mod K fixed point classes of f_b, that combine into the single fixed point class of f containing x.

(5.1) EXAMPLE (The Möbius pretzel). Define E to be the union $E_1 \cup E_2$, where

$$\begin{aligned} E_1 &= \{(z,w) \in \mathbb{C}^2 : |z| = 1,\ |w - \sqrt{z}| = 1 \text{ or } |w + \sqrt{z}| = 1\}, \\ E_2 &= \{(z,w) \in \mathbb{C}^2 : |z-2| = 1,\ |w - 1| = 1 \text{ or } |w + 1| = 1\}. \end{aligned}$$

The space E_1 can be thought of as a quotient of $\mathbf{8} \times I$, the figure eight cross the unit interval, where the ends of the figure eight cylinder are joined after twisting through π radians (rather like the Möbius stip with a figure eight cross section). On the other hand, E_2 (Midas man) can be thought of as revolving the figure eight about a central axis. Now E cannot be drawn in three space, without self intersection, but it can be obtained from Figure 2 by identifying the figure eight cross sections which are closest to each other in the diagram. E fibres over the figure eight, by projecting onto the first factor. The fibre of course is also a figure eight. We define a self map f of E over the identity. In particular each fibre is taken to itself by this map which in each fibre in E_2, flips the figure eight about the vertical axis through the wedge point, and about this axis gradually rotated through π radians in E_1 in the obvious way. Thus in each fibre there are three fixed points, each in its own class when considered to be in the fibre.

In the diagram we have drawn some lines of fixed points. In E_1 the line drawn can be thought of as starting at the top intersection point of E_1 and E_2 and, by

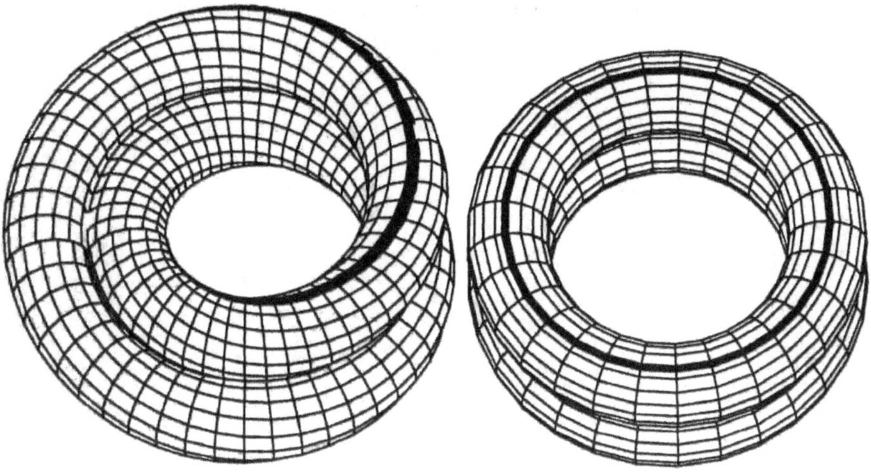

E_1 E_2

FIGURE 2. The Möbius pretzel

going round twice traces a continuous loop which joins the two outside fixed points in each fibre. Thus in E there are exactly two fixed point classes of f given by $\{(1,0)\}$ and $\{(1,\pm 2)\}$, while in each fibre there are three distinct classes. By the observation immediately before this example, in the fibre over the wedge point in the base, if x is either $(1,2)$ or $(1,-2)$, then $[\operatorname{Fix} \overline{f}_*^{p(x)}; p_*(\operatorname{Fix} f_*^x)] = 2$. Whereas if $x = (1,0)$ then $[\operatorname{Fix} \overline{f}_*^{p(x)}; p_*(\operatorname{Fix} f_*^x)] = 1$.

Recall from the proof of Theorem (2.14), that for each $b \in \Phi(\overline{f})$, the number c_b denotes the cardinality of the set of equivalence classes of $\mathcal{E}_K(f_b)$ which are identified under $j_{b*}\colon \mathcal{E}_K(f_b) \to \mathcal{E}(f)$ (which by (2.9) is the cardinality of $p_{\mathcal{E}}^{-1}(\overline{F}_b)$). We would like conditions under which this number is simply $N_K/[\operatorname{Fix} \overline{f}_*^{p(x)}; p_*(\operatorname{Fix} f_*^x)]$. For this we need that $[\operatorname{Fix} \overline{f}_*^{p(x)}; p_*(\operatorname{Fix} f_*^x)]$ is independent of appropriate x.

(5.2) DEFINITION. Let $f\colon E \to E$ be a fibre preserving map of a fibration $p\colon E \to B$. We say that f is *locally essentially* Fix *group uniform* if for any essential fixed point class $\overline{\mathbf{F}}$ of $\overline{f}$, and for any $x \in F_b$ belonging to an essential class of f, that the local Fix group index $[\operatorname{Fix} \overline{f}_*^{p(x)}; p_*(\operatorname{Fix} f_*^x)]$ is independent of the choice of x and b (but not necessarily of $\overline{\mathbf{F}}$).

Example (5.1) illustrates when that this does not always hold. From [HKW] we have that $[\operatorname{Fix} \overline{f}_*^{p(x)}; p_*(\operatorname{Fix} f_*^x)]$ is independent of the choice of x within the homotopy class of f in E. Furthermore f is locally essentially Fix group uniform if: $f_*(\pi_1(E))$ commutes with the image of $i_*(\pi_1(F))$, or f is essentially fix group trivial (in particular if $B = T^k$, a k torus).

(5.3) LEMMA. *If f is locally essentially* Fix *group uniform, then for each b in an essential fixed point class of $\overline{f}$, we have that*

$$\#(p_*^{-1}(\overline{\mathbf{F}}_b)) = N_K/[\operatorname{Fix}\overline{f}_*^{p(x)}; p_*(\operatorname{Fix} f_*^x)],$$

for any x in an essential fixed point class of f_b.

The equation $N(f) = \sum_i c_i$ in the proof of Theorem (2.14) now allows us to prove:

(5.4) THEOREM ([HKW]). *Let $f\colon E \to E$ be a fibre preserving map of a Hurewicz fibration $p\colon E \to B$, if f is locally essentially* Fix *group uniform, then*

$$N(f) = \sum_{b \in \chi} \frac{N_K(f_b)}{[\operatorname{Fix}\overline{f}_*^{p(x)}; p_*(\operatorname{Fix} f_*^y)]},$$

where y is chosen in an essential class of f_b (if no such y exists, we interpret that summand to be zero), and where χ is a set that has exactly one representative of each essential fixed point class of $\overline{f}$.

(5.5) COROLLARY ([HKW]). *If under the conditions of Theorem (5.4) we have that f is globally essentially* Fix *group uniform, then*

$$[\operatorname{Fix}\overline{f}_*^{p(x)}; p_*(\operatorname{Fix} f_*^y)]N(f) = \sum_{b \in \chi} N_K(f_b).$$

Furthermore, f satisfies the naïve addition formula

$$N(f) = \sum_{b \in \chi} N(f_b),$$

if and only if $N_K(f_b) = N(f_b)$ and for each $x \in \mathbf{F} \in \mathcal{E}(f)$ we have

$$[\operatorname{Fix}\overline{f}^b{}_*; p_*(\operatorname{Fix} f_*^x)] = 1.$$

In particular, from Corollary (3.9), the naïve addition formula holds for all fibre-preserving maps of solvmanifolds.

From now on will will refer to the necessary and sufficient conditions given in the second part of Corollary (5.5) the *naïve addition conditions.* We close this section with Chris McCord's algorithm for computing $N(f)$ for self maps f of solvmanifolds.

(5.6) ALGORITHM. Let $f: S \to S$ be a map of a solvmanifold S. The computation of $N(f)$ can be accomplished by the following algorithmic procedure:

(5.6.1) Form the minimal Mostow fibration $N^u \to S \xrightarrow{\pi} T^k$, where N^u is a nilmanifold with dimension u.

(5.6.2) Linearize the gluing data of π to obtain $A: \mathbb{Z}^k \to \mathrm{Aut}(\mathbb{Z}^u)$.

(5.6.3) Obtain the linearization X of the restriction f_0 of f to the principal fibre over $\overrightarrow{0}$.

(5.6.4) Determine the Redeimeister group $\mathcal{R}(\overline{f})$.

(5.6.5) Determine a set $\mathcal{T} \subset \mathbb{Z}^k$ of Redeimeister representatives (one element for each class in $\mathcal{R}(\overline{f})$). Then

$$N(f) = \sum_{\nu \in \mathcal{T}} |\det(A(\nu)X - I)|.$$

(5.7) EXAMPLE. Let $A: \mathbb{Z}^2 \to \mathrm{Aut}(\mathbb{Z}) = \{\pm 1\}$ be given by

$$A(l_1, l_2) = (-1)^{(l_1 + l_2)},$$

and consider the quotient S of $\mathbb{R} \times \mathbb{R}^2$ generated by having

$$(x + k, \overrightarrow{y}) \sim (A(\overrightarrow{l})(\overrightarrow{x}), \overrightarrow{y} + \overrightarrow{l}),$$

for any $\overrightarrow{l} \in \mathbb{Z}^p$, and any $k \in \mathbb{Z}$. Then S is a solvmanifold with associated Mostow fibration $S^1 \to S \to T^p$, and gluing data A (see Theorem (7.1)). Consider the matrix Y with columns $(0,3)$ and $(3,0)$ then the map $f : \mathbb{R} \times \mathbb{R}^2 \to \mathbb{R} \times \mathbb{R}^2$ induced by $(-1, Y)$ determines a well defined fibre preserving map f of S (Theorem (7.1) again). Then the linearization of $\overline{f}$ is Y, and that of f_0 is -1. So $\mathcal{R}(\overline{f}) = \mathbb{Z}^2 \times \mathbb{Z}^2$ (Theorem (2.2), and $\mathcal{T} = \{(0,0), (1,0), (1,1), (0,1)\}$ is a set of Redeimeister representatives for $\overline{f}$. From (5.6) we have $N(f) = \sum_{\nu \in \mathcal{T}} |-A(\nu) - 1| = 4$.

(5.8) REMARK. We note that once again is it not necessary to know which fixed point in the base corresponds to which Nielsen number in the fibre. We simply add the Nielsen number of the fibres over a set of Reidemeister representatives.

(5.9) REMARK. The algorithmic nature of the calculation of Nielsen numbers on solvmanifolds has been taken one step further by Albian Raviston (see [R]), who has turned the algorithm into a computer program.

(5.10) REMARK. As already mentioned in Section 4, Fernanda Cardona and Peter Wong, using the ideas of this section, have extended Schusteff's relative fibre space results ([CW2]). Thus [CW2] may be thought of as a kind of "pushout" of [Sct] and [HKW], i.e. [CW2] = [Sct] $\sqcup$ [HKW] (actually we should also include the work of Zhao [Z] in this pushout).

6. Nielsen type numbers for fibre preserving maps

This section deals with a Nielsen type number $N_{\mathcal{F}}(f,p)$ for fibre preserving maps f of a fibration p. The number $N_{\mathcal{F}}(f,p)$ is dual (in the sense of Eckmann Hilton duality) to the relative Nielsen number $N(f;X,A)$. These numbers can be thought of as examples of restricted Nielsen theories, where one restricts both maps and homotopies. The idea is that by restricting to such maps and their respective homotopies, the minimum number of fixed points can be greater than in the unrestricted case. The corresponding Nielsen type number should then provide a (hopefully) sharp lower bound for the number of fixed points within the restricted homotopy class. In the relative case of maps $f\colon (X,A) \to (X,A)$, one must take into account the fact that maps f and homotopies restrict to maps and homotopies of A. In the situation considered here, one must take account of the fact that the maps and homotopies are fibre preserving. The main reference for this section ([H3]) was written while we were still mesmerized by the product formula. In [H3] then, under the hypothesis of fibre uniformity, the number was defined by $N_{\mathcal{F}}(f,p) := N(\overline{f})N(f_b)$. The natural extension, which defines $N_{\mathcal{F}}(f,p) := \sum_{b\in\chi} N(f_b)$, came in [HKW], as our attention and interests were turned to fibrations with non-fibre uniformity. Let f be a fibre preserving map of a fibration $p\colon E \to B$, a set $\chi \subseteq \Phi(\overline{f})$ is said to be *a set of essential representatives for* $\overline{f}$, if χ contains exactly one point of each essential class of $\overline{f}$.

(6.1) Definition. Let f be a fibre preserving map of a fibration $p\colon E \to B$, and let χ be a set of essential representatives for $\overline{f}$. Then $N_{\mathcal{F}}(f,p)$ is defined by

$$N_{\mathcal{F}}(f,p) := \sum_{b\in\chi} N(f_b) \quad (= N(\overline{f})N(f_b) \text{ if } f \text{ is fibre uniform}).$$

To make our definition of fibre homotopy absolutely clear (and so also the category in which we are working), we say that a homotopy $H\colon E\times I \to E$ is a *fibre homotopy* if it induces a homotopy $\overline{H}\colon B\times I \to B$. Note that for each $t \in I$ the map $H_t := H(\,\cdot\,,t)$ is a fibre preserving map. Let $M_{\mathcal{F}}(f,p)$ be the least number of fixed points of any map g which is fibre homotopic to f.

(6.2) Theorem. *The number* $N_{\mathcal{F}}(f,p) \le M_{\mathcal{F}}(f,p)$.

Proof. Let g be a fibre preserving map which is fibre homotopic to f, and let χ' be a set of essential representatives for $\overline{g}$. Then

$$\#(\Phi(g)) \ge N(g;E,p^{-1}(\chi')) \ge N(g_{\chi'}) = \sum_{d\in\chi'} N(g_d) = \sum_{b\in\chi} N(f_b) = N_{\mathcal{F}}(f,p),$$

where $N(g;E,p^{-1}(\chi'))$ is the relative Nielsen number of

$$g\colon (E,p^{-1}(\chi')) \to (E,p^{-1}(\chi')),$$

and where $g_{\chi'}$ denotes the restriction of g to $p^{-1}(\chi')$. The two inequalities follow from relative Nielsen theory, and the last equality by definition. To see the first equality, observe that g_χ is the disjoint union of $N(\overline{g})$ maps between fibres, with respective Nielsen numbers $N(g_b)$. Finally the middle equality follows from the homotopy invariance of $N(\overline{g})$ and the homotopy type invariance of the $N(g_d)$. □

We remark, in the context of fibre preserving maps, that this is essentially Bob Brown's insight (see his "proof" in our introduction) of the naïve product theorem. Theorem (6.2) shows that $N_{\mathcal{F}}(f,p)$ is a fibre homotopy type invariant. It is also fibre homotopy type invariant, and it satisfies the commutative property.

(6.3) THEOREM. *Let p be an arbitrary fibration, and f a self-fibre preserving map of p.*

(6.3.1) If χ is a set of essential representatives of $\overline{f}$, then

$$N_{\mathcal{F}}(f,p) = N(f; E, p^{-1}(\chi)),$$

where $N(f; E, p^{-1}(\chi))$, is the relative Nielsen number of the map

$$f\colon (E, p^{-1}(\chi)) \to (E, p^{-1}(\chi)).$$

(6.3.2) $N_{\mathcal{F}}(f,p) \geq N(f)$. Equality holds if and only if the naïve addition conditions hold.

(6.3.3) If $N(\overline{f}) \neq 0$, then for any b in an essential class of $\overline{f}$, we have that

$$N_{\mathcal{F}}(f,p) \geq N(f_b).$$

In (6.3.3) the equality may or may not be strict. This is illustrated by taking b first to be "b", then to be d, in Example (2.10). That $N(f) = N(f, f_\chi)$ (and so (6.3.1) above), is a simple consequence of Lemma (2.8), while (6.3.2) and (6.3.3) are simple relative Nielsen theory properties.

(6.4) THEOREM (Minimum Theorem). *Let f be a self fibre preserving map of the fibration $p\colon E \to B$. If B and all fibres are Wecken spaces, then there is a fibre preserving map g that is fibre homotopic to f with the property that*

$$N_{\mathcal{F}}(g,p) = M_{\mathcal{F}}(f,p).$$

The following corollary is immediate from (6.3.2) and from Theorem (6.4).

(6.5) COROLLARY (Solvmanifolds are Wecken). *If in addition to the conditions of Theorem (6.4), we have that any self map is homotopic to a fibre preserving map f which satisfies the naïve addition conditions, then E is (ordinary) Wecken.*

In particular all nilmanifolds and solvmanifolds are Wecken, and this of course includes the Klein bottle.

The last part of the corollary was a throw away remark in [HKW], in the paragraph following 8.3. It does however seem to be the first published proof that solvmanifolds, and in particular the Klein bottle, is Wecken. Of course for the proof of the Wecken property for nilmanifolds, one has to use induction. The proof for solvmanifolds sketched below is then straighforward. Details can be found in [H3] by simply ignoring fibre uniformity there.

PROOF OF THEOREM (6.4). Since p satisfies the homotopy lifting property, we may assume without loss of generality that $\overline{f}$ has exactly $N(\overline{f})$ fixed points. Let $b \in \Phi(\overline{f}) := \chi$, and for each such b, let $H_b: p^{-1}(b) \times I \to p^{-1}(b)$ be a homotopy from f_b to g_b where g_b has exactly $N(f_b)$ fixed points. Let $H_\chi: p^{-1}(\chi) \times I \to p^{-1}(\chi)$ be the union of the homotopies H_b. It should be clear that $g_\chi := H_\chi(\,\cdot\,, 1)$ has exactly $N_{\mathcal{F}}(f,p)$ fixed points. Our task is to extend g_χ to a fibre preserving map $g: E \to E$ that is fibre homotopic to f, and without picking up any more fixed points. We do this by using Strøm's relative extension theorem ([St, Theorem 12]), which allows us to extend the map $f \cup H_\chi \to E$ over the constant homotopy $\overline{F}$ of $\overline{f}$ with itself, to a map $F: E \times I \to E$ which preserves the commutativity of the diagram.

$$\begin{array}{ccc} E \times \{0\} \cup p^{-1}(\chi) \times I & \xrightarrow{f \cup H_\chi} & E \\ {\scriptstyle j}\big\downarrow & & \big\downarrow{\scriptstyle p} \\ E \times I & \xrightarrow[\overline{F}]{} & B \end{array}$$

Here j is the inclusion. In particular the map $g := F(\,\cdot\,, 1)$ extends g_χ and is over $\overline{f}$. To see that g has only $N_{\mathcal{F}}(f,p)$ fixed points, we observe that if x is a fixed point of g, which is in the complement of $p^{-1}(\chi)$ in E, then $p(x)$ must be a fixed point of $\overline{f}$ in the complement of χ in B, a contradiction. □

In their paper on the computation of the relative Nielsen number Fernanda Cardona and Peter Wong ([CW2]) consider relative Reidemeister numbers (with the obvious definitions) under Jiang type conditions. They then give the following theorem for the calculation of $N_{\mathcal{F}}(f,p)$ and/or $N(f; E, p^{-1}(\chi))$.

(6.6) THEOREM ([CW2]). *Let f be a fibre preserving map of the fibration $p: E \to B$. If B, E and all fibres are 0-connected compact Jiang type ANR's, and $L(f) \cdot \Pi_{b \in \chi} L(f_b) \neq 0$. Then*

$$N_{\mathcal{F}}(g,p) = R(f; E, p^{-1}(\chi)).$$

If in addition f satisfies the naïve addition condition for all x and b, then

$$N(f; E, p^{-1}(\chi)) = N_{\mathcal{F}}(f,p) = R(f; E, p^{-1}(\chi)) = R(f).$$

7. Model solvmanifolds: applications of naïve addition

About the time [HKW] was in its final stages of preparation, the need for a variety of concrete examples of solvmanifolds and their maps had become clear. I remember teasing Ed Keppelmann about his universal theorems on solvmanifolds (see [K]), and asking him for examples. "The Klein bottle and nilmanifolds", he replied confidently. 'So you have a theorem about the Klein bottle and nilmanifolds?' I teased. The problem was that everyone we asked for examples of solvmanifolds gave us the same answer "The Klein bottle and nilmanifolds". This paucity of readily available examples of solvmanifolds and their maps, lead us to the theorem below, in which we proved that this construction given there really does give rise to a solvmanifold; that the given matrix theoretical equation is both necessary and sufficient (as opposed to earlier descriptions on arbitrary solvmanifolds that showed it was mearly necessary); and that calculations of Nielsen theory on arbitrary solvmanifolds never got harder than they do on these models.

(7.1) THEOREM (Model solvmanifolds and their maps, [HK4]). *Let p and u be positive integers, and let $A\colon \mathbb{Z}^p \to \mathrm{Aut}(\mathbb{Z}^u)$ be a homomorphism of groups. Consider the equivalence relation on $\mathbb{R}^u \times \mathbb{R}^p$ generated by*

(7.1.1) $(\overrightarrow{x}, \overrightarrow{y}) \sim (\overrightarrow{x} + \overrightarrow{k}, \overrightarrow{y})$, for any $\overrightarrow{k} \in \mathbb{Z}^u$, and
(7.1.2) $(\overrightarrow{x}, \overrightarrow{y}) \sim (A(\overrightarrow{l})(\overrightarrow{x}), \overrightarrow{y} + \overrightarrow{l})$, for any $\overrightarrow{l} \in \mathbb{Z}^p$.

Then the quotient space $S = (\mathbb{R}^u \times \mathbb{R}^p)/\sim$ is a solvmanifold, and any self map f of S is, up to homotopy fibre-preserving with respect to the (Mostow) fibration $T^u \hookrightarrow S \xrightarrow{\pi} T^p$ where π is induced by projection $\mathbb{R}^u \times \mathbb{R}^p \to \mathbb{R}^p$. Furthermore, let X and Y be $u \times u$ and $p \times p$ matrices respectively over $\mathbb{Z}$. Then the correspondence on $\mathbb{R}^u \times \mathbb{R}^p$ given by $\widetilde{f}(\overrightarrow{x}, \overrightarrow{y}) = (X\overrightarrow{x}, Y\overrightarrow{y})$ induces a well defined map f on S if and only if for any $\overrightarrow{v} \in \mathbb{Z}^p$ we have that

(7.1.3) $$XA(\overrightarrow{v}) = A(Y\overrightarrow{v})X.$$

We call maps induced by X and Y *maps of type* X, Y, or simply diagonal maps. We also abuse notation, and write $f = (X, Y)$ for short. Since these models are indeed solovmanifolds, then we can use the linearization lemmas from section 2 (i.e. (3.11)) as well as the naïve addition formulae of (5.5), to calculate Nielsen numbers. In particular we have:

(7.2) PROPOSITION. *Let $f = (X, Y)$ be a diagonal self map of a solvmanifold S with Mostow fibration as above. Then*

$$N(f) = \sum_{[\overline{\alpha}] \in \mathcal{R}(\overline{f}^b_*)} (|\det(A^{\overline{\alpha}}X - I)|).$$

Note again that we do not at this point, need to know which fibre has which linearization, since we can simply run through a set of representatives of the elements of $\mathcal{R}(\overline{f}^b_*)$ (see Chapter I.3). Our first example gives the calculation Cheng-ye You showed me in the late eighties.

(7.3) EXAMPLE ([Y3]). As can be seen by examining Example (3.13), the Klein bottle is an example of the construction given in Theorem (7.1). This same theorem allows us to determine *all* diagonal maps of K^2. The gluing data $A: \mathbb{Z} \to \mathrm{Aut}(\mathbb{Z})$ takes an integer n to $(-1)^n$, and maps of B and F_0 are denoted by integers (one by one matrices). Let f be a self map of K^2 of type (q, r). Then from Theorem (7.1) f is well defined if and only if $q(-1) = (-1)^r q$, that is if and only if either r is odd, or $q = 0$.

As is shown in Chapter I.3 of this handbook, linearization is functorial, so if $f = (q, r)$ is a map of type (q, r), then the nth iterate $f^n = (q^n, r^n)$ is a map of type (q^n, r^n). The Nielsen numbers of these iterates on the fibers are $|(-1)^v q^n - 1|)$ for $v \in \mathcal{R}((\overline{f}^b)^n_*) \cong \mathbb{Z}_{|r^n-1|}$. If $q = 0$, then all fibre maps have Nielsen number 1, and so from (7.2) we have that $N(f^n) = |(r^n - 1)|$. If r is odd, then $|r^n - 1|$ is even, and there is equal distribution of maps between the fibres with Nielsen numbers $|q^n - 1|$ and $|-q^n - 1| = |q^n + 1|$. From (7.2) we get $N(f^n) = |r^n - 1|(|q^n - 1| + |q^n + 1|)/2$. Thus

$$N(f^n) = \begin{cases} |q^n(r^n - 1)| & \text{if } q \neq 0, \\ |(r^n - 1)| & \text{if } q = 0. \end{cases}$$

These calculations for $N(f^n)$ (which Halpern equated not always correctly with the minimum number $M\Phi_n(f)$ — see Section 8) agree with those of [Ha] obtained by entirely different means. The reader may have noticed that the maps (r, q) do not exhaust all possibilities for maps of K^2. If k is arbitrary, then the maps (q, r) and the maps which take (s, t) to $(qt + ks, rs)$, all have the same Nielsen theory. This is because for fixed s the map which takes t to $qt + ks$ is homotopic to the map which takes t to qt. Thus the maps (r, q) *do* exhaust all the possibilities for calculations, and so we need only consider the standard (r, q) maps defined above.

For maps of $\mathcal{NR}$ solvmanifolds (i.e. Example (3.15)) we specifically excluded roots of unity other than 1. In the next example, which comes in several parts, we do allow roots of unity other than 1. When this occurs, the gluing data is often periodic (see [HK3]) and, as we shall see, when this is the case it simplifies calculations of all Nielsen type numbers, and in particular it simplifies algorithm (5.6).

(7.4) EXAMPLE (Circle base and two torus fibre). Consider the following example of the construction in Theorem (7.1) with $u = 2$, $p = 1$, and with gluing data given by $A\colon \mathbb{Z} \to \mathrm{Aut}(\mathbb{Z}^2)$ where $A(k) = D^k$ with

$$D = \begin{bmatrix} 0 & 1 \\ -1 & 1 \end{bmatrix}.$$

We denote the resulting solvmanifold by S with Mostow fibration $T^2 \to S \to S^1$. Note that $D^6 = I$, and for $0 \leq i < j \leq 5$ that $D^i \neq D^j$, so D is periodic of period 6.

Let $f = (X, r)$, where X is a 2×2 matrix with entries in $\mathbb{Z}$, and r is an integer. The required compatibility condition (7.1.3) is easily seen to be equivalent to $XD = D^r X$. This time however, since D is periodic with period 6, when $r \equiv s(\mathrm{mod}\ 6)$ then $XD = D^r X$ if and only if $XD = D^s X$. Thus once we find a single X and suitable r, there are many other r's with (X, r) a well defined map. We divide the the computation of the Nielsen numbers of iterates of maps given in Example (7.4) into mod 6 classes for the various r. Details of these and other calculations are given in Chapter I.3.

Example (7.4) for $r \equiv 0, 2, 3, 4 \bmod (6)$.

(7.5) PROPOSITION. *If $r \equiv 0, 2, 3, 4 \pmod 6$, then $f = (X, r)$ is well defined map only when $X = 0$. The maps $f^n = (X^n, r^n) = (0, r^n)$ are all fibre uniform with $N(f_b) = 1$, so*

$$N(f) = 1 \cdot N(\overline{f}^n) = |1 - r^n|.$$

Example (7.4) for $r \equiv 1 \pmod 6$.

(7.6) PROPOSITION. *Let $r \equiv 1 \pmod 6$. Then (X, r) is a well defined if and only if X is of the form shown for arbitrary integers a and b. For such an X we have that $N(f^n)$ is as displayed.*

$$X = \begin{bmatrix} a & b \\ -b & a+b \end{bmatrix}, \qquad N(f^n) = \frac{|1 - r^n|}{6} \sum_{j=0}^{5} |\det(D^j X^n - I)|.$$

PROOF. Since $D^r = D$, the first part of the proposition is proved by solving the matrix equation $XD - DX = 0$. For the second part, note that when $r \equiv 1$ mod 6, then $r^n \equiv 1$ mod 6 too, so $N(\overline{f}^n) = |1 - r^n|$ is divisible by 6, and each linearization on the fibres occurs exactly $|1 - r^n|/6$ times. □

Example (7.4) with $r \equiv 5 \pmod 6$. Note that $5^{2n} \equiv +1 \pmod 6$, while $5^{2n+1} \equiv -1 \pmod 6$. So for even iterates calculation of the Nielsen numbers are covered by (7.6). Of course we also need the X to satisfy the condition of our next proposition.

(7.7) PROPOSITION. *Suppose that* $r \equiv 5 \pmod 6$. *Then* $XD = D^r X$ *if and only if for any integers* a *and* b *the matrix* X *is of the form shown*

$$X = \begin{bmatrix} a & b \\ a+b & -a \end{bmatrix}, \qquad Z = \begin{bmatrix} 1 & 1 \\ 2 & -1 \end{bmatrix}.$$

Here Z *is simply an example of such an* X *with* $a = b = 1$. *The resulting maps* f^n *of type* (X^n, r^n) *are fibre uniform (and so* f *is weakly Jiang) for all* X *if and only if* n *is odd. Furthermore* $N(f^n)$ *is given by*

$$\begin{aligned} N(f^{2n+1}) &= |(1 - r^{2n+1})\det(X^{2n+1} - I)|, \\ N(f^{2n}) &= \frac{|1 - r^{2n}|}{6} \sum_{j=0}^{5} |\det(D^j X^{2n} - I)|. \end{aligned}$$

We note in passing that for the X in Proposition (7.7), that all even powers of X are diagonal.

(7.8) EXAMPLE. As an example of (7.7), for $n = 12$ and $X = Z$ there, if we let $s_j = |\det(D^j Z^{12} - I)|$, then $s_0 = 529,984$, $s_1 = 530,713$, $s_2 = 532,171$, $s_3 = 532,900$, $s_4 = 532,171$, and $s_5 = 530,713$. So

$$N(f^{12}) = \frac{1}{6}|5^{12} - 1| \sum_{j=0}^{5} s_j = 129,746,581,499,808.$$

In view of the emphasis in Nielsen coincidence theory, of the equality $N(f,g) = R(f,g)$ for homogeneous spaces (see the last part of Section 9), the question arrises as to what light the analysis here sheds on the fixed point version of this question. While we have no meta theorems, the following proposition shows that there are likely to be patterns around that bear investigation.

(7.9) PROPOSITION. *Let* $T^u \to S \to S^1$ *be a solvmanifold with gluing data* $A\colon \mathbb{Z} \to \mathrm{Aut}(\mathbb{Z}^u)$ *generated by* $A(1) = G$, *where* G *is a periodic matrix of least period* d. *If the only maps* (W, r) *of* S *with* $r \equiv 1 \operatorname{mod}(d)$, *and with* W *non singular are the zero matrices* $W = 0$, *then a map* $f = (X, r)$ *is weakly Jiang if and only if* X *is not a power of* G. *In particular the a map* (X, r) *in Proposition* (7.6) *is non weakly Jiang if and only if* $X = D^k$ *for some* k.

PROOF. Suppose that $X = D^k$ for some k. Clearly $DX = XD$, so (X, r) is a self map of S. It is also clear that there are integers s and t with $\det(D^t X - I) = 0$ and $D^t(X - I) \neq 0$, so (X, r) is non weakly Jiang by (3.17). Next suppose that (X, r) is a self map of S which is non weakly Jiang and with $r \equiv 1$ mod d. Then $XD = DX$, and by (3.17) again, there is an integer t with $det(D^t X - I) = 0$. Let

$W = D^tX - I$, then W is non-singular, and $WD = D^tXD - D = D^{t+1}X - D = DW$, so (W, r) is a self map. By hypothesis $W = 0$, so $X = D^{d-t}$ as required.

For the application to Proposition (7.6), we must show that if (W, r) is a map of S there with $\det(W) = 0$, then $W = 0$. Since (W, r) is a map, then W must be of the the form of X in (7.6) so $\det(W) = a^2 + ab + b^2 = 0$ for some integers a and b. So $a = b(-1/2 \pm i\sqrt{3}/2)$. Since $a \in \mathbb{Z}$ then $b = 0$ which then gives that both a and b and so W are zero as required. □

(7.10) REMARK. As can be seen from this section, the role of non-weakly Jiang maps in the calculation of ordinary Nielsen numbers of a self map f of a solvmanfifold is relatively uninteresting (it simply means that certain summands in the formulas (7.2) and (7.6), for example, are zero). On the other hand, as we shall see in the next section, it is the non-weakly Jiang maps that have the more interesting Nielsen periodic theory.

Now let $f = (X, Y)$ be a map of a model solvmanifold $T^u \hookrightarrow S \to T^p$, with gluing data $A\colon \mathbb{Z}^p \to \mathrm{Aut}(\mathbb{Z}^u)$, and let ε_i be the ith standard basis element in $\mathbb{Z}^p$. Then A is generated by p invertible (over $\mathbb{Z}$) matrices $D_i := A(\varepsilon_i)$. Moreover, since $\varepsilon_i + \varepsilon_j = \varepsilon_j + \varepsilon_i$, then for $1 \le i, j \le p$ we must have $D_iD_j = D_jD_i$. If $Y = (y_{ij})$, the compatibility condition requires for each $i = 1, \ldots, p$ that

$$XD_i = D_1^{y_{1i}} \ldots D_p^{y_{pi}} X.$$

(7.11) EXAMPLE. Let $u = p = 2$ in Theorem (7.1), thus $A\colon \mathbb{Z}^2 \to \mathrm{Aut}(\mathbb{Z}^2)$ is generated by 2 invertible matrices D_1 and D_2. Let D_1 be as shown below, and $D_2 = D_1^2$.

$$D_1 = \begin{bmatrix} -1 & 3 \\ -1 & 2 \end{bmatrix}, \quad Y_1 = \begin{bmatrix} 7 & 0 \\ 0 & 4 \end{bmatrix}, \quad Y_2 = \begin{bmatrix} 5 & 4 \\ 1 & 5 \end{bmatrix}, \quad X = \begin{bmatrix} a & -3c \\ c & a-3c \end{bmatrix}.$$

Note that D_1 has period 6. Since $D_2 = D_1^2$ then the following relationships $D_2 = D_2^4 = D_1^4D_2^5$, and $D_1^7 = D_1^5D_2$ are easily verified. In particular if $XD_1 = D_1^7X$, then we must also have that $XD_2 = XD_1^2 = D_1XD_1 = D_2X = D_2^4X$, so for such an X, and Y_1 as shown above we have that (X, Y_1) is a map. In addition, for this same X, we must also have that $XD_1 = D_1^5D_2X$ and $XD_2 = D_1^4D_2^5X$, so (X, Y_2) is a map too. Therefore, for $k = 1, 2$, in order to find compatible maps (X, Y_k) in this example, we need exactly those X for which $XD_1 = D_1X$. That the form of such X is as shown follows a by now familiar argument.

For the diagonal matrix Y_1 we have for each n, from (2.1) that $\mathcal{R}(\overline{f}_*^{0n}) \cong \mathbb{Z}_{7^n-1} \times \mathbb{Z}_{4^n-1}$. Now the linearization of the restriction of f to the fibre over (i, j) is $D_1^iD_2^jX^n$. So if $(i, j) \equiv (s, t) \bmod (6, 3)$ (i.e. $i \equiv s \bmod 6$, and $j \equiv t \bmod 3$) then the Nielsen numbers $|\det(D_1^iD_2^jX - I)|$, and $|\det(D_1^sD_2^tX - I)|$ are equal.

Thus by analogy with Example (7.4), $\mathcal{R}(\overline{f}_*^{0n})$ can be divided up according to fibre types, with $6 \cdot 3 = |\Phi(\overline{f})| = |\det(Y_1 - I)|$ equivalence classes, and each class has the same number of elements. It is not hard to see (putting $Y = Y_1$) that we have the following analogue of the formula for $N(f^n)$ in (7.7).

$$N(f^n) = \frac{|\det(Y^n - I)|}{|\det(Y - I)|} \sum_{[\alpha] \in \mathcal{R}(\overline{f}_*^{0n})} |\det(A^\alpha X^n - I)|, \tag{7.11.1}$$

where for $\alpha = (i_1, \ldots, i_p) \in \mathcal{R}(\overline{f}) \cong \mathbb{Z}^p/(Y - I)(\mathbb{Z}^p)$, we have $A^\alpha = D_1^{i_1} \ldots D_p^{i_p}$.

(7.12) REMARK. It should be clear that formula (7.11.1) works in a much broader context than that of the single map $f_1 = (X, Y_1)$. In fact, whenever each of the D_i (as defined by an A as above) are periodic, this formula will work. In particular it works for $f_2 = (X, Y_2)$ above. However the formula does not (as it stands) particularize to (7.7) (unless $|r - 1| = 6$). In other words, the calculation of $N(f^n)$ from formula (7.11.1) can, in individual cases, often be done even more efficiently than considering $|\det(Y - I)|$ fibre types (especially since this can be huge). However, care must be taken in determining the minimum number of fibre types. In this example, there are a total of only four different Nielsen numbers on the fibre. On the other hand the minimum number of fibre types is 6 (since two are repeated). The point is that we need 6 fibre types in order to maintain uniformity of distribution.

8. Fibre techniques in Nielsen periodic point theory

This section reports briefly on four papers [HK1]–[HK4], which do Nielsen periodic point theory on nil- and solvmanifolds. For a self map $f: X \to X$ and a fixed positive integer n, there are two Nielsen type numbers $NP_n(f)$ and $N\Phi_n(f)$, which are lower bounds with respect to maps homotopic to f, for the cardinality of the sets $P_n(f) := \{x \in \Phi(f^n) : f^n(x) = x, \text{ and } f^m(x) \neq x \text{ for any } m < n\}$ and $\Phi(f^n)$, respectively. Of course $\Phi(f^n) = \bigcup_{m|n} P_m(f)$, and a first expectation might be that $N\Phi_n(f) = N(f^n)$, and $N\Phi_n(f) = \sum_{m|n} NP_m(f)$. Unfortunately in general, periodic point theory is much more subtle than this, and both equalities have counterexamples, the second even for very nice spaces (consider $n = 2$ for the flip map on S^1). Of course we would like reasonable conditions to know when these (and other) equalities hold. In this regard, once again nil- and solvmanifolds show themselves to be ideal spaces for applying fibre space techniques, in that they exhibit many useful properties (including $N\Phi_n(f) = \sum_{m|n} NP_m(f)$, and conditions under which $N\Phi_n(f) = N(f^n)$), that allow for calculations. We review briefly the definitions of these numbers then give and illustrate three main theorems. For details of Nielsen periodic point theory see [J], [HPY], [HY].

The definitions of the Nielsen type periodic numbers $N\Phi_n(f)$ and $NP_n(f)$ are based on orbits and reducibility. Let $m|n$ be positive integers. Let $\mathbf{F}^m$ be a fixed point class of f^m, and $y \in \mathbf{F}^m$, then $f^n(y) = y$, so y is also a fixed point of f^n, and in this induces a well defined function $\gamma_{m,n}\colon \Phi(f^m)/\sim\ \to \Phi(f^n)/\sim$. On the other hand, if $y \in \mathbf{F}^m$, then $f^m(f(y)) = f(f^m(y)) = f(y)$ is also a fixed point of f^m, and the designation $y \to f(y)$ yields a well defined index preserving bijection $f\colon \Phi(f^m)/\sim\ \to \Phi(f^m)/\sim$. Corresponding to all of this, if x is a fixed point of f, there is a "boosting function" $\iota_{m,n}\colon \mathcal{R}((f^m)^x_*) \to \mathcal{R}((f^n)^x_*)$ defined on a Reidemeister class $[\alpha]^m$ by

$$\iota_{m,n}([\alpha]^m) = [\alpha f^m_*(\alpha) f^{2m}_*(\alpha) \dots f^{(n-m)}_*(\alpha)]^n.$$

Note that the $\gamma_{m,n}$ and the $\iota_{m,n}$ may not be injective, for example the flip map in S^1 from level 1 to level 2. Note also the superscritps on both geometric and Reidemiester classes. These superscripts keep track of the various levels. With regard to algebraic orbits, for each m there is a well defined index preserving bijection $f^x_*\colon \mathcal{R}((f^m)^x_*) \to \mathcal{R}((f^m)^x_*)$ defined by $f^x_*([\alpha]^m) = [f^x_*(\alpha)]^m$. Because of the index preserving nature of both geometric and algebraic bijections, the orbits (written $\langle \mathbf{F}^m \rangle = \{\mathbf{F}^m, f(\mathbf{F}^m), \dots\}$, respectively $\langle [\alpha]^m \rangle = \{[\alpha]^m, [f^x_*(\alpha)]^m, \dots\}$) may be spoken about as being essential or not.

(8.1) LEMMA. *The following diagrams are commutative.*

$$\begin{array}{ccc} \Phi(f^m)/\sim & \xrightarrow{\rho^m} & \mathcal{R}((f^m)^x_*) \\ {\scriptstyle \gamma_{m,n}}\downarrow & & \downarrow{\scriptstyle \iota_{m,n}} \\ \Phi(f^n)/\sim & \xrightarrow[\rho^n]{} & \mathcal{R}((f^n)^x_*) \end{array} \qquad \begin{array}{ccc} \mathcal{R}((f^m)^x_*) & \xrightarrow{f^x_*} & \mathcal{R}((f^m)^x_*) \\ {\scriptstyle \iota_{m,n}}\downarrow & & \downarrow{\scriptstyle \iota_{m,n}} \\ \mathcal{R}((f^n)^x_*) & \xrightarrow[f^x_*]{} & \mathcal{R}((f^n)^x_*), \end{array}$$

and for all $k|m|n$, we have $\gamma_{k,n} = \gamma_{m,n}\gamma_{k,m}$ and $\iota_{k,n} = \iota_{m,n}\iota_{k,m}$. Furthermore $\iota_{m,m} = 1$.

We say that a class $[\alpha]^n$ (respectively, $\mathbf{A}^n$) is reducible, provided there exists an $m|n$ and a class $[\beta]^m$ (respectively, $\mathbf{B}^m$) for which $\iota_{m,n}([\beta]^m) = [\alpha]^n$ (respectively, $\gamma_{m,n}(\mathbf{B}^m) = \mathbf{A}^n$). The depth $d([\alpha]^n)$ (or $d(\mathbf{A}^n)$) is the smallest integer $d|n$ such that the class $[\alpha]^n$ (respectively, $\mathbf{A}^n$) reduces to level d. The class $[\alpha]^n$ (respectively, $\mathbf{A}^n$) is *irreducible* provided it has depth n. From the commutativity of the diagrams in (8.1) we have

(8.2) COROLLARY. *The geometric depth of a class is greater than or equal to the depth of its ρ image. (i.e. $d(A^n) \geq d([\rho(A^n)])$). The length of an algebraic orbit is preserved under the $\iota_{m,n}$. If an orbit $\langle [\alpha]^m \rangle$ is irreducible and essential, it represents at least m periodic points.*

The last statement bears further explanation. It means that if there is one periodic point y of least period m, then there are at least m of them $y, f(y), f^2(y), \ldots, f^{m-1}(y)$ in the same orbit. Note that the orbit we are thinking of here is at the geometric level of points, and not classes. The orbits of classes (both geometric and algebraic) can have lengths which are less than m in this context. In fact this is key to understanding the definitions below. We refer the reader to [J] and [HPY] for an example where for certain n, there are irreducible essential orbits of length 1. By ordinary Nielsen theory, such single classes reveal the existence of a fixed point x (say) of f^n. However the irreducibility of the class of this x, from the left hand diagram above, shows that the elements of set $\{x, f(x), \ldots, f^{n-1}(x)\}$ are distinct. Thus the single Nielsen class must contain at least n periodic points of least period n. With this insight, we make the following definition. For each positive integer n define

$$NP_n(f) := n\#(\mathrm{IEO}_n),$$

where $\#(\mathrm{IEO}_n)$ is the number of irreducible essential algebraic *orbits* of f at level n.

(8.3) EXAMPLE. Let $f\colon S^1 \to S^1$ be the map $f(e^{i\theta}) = e^{3i\theta}$. Then

$$\Phi(f^n) = \{e^{2\pi ik/(3^n-1)} : k = 0, \ldots, 3^n - 2\},$$

in particular $\#(\Phi(f^n)) = 3^n - 1$. Note that $n = 1$, $k = 1$, and $n = 2$, $k = 4$, and $n = 4$, $k = 40$ determine the same point $e^{\pi i} = -1$, similarly $n = 2$, $k = 1$ and $n = 4$, $k = 10$ determine $e^{\pi i/4}$. In the latter case $P_4(f) = \Phi(f^4) - \Phi(f^2)$, and $\#(P_4(f)) = 72$.

At the algebraic level $\mathcal{R}(f^1_*) \cong \mathbb{Z}_2$, $\mathcal{R}(f^2_*) \cong \mathbb{Z}_8$ and $\mathcal{R}(f^4_*) \cong \mathbb{Z}_{80}$. Since we are working with abelian groups we denote the binary operation by $+$. Note then that $\iota_{1,2}(\alpha) = f(\alpha) + \alpha = 3\alpha + \alpha$, that is $\iota_{1,2}$ is multiplication by $3 + 1 = 4$. Similarly $\iota_{2,4}$ is multiplication by $3^2 + 1 = 10$, while $\iota_{1,4}$ is multiplication by $3^3 + 3^2 + 3^1 + 1 = 40$. This is illustrated in the diagram below

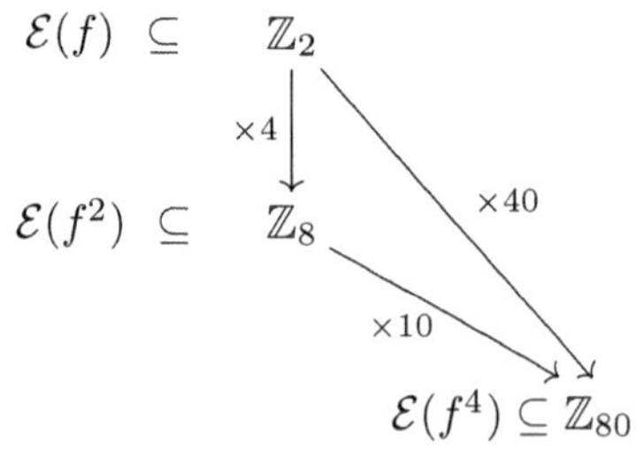

The inclusions are of course given by the various ρ. Since S^1 is a Jiang space and $L(f^j) \neq 0$ for all j, the inclusions are in fact bijections, so all the algebraic classes are essential, and we need only deal with the algebra. The fact that $4 \times 10 = 40$

illustrates that part of Lemma (8.1), which concerns the relationship between the ι $(\iota_{k,n} = \iota_{m,n}\iota_{k,m})$.

Using the fact that $f_*\colon \mathcal{R}(f_*^n) \to \mathcal{R}(f_*^n)$ is multiplication by 3 mod $3^n - 1$, we see at level 4 that there are 18 irreducible essential orbits, at level 2 there are three, and at level 1 just 2, so $NP_1(f) = 2$, $NP_2(f) = 2 \cdot 3$ and $NP_4(f) = 4 \cdot 18$.

The other periodic point number $N\Phi_n(f)$, is defined by computing the *minimal height* (that is the sum of the depths of its members) of a set of *n-representatives for f* (that is a collection of algebraic *orbits* at various levels with the property that any essential orbit $\langle \alpha^m \rangle$ for any $m|n$ reduces to some element in this set). The numbers $NP_n(f)$ and $N\Phi_n(f)$ are homotopy invariants. While the definition for $N\Phi_n(f)$ is rather complicated it turns out that for nil and solvmanifolds we do not need to compute it in this way. From the fact that the set $\bigcup_{m|n} \mathrm{IEO}_m(f)$ is contained in any set of n-representatives we have

(8.4) THEOREM ([J], [HY]). *$NP_n(f) \le MP_n(f)$, and*

$$\sum_{m|n} NP_m(f) \le N\Phi_n(f) \le M\Phi_n(f).$$

The following theorem is one of the main ones from [HK1]. The class of weakly Jiang maps include all maps on nil, and many maps on arbitrary solvmanifolds (see e.g. (7.9)).

(8.5) THEOREM. *Let $f\colon S \to S$ be a self map of an arbitrary solvmanifold, then*

$$N\Phi_n(f) = \sum_{m|n} NP_m(f).$$

If S is a nilmanifold or $\mathcal{NR}$ solvmanifold, or if S is an arbitrary solvmanifold with f^n weakly Jiang, then for all $m|n$

(8.5.1) *if $N(f^n) \neq 0$ then $N\Phi_m(f) = N(f^m)$, and*

(8.5.2) $$NP_m(f) = \sum_{\tau \subseteq \mathbf{p}(m)} (-1)^{\#\tau} N(f^{m:\tau}),$$

where $\mathbf{p}(m)$ denotes the set of prime divisors of m and $m : \tau = m \prod_{p\in\tau} p^{-1}$.

Note that (8.5.2) is not simply Möbius inversion applied to the displayed equation, since (8.5.2) involves $N(f^{m:\tau})$ and not $N\Phi_{m:\tau}(f)$.

(8.6) EXAMPLE. Taking (X, r) as in Proposition (7.7), we have for $n = 12$ that $\mathbf{p}(12) = \{2, 3\}$, that $N\Phi_{12}(f) = N(f^{12})$, and $NP_{12}(f) = N(f^{12}) - N(f^6) - N(f^4) + N(f^2)$.

Sketch of the Proof of Theorem (8.5). The technique of the proof is to mimic "essential" properties (pun intended) which hold for maps of tori. These properties were used to prove similar theorems in that context in [HY]. Such properties among several others, include firstly the fact that for a map f of a torus, if $N(f^n) \neq 0$, then the boosting functions $\iota_{m,n}$ are injective when boosted to essential classes, and secondly that all maps are essentially reducible. A map $f: X \to X$ is said to be *essentially reducible* provided that for any class $[\alpha]^n$ of f^n, and any $[\beta]^m$ with $\iota_{m,n}([\beta]^m) = [\alpha]^n$, if $[\alpha]^n$ is essential, then so is $[\beta]^m$. If for any $f: X \to X$ we have that f is is essentially reducible, then we say that X itself is essentially reducible. A consideration of roots of unity on the linearizations of maps of tori and nilmanifolds, easily implies that tori and nilmanifolds are essentially reducible (see [HK1, 4.14]). Theorem (8.5), a complete proof of which we cannot give here in detail, uses fibre techniques and induction to deduce results on nil and solvmanifolds from the corresponding properties on the fiber and base components (note again that the towers of Theorem (3.1) start out with tori in base and fibre). With regard to the two properties mentioned above, after changing basepoints as in the proof of (3.16), we construct diagrams of the form

$$\begin{array}{ccccc}
 & \mathcal{R}((f^q)_{b*}^{m\mu/q}) & \xrightarrow{\iota_{m/q,n/q}} & \mathcal{R}((f^q)_{b*}^{ny/q}) & \\
 & \downarrow j_* & & \downarrow j_* & \\
[\beta'], [\beta] \in & \mathcal{R}((f^m)_*^{\mu}) & \xrightarrow{\iota_{m,n}} & \mathcal{R}((f^n)_*^{y}) & \ni [\alpha'] \\
 & \downarrow & & \downarrow & \\
[0_b] \in & \mathcal{R}((\overline{f}^q)_*^{mb/q}) & \xrightarrow[\iota_{m,n}]{} & \mathcal{R}((\overline{f}^n)_*^{p(y)}) & \ni [\overline{\alpha}']
\end{array}$$

Here b is a periodic point of least period of q of $\overline{f}$ in the base B of the fibration. Such a b exists, either because either B is a torus, or by induction hypothesis (essential reducibility). To show the injectivity of the middle $\iota_{m,n}$ (under the assumptions that the other two are injective), we note that if $\iota_{m,n}([\beta']) = \iota_{m,n}([\beta]) = [\alpha']$, then $p_*([\beta']) = p_*([\beta]) = [0_b]$ since the lower $\iota_{m,n}$ is injective. By the exactness of the appropriate version of the ammended sequence in the proof of Theorem (3.17), there are $[\gamma]$ and $[\gamma']$ in $\mathcal{R}((f^q)_{b*}^{m\mu/q})$ with $j_*([\gamma])) = [\beta]$ and $j_*([\gamma']) = [\beta']$. Now $j_* \circ \iota_{m/q,n/q}([\gamma]) = j_* \circ \iota_{m/q,n/q}([\gamma']) = [\alpha]$, and the injectivity of j_* and $\iota_{m/q,n/q}$ imply that $[\gamma] = [\gamma']$, so that $[\beta'] = [\beta]$ as required.

A similar argument shows that the middle $\iota_{m,n}$ is essentially reducible when the other two are. □

The equation $N\Phi_n(f) = N(f^n)$ in Theorem (8.5) may not hold if $N(f^n) = 0$ as can be seen from the flip map on S^1. Here $N(f^2) = 0$ but there are clearly 2 fixed points of f, and in fact $N\Phi_2(f) = 2$. The next theorem of which Theorem (8.5) is in fact a special case, allows us to compute $N\Phi_n(f)$ for weakly Jiang maps when $N(f^n) = 0$. Like (8.5), this theorem is also a generalization of one given originally mainly for tori (see [HY, 4.17]).

For a given $f\colon X \to X$ and a fixed natural number n, the set $M(f,n)$ is defined by $m \in M(f,n)$ if and only if $m|n$, $N(f^m) \neq 0$ and for k with $m|k|n$ and $m \neq k$, that $N(f^k) = 0$.

(8.7) EXAMPLE. Let $f\colon T^2 \to T^2$ be the map with linearization D, where D is the gluing data in Example (7.4). As we noticed there D has period 6, so in particular if $6|n$, then $N(f^n) = 0$. Let $n = 2^p 3^q k$, where $p, q \geq 1$ and k is odd and not a multiple of 3, then $M(f,n) = \{2^p k, 3^q k\}$.

The hypothesis of the next theorem includes all maps on nil and $\mathcal{N}R$ solvmanifolds, and many maps on arbitrary solvmanifolds.

(8.8) THEOREM. *Let $f\colon S \to S$ be a self map of a solvmanifolds S, and suppose that n is a fixed positive integer, and that for all $m \in M(f,n)$, the map f^m is weakly Jiang. Then*

$$N\Phi_n(f) = \sum_{\emptyset \neq \mu \subseteq M(f,n)} (-1)^{\#\mu - 1} N(f^{\xi(\mu)}),$$

where $\xi(\mu)$ is the greatest common divisor of all elements of μ. Furthermore, the Möbius inversion of (8.5.2) *can be used to obtain the $NP_n(f)$.*

We note in the theorem that if $N(f^n) \neq 0$, then $M(f,n) = \emptyset$, so Theorem (8.8) contains (8.5.1) as a special case. In Example (8.7) if $n = 180 = 2^2 \cdot 3^2 \cdot 5$, then $N\Phi_{180} = N(f^{20}) + N(f^{45}) - N(f^5)$. As a non-toral example we give:

(8.9) EXAMPLE. Consider the matrices X and D over $\mathbb{Z}$

$$X = \begin{bmatrix} a & 0 \\ b & ra \end{bmatrix}, \quad D = \begin{bmatrix} 1 & 0 \\ -1 & 1 \end{bmatrix}, \quad D^r = \begin{bmatrix} 1 & 0 \\ -r & 1 \end{bmatrix},$$

where a, b and r are integers. Let S be the solvmanifold with Mostow fibration $T^2 \to S \to S^1$, obtained by putting $A(k) := D^k$ for $k \in \mathbb{Z}$ in Theorem (7.1). A map $f = (X, r)$ is well defined if and only if X is of the form shown. In each case, f^n is fibre uniform and hence weakly Jiang for all n, and $N(f^n) = |r^n - 1||a^n - 1||r^n a^n - 1|$. Furthermore, if $N(f^n) \neq 0$, then $N\Phi_n(f) = N(f^n)$. If $r = 1$, then $N\Phi_n(f) = NP_n(f) = N(f^n) = 0$ for all n. If $a \in \{\pm 1\}$ then

$$N\Phi_n(f) = N\Phi_m(f) = N(f^m) = |r^m - 1||a^m - 1||r^m a^m - 1|,$$

where m is the largest odd divisor of n.

We come now to the place where we deal with non-weakly Jiang maps. In this context, Theorems (8.5) and (8.8) do not apply, and in fact counter-examples to their conclusions can be given (see for example Corollary (8.15)). Let $f\colon S \to S$ be a map of a solvmanifold with linearization (X, Y). If $\det(Y - I) \neq 0$ then we can assume that $\overline{f}$ is a linear map with $\#(\Phi(\overline{f}^n)) = N(\overline{f}^n)$ (for details see Chapter I.3). In such cases, let $\Phi(\overline{f}^n) = \{x_j : j = 0, \dots, N(\overline{f}^n) - 1\}$. Suppose we want to calculate $NP_6(f)$, the minimum number of periodic points of minimal period 6. Because the map is fibre preserving, each periodic point of f lies in the fibre over some periodic point in the base. To see this, let x be such that $f^m(x) = x$, then $\overline{f}^m(p(x)) = p(f^m(x)) = p(x)$, so $p(x)$ is a periodic point of $\overline{f}$. Furthermore, as in the fixed point case, it turns out that when the fibration satisfies the naïve addition conditions, then we can count the number of n periodic classes of f, by counting them as the sum of the periodic classes in the fibres over the x_j in the base. At this point unlike the fixed point case, it becomes very important to take account the period $\mathrm{per}(x_j)$ of the x_j. For $n = 6$, the possibilities for $\mathrm{per}(x_j)$ are 1, 2, 3 or 6. If x_j has minimum period 3 say, then we must iterate f three times before its restriction $(f^3)_{x_j}$ is a self map of the fibre F_{x_j} over x_j. Thus the 6th iterate of f, when restricted to the fibre over this x_j will be the second iterate of $(f^3)_{x_j}$, and so periodic points of minimum period 2 of $(f^3)_{x_j}$ are periodic points of minimum period 6 of f. Of course there are at least $NP_2((f^3)_{x_j})$ of them sitting in the fibre over x_j. Similar calculations can then be made for each point in the base with periods 1, 2, 3 and 6, and so we deduce that there are at least $\sum_{b \in \Phi(\overline{f}^6)} NP_{6/\mathrm{per}(b)}((f^{\mathrm{per}(b)})_b)$ periodic points of f of least period 6. The question of course is if this is $NP_6(f)$ (compare the discussion of $N(f)$ as the sum of the $N(f_b)$ in the introduction). Because the Mostow fibrations on solvmanifolds satisfy the naïve addition conditions the answer is yes, as the next theorem states.

(8.10) THEOREM ([HK2]). *Let S be a solvmanifold, and $N \to S \to T$ its Mostow fibration. Let n be a positive integer and $f\colon S \to S$ a fibre preserving map inducing a linear map $\overline{f}\colon T \to T$. If $N(\overline{f}^n) = 0$, then $NP_n(f) = 0$, if $N(\overline{f}^n) \neq 0$, then*

$$NP_n(f) = \sum_{b \in \Phi(\overline{f}^n)} NP_{n/\mathrm{per}(b)}((f^{\mathrm{per}(b)})_b).$$

By Theorem (8.5) we have on solvmanifolds that $N\Phi_n(f) = \sum_{m|n} NP_m(f)$, so when the $NP_m(f)$ have been calculated from (8.10), then the $N\Phi_n(f)$ can be calculated from this formula. On the other hand, when we know the $N\Phi_n(f)$, we can calculate the $NP_n(f)$ by the formula $NP_n(f) = \sum_{\tau \subseteq \mathbf{P}(n)} (-1)^{|\tau|} N\Phi_{n:\tau}(f)$. Therefore for a given f we only need fibre techniques to calculate either the $N\Phi_m(f)$, or the $NP_m(f)$. Typically, for weakly Jiang maps, it is easier to compute the $N\Phi_n(f)$

using Theorem (8.5), while for non-weakly Jiang maps it is easier to compute the $NP_n(f)$, using (8.10).

(8.11) EXAMPLE. Let K^2 be the Klein bottle, $r \neq 1 \in \mathbb{Z}$ be odd, and let $f = (-1, r)$ be the self map of K^2. Then f induces the standard map $\overline{f}$ of degree r on the base of the Mostow fibration $S^1 \hookrightarrow K^2 \xrightarrow{p} S^1$, and $\overline{f}^n$ has exactly $N(\overline{f}^n)$ periodic points, which we denote by $\Phi(\overline{f}^n) = \{x_j : j = 0, \ldots, |r^n - 1| - 1\}$, where $x_0 = \{0, 1\}$ in $S^1 = I/[0 \sim 1]$, and the other $x_j = j/|1 - r^n|$ consist of points equally spaced on the circle (see [HK1]).

It is not hard to see that $\rho\colon \Phi(\overline{f}^m) \to \mathcal{R}(\overline{f}^m_*)$ takes x_j to the class of $j \in \mathbb{Z}_{|1-r^m|}$. When $m|n$ and $x_j \in \Phi(\overline{f}^m)$, the restriction $(f^m)_{x_j}$ of f^m to the fibre over x_j has degree $(-1)^j(-1)^m$ (see [HK1, 5.5] or Chapter I.3). Since $N((f^m)_{x_j}) = 1 - (-1)^{j+m}$, and $|r^m - 1|$ is even, then exactly half of these numbers are zero, and the other half 2. In particular no f^n is weakly Jiang.

We discuss cases $n = 6$, and $n = 2^t$ with $t \geq 1$. Since $(f^t)_b = \pm 1$ we have in both cases, that $NP_{n/\text{per}(b)}((f^{\text{per}(b)})_b) = 0$ for all b with $n/\text{per}(b) > 1$. In other words the only possible non-zero contribution to $NP_n(f)$ from Theorem (8.10) occurs when $\text{per}(b) = n$, that is on the irreducible elements of $\Phi(\overline{f}^n)$. Furthermore, for $n = 6$, and $n = 2^t$, we have $(f^n)_{x_j}$ is of degree $(-1)^{n+j} = (-1)^j$, so it is only the odd values of j where the $(f^n)_{x_j}$ contribute to $NP_n(f)$. So we need to exclude those odd indices for $\overline{f}$ which come up from iterates less than n. Now on the S^1 in the base, $\iota_{m,n} = 1 + r^m + r^{2m} + \ldots + r^{n-m}$, and this simply multiplies the index (j) of geometric classes by this number (see [HK3] for a more detailed discussion). Because r is odd we know that $\iota_{m,n}$ consists of n/m odd terms. Thus $r_{m,n} \equiv (n/m) \pmod 2$. Multiplication of each element from $\{0, \ldots, |r^m - 1| - 1\}$ by an odd number will give numbers, half of which are odd, and the other half even. Obviously multiplying by an even number will give only even numbers.

Now when $n = 2^k$ then n/m is always even. Therefore in this case all the odd indices are irreducible, and as we remarked above, all the even indices (irreducible or not) contribute nothing. Thus

$$\begin{aligned} NP_n(f) &= \sum_{j\text{odd}} NP_1((f^n)_{x_j}) \\ &= \sum_{x_j \in \Phi(\overline{f}^n)} NP_1((f^n)_{x_j}) = \sum_{x_j \in \Phi(\overline{f}^n)} N((f^n)_{x_j}) = N(f^n) \end{aligned}$$

where the last step is from the naïve addition formula. Since on the odd indices $N((f^n)_{x_j}) = 2$ and zero elsewhere, the naïve addition formula also gives that $N(f^n) = N(\overline{f}^n) = r^n - 1$. So for $q = -1$ on K^2 when $n = 2^k$ we get the surprising result that $NP_n(f) = N(f^n)$. Surprising that is, when we contrast this with the more usual result that $N\Phi_n(f) = N(f^n)$ (see (8.5.1)). We also have that $N(f^{2^i}) = |r^{2^i} - 1|$ and $N\Phi_n(f) = \sum_{i=0}^{k} N(f^{2^i}) = \sum_{i=0}^{k} |r^{2^i} - 1|$.

For $n = 6$ we need to know which of the $(r^6 - 1)/2$ odd indices are irreducible. In this way we can exclude those indices coming up from $\Phi(\overline{f}^3)$, $\Phi(\overline{f}^2)$, and $\Phi(\overline{f})$. Since $6/3$ and $6/1$ are even, the boosts from $\Phi(\overline{f}^3)$ and $\Phi(\overline{f})$ give only even indices, as do the $(r^2 - 1)/2$ even indices from $\Phi(\overline{f}^2)$. Thus we need to exclude from our sum only the $(r^2 - 1)/2$ odd indices coming up injectively from $\Phi(\overline{f}^2)$. So there are

$$\frac{1}{2}(r^6 - 1) - \frac{1}{2}(r^2 - 1) = \frac{1}{2}(r^6 - r^2)$$

irreducible indices which each contribute 2 to $NP_n(f)$, we therefore have $NP_n(f) = r^6 - r^2$.

We wish to draw attention to a number of features of Example (8.11). Firstly we only needed to consider those b for which $n/\mathrm{per}(b) = 1$. As we saw from Proposition (7.9) for D in (7.4) with $r \equiv 1 \bmod (6)$ $f = (X, Y)$ is non-weakly Jiang if and only if $X = D^t$ for some t. Clearly then f^n is non-weakly Jiang under the same conditions. Consider the following table $NP_s(D^\ell)$ which illustrates the possibilities for $NP_{n/\mathrm{per}(b)}((f^{\mathrm{per}(b)})_b)$ in this situation.

D^ℓ	$D^0 = I$	D	D^2	D^3	D^4	D^5
$NP_1(D^\ell)$	0	1	3	4	3	1
$NP_2(D^\ell)$	0	2	0	0	0	2
$NP_3(D^\ell)$	0	3	0	0	0	3
$NP_k(D^\ell)$, $(k > 3)$	0	0	0	0	0	0

In particular the $NP_{n/\mathrm{per}(b)}$ summands are zero in (8.10) except for $\mathrm{per}(b) \notin \{n, n/2, n/3\}$. Arguing the same way as we did in the Example (8.11), we see that Theorem (8.10)simplifies to give the following sum for $NP_n(f)$

$$NP_n(f) = \sum_{\{b:\mathrm{per}(b)\in\{n,n/2,n/3\}\}} NP_{n/\mathrm{per}(b)}((f^{\mathrm{per}(b)})_b).$$

With the aid of a modular version of Möbius inversion for circle maps ([HK3; 4.10]), (which helps in this type of situation to determine the distribution of the NP numbers on the fibres) we have the following consequence of these considerations (see [HK3] for details).

(8.12) PROPOSITION. *Let f be a map of type (D^k, r), with D as in (7.4), with $r \equiv 1 \bmod (6)$, and k arbitrary. If n is not divisible by 2 or 3, then*

$$N\Phi_n(f) = N(f^n) = 2|r^n - 1|, \quad \text{and} \quad NP_n(f) = 2NP_n(\overline{f}^n).$$

If $n = 6^s$ $(s \geq 1)$ *then* $NP_n(f) = 2|r^n - 1| - 4|r^{n/2} - 1|/3 - |r^{n/3} - 1|$.

(8.13) REMARK. In the non-weakly Jiang Example (8.11), we can think of the Nielsen numbers on the fibres as averaging out to be 1 over each point in the base. In Proposition (8.12), after recalling from Proposition (7.9) that the maps of type (D^k, r) above, are non-weakly Jiang, we see that there are similar averagings, one average for each of the three summands in the formula for $NP_n(f)$. For example when $n/\text{per}(b) = 1$, there are 6 fibre types, with $(0+1+3+4+3+1)/6 = 2$ as the average over each point in the base (explaining the 2 in the products $2|r^n - 1|$ and $2NP_n(\overline{f}^n)$ in Proposition (8.12)).

A second feature of Example (8.11) is the unusual equality $NP_n(f) = N(f^n)$ when $n = 2^t$. The obvious question asks if this is an isolated phenomena, or is if it widespread in some sense. The corollary following our next theorem (both from [HK3]) gives an indication of the answer. The theorem also determines the NP for all none weakly Jiang maps of K^2.

(8.14) THEOREM. *Let* $A\colon \mathbb{Z}^p \to \text{Aut}(\mathbb{Z}) = \{\pm 1\}$ *be the parity homomorphism given by* $A(l_1, \dots, l_p) = (-1)^{\sum_{i=1}^p l_i}$, *and let* S *the model constructed from* A *with its associated Mostow fibration* $S^1 \to S \to T^p$. *Let* Y *be a* $p \times p$ *matrix with odd column sums and* $q = \pm 1$. *Then* $f = (q, Y)$ *is a map of* S. *If* $N(\overline{f}^n) = 0$, *then* $NP_n(f) = 0$. *Otherwise*

$$NP_n(f) = \begin{cases} \displaystyle\sum_{\tau \subseteq \mathbf{P}(n) - \{2\}} (-1)^{|\tau|} N(\overline{f}^{n:\tau}) & \text{if } n \text{ is even,} \\ NP_n(\overline{f}) & \text{if } n \text{ is odd.} \end{cases}$$

In particular, if $n = 2^t$, *then* $NP_n(f) = N(\overline{f}^n) = |\det(Y^n - I)| = N(f^n)$.

The last part follows from the first, since when $n = 2^t$ then $\mathbf{P}(n) = \{2\}$, and $\mathbf{P}(n) - \{2\} = \emptyset$.

(8.15) COROLLARY. *For each* $m > 1$, *there is a solvmanifold* S *of dimension* m, *and self maps* f *on* S *for which, for an infinite number of* n, *we have* $NP_n(f) = N(f^n)$.

In work in progress the current author and Ed Keppelmann are investigating the periodic numbers on fibrations that do not satisfy the naïve addition conditions (see [HK6]).

9. Product and addition formulae for coincidences

Traditionally fixed point theory has been regarded as the special case of coincidence theory in which g is taken to be the identity. While this statement has also been regarded as overly simplistic because of the need to impose conditions (i.e. that X and Y be orientable manifolds), it is becoming increasingly cleary that even this is a somewhat naïve view. Part of this is because of the difficulty in generalizing fixed point theory results to coincidences. This manifests itself for example in the length of time it took to generalize the result $|L(f)| \leq N(f)$ for maps of solvmanifolds (see Remark (9.13)), and for example of the fact that while all self maps of $\mathcal{NR}$ solvmanifolds are fibre uniform, at the time of writing, the result is an open question for coincidences. In this section we follow the progress made in generalizing fibered Nielsen theory to coincidences. We extend the modified fundamental group approach to coincidences (see [GH]), and of course to the results of Jurek Jezierski (see [Jz1]).

Throughout this section $f, g\colon X \to Y$ will be maps, with X and Y closed oriented manifolds of the same dimension. We donote the set of coincidences of f and g, by $\Phi(f,g) = \{x \in X : f(x) = g(x)\}$. As usual $\Phi(f,g)$ is divided into Nielsen classes defined by requiring that x and y are equivalent, if there is a path $c\colon x \to y$ in X with $f(c) \sim g(c)$. A generic class is denoted by $\mathbf{C}$. Suppose that K and H are normal subgroups ([12]) of $\pi_1(X)$ and $\pi_1(Y)$, respectively, that $x \in \Phi(f,g)$, and that $f_*^x(K) \subseteq H$ and $g_*^x(K) \subseteq H$, then f and g induce homomorphisms $f_*^x, g_*^x\colon \pi_1(X)/K \to \pi_1(Y)/H$. These homomorphisms in turn define the Reidemeister classes given by the relation where $\alpha, \beta \in \pi_1(Y)/H$ are equivalent, if there is a class $\gamma \in \pi_1(X)/K$ such that $\alpha = g_*^x(\gamma)\beta f_*^x(\gamma^{-1})$. The set of such classes is denoted by $\mathcal{R}(f_*^x, g_*^x)$, with cardinality $R(f,g)$. The following is an exact sequence of based sets and basepoint preserving functions (with the obvious basepoints).

$$(9.1)\quad 1 \to \mathrm{Coin}_K(f_*^x, g_*^x) \to \pi_1(X,x)/K \xrightarrow{g_*^x \cdot f_*^{x-1}} \pi_1(Y, f(x))/H \to \mathcal{R}_H(f_*^x, g_*^x) \to 1,$$

where $g_*^x \cdot f_*^{x-1}(\alpha) = g_*^x(\alpha) f_*^x(\alpha^{-1})$, and $\mathrm{Coin}_K(f_*^x, g_*^x)$, the mod K subgroup of algebraic coincidences, is the kernel of the given function. When $\pi_1(Y, f(x))/H$ is abelian then $g_*^x \cdot f_*^{x-1}$ is a homomorphism which we write additively as $g_*^x - f_*^x$. In this case $\mathcal{R}_H(f_*^x, g_*^x)$ inherits a canonical abelian group structure as the cokernel of $g_*^x - f_*^x$.

(9.2) PROPOSITION ([GH]). *Let $f, g\colon X \to Y$ be maps as given above, with $\pi_1(Y, f(x))$ abelian, then sequence* (9.1) *is an exact sequence of groups and homomorphisms.*

The notion of index on coincidence Reidemeister classes, mimics the idea of index for the Reidemeister classes in the fixed point case. Let $[\alpha] \in \mathcal{R}_H(f_*^x, g_*^x)$,

([12]) As in the fixed point case, we are really thinking of normal subgroupoids of $\pi(X)$ and $\pi(Y)$. See [Jz1, p. 184].

define the index $i([\alpha])$ by

$$i([\alpha]) = \begin{cases} \mathrm{ind}(\mathbf{C}) & \text{if } [\alpha] = \rho(\mathbf{C}), \\ 0 & \text{otherwise.} \end{cases}$$

where this time $\mathrm{ind}(\mathbf{C})$ is the coincidence point index defined for example in [V]. As with the fixed point case, a class $[\alpha]$ is said to be *essential* if $i([\alpha]) \neq 0$.

The equality $N(f,g) = |\det(B-A)|$ ($= R(f,g)$ when $L(f,g) \neq 0$) in the next proposition is immediate since tori are Jiang spaces. The more subtle part (which generalizes the result in [BBPT]) is proved in [Jz1, p. 205].

(9.3) THEOREM (Jezierski, [J1]). *Let $f, g: T \to T$ be maps of a torus T, with linearization matrices A and B respectively, then*

$$N(f,g) = |L(f,g)| = |\det(B-A)|,$$

and if $L(f,g) \neq 0$, then $N(f,g) = R(f,g)$.

With respect to fibre preserving maps, we work in the context of two fibrations $p: E \to B$ and $q: E' \to B'$, and of fibre preserving maps $f, g: E \to E'$. That is we have a commutative diagram (or is it two?).

$$(9.4) \qquad \begin{array}{ccc} E & \xrightarrow{f,g} & E' \\ {\scriptstyle p}\downarrow & & \downarrow{\scriptstyle q} \\ B & \xrightarrow[\overline{f},\overline{g}]{} & B' \end{array}$$

Our exposition follows a modified fundamental group approach generalizing the fixed point case in [H1]. It leads to both Reidemeister and Nielsen number formulas for coincidences, as well as the following generalization of Theorem (5.4).

(9.5) THEOREM. *Let $f, g: E \to E'$ be fibre preserving maps of a Hurewicz fibrations $p: E \to B$ and $p': E' \to B'$, $x \in \Phi(f,g)$, $b := p(x)$ and $K = \ker j_*: \pi_1(F_b, x) \to \pi_1(E, x)$. Then the sequence of groups homomorphisms, and based sets and basepoint preserving functions*

$$\begin{aligned} 1 \to \mathrm{Coin}_K(f^x_{b*}, g^x_{b*}) &\xrightarrow{j_{K*}} \mathrm{Coin}(f^x_*, g^x_*) \xrightarrow{p_*} \mathrm{Coin}(\overline{f}^b_*, \overline{g}^b_*) \\ &\xrightarrow{\delta} \mathcal{R}_K(f^x_{b*}, g^x_{b*}) \xrightarrow{j_*} \mathcal{R}(f^x_*, g^x_*) \xrightarrow{p_*} \mathcal{R}(\overline{f}^b_*, \overline{g}^b_*) \to 1 \end{aligned}$$

is exact. So, if $\pi_1(E')$ is abelian, then

$$[\mathrm{Coin}(\overline{f}^b_*, \overline{g}^b_*); p_*(\mathrm{Coin} f^x_*, g^x_{b*})] R(f,g) = R(\overline{f}, \overline{g}) \cdot R_K(f_b, g_b). \qquad \square$$

Because of complications in the definition of coincidence index, and the coincidence version of the product theorem for index ([J1; Theorem 5.5]), we make the following assumptions implicitly throughout the rest of this section. We assume *that* $f, g\colon E \to E'$ *are fibre preserving maps of locally trivial fibre bundles* $p\colon E \to B$ *and* $p'\colon E' \to B'$; *that* E, E', B, B', *and all* F_b, F'_b *are closed compact connected orientable manifolds which as listed pairs, have the same dimensions; that* $\Phi(f, g)$ *is finite* (see [Jz1, Theorem 3.1]).

(9.6) LEMMA ([Jz1, 5.7]). *Let* $f, g\colon E \to E'$ *be as in Theorem* (9.5). *Suppose that* b *is an isolated coincidence point of* $\overline{f}$ *and* $\overline{g}$, *and that* $A \subseteq \Phi(f_b, g_b)$ *be an isolated set of coincidence points of* f_b *and* g_b, (*i.e.* A *is open and compact in* $\Phi(f_b, g_b)$). *Then* A *is also an isolated set of fixed points of* f *and* g, *and*

$$\operatorname{ind}(f, g : A) = \operatorname{ind}(f_b, g_b : A) \cdot \operatorname{ind}(\overline{f}, \overline{g} : b).$$

The next result is a global product theorem for index of coincidence points.

(9.7) THEOREM ([Jz1, Theorem 5.10]). *Suppose that* $N(\overline{f}, \overline{g}) \neq 0$, *and that* $L(f_b, g_b)$ *is independent of* $b \in \Phi(\overline{f}, \overline{g})$, *then* $L(f, g) = L(f_b, g_b) \cdot L(\overline{f}, \overline{g})$. *In particular the result is true if* $p'\colon E' \to B'$ *is orientable.*

PROOF. As indicated earlier, we may assume that $\Phi(\overline{f}, \overline{g})$ is finite. Then by normality and (9.6) we have

$$\begin{aligned} L(f, g) &= \sum_{b \in \Phi(\overline{f}, \overline{g})} \operatorname{ind}(f, g : F_b) = \sum_{b \in \Phi(\overline{f}, \overline{g})} L(f_b, g_b) \cdot \operatorname{ind}(\overline{f}, \overline{g} : b) \\ &= L(f_b, g_b) \sum_{b \in \Phi(\overline{f}, \overline{g})} \operatorname{ind}(\overline{f}, \overline{g} : b) = L(f_b, g_b) \cdot L(\overline{f}, \overline{g}). \end{aligned}$$ □

The next corollary of Lemma (9.6) is the analogue of Lemma (2.8).

(9.8) LEMMA. *Let* p *and* p' *be locally trivial fibre bundles, and let* $[\alpha K] \in \mathcal{R}_K(f^x_{b*}, g^x_{b*})$, $[\beta] \in \mathcal{R}(f^x_*, g^x_*)$ *and* $[\overline{\beta}] \in \mathcal{R}(\overline{f}^b_*, \overline{g}^b_*)$ *be such that* $j_{b*}([\alpha K]) = [\beta]$, *and* $p_*([\beta]) = [\overline{\beta}]$. *Then* $[\beta]$ *is essential, if and only if both* $[\alpha K]$ *and* $[\overline{\beta}]$ *are essential.*

The results and techniques for product/addition formula for fixed points, can now be mimicked. In particular we have exact sequences

$$\mathcal{E}(f_b, g_b) \xrightarrow{j_{b*}} \mathcal{E}(f, g) \xrightarrow{p_*} \mathcal{E}(\overline{f}, \overline{g}),$$

from which we deduce that

$$N(f, g) = \#(\mathcal{E}(f, g)) = \#\Bigg(\bigcup_{\overline{\mathbf{C}}_b \in \mathcal{E}(\overline{f}, \overline{g})} p_*^{-1}(\overline{\mathbf{C}}_b) \Bigg) = \sum_{\overline{\mathbf{C}}_b \in \mathcal{E}(\overline{f}, \overline{g})} \#(p_*^{-1}(\overline{\mathbf{C}}_b)),$$

where $\overline{\mathbf{C}}_b$ denotes the coincidence class of b. The following theorem does not appear elsewhere in the literature, but it makes for a smoother exposition, and it is too hard to resist.

(9.9) THEOREM. *Let $f, g: E \to E'$ be fibre preserving maps, and suppose for each $\overline{\mathbf{C}}_b$, that $[\mathrm{Coin}(\overline{f}_*^b, \overline{g}_*^b); p_*(\mathrm{Coin}(f_*^x, g_{b*}^x))]$ is independent of $x \in \mathbf{F} \in p_*^{-1}(\overline{\mathbf{C}}_b)$, and $b \in \overline{\mathbf{C}}_b$. Then*

$$N(f,g) = \sum_{b \in \chi} \frac{N_K(f_b, g_b)}{[\mathrm{Coin}(\overline{f}_*^b, \overline{g}_*^b); p_*(\mathrm{Coin}(f_*^x, g_*^x))]},$$

where the x are chosen in an essential class of f_b and g_b, (if no such x exists, we interpret that summand to be zero), and where χ is a set that has exactly one representative of each essential coincidence class of $\overline{f}$ and $\overline{g}$.

From this we easily have:

(9.10) COROLLARY ([Jz1, Theorem 6.9]). *If $N_K(f_b, g_b)$ is independent of b in any essential coincidence class of $\overline{f}$ and $\overline{g}$ and if $[\mathrm{Coin}(\overline{f}_*^b, \overline{g}_*^b); p_*(\mathrm{Coin}(f_*^x, g_*^x))]$ is independent of x in any essential class of f and g. Then*

$$(9.10.1) \qquad [\mathrm{Coin}(\overline{f}_*^b, \overline{g}_*^b); p_*(\mathrm{Coin}(f_*^x, g_{b*}^x))] N(f,g) = N_K(f_b, g_b) \cdot N(\overline{f}, \overline{g}).$$

In particular the formula holds if $\pi_1(Y)$ is abelian, and p' is orientable. Furthermore

$$N(f,g) = \sum_{b \in \chi} N(f_b, g_b)$$

($= N(\overline{f}, \overline{g}) N(f_b, g_b)$ in the presence of fibre uniformity), if and only if

(9.10.2) for each $x \in \mathbf{C} \in \mathcal{E}(f,g)$ we have $[\mathrm{Coin}(\overline{f}_*^b, \overline{g}_*^b); p_*(\mathrm{Coin}(f_*^x, g_*^x))] = 1$,
(9.10.3) and $N_K(f_b, g_b) = N(f_b, g_b)$.

We will refer to conditions (9.10.2) and (9.10.3) as the naïve addition conditions for coincidences. The proof of the next lemma is the analogue of the proof of (3.9).

(9.11) LEMMA. *Nilmanifolds and solvmanifolds are aspherical and satisfy the naïve addition conditions for coincidences for all pairs of self maps f and g for which $L(\overline{f}, \overline{g}) \neq 0$.*

(9.12) COROLLARY ([Jz1]). *Let $f, g: N \to N$ be self maps of a nilmanifold N, then $N(f,g) = |L(f,g)|$.*

PROOF. The proof follows those of (3.4) and (3.9), but this time using Theorem (9.3). The induction step uses

$$N(f,g) = N(\overline{f}, \overline{g}) N(f_b, g_b) = |L(\overline{f}, \overline{g})| L(f_b, g_b)| = |L(f,g)|. \qquad \square$$

(9.13) REMARK. In light of Lemma (9.11), the reader may be anticipating further generalizations of fixed point versions of theorems in Section 7, and in particular of the inequality $N(f) \geq |L(f)|$ for solmanifolds. There was however a wrinkle in the process. Clearly in order to generalize the proofs of the inductive fixed point results, one expects to use induction on the dimensions involved. Let $f, g: S_1 \to S_2$ be maps of solvmanifolds inducing maps $\overline{f}, \overline{g}: T_1 \to T_2$ on the tori bases of the Mostow fibrations for S_1 and S_2, respectively. Chris McCord in [Mc4] writes "Of course, in order for induction to work, the dimensions must match: we need $\dim(T_1) = \dim(T_2)$. But if S_1 and S_2 are different solvmanifolds, there is no guarantee *a proir* that $\dim(T_1) = \dim(T_2)$." Chris writing this in [Mc4, p. 82] is writing a correction to [Mc3]. The goal in the latter paper, when $\dim(T_1) \neq \dim(T_2)$, was to show either that $N(f,g) = |L(f,g)| = 0$, or to construct a new fibration $N' \to S_1 \to T_1'$ with $\dim(T_1') = \dim(T_2)$ so that f and g were still fibre preserving. However a counter example to the validity of one of the building blocks in this construction was discovered by Daciberg Gonçalves. The counter example did not show the inequality $N(f,g) \geq |L(f,g)|$ was invalid for solvmanifolds, but it did mean the generality of some of the theorems in [Mc3] were not yet sure. In [Mc4] Chris recovered the first part of the next result from [Mc3]. However the recovered proof in [Mc4] does not use fibre techniques. Since the counter example did not show that that the fibre techniques employed in [Mc3] were invalid when $\dim(T_1) = \dim(T_2)$ (for example when $S_1 = S_2$), then the second part of (9.14) (which is an easy generalization of Theorem (3.10) — see [Mc3, p. 260]) can also be recovered from [Mc3]. This left the inequality in doubt when $\dim T_1 > \dim T_2$ (but see Theorem (9.19)).

(9.14) THEOREM ([Mc3], [Mc4]). *Let $f, g: S_1 \to S_2$ be maps of solvmanifolds of the same dimension. If S_2 and S_2 are nilmanifolds, then $N(f,g) = |L(f,g)|$. If for $i = 1, 2$ the Mostow fibrations $N_i \to S_i \to T_i$ are such that* $\dim T_1 \leq \dim T_2$, *then*

$$|L(f,g)| \leq N(f,g).$$

After the initial success of calculating $N(f)$ by comparing it with $|L(f)|$ (see Remark (9.13)) was not mirrored in Nielsen coincidence theory, it was perhaps natural that attention be turned to the so to speak "next" algebraic (but not so easily computable) invariant $R(f,g)$. In particular attention was focused on conditions for the equality $N(f,g) = R(f,g)$ to hold. Consider the following four properties which, among other things, we relate in the next theorem.

(9.15.1) $$N(f,g) \neq 0,$$

(9.15.2) $$\mathrm{Coin}(f_*, g_*) = 1,$$

(9.15.3) $$R(f,g) < \infty,$$

(9.15.4) $$N(f,g) = R(f,g) = |L(f,g)|.$$

For the latter part of the theorem, we need a new definition. A space Y is said to be *Jiang type space for coincidences* if for any $f, g\colon X \to Y$ either $L(f,g) = 0$ in which case $N(f,g) = 0$ too, or if $L(f,g) \neq 0$, then $N(f,g) = R(f,g)$. Clearly if Y is a Jiang space, then Y is a Jiang type space for coincidences. In fact there is a generalization of this to pseudo Jiang maps in [GH, Theorem 2.7], but neither result uses fibre-techniques. Some of our earlier results showed that certain spaces were Jiang type spaces for fixed points.

(9.16) THEOREM. *Let $f, g\colon X \to Y$ be a pair of maps of compact manifolds.*

(9.16.1) *If $X = Y$ is a torus then* (9.15.1) $\Leftrightarrow$ (9.15.2) $\Leftrightarrow$ (9.15.3) $\Rightarrow$ (9.15.4).

(9.16.2) (see [Mc4]) *If X and Y, and are nilmanifolds and* $\dim(X) = \dim(Y)$, *then* (9.15.1) $\Leftrightarrow$ (9.15.2).

(9.16.3) *If $X = Y$ belongs to a nilpotent class* ([G1, Theorem 2.5]), *or if $X = N_1$ and $Y = N_2$ are compact nilmanifolds of the same dimension* ([G2, Theorem 2.6]), *then* (9.15.1) $\Leftrightarrow$ (9.15.2) $\Leftrightarrow$ (9.15.3), *in each case these equivalent conditions imply* (9.15.4).

(9.16.4) ([GW2, Theorem 3]) *If X is a finite complex, and $Y = N$ a nilmanifold,* (9.15.1) $\Rightarrow$ (9.15.3).

(9.16.5) ([G1; Theorem 3.7]) *If X and Y are the total spaces of fibrations p_1 and p_2 respectively with matching dimensions, if f and g are fibre preserving, p_2 is nilpotent, and either f_* or g_* are surjective at the π_1 level, then* (9.15.1) $\Leftrightarrow$ (9.15.2$'$) $\Leftrightarrow$ (9.15.3$'$), *where where* (9.15.2$'$) *and* (9.15.3$'$) *are the same as* (9.15.2) *and* (9.15.3), *respectively but add the condition that $L(f_b, g_b) \neq 0$. Furthermore each condition implies*

$$N(f,g) = N(f_b, g_b)N(\overline{f}, \overline{g}) = R(f_b, g_b)R(\overline{f}, \overline{g}) = R(f,g) = |L(f,g)|.$$

(9.16.6) ([GW3, Theorem 2.2]) *If X and Y are as in* (9.16.5), *and if $N(\overline{f}, \overline{g}) = N(\overline{f}, \overline{g})$ and $N(f_b, g_b) = R(f_b, g_b)$ for all $b \in \Phi(\overline{f}, \overline{g})$, then $N(f,g) = R(f,g)$.*

(9.16.7) *If X is a solvmanifold, and Y is a nilmanifold with* $\dim(X) \geq \dim(Y)$, *then* (9.15.1) $\Rightarrow$ (9.15.3) ([W1, Theorem 4]). *If* $\dim(X) = \dim(Y)$, *then* (9.15.1) $\Leftrightarrow$ (9.15.2) $\Leftrightarrow$ (9.15.3) $\Rightarrow$ (9.15.4) ([W3; Theorem 5]).

(9.16.8) ([W1, Lemma 4]) *If X and Y are solvmanifolds of the same dimension,* (9.15.3) $\Rightarrow$ (9.15.1).

(9.16.9) ([GW3; Corollary 2.3]) *If $X = Y = M$ is a solvmanifold, then* (9.15.3) $\Rightarrow N(f,g) = R(f,g)$.

(9.16.10) ([W2, Theorem 3.1]) *If X and Y are finite complexes, and $p\colon Y \to N$ is a fibering over a nilmanifold N, then then $N(f,g) > 0$ implies that $R(pf, pg) < \infty$.*

Furthermore Y is a Jiang type space for coincidences under the following three conditions.

(9.16.11) ([GW1, Theorem 4]) $Y = G/K$, *where G is a compact connected Lie group, and K is a finite subgroup.*

(9.16.12) ([GW1]) *Y is a $\mathcal{C}$-nilpotent space, whose fundamental group has center of finite index.*

(9.16.13) ([GW2]) *Y is a nilpotent space, and $R(f, g) < \infty$.*

The proof of (9.16) use the combination of a number of different and often subtle techniques, which seem to have parallels with localization proofs. One recurring theme however, has to do with sequences of Reidemeister classes which are associated with certain maps of short exact sequences of groups. The said sequences appear in the proofs of several parts of Theorem (9.16). One could, I suppose, debate whether these are fibre techniques or not. There are two things to say in this regard. Firstly, since for the most part we are dealing with $K(\pi, 1)$'s, we can always chant the McCord mantra over the algebra, and so produce fibre preserving maps of fibrations which reflect such sequences. Secondly as we shall soon see, there is a more general sequence (than the ones to be found in the references quoted thus far). This sequence is, in fact, the coincidence analogue of one given in [H1] in the fixed point case. The idea there was to separate the algebra from the geometry. Lurking behind the theorem in [H1], and therefore behind the one given below, lies the idea of a fibration of groupoids. Actually the theory of fibrations of groupoids gives a more subtle version than the one given below (see [H4]). The theorem then concerns maps of exact sequences of groups and homomorphisms of the form

(9.17)
$$\begin{array}{ccccccccc} 1 & \longrightarrow & H_1 & \xrightarrow{i_1} & G_1 & \longrightarrow & Q_1 & \longrightarrow & 1 \\ & & f' \big\downarrow \big\downarrow g' & & f \big\downarrow \big\downarrow g & & \overline{f} \big\downarrow \big\downarrow \overline{f} & & \\ 1 & \longrightarrow & H_2 & \xrightarrow[i_2]{} & G_2 & \longrightarrow & Q_2 & \longrightarrow & 1 \end{array}$$

(9.18) THEOREM ([H4]). *In the context of* (9.17) *there is an exact sequence of groups and homomorphisms, and based sets.*

$$1 \to \mathrm{Coin}(f', g') \xrightarrow{i_1} \mathrm{Coin}(f, g) \xrightarrow{p_1} \mathrm{Coin}(\overline{f}, \overline{g}) \xrightarrow{\delta} \mathcal{R}(f', g') \xrightarrow{i_*} \mathcal{R}(f, g) \xrightarrow{p_{2*}} \mathcal{R}(\overline{f}, \overline{g}) \to 1.$$

This sequence generalizes ones in several of the references given in Theorem (9.16) above. The generalization concerns the continuation of the sequences to the left of $\mathcal{R}(f', g')$ (such continuations do not appear in the indicated references).

In each case however, the respective authors are interested in the injectivity of $i_*\colon \mathcal{R}(f',g') \to \mathcal{R}(f,g)$, and hence implicitly in the continuation of their sequences to the left. Of course Theorems (9.5) and (2.5) are special cases of (9.18), where the horizontal sequences to which Theorem (9.18) must be applied, are truncated versions of the homotopy sequences of the fibrations involved (i.e. the induced self homomorphisms of sequence $\pi_1 i(F/K) \to \pi_1(B)$, in the fixed point case).

Individual methods are used in the many steps in proving (9.16). However a common step is in the discovery of appropriate algebraic sequences and maps thereof, to which (from our point of view) Theorem (9.18) may be applied. In [GW2] for example, given a homomorphism $\phi\colon \pi \to G$, where G is finitely generated torsion free nilpotent, and π is finitely generated, the authors derive the isolators of the kth stage of the lower central series of π and G, then by inclusion and quotient operations form short exact sequences. Maps $f, g\colon X \to N$, where $\pi_1(X) = \pi$, and $\pi_1(N) = G$ then naturally induced maps of these exact sequences, giving diagrams of the of the form shown in (9.17). In other cases the key involves the question of correctly factorizing induced homomorphisms, while in [GW1], the fact that the sequences in question are central extensions, yields (like eventual commutativity in the fixed point case), enough commutativity to produce product formulae. Finally in terms of intuitive insight into these proofs, we remark that it is not hard to see (since the last term in the sequence in (9.18) is surjective), that the finiteness (or not) of $\mathcal{R}(f,g)$ is closely related to those of $\mathcal{R}(\overline{f},\overline{g})$ and $\mathcal{R}(f',g')$ (especially when it is known that $\mathrm{Coin}(\overline{f},\overline{g}) = 1$).

We close this section with the completion of Theorem (9.14), which left the case $\dim T_1 > \dim T_2$ open. However, since the proof does not use fibre techniques, we omit it.

(9.19) THEOREM ([W1, Theorem 6]). *Let $f, g\colon M_1 \to M_2$ be maps of compact orientable solvmanifolds of the same dimension. Then*

$$|L(f,g)| \leq N(f,g).$$

10. The coincidence semi-index product formulae

We come now to the semi-index product formula due to J. Jezierski (see [Jz2]). We recall the version of semi-index theory given in [Jz2], which is a minor variation of that given originally in [DJ]. Let M, M' be smooth closed m-manifolds, the pair $f, g\colon M \to M'$ smooth transverse maps, and let x_0, $x_1 \in \Phi(f,g)$.

We say that x_0 and x_1 are *R-related* (and write $x_0 R x_1$) if and only if there is a path $u\colon x_0 \to x_1$ such that $fu \sim gu$, and exactly one of the paths u or fu is orientation preserving. It is evident that R need not be an equivalence relation.

Let $A \subset \Phi(f,g)$. If $A = \{a_1, b_1, \dots, a_k, b_k, c_1, \dots, c_s\}$ with $a_i R b_i$ for $i = 1, \dots, k$, and no c_i R-related to c_j for $i \neq j$, and $i, j = 1, \dots, s$. Then $A = \{a_1, b_1, \dots, a_k, b_k, c_1, \dots, c_s\}$ is said to be *a decomposition of* A, and the elements $c_1, \dots, c_s$ are said to be *free* in this decomposition. The *semi-index of* A denoted $|\text{ind}|(f, g : A)$, is the number of free elements in a decomposition of A. If A is a Nielsen class, then A is said to be essential if $|\text{ind}|(f, g : A) \neq 0$. It is clear that this notion of essentiality extends to Riedemiester classes as in Section 9.

Finally the *semi-index Nielsen number* $N_S(f,g)$ is defined to be the number of such essential Nielsen classes. The number $N_S(f,g)$ is independent of the decompositions of the Nielsen classes, it is a homotopy and homotopy type invariant lower bound for the number of coincidences of f and g, furthermore in the context of fibre preserving maps, all the T-type transformations used in the proof of addition and product formula are semi-index preserving. There is however, an obstruction to the semi-index product formula which we now proceed to elucidate.

Let $x \in \Phi(f,g)$ if xRx, then x is said to be *self reducing*. Evidently x is self reducing if there is a loop u at x with $f(u) \sim g(u)$, and exactly one of u and $f(u)$ orientation-preserving. Let A be a Nielsen class of f and g, we say that A *is defective* if it contains a self reducing coincidence point. The following lemma is a rewrite of the ideas of [Jz2, Lemma 2.3].

(10.1) LEMMA ([Jz2]). *Let A be a defective Nielsen class, then for any $x, y \in A$ we have xRx, xRy and yRy. Thus $|\text{ind}|(f, g : A)$ is 1 if $\#(A)$ is odd, and 0 if $\#(A)$ is even. Furthermore, in the presence of orientability*

$$|\text{ind}|(f, g : A) = |\text{ind}(f, g : A)|.$$

Note in particular, that in a defective class every point is self reducing. The next example demonstrates that the semi-index product formula can fail in the presence of defective classes.

(10.2) EXAMPLE ([Jz2, Example 2.4]). Let K^2 be the Klein bottle regarded as in Example (2.10), as the quotient of a rectangle. Let f' be the map which takes the boundary of the rectangle to the north pole N of S^2, and the rest of K^2 diffeomorphically onto $S^2 - N$. Let $g' \colon K^2 \to S^2$ be a constant map to $y \neq N$. The pair f', g' is transverse and $\Phi(f', g') = \{z\}$ say, a singleton. So $|\text{ind}|(f', g' : \{z\}) = 1$ and $N_S(f,g) = 1$ also.

Now z is self reducing, since there is an orientation reversing loop at z whose images by f' and g' are nul-homotopic. Consider next the maps $\overline{f}$, $\overline{g} \colon S^2 \to S^2$ with $\overline{f}$ the identity, and $\overline{g}(x,y,z) = (-x,-y,z)$. This pair is transverse and $\Phi(\overline{f}, \overline{g}) = \{(0,0,1), (0,0,-1)\} := \{b_1, b_2\}$. Since S^2 is simply connected, there is a single Nielsen class denoted $\overline{b}_1 = \{b_1, b_2\}$. Since S^2 is orientable, the points b_1

and b_2 are not R-related and so $|\mathrm{ind}|(\overline{f},\overline{g} : \overline{b}_1) = 2$ and $N_S(\overline{f},\overline{g}) = 1$. Consider the following diagram

$$\begin{array}{ccc} S^2 \times K^2 & \xrightarrow{f,g} & S^2 \times S^2 \\ {\scriptstyle p_1}\downarrow & & \downarrow{\scriptstyle p_1} \\ S^2 & \xrightarrow[\overline{f},\overline{g}]{} & S^2 \end{array}$$

where $f = \overline{f} \times f'$, $g = \overline{g} \times g'$ and in each case p_1 denotes the projection on the first factor. Note that $\Phi(f,g) = \{(b_1,z),(b_2,z)\}$ and these two points are Nielsen equivalent. Since z is self reducing, then so are (b_1,z) and (b_2,z), and the single Nielsen class $\overline{(b_1,z)}$ is defective. From Lemma (10.1) we have $|\mathrm{ind}|(f,g : \overline{(b_1,z)}) = 0$, so $N_S(f,g) = 0$. Then

$$|\mathrm{ind}|(f,g : \overline{(b_1,z)}) = 0 \neq 2 \cdot 1 = |\mathrm{ind}|(f',g' : z) \cdot |\mathrm{ind}|(\overline{f},\overline{g} : \overline{b}_1) = 2,$$

so that the *semi-index product theorem* (SIPT) is seen to fail in the presence of defective classes. Note also that the naïve addition and product formulas fail too, and this even for product maps between product fibrations (compare Example (2.20)).

The analysis in [Jz2] that allows the (SIPT) is long, technical and beyond the scope of this chapter (as in fact was the proof of the ordinary index product formula). We state the main result and refer the reader to [Jz2, Theorem 3.13] for details.

(10.3) THEOREM (The Semi-Index Product Theorem (SIPT), [Jz2]). *In the context of diagram* (9.4), *let* E, E', B, B' *and all fibres be smooth topological manifolds, with* $\dim(E_1) = \dim(E_2)$, *and* $\dim(B_1) = \dim(B_2)$. *Let* f, g *be fibre preserving maps,* A *be a Nielsen class of* f, g, *and let* $p(A) \subset \overline{A}$, *where* $\overline{A}$ *is a Nielsen class of* $\overline{f}$, $\overline{g}$. *Suppose that* $p\colon E \to B$ *and* $p'\colon E' \to B'$ *are locally trivial fibre bundles, then the equality*

$$|\mathrm{ind}|(f,g : A) = |\mathrm{ind}|(f_b,g_b : F_b \cap A) \cdot |\mathrm{ind}|(\overline{f},\overline{g} : \overline{A})$$

holds if and only if at least one of the following conditions is satisfied:

(10.3.1) $|\mathrm{ind}|(f_b,g_b : F_b \cap A) \cdot |\mathrm{ind}|(\overline{f},\overline{g} : \overline{A}) \leq 1$,
(10.3.2) the class $\overline{A}$ is not defective, and neither is any class in $F_b \cap A$.

Note that the product given in (10.3.1) is 2 in Example (10.2), that the left hand side of (SIPT) is always less than or equal to the right hand side (see [Jz2]), and that condition (10.3.1) does not exclude the possibility (SIPT) holding in the presence of defective classes. We have the following analogue of Lemma (9.8), which as there, we give in terms of Riedemeister classes.

(10.4) LEMMA. *In the context of Theorem* (10.3), *let A be a Nielsen class of f and g, $x \in A$ and $[\beta] = \rho(A) \in \mathcal{R}(f_*^x, g_*^x)$. If $[\beta]$ is essential, then so is $[\overline{\beta}] = p_*([\beta]) \in \mathcal{R}(\overline{f}_*^b, \overline{g}_*^b)$, and any class $[\alpha K] \in \mathcal{R}_K(f_{b*}^x, g_{b*}^x)$ with $j_*^x([\alpha K]) = [\beta]$. If A satisfies* (SIPT), *the converse also holds.*

We now have all the ingredients needed to replicate the proof of Theorem (9.10) to give:

(10.5) THEOREM. *In the context of Theorem* (10.3), *if every non-empty class of f and g satisfies the* (SIPT), *then*

$$N_{\mathcal{S}}(f,g) = \sum_{b \in \chi} N_{\mathcal{S}}(f_b, g_b),$$

if and only if

(10.5.1) for each $x \in \mathbf{C} \in \mathcal{E}(f,g)$ we have $[\mathrm{Coin}(\overline{f}_*^b, \overline{g}_*^b); p_*(\mathrm{Coin}(f_*^x, g_*^x))] = 1$,
(10.5.2) and $N_K(f_b, g_b) = N(f_b, g_b)$.

(10.6) REMARK. It seems to this author, that results along the lines of formulas (1.3) and (9.10.1) are likely for semi-index. Particularly useful for this is the fact (shown by Jurek Jezierski), that the property of Nielsen class A being defective or not, can be expressed algebraically. In this way it can be shown to be a homotopy invariant. More precisely, if for $f, g\colon M \to N$ we let $H \subset \pi_1(M)$ denote $H' \subset \pi_1(N)$ denote the subgroups of orientation preserving elements, then a point $x \in \Phi(f,g)$ is defective if and only if

$$\mathrm{Coin}(f_*^x, g_*^x) \cap H \neq \mathrm{Coin}(f_*^x, g_*^x) \cap (f_*^x)^{-1}(H')).$$

Thus even when $\mathrm{Coin}(f_*^x, g_*^x) \neq 1$ then equality of the above sets for all appropriate x, together with other appropriate conditions, should ensure the desired results.

(10.7) COROLLARY. *If in the context of Theorem* (10.5) *we have that $B = B' = T^k$ the k torus for some k, then the naïve addition formula in that theorem holds.*

PROOF. If $N(\overline{f}, \overline{g}) = 0$, then $\overline{f}$, $\overline{g}$ (and hence f, g) are homotopic to coincidence free maps, and so both sides of the equation are zero. Next, suppose that $N(\overline{f}, \overline{g}) \neq 0$, then by Theorem (2.3) any Reidemeister class of $\overline{f}$, $\overline{g}$ must have index ± 1. Since B is a torus we have that $K = 1$, and $\mathrm{Coin}(f_*^x, g_*^x) = 1$ for all essential x, so the essential classes of f_b, g_b inject into the essential classes of f and g.

Now let A be a Nielsen class of f and g. First suppose for any $b \in p(A)$, that no defective class of f_b and g_b is contained in A. Since T^k is orientable, then no class of $\overline{f}$ and $\overline{g}$ is defective, and the (SIPT) follows from (10.3.2). Suppose, on the other hand, that $A \cap F_b$ contains a self reducing point y say for some b. But the classes of f_b and g_b inject into those of f and g, so $\mathbf{C}_b := A \cap F_b$ is a single class of f_b and g_b. Now $\mathbf{C}_b$ is defective, since it contains the self reducing point y. From Lemma (10.1) $|\text{ind}|(f_b, g_b : F_b \cap A) \leq 1$. Putting this together with the fact that $|\text{ind}|(\overline{f}, \overline{f} : p(A)) = \pm 1$, we have this time, that assumption (10.3.1) is satisfied, and so for this A too we have the (SIPT) as required. □

We illustrate these results from ongoing work from [HK5], from which we also take the following generalizations of the last part of Theorem (7.1), and of Proposition (7.2), extending those results to include maps between different solvmanifolds.

(10.8) THEOREM ([HK5]). *Let t, s and u, p be given positive integers, and let $A: \mathbb{Z}^s \to \text{Aut}(\mathbb{Z}^t)$, and $B: \mathbb{Z}^p \to \text{Aut}(\mathbb{Z}^u)$ be homomorphisms of groups. Let S_1 and S_2 be the solvmanifolds defined using Theorem (7.1) with gluing data A respectively B, and let $T^t \hookrightarrow S \xrightarrow{\pi_1} T^s$ and $T^u \hookrightarrow S \xrightarrow{\pi_2} T^p$ be the respective Mostow fibrations. Let X and Y be respectively $u \times t$ and $p \times s$ matrices over $\mathbb{Z}$. Then the correspondence $\widetilde{f}(\overrightarrow{x}, \overrightarrow{y}) = (X\overrightarrow{x}, Y\overrightarrow{y})$ for $(\overrightarrow{x}, \overrightarrow{y}) \in \mathbb{R}^t \times \mathbb{R}^s$, induces a well defined map $f: S_1 \to S_2$ (denoted (X, Y)), if and only if*

$$XA(\overrightarrow{v}) = B(Y\overrightarrow{v})X$$

for any $\overrightarrow{v} \in \mathbb{Z}^s$. Furthermore, if $t = u$, $s = p$ and $f = (X, Y)$ and $g = (W, Z)$ are maps from S_1 to S_2, with $\det(Y - Z) \neq 0$, then

$$N_S(f, g) = \sum_{[\overline{\alpha}] \in \mathcal{R}(\overline{f}_*, \overline{g}_*)} (|\det(B([\overline{\alpha}])X - W)|).$$

This theorem is illustrated in [HK5] by the following example.

(10.9) EXAMPLE. Let $p = s = 1$, and $t = u = 2$ in Theorem (10.8), and let S_1 and S_2 be the solvmanifolds defined using gluing data determined by the matrices A and B, respectively.

$$A = \begin{bmatrix} 2 & 1 \\ 5 & 3 \end{bmatrix}, \qquad B = \begin{bmatrix} 3 & -5 \\ -1 & 2 \end{bmatrix},$$

$$X = \begin{bmatrix} 5a + 5b & a \\ b & -a - b \end{bmatrix}, \qquad W = \begin{bmatrix} 5x - y & y \\ y & x \end{bmatrix}.$$

The Mostow fibrations of S_i are of the form $T^2 \to S_i \xrightarrow{\pi_i} S^1$ for $i = 1, 2$. Using the techniques of Section 7 together with Theorem (10.8), we see that the pairs $f = (X, 1)$ and $g = (W, -1)$ are well defined fibre preserving maps from S_1 to S_2, if and only if X and W are of the from given above. Since $R(\overline{f}, \overline{g}) = |1-(-1)| = 2$, we have

$$N_S(f,g) = \sum_{i=0}^{1}(|\det(B^i X - W)|) = |\det(X - W)| + |\det(BX - W)|.$$

Now $|\det(X - W)| = |\det(BX - W)| = |-5a^2 - 112 + 5x^2 - xy - y^2|$, so we have that $N_S(f,g) = |-10a^2 - 22ab - 10b^2 + 10x^2 - 2xy - 2y^2|$. Note that the pair f and g here, are fibre uniform for coincidences.

11. Nielsen Root numbers of fibre preserving maps

Nielsen root theory can be thought of as a special case of Nielsen coincidence theory, obtained by putting g equal to the constant map (at $c \in Y$ say). Thus if $f: X \to Y$, then *a root of f at c* is simply a point $x \in f^{-1}(c)$. Nielsen root classes then can be identified with Nielsen coincidence classes of f with the constant map. A special feature of Nielsen root theory, however is the fact that if X is a manifold, then Y is a Jiang type space for root classes (see for example [Kg]). In other words under this simple assumption on X, either every root class is inessential, or every root class is essential. In the latter case, we say that *f is root essential.* Now $NR(f)$, the *Nielsen root number of f*, is the number or essential Nielsen root classes of f. Since for any $x \in f^{-1}(c)$, the homomorphism $g_*^x = 0$, then we have from Proposition (2.1) that

$$\mathrm{Coin}(f^x, *) = \mathrm{Ker}\,(f_*).$$

From the exactness of sequence (9.1) we have

(11.1) THEOREM. *If $f: X \to Y$ is a map, with X a manifold. If f is root inessential, then $NR(f) = 0$, else*

$$NR(f) = [\pi_1(Y) : f_*(\pi_1(X))].$$

Motivated by the multiplicative property of both the Hopf degree and the Brouwer degree for the composition of maps, together with the connection of these numbers to Nielsen root theory, Robin Brooks, Bob Brown and Helga Schirmer in [BBS1, Theorem 4.3] produced a multiplicative formula with a certain correction factor $\kappa(f,g)$ giving

$$\kappa(f,g)NR(f \circ g) = NR(f)NR(g)$$

for maps $f: X \to Y$ and $g: Y \to Z$. As a corollary, by factoring a product $f_1 \times f_2: X_1 \times X_2 \to Y_1 \times Y_2$ as $(f_1 \times 1) \circ (1 \times f_2)$, by equating $NR(f_1 \times 1)$ with $NR(f_1)$, $NR(1 \times f_2)$ with $NR(f_2)$ and showing for this case, that $\kappa(f_1 \times 1, 1 \times f_2) = 1$ they obtained the following cartesian product formula

$$NR(f_1 \times f_2) = NR(f_1)NR(f_2).$$

This is then generalized in [BBH2, Theorem 6.1] to give:

(11.2) THEOREM. *Let $p: E \to B$ and $q: E' \to B'$ be locally trivial fibre bundles with fibres F and G, where* $\dim(F) = \dim(G)$, $\dim(E) = \dim(E')$ *and* $\dim(B) = \dim(B')$. *Let $f: E \to E'$ be a fibre preserving map, $x \in E$, and $c = f(x)$. If f is root essential, then*

$$\kappa(f)NR(f) = NR(f_x)NR(\overline{f}),$$

where $\kappa(f)$ is the number of root classes of f_x contained in a root class of f at c.

The proof of this theorem in [BBS2] is a generalization of the cartesian product proof mentioned above. Insofar as it uses fibre-techniques it factors f via the pullback r of q and $\overline{f}$ as shown in the diagram (where $g(y) = (p(y), f(y))$)

$$\begin{array}{ccccc}
F & \xrightarrow{f_x} & G & \xrightarrow{1} & G \\
\downarrow & & \downarrow & & \downarrow \\
E & \xrightarrow{g} & B \sqcap E' & \xrightarrow{f} & E' \\
{\scriptstyle p}\downarrow & & \downarrow{\scriptstyle r} & & \downarrow{\scriptstyle q} \\
B & \xrightarrow[1]{} & B & \xrightarrow[\overline{f}]{} & B'
\end{array}$$

From [BBH2, Theorem 4.3] (stated above) we have

$$\kappa(f, g)NR(f \circ g) = NR(f)NR(g).$$

Next the results $NR(f_1 \times 1) = NR(f_1)$ and $NR(1 \times f_2) = NR(f_2)$ above, are generalized to give $NR(g) = NR(f_x)$ and $NR(f) = NR(\overline{f})$ in separate theorems by non-fibre techniques. Finally it is shown for this f and g, that $\kappa(f, g) = \kappa(f)$ which then gives the desired result.

Using the ideas of this chapter, we sketch an alternative proof of Theorem (11.2), together with a mod K version of it (formula (11.3) below). In fact when f is root essential, simply by putting $g = *$ in the sequence of Theorem (9.5), we obtain the exact sequence

$$\begin{aligned} 1 \to \operatorname{Ker}_K(f^x_{p(x)*}) \xrightarrow{j_{K*}} \operatorname{Ker}(f^x_*) \xrightarrow{p_*} \operatorname{Ker}(\overline{f}^{p(x)}_*) \\ \xrightarrow{\delta} \mathcal{ER}_K(f^x_{p(x)*}) \xrightarrow{j_*} \mathcal{ER}(f^x_*) \xrightarrow{p_*} \mathcal{ER}(\overline{f}^{p(x)}_*) \to 1, \end{aligned}$$

where the symbol $\mathcal{ER}$ denotes the essential root classes of f. It is easy to see that for different x, the sequences are conically bijective, and we deduce the mod K version

(11.3) $$[\operatorname{Ker} \overline{f}^{p(x)}_* : p_*(\operatorname{Ker} f^x_*)]NR(f) = NR_K(f_x)NR(\overline{f}).$$

(11.4) COROLLARY (The Naïve Root Product Theorem). *Under the above hypothesis, we have $NR(f) = NR(f_b)NR(\overline{f})$, if and only if*

(11.4.1) $[\text{Ker } \overline{f}_*^{p(x)} : p_*(\text{Ker } f_*^x)] = 1$, and

(11.4.2) $NR_K(f_b) = NR(f_b)$.

In particular $NR(f_1 \times f_2) = NR(f_1)NR(f_2)$.

The first part of the corollary is obvious, while the last part follows along the lines of the proof of Examples (2.20.4) (or one could use the ideas of the proof of (2.18) in the coincidence context).

To get the non-mod K version, we refer back to the details of the proof of Theorem (2.14). This gives us by analogy that

$$\#(\mathcal{ER}(f_*^x)) = \#\Big(\bigcup_{b\in\chi} p_*^{-1}(\overline{\mathbf{N}}_b)\Big)$$

where $\overline{\mathbf{N}}_b$ is the root class of b where $p(x) = b$, and χ is a set of essential representatives. So $NR(f)$ is the sum of the cardinalities of the $p_*^{-1}(\overline{\mathbf{N}}_b))$. These sets are the kernels of p_* for the various choices of b and x in the Nielsen root theory sequence above. From this exact sequence, we see that the kernel of p_* is the image of $j_*: \mathcal{ER}_K(f_{b*}^x) \to \mathcal{ER}(f_*^x)$. Next this $\text{Im}(j_*)$ is the same as the image of the composite

$$j_*': \mathcal{ER}_K(f_{p(x)*}^x) \to \mathcal{ER}_K(f_{p(x)*}^x) \to \mathcal{ER}(f_*^x).$$

Thus $\#(p_*^{-1}(\overline{\mathbf{N}}_b)$ is the same as the cardinality of the set $\mathcal{ER}(f_{b*}^x)/\sim$, where $\sim$ is the equivalence relation which identifies points of $\mathcal{ER}(f_{b*}^x)/\sim$ under the image of j_*'. That is

$$\#(p_*^{-1}(\overline{\mathbf{N}}_b)) = NR(f_b)/\kappa(f)$$

as required. The reader will note the similarity of this sketch, to the sketch we gave of the Pak version of the fixed point formula.

References

[A] D. V. Anosov, *The Nielsen number of maps of nil-manifolds*, Russian Math. Surveys **40** (1985), 149–150.

[BBS1] R. Brooks, R. F. Brown and H. Schirmer, *The absolute degree and the Nielsen root number of compositions and Cartesian Products of maps*, Topology Appl. **116** (2001), 5–27.

[BBS2] ———, *The absolute degree and the Nielsen root number of a fibre-preserving map*, Topology Appl. **125** (2002), 1–46.

[BBPT] R. Brooks, R. F. Brown, J. Pak and D. G. Taylor, *Nielsen numbers of maps of tori*, Proc. Amer. Math. Soc. **52** (1975), 398–400.

[B1] R. F. Brown, *The Nielsen number of a fibre map*, Ann. Math. **85** (1967), 483–493.

[B2] ———, *The Lefschetz Fixed Point Theorem*, Scott, Foresman and Co., Glenview, Ill., 1971.

[BF] R. F. Brown and E. Fadell, *Corrections to "The Nielsen number of a fibre map"*, Ann. of Math. **95** (1972), 365–367.

[CW1] F. S. P. Cardona and P. N.-S. Wong, *On the computation of the relative Nielsen number*, Theory of Fixed Points and its Applications (São Paulo, 1999), Topology Appl. **116** (2001), 29–41.

[CW2] ———, *The relative Reidemeister numbers of fiber map pairs*, Topol. Methods Nonlinear Anal. **21** (2003), 131–145.

[DJ] R. Dobreńko nd J. Jezierski, *The coincidence Nielsen number on nonorientable manifolds*, Rocky Mountain J. Math. **23** (1993), 67–85.

[F] E. Fadell, *Natural fiber splittings and Nielsen numbers*, Houston J. Math. **2** (1976), 71–84.

[FH] E. Fadell and S. Husseini, *On a theorem of Anosov on Nielsen numbers for nilmanifolds*, Funct. Anal. Appl. (1985), 47–53; NATO Adv. Sci. Inst. Ser. C Math. Phys. Sci., vol. 173, Reidel, Dordrecht–Boston, Mass., 1986.

[G1] D. L. Gonçalves, *Coincidence Reidemeister classes on nilmanifolds and nilpotent fibrations*, Topology Appl. **83** (1998), 169–186.

[G2] ———, *The coincidence Reidemeister classes of maps on nilmanifolds*, Topol. Methods Nonlinear Anal. **12** (1998), 375–386.

[GW1] D. L. Gonçalves and P. N.-S. Wong, *Homogeneous spaces in coincidence theory*, 10th Brazilian Topology Meeting (São Carlos, 1996), Mat. Contemp. **13** (1997), 143–158.

[GW2] ———, *Nilmanifolds are Jiang type spaces for coincidences*, Forum Math. **13** (2001), 133–141.

[GW3] ———, *Homogeneous spaces in coincidence theory* II, preprint.

[GH] J. Guo and P. R. Heath, *Coincidence theory on the complement*, Topology Appl. **95** (1999), 229–250.

[Ha] B. Halpern, *Periodic points on the Klein bottle*, never published.

[H1] P. R. Heath, *Product formulae for Nielsen numbers of fibre maps*, Pacific J. Math. **117** (1985), 267–289.

[H2] ———, *A pullback theorem for locally-equiconnected spaces*, Manuscripta Math. **55** (1986), 233–237.

[H3] *A Nielsen type number for fibre preserving maps*, Topology Appl. **53** (1993), 19–35.

[H4] ———, *Groupoids and fibred addition and product theorems for Nielsen theory*, Talk delivered at special session on Fixed point theory. Pheonix 2004.

[HK1] P. R. Heath and E. Keppelmann, *Fibre techniques in Nielsen periodic point theory on nil and solvmanifolds* I, Topology Appl. **76** (1997), 217–247.

[HK2] ———, *Fibre techniques in Nielsen periodic point theory on nil- and solvmanifolds* II, Topology Appl. **106** (2000), 149–167.

[HK3] Fibre techniques in Nielsen periodic point theory on solvmanifolds III: Calculations, Quaest. Math. **25** (2002), 177–208.

[HK4] ———, *Model solvmanifolds for Lefschetz and Nielsen theories*, Quaest. Math. **25** (2002), 483–501.

[HK5] ———, *Model solvmanifolds* II, *Nielsen coincidence and root theory on solvmanifolds*, in preparation.

[HK6] ———, *Fibre techniques in Nielsen periodic point theory for arbitrary fibre preserving maps*, in preparation.

[HPY] P. R. Heath, R. Piccinini and C. You, *Nielsen type numbers for periodic points* I, Topological Fixed Point Theory and Applications, Proceedings, Tianjin 1988, Lecture Notes in Math., vol. 1411, Springer–Verlag, 1989, pp. 88–106.

[HMP] P. R. Heath, C. Morgan nd R. A. Piccinini, *Nielsen numbers and pullbacks*, Topology Appl. **26** (1987), 65–82.

[HY] P. R. Heath and C. You, *Nielsen type numbers for periodic points* II, Topology Appl. **43** (1992), 219–236.

[J] B. Jiang, *Lectures on Nielsen fixed point theory*, Contemp. Math. **14** (1983), Amer. Math. Soc., Providence, Rhode Island.

[Jz1] J. Jezierski, *The Nielsen number product formula for coincidences*, Fund. Math. **134** (1990), 183–212.

[Jz2] *The semi-index product formula*, Fund. Math. **140** (1992), 99–120.

[K] E. C. Keppelmann, *Periodic points on nilmanifolds and solvmanifolds*, Pacific J. Math. **164** (1994), 105–128.

[KMc] E. C. Keppelmann nd C. McCord, *The Anosov theorem for exponential solvmanifolds*, Pacific J. Math. **170** (1995), 143–159.

[Kg] T.-h. Kiang, *The theory of fixed point classes*; from the second Chinese edition (1989), Springer–Verlag, Berlin; Science Press, Beijing,.

[L] J. Lillig, *A union theorem for cofibrations*, Arch. Math. **24** (1973), 410–415.

[M] A. I. Mal'cev, *On a class of homogeneous spaces*, Amer. Math. Soc. Transl. (1) **9** (1962), 276–307; English transl..

[Mc1] C. K. McCord, *Nielsen numbers and Lefschetz numbers on solvmanifolds*, Pacific J. Math. **147** (1991), 153–164.

[Mc2] ———, *Computing Nielsen numbers*, Nielsen theory and dynamical systems (South Hadley, MA, 1992), pp. 249–267; Contemp. Math. **152** (1993), Amer. Math. Soc., Providence, RI.

[Mc3] ———, *Lefschetz and Nielsen coincidence numbers on nilmanifolds and solvmanifolds*, Topology Appl. **43** (1992), 249–261.

[Mc4] ———, *Lefschetz and Nielsen coincidence numbers on nilmanifolds and solvmanifolds* II, Topology Appl. **75** (1997), 81–92.

[Mo] G. D. Mostow, *Factor spaces of solvable groups*, Ann. of Math. **60** (1954), 1–27.

[N-O] B. Norton-Odenthal, *A product formula of the generalized Lefschetz number*, Ph.D. dissertation (1991), University of Wisconsin-Madison.

[P] J. Pak, *On the fixed point indices and Nielsen numbers of fiber maps on Jiang spaces*, Trans. Amer. Math. Soc. **212** (1975), 403–415.

[R] A. H. Rarivoson, *Calculation of the Nielsen numbers on nilmanifolds and solvmanifolds*, Ph.D. dissertation (1993), University of Cincinnati.

[S1] H. Schirmer, *Mindestzahlen von Koinzidenzpunkten*, J. Reine Angew. Math. **194** (1955), 21–39.

[S2] ———, *A relative Nielsen number*, Pacific J. Math. **122** (1986), 459–473.

[Sct] A. L. Schusteff, *Product formulas for relative Nielsen Numbers of Fibre map pairs*, Ph.D. thesis (1990), University of California, Los Angeles.

[St] A. Strøm, *Note on cofibrations*, Math. Scand. **19** (1966), 11–14.

[V] J. Vick, *Homology Theory*, Academic Press, Inc., New York, 1973.

[W1] P. N.-S. Wong, *Reidemeister number, Hirsch rank, coincidences on polycyclic groups and solvmanifolds*, J. Reine Angew. Math. **524** (2000), 185–204.

[W2] ———, *Coincidence theory for spaces which fiber over a nilmanifold*, preprint.

[Y1] C. Y. You, *Fixed point classes of a fiber map*, Pacific J. Math. **100** (1982), 217–241.

[Y2] ———, *A Nielsen number product formula for pullbacks*, Beijing Daxue Xuebao Ziran Kexue Ban **29** (1993), 21–25.

[Y3] ———, Private communication.

[Z] X. Zhao, *A relative Nielsen number for the complement*, Topological Fixed Point Theory and Applications, Lecture Notes in Mathematics, vol. 1411, Springer–Verlag, 1989, pp. 189–199.

15. WECKEN THEOREM FOR FIXED AND PERIODIC POINTS

Jerzy Jezierski

The Nielsen number $N(f)$ has two basic properties which follow straight from its definition:

(1) $N(f)$ is a homotopy invariant,

(2) $N(f)$ is the lower bound of the number of fixed points.

It is natural to ask if $N(f)$ is the best such lower bound being a homotopy invariant i.e. is f homotopic to a map g such that $\#\mathrm{Fix}(g) = N(f)$?

In fact Jakob Nielsen just after defining this invariant (for self maps of surfaces) conjectured that the answer is positive. Two decades later Franz Wecken proved that it is true for self-maps of manifolds of dimension at least three (see [W]) and since this time all results of this type are called Wecken theorems. In 1966 Shi in [Shi] extended this result onto a very large class of polyhedra. In 1979 Boju Jiang extended this class to all finite polyhedra who do satisfy two conditions. For the first there is no local separating point and this assumption seems natural since separating points make the space too "wild" for such nice theorem. If we restrict to the polyhedra without local separating points the second assumption is that the polyhedron is not a surface (of negative Euler characteristic) — the original domain of Nielsen's research! In 80' (forty years after Wecken) it turned out that Wecken theorem is not true for surfaces (of negative Euler characteristic). The counter-examples were given by Fadell and Husseini (1982, [FH]) in noncompact case and Boju Jiang ([Ji4], [Ji5]). Then there appeared many papers proving that in the case of surfaces the cardinality of the fixed point set is usually much larger than the Nielsen number (see [Ke], [Z]).

Thus one can say that the problem of realizing the Nielsen number is almost solved for polyhedra. Surprisingly almost no progress was done in extending this theorem onto other ANRs. It is still open whether the fixed point set can be made finite after a deformation of a self-map of an arbitrary ANR.

Unfortunately Wecken property holds (for manifolds) only in dimension ≥ 3. It turned out that it is not true in the case of the maps of surfaces. In 1984 Boju Jiang gave first counter example and soon after Michael Kelly and independently Zhang showed that in the case of self maps of surfaces the gap between the least number of fixed points and the Nielsen number may be arbitrarily large.

In this chapter we start with the proof of the classical Wecken theorem (for manifolds of dimension $\neq 2$). Then we present the proof of the extension of this theorem to polyhedra. In the second part we consider the periodic points of a map $f: M \to M$. We give some Wecken type results in this category. First we give a simple (necessary and sufficient) condition for the map to be homotopic to a map with no n-periodic points ($n \in \mathbb{N}$ a fixed number) which we call Weak Wecken's Theorem for Periodic Points. Then we define the invariant $NF_n(f)$ introduced in 1983 by Boju Jiang (see [Ji1]) which plays the role of the Nielsen number as a lower bound of the cardinality of $\text{Fix}(f^n)$. We show that also here a Wecken Theorem holds.

1. Classical Wecken theorem for manifolds

Here we will prove Wecken theorem for manifolds. An elegant proof of this Theorem is given in [Ji2] where the problem is reduced to minimizing the number of the intersections of the graph $\Gamma_f = \{(x, fx) : x \in M\} \subset M \times M$ with the diagonal $\Delta M = \{(x, x) : x \in M\}$. The intersections are being removed with the Whitney trick. We decided to present a more elementary, although longer, proof since we will use its modifications to get a similar result about periodic points in the next sections.

(1.1) THEOREM ([W]). *Every self map $f: M \to M$ of a compact manifold of dimension $m \neq 2$ is homotopic to a map g realizing the Nielsen number $\#\text{Fix}(g) = N(f)$.*

PROOF. In dimension $m = 1$ there are only two compact manifolds: interval $[a, b]$ and the sphere S^1. Every self-map of the interval is homotopic to the constant which realizes $N(f) = 1$. On the other hand any self map $f: S^1 \to S^1$ is homotopic to a map $f(z) = z^k$, $k \in \mathbb{Z}$, which has exactly $N(f) = |k-1|$ fixed points for $k \neq 1$. If $k = 1$ then we get the identity map which is homotopic, after a small twist, to a fixed point free map.

Let $\dim M \geq 3$. If M has the boundary then it may be deformed into $\text{int}M$ so we may assume that the image of f hence all fixed points are contained in $\text{int}M$. By transversality we may assume that $\text{Fix}(f)$ is finite. Then we use successively Lemma (1.3) to reduce each Nielsen class to a single point. At last Lemma (1.4) removes inessential classes. Thus all Nielsen classes of the obtained map are essential and each of them contains exactly one point hence the cardinality of the fixed point set equals $N(f)$.

The proof is complete once the below lemmas are proved. □

(1.2) LEMMA. *Let $A \subset M$ be a compact subset disjoint from* $\mathrm{Fix}(f)$. *Then there is a number $\varepsilon > 0$ such that* $\mathrm{d}(f, f') < \varepsilon$ *and* $f'(x) = f(x)$ *(for all $x \notin A$) imply* $\mathrm{Fix}(f') = \mathrm{Fix}(f)$.

PROOF. Since A is compact, $\eta = \inf\{\mathrm{d}(x, f(x)) : x \in A\} > 0$. Let $\varepsilon = \frac{1}{2}\eta$ Assume that $\mathrm{d}(f(a), f'(a)) < \varepsilon$ for $a \in A$ and $f(x) = f'(x)$ for $x \notin A$. It is evident that, for any $x \notin A$, $x \in \mathrm{Fix}(f)$ iff $x \in \mathrm{Fix}(f')$. On the other hand for $x \in A$

$$\mathrm{d}(x, f'(x)) \geq \mathrm{d}(x, f(x)) - \mathrm{d}(f(x), f'(x)) \geq \eta - \frac{1}{2}\eta = \frac{1}{2}\eta > 0$$

hence no new fixed point in A appears. □

The next lemma shows how to unite fixed points from the same Nielsen class.

(1.3) LEMMA. *Let* $\dim M \geq 3$, *let* $\mathrm{Fix}(f)$ *be finite and contained in* $\mathrm{int} M$. *Suppose that $x_0 \neq x_1 \in \mathrm{Fix}(f)$ are Nielsen related. Then there is a homotopy f_t constant in a neighbourhood of* $\mathrm{Fix}(f) \setminus \{x_0, x_1\}$ *satisfying $f_0 = f$ and* $\mathrm{Fix}(f_1) = \mathrm{Fix}(f_0) \setminus \{x_1\}$.

PROOF. Since $\dim M \geq 3$, we may assume that a path ω realizing the Nielsen relation is a flat arc in M. We may also fix a Euclidean neighbourhood $V \subset M$ satisfying $\mathrm{Fix}(f) \cap V = \{x_0, x_1\}$ and ω is a segment in $V = \mathbb{R}^m$.

First we consider a special case: we assume moreover, that $f\omega \subset V$ i.e. ω and $f\omega$ are close.

Since $f(\omega) \subset V$, there is a smaller convex neighbourhood $V_0 \subset V = \mathbb{R}^m$ satisfying $\omega[0,1] \subset V_0 \subset \mathrm{cl}\, V_0 \subset V$ and $f(\mathrm{cl}\, V_0) \subset V$. Since $V_0 \subset V$ are convex, the restriction $f\colon \mathrm{cl}\, V_0 \to V$ is homotopic (rel. boundary) to the map $f'((1-t)x + tx_0) = (1-t)f(x) + tx_0$ $(x \in \mathrm{bd}\, V_0)$. The homotopy may be extended, by the constant, onto whole M and we obtain the map $f'\colon M \to M$ satisfying $\mathrm{Fix}(f') = \mathrm{Fix}(f) \setminus \{x_1\}$. In fact since $x \neq f'(x)$ for all $x \in \mathrm{bd}\, V_0$, x_0 is the only fixed point of f' in $\mathrm{cl}\, V_0$.

Now we consider the general case. The paths ω and $f\omega$ may not be contained in a Euclidean neighbourhood. We will show that there is a homotopy reducing this case to the previous one. The main difficulty is the construction of this homotopy is to avoid producing new fixed points. In fact this is the only place when the assumption on dimension (≥ 3) is essential.

Let us fix a Euclidean neighbourhood $V \supset \omega[0,1]$. Then there is an $r_0 > 0$ such that $f\omega[0, r_0] \cup f\omega[1 - r_0, 1] \subset V$. We may assume that $f\omega[r_0, 1 - r_0] \cap \omega[0,1] = \emptyset$. In fact, a small homotopy with the carrier in a small neighbourhood of $\omega[r_0, 1-r_0]$ (set disjoint from $\mathrm{Fix}(f)$; Lemma (4.4)) takes $f\omega[r_0, 1-r_0]$ off the segment $\omega[0,1]$. Let $\overline{\omega}\colon [r_0, 1-r_0] \to V \setminus \omega[0,1]$ be a path joining the point $f\omega(r_0)$ with $f\omega(1-r_0)$. Now the homotopy between $f\omega$ and ω implies a fixed end point homotopy $h\colon [r_0, 1-r_0] \times I \to M$ between the restriction $f\omega_{|[r_0,1-r_0]}$ and $\overline{\omega}$. Since $h(t,0), h(t,1) \notin \omega[0,1]$ and $\dim M \geq 3$, the homotopy h may be deformed rel. $\mathrm{bd}([r_0, 1-r_0] \times [0,1])$ to a homotopy into $M \setminus \omega[0,1] \approx M \setminus *$.

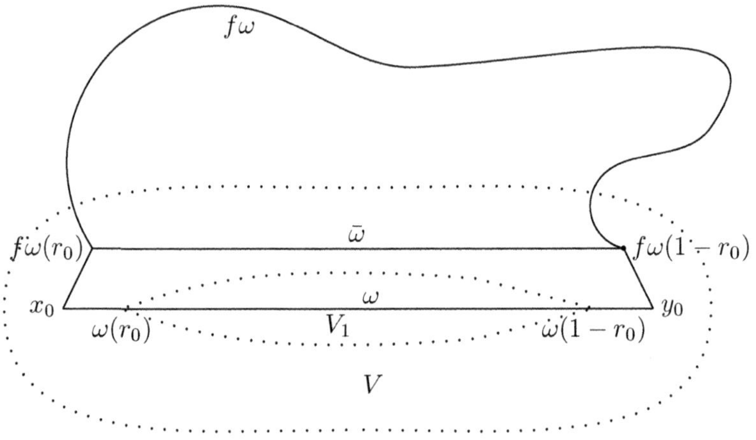

FIGURE 1

Since the compact sets $h([r_0, 1-r_0] \times [0,1])$ and $\omega[0,1]$ are disjoint, there is a convex neighbourhood V_1 satisfying $\mathrm{cl}\, V_1 \subset V$, $\omega[0,1] \cap V_1 = \omega(r_0, 1-r_0)$, $f(\mathrm{cl}\, V_1) \cap \mathrm{cl}\, V_1 = \emptyset$ and $h([r_0, 1-r_0] \times [0,1]) \cap \mathrm{cl}\, V_1 = \emptyset$. The map $f\colon M \to M$, the partial homotopy $h\colon [r_0, 1-r_0] \times I \to M$ and the Homotopy Extension Property give a homotopy $f_t\colon M \to M$ satisfying: $f_0 = f$, $f_t(\omega(s)) = h(s,t)$, the carrier of the homotopy f_t is contained in V_1 and $f_t(\mathrm{cl}\, V_1) \cap \mathrm{cl}\, V_1 = \emptyset$. It remains to notice that

(1.3.1) $\mathrm{Fix}(f_t)$ does not depend on t,
(1.3.2) f_1 satisfies the special case i.e. $f_1(\omega[0,1]) \subset V$.

In fact for all $x \in \mathrm{cl}\, V_1$ (carrier of the homotopy), $f_t(x) \notin \mathrm{cl}\, V_1$, which implies (1.3.1). On the other hand $f_1(\omega(s)) = h(s,1) \in V$ hence f_1 is the special case. □

The next lemma removes inessential fixed points.

(1.4) LEMMA (Hopf). *Let $U \subset \mathbb{R}^m$ be an open subset, $f\colon U \to \mathbb{R}^m$ a map with an isolated fixed point x_0 and $\mathrm{ind}(f; x_0) = 0$. Then there is a homotopy $f_t\colon U \to \mathbb{R}^m$ starting from $f_0 = f$, with the carrier in a prescribed neighbourhood of x_0 and satisfying $\mathrm{Fix}(f_1) = \mathrm{Fix}(f) \setminus x_0$. Moreover, the homotopy may be arbitrarily small.*

PROOF. We may assume that $x_0 = 0 \in U \subset \mathbb{R}^m$. Let $V_0 \subset \operatorname{cl} V_0 \subset U$ be a ball centered at 0. Since $\operatorname{ind}(f; V_0) = \operatorname{ind}(f; 0) = 0$, the homomorphism

$$H_m(\operatorname{cl} V_0, \operatorname{cl} V_0 \setminus 0) \xrightarrow{(i-f)_*} H_m(\mathbb{R}^m, \mathbb{R}^m \setminus 0)$$

is trivial. From the long homology sequence the homomorphism

$$H_{m-1}(\operatorname{cl} V_0 \setminus 0) \xrightarrow{(i-f)_*} H_{m-1}(\mathbb{R}^m \setminus 0)$$

is also trivial.

Since $\operatorname{cl} V_0 \setminus 0$ and $\mathbb{R}^m \setminus 0$ have the homotopy type of S^{m-1} and $H_{m-1}(S^{n-1}) = \pi_{m-1}(S^{m-1}) = \mathbb{Z}$, the map $i - f \colon \operatorname{bd} V_0 \to \mathbb{R}^m \setminus 0$ is homotopy trivial. Thus $i - f$ admits an extension $h \colon \operatorname{cl} V_0 \to \mathbb{R}^m \setminus 0$. Then the map $f' \colon \operatorname{cl} V_0 \to S^{n-1}$ given by $f'(x) = x - h(x)$ is the desired fixed point free map. In fact $x = f'(x)$ implies $x = f'(x) = x - h(x)$ hence $h(x) = 0$ contradicting to $h(\operatorname{cl} V_0) \subset V \setminus 0$.

If the ball V_0 is chosen small then the deformation can be made small. □

We will end this section by a small correction of Wecken theorem. It shows that the map realizing the least number of periodic points may be given near each fixed point by a prescribed formula.

Let us suppose that we are given some (model) maps $S_r \colon \mathbb{R}^m \to \mathbb{R}^m$ where 0 is the only fixed point and $\operatorname{ind}(S_r; 0) = r$ ($r \in \mathbb{Z}$). In the next corallary we will use the method of the proof of the Creating Formula from [Ji2]

(1.5) COROLLARY. *Any self map of a compact manifold of dimension* ≥ 3 *is homotopic to a map* g *with* $\#\operatorname{Fix}(g) = N(f)$ *and moreover, near each point* $x_0 \in \operatorname{Fix}(g)$ *the map* g *is given by the formula* S_r *where* $r = \operatorname{ind}(f; x_0)$.

PROOF. We introduce a Euclidean neighbourhood $U = \mathbb{R}^m$ near the point x_0 and we fix a smaller U_0 of x_0 such that $f(\operatorname{cl} U_0) \subset U$. We fix another point $y_0 \in U_0$ and its ball neighbourhood U_1 such that $x_0 \notin \operatorname{cl} U_1 \subset U_0$. Let $f'_t \colon \operatorname{cl} U_1 \to U$ be a partial homotopy, constant on the boundary of $\operatorname{cl} U_1$, from $f'_0 = f_{\operatorname{cl} U_1}$ to the map f'_1 which is given by formula S_r near y_0. Such homotopy exists since any two maps into $U = \mathbb{R}^m$ are homotopic. Then we take an extension of this homotopy with the carier in U_0. Since the fixed point index is the homotopy invariant, the creation of a new fixed point of index r at y_0 implies new fixed point of total index $-r$. Now we make the number of these points finite by the homotopy with the carrier in U_1. Then we may use Lemma (1.3) to coalesce these extra points to the point x_0. Now the index at x_0 must be zero hence we may use the Hopf lemma to get rid of this point. Thus the fixed point x_0 is replaced by y_0 and f is given near this point by the formula S_r. □

2. Wecken theorem for polyhedra

We will prove that the Nielsen number can be also realized in the homotopy class of selfmaps of a large class of polyhedra. The content of this section is the mixture of the proofs given by H. G. Shi in 1966 ([Shi]) and Boju Jiang in 1980 ([Ji3], compare [B]).

(2.1) DEFINITION. A point $x \in X$ is called *local separating point*if there is an open connected subset $U \subset X$ such that $U \setminus x$ is not connected.

It turns out that to get a Wecken theorem it is reasonable to assume that the considered polyhedron has no local separating points. In general such points may be an obstruction to Wecken theorem, see [Ki, Chapter IV, Section 4].

Our aim is to prove

(2.2) THEOREM (Wecken Theorem for polyhedra). *Let X be a compact connected polyhedron without local separating points. We moreover assume that X is not a 2-manifold. Then each self map $f\colon X \to X$ is homotopic to a map g satisfying* $\#\mathrm{Fix}(g) = N(f)$.

SCHEME OF THE PROOF. By Lemma (2.3) we may assume that $\mathrm{Fix}(f)$ is finite and each fixed point is lying inside a maximal simplex. As in the case of manifolds it is enough to show that each Nielsen class may be reduced to a point. As in the case of manifolds we will be coalescing successfully two Nielsen related fixed points. First we will show that the last is possible if a path ω, establishing the Nielsen relation, and its image $f\omega$ are near. Here we will use only the assumption that X has no local separating points. Then we will show that the general case can be reduced to the last one. Now the assumption that X is not a 2-manifold will be necessary. □

(2.3) LEMMA. *Each self map of a compact polyhedron $f\colon X \to X$ is homotopic to a map g such that* $\mathrm{Fix}(g)$ *is finite, each fixed point is containted inside a maximal simplex of a fixed triangulation. Moreover, the homotopy may be arbitrarily small.*

The proof will be given at the end of the section.

We start by reformulating the assumptions on the polyhedron X.

(2.4) LEMMA. *Let X be a polyhedron. A point $x \in X$ is a local separating point if and only if* $H_1(X, X \setminus x) \neq 0$.

PROOF. The lemma is obvious for X= a point, hence we may assume that the dimension of each component is ≥ 1.

$\Rightarrow$ Let $U \subset X$ be a connected neighbourhood with $U \setminus x$ disconnected. From the homology exact sequence of the pair $(U, U \setminus x)$ we have

$$H_1(U, U \setminus x) \xrightarrow{\partial_*} H_0(U \setminus x) \xrightarrow{i_*} H_0(U)$$

where $H_0U = \mathbb{Z}$, $H_0(U \setminus x) = \mathbb{Z} \oplus \ldots \oplus \mathbb{Z}$ with the number of components ≥ 2. Thus $0 \neq \ker i_* = \operatorname{im} \partial_*$ hence $H_1(U, U \setminus x) \neq 0$ and the excision isomorphism gives $H_1(X, X \setminus x) \neq 0$.

$\Leftarrow$ Let $H_1(X, X \setminus x) \neq 0$. We will show that for any contractible neighbourhood U of x, (for example for the open star: $U = \operatorname{st} x$) $U \setminus x$ is disconnected. We prove by a contradiction: we assume that $U \setminus x$ is connected.

By the assumption $H_1(U, U \setminus x) = H_1(X, X \setminus x) \neq 0$. We consider again the exact sequence

$$H_1U \xrightarrow{j_*} H_1(U, U \setminus x) \xrightarrow{\partial_*} H_0(U \setminus x) \xrightarrow{i_*} H_0U.$$

By connectedness of $U \setminus x$, $i_*\colon H_0(U \setminus x) \to H_0U$ is the isomorphism hence $\operatorname{im} \partial_* = \ker i_* = 0$. On the other hand $H_1U = 0$ implies ∂_* mono hence $H_1(U, U \setminus x) = 0$ and we get a contradiction. □

The next lemma gives a useful characterization of polyhedra without local separating points. Let X again denote a compact polyhedron with a fixed triangulation.

(2.5) LEMMA. *X has no local separating points if and only if each maximal simplex has dimension ≥ 2 and the boundary of the star of each vertex is connected.*

PROOF. $\Rightarrow$ X can not have a maximal simplex of dimension 1 since each point of a maximal 1-simplex is the local separating point. Let v be a vertex. Since v is not a local separating point,

$$0 = H_1(X, X \setminus v) = H_1(\operatorname{st} v, \operatorname{st} v \setminus v) = H_1(\operatorname{cl}(\operatorname{st} v), \operatorname{bd}(\operatorname{st} v))$$

hence from the exact homology sequence of $(\operatorname{cl}(\operatorname{st} v), \operatorname{bd}(\operatorname{st} v))$ the natural homomorphism $H_0(\operatorname{bd}(\operatorname{st} v)) \to H_0(\operatorname{cl}(\operatorname{st} v)) = \mathbb{Z}$ is mono hence the free group $H_0(\operatorname{bd}(\operatorname{st} v))$ may not have more than one generator hence $\operatorname{bd}(\operatorname{st} v)$ is connected.

$\Leftarrow$ Let $x \in X$ be not a vertex and let the simplex σ be the carrier of x. Let U be an open connected set containing x. Since there is a (possibly non proper) face σ of a simplex σ' of dimension ≥ 2, each path in U joining two points different than x can be deformed near x to a path avoiding x which proves that $U \setminus x$ is connected.

Let $v \in X$ be a vertex. Then, by the above lemma, $\operatorname{bd}(\operatorname{st} v)$ is connected $\Leftrightarrow$ $H_1(X, X \setminus v) = 0 \Leftrightarrow v$ is not a local separating point. □

(2.6) LEMMA. *Let X be a polyhedron without local separating points which is different than a point. Then the following three conditions are equivalent:*

(2.6.1) *X is not a 2-manifold.*

(2.6.2) *There is a 1-simplex which is a common face of at least three 2-simplices (in a triangulation).*

(2.6.3) *The space $T \times I$ (tree of three 1-simplices times the interval) can be imbedded into X.*

PROOF. Recall that each maximal simplex in a polyhedron without local separating points must be at least two dimensional.

(2.6.1) $\Rightarrow$ (2.6.2) is obvious if $\dim X \geq 3$. Now let $\dim X = 2$. If each 1-simplex is a face of at most two 2-simplices then X is a 2-pseudomanifold hence a 2-manifold.

(2.6.2) $\Rightarrow$ (2.6.3) obvious.

(2.6.3) $\Rightarrow$ (2.6.1) Let us fix the point $x = (v, 1/2) \in T \times I$ where v is the common point of three segments in T. Suppose that $j: T \times I \to X$ is an imbedding into a 2-manifold. This would induce the isomorphism between $H_1(T \times I, T \times I \setminus x) = \mathbb{Z} \oplus \mathbb{Z}$ and $H_1(j(T \times I), j(T \times I) \setminus j(x)) = H_1(X, X \setminus j(x)) = \mathbb{Z}$ which gives a contradiction. □

Let us fix the notation. We consider a compact connected polyhedron X with a finite triangulation K. Let $\{v_i\}_{i \in J}$ be the family of all vertices of K. Then each point $x \in X$ can be written uniquely as $x = \sum_{i \in J} \lambda_i v_i$ where $\lambda_i \geq 0$ and $\sum_{i \in J} \lambda_i = 1$. Now $\{i \in J; \lambda_i \neq 0\}$ is a simplex in K. When the triangulation K is fixed we may identify a point $x \in X$ with the sequence of its coordinates: $x = (\lambda_i)_{i \in J}$ or $x = \sum_{i \in J} \lambda_i v_i$. Let

$$\sigma(x) = \left\{ \sum_{i \in J} \lambda'_i v_i \in X : \lambda'_i \neq 0 \Leftrightarrow \lambda_i \neq 0 \right\}$$

denote the (open) simplex containing x. Then its closure equals

$$\mathrm{cl}(\sigma(x)) = \left\{ \sum_{i \in J} \lambda'_i v_i \in X : \lambda'_i \neq 0 \Rightarrow \lambda_i \neq 0 \right\}.$$

The star

$$\operatorname{st} x = \{y \in X : \sigma(x) \subset \mathrm{cl}(\sigma(y))\} = \left\{ \sum_{i \in J} \lambda'_i v_i \in X : \lambda_i \neq 0 \Rightarrow \lambda'_i \neq 0 \right\}.$$

Let us denote subsets

$$\begin{aligned} V(x) &= \{y \in X : \mathrm{cl}(\sigma(x)) \cap \mathrm{cl}(\sigma(y)) \neq \emptyset\} \\ &= \left\{ \sum_{i \in J} \lambda'_i v_i : \lambda_i \neq 0 \text{ and } \lambda'_i \neq 0 \text{ for an } i \in J \right\}. \end{aligned}$$

We notice that for $x, y \in \sigma$ we have the equality $V(x) = V(y)$ which allows to define $V(\sigma) = V(x)$ for an $x \in \sigma$.

For a subset A of a polyhedron X and $\eta > 0$ we denote $U(A, \eta) = \{x \in X : \operatorname{dist}(x, A) < \eta\}$ where dist is taken with the respect to the Euclidean norm $\mathrm{d}(x, y) = \sqrt{\sum_i (x_i - y_i)^2}$ and $\operatorname{dist}(x, A) = \inf\{d(x, a) : a \in A\}$.

(2.7) LEMMA. *Let σ be a maximal simplex in X. Then there is a map $\alpha\colon \sigma \times V(\sigma) \times I \to V(\sigma)$ satisfying:*

(2.7.1) $\alpha(x, y, 0) = x$, $\alpha(x, y, 1) = y$, $\alpha(x, x, t) = x$,
(2.7.2) *if $x \neq y$ then the map $\alpha(x, y, \cdot\,)\colon I \to X$ is mono,*
(2.7.3) *if $x, y \in \sigma(x), z \in V(\sigma)$ and $0 < t \leq 1$ then $\alpha(x, y, t) = \alpha(x, z, t)$ implies $y = z$.*

PROOF. We start with defining a natural deformation retraction $r_t\colon V(\sigma) \to V(\sigma)$. We identify the point $x \in X$ with the sequence of its coordinates $x = (x_i)_{i\in J}$. Let $J_\sigma \subset J$ be the set of vertices spanning the simplex σ. We put $r_t(x_i) = (x'_i)$ where

$$x'_i = \begin{cases} \dfrac{x_i}{\sum_{j\in J_\sigma} x_j + \sum_{j\notin J_\sigma} t\cdot x_j} & \text{for } i \in J_\sigma, \\ \dfrac{t\cdot x_i}{\sum_{j\in J_\sigma} x_j + \sum_{j\notin J_\sigma} t\cdot x_j} & \text{for } i \notin J_\sigma. \end{cases}$$

Now $r_1 = \mathrm{id}_{V(\sigma)}$ and r_0 is a projection onto $\mathrm{cl}\,\sigma$.

For $x, y \in \mathrm{cl}\,\sigma$ we define $\alpha(x, y, t) = (1-t)x + ty$. If $y \in V(\sigma) \setminus \sigma$, we define $\alpha(x, y, \cdot\,)$ as the path running through the broken line $[x, r_0(y), y]$ with the constant velocity. Since $r_0(y) = y$ for $y \in \mathrm{bd}\,\sigma$, α is well defined and continuous. The properties are evident. □

We say that a self map $f\colon X \to X$ is a *proximity map* on a subset $A \subset X$ if $f(a) \in V(a)$ for all $a \in A$.

Before we start to unite the points from the same Nielsen class we show that if f is a proximity map on a convex set lying inside a maximal simplex then all the fixed points in this set can be replaced by one single fixed point. For a subset $A \subset X$ of a compact polyhedron and $\eta > 0$ we denote

$$U(A, \eta) = \{x \in X : \mathrm{dist}(x, A) < \eta\}$$

where the distance is taken with the respect with the metric

$$\mathrm{d}(x, y) = \sqrt{\sum (x_i - y_i)^2}.$$

We will say the the maps $f, g\colon X \to Y$ are *homotopic on* $A \subset X$ if they can be connected with a homotopy constant outside A.

(2.8) LEMMA ([Shi]). *Let $a \in X$ be a point lying in a maximal simplex $\sigma(a)$ and let $\eta > 0$ satisfy:* $\mathrm{cl}(U(a, \eta)) \subset \sigma(a)$, $\mathrm{Fix}(f) \cap \mathrm{bd}\, U(a, \eta) = \emptyset$, *$f$ is a proximity map on* $\mathrm{cl}\, U(a, \eta)$. *Then there is a map $f'\colon X \to X$ satisfying:*

(2.8.1) *f, f' are homotopic on* $\mathrm{cl}(U(a, \eta))$,

(2.8.2) $\mathrm{Fix}(f') \cap \mathrm{cl}(U(a,\eta)) = \{a\}$
(2.8.3) f' *is a proximity map on* $\mathrm{cl}(U(a,\eta))$.

PROOF. We start with a special case assuming that $f(\mathrm{bd}U(a,\eta)) \subset \sigma(a)$. Then we may define the map $f' \colon X \to X$ by the formula

$$f'(x) = \begin{cases} (1-t)a + tf(y) & \text{for } x = (1-t)a + ty \in \mathrm{cl}\, U(a,\eta)), \\ & \text{and } y \in \mathrm{bd}\, U(a,\eta), \\ f(x) & \text{for } x \notin U(a,\eta). \end{cases}$$

Then the restriction of f and f' to $\mathrm{cl}\, U(a,\eta)$ as maps into $\sigma(x)$ are homotopic rel. boundary. Since $f(y) \neq y$ for $y \in \mathrm{bd}\, U(a,\eta)$, (2.8.2) follows. Since $f'(\mathrm{cl}\, U(a,\eta)) \subset \sigma(a)$, (2.8.3) is satisfied.

Now we reduce the general situation to the above special case. Since

$$f(\mathrm{cl}\, U(a,\eta)) \subset V(a), \quad f(x) \in V(a) = V(x) \quad \text{for each } x \in \mathrm{cl}\, U(a,\eta)$$

hence $\alpha(x, f(x), t)$ is defined. The compactness of $\mathrm{cl}\, U(a,\eta) \subset \sigma(a)$ implies the existence of an $\varepsilon > 0$ such that $\alpha(x, f(x), t) \in \sigma(a)$ for $0 \leq t \leq \varepsilon$ and $x \in \mathrm{cl}\, U(a,\eta)$. Since $\mathrm{Fix}(f) \cap (\mathrm{bd}\, U(a,\eta)) = \emptyset$ there exists $0 < \eta' < \eta$ such that $\mathrm{Fix}(f) \cap (\mathrm{cl}\, U(a,\eta) \setminus U(a,\eta')) = \emptyset$. Let $\lambda \colon [0,\eta] \to \mathbb{R}$ be an increasing function satisfying $\lambda(0) = 0$, $\lambda(\eta') < \varepsilon$, $\lambda(\eta) = 1$. Now we consider the map

$$f'(x) = \begin{cases} f(x) & \text{for } x \notin U(a,\eta), \\ \alpha(x, f(x), \lambda(\mathrm{d}(x,a)) & \text{for } \mathrm{cl}\, U(a,\eta)). \end{cases}$$

Now f, f' are homotopic on $\mathrm{cl}\, U(a,\eta)$ (as the maps into the contractible subspace $V(\sigma)$). Moreover, f' is a proximity map on $\mathrm{cl}\, U(a,\eta)$ and $f'(\mathrm{cl}\, U(a,\eta')) \subset \sigma(a)$. Moreover, for $x \in \mathrm{cl}\, U(a,\eta) \setminus U(a,\eta')$, we have $\mathrm{d}(a,x) > 0$ and $f(x) \neq x$ which implies $f'(x) = \alpha(x, f(x), \lambda(\mathrm{d}(a,x)) \neq x$ (property (2.7.3)) hence there are no fixed points of f' in $\mathrm{cl}\, U(a,\eta) \setminus U(a,\eta')$. Now we may apply the first part to $\mathrm{cl}\, U(a,\eta')$. □

(2.9) REMARK. The above lemma remains valid if we replace the ball $U(a,\eta)$ by an open convex subset C whose closure is contained in $\sigma(a)$. We define then $D \colon C \to \mathbb{R}$ by the formula

$$D(x) = \frac{\mathrm{d}(x,a)}{\mathrm{d}(y,a)}$$

where $y \in \mathrm{bd}\, C$ is the intersection point of the half-line starting from a and passing through x. Then we follow the above proof using $D(x)$ instead of $\mathrm{d}(a,x)$.

(2.10) LEMMA ([Shi]). *Let σ_1, σ_2 be closed maximal simplices (of dimension ≥ 2), $\dim(\mathrm{cl}\,\sigma_1 \cap \mathrm{cl}\,\sigma_2) \geq 1$, $a \in \sigma_1$, $b \in \mathrm{int}(\mathrm{cl}\,\sigma_1 \cap \mathrm{cl}\,\sigma_2)$. Let $f\colon X \to X$ be a continuous map satisfying:*

(2.10.1) *a is an isolated fixed point, $\mathrm{Fix}(f) \cap [a,b] = \{a\}$, $\mathrm{ind}(f;a) = j \neq 0$,*
(2.10.2) *f is a proximity map on $[a,b]$.*

Then there exist $\varepsilon > 0$ and a map $f'\colon X \to X$ satisfying:

(2.10.3) *f, f' are homotopic on $U([a,b],\varepsilon)$,*
(2.10.4) *$\mathrm{Fix}(f') \cap \mathrm{cl}(U([a,b])) = \{c\}$ a point in σ_2,*
(2.10.5) *f' is a proximity map on $\mathrm{cl}(U[a,b],\varepsilon)$,*
(2.10.6) *the Nielsen classes of the fixed points $a \in \mathrm{Fix}(f)$ and $c \in \mathrm{Fix}(f')$ correspond by the above homotopy.*

PROOF. We notice that for a sufficiently small $\varepsilon > 0$:

(2.11.1) $\mathrm{cl}(U(b,\varepsilon)) \subset \mathrm{st}\, b$,
(2.11.2) $f(\mathrm{cl}(U(b,\varepsilon))) \cap \mathrm{cl}(U(b,\varepsilon)) = \emptyset$,
(2.11.3) $\mathrm{Fix}(f) \cap \mathrm{cl}(U(a,\varepsilon)) = a$,
(2.11.4) f is a proximity map on $\mathrm{cl}(U[a,b],\varepsilon)$.

Let us fix points $p \in (a,b] \setminus \mathrm{cl}(U(b,\varepsilon))$, $q \in \sigma_2 \setminus U(b,\varepsilon)$.

Since $\mathrm{st}\,(b) \setminus U(b,\varepsilon)$ is path-connected, there is a broken line Q joining the point q with p in this set. Moreover, there is a path P joining $f(b)$ and p in $V(b)$. We define the path $r = P * ([p,b,q] * Q)^j$ and the map

$$f'(x) = \begin{cases} f(x) & \text{for } x \notin U(b,\varepsilon), \\ f\left(\left(\dfrac{2}{\varepsilon}\mathrm{d}(x,b) - 1\right)x\right) & \text{for } U(b,\varepsilon) \setminus U\left(b, \dfrac{\varepsilon}{2}\right), \\ r\left[1 - \dfrac{2}{\varepsilon}\mathrm{d}(x,b)\right] & \text{for } x \in U\left(b, \dfrac{\varepsilon}{2}\right). \end{cases}$$

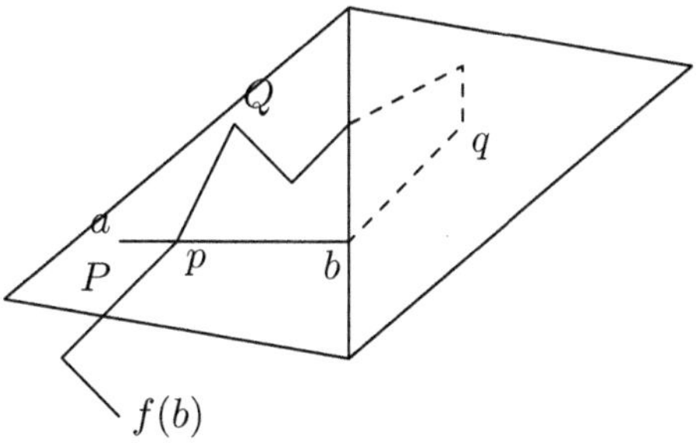

FIGURE 2

We notice that this map has no fixed point in $U(b,\varepsilon)\setminus U(b,\varepsilon/2)$ since $f'(U(b,\varepsilon)\setminus U(b,\varepsilon/2))\subset f(U(b,\varepsilon))$ and the last set is disjoint from $U(b,\varepsilon)$.

It remains to consider $U(b,\varepsilon/2)$. We notice that $f'(U(b,\varepsilon/2))\cap U(b,\varepsilon/2)\subset [a,b,q]\cap U(b,\varepsilon/2)$. The last set consists of two segments $S=(S\cap[a,b])\cup(S\cap[b,q])$. The analysis of the map r shows that the restriction of f' to the segment $S\cap(a,b]$ runs $|j|$ times through it. Each time we have an extending map hence each time we have exactly one fixed point. Moreover, the fixed point index (of the restriction of f to the segment) at this point equals $-\operatorname{sgn}(j)$ (for $j>0$ the restriction near the fixed point is reversing the orientation). By the Commutativity Property the index of the map $f'\colon X\to X$ near the fixed point is the same. An analysis of the restriction of f' to $[b,q]$ gives $|j|$ fixed points in $[b,q]$ each of index $\operatorname{sgn}(j)$ (here f' preserves the orientation of the segment).

Now we apply Remark (2.9) to the convex set contained in $U([a,b],\varepsilon)\cap\sigma_1$ containing inside a and $|j|$ new fixed points.Then we get a single fixed point of index $j-\operatorname{sgn}(j)|j|=0$ hence this point may be removed by a small local deformation. Similarly we find a convex set in σ_2 containing new fixed points which can be united to a point $c\in\sigma_2$ (of index $=j$).

The maps f, f' are homotopic on $U(b,\varepsilon)$ and are proximity maps there, since $f(U(b,\varepsilon)\cup f'(U(b,\varepsilon))\subset V(b)$ which is contractible. □

The above lemma allows to shift a fixed point from a maximal simplex to a neighbour one along a path on which f is a proximity map. As a consequence we get the main result of this subsection.

(2.12) LEMMA. *Let X be a compact polyhedron without local separating points. Let $f\colon X\to X$ be a self map with* $\operatorname{Fix}(f)$ *finite and let $a,b\in\operatorname{Fix}(f)$ lie inside maximal simplices. We assume moreover, that there is a path ω joining these points such that the restriction of f to $\omega[0,1]$ is a proximity map. Then there is a homotopy f_t starting from $f_0=f$ and satisfying* $\operatorname{Fix}(f_1)=\operatorname{Fix}(f)\setminus a$. *Moreover, the carrier of the homotopy can be contained in a prescribed neighbourhood of ω.*

PROOF. We may approximate ω with a broken line avoiding vertices and other fixed points and such that $\omega(t)$ belongs to maximal simplices for all but a finite number of $t\in I$. Moreover, in these exceptional t, ω is passing from one maximal simplex into another. Now may use the above lemma to shift the fixed point a along ω to the simplex $\sigma(b)$. Finally we use Lemma (2.8) to unite the two fixed points. □

Thus Wecken theorem will be proved once we show that for any two fixed points from the same Nielsen class there is a homotopy $\{f_t\}$ such that $\operatorname{Fix}(f_t)$ is constant and that f_1 is a proximity map on a path joining these two points.

(2.13) COROLLARY. *Let X be a compact polyhedron without local separating points. Then the identity map id_X is homotopic to a map with exactly one fixed point. If moreover, $\chi(X) = 0$ then id_X is homotopic to a fixed point free map.*

PROOF. By Lemma (2.3) we get a small homotopy from the identity to a map f with $\mathrm{Fix}(f)$ finite and each fixed point is lying inside a maximal simplex. Then we find a broken line running over $\mathrm{Fix}(f)$. Now we may apply successfully Lemma (2.10) to shift these points to one point. If moreover, $\chi(X) = 0$ then the index at this point is zero hence the fixed point can be removed by a small homotopy. □

Now we come back to the proof of Lemma (2.3). We start with an auxiliary lemma.

(2.14) LEMMA. *Each self-map $f: X \to X$ of a compact cw-complex is homotopic to a map g satisfying*

(2.14.1) *for each $k = 1, \dots, \dim X$ there is a neighbourhood $U_k \subset X$ of the skeleton $X^{(k)}$ such that $g(U_k) \subset X^{(k)}$,*

(2.14.2) *the fixed point set splits into the mutually disjoint closed-open subsets*

$$\mathrm{Fix}(g) = \bigcup_{\sigma} \mathrm{Fix}(g) \cap \sigma$$

where the summation runs the family of (open) cells in X,

(2.14.3) *there exist open sets $\{V_\sigma\}$ satisfying $\mathrm{cl}\, V_\sigma \cap \mathrm{cl}\, V_{\sigma'} = \emptyset$ (for $\sigma \neq \sigma'$), $\mathrm{Fix}(g) \cap \sigma \subset V_\sigma$, $g(V_\sigma) \subset \sigma$.*

PROOF. We may assume that f is a cellular map.

(2.14.1) We will show that each finite cw-complex (with a fixed triangulation) admits a cellular deformation $R_t: X \to X$ such that $R_0 = \mathrm{id}$ and, for each $k = 1, \dots, \dim X$, there is an open subset $U_k \subset X$ mapped into k-skeleton $R_1(U_k) \subset X^{(k)}$. Then we put $g = fR_1$ and we see that $g(U_k) = fR_1(U_k) \subset f(X^{(k)}) \subset X^{(k)}$.

It remains to show the existence of the deformation of a cw-complex X. We will do it inductively with the respect to the dimension: we will define the deformations $R_t^{(k)}: X^{(k)} \to X^{(k)}$ satisfying:

- $R_0^{(k)} = \mathrm{id}$,
- $R_t^{(k)}(x) = R_t^{(k-1)}(x)$ for $x \in X^{(k-1)}$,
- for each $l < k$ there is an open subset $U_l^k \subset X^{(k)}$ such that $R_1^{(k)}(U_l^k) \subset X^{(l)} \subset U_l^k$.

Then R_t^d, where $d = \dim X$ will satisfy our claim.

We start by defining deformations of the standard unit balls. Let $r_t^k: D^k \to D^k$ be given by:

$$r_t^k(x) = \begin{cases} (1+t)x & \text{for } \|x\| \le \dfrac{1}{1+t}, \\ \dfrac{x}{\|x\|} & \text{for } \|x\| \ge \dfrac{1}{1+t}. \end{cases}$$

For $k = 1$ we define

$$R_t^1(x) = \begin{cases} x & \text{for } x \in X^{(0)}, \\ \phi_\sigma(r_t^1(y)) & \text{for } x = \phi_\sigma(y). \end{cases}$$

Now $X^{(0)} \cup \bigcup_\sigma \phi_\sigma(\{x \in D^1 : \|x\| > 1/2\})$, where σ runs the family on all 1-cells in X, is the open neighbourhood of $X^{(0)}$ in $X^{(1)}$ which is sent by $R_1^{(1)}$ into $X^{(0)}$.

Now we assume that we have a map $R_t^{(k)}: X^{(k)} \to X^{(k)}$ satisfying the above three conditions. We define $R_t^{(k+1)}: X^{(k+1)} \to X^{(k+1)}$ by

$$R_t^{(k+1)}(x) = \begin{cases} R_t^{(k)}(x) & \text{for } x \in X^{(k)}, \\ R^{(k)}_{((1+t)\|y\|-1)}(\phi_\sigma(r_t^{k+1}(y))) & \text{for } x = \phi_\sigma(y) \text{ and } \|y\| \ge \dfrac{1}{1+t}, \\ \phi_\sigma(r_t^{k+1}(y)) & \text{for } x = \phi_\sigma(y) \text{ and } \|y\| \le \dfrac{1}{1+t}. \end{cases}$$

Now $U_k = X^{(k)} \cup \bigcup_\sigma \phi_\sigma(\{x \in D^{k+1} : \|x\| > 1/2\})$, where σ runs the family on all (k+1)-cells in X, is the open neighbourhood of $X^{(k)}$ in $X^{(k+1)}$ satisfying $R_1^{k+1}(U_k) \subset X^{(k)}$. For $l < k$ we denote by U_l' the neighbourhood $X^{(l)} \subset U_l' \subset X^{(k)}$ satisfying the inductive assumption. Then $U_l = (R^{(k+1)}_{1|U_k})^{-1}(U_l') \subset X^{(k+1)}$ is the desired neighbourhood of $X^{(l)}$ in $X^{(k+1)}$.

(2.14.2) Let σ be a cell of dimension s. We show that $\mathrm{Fix}(g) \cap \sigma$ is closed and open in $\mathrm{Fix}(g)$. By the above $\mathrm{Fix}(g) \cap \sigma \subset \sigma \setminus U_{s-1}$ implies that $\mathrm{Fix}(g) \cap \sigma = \mathrm{Fix}(g) \cap (\sigma \setminus U_{s-1})$ is closed in X. Now $\mathrm{Fix}(g) = \bigcup_\sigma \mathrm{Fix}(g) \cap \sigma$ is the the sum of mutually disjoint closed summands. Since the sum is finite, all summands are also open in $\mathrm{Fix}(g)$.

(2.14.3) since the sets $\mathrm{Fix}(f) \cap \sigma$ are isolated, we can take neighbourhoods V_σ' satisfying $\mathrm{Fix}(f) \cap \sigma \subset V_\sigma'$, $f(V_\sigma') \subset X^{(s)}$ and $\mathrm{cl}\, V_\sigma' \cap \mathrm{cl}\, V_{\sigma'}' = \emptyset$ for $\sigma \neq \sigma'$. We put $V_\sigma = V_\sigma' \cap f^{-1}\sigma$. □

Proof of Lemma (2.3). Let $\{V_\sigma\}$ be the family of open subsets from Lemma (2.14) (we regard the polyhedron X as a cw-complex). Then $\mathrm{Fix}(f) = \bigcup_\sigma \mathrm{Fix}(f) \cap V_\sigma$ is the mutually disjoint sum of closed open subsets (σ runs the family of all cells in X). For each non-maximal simplex σ:

- we define an Urysohn function $\lambda_\sigma: X \to [0,1]$ satisfying

$$\lambda_\sigma(\mathrm{Fix}(f) \cap \mathrm{cl}\, V_\sigma) = 1, \quad \lambda_\sigma(X \setminus V_\sigma) = 0,$$

- we fix a maximal simplex $\sigma_{\max}$ such that $\sigma \subset \operatorname{cl} \sigma_{\max}$,
- we fix a point $v_\sigma \in \sigma_{\max}$.

Now we define the map $g\colon X \to X$ putting

$$g(x) = \begin{cases} (1 - \lambda_\sigma(x))f(x) + \lambda_\sigma(x)v_\sigma & \text{for } x \in \operatorname{cl} V_\sigma (\sigma \text{ non-maximal}), \\ f(x) & \text{otherwise.} \end{cases}$$

To see that the map is well defined we notice that for $x \in \operatorname{bd} V_\sigma$ we have $\lambda_\sigma(x) = 0$ which implies $g(x) = f(x)$. Moreover, $g(\operatorname{cl} V_\sigma) \cup f(\operatorname{cl} V_\sigma) \subset \operatorname{cl}(\sigma_{\max})$ implies that f, g are homotopic. We notice that if $\lambda_\sigma(x) = 0$ (for all non-maximal σ) then $g(x) = f(x)$. But then $x \notin \bigcup_\sigma V_\sigma \supset \operatorname{Fix}(f)$ hence $g(x) = f(x) \neq x$. Thus $g(x) = x$ implies $x \in V_\sigma$ and $\lambda_\sigma > 0$ for a σ. But then $x = g(x) \in \sigma_{\max}$ hence each fixed point of g is lying inside a maximal simplex.

Thus $\operatorname{Fix}(g)$ splits into mutually disjoint sum $\operatorname{Fix}(g) = \bigcup_\sigma \operatorname{Fix}(g) \cap \operatorname{cl} V_\sigma$ where now σ runs the family of all maximal simplices. In each maximal simplex we find an open set U_σ satisfying: $\operatorname{Fix}(g) \cap \sigma \subset U_\sigma \subset \operatorname{cl} U_\sigma \subset \sigma$ and $f(U_\sigma) \subset \sigma = \mathbb{R}^m$. Now the transversality gives a homotopy rel. $X \setminus U_\sigma$ to a map with $\operatorname{Fix}(g) \cap U_\sigma$ finite. Repeating this in all maximal simplices we get Lemma (2.3).

The homotopy can be small: let us notice that the image of a point x during the homotopy does not leave the star of $f(x)$. It remains to take the triangulation sufficiently fine. □

2.1. Reduction to a proximity map. Shi proved Wecken theorem for connected polyhedra which are at least three dimensional and have no local separating points. The idea of this proof is to shift all fixed points of a Nielsen class to a 3-simplex and then to unite them there (see [Shi], [B]). In 1979 Boju Jiang gave an alternative proof of this result extending it by the way onto a larger class of polyhedra (see [Ji3]). Here we will present this proof.

We start with some notation inspired by the geometry of surfaces. We consider a polyhedron X without local separating points. A path $p\colon I \to X$ is called PL-*path* if it is affine for some subdivisions of I and X. The image of each vertex of I is called a *corner*. A path p is called an *normal* PL-*arc* if

(2.15.1) p avoids vertices of X,

(2.15.2) p has only double intersections and no intersection is a corner,

(2.15.2) $p(s)$ is in maximal simplices for all but a finite number of points. In these exceptional points, $p(s)$ leaves a maximal simplex and enters another.

A PL-path with different end points and without self-intersections is called PL-*arc*.

(2.16) LEMMA. *Any path in X with different end points which are not vertices of X is ε-homotopic (rel. ends) to a normal* PL-*path.*

PROOF. We approximate the path with a broken-line. Since there are no local separating points, a small shift of corners makes (2.15.1)–(2.15.3) hold. □

(2.17) LEMMA. *Let $q\colon I \to X$ be a normal* PL*-arc and let $S = [0,1] \times [-\eta, \eta]$ be a rectangular strip. Then there is a* PL*-imbedding $\widetilde{q}\colon S \to X$ extending q, i.e. $\widetilde{q}(t,0) = q(t)$.*

PROOF. This is evident for X = a manifold (of dimension ≥ 2). In general we imbed successively the piece $[t_i, t_{i+1}] \times [-\eta, \eta]$ where t_i, t_{i+1} are consecutive corners. If q passes at the point t_{i+1} from one simplex to another one we choose $\widetilde{q}(t_{i+1} \times [-\eta, \eta])$ as a segment in the common face of these simplices, see Figure 4.□

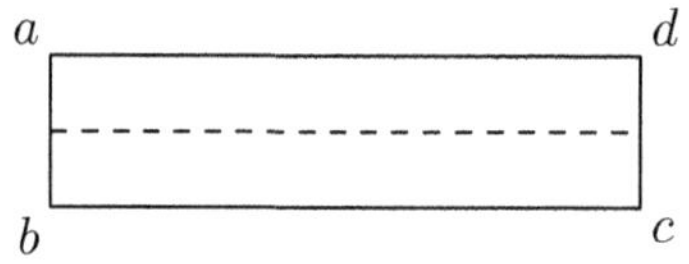

FIGURE 3

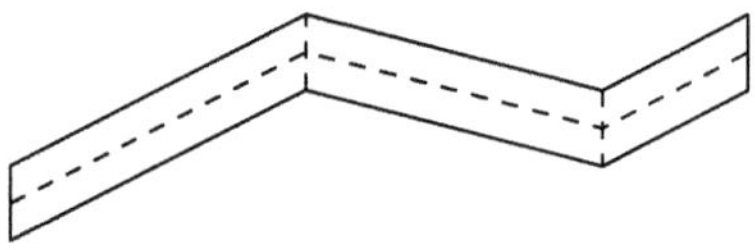

FIGURE 4

(2.18) LEMMA. *Each path whose end points are contained inside maximal simplices and are different is homotopic, rel. ends, to a normal* PL*-arc.*

PROOF. By the above we may replace the path $p\colon I \to X$ by a normal PL-path. It remains to remove the intersections. If an intersection takes place in a maximal simplex of dimension ≥ 3 then a local small deformation allows to remove it. In general case we proceed as follows. Let $p(s') = p(s)$ where $s \in I$ is the greatest parameter such that $p(s)$ is the intersection point. We consider a strip Σ along $[p(s), p(1)]$ as in Lemma (2.17). We may assume that $[p(s'-\varepsilon), p(s'+\varepsilon)] = [a,b]$ a side of this strip. If moreover, $\varepsilon > 0$ is small enough, Σ has no other intersections with $p([0,1])$. Now we may homotope the segment $[a,b]$ into $[a,d,c,b]$ through the strip, see Figure 3. Since by the assumption $p(s)$ is the last intersection point, no new intersections appear. Finally we can remove the intersection point $p(1)$, since $p(1)$ is lying inside a maximal simplex. Following the procedure we eliminate all intersection points. □

(2.19) LEMMA. *Let X be a polyhedron without local separating points. Then each path $p\colon I \to X$ whose ends are different and are lying inside maximal simplices is homotopic to a normal* PL*-arc $q\colon I \to X$ satisfying:*

(2.19.1) *there is a maximal simplex τ and numbers $0 < s_1 < s_3 < s_2 < 1$ such that $q(s) \in \operatorname{bd}\tau$ if and only if $s = s_1$ or s_2, $q(s_3) \in \tau$ and q is affine on $[s_1, s_2]$ and $[s_3, s_2]$ (τ is a simplex in a suitable subdivision).*

(2.19.2) $q(s_1)$, $q(s_2)$ *may belong to a prescribed simplex* σ_1 *of dimension* ≥ 1 *which is assumed to be a commmon face of at least two maximal simplices of* X.

PROOF. By Lemma (2.18) we may assume that p is a normal PL-arc q'. Let σ_1 be a simplex of dimension ≥ 1 which is a common face of at least two maximal simplices. Passing to a subdivision, if necessary, we may assume that the arc q avoids the star of σ_1. Now we fix numbers $0 < s'_1 < s'_3 < s'_2 < 1$ such that there are no corners in $q'[s'_1, s'_2]$. We find a normal PL-arc from $q'(s'_3)$ to a point in σ_1 which has no other common points with $q'[0,1]$ but $q'(s'_3)$. Then we take a PL-imbedding of the strip along this arc (see Figure 5).

Now we may homotope $q'[s'_1, s'_2]$ along the strip to get an arc q, $q(s) = q'(s)$ for $s \notin [s_1, s_2]$ and q satisfies the lemma for some $s_1 < s_2 \in (s'_1, s'_2)$ and $s'_3 = s_3$. □

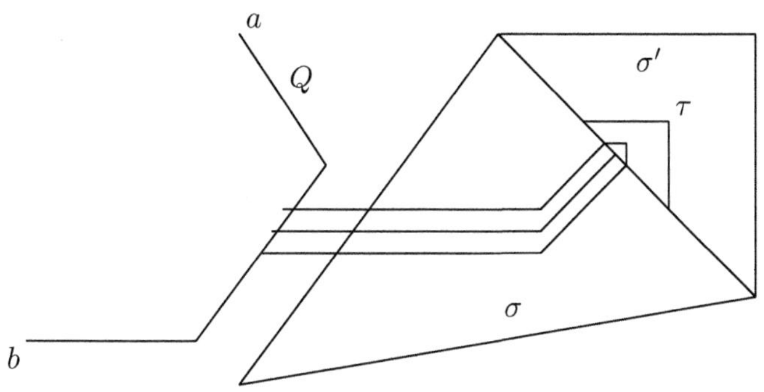

FIGURE 5

Let us fix a normal PL-arc $q\colon I \to X$. We say that a path $p\colon I \to X$ is *q-special* if $p(0) = q(0)$, $p(1) = q(1)$ and $p(s) \neq q(s)$ for $0 < s < 1$. We say that two paths p_0, p_1 are *q-specially homotopic* if they can be connected by a homotopy $\{p_t\}$ where each p_t is q-special.

On the other hand let $A \subset X$ and let $h_t\colon A \to X$ be a homotopy. This homotopy is called *special* if $\mathrm{Fix}(h_s) = \{a \in A; h_s(a) = a\}$ does not depend on s.

(2.20) REMARK. Let us notice that if q is a normal PL-arc then the path fq is q-specially homotopic to p if and only if the restriction $f\colon q(I) \to X$ is specially homotopic to $pq^{-1}\colon q(I) \to X$.

(2.21) LEMMA (Special Homotopy Extension Property). *Let* X *be a metric* ANR *and* $A \subset X$ *its closed subset. Let* $f_0\colon X \to X$ *be a homotopy and* $h_t\colon A \to X$ *a special homotopy starting from* $h_0 = f_{0|A}$. *Then* h_t *can be extended to a special homotopy* $f_t\colon X \to X$ *of* f_0.

PROOF. We define a map $H'\colon X \times 0 \cup A \times I \to X$ by the formula:

$$H'(x,t) = \begin{cases} f_0(x) & \text{for } t = 0, \\ h_t(x) & \text{for } x \in A. \end{cases}$$

Since X is ANR, there is an extension $H\colon X \times I \to X$. The set $C = \{x \in X : x \in H(x \times I)\}$ is closed since H is continuous and I is compact. We define

$$F(x,t) = \begin{cases} H(x,0) & \text{for } \mathrm{d}(x,A) \geq \mathrm{d}(x,C), \\ H(x, t - t\mathrm{d}(x,A))/\mathrm{d}(x,C)) & \text{for } \mathrm{d}(x,A) \leq \mathrm{d}(x,C) > 0. \end{cases}$$

The homotopy is continuous since $\mathrm{d}(x,A) = \mathrm{d}(x,C) = 0$ means $x \in A \cap C$ is a fixed point of the special homotopy and $H(x,t)$ does not depend on t. This and the compactness of I imply the continuity of F for x satisfying $\mathrm{d}(x,A) = \mathrm{d}(x,C) = 0$. Now we check that the homotopy F is special. Since F is an extension of the special homotopy $\{h_t\}$, $\mathrm{Fix}F(\,\cdot\,,t) \cap A$ does not depend on t. Let $x \notin A$. If moreover, $\mathrm{d}(x,A) \geq \mathrm{d}(x,C)$ then $F(x,t) = H(x,0) = f_0(x)$ does not depend on t hence x is a fixed point for all t or for none of them. On the other hand $\mathrm{d}(x,A) < \mathrm{d}(x,C)$ implies $F(x,t) \in H(x \times I)$ but the last set does not contain x since $x \notin C$. □

(2.22) LEMMA. *Each q-special path $p\colon I \to X$ is q-specially homotopic to a normal* PL*-path p'. The homotopy may be arbitrarily small.*

PROOF. Let us denote the ends $p(0) = q(0) = a$, $p(1) = q(1) = b$. Since q is a normal PL-arc, a and b are not vertices. Let $\varepsilon > 0$ be so small that q is linear on $[0,\varepsilon]$, $[1-\varepsilon, 1]$ and that $p[0,\varepsilon] \subset \mathrm{st}\,(a)$, $p[1-\varepsilon,1] \subset \mathrm{st}\,(b)$. First we q-specially homotope p to an affine map on $[0,\varepsilon]$. We define $h\colon [0,\varepsilon] \to X$

$$h(s,t) = \begin{cases} \left(1 - \dfrac{s}{\varepsilon t}\right)a + \dfrac{s}{\varepsilon t} \cdot p(\varepsilon t) & \text{for } s \leq \varepsilon t \text{ and } t > 0, \\ p(s) & \text{for } s \geq \varepsilon t. \end{cases}$$

Let us notice that $h(s,t) \neq q(s)$ for $0 < s \leq \varepsilon$, $t \in I$. In fact since $q(s) \neq p(s)$ for $0 < s < 1$,

$$h(s,t) = \left(1 - \frac{s}{\varepsilon t}\right)a + \frac{s}{\varepsilon t} \cdot p(\varepsilon t) \neq \left(1 - \frac{s}{\varepsilon t}\right)q(0) + \frac{s}{\varepsilon t} \cdot q(\varepsilon t) = q(s).$$

Similarly we define $H(s,t)$ for $1 - \varepsilon \leq s \leq 1$. Since the above homotopies are constant for $s = \varepsilon$ and $1-\varepsilon$, we may assume that p is affine on $[0,\varepsilon]$ and $[1-\varepsilon,1]$. Now $\inf\{\mathrm{d}(p(s),q(s)) : \varepsilon \leq s \leq 1-\varepsilon\} > 0$, hence we may extend the homotopy h onto $I \times I$ from $p = h(\,\cdot\,,x)$ to a PL-path and $h(s,t) \neq q(s)$ for $0 < s < 1$. □

Consider again $f: X \to X$ with $\mathrm{Fix}(f)$ finite and Nielsen related points $a_0, a_1 \in \mathrm{Fix}(f)$ lying in maximal simplices. By Lemma (2.18) we may assume that there is a normal PL-arc q joining these points avoiding other fixed points and $fq \sim q$. Suppose that the path $fq: I \to X$ is q-specially homotopic to a path close to q. Now the restriction $f: q(I) \to X$ is specially homotopic to a proximity map (Remark (2.20)). The last homotopy extends (Lemma (2.21)) to a special homotopy on X. The obtained map f_1 has the same fixed points as f but $f_1: q(I) \to X$ is a proximity map. Now Lemma (2.12) allows to join these points.

Thus to prove Wecken theorem it is enough to show that for each pair of periodic points from the same Nielsen class there is a normal arc q such that fq is q-specially homotopic to a map close to q. We will show that this is possible for polyhedra satisfying the assumption of Wecken Theorem (2.2).

(2.23) LEMMA. *Suppose that a normal* PL-*arc* q *satisfies condition* (2.19.1). *Then any* q*-special path* $p: I \to X$ *is* q*-specially homotopic to a path* p' *with* $p'(0,1) \cap (q(0,1) \cup \tau) = \emptyset$.

PROOF. By Lemma (2.16) we may assume that p is a normal PL-arc. By a small correction we may reduce the intersections with $q(I)$ to a finite number (none at a corner). Now we will eliminate the intersections. Let $p(s') = q(s)$ for $s' < s$. As in the proof of Lemma (2.18) we homotope $p(s'-\delta, s'+\delta)$ along the arc $q[s,1]$ and we remove the intersection. If $s' > s$ we do the same along $q[0,s]$. If p does not avoid τ we compose it with the radial projection $\tau \setminus q(s_3) \to \mathrm{bd}\,\tau$. Then $p(0,1) \cap \tau = \emptyset$. □

(2.24) LEMMA. *Suppose that the normal* PL-*arc* q *satisfies condition* (2.19.1) *Let* $p_0, p_1: I \to X$ *be special paths whose images* $p_0(I)$ *and* $p_1(I)$ *do not meet* $q(0,1) \cup \tau$. *If* p_0 *and* p_1 *are homotopic in* $X \setminus \tau$ *then they are* q*-specially homotopic in* X.

PROOF. Let $\lambda: I \to I$ be given by Figure 6 (i.e. $\lambda[0, s_1] = 0$, $\lambda[s_2, 1] = 1$).

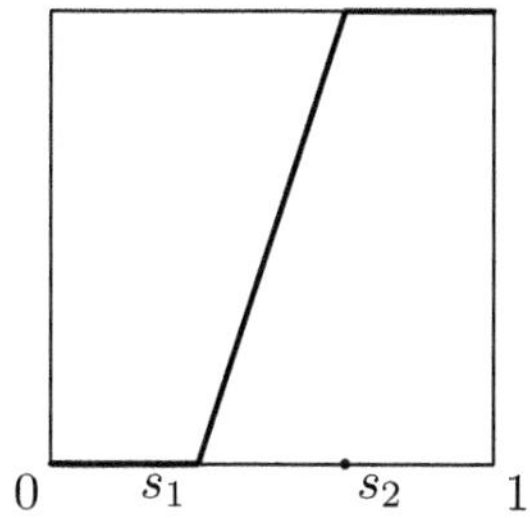

FIGURE 6

The paths p_0 and $p_0 \cdot \lambda$ are connected by the homotopy $\{h_t\} = p_0((1-t)s + t\lambda(s))$ which is q-special. Similarly p_1 and $p_1 \cdot \lambda$ are q-specially homotopic.

Let $\{p_t\}$ be a homotopy connecting p_0 and p_1 as the paths in $X \setminus \tau$. Then $\{p_t \cdot \lambda\}$ is a homotopy connecting $p_0 \cdot \lambda$ and $p_1 \cdot \lambda$. This homotopy is q-special: $p_t \cdot \lambda(s) = q(0) \neq q(s)$ for $0 < s \leq s_1$, $p_t \cdot \lambda(s) = q(1) \neq q(s)$ for $s_2 \leq s < 1$ and for $s_1 < s < s_2$ we have $p_t \cdot \lambda(s) \in X \setminus \tau$ while $q(s) \in \tau$. □

(2.25) LEMMA. *Let the normal* PL-*arc q satisfy Lemma* (2.19) *where σ_1 is a 1-dimensional face of at least three 2-simplices of X. If two special paths $p_0, p_1 \colon I \to X$ are homotopic then they are q-specially homotopic.*

PROOF. In view of Lemma (2.23), we may assume that $p_0[0,1]$ and $p_1[0,1]$ are disjoint from $q(0,1) \cup \tau$. We choose a point $a = q(0)$ as the base point in X and in $X \setminus \tau$. Notice that $p_0 \sim p_1$ means $p_1 * p_0^{-1}$ is nullhomotopic in X.

Suppose that moreover $\dim \tau \geq 3$. Then the inclusion $i \colon X \setminus \tau \subset X$ induces the isomorphism $i_\# \colon \pi_1(X \setminus \tau) \to \pi_1 X$. Now $p_0 \sim p_1$ in X implies $p_0 \sim p_1$ in $X \setminus \tau$ hence by Lemma (2.24) p_0 and p_1 are q-specially homotopic.

Now we consider the case $\dim X = 2$.

We take a point $c \in \operatorname{bd} \tau$ different than $q(s_1)$, $q(s_2)$ and the vertices of X. Let u be a loop on $\operatorname{bd} \tau$, based in c, running once around $\operatorname{bd} \tau$. Then the group $\pi_1 X$ is the factor group of $\pi_1(X \setminus \tau)$ modulo the normal subgroup generated by u. Now $p_1 p_0^{-1} \sim 1$ in X implies $p_1 p_0^{-1}$ is homotopic in $X \setminus \tau$ to a loop of the form $(\nu_1 u^{k_1} \nu_1^{-1})(\nu_2 u^{k_2} \nu_2^{-1}) \ldots (\nu_m u^{k_m} \nu_m^{-1})$, where $k_1, \ldots, k_m$ are integers, $\nu_1, \ldots, \nu_m$ are paths in $X \setminus \tau$ from a to c.

For every $j = 1, \ldots, m$ we can find a PL-path ν_j' having only a finite number of intersections with Q such that $\nu_j' \sim \nu_j$ in $X \setminus \tau$. Then we can eliminate the intersections of ν_i' with $q(0,1)$ just as we did in the proof of Lemma (2.18). Thus we get a path w_j in $X \setminus (\tau \cup q(0,1))$ such that $w_j \sim \nu_j$ in $X \setminus \tau$.

Take a neighbourhood W of a such that $\operatorname{cl} W \cap \operatorname{cl} \tau = \emptyset$. The condition (2.19.2) and the assumption that σ_1 is the common face of at least three 2-simplices make it possible to modify u into a loop μ in $X \setminus (\tau \cup q(0,1))$, based at c satisfying $\mu \sim u$ in $X \setminus (\tau \cup W)$.

Let $p_0' = (w_1 u^{k_1} w_1^{-1}) \ldots (w_m u^{k_m} w_m^{-1}) p_0$. Now p_1 and p_0' are in $X \setminus (\tau \cup q(0,1))$ and they are homotopic in $X \setminus \tau$. By Lemma (2.24), p_1 and p_0' are q-specially homotopic. It remains to show that p_0 and p_0' are q-specially homotopic

Let $\varepsilon > 0$ be so small that $q[0, \varepsilon]$ is contained in the neighbourhood $W \ni a$ chosen above. In view of the reparametrization used in the proof of Lemma (2.24), we may assume that p_0 is constant on $[0, \varepsilon]$ and that p_0' makes up $(w_1 u^{k_1} w_1^{-1}) \ldots (w_m u^{k_m} w_m^{-1})$ on $[0, \varepsilon]$ and then goes on p_0 on $[\varepsilon, 1]$. Since $\mu \sim u$ in $X \setminus (\tau \cup W)$ and $u \sim 1$ in $\operatorname{cl} \tau \subset X \setminus W$, $\mu \sim 1$ in $X \setminus W \subset X \setminus q(0, \varepsilon]$. But w_j is in $X \setminus q(0, \varepsilon]$,

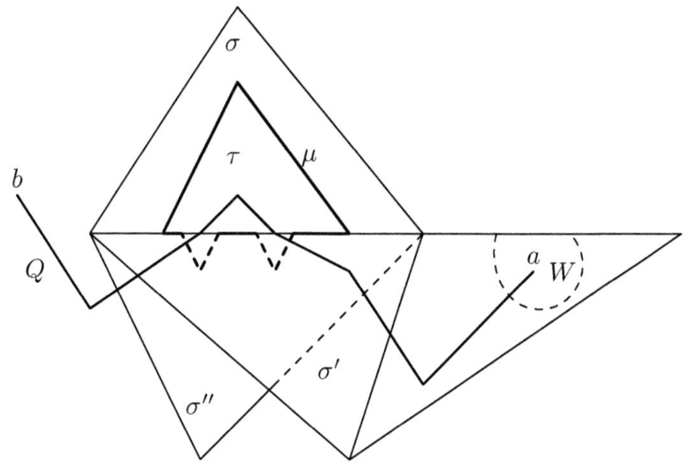

FIGURE 7

so $(w_j \mu^{k_j} w_j^{-1}) \sim (w_j w_j^{-1}) \sim 1$ in $X \setminus q(0, \varepsilon]$. Hence, the part of p_0' on $[0, \varepsilon]$ can be homotoped to constant without touching $q(0, \varepsilon]$. Thus p_0 and p_0' are q-specially homotopic. □

PROOF OF WECKEN THEOREM (2.2). As we noticed in the scheme of the proof of theorem it is enough to show that f is homotopic to a map with all fixed points lying in different Nielsen classes. We may assume that $\mathrm{Fix}(f)$ is finite and each fixed point is lying inside a maximal simplex. Let a, b be two fixed points lying in the same Nielsen class. We will show that there is a homotopy $\{f_t\}$ starting from f, $\{f_t\}$ is constant in a neighbourhood of $\mathrm{Fix}(f) \setminus \{a, b\}$ and $\mathrm{Fix}(f_1) = \mathrm{Fix}(f) \setminus a$.

Since a and b are Nielsen related, there is a path p joining these points and satisfying $fp \sim p$. By Lemma (2.24), p may be replaced by a normal PL-arc q satisfying moreover, the assumption of Lemma (2.24) and that a, b are the only fixed points of f on $Q = q(I)$. We are going to apply Lemma (2.25) for this q.

Let $p_\varepsilon' = q \cdot h$ where $h: I \to I$ is a homeomorphism such that $h(0) = 0$, $h(1) = 1$ and $0 < |h(s) - s| < \varepsilon$ for $0 < s < 1$. Then, for sufficiently small $\varepsilon > 0$, $p_\varepsilon'(s)$ and $q(s)$ are close, so that the map $\phi = p_\varepsilon' \cdot q^{-1}: Q \to X$ is a proximity map.

Now both p_ε' and $f \cdot q$ are q-special paths and they are homotopic, since they both are homotopic to q. By Lemma (2.25) they are q-specially homotopic. Now $f, p_\varepsilon' \cdot q^{-1}: Q \to X$ are specially homotopic maps (Remark (2.21)). Then Lemma (2.21) extends this special homotopy to a special homotopy $f_t: X \to X$. Now $\mathrm{Fix}(f_1) = \mathrm{Fix}(f_0)$ and f_1 is a proximity map on q hence Lemma (2.12) yields a homotopy $\{f_t : 1 \le t \le 2\}$ with $\mathrm{Fix}(f_2) = \mathrm{Fix}(f) \setminus a$. □

3. Nielsen relation for periodic points

Now we ask about the least number of periodic points. We are given a self map $f: X \to X$ and a natural number $n \in \mathbb{N}$. How large must be $\mathrm{Fix}(g^n)$ for any g homotopic to f?

We will consider a self map $f: X \to X$, a number $n \in \mathbb{N}$ and we will be trying to find lower bounds of the cardinality of sets $\mathrm{Fix}(g^n)$ and

$$P_n(g) = \{x \in \mathrm{Fix}(g^n) : g^k(x) \neq x \text{ for all } k|n,\ k \neq n\}$$

where g runs the set of all maps homotopic to f? Let us emphasis that we deform only the map f while we are investigating the fixed points of the iterate f^n. The first lower bounds of these cardinalities appeared in an (unpublished) paper [H4] of Halpern. In [Ji1] Boju Jiang introduced general homotopy invariants which are estimates of the number of periodic points. These invariants were intensively explored in 80' and 90' (see [HK1]–[HK4], [Mc]).

Since $\#\mathrm{Fix}(g^n) \geq N(f^n)$ for every map g homotopic to the given f, the number $N(f^n)$ is a lower bound of the cardinality of $\mathrm{Fix}(f^n)$. But this estimation is not the best one.

(3.1) Example (Flip map). Let $f: S^1 \to S^1$ be given by $f(z) = \overline{z}$ (conjugation of the complex number $z \in \mathbb{C}$) and let $n = 2$. Then $f^2 = \mathrm{id}$ hence $N(f^2) = N(\mathrm{id}) = 0$. But every map homotopic to f must have at least two fixed (hence also 2-periodic) points since $N(f) = |L(f)| = |1 - (-1)| = 2$.

The above example shows that we must not forget about the periodic points of periods smaller than n. On the other hand the sum $\sum_{k|n} N(f^k)$ may be too large as it is in the case of the constant map (since then the unique fixed point is counted in each period). Thus we must consider more carefully the relations between Reidemeister sets $\mathcal{R}(f^k)$ for $k|n$.

4. Weak Wecken's theorem for periodic points

The above discussion shows that it is not so easy to propose a sensible lower bound of the number of periodic points so we will confine, for the beginning, to an easier question. When a given map f is homotopic to a map with no periodic points of period n?

Let us notice that if a map $f: X \to X$ is homotopic to a map g with $\mathrm{Fix}(g^n)$ empty then also $\mathrm{Fix}(g^k) = \emptyset$ for each divisor $k|n$. This implies $N(f^k) = 0$ for all $k|n$. We will show that in the case of manifolds of dimension at least three this is also the sufficient condition for the deformation of the map f to a map with no n-periodic points. The rest of this section is the proof of the following Weak Wecken's Theorem for Periodic Points. In fact the proof is quickly reduced to the Cancelling Procedure (4.3).

(4.1) THEOREM (Weak Wecken's Theorem for Periodic Points). *Any self map $f\colon M \to M$ of a compact PL-manifold of dimension ≥ 3 is homotopic to a map g without periodic points of period n (i.e. $g^n(x) \neq x$) if and only if for any divisor k of n the Nielsen number $N(f^k) = 0$.*

PROOF. $\Rightarrow$ is evident. The rest of this section is the proof of $\Leftarrow$. We use the induction with the respect to the divisors of the given $n \in \mathbb{N}$: for every $k|n$ we show that

f is homotopic to a map g satisfying $g^l(x) \neq x$ for all $l|n$, $l \leq k$.

For $k = 1$ the theorem follows from the classical Wecken theorem for fixed points. Now we assume that the induction assumption holds for all divisors of n which are less than k. We will show how to remove k-periodic points. We will base on the following technical result (for the proof we refer to [Je1]).

(4.2) THEOREM [Je1]. *Let $M \subset \mathbb{R}^N$ be a compact PL-submanifold with the metric inherited after the Euclidean metric in $\mathbb{R}^N$. Let $n \in \mathbb{N}$ be a fixed number. Then any continuous map $f\colon M \to M$ is homotopic to a map g such that $\operatorname{Fix}(g^n)$ is finite and g is a PL-homeomorphism near any point $x \in \operatorname{Fix}(g^n)$. Moreover, for any $\varepsilon > 0$ we may choose a g satisfying $d(f,g) < \varepsilon$.* □

By the above theorem we may assume that $\operatorname{Fix}(f^n)$ is finite and f is a linear homeomorphism near each $x \in \operatorname{Fix}(f^n)$. In particular $\operatorname{ind}(f^k, x) = \pm 1$ at these points. Consider an orbit of Nielsen classes $A \subset \operatorname{Fix}(f^k)$. Since by the induction assumption $f^l(x) \neq x$ ($l|n$, $l < k$, $x \in M$), all orbits of points in $\operatorname{Fix}(f^k)$ have length k. Since $N(f^k) = 0$, $\operatorname{ind}(f^k; A) = 0$ hence A splits into finite pairs of orbits $\{x_1, \ldots, x_k\}$, $\{y_1, \ldots, y_k\}$ such that there is a path $\omega\colon [-1, +1] \to M$ establishing the Nielsen relation between x_1, y_1 and $\operatorname{ind}(f^k; \{x_1\}) + \operatorname{ind}(f^k; \{y_1\}) = 0$.

Thus the induction step will be done once we prove that the orbits $\{x_1, \ldots, x_k\}$, $\{y_1, \ldots, y_k\}$ can be removed by a homotopy which is constant in a neighbourhood of the other periodic points and which does not produce new periodic points. In other words it remains to show the following.

(4.3) THEOREM (Cancelling Procedure). *Let $f\colon M \to M$ be a map with $\operatorname{Fix}(f^k)$ finite ($\dim M \geq 3$). Assume that*

(4.3.1) *$\{x_0, \ldots, x_{k-1}\}$, $\{y_0, \ldots, y_{k-1}\}$ are disjoint orbits of length k which are Nielsen related i.e. there is a path $\omega\colon [-1, +1] \to M$ from $f(-1) = x_0$ to $f(+1) = y_0$ such that $f^k\omega$ and ω are fixed end point homotopic,*

(4.3.2) *f is a PL-homeomorphism near each point in $\{x_0, \ldots, x_{k-1}; y_0, \ldots, y_{k-1}\}$.*

(4.3.3) $\operatorname{ind}(f^k; x_0) + \operatorname{ind}(f^k; y_0) = 0$

Then there is a homotopy $\{f_t\}$ *starting from* $f_0 = f$ *constant in a neighbourhood of* $\text{Fix}(f^k) \setminus \{x_0, \dots, x_{k-1}; y_0, \dots, y_{k-1}\}$ *and satisfying*

$$\text{Fix}(f_1^k) = \text{Fix}(f^k) \setminus \{x_0, \dots, x_{k-1}; y_0, \dots, y_{k-1}\}.$$

The rest of this section is the proof the above Cancelling Procedure. Here is the outline of the proof.

Some technical lemmas allow to make ω and their images $f\omega, \dots, f^{k-1}\omega$ flat arcs and f the homeomorphism in neighbourhoods of these arcs (more exactly in neighbourhoods of $f^i(\omega[-1,0))$ and $f^i(\omega(0,1])$ for $i = 0, \dots, k-2$).

If it moreover, happens that $f^k\omega$ is close to ω then the use of a modified Hopf lemma allows to remove the two orbits. Thus the main difficulty in the general case is to make $f^k\omega$ close to ω without adding new periodic points.

4.1. How to control the periodic point set during a homotopy. In all deformations $\{f_t\}$ we will have either not to change the periodic point set $\text{Fix}(f_t^k)$ or to control its changes. We will use (often implicitly)the following three methods.

(1) The support of the homotopy

$$\text{Supp}\{f_t\} = \{x \in X : f_t(x) \neq f(x) \text{ for } t \in [0,1]\}$$

is isolated from $\text{Fix}(f_0^k)$ and the homotopy $\{f_t\}$ is sufficiently small.

(4.4) LEMMA. *Let* (M, d) *be a metric space,* $f\colon M \to M$ *a continuous map,* $C \subset M$ *a compact subset disjoint from* $\text{Fix}(f^n)$. *Then there exists an* $\varepsilon > 0$ *such that for any map* $g\colon M \to M$ *satisfying* $g(x) = f(x)$ *for* $x \notin C$ *and* $d(f(x), g(x)) < \varepsilon$ *for* $x \in C$, *the equality* $\text{Fix}(g^n) = \text{Fix}(f^n)$ *holds.*

PROOF. It is enough to notice that $d(x, f^n(x)) > 0$, for $x \in C$ (compact set), implies $\inf\{d(x, f^n(x)) : x \in C\} > 0$. □

(2) The homotopy does not send its carrier back to itself.

(4.5) LEMMA. *Suppose that there exist sets* $A_0, \dots, A_k \subset X$ *such that* $f(A_i) \subset A_{i+1}$ $(i = 0, \dots, k-1)$ *and moreover,* $A_1 \cup \dots \cup A_k$ *is disjoint from* A_0. *Suppose that* $\text{Supp}\{f_t\} \subset A_0$, $f_0 = f$ *and* $f_t(A_0) \subset A_1$. *Then* $\text{Fix}(f_0^k) = \text{Fix}(f_1^k)$.

PROOF. Since $A_1 \cup \dots \cup A_k$ is disjoint from $\text{Supp}\{f_t\} \subset A_0$,

$$f_t^k(a) = f^{k-1}(f_t(a)) \in f^{k-1}A_1 \subset A_k \quad \text{for any } a \in A_0.$$

This implies $f_t^k(a) \neq a$ for any point from the support of the homotopy. Thus the k-periodic points of both maps are lying outside the carrier of the homotopy hence they must be equal. □

(3) If we have to change f near the periodic points we will need a formula to control the set of periodic points. Since majority of homotopies $\{f_t\}$ will be constant on $\text{Fix}(f_0^k)$ we will use the following lemma.

(4.6) LEMMA. *Let $f_t: X \to X$ be a homotopy constant on* $\mathrm{Fix}(f_0^k)$. *Then*

$$\mathrm{Fix}(f_1^k) = \mathrm{Fix}(f_0^k) \cup \{\textit{orbits of } f_1^k \textit{ cutting } \mathrm{Supp}\{f_t\}\}.$$

PROOF. $\supset$ is obvious since $f_1^k(a) = f_0^k(a) = a$ for any $a \in \mathrm{Fix}(f_0^k)$.

$\subset$ Let $\{x_0, \dots, x_{k-1}\}$ be an orbit in $\mathrm{Fix}(f_1^k)$. If the orbit does not cut $\mathrm{Supp}\{f_t\}$ then $f_1(x_i) = f_0(x_i)$ for all $i = 0, \dots, k-1$ hence it is also the orbit in $\mathrm{Fix}(f_0^k)$.□

We will also need a lemma making the inverse images of a neighbourhood of a point small in the homotopy sense explained in the next lemma.

(4.7) LEMMA. *Let $f: M \to M$ be a self map of a compact manifold and let the iterations $x_0, fx_0, \dots, f^k x_0$ be different. Then f is homotopic to a map f_1 whose iterations $f_1, f_1^2, \dots, f_1^k$ are transverse to x_0. Moreover,*

(4.7.1) *the homotopy may be arbitrarily small,*

(4.7.2) *if the iterations of f are already transverse to x_0 on a closed subset $A \subset M$ then the homotopy may be constant in a neighbourhood of A,*

(4.7.3) *the support of the homotopy may be contained in a prescribed neighbourhood of $x_0 \cup f^{-1}(x_0) \cup \ldots \cup f^{-k}(x_0)$,*

(4.7.4) *in particular the homotopy may be constant in a neighbourhood of* $\mathrm{Fix}(f^k)$.

PROOF. We notice that the sets x_0, $f^{-1}(x_0), \dots, f^{-k}(x_0)$ are mutually disjoint. If fact if $0 \le i < j \le k$ and $y \in f^{-i}(x_0) \cap f^{-j}(x_0)$ then $x_0 = f^j(y) = f^{j-i}(f^i(y)) = f^{j-i}(x_0)$ which contradicts that the first k elements of the orbit of x_0 are different. We will denote $x_i = f^i(x_0)$. Since the below deformation can be arbitrarily small we may assume that the values $x_0, \dots, x_k$ remain different hence the sets x_0, $f_t^{-1}(x_0), \dots, f_t^{-k}(x_0)$ are disjoint in each moment of this homotopy.

A small deformation constant outside a prescribed neighbourhood of $f^{-1}(x_0)$ makes f transverse to x_0.

Assume that $f, \dots, f^{l-1}$ are transverse to x_0. We may correct this map (after a homotopy with the support in a prescribed neighbourhood of $f^{-l}(x_0)$) to a map transverse to $x_0, f^{-1}(x_0), \dots, f^{-(l-1)}(x_0)$. This is possible since the considered inverse images are mutually disjoint. Now the obtained map satisfies the main claim of the lemma for $k = l$ hence the inductive step is done.

The points (4.7.1)–(4.7.3) can be satisfied since any correction to a transverse map may be arbitrarily small and its image may be constant outside a prescribed neighbourhood of the corresponding inverse image. □

(4.8) COROLLARY. *Let the first k iterations of the map $f: M \to M$ be transverse to x_0 and let the sets x_0, $f^{-1}(x_0), \dots, f^{-k}(x_0)$ be mutually disjoint. Then for a sufficiently small ball neighbourhood of $U_0 \ni x_0$ the sets $U_0, f^{-1}(U_0), \dots, f^{-k}(U_0)$ are* *disjoint and each $f^{-i}(U_0)$ splits into the finite sum of connected components each of them mapped by f^i homeomorphically onto U_0.*

PROOF. By the compactness of M and the transversality, the sum $f^{-1}(x_0) \cup \ldots \cup f^{-k}(x_0)$ is finite and each its point $z \in f^{-i}(x_0)$ admits a neighbourhood U_z which is mapped by f^i homeomorphically onto a neighbourhood of x_0. Thus the finite intersection $V = \bigcap_{i=0}^{k} \bigcap_{z \in f^{-i}(x_0)} f^i(U_z)$ is a neighbourhood of x_0. On the other hand x_0 does not belong to the compact set $\bigcup_{i=0}^{k} f^i(M \setminus \bigcup_{z \in f^{-i}(x_0)} U_z)$ hence there is a ball neighbourhood U_0 ($x_0 \in U_0 \subset V$) disjoint from this sum. It remains to prove that U_0 satisfies our corollary. Let $z' \in f^{-i}(U_0)$. Then $f^i(z') \in U_0$ hence $f^i(z') \notin f^i(M \setminus \bigcup_{z \in f^{-i}(x_0)} U_z)$. Now $z' \notin M \setminus \bigcup_{z \in f^{-i}(x_0)} U_z$ so $z' \in U_{z''}$ for a $z'' \in f^{-i}(x_0)$. By the above $f^i \colon U_{z''} \to f^i(U_{z''})$ is a homeomorphism, $f^i(U_{z''}) \supset V \supset U_0$ and $U \backslash 0$ is connected. Let U_0'' denote the connected component of $z'' \in f^{-i}$. Then the restriction of f^i to $U_0^{''} \subset U_{z''}$ is a homeomorphism onto $U_0 \subset f^i(U_{z''})$. □

(4.9) COROLLARY. *Let $f \colon M \to M$ be a map with $\mathrm{Fix}(f^l)$ finite and let $x_0, f(x_0), \ldots, f^l(x_0)$ be different. Then f is homotopic to a map g and there is a neighbourhood $U \ni x_0$ such that $\mathrm{cl}\, U, g^{-1}(\mathrm{cl}\, U), \ldots, g^{-l}(\mathrm{cl}\, U)$ is contained inside a finite number of mutually disjoint closed balls in M. Moreover,*

(4.9.1) *the homotopy may be arbitrarily small,*
(4.9.2) *the support of the homotopy may be contained inside a prescribed neighbourhood of $x_0, f^{-1}(x_0), \ldots f^{-l}(x_0)$,*
(4.9.3) $\mathrm{Fix}(f^k)$ *does not change during the homotopy,*
(4.9.4) *the corollary is stable in the following sense: if g satisfies the corollary then so also does any map sufficiently close to g.*

PROOF. Let U_0 be the neighbourhood from Corollary (4.8). Then we may take as U any ball satisfying $x_0 \in U \subset \mathrm{cl}\, U \subset U_0$. Now the property (4.9.4) follows from the compactness of M. □

4.2. Making $f^k(\omega)$ close to ω. The crucial step in the proof of the Cancelling Procedure (Theorem (4.3)) is the following theorem which makes the path $f^k\omega$ close to ω. This step corresponds to the reduction to the Euclidean case in the proof of the classical Wecken Theorem. For $V \subset \mathbb{R}^{m-1} \times \mathbb{R}$ we denote

$$V^+ = \{(x,t) \in V : t > 0\},\ V^- = \{(x,t) \in V : t < 0\},\ V^0 = \{(x,t) \in V : t = 0\}.$$

(4.10) THEOREM. *Under the assumption of Cancelling Procedure (Theorem (4.3)) there exists a homotopy which does not change $\mathrm{Fix}(f^k)$, is constant in a neighbourhood of $\mathrm{Fix}(f^k)$ and after which the map f satisfies:*

(4.10.1) *a path* $\omega_0\colon [-1,+1] \to M$ *establishing the Nielsen relation between* x_0 *and* y_0 *in* $\mathrm{Fix}(f^k)$ *is a* PL-*arc avoiding other periodic points,*

(4.10.2) *there exist mutually disjoint Euclidean neighbourhoods* V_0' $'_{k-1}$ *such that, for* $i = 0, \dots, k-2$,

$$f^i\omega(t) = (0,t) \in V_i' = \mathbb{R}^{m-1} \times \mathbb{R} \quad (i = 0, \dots, k-1)$$

$$f(V_i'^{+}) \subset (V_{i+1}'^{+}), \quad f(V_i'^{-}) \subset (V_{i+1}'^{-}), \quad f((V_i')^0) = 0 \in V_{i+1}' = \mathbb{R}^m.$$

(4.10.3) *the restriction of* f *to* $V_i' \setminus (V_i')^0$ *is a homeomorphism on its image* ($i = 0, \dots, k-2$).

(4.10.4) *there exists a Euclidean neighbourhood* W *such that* $V_0' \subset W$, $f(V_{k-1}') \subset W$ *and the restriction of* f^{k-1} *to* $W \setminus W^0$ *is a homeomorphism.*

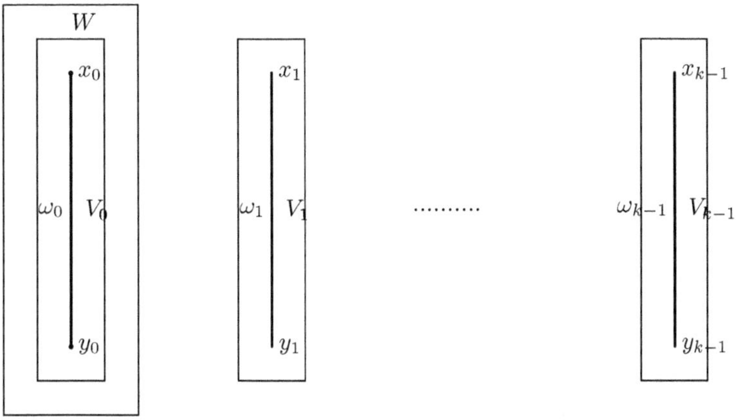

FIGURE 8

The proof of Theorem (4.10) will be given after Lemma (4.16) and will follow from a sequence of lemmas. Let $f\colon M \to M$ satisfy the assumptions of Theorem (4.10) (i.e. of the Cancelling Procedure).

(4.11) LEMMA. *Under the assumption of Cancelling Procedure, there is a homotopy* $\{f_t\}$ *starting from* $f_0 = f$ *and an arc* $\omega_0\colon [-1,+1] \to M$ *from* $\omega_0(-1) = x_0$ *to* $\omega_0(+1) = y_0$ *satisfying* $f^k\omega_0 \sim \omega_0$ *and:*

(4.11.1) $\omega_0, \omega_1 = f\omega_0, \dots, \omega_S = f^S\omega_0$ $(S = 3k-1)$ *are* PL-*arcs whose interiors,* $\{\omega_i(-1,+1)\}$ *are mutually disjoint and disjoint from* $\mathrm{Fix}(f^k)$,

(4.11.2) $\{f_t\}$ *is constant in a prescribed neighbourhood of* $\mathrm{Fix}(f^k)$,

(4.11.3) $\{f_t\}$ *can be arbitrarily small,*

(4.11.4) $\mathrm{Fix}(f_1^k) = \mathrm{Fix}(f^k)$.

PROOF. Let us concentrate on proving (4.11.1). Since $\mathrm{Fix}(f^k)$ is finite, ω_0 may be chosen a PL-arc such that $\omega_0(t) \notin \mathrm{Fix}(f^k)$ for $-1 < t < +1$. Since f is a linear homeomorphism in neighbourhoods of x_0 and y_0, there exists an $\varepsilon > 0$ such that $f\omega_0[-1, -1+\varepsilon]$, $f\omega_0[1-\varepsilon, 1]$ are segments in the corresponding Euclidean neighbourhoods. Since $\dim M \geq 3$, there exists a small homotopy (rel. ends) on $\omega_0[-1+\varepsilon, 1-\varepsilon]$ after which the path $f(\omega_0[-1+\varepsilon, 1-\varepsilon])$ becomes also PL-arc and moreover, the homotopy may be extended on the whole M by a homotopy with the carrier in a prescribed neighbourhood of $\omega_0[-1+\varepsilon, 1-\varepsilon]$. Thus we may assume that $\omega_1 = f(\omega_0)$ is also an arc and is disjoint from $\omega_0[-1, +1]$ (Lemma (4.4)).

We repeat the above construction to the arc ω_1 and we get that ω_2 is also an arc disjoint from ω_0 and ω_1. Thus we may assume that $\omega_0, \dots, \omega_{k-1}$ are mutually disjoint arcs. Let us recall that f is still a PL-homeomorphism in neighbourhoods of the points $\{x_0, \dots, x_{k-1}; y_0, \dots, y_{k-1}\}$. We may continue this procedure to make the arcs $f^k[-1+\varepsilon, 1-\varepsilon], \dots, f^S[-1+\varepsilon, 1-\varepsilon]$ mutually disjoint. Since f is a PL-homeomorphism near the points $\{x_i; y_j\}$ and $\varepsilon > 0$ may be arbitrarily small, we may assume that $i = j$ and $f^i\omega_0(t) = f^j\omega_0(s)$ imply $k|(j-i)$ and $t = s = \pm 1$ for $i, j = 0, \dots, S = 3k-1$, see Figure 9.

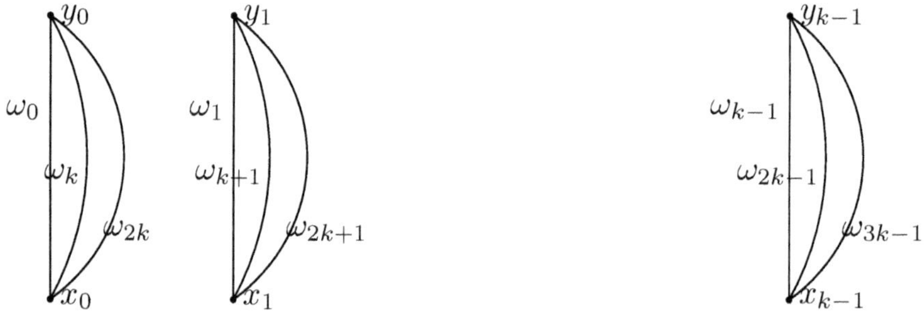

FIGURE 9

It remains to notice that all the above deformations have the carrier isolated from $\mathrm{Fix}(f^k)$. Thus we have (4.11.2). On the other hand these deformations may be arbitrarily small. Now (4.11.3) implies (4.11.4) — see Lemma (4.4). □

Let us denote $z_i = \omega_i(0)$ for $i = 0, \dots, k-1$.

(4.12) LEMMA. *Under the assumptions of Cancelling Procedure. There is a homotopy $\{f_t\}$ starting from $f_0 = f$ satisfying:*

(4.12.1) *$\{f_t\}$ is constant in a prescribed neighbourhood of* $\mathrm{Fix}(f^k)$,

(4.12.2) *$\{f_t\}$ can be arbitrarily small,*

(4.12.3) *there exists a neighbourhood $U_{k-1} \ni z_{k-1}$ such that the sum*

$$\mathrm{cl}\, U_{k-1} \cup f_1^{-1}(\mathrm{cl}\, U_{k-1}) \cup \ldots \cup f_1^{-(k-1)}(\mathrm{cl}\, U_{k-1})$$

is contained inside a finite sum of mutually disjoint m-balls,

(4.12.4) *the interiors of the arcs ω_i are mutually disjoint for $i = 0, \dots, 2k-1$.*

PROOF. We may assume that $f = f_1$ from Lemma (4.11). Since ω_i are mutually disjoint arcs, we may choose mutually disjoint Euclidean neighbourhoods V_i where $\omega_i(t) = (0,t) \in V_i = \mathbb{R}^{m-1} \times \mathbb{R}$ $(i = 0, \dots, k-1)$. Since the points $z_i = \omega_i(0)$ $(i = 0, \dots, k-1)$ do not belong to $\mathrm{Fix}(f^k)$, we may deform f in small neighbourhoods U_i $(z_i \in U_i \subset V_i = \mathbb{R}^m)$ so that f takes there the form

$$U_i \ni (x,t) \to (x,t) \in V_{i+1} \quad \text{for } (x,t) \in \mathbb{R}^{m-1} \times \mathbb{R} \text{ and } i = 0, \dots, k-2.$$

Now lemma follows from Corollary (4.9) (for $x_0 = z_{k-1}$ and $l = k-1$). □

(4.13) LEMMA. *Under the assumptions of Cancelling Procedure. There is a homotopy $\{f_t\}$ starting from $f_0 = f$ and satisfying:*

(4.13.1) *there exist mutually disjoint Euclidean neighbourhoods $V_0, \dots, V_{k-1} \subset M$ where $\omega_i(t) = (0,t) \in V_i = \mathbb{R}^{m-1} \times \mathbb{R}$ for $-1 \le t \le +1$ and $i = 0, \dots, k-1$,*

(4.13.2) *$f_1(V_i^+) \subset V_{i+1}^+$, $f_1(V_i^-) \subset V_{i+1}^-$, $f_1(V_i^0) = z_{i+1} \in V_{i+1}^0$ for $i = 0, \dots, k-2$,*

(4.13.3) *the restriction of f_1 to $V_i \setminus V_i^0$ is a homeomorphism on the image,*

(4.13.4) *$\{f_t\}$ is constant in a prescribed neighbourhood of* $\mathrm{Fix}(f^k)$,

(4.13.5) *$\{f_t\}$ can be arbitrarily small,*

(4.13.6) $\mathrm{Fix}(f_t^k)$ *does not depend on t,*

(4.13.7) *there exists a neighbourhood $U_{k-1} \ni z_0$ such that the sum*

$$\mathrm{cl}\, U_{k-1} \cup f_1^{-1}(\mathrm{cl}\, U_{k-1}) \cup \dots \cup f_1^{-(k-1)}(\mathrm{cl}\, U_{k-1})$$

is contained inside a finite sum of mutually disjoint m-balls.

PROOF. We may assume that $f = f_1$ from Lemma (4.12) We will correct f to make it a homeomorphism near $\omega_i[-1,0)$ and $\omega_i(0,+1]$ for $i = 0, \dots, k-2$. We fix disjoint Euclidean neighbourhoods of ω_0 and $\omega_1 = f\omega_0$ where $\omega_0(t) = (0,t) \in \mathbb{R}^{m-1} \times \mathbb{R}$, $f\omega_0(t) = (0,t) \in \mathbb{R}^{m-1} \times \mathbb{R}$, respectively.

We fix m-simplices σ_x, σ_y containing x_0 and y_0 respectively on which f is a homeomorphism. Moreover, we assume that $(m-1)$-dimensional faces $\sigma_{-1+\varepsilon} \subset \sigma_x$, $\sigma_{1-\varepsilon} \subset \sigma_y$ belong to the hyperplanes $x_m = -1+\varepsilon$ and $x_m = 1-\varepsilon$, respectively (see the Figure 10).

Let $p' \colon \mathbb{R}^m \to \mathbb{R}^{m-1}$, $p'' \colon \mathbb{R}^m \to \mathbb{R}$ denote the projections $p'(x,t) = x$, $p''(x,t) = t$ for $(x,t) \in \mathbb{R}^{m-1} \times \mathbb{R} = \mathbb{R}^m$. Let σ_0 be an $(m-1)$-simplex contained in $p'(\sigma_{-1+\varepsilon}) \cap p'(\sigma_{1-\varepsilon})$ and let $0 \in \mathrm{int}\sigma_0$. If σ_0 is chosen small enough then $p''(f(x, -1+\varepsilon)) <$

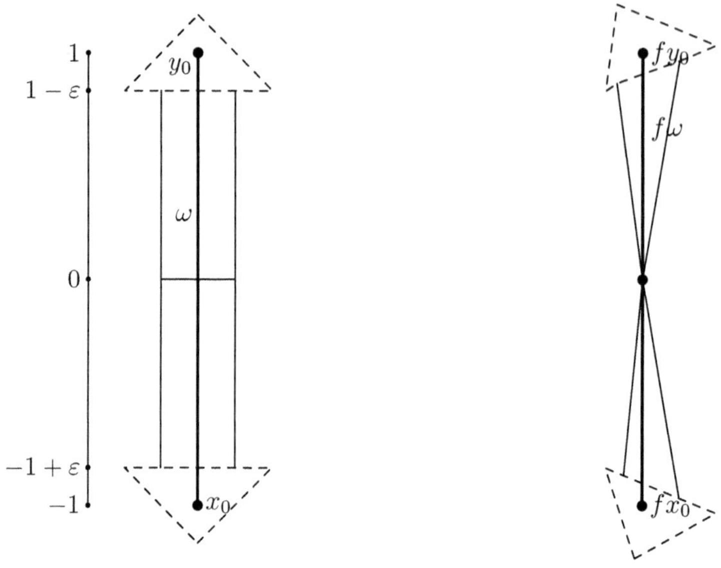

FIGURE 10

$0 < p''(f(x, 1-\varepsilon))$ for $x \in \sigma_0$. We define the map $f'\colon \sigma_0 \times [-1+\varepsilon, 1-\varepsilon] \to \mathbb{R}^m$ by the formula

$$f'(x,t) = \begin{cases} \dfrac{t}{1-\varepsilon} f(x, 1-\varepsilon) & \text{for } 0 \leq t \leq 1-\varepsilon, \\ \dfrac{t}{-1+\varepsilon} f(x, -1+\varepsilon) & \text{for } -1+\varepsilon \leq t \leq 0. \end{cases}$$

Then the restrictions of f' to $\sigma_0 \times [-1+\varepsilon, 0)$ and to $\sigma_0 \times (0, 1-\varepsilon]$ are homeomorphisms. Let us fix a number $\eta > 0$. If the simplex σ_0 is small enough then $d(f'(x,t), f(x,t)) < \eta$ for $(x,t) \in \sigma_0 \times [-1+\varepsilon, 1-\varepsilon]$. Moreover, the homotopy from f to f' (by segments) is still η-homotopy and admits an extension onto M which is constant outside a prescribed neighbourhood of $\sigma_0 \times [-1+\varepsilon, 1-\varepsilon]$. Thus we may assume that after this homotopy no new periodic point (of minimal period $l \leq k$, $l|n$) appears (Lemma (4.4)).

On the other hand $f'(x, 1-\varepsilon) = f(x, 1-\varepsilon)$, $f'(x, -1+\varepsilon) = f(x, -1+\varepsilon)$ for $x \in \sigma_0$ so we may assume that the homotopy is constant on σ_x, σ_y. For a sufficiently small number $\varepsilon' > 0$; $\sigma_0 \times [1-\varepsilon', 1+\varepsilon'] \subset \sigma_y$, $\sigma_0 \times [-1-\varepsilon', -1+\varepsilon'] \subset \sigma_x$ hence the restriction of f to $\sigma_0 \times (0, 1+\varepsilon']$ and to $\sigma_0 \times [-1-\varepsilon', 0)$ is a homeomorphism. Finally $V_0 = \text{int}\sigma_0 \times (-1-\varepsilon', 1+\varepsilon')$ is the desired Euclidean neighbourhood.

In general we proceed as follows: we fix a Euclidean neighbourhood V_{k-1} of ω_{k-1} (in which $\omega_{k-1}(t) = (0,t)$). The above construction gives a neighbourhood $V_{k-2} \supset \omega_{k-2}$ such that $f\colon V_{k-2} \backslash V^0_{k-2} \to V_{k-1}$ is the homeomorphism on the image. Then we choose V_{k-3} and so on. The obtained map satisfies the points (4.13.1) and (4.13.2). The points (4.13.3) and (4.13.4) follow from the construction of V_i. Since these sets may be arbitrarily thin, (4.13.5) is satisfied which in turn implies (4.13.6). Since the homotopy may be small, (4.13.7) follows from (4.9.4). □

For a fixed number $\varepsilon > 0$ we put $V_{k-1}^{\varepsilon} = \{(x,t); -1+\varepsilon \le t \le 1-\varepsilon\}$.

(4.14) REMARK. $f(V_{k-1}^{\varepsilon}), \dots, f^k(V_{k-1}^{\varepsilon})$ are mutually disjoint for a sufficiently thin V_{k-1}, because $\omega_k, \dots, \omega_{2k-1}$ are mutually disjoint.

The assumption of the next lemma differs from the earlier ones; the path ω_0 exceptionally does not join periodic points. This lemma will be applied for $a = 1-\varepsilon$ at the end of the proof of Theorem (4.10).

(4.15) LEMMA. *Let $f\colon M \to M$ be a self-map of a compact* PL*-manifold of dimension* ≥ 3. *We assume that:*

(4.15.1) $\mathrm{Fix}(f^k)$ *is finite,*
(4.15.2) $\omega_0\colon [-a,a] \to M$ *is a* PL*-arc such that* $\omega_i = f^i\omega_0$ *for* $i = 0, \dots, 2k-1$ *are mutually disjoint arcs, all disjoint from* $\mathrm{Fix}(f^k)$,
(4.15.3) $V_0 \subset M$ *is a Euclidean neighbourhood such that* $\omega_0(t) = (0,t) \in \mathbb{R}^{n-1} \times \mathbb{R} = V_0$,
(4.15.4) $\omega_k(-a), \omega_k(+a) \in V_0$ *and* ω_k *is homotopic (in* M, *rel. end points) to a path lying in* V_0,
(4.15.5) *there exists a neighbourhood* $U_{k-1} \ni z_{k-1} = \omega_{k-1}(0)$ *such that the sum*

$$\mathrm{cl}\, U_{k-1} \cup f_1^{-1}(\mathrm{cl}\, U_{k-1}) \cup \ldots \cup f_1^{-(k-1)}(\mathrm{cl}\, U_{k-1})$$

is contained inside a finite sum of mutually disjoint m-balls.

Then there is a partial homotopy $h_s\colon \omega_{k-1}[-a,a] \to M$ $(0 \le s \le 2)$ *satisfying:*

(4.15.6) $h_0 = f_{|\omega_{k-1}}$, h_2 *is a path in* $V_0 \setminus \omega[-a,a]$,
(4.15.7) h_s *is constant at the ends* $\omega_{k-1}(-a)$ *and* $\omega_{k-1}(-a)$,
(4.15.8) $f^i h_s \omega_{k-1}[-a,a]$ *is disjoint from* $\omega_{k-1}[-a,+a]$ *for* $i = 0, \dots, k-1$,
(4.15.9) *in particular* $f^i h_s(x) \ne x$ *for* $x \in \omega_{k-1}[-a,+a]$ *for* $0 \le s \le 2$ *and* $i = 0, \dots, k-1$.

PROOF. The partial homotopy $\{h_s\}$ will be obtained in two steps: reparametrization (for $0 \le t \le 1$) and the contraction of the path $f\omega_{k-1}$ to V_0 (for $1 \le t \le 2$).

Step 1. Let $\eta > 0$ be so small that $0 \times [-\eta, \eta] \subset U_{k-1} \subset \mathbb{R}^{m-1} \times \mathbb{R} = V_{k-1}$ (U_{k-1} is the neighbourhood from assumption (4.15.5)). Let $r\colon [-a,a] \to [-a,a]$ be the map given by Figure 11 and let $r_t\colon \omega_{k-1}[-a,a] \to \omega_{k-1}[-a,a]$ be the segment homotopy from $r_0 = \mathrm{id}_{\omega_{k-1}[-a,a]}$ to $r_1(\omega_{k-1}(t)) = \omega_{k-1}(r(t))$ (*reparametrization of the path* ω_{k-1}). We define the partial homotopy $h_t\colon \omega_{k-1}[-a,a] \to M$ putting

$$h_t(x) = f r_t(x) \quad \text{for } x \in \omega_{k-1}[-a,a].$$

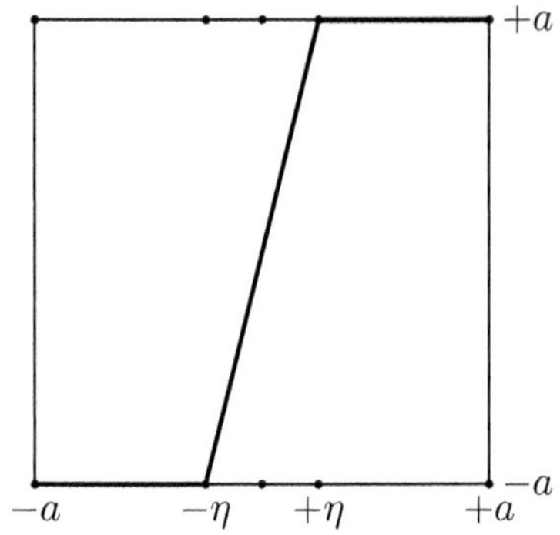

FIGURE 11

Since $f^i h_s \omega_{k-1}(x) \in \omega_{k+i}[-a,+a]$ and the arcs ω_j are mutually disjoint for $j = 0 \quad 2k-1$, $f^i h_s \omega_{k-1}(x) = x$ for $x \in \omega_{k-1}$, $i = 0, \dots, k-1$. Notice that $r_1(\omega_{k-1}[-a,-\eta])$, $r_1(\omega_{k-1}[\eta,a])$ are points hence so are $h_1(\omega_{k-1}[-a,-\eta])$, $h_1(\omega_{k-1}[\eta,a])$.

Step 2. The homotopy h_s (for $1 \le s \le 2$) will be constant on $\omega_{k-1}[-a,-\eta]$ and on $\omega_{k-1}[\eta,a]$. We are going to define this homotopy on $\omega_{k-1}[-\eta,\eta]$. Let us fix a path $\overline{\omega}\colon [-a,+a] \to V_0 \setminus \omega_0[-a,a]$.

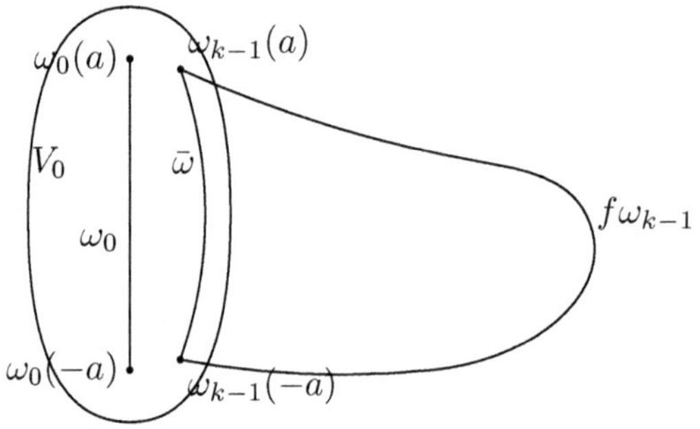

FIGURE 12

Since $f\omega_{k-1}$ is homotopic to a path $\overline{\omega}\colon \omega_{k-1}[-a,a] \to V_0$, there is a homotopy between the maps $h_1, \overline{\omega} r_1$ considered as the maps from $\omega_{k-1}[-\eta,\eta]$ to M. We will show that the last homotopy may avoid the set

$$\operatorname{cl} U_{k-1} \cup f_1^{-1}(\operatorname{cl} U_{k-1}) \cup \ldots \cup f_1^{-(k-1)}(\operatorname{cl} U_{k-1}).$$

In fact by the assumption (4.15.5) the sum is contained in a finite number of the disjoint balls $\bigcup_\alpha K_\alpha$. Since $\dim M \ge 3$,

$$\pi_2\Big(M, M \setminus \bigcup_\alpha K_\alpha\Big) = \pi_2(M, M \setminus \text{finite set}) = 0$$

so the image of the (two dimensional homotopy) $h_s\colon \omega_{k-1}[-\eta,+\eta]\times[1,2]\to M$ may be deformed (rel. boundary of $[-\eta,\eta]\times[1,2]$) outside $\bigcup_\alpha K_\alpha$. Thus we may assume that $h_s(\omega_{k-1}(t))\notin\bigcup_{i=0}^{k-1}f^{-i}(\operatorname{cl}U_{k-1})$. Since $\omega_{k-1}[-\eta,+\eta]\subset U_{k-1}$ we get (4.15.8) (for $x\in\omega_{k-1}[-\eta,+\eta]$). We extend the homotopy h_t onto $\omega_{k-1}[-a,+a]$ by the constant homotopy outside $\omega_{k-1}[-\eta,+\eta]$ and $1\le t\le 2$. Now the homotopy $\{h_t\}$ for $0\le t\le 2$ satisfies the lemma. □

(4.16) LEMMA. *The partial homotopy from Lemma* (4.15) *can be extended to* $f_t\colon M\to M$ *where*

(4.16.1) *the carrier of* $\{f_t\}$ *is contained in* $D\times[-a,+a]\subset V_{k-1}=\mathbb{R}^{m-1}\times\mathbb{R}$ *where* D *is any prescribed neighbourhood of* $0\in\mathbb{R}^{m-1}$,

(4.16.2) $\operatorname{Fix}(f_1^k)=\operatorname{Fix}(f^k)$.

PROOF. Let $D\subset\mathbb{R}^{m-1}$ be a closed ball centered in 0. Take an arbitrary extension f'_t of the partial homotopy $\{h_t\}$ on M. We consider the metric space $X=M\setminus\{\omega_{k-1}(-a),\omega_{k-1}(+a)\}$. Then the sets $X\setminus D\times[-a,+a]$ and $0\times(-a,+a)$ are disjoint closed subsets of X. Let $\lambda\colon X\to[0,1]$ be an Urysohn function satisfying $\lambda(X\setminus D\times[-a,+a])=0$, $\lambda(0\times(-a,+a))=1$.

Then then map $f_t\colon M\to M$

$$f_t(x)=\begin{cases} f'_{\lambda(x)t}(x) & \text{for } x\neq\omega_{k-1}(\pm a),\\ f(x) & \text{for } x=\omega_{k-1}(\pm a),\end{cases}$$

gives a homotopy satisfying (4.16.1) (f_t is continuous in the points $\omega_{k-1}(-a)$ and $\omega_{k-1}(+a)$ since the homotopy f'_t is constant there).

It remains to show that if D is small enough then $\operatorname{Fix}(f_2^k)=\operatorname{Fix}(f^k)$. Suppose otherwise. Let D_n be a ball of radius $1/n$. Now we have $x_n\in D_n$ and $0\le t_n\le 2$ such that $f'^k_{t_n}(x_n)=x_n$. The compactness gives subsequences convergent to $x_0\in 0\times[-a,+a]$ and $0\le t_0\le 2$. Then $f^k_{t_0}(x_0)=x_0$ contradicts (4.15.4). □

END OF THE PROOF OF THEOREM (4.10). We may assume that f satisfies Lemma (4.13). Then (4.10.1)–(4.10.3) are satisfied. We will get (4.10.4) changing f only near $\omega_{k-1}[-1+\varepsilon,1-\varepsilon]$. Let $\varepsilon>0$ be so small that $f^k(\omega_0([-1,-1+\varepsilon]\cup[1-\varepsilon,1]))\subset V_0$. We may apply Lemmas (4.15) and (4.16) to $a=1-\varepsilon$ and we get $f^k(\omega_0)\subset V_0$. Now we put $W=V_0$ and we find $V'_{k-1}\subset V_{k-1}$ satisfying Lemma (4.13) and so thin that $fV'_{k-1}\subset W$. Then we find $V'_{k-2}\subset V_{k-2}$ satisfying Lemma (4.13) and $fV'_{k-2}\subset V'_{k-1}$ and so on until we get $V'_0\subset V_0$.

(4.17) REMARK. Since the only condition on the path $\overline{\omega}$ in Lemma (4.15) was $\overline{\omega}\subset V_0\setminus\omega_0[-a,+a]$, we may assume that $\overline{\omega}(0)\in V_0^-$. Then $f_2(z_{k-1})=\overline{\omega}(0)\in V_0^-$.

This ends the first step of the proof of the Cancelling Procedure: reduction to the local situation. □

4.3. Extension of the partial homotopy. We will show that the map f satisfying Theorem (4.10) may be deformed so that f^k is given, near the arc $\omega_0[-1,+1]$, by a prescribed formula. We will do it first locally and then we will need an extension of this local deformation onto whole M without new periodic points.

We may assume that $f: M \to M$ satisfies Theorem (4.10). We consider the orbits $\{x_0, \dots, x_{k-1}\}$, $\{y_0, \dots, y_{k-1}\}$. Since f is a PL-homeomorphism in a neighbourhood of each periodic point and $\operatorname{ind}(f^k; y_0) + \operatorname{ind}(f^k; x_0) = 0$, we may assume that $\operatorname{ind}(f^k; x_0) = +1$, $\operatorname{ind}(f^k; y_0) = -1$.

Let us denote $P = [-2,+2]^m$ and consider the map: $h: P \to \mathbb{R}^m$ given by the formula

$$h(x,t) = \left(\frac{|t|}{3} \cdot x, \eta(t)\right)$$

where $(x,t) \in \mathbb{R}^{m-1} \times \mathbb{R}$ and $\eta: [-2,2] \to \mathbb{R}$ is a function satisfying: $\eta(t) = t$ if and only if $t = \pm 1$, $\eta(-2) > -2$, $\eta(0) < 0$, $\eta(2) > 2$.

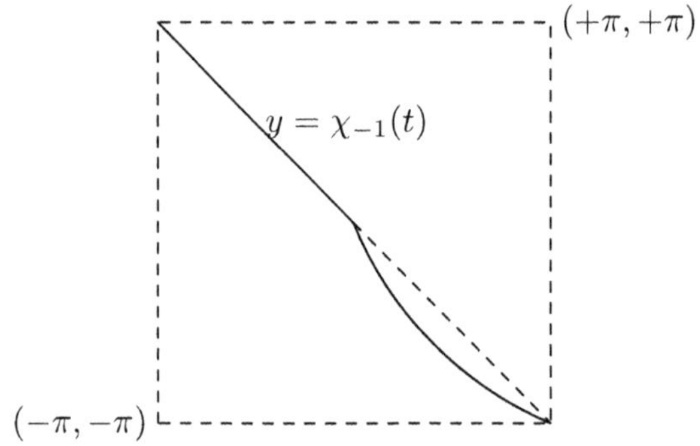

FIGURE 13

Now $\operatorname{Fix}(h) = \{(0,-1), (0,+1) \in \mathbb{R}^{m-1} \times \mathbb{R}\}$. We identify $P = [-2,2]^m$ with a subset of $V_0 = \mathbb{R}^m$. Then $\operatorname{ind}(h, x_0) = +1 = \operatorname{ind}(f^k; x_0)$, $\operatorname{ind}(h, y_0) = -1 = \operatorname{ind}(f^k; y_0)$.

Let us notice that putting $P_0 = [-2,2]^{m-1} \times 0$ both $h(P_0)$, $f^k(P_0)$ are points. We are going to show that f is homotopic to a map such that f^k equals h near the path ω_0.

We start by recalling a classical result. Let us denote $Q = [0,1]^m$ and $Q_0 = \{(x_1, \dots, x_m) \in Q : x_m = 0\}$

(4.18) THEOREM (Hopf). *Let $f_0, f_1: Q \to \mathbb{R}^m$ be maps satisfying $f_0(z) \neq z \neq f_1(z)$ for $z \in \operatorname{bd} Q$ and $\operatorname{ind}(f_0) = \operatorname{ind}(f_1)$. Then there is a homotopy $h: Q \times I \to \mathbb{R}^m$ satisfying $h(z,i) = f_i(z)$ for $i = 0,1$, $z \in Q$ and $h(z,t) \neq z$ for $t \in I$, $z \in \operatorname{bd} Q$.* □

We will need its modification:

(4.19) LEMMA. *Let us add to the assumptions of the above Hopf Theorem* (4.18) *the following three conditions:*

(4.19.1) $f_0(Q_0)$ *and* $f_1(Q_0)$ *are points,*
(4.19.2) $\gamma\colon [0,1] \to \mathbb{R}^m \setminus Q_0$ *is a path from* $\gamma(0) = f_0(Q_0)$ *to* $\gamma(1) = f_1(Q_0)$,
(4.19.3) $\dim M = m \geq 3$.

Then the homotopy h *in Hopf Lemma* (4.18) *may be chosen to satisfy* $h(z,t) = \gamma(t)$ *for* $z \in P_0$, $t \in I$.

PROOF. We define the map $H_1\colon Q \times 0 \cup Q \times 1 \cup Q_0 \times I \to \mathbb{R}^m$ by the formula:

$$H_1(z,t) = \begin{cases} z - f_0(z) & \text{for } t = 0, \\ z - f_1(z) & \text{for } t = 1, \\ z - \gamma(t) & \text{for } z \in Q_0. \end{cases}$$

Then $\deg H_1 = \mathrm{ind}(f_0) - \mathrm{ind}(f_1) = 0$ hence there is an extension $H_2\colon \mathrm{bd}(Q \times I) \to \mathbb{R}^m$ such that $H_2^{-1}(0) = H_1^{-1}(0)$. Let $H_3\colon Q \times I \to \mathbb{R}^m$ be an arbitrary extension of H_2. Now $h(z,t) = z - H_3(z,t)$ is the desired homotopy. □

We will use the following notation:

$$P = [-2,+2]^m \subset \mathbb{R}^m, \qquad P_0 = \{(x_1,\dots,x_m) \in P : x_m = 0\},$$
$$P_+ = \{(x_1,\dots,x_m) \in P : x_m \geq 0\}, \qquad P_- = \{(x_1,\dots,x_m) \in P : x_m \leq 0\},$$
$$P_{[a,b]} = \{(x_1,\dots,x_m) \in P : a \leq x_m \leq b\}.$$

(4.20)) LEMMA. *Let the maps* $f, g\colon P \to \mathbb{R}^m$ *satisfy:*

(4.20.1) $f(z) \neq z \neq g(z)$ *for* $z \in P_0 \cup \mathrm{bd}\, P$,
(4.20.2) $f(P_0)$, $g(P_0)$ *are points,*
(4.20.3) $\mathrm{ind}(f; P_+) = \mathrm{ind}(g; P_+)$, $\mathrm{ind}(f; P_-) = \mathrm{ind}(g; P_-)$.

Let $\gamma\colon [0,1] \to (\mathbb{R}^m \setminus P_0)$ *be a path from* $f(P_0)$ *to* $g(P_0)$. *Then there is a homotopy* $h\colon P \times I \to \mathbb{R}^m$ *satisfying:*

(4.20.4) $h(z,0) = f(z)$, $h(z,1) = g(z)$ *for* $z \in P$,
(4.20.5) $h(z,t) \neq z$ *for* $z \in P_0 \cup (\mathrm{bd}\, P)$,
(4.20.6) $h(z,t) = \gamma(t)$ *for* $z \in P_0$.

PROOF. By Lemma (4.19) there is a homotopy $h_+\colon P_+ \times I \to \mathbb{R}^m$ satisfying:

(4.21.1) $h_+(z,0) = f(z)$, $h_+(z,1) = g(z)$ for $z \in P_+$,
(4.21.2) $h_+(P_0,t) = \gamma(t)$, $h_+(z,t) \neq z$ for $z \in \mathrm{bd}\, P_+$.

Then we may define a similar homotopy $h_-\colon P_- \times I \to \mathbb{R}^m$ and $h = h_+ \cup h_-$ will satisfy the lemma. □

The above Lemma implies a homotopy $\{h_t\}$ from $h_0 = f^k$ (more precisely from its restriction to P) to $h_1 = h$ (given by a formula at the beginning of the section) such that P_0 is sent into a point and no point from $\operatorname{bd} P \cup P_0$ is fixed in any moment of this homotopy. This induces the partial homotopy $f_t'\colon f^{k-1}(P) \to M$ by the formula

$$f_t'(x) = h_t(y) \quad \text{for } x = f^{k-1}(y),\ y \in P.$$

In other words

$$f_t'(x) = h_t((f_|^{k-1})^{-1}(x))$$

where $f_|^{k-1}$ denotes the restriction $f^{k-1}\colon P \to V_{k-1} = \mathbb{R}^m$. Since $f_|^{k-1}$ is mono on $P - P_0$ and $h_t(P_0)$ is a point (for any fixed t), the definition is correct. Moreover, we recall that the points $f(z_{k-1}) = f^k(z_0)$ and $h(z_0)$ belong to V_0^- (see Remark (4.17) and recall that $h(0,0) = (0, \eta(0))$ where $\eta(0) < 0$). Thus we may join these two points with a path $\gamma\colon [0,1] \to V_0^-$ and we may assume that$f_t'(f^{k-1}P_0) = \gamma(t)$.

(4.22) Lemma. *The partial homotopy $f_t'\colon f^{k-1}(P) \to M$ admits an extension $f_t\colon M \to M$ with the carrier contained inside any prescribed neighbourhood of $f^{k-1}(P)$ and such that* $\operatorname{Fix}((f_1)^k) = \operatorname{Fix}((f_0)^k)$.

Proof. The Lemma follows from Lemma (4.23) applied for $X = M$, $A = f^{k-1}(P)$. It remains to notice that then $\operatorname{Fix}(f_0^k) \cap f^{k-1}(P) = \operatorname{Fix}(f_1^k) \cap f^{k-1}(P) = \{x_{k-1}, y_{k-1}\}$. □

(4.23) Lemma. *Let X be a compact metric* ANR *and let $A \subset X$ be its closed subset. Let $f\colon X \to X$ be a selfmap let $f_t'\colon A \to X$ be a partial homotopy satisfying:*

(4.23.1) $f_0'(a) = f(a)$ *for* $a \in A$,
(4.23.2) $f^{l-1}(f_t'(A)) \cap A = \emptyset$ *for* $l = 1, \dots, k-1$,
(4.23.3) $f^{k-1}(f_t'(x)) \neq x$ *for* $x \in \operatorname{bd} A$.

Then for arbitrary neighbourhood $U \supset A$ there exists a homotopy $f_t\colon X \to X$ satisfying:

(4.23.4) $f_t(a) = f_t'(a)$ *for* $a \in A$,
(4.23.5) $f_0(x) = f(x)$ *for* $x \in X$,
(4.23.6) $f_t(x) = f(x)$ *for* $x \notin U$,
(4.23.7) *for any $t_0 \in [0,1]$ the orbits of* $\operatorname{Fix}(f_{t_0}^k)$ *avoiding A coincide with the orbits of* $\operatorname{Fix}(f^k)$ *avoiding A.*

In particular if $\operatorname{Fix}(f_{t_0}^k) \cap A = \emptyset$, *for a $t_0 \in [0,1]$, then*

$$\operatorname{Fix}(f_{t_0}^k) = \operatorname{Fix}(f^k) \setminus \{\textit{orbits of } f^k \textit{ cutting } A\}.$$

Proof. Since X is an ANR, the partial homotopy $f_t'\colon A \to A$ with $f_0' = f$ admits an extension $f_t'\colon X \to X$ Then by the assumption (4.23.2) there exists an open neighbourhood U' such that $A \subset U' \subset U$ and $f^{l-1}(f'_t(\mathrm{cl}\, U')) \cap \mathrm{cl}\, U' = \emptyset$ for $l = 1, \dots, k-1$. Let $\mu: X \to [0,1]$ be an Urysohn function satisfying $\mu(A) = 1$, $\mu(M - U') = 0$. Then the map $f''_t(x) = f'_{\mu(x)t}(x)$ satisfies $(f''_t)^l(\mathrm{cl}\, U')) \cap \mathrm{cl}\, U' = \emptyset$ for $l = 1, \dots, k-1$ since for any $x \in \mathrm{cl}\, U'$, $(f''_t)^l(x) = f^{l-1}(f'_{\mu(x)t})(x)$. Let $C = \{x \in X : (f''_t)^k(x) = x \text{ for a } t \in [0,1]\}$. This is a compact subset disjoint from $\mathrm{bd}\, A$ hence $C - A$ is also compact. Let $\lambda: X \to [0,1]$ be an Urysohn function $\lambda(A) = 1$, $\lambda((M - U') \cup ((C - A)) = 0$. We put $f_t(x) = f''_{t\lambda(x)}(x)$. Properties (4.23.4)–(4.23.6) are easy to check. Now we prove (4.23.7). Consider an orbit $\mathcal{O} \subset \mathrm{Fix}(f^k_{t_0})$ disjoint from A. Then $\mathcal{O} \subset C \setminus A$ hence $\lambda(x) = 0$ and the homotopy $\{f_t(x)\}$ is constant for any $x \in \mathcal{O}$. Thus $f_{t_0}(x) = f(x)$ and $\mathcal{O}$ turns out to be an orbit of f. Reversing this argument we prove that any orbit of f avoiding A is also the orbit of in $\mathrm{Fix}(f^k_1)$ avoiding A. □

4.4. End of the proof of the Cancelling Procedure. Thus we may assume that $f: M \to M$ satisfies Theorem (4.10) and moreover, $f^k(x,t) = (|t| \cdot x/3, \eta(t))$ for $(x,t) \in P = [-2,+2]^m \subset W = \mathbb{R}^m$.

(4.24) LEMMA. *There is a homotopy $h_s: P \to \mathbb{R}^m$ $(0 \le s \le 1)$ satisfying:*

(4.24.1) $h_0(x,t) = (|t| \cdot x/3, \eta(t))$,
(4.24.2) $h_s(P_0)$ *is a point, for each fixed* s,
(4.24.3) $h_s(z) \neq z$ *for* $z \in \mathrm{bd}\, P$,
(4.24.4) $h_s(P_0) \in \mathrm{int} P$,
(4.24.5) $h_1(z) \neq z$ *for all* $z \in P$.

PROOF. Let $\eta_s: I \to \mathbb{R}$ be a homotopy constant on the ends $\eta(-2) > -2$, $\eta(2) > 2$ from $\eta_0 = \eta$ to a fixed point free map η_1. Then the required homotopy is given by

$$h_s(x,t) = \left(\frac{|t|}{3} \cdot x, \eta_s(t)\right).$$ □

The homotopy from Lemma (4.24) induces a partial homotopy $h'_s: f^{k-1}(P) \to M$ by the formula: $h'_s(x) = h_s((f^{k-1}_|)^{-1}(x))$. Then the Cancelling Procedure (Theorem (4.3)) follows from the application of Lemma (4.25) to

$$X = M, \quad A = f^{k-1}(P), \quad A' = f^{k-1}(P_0) = \{z_{k-1}\}.$$

It remains to show that the assumptions of Lemma (4.25) are satisfied. To see that the above $A' = \{z_{k-1}\}$ corresponds to A' in Lemma (4.25) we show that $f^{k-1}h'_t(v) \neq v$ for any $v \in \mathrm{bd}\, f^{k-1}(P) \setminus z_{k-1}$. Then $v = f^{k-1}(z)$ for a $z \in \mathrm{bd}\, P$, $z \notin P_0$. Suppose that $f^{k-1}h'_t(v) = v$. Then $f^{k-1}h'_t(v) = f^{k-1}(z)$. Since the restriction of f^{k-1} to $P \setminus P_0$ is mono, $h'_t(v) = z$. Thus $z = h'_t(v) = h'_t f^{k-1}(z) = h_t(z)$ contradicting to (4.24.3).

Now we are in a position to prove that the assumptions of Lemma (4.25) are satisfied.

(4.25.1) is evident.

(4.25.2) Since $h'_t(A) = A'(f^{k-1}(A)) \subset W$, $f^i(h'(A)) \subset f^i W$ are disjoint from $W \supset h'_t(A)$ for $i = 0, \dots, k-2$.

(4.25.3) Let $U' = \operatorname{int} P$. By (4.24.4) $h_s(P_0) \subset U'$ hence $h'_s(z_0) = h_s(P_0) \subset U'$. This proves (a). At last the equality $f^{k-1}(\operatorname{cl} U') = f^{k-1}(P) = A$ proves (b).

(4.25) LEMMA. *Let X be a compact* ANR, *A its closed subset, $k \in \mathbb{N}$, $f\colon X \to X$ a continuous map, $h'_t\colon A \to X$ a partial homotopy and let $A' = \{x \in \operatorname{bd} A : x = f^{k-1}h'_t(x)$ for some $t \in [0,1]\}$. Moreover, we assume that:*

(4.25.1) *$h'_0(a) = f(a)$ for $a \in A$,*

(4.25.2) *the sets $A_0 = \{h'_t(a) : a \in A,\ 0 \le t \le 1\}$, $A_1 = f(A_0), \dots, A_{k-2} = f^{k-2}(A_0)$ are disjoint from A,*

(4.25.3) *there exists an open subset $U' \subset X$ satisfying:*
(a) *$h'_t(x) \in U'$ for $x \in A'$ and $0 \le t \le 1$,*
(b) *$f^{k-1}(\operatorname{cl} U') \subset A$.*

Then there exists an extension of the partial homotopy $\{h'_t\}$ to $h_t\colon X \to X$ satisfying:

(4.25.4) *$h_0 = f$,*

(4.25.5) *the carrier of the homotopy $\{h_t\}$ is contained in an arbitrarily prescribed neighbourhood of A,*

(4.25.6) *the set* $\operatorname{Fix}(h_t^k) \setminus$ *(the orbits cutting A) does not depend on $t \in [0,1]$.*

PROOF. Let $H'\colon X \times I \to X$ be an arbitrary extension of the partial homotopy h'_t onto the ANR X. We will write $H'(x,t) = h'_t(x)$ for all $x \in X$.

Since $H(A' \times I) \subset U'$, the compactness implies the existence of an open subset $U \subset X$ containing A' and satisfying $H'(\operatorname{cl} U \times I) \subset U'$. Moreover, if U is sufficiently small then, by assumption (4.25.2) the sets

$$B_0 = H'((\operatorname{cl} U \cup A) \times I), B_1 = f(B_0), \dots, B_{k-2} = f^{k-2}(B_0)$$

are disjoint from $\operatorname{cl} U \cup A$. Then we have $(h'_t)^k(x) = f^{k-1}h'_t(x)$ for all $x \in \operatorname{cl} U$.

We will show that $(h'_t)^k(x) \neq x$ for all $x \in \operatorname{cl} U \setminus A$, $t \in [0,1]$. In fact $x \in \operatorname{cl} U \setminus A$ implies $x \in \operatorname{cl} U$ hence $h'_t(x) \in U'$. Now $(h'_t)^k(x) = f^{k-1}h'_t(x) \in f^{k-1}U' \subset A$ implies $(h'_t)^k(x) \neq x$ since $x \notin A$.

Now we show that $(h'_t)^k(x) \neq x$ for all $x \in \operatorname{bd}(A \cup \operatorname{cl} U)$. Suppose contrary i.e. $(h'_t)^k(x) = x$. We notice that $\operatorname{bd}(A \cup \operatorname{cl} U) \subset \operatorname{bd} A \cup \operatorname{bd}(\operatorname{cl} U)$. First we assume that $x \in \operatorname{bd} A$. Then the equality $(h'_t)^k(x) = x$ implies $x \in A'$ so $x \in A' \subset U \subset \operatorname{int}(A \cup \operatorname{cl} U)$ hence $x \notin \operatorname{bd}(A \cup \operatorname{cl} U)$ contradicting to the assumption. If $x \in \operatorname{bd}(\operatorname{cl} U) \setminus \operatorname{bd} A$ then $x \in \operatorname{cl} U \setminus A$ and $(h'_t)^k(x) \in A$ as above hence $h'_t)^k(x) \neq x$.

Let $B = \{x \in X \setminus A : (h'_t)^k(x) = x \text{ for a } t \in I\}$. By the above, B is a closed subset of X disjoint from A. Let V be a neighbourhood of A disjoint from B. Let $\lambda: X \to [0,1]$ be an Urysohn function: $\lambda(A) = 1$, $\lambda(X \setminus V) = 0$. We will show that $h_t(x) = h'_{\lambda(x)t}(x)$ is the desired homotopy. In fact (4.25.4) is evident. To see that (4.25.5) holds we notice that $h_t(x) = h'_{\lambda(x)t}(x) = h'_0(x) = f(x)$ for all $x \notin V$. It remains to show (4.25.6). Let us fix $t \in [0,1]$ and consider an orbit of h_t avoiding A. This orbit must belong to B. Now $h_t(x) = h'_{\lambda(x)t}(x) = h'_0(x) = f(x)$ for any x from the orbit which implies that this is also the orbit of f. The same arguments show that each orbit of $f = h_0$ avoiding A belongs to B and is the orbit of h_t which gives (4.25.6). □

This ends the proof of Cancelling Procedure (Theorem (4.3)) which implies the Weak Wecken's Theorem for periodic points — see the scheme of the proof. □

5. Wecken's theorem for periodic points

We consider a self map of a compact ANR $f: X \to X$ and a natural number n. We will define a non-negative integer $NF_n(f)$ such that

(5.1.1) $NF_n(f)$ is a homotopy invariant,

(5.1.2) $NF_n(f)$ is a lower bound of the number of n-periodic points of f

$$NF_n(f) \leq \#\mathrm{Fix}(f^n).$$

This number was introduced by Boju Jiang in 1983 in [Ji1]. The definition requires an analysis of the relations among the Reidemeister classes of the iterations of f. We will show that $NF_n(f)$ satisfies the above two conditions and then we will show that, in the case of manifolds of dimension ≥ 3, $NF_n(f)$ is the best such lower bound (Wecken's Theorem for Periodic Points (5.13)).

We recall the definition of the set of Reidemeister classes of a self map $f: X \to X$. We suppose that the space X is connected and admits a universal covering $p: \widetilde{X} \to X$ (this is satisfied for $X=$ ANR). We define the group of natural transformations

$$\mathcal{O}_X = \{\alpha: \widetilde{X} \to \widetilde{X} : p\alpha = p\}$$

and the set of lifts

$$\mathrm{lift}(f) = \{\widetilde{f}: \widetilde{X} \to \widetilde{X} : p\widetilde{f} = fp\}.$$

Then $\mathcal{O}_X$ is a group isomorphic to the fundamental group $\pi_1 X$. Moreover, $\mathcal{O}_X$ is acting on lift(f) by the formula $\alpha \circ \widetilde{f} = \alpha\widetilde{f}\alpha^{-1}$.

The quotient set $\mathcal{R}(f)$ will be called *set of Reidemeister classes*. For any lift $\widetilde{f}$ the set $p(\mathrm{Fix}(\widetilde{f}))$ is a Nielsen class (or is empty). Moreover, each Nielsen class is of the above form. It turns out that the subordination to each Reidemeister class

$A \subset \mathrm{Fix}(f)$ a lift $\widetilde{f}$ satisfying $p(\mathrm{Fix}(\widetilde{f})) = A$ induces a map $j\colon \mathcal{N}(f) \to \mathcal{R}(f)$ which is an injection. Thus we may identify each Nielsen class with a Reidemeister class.

We consider a self map of a compact ANR $f\colon X \to X$ and its iteration f^n. Then $\mathrm{Fix}(f) \subset \mathrm{Fix}(f^n)$. Let us notice that this inclusion preserves the Nielsen relation: if $x_0, x_1 \in \mathrm{Fix}(f)$ and the path ω satisfies $f\omega \sim \omega$ then also $f^n\omega \sim \omega$. Thus we get the map $\gamma\colon \mathcal{N}(f) \to \mathcal{N}(f^n)$. Notice that γ may be no longer an inclusion (as in Example (3.1): the flip map and $n = 2$).

(5.2) DEFINITION. We define $\gamma\colon \mathcal{R}(f) \to \mathcal{R}(f^n)$ putting $\gamma[\widetilde{f}] = [\widetilde{f}^n]$.

This definition is correct since $(\alpha\widetilde{f}\alpha^{-1})^n = \alpha\widetilde{f}^n\alpha^{-1}$. Now we show that the above map is really an extension of the map $\gamma\colon \mathcal{N}(f) \to \mathcal{N}(f^n)$.

(5.3) LEMMA. *The diagram*

$$\begin{array}{ccc} \mathcal{N}(f) & \xrightarrow{\ \gamma\ } & \mathcal{N}(f^n) \\ {\scriptstyle j}\big\downarrow & & \big\downarrow{\scriptstyle j} \\ \mathcal{R}(f) & \xrightarrow[\ \gamma\ ]{} & \mathcal{R}(f^n) \end{array}$$

is commutative. Here the vertical arrows denote the canonical inclusion of the set of the Nielsen classes into the set of the Reidemeister classes.

PROOF. Let us fix a Nielsen class $A \in \mathcal{N}(f)$ and a point $x_0 \in A$. Then $j(A) = [\widetilde{f}]$ for a lift $\widetilde{f}$ satisfying $p(\mathrm{Fix}(\widetilde{f})) = A$. Now $\gamma j(A) = \gamma[\widetilde{f}] = [\widetilde{f}^n]$. On the other hand $\gamma(A) = A' \in \mathcal{N}(f^n)$ means $A \subset A'$ and $j\gamma(A) = j(A')$. It remains to show that $j(A') = [\widetilde{f}^n]$. But the Nielsen classes $A', p(\mathrm{Fix}(\widetilde{f}^n)) \in \mathcal{N}(f^n)$ contain x_0 hence are equals. Now $j(A') = j(p(\mathrm{Fix}(\widetilde{f}^n))) = [\widetilde{f}^n]$. $\square$

Since $f^n = (f^m)^{n/m}$ (for $m|n$), the above lemma gives the map $\gamma_{nm}\colon \mathcal{R}(f^m) \to \mathcal{R}(f^n)$ such that the diagram

$$\begin{array}{ccc} \mathcal{N}(f^m) & \xrightarrow{\ \gamma_{nm}\ } & \mathcal{N}(f^n) \\ {\scriptstyle j}\big\downarrow & & \big\downarrow{\scriptstyle j} \\ \mathcal{R}(f) & \xrightarrow[\ \gamma_{nm}\ ]{} & \mathcal{R}(f) \end{array}$$

is commutative, i.e. γ_{nm} sends the Nielsen class $A \in \mathcal{N}(f^m)$ into the unique class $A' \in \mathcal{N}(f^n)$ containing A.

The map γ_{nm} is sometimes called *boosting* function (see [HaK]). We notice that $\gamma_{nm}\gamma_{mk} = \gamma_{nk}$ for $k|m|n$ and $\gamma_{mm} = \mathrm{id}$.

(5.4) DEFINITION. We define the depth of a Reidemeister class $A \in \mathcal{R}(f^n)$ as the smallest divisor k of n satisfying $A \in \operatorname{im}\gamma_{nk}$.

5.1 Orbits of Reidemeister classes. We notice that the restriction of the map $f\colon X \to X$ defines the homeomorphism $f_|\colon \mathrm{Fix}(f^n) \to \mathrm{Fix}(f^n)$ satisfying ${f_|}^n = \mathrm{id}$. In other words we have the action of the group $\mathbb{Z}_n$ on $\mathrm{Fix}(f^n)$. Moreover, this map preserves the Nielsen relation: if ω is joining the points $x, y \in \mathrm{Fix}(f^n)$ and $f^n\omega \sim \omega$ then $f\omega$ is joining the points $fx, fy \in \mathrm{Fix}(f^n)$ and $f^n(f\omega) \sim f(\omega)$. This yields the map $\mathcal{N}_f\colon \mathcal{N}(f^n) \to \mathcal{N}(f^n)$ satisfying $(\mathcal{N}_f)^n = \mathrm{id}$. Thus we get the action of $\mathbb{Z}_n$ on the set of Nielsen classes $\mathcal{N}(f^n)$. We will show that the map $\mathcal{N}_f$ extends onto the sets of Reidemeister classes. We start with a lemma about the set $\mathrm{lift}(f^n)$.

(5.5) LEMMA ([Ji1]). *Let $\widetilde{f}_1, \dots, \widetilde{f}_n$ be lifts of the given map f. Then the composition $\widetilde{f}_1 \dots \widetilde{f}_n$ is also a lift of f^n. Conversely each lift of f^n is of the above form.*

PROOF. The first part is obvious. Now for any lift $\widetilde{h}^{(n)} \in \mathrm{lift}(f^n)$ there exists a deck transformation $\alpha \in \mathcal{O}_X$ satisfying $\widetilde{h}^{(n)} = (\alpha\widetilde{f}_1)\widetilde{f}_2 \dots \widetilde{f}_n$. □

The commutative diagram

$$\begin{array}{ccc} X & \xrightarrow{f^n} & X \\ {\scriptstyle f}\downarrow & & \downarrow{\scriptstyle f} \\ X & \xrightarrow[f^n]{} & X \end{array}$$

defines the map $\mathcal{R}_f\colon \mathcal{R}(f^n) \to \mathcal{R}(f^n)$ as follows. We fix a lift $\widetilde{f}\colon \widetilde{X} \to \widetilde{X}$ of f. Then for any lift $\widetilde{h}^{(n)}$ of f^n there is a unique lift $\widetilde{h}'^{(n)}$ such that the following diagram commutative.

$$\begin{array}{ccc} X & \xrightarrow{\widetilde{h}^{(n)}} & X \\ {\scriptstyle \widetilde{f}}\downarrow & & \downarrow{\scriptstyle \widetilde{f}} \\ X & \xrightarrow[\widetilde{h}'^{(n)}]{} & X \end{array}$$

We define $\mathcal{R}_{\widetilde{f}}\colon \mathrm{lift}(f^n) \to \mathrm{lift}(f^n)$ which induces the map $\mathcal{R}_f\colon \mathcal{R}(f^n) \to \mathcal{R}(f^n)$. It turns out that $\mathcal{R}_f$ does not depend on the choice of the lift $\widetilde{f}$.

(5.6) LEMMA ([Ji1]).

(5.6.1) $\mathcal{R}_f[\widetilde{f}_1 \dots \widetilde{f}_n] = [\widetilde{f}_n\widetilde{f}_1 \dots \widetilde{f}_{n-1}]$,
(5.6.2) $(\mathcal{R}_f)^n = \mathrm{id}$,
(5.6.3) $f(p(\mathrm{Fix}(\widetilde{f}_1 \dots \widetilde{f}_n))) = p(\mathrm{Fix}(\widetilde{f}_n\widetilde{f}_1 \dots \widetilde{f}_{n-1}))$,
(5.6.1) $\mathrm{ind}(f^n; p(\mathrm{Fix}(\widetilde{f}_1 \dots \widetilde{f}_n))) = \mathrm{ind}(f^n; p(\mathrm{Fix}(\widetilde{f}_n\widetilde{f}_1 \dots \widetilde{f}_{n-1})))$.

PROOF. (5.6.1) follows from the commutativity of the diagram

$$\begin{array}{ccc} \widetilde{X} & \xrightarrow{\widetilde{f}_1 \dots \widetilde{f}_n} & \widetilde{X} \\ {\scriptstyle \widetilde{f}_n}\downarrow & & \downarrow{\scriptstyle \widetilde{f}_n} \\ \widetilde{X} & \xrightarrow[\widetilde{f}_n \widetilde{f}_1 \dots \widetilde{f}_{n-1}]{} & \widetilde{X} \end{array}$$

(5.6.2) We apply n-times (5.6.1).

(5.6.3) Notice that $\widetilde{f}_n(\mathrm{Fix}(\widetilde{f}_1 \dots \widetilde{f}_n)) = \mathrm{Fix}(\widetilde{f}_n \widetilde{f}_1 \dots \widetilde{f}_{n-1})$. Then we apply p to the both sides.

(5.6.4) Let us denote $A = p(\mathrm{Fix}(\widetilde{f}_1 \dots \widetilde{f}_n))$ and $B = p(\mathrm{Fix}(\widetilde{f}_n \widetilde{f}_1 \dots \widetilde{f}_{n-1}))$. Then by the above $f(A) = B$ and $f^{n-1}B = A$. The commutativity of the fixed point index implies $\mathrm{ind}(f^n; A) = \mathrm{ind}(f^n; B))$. □

(5.7) COROLLARY. *The map $\mathcal{R}_f: \mathcal{R}(f^n) \to \mathcal{R}(f^n)$ defines the action of the group $\mathbb{Z}_n$ on $\mathcal{R}(f^n)$ (by (5.6.2)). The restriction of this action coincides with the earlier action of $\mathbb{Z}_n$ on the set of Nielsen classes (by (5.6.1) and (5.6.3)). The classes in the same orbit have the same fixed point index (by (5.6.4)).*

Let $\mathcal{OR}(f^n)$ denote the quotient set of the above action. We will call its elements *orbits of Reidemeister classes.* Let us notice that $\gamma_{nk}: \mathcal{R}(f^k) \to \mathcal{R}(f^n)$ sends an orbit into an orbit hence we get the map $\gamma_{nk}: \mathcal{OR}(f^k) \to \mathcal{OR}(f^n)$. We may extend the notion of the depth onto the Reidemeister classes putting for an $\mathcal{A} \in \mathcal{OR}(f^n)$

$$\mathrm{d}(\mathcal{A}) = \text{the least divisor k of n satisfying } \mathcal{A} \in \mathrm{im}\, \gamma_{nk}.$$

The Reidemeister class $A \in \mathcal{R}(f^n)$ is called *reducible* if there is a $k|n$, $k < n$ and $B \in \mathcal{R}(f^k)$ satisfying $\gamma_{nk}(B) = A$. We notice that then each class in this orbit is reducible.

The orbit of Reidemeister classes is called *reducible* if it contains a reducible Reidemeister class. Otherwise the orbit is called irreducible.

We notice that the orbit $\mathcal{A} \in \mathcal{OR}(f^n)$ is irreducible if and only if $\mathrm{d}(\mathcal{A}) = n$.

By the commutativity of the fixed point index any two Reidemeister classes A, B in the same orbit have the same index $\mathrm{ind}(f^n; A) = \mathrm{ind}(f^n; B)$. We call an orbit $\mathcal{A} \in \mathcal{OR}(f^n)$ *essential* if a (hence any) Reidemeister class in $\mathcal{A}$ is essential.

(5.8) LEMMA. *If the orbit $\mathcal{A} \in \mathcal{OR}(f^n)$ is essential and irreducible then it contains at least n periodic points.*

PROOF. Since the orbit is essential, all its classes are nonempty. Let us fix a point $a \in A \in \mathcal{A}$. Then the length of the orbit of points $a, fa, \dots$ must be n. In fact $f^n(a) = a$ since $a \in \mathrm{Fix}(f^n)$ and if moreover, $f^k a = a$ for an $k < n$ then

$f^l(a) = a$ for $l = \gcd(n,k)$ would imply $a \in \mathrm{Fix}(f^l)$ and the Nielsen class $B \subset \mathrm{Fix}(f^l)$ containing a would satisfy $\gamma_{nl}(B) = A$ contradicting to the irreducibility of $\mathcal{A}$. □

5.2. Points of pure period n. Recall that $P_n(f) = \{x \in X; f^n(x) = x$ but $f^k(x) = x$ for no $k|n, k < n\}$ denotes the set of points of the pure period n. We are going to define a homotopy invariant being the lower bound of the cardinality of $P_n(f)$. We define

$$\begin{aligned} IEC_n(f) &= \text{number of irreducible essential Reidemeister classes in } \mathcal{R}(f^n), \\ NP_n(f) &= (\text{number of irreducible essential orbits} \\ &\qquad \text{of Reidemeister classes in } \mathcal{OR}(f^n)) \times n. \end{aligned}$$

The factor n is explained in Lemma (5.8).

(5.9) THEOREM. *$NP_n(f)$ is the homotopy invariant satisfying*

$$NP_n(f) \le \#P_n(f).$$

PROOF. The first part is evident since $NP_n(f)$ is defined by the homotopy invariants. The inequality follows from Lemma (5.8). □

In Section 6 we will show that, in dimensions ≥ 3, $NP_n(f)$ is the best such lower bound.

5.3. The estimation of $\#\mathrm{Fix}(f^n)$. We started the study of periodic points by presenting some naive bounds of the number $\#\mathrm{Fix}(f^n)$. Now we are going to define a Nielsen type number of periodic points introduced by Boju Jiang in 1983 in [Ji1].

Let us consider the disjoint sum $\bigcup_{k|n} \mathcal{OR}(f^k)$. A subset $\mathfrak{A} \subset \bigcup_{k|n} \mathcal{OR}(f^k)$ is called *preceding system* if every essential orbit in $\bigcup_{k|n} \mathcal{OR}(f^k)$ is preceded by an element from $\mathfrak{A}$. The preceding system $\mathfrak{A}$ is called *minimal preceding system* (MPS for the shorthand) if the number $\sum_{\mathcal{A}\in\mathfrak{A}} \mathrm{d}(\mathcal{A})$ is minimal. We denote this minimal number by $NF_n(f)$.

(5.10) REMARK. Each preceding system contains all essential irreducible orbits of Reidemeister classes.

PROOF. Let $\mathcal{S}$ be a preceding system and let $\mathcal{A}$ be an irreducible essential orbit of Reidemeister classes. Since $\mathcal{A}$ is essential, it must be preceded by an orbit from $\mathcal{S}$. But $\mathcal{A}$ as irreducible is preceded only but itself. Thus $\mathcal{A} \in \mathcal{S}$. □

(5.11) THEOREM. *$NF_n(f)$ is the homotopy invariant and a lower bound of the number of n-periodic points: $NF_n(f) \le \#\mathrm{Fix}(f^n)$.*

PROOF. The homotopy invariance is obvious. We prove the inequality. If $\mathrm{Fix}(f^n)$ is infinite the inequality is evident. Assume that $\mathrm{Fix}(f^n)$ is finite and let us split it into the orbits of points

$$\mathrm{Fix}(f^n) = \{x_1^1, \dots, x_{s_1}^1; \dots, x_1^l, \dots, x_{s_l}^l\}$$

The orbit $\{x_1^i, \dots, x_{s_i}^i\}$ determines the orbit of Reidemeister classes $\mathcal{A}_i \in \mathcal{OR}(f^{s_i})$. Then $\mathrm{d}(\mathcal{A}_i) \le s_i$ and let $\mathcal{B}_i \in \mathcal{OR}(f^{\mathrm{d}(\mathcal{A}_i)})$ denote the orbit preceding $\mathcal{A}_i$. Then

$$\sum_{i=1}^{l} \mathrm{d}(\mathcal{B}_i) = \sum_{i=1}^{l} \mathrm{d}(\mathcal{A}_i) \le \sum_{i=1}^{l} s_i = \#\mathrm{Fix}(f^n).$$

It remains to show that the orbits $\mathcal{B}_1, \dots, \mathcal{B}_l$ form a Preceding System. In fact if $\mathcal{C} \in \mathcal{OR}(f^k)$ is essential then it contains an orbit $\{x_1^i, \dots, x_{s_i}^i\}$ hence $\mathcal{B}_i$ is preceding $\mathcal{C}$. □

The next example shows that we may not restrict in the definition of MPS to the essential orbits only.

(5.12) EXAMPLE. Let $f\colon S^{2m} \to S^{2m}$ be the antipodal map $f(x) = -x$ and $n = 2$. Since S^{2m} is simply connected, $\mathcal{R}(f^k)$ consists of one class for each $k \in \mathbb{N}$. In particular the unique orbit in $\mathcal{R}(f^2)$ reduces to the unique class in $\mathcal{R}(f^1)$. But the last one is inessential since $\mathrm{Fix}(f) = \emptyset$. Thus there is no essential irreducible orbit. Nevertheless $\mathrm{Fix}(g^2) \ne \emptyset$ for each $g \sim f$ since $g^2 \sim f^2 = \mathrm{id}$ but $L(\mathrm{id}) = \chi_{S^{2m}} = 2$. Thus MPS consists of the unique (inessential) class in $\mathcal{R}(f^1)$. □

Since any Preceding System contains all essential irreducible orbits, we have

$$\sum_{k|n} NP_k(f) \le NF_n(f)$$

The above example of antipodism shows that this inequality may be sharp.

Fortunately in many situation the equality $\sum_{k|n} NF_k(f) = NP_n(f)$ holds which allowed to make some computations of this invariant. Such computations were done for self maps of tori, nilmanifolds and some solvmanifolds in the papers of P. Heath and E. Keppelmann, see [HK2]–[HK4].

The aim of this section is to prove that $NF_n(f)$ is the best lower bound of the number of periodic points.

(5.13) THEOREM (Wecken Theorem for Periodic Points). *If $\dim M \ge 3$ then f is homotopic to a map g satisfying $\#\mathrm{Fix}\,(g^n) = NF_n(f)$.*

5.4. Scheme of the proof of Wecken's theorem for periodic points. Let $f\colon M \to M$, $\dim M = m \geq 3$ and let n be a fixed natural number. We fix a minimal preceding system of Reidemeister classes and we denote it MPS. Let us notice that it is enough to show that f realizes the number $NF_n(f)$ if and only if each orbit of periodic points $\{x_0, \dots, x_{l-1}\}$ (of length $l|n$) represents an orbit of Reidemeister classes from MPS of depth l which contains only the points $\{x_0, \dots, x_{l-1}\}$. This gives an injection from the set of orbits of points into the orbits in MPS. Now

$$\#\mathrm{Fix}(f^n) = \sum_A \#\, A \leq \sum_B \mathrm{d}(B) = NF_n(f)$$

where A runs through the set of all orbits of points in $\mathrm{Fix}(f^n)$ and B runs through MPS. The opposite inequality follows from Theorem (5.11).

We will construct the homotopy from a given map f to a map realizing $NF_n(f)$ in two steps. First we show how to deform f to get $\mathrm{Fix}(f^n)$ as above, plus some periodic points representing orbits of Reidemeister classes which are preceded by the classes from the MPS. In the second step we will coalesce these extra points to the MPS.

(5.14) LEMMA. *Any continuous map $f\colon M \to M$ is homotopic to a map such that for any $k|n$ each orbit of points $\{x_0, \dots, x_{k-1}\} \in \mathrm{Fix}(f^k)$ (of length $k|n$) either represents an orbit of Reidemeister classes which is preceded by an orbit from* MPS *or represents an orbit of Reidemeister classes of depth k and this class contains only these k points and belongs to* MPS. *Moreover, S^k_r is satisfied at any point x_i for an $r \in \mathbb{Z}$ (see Definition* (5.15)*).*

PROOF. We use induction on all divisors $k|n$. Let $k = 1$. By Corollary (1.5) we may assume that f has exactly $N(f)$ fixed points and near each of them f is given by the formula S^1_r. If an inessential Nielsen class from $\mathcal{R}(f)$ belongs to MPS then the Procedure (5.19) allows us to produce a point in this class and S^1_0 is satisfied at this point.

Now we fix a $k|n$ and we assume that induction assumption holds for all $l < k$, $l|n$. We will homotope f to make the lemma hold for k.

We use Theorem (4.2) to make

$$P_k(f) = \{x \in \mathrm{Fix}(f^k) : f^l(x) \neq x, \text{ for } l|k,\ l < k\}$$

finite and a local homeomorphism near each point from $P_k(f)$ without changing $\mathrm{Fix}(f^k) \setminus P_k(f)$.

Let $A \in \mathcal{O}R(f^k)$ be an orbit of Reidemeister classes. If A is not preceded by an orbit from MPS then each orbit of points from A is of length k. By Theorem (4.2) we may replace the orbit by a finite number of orbits each of index ± 1. After the

Canceling Procedure only a finite number of orbits of the same index remain. If A is inessential it disappears. If A is an essential irreducible orbit, it belongs to MPS. Then Creating Procedure (Theorem (5.16)) allows us to replace these orbits by a single orbit satisfying S_r^k. If A is an inessential orbit belonging to MPS then the Addition Procedure (5.19) allows us to create a k-point orbit representing A and satisfying S_0^k. This ends the inductive step. □

The main theorem will be proved when all the classes which do not belong to MPS are removed. We apply Theorem (4.2) to make each such orbit finite. Let $\{z_0, \dots, z_{k-1}\}$ be an orbit of points representing an orbit of Reidemeister classes proceeded by an orbit from MPS. Then the Coalescing Procedure (5.23) allows us to get rid of $\{z_0, \dots, z_{k-1}\}$. Finally we will show that if dimension ≥ 3 then there is still enough space to perform all the deformations coalescing orbits simultaneously with mutually disjoint carriers.

5.5. Procedures. We will enlist several formulae for maps with an isolated fixed point and these maps will be called standard. Then we will show that near each periodic point the considered map may have the standard form.

(5.15) DEFINITION. We say that a self-map $f\colon \mathbb{R}^m \to \mathbb{R}^m$ satisfies S_k ($k \in \mathbb{Z}$) if it is given by the formula $f(z,v) = \lambda(\rho_{k'}(z), v)$ where $(z,v) \in \mathbb{C} \times \mathbb{R}^{m-2} = \mathbb{R}^m$, $k \in \mathbb{Z}$, $k' = (-1)^m k$, $\lambda > 1$ and $\rho_{k'}\colon \mathbb{C} \to \mathbb{C}$ is given by (in polar coordinates: $r \geq 0$ and $\phi \in \mathbb{R}$ regarded modulo $2\pi\mathbb{Z}$).

$$\rho_{k'}(r,\phi) = \begin{cases} (r, k'\phi) & \text{for } |k'| \geq 2, \\ (r, \phi + \alpha_0) & \text{for } k' = 1, \\ (r, \chi_{-1}(\phi)) & \text{for } k' = -1, \\ (r, \chi_0(\phi)) & \text{for } k' = 0, \end{cases}$$

where

(5.15.1) for $k' = 1$: $\alpha_0 > 0$ denotes an irrational angle i.e. α_0/π is irrational,

(5.15.2) for $k' = -1$ we define $\chi_{-1}\colon [-\pi,\pi] \to [-\pi,\pi]$ as a homeomorphism satisfying: $\chi_{-1}(-\pi) = +\pi$, $\chi_{-1}(+\pi) = -\pi$, $\chi_{-1}(t) < -t$ for $0 < t < \pi$, $\chi_{-1}(t) = t$ for $-\pi \leq x \leq 0$, i.e. $\rho_{-1}(1,\phi)$ is a small deformation of the flip map (see Figure 14),

(5.15.3) for $k' = 0$: we define $\chi_0\colon [-\pi,\pi] \to [-\pi,\pi]$ by the formula $\chi_0(t) = \chi_{-1}(|t|)$.

If $f^l(x) = x$ and f^l satisfies S_k in x then we will write S_k^l. Since $\lambda > 1$, 0 is the only fixed point and $\operatorname{ind}(\rho_k) = \deg(\rho_{k'}) = (-1)^m \cdot k' = k$ which explains the factor $(-1)^m$.

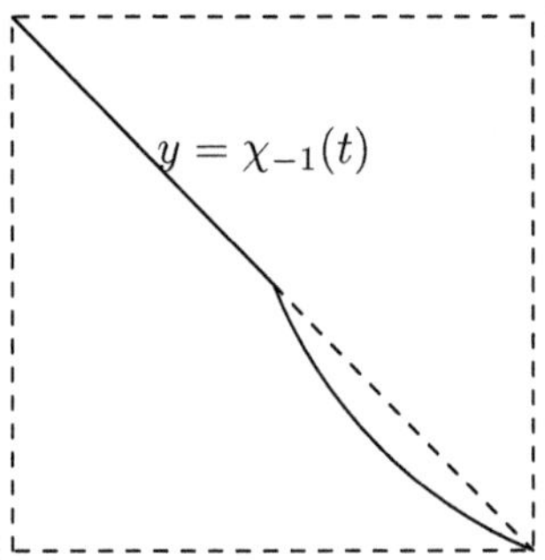

FIGURE 14

Let us notice that the restriction $\rho_k: \mathbb{C} \setminus 0 \to \mathbb{C} \setminus 0$ is a covering map for $k \neq 0$. Thus f is a ramified covering. Moreover, for $k = \pm 1$ this restriction is a homeomorphism. In any case the fixed point set of the restriction $\rho_k: S^1 \to S^1$ is always finite.

The next procedure allows to create a new periodic point of the given index r. Since the index is the homotopy invariant, the new orbit will be balanced by $|r|$ new orbits of points each of index $= -\mathrm{sgn}\,(r)$.

(5.16) THEOREM (Creating Procedure). *Let $\{x_0, \dots, x_{k-1}\}$ be an isolated k-orbit of a map $f: M \to M$ which is a local homeomorphism near each x_i. We fix a Euclidean neighbourhood $U \ni x_0$ such that $f^i(\mathrm{cl}\, U) \cap f^j(\mathrm{cl}\, U) = \emptyset$ and f is a homeomorphism on each $f^i(\mathrm{cl}\, U)$ for $0 \le i < j \le k-1$. Let $V_0 \subset U$ be a Euclidean neighbourhood of x_0 satisfying $f^k(\mathrm{cl}\, V_0) \subset U$. We denote $V_i = f^i(V_0)$ and suppose that $0 \in V_0 \subset U = \mathbb{R}^m$ and $x_0 \neq 0$. If we take V_0 small enough then $x_0 \notin V_0$. Then there exists a homotopy $\{f_t\}$ constant outside $V_{k-1} = f^{k-1}(V_0)$ satisfying $f_0 = f$, $f_1^k(0) = 0$ (hence $0 \in V_0$ becomes a new periodic point) and S_r^k is satisfied at this point for a prescribed number $r \neq 0$. Moreover, still $f_1(\mathrm{cl}\, V_{k-1}) \subset U$ and*

$$\begin{aligned}\mathrm{Fix}(f_1^k) &= \mathrm{Fix}(f^k) \cup \{\textit{the orbit of}\ 0\}\\ &\cup \{\textit{orbits of}\ |r|\ \textit{new periodic points in}\ V_{k-1}\ \textit{each of index} = -\mathrm{sgn}\,(r)\}.\end{aligned}$$

PROOF. Let us fix balls centered at $0 \in V_0$: $K(0,\rho) \subset K(0,\rho') \subset V_0$, $\rho < \rho'$. Since the restriction $f^{k-1}: \mathrm{cl}\, V_0 \to \mathrm{cl}\, V_{k-1}$ is a homeomorphism, we may define the map $f_1: f^{k-1}(K(0,\rho)) \to U = \mathbb{C} \times \mathbb{R}^{m-2}$ by the formula $f_1(f^{k-1}(z,x)) = \lambda(\rho_r(z), x)$ (see Definition (5.15)) and we extend it to $f_1: f^{k-1}(K(0,\rho')) \to U$, a map satisfying $f_1(y) = f(y)$ for $y \in f^{k-1}(\mathrm{bd}\, K(0,\rho'))$, and then by the same formula onto the rest of M. This is possible since U is contractible and any two maps into U are homotopic. Since $x_0 \notin V_0$, it still remains a fixed point of f_1^k. Thus $\mathrm{Fix}(f_1^k) = \mathrm{Fix}(f^k) \cup \{\text{orbits of } f_1 \text{ crossing } V_{k-1}\}$. Now we show that there is an arbitrarily small homotopy $f_t\colon M \to M$, $1 \le t \le 2$, constant outside $V_{k-1} \setminus f^{k-1}(K(0,\rho))$ such that $\mathrm{Fix}(f_2^k) \cap V_{k-1}$ is finite.

Since the restriction $f_1^k = f_1(f^{k-1})\colon \mathrm{cl}\, K(0,\rho') \setminus K(0,\rho) \to U$ has no fixed point on the boundary, there is an arbitrarily small homotopy $\phi_t\colon \mathrm{cl}\, K(0,\rho') \setminus K(0,\rho) \to U$ $(1 \le t \le 2)$ satisfying $\phi_1 = f_1(f^{k-1})$, ϕ_t is constant on the boundary, $\mathrm{Fix}(\phi_2)$ is finite and ϕ_2 is a linear isomorphism near each point in $\mathrm{Fix}(\phi_2)$. Then we define the homotopy of $f_t\colon M \to M$ by the formula

$$f_t(x) = \begin{cases} \phi_t(z) & \text{for } x = f^{k-1}(z), z \in \mathrm{cl}\, K(0,\rho') \setminus K(0,\rho), \\ f(x) & \text{otherwise.} \end{cases}$$

Since $\mathrm{ind}(f_2^k, 0) = r$, there must appear in V_{k-1}: $s + |r|$ new fixed points of f_1^k each of index $(-\mathrm{sgn}\, r)$ and s new fixed points of f_1^k each of index $(+\mathrm{sgn}\, r)$ (for a non-negative integer s).

Since all these points belong to the same Nielsen class (in $\mathrm{Fix}(f_2^k)$), we apply the Cancelling Procedure to remove all pairs of orbits with opposite indices. The map f_3 thus obtained satisfies the lemma. □

(5.17) Corollary. *Let* $\mathrm{Fix}(f^k)$ *be finite and let* $x_0 \notin \mathrm{Fix}(f^k)$. *Suppose that* A *is the sum of orbits, from the same Nielsen class, of period* k *and* $\mathrm{ind}(f^k; A) = rk$ *for an* $r \neq 0$. *Then there is a homotopy* f_t *starting from* $f_0 = f$, *constant in a neighbourhood of* $\mathrm{Fix}(f^k) \setminus A$ *satisfying*

$$\mathrm{Fix}(f_1^k) = (\mathrm{Fix}(f_1^k) \setminus A) \cup \{x_0, f(x_0), \dots, f^{k-1}(x_0)\}.$$

Moreover, the orbit $\{x_0, f(x_0), \dots, f^{k-1}(x_0)\}$ *satisfies* S_r^k.

Proof. We choose a point $a \in A$, a Euclidean neighbourhood U of a and a point $x_0 \neq 0$, $x_0 \in U$. We apply the Creating Procedure to get the orbit $\{x_0, f(x_0), \dots, f^{k-1}(x_0)\}$ of index rk and $|r|$ orbits of points of index $-\mathrm{sgn}\,(r)k$. Then we apply the Cancelling Procedure to remove all pairs of orbits of opposite indices. □

(5.18) Remark. Using the above corollary to each orbit of points we can replace each orbit by an orbit satisfying S_r^k condition.

The next procedure makes a non-periodic point a periodic one (of index zero) in an arbitrarily prescribed Nielsen class.

(5.19) Theorem (Addition Procedure). *Given numbers* $k, n \in \mathbb{N}$, $k|n$ *a map* $f\colon M \to M$ *such that* $\mathrm{Fix}(f^n)$ *is finite and a point* $x_0 \in M$ *such that the points*

$x_0, x_1 = f^1(x_0), \dots, x_{2n} = f^{2n}(x_0)$ *are different. Let moreover,* $\dim M \geq 3$. *Then there is a homotopy* $\{f_t\}_{0 \leq t \leq 2}$ *satisfying:*

(5.19.1) $f_0 = f$,
(5.19.2) $\{f_t\}$ *is constant in a neighbourhood of* $\mathrm{Fix}(f^n)$,
(5.19.3) $f_2^k(x_0) = x_0$ *and* $f_2^i(x_0) \neq x_0$ *for* $i = 1, \dots, k-1$,
(5.19.4) $\mathrm{Fix}(f_2^n) = \mathrm{Fix}(f^n) \cup \{x_0\}$,
(5.19.5) f_2 *satisfies* S_0^k *at the point* x_0.

PROOF. Since the points $x_0, x_1, \dots, x_{n-1}$ do not belong to $\mathrm{Fix}(f^n)$, we may deform f near them to make f a local homeomorphism there. The deformation may be arbitrarily small hence $\mathrm{Fix}(f^n)$ does not change. Let $V_0 = \mathbb{R}^m$ be a Euclidean neighbourhood of x_0 where $f, f^2, \dots, f^{n-1}$ are homeomorphisms. We define $V_i = f^i(V_0)$ $(i = 1, \dots, 2n)$. For $i = 1, \dots, k-1$ we get Euclidean neighbourhoods $V_i \ni x_i$ and if V_0 is chosen small enough, the sets $V_0, \dots, V_{2n}$ are mutually disjoint. Then the restriction of f near x_i is given in the coordinates by

$$V_i = \mathbb{R}^m \ni x \to x \in \mathbb{R}^m = V_{i+1} \quad \text{for } i = 0, \dots, k-2.$$

We correct f so that $f, f^2, \dots, f^{n-1}$ become transverse to x_{k-1}. The deformation may be arbitrarily small and constant in a neighbourhood of $\mathrm{Fix}(f^n)$ and the points still $x_0, \dots, x_{2n}$ remain different.

Let us fix a ball neighbourhood $U_0 = B(x_0, r_0) \subset V_0 = \mathbb{R}^m$. We will denote $U_i = f^i(U_0) \subset V_i$. By the above convention $U_i = B(x_i, r_0) \subset V_i = \mathbb{R}^m$ for $i = 1, \dots, k-1$. If $r_0 > 0$ is chosen small enough then U_{k-1} is also small and $\mathrm{cl}\, U_{k-1} \cup f^{-1}\mathrm{cl}\, U_{k-1} \cup \dots \cup f^{-(k-1)}\mathrm{cl}\, U_{k-1}$ is a finite number of mutually disjoint mutually disjoint balls in M, see Corollary (4.9). Let us fix a number $r_1 \in (0, r_0)$.

Let $\omega \colon [0, r_1] \to M$ be a path satisfying:

(5.20.1) $\omega(0) = x_0$, $\omega(r_1) = f(x_{k-1}) = x_k$,
(5.20.2) ω is avoiding $(\mathrm{cl}\, U_{k-1} \cup f^{-1}(\mathrm{cl}\, U_{k-1}) \cup \dots \cup f^{-(n-1)}(\mathrm{cl}\, U_{k-1})) \setminus \mathrm{cl}\, U_0$,
(5.20.3) $\omega(t) = (\lambda t, 0) \in \mathrm{cl}\, U_0 \subset \mathbb{R} \times \mathbb{R}^{n-1}$ for $0 \leq t \leq \varepsilon$ (where $\lambda > 1$ satisfies $r_1 > \varepsilon \overset{\text{def}}{=} r_0/\lambda > 0$),
(5.20.4) $\omega(t) \notin \mathrm{cl}\, U_0$ for $\varepsilon < t \leq r_1$.

We define the map $f_1 \colon M \to M$ deforming f only inside $U_{k-1} = B(0, r_0) \subset \mathbb{R}^m$ by

$$f_1(x) = \begin{cases} f(x) & \text{for } x \notin U_{k-1} = \mathrm{cl}\, B(0, r_0), \\ f\left(\dfrac{\|x\| - r_1}{r_0 - r_1} \cdot x\right) & \text{for } r_1 \leq \|x\| \leq r_0, \\ \omega(\|x\|) & \text{for } \|x\| \leq r_1. \end{cases}$$

We notice that f_1 is homotopic to f by a homotopy constant outside U_{k-1}, hence $\mathrm{Fix}(f^n) \cap U_{k-1} = \emptyset$ implies $\mathrm{Fix}(f^n) \subset \mathrm{Fix}(f_1^n)$. We will show that $\mathrm{Fix}(f_1^n) \cap B(0, r_0) = \{0\}$ $(= x_{k-1}$ a point of pure period $k)$.

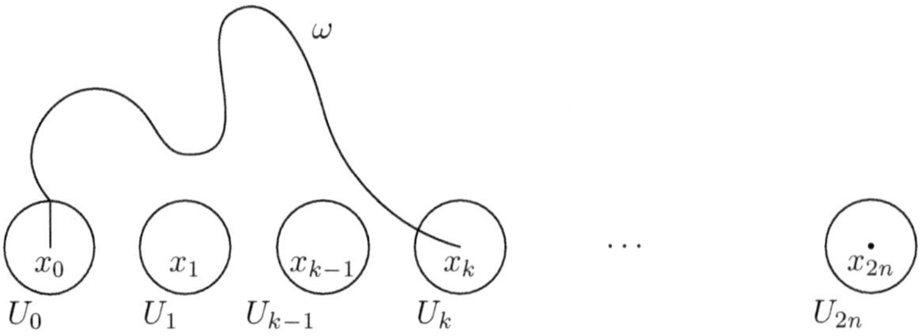

FIGURE 15

Assume that $x = f_1^n(x)$ for $r_1 \leq \|x\| \leq r_0$ i.e. x belongs to the annulus $A(0; r_1, r_0) = \{x \in \mathbb{R}^n : r_1 \leq \|x\| \leq r_0\}$. Now

$$x \in f_1^n(A(0; r_1, r_0)) = f^{n-1}(f(\operatorname{cl} B(0, r_0))) = f^n(\operatorname{cl} B(0, r_0)) = \operatorname{cl} U_{n+k-1}.$$

This contradicts to the assumption $x \in U_{k-1}$, since $\operatorname{cl} U_{k-1} \cap \operatorname{cl} U_{n+k-1} = \emptyset$.

Now we assume that $0 \leq \|x\| \leq r_1/\lambda^{n/k}$ (in U_{k-1}). Then $\|x\| \leq r_1$, so $f_1(x) = \omega(\|x\|) = (\lambda\|x\|, 0)$ (in U_0) hence $\|f_1(x)\| = \lambda\|x\|$. Since the restriction $f^{k-1}\colon U_0 \to U_{k-1}$ is an isometry in coordinates, $\|f_1^k(x)\| = \lambda\|x\| \leq r_1/\lambda^{n/k-1} \leq r_1$ (in U_{k-1}). We may repeat the above and get $\|f_1^{2k}(x)\| = \lambda^2\|x\|, \dots, \|f_1^n(x)\| = \lambda^{n/k}\|x\|$. Since $\lambda > 1$, $f_1^n(x) = x$ if and only if $x = 0 \in U_{k-1}$ in this case.

Now let $r_1/\lambda^{n/k} < \|x\| \leq r_1$ in U_{k-1}. Then the inclusion $f_1^{ik}(x) \in B(0, r_1) \subset U_{k-1}$ does not hold for all $i = 1, \dots, n/k$. Let i be the smallest number for which the inclusion does not hold. Then:

- either $f_1^{ik}(x) \in A(0, r_1, r_0) \subset U_{k-1}$ hence

  $$f_1^{ik+1}(x) = f_1(f_1^{ki}(x)) \subset f_1(A(0, r_1, r_0)) = f(\operatorname{cl} U_{k-1}) = \operatorname{cl} U_k.$$

  Then $f_1^{ik+2}(x) \in f(\operatorname{cl} U_k) = \operatorname{cl} U_{k+1}$ and so on. But the sets $U_k, \dots, U_{2n}$ are disjoint from U_{k-1} hence $f_1^n(x) \neq x \in U_{k-1}$,
- or $f_1^{ik}(x) \in \omega[\varepsilon, r_1]$. Since i was the smallest number for which the inclusion did not take place, $f_1^{(i-1)k}(x) = (\lambda^{i-1}\|x\|, 0) \in B(0, r_1) \subset U_{k-1}$ while $f_1^{(i-1)k+1}(x) = \omega(\lambda^{i-1}\|x\|, 0) = (\lambda^i\|x\|, 0) \notin B(0, r_1) \subset U_0$. We consider two subcases:
  If $f_1^{(i-1)k+1}(x) \in A(0, r_1, r_0) \subset U_0$ then $f_1^{ik}(x) \in A(0, r_1, r_0) \subset U_{k-1}$ hence $f_1^{ik+1}(x) \in f(A(0, r_1, r_0)) \subset \operatorname{cl} U_k$ and successively as above $f_1^{ik+2}(x) \in \operatorname{cl} U_{k+1}$ and so on. But these sets are disjoint from $\operatorname{cl} U_{k-1} \ni x$.
  If $f_1^{(i-1)k+1}(x) \notin A(0, r_1, r_0) \subset U_0$. Then it must belong to $\omega[\varepsilon, r_1] = \omega[0, r_1] \setminus U_0$. But the last set is disjoint from $(U_{k-1} \cup f^{-1}(\operatorname{cl} U_{k-1}) \cup \dots \cup f^{-(n-1)}(\operatorname{cl} U_{k-1}))$ hence no iteration of this point comes back to $U_{k-1} \ni x$.

Thus the map f_1 satisfies (5.20.1)–(5.20.4). It remains to modify f_1 in a neighbourhood of x_{k-1} to make it satisfy S_0^k there without varying $\mathrm{Fix}(f_1^n)$.

We recall that the restriction of f_1 is given near $0 = x_0 \in \mathbb{R}^n \subset M$ by the formula $f_1(x) = (\lambda\|x\|, 0) \in \mathbb{R} \times \mathbb{R}^{n-1}$ for $\|x\| \le r_1$ and a number $\lambda > 1$. We fix the numbers $0 < r_3 < r_2 < r_1$ where $r_2 \le r_1/\lambda^n$. We define the map $f_2 \colon M \to M$ putting

$$f_2(x) = \begin{cases} f_1(x) & \text{for } x \notin B(0, r_2) \subset U_{k-1}, \\ \chi_0(x) & \text{for } x \in \operatorname{cl} B(0, r_3) \subset U_{k-1}, \\ \text{by Lemma (5.22)} & \text{for } x \in A(0; r_3, r_2) \subset U_{k-1} \end{cases}$$

(where in the definition of χ_0 we use the same $\lambda > 1$ as above). Notice that we may apply Lemma (5.22) since the image of the map χ_0 is lying in the hemispace $\{(x_1, \dots, x_n) \in \mathbb{R}^n : x_1 \ge 0\}$. The map f_2 satisfies $\|f_2(x)\| = \lambda\|x\|$ for $x \in B(0, r_1) \subset U_{k-1}$. This implies (by the choice of r_2) $f_2^n(B(0, r_2)) \subset B(0, r_1)$ hence the only n-periodic point in $B(0, r_1)$ is 0. Moreover, S_0^k is satisfied in $B(x_0, r_3)$ and f_2 is homotopic to f_1 since the deformation is small. □

(5.21) REMARK. Since the path $\omega \colon [0, r_1] \to M$ may represent any homotopy class (we require only that it avoids the sum $\bigcup K_\alpha$), the obtained new fixed point may represent any prescribed Reidemeister class in $\mathcal{R}(f_2)$.

(5.22) LEMMA. *Suppose that we are given the numbers $0 < r_3 < r_2$ and functions $\psi \colon S(0, r_2) \to \mathbb{R}^n$, $\phi \colon S(0, r_3) \to \mathbb{R}^n$ such that the vectors $\psi(r_2 x)$, $\phi(r_3 x)$ (for $\|x\| = 1$) are not opposite i.e. $0 \notin [\psi(r_2 x), \phi(r_3 x)]$. Let moreover, $\|\psi(x)\| = \lambda\|x\|$, $\|\phi(x')\| = \lambda\|x'\|$ for $\|x\| = r_2$, $\|x'\| = r_3$. Then there is a continuous extension of these maps onto the annulus $h \colon A(0; r_3, r_2) \to \mathbb{R}^n$ satisfying $\|h(x)\| = \lambda\|x\|$.*

PROOF. For a point $x \in A(0; r_3, r_2)$ we denote $x_2 = r_2 \cdot x/\|x\|$, $x_3 = r_3 \cdot x/\|x\|$ and we define an extension sending the segment $[x_3, x_2]$ linearly onto $[\phi(x_3), \psi(x_2)]$ i.e.

$$h'(x) = \left(1 - \frac{\|x\| - r_2}{r_3 - r_2}\right)\psi(x_2) + \frac{\|x\| - r_2}{r_3 - r_2}\phi(x_3).$$

Since the directions of the vectors $\psi(x_2)$, $\phi(x_3)$ are not opposite, $h'(x) \ne 0$. Now $h(x) = \lambda\|x\|h'(x)/\|h'(x)\|$ is the desired extension. □

5.6. Coalescing procedure. The aim of this section is to prove the Coalescing Procedure (Theorem (5.23)) which permits us to join two orbits in the same Nielsen class. This involves some technical difficulties since, in contrast to the previous sections, we have to change the map f near the periodic points so we will need explicit formulae for the deformations to be sure that no other periodic points appear. Lemma (4.2) (saying that if a deformation on a subset disjoint from the set of periodic points is sufficiently small then $\mathrm{Fix}(f^n)$ does not vary) will be not always applicable.

(5.23) THEOREM (Coalescing Procedure). *Suppose that* $\mathrm{Fix}(f^k)$ *is finite. Let the orbits* $\{y_0, \ldots, y_{l-1}\} \in \mathrm{Fix}(f^l)$, $\{x_0, \ldots, x_{k-1}\} \in \mathrm{Fix}(f^k)$ *be disjoint and let the points* x_0, y_0 *be Nielsen related as fixed points of* f^k $(l|k)$. *Moreover, let* S^l_r *and* $S^k_{r'}$ *be satisfied at* y_0 *and* x_0, *respectively. Then there is a homotopy* f_t *constant in a neighbourhood of* $\mathrm{Fix}(f^k) \setminus \{x_0, \ldots, x_{k-1}; y_0, \ldots, y_{l-1}\}$ *satisfying* $f_0 = f$, $\mathrm{Fix}(f_1^k) = \mathrm{Fix}(f^k) \setminus \{x_0, \ldots, x_{k-1}\}$.

SCHEME OF THE PROOF. We will deform f in four steps to make it satisfy the following: (in the first two steps we do not change f in a neighbourhood of $\mathrm{Fix}(f^k)$).

(CP1) There is an arc $\omega_0\colon [-1, +1] \to M$ such that $\omega_0(-1) = y_0$, $\omega_0(+1) = x_0$ and the paths $f^k\omega_0$, ω_0 are homotopic (Nielsen relation). Moreover, $\omega_i = f^i\omega_0$ are also arcs $(i = 0, \ldots, 3k-1)$ and for $i < j$ we have: $\omega_i(t) = \omega_j(s)$ if and only if $[(t = s = -1)$ and $(j - i$ is a multiple of $l)]$ or $[(t = s = 1)$ and $(j - i$ is a multiple of $k)]$. Moreover, $f, \ldots, f^{k-1}$ are transverse to $z_{k-1} = \omega_{k-1}(0)$.

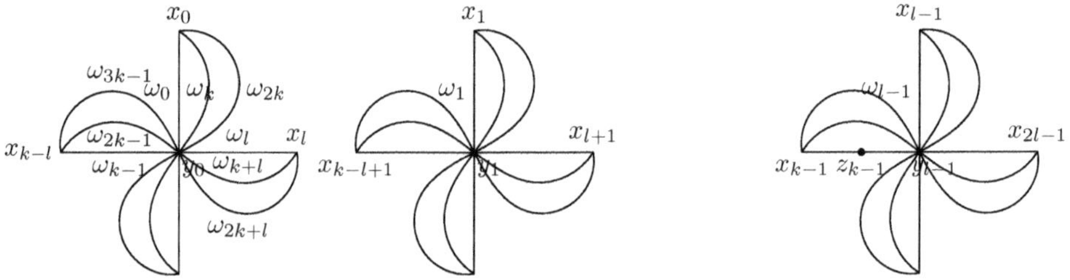

FIGURE 16

(CP2) Let us fix Euclidean neighbourhoods $U_i \supset \omega_i[-1, +1]$ where $\omega_i(t) = (t, 0) \in U_i = \mathbb{R} \times \mathbb{R}^{m-1}$ for $-1 \le t \le +1$ $(i = 0, \ldots, k-1)$. Then we may choose other, mutually disjoint convex, neighbourhoods V_i of $\omega_i(-1, +1) \subset V_i \subset U_i$ such that the restrictions of f to V_i^- and to V_i^+ are homeomorphisms and $f(V_i^0) = z_{i+1}$ where $V_i^- = \{(t, x) \in V_i : t < 0\}$, $V_i^+ = \{(t, x) \in V_i : t > 0\}$, $V_i^0 = \{(t, x) \in V_i : t = 0\}$ and $i = 0, \ldots, k-2$. Moreover, $z_{k-1} = \omega_{k-1}(0)$ admits a neighbourhood U such that $\mathrm{cl}\, U \cup f^{-1}(\mathrm{cl}\, U) \cup \ldots \cup f^{k-1}(\mathrm{cl}\, U)$ is contained inside a finite number of mutually disjoint balls.

Let us emphasize once more that all the above deformations take place on a subset isolated from $\mathrm{Fix}(f^k)$ and can be arbitrarily small hence $\mathrm{Fix}(f^k)$ remains invariant. In contrast, periodic points x_{k-1} and y_{l-1} belong to the boundary of the carrier of the next homotopy.

(CP3) We will show that there is a homotopy $f_t\colon M \to M$ with the carrier in V_{k-1} such that:

(5.24.1) $f_0 = f$ and $\mathrm{Fix}(f_1^k) = \mathrm{Fix}(f^k) \cup \{z_0, \ldots, z_{k-1}\}$ (where $z_i = \omega_i(0)$),

(5.24.2) $f_i(\omega_0[-1,+1]) = \omega_i[-1,+1]$ for $i = 0,\ldots k-1$ and $f_k(\omega_0[-1,+1]) = \omega_0[-1,+1]$.

(CP4) After a homotopy f_τ $(1 \le \tau \le 2)$ constant in a neighbourhood of $\omega_0 \cup \ldots \cup \omega_{k-1}$ we get $\mathrm{Fix}(f_2^k) = \mathrm{Fix}(f^k) \setminus \{x_0, \ldots, x_{k-1}\}$. The proof of each step will be given in a subsubsection.

5.6.1. Disjoint arcs. We assume that $\mathrm{Fix}(f^k)$ is finite, $x_0 \in \mathrm{Fix}(f^k)$ and $y_0 \in \mathrm{Fix}(f^l)$ (for an $l|k$) are Nielsen related as fix points of f^k. Moreover, S^k_r, $S^l_{r'}$ conditions are satisfied near these points.

Let us fix a Euclidean neighbourhood $\mathbb{R}^m = U \subset M$ of the fixed point $y_0 = 0$ where $S^l_{r'}$ is satisfied. In particular $|f^l(x)| = \lambda|x|$ in a ball neighbourhood B of 0 for a $\lambda > 1$. Similarly we fix a Euclidean neighbourhood of x_0 where S^k_r holds. We fix an arc $\omega_0\colon [-1,+1] \to M$ for which $\omega_0(-1) = y_0$, $\omega_0(+1) = x_0$, the paths $\omega_k = f^k\omega_0$ and ω_0 are homotopic (Nielsen relation). Now as in Lemma (4.11) we may assume that the arcs $\omega_0(-1,-1+\varepsilon], f(\omega_0(-1,-1+\varepsilon]), \ldots, f^{3k-1}(\omega_0(-1,-1+\varepsilon])$ are mutually disjoint. The same can be done near x_0. Then we may follow the proof of Lemma (4.11) to make $\omega_0, \ldots, \omega_{3k-1}$ arcs with mutually disjoint interiors.

Finally we may apply Lemma (4.7) to make $f, f^2, \ldots, f^{k-1}$ transverse to z_{k-1}.

5.6.2. Neighbourhoods of the arcs ω_i. To get (CP2) we follow the proof of Lemma (4.13), the main difficulty is that the neighbourhoods $V_i \supset \omega(-1,+1)$ should be mutually disjoint. For any $i = 0, \ldots, k-1$ we establish a neighbourhood $U_i = \mathbb{R}^m$ so that $\omega_i(t) = (t, 0) \in \mathbb{R} \times \mathbb{R}^{m-1}$ for $-1 \le t \le +1$.

Consider the arc $\omega_0[-1,+1] \subset U_0$. We fix positive numbers $\varepsilon_0 > 0$, $\delta_0 > 0$ and we define sets $C_- = \{(t,x) \in \mathbb{R} \times \mathbb{R}^{m-1} \colon t = -1 + \varepsilon_0,\ |x| \le \delta_0\}$, $C_+ = \{(t,x) \in \mathbb{R} \times \mathbb{R}^{m-1} \colon t = +1 - \varepsilon_0,\ |x| \le \delta_0\}$ and $D_- = \mathrm{conv}\{y_0 \cup C_-\}$, $D_+ = \mathrm{conv}\{x_0 \cup C_+\}$, $D_0 = \mathrm{conv}\{C_- \cup C_+\}$. Notice that $D_0 = \bigcup_{|x| \le \delta_0}[(-1+\varepsilon_0, x), (1-\varepsilon_0, x)]$. We denote $V_0 = \mathrm{int}(D_- \cup D_0 \cup D_+)$.

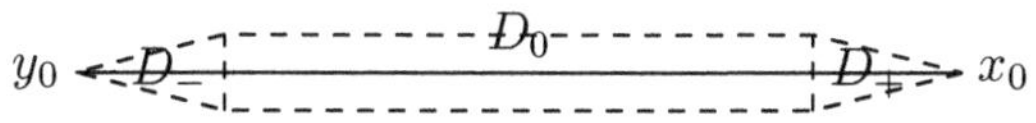

FIGURE 17

Then we fix numbers $\varepsilon_1, \delta_1 > 0$ and we define in a similar way a neighbourhood $V_1 \supset \omega_1(-1,+1)$. Let us notice that if the numbers δ_0, δ_1 are sufficiently small then

$$\mathrm{cl}\, V_0 \cap \mathrm{cl}\, V_1 = \begin{cases} \emptyset & \text{for } l > 1, \\ \{y_0\} & \text{for } l = 1. \end{cases}$$

Since $f\omega_0(-1,+1) = \omega_1(-1,+1)$, for fixed ε_1, δ_1 we may find ε_0, δ_0 such that $f(V_0) \subset V_1$.

Following this we may define neighbourhoods V_i ($i = 0, \dots, k-1$) of the above form and satisfying $\omega_i(-1,+1) \subset V_i \subset U_i$, $f(V_i) \subset V_{i+1}$, and

$$\operatorname{cl} V_i \cap \operatorname{cl} V_j = \begin{cases} \{y_i\} & \text{for } 0 \le i < j \le k-1,\ l | j-i, \\ \emptyset & \text{otherwise.} \end{cases}$$

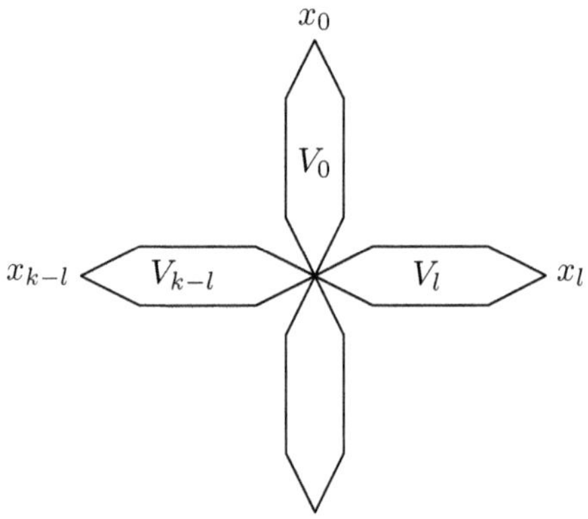

FIGURE 18

We define V_{2k-1} then we choose V_{2k-2} and so on. Now we are in a position to follow the proof of Lemma (4.13).

Let us emphasize that the neighbourhood $V_0 \subset U_0 = \mathbb{R} \times \mathbb{R}^{m-1}$ may be presented as

$$V_0 = \{(t,x) : -1 < t < +1,\ |x| < \rho(t)\}$$

for a continuous concave function $\rho\colon [-1,+1] \to \mathbb{R}$ satisfying $\rho(\pm 1) = 0$ and $\rho(x) > 0$ for $-1 < x < +1$. Moreover, we may assume that ρ is linear on intervals $[-1,-1+\varepsilon]$, $[1-\varepsilon, 1]$ and is constant on $[-1+\varepsilon, 1-\varepsilon]$ and ρ may be arbitrarily small.

To get the last statement of (CP2) we notice that the maps $f, f^2, \dots, f^{k-1}$ were transverse to z_{k-1} and the deformations could be arbitrarily small. It remains to apply Corollary (4.9).

5.6.3. Property (CP3). We consider a map satisfying (CP2). We will construct a deformation $\{f_s\}$ $0 \le s \le 3$ such that: $\operatorname{supp}\{f_s\} \subset \operatorname{cl} V_{k-1}$, $f_0 = f$, $\operatorname{Fix}(f_s)$ is constant. Moreover, there is a thinner convex neighbourhood $V'_{k-1} \supset \omega_{k-1}(-1,+1)$ and a number $\eta > 0$ such that for $(t,x) \in V'_{k-1} \subset \mathbb{R} \times \mathbb{R}^{m-1}$ we have:

(5.25.1) $f_1(t,x) = f(r(t),0)$ and moreover, $f_1(t,x) = y_0 \in V_0$ for $t \le -\eta$, $f_1(t,x) = x_0 \in V_0$ for $t \ge +\eta$ where $r\colon [-1,+1] \to [-1,+1]$ is given by the formula

$$r(t) = \begin{cases} -1 & \text{for } t \leq -\eta, \\ t/\eta & \text{for } -\eta \leq t \leq +\eta, \\ +1 & \text{for } +\eta \leq t. \end{cases}$$

(5.25.2) $f_2(x,t) = (r(t), 0, \rho_0(r(t))) \in \operatorname{bd} V_0' \subset \mathbb{R}\times\mathbb{R}^{m-2}\times\mathbb{R}$ where $\rho_0\colon [-1,1] \to \mathbb{R}$ is the concave function defining V_0.

In the two above steps $\operatorname{Fix}(f^k)$ does not change.

(5.25.3) Finally we give a formula for a homotopy from f_2 to a map f_3 mapping ω_{k-1} to ω_0.

Ad (5.25.1). Since the arcs ω_i $(i = 0, \dots, 2k-1)$ have mutually disjoint interiors, we may choose the neighbourhoods V_i satisfying: $\omega_i(-1,+1) \subset V_i$, $f(V_i) \subset V_{i+1}$ and

$$\operatorname{cl} V_i \cap \operatorname{cl} V_j = \begin{cases} \{x_i\} & \text{for } 0 \leq i < j \leq k-1,\ k|j-i, \\ \{y_i\} & \text{for } 0 \leq i < j \leq k-1,\ l|j-i, \\ \emptyset & \text{otherwise.} \end{cases}$$

Let us notice that if $h\colon M \to M$ is a map constant outside V_{k-1} and $h(\operatorname{cl} V_{k-1}) \subset \operatorname{cl} V_{k-1}$ then the composition $f_1 = fh$ is homotopic to f (since $\operatorname{cl} V_{k-1}$ is contractible) and $\operatorname{Fix}(f_1^k) = \operatorname{Fix}(f^k)$ (since $f_1(\operatorname{cl} V_{k-1}) \subset \operatorname{cl} V_k$, $f_2(\operatorname{cl} V_{k-1}) \subset \operatorname{cl} V_{k+1} \cdots$ and these sets are disjoint from V_{k-1} Lemma (4.5)).

Let $\eta > 0$ be so small that for $W_\eta = \{(t,x) \in V_{k-1}\colon |t| \leq \eta\}$ the sum $W_\eta \cup f^{-1}(W_\eta) \cup \ldots \cup f^{k-1}(W_\eta)$ is contained inside a finite number of mutually disjoint closed balls $\bigcup_\alpha K_\alpha$ (see (CP2)). Let

$$V_{k-1}' = \{(t,x) \in V_{k-1} : (t, 2x) \in V_{k-1}\}$$

be a thinner convex neighbourhood of $\omega_{k-1}(-1,+1)$.

We take as $h\colon M \to M$ a map satisfying: h is constant outside V_{k-1}, maps the convex set $\operatorname{cl} V_{k-1}$ into itself and for moreover, for $(t,x) \in \operatorname{cl} V_{k-1} \subset \mathbb{R}^{m-1} \times \mathbb{R}$,

$$h(t,x) = \begin{cases} x_{k-1} & \text{for } t \leq -\eta, \\ y_{k-1} & \text{for } t \geq +\eta, \\ (r(t), 0) & \text{for } (t,x) \in V_{k-1}', \end{cases}$$

where $r\colon [-1,+1] \to [-1,+1]$ is given above. Then the map $f_1 = fh$ satisfies condition (5.25.1) (on V_{k-1}').

Ad (5.25.2). Thus we may assume (replacing V_{k-1} by V_{k-1}') that $f_1(t,x) = f_1(t,0)$ for all $(t,x) \in V_{k-1}$.

Consider the neighbourhood $V_0 \subset \mathbb{R}^m$ of $\omega_0(-1,+1)$. We may assume that this neighbourhood is of the form $V_0 = \{(t,x) \in \mathbb{R}^{m-1} \times \mathbb{R} : \|x\| < \rho_0(t)\}$. Let us define a path $\overline{\omega}\colon [-1,+1] \to \operatorname{bd} V_0$ by the formula $\overline{\omega}(t) = (r(t), 0, \rho_0(r(t))) \in \mathbb{R} \times \mathbb{R}^{m-2} \times \mathbb{R}$. This path is homotopic to ω_0 (rel. end points). Since $\omega_0 \sim f^k(\omega_0) = f(\omega_{k-1})$ (Nielsen relation), there is a partial homotopy $f'_s\colon \omega_{k-1}[-1,+1] \to M$ satisfying: $f'_0(\omega_{k-1}(t)) = f(\omega_{k-1}(t))$, $f'_1(\omega_{k-1}(t)) = \overline{\omega}_{k-1}(t)$ and moreover $f'_s(\omega_{k-1}(t)) = y_0$ for $t \le -\eta$, $f_s(\omega_{k-1}(t)) = x_0$ for $t \ge +\eta$. We may correct this homotopy (as in the proof of Cancelling Property) to get $f'_s(t,x) \notin \bigcup K_\alpha$ since $\pi_2(M, M \setminus \bigcup K_\alpha) = 0$. The last implies $f^{k-1} f'_s(z) \ne z$ for any $z \in \omega_{k-1}([-1,+1])$.

We extend the partial homotopy: on $V'_{k-1} = \{(t,x) \in V_{k-1}; (t,2x) \in V_{k-1}\}$ putting $f_s(t,x) = f_s(t,0)$ and on $\operatorname{bd} V_{k-1}$ by the constant homotopy. By the Homotopy Extension Property we get an extension of the homotopy onto $\operatorname{cl} V_{k-1}$. Since the last homotopy is constant on the boundary, it extends by the constant homotopy into the whole M.

Let us denote $f\omega_{k-1}(t) = (r(t), x^*_t) \in \operatorname{bd} V_0 \subset U_0$ for $x^*_t = (0, \rho_0(r(t)))$. Moreover, we may compose the last homotopy with the retraction of a thinner neighbourhood onto $\omega_1(-1,+1)$ and assume that $f\omega_{k-1}(t,x) = (r(t), x^*_t)$ for (x,t) from this thinner neighbourhood. Replacing V_{k-1} by this thinner neighbourhood we may assume that the equality holds on the whole V_{k-1}.

Ad (5.25.3). The last step is the homotopy $H_s\colon M \to M$ constant outside V_{k-1} and mapping $\operatorname{cl} V_{k-1}$ into $\operatorname{cl} V_0$ by the formula

$$H_s(x,t) = \left(r(t), 0, \left((1-s) + s \cdot \frac{\|x\|}{\rho_{k-1}(t)}\right) \cdot \rho_0(r(t))\right) \in V_0 \subset \mathbb{R} \times \mathbb{R}^{m-2} \times \mathbb{R}.$$

It remains to notice that the last homotopy has no new fixed points in $\operatorname{cl} V_{k-1} \setminus \omega_{k-1}[-1,+1] = \{(t,x) \in V_{k-1} \subset \mathbb{R} \times \mathbb{R}^{m-1} \colon t \ne 0\}$. In fact since the map r is expanding for $|t| \le \eta$, $f_1^k(t,x) = (t^*, x^*)$ and $t \ne 0$ imply $|t| < |t^*|$. The map $f_3 = H_1$ satisfies (CP3).

5.6.4. Property (CP4). We assume that $f\colon M \to M$ satisfies (CP3). Then $T = \omega_0[-1,+1] \cup \ldots \cup f^{k-1}(\omega_0[-1,+1])$ splits into $T = T_0 \cup \ldots \cup T_{l-1}$ where $T_i = f^i\omega_0[-1,+1] \cup f^{i+l}(\omega_0[-1,+1]) \cup \ldots f^{i+(k-l)}(\omega_0[-1,+1])$. Then each T_i is a connected component of $\operatorname{Fix}(f^k)$. Moreover, $f(T_i) = T_{i+1}$ for $0 \le i \le l-2$ and $f(T_{l-1}) \subset T_0$. Since the manifolds are homogeneous we have the following

(5.26) Lemma. *There is a homotopy $h_t\colon M \to M$ satisfying:*

(5.26.1) $h_0 = \mathrm{id}$,

(5.26.2) $h_t(T) \subset T$,

(5.26.3) *h_t is a homeomorphism for $t < 1$ and the restriction $f_1\colon M \setminus T \to M \setminus \{y_0, \ldots, y_{l-1}\}$ is a homeomorphism,*

(5.26.4) *the carrier of the homotopy is contained in a prescribed neighbourhood of* $T \setminus \{y_0, \dots, y_{l-1}\}$.

PROOF. We define the homotopy $f_t(x) = h_t f h_t^{-1}(x)$. The definition is correct since $h_t^{-1}(x)$ is not a singleton only for $t = 1$ and $x \in \{y_0, \dots, y_{l-1}\}$. But then $h_1 f h_1^{-1}(y_i) = h_1 f(T_i) = h_1(T_{i+1}) = y_{i+1}$. Now we show that

$$\mathrm{Fix}(f_1^k) = (\mathrm{Fix}(f_0^k) \setminus T) \cup \{y_0, \dots, y_{l-1}\}.$$

$\supset$ is evident since the homotopy is constant on $(\mathrm{Fix}(f_0^k) \setminus T) \cup \{y_0, \dots, y_{l-1}\}$.

Now we prove $\subset$. Let $x \in \mathrm{Fix}(f_1^k)$. If $x \in \{y_0, \dots, y_{l-1}\}$ then x belongs to the right hand side. Now we assume that $x \notin \{y_0, \dots, y_{l-1}\}$. Then $h_1^{-1}(x)$ is a single point hence $x = f_1^k(x) = h_1 f^k h_1^{-1}(x)$ implies $h_1^{-1}(x) = f^k h_1^{-1}(x)$ hence $h_1^{-1}(x) \in \mathrm{Fix}(f^k)$. Thus $x \in h_1(\mathrm{Fix}(f^k)) = (\mathrm{Fix}(f_0^k) \setminus T) \cup \{y_0, \dots, y_{l-1}\}$.

Thus we have proved that each orbit of points can be shifted to an orbit of points preceding it. □

END OF THE PROOF OF THEOREM (5.13). Now we may end the proof of the Wecken Theorem for periodic points (see the scheme of the proof in Subsection 5.4). We may assume that f satisfies Lemma (5.14). It remains to remove the orbits of points which are preceded by the orbits from MPS. Coalescing Procedure allows to remove each orbit individually. It remains to notice that this operation can be done simultaneously for all such orbits (without producing new orbits).

This is possible since we are working in dimension ≥ 3. Majority of deformations are arbitrarily small with supports near 1-dimensional segments which can be made mutually disjoint. Only two moments of the construction should be explained.

Consider the orbits $\{x_0^\gamma, \dots, x_{k-1}^\gamma\}$ $\gamma \in \Gamma$ which should be coalesced to the orbit $\{y_0, \dots, y_{l-1}\}$. We should be able to make the arcs ω_0^γ and their iterations $f^i \omega_0^\gamma$ mutually disjoint. This can be achieved by choosing the segments $\omega_0^\gamma(1-\varepsilon, 1)$ on different "levels", i.e. we put $\omega_0^\gamma(t) = (0, (1-t)r_\gamma, t) \in \mathbb{R}^{n-2} \times \mathbb{R} \times \mathbb{R}$ for different r_γ. Because of the S_r conditions the iterations will be also on different levels (for $1 - \varepsilon < t < 1$).

The homotopies from $f\omega_{k-1}^\gamma$ to ω_0^γ (which avoid the finite sum of balls $\bigcup_\alpha K_\alpha^\gamma$) should have disjoint supports and all should avoid such sums. But all these sums $(\bigcup_\alpha K_\alpha^\gamma)$ can be made mutually disjoint, since we can make $f, \dots, f^{k-1}$ simultaneously transverse to the points z_{k-1}^γ $(\gamma \in \Gamma)$. □

6. The least number of periodic points of the given minimal periods

The procedures from the last two sections allow to answer another natural question. One of the basic problems in dynamical systems is the existence of periodic points of the given minimal period. As we have seen if the number $NP_n(f) \neq 0$

then such a point (and even an orbit of length n) must exist for each map homotopic to f. Is the above inequality the only obstruction to a deformation of f to a map with no periodic point of the minimal period n? We will show that in dimensions ≥ 3 the answer is positive.

(6.1) THEOREM. *Let $f\colon M \to M$ be a self-map of a compact* PL*-manifold of dimension $d \geq 3$. Then f is homotopic to a map g satisfying $P_n(g) = \emptyset$ if and only if $NP_n(f) = 0$.*

PROOF. Let us recall that $NP_n(f) = 0$ if and only if there is no essential irreducible Reidemeister class in $\mathcal{R}(f^n)$. Now $\Rightarrow$ is evident.

It remains to prove $\Leftarrow$. By Theorem (4.2) we may assume that $\mathrm{Fix}(f^n)$ is finite and moreover, f is a local PL-homeomorphism near each fixed point of f^n. Let us denote $\mathrm{Fix}(f^n) = A \cup B$ where A denotes the sum of all irreducible classes and B the sum of all reducible ones.

Lemma (6.2) gives a homotopy $\{f_t\}_{0\leq t\leq 1}$ which is constant in a neighbourhood of $\mathrm{Fix}(f^n) \setminus A = B$, $f_0 = f$ and $\mathrm{Fix}(f_1^n) = \mathrm{Fix}(f_0^n) \setminus A$. Since the homotopy is constant in $\mathrm{Fix}(f_1^n)$, all classes in $\mathrm{Fix}(f_1^n)$ are reducible. Now Lemma (6.4) yields a homotopy from f_1 to a map g satisfying $P_n(g) = \emptyset$.

The proof of Theorem (6.1) will be complete once Lemmas (6.2) and (6.4) are proved. □

(6.2) LEMMA. *Let $f\colon M \to M$ be a self-map of a compact* PL*-manifold satisfying $NP_n(f) = 0$ where* $\mathrm{Fix}(f^n)$ *is finite and f is a local PL-homeomorphism near each fixed point of f^n. Let A be the sum of all irreducible Nielsen classes of f^n i.e. classes of depth $= n$. Then there is a homotopy $\{f_t\}$ constant in a neighbourhood of* $\mathrm{Fix}(f^n) \setminus A$ *and satisfying $f_0 = f$,* $\mathrm{Fix}(f_1^n) = \mathrm{Fix}(f_0^n) \setminus A$.

PROOF. Since each orbit in A is irreducible, the length of each orbit of points in A is n. Since the orbits of Nielsen classes are inessential ($NP_n(f) = 0$), their sum splits into pairs of orbits of length n of opposite indices ± 1.

Consider a pair of orbits of points $\{x_1, \dots, x_n; y_1, \dots, y_n\} \subset A$ of opposite indices. The Cancelling Procedure yields a homotopy constant in a neighbourhood of $\mathrm{Fix}(f^n) \setminus \{x_1, \dots, x_n; y_1, \dots, y_n\}$ such that $f_0 = f$ and $\mathrm{Fix}(f_1^n) = \mathrm{Fix}(f^n) \setminus \{x_1, \dots, x_n; y_1, \dots, y_n\}$. Following this procedure we can reduce the number of such orbits to zero hence we get $\mathrm{Fix}(f_1^n) = \mathrm{Fix}(f_0^n) \setminus A$ as required. □

(6.3) REMARK. If f_1 satisfies the above Lemma then all the Nielsen classes in $\mathrm{Fix}(f^n)$ are reducible.

(6.4) LEMMA. *If all Nielsen classes in $P_n(f)$ are reducible then f is homotopic to a map g satisfying $P_n(g) = \emptyset$.*

PROOF. By Theorem (4.2) we may assume that $\mathrm{Fix}(f^n)$ is finite and f is a PL-homeomorphism near each fixed point of f^n. If each orbit of points in $\mathrm{Fix}(f^n)$ has length $< n$ then $P_n(f) = \emptyset$. Now we assume that $\mathrm{Fix}(f^n)$ contains exactly one orbit $\{a_0, \dots, a_{n-1}\}$ of length n. By the assumption the orbit of the Nielsen classes containing $\{a_0, \dots, a_{n-1}\}$ is reducible hence we may assume that there is also an orbit $\{b_0, \dots, b_k\}$ of length $k < n$ in this class. In general such orbit may not exist but then we may apply Addition Formula and create such orbit. Now we may apply Coalescing Procedure and we get a homotopy constant on $\mathrm{Fix}(f^n) \setminus \{a_0, \dots, a_{n-1}\}$ such that $f_0 = f$, $\mathrm{Fix}(f_1^n) = \mathrm{Fix}(f^n) \setminus \{a_0, \dots, a_{n-1}\}$. Then all orbits of points in $\mathrm{Fix}(f_1^n)$ have length $< n$ hence $P_n(f_1) = \emptyset$.

In general $\mathrm{Fix}(f^n)$ may contain several orbits of length n. Then we use the arguments from Subsection 5.7 and we may coalesce simultaneously all these orbit to shorter ones. □

One may expect that such operation of realizing $NP_k(f)$ can be done simultaneously for two or more periods. But the example of antipodism on S^{2m} (Example (5.12)) shows that this is not possible in general: we can not remove points of periods 1 and 2 simultaneously although there is no essential irreducible orbit.

The next theorem makes precise for which periods it is possible to remove periodic points simultaneously.

(6.5) THEOREM. *Let $f\colon M \to M$ be a self map of a compact PL-manifold of dimension ≥ 3. Let $N_0 \subset \mathbb{N}$ be finite. Then there is a homotopy $f_t\colon M \to M$ such that $f_0 = f$ and $\mathrm{P}_r(f_1) = \emptyset$ for all $r \in N_0$ if and only if for any $r \in N_0$ any essential Reidemeister class $A^r \in \mathcal{R}(f^r)$ reduces to a class $B^s \in \mathcal{R}(f^s)$ for an $s \notin N_0$.*

PROOF. $\Rightarrow$ Assume that $\mathrm{P}_r(f) = \emptyset$ for all $r \in N_0$. Consider an essential orbit of Reidemeister classes $A^r \in \mathcal{OR}(f^r)$ where $r \in N_0$. Since A^r is essential, it contains an orbit of points $\{x_0, \dots, x_{s-1}\}$ for an $s|r$. Then $P_s(f) \neq \emptyset$ hence $s \notin N_0$. Now the orbit of Reidemeister classes A^r reduces to the orbit $B^s \in \mathcal{OR}(f^s)$ represented by the points $\{x_0, \dots, x_{s-1}\}$.

$\Leftarrow$ We use the induction with the respect to the number $l = \#N_0$. For $l = 1$ the Theorem follows from Theorem (6.1). Now we assume that the Theorem holds for $< l$. Let $N_0 \subset \mathbb{N}$ be a subset of cardinality l. Let r be the greatest element in N_0. By inductive assumption f is homotopic to a map f_1 satisfying $\mathrm{P}_s(f_1) = \emptyset$ for all $s \in N_0 \setminus \{r\}$. It remains to remove all orbits from $\mathrm{P}_r(f_1)$. Let $A^r \in \mathcal{OR}(f_1^r)$ be nonempty.

Suppose that A^r does not reduce (as a Reidemeister class) to any class B^s with $s \notin N_0$. Then A^r is inessential and each orbit of points in A^r must be of length r.

Now A^r splits into the pairs of orbits of points of length of opposite indices. We may apply Cancelling Procedure to remove A^r.

Now we suppose that A^r reduces to an orbit of Reidemeister classes $B^s \in \mathcal{OR}(f_1^s)$, $s \notin N_0$. If $B^s \neq \emptyset$ then we may apply Coalescing Procedure and we match A^r to B^s. If $B^s = \emptyset$ then we apply Addition Property and we create a new orbit of length s representing B^s and we may apply Coalescing Procedure. Thus each $A^r \in \mathcal{OR}(f_1^r)$ may be either removed or coalesced to an orbit or points of length $s \notin N_0$. Following the arguments from the end of the proof of Coalescing Procedure we deduce that these operations can be done simultaneously to all orbits of Reidemeister classes in $\mathcal{OR}(f_1^r)$. After this $P_s(f_1) = \emptyset$ for all $s \in N_0$. □

References

[B] R. F. Brown, *The Lefschetz Fixed Point Theorem*, Glenview, New York, 1971.

[D] A. Dold, *Fixed point indices of iterated maps*, Invent. Math. **74** (1983), 419-435.

[FH] E. Fadell and S. Husseini, *Fixed point theory for non-simply connected manifolds*, Topology **20** (1981), 53–92.

[H1] B. Halpern, *Periodic points on tori*, Pacific J. Math. **83** (1979), 117–133.

[H2] ———, *The minimum number of periodic points*, Abstract 775-G8, Abstracts Amer. Math. Soc. **1** (1980), 269.

[H3] ———, *Periodic points on the Klein bottle*, preprint.

[H4] ———, *Nielsen type numbers for periodic points*, preprint.

[HaK] E. Hart and E. Keppelmann, *Nielsen periodic number for periodic maps on orientable sudfaces*, Pacific J. Math.

[HK1] Ph. Heath and C. Y. You, *Nielsen type numbers for periodic points*, Topology Appl. **43** (1992), 219–236.

[HK2] ———, *Fibre techniques in Nielsen periodic point theory on nil and solvmanifolds* I, Topology Appl. **76** (1997), 217–247.

[HK3] ———, *Fibre techniques in Nielsen periodic point theory on nil and solvmanifolds* II.

[HK4] Ph. Heath and E. Keppelmann, *Fibre techniques in Nielsen periodic point theory on nil and solvmanifolds* III.

[Je1] J. Jezierski, *Canceling periodic points*, Math. Ann. **321** (2001), 107–130.

[Je2] J. Jezierski, *Wecken's theorem for periodic points*, Topology **42** (2003), 1101–1124.

[Je3] ———, *Weak Wecken's theorem for periodic points in dimension* 3, Fund. Math. (to appear).

[Je4] ———, *Wecken's theorem for periodic points in dimension* 3, submitted.

[JM] J. Jezierski and W. Marzantowicz, *Homotopy minimal periods for nilmanifolds maps*, Math. Z. **239** (2002), 381–414.

[Ji1] B. J. Jiang, *Lectures on the Nielsen Fixed Point Theory*, Contemp. Math., vol. 14, Amer. Math. Soc., Providence, 1983.

[Ji2] ———, *Fixed point classes from a differential viewpoint*, Lecture Notes in Math., vol. 886, Springer, 1981, pp. 163–170.

[Ji3] ———, *The least number of fixed points*, American J. Math. **102** (1980), 749–763.

[Ji4] ———, *Fixed points and braids*, Invent. Math. **75** (1984), 69–74.

[Ji5] ———, *Fixed points and braids* II, Math. Ann. **272** (1985), 249–256.

[Ji6] ———, *Surface maps and braid equations*, Differential Geometry and Topology (Tianjin, 1988), Lecture Notes in Math., vol. 1411, Springer, Berlin, 1989, pp. 125–141.

[Ji7] ———, *Estimation of the number of periodic orbits*, Pacific J. Math. **172** (1966), 151–185.

[Ke] M. Kelly, *Minimizing the number of the fixed points for self-maps of compact surfaces*, Pacific J. Math. **126** (1987), 81–123.

[Ki] T. H. Kiang, *The Theory of Fixed Point Classes*, Springer–Verlag, Berlin–Heidelberg–New York, 1989.
[Mc] Ch. McCord, *Nielsen numbers and Lefschetz numbers on solvmanifolds*, Pacific J. Math. **147** (1991), 153–164.
[N] J. Nielsen, *Uber die Minimalzahl der Fixpunkte bei Abbildungstypen der Ringflachen*, Math. Ann. **82** (1921), 83–90.
[Shi] H. G. Shi, *On the least number of fixed points and Nielsen numbers*, Acta Math. Sinica **16** (1966), 223–232.
[Sch1] H. Schirmer, *Mindestzahlen von Koinzidenzpunkten*, J. Reine Angew. Math. **194** (1955), 21–39.
[Sch2] ———, *A relative Nielsen number*, Pacific J. Math. **122** (1986), 459–473.
[W] F. Wecken, *Fixpnktklassen* I, Math. Ann. **117** (1941), 659-671; II **118** (1942), 216–234; III **118** (1942), 544–477.
[Y1] C. Y. You, *The least number of periodic points on tori*, Adv. in Math. (China) **24** (1995), 155–160.
[Y2] ———, *A note on periodic points on tori*, Beijing Math. **1** (1995), 224–230.
[Z] X. G. Zhang, *The least number of fixed points can be arbitrarily larger than the Nielsen number*, Beijing Daxue Xuebao **3** (1986), 15–25.

16. A PRIMER OF NIELSEN FIXED POINT THEORY

Boju Jiang

1. Introduction

Let X be a space, and let $f: X \to X$ be a self-map. A fixed point of f is a solution of the equation $x = f(x)$. The set of all fixed points of f we will denote by $\mathrm{Fix}(f)$. Fixed point theory studies the nature of the fixed point set $\mathrm{Fix}(f)$ in relation to the space X and the map f, such as: existence (is $\mathrm{Fix}(f) \neq \emptyset$?); the number of fixed points $\#\mathrm{Fix}(f)$ (we will use the notation $\#S$ for the cardinality of a set S); the behavior under homotopy (how $\mathrm{Fix}(f)$ changes when f changes continuously); etc.

Fixed point theory started in the early days of topology, because of its close relationship with other branches of mathematics. Existence theorems are often proved by converting the problem into an appropriate fixed point problem. For example, in the qualitative theory of dynamical systems, a closed orbit in a flow can be viewed as a fixed point of the return map on a section of the flow. And in nonlinear elliptic partial differential equations, a solution can be formulated as a fixed point of a suitable map in a functional space. In many problems, however, one is not satisfied with the mere existence of a solution. One wants to know the number, or at least a lower bound for the number of solutions. But the actual number of fixed points of a self-map can hardly be the subject of an interesting theory, since it can be altered by an arbitrarily small perturbation of the map. So, in topology, one proposes to determine the minimal number of fixed points in a homotopy class. This is the motivation of Nielsen fixed point theory.

Perhaps the best known fixed point theorem in topology is the Lefschetz fixed point theorem.

(1.1) Theorem (Lefschetz 1923; Hopf 1929). *Let X be a compact polyhedron, and let $f: X \to X$ be a map. Define the Lefschetz number $L(f)$ of f to be*

$$L(f) := \sum_q (-1)^q \operatorname{trace}(f_{*q}\colon H_q(X;\mathbb{Q}) \to H_q(X;\mathbb{Q})),$$

Partially supported by a MOSTC and a MOEC grants.

where $H_q(X;\mathbb{Q})$ is the rational homology of X. If $L(f) \neq 0$, then every map homotopic to f has a fixed point.

The Lefschetz number is the total algebraic count of fixed points. It is a homotopy invariant and is easily computable. But it counts the fixed points "by multiplicity", just like what one does when one says an equation of degree n has n roots. So, the Lefschetz theorem, along with its special case, the Brouwer fixed point theorem, and its generalization, the widely used Leray–Schauder theorem in functional analysis, can tell of existence only. In contrast, the (chronologically) first result of Nielsen theory has set a beautiful example of a different type of theorem.

(1.2) THEOREM (Nielsen–Brouwer, 1921). *Let $f: T^2 \to T^2$ be a self-map of the torus. Suppose the endomorphism induced by f on the fundamental group $\pi_1(T^2) \cong \mathbb{Z} \oplus \mathbb{Z}$ is represented by the 2×2 integral matrix A. Then the least number of fixed points in the homotopy class of f equals the absolute value of the determinant of $E - A$, where E is the identity matrix; in symbols,*

$$\min\{\#\operatorname{Fix}(g) : g \simeq f\} = |\det(E - A)|.$$

It can be shown that $L(f) = \det(E - A)$ on the torus. The Nielsen–Brouwer theorem says much more than the Lefschetz theorem specialized to the torus, since it gives a lower bound for the number of fixed points, or it confirms the existence of a homotopic map which is fixed point free. The proof was via the universal covering space $\mathbb{R}^2$ of the torus. From this instance evolved the central notions of Nielsen theory — the fixed point classes and the Nielsen number. Self-maps of the circle present the simplest case for illustrating Nielsen theory, which we now describe following the exposition in the classic [AH, p. 533].

(1.3) PROPOSITION. *Let $f: S^1 \to S^1$ be a self-map of the circle. Suppose the degree of f is d. Then the least number of fixed points in the homotopy class of f is $|d - 1|$.*

PROOF. Let S^1 be the unit circle on the complex plane, i.e.

$$S^1 = \{z \in \mathbb{C} : |z| = 1\}.$$

Let $p: \mathbb{R} \to S^1$ be the exponential map $p(\theta) = z = e^{i\theta}$. Then θ is the argument of z, which is a multi-valued function of z. For every $f: S^1 \to S^1$ one can always find "argument expressions" (or liftings) $\widetilde{f}: \mathbb{R} \to \mathbb{R}$ such that $f(e^{i\theta}) = e^{i\widetilde{f}(\theta)}$, in fact a whole series of them, differing from each other by integral multiples of 2π. We shall pick an arbitrary lifting $\widetilde{f}_0$ as the "principal" one. Any lifting is of the form

$\widetilde{f_k} = \widetilde{f_0} - 2k\pi$, k being an integer. Since the degree of f is d, these functions are such that $\widetilde{f_k}(\theta + 2\pi) = \widetilde{f_k}(\theta) + 2d\pi$.

Suppose z is a fixed point of f, and θ is an argument of z. It is evident that $\widetilde{f_0}(\theta) = \theta + 2k\pi$, or, equivalently, $\theta \in \mathrm{Fix}(\widetilde{f_k})$, for some integer k. This integer k is determined by z only up to a multiple of $d-1$, because for any other argument $\theta' = \theta + 2q\pi$ of z,

$$\widetilde{f_0}(\theta') = \widetilde{f_0}(\theta + 2q\pi) = \widetilde{f_0}(\theta) + 2qd\pi = \widetilde{f_0}(\theta') + 2q(d-1)\pi.$$

Thus, if $k \equiv \ell \bmod (d-1)$, then fixed points of $\widetilde{f_k}$ and fixed points of $\widetilde{f_\ell}$ project to the same fixed points of f, i.e. $p\,\mathrm{Fix}(\widetilde{f_k}) = p\,\mathrm{Fix}(\widetilde{f_\ell})$; if $k \not\equiv \ell \bmod (d-1)$, then a fixed point of $\widetilde{f_k}$ and a fixed point of $\widetilde{f_\ell}$ never project to the same fixed point of f, i.e. $p\,\mathrm{Fix}(\widetilde{f_k}) \cap p\,\mathrm{Fix}(\widetilde{f_\ell}) = \emptyset$.

So, the liftings fall into equivalence classes (called lifting classes) by the relation $\widetilde{f_k} \sim \widetilde{f_\ell}$ if and only if $k \equiv \ell \bmod (d-1)$, and the fixed points of f split into $|d-1|$ classes (called fixed point classes) of the form $p\,\mathrm{Fix}(\widetilde{f_k})$. That is, two fixed points are in the same class if and only if they come from fixed points of the same lifting. Note that each fixed point class is by definition associated with a lifting class, so that the number of fixed point classes is $|d-1|$ if $d \neq 1$, and is ∞ if $d = 1$. Also note that a fixed point class need not be nonempty.

Now, to prove that a map f of degree d has at least $|d-1|$ fixed points, we only have to show that every fixed point class is nonempty, or equivalently, that every lifting has a fixed point, if $d \neq 1$. In fact, for each k, by means of the equality $\widetilde{f_k}(\theta+2\pi) - \widetilde{f_k}(\theta) = 2d\pi$, it is easily seen that the function $\theta - \widetilde{f_k}(\theta)$ takes different signs when θ approaches $\pm\infty$, hence $\widetilde{f_k}$ has at least one fixed point.

That $|d-1|$ is indeed the least number of fixed points in the homotopy class is seen by checking the special map $f(z) = -z^d$. □

The standard references of Nielsen fixed point theory are: [Br1], [Ji4], [Ki].

This chapter has two parts. The first part focuses on the central notion of Nielsen fixed point theory — the fixed point classes, and contains detailed proofs. The second part introduces the invariants and basic facts, some with a very sketchy outline of the proof. We assume that the reader is already familiar with the notion of fixed point index.

The aim of this primer is to help the reader understand and appreciate the rich mathematics in the other chapters. We will try to explain the mathematical notions and facts, rather than the historical development (the reader is referred to [Br2]). The choice of topics is in the "classical" range and limited by the personal interest of the author. No completeness is claimed.

FIXED POINT CLASSES

The Lefschetz Fixed Point Theorem emerged from the notion of fixed point index. The notion of fixed point class adds another structure to the set of fixed points. We could say

Nielsen fixed point theory = fixed point index + fixed point class.

There are several good expositions for the fixed point index, for example [D], [Br1]. So in this part we focus on the notion of fixed point class. We require our spaces X, Y etc. to have a universal covering space. This means that the spaces considered are assumed to be path-connected, locally path-connected and semilocally 1-connected, see [M].

As an important notion, there are several approaches toward fixed point classes, each has its merit and weakness. We shall focus on three approaches in the first few sections, then introduce the notion of coordinate. See Section 5.4 for a brief summary. The remaining sections are devoted to the behavior of fixed point classes under homotopy and other operations on the self-map. A comparison of the three approaches in the light of their relation to these operations is summarized in Section 7.4.

There are still other approaches to the fixed point classes. See [Ji3] for a view from Kervaire's intersection theory [Ke] in non-simply connected differential manifolds, and see [FH1] from obstruction theory on non-simply connected triangulated manifolds.

2. The covering space approach

This is the original approach of Nielsen in his seminal paper [N2], although he focused only on closed surfaces.

Let $p\colon \widetilde{X} \to X$ be the universal covering of X.

(2.1) DEFINITION. A *lifting* of a map $X \xrightarrow{f} X$ is a map $\widetilde{X} \xrightarrow{\widetilde{f}} \widetilde{X}$ such that $p \circ \widetilde{f} = f \circ p$. A *covering translation* is a map $\widetilde{X} \xrightarrow{\alpha} \widetilde{X}$ such that $p \circ \alpha = p$, i.e. a lifting of the identity map.

$$\begin{array}{ccc} \widetilde{X} & \xrightarrow{\widetilde{f}} & \widetilde{X} \\ {\scriptstyle p}\downarrow & & \downarrow{\scriptstyle p} \\ X & \xrightarrow[f]{} & X \end{array} \qquad\qquad \begin{array}{ccc} \widetilde{X} & \xrightarrow{\alpha} & \widetilde{X} \\ {\scriptstyle p}\downarrow & & \downarrow{\scriptstyle p} \\ X & = & X \end{array}$$

The covering translations of $\widetilde{X}$ form a group $\mathcal{D} = \mathcal{D}(\widetilde{X})$ which is isomorphic to the fundamental group $\pi_1(X)$.

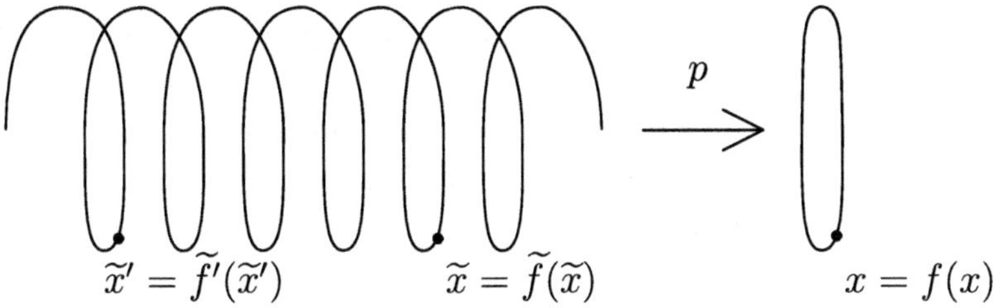

FIGURE 1

(2.2) LEMMA. *Suppose* $\widetilde{x} \in p^{-1}(x)$ *is a fixed point of a lifting* $\widetilde{f}$ *of* f, *and* $\alpha \in \mathcal{D}$ *is a covering translation on* $\widetilde{X}$. *Then, a lifting* $\widetilde{f}'$ *of* f *has* $\alpha(\widetilde{x}) \in p^{-1}(x)$ *as a fixed point if and only if* $\widetilde{f}' = \alpha \circ \widetilde{f} \circ \alpha^{-1}$.

PROOF. "If" is obvious: $\widetilde{f}'(\alpha(\widetilde{x})) = \alpha \circ \widetilde{f} \circ \alpha^{-1}(\alpha(\widetilde{x})) = \alpha \circ \widetilde{f}(\widetilde{x}) = \alpha(\widetilde{x})$.

"Only if": Both $\widetilde{f}'$ and $\alpha \circ \widetilde{f} \circ \alpha^{-1}$ are liftings of f and have $\alpha(\widetilde{x})$ as a fixed point. Since they agree at the point $\alpha(\widetilde{x})$, they are the same lifting. □

(2.3) DEFINITION. Two liftings $\widetilde{f}$ and $\widetilde{f}'$ of $f: X \to X$ are said to be *conjugate* if there exists $\alpha \in \mathcal{D}$ such that $\widetilde{f}' = \alpha \circ \widetilde{f} \circ \alpha^{-1}$. The equivalence classes under conjugacy are called *lifting classes* and written

$$[\widetilde{f}] = \{\alpha \circ \widetilde{f} \circ \alpha^{-1} : \alpha \in \mathcal{D}\}.$$

(2.4) THEOREM.

(2.4.1) $\mathrm{Fix}(f) = \bigcup_{\text{liftings } \widetilde{f}} p\,\mathrm{Fix}(\widetilde{f})$.

(2.4.2) $p\,\mathrm{Fix}(\widetilde{f}) = p\,\mathrm{Fix}(\widetilde{f}')$ if $[\widetilde{f}] = [\widetilde{f}']$.

(2.4.3) $p\,\mathrm{Fix}(\widetilde{f}) \cap p\,\mathrm{Fix}(\widetilde{f}') = \emptyset$ if $[\widetilde{f}] \neq [\widetilde{f}']$.

PROOF. (2.4.1) Suppose $x_0 \in \mathrm{Fix}(f)$. Let $\widetilde{f}_0$ be a given lifting of f. Then for any $\widetilde{x}_0 \in p^{-1}(x_0)$, there exists $\alpha \in \mathcal{D}$ such that $\widetilde{f}_0(\widetilde{x}_0) = \alpha\widetilde{x}_0$. Hence $x_0 \in p\,\mathrm{Fix}(\alpha^{-1}\widetilde{f}_0)$.

(2.4.2) If $\widetilde{f}' = \alpha \circ \widetilde{f} \circ \alpha^{-1}$, then by Lemma (2.2), $\mathrm{Fix}(\widetilde{f}') = \alpha\,\mathrm{Fix}(\widetilde{f})$, so that $p\,\mathrm{Fix}(\widetilde{f}') = p\,\mathrm{Fix}(\widetilde{f})$.

(2.4.3) If $x_0 \in p\,\mathrm{Fix}(\widetilde{f}) \cap p\,\mathrm{Fix}(\widetilde{f}')$, there are $\widetilde{x}_0, \widetilde{x}_0' \in p^{-1}(x_0)$ such that $\widetilde{x}_0 \in \mathrm{Fix}(\widetilde{f})$ and $\widetilde{x}_0' \in \mathrm{Fix}(\widetilde{f}')$. Suppose $\widetilde{x}_0' = \alpha\widetilde{x}_0$. By Lemma (2.2), $\widetilde{f}' = \alpha \circ \widetilde{f} \circ \alpha^{-1}$, hence $[\widetilde{f}] = [\widetilde{f}']$. □

(2.5) DEFINITION. The subset $p\,\mathrm{Fix}(\widetilde{f})$ of $\mathrm{Fix}(f)$ is called the *fixed point class* of f determined by the lifting class $[\widetilde{f}]$. Thus the fixed point set $\mathrm{Fix}(f)$ splits into a disjoint union of fixed point classes.

(2.6) REMARK. For nonempty fixed point classes, this definition essentially says: Two fixed points of f are in the same class if and only if there is a lifting $\widetilde{f}$

of f having fixed points above both of them. But there are also empty fixed point classes.

A fixed point class (empty or not) is always considered to carry a label — the lifting class determining it. Thus two empty fixed point classes are considered different if they are determined by different lifting classes.

3. The path approach

There is another way of defining a fixed point class, which does not use covering space explicitly. Nielsen [N2] mentioned it as an alternative description of fixed point class. It was taken by Wecken [W1] as the primary definition, and has been in popular use since then.

(3.1) DEFINITION. Two fixed points x_0 and x_1 of $f\colon X \to X$ belong to the same *fixed point class* if and only if there is a path c from x_0 to x_1 such that $c \simeq f \circ c$ (homotopy keeping the endpoints fixed).

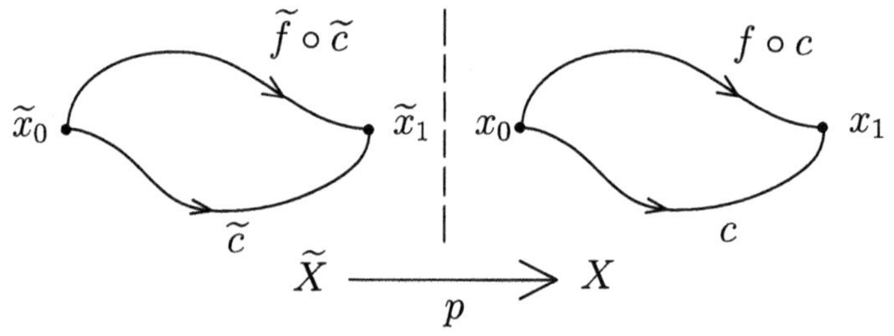

FIGURE 2

PROOF (Equivalence between Definitions (2.5) and (3.1)). Suppose fixed points x_0 and x_1 are in the same class according to Definition (2.5). Then there exist a lifting $\widetilde{f}\colon \widetilde{X} \to \widetilde{X}$ and points $\widetilde{x}_0 \in p^{-1}(x_0)$ and $\widetilde{x}_1 \in p^{-1}(x_1)$ such that $\widetilde{f}(\widetilde{x}_0) = \widetilde{x}_0$ and $\widetilde{f}(\widetilde{x}_1) = \widetilde{x}_1$. Take any path $\widetilde{c}$ in $\widetilde{X}$ from $\widetilde{x}_0$ to $\widetilde{x}_1$. Since $\widetilde{X}$ is simply-connected, $\widetilde{c} \simeq \widetilde{f} \circ \widetilde{c}$. Projecting down to X, we have $c \simeq f \circ c$ where $c = p \circ \widetilde{c}$.

Conversely, suppose fixed points x_0 and x_1 of f are in the same class according to Definition (3.1). Let $\widetilde{f}$ be a lifting of f such that $x_0 \in p\,\mathrm{Fix}(\widetilde{f})$. We want to prove $x_1 \in p\,\mathrm{Fix}(\widetilde{f})$.

Suppose $\widetilde{x}_0 \in p^{-1}(x_0)$ and $\widetilde{f}(\widetilde{x}_0) = \widetilde{x}_0$. Lift the path c of Definition (3.1) to get a path $\widetilde{c}$ in $\widetilde{X}$, starting from $\widetilde{x}_0 \in p^{-1}(x_0)$ and ending at some $\widetilde{x}_1 \in p^{-1}(x_1)$. Now $\widetilde{f} \circ \widetilde{c}$ is the lift of $f \circ c$ from $\widetilde{x}_0$. Since $c \simeq f \circ c$, their lifts from the same starting point $\widetilde{x}_0$ should have the same ending point. Hence $\widetilde{x}_1 = \widetilde{f}(\widetilde{x}_1)$, and $x_1 = p(\widetilde{x}_1) \in p\,\mathrm{Fix}(\widetilde{f})$. □

(3.2) REMARK. Definition (3.1) is a definition only for *nonempty* fixed point classes. The advantage of Definition (2.1) is that it works directly on X, hence is more convenient for geometric questions. The disadvantage is that empty fixed point classes are not mentioned, which are crucial to the homotopy invariance. Consequently in Wecken's treatment, a homotopy of self-maps gives rise to only a partial correspondence between fixed point classes, in contrast to the one-one correspondence in Definition 6.2 below.

(3.3) PROPOSITION. *Every fixed point class of $f: X \to X$ is an open subset of* $\mathrm{Fix}(f)$.

PROOF. Given a fixed point x_0 of f, we want to find a neighbourhood U of x_0 such that any fixed point $x_1 \in U$ belongs to the same class. Since X has a universal covering, X is locally path-connected and semilocally 1-connected. There is a neighbourhood W of x_0 such that every loop in W at x_0 is trivial in X. There also is a path-connected neighbourhood U of x_0 such that $U \subset W \cap f^{-1}(W)$.

Now, if $x_1 \in U \cap \mathrm{Fix}(f)$, take a path c in U from x_0 to x_1, then both c and $f \circ c$ are in W, hence $c \simeq f \circ c$. Thus x_0, x_1 are in the same class by Definition (3.1).□

4. The mapping torus approach

The notion of fixed point classes can be interpreted in terms of the mapping torus (see [Ji7]).

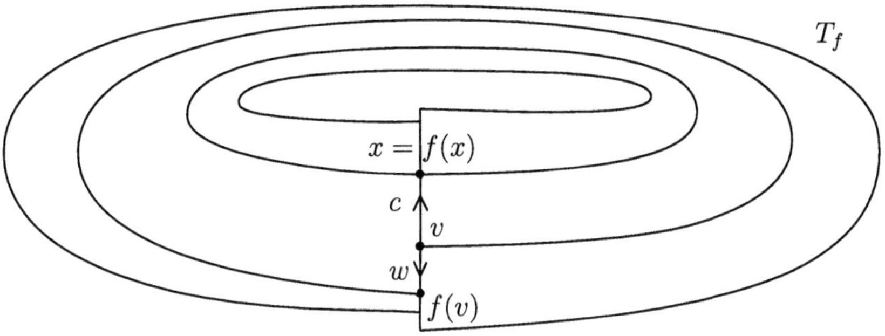

FIGURE 3

(4.1) DEFINITION. The *mapping torus* T_f of $f: X \to X$ is the space obtained from $X \times I$ by identifying $(x, 1)$ with $(f(x), 0)$ for all $x \in X$. We shall regard T_f as the space obtained from $X \times \mathbb{R}_+$, where $\mathbb{R}_+$ stands for the real interval $0 \leq s < \infty$, by identifying $(x, s+1)$ with $(f(x), s)$ for all $x \in X$ and $s \geq 0$. On T_f there is a natural semi-flow ("sliding along the rays")

$$\varphi: T_f \times \mathbb{R}_+ \to T_f, \quad \varphi_t([x, s]) = [x, s+t] \quad \text{for all } t \geq 0.$$

A point $x \in X$ and a positive number $\tau > 0$ determine the *time-τ orbit curve* $\mathrm{orb}_x^\tau := \{\varphi_t(x)\}_{0 \le t \le \tau}$ in T_f. We may identify X with the cross-section $X \times 0 \subset T_f$, then the map $f\colon X \to X$ is just the return map of the semi-flow φ.

(4.2) EXAMPLE. Let X be the circle $S^1 = \{z \in \mathbb{C} : |z| = 1\}$, and let $f\colon S^1 \to S^1$ be the reflection $z \mapsto \bar{z}$. Then the mapping torus T_f is a Klein bottle.

Now, a point $x \in X$ is a fixed point of f if and only if the time-1 orbit curve orb_x^1 is a closed curve.

(4.3) DEFINITION. Two fixed points x_0 and x_1 of $f\colon X \to X$ belong to the same *fixed point class* if and only if the time-1 closed orbit curves $\mathrm{orb}_{x_0}^1$ and $\mathrm{orb}_{x_1}^1$ are freely homotopic in T_f. (The term "freely homotopic" means homotopic as maps from the circle $S^1 = I/\{0,1\}$ into T_f.)

PROOF (Equivalence between Definitions (3.1) and (4.3)). Suppose fixed points x_0 and x_1 are in the same class according to Definition (3.1). Then there exists a homotopy $F\colon I \times I \to X$ with $F(s,0) = x_0$, $F(s,1) = x_1$, and $F(0,t) = f(F(1,t))$, for all $s,t \in I$. Define $H\colon I \times I \to T_f$ by $H(s,t) = [F(s,t),s]$. It is readily seen that $H(s,0) = \mathrm{orb}_{x_0}^1(s)$, $H(s,1) = \mathrm{orb}_{x_1}^1(s)$, and $H(0,t) = H(1,t)$, for all $s,t \in I$. Hence $\mathrm{orb}_{x_0}^1$ and $\mathrm{orb}_{x_01}^1$ are freely homotopic as closed curves in T_f.

Conversely, suppose fixed points x_0 and x_1 of f are in the same class according to Definition (4.3). Then there exists a homotopy $H\colon I \times I \to T_f$ with $H(s,0) = \mathrm{orb}_{x_0}^1(s)$, $H(s,1) = \mathrm{orb}_{x_1}^1(s)$, and $H(0,t) = H(1,t)$, for all $s,t \in I$.

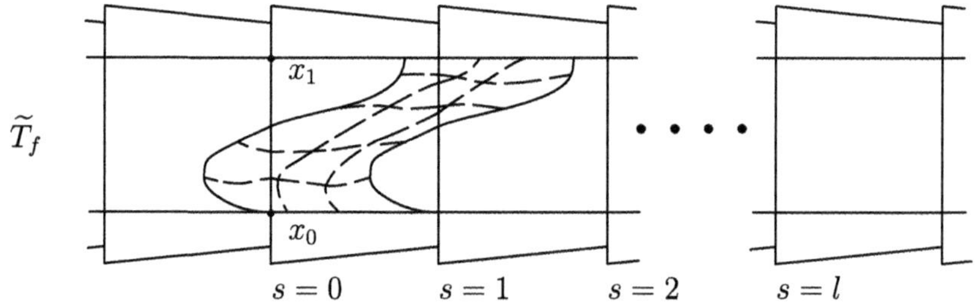

FIGURE 4

Consider the covering space $\widetilde{T}_f$ obtained from the disjoint union $\bigsqcup_{k\in\mathbb{Z}} X \times I_k$, where I_k stands for the real interval $k \le s \le k+1$, by identifying $(x,k) \in X \times I_{k-1}$ to $(f(x),k) \in X \times I_k$, for all $x \in X$ and $k \in \mathbb{Z}$. The covering map $\widetilde{T}_f \to T_f$ maps $(x,s) \in X \times I_k$ to $[x, s-k] \in T_f$. Note that the function $\sigma\colon (x,s) \mapsto s$ is continuous on $\widetilde{T}_f$.

Lift the homotopy H to a homotopy $\widetilde{H}\colon I \times I \to \widetilde{T}_f$ such that $\widetilde{H}(0,0) = (x_0, 0)$. By compactness the image of $\widetilde{H}$ is bounded from the right, i.e. lies in the subset

$A_\ell = \{(x,s) : \sigma(x,s) < \ell\}$ of $\widetilde{T}_f$ for some $\ell \in \mathbb{Z}$. Define a map $r: A_\ell \to X$ by sending $(x,s) \in X \times I_k$ to $f^{\ell-k}(x)$. It is well-defined and continuous. Now define a homotopy $F := r \circ \widetilde{H}: I \times I \to X$. Clearly $F(s,0) = x_0$ and $F(s,1) = x_1$ for all s.

Since $H(0,t) = H(1,t)$ for all $t \in I$, $\sigma(\widetilde{H}(1,t)) - \sigma(\widetilde{H}(0,t))$ is always an integer, hence is a constant. From $H(s,0) = \mathrm{orb}^1_{x_0}(s)$ and $\widetilde{H}(0,0) = (x_0, 0)$ we see $\sigma(\widetilde{H}(s,0)) = s$, so this constant is 1. Therefore $F(0,t) = f(F(1,t))$ for all t. This homotopy F shows that x_0 and x_1 are in the same class according to Definition (3.1). □

(4.4) REMARK. Although Definition (4.3) is also a definition for *nonempty* fixed point classes, it does imply the notion that fixed point classes are labelled by free homotopy classes of closed curves which wind around T_f once. A free homotopy class of closed curves (which wind around T_f once) determines an empty fixed point class if it does not contain any closed time-1 orbit curve in T_f.

Definition (4.3) has several advantages. Geometrically, it relates to the "suspension" semi-flow in dynamical systems theory, hence it allows generalization to periodic orbits. Algebraically, free homotopy classes of closed curves are nothing but conjugacy classes in the fundamental group. So it involves a familiar algebraic notion. In contrast, the conjugacy in Definition (2.3) is a strange one, because the set of liftings is only a set, not a group.

5. The coordinates of a fixed point class

5.1. The covering space approach and the path approach. These two approaches lead to the same coordinates for fixed point classes, because of the standard construction of universal covering space via path classes (cf. [M, proof of Theorem V.10.2]).

Suppose v is the base point in X. For brevity we shall often write the fundamental group $\pi_1(X,v)$ as π. It is well known that points of the universal covering space $\widetilde{X}$ can be identified with path classes in X which start at v. (Recall that two paths in X are in the same path class if and only if they have common end points and they are homotopic relative to end points. The path class of a path c will be denoted by $\langle c \rangle$.) The covering map $p: \widetilde{X} \to X$ sends the path class $\langle c \rangle$ to the terminal point of c. The group π acts on $\widetilde{X}$ by left multiplication, thus identifying the group $\mathcal{D}$ of covering translations with π. In symbols, the loop class $\alpha = \langle a \rangle \in \pi$ acts on $\widetilde{X} = \{\langle c \rangle : c \text{ starts at } v\}$ by

$$\alpha(\langle c \rangle) = \langle ac \rangle \quad \text{for all } \langle c \rangle \in \widetilde{X}.$$

The base point $\tilde{v}$ in $\widetilde{X}$ is taken to be the trivial path class $\langle e \rangle \in \pi_1(X,v)$.

(5.1) DEFINITION. Suppose $f: X \to X$ is a map, and suppose a path w from v to $f(v)$ has been chosen as the *reference path* for f. An endomorphism $f_\pi: \pi \to \pi$ is defined as the composition of

$$\pi_1(X, v) \xrightarrow{f_*} \pi_1(X, f(v)) \xrightarrow{w_*} \pi_1(X, v)$$

where f_* is the homomorphism induced by the map f and w_* is the isomorphism induced by the reference path w. In other words,

$$f_\pi(\langle a \rangle) := \langle w(f \circ a)w^{-1} \rangle \quad \text{for all } \langle a \rangle \in \pi.$$

And a lifting $\widetilde{f}: \widetilde{X} \to \widetilde{X}$ of f is defined by

$$\widetilde{f}(\langle c \rangle) := \langle w(f \circ c) \rangle \quad \text{for all } \langle c \rangle \in \widetilde{X}.$$

Clearly $\widetilde{f}(\widetilde{v}) = \langle w \rangle$.

(5.2) LEMMA. *For any $\alpha \in \pi$ regarded as a covering translation on $\widetilde{X}$, we have*

$$\widetilde{f} \circ \alpha = f_\pi(\alpha) \circ \widetilde{f}: \widetilde{X} \to \widetilde{X}.$$

PROOF. Suppose $\alpha = \langle a \rangle \in \pi$. Using the definitions above, it is easy to check that $\widetilde{f} \circ \alpha$ and $f_\pi(\alpha) \circ \widetilde{f}$ both map $\widetilde{x} = \langle c \rangle \in \widetilde{X}$ to the same point $\langle w(f \circ a)(f \circ c) \rangle$ in $\widetilde{X}$. □

(5.3) LEMMA. *Two liftings $\beta^{-1} \circ \widetilde{f}$ and $\beta'^{-1} \circ \widetilde{f}$ are conjugate if and only if there is $\alpha \in \pi$ such that*

$$\beta' = f_\pi(\alpha)\beta\alpha^{-1}.$$

PROOF. $[\beta^{-1} \circ \widetilde{f}] = [\beta'^{-1} \circ \widetilde{f}]$ if and only if there is $\alpha \in \pi$ such that $\beta'^{-1} \circ \widetilde{f} = \alpha \circ (\beta^{-1} \circ \widetilde{f}) \circ \alpha^{-1} = (\alpha \circ \beta^{-1} \circ f_\pi(\alpha^{-1})) \circ \widetilde{f}$. □

This lemma suggests a general definition:

(5.4) DEFINITION. Let G be a group, $\phi: G \to G$ an endomorphism. Two elements $\beta, \beta' \in G$ are said to be *ϕ-conjugate* if and only if there exists $\alpha \in G$ such that

$$\beta' = \phi(\alpha)\beta\alpha^{-1}.$$

We shall write $[\beta]_\phi$ for the ϕ-conjugacy class of $\beta \in G$. The set of ϕ-conjugacy classes in G will be called the *Reidemeister set* of ϕ, denoted by $\mathcal{R}(\phi)$. The number of ϕ-conjugacy classes in G will be called the *Reidemeister number* of ϕ, denoted by $R(\phi)$.

A fixed point class is labelled by a lifting class which corresponds to such a twisted conjugacy class. So the latter can serve as the coordinate of the fixed point class.

(5.5) DEFINITION. Suppose $\beta \in \pi$. The *π-coordinate* of the lifting class $[\beta^{-1} \circ \widetilde{f}]$, and also of the fixed point class $p\,\mathrm{Fix}(\beta^{-1} \circ \widetilde{f})$, is defined to be the f_π-conjugacy class $[\beta]_{f_\pi} \in \mathcal{R}(f_\pi)$.

Alternatively, the *π-coordinate* $\mathrm{cd}_\pi(x, f)$ of a fixed point $x \in \mathrm{Fix}(f)$ is defined to be the f_π-conjugacy class of $\langle w(f \circ c)c^{-1}\rangle \in \pi$, where c is any path from v to x. Clearly, two fixed points are in the same fixed point class if and only if they have the same π-coordinate. The *π-coordinate* $\mathrm{cd}_\pi(\mathbf{F}, f)$ of a nonempty fixed point class $\mathbf{F}$ is then defined to be the common π-coordinate of its members.

$$\mathrm{cd}_\pi(\mathbf{F}, f) = [\langle w(f \circ c)c^{-1}\rangle]_{f_\pi} \in \mathcal{R}(f_\pi),$$

where c is any path beginning at v and ending in $\mathbf{F}$.

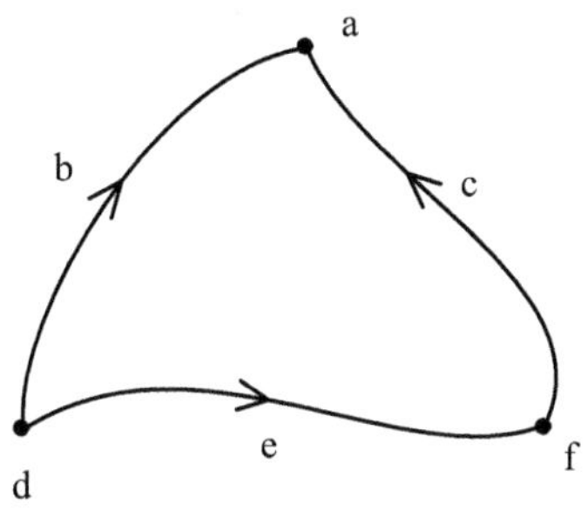

FIGURE 5

PROOF. We need to show the equivalence of the two definitions which are from the covering space approach and from the path approach, respectively.

A point in $p^{-1}(x)$ is nothing but a path class $\langle c\rangle$ from v to x. Since $\widetilde{f}(\langle c\rangle) = \langle w(f \circ c)\rangle$, we see $\widetilde{f}(\langle c\rangle) = \beta\langle c\rangle$ if and only if $\beta = \langle w(f \circ c)c^{-1}\rangle$. □

(5.6) REMARK. Definition (5.5) supplements Definition (3.1) in that we can now talk about the fixed point class (possibly empty) labelled by a f_π-conjugacy class in π. This resolves the problem mentioned in Remark (3.2).

(5.7) EXAMPLE. Let $X = T^2$ be the torus and $f\colon T^2 \to T^2$ be a self-map. Since the fundamental group π is commutative, we shall identify it with the first homology group $H_1(T^2) = \mathbb{Z} \oplus \mathbb{Z}$. Suppose the endomorphism f_π induced by f on $H_1(T^2)$ is represented by the 2×2 integral matrix A. Then the Reidemeister set $\mathcal{R}(f_\pi)$ is identified to the group $\mathrm{coker}(f_\pi - 1)$. The Reidemeister number $R(f_\pi) = |\det(A - E)|$ if $\det(A - E) \neq 0$, and $R(f_\pi) = \infty$ otherwise.

When $f = \mathrm{id}$ is the identity map, the Reidemeister set $\mathcal{R}(\mathrm{id}_\pi)$ is identified with the group $H_1(T^2)$ itself. The fixed point class with coordinate $0 \in H_1$ is the whole T^2. All the other classes are empty.

(5.8) REMARK. Definition (5.4) (and (5.5) accordingly) is in Reidemeister's original form [R]. Another form [Ji4, Definition II.1.5] has also been used in the literature which differs by an inversion in the group. (With respect to the reference lifting $\widetilde{f}$, the expression $\widetilde{x} \in \mathrm{Fix}(\beta^{-1} \circ \widetilde{f})$ and the equality $\widetilde{f}(\widetilde{x}) = \beta(\widetilde{x})$ are equivalent. Choosing the β in the latter or the β^{-1} in the former as coordinate is a matter of convention.) We now prefer the original form because it is more convenient in the important trace formula (Theorem (9.1)), and because it arises more naturally in the mapping torus approach (Proposition (5.11)).

5.2. The mapping torus approach. In this approach, the *coordinate* of a fixed point $x \in \mathrm{Fix}(f)$ is simply the free homotopy class of its closed orbit curve orb_x^1 in the mapping torus T_f. It is common sense that such a free homotopy class is a conjugacy class in the fundamental group $\pi_1(T_f)$, without referring to any base point. However, for comparison with the π-coordinate above, we need a bit more detail.

In the mapping torus T_f, take the base point v of X as the base point (recall that X is regarded as embedded in T_f). Let $\Gamma = \pi_1(T_f, v)$. By the van Kampen Theorem, Γ is obtained from π by adding a new generator z represented by the loop $(\mathrm{orb}_v^1)w^{-1}$, and adding the relations $z^{-1}\alpha z = f_\pi(\alpha)$ for all $\alpha \in \pi$:

$$\Gamma = \langle \pi, z : \alpha z = z f_\pi(\alpha) \text{ for all } \alpha \in \pi \rangle.$$

(5.9) REMARK. The homomorphism $\pi \to \Gamma$ induced by the inclusion $X \subset T_f$ is not necessarily injective. Its kernel equals $\bigcup_{n>0} \ker(f_\pi^n)$, the union of the kernels of all iterates of $f_\pi\colon \pi \to \pi$. Therefore, π is not a subgroup of Γ unless the endomorphism $f_\pi\colon \pi \to \pi$ is injective.

The *winding number* of an element $\gamma \in \Gamma$, denoted $\omega(\gamma)$, counts the number of times that γ winds around the mapping torus. It equals the total z-exponent of the word γ in the above presentation of Γ.

Let Γ_c denote the set of conjugacy classes in Γ. We shall use the bracket notation $\gamma \mapsto [\gamma]$ for the projection $\Gamma \to \Gamma_c$ to the set of conjugacy class. For $n \in \mathbb{Z}$, let $\Gamma_c^{(n)} \subset \Gamma_c$ denote the subset of conjugacy classes with winding number n. Thus $\Gamma_c^{(1)}$ is the set of free homotopy classes of closed curves winding around T_f once.

(5.10) DEFINITION. Suppose $x \in \mathrm{Fix}(f)$. The *Γ-coordinate* $\mathrm{cd}_\Gamma(x, f)$ of x is defined to be the free homotopy class of the closed orbit curve orb_x^1, i.e. the conjugacy class $[\mathrm{orb}_x^1] \in \Gamma_c^{(1)}$. Two fixed points are in the same fixed point class if and only if they have the same Γ-coordinate. The *Γ-coordinate* $\mathrm{cd}_\Gamma(\mathbf{F}, f)$ of a nonempty fixed point class $\mathbf{F}$ is defined to be the common Γ-coordinate of its members

$$\mathrm{cd}_\Gamma(\mathbf{F}, f) = [\mathrm{orb}_x^1] \in \Gamma_c^{(1)}$$

for any $x \in \mathbf{F}$. We shall also talk about the fixed point class (possibly empty) labelled by any conjugacy class in $\Gamma_c^{(1)}$.

5.3. Relationship between the π- and the Γ-coordinates.

(5.11) PROPOSITION. *Suppose* $x \in \mathrm{Fix}(f)$. *Then* $\mathrm{cd}_\pi(x, f) = [\alpha]_{f_\pi} \in \mathcal{R}_{f_\pi}$ *if and only if* $\mathrm{cd}_\Gamma(x, f) = [z\alpha] \in \Gamma_c^{(1)}$.

PROOF. Pick any path c from v to x and let $\alpha = \langle w(f \circ c)c^{-1} \rangle$ in π. Then $\mathrm{cd}_\pi(x, f) = [\alpha]_{f_\pi}$. On the other hand, $\mathrm{cd}_\Gamma(x, f)$, the conjugacy class of orb_x^1, is represented by $\langle c(\mathrm{orb}_x^1)c^{-1} \rangle = \langle (\mathrm{orb}_v^1)(f \circ c)c^{-1} \rangle = \langle (\mathrm{orb}_v^1)w^{-1} \rangle \langle w(f \circ c)c^{-1} \rangle = z\alpha$ in Γ. □

(5.12) REMARK. It can be proved algebraically that the rule $[\alpha]_{f_\pi} \leftrightarrow [z\alpha]$ establishes a one-one correspondence $\mathcal{R}_{f_\pi} \leftrightarrow \Gamma_c^{(1)}$. This means, by going from X to T_f, a twisted conjugacy problem is replaced by an ordinary conjugacy problem in a more complicated group.

5.4. Summary. We shall use the notation $\mathrm{FPC}(f)$ for the set of all fixed point classes (including empty ones) of a self-map f. It can be identified with the set of conjugacy classes of liftings of f (in the covering space approach), or to the set of free homotopy classes of closed curves in T_f with winding number 1 (in the mapping torus approach). By means of coordinates, it is labelled by the Reidemeister set $\mathcal{R}(f_\pi)$, or by the set of conjugacy classes $\Gamma_c^{(1)}$. The total number of fixed point classes of f equals the Reidemeister number $R(f_\pi)$.

6. Behavior under a homotopy

Given a homotopy $H = \{h_t\} : f_0 \simeq f_1 : X \to X$, we want to see its influence on the fixed point classes of f_0 and f_1.

(6.1) DEFINITION. A homotopy $H = \{h_t\}_{t \in I} : f_0 \simeq f_1 : X \to X$ between self-maps gives rise to a level-preserving map $H : X \times I \to X \times I$,

$$H(x, t) = (h_t(x), t),$$

called the *fat homotopy* of H (which piles up all the h_t's). Accordingly, $h_t : X \to X$ will be called the *t-slice* of the fat homotopy. Similarly, for a subset $A \subset X \times I$, the subset $A_t := \{x \in X : (x, t) \in A\} \subset X$ will be called the *t-slice* of A.

The advantage of the fat homotopy H is that it is a *self-map* of $X \times I$, so we may talk about its fixed point classes (as well as liftings, mapping torus, etc).

(6.2) DEFINITION. Let $H : f_0 \simeq f_1$ be a homotopy. Let $\mathbf{F}_0$ and $\mathbf{F}_1$ be fixed point classes of f_0 and f_1, respectively. We say that $\mathbf{F}_0$ *corresponds to* $\mathbf{F}_1$ *via* H,

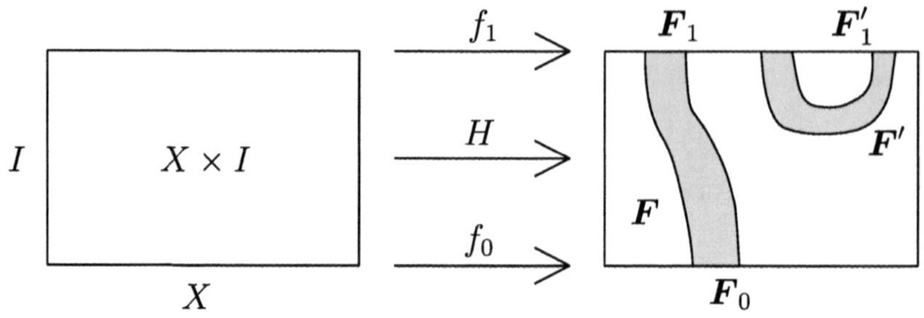

FIGURE 6

or $\mathbf{F}_0$ *is* H*-related to* $\mathbf{F}_1$, if and only if they are, respectively, the 0-slice and 1-slice of the same fixed point class $\mathbf{F}$ of the fat homotopy $H: X \times I \to X \times I$. Thus H gives rise to a one-one correspondence

$$H_{\mathrm{FPC}}: \mathrm{FPC}(f_0) \to \mathrm{FPC}(f_1)$$

from the fixed point classes of f_0 to the fixed point classes of f_1.

The last statement will become clear when we examine the manifestations of this definition in the three approaches. In particular, it implies that the Reidemeister number $R(f)$ is a homotopy invariant of the map f. But we must first point out some important phenomena:

- A nonempty fixed point class may disappear or emerge under a homotopy. The correspondence under homotopy may fail to be one-to-one unless empty fixed point classes are included in the consideration.

(6.3) EXAMPLE. Consider maps $S^1 \to S^1$. Let $H = \{h_t : e^{i\theta} \mapsto e^{i(\theta + t\sin\theta + 2\pi t)}\}$. Then $\mathrm{Fix}(h_0) = S^1$ is a fixed point class $\mathbf{F}_0$ of f_0, and $\mathrm{Fix}(h_1) = \{1, -1\}$ is a fixed point class $\mathbf{F}_1$ of f_1.

The fixed point set $\mathrm{Fix}(H)$ splits into two fixed point classes, one is the bottom circle $S^1 \times 0$, the other is a curve $\{(e^{i\theta}, t) \in S^1 \times I : t = 2\pi/(2\pi + \sin\theta)\}$ which hits the top circle $S^1 \times 1$ at two points. So $\mathbf{F}_0$ and $\mathbf{F}_1$ are not H-related. In the course of H, $\mathbf{F}_0$ disappeared immediately and $\mathbf{F}_1$ emerged later.

- The correspondence via H may depend on the homotopy H.

(6.4) EXAMPLE. Consider maps $S^1 \to S^1$. Let $f_0 = f_1: z \mapsto z^{-2}$. Then $\mathrm{Fix}(f_0) = \mathrm{Fix}(f_1) = \{1, e^{2\pi i/3}, e^{4\pi i/3}\}$, each point being a fixed point class. Consider two homotopies $H' = \{h'_t: z \mapsto z^{-2}\} : f_0 \simeq f_1$ and $H = \{h_t : z \mapsto z^{-2}e^{2\pi t i}\} : f_0 \simeq f_1$.

Then $1 \in \mathrm{Fix}(f_0)$ is H'-related to $1 \in \mathrm{Fix}(f_1)$, but H-related to $e^{2\pi i/3} \in \mathrm{Fix}(f_1)$.

In the covering space approach, a lifting $\widetilde{f_0}$ of f_0 (regarded as the 0-slice of the fat homotopy H) extends to a unique lifting $\widetilde{H}$ of H, and the 1-slice of $\widetilde{H}$ provides a lifting $\widetilde{f_1}$ of f_1. This correspondence $\widetilde{f_0} \mapsto \widetilde{f_1}$ is one-to-one and preserves the conjugacy relation, so it gives a one-one correspondence from lifting classes of f_0 to lifting classes of f_1. Explicitly, the correspondence is

$$p\,\mathrm{Fix}(\beta^{-1} \circ \widetilde{f_0}) \overset{H}{\longmapsto} p\,\mathrm{Fix}(\beta^{-1} \circ \widetilde{f_1}).$$

In the path approach, a fixed point $x_0 \in \mathrm{Fix}(f_0)$ is H-related to a fixed point $x_1 \in \mathrm{Fix}(f_1)$ if and only if $(x_0, 0)$ and $(x_1, 1)$ are in the same fixed point class of the fat homotopy $H: X \times I \to X \times I$; that is to say, if and only if there is a path c in $X \times I$ from $(x_0, 0)$ to $(x_1, 1)$ such that $c \simeq H \circ c$.

In the mapping torus approach, consider the mapping torus T_H of the fat homotopy $H: X \times I \to X \times I$. In an obvious sense we can regard the mapping torus T_{f_0} as the 0-slice of T_H, and T_{f_1} the 1-slice. Both slices are deformation retracts of T_H. A fixed point $x_0 \in \mathrm{Fix}(f_0)$ is H-related to a fixed point $x_1 \in \mathrm{Fix}(f_1)$ if and only if the closed orbit curves $\mathrm{orb}^1_{x_0}$ and $\mathrm{orb}^1_{x_1}$ are freely homotopic in T_H.

Finally, let us look at the correspondence via homotopy in terms of coordinates.

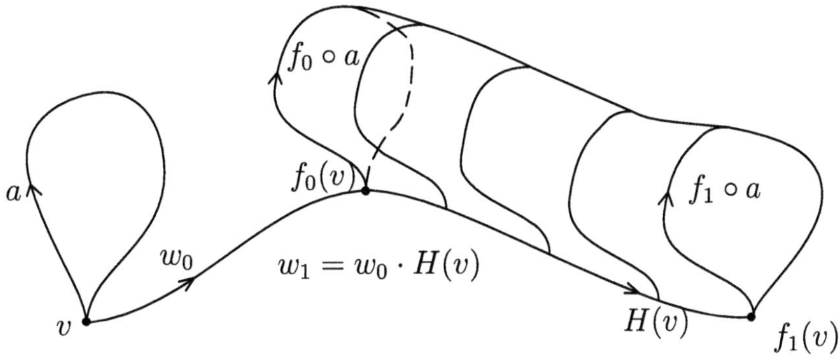

FIGURE 7

Let v be the base point of the space X. The *track* of a homotopy $H = \{h_t\}_{t \in I}$: $f_0 \simeq f_1: X \to X$ is the path $H(v) = \{h_t(v)\}_{t \in I}$ from $f_0(v)$ to $f_1(v)$. It follows directly from Definition (5.1) and other relevant definitions that

(6.5) PROPOSITION. *Given a homotopy $H: f_0 \simeq f_1: X \to X$. Assume the reference paths w_0 and w_1 for the maps f_0 and f_1, respectively, are such that*

$$w_1 \simeq w_0 \cdot H(v), \quad \textit{or equivalently,} \quad w_0 \simeq w_1 \cdot H^{-1}(v).$$

Then both maps f_0 and f_1 induce the same endomorphism on $\pi = \pi_1(X, v)$,

$$f_{0\pi} = f_{1\pi}: \pi \to \pi.$$

Hence they have the same Reidemeister set $\mathcal{R}(f_0) = \mathcal{R}(f_1)$, and the same mapping torus group $\Gamma_{f_0} = \Gamma_{f_1}$.

Under this assumption on reference paths, a fixed point class $\mathbf{F}_0$ of f_0 corresponds via H to a fixed point class $\mathbf{F}_1$ of f_1 if and only if they have the same π-coordinate $\mathrm{cd}_\pi(\mathbf{F}_0, f_0) = \mathrm{cd}_\pi(\mathbf{F}_1, f_1)$, and/or the same Γ-coordinate $\mathrm{cd}_\Gamma(\mathbf{F}_0, f_0) = \mathrm{cd}_\Gamma(\mathbf{F}_1, f_1)$.

7. Behavior under morphism and commutativity

7.1. Morphisms of self-maps.

(7.1) DEFINITION. Let $f: X \to X$ and $g: Y \to Y$ be two self-maps. A *morphism* from f to g is a map $\mu: X \to Y$ such that $\mu \circ f = g \circ \mu$, i.e. the diagram

$$\begin{array}{ccc} X & \xrightarrow{f} & X \\ \mu \downarrow & & \downarrow \mu \\ Y & \xrightarrow[g]{} & Y \end{array}$$

commutes. Self-maps and morphisms between them form a category — the *category of self-maps.*

(7.2) PROPOSITION. *In the above setting, the map $X \xrightarrow{\mu} Y$ restricts to a function*

$$\mathrm{Fix}(f) \xrightarrow{\mu} \mathrm{Fix}(g).$$

This function respects fixed point classes, hence induces a function

$$\mathrm{FPC}(f) \xrightarrow{\mu_{\mathrm{FPC}}} \mathrm{FPC}(g).$$

PROOF. The first part is obvious. We will show that if x_0 and x_1 are in the same fixed point class of f, then $\mu(x_0)$ and $\mu(x_1)$ are in the same fixed point class of g. In fact, if a path c from x_0 to x_1 is such that $c \simeq f \circ c$, then for the path $\mu \circ c$ from $\mu(x_0)$ to $\mu(x_1)$ we have

$$\mu \circ c \simeq \mu \circ (f \circ c) = g \circ (\mu \circ c),$$

which ties $\mu(x_0)$ and $\mu(x_1)$ together.

That the function is well defined even for empty fixed point classes can be seen via the covering space approach [Ji4, pp. 42–44]. However, it is easier via the mapping torus approach because μ induces a map $\mu_T: T_f \to T_g$, defined by $[x, s] \mapsto [\mu(x), s]$, which sends orbit curves in T_f to orbit curves in T_g. □

(7.3) EXAMPLE. The morphism

$$\begin{array}{ccc} X & \xrightarrow{f} & X \\ {\scriptstyle f}\downarrow & & \downarrow{\scriptstyle f} \\ X & \xrightarrow[f]{} & X \end{array}$$

induces the identity function $f_{\mathrm{FPC}} = \mathrm{id}\colon \mathrm{FPC}(f) \to \mathrm{FPC}(f)$.

7.2. Commutativity. Commutativity is a built-in symmetry of the fixed point problem. Suppose we have two maps $X \xrightarrow{\phi} Y$ and $Y \xrightarrow{\psi} X$. To find a fixed point of $\psi \circ \phi$ is equivalent to solving a system of equations

$$\begin{cases} x = \psi(y), \\ y = \phi(x). \end{cases}$$

This system is symmetric in x and y. Geometrically speaking, finding a fixed point of $\psi\circ\phi$ is equivalent to finding an intersection of the two graphs $\{(x,y) : y = \phi(x)\}$ and $\{(x,y) : x = \psi(y)\}$ in $X \times Y$. This is again a symmetric situation.

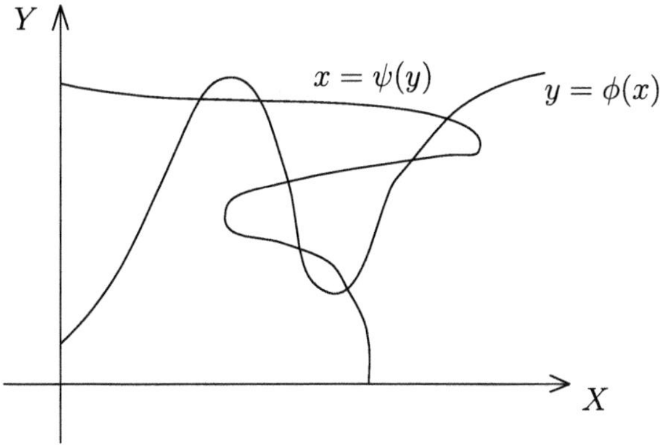

FIGURE 8

The notion of fixed point class respects this symmetry.

(7.4) PROPOSITION. *The maps* $X \xrightarrow{\phi} Y$ *and* $Y \xrightarrow{\psi} X$ *restrict to a pair of mutually inversive bijections*

$$\mathrm{Fix}(\psi \circ \phi) \xrightarrow{\phi} \mathrm{Fix}(\phi \circ \psi) \quad \textit{and} \quad \mathrm{Fix}(\phi \circ \psi) \xrightarrow{\psi} \mathrm{Fix}(\psi \circ \phi).$$

They respect fixed point classes, and induce a pair of mutually inversive bijections

$$\mathrm{FPC}(\psi \circ \phi) \xrightarrow{\phi_{\mathrm{FPC}}} \mathrm{FPC}(\phi \circ \psi) \quad \textit{and} \quad \mathrm{FPC}(\phi \circ \psi) \xrightarrow{\psi_{\mathrm{FPC}}} \mathrm{FPC}(\psi \circ \phi).$$

PROOF. Obviously we have morphisms

$$\begin{array}{ccc} X & \xrightarrow{\psi\circ\phi} & X \\ {\scriptstyle\phi}\downarrow & & \downarrow{\scriptstyle\phi} \\ Y & \xrightarrow[\phi\circ\psi]{} & Y \end{array} \qquad \text{and} \qquad \begin{array}{ccc} Y & \xrightarrow{\phi\circ\psi} & Y \\ {\scriptstyle\psi}\downarrow & & \downarrow{\scriptstyle\psi} \\ X & \xrightarrow[\psi\circ\phi]{} & X \end{array}$$

The functions ϕ_{FPC} and ψ_{FPC} are defined by Proposition (7.2). Since $\psi_{\mathrm{FPC}} \circ \phi_{\mathrm{FPC}} = (\psi \circ \phi)_{\mathrm{FPC}} = \mathrm{id}\colon \mathrm{FPC}(\psi \circ \phi) \to \mathrm{FPC}(\psi \circ \phi)$ and similarly $\phi_{\mathrm{FPC}} \circ \psi_{\mathrm{FPC}} = \mathrm{id}\colon \mathrm{FPC}(\phi \circ \psi) \to \mathrm{FPC}(\phi \circ \psi)$, they are inverses to each other. □

The commutativity property is very powerful. It permits us to switch from one space to another with a possibly different fundamental group.

(7.5) EXAMPLE. Suppose $A \subset X$ is a path-connected subspace, and $f\colon X \to X$ maps X into A. Let $i\colon A \to X$ denote the inclusion map, $f_A\colon A \to A$ denote the restriction of f to A, and $f'\colon X \to A$ the same as f except regarding A as the target space. Then we have $f = i \circ f'\colon X \to X$ and $f_A = f' \circ i\colon A \to A$. So, by commutativity, we have a bijection

$$i_{\mathrm{FPC}}\colon \mathrm{FPC}(f_A) \to \mathrm{FPC}(f).$$

When we interpret the commutativity property in the covering space approach, we see that a natural one-one correspondence between $\psi \circ \phi$ and $\phi \circ \psi$ can be established only at the level of lifting classes, not at the level of liftings.

In the path approach, the correspondence between nonempty fixed point classes is very clear. If a path c ties x_0, x_1 into the same fixed point class of $\psi \circ \phi$, then the path $\phi \circ c$ ties $\phi(x_0), \phi(x_1)$ into the same fixed point class of $\phi \circ \psi$. But when we look at the π-coordinate, $\pi = \pi_1(X)$ and $\pi' = \pi_1(Y)$ could be vastly different. The Reidemeister sets $\mathcal{R}((\psi \circ \phi)_\pi)$ and $\mathcal{R}((\phi \circ \psi)_\pi)$ describe partitions in very different sets, although algebraically we can show they are in one-one correspondence.

In the mapping torus approach, the situation is better. The natural maps $\phi_T\colon T_{\psi\circ\phi} \to T_{\phi\circ\psi}$ and $\psi_T\colon T_{\phi\circ\psi} \to T_{\psi\circ\phi}$ (defined in the proof of Proposition (7.2)) are in fact a pair of homotopy equivalences. Thus we can naturally identify the fundamental groups of the two mapping tori. The corresponding fixed point classes of $\psi \circ \phi$ and $\phi \circ \psi$ simply have the same Γ-coordinate.

7.3. Homotopy type of self-maps. The following notion was introduced in [KiJi] and [F].

(7.6) DEFINITION. Two self-maps $f\colon X \to X$ and $g\colon Y \to Y$ are said to be of the same *homotopy type* if there is a homotopy equivalence $h\colon X \to Y$ such that $h \circ f \simeq g \circ h$. Clearly this is an equivalence relation among self-maps. (Note that h is not required to be a morphism from f to g in the sense of Definition (7.1).)

(7.7) PROPOSITION. *If two self-maps $f: X \to X$ and $g: Y \to Y$ are of the same homotopy type, then they have the same number of fixed point classes, i.e.* $R(f) = R(g)$.

PROOF. Let $k: Y \to X$ be a homotopy inverse of the homotopy equivalence $h: X \to Y$. Then $f \simeq (k \circ h) \circ f$ and $g \simeq g \circ (h \circ k)$. Hence, by the homotopy invariance (6.2) and the commutativity (7.4) of the Reidemeister number, $R(f) = R((k\circ h)\circ f) = R(k\circ(h\circ f)) = R((h\circ f)\circ k) = R((g\circ h)\circ k) = R(g\circ(h\circ k)) = R(g)$.□

7.4. Comparison. Homotopy and commutativity are the most important operations in fixed point theory. It is the behavior under homotopy that makes us insist on an adequate notion of empty fixed point classes. We have seen in Section 5.4 that the covering space approach and the mapping torus approach are better on this count.

Commutativity, by its very nature, focuses on nonempty fixed point classes. Thus it is no surprise that the path approach and the mapping torus approach work better. However, for the extension to encompass empty fixed point classes, the mapping torus approach seems more natural. It is also conceptually simpler from a coordinate view point, since the Γ group respects the commutativity, while the π group does not.

INVARIANTS OF NIELSEN FIXED POINT THEORY

In this part we consider spaces that admit both a fixed point index and a universal covering space. The usual assumption is that the space is a compact connected ANR (= absolute neighbourhood retract). For simplicity we restrict ourselves to self-maps of compact connected polyhedra (= spaces decomposable into simplicial complexes). Extension to ANR's is done by the standard method of domination (cf. [Br1]).

8. The Nielsen number and the generalized Lefschetz number

Let X be a compact connected polyhedron, $f: X \to X$ be a map. By Proposition (3.3), every fixed point class $\mathbf{F}$ is an isolated (= both open and closed) subset of $\mathrm{Fix}(f)$. Hence by compactness f can have only a finite number of nonempty fixed point classes. By the theory of fixed point index, to any isolated set of fixed points there is associated an integer called its index.

(8.1) DEFINITION. The fixed point index $\mathrm{ind}(\mathbf{F}, f) \in \mathbb{Z}$ is called the *index of the fixed point class* $\mathbf{F}$. Thus the set $\mathrm{FPC}(f)$ of fixed point classes will be regarded as a weighted set, the weight of a fixed point class $\mathbf{F}$ being the index $\mathrm{ind}(\mathbf{F}, f)$.

A fixed point class is called *essential* if its index is non-zero. The number of essential fixed point classes is called the *Nielsen number* $N(f)$ of f.

Suppose a reference path w from the base point $v \in X$ to $f(v)$ is given. For every fixed point class $\mathbf{F}$, in Definitions (5.5) and (5.10) we have defined the π-coordinate $\mathrm{cd}_\pi(\mathbf{F}, f) \in \mathcal{R}(f_\pi)$ and the Γ-coordinate $\mathrm{cd}_{\Gamma}(\mathbf{F}, f) \in \Gamma_c^{(1)}$. For a set S we shall use the notation $\mathbb{Z}S$ to denote the free abelian group with basis S.

(8.2) DEFINITION. The *π-generalized Lefschetz number* of f is defined as the formal sum

$$L_\pi(f) := \sum_{\mathbf{F}} \mathrm{ind}(\mathbf{F}, f) \cdot \mathrm{cd}_\pi(\mathbf{F}, f) \in \mathbb{Z}\mathcal{R}(f_\pi),$$

the summation being over all fixed point classes $\mathbf{F}$ of f. Similarly the *Γ-generalized Lefschetz number* of f is defined as the formal sum

$$L_{\Gamma}(f) := \sum_{\mathbf{F}} \mathrm{ind}(\mathbf{F}, f) \cdot \mathrm{cd}_{\Gamma}(\mathbf{F}, f) \in \mathbb{Z}\Gamma_c^{(1)}.$$

Thus the generalized Lefschetz number is a way of organizing the index and coordinate information of all fixed point classes. The Nielsen number $N(f)$ is nothing but the number of non-zero terms in the generalized Lefschetz numbers $L_\pi(f)$ and $L_\Gamma(f)$.

The Nielsen number and the generalized Lefschetz number are the principal invariants of Nielsen fixed point theory. Their basic invariance properties are (cf. [Ji4, Sections I.4–5])

(8.3) THEOREM (Homotopy Invariance). *Suppose $f_0 \simeq f_1 : X \to X$ via a homotopy $H = \{h_t\}_{0 \le t \le 1}$. Then the bijection between fixed point classes*

$$\mathrm{FPC}(f_0) \xrightarrow{H_{\mathrm{FPC}}} \mathrm{FPC}(f_1)$$

in Definition (6.2) is index preserving. With the reference paths for f_0, f_1 described in Proposition (6.5), for the generalized Lefschetz numbers we have

$$L_\pi(f_0) = L_\pi(f_1) \quad \textit{and} \quad L_{\Gamma}(f_0) = L_{\Gamma}(f_1).$$

Hence the Nielsen number $N(f)$ is a homotopy invariant of the self-map f. Every self-map homotopic to f must have at least $N(f)$ fixed points.

(8.4) THEOREM (Commutativity). *Suppose $\phi: X \to Y$ and $\psi: Y \to X$. Then the bijection between fixed point classes*

$$\mathrm{FPC}(\psi \circ \phi) \xrightarrow{\phi_{\mathrm{FPC}}} \mathrm{FPC}(\phi \circ \psi)$$

given in Proposition (7.4) is index preserving. Moreover, the mapping tori $T_{\psi\circ\phi}$ and $T_{\phi\circ\psi}$ are homotopy equivalent in a standard way. If we identify their fundamental groups, then the generalized Lefschetz numbers are equal:

$$L_{\Gamma}(\psi \circ \phi) = L_{\Gamma}(\phi \circ \psi).$$

It follows that the Nielsen number has the commutativity property

$$N(\psi \circ \phi) = N(\phi \circ \psi).$$

(8.5) THEOREM (Homotopy Type Invariance). *Suppose $f: X \to X$ and $g: Y \to Y$ are self-maps of the same homotopy type. Then there is an index preserving bijection between* FPC(f) *and* FPC(g). *Hence the Nielsen number is a homotopy type invariant of self-maps.*

OUTLINE OF PROOF FOR THEOREMS (8.3)–(8.5). For Theorems (8.3) and (8.4), we only need to show that the two corresponding fixed point classes have the same index. This is true by the homotopy invariance and the commutativity property of the fixed point index, respectively.

Theorem (8.5) follows from (8.3) and (8.4), just as in the proof of Proposition (7.7). □

9. The Reidemeister trace formula

Assume that X is a finite cell complex and $f: X \to X$ is a cellular map. Pick a cellular decomposition $\{e_j^d\}$ of X, the base point v being a 0-cell. It lifts to a π-invariant cellular structure on the universal covering $\widetilde{X}$. Choose an arbitrary lift $\widetilde{e}_j^d$ for each e_j^d. These lifts constitute a free $\mathbb{Z}\pi$-basis for the cellular chain complex of $\widetilde{X}$. The lift $\widetilde{f}$ of f is also a cellular map. In every dimension d, the cellular chain map $\widetilde{f}$ gives rise to a $\mathbb{Z}\pi$-matrix $\widetilde{F}_d$ with respect to the above basis, i.e. $\widetilde{F}_d = (a_{ij})$ if $\widetilde{f}(\widetilde{e}_i^d) = \sum_j a_{ij}\widetilde{e}_j^d$, $a_{ij} \in \mathbb{Z}\pi$.

In this notation, the trace formula for generalized Lefschetz number ([R], [W2]; see [Ji8] for the mapping torus version) says:

(9.1) THEOREM (Reidemeister Trace Formula). *For the generalized Lefschetz number we have*

$$L_{\pi}(f) = \sum_d (-1)^d [\operatorname{trace} \widetilde{F}_d]_{f_\pi} \quad \textit{in } \mathbb{Z}\mathcal{R}(f_\pi),$$

where the brackets denote the linear map $\mathbb{Z}\pi \to \mathbb{Z}\mathcal{R}(f_\pi)$ extending the projection $\pi \to \mathcal{R}(f_\pi)$, $\alpha \mapsto [\alpha]_{f_\pi}$. Equivalently, the mapping torus version is

$$L_{\Gamma}(f) = \sum_d (-1)^d [\operatorname{trace}(z\widetilde{F}_d)] \quad \textit{in } \mathbb{Z}\Gamma_c,$$

where $z\widetilde{F}_d$ is regarded as a matrix in $\mathbb{Z}\Gamma$, and the brackets denote the linear map $\mathbb{Z}\Gamma \to \mathbb{Z}\Gamma_c$ extending the projection $\Gamma \to \Gamma_c$ defined by $\gamma \mapsto [\gamma]$.

OUTLINE OF PROOF. Firstly, on the algebraic level, given an endomorphism $\phi: \pi \to \pi$, consider a free $\mathbb{Z}\pi$-chain complex $\widetilde{C}$ with a finite $\mathbb{Z}\pi$-basis. Suppose $\widetilde{f}: \widetilde{C} \to \widetilde{C}$ is a ϕ-chain map, i.e. a $\mathbb{Z}$-chain map which satisfies the action rule $\widetilde{f}(\alpha\widetilde{c}) = \phi(\alpha)\widetilde{f}(\widetilde{c})$ for all $\alpha \in \pi$ and $\widetilde{c} \in \widetilde{C}$. Define the "trace invariant" of $\widetilde{f}$, belonging to $\mathbb{Z}\mathcal{R}(\phi)$, as an alternating sum similar to the right hand side of the trace formula. An algebraic theory can be easily developed to show that it is an invariant under ϕ-chain homotopy and ϕ-chain homotopy equivalence (comp. [H]).

Secondly, on the topological level, show that for a lifting of a map to the universal covering space, the trace invariant of the chain map is independent of the cellular decomposition of the polyhedron and the cellular approximation of the map.

Thirdly, by a suitable approximation theorem (cf. [Br1, p. 118] or [Ji4, p. 62]), the trace formula is verified for a special cellular approximation of the map. □

(9.2) EXAMPLE (Recipe for one-dimensional polyhedra). Let X be a bouquet of r circles with one 0-cell v and r 1-cells $a_1, \dots, a_r$, and let $f: X \to X$ be a cellular map. Then $\{a_1, \dots, a_r\}$ is a free basis for $\pi = \pi_1(X)$. The homomorphism $f_\pi: \pi \to \pi$ induced by f is determined by the images $a_i' := f_\pi(a_i)$, $i = 1, \dots, r$. The fundamental group of the mapping torus $\Gamma = \pi_1(T_f)$ has a presentation

$$\Gamma = \langle a_1, \dots, a_r, z : a_i z = z a_i', \ i = 1, \dots, r \rangle.$$

As pointed out in [FH2], the matrices of the lifted chain map $\widetilde{f}$ are

$$\widetilde{F}_0 = (1) \quad \text{and} \quad \widetilde{F}_1 = D := \left(\frac{\partial a_i'}{\partial a_j} \right),$$

where D is the Jacobian matrix in Fox calculus (see [Bi, Section 3.1] for an introduction). Then, in $\mathbb{Z}\mathcal{R}(f_\pi)$ and $\mathbb{Z}\Gamma_c$, we have

$$L_\pi(f) = [1] - \sum_{i=1}^{r} \left[\frac{\partial a_i'}{\partial a_i} \right]_{f_\pi} \quad \text{and} \quad L_\Gamma(f) = [z] - \sum_{i=1}^{r} \left[z \frac{\partial a_i'}{\partial a_i} \right].$$

10. The least number of fixed points

10.1. In a homotopy class of self-maps. Let X be a compact connected polyhedron, and let $f: X \to X$ be a map. Consider the number

$$\mathrm{MF}[f] := \mathrm{Min}\{\#\,\mathrm{Fix}(g) : g \simeq f\},$$

i.e. the least number of fixed points in the homotopy class $[f]$ of f. We know from Theorem (8.3) that $N(f)$ is a lower bound for $MF[f]$. The equality $N(f) = \mathrm{MF}[f]$, conceived in [N2], is known as the *Nielsen fixed point conjecture.* It turns out to be true under mild restrictions on the space X involved.

The equality $N(f) = \mathrm{MF}[f]$ means we can deform the map f so that each essential fixed point class is combined into a single fixed point and each inessential fixed point class is removed. This problem is geometric in nature and can be solved only by careful constructions.

The very first result is Wecken's theorem [W3] that the equality $N(f) = \mathrm{MF}[f]$ is always true if the polyhedron X is a compact manifold of dimension ≥ 3. Spaces on which the Nielsen conjecture holds true are sometimes called *Wecken spaces* in the literature.

The Nielsen conjecture is sensitive to local properties of the space.

(10.1) DEFINITION. A point x of a connected space X is a *global separating point* of X if $X - x$ is not connected. A point x of a space X is a *local separating point* if x is a separating point of some connected open subspace U of X.

(10.2) THEOREM ([Ji1]). *Let X be a compact connected polyhedron without local separating points. Suppose X is not a surface (closed or with boundary). Then* $\mathrm{MF}[f] = N(f)$ *for any map* $f\colon X \to X$.

OUTLINE OF PROOF. By the Hopf Approximation Theorem (cf. [Br1, p. 118]), we may assume that f has finitely many fixed points, each having a Euclidean neighbourhood.

We only need to find a procedure to coalesce two points x_0, x_1 in the same fixed point class. Repeating this procedure we will be able to combine each fixed point class into a single fixed point, and then to remove it if its index is 0.

Coalescing can be done in three steps.

Step 1. Find an arc (= path without self-intersections) c, connecting x_0 to x_1 but avoiding other fixed points, so that $f \circ c \simeq c$. Local separating points may prevent us from achieving this.

Step 2. Deform f in a regular neighbourhood of the arc c until $f(c) = c$ (using the homotopy from $f(c)$ to c), *without creating any new fixed points.* We would not have enough room to get rid of new fixed points if we are confined to a surface.

Step 3. Coalescing along the arc c. □

An immediate consequence is

(10.3) COROLLARY (A converse of the Lefschetz Fixed Point Theorem). *Let X be a compact connected and simply connected polyhedron, without global separating* *points. Then, for a self-map $f: X \to X$, the Lefschetz number $L(f) = 0$ if and only if f is homotopic to a fixed point free map.*

If the assumption of no local separating points is removed, the equality $N(f) = \mathrm{MF}[f]$ may break down even for the identity class $[\mathrm{id}_X]$.

For a general polyhedron X but a special homotopy class $f = \mathrm{id}_X$, a method of calculating $\mathrm{MF}[\mathrm{id}_X]$ has been found by Shi (see [Sh]). Instead of describing the rather complicated result, we only state a simpler consequence.

(10.4) THEOREM ([Sh]). *Let X be a compact connected polyhedron. Suppose X is not disconnected by removing all local separating points. Then $\mathrm{MF}[\mathrm{id}_X] = N(\mathrm{id}_X)$, which is 0 if the Euler characteristic $\chi(X) = 0$ and is 1 if $\chi(X) \neq 0$.*

10.2. In a homotopy type of self-maps. We can characterize the geometric significance of the Nielsen number.

(10.5) THEOREM (Geometric characterization of the Nielsen number). *In the category of compact connected polyhedra, the Nielsen number of a self-map equals the least number of fixed points among all self-maps having the same homotopy type.*

$$N(f) = \mathrm{Min}\{\#\,\mathrm{Fix}(g) : g \textit{ has the same homotopy type as } f\}.$$

PROOF. Let m stand for the number on the right hand side. Then $N(f) \leq m$ by Theorem (8.5). On the other hand, there always exists a compact manifold M (with boundary) of dimension ≥ 3 having X as deformation retract. (For example, M = the regular neighbourhood of X imbedded in a Euclidean space.) So, by Theorem (10.2), the lower bound $N(f)$ is realizable on M by a map having the same homotopy type as f. □

10.3. Self-maps on surfaces. Dimension 2 turns out to be the hard and rich dimension in the fixed point theory for self-maps. In a manifold M of dimension n, finding a fixed point of a map $f: M \to M$ is equivalent to finding, in the $2n$-dimensional manifold $M \times M$, an intersection of two n-dimensional submanifolds, namely the graph $\Gamma(f) = \{(x_1, x_2) \in M \times M : x_2 = f(x_1)\}$ and the diagonal $\Delta(M) = \{(x_1, x_2) \in M \times M : x_1 = x_2\}$. It is well known that the intersection theory does not work well when the dimension of the ambient manifold is less than 5. This is a typical difficulty in low dimensions. In our context, $2n < 5$ means $n \leq 2$, but since the case $n = 1$ is trivial, $n = 2$ is the only special dimension. We are at the intersection of fixed point theory with low-dimensional topology.

The following theorem uncovers the anomaly in dimension 2.

(10.6) THEOREM ([Ji6]). *Let M be a connected compact surface with Euler characteristic $\chi < 0$. Then there is a self-map $f: M \to M$ such that $N(f) = 0 <$* $\mathrm{MF}[f]$.

The first example [Ji5] is a very simple map:

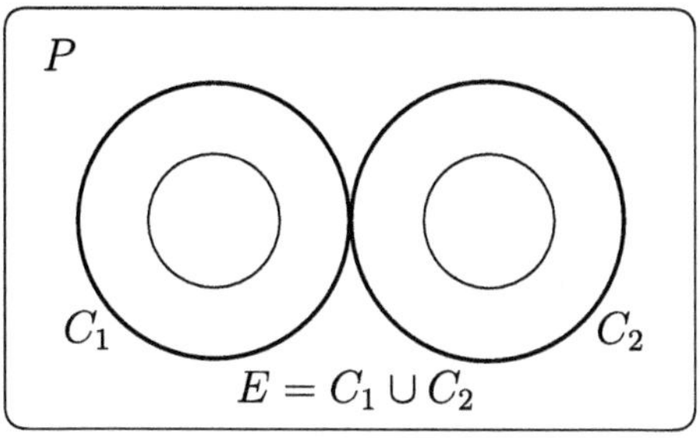

FIGURE 9

(10.7) EXAMPLE. Let P be the pants space (= a disk with two holes). Let C_1 and C_2 be two circles in P touching at a single point, each encircling a hole. Denote by E the figure eight space $C_1 \cup C_2$. Let $r: P \to E$ be a retraction.

Construct a map $\phi: E \to E$ that sends C_1 into itself with degree -1, and sends C_2 into itself with degree 2. Regard the composition $\phi \circ r$ as a self-map $f: P \to P$. Then we have $N(f) = 0$ but $\mathrm{MF}[f] = 2$.

The proof of $\mathrm{MF}[f] = 2$ makes use of an algebraic tool — Artin's pure braid group (see [Bi, p. 5] for the definition).

There are only seven compact surfaces with $\chi \geq 0$: the sphere, the projective plane, the torus, the Klein bottle, the disk, the annulus and the Möbius band. The equality $N(f) = \mathrm{MF}[f]$ can be verified directly. Therefore combining with Theorem (10.2) we have

(10.8) THEOREM (A characterization of Wecken spaces, [Ji6]). *Let X be a connected compact polyhedron with no local cut points. Then $N(f) = \mathrm{MF}[f]$ for every map $f: X \to X$ if and only if X is not a surface with negative Euler characteristic.*

11. The least number of fixed points in an isotopy class

We can also ask the following question. Let M be a compact manifold, and let $f: M \to M$ be a self-homeomorphism. Is it true that

$$N(f) = \min\{\#\,\mathrm{Fix}(g) : g \text{ is isotopic to } f\}?$$

Actually this is the original question studied by Nielsen, for orientable closed surfaces. The generalization to self-maps came later, due to Reidemeister and Wecken.

11.1. Dimension 2. The answer for the surface case relies on Thurston's classification of surface homeomorphisms:

(11.1) THEOREM (Nielsen–Thurston classification, [T]). *Assume that M is a compact surface such that each of its connected components has negative Euler characteristic. Then every homeomorphism $f\colon M \to M$ is isotopic to a diffeomorphism φ such that either*

(11.1.1) φ is an isometry with respect to some hyperbolic metric on M, or equivalently, φ is a periodic map, i.e. $\varphi^m = \mathrm{id}$; or

(11.1.2) φ is a pseudo-Anosov map, i.e. there is a number $\lambda > 1$ and a pair of transverse measured foliations $(\mathfrak{F}^s, \mu^s)$ and $(\mathfrak{F}^u, \mu^u)$ such that $\varphi(\mathfrak{F}^s, \mu^s) = (\mathfrak{F}^s, \mu^s/\lambda)$ and $\varphi(\mathfrak{F}^u, \mu^u) = (\mathfrak{F}^u, \lambda\mu^u)$; or

(11.1.3) φ is a reducible map, i.e. there is a system of disjoint simple closed curves $\Gamma = \{\Gamma_1, \dots, \Gamma_n\}$ in $\operatorname{int} M$ such that Γ is invariant by φ (but the Γ_i may be permuted) and Γ has a φ-invariant tubular neighbourhood $\mathcal{N}(\Gamma)$ such that each component of $M - \mathcal{N}(\Gamma)$ has negative Euler characteristic and on each (not necessarily connected) φ-component of $M - \mathcal{N}(\Gamma)$, φ satisfies (11.1.1) or (11.1.2).

The answer to the Nielsen conjecture is

(11.2) THEOREM ([Ji2], [JiG]). *Let M be a compact surface, closed or with boundary. Let $f\colon M \to M$ be a self-homeomorphism. Then f is isotopic to a self-embedding which has $N(f)$ fixed points. If, in addition, no boundary component of M is mapped onto itself by f in an orientation-reversing manner, then f is isotopic to a self-homeomorphism having $N(f)$ fixed points.*

(11.3) REMARK. It is necessary to allow self-embeddings in order to get as few as $N(f)$ fixed points when there are orientation-reversing invariant boundary components, because any self-homeomorphism would have at least two fixed points on each such boundary component. If we insist on homeomorphisms, the least number of fixed points in the isotopy class turns out to be the relative Nielsen number $N(f; M, \partial M)$ defined by Schirmer in [Sch].

OUTLINE OF PROOF OF THEOREM 11.2. Observe that the conclusion is true if φ is a periodic map or a pseudo-Anosov map. In the reducible case, along the reducing curves we cut M into periodic and/or pseudo-Anosov pieces. On each piece the least number of fixed points is realized by φ.

A major difficulty arises when we try to put the pieces back together. Two fixed points that are inequivalent in a piece may become equivalent in the whole manifold.

To overcome this difficulty, a standard form for φ is developed so that each fixed point class (in the whole manifold) is connected, hence can be coalesced to a point. □

11.2. Dimension 3. For three dimensional manifolds, we have the following results. References on topology and geometry of 3-manifolds: [J], [S].

(11.4) THEOREM ([JiWW]). *Suppose M is a closed orientable manifold which is either Haken or geometric (admits one of Thurston's eight 3-dimensional geometries), and $f\colon M \to M$ is an orientation-preserving homeomorphism. Then f is isotopic to a homeomorphism g with $\#\operatorname{Fix}(g) = N(f)$.*

If Thurston's Geometrisation Conjecture is true, the above theorem applies to all aspherical (i.e. $\pi_2 = 0$), orientable, closed 3-manifolds.

The proof in the Haken case uses the Jaco–Shalen–Johannson torus-decomposition into pieces which admit geometric structures (mainly Seifert fiber spaces and hyperbolic 3-manifolds). The given map is then deformed to a map which is of a certain standard form on each piece of the decomposition (in particular, fiber-preserving or isometric), and also on its boundary and around the fixed points. The proof follows the general pattern of cutting and pasting as in the proof of Theorem (11.2), but now along the decomposition tori instead of the reducing curves. Everything is more complicated though. For example, in the Seifert fibered case, this uses a generalization of Theorem (11.2) to the case of 2-dimensional orbifolds.

For many manifolds, a much stronger result holds: All homeomorphisms are isotopic to fixed point free homeomorphisms.

(11.5) THEOREM ([JiWW]). *Let M be a closed orientable 3-manifold which is either Haken or geometric. Then any orientation-preserving homeomorphism f on M is isotopic to a fixed point free homeomorphism, unless some component of the JSJ torus-decomposition of M is a Seifert fiber space with big orbifold.*

Here the basis orbifold of a Seifert fiber space is said to be *small* if it is a sphere with at most three holes or cone points, or a projective plane with at most two holes or cone points; otherwise it is *big*.

In particular, the theorem applies to all closed hyperbolic and spherical 3-manifolds, and four of the six Euclidean 3-manifolds.

11.3. Higher dimensions. Kelly gave in [K] a proof of the Nielsen conjecture for self-homeomorphisms on closed manifolds of dimension at least 5, using the fact that any null-homotopic simple closed curve in such a manifold bounds an embedded disc.

The dimension 4 is still a blank.

References

[AH] P. S. Alexandroff and H. Hopf, *Topologie*, Springer–Verlag, Berlin, 1935. (German)

[Bi] J. S. Birman, *Braids, links, and mapping class groups*, Princeton University Press, Princeton, N.J., 1974.

[Br] L. E. J. Brouwer, *Über die Minimalzahl der Fixpunkte bei den Klassen von eindeutigen stetigen Transformationen der Ringflächen*, Math. Ann. **82** (1921), 94–96. (German)

[Br1] R. F. Brown, *The Lefschetz Fixed Point Theorem*, Scott, Foresman and Co., Glenview, Ill.-London, 1971.

[Br2] ______, *Fixed point theory*, History of Topology (I. M. James, ed.), North-Holland, Amsterdam, 1999, pp. 271–299.

[D] A. Dold, *Lectures on Algebraic Topology*, Springer–Verlag, New York, 1972; second ed., 1980.

[F] E. Fadell, *Nielsen numbers as a homotopy type invariant*, Pacific J. Math. **63** (1976), 381–388.

[FH1] E. Fadell and S. Husseini, *Fixed point theory for non-simply-connected manifolds*, Topology **20** (1981), 53–92.

[FH2] ______, *The Nielsen number on surfaces*, Topological Methods in Nonlinear Functional Analysis (Toronto, Ont., 1982), Contemp. Math., vol. 21, Amer. Math. Soc., Providence, RI, 1983, pp. 59–98.

[H] S. Husseini, *Generalized Lefschetz numbers*, Trans. Amer. Math. Soc. **272** (1982), 247–274.

[J] W. Jaco, *Lectures on three-manifold topology*, CBMS Regional Conf. Ser. in Math. **43** (1980), Amer. Math. Soc., Providence, R.I..

[Ji1] B. Jiang, *On the least number of fixed points*, Amer. J. Math. **102** (1980), 749–763.

[Ji2] ______, *Fixed points of surface homeomorphisms*, Bull. Amer. Math. Soc. (N.S.) **5** (1981), 176–178.

[Ji3] ______, *Fixed point classes from a differential viewpoint*, Fixed Point Theory (Sherbrooke, Que., 1980), Lecture Notes in Math., vol. 886, Springer, Berlin, 1981, pp. 163–170.

[Ji4] ______, *Lectures on Nielsen fixed point theory*, Contemp. Math. **14** (1983), Amer. Math. Soc.

[Ji5] ______, *Fixed points and braids*, Invent. Math. **75** (1984), 69–74.

[Ji6] ______, *Fixed points and braids* II, Math. Ann. **272** (1985), 249–256.

[Ji7] ______, *A characterization of fixed point classes*, Fixed Point Theory and its Applications, Contemp. Math. (Berkeley, CA, 1986), vol. 72, Amer. Math. Soc., Providence, RI, 1988, pp. 157–160.

[Ji8] ______, *Estimation of the number of periodic orbits*, Pacific J. Math. **172** (1996), 151–185.

[JiG] B. Jiang and J. Guo, *Fixed points of surface diffeomorphisms*, Pacific J. Math. **160** (1993), 67–89.

[JiWW] B. Jiang, Sh. Wang and Y.-Q. Wu, *Homeomorphisms of 3-manifolds and the realization of Nielsen number*, Comm. Anal. Geom. **9** (2001), 825–877.

[K] M. R. Kelly, *The Nielsen number as an isotopy invariant*, Topology Appl. **62** (1995), 127–143.

[Ke] M. A. Kervaire, *Geometric and algebraic intersection numbers*, Comment. Math. Helv. **39** (1965), 271–280.

[Ki] T.-h. Kiang, *The Theory of Fixed Point Classes*, Springer–Verlag, Berlin, 1989.

[KiJi] T.-h. Kiang and B. Jiang, *The Nielsen numbers of self-mappings of the same homotopy type*, Sci. Sinica **12** (1963), 1071–1072.

[M] W. S. Massey, *Algebraic Topology: An Introduction*, Harcourt, Brace & World, Inc., New York, 1967.

[N1] J. Nielsen, *Über die Minimalzahl der Fixpunkte bei den Abbildungstypen der Ringflächen*, Math. Ann. **82** (1921), 83–93. (German)

[N2] ———, *Untersuchungen zur Topologie der geschlossenen zweiseitigen Flächen,* I, Acta Math. **50** (1927), 189–358 (German); English transl.V. L. Hansen (ed.), *Investigations in the topology of closed orientable surfaces,* I, Jakob Nielsen: Collected Mathematical Papers, vol. 1, Birkhäuser, Boston, 1986, pp. 223–341.

[R] K. Reidemeister, *Automorphismen von Homotopiekettenringen*, Math. Ann. **112** (1936), 586–593. (German)

[S] P. Scott, *The geometries of* 3*-manifolds*, Bull. London Math. Soc. **15** (1983), 401–487.

[Sch] H. Schirmer, *A relative Nielsen number*, Pacific J. Math. **122** (1986), 459–473.

[Sh] G. Shi, *Least number of fixed points of the identity class*, Acta Math. Sinica **18** (1975), 192–202 (Chinese); an English exposition can be found in Chapter 4 of [Ki]

[T] W. P. Thurston, *On the geometry and dynamics of diffeomorphisms of surfaces*, Bull. Amer. Math. Soc. (N.S.) **19** (1988), 417–431.

[W1] F. Wecken, *Fixpunktklassen,* I, Math. Ann. **117** (1941), 659–671. (German)

[W2] ———, *Fixpunktklassen,* II, *Homotopieinvarianten der Fixpunkttheorie*, Math. Ann. **118** (1941), 216–234. (German)

[W3] ———, *Fixpunktklassen,* III, *Mindestzahlen von Fixpunkten*, Math. Ann. **118** (1942), 544–577. (German)

17. NIELSEN FIXED POINT THEORY ON SURFACES

Michael R. Kelly

The purpose of this chapter is to give a survey of some of the research problems that arose out of the early work of Jacob Nielsen concerning the study of fixed points of surface mappings. One aspect of this work is the *Nielsen number* of a self-map on a topological space and its role as a lower bound for the number of fixed points. The primary focus of this chapter is to address the question of the sharpness of the Nielsen number as a lower bound for the number of fixed points. The chapter also considers a number of variations on this study and other problems that arose out of work on such questions. Outlines of the proofs of some of the results presented in this chapter are given. References are cited for the rest.

The chapter is not meant to be a comprehensive survey of results regarding fixed points of surface mappings. For instance, one area of research which has been influenced by Nielsen's work is the Birkhoff and Aubrey–Mather theories which study the dynamics of area-preserving diffeomorphisms of the 2-dimensional annulus. For references along these lines the reader is referred to the surveys given in [Bo], [L].

1. The classical Wecken problem

Let $f: X \to X$ be a self-mapping of the topological space X. We denote by $\mathrm{MF}[f]$ the minimal number of fixed points possible for a map in the homotopy class of f. It's value will be a non-negative integer when X is, for example, a finite polyhedron or a compact manifold. The Nielsen number $N(f)$, by its homotopy invariance, provides a lower bound for the value of $\mathrm{MF}[f]$. Thus it becomes an interesting object of study on its own and it is a fundamental problem to determine the sharpness of this bound. One would expect that an answer could easily depend on the category of spaces under consideration.

When Nielsen first formulated the notion of the Nielsen number [Ni1] it was done in context of a very specific calculation. The purpose was to compute the number of fixed points for certain maps on the 2-torus. Of course, they were the naturally defined linear maps. He also established that these maps had the feature (with a few exceptions) that they had the least number of fixed points among all maps in their respective homotopy class of maps.

In 1942, Wecken proved in [Wec] that if X is a compact, connected PL manifold, with or without boundary, of dimension $\dim(X) \geq 3$ and $f\colon X \to X$ is any map, then $N(f) = \mathrm{MF}[f]$. So the lower bound is sharp for most manifolds. Consequently, a space X such that $N(f) = \mathrm{MF}[f]$ for all maps $f\colon X \to X$ is said to have the *Wecken property*. Wecken's result was extended by Weier (see [Wei]) by removing the dependence on PL-topology methods for a proof. He gives a proof for topological manifolds of dimension at least three. Since then a number of authors gave results extending Wecken's theorem, culminating in the theorem of Jiang (see [J1]), which said that among compact polyhedra without local separating points surfaces with negative Euler characteristic were the only ones that were not known to have the Wecken property.

The natural question of whether or not surfaces with negative Euler characteristic also possess the Wecken property was not settled until long after Wecken's paper was published. In 1985, Jiang proved in [J5] that if X is a compact, connected surface, with or without boundary, then X has the Wecken property if and only if the Euler characteristic $\chi(X)$ of X is non-negative. Since there are only six surfaces X for which $\chi(X) \geq 0$, almost all surfaces fail to have the Wecken property.

We give here an outline of two different proofs of this result. One is the proof due to Jiang and the second is due to this author. The idea of the proof is to first find a counterexample on the *pants* surface which Jiang gives in [J4]. Next one embeds the pants surface in any other surface with negative Euler characteristic in such a way that the nature of the counterexample persists. Jiang's original example did not work for this, but in [J5] he produces another example which does the job.

Let P denote the pants surface, the 2-sphere with three open disks removed. Let $a \vee b$ be the wedge of two loops in the plane and identify P with a regular neighbourhood of this wedge. Orient a, b so that the loop ab corresponds to one of the boundary components of P. Prescribe a homotopy class of maps on P by sending a to a word U and b to a word V. These are actually maps on $\pi_1(P)$, but they determine a homotopy class of maps on P unique up to conjugation of the pair (U, V). The homotopy class that Jiang considers is obtained by setting $U = a^{-1}$, $V = a^{-1}b^2$. It is not difficult to define a particular map f which has exactly two fixed points, one with index 1 and the other with index -1 and such tht the two are in the same Nielsen class.

Jiang claims that any map homotopic to f must have a fixed point. Here is a brief outline of his argument.

Consider a map g (homotopic to f) with a finite fixed point set in the interior of P. Choose loops ℓ_i, $i = 0, 1, 2$ in P going from a basepoint to the boundary component ∂_i along an arc, around the boundary and back along the arc, and missing any fixed points. Orient the loops so that $\ell_1 \ell_2 \ell_0^{-1}$ is represented by the trivial element in $\pi_1(P)$. Let h_t denote the homotopy from this path to the constant path at the base point. Let Δ be the diagonal in $P \times P$. Then $(x, g(x))$ takes each ℓ_i to a loop in $P \times P - \Delta$, represented by an element τ_i in $\pi_1(P \times P - \Delta)$.

Now suppose there is a map g with no fixed points. Then $g \circ h_t$ is a homotopy ending in the trivial path, so $\tau_1 \tau_2 \tau_0^{-1} = 1$. Let σ_i denote the element obtained from ℓ_i using $(x, f(x))$. Then $\tau_i = u_i \sigma_i u_i^{-1}$ for some word u_i. Since the map f is given explicitly Jiang calculates each σ_i and obtains an equation in the braid group, which looks like

$$u_1 \sigma_1 u_1^{-1} u_2 \sigma_2 u_2^{-1} = u_0 \sigma_0 u_0^{-1},$$

with the u_i's being unknowns. In general, it is difficult to decide if such an equation in three unknowns has any solutions. Jiang's trick was to find a useful representation into the free product of the integers with the integers mod 2, where the equation becomes $vab^2abv^{-1} = b^3$, with $a^2 = 1$. The utility is that there is now only one unknown, namely v, and its not hard to see that ab^2ab and b^3 are not conjugate in this group. So we get a contradiction to the existence of a fixed point free map.

An alternative method is developed in [K1], which computes the value of MF[f] for any self-map of P. We outline how this method works on Jiang's original example given above. Included are some improvements of this method which can be found in [K9]. The viewpoint of this approach is to consider handle structures on the surface and to work with a weak form of transversality associated to the preimages of the attaching curves for the 1-handles. This then turns the problem into a combinatorial one as opposed to the algebraic problem in Jiang's argument.

Consider a handle decomposition of P consisting of one 0-handle and two 1-handles. Let A denote the set of attaching arcs for the 1-handles of the decomposition. We first have a weak transversality type result for maps that have exactly MF[f] fixed points. We deform our map (without adding fixed points) so that $f^{-1}(A)$ is a proper 1-manifold in P, with each component being essential in P and meeting A in a finite set of points. Moreover, the results of [K9] assert that for the pants surface one can choose the handle structure so that the 1-handles are mapped into the interior of the 0-handle. Thus, all of the arcs in $f^{-1}(A)$ lie in the 0-handle, which we denote by D.

The homeomorphism group for the pants surface is finite of order 12, and since we work with unoriented handle structures there will be only 6 such structures up to isotopy. Furthermore, a homeomorphism that interchanges the two 1-handles of a given handle structure essentially gives back the same structure. Thus we are reduced to considering 3 possible handle structures. One way to describe them is as follows. Up to isotopy in P there are exactly three distinct proper arcs which are essential and non-separating in P. Choose one arc in each class so that the choices are pairwise disjoint. Now choose two of the three and take a regular neighborhood of each. These will be the 1-handles for a decomposition, the complimentary disk is the 0-handle D.

We now focus on Jiang's example as described above, but the method outlined works for any self-map on P. Consider the map $a \mapsto a^{-1}$, $b \mapsto a^{-1}b^2$, and the handle structure given by choosing the two arcs which meet $a \vee b$ in exactly one point, and let H_a, H_b denote the corresponding 1-handles. The third isotopy class of arcs needed for the other handle structures meets both a and b.

Given this data we can now quantify the possible configurations for the curves in $f^{-1}(A)$. One basic configuration has 2 arcs in D each isotopic to the attaching arcs for H_a and 6 arcs in D each isotopic to those of H_b. These arcs and their images are determined by the two words a^{-1} and $a^{-1}b^2$ which prescribe the map. We assume that all of these curves lie in a neighbourhood of one attaching arc for each of H_a and H_b. All other possible configuations are obtained by *rearrangements* of this one. This involves sliding arcs around 1-handles and coalesing a pair of arcs that have the same image to form a new pair of arcs. We don't give the details here but simply note that there is a finite number of distinct configuations up to isotopy in D (with the endpoints of the arcs always staying in the boundary of P).

Now, each component of $D \setminus f^{-1}(A)$ which is mapped by f into D has a fixed point index which is determined by the images of the arcs of $f^{-1}(A)$ on its border. Given the initial configuration above we detect such a region between the two regions (mapping to 1-handles) formed by the b^2 given in the map, and this region has index -1. Checking all the possible rearrangements for this setting one sees that a fixed point with index -1 persists in each.

The proof is completed by repeating the process for the other two handle structures. In summary, we show that for each handle structure all maps that have $f^{-1}(A)$ contained in D must have a fixed point. Thus from [K9] any map in this homotopy class has a fixed point.

Now that one has an example on the pants surface it remains to see how useful the pants surface is in obtaining an example for an arbitrary surface with negative Euler characteristic. The idea here is simple, but does not work with any such example. The pants surface can be embedded in any surface with negative Euler characteristic so that the induced map on fundamental group is injective. One would like to embedd the example of Jiang in such a way that the non-Wecken property persists in the new surface. It is not at all clear how to do this with the above example, but Jiang [J6] produces a different example with Nielsen number zero for which the extension is clear, and the proof of the existence of a fixed point for any map in the homotopy class reduces to the case of the map defined on the pants surface. Here is the example: $a \mapsto a^2ba^{-1}b^{-1}$ and $b \mapsto 1$. What makes the computations feasible for this example is the fact that one of the generators is killed by the map.

Since we have examples of non-Wecken maps on all surfaces with negative Euler characteristic it is reasonable to investigate to what degree the Wecken property fails. With this in mind a manifold X is said to be *totally non-Wecken* if, for each integer $m \geq 1$, there exists a map $f_m\colon X \to X$ such that $\mathrm{MF}[f_m] - N(f_m) \geq m$. The second method described above applies to all self-maps of the pants surface and thus an algorithum for the calculation of $\mathrm{MF}[f]$ on this surface. By selecting a suitable family of maps on the pants surface, for example f_m determined by $a \mapsto (bab^{-1}a^{-1})^m ba$, $b \mapsto 1$, one calculates that $\mathrm{MF}[f_m] \geq 2m$ and $N(f_m) = 0$. So the pants surface is totally non-Wecken.

Extending earlier work in [K2], Jiang proves a commutativity type theorem in [J6] and as a consequence is able to prove the following: if X is a compact, connected surface, with or without boundary, such that $\chi(X) < 0$, then X is totally non-Wecken [J6].

Another way to consider how the Wecken property fails is to try to classify the maps f on a given surface that satisfy $\mathrm{MF}[f] = N(f)$. As noted above the value of $\mathrm{MF}[f]$ can be computed for all self-maps on the pants surface. It would then be useful to make a comparison of this value against that of the Nielsen number. This was partially done by J. Wagner [Wa1], [Wa2], where the value of $N(\phi)$ is computed for a large class of endomorphisms of the free group. For the pants surface, when compared to the value of $\mathrm{MF}[f]$ it is shown that equality occurs for virtually all maps in the class. It is also shown that this class of endomorphisms is large in that the probability that a randomly chosen self-map is in this class is one. It is not surprizing that maps of the form $a \mapsto W$, $b \mapsto 1$ are not in this class. As we have already seen this class generates the non-Wecken result for surfaces in general. Similarly, self-maps of the pants of the form $a \mapsto WX$, $b \mapsto WY$ can be used to generate a large difference between $\mathrm{MF}[f]$ and $N(f)$. Is it possible that such maps, and any conjugate of such, are the only ones that lead to the pants surface being totally non-Wecken?

2. Homeomorphisms and fixed point indices

These Wecken type problems, which we see tend to have negative answers for surfaces can also be considered from other viewpoints such as natural restrictions imposed on the maps or spaces. One such is to only consider the problem for maps that are homeomorphisms. In fact, in some of Nielsen's early work we see that his primary concern was with the structure of surface homeomorphisms, and one issue was what is now called the Wecken property. He conjectured in 1927 that this Wecken property should hold for surface homeomorphisms (see [Ni2]). Over fifty years later Jiang announced in [J2] a proof of this conjecture. The proof relies heavily on the Nielsen–Thurston classification of surface homeomorphisms, which did not appear in published form until 1988 (see [Th] and [CB]). The proof of the Wecken property for surface homeomorphims finally appears in 1993 (see [JG]).

Besides giving a detailed proof of the Wecken property in that paper the authors also establish two inequalities for the indices of the Nielsen classes of surface homeomorphisms. Consider a surface F with Euler characteristic $\chi(F) < 0$. From the Nielsen–Thurston classification ([Th]) we see that hyperbolic geometry plays an important role in the structure of the *pseudo-Anosov* homeomorphisms on F. One feature is that the fixed point indices obey the following bounds:

$$\sum_{\{x:\text{index}(x)\geq 1\}} (\text{index}(x) - 1) \leq 0 \quad \text{and} \quad \sum_{\{x:\text{index}(x)\leq -1\}} (\text{index}(x) + 1) \geq 2\chi(F).$$

The first inequality simply states that $+1$ is an upper bound for the index of a single fixed point. This had been established independently by Pelikan and Slaminka in the more general setting of area-preserving surface homeomorphisms (see [PS]). The second inequality is roughly the Euler–Poincaré formula, which follows from the existence of a singular foliation associated to a given pseudo-Anosov homeomorphism. These bounds also highlight the 1-dimensional nature of pseudo-Anosov's in that its easy to verify the two inequalities for the isolated fixed points of an arbitrary self-map $f\colon G \to G$, where G is a finite graph.

Jiang and Guo extend this result about pseudo-Anosov's by constructing, in an arbitrary homotopy class of homeomorphisms, a representative homeomorphism on F which has the minimal number of fixed points in its homotopy class and also satisfies both of these inequalities. But since surface homeomorphisms are Wecken, this also says that the indices of the Nielsen classes obey the same bounds. Consequently, any map in the homotopy class (of a homeomorphism) having the minimal number of fixed points possible satisfies the inequalities as well.

In that paper the authors raise the question as to the validity of these bounds on the indices of Nielsen classes when one considers arbitrary self-mappings of surfaces with negative Euler characteristic. This was settled independently by

Jiang (see [J7]) and this author (see [K7]). Both papers show that the same bounds are valid for all self-maps. The two papers give very different proofs of this result, but both work from the observation that the problem reduces to the setting of surfaces with non-empty boundary as follows. We can assume that $f\colon F \to F$ is not homotopic to a homeomorphism. Since $\chi(F) < 0$ there is a disk $D \subset F$ and a map g homotopic to f with image in $F \setminus D$. Due to the commutativity property in Nielsen theory (see [J3]) there is a one-to-one correspondence between the Nielsen classes of f and those of g restricted to $F \setminus D$ which respects indices.

Jiang makes further use of the commutativity property to reduce the problem to the case where the map is π_1-injective. Then the methods developed in [BH] and [DV] can be applied to produce a graph map which carries the Nielsen class data of the original map. This gives the desired bounds because these graph maps have exactly one fixed point corresponding to each essential Nielsen class. In [K7] an implicit construction of a representative map which carries information about the Nielsen classes is given. From this information one then deduces the two index bounds.

On the other hand the index bound problem for fixed point minimal maps on surfaces is more subtle. In particular, there does not exist a commutativity property for the minimal number of fixed points. Partial results given in [J6] and [K2] do not suffice for this problem, so one needs to work with the information given by the particular map defined on the surface. This is exactly the approach used in [K6]. Roughly, the idea is to deform a given fixed point minimal map into a representative map that carries a certain defined embedded graph, which in turn encodes the fixed point data for the original map. The two inequalities are then derived from this combinatorial data. No details are given here but we do point out that the method used requires that the surface have non-empty boundary. At this time the validity of these index bounds for fixed point minimal maps on closed surfaces remains an open problem.

3. The relative setting

We now return to considerations of the Wecken property, this time for the relative category. Here we consider compact surfaces which have nonempty boundary and all maps, including homotopies, map the boundary to itself. The relative Nielsen number due to Schirmer (see [Sch1]) is now compared against the minimal number of fixed points in a given homotopy class of maps.

The types of arguments used for this study are similar to those in the non-relative case. For small surfaces, that is either the disk, annulus or Möbius band, it is staightforward to verify the Wecken property. For almost all other surfaces a totally non-Wecken result is established by producing a family of maps where the difference between the minimal number of fixed points and the relative Nielsen number becomes large, just as in the absolute case. For some of these surfaces this was just a matter of embedding certian examples from [K1] into a relative setting (see [BS]). But this was not sufficient to handle all bounded surfaces. To take care of this problem the methods from [K1], [K2] were recast in the relative setting to take care of all surfaces obtained by removing two or more open disks from either the torus or the Klein bottle (see [K3]), while further improvements by J. Nolan in 1996 (see [No]) dealt with the planar surfaces. But after all this work there still remained one family of surfaces for which the relative Wecken problem was not resolved, the Möbius band with one or more open disks removed. The case where there is more than one puncture is resolved in [K8] and also a partial result is obtained for the special case of a single puncture.

That the once punctured Möbius band should present a difficulty was in a way not a surprize, as it is the non-orientable cousin of the pants surface. Of the four surfaces each having fundamental group being free on two generators two are modeled on spherical geometry (pants and punctured Möbius band) and two are hyperbolic (punctured torus and punctured Klein bottle). It was observed earlier in [BS] and [K3] that a surprizing feature emerges for the pants surface. Namely, that it is neither Wecken nor totally non-Wecken. What happens is that by classifying all of the boundary-preserving maps on the pants surface one can calculate a bound on the difference $|\mathrm{MF}_\partial(f) - N_\partial(f)|$ by using a suitable representative map for the homotopy class given by f. This turns out to be either 0 or 1 depending on the choice of f. Then by using a suitable choice for f its shown that the bound of 1 can be sharp. So there are relative maps on the pants which must have one more fixed point than that predicted by the relative Nielsen number. The analagous result for the once punctured Möbius band is resolved in [K8] and [BK].

At the time when this *almost-Wecken* phenomenon was discovered it was thought that something similar might happen for all the planar surfaces; that a bound on the difference $\mathrm{MF}_\partial(f) - N_\partial(f)$ dependent only on the number of punctures added to the annulus could also happen. This turns out not to be the case and can be explained by a fairly simple reason. The construction of a family of examples f_n of boundary-preserving maps such that the difference $\mathrm{MF}_\partial(f_n) - N_\partial(f_n) \to \infty$ can be done by choosing a two-sided simple closed curve which is not isotopic to a boundary curve. Outside an annular neighborhood A of the curve prescribe a single boundary-preserving map such that ∂A maps to a single point p in the interior, then generate a family of maps determined by mapping the annulus into the target surface so that an arc transveral to the boundary of A is sent to some loop based at p. For a suitable choice of such maps one can keep N_∂ constant.

The hard work is to show that the value of MF_∂ can be made arbitrarily large. Notice that there are only two surfaces which have negative Euler characteristic and do not contain a simple closed curve as above; the pants surface and the once punctured Möbius band.

4. Coincidences for maps between surfaces

Whenever one has a question or problem in Nielsen fixed point theory there is always the natural generalization to coincidence theory. Here we consider pairs of maps $f, g: X \to Y$ between two spaces and study the equation $f(x) = g(x)$. In the setting where both X and Y are manifolds there is a defined *semi-index* with which one gets a *Nielsen coincidence number* denoted by $N(f,g)$, and as with the Nielsen number it gives a lower bound for $MC(f,g)$, the minimal possible number of components for the coincidence set in the homotopy class of the pair (f,g). When X and Y are manifolds of the same dimension then each such component can be replaced by a single point.

Just as with the classical Wecken problem treated above there is also the analogous Wecken problem for coincidences. That is, when does $N(f,g) = MC(f,g)$? As in the self-mapping setting this was answered in the affirmative when both X and Y are manifolds of the same dimension $d \geq 3$ (see [Sch2]). So again the interest is in the case where both X and Y are surfaces. There has been a number of results published on this problem. In fact, the first known counterexample was obtained well before Jiang's example for the fixed point problem. In 1930, H. Hopf considers in [Ho] the setting where one of the maps is the constant map and produces an example for which this Wecken property fails. For orientable surfaces the Wecken problem where one of the maps is the constant map has been studied in detail in [BGZ] and [GZ]. When the target surface is the torus the authors also establish that the Wecken property holds in general. Another special case for which the Wecken problem has been resolved is the situation where X and Y are the same closed surface and f is the identity map. Then Fuller tells us in [Fu] that the coincidence Wecken problem gives nothing new. That is, $MC(\mathrm{id}, g) = \mathrm{MF}(g)$. Besides these special cases not much else has been done concerning the Wecken problem for coincidences between surfaces.

In the paper [GJ] the index bounds for Nielsen classes analogous to the inequalities for surface maps are established.

5. Open problems

We conclude this chapter we a brief discussion of some open problems and possible directions for further research in the area of Nielsen fixed point theory as it applies to surface mappings. Probably foremost is the following

(5.1) QUESTION. Is there an invariant which is essentially algebraic in nature and which measures the least number of fixed points for surface self-mappings?

Related to Question (5.1) one could pose the following

(5.2) QUESTION. Is there an algorithm for the calculation of $\mathrm{MF}[f]$ for a self-map f of a surface? Similarly, is there an algorithm for the calcuation of the Nielsen number?

An algorithm for $\mathrm{MF}[f]$ is clearly a problem about surface self-mappings, whereas the Nielsen number is more about groups, in particular free groups and surface groups, and their endomorphisms. The result in [K1] gives an algorithm for $\mathrm{MF}[f]$ for arbitrary f in the case that the surface is the pants surface. It is not at all clear how one might extend this to other surfaces. The problem of an algorithm for the Nielsen number is addressed in [Wa2] and in [Yi]. For surface homeomorphisms an algorithm for Nielsen numbers is presented in [K5], while one is implicitly present in the paper [BH]. In the preprint [K9] an alternative approach is taken. Instead of an algorithm one looks for a representative map in the homotopy class of a given map f that carries the information for the computation of $\mathrm{MF}[f]$. This leads to the following

(5.3) QUESTION. Can one find a 1-dimensional repsentative for the minimal fixed point problem? That is, an associated graph map which has exactly the same fixed point data as the given surface mapping.

(5.4) PROBLEM. Establish the index bounds given in Section 2 for fixed point minimal self-maps on closed surfaces.

(5.5) PROBLEM. Formulate and prove the coincidence analogs for the Wecken classification of surfaces, both in the classical and relative settings.

(5.6) PROBLEM. Formulate and prove an index bounds result for pairs of maps between surfaces which have the minimal number of coincidence points.

References

[BH] M. Bestvina and M. Handel, *Train-tracks for surface homeomorphisms*, Topology **34** (1995), 1–51.

[BGZ] S. Bogatyi, D. L. Gonçalves and H. Zieschang, *The minimal number of roots of surface mappings and quadratic equations in free groups*, Math. Z. **236** (2001), 419–452.

[Bo] P. Boyland, *Topological methods in surface dynamics*, Topology Appl. **58** (1994), 223–298.

[Br1] R. F. Brown, *The Lefschetz Fixed Point Theorem*, Scott–Foresman, 1971.

[Br2] ______, *Nielsen theory on manifolds*, Proceedings of the conference on Nielsen theory and Reidemeister torsion, vol. 49, Banach Center Publ., 1999, pp. 19–27.

[BK] R. F. Brown and M. Kelly, *The boundary-Wecken classification of surfaces*, Algebr, Geom. Topol. **4** (2004), 49–71.

[BS] R. F. Brown and B. Sanderson, *Fixed points of boundary preserving maps on surfaces*, Pacific J. Math. **158** (1993), 243–264.

[CB] A. Casson and S. Bleiler, *Automorphisms of surfaces after Nielsen and Thurston*, London Math. Soc. Stud. Texts **9** (1988), Cambridge Univ. Press.

[DV] W. Dicks and E. Ventura, *The group fixed by a family of injective endomorphisms of a free group*, Contemp. Math. **195** (1996).

[Fu] F. B. Fuller, *The homotopy theory of coincidences*, Doctoral Dissertation, Princeton University (1951).

[GJ] D. L. Gonçalves and B. Jiang, *The index of coincidence Nielsen classes of maps between surfaces*, Topology Appl. **116** (2001), 73–89.

[GZ] D. L. Gonçalves and H. Zieschang, *Equations in free groups and coincidence of mappings on surfaces*, Math. Z. **237** (2001), 1–29.

[Ho] H. Hopf, *Zur Topologie der Abbildungen von Mannigfalltigkeiten* II, Math. Ann. **102** (1930), 562–623.

[J1] B. Jiang, *On the least number of fixed points*, Amer. J. Math. **102** (1980), 749–763.

[J2] ______, *Fixed points of surface homeomorphisms*, Bull. Amer. Math. Soc. **5** (1981), 176–178.

[J3] ______, *Lectures on Nielsen Fixed Point Theory*, Contemporary Math., vol. 14, Amer. Math. Soc., Providence, RI, 1983.

[J4] ______, *Fixed points and braids*, Invent. Math. **75** (1984), 69–74.

[J5] ______, *Fixed points and braids* II, Math. Ann. **272** (1985), 249–256.

[J6] ______, *Commutativity and Wecken properties for fixed points of surfaces and 3-manifolds*, Topology Appl. **53** (1993), 221–228.

[J7] ______, *Bounds for fixed points on surfaces*, Math. Ann. **311** (1998), 467–479.

[JG] B. Jiang and J. Guo, *Fixed points of surface diffeomorphisms*, Pacific J. Math. **160** (1993), 67–89.

[K1] M. Kelly, *Minimizing the number of fixed points for selfmaps of surfaces with boundary*, Pacific J. Math. **126** (1987), 81–123.

[K2] ______, *Minimizing the cardinality of the fixed point set for selfmaps of surfaces with boundary*, Mich. Math. J. **39** (1992), 201–217.

[K3] ______, *The relative Nielsen number and boundary-preserving surface maps*, Pacific J. Math. **161** (1993), 139–153.

[K4] ______, *The Nielsen number as an isotopy invariant*, Topology Appl. **62** (1995), 127–143.

[K5] ______, *Computing Nielsen numbers of surface homeomorphisms*, Topology **35** (1996), 13–25.

[K6] ______, *A bound for the fixed-point index for surface mappings*, Ergodic Theory Dynam. Systems **17** (1997), 1393–1408.

[K7] ______, *A bound on the fixed point index for self-maps of certain simplicial complexes*, Topology Appl. **108** (2000), 179–196.

[K8] ______, *Fixed points of boundary-preserving maps on punctured projective planes*, Topology Appl. **124** (2002), 145–157.

[K9] ______, *Graph representatives for fixed point minimal self-maps on surfaces*, preprint.

[L] P. Le Calvez, *Dynamical properties of diffeomorphisms of the annulus and of the torus*; by P Mazaud, SMF/AMS Texts and Monographs, vol. 4, Amer. Math. Soc., Providence, RI, 2000.

[Ni1] J. Nielsen, *Ringfladen ag Planen*, Mat. Tidsskr. **B** (1924), 1–22; English transl., Jakob Nielsen Collected Mathematical Papers, Birkhaüser, Basel, 1986, pp. 130-146.

[Ni2] ______, *Untersuchungen zur Topologie der geschlossenen zweiseitigen Flächen*, Acta Math. **50** (1927), 189–358.

[No] J. Nolan, *Fixed points of boundary-preserving maps of punctured discs*, Topology Appl. **73** (1996), 57–84.

[PS] S. Pelikan and E. Slaminka, *A bound for the fixed point index of area-preserving homeomorphisms of two-manifolds*, Ergodic Theory and Dynam. Systems **7** (1987), 463–479.

[Sch1] H. Schirmer, *A relative Nielsen number*, Pacific J. Math. **122** (1986), 459–472.

[Sch2] H. Schirmer, *Mindestzahlen von Koinzidenzpunkten*, J. Reine Angew. Math. **194** (1955), 21–39.

[Th] W. Thurston, *On the geometry and dynamics of surface diffeomorphisms*, Bull. Amer. Math. Soc. **19** (1988), 417–431.

[Wa1] J. Wagner, *Classes of Wecken maps of surfaces with boundary*, Topology Appl. **76** (1997), 27–46.

[Wa2] ———, *An algorithm for calculating the Nielsen number on surfaces with boundary*, Trans. Amer. Math. Soc. **351** (1999), 41–62.

[Wec] F. Wecken, *Fixpunktklassen* III, Math. Ann. **118** (1942), 544–577.

[Wei] F. Weier, *Fixpunkttheorie in topologischen Mannigfaltigkeiten*, Math. Z. **59** (1953), 171–190.

[Yi] P. Yi, *An algorithm for computing the Nielsen number of maps on the pants surface*, preprint.

18. RELATIVE NIELSEN THEORY

XUEZHI ZHAO

Nielsen theory is concerned mainly with the description of the fixed point set of any given map $f: X \to X$. In 1986, H. Schirmer considered the same problem for the maps of the form $f: (X, A) \to (X, A)$ (see [S1]). Compare with the ordinary Nielsen theory, the maps of the form $f: (X, A) \to (X, A)$, which are called relative maps, have an extra invariant subset A. Such an invariant set will bring more fixed points. The location of the fixed points of such kind of relative maps will be not arbitrary in the total space X.

In this part, we shall show some basic problems and main results in this relative Nielsen theory, together with the special techniques which are different from those in ordinary Nielsen theory.

1. Basic definitions

Throughout this part, we always assume that (X, A) is a pair of compact polyhedra, i.e. a pair of topological spaces homotopic to a pair of finite simplicial complexes, where X is connected. The subpolyhedron A may be disconnected. Its component number is finite because of the compactness. The interior, boundary and closure of any subset B in X are written as $\mathrm{int}(B)$, $\mathrm{bd}(B)$ and $\mathrm{cl}(B)$, respectively.

Consider a relative map $f: (X, A) \to (X, A)$. The restriction of f on A is written as $\overline{f}: A \to A$. As self maps, $f: X \to X$ and $\overline{f}: A \to A$ have their own fixed point classes respectively. These are $\mathrm{FPC}(f)$ and $\mathrm{FPC}(\overline{f})$, where

$$\mathrm{FPC}(\overline{f}) = \bigsqcup_{f(A_k) \subset A_k} \mathrm{FPC}(f_{A_k})$$

is the set of the disjoint union of the sets of the fixed point classes of the restriction of $f_{A_k}: A_k \to A_k$ of f on all components of A which are mapped by f into themselves. For a component A_k of A, if $f(A_k) \subset A_k$, we have a morphism of self maps, i.e. a commutative diagram:

$$\begin{array}{ccc} A_k & \xrightarrow{f_{A_k}} & A_k \\ {\scriptstyle i_k}\downarrow & & \downarrow{\scriptstyle i_k} \\ X & \xrightarrow[f]{} & X \end{array}$$

where $i_k: A_k \to X$ is the inclusion. Note that X and each component A_k of A have universal coverings

$$p: \widetilde{X} \to X, \quad p_{A_k}: \widetilde{A}_k \to A_k.$$

Pick a map $\widetilde{i}_k: \widetilde{A}_k \to \widetilde{X}$ such that $p \circ \widetilde{i}_k = i_k \circ p_k$. Thus, for each lifting $\widetilde{f}_{A_k}$ of f_{A_k}, there is a unique lifting $\widetilde{f}$ of f such that

$$\begin{array}{ccc} \widetilde{A}_k & \xrightarrow{\widetilde{f}_{A_k}} & \widetilde{A}_k \\ {\scriptstyle \widetilde{i}_k}\downarrow & & \downarrow{\scriptstyle \widetilde{i}_k} \\ \widetilde{X} & \xrightarrow[\widetilde{f}]{} & \widetilde{X} \end{array}$$

This is a correspondence from the set of liftings of f_{A_k} to set of liftings of f, which is denoted $\widetilde{i}_{k,\text{lift}}$. As a special case of [J3, p. 43, Theorem 1.8], all liftings $\widetilde{i}_k$ of i_k induce the same correspondence $i_{k,\text{FPC}}: \text{FPC}(f_{A_k}) \to \text{FPC}(f)$.

Any fixed point of f_{A_k} is certainly a fixed point of f. In geometry, two fixed points x and y of a map g are said to be in the same fixed point classes if there is a path α from x to y such that $\alpha \doteq g \circ \alpha$, where $\doteq$ denotes the path homotopy keeping end points fixed. Such a path is said to be a *Nielsen path* between x and y. Using this geometric definition, it is easy to check that any non-empty fixed point class of $\overline{f}$ is totally contained in a unique fixed point class of f.

(1.1) PROPOSITION. *Let F_k be a non-empty fixed point class of $f_{A_k}: A_k \to A_k$. Then F_k is contained in a fixed point class F of f if and only if $i_{k,\text{FPC}}(F_k) = F$.*

In this part, fixed point classes should be taken in the lifting sense. In other words, two fixed point classes will be regarded as different if they are determined by different lifting classes. In geometry, an empty fixed point class of $\overline{f}$ is contained any fixed point class of f. But, we do not use such a relation. Instead, we define

(1.2) DEFINITION. Let F_k and F be fixed point classes of $f_{A_k}: A_k \to A_k$ and f, respectively. F_k is said to be contained in F if $i_{k,\text{FPC}}(F_k) = F$.

Thus, in this sense, any (empty or non-empty) fixed point class of $\overline{f}$ is contained in a unique fixed point class of f.

2. Fixed points on total space

In present section, we shall consider the estimation of the number of fixed points of a relative map $f\colon (X,A) \to (X,A)$. Clearly $N(f)$, the number of essential fixed point classes of f, is a lower bound for the number of fixed points of f. But, it is not a "good" one, because the invariant subset A was not taken into account. Here is an example:

(2.1) EXAMPLE ([S3, Example 1.1]). Let P be a disc with two holes, and let f be the homeomorphism obtained by a reflection which interchanges the boundary of the two holes.

This example originally comes from [J2], where the fixed points of homemorphisms on surfaces are in consideration. Clearly, any homeomorphism isotopic to f has at least two fixed points because the Nielsen number $N(f|\partial P)$ is 2. But, $N(f) = 1$.

On the other hand, the sum $N(f)+N(\overline{f})$ of the number of essential fixed point classes of f and essential fixed point classes of $\overline{f}\colon A \to A$ is not a lower bound of the number of fixed point of f, because there may be an overlap between the set of fixed point classes of f and that of $\overline{f}$.

(2.2) EXAMPLE. Let $X = \{(x,y) : x^2+y^2 \leq 1\}$ be a two-disc, and $A = \{(x,y) : x^2+y^2 = 1\}$ be the boundary of X. The relative $f\colon (X,A) \to (X,A)$ is defined by

$$f(x,y) = \left(-x, y + \frac{1-x^2-y^2}{\pi} \sin\left(\frac{\pi(y+1)}{2}\right)\right).$$

It is obvious that f has a unique fixed point class. It has index 1. And $\overline{f}$ has two fixed point classes. Both of the fixed point classes of $\overline{f}$ have index 1. So, $N(\overline{f}) + N(f) = 3$, but f has two fixed points $(0,1)$ and $(0,-1)$.

Any fixed point class of $\overline{f}$ is contained in a unique fixed point class of f, a fixed point class of f may or may not contain some fixed point classes of $\overline{f}$. So, we have the definition:

(2.3) DEFINITION ([S1, Definition 2.1]). Let $f\colon (X,A) \to (X,A)$ be a relative map. A fixed point class of f is said to be a *common* fixed point class if it contains an essential fixed point class of $\overline{f}$.

The number of common fixed point classes of f is written as $R(f,\overline{f})$, the number of common and essential fixed point classes of f is written as $N(f,\overline{f})$.

Note that any essential fixed point class of $\overline{f}$ is non-empty. We have

(2.4) PROPOSITION. *If F is a common fixed point class. Then $F \cap A \neq \emptyset$.*

(2.5) DEFINITION ([S1, Definition 2.4]). Let $f: (X, A) \to (X, A)$ be a relative map. We define

$$N(f; X, A) = N(\overline{f}) + N(f) - N(f, \overline{f}),$$

which is said to be the *relative Nielsen number* of f.

In Example (2.1), f has only one non-empty fixed point class with index 2, and $\overline{f}$ has two essential fixed point classes, either of them is contained in the unique fixed point class of f. So, the non-empty fixed point class of f is common. Thus, $N(f; X, A) = N(\overline{f}) + N(f) - N(f, \overline{f}) = 2 + 1 - 1 = 2$. Similarly, in Example (2.2), we have $N(f; X, A) = 2$.

(2.6) THEOREM (Lower Bound Theorem). *Any relative map $f: (X, A) \to (X, A)$ has at least $N(f; X, A)$ fixed points.*

PROOF. Since each essential fixed point class is non-empty, $\overline{f}$ has $N(\overline{f})$ distinct fixed points $a_1, \dots, a_n$, where $n = N(\overline{f})$. These are also fixed points of f. Similarly, $N(f) - N(f, \overline{f})$ essential fixed point classes of f which are not common produce distinct fixed point $b_1, \dots, b_m$, where $m = N(f) - N(f, \overline{f})$. If $a_i = b_j$ for some i and j with $1 \leq i \leq n$ and $1 \leq j \leq m$. Then the fixed point class F of f containing b_j would contain the fixed point a_i in A which lies in an essential fixed point class of $\overline{f}$. Thus, F would contain this essential fixed point class of $\overline{f}$. This contradicts to the fact that F is not common. Thus, two sets $\{a_1, \dots, a_n\}$ and $\{b_1, \dots, b_m\}$ are disjoint. Thus, the map f has at least $n + m = N(\overline{f}) + N(f) - N(f, \overline{f}) = N(f; X, A)$ fixed points. □

3. Fixed points on the closure of the complement

The consideration for the number of fixed points on the closure of the complement of a relative map $f: (X, A) \to (X, A)$ can be traced back to the C. Bowszyc's result, which is a natural generalization of Lefshetz fixed point theorem:

(3.1) PROPOSITION ([Bo, Theorem 3.1]). *Let $f: (X, A) \to (X, A)$ be a relative map. If $L(f) - L(\overline{f}) \neq 0$. Then f has at least one fixed point on the closure $\mathrm{cl}(X - A)$ of the complement $X - A$.*

PROOF. Suppose on the contrary that f has no fixed point on $\mathrm{cl}(X - A)$. Then the fixed point set $\mathrm{Fix}(f)$ of f is contained in $\mathrm{int}(A)$. Thus, $\mathrm{Fix}(f)$ and $\mathrm{cl}(X - A)$ are disjoint compact sets. There is an open neighbourhood U of $\mathrm{Fix}(f)$ with $\mathrm{cl}(U) \subset X - \mathrm{cl}(X - A) = \mathrm{int}(A)$. We have $L(f) = \mathrm{ind}(f, \mathrm{Fix}(f)) = \mathrm{ind}(f, U)$. Since $\overline{f}$ has no fixed point on $\mathrm{bd}(U)$, we have $L(\overline{f}) = \mathrm{ind}(f, \mathrm{Fix}(\overline{f})) = \mathrm{ind}(\overline{f}, U)$. It follows that $L(f) = L(\overline{f})$. □

The above property belongs to the Lefshetz fixed point theory. We need to look for the corresponding Nielsen theory, i.e. to consider the fixed points on $\mathrm{cl}(X-A)$ in fixed point class level. Using a similar argument as above, one can prove:

(3.2) LEMMA. *Let F be a fixed point class of $f\colon (X,A)\to(X,A)$. If $F\subseteq \mathrm{int}(A)$. Then* $\mathrm{ind}(f,F)=\mathrm{ind}(\overline{f},F\cap A)$.

In other word, a sufficient condition to ensure that a fixed point class F has non-empty intersection with $\mathrm{cl}(X-A)$ is that $\mathrm{ind}(f,F)\neq\mathrm{ind}(\overline{f},F\cap A)$. This lead to the following definition:

(3.3) DEFINITION ([S2]). A fixed point class F of $f\colon (X,A)\to(X,A)$ is said to assume its index in A, if $\mathrm{ind}(f,F)=\mathrm{ind}(\overline{f},F\cap A)$.

The number of fixed point classes of f which do not assume their indices in A is written as $N(f;\overline{X-A})$, which is said to be the relative Nielsen number of f on the closure of the complement.

(3.4) EXAMPLE. Let X be a circle and A a point in X. The relative $f\colon (X,A)\to (X,A)$ is the identity map.

Since $L(f)=0$, the unique non-empty fixed point class F of f has index zero. Clearly, $\mathrm{ind}(\overline{f},F\cap A)=\mathrm{ind}(\overline{f},A)=L(\overline{f})=1$. Hence F does not assume its index in A. Note that the other fixed point classes of f are empty and certainly assume their indices in A. Thus, $N(f;\overline{X-A})=1$.

For a fixed point class F, the difference $\mathrm{ind}(f,F)-\mathrm{ind}(\overline{f},F\cap A)$ indicates in some sense the "index" of the set $F\cap(X-A)$. As in the above example, such an "index" does not make sense in general because $F\cap(X-A)$ may be not an isolated fixed point set. But, we have

(3.5) PROPOSITION. *Let $f\colon (X,A)\to(X,A)$ be a relative map. If there is a neighbourhood V of A in X such that $f(V)\subset A$. Then, for any fixed point class F of f, $F\cap(X-A)$ is an isolated fixed point set of f with* $\mathrm{ind}(f,F\cap(X-A))=\mathrm{ind}(f,F)-\mathrm{ind}(\overline{f},F\cap A)$.

PROOF. Let F be a fixed point class of f. Pick a regular neighbourhood V' of A in X with $V'=\mathrm{cl}(V')\subset V$. Then f has no fixed point on $\mathrm{cl}(V')-A$. Since F is an isolated fixed point set, there is an open neighbourhood U of F in X with $\mathrm{cl}(U)\cap\mathrm{Fix}(f)=F$. Then $U-V'$ is an open neighbourhood of $F\cap(X-A)$ with $\mathrm{cl}(U-V')\cap\mathrm{Fix}(f)=F\cap(X-A)$. It follows that $F\cap(X-A)$ is an isolated fixed point set of f. Clearly, $U\cap\mathrm{int}(V')$ is an open neighbourhood of $F\cap A$ with $\mathrm{cl}(U\cap\mathrm{int}(V'))=F\cap A$. Thus, $F\cap A$ is also an isolated fixed point set of f. By the additivity of fixed point index, we have

$$\begin{aligned}\mathrm{ind}(f,F)=\mathrm{ind}(f,U)&=\mathrm{ind}(f,U\cap\mathrm{int}(V'))+\mathrm{ind}(f,U-V')\\&=\mathrm{ind}(f,U\cap\mathrm{int}(V'))+\mathrm{ind}(f,F\cap(X-A)).\end{aligned}$$

Notice that $f(U \cap \mathrm{int}(V')) \subset A$. By the restriction property of fixed point index ([J3, p. 16, Corollary 3.7]), we have that

$$\mathrm{ind}(f, U \cap \mathrm{int}(V')) = \mathrm{ind}(f|_A, (U \cap \mathrm{int}(V')) \cap A) = \mathrm{ind}(\overline{f}, F \cap A).$$

This proposition holds. □

From this proposition, we get that

(3.6) COROLLARY. *Let $f\colon (X, A) \to (X, A)$ be a relative map. If there is a neighbourhood V of A in X such that $f(V) \subset A$. Then any fixed point class which does not assume its index in A will contain a fixed point in $X - A$. Hence, f has at least $N(f; \overline{X - A})$ fixed points in $X - A$.*

It should be noticed that the above proposition is not true for general map. Here is an example, which is a modification of previous example:

(3.7) EXAMPLE. Let $X = \{e^{i\theta}\}_{0 \le \theta < 2\pi}$ be a circle, and let $A = \{e^0\}$ be a singleton. A relative map $f\colon (X, A) \to (X, A)$ is defined by $f(e^{i\theta}) = e^{i(\theta + \sin(\theta/2))}$.

Clearly, f has unique fixed point e^0. We write F for the fixed point class containing e^0. The fixed point set $F \cap (X - A)$ is empty, and is therefore an isolated fixed point set with zero fixed point index. Since f is homotopic to the identity map, we have $L(f) = \chi(X) = 0$. Thus, $\mathrm{ind}(f, F) = 0$, and hence $\mathrm{ind}(f, F) - \mathrm{ind}(\overline{f}, F \cap A) = 0 - 1 = -1$. This value is different from $\mathrm{ind}(f, F \cap (X - A)) = 0$.

So, the assumption of the existence of such a V is necessary. A fixed point class F of f which does not assume its index in A may have no fixed point $X - A$ even if $F \cap (X - A)$ is an isolated fixed point set. Hence, for general relative maps, $N(f; \overline{X - A})$ can not be used to estimate the number of fixed points of f on $X - A$.

By Lemma (3.2), we have

(3.8) THEOREM ([S2, Theorem 3.1]). *Any relative map $f\colon (X, A) \to (X, A)$ has at least $N(f; \overline{X - A})$ fixed points on the closure $\mathrm{cl}(X - A)$ of the complement $X - A$.*

If a component A_j of A has empty interior, then the fixed points of $f_{A_j}\colon A_j \to A_j$ lie automatically in $\mathrm{cl}(X - A)$. In order to give more precise estimation for the number of fixed points on $\mathrm{cl}(X - A)$, one must deal with the essential fixed point classes of $f_{A_j}\colon A_j \to A_j$ for the components A_j of A with $\mathrm{int}(A_j) = \emptyset$. The consideration to such fixed point classes lead to the following:

(3.9) DEFINITION (see [Z3, Definition 5.1]). Let $f\colon (X, A) \to (X, A)$ be a relative map. We write $N'(f; \overline{X - A})$ for the sum of the number of essential fixed point classes of all $f_{A_j}\colon A_j \to A_j$ with $\mathrm{int}(A_j) = \emptyset$ and the number of the fixed point classes which do not assume their indices in A and do not contain any essential fixed point classes of $f_{A_j}\colon A_j \to A_j$ with $\operatorname{int}(A_j) = \emptyset$.

When the two the Nielsen type numbers $N(f;\overline{X-A})$ and $N'(f;\overline{X-A})$ were introduced in [S2] and [Z3], respectively, they are written as $\widetilde{N}(f;X,A)$ and $\widetilde{N}'(f;X,A)$. Here, we follow the idea in [S3] and use more meaningful notations.

(3.10) PROPOSITION. *Let $f\colon (X,A) \to (X,A)$ be a relative map. Then*

$$N'(f;\overline{X-A}) \geq N(f;\overline{X-A}).$$

If each component of A has non-empty interior, then

$$N'(f;\overline{X-A}) = N(f;\overline{X-A}).$$

PROOF. Notice that for any component A_j of A, the number of fixed point classes of f containing essential fixed point classes of $f_{A_j}\colon A_j \to A_j$ is less than the number of essential fixed point classes of f_{A_j}. The number of fixed point classes of f which do not assume their indices in A and do contain an essential fixed point classes of $f_{A_j}\colon A_j \to A_j$ with $\operatorname{int}(A_j) = \emptyset$ will therefore less that $\sum_{\operatorname{int}(A_j)=\emptyset} N(f_{A_j})$. So, we have the first conclusion.

The second conclusion can be obtained directly from the definitions. □

(3.11) THEOREM (Lower Bound Theorem). *Any relative map $f\colon (X,A) \to (X,A)$ has at least $N'(f;\overline{X-A})$ fixed points on the closure $\operatorname{cl}(X-A)$ of the complement $X-A$.*

PROOF. Since any essential fixed point class is non-empty, there are m distinct fixed points, say $a_1, \dots, a_m$, of f on $\bigcup_{\operatorname{int}(A_j)=\emptyset} A_j$, where $m = \sum_{\operatorname{int}(A_j)=\emptyset} N(f_{A_j})$ These fixed points are in different essential fixed point classes of $f_{A_j}\colon A_j \to A_j$ with $\operatorname{int}(A_j) = \emptyset$. Since any component of A without interior is totally contained in $\operatorname{cl}(X-A)$, all a_i lie in $\operatorname{cl}(X-A)$.

Write $\mathcal{F}$ for the set of fixed point classes of f which do not assume their indices in A and do not contain any essential fixed point classes of a $f_{A_j}\colon A_j \to A_j$ with $\operatorname{int}(A_j) = \emptyset$. By Lemma (3.2), any fixed point class in $\mathcal{F}$ contains a fixed point on $\operatorname{cl}(X-A)$. Thus, there are $n = \sharp\mathcal{F}$ fixed points on $\operatorname{cl}(X-A)$, say $b_1, \dots, b_n$. They are in different fixed point classes in $\mathcal{F}$.

If $b_i = a_j$ for some i and j, then there a fixed point class in $\mathcal{F}$ contains a fixed point belonging to an essential fixed point class of $f_{A_k}\colon A_k \to A_k$ with $\operatorname{int}(A_j) = \emptyset$. This is a contradiction.

Thus, the two fixed point sets $\{a_1, \dots, a_m\}$ and $\{b_1, \dots, b_n\}$ have no intersection. It follows that f has at least $m + n = N'(f;\overline{X-A})$ fixed points on $\operatorname{cl}(X-A)$. □

4. Fixed points on the complement

In this section, we shall deal with the estimation for the number of the fixed points of a relative map $f\colon (X, A) \to (X, A)$ on the complement $X - A$ of the pair (X, A).

Consider an essential fixed point class F of f. If $F \cap (X - A) = \emptyset$, then F, as a subset of X, is contained in A because F is not empty. It follows that F contains a fixed point of $\overline{f}$ and hence contains a fixed point class of $\overline{f}$. This idea will lead to:

(4.1) DEFINITION. Let $f\colon (X, A) \to (X, A)$ be a relative map. A fixed point class of f is said to be a *weakly common* fixed point class if it contains a fixed point class of $\overline{f}$.

The number of essential fixed point classes of f which are not weakly common is written as $N(f; X - A)$.

It should be noticed that our "containing" relation is in the sense of lifting. A fixed point class F may be a weekly common one even if $F \cap A = \emptyset$. Following proposition gives a geometric method to identify the weekly common fixed point classes.

(4.2) PROPOSITION. *Let F be a non-empty fixed point class of relative map $f\colon (X, A) \to (X, A)$. Then the following three statements are equivalent:*

(4.2.1) *F contains a fixed point class of $f_{A_k}\colon A_k \to A_k$, i.e. F is a weakly common fixed point class.*

(4.2.2) *For any $a \in A_k$, there is a path $\zeta\colon (I, 0, 1) \to (X, F, a)$ such that $\zeta \simeq f\circ\zeta\colon (I, 0, 1) \to (X, \zeta(0), A_k)$.*

(4.2.3) *There is a point $a \in A_k$ and a path $\zeta\colon (I, 0, 1) \to (X, F, a)$ such that $\zeta \simeq f\circ\zeta\colon (I, 0, 1) \to (X, \zeta(0), A_k)$.*

PROOF. (1) $\Rightarrow$ (2). Suppose that F is determined by the lifting class $[\widetilde{f}]$. Pick a point $x_0 \in F$, there is a $\widetilde{x}_0 \in p^{-1}(x_0)$ with $\widetilde{f}(\widetilde{x}_0) = \widetilde{x}_0$. Since F contains a fixed point class of $f_{A_k}\colon A_k \to A_k$, there exists a lifting $\widetilde{f}_{A_k}$ of $f_{A_k}\colon A_k \to A_k$ so that $\widetilde{i}_{k,\text{lift}}(\widetilde{f}_{A_k}) = \widetilde{f}$.

For any given $a \in A_k$, we can find a point $\widetilde{a} \in p_{A_k}^{-1}(a)$. Thus, we have $\widetilde{i}_k(\widetilde{a}) \in \widetilde{i}_k(\widetilde{A}_k) \cap p^{-1}(a)$, and therefore $\widetilde{f}(\widetilde{i}_k(\widetilde{a})) = \widetilde{i}_k(\widetilde{f}_{A_k}(\widetilde{a})) \in \widetilde{i}_k(\widetilde{A}_k)$. Choose a path $\widetilde{\zeta}$ in $\widetilde{X}$ from $\widetilde{x}_0$ to $\widetilde{i}_k(\widetilde{a})$. Since the universal covering $\widetilde{X}$ is simply-connected and $i_k(\widetilde{A}_k)$ is connected, there is a homotopy of the form $\widetilde{\zeta} \simeq \widetilde{f}\circ\widetilde{\zeta}\colon (I, 0, 1) \to (\widetilde{X}, \widetilde{x}_0, \widetilde{i}_k(\widetilde{A}_k))$. Projecting down to X, we have $\zeta \simeq f\circ\zeta\colon (I, 0, 1) \to (X, x_0, A_k)$, where $\zeta = p\circ\widetilde{\zeta}$ is a path from x_0 to a.

(2) $\Rightarrow$ (3). It is trivial.

(3) $\Rightarrow$ (1). Write $x_0 = \zeta(0)$. Then $x_0 \in F$. Suppose that F is determined by lifting class $[\widetilde{f}]$. We may choose a point $\widetilde{x}_0 \in p^{-1}(x_0)$ with $\widetilde{f}(\widetilde{x}_0) = \widetilde{x}_0$. Lift the

path ζ from $\widetilde{x}_0$ to get a path $\widetilde{\zeta}$ in $\widetilde{X}$. Pick $\widetilde{a} \in p_k^{-1}(a)$, then there is a lifting $\widetilde{i}_k$ of the inclusion $i_k \colon A_k \to A$ such that

$$\begin{array}{ccc} (\widetilde{A}_k, \widetilde{a}) & \xrightarrow{\ \widetilde{i}_k\ } & (\widetilde{X}, \widetilde{\zeta}(1)) \\ {\scriptstyle p_k}\big\downarrow & & \big\downarrow{\scriptstyle p} \\ (A_k, a) & \xrightarrow[\ i_k\ ]{} & (X, a) \end{array}$$

commutes ([J3, p.42, Proposition 1.2(i)]). Let $H \colon I \times I \to X$ be the homotopy from ζ to $f \circ \zeta$, i.e. $H(t, 0) = \zeta(t)$, $H(t, 1) = f(\zeta(t))$ for all $t \in I$. Then $\widetilde{\zeta}$ determines a lifting $\widetilde{H} \colon I \times I \to \widetilde{X}$ of H. Define a path $\eta \colon I \to A_k$ by $\eta(s) = H(1, s)$. Lift the path $\eta \colon I \to A_k$ from $\widetilde{a}$ to get a path $\widetilde{\eta}$ in $\widetilde{A}_k$, then $\widetilde{i}_k \circ \widetilde{\eta} : I \to \widetilde{X}$ is a lifting from $\widetilde{\zeta}(1)$ in $\widetilde{X}$ of the path $i_k \circ \eta$. By the unique lifting property of covering spaces, we have $\widetilde{i}_k(\widetilde{\eta}(s)) = \widetilde{H}(1, s)$ for all $s \in I$. Especially, $\widetilde{i}_k \circ \widetilde{\eta}(1) = \widetilde{H}(1, 1) = \widetilde{f}(\widetilde{\zeta}(1))$. Write $\widetilde{f}_{A_k}$ for the unique lifting of $f_{A_k} \colon A_k \to A_k$ with $\widetilde{f}_{A_k}(\widetilde{a}) = \widetilde{\eta}(1)$. Thus, $\widetilde{i}_k \circ \widetilde{f}_{A_k}(\widetilde{a}) = \widetilde{f} \circ \widetilde{i}_k(\widetilde{a})$. By the unique lifting property of covering spaces, we have $\widetilde{i}_k \circ \widetilde{f}_{A_k} = \widetilde{f} \circ \widetilde{i}_k$, i.e. $\widetilde{f} = i_{k,\mathrm{lift}}(\widetilde{f}_{A_k})$. This implies $[\widetilde{f}] = i_{k,\mathrm{FPC}}[\widetilde{f}_{A_k}]$, i.e. F contains the fixed point class $p_{A_k}(\mathrm{Fix}(\widetilde{f}_{A_k}))$ of f_{A_k}. □

From this proposition, we get immediately

(4.3) COROLLARY. *A fixed point class of $f \colon (X, A) \to (X, A)$ containing a fixed point on A is a weakly common fixed point class.*

Since any essential fixed point class of $\overline{f}$ is non-empty, we have

(4.4) PROPOSITION. *Any common fixed point class of $f \colon (X, A) \to (X, A)$ is a weakly common fixed point class.*

(4.5) THEOREM (Lower Bound Theorem). *Any map $f \colon (X, A) \to (X, A)$ has at least $N(f; X - A)$ fixed points on $X - A$.*

PROOF. Recall that each essential fixed point class contains at least one fixed point. By the corollary of Proposition (4.2), the fixed points belonging to the fixed point classes which are not weakly common can not lie in A. Then, we are done. □

(4.6) EXAMPLE. Let $f \colon (X, A) \to (X, A)$ be the identity map, where X is a two disc and A is the boundary of the disc.

The unique fixed point class of f has index 1 and contains a fixed point class of $\overline{f} \colon A \to A$, because it contains fixed points on A. Thus, it is a weakly common fixed point class, we have $N(f; X - A) = 0$.

Note that $L(\overline{f}) = 0$. The unique non-empty fixed point class of $\overline{f}$ has index 0. The unique fixed point class of f is not common. So, $N(f; X, A) = N(\overline{f}) + N(f) - N(f, \overline{f}) = 0 + 1 - 0 = 1$.

(4.7) EXAMPLE. Let $X = \{re^{i\theta}\}_{0\le r\le 1, 0\le\theta<2\pi}$ be a two disc, and let $A = \{e^{i\theta}\}_{0\le\theta<2\pi}$ be the boundary of the disc. A map $f\colon (X,A) \to (X,A)$ is defined by $f(re^{i\theta}) = re^{i(\theta+\pi)}$.

Clearly, f has unique fixed point $x_0 = 0$. We have a path $\zeta\colon (I,0,1) \to (X,x_0,A)$ defined by $\zeta(t) = te^0$. Thus, the map $H\colon (I\times I, \{0\}\times I, \{1\}\times I) \to (X,x_0,A)$ defined by $H(t,s) = te^{i(\theta+s\pi)}$ is the homotopy from ζ to $f\circ\zeta$. Hence, x_0 is contained in a weekly common fixed point class of f, although f has no fixed point on A.

5. Properties of relative Nielsen numbers

We first consider the influence of homotopies between relative maps on the fixed point classes.

Let $H\colon (X\times I, A\times I) \to (X,A)$ be a relative homotopy between relative maps $f,g\colon (X,A) \to (X,A)$. We define a map $\mathbf{H}\colon (X\times I, A\times I) \to (X\times I, A\times I)$ by $\mathbf{H}(x,t) = (H(x,t),t)$, and call $\mathbf{H}$ the fat homotopy of H. For any $s\in I$, the map $h_s\colon (X,A) \to (X,A)$ defined by $h_s(x) = H(x,s)$ is said to be the s-slice of $\mathbf{H}$.

Let $p\colon \widetilde{X} \to X$ be the universal covering of X. If we define $\mathbf{P}\colon \widetilde{X}\times I \to X\times I$ by $\mathbf{P}(\widetilde{x},t) = (p(\widetilde{x}),t)$, then $\mathbf{P}$ is a universal covering of $X\times I$. Then, we can talk about the liftings of $\mathbf{H}$ with respect to $\mathbf{P}$. We shall say that $\widetilde{\mathbf{H}}\colon \widetilde{X}\times I \to \widetilde{X}\times I$ is a lifting of $\mathbf{H}$ if $\mathbf{P}\circ\widetilde{\mathbf{H}} = \mathbf{H}\circ\mathbf{P}$. Recall that a fixed point class F of f and a fixed point class G are say to be H-related if there is a lifting $\widetilde{f}$ of f and a lifting $\widetilde{g}$ of g which determine F and G, respectively, i.e. $F = p(\mathrm{Fix}(\widetilde{f}))$ and $G = p(\mathrm{Fix}(\widetilde{g}))$, such that $\widetilde{f}$ and $\widetilde{g}$ are 0-slice and 1-slice of a lifting of $\mathbf{H}$, respectively.

For a component A_k, since A_k is connected, $f(A_k)\subseteq A_k$ if and only if $g(A_k)\subseteq A_k$. Let A_k be a component of A with $f(A_k)\subseteq A_k$. The restriction of $H\colon (X\times I, A\times I) \to (X,A)$ on $A_k\times I$ gives a homotopy from $f|_{A_k}$ to $g|_{A_k}$. Hence, we can say "$H|_{A_k\times I}$-related" relation between the set of fixed point classes of f_{A_k} and that of g_{A_k}. Using [J3, p. 47, 1.16], we get

(5.1) PROPOSITION. *Let $H\colon (X\times I, A\times I) \to (X,A)$ be a relative homotopy between relative maps $f,g\colon (X,A) \to (X,A)$. Let F_k, G_k, F and G be fixed point classes of f_{A_k}, g_{A_k}, f and g respectively. Suppose that F_k and G_k are $H|_{A_k\times I}$-related, and that F and G are H-related, then F_k is contained in F if and only if G_k is contained in G.*

(5.2) LEMMA. *Let $H\colon (X\times I, A\times I) \to (X,A)$ be a relative homotopy between relative maps $f,g\colon (X,A) \to (X,A)$. Suppose that a fixed point class F of f and a fixed point class G of g are H-related. Then:*

(5.2.1) *F contains an essential fixed point class of f_{A_k} if and only if G contains an essential fixed point class of g_{A_k}.*

(5.2.2) *F contains an inessential fixed point class of f_{A_k} if and only if G contains an inessential fixed point class of g_{A_k}.*

(5.2.3) *F assumes its index in A if and only if G assumes its index in A.*

PROOF. (5.2.1) Let F contain an essential fixed point class F_k of f_{A_k}. Then F_k is $H|_{A_k \times I}$-related to a fixed point class G_k of g_{A_k}. By Proposition (5.1), G contains G_k. From [J3, p. 18, 4.5], we know that homotopy related fixed point classes have the same fixed point indices. Thus, G_k is essential.

The converse is the same.

(5.2.2) The proof is similar to (5.2.1).

(5.2.3) By Proposition (5.1), the $\overline{H}$-related relation gives an index-preserving one-to-one correspondence from the set of the fixed point classes of $\overline{f}$ contained in F to the set of the fixed point classes of $\overline{g}$ contained in G. Notice that

$$\begin{aligned} \operatorname{ind}(\overline{f}, F \cap A) &= \sum_{\overline{F} \in \mathrm{FPC}(\overline{f}), \overline{F} \subset F} \operatorname{ind}(\overline{f}, \overline{F}), \\ \operatorname{ind}(\overline{g}, G \cap A) &= \sum_{\overline{G} \in \mathrm{FPC}(\overline{g}), \overline{G} \subset G} \operatorname{ind}(\overline{g}, \overline{G}). \end{aligned}$$

We have $\operatorname{ind}(\overline{f}, F \cap A) = \operatorname{ind}(\overline{g}, G \cap A)$. As $\operatorname{ind}(f, F) = \operatorname{ind}(g, G)$, it follows that $\operatorname{ind}(f, F) = \operatorname{ind}(\overline{f}, F \cap A)$ if and only if $\operatorname{ind}(g, G) = \operatorname{ind}(\overline{g}, G \cap A)$. By definition of "assuming its index in A", we are done. □

For each fixed point F, one may ask:

(5.3.1) Is F essential?

(5.3.2) Does F contain an essential fixed point class of $f_{A_k}\colon A_k \to A_k$?

(5.3.3) Does F contain an inessential fixed point class of $f_{A_k}\colon A_k \to A_k$?

(5.3.4) Does F assume its index in A?

For a union $\bigcup_{j=1}^{k} A_j$ of some components of A, we write $N_{mnpq}(f; X, \bigcup_{j=1}^{k} A_j)$, where $m, n, p, q = 0, 1$, for the number of fixed point classes which have negative or positive answers to each of the four questions. For example, $N_{1010}(f; X, \bigcup_{j=1}^{k} A_j)$ is the number of fixed point classes of f which are essential, do not contain any essential fixed point class of $f_{A_j}\colon A_j \to A_j$ with $1 \le j \le k$, contain an inessential fixed point class of $f_{A_j}\colon A_j \to A_j$ with $1 \le j \le k$, and do not assume their indices in A. Especially, $N_{1010}(f; X, A)$ is the number of fixed point classes of f which are essential, not common, weekly common and do not assume their indices in A. These relative Nielsen type numbers are said to be basic relative Nielsen numbers because each relative Nielsen type numbers can be written as a sum of some of these "basic numbers" togother with $N(\overline{f})$ and $R(\overline{f})$ (see [Z2]).

From above lemma, we get immediately that

(5.4) LEMMA. *Let $f, g\colon (X, A) \to (X, A)$ be two homotopic relative map. Then for any components $A_1, \dots, A_k$ of A, we have*

$$N_{mnpq}\Big(f; X, \bigcup_{j=1}^{k} A_j\Big) = N_{mnpq}\Big(g; X, \bigcup_{j=1}^{k} A_j\Big),$$

for any $m, n, p, q = 0, 1$.

Notice that

$$\begin{aligned}
N(f; X, A) &= N(\overline{f}) + \sum_{p,q=0}^{1} N_{10pq}(f; X, A),\\
N(f; \overline{X - A}) &= \sum_{m,n,p=0}^{1} N_{mnp0}(f; X, A),\\
N'(f; \overline{X - A}) &= \sum_{m,n,p=0}^{1} N_{mnp0}(f; X, A) + \sum_{\operatorname{int}(A_k)=\emptyset} N(f_{A_k})\\
&\quad - \sum_{m,p=0}^{1} N_{m1p0}\Big(f; X, \bigcup_{\operatorname{int}(A_k)=\emptyset} A_k\Big),\\
N(f; X, A) &= N_{1000}(f; X, A)
\end{aligned}$$

(see [Z2, Theorems 3.6, 3.7] and [Z3, Definition 5.1]). We have

(5.5) THEOREM (Homotopy Invariance). *Let $f, g\colon (X, A) \to (X, A)$ be two homotopic relative maps. Then*

$$\begin{aligned}
N(f; X, A) &= N(g; X, A), &\qquad N(f; \overline{X - A}) &= N(g; \overline{X - A}),\\
N'(f; \overline{X - A}) &= N'(g; \overline{X - A}), &\qquad N(f; X - A) &= N(g; X - A).
\end{aligned}$$

By the lower bound property and the homotopy invariance of the three relative Nielsen type numbers, $N(f; X, A)$, $N'(f; \overline{X - A})$ and $N(f; X - A)$, we have

(5.6) THEOREM. *Let $f\colon (X, A) \to (X, A)$ be a relative map. Then any relative map which is relatively homotopic to f has at least $N(f; X, A)$ fixed points on X, at least $N'(f; \overline{X - A})$ fixed points on $\operatorname{cl}(X - A)$, and at least $N(f; X - A)$ fixed points on $X - A$.*

Other properties of relative Nielsen type numbers are commutativity and homotopy type invariance (See [S1, Theorems 3.4 and 3.5]). We omit them here.

6. Realization of Nielsen type numbers

As we have seen in previous sections, there are several lower bounds for the numbers of fixed points of relative maps on different parts of the pairs (X, A) for relative maps on (X, A). A natural question is that: whether these lower bounds can be realized in a given relative homotopy class.

In order to minimize the fixed point set of relative maps, we have to unite the fixed points in the same fixed point classes as more as possible. We need to adjust such basic techniques in ordinary fixed point theory (see [J1] and [Sh]) so that they will be suitable to the relative case here.

Now we make some conventions in notations. Throughout this section, K is assumed to be a finite simplicial complex and will be considered as a subset of a Euclidean space R^n, hence it admits a standard metric, say $d(\cdot\,,\cdot)$. By a simplex we mean an open simplex. For any $x \in K$, the unique simplex containing x is written as car_x, the carrier simplex of x. The corresponding closed simplex is written as $\overline{\mathrm{car}}_x$. We define

$$\mathrm{St}_x = \{y \in |K| : x \in \overline{\mathrm{car}}_y\}.$$

A map $g\colon |L| \to |K|$ is said to be *proximity at* x if $\overline{\mathrm{car}}_x \cap \overline{\mathrm{car}}_{f(x)} \neq \emptyset$. It is said to be *proximity* if it is proximity at each point.

A piece-wise linear path $\eta\colon I \to |K|$ in $|K|$ is said to be an *almost normal* PL-*arc* (see [S1, p. 469]) if

(6.1.1) $\eta((0,1))$ does not meet any vertex of K,

(6.1.2) $\eta(t_1) = \eta(t_2)$ only if $t_1 = t_2$,

(6.1.3) $\eta(t)$ lies in maximal simplices of K but for finitely many values of $t \in I$, at these exceptional values, $\eta(s)$ is either an end-point or a connecting points of two lie segments in two different maximal simplices.

Compare with the ordinary definition ([J1, p. 752]), our definition here has less limitation. The local cut points on end points are allowable.

Let $\eta\colon I \to |K|$ be an almost normal PL-arc. A path ζ is said to be *special* with respect to η if $\zeta(0) = \eta(0)$, $\zeta(1) = \eta(1)$ and $\zeta(t) \neq \eta(t)$ for $0 < t < 1$. Two special paths ζ_0 and ζ_1 are said to be *specially homotopic with respect to* η if they can be connected by a homotopy $\{\zeta_s\}_{s\in I}$ such that each $\zeta_s\colon I \to |K|$ is a special path with respect to η.

(6.2) LEMMA ([J1, Lemma 5.1]). *Let $\eta\colon I \to |K|$ be an almost normal* PL-*arc satisfying the following conditions:*

(6.2.1) *There is a simplex τ, being an open subset of $|K|$, and number $0 < s_1 < s_3 < s_2$, such that $\eta(s)$ lies in a face of τ if and only if $s = s_1$ or s_2, $\eta(s_3) \in \tau$, and η is linear on $[s_1, s_3]$ and $[s_3, s_2]$.*

(6.2.2) *The face* σ *of* τ *containing the points* $\eta(s_1)$ *and* $\eta(s_2)$ *has dimension* 1, *which is a common face of at least three* 2*-simplices.*

If special paths ζ_0 *and* ζ_1 *with respect to* η *are homotopic, then they are specially homotopic with respect to* η.

PROOF. See the proof of [J1, Lemma 5.1]. □

(6.3) LEMMA. *Let* $\zeta: I \to |K|$ *be a path with* $\zeta((0,1)) \subset |K| - |L|$, *where* L *is subcomplex of* K. *If* $\zeta(I) - \{\zeta(0), \zeta(1)\}$ *does not meet any local cut point and meets a simplex which either has dimension at least* 3 *or is a one-dimensional simplex being the face of at least three* 2*-simplices, then for any* $\varepsilon > 0$, *using some subdivision of* K *if necessary, there is an almost normal* PL*-arc* η *satisfying the conditions* (6.2.1) *and* (6.2.2) *such that*

(6.3.1) $\eta \doteq \zeta$,

(6.3.2) $\eta(I) \subset \{x \in |K| - |L| : d(x, \zeta(I)) < \varepsilon\} \cup \zeta(I)$.

PROOF. By a simple approximation and general position argument, we can modify ζ into a PL-path ζ' with $\zeta \doteq \zeta'$ satisfying the conditions of almost normal PL-arc but may have self-intersections. Note that $\zeta(I) - \{\zeta(0), \zeta(1)\}$ meets a simplex which either has dimension at least 3 or is a one-dimensional simplex being the face of at least three 2-simplices. In the first case, using some subdivision of K, we can make the ζ' meet transversally a new one-dimensional simplex σ, which is the face of three 2-simplices. In the second case, we write the one-dimensional simplex as σ.

If the resulting path ζ' has self intersections, these self intersections can be moved out in the following way, with the help of the three 2-simplices around σ.

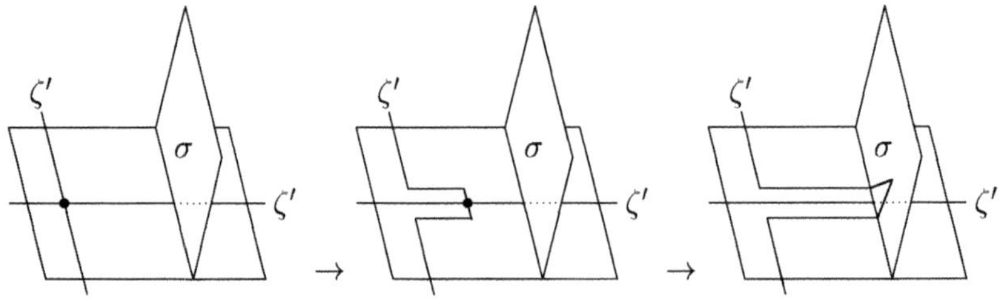

Clearly, the new path is in an arbitrary small neighbourhood of the original path. Using twice subdivision of K, a change at one of $\zeta'(I) \cap \sigma$ showing below and a suitable re-parameterring will lead to our desired almost mormal PL-arc, satisfying conditions (6.2.1) and (6.2.2).

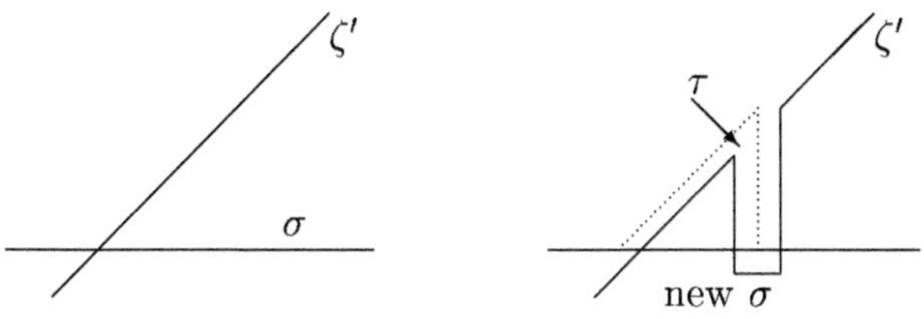

□

(6.4) LEMMA. *Let $f: |K| \to |K|$ be a self map on a compact polyhedron $|K| = X$, $|L|$ be a subpolyhedron of $|K|$. Suppose that x_0 and x_1 are two fixed points of f such that x_1 is an isolated fixed point of f in a maximal simplex and x_0 lies in a simplex in L which is the face of a simplex in $K - L$, i.e. $x_0 \in \mathrm{bd}(|L|)$. If there is a path $\zeta: I \to |K|$ from x_0 to x_1, such that:*

(6.4.1) $\zeta \stackrel{\cdot}{\simeq} f\zeta$,
(6.4.2) $\zeta(I) \cap \mathrm{Fix}(f) = \{\zeta(0), \zeta(1)\}$,
(6.4.3) $\zeta(I) - \{\zeta(0), \zeta(1)\}$ *does not meet any local cut point,*
(6.4.4) $\zeta(I) - \{\zeta(0), \zeta(1)\}$ *meets a simplex which either has dimension at least* 3 *or is a one-dimensional simplex being the face of at least three* 2*-simplices,*
(6.4.5) $\zeta(I) \cap |L| = \{x_0\}$.

Then, there is a map $f': |K| \to |K|$ such that:

(6.4.6) $f' \simeq f \operatorname{rel} |L|$,
(6.4.7) $\mathrm{Fix}(f') = \mathrm{Fix}(f) - \{x_1\}$.

PROOF. By Lemma (6.3), we have an almost normal PL-arc $\zeta': I \to |K|$ with $\zeta' \stackrel{\cdot}{\simeq} \zeta$ such that ζ' lies in a small neighbourhood and ζ' still satisfies the conditions (6.4.1)–(6.4.5) and (6.2.1)–(6.2.2). It follows that $\zeta' \stackrel{\cdot}{\simeq} f\zeta'$.

Let $H: I \times I \to |K|$ be a homotopy from ζ' to $f \circ \zeta'$, i.e. $H(t, 0) = \zeta'(t)$ and $H(t, 1) = f(\zeta'(t))$ for any $t \in I$, $H(0, s) = x_0$ and $H(1, s) = x_1$ for any $s \in I$. As ζ' is piece-wise linear and $H(0, I) = f(x_0) = x_0$, there is a value t_* in $(0, 1)$ such that $\zeta'((0, t_*])$ is a line segment lying in a maximal simplex in $K - L$, and that $H([0, t_*], I) \subset \mathrm{St}_{x_0}$. This implies that the restriction of f on $\zeta'([0, t_*])$ is a proximity map.

Denote $x_* = \zeta'(t_*)$. We have $f(x_*) \neq x_*$. Pick small positive number ε so that the open ball $B(x_*, \varepsilon) = \{x \in |K| : d(x, x_*) < \varepsilon\} \subset K - L$, $B(x_*, \varepsilon) \cap f(B(x_*, \varepsilon)) = \emptyset$ and $B(x_*, \varepsilon) \cup f(B(x_*, \varepsilon)) \subset \mathrm{St}_{x_0}$. Since f is proximity at x_*, we may pick a point $y_* \in \overline{\mathrm{car}}_{x_*} \cap \overline{\mathrm{car}}_{f(x_*)}$. There is a line segment from $f(x_*)$ to y_* and a line segment from y_* to x_*. Gluing the two line segments, we obtain an arc from $f(x_*)$ to x_*, denoted $\rho: I \to |K|$. Of course, ρ itself can be chosen as a straight line if $\overline{\mathrm{car}}_{f(x_*)} \subset \overline{\mathrm{car}}_{x_*}$.

Define a map $G: |K| \times I \to |K|$ by

$$G(x,t) = \begin{cases} f(x) & \text{if } x \notin B(x_*, \varepsilon t), \\ f\Big(\Big(\frac{2}{\varepsilon t}d(x,x_*) - 1\Big)x + \Big(2 - \frac{2}{\varepsilon t}d(x,x_*)\Big)x_*\Big) & \\ & \text{if } 0 < \frac{\varepsilon t}{2} < d(x,x_*) \le \varepsilon t, \\ \rho\Big(t - \frac{2}{\varepsilon}d(x,x_*)\Big) & \text{if } 0 \le d(x,x_*) \le \frac{\varepsilon t}{2}. \end{cases}$$

Such a homotopy is said to a point-tail homotopy (see [Ki, p. 82, Definition 1.2]).

Let us consider the fixed point set of $G_t: |K| \to |K|$ defined by $G_t(x) = G(x,t)$ for all $x \in |K|$. Clearly, $\mathrm{Fix}(G_t) \cap (|K| - B(x_*, \varepsilon)) = \mathrm{Fix}(f) \cap (|K| - B(x_*, \varepsilon))$. Since $G_t(B(x_*, \varepsilon) - B(x_*, \varepsilon t/2)) = f(B(x_*, \varepsilon))$ and $B(x_*, \varepsilon) \cap f(B(x_*, \varepsilon)) = \emptyset$, G_t has no fixed point on $B(x_*, \varepsilon) - B(x_*, \varepsilon t/2)$. Notice that $\rho(I) \cap B(x_*, \varepsilon)$ is a line segment, after re-arranging the parameter t of ρ, there is a positive c such that

$$d(\rho(t), x_*) = c(1-t), \quad \text{for } \rho(t) \in B(x_*, \varepsilon).$$

Let $x \in B(x_*, \varepsilon t/2) \cap \mathrm{Fix}(G_t)$. Then $x = G_t(x) = \rho(t - (2/\varepsilon)d(x,x_*))$. Thus, $d(x,x_*) = d(\rho(t - (2/\varepsilon)d(x,x_*)), x_*) = c(1 - t + (2/\varepsilon)d(x,x_*))$. It follows that $d(x,x_*) = c\varepsilon(1-t)/(\varepsilon - 2c)$. As different points in $\rho(I)$ has different distances to x_*, G_t has at most one fixed point on $B(x_*, \varepsilon t/2)$.

Denote $f'(x) = G(x,1)$. Then $f' \simeq f \mathrm{rel}\, |K| - B(x_*, \varepsilon)$. We have $f'(x_*) = x_*$. The above argument concerning with the fixed point set of G_t shows that G_1 has at most one fixed point on $B(x_*, \varepsilon)$. Thus, x_* is the unique fixed point of f' on $B(x_*, \varepsilon)$. So, we have that $\mathrm{Fix}(f') = \mathrm{Fix}(f) \cup \{x_*\}$.

Define a map $F: I \times I \to |K|$ by

$$F(t,s) = \begin{cases} H(t, 2s) & \text{if } 0 \le s \le 1/2, \\ G(\zeta'(t), 2s-1) & \text{if } 1/2 \le s \le 1. \end{cases}$$

It is a homotopy from ζ' to $f' \circ \zeta'$. We shall modify it into a homotopy which keeps the point $x_* = \zeta'(t_*)$. Notice that $F(\{t_*\} \times I) = H(\{t_*\} \times I) \cup G(\{\zeta(t_*)\} \times I) = H(\{t_*\} \times I) \cup \rho(I) \subset \mathrm{St}_{x_*}$, and that $F(t_*, 0) = F(t_*, 1) = x_*$. As St_{x_*} is contractible, the path $F(\{t_*\} \times I)$ is homotopic keeping end points fixed to the constant path at x_*. Thus, there is a map $Q: \{t_*\} \times I \times I \to |K|$ such that $Q(t_*, s, 0) = F(t_*, s)$ for all $s \in I$, $Q(t_*, s, 1) = x_*$ for all $s \in I$, and $Q(t_*, 0, p) = Q(t_*, 1, p) = x_*$ for all $p \in I$. Extend Q so that $Q(t, s, 0) = F(t, s)$ for all $t, s \in I$, $Q(0, s, q) = x_0$ and $Q(1, s, q) = x_1$ for all $s, q \in I$, and $Q(t, 0, q) = \zeta'(t)$ and $Q(t, 1, q) = f'(\zeta'(t))$ for all $t, q \in I$. We then have a map $Q: I \times I \times \{0\} \cup (\{0, t_*, 1\} \times I \cup I \times \{0,1\}) \times I \to |K|$. Applying the homotopy extension property, we can extend Q into $Q: I \times I \times I \to |K|$. Clearly, $Q(t,s,1)$ is a homotopy from ζ' to $f'\circ\zeta'$. Under this homotopy, we have $\zeta' \simeq f'\circ\zeta' \operatorname{rel}\{0, t_*, 1\}$.

Write $\zeta'_*: I \to |K|$ and $\zeta'_{**}: I \to |K|$ for the sub-arcs of ζ' from $x_0 = \zeta'(0)$ to $\zeta'(t_*)$ and from $\zeta'(t_*)$ to $x_1 = \zeta'(1)$ respectively, i.e. $\zeta'_*(t) = \zeta'(t_* t)$ and $\zeta'_{**}(t) = \zeta'((1-t_*)t + t_*)$ for all $t \in I$. Clearly, the arc ζ'_{**}, as well as ζ', still satisfies the conditions (6.4.1)–(6.4.5) which the path ζ satisfies. By Lemma (6.3), we can modify homotopically the normal PL-arc ζ'_{**} into an almost normal PL-arc η such that η satisfies the conditions (6.4.1)–(6.4.5) and (6.2.1)–(6.2.2).

Define an arc η'_ε by $\eta'_\varepsilon(t) = \eta(t - \varepsilon \sin t\pi)$ ($\varepsilon > 0$). When ε is small enough, we have $\overline{\mathrm{car}}_{\eta'_\varepsilon(t)} \cap \overline{\mathrm{car}}_{\eta(t)} \neq \emptyset$ for all $t \in I$. The condition (6.1.2) of almost normal PL-arc implies that η'_ε is a special path with respect to η. Since $\mathrm{Fix}(f') \cap \eta(I) = \{x_*, x_1\}$, $f'\circ\eta$ is also a special path with respect to η. As $\eta'_\varepsilon \doteq\!\!\simeq \eta$ and $\eta \doteq\!\!\simeq \zeta'_{**} \simeq f'\circ\zeta'_{**} \doteq\!\!\simeq f'\circ\eta$, we have $\eta'_\varepsilon \doteq\!\!\simeq f'\circ\eta$. By Lemma (6.2), η'_ε and $f'\circ\eta$ are specially homotopic.

Let $S: I\times I \to |K|$ be the above special homotopy from $f'\circ\eta$ to η'_ε. We can define a homotopy $T: \eta(I)\times I \to |K|$ by $T(x,s) = S(\eta^{-1}(x), s)$ for any $(x,s) \in \eta(I)\times I$. If $T(z,s) = z$ for some $(z,s) \in \eta(I) \times I$, then $S(\eta^{-1}(z), s) = z = \eta(\eta^{-1}(z))$. Because S is a special homotopy with respect to η, $S(t,s) \neq \eta(t)$ for $t \in (0,1)$ and $s \in I$. It follows that $z = x_*$ or x_1. Thus, $T: \eta(I)\times I \to |K|$ is special homotopy in the sense of [J1, Section 2]. Extend it to $T: (\eta(I) \cup |L|)\times I \to |K|$ by defining $T(x,t) = f'(x)$ for $x \in \zeta'_*(I) \cup |L|$ and $t \in I$. Using the "special homotopy extension property" ([J1, Lemma 2.1]), T can be extended to a special homotopy $T: |K| \times I \to |K|$ with $T(x,0) = f'(x)$ for any $x \in |K|$.

Define a map $f'': |K| \to |K|$ by $f''(x) = T(x,1)$ for all $x \in |K|$. Then $f'' \simeq f \operatorname{rel} |L|$, $\mathrm{Fix}(f'') = \mathrm{Fix}(f') = \mathrm{Fix}(f) \cup \{x_*\}$, and f'' is proximity at $\zeta_*(I) \cup \eta(I)$.

By successive use of [Sh, Lemma 1.3] (cf. [Br, Lemma VIII.C.3] or [Ki, p. 88, 2.2]), for any given $t' \in (0,1)$, there is a small $\varepsilon > 0$ and a map $f''': |K| \to |K|$ with $B(\eta([t',1]),\varepsilon) \in |K| - |L|$ such that $f''' \simeq f'' \operatorname{rel} |K| - B(\eta([t',1]),\varepsilon)$, f''' is proximity on $B(\eta([t',1]),\varepsilon)$, and that $\mathrm{Fix}(f''') = (\mathrm{Fix}(f'') - \{x_1\}) \cup \{\eta(t')\} = (\mathrm{Fix}(f) - \{x_1\}) \cup \{x_*, \eta(t')\}$. In other word, the fixed point x_1 is moved to $\eta(t')$.

We may choose $t' > 0$ small enough so that $\eta([0,t'])$ is contained in the maximal simplex car_{x_*}. Thus, we can find out an convex closed set U such that $U - \{x_0\} \subset \mathrm{car}_{x_*}$, $\zeta_*(I) \cup \eta([0,t']) \subset \mathrm{int}(U) \cup \{x_0\}$, and f''' is proximity at U.

Note that each point x in $U - \{x_0\}$ can be written uniquely as $x = t_x b_x + (1 - t_x)x_0$, where b_x is in $\mathrm{bd}(U)$ and is determined uniquely by x (See [S1, pp. 471–472]). We define a map $g: |K| \to |K|$ by

$$g(x) = \begin{cases} x_0 & \text{if } x = x_0, \\ t_x f'''(b_x) + (1-t_x)x_0 & \text{if } x = t_x b_x + (1-t_x)x_0,\ x \in U - \{x_0\}, \\ f'''(x) & \text{if } x \in |K| - U. \end{cases}$$

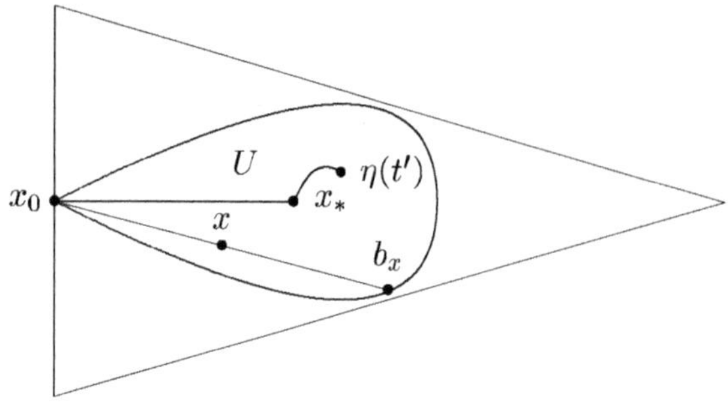

In fact, all simplexes we talk above may be assumed to be those of some subdivision of K and L. It is easy to check that the condition f''' is proximity at U with respect to a subdivision of K implies that $f(x) \in \mathrm{St}_{x_0}$ with respect to K itself. So, the addition in defining g makes sense.

Since St_{x_0} is contractiable, we have that $g \simeq f''' \mathrm{rel}\, |K| - \mathrm{int}(U)$. Note that g has unique fixed point x_0 on U. The map g is the desired map. □

This lemma is a generalization to the main technique in [J1], and can be used to case relative fixed point theory. Especially, the fixed point x_0 can be a local cut point and can be non-isolated.

When combining the fixed points on $X - A$, the homotopic change to the map happens to a neighbourhood of the Nielsen paths between the fixed points. This change can be very "big". In order to make the result map to be relative one, we have to assume that the Nielsen paths lie totally in $X - A$. So, a further restriction on the pair (X, A) to ensure the existence of such kinds of Nielsen paths is needed. That is the widely used "by-passed" condition.

(6.5) DEFINITION ([S1, Definition 5.1]). A subspace A of X is said to be by-passed in X if every path in X with end points in $X - A$ is homotopic keeping end points fixed to a path in $X - A$.

By this definition, one can prove

(6.6) PROPOSITION ([S1, Theorem 5.2]). *Let (X, A) be a pair of compact polyhedra. Then A can be by-passed in X if and only if $X - A$ is connected and $\pi_1(X - A) \xrightarrow{i_\pi} \pi_1(X)$ is onto.*

In order to minimize the fixed points of relative maps on (X, A), one naturally asks the subspace A to be "nice". So, Schirmer made such a definition: a space Y is said to be a *Nielsen space* if each map $g: Y \to Y$ is homotopic to a map with $N(g)$ fixed points which can lie anywhere in Y.

(6.7) THEOREM. *Let (X, A) be a pair of compact polyhedra satisfying:*

(6.7.1) *A is a Nielsen space,*

(6.7.2) *A can be by-passed in X,*

(6.7.3) *$X - A$ has no local cut point and is not a 2-manifold.*

Then every map $f\colon (X, A) \to (X, A)$ is relatively homotopic to a map $g\colon (X, A) \to (X, A)$ with $N(f; X, A)$ fixed points on X, $N'(f; \overline{X - A})$ fixed points on $\mathrm{cl}(X - A)$ and $N(f; X - A)$ fixed points on $X - A$.

PROOF. Let $\mathcal{F}_1 = \{F_{1,1}, \ldots, F_{1,n_1}\}$ be the set of the fixed point classes of f which do not their indices in A, are common and do not contain any essential fixed point class of $f_{A_k}\colon A_k \to A_k$ with $\mathrm{int}(A_k) = \emptyset$. Let $\mathcal{F}_2 = \{F_{2,1}, F_{2,2}, \ldots, F_{2,n_2}\}$ be the set of essential fixed point classes which are weakly common and not common. Thus, $n_1 = N'(f; \overline{X - A}) - \sum_{\mathrm{int}(A_k)=\emptyset} N(f_{A_k})$ and $n_2 = N(f) - N(f; X - A) - N(f, \overline{f})$. In the following proof, we shall not distinguish a fixed point class of f or $\overline{f}$ and those homotopy related to it.

Step 1. Minimize the fixed points on A.

Choose an essential fixed point class $\overline{F}_{1,i}$ of $\overline{f}$ for each $F_{1,i}$, where $i = 1, \ldots, n_1$. Clearly, each $F_{1,i}$ is an essential fixed point class of $f_{A_{k(i)}}\colon A_{k(i)} \to A_{k(i)}$ with $\mathrm{int}(A_{k(i)}) \neq \emptyset$. Since A is a Nielsen space, $\overline{f}$ is homotopic to a map $\overline{f}'$ with $N(\overline{f})$ fixed points. Moreover, we may arrange the $N(\overline{f})$ fixed points on A so that a fixed point of $\overline{f}'$ lies in $\mathrm{int}(A)$ if and only if it belongs to an essential fixed point class $\overline{F}$ of $f_{A_k}\colon A_k \to A_k$ with $\mathrm{int}(A_k) \neq \emptyset$ and $\overline{F} \notin \{\overline{F}_{1,1}, \overline{F}_{1,2}, \ldots, \overline{F}_{1,n_1}\}$. The fixed points in $\overline{F}_{1,i}$, $i = 1, \ldots, n_1$, lie on $\mathrm{bd}(A)$.

Step 2. Make map to have finitely many fixed points.

Let V and U be two regular neighbourhoods of A with $U \subset \mathrm{int}(V)$. Then A is a strong deformation retractor of V, i.e. there is map $r\colon V \times I \to A$ such that $r(x, 0) = x$ for each $x \in V$, $r(a, t) = a$ for any $a \in A$ and $t \in I$, and $r(x, 1) \in A$ for any $x \in V$. Let $H\colon A \times I \to A$ be a homotopy from $\overline{f}$ to $\overline{f}'$. Using homotopy extension property, H can be extended to a homotopy $H\colon (X \times I, A \times I) \to (X, A)$. Define a map $H^*\colon (V \times I, A \times I) \to (V, A)$ by $H^*(x, t) = H(r(x, t), 1)$, and use homotopy extension property again, we have a homotopy $H^*\colon (X \times I, A \times I) \to (X, A)$. Thus, $H^*(x, 1) \in A$ for any $x \in V$.

Since $H^*(x, 1)$ has no fixed point on ∂V, there is a $\delta > 0$ such that $d(H^*(v, 1), v) > \delta$ for any $v \in \partial V$. Apply Hopf simplical approximation theorem to the map $H^*|_{(X - \mathrm{int}(V)) \times \{1\}}$, there is homotopy $H^{**}\colon (X - \mathrm{int}(V)) \times I \to X$ from $H^*(x, 1)$ such that

(6.8.1) $H^{**}(x, t) \in B(H^*(x, 1), \delta/2) = B(H^{**}(x, 0), \delta/2)$ for any $x \in X - \mathrm{int}(V)$, and

(6.8.2) $H^{**}(x, 1)$ has finitely many fixed points, each of them lies in a maximal simplex.

Thus, $d(H^{**}(x,t),x) \geq d(H^{**}(x,0),x) - d(H^{**}(x,t),H^{**}(x,0)) \geq \delta/2$ for any $x \in \partial V$, and hence $H^{**}(x,t) \neq x$ for any $x \in \partial V$ and any $t \in I$. So, $H^{**}|_{\partial V \times I}$ is special homotopy. Extend H^{**} to

$$H^{**}: (U \cup (X - \operatorname{int}(V)) \times I) \to X$$

by $H^{**}(x,t) = H^{*}(x,1)$ for any $x \in U$. Using the "special homotopy extension property" ([J1, Lemma 2.1]) to $H^{**}|_{(U \cup \partial V)\times I}$, we can extend H^{**} to a special homotopy $H^{**}: V \times I \to X$. Combine the existed definition of H^{**} on $(X - V) \times I$, we get a homotopy $H^{**}: X \times I \to X$.

Define a map $f': X \to X$ by $f'(x) = H^{**}(x,1)$. It is obvious that f' is a map relatively homotopic to f such that

(6.9.1) the restriction of f' on A is just $\overline{f}'$, hence f' has $N(\overline{f})$ fixed points on A,
(6.9.2) $f'(U) \subset A$,
(6.9.3) the number of fixed points of f' is finite, each fixed point of f' on $X - A$ lies in a maximal simplex.

Step 3. Create $n_2 = N(f) - N(f; X - A) - N(f, \overline{f})$ fixed points on $\operatorname{bd}(A)$, which are in different fixed point classes in $\mathcal{F}_2$.

For each $F_{2,j}$ in $\mathcal{F}_2$, where $j = 1, \dots, n_2$, by Proposition (4.2), we may pick a point $b_j \in \operatorname{bd}(A_{k_j})$ and a path $\zeta_j: (I,0,1) \to (X, F_{2,j}, b_j)$ such that

$$\zeta_j \simeq f \circ \zeta_j: (I,0,1) \to (X, \zeta_j(0), A_{k_j}).$$

We claim that $\zeta_j(1)$'s can be chosen so that they are n_2 distinct points, and none of them is a fixed point of f'. In fact, if $\operatorname{bd}(A_{k_j})$ has infinitely many points, we shall have enough freedom in choosing $\zeta_j(1)$'s. If $\operatorname{bd}(A_{k_j})$ consists of finitely many points. By van Kampen theorem, $\pi_1(X)$ will contain $\pi_1(A_{k_j}) * \pi_1(X - A)$ as subgroup. Because A can be by-passed in X, the homomorphism $\pi_1(X-A) \xrightarrow{i_\pi} \pi_1(X)$ is onto. It follows that $\pi_1(A_{k_j})$ is a trivial group, i.e. A_{k_j} is simply-connected. (Furthermore, we can prove that $\operatorname{bd}(A_{k_j})$ is singleton.) Thus, $f'_{A_{k_j}}: A_{k_j} \to A_{k_j}$ has only one fixed point class. Since $F_{2,j}$ contains an inessential fixed point class of $f_{A_{k_j}}$, the unique fixed point class of $f_{A_{k_j}}$ is inessential. Note that f' has exactly $N(f) = N(f')$ fixed points on A. Any inessential fixed point class of $\overline{f}'$ contains no fixed point. Thus, f' has no fixed point on such a component A_{k_j}. Since each fixed point class of $\overline{f}'$ is contained in a unique fixed point of f', the simply-connected component of A will be appeared at most once in considering the terminal points $\zeta_j(1)$, $j = 1, \dots, n_2$. So, we prove the claim.

For $j = 1, \dots, n_2$, we write $H_j: (I \times I, \{0\} \times I, \{1\} \times I) \to (X, \zeta_j(0), A_{k_j})$ for the homotopy from ζ_j to $f \circ \zeta_j$. As $\zeta_j(1)$ is not a fixed point, we may choose a positive number ε so small that any two of $B(\zeta_j(1), \varepsilon)$'s are distinct and

$$f'(\operatorname{cl}(B(\zeta_j(1), \varepsilon))) \cap \operatorname{cl}(B(\zeta_j(1), \varepsilon)) = \emptyset.$$

Define a map $G: X \times I \to X$ by

$$G(x,t) = \begin{cases} f'(x) & \text{if } x \notin \bigcup_{j=1}^{n_2} B(\zeta_j(1), \varepsilon t), \\ f'\Big(\Big(\dfrac{2}{\varepsilon t} d(x, \zeta_j(1)) - 1\Big) x & \\ \qquad + \Big(2 - \dfrac{2}{\varepsilon t} d(x, \zeta_j(1))\Big) \zeta_j(1)\Big) & \text{if } 0 < \dfrac{\varepsilon t}{2} < d(x, \zeta_j(1)) \le \varepsilon t, \\ H_j\Big(1, 1 - t + \dfrac{2}{\varepsilon} d(x, \zeta_j(1))\Big) & \text{if } 0 \le d(x, \zeta_j(1)) \le \dfrac{\varepsilon t}{2}. \end{cases}$$

As ε can be arbitrary small, we may assume that each $B(\zeta_j(1), \varepsilon)$ lies in $\mathrm{St}_{\zeta_j(1)}$. Hence the second case in defining G makes sense. Since any two of $B(\zeta_j(1), \varepsilon)$'s are disjoint, G is well-defined at each point in $\bigcup_{j=1}^{n_2} B(\zeta_j(1), \varepsilon)$. Notice that we can modify H_j so that $\{H_j(1, s)\}_{0 \le s \le 1}$ is piece-wise linear. Thus, we can make $H_j(\{1\} \times I) \cap B(\zeta_j(1), \varepsilon)$ to be a line segment. Define $f'': X \to X$ by $f''(x) = G(x, 1)$. By the same argument as in the proof of Lemma (6.4), we have that $\mathrm{Fix}(f'') = \mathrm{Fix}(f') \cup \{\zeta_1(1), \dots, \zeta_{n_2}(1)\}$. Thus, $\sharp(\mathrm{Fix}(f'') \cap A) = N(\overline{f}) + n_2$. Using the Alexander trick to the homotopy, H_j can be modified to homotopy from ζ_j to $f'' \circ \zeta_j$. Hence, $\zeta_j \doteq f'' \circ \zeta_j$. Thus, we have that $\zeta_j(1) \in F_{2,j}$ for $j = 1, \dots, n_2$.

When ε is small enough, the set $\bigcup_{j=1}^{n_2} B(\zeta_j(1), \varepsilon)$ will be contained in the regular neighbourhood U of A, which has been chose in previous step. Since $f'(U) \subset A$, we have that $G(A \times I)) \subset f'(A \cup (\bigcup_{j=1}^{n_2} B(\zeta_j(1), \varepsilon))) \subset f'(U) \subset A$, i.e. G is a relative homotopy. Hence, $f'': (X, A) \to (X, A)$ is a relative map relatively homotopic to f such that

(6.10.1) f'' has $N(\overline{f}) + N(f) - N(f; X - A) - N(f, \overline{f})$ fixed points on A,
(6.10.2) each fixed point class in $\mathcal{F}_2$ contains a fixed point on $\mathrm{bd}(A)$,
(6.10.3) $f''(U) \subset A$,
(6.10.4) the number of fixed points of f'' is finite, each fixed point of f'' on $X - A$ lies in a maximal simplex.

Step 4. Combine the fixed points on $X - A$.

Let y_0 and y_1 be two fixed points in $X - A$ which are in the same fixed point class of f''. By definition, there is a path ζ from y_0 to y_1 such that $\zeta \doteq f'' \circ \zeta$. Since A can be by-passed in X, ζ is homotopic relative end points to a path ζ' in $X - A$. Thus, we have $\zeta' \doteq f'' \circ \zeta'$. Since $X - A$ has no local cut point, and since f'' has finitely many fixed points, we may modify ζ' so that it satisfies the conditions (6.4.2) and (6.4.3). Since $X - A$ is not a 2-manifold, there is a 1-simplex in $X - A$ which is the face of at least three 2-simplices. Pick a point z' in this 1-simplex with $z' \notin \mathrm{Fix}(f'')$ and a point z'' in $\zeta'(I)$ with $z'' \notin \{y_0, y_1\}$. As $X - A$ is path-connected, there is a path ρ in $X - A$ from z' to z''. Write ζ'_* and ζ'_{**}

for the sub-path of ζ' from y_0 to z'' and the sub-path from z'' to y_1 respectively. Then the path $\zeta'_*\rho^{-1}\rho\zeta'_{**}$ satisfies each of the conditions (6.4.1)–(6.4.5), where the subpolydedron $|L|$ is $U \cup \{y_0\}$. By Lemma (6.4), f'' is homotopic relative U to a map with the fixed point set $\mathrm{Fix}(f'') - \{y_1\}$.

Repeat the above procedure, we can unite the fixed points in the same fixed point class into one point. Remove the fixed points on $X - A$ of index zero (see [Br, p. 123, Theorem 4]), we have that, by Proposition (3.5), f''' has $N(f; \overline{X - A})$ fixed points on $X - A$, which are in different fixed point classes that do not assume their indices in A. Thus, we have a map $f''' \colon (X, A) \to (X, A)$ relatively homotopic to f with:

(6.11.1) f''' has $N(\overline{f}) + N(f) - N(f; X - A) - N(f, \overline{f})$ fixed points on A,
(6.11.2) each fixed point class in $\mathcal{F}_2$ contains a fixed point on $\mathrm{bd}(A)$,
(6.11.3) $f'''(U) \subset A$,
(6.11.4) f''' has $N(f; \overline{X - A})$ fixed points on $X - A$, each of them lies in a maximal simplex.

Step 5. Combine $N(f; \overline{X - A}) - N(f; X - A)$ fixed points on $X - A$ into $\mathrm{bd}(A)$.

Let y be a fixed point of f''' on $X - A$. Then y belongs to a fixed point class F of f''' which does not assume its index in A. There are four cases:

Case 1. F is not weekly common,

Case 2. F is weekly common and not common, i.e. $F \in \mathcal{F}_2$,

Case 3. F is common and contains a fixed point class of $f_{A_j} \colon A_j \to A_j$ with $\mathrm{int}(A_j) = \emptyset$,

Case 4. F is common and does not contains any essential of fixed point class of $f_{A_j} \colon A_j \to A_j$ with $\mathrm{int}(A_j) = \emptyset$, i.e. $F \in \mathcal{F}_1$.

In Case 1, we do nothing. In Case 2, we have that $F = F_{2,j}$ for a j with $1 \le j \le n_2$. Notice that fixed point $\zeta_j(1) \in F$. There is path η_j from $\zeta_j(1)$ to y such that $\eta_j \doteq f''' \circ \eta_j$. Since A can be by-passed in X, η_j can be chosen so that $\eta_j(I) \cap A = \eta_j(0)$. By a similar argument as in Step 4, we can combine the fixed point y into $\zeta_j(1) \in A$. In Case 3, since F contains an essential fixed point class of $f_{A_j} \colon A_j \to A_j$ with $\mathrm{int}(A_j) = \emptyset$, F contains a fixed point on $A_j \subset \mathrm{cl}(X - A)$. We then can combine y into the fixed point on A_j. In case (4), the fixed point y will be combined into the fixed point in $\overline{F}_{1,j}$, where $y \in F_{1,j}$.

Finally, we shall get the desired map $g \colon (X, A) \to (X, A)$. □

By above theorem and Proposition (3.10), we have

(6.12) THEOREM. *Let (X, A) be a pair of compact polyhedra satisfying:*

(6.12.1) *each component of A is a Nielsen space with non-empty interior,*
(6.12.2) *A can be by-passed in X,*
(6.12.3) *$X - A$ has no local cut point and is not a 2-manifold.*

Then every map $f: (X, A) \to (X, A)$ *is relatively homotopic to a map* $g: (X, A) \to (X, A)$ *with* $N(f; X, A)$ *fixed points on* X, $N(f; \overline{X - A})$ *fixed points on* $\mathrm{cl}(X - A)$ *and* $N(f; X - A)$ *fixed points on* $X - A$.

From the proof of above minimum theorem, the condition in realizing the three relative Nielsen type numbers, $N(f; X, A)$, $N(f; \overline{X - A})$ and $N(f; X - A)$, are very similar. The key point is to combine the fixed points in $X - A$. So, people focus on the realization of $N(f; X, A)$. A pair (X, A) is said to be Wecken if each map $f: (X, A) \to (X, A)$ is homotopic to a map with $N(f; X, A)$ fixed points. This includes the case $A = \emptyset$. The minimum theorems here give some answers in the case of polyhedral pair.

Consider some "nice" pairs, such as $(M, \partial M)$, where M is a compact manifold and ∂M is the boundary of M. It is trivial that $(M, \partial M)$ is Wecken if $\dim M = 1$. We know from above theorems that $(M, \partial M)$ is Wecken if $\dim M \geq 4$. When $\dim M = 3$ and M is orientable, B. Jiang ([J5]) proved that $(M, \partial M)$ is Wecken if and only if ∂M is Wecken. It is known from [J4] that a compact 2-dimensional manifold M is Wecken if and only if $\chi(M) > 0$. So, it easy to check that if $\dim M = 2$, $(M, \partial M)$ is Wecken if M is either a disc or a Möbius band or an annulus.

Since most surfaces are not Wecken, we ask a further question: what is the difference between $N(f; M, \partial M)$ and the minimal fixed point number of maps in the relative homotopy class of $f: (M, \partial M) \to (M, \partial M)$. Related results can be found in [BS], [J5], [K1] and [K2].

7. Minimal fixed point sets

For a relative map $f: (X, A) \to (X, A)$, the relative Nielsen number $N(f; X, A)$ is a lower bound for the number of the fixed points of maps in the relative homotopy class of f. In previous section, we have shown that under some assumptions on (X, A), this lower bound can be realized by a relative map in the relative homotopy class of f. It is natural to ask: what is the distribution of the fixed point set of f when f has $N(f; X, A)$ fixed points? Of course, f has at least $N'(f; \overline{X - A})$ fixed points on $\mathrm{cl}(X - A)$ and at least $N(f; X - A)$ fixed points on $X - A$. This is a restriction. In fact, there are still something more. Here is an example:

(7.1) EXAMPLE. Let (X, A) be a pair of compact polyhedra, where X is a circle and A is a small arc in X. A relative map $f: (X, A) \to (X, A)$ is the identity map.

Clearly, $\overline{f}$ has unique fixed point class, it is essential. And f has no essential fixed point class. Thus, we have $N(f; X, A) = N(\overline{f}) + N(f) - N(f, \overline{f}) = 1 + 0 - 0 = 1$. Moreover, the fixed point class of f containing the essential fixed point class of

$\overline{f}$ does not assume its index in A. Thus, any map relatively homotopic to f has a fixed point on $\mathrm{cl}(X - A)$.

Note that any map relatively hommotopic to f has at least one fixed point on A. If a relative map g in the relative homotopy class of has minimal fixed point set, i.e. $\sharp(\mathrm{Fix}(g)) = N(f; X, A)$, the unique fixed point of g must lie in $\mathrm{bd}(A)$.

This phenomenon was noticed in [S2]. From the proof of the lower bound property of $N(f; X, A)$, we may prove

(7.2) LEMMA ([Z3, Lemma 3.3]). *Let $f\colon (X, A) \to (X, A)$ be a relative map with $N(f; X, A)$ fixed points. Then each essential fixed point class of $\overline{f}$ contains exactly one fixed point, the number of fixed points in a fixed point class F of f:*

(7.2.1) *is the number of essential fixed point classes of $\overline{f}$ being contained in F, if F is common;*

(7.2.2) *is 1, if F is non-common and essential;*

(7.2.3) *is 0, otherwise.*

(7.3) LEMMA. *Let $f\colon (X, A) \to (X, A)$ be a relative map with $N(f; X, A)$ fixed points. If a fixed point class F of f contains an essential fixed point of $f_{A_k}\colon A_k \to A_k$, does not contain any essential fixed point class of $f_{A_j}\colon A_j \to A_j$ with $A_j \neq A_k$, and does not assume its index in A. Then $F \cap \mathrm{bd}(A_k) \neq \emptyset$.*

PROOF. Let F contain n essential fixed point classes of $\overline{f}$. Then F contains n fixed points $b_1, \dots, b_n$ which are in distinct essential fixed point classes of $\overline{f}$ being contained in F. Since F does not assume its index in A, F contains a fixed point y in $\mathrm{cl}(X - A)$. By Lemma (7.2), F contains n fixed points. Thus, $y = b_j$ for a j with $1 \leq j \leq n$. As F does not contain any essential fixed point class of $f_{A_j}\colon A_j \to A_j$ with $A_j \neq A_k$, we have $y = b_j$ is a fixed point of f_{A_k}. Thus, $y \in A_k \cap (F \cap \mathrm{cl}(X - A)) \subset \mathrm{bd}(A_k)$. So, $F \cap \mathrm{bd}(A_k) \neq \emptyset$. □

This lemma shows a kind of restriction for the fixed point of f on $\mathrm{bd}(A)$ if f has $N(f; X, A)$ fixed points.

(7.4) DEFINITION (see [S2, Section 2] and [Z3, Definition 3.1]). Let $f\colon (X, A) \to (X, A)$ be a relative map. For a union $S = \bigcup_{j=1}^{k} A_j$ of some components of A, we define $\widetilde{n}(f, X, S)$ for the number of fixed point classes of f which contain an essential fixed point of $f_{A_j}\colon A_j \to A_j$ with $A_j \subset S$ and do not assume their indices in A, and $n(f, X, S)$ for the number of non-common and essential fixed point classes of f which contain an inessential fixed point of $f_{A_j}\colon A_j \to A_j$ with $A_j \subset S$.

If an essential fixed point class F of f is non-common, weekly common, it will not assume its index in A. Such a F may lie in $X - A$ or a boundary $\mathrm{bd}(A_j)$ when F contains an (inessential) fixed point class of $f_{A_j}\colon A_j \to A_j$. By above lemmas, we have

(7.5) THEOREM. *Let $f\colon (X,A) \to (X,A)$ be a relative map with $N(f;X,A)$ fixed points. Then for any union $S = \bigcup_{j=1}^{k} A_j$ of some components of A,*

(7.5.1) *f has at most*

$$\sum_{A_j \subset S, \operatorname{int}(A_j)\neq\emptyset} N(f_{A_j}) + \widetilde{n}\Big(f;X, \bigcup_{A_j \not\subset S \text{ or } \operatorname{int}(A_j)=\emptyset} A_j\Big) - \widetilde{n}(f;X,A)$$

fixed points on $\operatorname{int}(S)$.

(7.5.2) *f has at least*

$$\sum_{A_j \subset S, \operatorname{int}(A_j)=\emptyset} N(f_{A_j}) - \widetilde{n}\Big(f;X, \bigcup_{A_j \not\subset S \text{ or } \operatorname{int}(A_j)=\emptyset} A_j\Big) + \widetilde{n}(f;X,A)$$

and at most

$$\sum_{A_j \subset S} N(f_{A_j}) + n\Big(f;X, \bigcup_{A_j \subset S} A_j\Big)$$

fixed points on $\operatorname{bd}(S)$.

(7.5.3) *f has at least $\sum_{A_j \subset S} N(f_{A_j})$ and at most*

$$\sum_{A_j \subset S} N(f_{A_j}) + n\Big(f;X, \bigcup_{A_j \subset S} A_j\Big)$$

fixed points on S.

It is obvious that

(7.6) THEOREM. *Let $f\colon (X,A) \to (X,A)$ be a relative map with $N(f;X,A)$ fixed points. Then f has at least $N(f;X-A)$ and at most $N(f;X-A)+n(f;X,A)$ fixed points on $X-A$, has at least $N'(f;\overline{X-A})$ and at most $N(f;X,A)$ fixed points on* $\operatorname{cl}(X-A)$.

These theorems give a necessary for the distribution of the minimal fixed point set. In fact, under the same assumptions on (X,A) as in Theorem (6.7), these conditions are sufficient (see [Z3] for the details).

References

[Bo] C. Bowszyc, *Fixed point theorem for the pairs of spaces*, Bull. Acad. Polon. Sci. Ser. Sci. Math. Astr. Phys. **16** (1968), 845–850.

[Br] R. F. Brown, *The Lefschetz Fixed Point Theorem*, Scott Foresman and Co., Glenview, IL, 1971.

[BS] R. F. Brown and R. Sanderson, *Fixed points of boundary-preserving maps of surfaces*, Pacific J. Math. **158** (1993), 243 – 264.

[J1] B. Jiang, *On the least number of fixed points*, Amer. J. Math. **102** (1980), 749–763.

[J2] ______, *Fixed point class from a differentable viewpoint*, Fixed Point Theory, Lecture Notes in Math., vol. 886, Springer Berlin, 1981, pp. 163–170.

[J3] ______, *Lectures on Nielsen Fixed Point Theory, Contemporary Mathematics*, vol. 14, AMS, Providence, RI, 1983.

[J4] ______, *Fixed point and braids* II, Math. Ann. **272** (1985), 249–256.

[J5] ______, *Commutativity and Wecken properties for fixed points on surfaces and* 3*-manifolds*, Topology Appl. **53** (1993), 221–228.

[K1] M. Kelly, *The relative Nielsen number and boundary-preserving maps of compact surfaces*, Pacific J. Math. **161** (1993), 139–153.

[K2] ______ Fixed points boundary-preserving maps on punctured projective planes, Topology Appl. **124** (2002), 145–157.

[Ki] T. H. Kiang, *The Theory of Fixed Point Classes*, Springer–Verlag, Berlin, Science Press, Beijing, 1989.

[S1] H. Schirmer, *A relative Nielsen number*, Pacific J. Math. **122** (1986), 459–473.

[S2] ______, *On the location of fixed points on pairs of spaces*, Topology Appl. **30** (1988), 253–266.

[S3] ______, *A survey of relative Nielsen fixed point theory*, Nielsen Theory and Dynamical Systems, Contemporary Mathematics, vol. 152, AMS, Providence, RI, 1993, pp. 291–310.

[Sh] G. Shi, *On the least number of fixed points and Nielsen numbers*, Acta Math. Sinica **16** (1966), 223–232.

[Z1] X. Zhao, *A relative Nielsen number for the complement*, Topological Fixed Point Theory and Applications, Lecture Notes in Math., vol. 1411, Springer, Berlin, 1990, pp. 189–199.

[Z2] ______, *Basic relative Nielsen numbers*, Topology — Hawaii, World Scientific, 1992, pp. 215–222.

[Z3] ______, *Minimal fixed point sets of relative maps*, Fund. Math. **162** (1999), 163–180.

CHAPTER IV

APPLICATIONS

Topological fixed point theory is very useful in many branches of mathematics, such as:

(a) control theory,
(b) nonlinear analysis,
(c) game theory,
(d) mathematical economics.

The first time a direct application of degree theory to differential equations was presented by J. Leray and J. P. Schauder in 1934.

In this chapter several important applications are briefly discussed.

19. APPLICABLE FIXED POINT PRINCIPLES

Jan Andres

1. Introduction

The main purpose of our contribution is to present, according to our opinion, the most powerful topological fixed point principles applicable in the theories of differential equations (inclusions) and (multivalued) dynamical systems. The solutions of given initial and boundary value problems can be usually represented as fixed points of related single-valued or multivalued operators in Banach or Fréchet spaces. These operators are mostly compact or condensing and admissible in the sense of L. Górniewicz, provided they are multivalued. Since harmonic (or subharmonic) periodic solutions can be also determined by fixed (periodic) points of the associated Poincaré operators which are again, in the lack of uniqueness, admissible, we decided to introduce topological tools (Lefschetz number, Nielsen number and fixed point index) just for admissible compact and, more generally, condensing maps in Fréchet spaces. More precisely, retracts of open subsets of (convex sets in) Fréchet spaces, i.e. in particular ANRs, are taken into account. Nevertheless, one can take directly ANRs or even Fréchet spaces can be replaced, under slight restrictions, by (nonmetric) locally convex spaces, as in our monograph [AG].

In this frame, the Conley index theory dealing mainly with invariant sets should also appear with no objections. Instead, we however decided to present here another invariant called the Conley type (integer-valued) index which was stimulated by our interest in (multivalued) fractals (see [AF], [AFGL], [AG]), considered as fixed points of special (so called the Hutchinson–Barnsley) induced operators in hyperspaces or, equivalently, as compact invariant sets of the inducing (Hutchinson–Barnsley) maps. Since this index is just defined by means of a fixed point index (Lefschetz number or Nielsen number) for induced single-valued maps in hyperspaces being ANRs, we prefered this elementary "transcription" to the standard Conley index theory, because we can express in this way formally extremely simple existence and multiplicity criteria for compact invariant and periodic sets. As an almost immediate application, continuation principles for fractals can be formulated.

Because of a given limit, applications of all developed principles are only mentioned in the form of remarks and comments at the end of single sections, jointly with some other relevant informations like the standard reference sources, etc. The material for our contribution was selected and elaborated w.r.t. the indicated applications from our monograph [AG] with L. Górniewicz and quoted joint papers [AGJ3], [AGJ4] with L. Górniewicz and J. Jezierski (Section 3); [An2] and [AGJ1]–[AGJ4], [AV2] with L. Górniewicz, J. Jezierski and M. Väth (Section 4); [ABa], [AGG1]–[AGG3], [AV1] with R. Bader, G. Gabor, L. Górniewicz and M. Väth (Section 5). Section 6 (partly motivated by the paper [DPS]) is formally new. The collaboration and fruitful discussions with all mentioned colleagues and friends are highly appreciated.

2. Preliminaries

In the entire text, all spaces are metric and by a (multivalued) map $\varphi\colon X \multimap Y$, i.e. $\varphi\colon X \to 2^Y \setminus \{\emptyset\}$, we mean the one with nonempty, closed values. The reader is supposed to be familiar with the elements of degree theory and algebraic topology. Otherwise, we recommend the monographs [AG], [Go2] and [Sp].

A *Fréchet space* is a complete, metrizable, locally convex space. Its topology can be generated by a family of seminorms. If it is normable, then it becomes Banach.

By AR (or ANR) we denote, as usual, the class of *absolute retracts* (or *absolute neighbourhood retracts*), namely X is an AR (or ANR) if each embedding $h\colon X \to Y$, i.e. $h\colon X \to h(X)$ is a homeomorphism, into a metrizable space Y, such that $h(X) \subset Y$ is closed, is a retract (or a neighbourhood retract) of Y. It is well-known that every ANR is a retract of some open subset of a normed space and that every retract of an open subset of a convex set in a Fréchet space is an ANR. Furthermore, every AR is *contractible*, i.e. homotopically equivalent to a one point space, and every ANR X is *locally contractible*, namely locally contractible in each of its points $x \in X$ which means that, for every $\varepsilon > 0$, there exists $\delta > 0$ ($\delta < \varepsilon$) such that the ball $B(x,\delta)$ is contractible in $B(x,\varepsilon)$. If there exists a decreasing sequence $\{X_n\}$ of compact, contractible sets X_n such that $X = \bigcap\{X_n : n = 1, 2\dots\}$, then X is called an *R_δ-set.* Let us note that any R_δ-set is *acyclic* w.r.t. any continuous theory of homology (e.g. the Čech homology), i.e. homologically equivalent to a one point space, and so any R_δ-set is nonempty, compact and connected. The following hierarchies hold for metric spaces:

$$\begin{array}{c} \text{contractible} \subset \text{acyclic} \\ \cup \\ \text{convex} \subset \text{AR} \subset \text{ANR} \end{array}$$

compact, convex $\subset$ compact AR $\subset$ compact, contractible $\subset R_\delta \subset$ compact, acyclic, and all the above inclusions are proper.

A map $\varphi\colon X \multimap Y$ is said to be *upper semicontinuous* (u.s.c.) if, for every open $U \subset Y$, the set $\{x \in X : \varphi(x) \subset U\}$ is open in X. It is said to be *lower semicontinuous* (l.s.c.) if, for every open $U \subset Y$, the set $\{x \in X : \varphi(x) \cap U \neq \emptyset\}$ is open in X. If it is both u.s.c. and l.s.c. then it is called *continuous.* A compact-valued map $\varphi\colon X \multimap Y$, or equivalently $\varphi\colon X \to \mathcal{K}(Y) := \{K \subset Y : K \text{ is compact}\}$, is continuous if and only if it is *Hausdorff-continuous*, i.e. continuous w.r.t. the metric d in X and the Hausdorff-metric d_H in $\mathcal{B}(Y) := \{D \subset Y : D \text{ is nonempty and bounded}\}$, where $d_H(A, B) := \inf\{\varepsilon > 0 : A \subset O_\varepsilon(B) \text{ and } B \subset O_\varepsilon(A)\}$ and $O_\varepsilon(D) := \{x \in X : \exists y \in D : d(x, y) < \varepsilon\}$. The *hyperspace* $(\mathcal{K}(Y), d_H)$ will be employed in Section 6. Observe that every single-valued u.s.c. or l.s.c. map is continuous in the usual sense. The class of *compact* maps $\varphi\colon X \multimap Y$, i.e. $\overline{\varphi(X)}$ is compact, will be denoted by $\mathbb{K}(X, Y)$, or simply by $\mathbb{K}(X)$, provided φ is a self-map (an endomorphism).

Let (X, d) be a metric space and $\mathcal{B}(X)$ be the set of nonempty, bounded subsets of X, defined above. The function $\alpha\colon \mathcal{B} \to [0, \infty)$, where $\alpha[B] := \inf\{\delta > 0 : B \in \mathcal{B}$ admits a finite covering by the sets of diameter $\leq \delta\}$ is called the *Kuratowski measure of noncompactness* and the function $\gamma\colon \mathcal{B} \to [0, \infty)$, where $\gamma[B] := \inf\{\varepsilon > 0 : B \in \mathcal{B}$ has a finite ε-net$\}$, is called the *Hausdorff measure of noncompactness.* These measures of noncompactness (MNC) are related as follows: $\gamma[B] \leq \alpha[B] \leq 2\gamma[B]$. Moreover, they satisfy the following properties:

(2.1.1) $\mu[B] = 0 \Leftrightarrow \overline{B}$ is compact,
(2.1.2) $B_1 \subset B_2 \Rightarrow \mu[B_1] \leq \mu[B_2]$,
(2.1.3) $\mu[\overline{B}] = \mu[B]$,
(2.1.4) if $\{B_n\}$ is a decreasing sequence of nonempty, closed sets $B_n \in \mathcal{B}$ with $\lim_{n\to\infty} \mu[B_n] = 0$, then $\cap\{B_n : n = 1, 2, \dots\} \neq \emptyset$,
(2.1.5) $\mu[B_1 \cup B_2] = \max\{\mu[B_1], \mu[B_2]\}$,
(2.1.6) $\mu[B_1 \cap B_2] = \min\{\mu[B_1], \mu[B_2]\}$,
(2.1.7) $\mu\colon \mathcal{B} \to [0, \infty)$ is a seminorm, i.e. $\mu[\lambda B] = |\lambda|\, \mu[B]$ and $\mu[B_1 \cup B_2] \leq \mu[B_1] + \mu[B_2]$, for every $\lambda \in \mathbb{R}$ and $B, B_1, B_2 \in \mathcal{B}$, provided (X, d) is linear,

where μ denotes either α or γ.

Letting $\mu := \alpha$ or $\mu := \gamma$, a bounded mapping $\varphi\colon X \supset U \multimap X$, i.e. $\varphi(B) \in \mathcal{B}$, for $\mathcal{B} \ni B \subset U$, is said to be μ*-condensing* (shortly, *condensing*) if $\mu[\varphi(B)] < \mu[B]$, whenever $\mathcal{B} \ni B \subset U$ and $\mu[B] > 0$, or equivalently, if $\mu[\varphi(B)] \geq \mu[B]$ implies $\mu[B] = 0$, whenever $\mathcal{B} \ni B \subset U$. Analogously, a bounded mapping $\varphi\colon X \supset U \multimap X$ is said to be a k*-set contraction* w.r.t. μ (shortly, a k*-contraction* or a *set-contraction*) if $\mu[\varphi(B)] \leq k\mu[B]$, for some $k \in [0, 1)$, whenever $\mathcal{B} \ni B \subset U$.

Obviously, any set-contraction is condensing. Furthermore, compact self-maps or contractions with compact values (in vector spaces, also their sum) are well-known to be set-contractions, and so condensing. The class of condensing maps (k-contractions) $\varphi\colon X \supset U \multimap X$ will be denoted by $\mathbb{C}(U,X)$ ($\mathbb{C}_k(U,X)$), or simply by $\mathbb{C}(X)$ ($\mathbb{C}_k(X)$), provided $U = X$.

An u.s.c. map with R_δ-(acyclic) values will be called an R_δ-(acyclic) map. R_δ-maps $\varphi\colon X \multimap Y$ will be identified here with J-maps, written $\varphi \in J(X,Y)$. A map which is a finite composition of compact-valued acyclic maps is called *admissible* (in the sense of L. Górniewicz). Admissible maps can be equivalently defined as follows:

Assume that we have a diagram $X \overset{p}{\Longleftarrow} \Gamma \overset{q}{\longrightarrow} Y$, where $p\colon \Gamma \Rightarrow X$ is a *Vietoris map*, namely

(2.2.1) p is onto, i.e. $p(\Gamma) = X$,
(2.2.2) p is proper, i.e. $p^{-1}(K)$ is compact, for every compact $K \subset X$,
(2.2.3) $p^{-1}(x)$ is an acyclic set, for every $x \in X$, where acyclicity is understood in the sense of the Čech homology functor with compact carriers and coefficients in the field $\mathbb{Q}$ of rationals (for more details, see [Go2], [AG]),

and $q\colon \Gamma \to Y$ is a continuous map. The *admissible map* $\varphi(p,q)\colon X \multimap Y$ is induced by $\varphi(x) = q(p^{-1}(x))$, for every $x \in X$. We, therefore, identify the admissible map $\varphi(p,q)$ with the pair (p,q) called an *admissible pair*.

The class of admissible maps contains u.s.c. maps with convex and compact values, u.s.c. maps with contractible and compact values, R_δ-maps, acyclic maps with compact values and their compositions. Moreover, the class of admissible maps is, unlike the mentioned subclasses, closed under composition, i.e. composition of admissible maps remains admissible. Compact and condensing admissible maps are, therefore, extremely important in the topological fixed point theory.

3. Lefschetz number

We start with the Lefschetz theory, because it is a base for our further investigation. More precisely, the generalized Lefschetz number can be used for the definition of essential classes in the Nielsen theory as well as the possible normalization property of the fixed point index. Since, however, this theory is studied in detail in the contribution of L. Górniewicz in this handbook, we restrict ourselves only to the presentation of necessary facts.

3.1. Lefschetz number for compact maps. At first, we recall some algebraic preliminaries (for details, see [AG], [Sp]).

In what follows, all vector spaces are taken over $\mathbb{Q}$. Let $f\colon E \to E$ be an endomorphism of a finite-dimensional vector space E. If $v_1, \dots, v_n$ is a basis for E, then we can write

$$f(v_i) = \sum_{j=1}^{n} a_{ij} v_j, \quad \text{for all } i = 1, \ldots, n.$$

The matrix $[a_{ij}]$ is called the matrix of f (with respect to the basis $v_1, \ldots, v_n$). Let $A = [a_{ij}]$ be an $(n \times n)$-matrix; then the trace of A is defined as $\sum_{i=1}^{n} a_{ii}$. If $f\colon E \to E$ is an endomorphism of a finite-dimensional vector space E, then the trace of f, written $\mathrm{tr}(f)$, is the trace of the matrix of f with respect to some basis for E. If E is a trivial vector space then, by definition, $\mathrm{tr}(f) = 0$. It is a standard result that the definition of the trace of an endomorphism is independent of the choice of the basis for E. For more properties of the trace, see [AG].

Hence, let $E = \{E_q\}$ be a graded vector space of a finite type. If $f = \{f_q\}$ is an endomorphism of degree zero of such a graded vector space, then the (*ordinary*) *Lefschetz number* $\lambda(f)$ of f is defined by

$$\lambda(f) = \sum_q (-1)^q \mathrm{tr}(f_q).$$

Let $f\colon E \to E$ be an endomorphism of an arbitrary vector space E. Denote by $f^{(n)}\colon E \to E$ the nth iterate of f and observe that the kernels

$$\ker f \subset \ker f^{(2)} \subset \ldots \subset \ker f^{(n)} \subset \ldots$$

form an increasing sequence of subspaces of E. Let us now put

$$N(f) = \bigcup_n \ker f^{(n)} \quad \text{and} \quad \widetilde{E} = E/N(f).$$

Clearly, f maps $N(f)$ into itself and, therefore, induces the endomorphism

$$\widetilde{f}\colon \widetilde{E} \to \widetilde{E}$$

on the factor space $\widetilde{E} = E/N(f)$.

Let $f\colon E \to E$ be an endomorphism of a vector space E. Assume $\dim \widetilde{E} < \infty$. In this case, we define the generalized trace $\mathrm{Tr}(f)$ of f by putting $\mathrm{Tr}(f) = \mathrm{tr}(\widetilde{f})$.

(3.1) LEMMA. *Let $f\colon E \to E$ be an endomorphism. If $\dim E < \infty$, then $\mathrm{Tr}(f) = \mathrm{tr}(f)$.*

For the proof, see [AG].

Let $f = \{f_q\}$ be an endomorphism of degree zero of a graded vector space $E = \{E_q\}$. We say that f is a *Leray endomorphism* if the graded vector space $\widetilde{E} = \{\widetilde{E}_q\}$ is of finite type. For such an f, we define the (*generalized*) *Lefschetz number* $\Lambda(f)$ of f by putting

$$\Lambda(f) = \sum_q (-1)^q \mathrm{Tr}(f_q).$$

It is immediate from (3.1) that

(3.2) Lemma. *Let $f: E \to E$ be an endomorphism of degree zero, i.e. $f = \{f_q\}$ and $f_q: E_q \to E_q$ is a linear map. If E is a graded vector space of finite type, then $\Lambda(f) = \lambda(f)$.*

The following property of the Leray endomorphism is of an importance.

(3.3) Property. *Let*

$$\begin{array}{ccccccccccc} \cdots & \longrightarrow & E'_q & \longrightarrow & E_q & \longrightarrow & E''_q & \longrightarrow & E'_{q-1} & \longrightarrow & \cdots \\ & & \downarrow{\scriptstyle f'_q} & & \downarrow{\scriptstyle f_q} & & \downarrow{\scriptstyle f''_q} & & \downarrow{\scriptstyle f'_{q-1}} & & \\ \cdots & \longrightarrow & E'_q & \longrightarrow & E_q & \longrightarrow & E''_q & \longrightarrow & E'_{q-1} & \longrightarrow & \cdots \end{array}$$

be a commutative diagram of vector spaces in which the rows are exact. If two of the following endomorphisms $f = \{f_q\}$, $f' = \{f'_q\}$, $f'' = \{f''_q\}$ are the Leray endomorphisms, then so is the third, and, moreover, in that case we have:

$$\Lambda(f'') + \Lambda(f') = \Lambda(f).$$

For the proof, see [AG].

Besides the above property of the Leray endomorphisms, we also point out some information about weakly nilpotent endomorphisms.

(3.4) Definition. A linear map $f: E \to E$ of a vector space E into itself is called *weakly nilpotent* if, for every $x \in E$, there exists n_x such that $f^{n_x}(x) = 0$.

Observe that if $f: E \to E$ is weakly nilpotent, then $N(f) = E$, and so, we have:

(3.5) Lemma. *If $f: E \to E$ is weakly nilpotent, then* $\mathrm{Tr}(f)$ *is well-defined and* $\mathrm{Tr}(f) = 0$.

Assume that $E = \{E_q\}$ is a graded vector space and $f = \{f_q\}: E \to E$ is an endomorphism. We say that f is *weakly nilpotent* if f_q is weakly nilpotent, for every q.

From (3.5), we deduce:

(3.6) Lemma. *Any weakly nilpotent endomorphism $f: E \to E$ is a Leray endomorphism, and $\Lambda(f) = 0$.*

Now, the Lefschetz number will be defined for admissible mappings. Namely, let $\varphi: X \multimap X$ be an admissible map and $(p, q) \subset \varphi$ be a selected pair of φ. Then the induced homomorphism $q_* \circ p_*^{-1}: H_*(X) \to H_*(X)$ is an endomorphism of the graded vector space $H_*(X)$ into itself. So, we can define the Lefschetz number $\Lambda(p, q)$ of the pair (p, q) by putting $\Lambda(p, q) = \Lambda(q_* \circ p_*^{-1})$, provided the Lefschetz number $\Lambda(q_* \circ p_*^{-1})$ is well-defined.

It allows us to define the Lefschetz set Λ of φ as $\Lambda(\varphi) = \{\Lambda(p,q) : (p,q) \subset \varphi\}$. In what follows, we say that the Lefschetz set $\Lambda(\varphi)$ of φ is *well-defined* if, for every $(p,q) \subset \varphi$, the Lefschetz number $\Lambda(p,q)$ of (p,q) is defined.

Moreover, from the homotopy property, we get:

(3.7) LEMMA.

(3.7.1) *If $\varphi, \psi: X \multimap X$ are homotopic ($\varphi \sim \psi$), then: $\Lambda(\varphi) \cap \Lambda(\psi) \neq \emptyset$.*

(3.7.2) *If $\varphi: X \multimap X$ is an admissible map and X is acyclic, then the Lefschetz set $\Lambda(\varphi)$ is well-defined and $\Lambda(\varphi) = \{1\}$.*

It is useful to formulate

(3.8) THEOREM (Coincidence Theorem, see [AG, (6.12)]). *Let U be an open subset of a finite dimensional normed space E. Consider the following diagram:*

$$U \overset{p}{\Longleftarrow} \Gamma \overset{q}{\longrightarrow} U$$

in which q is a compact map. Then the Lefschetz number $\Lambda(p,q)$ of the pair (p,q), given by the formula $\Lambda(p,q) = \Lambda(q_* \circ p_*^{-1})$, is well-defined, and $\Lambda(p,q) \neq 0$ implies that $p(y) = q(y)$.

Theorem (3.8) can be reformulated in terms of multivalued mappings as follows.

Let $U \subset E$ be the same as in (3.8) and let $\varphi: U \multimap U$ be a compact, admissible map, i.e. $\varphi \in \mathbb{K}(U)$. We let $\Lambda(\varphi) = \{\Lambda(p,q) : (p,q) \subset \varphi\}$, where $\Lambda(p,q) = \Lambda(q_* p_*^{-1})$. Then we have:

(3.9) THEOREM.

(3.9.1) *The set $\Lambda(\varphi)$ is well-defined, i.e. for every $(p,q) \subset \varphi$, the generalized Lefschetz number $\Lambda(p,q)$ of the pair (p,q) is well-defined, and*

(3.9.2) *$\Lambda(\varphi) \neq \{0\}$ implies that the set* $\mathrm{Fix}(\varphi) = \{x \in U : x \in \varphi(x)\}$ *is nonempty.*

Now, we shall generalize (3.8) to the case, when $\varphi \in \mathbb{K}(U)$ and U is an open subset of a Fréchet space E. To do it, we recall some additional notions and facts.

Two maps $\varphi, \psi: X \multimap Y$ are called *α-close*, $\alpha \in \mathrm{Cov}(Y)$, i.e. an open covering of Y, if, for every $x \in X$, there exists $U_x \in \alpha$ such that $\varphi(x) \cap U_x \neq \emptyset$ and $\psi(x) \cap U_x \neq \emptyset$.

Assume that $X \subset Y$ and $\alpha \in \mathrm{Cov}(Y)$. A point $x \in X$ is called an *α-fixed point* of $\varphi: X \multimap Y$ if there exists $U_x \in \alpha$ such that $x \in U_x$ and $\varphi(x) \cap U_x \neq \emptyset$.

We prove:

(3.10) LEMMA. *Let $\varphi \in \mathbb{K}(X)$ and $D \subset \mathrm{Cov}(X)$ be a cofinal family of open coverings. If, for every $\alpha \in D$, there exists an α-fixed point of φ, then* $\mathrm{Fix}(\varphi) =$ *$\{x \in X : x \in \varphi(x)\}$ is a nonempty set.*

PROOF. Assume, on the contrary, that $\mathrm{Fix}(\varphi) = \emptyset$. Then $x \notin \varphi(x)$, for every $x \in X$. Since we consider X to be a vector space, i.e. a $T_{3\frac{1}{2}}$-space, there are U_x and V_x such that $x \in U_x$, $\varphi(x) \subset V_x$ and $U_x \cap V_x = \emptyset$. Observe that the set $\{x \in X : \varphi(x) \subset V_x\}$ is an open neighbourhood of x (we have assumed that φ is u.s.c.).

Let $W_x = U_x \cap \{x \in X : \varphi(x) \subset V_x\}$. Then $\beta = \{W_x\}_{x \in X}$ is an open covering of X. Since D is cofinal in $\mathrm{Cov}(X)$, there exists $\gamma = \{O_x\} \in D$ such that $\gamma \geq \beta$. Then φ has no γ-fixed point, and we obtain a contradiction. □

(3.11) LEMMA. *Let U be an open subset of a Fréchet space E and let $\varphi \in \mathbb{K}(U)$. Then:*

(3.11.1) *the set $\Lambda(\varphi)$ is well-defined, and*

(3.11.2) $\Lambda(\varphi) \neq \{0\}$ *implies* $\mathrm{Fix}(\varphi) \neq \emptyset$.

PROOF. Let $(p, q) \subset \varphi$ and assume that: $U \overset{p}{\Leftarrow} \Gamma \overset{q}{\longrightarrow} U$, where p is a Vietoris map and q is compact.

Therefore, there exists a compact $K \subset U$ such that $q(\Gamma) \subset K$. Let $\alpha \in \mathrm{Cov}_U(K)$, i.e. an open covering of K in U. We choose $\beta \in \mathrm{Cov}_U(K)$, according to Lemma (3.10). Now, applying the Schauder-like approximation theorem (I.1.37) in [AG] (cf. Lemma (5.24) below), we obtain a finite dimensional map $\pi_\beta\colon K \to U$ which is β-close to the inclusion map $i\colon K \to U$. Then the map $q_\beta = \pi_\beta \circ q\colon \Gamma \to U$ satisfies the following conditions (for more details, see [AG]):

(3.11.3) q_β is compact,

(3.11.4) q and q_β are β-close,

(3.11.5) q and q_β are stationary α-homotopic,

(3.11.6) $\overline{q_\beta(\Gamma)} \subset U_\beta$, where $U_\beta = U \cap E^n$ and E^n is a finite dimensional subspace of E chosen according to the approximation theorem.

Summing up the above, we get the following commutative diagram:

$$\begin{array}{ccc} H_*(U_\beta) & \xrightarrow{\;j_*\;} & H_*(U) \\ {\scriptstyle (\overline{q}_\beta)_*(p_\beta)_*^{-1}} \Big\uparrow & \nwarrow {\scriptstyle (q'_\beta)_* p_*^{-1}} & \Big\uparrow {\scriptstyle q_{\beta*} p_*^{-1}} \\ H_*(U_\beta) & \xrightarrow[\;j_*\;]{} & H_*(U) \end{array}$$

in which $j\colon U_\beta \to U$ is the inclusion map and $\overline{q}_\beta$, p_β, q'_β are the respective restrictions of q_β and p.

Now, from the commutativity of the above diagram and (3.8), we get

$$\Lambda((\overline{q}_\beta)_*(p_\beta)_*^{-1}) = \Lambda(q_* p_*^{-1})$$

which proves assertion (3.11.1).

If we assume that $\Lambda(q_* p_*^{-1}) \neq 0$ for some $(p,q) \subset \varphi$, then $\Lambda((\overline{q}_\beta)_*(p_\beta)_*^{-1}) \neq 0$ and, in view of (3.11.1), we get that φ has an α-fixed point. Consequently, our claim follows from (3.10), and the proof is completed. □

We are ready to give the main theorem of this section.

(3.12) THEOREM (The Lefschetz Fixed Point Theorem). *Let X be a retract of an open subset U of (a convex set in; cf. [AG]) a Fréchet space E. Assume, furthermore, that $\varphi \in \mathbb{K}(X)$. Then:*

(3.12.1) *the Lefschetz set $\Lambda(\varphi)$ of φ is well-defined,*
(3.12.2) *if $\Lambda(\varphi) \neq \{0\}$, then* $\mathrm{Fix}(\varphi) \neq \emptyset$.

PROOF. Let $r: U \to X$ be the retraction map and $i: X \to U$ the inclusion map. We have the following commutative diagram:

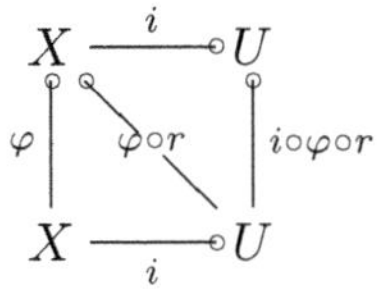

Consequently, by using (3.11), we obtain our theorem (observe that $i\circ\varphi\circ r: U \multimap U$ is a compact map); the proof is completed. □

3.2. Relative Lefschetz number. In this section, by X and A we shall denote retracts of some open subsets of (convex sets in) a Fréchet space E. We shall also assume that $A \subset X$. By a multivalued mapping, we shall again understand a pair (p,q), where:

$$X \overset{p}{\Longleftarrow} \Gamma \overset{q}{\longrightarrow} Y,$$

but we shall also consider multivalued maps of pairs of spaces, i.e. the diagrams of the following type:

$$(X,A) \overset{p}{\Longleftarrow} (\Gamma,\Gamma_0) \overset{q}{\longrightarrow} (Y,B)$$

and, as before, we shall use the following notation

$$(p,q): (X,A) \multimap (Y,B) \quad \text{or} \quad \varphi = \varphi(p,q): (X,A) \multimap (Y,B).$$

For such a pair, we let $(p,q)_X: X \multimap Y$ and $(p,q)_A: A \multimap B$, where $(p,q)_X = (p_\Gamma, q_\Gamma)$ and $(p,q)_A = (p_{\Gamma_0}, q_{\Gamma_0})$.

For a multivalued map $(p,q): (X,A) \multimap (X,A)$, we define the *generalized Lefschetz number* $\Lambda((p,q))$ by putting

$$\Lambda((p,q)) = \Lambda(q_* \circ p_*^{-1});$$

see (3.8) in the foregoing section.

(3.13) DEFINITION. A multivalued map $(p,q)\colon (X,A) \multimap (X,A)$ is called *compact* if both $(p,q)_X$ and $(p,q)_A$ are compact.

As a direct consequence of (3.3), we get:

(3.14) PROPOSITION. *Let $(p,q)\colon (X,A) \multimap (X,A)$. If for any two maps of: (p,q), $(p,q)_X$, $(p,q)_A$, the generalized Lefschetz number is well-defined, then so is the third, and in that cases we have:*

$$\Lambda((p,q)) = \Lambda((p,q)_X) - \Lambda((p,q)_A).$$

Now, we are able to formulate the relative version of the Lefschetz fixed point theorem:

(3.15) THEOREM (Relative Version of the Lefschetz Fixed Point Theorem). *If $(p,q)\colon (X,A) \multimap (X,A)$ is a compact admissible map, then:*

(3.15.1) *is well-defined, and*
(3.15.2) *$x \in \mathrm{Fix}((p,q))$ such that $x \in \overline{X \setminus A}$.*

For the proof, see [AG].

Observe that from (3.15) we get not only the existence of a fixed point, but also its localization.

3.3. Lefschetz number for periodic points. Let $E = \{E_q\}$ be a graded vector space over $\mathbb{Q}$ and let $f\colon E \to E$ be a Leray endomorphism. For such f, we have defined the generalized Lefschetz number $\Lambda(f)$. Now, we shall define, for such f, the *generalized Euler characteristic* $\chi(f)$ by putting:

$$\chi(f) = \sum_q (-1)^q \dim(\widetilde{E}_q),$$

where the graded vector space $\widetilde{E} = \{\widetilde{E}_q\}$ is constructed in Section 3.1.

The following proposition is a direct consequence of our definitions:

(3.16) PROPOSITION. *Let $f\colon E \to E$ be an endomorphism and $f^n\colon E \to E$ denote its n-iterate, i.e.*

$$f^n = \underbrace{f \circ \ldots \circ f}_{n\text{-times}}.$$

We have: f is a Leray endomorphism if and only if f^n is a Leray endomorphism and, in this case, we obtain: $\chi(f) = \chi(f^n)$. Let $\mathbb{Q}\{x\}$ denote the integral domain of all formal power series:

$$S = a_0 + a_1 x + a_2 x^2 + \ldots = \sum_{n=0}^{\infty} a_n x^n$$

with coefficients $a_n \in \mathbb{Q}$. Then $\mathbb{Q}\{x\}$ contains the polynomial ring $\mathbb{Q}[x]$.

(3.17) DEFINITION. Let $f\colon E \to E$ be a Leray endomorphism. The *Lefschetz power series* $L(f)$ of f is an element of $\mathbb{Q}\{x\}$ defined by:

$$L(f) = \chi(f) + \sum_{n=0}^{\infty} \Lambda(f^n)x^n.$$

It is well-known (see [Bo2], [AGJ4]) that $L(f) = u \cdot v^{-1}$, where u and v are relatively prime polynomials with $\deg u < \deg v$ ($u \neq 0$). We let

$$P(f) = \deg v, \quad \text{where } L(f) = u \cdot v^{-1}. \tag{3.18}$$

The above can be summarized as follows:

(3.19) PROPOSITION. *Let $f\colon E \to E$ be a Leray endomorphism. Then we have:*

(3.19.1) $\chi(f) \neq 0$ *implies* $P(f) \neq 0$,
(3.19.2) $P(f) \neq 0 \Leftrightarrow \Lambda(f^n) \neq 0$, *for some* n,
(3.19.3) $P(f) \neq 0$, *then for any natural* m *one of the numbers:*

$$\Lambda(f^{m+1}), \Lambda(f^{m+2}), \ldots, \Lambda(f^{m+P(f)})$$

is different from zero.

We recall that an admissible map $(p,q)\colon X \multimap X$ ($(p,q)\colon (X,A) \multimap (X,A)$) is called a Lefschetz map if $(p,q)_*\colon H(X) \to H(X)$ ($(p,q)_*\colon H(X,A) \to H(X,A)$) is a Leray endomorphism (see Sections 3.1 and 3.2). We shall consider, for the sake of simplicity, admissible maps of the form $(p,q)\colon X \multimap X$, but all formulations remain true for pairs of spaces. We let:

$$\begin{aligned} \Lambda(p,q) &= \Lambda((p,q)_*), & \chi(p,q) &= \chi((p,q)_*), \\ L(p,q) &= L((p,q)_*), & P(p,q) &= P((p,q)_*), \end{aligned}$$

whenever (p,q) is a Lefschetz map. Then Proposition (3.16) can be reformulated as follows:

(3.20) PROPOSITION. *An admissible map $(p,q)\colon X \multimap X$ is a Lefschetz map if and only if so is any its iterate. In this case, we have:*

$$\chi(p,q) = \chi((p,q)^n), \quad \text{where } (p,q)^n = \underbrace{(p,q) \circ \ldots \circ (p,q)}_{n\text{-times}}.$$

Finally, the preceding discussion can be summarized as follows (cf. Proposition (3.19)).

(3.21) THEOREM. *Let* $(p,q)\colon X \multimap X$ *be a Lefschetz map. We have:*

(3.21.1) $\chi(p,q) \neq 0$ *implies* $P(p,q) \neq 0$;

(3.21.2) $P(p,q) \neq 0$ *if and only if* $\Lambda((p,q)^n) \neq 0$, *for some natural* n;

(3.21.3) $P(p,q) = k \neq 0$ *implies that, for any natural* m, *at least one of the coefficients*

$$\Lambda((p,q)^{m+1}), \Lambda((p,q)^{m+2}), \ldots, \Lambda((p,q)^{m+k})$$

of the series $L(p,q)$ *must be different from zero.*

For a multivalued map $(p,q)\colon (X,A) \multimap (X,A)$, a point $x \in X$ is called n-*periodic* if $x \in \mathrm{Fix}((p,q)^n)$.

According to the foregoing Section 3.2, by X and A we shall denote retracts of some open subsets of (convex sets in) a Fréchet space E. We shall also assume that $A \subset X$.

We are able to prove:

(3.22) THEOREM (Periodic Point Theorem). *Let*

$$(p,q)\colon X \multimap X \quad ((p,q)\colon (X,A) \multimap (X,A))$$

be a compact map. If $\chi(p,q) \neq 0$ *or* $P(p,q) \neq 0$, *then* (p,q) *has an* n-*periodic point* (*an* n-*periodic point in* $\overline{X \setminus A}$), *for some* $m+1 \leq n \leq m + P(p,q)$ *and arbitrary* $m \geq 0$.

PROOF. It follows from (3.12) (resp. (3.15)) that $\Lambda(p,q)$ is well-defined. In view of (3.21), it is sufficient to assume that $P(p,q) \neq 0$. Then, using (3.12) (resp. (3.15)), we deduce that $\mathrm{Fix}((p,q)^n) \neq \emptyset$ (and $\mathrm{Fix}((p,q)^n) \cap \overline{X \setminus A} \neq \emptyset$). Observe that every point $x \in \mathrm{Fix}((p,q)^n)$ is n-periodic for (p,q). The proof is completed. □

3.4. Lefschetz number for condensing maps. The definition of the Lefschetz number for condensing maps is far from to be obvious. We indicate one construction requiring the notion of a special neighbourhood retract.

(3.23) DEFINITION. A closed bounded subset X of a Fréchet space E is called a *special neighbourhood retract* (written, $X \in \mathrm{SNR}(E)$) if there exists an open subset U of (a convex set in) E such that $X \subset U$ and a (continuous) retraction $r\colon U \to X$ such that: $\mu(r(A)) \leq \mu(A)$, for every $A \subset U$, where μ is a measure of noncompactness.

(3.24) THEOREM. *Assume that* $X \in \mathrm{SNR}(E)$ *and* $\varphi\colon X \multimap X$ *is a condensing admissible map. Then:*

(3.24.1) *the Lefschetz set* $\Lambda(\varphi)$ *of* φ *is well-defined,*

(3.24.2) *if* $\Lambda(\varphi) \neq \{0\}$, *then* $\mathrm{Fix}(\varphi) \neq \emptyset$.

For the proof, see [AG].

(3.25) COROLLARY. *Let $X \in \mathrm{SNR}(E)$ and φ be a k-set contraction. Then:*

(3.25.1) *the Lefschetz set $\Lambda(\varphi)$ of φ is well-defined,*
(3.25.2) *if $\Lambda(\varphi) \neq \{0\}$, then $\mathrm{Fix}(\varphi) \neq \emptyset$.*

Corollary (3.25) can be obtained from (3.24), because every k-set contraction mapping is condensing.

(3.26) COROLLARY. *Let $X \in \mathrm{SNR}(E)$, for $U = E$ (see Definition (3.23)). If φ is condensing or, in particular, a k-set contraction, then φ has a fixed point.*

In fact, in the considered case, we obtain that $\Lambda(\varphi) \neq \{0\}$, and (3.26) follows from (3.24).

3.5. Remarks and comments. The standard references for the Lefschetz fixed point theory are, in the single-valued case [Br1], [Gr] and, in the multivalued case, [Go1], [Go2]. In [AG], we have generalized it for (not necessarily metric) locally convex spaces. The origins of the relative and the periodic Lefschetz theories are connected with the name of C. Bowszyc (see [Bo1], [Bo2]). A suitable definition of the Lefschetz number for condensing maps on nonsimply connected sets remains an open problem (cf. [Go3]).

Many applications of the Lefschetz type theorems to differential equations and inclusions can be found in [AG]. Since the celebrated Schauder fixed point theorem is a special case of the Lefschetz theorem on ARs, its huge amount of applications to boundary value problems in standard text-books and monographs can be also regarded as the one of the Lefschetz theory. Less traditional applications, e.g. to the existence problems of bounded and almost-periodic solutions of differential inclusions, can be found in [An1], [ABe], [Sr].

4. Nielsen number

The standard Nielsen number provides a lower estimate of the number of fixed points of a given (single-valued) self-map. On the other hand, we have constructed in [AGJ1] (cf. [AG]) the example showing that, in the multivalued case, it provides rather a lower estimate of the number of coincidences than the fixed points.

4.1. Nielsen number for compact maps. Let

$$X \overset{p_0}{\Longleftarrow} \Gamma_0 \overset{q_0}{\longrightarrow} Y \qquad \text{and} \qquad X \overset{p_1}{\Longleftarrow} \Gamma_1 \overset{q_1}{\longrightarrow} Y$$

be two maps. We say that (p_0, q_0) is *homotopic* to (p_1, q_1) (written $(p_0, q_0) \sim (p_1, q_1)$) if there exists a multivalued map

$$X \times I \overset{p}{\longleftarrow} \overline{\Gamma} \overset{q}{\longrightarrow} Y, \quad I = [0, 1],$$

such that the following diagram is commutative:

$$\begin{array}{ccccc} X & \stackrel{p_i}{\Longleftarrow} & \Gamma_i & \stackrel{q_i}{\longrightarrow} & Y \\ {\scriptstyle k_i}\downarrow & & \downarrow{\scriptstyle f_i} & \nearrow{\scriptstyle q} & \\ X \times I & \underset{p}{\Longleftarrow} & \overline{\Gamma} & & \end{array}$$

for $k_i(x) = (x, i)$, $i = 0, 1$, and $f_i\colon \Gamma_i \to \overline{\Gamma}$ is a homeomorphism onto $p^{-1}(X \times i)$, $i = 0, 1$, i.e. $k_0 p_0 = p f_0$, $q_0 = q f_0$, $k_1 p_1 = p f_1$ and $q_1 = q f_1$.

If $(p_0, q_0) \sim (p_1, q_1)$ and $h\colon Y \to Z$ is a continuous map, then we write $(p_0, hq_0) \sim (p, hq)$. We say that a multivalued map $X \stackrel{p}{\Longleftarrow} \Gamma \stackrel{q}{\longrightarrow} Y$ *represents a single-valued map* $\rho\colon X \to Y$ if $q = p\rho$. Now, we assume that $X = Y$ and we are going to estimate the cardinality of the coincidence set $C(p, q) := \{z \in \Gamma : p(z) = q(z)\}$. We begin by defining a Nielsen type relation on $C(p, q)$. This definition requires the following conditions on $X \stackrel{p}{\Longleftarrow} \Gamma \stackrel{q}{\longrightarrow} Y$:

(4.1) X, Y are connected, locally contractible metric spaces (observe that then they admit universal coverings),

(4.2) $p\colon \Gamma \Rightarrow X$ is a Vietoris map,

(4.3) for any $x \in X$, the restriction $q_1 = q|_{p^{-1}(x)}\colon p^{-1}(x) \to Y$ admits a lift $\widetilde{q}_1$ to the universal covering space $(p_Y\colon \widetilde{Y} \to Y)$:

$$\begin{array}{ccc} & & \widetilde{Y} \\ & \stackrel{\widetilde{q}_1}{\nearrow} & \Big\Downarrow{\scriptstyle p_Y} \\ p^{-1}(x) & \underset{q_1}{\longrightarrow} & Y \end{array}$$

Consider a single-valued map $\rho\colon X \to Y$ between two spaces admitting universal coverings $p_X\colon \widetilde{X} \Rightarrow X$ and $p_Y\colon \widetilde{X} \Rightarrow Y$. Let $\theta_X = \{\alpha : \widetilde{X} \to \widetilde{X} : p_X \alpha = p_X\}$ be the group of natural transformations of the covering p_X. Then the map ρ admits a lift $\widetilde{\rho}\colon \widetilde{X} \to \widetilde{Y}$. We can define a homomorphism $\widetilde{\rho}_!\colon \theta_X \to \theta_Y$ by the equality

$$\widetilde{q}(\alpha \cdot \widetilde{x}) = \widetilde{q}_!(\alpha)\widetilde{q}(\widetilde{x}) \quad (\alpha \in \theta_X,\ \widetilde{x} \in \widetilde{X}).$$

It is well-known (e.g. [Sp]) that there is an isomorphism between the fundamental group $\pi_1(X)$ and θ_X which may be described as follows. We fix points $x_0 \in X$, $\widetilde{x} \in \widetilde{X}$ and a loop $\omega\colon I \to X$ based at x_0. Let $\widetilde{\omega}$ denote the unique lift of ω starting from $\widetilde{x}_0$. We subordinate to $[\omega] \in \pi_1(X, x_0)$ the unique transformation from θ_X sending $\widetilde{\omega}(0)$ to $\widetilde{\omega}(1)$. Then the homomorphism $\widetilde{\rho}_!\colon \theta_X \to \theta_Y$ corresponds to the induced homomorphism between the fundamental groups $\rho_{\#}\colon \pi_1(X, x_0) \to \pi_1(Y, \rho(x_0))$.

We will show that, under the assumptions (4.1)–(4.3), a multivalued map (p, q) admits a lift to a multivalued map between the universal coverings. These lifts will split the coincidence set $C(p, q)$ into Nielsen classes. Besides that, we will also show that the pair (p, q) induces a homomorphism $\theta_X \to \theta_Y$ giving the Reidemeister set in this situation.

We start with the following lemma.

(4.4) LEMMA. *Suppose we are given Y, a locally contractible metric space, Γ a metric space, $\Gamma_0 \subset \Gamma$ a compact subspace, $q: \Gamma \to Y$, $\widetilde{q}_0: \Gamma_0 \to Y$ continuous maps for which the diagram*

$$\begin{array}{ccc} \Gamma_0 & \xrightarrow{\widetilde{q}_0} & \widetilde{Y} \\ {\scriptstyle i}\big\downarrow & & \big\downarrow{\scriptstyle p_Y} \\ \Gamma & \xrightarrow[q]{} & Y \end{array}$$

commutes (here, $p_Y: \widetilde{Y} \to Y$ denotes the universal covering). In other words, $\widetilde{q}_0$ is a partial lift of q. Then $\widetilde{q}_0$ admits an extension to a lift onto an open neighbourhood of Γ_0 in Γ.

For the proof, see [AGJ1] (cf. [AG]).

Consider again a multivalued map $X \xleftarrow{p} \Gamma \xrightarrow{q} Y$ satisfying (4.1). Define (a pullback)

$$\widetilde{\Gamma} = \{(\widetilde{x}, z) \in \widetilde{X} \times \Gamma : p_X(\widetilde{x}) = p(z)\}.$$

Now, we can apply Lemma (4.4) to the multivalued map $\widetilde{X} \overset{\widetilde{p}}{\Longleftarrow} \widetilde{\Gamma} \xrightarrow{q p_\Gamma} Y$, and so we get a lift $\widetilde{q}: \widetilde{\Gamma} \to \widetilde{Y}$ such that the diagram

$$\begin{array}{ccccc} \widetilde{X} & \overset{\widetilde{p}}{\Longleftarrow} & \widetilde{\Gamma} & \xrightarrow{\widetilde{q}} & \widetilde{Y} \\ {\scriptstyle p_X}\big\downarrow & & {\scriptstyle p_\Gamma}\big\downarrow & & \big\downarrow{\scriptstyle p_Y} \\ X & \underset{p}{\Longleftarrow} & \Gamma & \xrightarrow[q]{} & Y \end{array}$$

is commutative, where $\widetilde{p}(\widetilde{x}, z) = \widetilde{x}$ and $p_\Gamma(\widetilde{x}, z) = z$. Let us note that the lift $\widetilde{p}$ is given by the above formula, but $\widetilde{q}$ is not precised. We fix such a $\widetilde{q}$.

Observe that $p: \Gamma \Longrightarrow X$ and the lift $\widetilde{p}$ induce a homomorphism $\widetilde{p}^!: \theta_X \to \theta_\Gamma$ by the formula $\widetilde{p}^!(\alpha)(\widetilde{x}, z) = (\alpha\widetilde{x}, z)$. It is easy to check that the homomorphism $\widetilde{p}^!$ is an isomorphism (any natural transformation of $\widetilde{\Gamma}$ is of the form $\alpha \cdot (\widetilde{x}, z) = (\alpha\widetilde{x}, z)$) and that $\widetilde{p}^!$ is inverse to $\widetilde{p}_!$. Recall that the lift $\widetilde{q}$ defines a homomorphism $\widetilde{q}_!: \theta_\Gamma \to \theta_Y$ by the equality $\widetilde{q}(\lambda) = \widetilde{q}_!(\lambda)\widetilde{q}$.

In the sequel, we will consider the composition $\widetilde{q}_!\widetilde{p}^!: \theta_X \to \theta_Y$.

(4.5) LEMMA. *Let a multivalued map (p, q) satisfying (4.1) represent a single-valued map ρ, i.e. $q = \rho p$. Let $\widetilde{\rho}$ be the lift of ρ which satisfies $\widetilde{q} = \widetilde{\rho}\widetilde{p}$. Then $\widetilde{\rho}_!\widetilde{p}^! = \rho_!$.*

PROOF. $\widetilde{\rho}_!\widetilde{p}^! = (\widetilde{\rho}\widetilde{p})_!\widetilde{p}^! = \widetilde{\rho}_!\widetilde{p}_!\widetilde{p}^! = \rho_!$. □

Now, we are in a position to define the Nielsen classes. Consider a multivalued self-map $X \xleftarrow{p} \Gamma \xrightarrow{q} X$ satisfying (4.1). By the above consideration, we have a commutative diagram

$$\begin{array}{ccc} \widetilde{\Gamma} & \xrightarrow{\widetilde{p},\widetilde{q}} & \widetilde{X} \\ {\scriptstyle p_\Gamma}\downarrow & & \downarrow{\scriptstyle p_X} \\ \Gamma & \xrightarrow[p,q]{} & X \end{array}$$

Following the single-valued case (see e.g. [J]), we can prove (see [AG])

(4.6) LEMMA.

(4.6.1) $C(p, q) = \bigcup_{\alpha \in \theta_X} p_\Gamma C(\widetilde{p}, \alpha\widetilde{q})$,

(4.6.2) *if $p_\Gamma C(\widetilde{p}, \alpha\widetilde{q}) \cap p_\Gamma C(\widetilde{p}, \beta\widetilde{q})$ is not empty, then there exists a $\gamma \in \theta_X$ such that $\beta = \gamma \circ \alpha \circ (\widetilde{q}_!\widetilde{p}^!\gamma)^{-1}$,*

(4.6.3) *the sets $p_\Gamma C(\widetilde{p}, \alpha\widetilde{q})$ are either disjoint or equal.*

Define an action of θ_X on itself by the formula $\gamma \circ \alpha = \gamma\alpha(\widetilde{q}_!p^!\gamma)$. The quotient set will be called the *set of Reidemeister classes* and will be denoted by $R(p, q)$. The above lemma defines an injection: set of Nielsen classes $\to R(p, q)$, given by $A \to [\alpha] \in R(p, q)$, where $\alpha \in \theta_X$ satisfies $A = p_\Gamma(C(\widetilde{p}, \alpha\widetilde{q}))$.

Now, we are going to prove that our definition does not depend on $\widetilde{q}$.

Let us recall that the homomorphism $\widetilde{q}_!\colon \theta_\Gamma \to \theta_Y$ is defined by the relation $\widetilde{q}\alpha = \widetilde{q}_!(\alpha)\widetilde{q}$, for $\alpha \in \theta_\Gamma$. If $\widetilde{q}' = \gamma\widetilde{q}$ is another lift of q ($\widetilde{\gamma} \in \theta_\Gamma$), then the induced homomorphism $\widetilde{q}'_!\colon \theta_\Gamma \to \theta_Y$ is defined by the relation $\widetilde{q}'\alpha = \widetilde{q}'_!(\alpha)\widetilde{q}'$.

(4.7) LEMMA. *If $\widetilde{q}' = \gamma \cdot \widetilde{q}$ is another lift of q, then $\gamma \cdot \widetilde{q}'_!(\alpha) \cdot \gamma^{-1} = \widetilde{q}'_!(\alpha)$, for all $\alpha \in \theta_\Gamma$.*

PROOF. The equalities $\widetilde{q}' = \gamma \cdot \widetilde{q}$ and $\widetilde{q}'(\alpha\widetilde{u}) = \widetilde{q}'_!(\alpha)\widetilde{q}'(\widetilde{u})$ imply $\gamma \cdot \widetilde{q}(\alpha\widetilde{u}) = \widetilde{q}'_!(\alpha) \cdot \gamma \cdot \widetilde{q}(\widetilde{u})$, by which $\gamma \cdot \widetilde{q}_!(\alpha) \cdot \widetilde{q}(\widetilde{u}) = \widetilde{q}'_!(\alpha) \cdot \gamma \cdot \widetilde{q}'(\widetilde{u})$. Thus, $\gamma \cdot \widetilde{q}_!(\alpha) = \widetilde{q}'_!(\alpha) \cdot \gamma$ and finally $\widetilde{q}'_!(\alpha) \cdot \widetilde{q}(\widetilde{u}) = \gamma \cdot \widetilde{q}_!(\alpha) \cdot \gamma^{-1}$. □

(4.8) PROPOSITION. *Let us fix two lifts $\widetilde{q}$ and $\widetilde{q}'$. Let $\gamma \in \theta_X$ denote the unique transformation satisfying $\widetilde{q} = \gamma \cdot \widetilde{q}'$. Then $\alpha, \beta \in \theta_X$ are in the Reidemeister relation with respect to $\widetilde{q}$ if and only if so are $\alpha \cdot \gamma^{-1}, \beta \cdot \gamma^{-1}$ with respect to $\widetilde{q}'$.*

PROOF. Suppose that $\beta = \delta \cdot \alpha \cdot \widetilde{q}_!\widetilde{p}^!(\delta^{-1})$. Then $\beta \cdot \gamma^{-1} = \delta \cdot \alpha \cdot \widetilde{q}_!\widetilde{p}^!(\delta^{-1})\gamma^{-1} = \delta \cdot \alpha \cdot \gamma^{-1}(\gamma \cdot \widetilde{q}_!\widetilde{p}^!(\delta^{-1})\gamma^{-1}) = \delta(\alpha \cdot \gamma^{-1})\widetilde{q}_!\widetilde{p}^!(\delta^{-1})\gamma^{-1})\delta \cdot \alpha \cdot \gamma^{-1} \cdot \gamma \cdot \widetilde{q}_!\widetilde{p}^!(\delta^{-1})$. Then $\delta \cdot \alpha \cdot \gamma^{-1} \cdot \gamma \cdot \widetilde{q}_!\widetilde{p}^!(\delta^{-1}) = \delta(\alpha \cdot \gamma^{-1})\widetilde{q}'_!\widetilde{p}^!(\delta^{-1})$. □

The above consideration shows that the Reidemeister sets obtained by different lifts of q are canonically isomorphic. That is why we write $R(p,q)$ omitting tildes.

(4.9) PROPOSITION. *If $X \times [0,1] \overset{p}{\Leftarrow} \Gamma \overset{q}{\longrightarrow} Y$ is a homotopy satisfying (4.1)–(4.3), then the homomorphism $\widetilde{q}_{t!}\widetilde{p}_t^!\colon \theta_X \to \theta_Y$ does not depend on $t \in [0,1]$, where the lifts used in the definitions of these homomorphisms are restrictions of some fixed lifts p, q of the given homotopy.*

For the proof, see [AGJ1] (cf. [AG]).

(4.10) REMARK. If (p,q) represents a single-valued map $\rho\colon X \to Y (q = \rho p)$, then $\widetilde{q}_!\widetilde{p}^!$ equals $\widetilde{\rho}_!$ (here the chosen lifts satisfy $\widetilde{q} = \widetilde{\rho}\widetilde{p}$).

Below we shall define the Nielsen relation modulo a subgroup.

Let us point out that the above theory can be modified onto the relative case. Consider again a multivalued pair (p,q) satisfying (4.1)–(4.3). Let $H \subset \theta_X$, $H' \subset \theta_Y$ be normal subgroups. Then the action of H on $\widetilde{X}$ gives the quotient space $\widetilde{X}_H$ and the map $p_{XH}\colon \widetilde{X}_H \to X$ is also a covering. Similarly, we get $p_{YH'}\colon \widetilde{Y}'_H \to Y$. On the other hand, the action of H on $\widetilde{\Gamma}$ given by $h \circ (\widetilde{x}, z) = (h\widetilde{x}, z)$ determines the quotient space $\widetilde{\Gamma}_H$ with the natural map $\widetilde{p}_H\colon \widetilde{\Gamma}_H \to \widetilde{X}_H$ induced by $\widetilde{p}$. Assume that $\widetilde{q}_!\widetilde{p}^!(H) \subset H'$. Observe that this condition does not depend on the choice of the lifts $\widetilde{p}$, $\widetilde{q}$, because the subgroups H, H' are the normal divisors. Thus, $\widetilde{q}\colon \widetilde{\Gamma} \to \widetilde{Y}$ induces a map $\widetilde{q}_H\colon \widetilde{\Gamma}_H \to \widetilde{Y}_{H'}$ and the diagram

$$\begin{array}{ccccc}
\widetilde{X}_H & \overset{\widetilde{p}_H}{\Longleftarrow} & \widetilde{\Gamma}_H & \overset{\widetilde{q}_H}{\longrightarrow} & \widetilde{Y}_H \\
{\scriptstyle p_{XH}}\downarrow & & {\scriptstyle p_{\Gamma H}}\downarrow & & \downarrow{\scriptstyle p_{YH'}} \\
X & \underset{p}{\Longleftarrow} & \Gamma & \underset{q}{\longrightarrow} & Y
\end{array}$$

commutes. Now, we can get the homomorphisms $\widetilde{q}_{H!}\widetilde{p}_H^!\colon \theta_{XH} \to \theta_{YH'}$, where θ_{XH}, $\theta_{YH'}$ denote the groups of natural transformations of $\widetilde{X}_H$ and $\widetilde{Y}_{H'}$, respectively.

Assuming $X = Y$ and $H = H'$. We can give

(4.11) LEMMA.

(4.11.1) *$C(p,q) = \bigcup_{\alpha \in \theta_{XH}} p_{\Gamma H} C(\widetilde{p}_H, \alpha\widetilde{q}_H)$, if $p_{\Gamma H} C(\widetilde{p}_H, \alpha\widetilde{q}_H) \cap p_{\Gamma H} C(\widetilde{p}_H, \beta\widetilde{q}_H)$ is not empty, then there exists a $\gamma \in \theta_{XH}$ such that $\beta = \gamma \circ \alpha \circ (\widetilde{q}_{H!}\widetilde{p}_H^!\gamma)^{-1}$,*

(4.11.2) *the sets $p_{\Gamma H} C(\widetilde{p}_H, \alpha\widetilde{q}_H)$ are either disjoint or equal.*

Hence, we get the splitting of $C(p,q)$ into the H-Nielsen classes and the natural injection from the set of H-Nielsen classes into the set of Reidemeister classes modulo H, namely, $R_H(p,q)$.

Now, we would like to exhibit the classes which do not disappear under any (admissible) homotopy. For this, we need however (besides (4.1)–(4.3)) the following two assumptions on the pair $X \xLeftarrow{p} \Gamma \xrightarrow{q} Y$.

(4.12) Let X be a connected retract of an open subset of (a convex set in) a Fréchet space, p is a Vietoris map and $\mathrm{cl}(q(\Gamma)) \subset X$ is compact, i.e. q is a compact map.

(4.13) There exists a normal subgroup $H \subset \theta_X$ of a finite index satisfying

$$\widetilde{q}_! \widetilde{p}^!(H) \subset H.$$

(4.14) DEFINITION. We call a pair (p,q) *N-admissible* if it satisfies (4.1)–(4.3), (4.12) and (4.13).

(4.15) REMARK. The pairs satisfying (4.12) are called *admissible.*

Let us recall that, under the assumption (4.12), the Lefschetz number $\Lambda(p,q) \in \mathbb{Q}$ is defined (see Section 3.1). This is a homotopy invariant (with respect to the homotopies satisfying (4.12)) and $\Lambda(p,q) \neq 0$ implies $C(p,q) \neq \emptyset$ (cf. Section 3.1).

The assumption (4.13) gives rise to the last commutative diagram, where the coverings p_{XH}, $p_{\Gamma H}$, p_{YH} are finite, because the subgroup $H \in \theta_{XH}$ has a finite index. Now, we can observe that the pair $(\widetilde{p}_H, \alpha\widetilde{q}_H)$, for any $\alpha \in \theta_{XH}$, also satisfies (4.12) ($\widetilde{p}^{-1}(\widetilde{x}) = p^{-1}(x)$, $\mathrm{cl}(\alpha\widetilde{q}_H(\widetilde{\Gamma}_H)) \subset p_{XH}^{-1}(\mathrm{cl}\, q(\Gamma))$ and the last set is compact, because the covering p_{XH} is finite).

Let $A = p_{\Gamma H} C(\widetilde{p}, \alpha\widetilde{q})$ be a Nielsen class of an N-admissible pair (p,q). We say that (the N-Nielsen class) A is *essential* if $\Lambda(\widetilde{p}, \alpha\widetilde{q}) \neq 0$. The following lemma explains that this definition it correct, i.e. does not depend on the choice of α.

(4.16) LEMMA. *If $p_{\Gamma H} C(\widetilde{p}, \alpha\widetilde{q}) = p_{\Gamma H} C(\widetilde{p}, \alpha'\widetilde{q}) \neq \emptyset$, for some $\alpha, \alpha' \in \theta_{XH}$, then $\Lambda(\widetilde{p}, \alpha\widetilde{q}) = \Lambda(\widetilde{p}, \alpha'\widetilde{q})$.*

PROOF. Since α, α' represent the same element in $R_H(p,q)$, there exists $\gamma \in \theta_{XH}$ such that $\alpha' = \gamma \circ \alpha \circ \widetilde{q}_! \widetilde{p}^!(\gamma^{-1})$. Thus,

$$\begin{aligned}
\Lambda(\widetilde{p}, \alpha'\widetilde{q}) &= \mathrm{Tr}((\widetilde{p}^*)^{-1}(\alpha' \cdot \widetilde{q})^*) = \mathrm{Tr}((\widetilde{p}^*)^{-1}(\gamma \circ \alpha \circ (\widetilde{q}_! \widetilde{p}^!(\gamma^{-1})) \circ \widetilde{q})^*) \\
&= \mathrm{Tr}((\widetilde{p}^*)^{-1}(\gamma \circ \alpha \circ \widetilde{q}_!(\widetilde{p}^!(\gamma^{-1})))^*) = \mathrm{Tr}((\widetilde{p}^*)^{-1}((\widetilde{p}^!(\gamma^{-1})))^* \circ (\alpha\widetilde{q})^* \circ \gamma^*) \\
&= \mathrm{Tr}((\widetilde{p}^* \widetilde{p}^!(\gamma))^*)^{-1} \circ (\alpha\widetilde{q})^* \circ \gamma^*) = \mathrm{Tr}((\widetilde{\gamma}^*)^{-1}(\widetilde{p}^*)^{-1} \circ (\alpha\widetilde{q})^* \circ \gamma^*) \\
&= \mathrm{Tr}((\widetilde{p}^*)^{-1} \circ (\alpha\widetilde{q})^*) = \Lambda(\widetilde{p}, \alpha\widetilde{q}).
\end{aligned}$$

□

(4.17) DEFINITION. Let (p,q) be an N-admissible multivalued map (for a subgroup $H \subset \theta_X$). We define the *Nielsen number modulo H* as the number of essential classes in θ_{X_H}. We denote this number by $N_H(p,q)$.

(4.18) REMARK. Observe that the above method allows us to define only essential classes (and the Nielsen number) modulo a subgroup of a finite index in $\theta_X = \pi_1 X$. The problem how to get similar notions in an arbitrary case we leave open.

The following theorem is an easy consequence of the homotopy invariance of the Lefschetz number.

(4.19) THEOREM. *$N_H(p,q)$ is a homotopy invariant (with respect to N-admissible homotopies) $X \times [0,1] \overset{p}{\Longleftarrow} \Gamma \overset{q}{\longrightarrow} X$. Moreover, (p,q) has at least $N_H(p,q)$ coincidences.*

The following theorem shows that the above definition is consistent with the classical Nielsen number for single-valued maps.

(4.20) THEOREM. *If an N-admissible map (p,q) is N-admissibly homotopic to a pair (p',q'), representing a single-valued map p (i.e. $q' = \rho p'$), then (p,q) has at least $N_H(\rho)$ coincidences (here H denotes also the subgroup of $\pi_1 X$ corresponding to the given $H \subset \theta_X$ in (4.13)).*

For the proof, see [AGJ1] (cf. [AG]).

Although in the general case the theory, presented in the previous sections, requires special assumptions on the considered pair (p,q), we shall see that in the case of multivalued self-maps on a torus it is enough to assume that this pair satisfies only (4.12), i.e. it is admissible. We will do this by showing that in the case of any pair satisfying (4.12), it is homotopic to a pair representing a single-valued map.

(4.21) LEMMA. *For any compact space X, if $\widetilde{H}^1(X;\mathbb{Q})=0$, then $\widetilde{H}^1(X;\mathbb{Z})=0$.*

For the proof, see [AGJ1] (cf. [AG]).

(4.22) THEOREM. *Any multivalued self-map (p,q) on the torus satisfying (4.12) is admissibly homotopic to a pair representing a single-valued map.*

For the proof, see [AGJ1] (cf. [AG]).

(4.23) THEOREM. *Let $\mathbb{T}^n \overset{p}{\Longleftarrow} \Gamma \overset{q}{\longrightarrow} \mathbb{T}^n$ be such that p is a Vietoris map. Let $\rho\colon \mathbb{T}^n \to \mathbb{T}^n$ be a single-valued map representing a multivalued map homotopic to (p,q) (according to Theorem (4.22), such a map always exists). Then (p,q) has at least $N(\rho)$ coincidences.*

PROOF. Let us recall that $N(\rho) = |\Lambda(\rho)| = |\det(I-A)|$, where A is an integer $(n \times n)$-matrix representing the induced homotopy homomorphism $\rho_{\#}\colon \pi_1\mathbb{T}^n \to \pi_1\mathbb{T}^n$. Moreover, if $\det(I-A) \neq 0$, then

$$\operatorname{card}(\pi_1(\mathbb{T}^n)/\operatorname{Im}(\rho_{\#})) = |\det(I-A)|.$$

The case $N(\rho) = 0$ is obvious. Assume that $N(\rho) \neq 0$. By Theorem (4.22), it is enough to find a subgroup (of a finite index) $H \subset \pi_1\mathbb{T}^n = \mathbb{Z}^n$ satisfying

(4.23.1) $\rho_\#(H) \subset H$, and
(4.23.2) $N_H(\rho) = N(\rho)$.

We define $H = \{z - \rho_\#(x) : x \in \pi_1 X\}$. Then (4.23.1) is clear. Recall that, for any endomorphism of an abelian group $\rho: G \to G$, the Reidemeister set is the quotient group $R(\rho) = G/(\mathrm{Im}(\mathrm{id}-\rho)$. In our case, $H = \mathrm{Im}(\mathrm{id}-\rho)$ and the natural map $G \to G/H$ induces the bijection between $R(\rho) = G/H$ and $R_H(\rho) = (G/H)/(H/H)$. Thus, we get the bijection $R_H(\rho) = R(\rho)$. Finally, we notice that all the Nielsen classes of ρ have the same index ($=\mathrm{sign}(\det(I - A))$. Thus, all involving classes in $R_H(\rho)$ in $R(\rho)$ are essential, which proves $N_H(\rho) = N(\rho)$. □

4.2. Relative Nielsen numbers. Now, we are going to present the relative Nielsen theory.

Assuming that X is a retract of an open subset of (a convex set in) a Fréchet space E (i.e. X is an ANR), it admits a universal covering $p_X: \widetilde{X} \to X$.

Moreover, we consider all admisible maps $(p,q): X \multimap X$ for which the following conditions are satisfied (cf. (4.3), (4.13)):

(A) For any $x \in X$, the restriction $q|_{p^{-1}(x)}: p^{-1}(x) \to X$ admits a lift $\widetilde{q}$ to the universal covering space, i.e. , $p_X \circ \widetilde{q} = q$.
(B) There exists a normal subgroup $H \subset \pi_1(X)$ of a finite index such that $\widetilde{q}_!\widetilde{p}^!(H) \subset H$. Here, $\widetilde{q}_!\widetilde{p}^!$ denotes a homomorphism of the fundamental group $\pi_1(X)$ induced by $(p,q): X \multimap X$ (see the foregoing Section 4.1); if (p,q) represents a single-valued map, i.e. , $q(p^{-1}(x))$ is a singleton, for every $x \in X$, then $\widetilde{q}_!\widetilde{p}^!$ coincides with the induced homomorphism $f_\#: \pi_1(X) \to \pi_1(X)$.

Finally, in what follows, we shall assume that $A \subset X$ is, moreover, closed and connected.

Assume that $(p,q): X \multimap X$ is a compact map of the form $X \xLeftarrow{p} \Gamma \xrightarrow{q} X$. Denote $\Gamma_A = p^{-1}(A)$ and consider the restriction $A \xLeftarrow{p_|} \Gamma_A \xrightarrow{q_|} A$. If $U \subset X$ is an open subset in the definition of a compact mapping, for (p,q), then $U \cap A$ can be associated to the compact mapping of $(p_|, q_|)$. Let $H_0 = i_\#^{-1}(H) \subset \pi_1 A$. Since the induced homomorphism $i_\#: (\pi_1 A)/H_0 \to (\pi_1 X)/H$ is mono, H_0 is also a normal subgroup of a finite order. Hence, $A \xLeftarrow{p_|} \Gamma_A \xrightarrow{q_|} A$ (where $p_|$, $q_|$ denote the natural restrictions) also satisfies the assumptions (A) and (B).

Let us recall that a pair (p,q) satisfying condition (A) induces a homomorphism $\widetilde{q}_!\widetilde{p}^!: O_X \to O_X$, where $O_X = \{\alpha: \widetilde{X} \to \widetilde{X};\ p_X\alpha = p_X\}$, $p_X: \widetilde{X} \to X$ is a fixed covering related to the subgroup H, corresponding, in the single-valued case ($\rho = qp^{-1}$), to $\rho: \pi_1 X \to \pi_1 X$. We define the action of O_X on itself $\gamma \circ \alpha = \gamma\alpha\widetilde{q}_!\widetilde{p}^!\gamma^{-1}$.

By an analogy with the single-valued case, we define the quotient set, the *set of Reidemeister classes*, and we denote it by $\mathcal{R}_H(p,q)$. We define the *Nielsen class* corresponding to a class $[\alpha] \in \mathcal{R}(p,q)$, as $p_\Gamma(C(\widetilde{p}, \alpha\widetilde{q}))$. This splits $\mathrm{C}(p,q)$ into disjoint Nielsen classes and defines the natural injection $\eta\colon \mathcal{N}_H(p,q) \to \mathcal{R}_H(p,q)$. If $\Lambda(\widetilde{p}, \alpha\widetilde{q}) \neq 0$, then $\mathrm{C}(\widetilde{p}, \alpha\widetilde{q}) \neq \emptyset$, and the Nielsen class corresponding to $[\alpha] \in \mathcal{R}_H(p,q)$ is called *essential.* We define the *H-Nielsen number* as the number of essential classes and denote it by $N(p,q)$.

(4.24) LEMMA. *The following diagram commutes*

$$\begin{array}{ccc} \mathcal{N}_{H_0}(p_|, q_|)(i) & \xrightarrow{\mathcal{N}} & \mathcal{N}_H(p,q) \\ \eta \big\downarrow & & \big\downarrow \eta \\ \mathcal{R}_{H_0}(p_|, q_|) & \xrightarrow[\mathcal{R}(i)]{} & \mathcal{R}_H(p,q) \end{array}$$

For the proof, see [AGJ3] (cf. [AG]).

Let $S_H(p,q;A) \subset \mathcal{R}_H(p,q)$ denote the set of essential Reidemeister classes which contain no essential class from $\mathcal{R}_{H_0}(p_|, q_|)$.

(4.25) THEOREM. *Under the assumptions* (A) *and* (B), *a compact pair* (p,q) *has at least* $N_H(p,q) + (\#S_H(p,q;A))$ *coincidences.*

PROOF. We choose a point $z_1, \dots, z_k \in C(f_|, g_|)$, from each essential class of $f_|, g_|$, and a point $w_1, \dots, w_l \in C(f,g)$ from each one in $S_H(p,q;A)$. It remains to show that $i(z) \neq w$, for any $z = z_1, \dots, z_k$, $w = w_1, \dots, w_l$. Suppose the contrary. Then, by Lemma (4.24), the essential Nielsen class of $(p_|, q_|)$ containing z is involved in the class of (p,q) containing w. Since the last class belongs to $S_H(p,q;A)$, we get a contradiction. □

(4.26) REMARK. Since any essential Reidemeister class corresponds always to a nonempty Nielsen class, in the definition of $S_H(p,q;A)$, the name of Reidemeister can be replaced by Nielsen: $S_H(p,q;A)$ becomes the set of Nielsen classes (from $\mathcal{N}_H(p,q)$) which contains no essential Nielsen class (from $\mathcal{N}_{H_0}(p_|, q_|)$).

The following theorem gives a lower bound for the number of coincidences of the pair (p,q) lying outside Γ_A. Let $SN_H(p,q;A)$ be the cardinality of the set of essential classes in $\mathcal{R}_H(p,q) \setminus \operatorname{Im} \mathcal{R}(i)$.

Finally, we have:

(4.27) THEOREM. *Under the assumptions* (A) *and* (B), *a compact pair* (p,q) *has at least* $SN_H(p,q;A)$ *coincidences in* $\Gamma \setminus \Gamma_A$.

For the proof, it is sufficient to observe that each essential class from $\mathcal{R}_H(p,q) \setminus \operatorname{Im} \mathcal{R}(i)$ is nonempty and disjoint from Γ_A.

4.3. Nielsen number for periodic points. Consider a map:

$$X \overset{p}{\Longleftarrow} \Gamma \overset{q}{\longrightarrow} X.$$

A sequence of points $(z_1, \dots, z_k)$, satisfying $z_i \in \Gamma$, $i = 1, \dots, k$, $q(z_i) = p(z_{i+1})$, $i = 1, \dots, k-1$ and $q(z_k) = p(z_1)$, will be called a *k-periodic orbit of coincidences*, for (p, q).

Let us note that, for $(p, q) = (\mathrm{id}_X, f)$, a k-periodic orbit of coincidences equals the orbit of periodic points for f. In what follows, we shall consider periodic orbits of coincidences with the fixed first element $(z_1, \dots, z_k)$, unless otherwise stated. Then $(z_2, z_3, \dots, z_k, z_1)$ is considered as another periodic orbit. We say that two orbits $(z_1, \dots, z_k)$, $(z_1', \dots, z_k')$ are *cyclically equal* if

$$(z_1', \dots, z_k') = (z_l, \dots, z_k; z_1, \dots, z_{l-1}), \quad \text{for an } l = 1, \dots, k.$$

Otherwise, we call them *cyclically different*. Let us also note that, in the single-valued case, a k-periodic point x determines the whole orbit $\{x, fx, f^2x, \dots, f^{k-1}x\}$, but for a multivalued map $X \overset{p}{\Longleftarrow} \Gamma \overset{q}{\longrightarrow} X$, there can be distinct orbits starting from a given z_1 (the second element z_2 satisfies only $z_2 \in q^{-1}(pz_1)$, so it need not be uniquely determined).

Let us denote $\Gamma_k = \{(z_1, \dots, z_k) : z_i \in \Gamma,\ qz_i = pz_{i+1},\ i = 1, \dots, k-1\}$. We define the maps $p_k, q_k \colon \Gamma_k \to X$ by the formulae $p_k(z_1, \dots, z_k) = p(z_1)$, $q_k(z_1, \dots, z_k) = q(z_k)$.

(4.28) Remark. A sequence of points $(z_1, \dots, z_k) \in \Gamma_k$ is an orbit of coincidences if and only if $(z_1, \dots, z_k) \in C(p_k, q_k)$.

Thus, the study of k-periodic orbits of coincidences reduces to the coincidences of the pair $X \overset{p_k}{\longleftarrow} \Gamma_k \overset{q_k}{\longrightarrow} X$. Here, we shall try to find an estimation of the number of k-orbits of coincidences of the pair (p, q). However, to lift the multivalued map $X \overset{p}{\Longleftarrow} \Gamma \overset{q}{\longrightarrow} X$ to covering spaces, we need the following assumption. Let X be compact space which admits a universal covering $p_X \colon \widetilde{X} \to X$.

(C) $p \colon \Gamma \Rightarrow X$ is a Vietoris map and there exists a map $\widetilde{q}$ making the following commutative diagram:

$$\begin{array}{ccccc}
\widetilde{X} & \overset{\widetilde{p}}{\longleftarrow} & \widetilde{\Gamma} & \overset{\widetilde{q}}{\longrightarrow} & \widetilde{X} \\
{\scriptstyle p_X}\big\downarrow & & {\scriptstyle p_\Gamma}\big\downarrow & & \big\downarrow{\scriptstyle p_X} \\
X & \underset{p}{\Longleftarrow} & \Gamma & \underset{q}{\longrightarrow} & X
\end{array}$$

($\widetilde{\Gamma} = \{(\widetilde{x}, z) \in \widetilde{X} \times \Gamma : p_X(\widetilde{x}) = p(z)\}$ is the pullback with natural projections $\widetilde{p}(\widetilde{x}, z) = \widetilde{x}$, $p_\Gamma(\widetilde{x}, z) = z$).

We assume that (p, q) satisfies (C) and we fix such a lift $\widetilde{q}$.

(4.29) LEMMA. *If (p,q) satisfies (C), then so does (p_k, q_k).*

For the proof, see [AGJ4] (cf. [AG]).

(4.30) LEMMA. *For any point $(\widetilde{x}, (z_1, \ldots, z_k)) \in \widetilde{\Gamma}_k$, there exists a unique point $(\overline{z}_1, \ldots, \overline{z}_k) \in \overline{\Gamma}_k$ satisfying $\widetilde{p}(\overline{z}_1) = \widetilde{x}$ and $p_\Gamma(\overline{z}_i) = z_i$ $(i = 1, \ldots, k)$. The map $\varphi: \widetilde{\Gamma}_k \to \overline{\Gamma}_k$ given by this correspondence $(\varphi(\widetilde{x}; z_1, \ldots, z_k)) = (\overline{z}_1, \ldots, \overline{z}_k))$ is a bijection for which the second diagram commutes. The opposite map $\psi: \overline{\Gamma}_k \to \widetilde{\Gamma}_k$ is given by the formula $\psi(\overline{z}_1, \ldots, \overline{z}_k) = (\widetilde{p}\overline{z}_1; p_\Gamma \overline{z}_1, \ldots, p_\Gamma \overline{z}_k)$.*

For the proof, see [AGJ4] (cf. [AG]).

Let $\mathcal{O}_X, \mathcal{O}_\Gamma, \mathcal{O}_{\overline{\Gamma}_k}$ denote the groups of transformations of coverings: $\widetilde{X} \to X$, $\widetilde{\Gamma} \to \Gamma$ and $\overline{\Gamma}_k \to \Gamma_k$, respectively. By the arguments in Section 4.1, the lifts $\overline{p}_k$, $\overline{q}_k$ define homomorphisms $\overline{p}_k^!: \mathcal{O}_X \to \mathcal{O}_{\Gamma_k}$ and $\overline{q}_{k!}: \mathcal{O}_{\overline{\Gamma}_k} \to \mathcal{O}_X$ by the commutative diagrams:

$$
\begin{array}{ccc}
\widetilde{X} & \xleftarrow{\ \overline{p}_k\ } & \overline{\Gamma}_k \\
{\scriptstyle\alpha}\downarrow & & \downarrow{\scriptstyle \overline{p}_k^!(\alpha)} \\
\widetilde{X} & \xleftarrow[\ \overline{p}_k\ ]{} & \overline{\Gamma}_k
\end{array}
\qquad\qquad
\begin{array}{ccc}
\overline{\Gamma}_k & \xrightarrow{\ \overline{q}_k\ } & \widetilde{X} \\
{\scriptstyle\alpha}\downarrow & & \downarrow{\scriptstyle \overline{q}_{k!}(\alpha)} \\
\overline{\Gamma}_k & \xrightarrow[\ \overline{q}_k\ ]{} & \widetilde{X}
\end{array}
$$

Hence, we can define the action of $\mathcal{O}_X$ on itself by the formula $\alpha \circ \beta = \alpha \cdot \beta \cdot [(\overline{q}_k)_!(\overline{p}_k)^!(\alpha^{-1})]$. The quotient set will be called the *set of Reidemeister classes* and will be denoted by $\mathcal{R}(\overline{p}_k, \overline{q}_k)$ (cf. Section 4.1). Notice that the lifts $\overline{p}_k, \overline{q}_k$, are fixed, because the lifts $\widetilde{p}_k, \widetilde{q}_k$ were fixed. The composition $(\overline{q}_k)_!(\overline{p}_k)^!$ is a generalization of the homomorphism $\rho_\#: \mathcal{O}_X \to \mathcal{O}_X$ induced by a (single-valued) map $\rho: X \to X$ (or, equivalently, to the homomorphism of the fundamental group $\rho_\#: \pi_1 X \to \pi_1 X$). In fact, if (p,q) represents a single-valued map ρ (i.e. $\rho(x) = qp^{-1}(x)$), then $\rho_\# = q_! p^!$ (see Section 4.1).

(4.31) LEMMA. *Let $\alpha \in \mathcal{O}_X$ and let $(\overline{z}_1, \ldots, \overline{z}_k) \in \overline{\Gamma}_k$, $\overline{z}_i = (\widetilde{x}_i, z_i)$. Let us denote $\overline{p}_k^!(\alpha)(\overline{z}_1, \ldots, \overline{z}_k) = (\overline{z}'_1, \ldots, \overline{z}'_k)$, where $\overline{z}'_i = (\widetilde{x}'_i, z_i)$. Then*

$$\widetilde{x}'_i = [\widetilde{q}_! \widetilde{p}^!]^{i-1}(\alpha)(\widetilde{x}_i).$$

Conversely, if $(\overline{z}_1, \ldots, \overline{z}_k)$, $(\overline{z}'_1, \ldots, \overline{z}'_k) \in \overline{\Gamma}_k$ satisfy the above equality, then

$$\overline{p}_k^!(\alpha)(\overline{z}_1, \ldots, \overline{z}_k) = (\overline{z}'_1, \ldots, \overline{z}'_k).$$

For the proof, see [AGJ4] (cf. [AG]).

(4.32) COROLLARY. $(\overline{q}_k)_!(\overline{p}_k)^! = [(\widetilde{q}_l)_!(\widetilde{p}_l)^!]^{k/l}$ *for $l | k$.*

In this part, we show that, as in the single-valued case, the set of periodic coincidences splits into disjoint closed and open subsets (Nielsen classes). Since periodic coincidences are exactly coincidences of the map $X \overset{p_k}{\Longleftarrow} \Gamma_k \xrightarrow{q_k} X$, we can apply Lemma (4.6) in Section 4.1 to this map (and the covering $\widetilde{X} \overset{\widetilde{p}_k}{\Longleftarrow} \overline{\Gamma}_k \xrightarrow{\widetilde{q}_k} \widetilde{X}$).

(4.33) LEMMA. *Let the multivalued map* $X \xleftarrow{p} \Gamma \xrightarrow{q} X$ *satisfy* (A). *Then*

(4.33.1) $C(p_k, q_k) = \bigcup_{\alpha \in \mathcal{O}_X} p_{\overline{\Gamma}_k}(C(\overline{p}_k, \alpha\overline{q}_k))$.
(4.33.2) *If* $p_{\overline{\Gamma}_k}(C(\overline{p}_k, \alpha\overline{q}_k)) \cap p_{\overline{\Gamma}_k}(C(\overline{p}_k, \alpha'\overline{q}_k)) \neq \emptyset$, *then* $[\alpha] = [\alpha'] \in \mathcal{R}(\overline{p}_k, \overline{q}_k)$.
(4.33.3) *If* $[\alpha] = [\alpha'] \in \mathcal{R}(\overline{p}_k, \overline{q}_k)$, *then* $p_{\overline{\Gamma}_k}(C(\overline{p}_k, \alpha\overline{q}_k)) = p_{\overline{\Gamma}_k}(C(\overline{p}_k, \alpha'\overline{q}_k))$.

Thus, $C(p_k, q_k) = \bigcup p_\Gamma(C(\overline{p}_k, \alpha\overline{q}_k))$ *is a disjoint sum, where the summation runs one* α *from each Reidemeister class in* $\mathcal{R}(\overline{p}_k, \overline{q}_k)$.

Thus, any Reidemeister class $[\alpha]$ determines a subset $p_{\overline{\Gamma}_k}(C(\overline{p}_k, \alpha\overline{q}_k))$ which will be called the *Nielsen class*. The set of all (nonempty) Nielsen classes will be denoted by $\mathcal{N}(p_k, q_k)$. Thus, we have a natural injection $\mathcal{N}(p_k, q_k) \to \mathcal{R}(p_k, q_k)$.

Let us notice that following the arguments in Section 4.1, we can reformulate all the above, when replacing the universal covering $p_X: \widetilde{X} \to X$ by any finite regular covering; i.e. for any normal subgroup $H \lhd \mathcal{O}_X$ of a finite index we take a covering $p_{XH}: \widetilde{X}_H \to X$ determined by this group. This allows us to define the spaces $\widetilde{\Gamma}_H$, $\widetilde{\Gamma}_{kH}$, $\overline{\Gamma}_{kH}$ and the homomorphisms $\widetilde{p}_H^!, \widetilde{q}_{!H}, \overline{p}_H^!, \overline{q}_{!H}$. This generalization works once we assume the condition

(AH) $p: \Gamma \Rightarrow X$ is a Vietoris map and there exists a map $\widetilde{q}_H$ making the following diagram commutative

$$\begin{array}{ccccc} \widetilde{X}_H & \xleftarrow{\widetilde{p}_H} & \widetilde{\Gamma}_H & \xrightarrow{\widetilde{q}_H} & \widetilde{X}_H \\ {\scriptstyle p_{XH}}\downarrow & & \downarrow{\scriptstyle p_{\Gamma H}} & & \downarrow{\scriptstyle p_{XH}} \\ X & \xleftarrow[p]{} & \Gamma & \xrightarrow[q]{} & X \end{array}$$

Observe that (C) implies (AH) for any normal subgroup of a finite index $H \lhd \mathcal{O}_X$. In particular, Lemma (4.33) becomes

(4.34) LEMMA. *Let the multivalued map* $X \xleftarrow{p} \Gamma \xrightarrow{q} X$ *satisfy* (AH). *Then*

(4.34.1) $C(p_k, q_k) = \bigcup_{\alpha \in \mathcal{O}_{XH}} p_{\overline{\Gamma}_{kH}}(C(\overline{p}_{kH}, \alpha\overline{q}_{kH}))$.
(4.34.2) *If* $p_{\overline{\Gamma}_{kH}}(C(\overline{p}_{kH}, \alpha\overline{q}_{kH})) \cap p_{\overline{\Gamma}_{kH}}(C(\overline{p}_{kH}, \alpha'\overline{q}_{kH})) \neq \emptyset$, *then*

$$[\alpha] = [\alpha'] \in \mathcal{R}(\overline{p}_{kH}, \overline{q}_{kH}).$$

(4.34.3) *If* $[\alpha] = [\alpha'] \in \mathcal{R}(\overline{p}_{kH}, \overline{q}_{kH})$, *then* $p_{\overline{\Gamma}_k}(C(\overline{p}_k, \alpha\overline{q}_k)) = p_{\overline{\Gamma}_k}(C(\overline{p}_{kH}, \alpha'\overline{q}_{kH}))$.

Thus, $C(p_k, q_k) = \bigcup p_\Gamma(C(\overline{p}_{kH}, \alpha\overline{q}_{kH}))$ *is a disjoint sum, where the summation runs one* α *from each Reidemeister class from* $\mathcal{R}(\overline{p}_{kH}, \overline{q}_{kH})$.

Since the spaces $\widetilde{\Gamma}$, $\overline{\Gamma}_k$ can be quite arbitrary, we are not able to follow the case of polyhedra or ANR's to define essential classes. We will follow the ideas presented in Section 4.1. We fix a normal subgroup $H \lhd \mathcal{O}_X$ of a finite index satisfying $\widetilde{q}_!\widetilde{p}^!(H) \subset H$ and we say that an H-Nielsen class $p_\Gamma(C(\widetilde{p}, \alpha\widetilde{q}))$ is essential if $\Lambda(\widetilde{p}, \alpha\widetilde{q}) \neq 0$. The last (Lefschetz) number is defined, because the map is compact as the finite covering of (p, q).

Observe that we have a natural map $\nu\colon \mathcal{R}(p_k, q_k) \to \mathcal{R}(p_k, q_k)$ given by $\nu[\alpha] = [\widetilde{q}_!\widetilde{p}^!(\alpha)]$. One can check that this is well-defined and, moreover, $\nu^k[\alpha] = [\widetilde{q}_!\widetilde{p}^!(\alpha)] = [\alpha]$. Hence, we get an action of $\mathbb{Z}_k$ on $\mathcal{R}(p_k, q_k)$. We can talk about *orbits of Reidemeister classes*. On the other hand, we define a map $\nu'\colon C(p_k, q_k) \to C(p_k, q_k)$ by $\nu'(z_1, \dots, z_k) = (z_2, \dots, z_n, z_1)$.

(4.35) LEMMA. *The following diagram commutes:*

$$\begin{array}{ccc} \mathcal{N}(p_k, q_k) & \xrightarrow{\nu'} & \mathcal{N}(p_k, q_k) \\ \mu\downarrow & & \downarrow\mu \\ \mathcal{R}(p_k, q_k) & \xrightarrow[\nu]{} & \mathcal{R}(p_k, q_k) \end{array}$$

where μ denotes the natural inclusion.

(4.36) COROLLARY. *The map $j_{kl}\colon C(p_l, q_l) \to C(p_k, q_k)$ sends the Nielsen class corresponding to $[\alpha] \in \mathcal{R}(\widetilde{p}_l, \widetilde{q}_l)$ to the Nielsen class corresponding to $[i_{kl}(\alpha)] \in \mathcal{R}(\widetilde{p}_k, \widetilde{q}_k)$. In other words, the following diagram commutes:*

$$\begin{array}{ccc} \mathcal{N}(p_l, q_l) & \xrightarrow{j_{kl}} & \mathcal{N}(p_k, q_k) \\ \downarrow & & \downarrow \\ \mathcal{R}(\overline{p}_l, \overline{q}_l) & \xrightarrow[i_{kl}]{} & \mathcal{R}(\overline{p}_k, \overline{q}_k) \end{array}$$

(4.37) COROLLARY. *If the Nielsen classes A, $A' \subset C(p_k, q_k)$ contain cyclically equal orbits $(x_1, \dots, x_k) \in A$, $(x_{i+1}, \dots, x_k; x_1, \dots, x_i) \in A'$, then $\mu(A)$, $\mu(A')$ belong to the same orbit of the Reidemeister classes.*

PROOF. Since $(x_{i+1}, \dots, x_k; x_1, \dots, x_i) = {\nu'}^i(x_1, \dots, x_k)$, we have $A' = \nu^i(A)$. Now, $\mu(A') = \mu{\nu'}^i(A) = \nu^i\mu(A)$. □

An orbit of the Reidemeister classes will be called *essential* if $\Lambda(\overline{p}_k, \alpha\overline{q}_k) \neq 0$, for an α from this orbit. By the above lemma, the coincidence set is nonempty for the lift corresponding to *any* Reidemeister class from an essential orbit.

Corollary (4.37) makes the following definitions consistent.

(4.38) DEFINITION. A k-orbit of coincidences $(z_1, \dots, z_k)$ is called *reducible* if $(z_1, \dots, z_k) = j_{kl}(z_1, \dots, z_l)$, for an $l < k$ dividing k.

(4.39) DEFINITION. A Reidemeister class $[\alpha] \in \mathcal{R}(\widetilde{p}_k, \widetilde{q}_k)$ is called *reducible* if it lies in the image of $i_{kl}\colon \mathcal{R}(\overline{p}_l, \widetilde{q}_l) \to \mathcal{R}(\overline{p}_k, \widetilde{q}_k)$, for an $l < k$ dividing k.

(4.40) DEFINITION. An orbit of Reidemeister classes in $\mathcal{R}(\overline{p}_k, \widetilde{q}_k)$ is called *reducible* if it contains a reducible element.

(4.41) LEMMA. *Let* $[\alpha] \in \mathcal{R}(\overline{p}_k, \overline{q}_k)$ *represent an essential irreducible orbit of Reidemeister classes. Then* $p_{\overline{\Gamma}_k}(C(\overline{p}_k, \alpha\overline{q}_k) \neq \emptyset$. *Moreover, if*

$$(z_1, \dots, z_k) \in p_{\overline{\Gamma}_k}(C(\overline{p}_k, \alpha\overline{q}_k)),$$

then $(z_1, \dots, z_k) \neq (z_1, \dots, z_l; z_1, \dots, z_l; z_1, \dots, z_l)$, *for any* $l \mid k$, $l < k$.

For the proof, see [AGJ4] (cf. [AG]).

Let $S_k(\widetilde{p}, \widetilde{q})$ denote the number of irreducible and essential orbits in $\mathcal{R}(\overline{p}_k, \overline{q}_k)$.

(4.42) THEOREM. *The (multivalued) compact map* (p, q) *satisfying condition* (C) *has at least* $S_k(\widetilde{p}, \widetilde{q})$ *irreducible cyclically different orbits of coincidences.*

PROOF. We choose an orbit of points from each essential irreducible orbit of the Reidemeister classes. By Corollary (4.37), they are cyclically different. By Lemma (4.41), they are also irreducible. □

(4.43) REMARK. Since the essentiality is a homotopy invariant and irreducibility is defined in terms of Reidemeister classes, $S_k(\widetilde{p}, \widetilde{q})$ is a homotopy invariant.

4.4. Nielsen number for condensing maps. For single-valued (continuous) self-maps in metric (e.g. Fréchet) spaces, including condensing maps, the Nielsen theory was developed in [Scho], provided only that (i) the set of fixed points is compact, (ii) the space is a metric ANR, and (iii) the related generalized Lefschetz number is well-defined. On the other hand, we could see in Section 3.4 (cf. also Section 5.4) that to define the Lefschetz number for condensing maps on non-simply connected sets is a difficult task. Roughly speaking, once we have so defined the generalized Lefschetz number, the Nielsen number can be defined as well.

In the multivalued case, the situation becomes still more delicate, but the main difficulty related to definition of the generalized Lefschetz number remains actual. Therefore, in order to define essential classes without an explicit usage of the generalized Lefschetz number (or a fixed point index), we must proceed in another (recursive) way. More precisely, algebraic (homologic) requirements related to the underlying Lefschetz theory will be replaced by topological arguments related to stable homotopies. Of course, once the generalized Lefschetz number is well-defined, the essentiality should immediately follow. For more details, see [AV2].

We can restrict ourselves to the definition of essential classes, because H-Nielsen classes can be defined in the same way as in Section 4.1.

Hence, let a fixed class $\mathcal{H}$ of homotopies (P,Q) on X be given, where $P:\Gamma \to [0,1]\times X$ and $Q:\Gamma \to X$ satisfy conditions (4.1)–(4.3) (for P, with $[0,1]\times X$, instead of X). Such homotopies induce families of pairs (P_t,Q_t) $(0\le t\le 1)$, where $Q_t:\Gamma_t \to X$ is the restriction of Q to $\Gamma_t := P^{-1}(\{t\}\times X)$, and $P_t:\Gamma_t \to X$ is defined by $(t,P_t(x)) = P(x)$ $(x\in\Gamma_t)$. We write $(P_{t_0},Q_{t_0}) \sim (P_{t_1},Q_{t_1})$, for each $t_0,t_1\in[0,1]$. By $\mathcal{P}$, we denote the class of all pairs $(P_t,Q_t)\in\mathcal{P}$ (for all homotopies and all $t\in[0,1]$).

Each lifting of a homotopy $(P,Q)\in\mathcal{H}$ is itself of a homotopy $(\widetilde{P},\widetilde{Q})$ (only for $\widetilde{Q}$ there is a choice of the lifting), and to this homotopy there correspond pairs $(\widetilde{P}_t,\widetilde{Q}_t)$ $(0\le t\le 1)$ which are liftings of the pairs (P_t,Q_t) associated to the original homotopy (P,Q). In this situation, we call, for each $t_1,t_2\in[0,1]$, $(\widetilde{Q}_{t_1},\widetilde{Q}_{t_2})$ a (P,Q)-admissible lifting for the two pairs (P_{t_1},Q_{t_1}) and (P_{t_2},Q_{t_2}).

(4.44) DEFINITION. We say that a pair $(\widetilde{q}_0,\widetilde{q}_1)$ is an *$\mathcal{H}$-admissible lifting* for two pairs $(p_0,q_0),(p_1,q_1)\in\mathcal{P}$ if it is (P,Q)-admissible, for some $(P,Q)\in\mathcal{H}$.

Note that an $\mathcal{H}$-admissible lifting exists if and only if $(p_0,q_0)\sim(p_1,q_1)$.

Now, we define essential H-Nielsen classes. Fix a normal subgroup H (see (4.13)) and denote $\mathcal{N}_H(p,q) := \{C_\alpha : \alpha\in\theta_{XH}\}$ where $C_\alpha := C_\alpha(p,q,\widetilde{q}) := \mathrm{pr}_{\Gamma_H}(C(\widetilde{p}_H,\alpha\widetilde{q}_H))$, and the used symbols have the same meaning as in Section 4.1.

(4.45) DEFINITION. We call an H-Nielsen class $C\in\mathcal{N}_H(p,q)$ *0-essential*, for a pair $(p,q)\in\mathcal{P}$, if $C\neq\emptyset$.

For $n=1,2,\ldots$, we call C *n-essential* if, for each $(p_1,q_1)\in\mathcal{P}$ with $(p,q)\sim(p_1,q_1)$ and for each corresponding $\mathcal{H}$-admissible lifting $(\widetilde{q},\widetilde{q}_1)$, the following holds: There is some $\alpha\in\theta_{XH}$ with $C=C_\alpha(p,q,\widetilde{q})$ such that $C_\alpha(p_1,q_1,\widetilde{q}_1)$ is $(n-1)$-essential, for (p_1,q_1).

We call C *essential* if C is n-essential, for each n. We denote the cardinality of the set of n-essential H-Nielsen classes by $N_H^{\sim,(n)}(p,q)$. The cardinality of the set of essential H-Nielsen classes is denoted by $N_H^{\sim}(p,q)$.

Before we discuss this definition, let us note that it is equivalent to replace "some α" by "every α" in Definition (4.45). More precisely, the class $C_\alpha(p_1,q_1,\widetilde{q}_1)$ in the above definition is actually independent of the particular choice of α in the following sense.

(4.46) PROPOSITION. *If C is n-essential and $(p,q)\sim(p_1,q_1)$ with a corresponding $\mathcal{H}$-admissible lifting $(\widetilde{q},\widetilde{q}_1)$, then, for every $\alpha,\beta\in\theta_{XH}$ with $C=C_\alpha(p,q,\widetilde{q})=C_\beta(p,q,\widetilde{q})$, we have $C_\alpha(p_1,q_1,\widetilde{q}_1)=C_\beta(p_1,q_1,\widetilde{q}_1)$, and, in case $n\ge 1$, this class is $(n-1)$-essential, for (p_1,q_1).*

For the proof, see [AV2].

Let us now formulate the main properties of the Nielsen number $N_H^{\sim}(p,q)$.

(4.47) THEOREM. *Each n-essential H-Nielsen class is m-essential, for each $m \leq n$. In particular, n-essential H-Nielsen classes are nonempty. Moreover,*

$$N_H^{\sim,(0)}(p,q) \geq N_H^{\sim,(1)}(p,q) \geq N_H^{\sim,(2)}(p,q) \geq \ldots \geq N_H^{\sim}(p,q),$$

and all these numbers are lower bounds for the cardinality of $C(p,q)$.

For the proof, see [AV2].

(4.48) THEOREM (Homotopy Invariance). *If $(p_0,q_0) \sim (p_1,q_1)$, then there exists a one-to-one map of the set of 1-essential H-Nielsen classes of (p_0,q_0) into the set of 0-essential H-Nielsen classes of (p_1,q_1), and this mapping sends $(n+1)$-essential H-Nielsen classes of (p_0,q_0) into n-essential H-Nielsen classes and essential classes into essential classes. In particular, $N_H^{\sim,(1)}(p,q)$ is a lower bound for the cardinality of $C(p_1,q_1)$. Moreover, $N_H^{\sim,(n)}(p_1,q_1) \geq N_H^{\sim,(n+1)}(p_0,q_0)$ and $N_H^{\sim}(p_1,q_1) = N_H^{\sim}(p_0,q_0)$.*

For the proof, see [AV2].

Theorem (4.48) shows impressively that, for most applications, it suffices to calculate $N_H^{\sim,(1)}(p_0,q_0)$, i.e. it usually suffices to verify that a class is 1-essential (and so actually the complicated recursion in Definition (4.45) is not needed). On the other hand, it only guarantees the homotopy invariance of the Nielsen number $N_H^{\sim}(p,q)$.

Definition (4.45) differs from the "homologic" definitions in Section 4.1, where an (under additional assumptions defined) Lefschetz number was used to define the notion of essentiality. Of course, if such a Lefschetz number is defined (such that the Lefschetz fixed point theorem holds) and is nonzero, then the corresponding class is nonempty, i.e. 0-essential in the sense of Definition (4.45). Moreover, if this Lefschetz number is stable under the homotopies from $\mathcal{H}$ (which obviously holds for the classes of homotopies considered in Section 4.1), then this class is essential in the sense of Definition (4.45). Hence, $N_H^{\sim}(p,q)$ is always at least as large as the Nielsen numbers defined in Section 4.1, when $\mathcal{H}$ corresponds to the class of homotopies considered there. But $N_H^{\sim}(p,q)$ might be larger (and is defined for a richer class of maps), so it is — at least from a theoretical point of view — a better homotopy invariant number which estimates the number of coincidences.

4.5. Remarks and comments. The standard references for the Nielsen fixed point theory are, in the single-valued case, [Br1], [J], [KT]. Its further development with some applications are described in [McC]. One can also learn about its significant relationship with the Thurston theory concerning the classification of surfaces in [CB]. The multivalued Nielsen theory was developed rather recently in [An2], [AGJ1]–[AGJ4] (see also the references therein). In [AG2], for (not necessarily metric) locally convex spaces.

The relative Nielsen theory was initiated by H. Schirmer (see [Schi], where further directions are also indicated). The multivalued case is treated in [AGJ3]. The origins of the Nielsen theory for periodic points are connected with the names of Boju Jiang and B. Halpern. For further extensions, see the survey paper [He] and the monograph [Fel], where Zeta functions are employed. The multivalued case was studied in [AGJ4]. The scheme for noncompact versions of the Nielsen theory, including the one for condensing maps, was given in [Scho], but provided the troublesome Lefschetz number is well-defined (cf. Section 3.5). In [AV2], we introduced two recursive (Wecken type and Nielsen type) definitions for coincidences which are not necessarily based on a Lefschetz number.

Application of the Nielsen theory to multiplicity results for differential equations is sometimes called a *problem of J. Leray* (cf. e.g. the paper of R. F. Brown in [McC]). Because of its difficulty, these applications are rather rare (see e.g. [An2], [An3], [AGJ1]–[AGJ4], [Br2], [Br3], [Fec], [Mat1], [Mat2], [McC], and the references therein).

5. Fixed point index

In this section, fixed point indices will be defined separately for compact and condensing admissible maps. The continuation principles, suitable for applications, will be formulated for particular classes of J-maps.

5.1. Fixed point index for compact maps. The aim of this section is to define the fixed point index for compact admissible mappings on retracts of open subsets of (convex sets in) Fréchet spaces.

At first, following [AG], we recall the definition and properties of the fixed point index for admissible maps on open sets in the Euclidean space $\mathbb{R}^n$.

Consider the diagram:

$$X \overset{p}{\Longleftarrow} \Gamma \overset{q}{\longrightarrow} Y.$$

The above diagram induces a multivalued mapping $\varphi(p,q)\colon X \multimap Y$ defined by the formula:

$$\varphi(p,q)(x) = q(p^{-1}(x)), \quad \text{for every } x \in X.$$

In what follows, we shall identify the map $\varphi(p,q)$ with the pair (p,q). Of course, $\varphi(p,q)$ is admissible, but we keep only one selected pair (p,q) of this map. Moreover, $\varphi(p,q)$ is compact if and only if q is compact.

For a multivalued map $(p,q)\colon X \multimap Y$, we denote the set of coincidence points and fixed points as:

$$C(p,q) = \{z \in \Gamma : p(z) = q(z)\}, \quad \mathrm{Fix}(p,q) = \{x \in X : x \in q(p^{-1}(x))\}.$$

Evidently, $p(C(p,q)) = \mathrm{Fix}(p,q)$, and so:

$$\mathrm{Fix}(p,q) \neq \emptyset \Leftrightarrow C(p,q) \neq \emptyset.$$

By S^n, we shall denote the unit sphere in $\mathbb{R}^{n+1}$, i.e. $S^n = \{x \in \mathbb{R}^{n+1} : |x| = 1\}$. It is well-known that the one point compactification of $\mathbb{R}^n$ is S^n, i.e. $S^n = \mathbb{R}^n \cup \{\infty\}$.

Let U be an open subset of $\mathbb{R}^n$ and assume that $(p,q)\colon U \multimap \mathbb{R}^n$ is an admissible map such that $\mathrm{Fix}(p,q)$ is compact. We have the following diagram:

$$U \stackrel{p}{\Longleftarrow} \Gamma \stackrel{q}{\longrightarrow} \mathbb{R}^n$$

and its induced one:

$$(U, U \setminus \mathrm{Fix}(p,q)) \stackrel{\overline{p}}{\Longleftarrow} (\Gamma, p^{-1}(U \setminus \mathrm{Fix}(p,q)) \stackrel{\overline{q}}{\longrightarrow} (\mathbb{R}^n, \mathbb{R}^n \setminus \{0\}), \tag{5.1}$$

where $\overline{p}(z) = p(z)$ and $\overline{q}(z) = p(z) - q(z)$, for every $z \in \Gamma$. Observe that $\overline{p}$ and $\overline{q}$ are well-defined and $\overline{p}$ is a Vietoris map.

Now, we can extend the diagram (5.1) to the following one:

$$S^n \stackrel{i}{\longrightarrow} (S^n, S^n \setminus \mathrm{Fix}(p,q)) \stackrel{j}{\longleftarrow} (U, U \setminus \mathrm{Fix}(p,q)) \stackrel{\overline{p}}{\Longleftarrow}$$
$$(\Gamma, p^{-1}(U \setminus \mathrm{Fix}(p,q)) \stackrel{\overline{q}}{\longrightarrow} (\mathbb{R}^n, \mathbb{R}^n \setminus \{0\}),$$

where i, j are the respective inclusions. We would like to apply the Čech homology functor with compact carriers and coefficients in integers $\mathbb{Z}$ (for more details, see [Go1].

Note that, by the excision axiom, j_* is an isomorphism. Moreover,

$$H_n(S^n) = H_n(\mathbb{R}^n, \mathbb{R}^n \setminus \{0\}) = \mathbb{Z}.$$

By applying the functor H_n to (5.1), we can define the *fixed point index* $\mathrm{ind}(p,q)$ of (p,q) by putting (cf. [Go1]):

$$\mathrm{ind}(p,q) = ((\overline{q}_{*n} \circ (\overline{p}_{*n})^{-1} \circ (j_{*n})^{-1} \circ i_{*n}))(1) \in \mathbb{Z}. \tag{5.2}$$

It is useful to consider the homology class

$$O_{\mathrm{Fix}(p,q)} = ((j_{*n})^{-1} \circ i_{*n})(1) \in H_n(U, U \setminus \mathrm{Fix}(p,q))$$

called the *fundamental class* of $\mathrm{Fix}(p,q)$ in U. Using the notion of the fundamental class, one can define:

$$\mathrm{ind}(p,q) = (\overline{q}_{*n} \circ (\overline{p}_{*n})^{-1})(O_{\mathrm{Fix}(p,q)}). \tag{5.3}$$

Below we shall collect the most important properties of the fixed point index (cf. [Go1]):

(5.4) PROPOSITION. *Assume that* $(p,q)\colon U \multimap \mathbb{R}^n$ *is a multivalued map and* $\mathrm{Fix}(p,q)$ *is compact.*

(5.4.1) (Existence) *If* $\mathrm{ind}(p,q) \neq 0$, *then* $\mathrm{Fix}(p,q) \neq \emptyset$.

(5.4.2) (Localization) *If* V *is an open subset of* $\mathbb{R}^n$ *such that* $\mathrm{Fix}(p,q) \subset V \subset U$, *then* $\mathrm{ind}(p,q) = \mathrm{ind}(p_1,q_1)$, *where*

$$V \xleftarrow{p_1} p^{-1}(V) \xrightarrow{q_1} \mathbb{R}^n, \quad p_1(z) = p(z), q_1(z) = q(z),$$

i.e. $(p_1,q_1)\colon V \multimap \mathbb{R}^n$ *is the restriction of* (p,q).

(5.4.3) (Additivity) *Assume that* $U = U_1 \cup U_2$, *where* U_1, U_2 *are open in* $\mathbb{R}^n$. *Assume, furthermore, that* $(p_1,q_1)\colon U_1 \multimap \mathbb{R}^n$, $(p_2,q_2)\colon U_2 \multimap \mathbb{R}^n$ *are respective restrictions of* (p,q), $\mathrm{Fix}(p_1,q_1)$, $\mathrm{Fix}(p_2,q_2)$ *are compact and* $\mathrm{Fix}(p_1,q_1) \cap \mathrm{Fix}(p_2,q_2) = \emptyset$, *then*

$$\mathrm{ind}(p,q) = \mathrm{ind}(p_1,q_1) + \mathrm{ind}(p_2,q_2).$$

(5.4.4) (Homotopy) *If* $(p_1,q_1), (p_2,q_2)\colon U \multimap \mathbb{R}^n$ *are homotopic and the joining homotopy of* (p_1,q_1) *with* (p_2,q_2) *has a compact set of fixed points, then*

$$\mathrm{ind}(p_1,q_1) = \mathrm{ind}(p_2,q_2).$$

(5.4.5) (Contraction) *Assume that* $q(p^{-1}(U)) \subset \mathbb{R}^{n-k}$ *and let* $\widetilde{U} = U \cap \mathbb{R}^{n-k}$, $(p_1,q_1)\colon \widetilde{U} \multimap \mathbb{R}^{n-k}$ *be the respective contraction of* (p,q). *Then:*

$$\mathrm{ind}(p,q) = \mathrm{ind}(p_1,q_1).$$

(5.4.6) (Multiplicity) *Let* $U \subset \mathbb{R}^n$, $U' \subset \mathbb{R}^{n'}$ *be open sets and* $(p,q)\colon U \multimap \mathbb{R}^n$, $(p',q')\colon U' \multimap \mathbb{R}^{n'}$ *be two maps such that* $\mathrm{Fix}(p,q)$ *and* $\mathrm{Fix}(p',q')$ *are compact sets of fixed points and*

$$\mathrm{ind}(p \times p', q \times q') = \mathrm{ind}(p,q) \cdot \mathrm{ind}(p',q').$$

(5.4.7) (Normalization) *Assume that* $(p,q)\colon U \multimap U$ *is compact. Then*

$$\mathrm{ind}(p,\widetilde{q}) = \Lambda(p,q),$$

where $\widetilde{q}\colon \Gamma \to \mathbb{R}^n$, $\widetilde{q}(z) = q(z)$, *for every* $z \in \Gamma$.

Instead of calculating the fixed points of a map (p,q), we can calculate the points for which zero is in its image. Such points are usually called *equilibrium points*. A good tool to do it is the topological degree. We shall explain below what the topological degree is and what is its connection with the fixed point index.

Let U be an open subset of $\mathbb{R}^n$ and let $(p,q)\colon \overline{U} \multimap \mathbb{R}^n$ be a map such that:

$$0 \notin (x - \varphi(p,q))(x), \quad \text{for every } x \in \partial U. \tag{5.5}$$

Assume, furthermore, that (p,q) is a compact map and the map $\Phi(p,q)\colon \overline{U} \multimap \mathbb{R}^n$ is defined by the formula:

$$\Phi(p,q)(x) = x - \varphi(p,q)(x), \quad \text{for every } x \in \overline{U}.$$

Then $\Phi(p,q)$ is called the *vector field* associated with $\varphi(p,q)$, and Φ is called a *compact vector field* if $\varphi(p,q)$ is compact.

(5.6) DEFINITION. Let $\Phi(p,q)\colon \overline{U} \multimap \mathbb{R}^n$ be a compact vector field such that:

$$\{x \in \overline{U},\ 0 \in \Phi(p,q)(x)\} \cap \partial U = \emptyset.$$

We define the *topological degree* $\deg(\Phi(p,q);U)$ of $\Phi(p,q)$ with respect to U by the formula:

$$\deg(\Phi(p,q),U) = \operatorname{ind}(p,q).$$

Let us observe that, under the above assumptions, the set $\operatorname{Fix}(p,q)$ is compact.

Now, the properties of the topological degree analogous to those presented in (5.4) can be formulated as follows.

(5.7) PROPOSITION. *Let $\Phi(p,q)\colon \overline{U} \multimap \mathbb{R}^n$ be a compact vector field such that:*

$$\{x \in \overline{U} : 0 \in \Phi(p,q)(x)\} \cap \partial U = \emptyset.$$

(5.7.1) (Existence) *If $\deg(\Phi(p,q);U) \neq 0$, then there is $x \in U$ such that $0 \in \Phi(p,q)(x)$.*

(5.7.2) (Localization) *If $\{x \in \overline{U} : \Phi(p,q)(x)\} \subset V \subset U$, then*

$$\deg(\Phi(p,q),U) = \deg(\Phi(p,q),V).$$

(5.7.3) (Additivity) *Let U, U_1, U_2 be open subset of $\mathbb{R}^n$ such that $U_1 \cup U_2 \subset U$, $U_1 \cap U_2 = \emptyset$ and $0 \notin \phi(p,q)(\overline{U} \setminus (U_1 \cup U_2))$, then*

$$\deg(\phi(p,q),U) = \deg(\phi(p,q),U_1) + \deg(\phi(p,q),U_2).$$

(5.7.4) (Homotopy) (1) *If $\phi(p,q) \sim \psi(p',q')$, then*

$$\deg(\varphi(p,q),U) = \deg(\psi(p',q'),U).$$

(1) Here, the homotopy is understood as a homotopy of compact vector fields with no zero, for every $x \in \partial U$ and every $t \in [0,1]$.

(5.7.5) (Multiplicity) *Assume that* $U \subset \mathbb{R}^n$ *and* $V \subset \mathbb{R}^m$ *are open and* $\phi_1 : \overline{U} \multimap \mathbb{R}^n$, $\phi_2 : \overline{V} \multimap \mathbb{R}^m$ *are admissible compact vector fields with no zero on the boundaries* ∂U *and* ∂V, *respectively. Then*

$$\deg(\phi \times \psi, U \times V) = \deg(\phi, U) \cdot \deg(\psi, V).$$

(5.7.6) (Contraction) *Let* $\phi(p,q) : \overline{U} \multimap \mathbb{R}^n$ *and assume that* $\phi(p,q)(\overline{U}) \subset \mathbb{R}^{n-k}$, *for some* $k = 1, 2, \ldots$ *Then*

$$\deg(\phi(p,q), U) = \deg(\phi(p,q)|_{U \cap \mathbb{R}^{n-k}}, U \cap \mathbb{R}^{n-k}).$$

Now, we are going to define the topological degree and the fixed point index for admissible compact vector fields (admissible mappings) in Fréchet spaces. We shall start by the topological degree. As above, (p,q) is a compact multivalued map defined on $\overline{U}$ with values in E, where U is an open subset of a Fréchet space E. We shall put $\phi = \phi(p,q) : \overline{U} \multimap E$ to be a compact vector field defined as follows:

$$\phi(x) = x - q(p^{-1}(x)) = \{x - y : y \in q(p^{-1}(x))\}.$$

We shall also assume that:

$$0 \notin \phi(x), \quad \text{for every } x \in \partial U. \tag{5.8}$$

We prove the following

(5.9) Proposition. *If* $\phi : \overline{U} \multimap E$ *is a compact vector field, then* ϕ *is a closed map, i.e. for every closed* $A \subset \overline{U}$, *the set* $\phi(A)$ *is closed in* E.

Proof. Assume that A is a closed subset of $\overline{U}$ and $y \in \overline{\phi(A)}$. It is enough to show that $y \in \phi(A)$. Since $\phi = \phi(p,q)$ and q is compact, there is a compact set $K \subset E$ such that $qp^{-1}(\overline{U}) \subset K$. Since $y \in \overline{\phi(A)}$, it follows that every open neighbourhood V of the zero point in E, we have $(y + V) \cap \phi(A) \neq \emptyset$.

Let $y_V \in (y+V) \cap \phi(A)$ and let $x_V \in A$ be such that $y_V \in (x_V - q(p^{-1}(x_V))$. Let us consider the generalized sequence $\{x_V - y_V\}_V$, where V is an neighbourhood of the zero point in E. Since $\{x_V - y_V\} \subset K$, we can assume that it converges to a point $x \in K$.

Now, it is easy to observe that $\{x_V\}$ converges to $x + y$. Since A is closed, we get $(x + y) \in A$ and the u.s.c. of ϕ implies that the graph Γ_ϕ of ϕ is closed. Consequently, $y \in \phi(x + y)$, and so $y \in \phi(A)$. The proof is completed. □

We let $\mathcal{A}(U, \partial U) = \{\phi = \phi(p,q) : \phi : \overline{U} \to E,\ \phi \text{ satisfies } (5.8)\}$. We would like to define a function Deg: $\mathcal{A}(U, \partial U) \to \mathbb{Z}$, which satisfies the analogous properties to (5.7.1)–(5.7.6), where $\mathbb{Z}$ denotes the set of integers.

Let $\phi = \phi(p,q) \in \mathcal{A}(U, \partial U)$. Since $\phi(\partial U)$ is a closed subset of E and $0 \notin \phi(\partial U)$, there exists an open neighbourhood V_0 of the zero point in E such that $V_0 \cap \phi(\partial U) = \emptyset$. Let $V = V_0/2$ and $K = \overline{q(p^{-1}(\overline{U})}$. Since K is compact, there exists a finite set $\{y_1, \dots, y_k\} \subset K$ such that $K \subset \bigcup_{i=1}^{k}\{y_i + V\}$. We let $\alpha = \{y_i + V\}_{i=1,\dots,k}$. Now, let β be a refinement of α such that any two β-close mappings are α-stationary homotopic (for more details, see [AG]). By using the Schauder-like approximation Theorem (I.1.37) in [AG] (cf. Lemma (5.24) below), we get a β-approximation $\pi_\beta \colon K \to E$ of the inclusion $i \colon K \to E$ such that $\pi_\beta(K) \subset E^{n=n(\beta)} \subset E$. We let $q_\beta = \pi_\beta \circ q$ and let $\phi_\beta = \phi(p, q_\beta) \colon \overline{U} \to E$.

By a standard calculation, it is easy to see that:

$$\phi_\beta \in \mathcal{A}(U, \partial U). \tag{5.10}$$

Now, we let $\phi_{\beta,n(\beta)}$ to be the contraction of ϕ_β to the pair $(\overline{U} \cap E^{n(\beta)}, E^{n(\beta)})$. Then we have

$$\phi_{\beta,n(\beta)} \in (U \cap E^{n(\beta)}, E^n(\beta)). \tag{5.11}$$

Consequently, the topological degree $\deg(\phi_{\beta,n(\beta)}, U \cap E^{n(\beta)})$ is well-defined.

(5.12) Definition. $\operatorname{Deg}(\phi, U) = \deg(\phi_{\beta,n(\beta)}, U \cap E^{n(\beta)})$.

Using the properties (5.7.1)–(5.7.6), it is easy to see that Definition (5.12) is correct.

We would like to add that properties (5.7.1)–(5.7.6) can be reformulated for the topological degree Deg defined above. We left the respective formulations to the reader.

Now, we shall show that the topological degree $\operatorname{Deg} \colon \mathcal{A}(U, \partial U) \to \mathbb{Z}$ will help us to define the fixed point index for compact admissible mappings in Fréchet spaces.

Let $\varphi \in \mathbb{K}(U, E)$, where U is an open subset of a Fréchet space E. Moreover, we shall assume:

$$\operatorname{Fix}(\varphi) \text{ is a compact subset of } U. \tag{5.13}$$

We put $\mathbb{K}_0(U, E) = \{\varphi \in \mathbb{K}(U, E) : (5.13) \text{ is satisfied}\}$.

At first, we are going to define the fixed point index ind on $\mathbb{K}_0(U, E)$ as a function $\operatorname{ind} \colon \mathbb{K}_0(U, E) \to \mathbb{Z}$. To get $\operatorname{ind}(\varphi)$ as a singleton, we will assume that $(p, q) \subset \varphi$ is fixed, i.e. $\varphi = \varphi(p, q) = q \circ p^{-1}$.

Let $\varphi = \varphi(p, q) \in \mathbb{K}_0(U, E)$. Then there is an open subset U_0 of E such that

$$\operatorname{Fix}(\varphi) \subset U_0 \subset \overline{U}_0 \subset U.$$

Observe that then we have $\phi = \phi(p,q) \in \mathcal{A}(U_0, \partial U_0)$. Consequently, we let:

$$\operatorname{ind}(\varphi) = \operatorname{Deg}(\phi, U_0). \tag{5.14}$$

The additivity of the topological degree Deg implies that (5.14) does not depend on the choice of U_0.

Assume that X is a retract of an open subset U of E and $r\colon U \to X$ is a retraction map. Assume, furthermore, that: $\varphi = \varphi(p,q)\colon V \multimap X$ is a compact map with $\operatorname{Fix}(\varphi)$ being compact and V to be an open subset of X. Then we have the diagram

$$r^{-1}(V) \xrightarrow{\ r\ } V \overset{\varphi}{\multimap} X \xrightarrow{\ i\ } E.$$

It is easy to see that $\varphi \circ r \in \mathbb{K}_0(r^{-1}(V), E)$. We define

$$\operatorname{ind}(\varphi) = \operatorname{ind}(\varphi \circ r). \tag{5.15}$$

Note that the above definition depends on the choice of the retraction map.

Nevertheless, we have:

(5.16) THEOREM. *Assume that X is a neighbourhood retract of (a convex set in; cf.* [AG]) *E and $r\colon U \to X$ is a given retraction map.*

(5.16.1) (Additivity) *Let V_1, V_2, V be open subsets of X such that $V_1 \cap V_2 = \emptyset$, $V_1 \cup V_2 \subset V$ and let $\varphi \in \mathbb{K}_0(V, X)$, then* $\operatorname{ind}(\varphi) = \operatorname{ind}(\varphi_1) + \operatorname{ind}(\varphi_2)$, *where φ_i, $i = 1,2$ are restrictions of φ to V_1 and V_2, respectively.*

(5.16.2) *If* $\operatorname{ind}(\varphi) \neq 0$, *then* $\operatorname{Fix}(\varphi) \neq \emptyset$.

(5.16.3) (Homotopy) *If $\varphi(p,q) \sim \varphi(p_1,q_1)$ with the homotopy having the compact fixed point set, then* $\operatorname{ind}(\varphi(p,q)) = \operatorname{ind}(\varphi(p_1,q_1))$.

(5.16.4) *are neighbourhood retracts of (convex sets in; cf.* [AG]) *E_1 and E_2, respectively, and r_i, $i = 1,2$ are fixed retractions. Assume $\varphi_i \in \mathbb{K}(V_i, E_i)$, $i = 1,2$ and $\varphi = \varphi_1 \times \varphi_2$. Then* $\operatorname{ind}(\varphi) = \operatorname{ind}(\varphi_1) \cdot \operatorname{ind}(\varphi_2)$.

(5.16.5) (Normalization) *If $\varphi = \varphi(p,q) \in \mathbb{K}_0(X,X)$, then* $\operatorname{ind}(\varphi) = \Lambda(p,q)$.

The proof of (5.16) is straighforward and follows from the respective properties of the topological degree Deg.

5.2. Continuation principle for compact maps. Because of possible applications, it is very useful to formulate sufficiently general continuation principles.

For compact admissible maps from open subsets of a neighbourhood retract of a Fréchet space E into E, the definitions of a fixed point index were already given in the foregoing part.

Now, we will apply these results to formulating the appropriate continuation principles. We restrict ourselves only to a particular class of admissible maps, namely to R_δ-maps $\Phi\colon D \multimap E$, (written $\Phi \in J(D,E)$), i.e. u.s.c. maps with R_δ-values.

We often need to study fixed points for maps defined on sufficiently fine sets (possibly with an empty interior), but with values out of them. Making use of the previous results, we are in position to make the following construction.

Assume that X is a retract of a Fréchet space E and D is an open subset of X. Let $\Phi \in J(D,E)$ be locally compact, $\mathrm{Fix}(\Phi)$ be compact and let the following condition hold:

$$\text{(A)} \qquad \forall x \in \mathrm{Fix}(\Phi)\ \exists U_x \ni x,\ U_x \text{ is open in } D \text{ such that } \Phi(U_x) \subset X.$$

The class of locally compact J-maps from D to E with the compact fixed point set and satisfying (A) will be denoted by the symbol $J_A(D,E)$. We say that $\Phi, \Psi \in J_A(D,E)$ are homotopic in $J_A(D,E)$ if there exists a homotopy $H \in J(D \times [0,1], E)$ such that $H(\,\cdot\,,0) = \Phi$, $H(\,\cdot\,,1) = \Psi$, for every $x \in D$, there is an open neighbourhood V_x of x in D such that $H|_{V_x \times [0,1]}$ is compact, and

$$(\mathrm{A}_H) \quad \forall x \in D\ \forall t \in [0,1]\ [x \in H(x,t) \Rightarrow \exists U_x \ni x,\ U_x \text{ is open in } D \quad H(U_x \times [0,1]) \subset X].$$

Note that the condition (A_H) is equivalent to the following one:

- If $\{x_j\}_{j\geq 1} \subset D$ converges to $x \in H(x,t)$, for some $t \in [0,1]$, then $H(\{x_j\} \times [0,1]) \subset X$, for j sufficiently large.

Let $\Phi \in J_A(D,E)$. Then $\mathrm{Fix}(\Phi) \subset \bigcup\{U_x : x \in \mathrm{Fix}(\Phi)\} \cap V =: D' \subset D$ and $\Phi(D') \subset X$, where V is a neighbourhood of the set $\mathrm{Fix}(\Phi)$ such that $\Phi|_V$ is compact (by the compactness of $\mathrm{Fix}(\Phi)$ and local compactness of Φ) and U_x is a neighbourhood of x as in (A). Define

$$(5.17) \qquad \mathrm{Ind}_A(\Phi, X, r, D) = \mathrm{ind}(\Phi|_{D'}, X, r, D'),$$

where $\mathrm{ind}(\Phi|_{D'}, X, r, D')$ is defined as in (5.15). This definition is independent of the choice of D'.

In the following theorem, we give some properties of Ind_A which will be used in the proof of the continuation Theorem (5.19). The simple proof is omitted.

(5.18) THEOREM.

(5.18.1) (Existence) *If* $\mathrm{Ind}_A(\Phi, X, r, D) \neq 0$, *then* $\mathrm{Fix}(\Phi) \neq \emptyset$.

(5.18.2) (Localization) *If* $D_1 \subset D$ *are open subsets of a retract* X *of a space* E, $\Phi \in J_A(D,E)$ *is compact, and* $\mathrm{Fix}(\Phi)$ *is a compact subset of* D_1, *then*

$$\mathrm{Ind}_A(\Phi, X, r, D) = \mathrm{Ind}_A(\Phi, X, r, D_1).$$

(5.18.3) (Homotopy) *If H is a homotopy in $J_A(D,E)$, then*

$$\mathrm{Ind}_A(H(\cdot,0),X,r,D) = \mathrm{Ind}_A(H(\cdot,1),X,r,D).$$

(5.18.4) (Normalization) *If $\Phi \in J(X)$ is a compact map, then*

$$\mathrm{Ind}_A(\Phi,X,r,X) = 1.$$

(5.19) THEOREM (Continuation Principle). *Let X be a retract of a Fréchet space E, D be an open subset of X and H be a homotopy in $J_A(D,E)$ such that*

(5.19.1) $H(\cdot,0)(D) \subset X$,

(5.19.2) *there exists $H' \in J(X)$ such that $H'|_D = H(\cdot,0)$, H' is compact and* $\mathrm{Fix}(H') \cap (X \setminus D) = \emptyset$.

Then there exists $x \in D$ such that $x \in H(x,1)$.

PROOF. Applying the localization property, we obtain

$$\mathrm{Ind}_A(H(\cdot,0),X,r,D) = \mathrm{Ind}_A(H(\cdot,0),X,r,X).$$

By the normalization property, $\mathrm{Ind}_A(H(\cdot,0),X,r,X) = 1$. Thus, by the homotopy property, $\mathrm{Ind}_A(H(\cdot,0),X,r,D) = \mathrm{Ind}_A(H(\cdot,1),X,r,D) = 1$, which implies that $H(\cdot,1)$ has a fixed point. □

(5.20) COROLLARY. *Let X be a retract of a Fréchet space E and H be a homotopy in $J_A(X,E)$ such that $H(x,0) \subset X$, for every $x \in X$, and $H(\cdot,0)$ is compact. Then $H(\cdot,1)$ has a fixed point.*

(5.21) COROLLARY. *Let X be a retract of a Fréchet space E, D be an open subset of X and H be a homotopy in $J_A(D,E)$. Assume that $H(x,0) = x_0$, for every $x \in D$. Then there exists $x \in D$ such that $x \in H(x,1)$.*

PROOF. It is sufficient to define $H' \in J(X)$, $H'(x) = x_0$ and to use Theorem (5.19). □

The following result generalizes the well-known Ky Fan theorem in the case of Fréchet spaces.

(5.22) COROLLARY. *Let X be a retract of a Fréchet space E and $\Phi \in J(X)$ be compact. Then Φ has a fixed point.*

Some applications motivate us to consider weaker than (A_H) condition on H. Unfortunately, we cannot use the fixed point index technique described above.

(5.23) THEOREM (Continuation Principle). *Let X be a closed convex subset of a Fréchet space E and let $H \in J(X \times [0,1], E)$ be compact. Assume that*

(5.23.1) *$H(x,0) \subset X$, for every $x \in X$,*
(5.23.2) *for any $(x,t) \in \partial X \times [0,1)$ with $x \in H(x,t)$, there exist open neighbourhoods U_x of x in X and I_t of t in $[0,1)$ such that $H((U_x \cap \partial X) \times I_t) \subset X$.*

Then there exists a fixed point of $H(\cdot\,,1)$.

For the proof, let us only note that it is based on Lemma (5.24) below; for more details, see [AGG1] (cf. [AG, Theorem II.10.7]).

(5.24) LEMMA. *Let X be a convex closed subset of a Fréchet space E and $K \subset E$ be a compact subset such that $K \cap X \neq \emptyset$. Then, for every $\varepsilon > 0$, there exists a map $\pi_\varepsilon : K \to E$ whose image is contained in a finite dimensional space and such that*

(5.24.1) *$\pi_\varepsilon(K \cap X) \subset X$,*
(5.24.2) *$d(\pi_\varepsilon(x), x) < \varepsilon$, for all $x \in K$.*

(5.25) REMARK. Note that the convexity of X in Theorem (5.23) is essential only in the infinite dimensional case. For the proof, we have to intersect X with a finite dimensional subspace L.

5.3. Fixed point index for condensing maps. In this section, E is again a Fréchet space, C a convex subset of E and U an open subset of C. At first, we define the fixed point index for (compact) mappings in $\mathbb{K}(U,C)$ (cf. Section 5.1). Then, we do this for mappings in $\mathbb{C}^0(U,C) = \{\varphi \in \mathbb{C}(U,C) : \mathrm{Fix}(\varphi)$ is a compact subset of $U\}$ (cf. Section 2). Of course, $\mathbb{K}_0(U,C) := \{\varphi \in \mathbb{K}(U,C) : \mathrm{Fix}(\varphi)$ is a compact subset of $U\} \subset \mathbb{C}^0(U,C)$. In fact, we have more, namely:

$$\mathbb{K}_0(U,C) \subset \mathbb{C}^0_p(U,C) \subset \mathbb{C}^0(U,C), \quad \text{for every } p \in [0,1),$$

where $\mathbb{C}^0_p(U,C) = \{\varphi \in \mathbb{C}_p(U,C) : \mathrm{Fix}(\varphi)$ is a compact subset of $U\}$. In what follows, an admissible map $\varphi : U \multimap C$ will be identified with the pair (p,q), as above.

Note that, for such a pair (p,q), we are able to define a unique number as the fixed point index, Lefschetz number, etc.

Let $(p,q) \in \mathbb{K}_0(U,C)$. We denote $K = \overline{qp^{-1}(U)}$. By the hypothesis, K is a compact subset which contains the set $\mathrm{Fix}(p,q)$.

(5.26) PROPOSITION. *There exists a neighbourhood $V \in \mathcal{N}_E(0)$ ($\mathcal{N}_E(0)$ denotes the basis of topology at $0 \in E$) such that, for every V-approximation $\pi_V : K \to C$ of the inclusion $i: K \to C$, i.e. , a continuous map $\pi_V : K \to C$ such that $\pi_V(x) \in$*

$(x+V)$ and $\pi_V(K)\subset L$, where L is a finite dimensional subspace of E, we have that $\mathrm{Fix}(p,\pi_V\circ q)$ *is a compact subset of U.*

For the proof, see [AG].

Note that if (5.26) holds true in some $V\in\mathcal{N}_E(0)$ and $V_1\in\mathcal{N}_E(0)$ is a subset of V, then (5.26) is also true for V_1.

Now, for given $(p,q)\in\mathbb{K}_0(U,C)$, we choose V and $\pi_V\colon K\to C$ such that $\pi_V(x)\in(x+V)$ and $\pi_V(K)\subset L$, where L is a finite dimensional subspace of E.

We let $U^L=U\cap L$, $C^L=C\cap L$ and

$$U^L \xLeftarrow{p^L} p^{-1}(U^L)\xrightarrow{q^L} C^L,\quad \text{where } q^L=\pi_V\circ q.$$

Then the set $\mathrm{Fix}(p^L,q^L)$ is a compact subset of U^L and $(p^L,q^L)\in\mathbb{K}_0(U^L,C^L)$. Since C^L, as a convex subset of L, is an ANR-space, the fixed point index $\mathrm{ind}((p^L,q^L),U^L)$ is well-defined (cf. [Go2]). We define

$$\mathrm{ind}((p,q),U)=\mathrm{ind}((p^L,q^L),U^L). \tag{5.27}$$

Now, it is a standard procedure (cf. Section 5.1) to verify that (5.27) is correct.

Moreover, Theorem (5.16) can be reformulated in the considered case. So, its properties (5.16.1)–(5.16.5) hold true.

(5.28) Remark. Observe that, if $(p,q)\in\mathbb{K}_0(C,C)$, i.e. $U=C$, then from the property (5.16.5), we have $\mathrm{ind}((p,q),C)=\Lambda(p,q)$, but convexity of C implies that $\Lambda(p,q)=1$, for every $(p,q)\colon C\multimap C$. So, we have $\mathrm{ind}((p,q),C)=1$. Consequently, from the existence property (5.16.2), (ii), we get that $\mathrm{Fix}(p,q)\neq\emptyset$.

(5.29) Corollary. *If $(p,q)\in\mathbb{K}(C,C)$, then* $\mathrm{Fix}(p,q)\neq\emptyset$.

Starting from now, we shall assume that E is a Fréchet space, C is a convex subset of E, U is an open subset of C and μ is a measure of noncompactness defined in Section 2.

(5.30) Proposition. *If $\varphi\in\mathbb{C}(U,C)$, then* $\mathrm{Fix}(\varphi)$ *is relatively compact.*

Proof. In fact we have $\mathrm{Fix}(\varphi)\subset\varphi(\mathrm{Fix}(\varphi))$. Consequently, from the monotonicity property of μ, we have

$$\mu(\mathrm{Fix}(\varphi))\leq\mu(\varphi(\mathrm{Fix}(\varphi))).$$

But φ is condensing, so it implies that $\mu(\mathrm{Fix}(\varphi))=0$, and the proof is completed. □

We shall also use the notion of a fundamental set.

(5.31) Definition. Let $(p, q) \in \mathbb{C}(U, C)$. A compact, convex, nonempty set $S \subset C$ is called a *fundamental set*if:

(5.31.1) $q(p^{-1}(U \cap S)) \subset S$,
(5.31.2) if $x \in \overline{\mathrm{conv}}(\varphi(x) \cup S)$, then $x \in S$.

A set S is called *ω-fundamental* for (p, q) if S satisfies all conditions of (5.31), but it is not necessarily compact.

Below, we collect some important properties of fundamental sets.

(5.32) Properties. *Assume* $(p, q) \in \mathbb{C}(U, C)$.

(5.32.1) *If S is a fundamental set, for (p, q), then* $\mathrm{Fix}(p, q) \subset S$,
(5.32.2) *if S is an ω-fundamental set, for (p, q), and $P \subset S$, then the set $S' = \overline{\mathrm{conv}}(qp^{-1}(S \cap U) \cup P)$ is ω-fundamental, for (p, q), too,*
(5.32.3) *intersection of any family of fundamental sets (ω-fundamental sets), for (p, q), is also fundamental, for (p, q),*
(5.32.4) *if S is the intersection of all ω-fundamental sets, for (p, q), then we have:*

$$S = \overline{\mathrm{conv}}(qp^{-1}(S \cap U)),$$

(5.32.5) *for every compact $K \subset C$, there exists an ω-fundamental set such that*

$$S = \overline{\mathrm{conv}}(qp^{-1}(S \cap U) \cup K),$$

(5.32.6) *the family of all fundamental sets, for (p, q), is nonempty.*

For the proof, see [AG].

Now, we are able to extend the fixed point index to the class $\mathbb{C}^0(U, C)$. In what follows, we shall additionally assume that C is a closed convex subset of a Fréchet space. From the above assumption and (5.32), it follows, for any $(p, q) \in \mathbb{C}(U, C)$ that there exists a fundamental set S such that $S \subset C$.

Let $(p, q) \in \mathbb{C}^0(U, C)$ and $S = S(p, q)$ be a fundamental set for (p, q). Then $U_S = U \cap S$ is an open subset of S and the compact set $\mathrm{Fix}(p, q)$ is contained in U_S.

For $(p, q)\colon U \overset{p}{\Longleftarrow} \Gamma \overset{q}{\longrightarrow} C$, we consider $(p_S, q_S)\colon U_S \overset{p_S}{\Longleftarrow} p^{-1}(U_S) \overset{q_S}{\longrightarrow} S$, where p_S, q_S are the respective restrictions of p and q.

Therefore, $(p_S, q_S) \in \mathbb{K}_0(U_S, S)$ and $\mathrm{ind}((p_S, q_S), U_S)$ is defined as in (5.27). We define

$$\mathrm{ind}((p, q), U) = \mathrm{ind}((p_S, q_S), U_S). \tag{5.33}$$

It is straigtforward that Definition (5.33) is correct.

Moreover, properties (5.16.1)–(5.16.4) can be reformulated for our index defined in (5.33).

The problem of the normalization property is more complicated. It is related to the obstructions with the appropriate definition of the generalized Lefschetz number (see Section 3.4).

(5.34) THEOREM. *If* $(p,q) \in \mathbb{C}(C,C)$ *(in particular,* $(p,q) \in \mathbb{C}_\lambda(C,C)$, $\lambda \in [0,1]$*) with* C *to be a closed convex subset of a Fréchet space* E*, then* $\mathrm{Fix}(p,q) \neq \emptyset$.

PROOF. Let $(p,q)\colon C \overset{p}{\Longleftarrow} \Gamma \overset{q}{\longrightarrow} C$ and let S be a fundamental set for (p,q). We define:

$$(p_S, q_S)\colon S \overset{p_S}{\Longleftarrow} p^{-1}(S) \overset{q_S}{\longrightarrow} C$$

(as in (5.33)). Then $\mathrm{Fix}(p,q) = \mathrm{Fix}(p_S,q_S)$, but, in view of (3.12) in Section 3.1, we get $\mathrm{Fix}(p_S,q_S) \neq \emptyset$, and the proof is completed. □

Now, assume that U is an open subset of C and $(p,q)\colon U \multimap C$ is condensing (or, in particular, a λ-set contraction) and $\mathrm{Fix}(p,q)$ is compact.

Thus, instead of a (strong) normalization property, we can give the weak one, namely

(5.35) (Weak normalization) *Assume that* $(p,q) \in \mathbb{C}(U,C)$ *is a constant map, i.e.* $q(p^{-1}(x)) = a \notin \partial U$*, for each* $x \in U$*. Then*

$$\mathrm{Ind}((p,q),U) = \begin{cases} 1 & \text{for } a \in U, \\ 0 & \text{for } a \notin U. \end{cases}$$

5.4. Continuation principle for condensing maps. As already pointed out in Section 5.2, in the applications of the fixed point theory, we often need to consider maps with values in a Fréchet space and not in a closed convex set. We will extend our theory to this case, similarly as in Section 5.2.

Again, let E be a Fréchet space and X be a closed and convex subset of E. Let $U \subset X$ be open and consider the map $\varphi \in J_A(U,F)$, where the symbol $J_A(U,F)$ is again reserved for J-maps from U to E satisfying condition (A) (see Section 5.2). The notion of homotopy in J_A will be understood analogously. Thus, $\mathrm{Fix}(\varphi)$ is compact and φ has a compact fundamental set T. Set $\mathrm{Ind}_X^A(\varphi,U) := 0$, whenever $\mathrm{Fix}(\varphi) = \emptyset$. Otherwise, let $x_1,\dots,x_n \in \mathrm{Fix}(\varphi)$ such that $\mathrm{Fix}(\varphi) \subset \bigcup_{i=1}^n U_{x_i} =: V$, where U_{x_i} are neighbourhoods of x_i such that $\overline{U}_{x_i} \subset U$ and satisfy condition (A). Then $\varphi|_V\colon V \multimap X$ is a J-map with compact fundamental set T and satisfies $\mathrm{Fix}(\varphi) \cap \partial V = \emptyset$. Thus, we can define

$$\mathrm{Ind}_X^A(\varphi,U) := \mathrm{Ind}_X(\varphi|_{\overline{V}}, V). \tag{5.36}$$

The independence of this definition of the chosen set V follows from the additivity property. Furthermore, if $\varphi\colon \overline{U} \multimap X$ has a compact fundamental set and $\mathrm{Fix}(\varphi) \cap \partial U = \emptyset$, then $\mathrm{Ind}^A_X(\varphi, U)$ is defined and

$$\mathrm{Ind}^A_X(\varphi, U) = \mathrm{Ind}_X(\varphi, U).$$

The following proposition follows easily from the above argumentation.

(5.37) PROPOSITION.

(5.37.1) (Existence) *If* $\mathrm{Ind}^A_X(\varphi, U) \neq 0$, *then* $\emptyset \neq \mathrm{Fix}(\varphi)$.

(5.37.2) (Additivity) *Let* $\mathrm{Fix}(\varphi) \subset U_1 \cup U_2)$, *where* U_1, U_2 *are open disjoint subsets of* U. *Then*

$$\mathrm{Ind}^A_X(\varphi, U) = \mathrm{Ind}^A_X(\varphi|_{U_1}, U_1) + \mathrm{Ind}^A_X(\varphi|_{U_2}, U_2).$$

(5.37.3) (Homotopy) *Let* $\psi\colon U \multimap F$ *be homotopic in* J_A *to the map* φ. *Assume that the homotopy* $\chi\colon U \times [0,1] \multimap F$ *has a compact fundamental set and the set*

(5.37.3.1) $$\Sigma := \{(x,t) \in U \times [0,1] : x \in \chi(x,t)\}$$

is compact. Then $\mathrm{Ind}^A_X(\varphi, U) = \mathrm{Ind}^A_X(\psi, U)$.

(5.37.4) (Weak normalization) *Assume that* $\varphi\colon U \to F$ *is a constant map* $\varphi(x) = a \in F$, *for all* $x \in U$. *Then*

$$\mathrm{Ind}^A_X(\varphi, U) = \begin{cases} 1 & \text{for } a \in U, \\ 0 & \text{for } a \notin U. \end{cases}$$

Let us now formulate a continuation principle which is convenient for various applications.

(5.38) THEOREM (Continuation Principle). *Let* X *be a closed, convex subset of a Fréchet space* E, *let* $U \subset X$ *be open and let* $\chi\colon U \times [0,1] \multimap F$ *be a homotopy in* J_A *such that* Σ *(see (5.37.3.1)) is compact. Let* χ *be condensing and assume that there is a condensing* $\varphi \in J(X)$ *such that* $\varphi|_U = \chi(\cdot\,, 0)$ *and* $\mathrm{Fix}(\varphi) \cap (X \setminus U) = \emptyset$. *Then* $\chi(\cdot\,, 1)$ *has a fixed point.*

PROOF. The proof follows, in view of the existence property (5.37.1), from the following equations:

$$\mathrm{Ind}^A_X(\chi(\cdot\,,1), U) = \mathrm{Ind}^A_X(\chi(\cdot\,,0), U),$$

by the homotopy property (4.57.3),

$$\mathrm{Ind}^A_X(\chi(\cdot\,,0), U) = \mathrm{Ind}^A_X(\varphi|_U, U) = \mathrm{Ind}^A_X(\varphi, X),$$

by the additivity property (3.57.2). Finally, we see (cf. (5.34), (5.35)) that

$$\mathrm{Ind}^A_X(\varphi, X) = \mathrm{Ind}_X(\varphi, X) = 1. \qquad \square$$

(5.39) COROLLARY. *Let $\chi: X \times [0,1] \multimap F$ be a condensing homotopy in J_A such that $\chi(x,0) \subset X$, for every $x \in X$. Then $\chi(\,\cdot\,,1)$ has a fixed point.*

5.5. Remarks and comments. There is a lot of definitions of a topological degree and a fixed point index (see e.g. [GM], [Kz], [Mw], [Nu], [Ro], in the single-valued case, and [AG2], [BGMO], [Go2], [Kr], [Ma], in the multivalued case). Many interesting informations and references can be found in the survey paper [Mw], but we should mention with this respect at least the names of J. P. Schauder, J. Leray, F. E. Browder, A. Dold, A. Granas, M. A. Krasnosel'skiĭ, J. Mawhin, R. D. Nussbaum, ... A definition of an index for (especially condensing) noncompact maps on (especially nonsimply connected) nonconvex sets is of the same sort as in the Lefschetz theory (cf. [AV1]).

Degree arguments are perhaps the most frequent tools in nonlinear analysis at all and, in particular, in boundary value problems (cf. again [Mw]). It is almost impossible to indicate at least the most important applications in this field, and so we restrict ourselves (besides the quoted monographs containing plenty of applications) to the papers [ABa], [AGG1]–[AGG3], which are the most strictly related to our presentation here.

6. Conley type (integer-valued) index

Unlike the standard definitions of the Conley index, see e.g. [Co], [Ry2], [Sm], we can define an integer-valued Conley type index as the fixed point index, or other topological invariants, of induced maps in the related hyperspaces. This idea appeared independently in the paper [RPS] for single-valued continuous maps in locally compact ANR-spaces and in [AG], [AFGL], where it has been applied to develop a continuation principle for fractals generated by multivalued Hausdorff-continuous compact maps in spaces such that the related hyperspaces are ANRs.

6.1. Conley type index for compact maps.

(6.1) DEFINITION. We say that a metric space (X,d) is a *hyper absolute neighbourhood retract* (a *hyper absolute retract*) if $(\mathcal{K}(X), d_H)$, i.e. the hyperspace of compact subsets of X endowed with the induced Hausdorff metric, is an absolute neighbourhood retract (an absolute retract); written $X \in \mathrm{HANR}$ ($X \in \mathrm{HAR}$).

The best candidates for HANRs (HARs) are, to our knowledge, those in the following

(6.2) LEMMA ([Cur, p. 141]). *If X is a locally continuum-connected, i.e. every point has a basis of neighbourhoods such that every two points of each can be joined by a continuum inside, (a connected and locally continuum-connected) metric space, then $X \in \mathrm{HANR}$ ($X \in \mathrm{HAR}$).*

(6.3) REMARK. Obviously, AR $\Rightarrow$ ANR $\Rightarrow$ locally continuum-connected $\Rightarrow$ locally connected, where ANR (AR) denotes the class of absolute neighbourhood retracts (absolute retracts).

Now, we recall several notions which are typical in the frame of the Conley index theory (see e.g. [KM]).

Let $F\colon X \multimap X$ be a compact Hausdorff-continuous map with nonempty closed values. Defining the *(semi) invariant parts* of $N \subset X$ w.r.t. $F\colon X \multimap X$ as

- $\mathrm{Inv}^+(N,F) := \{x \in N : \sigma(i+1) \in F(\sigma(i))$, for all $i \in \mathbb{N} \cup \{0\}$, where $\sigma\colon \mathbb{N} \cup \{0\} \to N$ is a single-valued map with $\sigma(0) = x\}$,
- $\mathrm{Inv}^-(N,F) := \{x \in N : \sigma(i+1) \in F(\sigma(i))$, for all $i \in \mathbb{Z} \setminus \mathbb{N}$, where $\sigma : \mathbb{Z} \setminus \mathbb{N} \to N$ is a single-valued map with $\sigma(0) = x\}$,
- $\mathrm{Inv}(N,F) := \mathrm{Inv}^+(N,F) \cap \mathrm{Inv}^-(N,F)$,

we say that a compact invariant set $K \subset U \subset X$, where U is locally compact, (i.e. $F(K) = K$) is *isolated* w.r.t. F if there exists a compact neighbourhood N of K such that

$$B_{\mathrm{diam}(N,F)}(\mathrm{Inv}(N,F)) \subset \mathrm{int}K, \tag{6.4}$$

where $B_\varepsilon(A) := \{x \in X : d(x,A) < \varepsilon\}$ and $\mathrm{diam}(N,F) := \sup\{\mathrm{diam}\, F(x) : x \in N\}$, or, equivalently, $\mathrm{dist}(\mathrm{Inv}(N,F), \partial N) > \mathrm{diam}(N,F)$, where $\mathrm{dist}(A,B) := \min\{d(x,y) : x \in A,\ y \in B\}$, $A, B \subset X$, and ∂N stands for the boundary of N.

The neighbourhood N is then called an *isolating neighbourhood* of K.

A compact isolated invariant subset K of an open set $U \subset X$ is, furthermore, said to be an *attractor* if there exists an open neighbourhood $U_0 \subset U$ of K such that

$$F^m(U_0) = \underbrace{F \circ \ldots \circ F}_{m\text{-times}}(U_0) \subset U, \quad \text{for every } m \in \mathbb{N}, \tag{6.5}$$

and

(6.6) for every open neighbourhood V of K, there is $m(V) \in \mathbb{Z}$ such that, for all $n \geq m(V)$, $F^n(U_0) \subset V$.

We have proved in [AF] (cf. [AG]) that a compact Hausdorff-continuous map $F\colon X \subset U \multimap X$ induces in a natural way the compact (continuous) single-valued map F^* in the hyperspace $\mathcal{K}(X)$, i.e.

$$F^*|_{\mathcal{K}(U)}\colon \mathcal{K}(U) \to \mathcal{K}(X).$$

Hence, let $K \subset U$ be a compact isolated invariant set and N be an isolating neighbourhood of K. Considering an open set W such that $K \subset W \subset N$, we have defined a compact (continuous) single-valued map

$$F^*|_{\mathcal{K}(W)}\colon \mathcal{K}(W) \to \mathcal{K}(X).$$

Since $\mathrm{Fix}(F^*|_{\mathcal{K}(W)}) \subset \mathcal{K}(K)$, the set of fixed points of $F^*|_{\mathcal{K}(W)}$ is a compact subset of $\mathcal{K}(K)$. Moreover, if $X \in$ HANR (e.g. locally continuum-connected metric space, see Lemma (6.2), then $\mathcal{K}(W)$ is obviously an open subset of the ANR-space $\mathcal{K}(X)$, and so the fixed point index $\mathrm{ind}(F^*|_{\mathcal{K}(W)}, \mathcal{K}(W)) \in \mathbb{Z}$ of $F^*|_{\mathcal{K}(W)}\colon \mathcal{K}(W) \to \mathcal{K}(X)$ in $\mathcal{K}(W)$ is well-defined (see [AG] and cf. Section 5.1, where, instead of ANRs, retracts of open subsets of (convex sets in) Fréchet spaces were considered, but completeness was not employed). Let us point out that every ANR is a retract of an open subset of a normed space. Thus, we can also define the *Conley type* (integer-valued) *index* $I_X(K,F)$ of the pair (K,F) just by identifying

$$I_X(K,F) := \mathrm{ind}(F^*|_{\mathcal{K}(W)}, \mathcal{K}(W)). \tag{6.7}$$

(6.8) REMARK. Because of the definition, the Conley type index has all usual properties as a standard fixed point index like (see Section 5.1) the existence (Ważewski's property), homotopy, additivity, localization (excision), contraction (restriction), multiplicity and normalization properties. In particular, the additivity property reads as follows (see [RPS]):

$$I_X(K,F) = I_X(K_1,F) + I_X(K_2,F) + I_X(K_1,F)\, I_X(K_2,F), \tag{6.9}$$

where K is a compact isolated invariant set which is a disjoint union of two compact isolated invariant sets K_1 and K_2, i.e. $K = K_1 \cup K_2$, $K_1 \cap K_2 = \emptyset$. Moreover, it follows from the excision property of the related fixed point index that $I_X(K,F)$ depends neither of the choice of the isolating neighbourhood N of K, nor on the open set W.

The following slightly generalized (for multivalued maps) version of Corollary 8 in [RPS] can be used for computing the Conley type indices. We omit its proof, because it is strictly analogous to the one in [RPS].

(6.10) THEOREM (cf. [RPS]). *Let X be a locally continuum-connected metric space (see Lemma (6.2)), U its locally compact subset, and let $F\colon X \supset U \multimap X$ be a (locally defined) Hausdorff-continuous compact map. Let K be a compact isolated invariant set w.r.t. F which is a disjoint union of $p \in \mathbb{N}$ connected attractors. Then $I_X(K,F) = 2^{p-1}$.*

(6.11) REMARK. For one attractor (i.e. for $p = 1$), we get $I_X(K,F) = 1$.

Now, let K be a compact invariant set w.r.t. a (locally defined) Hausdorff-continuous compact map $F\colon X \supset U \multimap X$. Assume that K has a finite number of components $K_1, \ldots, K_p$. Since $F(K) = K$, F produces a permutation of the elements of this decomposition of K which can be expressed as follows:

$$K = K_{1,1} \cup K_{1,2} \cup \ldots \cup K_{1,k_1} \cup K_{2,1} \cup K_{2,2} \cup \ldots \cup K_{2,k_2} \cup \ldots \cup K_{r,1} \cup K_{r,2} \cup \ldots \cup K_{r,k_r},$$

where $k_1 + \ldots + k_r = p$, and for $i \in \{1, \ldots, r\}$ and $j \in \{1, \ldots, k_i - 1\}$, we have $F(K_{i,j}) = K_{i,j+1}$ and $F(K_{i,k_i}) = K_{i,1}$. Then we say that F *decomposes* K *in* r *cycles* and, for each $i \in \{1, \ldots, r\}$, the corresponding k_i is called the length of the cycle i.

Similarly as Theorem 4 in [RPS] can be proved the following

(6.12) THEOREM (cf. [RPS]). *Let X be a locally continuum-connected metric space (see Lemma (6.2)), U its locally compact subset, and let $F: X \supset U \multimap X$ be a (locally defined) Hausdorff-continuous compact map. Let K be an attractor. Then $I_X(K, F) = 2^r - 1$, where r is the number of cycles of K.*

6.2. Relative Conley type index. Let $A \subset X$ be such that both $A, X \in$ HANR (e.g. locally continuum-connected metric spaces, see Lemma (6.2), i.e. $\mathcal{K}(A) \in$ ANR and $\mathcal{K}(X) \in$ ANR. Keeping the notation in Section 3.2, let a compact multivalued self-map $F: (X, A) \multimap (X, A)$ be Hausdorff-continuous, i.e. let both F_X and F_A be Hausdorff-continuous, compact self-maps.

We have proved in [AF] (cf. [AG]) that a Hausdorff-continuous, compact map $F: (X, A) \multimap (X, A)$ induces in a natural way the compact (continuous) single--valued map F^* in the hyperspace $(\mathcal{K}(X), \mathcal{K}(A))$, i.e.

$$F^*: (\mathcal{K}(X), \mathcal{K}(A)) \to (\mathcal{K}(X), \mathcal{K}(A)).$$

Theorem (3.21), more precisely its slightly modified version, where $X, A \in$ ANR (see [AGJ3] or the comments in the foregoing Section 6.1), implies that the relative Lefschetz number $\Lambda(F^*)$ of F^* is well-defined, and $\Lambda(F^*) \neq 0$ yields that there exists a point $x \in \mathrm{Fix}(F^*)$ such that $x \in \overline{\mathcal{K}(X) \setminus \mathcal{K}(A)}$.

Hence, defining the Conley type index as

$$I_{(X,A)}(F) := \Lambda(F^*), \tag{6.13}$$

there existence of a compact invariant set $K \subset X$ w.r.t. F such that $K \not\subset \mathrm{int}A$ (because $\partial\mathcal{K}(A)) = \partial\mathcal{K}(\mathrm{int}A)$ and $\mathcal{K}(\mathrm{int}A)$ is open) follows from the nontriviality of the index $I_{(X,A)}(F)$. The nontriviality of the index means, according to a slightly modified (in the above spirit) Proposition (3.14), that $\Lambda(F_X^*) \neq \Lambda(F_A^*)$ or, in terms of the Conley type indices, that

$$I_X(F_X) := \Lambda(F_X^*) \neq \Lambda(F_A^*) =: I_A(F_A). \tag{6.14}$$

(6.15) COROLLARY. *Let $A \subset X$ be locally continuum-connected metric spaces (see Lemma (6.2)) and let $F: (X, A) \multimap (X, A)$ be a Hausdorff-continuous, compact map such that the relative Conley type index for F, defined in (6.13), is nontrivial, i.e. (6.14) holds. Then there exists a compact invariant set $K \subset X$ w.r.t. F such that $K \not\subset \mathrm{int}A$.*

6.3. Conley type index for periodic invariant sets. As in the foregoing Section 6.2, let $F\colon (X,A) \multimap (X,A)$ be a Hausdorff-continuous, compact map, where both $A, X \in \mathrm{HANR}$ and $A \subset X$. This induces the compact (continuous) single-valued map F^* in the hyperspace $(\mathcal{K}(X), \mathcal{K}(A))$, i.e.

$$F^*\colon (\mathcal{K}(X), \mathcal{K}(A)) \to (\mathcal{K}(X), \mathcal{K}(A)).$$

A slightly modified Theorem (3.22) (see [AGJ4]) or the comments in Section 6.1), therefore, guarantees that the relative homological invariants $\chi(F^*)$ and $P(F^*)$ are well-defined, and if $\chi(F^*) \neq 0$ or $P(F^*) \neq 0$, then there exists an n-periodic point in $\overline{\mathcal{K}(X) \setminus \mathcal{K}(A)}$, for some $m+1 \leq n \leq m + P(F)$ and an arbitrary $m \geq 0$. Of course, the absolute version for $F\colon X \multimap X$ with the induced map $F^*\colon \mathcal{K}(X) \to \mathcal{K}(X)$ holds all the better as a particular case, when $A = \emptyset$.

Hence, defining the relative Conley type index as

$$I_{\mathrm{per}}(F) := \chi(F^*), \tag{6.16}$$

$$\text{or} \quad I^{\mathrm{per}}(F) := P(F^*), \tag{6.17}$$

the existence of a compact n-periodic set $K \subset X$ and $K \not\subset \mathrm{int}A$ w.r.t. F, i.e.

$$F^n(K) = \underbrace{F \circ \ldots \circ F}_{m\text{-times}}(K) = K,$$

for some $m+1 \leq n \leq m + I^{\mathrm{per}}(F)$ and an arbitrary $m \geq 0$, follows from the nontriviality of one of the indices $I_{\mathrm{per}}(F)$ or $I^{\mathrm{per}}(F)$.

(6.18) COROLLARY. *Let $A \subset X$ be locally continuum-connected metric spaces (see Lemma* (6.2)) *and let $F\colon (X,A) \multimap (X,A)$ be a Hausdorff-continuous, compact map such that one of the relative Conley type indices for F, defined in* (6.16) *and* (6.17), *is nontrivial, i.e. $I_{\mathrm{per}}(F) \neq 0$ or $I^{\mathrm{per}}(F) \neq 0$. Then there exists an n-periodic compact set $K \subset X$ w.r.t. F such that $K \not\subset \mathrm{int}A$.*

(6.19) REMARK. As already pointed out, the absolute version of Corollary (6.18) holds for $F\colon X \multimap X$, when just putting $A = \emptyset$.

6.4. Conley type indices for number of invariant and periodic sets. Assuming that $X \in \mathrm{HANR}$ and $F\colon X \multimap X$ is a Hausdorff-continuous, compact map, the Nielsen number $N(F^*)$ is well-defined for the induced compact (continuous) single-valued map $F^*\colon \mathcal{K}(X) \to \mathcal{K}(X)$, because

(i) $\mathcal{K}(X) \in \mathrm{ANR}$,
(ii) $\mathrm{Fix}(F^*)$ is compact, and
(iii) the generalized Lefschetz number $\Lambda(F^*)$ is well-defined (see e.g. [Scho]).

Hence, defining the Conley type index as

$$I_N(F) := N(F^*), \tag{6.20}$$

$N(F^*)$ is a homotopy invariant for a lower estimate of the number of compact invariant sets K w.r.t. F, i.e. $I_N(F) \leq \operatorname{card}\{K \subset X \text{ is compact} : F(K) = K\} = \#\{K \subset X \text{ is compact} : F(K) = K\}$.

For the relative cases, one can easily check (cf. [Schi]) that the relative Nielsen numbers, studied in Section 4.2, are well-defined for single-valued compact self-maps, provided again only the analogous conditions to (i)–(iii). In fact, for single-valued compact self-maps, only the first analogous condition is enough.

Thus, let $F\colon (X, A) \multimap (X, A)$ be a Hausdorff-continuous, compact map, where both $A, X \in \mathrm{HANR}$ and $A \subset X$. This induces the compact (continuous) single--valued map $F^*\colon (\mathcal{K}(X), \mathcal{K}(A)) \to (\mathcal{K}(X), \mathcal{K}(A))$. Appropriately modified (in the spirit of comments in Section 6.1 and the above discussion) Theorems (4.25) and (4.27) guarantee that the relative Nielsen numbers $[N(F^*) + (\#S(F^*; A))]$ and $SN(F^*; A)$ (the index H can be omitted here) are well-defined, estimating from below $\#\operatorname{Fix}(F^*)$, on the total space $\mathcal{K}(X)$ and on the complement $\mathcal{K}(X) \setminus \mathcal{K}(A)$, respectively.

Defining, therefore, the relative Conley type indices as

$$I_X^N(F) := N(F^*) + (\#S(F^*; A)), \tag{6.21}$$

$$I_{X\setminus A}^N(F) := SN(F^*; A), \tag{6.22}$$

the following lower estimates are satisfied:

$$I_X^N(F) \leq \#\{K \subset X \text{ is compact} : F(K) = K\}, \tag{6.23}$$

$$I_{X\setminus A}^N(F) \leq \#\{K \subset X \text{ is compact} : F(K) = K \not\subset A\}, \tag{6.24}$$

respectively.

(6.25) COROLLARY. *Let $A \subset X$ be locally continuum-connected metric spaces (see Lemma* (6.2)) *and let $F\colon (X, A) \multimap (X, A)$ be a Hausdorff-continuous, compact map. Then the relative Conley type indices for F, defined in* (6.21) *and* (6.22), *are homotopic invariants satisfying inequalities* (6.23) *and* (6.24), *respectively.*

One can similarly check that the Nielsen number for periodic points, studied in Section 4.3, is well-defined for single-valued continuous self-maps on compact ANRs. Actually, compactness imposed on ANRs (as in Section 4.3), can be replaced by compactness of given maps.

Nevertheless, let $F\colon X \multimap X$ be a Hausdorff-continuous map on a compact $X \in \mathrm{HANR}$. This induces the continuous, single-valued map $F^*\colon \mathcal{K}(X) \to \mathcal{K}(X)$

on a compact hyperspace $\mathcal{K}(X) \in \mathrm{ANR}$. Appropriately modified (in the sense of the above discussion) Theorem (4.42) guarantees that the Nielsen number $S_k(\widetilde{F}^*)$ is well-defined, estimating from below $\#\,\mathrm{Per}_k(F^*)$, where

$$\mathrm{Per}_k(F^*) := \mathrm{Per}^k(F^*) \setminus \bigcup_{m<k} \mathrm{Per}^m(F^*)$$

and $\mathrm{Per}^n(F^*) := \mathrm{Fix}(F^{*n}) = \{x \in \mathcal{K}(X) : x = F^{*n}(x)\}$, for $k, m, n \in \mathbb{N}$.

Defining, therefore, the Conley type index as

$$I^{\mathrm{Per}}_{S_k}(F) := S_k(\widetilde{F}^*), \tag{6.26}$$

the following lower estimate, for compact k-periodic invariant sets w.r.t. F, is satisfied:

$$I^{\mathrm{Per}}_{S_k}(F) \leq \#\Big(P_k(F) := P^k(F) \setminus \bigcup_{m<k} P^m(F)\Big), \tag{6.27}$$

where $P^n(F) := \{K \subset X \text{ is compact} : F^n(K) = K\}$, for $k, m, n \in \mathbb{N}$.

(6.28) COROLLARY. *Let $F\colon X \multimap X$ be a Hausdorff-continuous map on a compact, locally connected metric space (see Lemma* (6.2)*). Then the Conley type index for F, defined in* (6.26)*, is a homotopic invariant satisfying inequality* (6.27)*, i.e. estimating from below, for every $k \in \mathbb{N}$, the number of compact k-periodic invariant sets w.r.t. F*

(6.29) REMARK. Similar corollaries can be also given for relative Conley type indices for the number of compact periodic sets, on the basis of results concerning the relative Nielsen numbers for periodic points (see e.g. the survey paper [He]).

(6.30) REMARK. Deeper results for compact invariant and periodic sets can be obtained by Conley type (integer-valued) indices defined by means of index pairs and, in particular, regular index pairs (see [DPS], where only single-valued maps on locally compact ANRs were considered with this respect). In [KM], the existence of an index pair, which is crucial for the definition of a cohomology (or a homotopy) index, was deduced from an isolating neighbourhood for an admissible (multivalued) map.

6.5. Remarks and comments. The standard references for the Conley index theory and its applications are [Co], [Ry2], [Sm]. Recently, the Conley index was also defined for multivalued maps in [Ga], [KM], [Ku], [Mr]. It allows us to detect, besides another, invariant sets.

Although we also investigate here, under similar assumptions, compact invariant sets, our approach is completely different. The Conley type (integer-valued) indices are namely defined, by means of topological invariants (Lefschetz number, Nielsen number, fixed point index), studied in the foregoing sections, for induced (single-valued) maps in hyperspaces. This way was stimulated by the similar idea in [DPS], but mainly by the possibility to construct so continuation principles for fractals (see [AF], [AFGL], [AG]). One can check that the appropriate elaboration of the ideas (for single-valued maps) in [DPS] would allow us to introduce the Conley type (integer-valued) index by means of index pairs (for multivalued maps) in [KM]. Nevertheless, such an index would require some additional restrictions, and it would be no longer applicable to fractals, because the related Hutchinson–Barnsley maps are not admissible (in particular, not with connected values), as required in [KM].

7. Concluding remarks

Although the notion of a degree is apparently the most frequent of all invariants applied to boundary value problems, it is not the most general one. Since the fixed point index is, in fact, a relative degree, it can be defined, unlike degree, on sets with an empty interior (e.g. on bounded, open sets in Fréchet spaces). Even in a finite dimensional case, there are situations when the Conley index can be computed, but not the degree (see e.g. [Ry1]). The Lefschetz number is usually used for the normalization property of a fixed point index. Nevertheless, it can apply to self-maps for which a fixed point index is not defined (cf. [Go1]). The simplest degree is (mod 2)-degree, sometimes also called topological essentiality or transversality or spectral theory for 0-epi maps (see e.g. [AG], [Go2]). It often makes sense to speak about 0-epi maps when no degree is defined (see e.g. [GV], [Vae]), but this is due to some loss of properties (e.g. only a weak form of the additivity property takes place). On the other hand, the Leray–Schauder degree can define a rotation of a vector field (cf. [KZ]) as well as a winding (rotation) number (cf. [Ro]). In [AV1], we defined, in a rather axiomatic way, a general coincidence index suitable for noncompact condensing maps on nonconvex sets and we also discussed its relation to well-known Mawhin's coincidence degree (see e.g. [GM], [Mw]).

The Nielsen theory (of fixed point classes) can be regarded as an extension of the Lefschetz theory, but in a multivalued case some additional restrictions must be imposed on the space. Moreover, the Wecken type approach is not very suitable for multivalued maps (cf. [AV2]). On the other hand, we employed in [AV2] (instead of an index or the Lefschetz number) topological essentiality for a Wecken type recursive definition. In applications, additivity property of a degree is often prefered for multiplicity results.

The spaces for our Conley type (integer-valued) index are locally connected (or even locally continuum-connected) and (for $I_X(K, F)$ in (6.7)) locally compact, and the maps are Hausdorff-continuous and compact-valued (i.e. both u.s.c. and l.s.c.), while the spaces for the (classical) Conley index are locally compact (or satisfying the Rybakowski condition [Ry2]) and (multivalued) maps are in [KM] admissible, i.e. in particular, u.s.c. with connected values. In certain situations, the Conley index can be connected, via the Euler–Poincaré characteristic, with the fixed point index (see e.g. [Sr] and the references therein).

For more comparisons, relationships, extensions and historical remarks, we recomended our monograph [AG].

References

[An1] J. Andres, *Almost-periodic and bounded solutions of Carathéodory differential inclusions*, Differential Integral Equations **12** (1999), 887–912.

[An2] J. Andres, *Multiple bounded solutions of differential inclusions. The Nielsen theory approach*, J. Differential Equations **155** (1999), 285–310.

[An3] ———, *A nontrivial example of application of the Nielsen fixed-point theory to differential systems: problem of Jean Leray*, Proc. Amer. Math. Soc. **128** (2000), 2921–2931.

[ABa] J. Andres and R. Bader, *Asymptotic boundary value problems in Banach spaces*, J. Math. Anal. Appl. **247** (2002), no. 1, 437–457.

[ABe] J. Andres and A. M. Bersani, *Almost-periodicity problem as a fixed-point problem for evolution inclusions*, Topol. Methods Nonlinear Anal. **18** (2001), 337–350.

[AF] J. Andres and J. Fišer, *Metric and topological multivalued fractals*, Internat. J. Bifur. Chaos Appl. Sci. Engrg. **14** (2004), no. 4, 1277–1289.

[AFGL] J. Andres, J. Fišer, G. Gabor and K. Leśniak, *Multivalued fractals*, Chaos Solitions Fractals **24** (2005), no. 3, 665–700.

[AGG1] J. Andres, G. Gabor and L. Górniewicz, *Boundary value problems on infinite intervals*, Trans. Amer. Math. Soc. **351** (1999), 4861–4903.

[AGG2] ———, *Topological structure of solution sets to multivalued asymptotic problems*, Z. Anal. Anwendungen **18** (1999), no. 4, 1–20.

[AGG3] ———, *Acyclicity of solution sets to functional inclusions*, Nonlinear Anal. **49** (2002), 671–688.

[AG] J. Andres and L. Górniewicz, *Topological Fixed Point Principles for Boundary Value Problems*, Kluwer, Dordrecht, 2003.

[AGJ1] J. Andres, L. Górniewicz and J. Jezierski, *A generalized Nielsen number and multiplicity results for differential inclusions*, Topology Appl. **100** (2000), 193–209.

[AGJ2] ———, *Noncompact version of the multivalued Nielsen theory and its application to differential inclusions*, Differential Inclusions and Optimal Control (J. Andres, L. Górniewicz and P. Nistri, eds.), vol. 2, Lecture Notes Nonlinear Anal., Nicolaus Copernicus Univ., Toruń, 1998, pp. 33–50.

[AGJ3] ———, *Relative versions of the multivalued Lefschetz and Nielsen theorems and their application to admissible semi-flows*, Topol. Methods Nonlinear Anal. **16** (2000), 73–92.

[AGJ4] ———, *Periodic points of multivalued mappings with applications to differential inclusions on tori*, Topology Appl. **127** (2003), no. 3, 447–472.

[AV1] J. Andres and M. Väth, *Coincidence index for noncompact mappings on nonconvex sets*, Nonlinear Funct. Anal. Appl. **7** (2002), no. 4, 619–658.

[AV2] ———, *Two topological definitions of a Nielsen number for coincidences of noncompact maps*, Fixed Point Theory Appl. **1** (2004), 49–69.

[BGMO] Yu. G. Borisovich, B. D. Gel'man, A. D. Myshkis and V. V. Obukhovskiĭ, *Topological methods in the fixed-point theory of multi-valued maps*, Uspekhi Mat. Nauk **35** (1980), no. 1, 59–126 (in Russian); Russian Math. Surveys **35** (1980), no. 1, 65–143. (Engl. transl.)

[Bo1] C. Bowszyc, *Fixed points theorem for the pairs of spaces*, Bull. Polish Acad. Sci. Math. **16** (1968), 845–851.

[Bo2] ———, *On the Euler–Poincaré characteristic of a map and the existence of periodic points*, Bull. Acad. Polon. Sci. **17** (1969), 367–372.

[Br1] R. F. Brown, *The Lefschetz Fixed Point Theorem*, Scott, Foresman and Co., Glenview, Ill., 1971.

[Br2] ———, *Multiple solutions to parametrized nonlinear differential systems from Nielsen fixed point theory*, Nonlinear Analysis (World Scientific, Singapore) (1987), 89–98.

[Br3] ———, *Nielsen fixed point theory and parametrized differential equations*, Contemp. Math., vol. 72, Amer. Math. Soc., Providence, R. I., 1988, pp. 33–45.

[CB] A. J. Casson and S. A. Bleiler, *Automorphisms of Surfaces after Nielsen and Thurston*, vol. 9, London Math. Soc. Student Texts, Cambridge Univ. Press, Cambridge, 1988.

[Co] C. C. Conley, *Isolated Invariant Sets and the Morse Index*, CBMS Regional Conference Ser. Math., vol. 38, Amer. Math. Soc., Providence, R. I., 1978.

[Cur] D. W. Curtis, *Hyperspaces of noncompact metric spaces*, Compositio Math. **40** (1980), 139–152.

[Fec] M. Fečkan, *Multiple solutions of nonlinear equations via Nielsen fixed-point theory: a survey*, Nonlinear Anal. in Geometry and Topology (Th. M. Rassias, eds.), Hadronic Press, Palm Harbor, 2000, pp. 77–97.

[Fel] A. Fel'shtyn, *Dynamical Zeta Function*, Nielsen Theory and Reidemeister Torsion, Memoirs of the Amer. Math. Soc., vol. 147, Amer. Math. Soc., Providence, R. I., 2000.

[Ga] G. Gabor, *Homotopy index for multivalued flows on sleek sets*, Preprint (2002).

[GM] R. G. Gaines and J. Mawhin, *Coincidence Degree and Nonlinear Differential Equations*, vol. 568, Lecture Notes in Math., Springer-Verlag, Berlin, 1977.

[GV] E. Giorgieri and M. Väth, *A characterization of* 0*-epi maps with a degree*, J. Funct. Anal. **187** (2001), 183–199.

[Go1] L. Górniewicz, *Homological Methods in Fixed Point Theory of Multivalued Maps*, Diss. Math., vol. 129, Warsaw, 1976.

[Go2] ———, *Topological Fixed Point Theory of Multivalued Mappings*, Kluwer, Dordrecht, 1999.

[Go3] ———, *On the Lefschetz fixed point theorem*, Math. Slovaca **52** (2002), no. 2, 221–233.

[Gr] A. Granas, *Points fixes pour les applications compactes: Espaces de Lefschetz et la theorie del'indice*, Sem. de Math. Sup., vol. 68, Montreal, 1980.

[He] P. R. Heath, *A survey of Nielsen periodic point theory (fixed n)*, Nielsen Theory and Reidemeister Torsion, vol. 49, Banach Center Publ., 1999, pp. 159–188; Inst. of Math, Polish Acad. Sci., Warsaw.

[J] B. Jiang, *Lectures on Nielsen Fixed Point Theory*, Contemp. Math., vol. 14, Amer. Math. Soc., Providence, R. I., 1983.

[KM] T. Kaczyński and M. Mrozek, *Conley index for discrete multivalued dynamical systems*, Topology Appl. **65** (1995), 83–96.

[KT] Kiang Tsai-han, *The Theory of Fixed Point Classes*, Springer, Berlin, 1989.

[KZ] M. A. Krasnosel'skiĭ and P. P. Zabreĭko, *Geometrical Methods of Nonlinear Analysis*, Springer–Verlag, Berlin, 1984; Nauka, Moscow, 1975. (in Russian)

[Kr] W. Kryszewski, *Topological and Approximation Methods in the Degree Theory of Set-Valued Maps*, Dissert. Math., vol. 336, Polish Acad. Sci., Warsaw, 1994.

[Ku] M. Kunze, *Non-Smooth Dynamical Systems*, Lecture Notes in Math., vol. 1744, Springer, Berlin, 2000.

[Ma] T.-W. Ma, *Topological Degrees of Set-Valued Compact Fields in Locally Convex Spaces*, Dissert. Math., vol. 92, Polish Acad. Sci., Warsaw, 1972.

[Mat1] T. Matsuoka, *The number and linking of periodic solutions of periodic systems*, Invent. Math. **70** (1983), 319–340.

[Mat2] ———, *Waveform in the dynamical study of ordinary differential equations*, Japan J. Appl. Math. **1** (1984), 417–434.

[Mr] M. Mrozek, *A cohomological index of Conley type for multivalued admissible flows*, J. Differential Equations **84** (1990), 15–51.

[Mw] J. Mawhin, *Leray–Schauder degree: A half century of extensions and applications*, Topol. Methods Nonlinear Anal. **14** (1999), no. 2, 195–228.

[McC] C. K. McCord (ed.), *Nielsen Theory and Dynamical Systems*, Contenp. Math., vol. 152, Amer. Math. Soc., Providence, R. I., 1993.

[Nu] R. D. Nussbaum, *The Fixed Point Index and Some Applications*, Séminaire de Mathematiques Supérieures, vol. 94, Les Presses de l'Université Montreal, C. P. 6128, succ. A, Montreal, Quebec, Canada, 1985.

[Ro] E. H. Rothe, *Introduction to Various Aspects of Degree Theory in Banach Spaces*, Math. Surveys Monogr., vol. 23, AMS, Providence, R. I., 1986.

[Ry1] K. P. Rybakowski, *On a relation between the Brouwer degree and the Conley index for gradient flows*, Bull. Soc. Math. Belg. (B) **37** (1985), no. (II), 87–96.

[Ry2] ———, *The Homotopy Index and Partial Differential Equations*, Springer, Berlin, 1987.

[RPS] F. R. Ruiz del Portal and J. M. Salazar, *Fixed point index in hyperspaces: a Conley type index for discrete semidynamical systems*, J. London Math. Soc. **64** (2001), 191–204.

[Schi] H. Schirmer, *A survey of relative Nielsen fixed point theory*, Nielsen Theory and Dynamical Systems, Contemp. Math. (C. K. McCord, ed.), vol. 152, Amer. Math. Soc. Colloq. Publ., Providence, R. I., 1993, pp. 291–309.

[Scho] U. K. Scholz, *The Nielsen fixed point theory for noncompact spaces*, Rocky Mountain J. Math. **4** (1974), 81–87.

[Sm] J. Smoller, *Shock Waves and Reaction Diffusion Equations*, Springer–Verlag, Berlin, 1980.

[Sp] E. H. Spanier, *Algebraic Topology*, McGraw-Hill, New York, 1966.

[Sr] R. Srzednicki, *Periodic and bounded solutions in blocks for time-periodic nonautonomous ordinary differential equations*, Nonlin. Anal. **22** (1994), no. 6, 707–737.

[Vae] M. Vaeth, *On the connection of degree theory and* 0*-epi maps*, J. Math. Anal. Appl. **257** (2001), 321–343.

20. THE FIXED POINT INDEX OF THE POINCARÉ TRANSLATION OPERATOR ON DIFFERENTIABLE MANIFOLDS

MASSIMO FURI — MARIA PATRIZIA PERA — MARCO SPADINI

1. Introduction

The fixed point index of the Poincaré translation operator associated to an ordinary differential equation is a very useful tool for establishing the existence of periodic solutions.

In this chapter we focus on ODEs on differentiable manifolds embedded in Euclidean spaces. Our purpose is twofold: on the one hand we aim to provide a short and accessible introduction to some topological tools (such as the topological degree, the degree of a tangent vector field and the fixed point index) that are useful in Nonlinear Analysis; on the other hand we offer a unifying approach to several results about the fixed point index of the Poincaré translation operator that were previously scattered among a number of publications.

Our main concern will be a formula for the computation of the fixed point index of the flow operator induced on a manifold by a first order autonomous ordinary differential equation. Other formulas for the fixed point index of the translation operator associated with non-autonomous equations will be deduced as consequences. We emphasize that other results, unrelated to our approach, but still involving the fixed point index of the Poincaré translation operator have been successfully exploited, for instance, by Srzednicki (see e.g. [Srz2], [Srz3]).

The chapter is organized as follows. In Section 2 we recall first some elements of calculus on finite dimensional manifolds. Then, in this context, we introduce the notions of fixed point index of a map and of degree of a tangent vector field. In particular, we show a simple axiomatic approach to the fixed point index theory for maps on a manifold based on just three axioms (see Theorem (2.32) below). In the same section we discuss some basics regarding first order differential equations on differentiable manifolds.

In Section 3, we consider the flow on a manifold M induced by an autonomous differential equation of the form

$$\dot{x} = g(x),$$

where g is a vector field tangent to M. Then, in the spirit of an earlier result by Krasnosel'skiĭ [Kra], we provide a relation, valid for sufficiently short times, between the degree of a tangent vector field and the fixed point index of its associated flow. From this result, we deduce a formula (see Theorem (3.9) below) for the computation, in terms of the degree of g, of the fixed point index of the flow operator associated with the above equation. This formula is valid whenever the fixed point index of the flow is well defined (and not merely for "sufficiently short" times). The idea behind the proof stems, besides the quoted result by Krasnosel'skiĭ, from more recent results by Capietto, Mawhin and Zanolin ([Maw], [CaMaZa]). Later, in the same section, we consider non-autonomous equations and derive some other formulas for the fixed point index of the (Poincaré) T-translation operator ($T > 0$ given) associated with the equations

$$\dot{x} = \lambda f(t,x), \quad \dot{x} = g(x) + \lambda f(t,x),$$

for $\lambda > 0$ sufficiently small, where g and f are tangent vector fields on M, and f is T-periodic. Similar results can be deduced for an equation of the form

$$\dot{x} = a(t)h(x) + \lambda f(t,x).$$

where h is a tangent vector field on M, and $a\colon \mathbb{R} \to \mathbb{R}$ is a T-periodic function with nonzero average. However, as far as we know, there is no general result in the literature encompassing all the different situations reflected by the above three equations.

Section 4, finally, collects a number of simple consequences of the formulas obtained in the previous section. Particular emphasis is given to results describing the structure of pairs (λ, p), where p is an initial point of a T-periodic solution to the equations considered in Section 3. As an illustrative application, we prove a continuation result for the T-periodic solutions of the equation $\dot{x} = f(t,x)$ when f is a T-periodic tangent vector field.

It should be remarked that the selection of applications collected in this chapter is very restrictive and made with the sole purpose of highlighting the role of the fixed point index of the translation operator. In fact, in order to remain within reasonable space limits, the most general situations are not sought for, and many interesting applications have been omitted. Among these, we mention multiplicity results for T-periodic solutions, guiding-functions-like and continuation-like existence results. Moreover, since a second order equation on a manifold can be seen as a particular first order equation on the tangent bundle (see e.g. [Fur]), a number of results could also be deduced for second order ODEs.

2. Notation and preliminaries

2.1. Tangent cones and tangent spaces to subsets of $\mathbb{R}^k$. In this subsection we introduce the notions of tangent cones and tangent spaces to arbitrary subsets of $\mathbb{R}^k$. We also recall the concept of C^r map, $r \in \mathbb{N} \cup \{\infty\}$, between arbitrary subsets of Euclidean spaces and discuss the notion of Fréchet derivative in this context. These concepts, which are well-known for maps defined on open sets, need an extended definition in the general case (see e.g. [Mil]). Roughly speaking, the extension of the notion of C^r map is obtained by forcing down the *hereditary property* of C^r maps on open sets, i.e. by requiring that the restriction of a C^r map to any subset of its domain is still a C^r map. The following definition also preserves the *local property* of C^r maps, i.e. any map which is locally C^r is C^r.

(2.1) DEFINITION. A map $f\colon X \to Y$, from a subset of $\mathbb{R}^k$ into a subset of $\mathbb{R}^s$, is said to be C^r, $r \in \mathbb{N} \cup \{\infty\}$, if for any $p \in X$ there exists a C^r map $g\colon U \to \mathbb{R}^s$, defined on an open neighbourhood of p, such that $f(x) = g(x)$ for all $x \in U \cap X$.

In other words, $f\colon X \to Y$ is C^r if it can be locally extended as a map into $\mathbb{R}^s$ (and not merely into Y) to a C^r map defined on an open subset of $\mathbb{R}^k$. To understand why one must seek the extension of f as a map into $\mathbb{R}^s$, observe that the identity $i\colon [0,1] \to [0,1]$ is not the restriction of any C^1 function $g\colon U \to [0,1]$ defined on an open neighbourhood U of $[0,1]$.

(2.2) REMARK. Using the well-known fact that any family of open subsets of $\mathbb{R}^k$ admits a subordinate smooth partition of unity, it is easy to show that any C^r map on $X \subseteq \mathbb{R}^k$ is actually the restriction of a C^r map defined on an open neighbourhood of X.

As a straightforward consequence of the definition one gets that, given $X \subseteq \mathbb{R}^k$, the identity $i\colon X \to X$ is a smooth map. Moreover, we observe that the composition of C^r maps between arbitrary Euclidean sets is again a C^r map, since the same is true for maps defined on open sets. Thus, one can view Euclidean sets as objects of a category, whose morphisms are C^r maps.

(2.3) DEFINITION. A C^r map $f\colon X \to Y$, from a subset X of $\mathbb{R}^k$ into a subset Y of $\mathbb{R}^s$, is said to be a *C^r-diffeomorphism* if it is bijective and f^{-1} is C^r. In this case X and Y are said to be *C^r-diffeomorphic*.

A straightforward consequence of the definition of diffeomorphism (and of the hereditary property of C^r maps) is that the restriction of a C^r-diffeomorphism is again a C^r-diffeomorphism onto its image.

(2.4) REMARK. An example of a C^r-diffeomorphism is given by the graph map associated with a C^r map $f\colon X \to \mathbb{R}^s$ defined on an arbitrary subset of $\mathbb{R}^k$. In fact, let $G_f = \{(x,y) \in \mathbb{R}^k \times \mathbb{R}^s : x \in X,\ y = f(x)\}$ denote the graph of f. The map $\widehat{f}: X \to G_f$, defined by $\widehat{f}(x) = (x, f(x))$, is clearly C^r and bijective. Observe now that $\widehat{f}^{-1}$ is just the restriction to G_f of the projection $(x,y) \mapsto x$ of $\mathbb{R}^k \times \mathbb{R}^s$ onto the first factor, which is a linear map (and, consequently, smooth). This proves that *the graph of a C^r map is C^r-diffeomorphic to its domain.*

We are ready to give the definitions of tangent vector, tangent cone, and tangent space to an arbitrary subset $X \subseteq \mathbb{R}^k$ at a point $p \in X$.

In the sequel $|\cdot|$ denotes the canonical Euclidean norm on $\mathbb{R}^k$.

(2.5) Definition. Let X be a subset of $\mathbb{R}^k$ and take $p \in X$. A unit vector $v \in S^{k-1} = \{x \in \mathbb{R}^k : |x| = 1\}$ is said to be *tangent* to X at p if there exists a sequence $\{p_n\}$ in $X \setminus \{p\}$ such that $p_n \to p$ and $(p_n - p)/|p_n - p| \to v$.

If p is isolated in X, then the *tangent cone* of X at p, C_pX, is just the trivial subspace $\{0\}$ of $\mathbb{R}^k$. If p is an accumulation point of X, then C_pX is the cone generated by the set of tangent unit vectors, i.e.

$$C_pX = \{\lambda v : \lambda \geq 0,\ v \in S^{k-1} \text{ is tangent to } X \text{ at } p\}.$$

The *tangent space* of X at p, T_pX, is the vector subspace of $\mathbb{R}^k$ spanned by C_pX.

It is fairly easy to check that the above definition of tangent cone is equivalent to the classical one introduced by Bouligand in [Bou] (see also [Sev, p. 149], for a precursor of this notion).

Observe that, because of the compactness of the unit sphere S^{k-1}, if p is an accumulation point of X, there exists at least one unit vector tangent to X at p. Moreover, the notion of tangent cone is local; that is, if two sets X and Y coincide in a neighbourhood of a common point p, they have the same tangent cone. Another important property is the translation invariance: $C_pX = C_{x+p}(x+X)$, for all $x \in \mathbb{R}^k$.

The following result is useful for the computation of the tangent cone to a set defined by inequalities. The easy proof, based on the Inverse Function Theorem, is left to the reader.

(2.6) Theorem. *Let $f: U \to \mathbb{R}^s$ be a C^1 map defined on an open subset of $\mathbb{R}^k$. Let $Y \subseteq \mathbb{R}^s$ and $p \in f^{-1}(Y)$. Assume that p is a regular point of f; i.e. the derivative $f'(p): \mathbb{R}^k \to \mathbb{R}^s$ of f at p is surjective. Then*

$$C_p(f^{-1}(Y)) = \{v \in \mathbb{R}^k : f'(p)v \in C_{f(p)}Y\} = f'(p)^{-1}(C_{f(p)}Y).$$

Given a C^1 map $f: X \to Y$ and a point $p \in X$, we shall define a linear operator $f'(p)$ from T_pX into $T_{f(p)}Y$, called the derivative of f at p, which maps the tangent cone of X at p into the tangent cone of Y at $f(p)$. This derivative will turn out to satisfy the usual functorial properties of the Fréchet derivative. To achieve this, we need the following three lemmas. The first one extends the well-known fact that the Fréchet derivative can be computed as a directional derivative. Its elementary proof is left to the reader.

(2.7) LEMMA. *Let $f: U \to \mathbb{R}^s$ be defined on an open subset of $\mathbb{R}^k$ and differentiable at $p \in U$. If $v \in S^{k-1}$ is a unit vector, then*

$$f'(p)v = \lim_{n\to\infty} \frac{f(p_n) - f(p)}{|p_n - p|},$$

where $\{p_n\}$ is any sequence in $U \setminus \{p\}$ such that $p_n \to p$ and $(p_n - p)/|p_n - p| \to v$.

(2.8) LEMMA. *Let $f: U \to \mathbb{R}^s$ be defined on an open subset of $\mathbb{R}^k$ and differentiable at $p \in U$. If f maps a subset X of U containing p into a subset Y of $\mathbb{R}^s$, then $f'(p)$ maps C_pX into $C_{f(p)}Y$. Consequently, because of the linearity of $f'(p)$, it also maps T_pX into $T_{f(p)}Y$.*

PROOF. It is sufficient to show that if $v \in S^{k-1}$ is tangent to X at p, then $f'(p)v$ is tangent to Y at $f(p)$. For this, let $\{p_n\}$ be a sequence in $X \setminus \{p\}$ such that $p_n \to p$ and $(p_n - p)/|p_n - p| \to v$. By Lemma (2.7), we have $(f(p_n) - f(p))/|p_n - p| \to f'(p)v$. If $f'(p)v = 0$ there is nothing to prove since $0 \in C_{f(p)}Y$ by the definition of tangent cone. On the other hand, if $f'(p)v \neq 0$, we have $f(p_n) \neq f(p)$, for n large enough. Thus, for such n's, we can write

$$\frac{f(p_n) - f(p)}{|f(p_n) - f(p)|} = \frac{|p_n - p|}{|f(p_n) - f(p)|} \frac{f(p_n) - f(p)}{|p_n - p|}.$$

Therefore,

$$\lim_{n\to\infty} \frac{f(p_n) - f(p)}{|f(p_n) - f(p)|} = \frac{f'(p)v}{|f'(p)v|}.$$

And this shows that $f'(p)v = \lambda w$, where $\lambda > 0$ and $w \in S^{k-1}$ is tangent to Y at $f(p)$. □

(2.9) LEMMA. *Let $f, g: U \to \mathbb{R}^s$ be defined on an open subset of $\mathbb{R}^k$ and differentiable at $p \in U$. Assume that f and g coincide on some subset of X containing p. Then $f'(p)$ and $g'(p)$ coincide on C_pX and, consequently, on T_pX.*

PROOF. Let $\varphi: U \to \mathbb{R}^s$ be defined by $\varphi(x) = f(x) - g(x)$; so that φ maps X into the trivial subspace $Y = \{0\}$ of $\mathbb{R}^s$. Thus, by Lemma (2.8), we obtain

$$\varphi'(p)v = f'(p)v - g'(p)v = 0, \quad \text{for all } v \in C_pX,$$

and the assertion is proven. □

Lemma (2.9) ensures that if $f: X \to \mathbb{R}^s$ is a C^1 map on a subset X of $\mathbb{R}^s$ and p is a point in X, then the restriction to T_pX of the derivative at p of any C^1 local extension of f to a neighbourhood of p does not depend on the chosen extension. In other words, all the C^1 extensions of f to an open neighbourhood of p have the same directional derivative along the vectors of the subspace T_pX. Moreover, Lemma (2.8) implies that if g is such an extension and f maps X into Y, then $g'(p)$ maps T_pX into $T_{f(p)}Y$. These two facts justify the following definition.

(2.10) DEFINITION. Let $f: X \to Y$ be a C^1 map from a subset X of $\mathbb{R}^k$ into a subset Y of $\mathbb{R}^s$. The *derivative* of f at p, $f'(p): T_pX \to T_{f(p)}Y$, is the restriction to T_pX of the derivative at p of any C^1 extension of f to a neighbourhood of p in $\mathbb{R}^k$.

We point out that this extended derivative inherits the two functorial properties of the classical derivative (the easy proof of this fact is left to the reader). As a consequence of this and Lemma (2.8) one gets the following result.

(2.11) THEOREM. *Let $f: X \subseteq \mathbb{R}^k \to Y \subseteq \mathbb{R}^s$ be a C^1-diffeomorphism. Then for any $p \in X$, $f'(p): T_pX \to T_{f(p)}Y$ is an isomorphism mapping C_pX onto $C_{f(p)}Y$.*

PROOF. To simplify the notation, put $q = f(p)$. By the definition of diffeomorphism we have $f^{-1} \circ f = i_X$ and $f \circ f^{-1} = i_Y$, where i_X and i_Y denote the identity on X and Y, respectively. Therefore, by the functorial properties of the extended derivative, the two compositions $(f^{-1})'(q)f'(p)$ and $f'(p)(f^{-1})'(q)$ coincide, respectively, with the identity on T_pX and T_qY. This means that $f'(p)$ is invertible and $f'(p)^{-1} = (f^{-1})'(q)$. The fact that C_pX and C_qY correspond to each other under $f'(p)$ is a direct consequence of Lemma (2.8). □

Let X be a subset of $\mathbb{R}^k$. We say that a point $p \in X$ is *singular* for X if $T_pX \neq C_pX$. In other words, since T_pX is the space spanned by C_pX, saying that p is a non-singular point for X means that C_pX is a vector space. The set of singular points of X will be denoted by δX.

For example, if X is an n-simplex in $\mathbb{R}^k$, δX is just the union of all the $(n-1)$-faces of X, $\delta\delta X$, denoted by $\delta^2 X$, is the union of all the $(n-2)$-faces of X, and so on.

Observe also that if X is an open subset of $\mathbb{R}^k$, then $\delta X = \emptyset$.

The following straightforward consequence of Theorem (2.11) shows that the concept of singular point is invariant under diffeomorphisms.

(2.12) THEOREM. *If $f: X \to Y$ is a C^r-diffeomorphism, then it maps δX onto δY. Consequently, for any $n \in \mathbb{N}$, $\delta^n X$ and $\delta^n Y$ are C^r-diffeomorphic.*

2.2. Differentiable manifolds in Euclidean spaces. A subset M of $\mathbb{R}^k$ is called a (boundaryless) m-dimensional (differentiable) manifold of class C^r, $r \in \mathbb{N} \cup \{\infty\}$, if it is locally C^r-diffeomorphic to $\mathbb{R}^m$; meaning that any point p of M admits a neighbourhood (in M) which is C^r-diffeomorphic to an open subset of $\mathbb{R}^m$. A C^r-diffeomorphism $\varphi: W \to V \subseteq M$ from an open subset W of $\mathbb{R}^m$ onto an open subset V of M is called a *parametrization* (of class C^r of V). The inverse $\varphi^{-1}: V \to W$ of φ is called a *chart* or a *coordinate system* on V, and its component functions, $x_1, \dots, x_m$, are the *coordinate functions* of φ^{-1} on V.

As a straightforward consequence of the definition of differentiable manifold and Theorem (2.12), any point p of an m-dimensional C^1-manifold M is non-singular (i.e. $C_pM = T_pM$). Moreover, $\dim T_pM = m$. In fact, since this property is true for open subsets of $\mathbb{R}^m$, according to Theorem (2.11), it holds true for m-dimensional C^1-manifolds. Incidentally, observe that Theorem (2.11) provides a practical method for computing T_pM. That is, if $\varphi: W \to V$ is a any C^1-parametrization of a neighbourhood V of p in M, then $T_pM = \operatorname{Im}\varphi'(w)$, where $\varphi(w) = p$.

The following direct consequence of the Implicit Function Theorem can be used to produce a large variety of examples of differentiable manifolds. It gives also a useful tool to compute the tangent space at any given point of a manifold. We recall first that if $f: U \to \mathbb{R}^s$ is a C^1 map on an open subset U of $\mathbb{R}^k$, an element $p \in U$ is said to be a *regular point* of f if the derivative $f'(p)$ of f at p is surjective. Non-regular points are called *critical (points)*. The *critical values* of f are those points of the target space $\mathbb{R}^s$ which lie in the image $f(C)$ of the set C of critical points. Any $y \in \mathbb{R}^s$ which is not in $f(C)$ is a regular value. Therefore, in particular, any element of $\mathbb{R}^s$ which is not in the image of f is a regular value. Notice that, in this terminology, the words "point" and "value" refer to the source and target spaces, respectively.

(2.13) THEOREM (Regularity of the level set). *Let $f: U \to \mathbb{R}^s$ be a C^r map of an open subset of $\mathbb{R}^k$ into $\mathbb{R}^s$. If $0 \in \mathbb{R}^s$ is a regular value for f, then $f^{-1}(0)$ is a C^r-manifold of dimension $k - s$. Moreover, given $p \in f^{-1}(0)$, we have*

$$T_p(f^{-1}(0)) = \operatorname{Ker} f'(p).$$

PROOF. Choose a point $p \in f^{-1}(0)$ and split $\mathbb{R}^k$ into the direct sum $\operatorname{Ker} f'(p) \oplus (\operatorname{Ker} f'(p))^\perp$. Since, by assumption, $f'(p): \mathbb{R}^k \to \mathbb{R}^s$ is onto, the restriction of $f'(p)$ to $(\operatorname{Ker} f'(p))^\perp$ is an isomorphism. Observe that this restriction is just the (second) partial derivative, $\partial_2 f(p)$, of f at p with respect to the given decomposition. It follows, by the Implicit Function Theorem, that in a neighbourhood of p, $f^{-1}(0)$ is the graph of a C^r map $\varphi: W \to \operatorname{Ker} f'(p)^\perp$ defined on an open subset W of $\operatorname{Ker} f'(p)$.

Remark (2.4) implies that in this neighbourhood $f^{-1}(0)$ is C^r-diffeomorphic to W. Thus $f^{-1}(0)$ is a C^r-manifold whose dimension is $\dim \mathrm{Ker} f'(p) = k - s$.

To prove that $T_p(f^{-1}(0)) = \mathrm{Ker} f'(p)$ observe first that $T_p(f^{-1}(0)) \subseteq \mathrm{Ker} f'(p)$. In fact, f maps $f^{-1}(0)$ into $\{0\}$ and, consequently, $f'(p)$ maps $T_p(f^{-1}(0))$ into $T_0(\{0\}) = \{0\}$. The equality follows by computing the dimensions of the two spaces. □

Theorem (2.13) can be partially inverted, in the sense that any C^r differentiable manifold in $\mathbb{R}^k$ can be locally regarded as a regular level set (i.e. as the inverse image of a regular value of a C^r map on an open subset of $\mathbb{R}^k$). In fact, the following theorem holds.

(2.14) THEOREM. *Let M be an m-dimensional manifold of class C^r in $\mathbb{R}^k$. Then, given $p \in M$, there exists a map $f: U \to \mathbb{R}^{k-m}$, C^r on a neighbourhood U of p in $\mathbb{R}^k$, which defines $M \cap U$ as a regular level set.*

PROOF. Let $\varphi: W \to \mathbb{R}^k$ be a C^r-parametrization of M around p and let $w = \varphi^{-1}(p)$. Consider any linear map $L: \mathbb{R}^{k-m} \to \mathbb{R}^k$ such that $\mathrm{Im} L \oplus T_pM = \mathbb{R}^k$ (this is clearly possible since $\dim T_pM = m$), and define $g: W \times \mathbb{R}^{k-m} \to \mathbb{R}^k$ by setting $g(x,y) = \varphi(x) + Ly$. The derivative of g at $(w,0) \in W \times \mathbb{R}^{k-m}$ is given by

$$g'(w,0)(h,k) = \varphi'(w)h + Lk,$$

which is surjective (therefore an isomorphism), since $\mathrm{Im}\, \varphi'(0) = T_pM$. By the Inverse Function Theorem, g is a C^r-diffeomorphism of a neighbourhood of $(w,0)$ in $W \times \mathbb{R}^{k-m}$ onto a neighbourhood U of p in $\mathbb{R}^k$. Let ψ be the inverse of such a diffeomorphism and define $f: U \to \mathbb{R}^{k-m}$ as the composition $\pi_2 \circ \psi$ of ψ with the projection $\pi_2: W \times \mathbb{R}^{k-m} \to \mathbb{R}^{k-m}$ of $W \times \mathbb{R}^{k-m}$ onto the second factor. We see that f satisfies the assertion. □

We point out that there are differentiable manifolds in $\mathbb{R}^k$ which cannot be globally defined as regular level sets. One can prove, in fact, that when this happens, the manifold must be orientable (the definition of orientability and the proof of this assertion would carry us too far away). As an intuitive example consider a Möbius strip M embedded in $\mathbb{R}^3$ and assume $M = f^{-1}(0)$, where $f: U \to \mathbb{R}$ is a C^1 map on an open subset of $\mathbb{R}^3$. If $0 \in \mathbb{R}$ were a regular value for f, the gradient of f at any point $p \in f^{-1}(0)$, $\nabla f(p)$, would be nonzero. Therefore, the map $\nu: M \to \mathbb{R}^3$, given by $\nu(p) = \nabla f(p)/|\nabla f(p)|$, would be a continuous normal unit vector field on M, and this is well-known to be impossible on the Möbius strip (a one-sided surface).

Now we want to define an "embedded" notion of tangent bundle TM associated with a C^r manifold M in $\mathbb{R}^k$. We will prove that if $r \geq 2$, TM is a C^{r-1}

differentiable manifold in $\mathbb{R}^k \times \mathbb{R}^k$. In order to do this, we shall define the concept of tangent bundle for any subset of $\mathbb{R}^k$, and prove that when two sets X and Y are C^r-diffeomorphic, the corresponding tangent bundles are C^{r-1}-diffeomorphic.

(2.15) Definition. Given $X \subseteq \mathbb{R}^k$, the subset

$$TX = \{(x, y) \in \mathbb{R}^k \times \mathbb{R}^k : x \in X,\ y \in T_x X\}$$

of $\mathbb{R}^k \times \mathbb{R}^k$ is called the *tangent bundle* of X. The canonical projection $\pi: TX \to X$ is the restriction to TX of the projection of $\mathbb{R}^k \times \mathbb{R}^k$ onto the first factor (thus, π is always a smooth map).

(2.16) Definition. Let $f: X \to Y$ be a C^r map from a subset X of $\mathbb{R}^k$ into a subset Y of $\mathbb{R}^s$ and assume $1 \le r \le \infty$. The *tangent map* of f, $Tf: TX \to TY$, is given by

$$Tf(x, y) = (f(x), f'(x)y).$$

As pointed out in Remark (2.2), one may regard a C^r map $f: X \to Y$ as the restriction of a C^r map $g: U \to \mathbb{R}^s$ defined on an open neighbourhood U of X. Consequently, if $r \ge 1$, $Tg: TU \to T\mathbb{R}^s$, given by $(x, y) \mapsto (g(x), g'(x)y)$, is a C^{r-1} map from the open neighbourhood $TU = U \times \mathbb{R}^k$ of TX into $T\mathbb{R}^s = \mathbb{R}^s \times \mathbb{R}^s$. This proves that Tf, which is just the restriction to TX of Tg, is a C^{r-1} map.

Clearly, if $f: X \to Y$ and $g: Y \to Z$ are C^r maps, one has $T(g \circ f) = Tg \circ Tf$. Moreover, if $i: X \to X$ is the identity on X, then $Ti: TX \to TX$ is the identity on TX. Therefore, one may regard T as a covariant functor from the category of Euclidean sets with C^r maps into the category of Euclidean sets with C^{r-1} maps. This implies that if $f: X \to Y$ is a C^r-diffeomorphism with $r \ge 2$, then $Tf: TX \to TY$ is a C^{r-1}-diffeomorphism. Therefore, if M is a C^r manifold of dimension m, since it is locally C^r-diffeomorphic to the open subsets of $\mathbb{R}^m$, its tangent bundle TM is a C^{r-1} manifold of dimension $2m$. Moreover, if $\varphi: W \to V \subseteq M$ is a parametrization of an open set V in M, $T\varphi: W \times \mathbb{R}^m \to TV \subseteq TM$ is a parametrization of the open set $TV = \pi^{-1}(V)$ of TM.

(2.17) Definition. Let X be a subset of $\mathbb{R}^k$. A *tangent vector field* on X is a continuous map $g: X \to \mathbb{R}^k$ with the property that $g(x) \in T_x X$ for all $x \in X$. The tangent vector field g on X is said to be *inward* if $g(x) \in C_x X$ for all $x \in X$.

Usually, in differential geometry, a tangent vector field on a differentiable manifold M is defined as a section of the tangent bundle TM. That is, a map $w: M \to TM$ with the property that the composition $\pi \circ w: M \to M$ of w with the bundle projection π is the identity on M. However, in our "embedded" situation (i.e. M in $\mathbb{R}^k$) this "abstract" definition turns out to be redundant. In fact, observe that, for M embedded in $\mathbb{R}^k$, a map $w\colon M \to \mathbb{R}^k \times \mathbb{R}^k$ is a section of TM if and only if for all $x \in M$ one has $w(x) = (x, g(x))$, with $g(x) \in T_xM$. Therefore, forgetting x in the pair $(x, g(x))$, one may accept the simpler definition given above.

An important example of a tangent vector field to a differentiable manifold $M \subseteq \mathbb{R}^k$ is the *gradient* of a C^1 function $f\colon M \to \mathbb{R}$. This is defined by assigning to any point $x \in M$ the unique vector $\nabla f(x) \in T_xM$ such that

$$\langle \nabla f(x), v \rangle = f'(x)v, \quad \text{for all } v \in T_xM,$$

where $\langle \cdot\,, \cdot \rangle$ denotes the (canonical) inner product on $\mathbb{R}^k$.

We now recall a classical theorem that will be essential for our definition of fixed point index. For its proof see e.g. [Hir, Chapter 4, §5].

(2.18) THEOREM (Tubular neighbourhoods). *Let $M \subseteq \mathbb{R}^k$ be a smooth manifold. Then there exists a neighbourhood W of M such that any point $x \in W$ possesses a unique closest point $r(x)$ in M. Moreover, the map $r\colon W \to M$ is a smooth submersion (i.e. it has surjective derivative at any point).*

Note that the map r in the above theorem is a *retraction*, i.e. it has the property that $r(x) = x$ for all $x \in M$. Thus any manifold in $\mathbb{R}^k$ is a retract of one of its neighbourhoods.

2.3. The Brouwer degree and the fixed point index. For the sake of simplicity, from now on "differentiable manifold" or, briefly, "manifold" will mean "smooth manifold" embedded in some Euclidean space. Moreover, any map between manifolds is assumed to be (at least) continuous.

Before introducing the fixed point index on manifolds, we need to discuss briefly a slightly extended notion of Brouwer degree in Euclidean spaces. For more details about degree, the reader is referred to [Llo], [Mil], [Nir].

Let V be an open subset of $\mathbb{R}^k$. A pair (g, V) is *admissible (for the Brouwer degree)* if g is an $\mathbb{R}^k$-valued (continuous) map whose domain $\mathcal{D}(g)$ contains V and such that $g^{-1}(0) \cap V$ is compact. In particular this holds if (g, V) is *strongly admissible*; that is, if V is bounded, g is defined at least on the closure $\overline{V}$ of V, and $g(x) \neq 0$ for all x in the boundary ∂V of V. On the set of admissible pairs there is defined an integer-valued function $\deg_B$, called *Brouwer degree*, satisfying the following three fundamental properties.

(2.19.1) (Normalization) For the identity I on $\mathbb{R}^k$ one has $\deg_B(I, \mathbb{R}^k) = 1$.

(2.19.2) (Additivity) Given an admissible pair (g, V), if V_1 and V_2 are two disjoint open subsets of V such that $g^{-1}(0) \cap V \subseteq V_1 \cup V_2$, then

$$\deg_B(g, V) = \deg_B(g|_{V_1}, V_1) + \deg_B(g|_{V_2}, V_2).$$

(2.19.3) (Homotopy invariance) If $V \subseteq \mathbb{R}^k$ is an open subset of $\mathbb{R}^k$, and $h\colon V \times [0,1] \to \mathbb{R}^k$ is an *admissible homotopy*, i.e. $h^{-1}(0)$ is compact, then

$$\deg_B(h(\cdot\,,0),V) = \deg_B(h(\cdot\,,1),V).$$

As a consequence of a well known result of Amann and Weiss (see [AmWa]), the function $\deg_B$ is uniquely determined by the above properties. This is because these properties imply the Amann–Weiss axioms, which are stated for strongly admissible pairs. Thus, henceforth, the three fundamental properties will be still referred to as *Amann–Weiss axioms* (for admissible pairs).

Below, we list some other important properties of the Brouwer degree, which can be easily derived from the Amann–Weiss axioms.

(2.20) (Excision) Given an admissible pair (g,V) and an open subset V_1 of V containing $g^{-1}(0) \cap V$, one has $\deg_B(g,V) = \deg_B(g|_{V_1}, V_1)$.

By excision, taking $V_1 = V$, we get the following property which shows that the degree is independent of the behavior of g outside V.

(2.21.1) (Localization) If (g,V) is admissible, then $\deg_B(g,V) = \deg_B(g|_V, V)$.

(2.21.2) (Solution) If (g,V) is admissible and $\deg_B(g,V) \neq 0$, then $g^{-1}(0) \cap V$ in nonempty.

(2.21.3) (Boundary dependence) If (g_1,V) and (g_2,V) are strongly admissible and $g_1|_{\partial V} = g_2|_{\partial V}$, then $\deg_B(g_1,V) = \deg_B(g_2,V)$.

(2.22) Remark. If (g,V) is admissible, g is C^1 on V and 0 is a regular value for g in V, then $g^{-1}(0) \cap V$ is a finite set. It can be shown (see e.g. [Llo]) that in this case

$$\deg_B(g,V) = \sum_{x \in g^{-1}(0) \cap V} \operatorname{sign} \det g'(x).$$

We now present the fixed point index in the context of differentiable manifolds. Let $M \subseteq \mathbb{R}^k$ be a manifold, U an open subset of M, and f an M-valued (continuous) map whose domain $\mathcal{D}(f) \subseteq M$ contains U. The fixed point index of f in U, $\operatorname{ind}(f,U)$, is a well defined integer whenever the set $\operatorname{fix}(f,U)$ of fixed points of f in U is compact. When this holds, the pair (f,U) is said to be *admissible* (*for the fixed point index on* M). Loosely speaking, $\operatorname{ind}(f,U)$ is an algebraic count of the elements of $\operatorname{fix}(f,U)$. The following is the precise definition.

(2.23) Definition. Let M, U and f be as above. If the pair (f,U) is admissible, the *fixed point index* of f in U is the integer

$$\operatorname{ind}(f,U) = \deg_B(I - f \circ r, r^{-1}(U)), \tag{2.24}$$

where I is the identity in $\mathbb{R}^k$ and $r: W \to M$ is any retraction defined on an open neighbourhood of M.

In the above definition, the existence of a retraction r is ensured by Theorem (2.18). Moreover, one can prove that $\deg_B(I - f \circ r, r^{-1}(U))$ does not depend on the choice of r (see e.g. [DuGr], [Gra], [Nus]). Thus, $\operatorname{ind}(f, U)$ is well defined. Note that, in particular, when $M = \mathbb{R}^k$, the fixed point index of a map $f: U \to \mathbb{R}^k$ coincides with the Brouwer degree of $I - f$ in U. Namely,

$$\operatorname{ind}(f, U) = \deg_B(I - f, U). \tag{2.25}$$

The fixed point index has a number of useful properties. Below we list some of the most important ones. For proofs and more details we refer to [Ama], [DuGr], [Gra], [Nus].

(2.26.1) (Normalization) Let $f: M \to M$ be constant. Then $\operatorname{ind}(f, M) = 1$.

(2.26.2) (Additivity) Given an admissible pair (f, U), if U_1 and U_2 are two disjoint open subsets of U such that $\operatorname{fix}(f, U) \subseteq U_1 \cup U_2$, then

$$\operatorname{ind}(f, U) = \operatorname{ind}(f|_{U_1}, U_1) + \operatorname{ind}(f|_{U_2}, U_2).$$

(2.26.3) (Homotopy invariance) A map $h: U \times [0,1] \to M$ with the property that the set $\{(x, \lambda) \in U \times [0,1] : x = h(x, \lambda)\}$ is compact is called an *admissible homotopy*. In this case,

$$\operatorname{ind}(h(\,\cdot\,, 0), U) = \operatorname{ind}(h(\,\cdot\,, 1), U).$$

(2.26.4) (Commutativity) Let U_1 and U_2 be open subsets of two manifolds M_1 and M_2, respectively. Given $f_1: U_1 \to M_2$ and $f_2: U_2 \to M_1$, if one of the pairs $(f_2 \circ f_1, f_1^{-1}(U_2))$ or $(f_1 \circ f_2, f_2^{-1}(U_1))$ is admissible, then so is the other and

$$\operatorname{ind}(f_2 \circ f_1, f_1^{-1}(U_2)) = \operatorname{ind}(f_1 \circ f_2, f_2^{-1}(U_1)).$$

(2.26.5) (Solution) If $\operatorname{ind}(f, U) \neq 0$, then the fixed point equation $f(x) = x$ has a solution in U.

(2.26.6) (Multiplicativity) Let U_1 and U_2 be open subsets of two manifolds M_1 and M_2, respectively. Assume that (f_1, U_1) and (f_2, U_2) are admissible. Consider the map $f_1 \times f_2: U_1 \times U_2 \to M_1 \times M_2$ given by $(x_1, x_2) \mapsto (f_1(x_1), f_2(x_2))$. Then the pair $(f_1 \times f_2, U_1 \times U_2)$ is admissible and

$$\operatorname{ind}(f_1 \times f_2, U_1 \times U_2) = \operatorname{ind}(f_1, U_1) \cdot \operatorname{ind}(f_2, U_2).$$

(2.27) REMARK. Given any M-valued map f defined on a subset $\mathcal{D}(f)$ of M, the pair $(f, \emptyset)$ is admissible. This includes the case when $\mathcal{D}(f)$ is the empty set (it is coherent with the notion of a map as a triple (A, B, R) where A is the domain, B is the codomain and $R \subseteq A \times B$ is such that there exists exactly one $(a, b) \in R$ for any $a \in A$). A simple application of the additivity property shows that $\operatorname{ind}(f|_\emptyset, \emptyset) = 0$ and $\operatorname{ind}(f, \emptyset) = 0$.

As a consequence of the additivity property and of Remark (2.27) one easily gets the following often-neglected property, which shows that the index of an admissible pair (f, U) does not depend on the behavior of f outside U.

(2.28) (Localization) If (f, U) is admissible, then $\operatorname{ind}(f, U) = \operatorname{ind}(f|_U, U)$.

Another consequence of the additivity is the following important property.

(2.29) (Excision) Given an admissible pair (f, U) and an open subset U_1 of U containing $\operatorname{fix}(f, U)$, one has $\operatorname{ind}(f, U) = \operatorname{ind}(f, U_1)$.

A stronger form of the homotopy property is often useful (see e.g. [Ama], [Nus]).

(2.30) (Generalized homotopy invariance) Let M be a manifold and let $W \subseteq M \times [0, 1]$ be open. Suppose $h: W \to M$ is such that the set

$$\{(x, \lambda) \in W \times [0, 1] : x = h(x, \lambda)\}$$

is compact. For any $\lambda \in [0, 1]$, put $W_\lambda = \{x \in M : (x, \lambda) \in W\}$ and $h_\lambda = h(\cdot, \lambda)$. Then $\operatorname{ind}(h_\lambda, W_\lambda)$ is well defined and independent of $\lambda \in [0, 1]$.

In the case when M is a compact manifold, it is well known that $\operatorname{ind}(f, M)$ coincides with the Lefschetz number $\Lambda(f)$ of f. This is often called the *strong normalization property* of the fixed point index. A discussion of the Lefschetz number, that would require homology theory, is beyond the scope of this chapter; the interested reader is referred to, e.g. [DuGr], [Spa].

It is well known that some of the above properties can be used as axioms for a fixed point index theory. For instance, it can be deduced from [Bro] that the first four determine uniquely the fixed point index. Actually the result of [Bro] is more general: it holds in the framework of metric ANRs. In this more general setting, other uniqueness results based on a stronger version of the normalization property are available for the (more restrictive) class of compact maps (see e.g. [DuGr, §16, Theorem 5.1]). Below, using the uniqueness result for the Brouwer degree given in [AmWa] we shall prove that the properties of normalization, additivity and homotopy invariance are sufficient to determine uniquely the fixed point index on manifolds.

An admissible pair for the index (f, U) will be called *regular* if f is smooth on U and any fixed point of f in U is nondegenerate; that is, 1 does not belong to the spectrum of the endomorphism $f'(x): T_xM \to T_xM$ for any $x \in \text{fix}(f, U)$. Note that, in this case, $\text{fix}(f, U)$ is necessarily a discrete set, therefore finite, being compact.

The following remark shows that the computation of the fixed point index of an admissible pair can always be reduced to that of a regular pair.

(2.31) REMARK. Let (f, U) be admissible (for the index) and let U_1 be a relatively compact neighbourhood of $\text{fix}(f, U)$ such that $\overline{U}_1 \subseteq U$. By excision, $\text{ind}(f, U_1)$ coincides with $\text{ind}(f, U)$. It can be shown, via standard transversality arguments, that f is admissibly homotopic on U_1 to some smooth approximation φ of it such that (φ, U_1) is regular.

The following proposition shows that the properties of normalization, additivity and homotopy invariance enforce a formula for the computation of the fixed point index that is valid for any regular pair (f, U). Thus, by Remark (2.31) and by the homotopy invariance property, we see that there exists a unique integer-valued function on the set of admissible pairs that satisfies the normalization, additivity and homotopy invariance properties. In other words, *the fixed point index is uniquely determined by these three properties.*

(2.32) THEOREM (Uniqueness of the fixed point index). *Let $M \subseteq \mathbb{R}^k$ be an m-dimensional manifold and let "ind" be an integer-valued function on the set of admissible pairs satisfying the properties of normalization, additivity and homotopy invariance. Then, given any regular pair (f, U), one has*

$$\text{ind}(f, U) = \sum_{x \in \text{fix}(f, U)} \text{sign}(\det(I_x - f'(x))),$$

where I_x denotes the identity of T_xM.

PROOF. Let W be an open subset of M which is diffeomorphic to the whole space $\mathbb{R}^m$ and let $\psi: W \to \mathbb{R}^m$ be any diffeomorphism (onto $\mathbb{R}^m$). Denote by $\mathcal{U}$ the set of all pairs (f, U) which are admissible for the fixed point index in M and such that $U \subseteq W$, $f(U) \subseteq W$. These pairs may be regarded as admissible for the fixed point index in W, and the restriction of the index function to $\mathcal{U}$ still satisfies the properties of normalization, additivity and homotopy invariance. We claim that for any $(f, U) \in \mathcal{U}$ one necessarily has

$$\text{ind}(f, U) = \deg_B(I - \psi \circ f \circ \psi^{-1}, \psi(U)), \tag{2.33}$$

where I is the identity in $\mathbb{R}^m$. To show this, denote by $\mathcal{V}$ the set of pairs (g, V) which are admissible for the degree in $\mathbb{R}^m$ and consider the one-to-one correspondence $\omega: \mathcal{U} \to \mathcal{V}$ defined by $\omega(f, U) = (I - \psi \circ f \circ \psi^{-1}, \psi(U))$. We need to prove that ind $= \deg_B \circ\, \omega$. Observe that $\omega^{-1}(g, V) = (\psi^{-1} \circ (I - g) \circ \psi, \psi^{-1}(V))$, and if two pairs $(f, U) \in \mathcal{U}$ and $(g, V) \in \mathcal{V}$ correspond under ω, then the sets fix(f, U) and $g^{-1}(0) \cap U$ correspond under ψ. It is also evident that the function $d: \mathcal{V} \to \mathbb{Z}$ defined by the composition $d = \text{ind} \circ \omega^{-1}$ satisfies the Amann–Weiss axioms. Thus, $\deg_B$ and d coincide on $\mathcal{V}$, and this implies ind $= \deg_B \circ \omega$, as claimed.

Assume now that (f, U) is a regular pair for the fixed point index in M. Let fix$(f, U) = \{x_1, \dots, x_n\}$ and let $W_1, \dots, W_n$ be n pairwise disjoint open subsets of U such that $x_i \in W_i$, for $i = 1, \dots, n$. Since M is locally diffeomorphic to the whole space $\mathbb{R}^m$, we may assume that each W_i is diffeomorphic to $\mathbb{R}^m$ under a diffeomorphism ψ_i. For any i, let U_i be an open subset of W_i such that $f(U_i) \subseteq W_i$. The additivity property yields

$$\text{ind}(f, U) = \sum_{i=1}^{n} \text{ind}(f, U_i),$$

and, by the previous argument, we get

$$\sum_{i=1}^{n} \text{ind}(f, U_i) = \sum_{i=1}^{n} \deg_B(I - \psi_i \circ f \circ \psi_i^{-1}, \psi_i(U_i)).$$

By Remark (2.22) and the chain property of the derivative, for any i one has

$$\deg_B(I - \psi_i \circ f \circ \psi_i^{-1}, \psi_i(U_i)) = \text{sign}(\det(I_{x_i} - f'(x_i))).$$

Thus

$$\text{ind}(f, U) = \sum_{i=1}^{n} \text{sign}(\det(I_{x_i} - f'(x_i))),$$

and this concludes the proof. □

2.4. The degree of a tangent vector field. Recall that, for the sake of simplicity, unless otherwise specified, all the manifolds are supposed smooth and all the maps are assumed continuous.

Let U be an open subset of a manifold $M \subseteq \mathbb{R}^k$ and let g be a tangent vector field defined at least on U. We say that (g, U) is an *admissible pair of M (for the degree of a vector field)* if $g^{-1}(0) \cap U$ is compact. When context precludes confusion about the universe M containing U, we will simply say that (g, U) is admissible (or, equivalently, that g is admissible on U). In this case (see e.g. [GuPo], [Hir], [Mil], [Tro] and references therein) one can assign to g an integer, $\deg(g, U)$, called the degree (or index, or Euler characteristic, or rotation) of the tangent vector field g on U.

To avoid any possible confusion, we point out that in the literature there exists a different extension of the Brouwer degree to the context of differentiable manifolds (see e.g. [Mil] and references therein), called the Brouwer degree for maps between manifolds. This second extension, roughly speaking, counts the (algebraic) number of solutions of an equation of the form $h(x) = y$, where $h: M \to N$ is a map between oriented manifolds of the same dimension and $y \in N$ is such that $h^{-1}(y)$ is compact. This dichotomy of notions in the context of manifolds arises from the fact that counting the solutions of an equation of the form $h(x) = y$ cannot be reduced to the problem of counting the zeros of a vector field, as one can do in $\mathbb{R}^k$ by defining $g(x) = h(x) - y$. Therefore, from the point of view of global analysis, the degree of a vector field and the degree of a map are necessarily two separated notions. The first one, which we are interested in, does not require any orientability and is particularly important for differential equations, since a tangent vector field on a manifold can be regarded as an autonomous differential equation.

We give here a brief idea of how this degree can be defined. We need first the following result (see e.g. [Mil]).

(2.34) THEOREM. *Let $g: M \to \mathbb{R}^k$ be a C^1 tangent vector field on a manifold $M \subseteq \mathbb{R}^k$. If g is zero at some point $p \in M$, then the derivative $g'(p): T_pM \to \mathbb{R}^k$ maps T_pM into itself. Therefore, $g'(p)$ can be regarded as an endomorphism of T_pM and, consequently, its determinant* $\det(g'(p))$ *is well defined.*

PROOF. It suffices to show that $g'(p)v \in T_pM$ for any $v \in T_pM$ such that $|v| = 1$. Given such a vector v, consider a sequence in $M \setminus \{p\}$ such that $p_n \to p$ and $(p_n - p)/|p_n - p| \to v$. By Lemma (2.7) we have

$$g'(p)v = \lim_{n\to\infty} \frac{g(p_n) - g(p)}{|p_n - p|} = \lim_{n\to\infty} \frac{g(p_n)}{|p_n - p|}.$$

Observe that for all $n \in \mathbb{N}$, the vector $w_n = g(p_n)/|p_n - p|$ is tangent to M at p_n. Let us show that this implies that the limit $w = g'(p)v$ of $\{w_n\}$ is in T_pM. In fact, because of Theorem (2.14), we may assume that M (around p) is a regular level set of a smooth map $f: V \to \mathbb{R}^s$ defined on some open subset V of $\mathbb{R}^k$. Thus, by Theorem (2.13), $f'(p_n)w_n = 0$, and passing to the limit we get $f'(p)w = 0$, which means $w \in T_pM$, as claimed. □

Let $g: M \to \mathbb{R}^k$ be a C^1 tangent vector field on a manifold $M \subseteq \mathbb{R}^k$. A zero $p \in M$ of g is said to be *nondegenerate* if $g'(p)$, as a map from T_pM into itself, is an isomorphism. In this case, its *index* at p, $\mathrm{i}(g,p)$, is defined to be 1 or -1 according to the sign of the determinant $\det(g'(p))$.

In the particular case when an admissible pair (g, U) is *regular* (i.e. g is smooth with only nondegenerate zeros), its degree, $\deg(g, U)$, is defined just summing up the indices at its zeros. This makes sense, since $g^{-1}(0) \cap U$ is compact (g being admissible in U) and discrete. Therefore, the sum is finite. Using transversality arguments (see e.g. [Hir]) one can show that if two such tangent vector fields can be joined by a smooth homotopy, then they have the same degree, provided that this homotopy is admissible (i.e. the set of zeros remains in a compact subset of U). Moreover, it is clear that given g as above, if V is any open subset of U containing $g^{-1}(0) \cap U$, then $\deg(g, U) = \deg(g, V)$.

The above "homotopy invariance property" for regular pairs gives an idea of how to proceed in the general case. If the pair (g, U) is admissible, consider any relatively compact open subset V of U containing $g^{-1}(0) \cap U$ and observe that, since the boundary ∂V of V (in M) is compact, we have $\min\{|g(x)| : x \in \partial V\} = \delta > 0$ (recall that M is embedded in $\mathbb{R}^k$). Let (g_1, V) be any regular pair with g defined (at least) on $\overline{V}$ such that

$$\max\{|g(x) - g_1(x)| : x \in \partial V\} < \delta.$$

Then $\deg(g, U)$ is defined as $\deg(g_1, V)$. To see that this definition does not depend on the approximating map, observe that if (g_2, V) is a different regular pair with g_2 satisfying the same inequality, we get $(1 - \lambda)g_1(x) + \lambda g_2(x) \neq 0$ for all $\lambda \in [0, 1]$ and $x \in \partial V$. Therefore $(x, \lambda) \mapsto (1 - \lambda)g_1(x) + \lambda g_2(x)$ is an admissible homotopy of tangent vector fields on V. This proves that $\deg(g_1, V) = \deg(g_2, V)$. The fact that this definition does not depend on the open set V containing $g^{-1}(0)$ is very easy to check and left to the reader.

The following are the main properties of the degree for admissible tangent vector fields on open subsets of a manifold.

(2.35.1) (Solution) If $\deg(g, U) \neq 0$, then g has a zero in U.

(2.35.2) (Additivity) Let (g, U) be admissible. If U_1 and U_2 are two disjoint open subsets of U whose union contains $g^{-1}(0) \cap U$, then

$$\deg(g, U) = \deg(g|_{U_1}, U_1) + \deg(g|_{U_2}, U_2).$$

(2.35.3) (Homotopy invariance) Let $h \colon U \times [0, 1] \to \mathbb{R}^k$ be an admissible homotopy of tangent vector fields; that is, $h(x, \lambda) \in T_x M$ for all $(x, \lambda) \in U \times [0, 1]$ and $h^{-1}(0)$ is compact. Then $\deg(h(\,\cdot\,, \lambda), U)$ does not depend on $\lambda \in [0, 1]$.

The above definition of degree implies that if two vector fields $g_1\colon M_1 \to \mathbb{R}^k$ and $g_2\colon M_2 \to \mathbb{R}^s$ correspond under a diffeomorphism $\psi\colon M_1 \to M_2$ (i.e. $\psi'(x)g_1(x) = g_2(\psi(x))$ for any $x \in M_1$), then, if one is admissible, so is the other, and they have the same degree (on M_1 and M_2, respectively). More precisely, the following property holds.

(2.36) (Topological invariance) Let ψ be a diffeomorphism of a manifold $M_1 \subseteq \mathbb{R}^k$ onto a manifold $M_2 \subseteq \mathbb{R}^s$. Let (g_1, U) be an admissible pair of M_1, and assume that $g_2\colon \psi(U) \to \mathbb{R}^s$ corresponds to g_1 under ψ, then $(g_2, \psi(U))$ is an admissible pair of M_2 and

$$\deg(g_1, U) = \deg(g_2, \psi(U)).$$

Note also that, in the particular case when U is an open subset of $M = \mathbb{R}^m$ and g is an admissible vector field on U (i.e. $g^{-1}(0) \cap U$ is compact), then $\deg(g, U)$ coincides with the Brouwer degree $\deg_B(g, U)$.

Another immediate consequence of the definition of degree is the relation between the degree of a vector field and that of its opposite. If (g, U) is an admissible pair of M, then

$$\deg(g, U) = (-1)^m \deg(-g, U), \tag{2.37}$$

where m is the dimension of M. More generally, for a given constant $\alpha \neq 0$, one has

$$\deg(g, U) = (\operatorname{sign} \alpha)^m \deg(\alpha g, U) \tag{2.38}$$

The additivity property, analogously to what happens for the fixed point index, implies the following property.

(2.39) (Excision) Given an admissible pair (g, U) and an open subset U_1 of U containing $g^{-1}(0) \cap U$, one has $\deg(g, U) = \deg(g|_{U_1}, U_1)$.

From the excision, taking $U_1 = U$, one gets the next property, which shows that the degree of an admissible pair (g, U) is not influenced by the behavior of g outside the open set U.

(2.40) (Localization) If (g, U) is admissible, then so is $(g|_U, U)$ and

$$\deg(g, U) = \deg(g|_U, U).$$

Given a relatively compact open set $U \subseteq M$, assume that $g_1\colon M \to \mathbb{R}^k$ and $g_2\colon M \to \mathbb{R}^k$ are tangent vector fields such that $g_1|_{\partial U} = g_2|_{\partial U}$. Clearly, (g_1, U) is admissible if and only if so is (g_2, U) and the map $(x, \lambda) \mapsto \lambda g_1(x) + (1-\lambda)g_2(x)$ is an admissible homotopy in U. By the homotopy invariance property, one gets the following property.

(2.41) (Boundary dependence) Assume the open set $U \subseteq M$ is relatively compact and that (g_1, U) or (g_2, U) is admissible. If $g_1|_{\partial U} = g_2|_{\partial U}$, then $\deg(g_1, U) = \deg(g_2, U)$.

Actually, for a relatively compact open set $U \subseteq M$, in [Pug] there is given a formula, valid for "most" tangent vector fields, that relates $\deg(g, U)$ to the topology of U and the behavior of g along ∂U.

It is known that, unless the target manifold is flat, the boundary dependence property does not hold for the degree of maps between oriented manifolds.

Another consequence of the homotopy invariance property is that the degree of a tangent vector field g on a compact manifold $M \subseteq \mathbb{R}^k$ is independent of g. In fact, if g_1 and g_2 are two tangent vector fields on M, then $h\colon M \times [0,1] \to \mathbb{R}^k$, given by

$$h(x, \lambda) = (1-\lambda)g_1(x) + \lambda g_2(x),$$

is an admissible homotopy. This permits the assignment of an integer $\chi(M)$, called the *Euler–Poincaré characteristic of* M, by setting

$$\chi(M) = \deg(g, M), \tag{2.42}$$

where $g\colon M \to \mathbb{R}^k$ is any tangent vector field on M. Thus, if $\chi(M) \neq 0$, then any tangent vector field on M must vanish at some point. Moreover, the topological invariance implies that if two compact manifolds M and N are diffeomorphic, then $\chi(M) = \chi(N)$.

Actually, there are other equivalent (and better) ways to define the Euler–Poincaré characteristic of a compact manifold. One of these is via homology theory, where $\chi(M)$ coincides with the Lefschetz number of the identity (see for example [DuGr], [Spa]). The powerful homological method has the advantage of being applicable to a large class of topological spaces, which includes those of the same homotopy type as compact polyhedra (such as compact manifolds with boundary). The celebrated Poincaré–Hopf theorem asserts that *the above definition of the Euler–Poincaré characteristic coincides with the homological one* (see e.g. [DuGr], [Hir], [Mil]).

Observe that formulas (2.37) and (2.42) imply that if M is an odd dimensional compact manifold, then $\chi(M) = 0$.

2.5. First order ordinary differential equations on manifolds. An autonomous first order differential equation on a manifold $M \subseteq \mathbb{R}^k$ (or, more generally, on a subset of $\mathbb{R}^k$) is determined by a tangent vector field $g\colon M \to \mathbb{R}^k$ on M.

The first order (autonomous) differential equation associated with g will be written in the form

$$\dot{x} = g(x), \quad x \in M. \tag{2.43}$$

However, the important fact about a differential equation is not the way this is written: what counts is the exact definition of what we mean by a solution (and this implicitly defines the notion of equation). By a solution of (2.43) we mean a C^1 curve $x\colon J \to \mathbb{R}^k$, defined on a (nontrivial) interval $J \subseteq \mathbb{R}$, which satisfies the conditions $x(t) \in M$ and $\dot{x}(t) = g(x(t))$, identically on J. Thus, even if, according to Remark (2.2), the map g may be thought of as defined on an open set U containing M, a solution $x\colon J \to \mathbb{R}^k$ of

$$\dot{x} = g(x), \quad x \in U \tag{2.44}$$

is a solution of (2.43) if and only if its image lies in M. However, if M is closed in U, under the uniqueness assumption of the Cauchy problem for (2.44), one can check that any solution of (2.44) starting from a point of M must lie entirely in M.

If $\varphi\colon M \to N$ is a diffeomorphism between two manifolds and g is a tangent vector field on M, one gets a tangent vector field h on N by setting $h(z) = \varphi'(\varphi^{-1}(z))g(\varphi^{-1}(z))$. In this way, if $x \in M$ and $z \in N$ correspond under φ, the two vectors $h(z)$ and $g(x)$ correspond under the isomorphism $\varphi'(x)\colon T_xM \to T_zN$. For this reason we say that the two vector fields g and h correspond under φ (or they are φ-related). We observe that in this case, as an easy consequence of the chain rule for the derivative (and the definition of solution of a differential equation), equation (2.43) is equivalent to

$$\dot{z} = h(z), \quad x \in N, \tag{2.45}$$

in the sense that $x\colon J \to M$ is a solution of (2.43) if and only if the composition $z = \varphi \circ x$ is a solution of (2.45). That is, the solutions of (2.43) and (2.45) correspond under the diffeomorphism φ.

A non-autonomous first order differential equation on a manifold $M \subseteq \mathbb{R}^k$ is given by assigning, on an open subset V of $\mathbb{R} \times M$, a non-autonomous (continuous) vector field $f\colon V \to \mathbb{R}^k$ such that $f(t,x) \in T_xM$ for all $(t,x) \in V$. That is, for any $t \in \mathbb{R}$, the map $f_t\colon V_t \to \mathbb{R}^k$, given by $f_t(x) = f(t,x)$, is a tangent vector field on the (possibly empty) open subset $V_t = \{x \in M : (t,x) \in V\}$ of M. In other words, $f(t,x) \in T_xM$ for each $(t,x) \in V$.

The first order differential equation associated with f is denoted as follows:

$$\dot{x} = f(t,x), \quad (t,x) \in V. \tag{2.46}$$

A solution of (2.46) is a C^1 map $x: J \to M$, on an interval $J \subseteq \mathbb{R}$, such that, for all $t \in J$, $(t, x(t)) \in V$ and $\dot{x}(t) = f(t, x(t))$.

We point out that (2.46) can be thought of as a special autonomous equation on the open submanifold V of $\mathbb{R} \times M \subseteq \mathbb{R}^{k+1}$. In fact (2.46) is clearly equivalent to the system

$$\begin{cases} \dot{t} = 1, \\ \dot{x} = f(t, x) \quad \text{for } (t, x) \in V, \end{cases} \tag{2.47}$$

and the vector field $(t, x) \mapsto (1, f(t, x))$ is tangent to V. By "equivalent" we mean that the solutions (2.46) and (2.47) are in a one-to-one correspondence.

As pointed out before, any differential equation on a manifold M is transformed into an equivalent one by a diffeomorphism $\varphi: M \to N$. Thus, since manifolds are locally diffeomorphic to open subsets of Euclidean spaces, the classical results about local existence and uniqueness for differential equations apply immediately to this more general context. Therefore, given $(t_0, x_0) \in V$, the continuity of the vector field $f: V \to \mathbb{R}^k$ is sufficient to ensure the existence, on an open interval J, of a solution $x: J \to M$ of (2.46) satisfying the Cauchy condition $x(t_0) = x_0$. If f is C^1 (or, more generally, locally Lipschitz), two solutions satisfying the same Cauchy condition coincide in their common domain. Moreover, by considering the partial ordering associated with graph inclusion, one gets that any solution of (2.46) can be extended to a maximal one (i.e. to a solution which is not the restriction of any different solution).

As in Euclidean spaces, one con prove that the domain of any maximal solution $x(\cdot)$ of (2.46) is an open interval (α, β), with $-\infty \leq \alpha < \beta \leq \infty$. Moreover, given any $t_0 \in (\alpha, \beta)$ and any compact set K in the domain V of $f: V \to \mathbb{R}^k$, both the graphs of the restrictions of $x(\cdot)$ to $(\alpha, t_0]$ and to $[t_0, \beta)$ are not contained in K. This is referred as the Kamke property of the maximal solution (in a manifold). In particular, if M is a compact manifold and $V = \mathbb{R} \times M$, any maximal solution of (2.46) is defined on the whole real axis.

As in Euclidean spaces (see e.g. [Cop]), one has the following result regarding the continuous dependence on data.

(2.48) THEOREM. *Let $M \subseteq \mathbb{R}^k$ be a manifold and $\{f_n\}$ a sequence of C^1 non-autonomous tangent vector fields on M defined on an open subset V of $\mathbb{R} \times M$. Assume that f_n converges uniformly on compact sets to a C^1 tangent vector field f_0. Given $(\tau, p) \in V$, denote (when defined) by $x_n(t, \tau, p)$ the value at t of the maximal solution of*

$$\begin{cases} \dot{x} = f_n(t, x), \\ x(\tau) = p. \end{cases}$$

Let $\{(\tau_n, p_n)\}$ be a sequence in V converging to $(\tau_0, p_0) \in V$ and $[a, b]$ a compact interval contained in the domain of $x_0(\cdot, \tau_0, p_0)$. Then, for n sufficiently large, $x_n(\cdot, \tau_n, p_n)$ is defined on $[a, b]$ and

$$x_n(t, \tau_n, p_n) \to x_0(t, \tau_0, p_0)$$

uniformly on $[a, b]$. In particular, the set of all (t, τ, p) such that $x(t, \tau, p)$ is well defined is an open subset of $\mathbb{R} \times V$ (obviously containing any (τ, τ, p) with $(\tau, p) \in V$).

For autonomous tangent vector fields, the following consequence of the well-known Kupka–Smale theorem (see e.g. [Pei]) will be crucial in the sequel.

(2.49) PROPOSITION. *Let $g: M \to \mathbb{R}^k$ be a tangent vector field. Then, there exists a sequence $\{g_n\}$ of C^1 tangent vector fields on M, uniformly converging to g on compact sets and such that, for any $n \in \mathbb{N}$, any $T > 0$ and any compact set $C \subseteq M$, the equation $\dot{x} = g_n(x)$ admits finitely many periodic orbits contained in C and with period in $(0, T]$.*

3. The fixed point index of the Poincaré translation operator

3.1. The autonomous case. Let $g: M \to \mathbb{R}^k$ be a given C^1 vector field tangent to $M \subseteq \mathbb{R}^k$. For $p \in M$ and $t \in \mathbb{R}$, let $\Phi_t(p)$ be the value at t (if defined) of the maximal solution of (2.43) starting from p at time $t = 0$. We shall also use the (more cumbersome) notation $\Phi_t^g(p)$ whenever it will be necessary to emphasize the dependence on g. The map $p \mapsto \Phi_t(p)$, when (and where) defined, is called *flow operator at time t* (associated with g). Obviously, if M is compact, $\Phi_t(p)$ is defined for all $(t, p) \in \mathbb{R} \times M$. Moreover, given a relatively compact subset U of M, $\Phi_t(p)$ is defined for any $p \in \overline{U}$ and $|t|$ small enough. In fact, Theorem (2.48) implies that the domain of the map $(t, p) \mapsto \Phi_t(p)$ is an open subset of $\mathbb{R} \times M$ containing the section $\{0\} \times M$.

(3.1) REMARK. Let ψ be a diffeomorphism from a manifold $M \subseteq \mathbb{R}^k$ onto a manifold $N \subseteq \mathbb{R}^s$, let $g: M \to \mathbb{R}^k$ be a C^1 tangent vector field on M, and let Φ_t be its associated flow operator at time t. Then, the composition $\psi \circ \Phi_t \circ \psi^{-1}$ coincides with the flow operator at time t associated with the tangent vector field $\widehat{g}: N \to \mathbb{R}^s$ that corresponds to g under ψ.

(3.2) LEMMA. *Let $g: M \to \mathbb{R}^k$ be a C^1 tangent vector field. For any compact subset K of M such that $g(p) \neq 0$ for $p \in K$, there exists a positive constant $\tau = \tau(K)$ such that*

$$\Phi_t(p) \neq p, \quad \textit{for } 0 < |t| \leq \tau \textit{ and all } p \in K.$$

PROOF. Assume, by contradiction, that there exist two sequences $\{t_n\}_{n\in\mathbb{N}}$ with $t_n \neq 0$ and $t_n \to 0$, and $\{p_n\}_{n\in\mathbb{N}} \subseteq K$ such that $p_n = \Phi_{t_n}(p_n)$.

Without loss of generality we may assume $p_n \to p_0 \in K$. Denote by $x_n(\cdot)$ the solution of the initial value problem

$$\begin{cases} \dot{x}(t) = t_n g(x(t)) & \text{for } t \in [0,1], \\ x(0) = p_n. \end{cases}$$

Clearly $x_n(1) = \Phi_{t_n}(p_n)$. As $M \subseteq \mathbb{R}^k$, the integrals $\int_0^1 g(x_n(t))\,dt$ make sense. Hence,

$$0 = x_n(1) - x_n(0) = t_n \int_0^1 g(x_n(t))\,dt,$$

so, as $t_n \neq 0$,

$$\int_0^1 g(x_n(t))\,dt = 0 \quad \text{for all } n \in \mathbb{N}. \tag{3.3}$$

Using Theorem (2.48) and the fact that any $x_n(\,\cdot\,)$ is defined on the whole interval $[0,1]$ one can easily prove that $x_n(\,\cdot\,)$ converges uniformly to the (constant) solution $x_0(t) \equiv p_0$.

Taking the limit in (3.3), we get $0 = \int_0^1 g(x_0(t))\,dt = g(p_0)$. This is a contradiction. □

In the "flat case", i.e. when M is an open subset of $\mathbb{R}^k$, there exists a simple relation between the degree of a vector field and the fixed point index of the associated flow operator. Namely, the following result holds.

(3.4) PROPOSITION. *Let U be a relatively compact open subset of $\mathbb{R}^k$ and let g be a C^1 vector field in $\mathbb{R}^k$ defined at least on $\overline{U}$ and such that $g^{-1}(0) \cap \partial U = \emptyset$. Then*

$$\deg(g,U) = \lim_{t\to 0^-} \deg(I - \Phi_t^g, U),$$

where I denotes the identity in $\mathbb{R}^k$ and Φ_t^g is the flow operator associated with g.

PROOF. From the proof of Lemma 6.1 in [Kra, §2] (or from Corollary 3.4 in [FuPe4]) it follows that

$$\deg(-g,U) = \deg(I - \Phi_t^g, U)$$

when $t > 0$ is sufficiently small. We see that if $\Phi_t^{-g}(p)$ is defined, so is $\Phi_{-t}^g(p)$ and $\Phi_t^{-g}(p) = \Phi_{-t}^g(p)$. Thus the assertion follows. □

A similar result holds on manifolds.

(3.5) THEOREM. *Let $g\colon M \to \mathbb{R}^k$ be a C^1 vector field tangent to a manifold $M \subseteq \mathbb{R}^k$. Let U be a relatively compact, open subset of M, and assume $g^{-1}(0) \cap \partial U = \emptyset$. Then*

$$\deg(g, U) = \lim_{t \to 0^-} \operatorname{ind}(\Phi_t^g, U),$$

where Φ_t^g denotes the flow operator associated with g.

PROOF. By Lemma (3.2), (Φ_t^g, U) is admissible for $|t| > 0$ sufficiently small. By standard approximation results on manifolds, one can find a smooth approximation γ of g with the following properties:

(3.5.1) γ has only nondegenerate zeros;

(3.5.2) for $|t| > 0$ small enough, the flow operator Φ_t^γ associated with γ is admissibly homotopic to Φ_t^g in U.

By the homotopy invariance property of both the degree and the fixed point index, it is not restrictive to assume that properties (3.5.1) and (3.5.2) hold true for g.

Since the zeros of g are nondegenerate, $g^{-1}(0)$ is a finite set, say $\{p_1, \ldots, p_n\}$. Let $V_1, \ldots, V_n$ be pairwise disjoint open neighbourhoods in U of $x_1, \ldots, x_n$, respectively.

By Lemma (3.2) one can take $\tau > 0$ so small that Φ_t^g has no fixed points in the compact set $\overline{U} \setminus \bigcup_{i=1}^n V_i$ for $0 < |t| \leq \tau$. Therefore, by the additivity property (both of the degree and of the index), it is enough to prove the assertion for the particular case when $g^{-1}(0) \cap U$ consists of just one nondegenerate zero p and $\overline{U}$ is contained in an open set W diffeomorphic to $\mathbb{R}^m$.

Let ψ be the diffeomorphism of W onto $\mathbb{R}^m$, and let $\widehat{g}$ be the vector field in $\mathbb{R}^m$ corresponding to g under ψ. By Remark (3.1) and Proposition (3.4), for $t \neq 0$ sufficiently small, one has

$$\deg(\widehat{g}, \psi(U)) = \operatorname{ind}(\psi \circ \Phi_t^{\hat{g}} \circ \psi^{-1}, \psi(U)).$$

The assertion follows from the topological invariance property of the degree and from the commutativity property of the fixed point index. □

(3.6) REMARK. Let U be a relatively compact open subset of M and let $\Phi\colon \mathbb{R} \times M \to M$ be a dynamical system with no *rest points* (i.e. points $x \in M$ with the property that $\Phi(t, x) = x$ for all $t \in \mathbb{R}$) on ∂U. In this case the *rest point index*

$$\mathrm{I}(\Phi, U) = \lim_{t \to 0^+} \operatorname{ind}(\Phi(t, \cdot\,), U)$$

introduced in [Srz1] in the more general settings of ENRs is well defined (because of the homotopy invariance property of the fixed point index). Assume that (2.43) induces a dynamical system $\Phi^g\colon \mathbb{R} \times M \to M$. Thus, taking into account that $\Phi_t^{-g} = \Phi_{-t}^g$, from Theorem (3.5) and formula (2.37) one gets the following relation between the degree of a vector field and the rest point index of the associated dynamical system:

$$\mathrm{I}(\Phi^g, U) = (-1)^m \deg(g, U),$$

where m is the dimension of M.

Below, following [FuSp1], we shall extend to the manifold setting a formula proved in [Maw], [CaMaZa] for the computation of the fixed point index of the flow operator associated with autonomous differential equations in Euclidean spaces (Theorem (3.9) below). A similar formula had been previously proved by Krasnosel'skiĭ [Kra], still in the "flat context", in the (more general) non-autonomous case under an additional T-irreversibility assumption (see below).

Below, by an *orbit* we mean the image of a periodic solution of (2.43). Given $T > 0$, by A_T we denote the union of all τ-periodic orbits with $0 < \tau \leq T$. Note that $g^{-1}(0) \subseteq A_T$ for all $T > 0$.

(3.7) LEMMA. *Given $T > 0$, let $O \subseteq M$ be a nontrivial isolated orbit of* (2.43) *in A_T. Then, there exists an open neighbourhood W of O such that, for all $0 < \tau \leq T$, Φ_τ is defined on $\overline{W}$, admissible on W, and* $\operatorname{ind}(\Phi_\tau, W) = 0$.

PROOF. Since O is a periodic orbit, Φ_t is defined on O for all $t \in \mathbb{R}$. Thus, O being compact, Φ_T is defined on some relatively compact, open neighbourhood W of O. Observe that, since O is isolated in A_T, one can choose W such that Φ_τ is fixed point free on ∂W for all $\tau \in (0, T]$. This implies, by the homotopy invariance property, that $\operatorname{ind}(\Phi_\tau, W)$ is independent of $\tau \in (0, T]$. Moreover, by the nontriviality of O, there exists a positive minimal period σ of O. Thus $\operatorname{ind}(\Phi_{\sigma/2}, W) = 0$ since $\Phi_{\sigma/2}$ is fixed point free on $\overline{W}$. □

(3.8) LEMMA. *Assume that Φ_T is defined on a relatively compact open subset U of M. Suppose there exist only finitely many orbits with period in $(0, T]$ which meet $\overline{U}$. Then, given $\tau, \sigma \in (0, T]$ such that Φ_τ and Φ_σ are fixed point free on ∂U, we have*

$$\operatorname{ind}(\Phi_\tau, U) = \operatorname{ind}(\Phi_\sigma, U).$$

PROOF. If all the orbits with period in $(0, T]$ that meet $\overline{U}$ are trivial, then Φ_τ and Φ_σ are admissibly homotopic on U. Otherwise, let $O_1, \dots, O_n$ be all the nontrivial ones. Lemma (3.7) implies the existence of n open subsets of M, $W_1, \dots, W_n$, such that $O_i \subseteq W_i$ and

$$\operatorname{ind}(\Phi_\tau, W_i) = \operatorname{ind}(\Phi_\sigma, W_i) = 0,$$

for all $i = 1 \dots n$. We can assume $\overline{W}_i \cap \overline{W}_j = \emptyset$ when $i \neq j$. Define

$$U_1 = U \setminus \bigcup_{i=1}^{n} \overline{W}_i.$$

By the additivity and the excision properties,

$$\operatorname{ind}(\Phi_\tau, U) = \operatorname{ind}(\Phi_\tau, U_1) \quad \text{and} \quad \operatorname{ind}(\Phi_\sigma, U) = \operatorname{ind}(\Phi_\sigma, U_1).$$

Using the homotopy invariance, we can write $\operatorname{ind}(\Phi_\tau, U_1) = \operatorname{ind}(\Phi_\sigma, U_1)$, and the claim follows. □

(3.9) THEOREM. *Let $g\colon M \to \mathbb{R}^k$ be a C^1 tangent vector field on a manifold $M \subseteq \mathbb{R}^k$ and let U be a relatively compact open subset of M. Let $T > 0$ and assume that, for any $p \in \overline{U}$, the (maximal) solution of the Cauchy problem*

$$\begin{cases} \dot{x} = g(x), \\ x(0) = p, \end{cases}$$

is defined on $[0, T]$. If Φ_T^g is fixed point free on ∂U, then

$$\operatorname{ind}(\Phi_T^g, U) = \deg(-g, U).$$

PROOF. Consider a sequence $\{g_n\}$ of C^1 tangent vector fields as in Proposition (2.49). As usual, for $n \in \mathbb{N}$, we will denote by $\{\Phi_t^{g_n}\}_{t \in \mathbb{R}}$ the local flow associated with the equation $\dot{x} = g_n(x)$. Since the flow is a continuous map of the twofold variable $(t, x) \in \mathbb{R} \times M$, the "attainable set" $\widehat{U}_T = \Phi_{[0,T]}(\overline{U})$ is a compact subset of M. Let B be a relatively compact open set containing $\widehat{U}_T$. Let c be the distance (in $\mathbb{R}^k$) between $\widehat{U}_T$ and ∂B. One can choose a sufficiently large $\overline{n}$ such that $\|\Phi_t(x) - \Phi_t^{g_n}(x)\| \leq c/2$ for all $x \in \overline{U}$, $t \in [0, T]$ and $n > \overline{n}$. This implies that, if $n > \overline{n}$, any solution of $\dot{x} = g_n(x)$ which meets $\overline{U}$ is contained in $\overline{B}$. By the choice of the sequence $\{g_n\}$, $\overline{B}$ contains only finitely many periodic orbits of $\dot{x} = g_n(x)$ with period in $(0, T]$.

Since $\overline{U}$ is compact and $g^{-1}(0) \cap \partial U = \emptyset$, by Theorem (3.5) there exists $\varepsilon > 0$ such that $\operatorname{ind}(\Phi_{-\varepsilon}^g, U) = \deg(g, U)$. Since $\Phi_\varepsilon^{-g} = \Phi_{-\varepsilon}^g$, one has

$$\operatorname{ind}(\Phi_\varepsilon^g, U) = \deg(-g, U). \tag{3.10}$$

Using the continuous dependence on data and the compactness of ∂U we can assume $\Phi_T^{g_n}(x) \neq x$ and $\Phi_\varepsilon^{g_n}(x) \neq x$ for all $x \in \partial U$. Moreover, by the homotopy invariance property of the index, we get

$$\operatorname{ind}(\Phi_T^{g_n}, U) = \operatorname{ind}(\Phi_T^g, U), \tag{3.11}$$

$$\operatorname{ind}(\Phi_\varepsilon^{g_n}, U) = \operatorname{ind}(\Phi_\varepsilon^g, U), \tag{3.12}$$

provided that n is large enough. Applying Lemma (3.8), we have

$$\operatorname{ind}(\Phi_T^{g_n}, U) = \operatorname{ind}(\Phi_\varepsilon^{g_n}, U).$$

Thus, by (3.10)–(3.12), we obtain $\operatorname{ind}(\Phi_T^g, U) = \operatorname{ind}(\Phi_\varepsilon^g, U) = \deg(-g, U)$, and this completes the proof. □

In spite of the fact that the restriction to ∂U of the flow operator Φ_T may be strongly influenced by the behavior of g outside ∂U, this is not so for its fixed point index. In fact, in some sense, the fixed point index of the flow depends only on how points are shot along the boundary of U. More precisely, we have the following result.

(3.13) COROLLARY. *Let M, g, U and T be as in the Theorem* (3.9). *Let $h\colon M \to \mathbb{R}^k$ be a C^1 tangent vector field and Φ_t^h its associated flow operator. If $g|_{\partial U} = h|_{\partial U}$, then*

$$\operatorname{ind}(\Phi_T^g, U) = \operatorname{ind}(\Phi_T^h, U),$$

provided that they are both fixed point free on the boundary of U.

PROOF. The assertion follows immediately from Theorem (3.9) and the boundary dependence property of the degree. □

Note also that Theorem (3.9) is not a trivial consequence of the homotopy property because, in general, the map $(p,t) \mapsto \Phi_t^g(p)$ is not an admissible homotopy on U. For example, consider in $M = \mathbb{R}^2$ the differential equation

$$(\dot{x}, \dot{y}) = (y, -x)$$

and let U be the unit open disk in $\mathbb{R}^2$. A direct computation shows that $\operatorname{ind}(\Phi_t, U)$ is well defined and equal to 1 for any $t \neq 2k\pi$, and it is not defined for $t = 2k\pi$ ($k \in \mathbb{Z}$). Therefore if t is considered in an interval containing one of these values, the flow does not give an admissible homotopy.

Let g, U and T satisfy the assumptions of Theorem (3.9). Consider the following differential equation depending on a parameter $\lambda \geq 0$:

$$\dot{x} = \lambda g(x). \tag{3.14}$$

Clearly, if $\lambda > 0$, any T-periodic solution of $\dot{x} = g(x)$ corresponds to a (T/λ)-periodic one of (3.14). Denote by $\Phi_t^{\lambda g}$ the flow operator associated with this equation. Observe that, for $\lambda \in [0,1]$ and $p \in \overline{U}$, $\Phi_T^{\lambda g}(p)$ and $\Phi_{\lambda T}^g(p)$ are both defined on $\overline{U}$ and

$$\Phi_T^{\lambda g}(p) = \Phi_{\lambda T}^g(p). \tag{3.15}$$

Given $\lambda_1, \lambda_2 \in (0,1]$, assume that $\Phi_T^{\lambda_1 g}$ and $\Phi_T^{\lambda_2 g}$ are fixed point free on ∂U. Then, by Theorem (3.9), we have

$$\mathrm{ind}(\Phi^g_{\lambda_1 T}, U) = \mathrm{ind}(\Phi^g_{\lambda_2 T}, U).$$

Therefore, by (3.15), we get

$$\mathrm{ind}(\Phi_T^{\lambda_1 g}, U) = \mathrm{ind}(\Phi_T^{\lambda_2 g}, U), \tag{3.16}$$

and this equality holds true no matter whether or not the homotopy

$$H : \overline{U} \times [\lambda_1, \lambda_2] \to M$$

given by $H(x,s) = \Phi_T^{sg}(x)$ is admissible.

3.2. The periodic case. Let $f \colon \mathbb{R} \times M \to \mathbb{R}^k$ be a T-periodic C^1 tangent vector field on M. For $p \in M$, let $P_T^f(p)$ be the value at T (if defined) of the maximal solution of the Cauchy problem

$$\dot{x} = f(t,x), \tag{3.17.1}$$

$$x(0) = p. \tag{3.17.2}$$

By Theorem (2.48), the domain of the map $p \mapsto P_T^f(p)$ is an open subset of M. This map is called the *(Poincaré) T-translation operator* associated with (3.17.1). Observe that for an autonomous equation $\dot{x} = g(x)$ one has $\Phi_T^g = P_T^g$.

Let U be a relatively compact open subset of M. Assume that P_T^f is defined on $\overline{U}$. One could ask if a formula like (3.16) is still valid replacing the flow with the Poincaré operator.

Consider the differential equation

$$\dot{x} = \lambda f(t,x), \quad \lambda \geq 0. \tag{3.18}$$

Given $\lambda_1, \lambda_2 \in (0,1]$, assume that $P_T^{\lambda_1 f}$ and $P_T^{\lambda_2 f}$ are fixed point free on ∂U. The question is whether the following equality holds:

$$\mathrm{ind}(P_T^{\lambda_1 f}, U) = \mathrm{ind}(P_T^{\lambda_2 f}, U). \tag{3.19}$$

The answer is affirmative in the case when $U = M$ is a compact manifold (this is an easy consequence of the homotopy invariance property of the fixed point index), but it is false in general. To see this, let U be the open unit disk in $M = \mathbb{R}^2$, and consider the following system:

$$\begin{cases} \dot{x}_1 = \lambda x_2, \\ \dot{x}_2 = -\lambda x_1 + \lambda \sin t. \end{cases} \tag{3.20}$$

This is a differential equation of the form (3.18) with $x = (x_1, x_2)$ and

$$f(t,x) = (x_2, -x_1 + \sin t).$$

For $\lambda = 1$, (3.20) does not admit 2π-periodic solution. Thus,

$$\operatorname{ind}(P^f_{2\pi}, U) = 0.$$

On the other hand, for λ sufficiently small, from Theorem (3.23) below (see also [FuPe3]) it follows that

$$\operatorname{ind}(P^{\lambda f}_{2\pi}, U) = \deg(-w, U) = 1,$$

where $w\colon \overline{U} \to \mathbb{R}^2$ is defined by

$$w(x_1, x_2) = \frac{1}{2\pi}\int_0^{2\pi} (x_2, -x_1 + \sin t)\, dt = (x_2, -x_1),$$

contradicting (3.19).

Let f and T be as above. Following Krasnosel'skiĭ (see [Kra]), a point $p \in M$ is said to be of *T-irreversibility* for equation (3.17.1) if the (maximal) solution $x(\,\cdot\,, p)$ of (3.17) is defined on $[0, T]$ and $x(t,p) \neq p$ for any $t \in (0, T]$. Using the homotopy property of the degree, Krasnosel'skii proves a formula for computing the fixed point index of the operator of translation along trajectories of a non-autonomous differential equation. His result (reformulated in the framework of differentiable manifolds) is the following.

(3.21) THEOREM. *Let $f\colon \mathbb{R} \times M \to \mathbb{R}^k$ be a C^1 tangent vector field that is T-periodic in the first variable. Let U be a relatively compact open subset of M. Assume that P^f_T is defined on $\overline{U}$. Suppose that all points of ∂U are of T-irreversibility for (3.17.1) and that $f(0,x) \neq 0$ on ∂U. Then*

$$\operatorname{ind}(P^f_T, U) = \deg(-f(0,\,\cdot\,), U).$$

Theorem (3.9) shows that, at least in the case of autonomous differential equations, the assumption of T-irreversibility can be removed: the essential fact is the absence of fixed points for P^f_T on ∂U (i.e. the admissibility on U of the Poincaré T-translation operator). Now, the question is if one can eliminate the T-irreversibility hypothesis also for the non-autonomous case. Equation (3.20), with $\lambda = 1$, shows that this is not possible. In fact, let U be the open unit disk in $\mathbb{R}^2$. A direct computation gives $\deg(-f(0,\,\cdot\,), U) = 1$ and $\operatorname{ind}(P^f_{2\pi}, U) = 0$ (since (3.20) has no 2π-periodic orbits for $\lambda = 1$).

Despite this limitation, it is possible to give a formula for the fixed point index of the T-translation operator associated with the equation

$$\dot{x} = \lambda f(t,x), \quad \lambda \geq 0, \tag{3.22}$$

where $f: \mathbb{R} \times M \to \mathbb{R}^k$ is a C^1 tangent vector field, T-periodic in the first variable.

Define the *average wind* vector field

$$w_f(p) = \frac{1}{T}\int_0^T f(t,p)\,dt,$$

which is clearly tangent to M.

The following theorem (compare the proof of Theorem 2.1 in [FuPe3]) provides a formula for the computation of the fixed point index of the T-translation operator $P_T^{\lambda f}$ associated with equation (3.22) for λ sufficiently small.

(3.23) THEOREM. *Let $f: \mathbb{R} \times M \to \mathbb{R}^k$ be C^1 tangent vector field that is T-periodic in the first variable. Consider a relatively compact open subset U of the manifold $M \subseteq \mathbb{R}^k$. Assume that (w_f, U) is admissible for the degree. Then, there exists $\lambda_0 > 0$ such that, for $0 < \lambda \leq \lambda_0$, $P_T^{\lambda f}$ is defined on $\overline{U}$, fixed point free on ∂U and*

$$\operatorname{ind}(P_T^{\lambda f}, U) = \deg(-w_f, U).$$

PROOF. Consider the equation

$$\dot{x} = \lambda(\mu f(t,x) + (1-\mu)w_f(x)), \quad \lambda \geq 0,\ \mu \in [0,1]. \tag{3.24}$$

Denote by H_T the translation operator that associates to any (λ, p, μ) the value at time T (if defined) of the solution of (3.24) starting from p at time 0. One can show that for $\lambda \geq 0$ small enough $H_T(\lambda, p, \mu)$ is defined for $p \in \overline{U}$ and $\mu \in [0,1]$.

We claim that there exists $\lambda_0 > 0$ such that $H_T(\lambda, p, \mu) \neq p$ for $0 < \lambda \leq \lambda_0$, $p \in \partial U$ and $\mu \in [0,1]$. Assume this is not the case. Thus there exist sequences $\lambda_n \to 0$, $\mu_n \in [0,1]$ and $p_n \in \partial U$ such that $\lambda_n > 0$ and

$$0 = H_T(\lambda_n, p_n, \mu_n) - p_n = \lambda_n \int_0^T (\mu_n f(t, x_n(t)) + (1-\mu_n) w_f(x_n(t)))\,dt,$$

where x_n denotes the solution of

$$\begin{cases} \dot{x} = \lambda_n(\mu_n f(t,x) + (1-\mu_n)w_f(x)), \\ x(0) = p_n. \end{cases}$$

Since $\lambda_n > 0$, one has

$$0 = \int_0^T (\mu_n f(t, x_n(t)) + (1-\mu_n) w_f(x_n(t)))\,dt. \tag{3.25}$$

Without loss of generality we can assume $\mu_n \to \mu_0 \in [0,1]$ and $p_n \to p_0 \in \partial U$. Thus, by Theorem (2.48), x_n converges uniformly on $[0,T]$ to a (necessarily constant) solution $x_0(t) \equiv p_0$. Hence, passing to the limit in (3.25), we get

$$0 = \int_0^T (\mu_0 f(t,p_0) + (1-\mu_0)w_f(p_0))\,dt,$$

that implies $w_f(p_0) = 0$. This contradicts the assumption.

Thus, there exists $\lambda_0 > 0$ such that, when $0 < \lambda \leq \lambda_0$, the map

$$H_T(\lambda, \cdot\,, \cdot\,): \overline{U} \times [0,1] \to M$$

given by $(p,\mu) \mapsto H_T(\lambda,p,\mu)$ is an admissible homotopy. The homotopy invariance property of the fixed point index shows that for such λ's

$$\text{(3.26)} \qquad \operatorname{ind}(\Phi_T^{\lambda w_f}, U) = \operatorname{ind}(P_T^{\lambda f}, U).$$

By Theorem (3.9) and formula (2.38), one has

$$\text{(3.27)} \qquad \operatorname{ind}(\Phi_T^{\lambda w_f}, U) = \deg(-\lambda w_f, U) = \deg(-w_f, U).$$

The assertion follows from equations (3.26) and (3.27). □

The following result is a well known consequence of the homotopy invariance property of the Lefschetz number (see e.g. [DuGr]) since, when the manifold M is compact, P_T^f is admissibly homotopic to the identity. We give here a different proof based on Theorem (3.23).

(3.28) COROLLARY. *Let $M \subseteq \mathbb{R}^k$ be a compact manifold. Consider a C^1 tangent vector field $f: \mathbb{R} \times M \to \mathbb{R}^k$, T-periodic in the first variable. Then* $\operatorname{ind}(P_T^f, M)$ *is well defined and*

$$\operatorname{ind}(P_T^f, M) = \chi(M).$$

PROOF. By Theorem (3.23), taking $U = M$ one can find $\lambda > 0$ such that $\operatorname{ind}(P_T^{\lambda f}, M) = \deg(-w_f, M)$. Moreover, the Poincaré–Hopf theorem implies

$$\deg(-w_f, M) = \chi(M).$$

Since M is compact, $P_T^{\lambda f}$ is admissibly homotopic to P_T^f. The assertion follows from the homotopy invariance property of the fixed point index. □

When the manifold M is not compact, Theorem (3.23) allows the computation of the fixed point index of P_T^f only for small values of λ. There is a case, however, when this limitation is not necessary: that is, when $f(t,x) = a(t)h(x)$.

Namely, consider the equation

$$\dot{x} = \lambda a(t)h(x), \quad \lambda \geq 0, \tag{3.29}$$

where $a: \mathbb{R} \to \mathbb{R}$ is a T-periodic continuous function and $h: M \to \mathbb{R}^k$ is a C^1 tangent vector field. Assume that the average

$$\overline{a} = \frac{1}{T}\int_0^T a(t)\,dt \neq 0.$$

As in the autonomous case, the fixed point index of the translation operator $P_T^{\lambda ah}$, when defined, does not depend on λ. In fact, the following result holds (compare [Spd3]).

(3.30) THEOREM. *Let $h: M \to \mathbb{R}^k$ be a C^1 tangent vector field, and let $a: \mathbb{R} \to \mathbb{R}$ be continuous, T-periodic, with*

$$\overline{a} = \frac{1}{T}\int_0^T a(t)\,dt \neq 0.$$

Given a relatively compact open subset U of M, assume that $P_T^{\lambda ah}$ is defined on $\overline{U}$ and fixed point free on ∂U. Then $\operatorname{ind}(P_T^{\lambda ah}, U) = (\operatorname{sign}\overline{a})^m \deg(-h, U)$, *where m is the dimension of M.*

PROOF. Without loss of generality, we may assume $\lambda = 1$ and $\overline{a} = 1$. Take any $p \in M$ and consider the Cauchy problems

$$\dot{x} = h(x), \qquad x(0) = p, \tag{3.31.1}$$

$$\dot{x} = a(t)h(x), \quad x(0) = p. \tag{3.31.2}$$

Denote by $x: I \to M$ and $\xi: J \to M$ the (unique) maximal solutions of (3.31.1) and of (3.31.2), respectively. Clearly, if $\int_0^\tau a(s)ds \in I$ for all $\tau \in [0, t]$, then

$$\xi(t) = x\Big(\int_0^t a(s)\,ds\Big).$$

Hence $t \in J$. Moreover, by a standard maximality argument, one can prove that $t \in J$ implies $\int_0^t a(s)ds \in I$. In particular, if $T \in J$, then $\int_0^T a(s)\,ds = T \in I$. When this happens, one has $\xi(T) = x(T)$. In other words, if $P_T^{ah}(p)$ is defined, then so is $P_T^h(p)$, and $P_T^h(p) = P_T^{ah}(p)$. Theorem (3.9) implies

$$\operatorname{ind}(P_T^{ah}(p), U) = \operatorname{ind}(P_T^h(p), U) = \deg(-h, U).$$

This proves the assertion. □

Let us now consider the parametrized equation

$$\dot{x} = g(x) + \lambda f(t, x), \quad \lambda \geq 0, \tag{3.32}$$

where $f: \mathbb{R} \times M \to \mathbb{R}^k$ and $g: M \to \mathbb{R}^k$ are C^1 tangent vector fields on $M \subseteq \mathbb{R}^k$, and f is T-periodic in the first variable. As before $P_T^{g+\lambda f}$ denotes the Poincaré T-translation operator associated with (3.32). Note that for $\lambda = 0$ one has $P_T^g = \Phi_T^g$.

By a *starting point* of (3.32) we mean a pair $(\lambda, p) \in [0, \infty) \times M$ such that $P_T^{g+\lambda f}(p) = p$. Clearly, (λ, p) is a starting point of (3.32) if and only if the unique solution of $\dot{x} = g(x) + \lambda f(t, x)$ starting at p for $t = 0$ is T-periodic. Observe that $p \in M$ belongs to a T-periodic orbit of (3.32) for $\lambda = 0$ if and only if $(0, p)$ is a starting point. In particular, the set $\{0\} \times g^{-1}(0)$ is made up of starting points, and will be referred to as the set of *trivial starting points* (of (3.32)). Of course there may exist starting points $(0, p)$ which are nontrivial (this happens when p belongs to a nontrivial T-periodic orbit of (3.32) for $\lambda = 0$).

The following results holds.

(3.33) THEOREM. *Let $f: \mathbb{R} \times M \to \mathbb{R}^k$ and $g: M \to \mathbb{R}^k$ be C^1 tangent vector fields on a manifold $M \subseteq \mathbb{R}^k$, and let f be T-periodic in the first variable. Let U be a relatively compact open subset of M such that $P_T^{g+\lambda f}$ is defined on $\overline{U}$ for any $\lambda \in [0, 1]$, and assume there are no starting points on $[0, 1] \times \partial U$ of (3.32). Then*

$$\operatorname{ind}(P_T^{g+f}, U) = \deg(-g, U).$$

PROOF. Since there are no fixed points of $P_T^g = \Phi_T^g$ on ∂U, (Φ_T^g, U) is admissible for the fixed point index. Thus, by Theorem (3.9),

$$\operatorname{ind}(P_T^g, U) = \operatorname{ind}(\Phi_T^g, U) = \deg(-g, U). \tag{3.34}$$

By assumption, there are no fixed points of $P_T^{g+\lambda f}$ on ∂U for any $\lambda \in [0, 1]$. Consequently, the map $(p, \lambda) \mapsto P_T^{g+\lambda f}(p)$ is an admissible homotopy. The homotopy invariance property of the fixed point index yields

$$\operatorname{ind}(P_T^g, U) = \operatorname{ind}(P_T^{g+f}, U). \tag{3.35}$$

The assertion follows from (3.34) and (3.35). □

(3.36) THEOREM. *Let $f: \mathbb{R} \times M \to \mathbb{R}^k$ and $g: M \to \mathbb{R}^k$ be C^1 tangent vector fields on a manifold $M \subseteq \mathbb{R}^k$, and let f be T-periodic in the first variable. Let U be a relatively compact open subset of M such that Φ_T^g is defined on $\overline{U}$ and fixed point free on ∂U. Then, there exists $\lambda_0 > 0$ such that $\operatorname{ind}(P_T^{g+\lambda f}, U)$ is well defined for $0 \leq \lambda \leq \lambda_0$, and*

$$\operatorname{ind}(P_T^{g+\lambda f}, U) = \deg(-g, U).$$

PROOF. Because of Theorem (2.48), the set

$$\{(\lambda, p) \in [0, \infty) \times M : P_T^{g+\lambda f}(p) \text{ is defined}\}$$

is open in $[0, \infty) \times M$. Consequently, since by assumption $P_T^g = \Phi_T^g$ is defined for any p in the compact set $\overline{U}$, so is the operator $P_T^{g+\lambda f}$ for small values of λ. As the set of starting points is closed and P_T^g is fixed point free on the compact set ∂U, there are no starting points on $[0, \lambda_0] \times \partial U$ for some $\lambda_0 > 0$. The assertion now follows from Theorem (3.33) replacing f by f/λ_0. □

Since in the above result we have assumed that (Φ_T^g, U) is fixed point free on ∂U, as a consequence one has $g(p) \neq 0$ for all $p \in \partial U$. Thus Theorem (3.36) cannot be seen as an extension of Theorem (3.23). As far as we know a result which includes both these cases is still unknown.

An argument as in the proof of Theorem (3.36) yields a similar result for the parametrized equation

$$\dot{x} = a(t)g(x) + \lambda f(t, x), \quad \lambda \geq 0, \tag{3.37}$$

where $a: \mathbb{R} \to \mathbb{R}$ is a T-periodic continuous function, $g: M \to \mathbb{R}^k$ and $f: \mathbb{R} \times M \to \mathbb{R}^k$ are C^1 tangent vector fields, and f is T-periodic in the first variable. As before, a pair $(\lambda, p) \in [0, \infty) \times M$ is a *starting point for* (3.37) if p is a fixed point of $P_T^{ag+\lambda f}$.

(3.38) THEOREM. *Assume that $a: \mathbb{R} \to \mathbb{R}$ is a T-periodic continuous function with nonzero average $\overline{a}$, $g: M \to \mathbb{R}^k$ and $f: \mathbb{R} \times M \to \mathbb{R}^k$ are C^1 tangent vector fields on a manifold $M \subseteq \mathbb{R}^k$, and f is T-periodic in the first variable. Let U be a relatively compact open subset of M such that P_T^{ag} is defined on $\overline{U}$ and fixed point free on ∂U. Then, there exists $\lambda_0 > 0$ such that* $\operatorname{ind}(P_T^{ag+\lambda f}, U)$ *is well defined for $0 \leq \lambda \leq \lambda_0$, and*

$$\operatorname{ind}(P_T^{ag+\lambda f}, U) = (\operatorname{sign} \overline{a})^m \deg(-g, U),$$

where m is the dimension of the manifold M.

The C^1 assumption on the vector fields f and g made throughout this section can be clearly relaxed. The important requirement is the uniqueness of the solutions of the initial value problems. Actually, using the techniques described in [FuPe5], [Spd2], the results presented in this section could be extended to the Carathéodory case. Namely, when g is locally Lipschitz and f is assumed to satisfy the following hypotheses:

(3.39.1) for each $p \in M$, the map $t \mapsto f(t, p)$ is Lebesgue measurable on $\mathbb{R}$;
(3.39.2) for a.a. $t \in \mathbb{R}$, the map $p \mapsto f(t, p)$ is continuous on M;

(3.39.3) for any compact set $K \subseteq M$, there exists $\gamma_K \in L^1([0,T])$ such that $|f(t,p)| \leq \gamma_K(t)$ for a.a. $t \in [0,T]$ and all $p \in K$;

(3.39.4) for any $p \in M$, $f(t+T,p) = f(t,p) \in T_pM$ a.e. in $\mathbb{R}$;

(3.39.5) for any compact subset K of M, there exists $\alpha_K \in L^1([0,T])$ such that

$$|f(t,p_1) - f(t,p_2)| \leq \alpha_K(t)|p_2 - p_1|,$$

for a.a. $t \in \mathbb{R}$ and for any $p_1, p_2 \in K$.

Conditions (3.39.1)–(3.39.3) are the so-called Carathéodory type assumptions while (3.39.4) says that f is a time-dependent T-periodic tangent vector field on M. The assumption (3.39.5) ensures the uniqueness of the solutions of the initial value problems (compare, e.g. [CoLe]).

In this framework, a solution to (3.32) is an absolutely continuous function $x\colon J \to M \subseteq \mathbb{R}^k$ defined on a (nontrivial) interval and satisfying the condition

$$\dot{x}(t) = g(x(t)) + \lambda f(t, x(t)), \quad \text{for a.a. } t \in J.$$

4. Applications and examples

Below, we shall use the results of the previous section to investigate the structure of the set of T-periodic solutions of equations (3.22) and (3.32). The possible field of application ranges from continuation theorems for periodic solutions to multiplicity results. However, we shall confine ourselves to the simplest applications and only present those that seem most appropriate to shed light on the results discussed in the previous section.

We will need the following global connectivity result.

(4.1) LEMMA ([FuPe6]). *Let X be a locally compact metric space and let $K \subseteq X$ be nonempty and compact. Assume that any compact subset of X containing K has nonempty boundary. Then $X \setminus K$ contains a connected set whose closure intersects K and is not compact.*

By Theorem (2.48), the set $\Omega \subseteq [0,\infty) \times M$ given by

$$\Omega = \{(\lambda, p) : \text{the solution } x(\,\cdot\,) \text{ of (3.32) satisfying } x(0) = p \text{ is defined on } [0,T]\},$$

is open. Thus it is locally compact. Clearly Ω contains the set S of all starting points of (3.32). Observe that S is closed in Ω, though not necessarily closed in $[0,\infty) \times M$. Therefore it is locally compact.

In the sequel, given any subset A of $[0,\infty) \times M$ and $\lambda \geq 0$, the symbol A_λ denotes the slice $\{x \in M : (\lambda, x) \in A\}$ of A.

(4.2) THEOREM ([FuSp1]). *Let $f:\mathbb{R}\times M\to\mathbb{R}^k$ and $g: M\to\mathbb{R}^k$ be two C^1 tangent vector fields on a manifold $M\subseteq\mathbb{R}^k$, and let f be T-periodic in the first variable. Denote by S the set of starting points of (3.32) and let V be an open subset of Ω. If $g^{-1}(0)\cap V_0$ is compact and $\deg(g,V_0)$ is nonzero, then the set $(S\cap V)\setminus(\{0\}\times g^{-1}(0))$ of nontrivial starting points in V admits a connected subset whose closure in V meets $\{0\}\times g^{-1}(0)$ and is not compact.*

PROOF. To prove the assertion it is enough to show that the topological pair

$$(X,K)=(S\cap V,\{0\}\times(g^{-1}(0)\cap V_0))$$

satisfies the assumptions of Lemma (4.1). Since the set $S\cap V$ is open in the locally compact set S, it is locally compact as well. Moreover, as $\deg(g,V_0)$ is nonzero, the compact set $\{0\}\times(g^{-1}(0)\cap V_0)$ is nonempty. Assume, by contradiction, that there exists a compact subset C of $S\cap V$ containing $\{0\}\times(g^{-1}(0)\cap V_0)$ and with empty boundary in the space $S\cap V$. Thus C is open in $S\cap V$ (in fact it is clopen). As V is open in $[0,\infty)\times M$, C is actually open as a subspace of S. Thus there exists an open subset W of $[0,\infty)\times M$ such that $S\cap W=C$. Because of the compactness of the slice C_0 of C, we may choose W is such a way that the neighbourhood W_0 of C_0 turns out to be relatively compact in M. Moreover, without loss of generality, we may assume that the boundary of W_0 in M does not contain points of S_0 (i.e. fixed points of $P^g_T=\Phi^g_T$). Thus, applying the excision property of the degree, Theorem (3.9) and formula (2.37), one gets

$$\operatorname{ind}(P^g_T,W_0)=\deg(-g,W_0)=(-1)^m\deg(g,W_0)=(-1)^m\deg(g,V_0)\neq 0,$$

where m is the dimension of M. As C is compact, there exists $\mu>0$ such that the Poincaré operator $P^{g+\mu f}_T$ is fixed point free on the slice W_μ. Then, from the generalized homotopy property and the solution property of the index, we get

$$\operatorname{ind}(P^g_T,W_0)=\operatorname{ind}(P^{g+\mu f}_T,W_\mu)=0,$$

and this is a contradiction. □

A similar (but more general) result can be proved for equation (3.37). The proof is analogous to that of Theorem (4.2) and, therefore, it will be omitted.

(4.3) THEOREM ([Spd3]). *Let $a:\mathbb{R}\to\mathbb{R}$ be a continuous function, and let $f:\mathbb{R}\times M\to\mathbb{R}^k$ and $g: M\to\mathbb{R}^k$ be two C^1 tangent vector fields on $M\subseteq\mathbb{R}^k$. Assume also that f and a are T-periodic, and that the average of a is nonzero. Denote by S the set of starting points of (3.37) and let V be an open subset of $[0,\infty)\times M$ such that $P^{ag+\lambda f}_T(p)$ is defined for any $(\lambda,p)\in V$. Assume that*

$\deg(g, V_0)$ is well defined and nonzero. Then the set $(S \cap V) \setminus (\{0\} \times g^{-1}(0))$ of nontrivial starting points (in V) of (3.37) admits a connected subset whose closure in V meets $\{0\} \times g^{-1}(0)$ and is not compact.

The following result regarding equation (3.22) can be proved similarly to Theorem (4.2).

(4.4) THEOREM ([FuPe3]). *Let $f : \mathbb{R} \times M \to \mathbb{R}^k$ be a C^1 tangent vector field which is T-periodic in the first variable. Denote by S the set of starting points of (3.22) and let $V \subseteq [0, \infty) \times M$ be open and such that $P_T^{\lambda f}(p)$ is defined for any $(\lambda, p) \in V$. Assume that $\deg(w_f, V_0) \neq 0$. Then the set $(S \cap V) \setminus (\{0\} \times M)$ of nontrivial starting points (in V) of (3.22) admits a connected subset whose closure in V meets $\{0\} \times w_f^{-1}(0)$ and is not compact.*

Below, we give some simple consequences of Theorems (4.2)–(4.4) which illustrate their usefulness in describing the structure of the starting point sets.

(4.5) COROLLARY. *Let M, a, g and f be as in Theorem (4.3). Assume that M is closed as a subset of $\mathbb{R}^k$ and*

$$\|f(t, x)\| \leq \alpha + \beta\|x\|, \quad \|g(x)\| \leq \alpha + \beta\|x\|$$

for some $\alpha, \beta \geq 0$ and all $(t, x) \in \mathbb{R} \times M$. If $g^{-1}(0)$ is compact and $\deg(g, M) \neq 0$, then there exists an unbounded connected set of starting points for (3.37) which meets $\{0\} \times g^{-1}(0)$.

PROOF. Since M is closed in $\mathbb{R}^k$, the assumptions on f and g imply that any (maximal) solution of (3.37) is defined on the whole real line. Thus, taking $V = [0, \infty) \times M$, Theorem (4.3) implies the existence a connected set Σ of starting points for the equation (3.37) whose closure (in $[0, \infty) \times M$ or, equivalently, in $[0, \infty) \times \mathbb{R}^k$) is not compact and meets $\{0\} \times g^{-1}(0)$. This implies that Σ is unbounded. □

We point out that the unbounded set of starting points ensured by Corollary (4.5) may be "completely vertical"; that is, contained in the slice $\{0\} \times M$ of $[0, \infty) \times M$. Of course this may happen only if M is not compact. The following example with $M = \mathbb{R}^2$ and $T = 2\pi$ illustrates this phenomenon:

$$\begin{cases} \dot{x} = y, \\ \dot{y} = -x + \lambda \sin t. \end{cases}$$

(4.6) COROLLARY. *Assume that M is a compact manifold with $\chi(M) \neq 0$. Let a, f and g be as in Theorem (4.3). Then, there exists a connected set of starting* *points* Σ *for* (3.37) *which meets* $\{0\} \times g^{-1}(0)$ *and such that* $\pi_1(\Sigma) = [0, \infty)$, *where* π_1 *denotes the projection on the first factor of* $[0, \infty) \times M$.

PROOF. By the compactness of M, any solution of (3.37) is globally defined. We apply Theorem (4.3) to the open set $V = [0, \infty) \times M$. By the Poincaré–Hopf Theorem, we have $\deg(g, V_0) = \chi(M) \neq 0$. Therefore there exists a connected set Σ of starting points of (3.37) which meets $\{0\} \times g^{-1}(0)$ and is not contained in any compact subset of V. This implies that Σ is unbounded and, M being compact, its projection on $[0, \infty)$ must be unbounded, connected and containing 0. □

An analogous argument proves (see e.g. [FuPe3]) the following consequence of Theorem (4.4).

(4.7) COROLLARY. *Assume that* M *is a compact manifold with* $\chi(M) \neq 0$. *Let* f *be as in Theorem* (4.4). *Then there exists an unbounded connected set of starting points which meets* $\{0\} \times w_f^{-1}(0)$. *In particular the equation* $\dot{x} = f(t, x)$ *admits a* T*-periodic solution.*

The fact that the global branch ensured by Theorem (4.2) emanates from the set of zeros of g, and not merely from the set of all T-periodic orbits of $\dot{x} = g(x)$, allows us to obtain information about the starting point set of equation (3.37) also in the case of a compact manifold with zero Euler-Poincaré characteristic, as in the following multiplicity result. Here the *index* at an isolated zero z of a tangent vector field $g: M \to \mathbb{R}^k$ is defined as $\deg(g, U)$, where U is an isolating neighbourhood of z. This makes sense because of the excision property of the degree of a vector field. In particular, when z is a nondegenerate zero, its index is either $+1$ or -1.

(4.8) COROLLARY. *Let* $M \subseteq \mathbb{R}^k$ *be a compact manifold. Assume that* f *and* g *are as in Theorem* (4.2) *and, in addition, that* g *has exactly two distinct zeros* z_1 *and* z_2 *with nonzero index. Denote by* S_1 *and* S_2 *the connected components of the set of starting points of* (3.32) *which contain respectively* z_1 *and* z_2. *Then just one of the following two possibilities holds:*

(4.8.1) $S_1 = S_2$,

(4.8.2) S_1 *and* S_2 *are disjoint and both unbounded* (*in* $[0, \infty) \times M$).

In particular, if (4.8.2) *holds, there exist at least two distinct* T*-periodic solutions of* (3.37) *for each* $\lambda \in [0, \infty)$.

PROOF. Since M is compact, any solution of (3.32) is globally defined. Take

$$V_1 = [0, \infty) \times M \setminus \{(0, z_2)\}, \quad V_2 = [0, \infty) \times M \setminus \{(0, z_1)\}.$$

Obviously $(0, z_i) \in V_i$ and, by the excision property, $\deg(g, V_i) \neq 0$ for $i \in 1, 2$. We may assume $S_1 \neq S_2$. In this case S_1 and S_2, being connected components, are clearly disjoint and, consequently, $S_1 \subseteq V_1$ and $S_2 \subseteq V_2$. Because of Theorem (4.2), S_1 and S_2 are not contained in any compact subset of V_1 and V_2 respectively and, in particular, they are not compact. Now S_1 and S_2 are closed in the set S of all starting points of (3.32). Since S is closed in $V = [0, \infty) \times M$, which is closed in $\mathbb{R}^{k+1}$, the two components S_1 and S_2 must be unbounded. □

A remarkable consequence of Theorem (4.4) is the following continuation result regarding T-periodic solutions (see [FuPe3] for a more general version). Observe that in Theorems (4.2)–(4.4) we have assumed the global existence in $[0, T]$ of the solutions of the initial value problem. This hypothesis is not explicitly stated in Theorem (4.9) below, since it is ensured by suitable a priori bounds on the T-periodic orbits of the equation (see e.g. [CaMaZa], [Kra], [Maw]).

(4.9) THEOREM. *Let $f: \mathbb{R} \times M \to \mathbb{R}^k$ be a C^1 tangent vector field which is T-periodic in the first variable. Assume that:*

(4.9.1) *$\deg(w_f, M)$ is well defined and nonzero;*
(4.9.2) *the T-periodic orbits of (3.22) for $\lambda \in (0, 1]$ lie in a compact subset of M.*

Then, the equation

$$\dot{x} = f(t, x) \tag{4.9.3}$$

has a T-periodic solution.

PROOF. Let K be a compact subset of M containing all the zeros of w_f and all the T-periodic orbits of (3.22) for $\lambda \in (0, 1]$, and let $U \subseteq M$ be a relatively compact open set containing K. Let $\sigma: M \to [0, 1]$ be a C^1 function with compact support in M and such that $\sigma(p) = 1$ for each $p \in U$. Since the tangent vector field σf has compact support, for each $(\lambda, p) \in [0, \infty) \times M$ the solution of the initial value problem

$$\begin{cases} \dot{x} = \lambda \sigma(x) f(t, x), \\ x(0) = p, \end{cases} \tag{4.10}$$

is globally defined on $\mathbb{R}$. We have $\sigma w_f = w_{\sigma f}$. Thus, from the excision property it follows that

$$\deg(w_{\sigma f}, U) = \deg(w_f, U) = \deg(w_f, M) \neq 0.$$

Hence, taking $V = [0, 1) \times U$, Theorem (4.4) implies the existence of a connected set $\Sigma \subseteq V$ of starting points of (4.10) which is closed in V, meets $w_{\sigma f}^{-1}(0) \cap U = w_f^{-1}(0)$ and is not compact. Consider the following subset of Σ:

$$\widetilde{\Sigma} = \{(\lambda, p) \in \Sigma : \text{the solution } x \text{ of (4.10) corresponding to } (\lambda, p) \\ \text{is such that } x(t) \in U \text{ for all } t\}.$$

As Σ meets $w_f^{-1}(0)$, $\tilde{\Sigma}$ is nonempty. Moreover, it is easy to check that $\tilde{\Sigma}$ is open and closed in Σ. Thus, $\tilde{\Sigma} = \Sigma$. Since $\sigma(p) \equiv 1$ in U, this implies that any $(\lambda, p) \in \Sigma$ is, in fact, a starting point of (3.22).

Observe that, Σ being noncompact and closed in $[0,1) \times U$, the closure of Σ in the compact set $[0,1] \times \overline{U}$ must intersect the boundary of $[0,1) \times U$ in $[0,\infty) \times M$, which is the union of $[0,1) \times \partial U$ and $\{1\} \times \overline{U}$.

By the continuous dependence on data of the solutions, if (λ, p) is in the closure of Σ, then it is a starting point. Therefore, by the choice of the compact set K, one has $p \in K$. As $K \subseteq U$, the closure of Σ does not intersect the set $[0,1) \times \partial U$. Thus, there exists a starting point of (3.22) of the form $(1, p)$, and this proves the assertion. □

In this section we have introduced only a small number of consequences of the formulas obtained in Section 3. Among the applications to T-periodic solutions of differential equations not presented here, we mention multiplicity results, guiding function-like existence results and other continuation results (see e.g. [FuPe2], [Spd1], [Spd3]). It should also be remarked that the results of this section, mainly Theorems (4.2)–(4.4), can be reformulated somewhat more elegantly in an infinite-dimensional framework (see e.g. [FuPe1], [FuSp2] and also [FuPe5], [Spd2] for the Carathéodory case).

Since second order ordinary differential equations on a manifold M can be written as first order equations on its tangent bundle TM (see e.g. [Fur]), one can apply the results of Section 3 to this kind of equation (see e.g. [FuPe6], [FuSp3], see also [FuPeSp1], [FuPeSp2] and references therein for a discussion of the multiplicity results obtainable using this methods).

References

[Ama] H. Amann, *Fixed point equations and nonlinear eigenvalue problems in ordered Banach spaces*, SIAM Rev. **18** (1976), 620–709.

[AmWa] H. Amann H and S. Weiss, *On the uniqueness of the topological degree*, Math. Z. **130** (1973), 37–54.

[Bou] G. Bouligand, *Introduction à la géométrie infinitesimale directe*, Gauthier–Villard, Paris, 1932.

[Bro] R. F. Brown, *An elementary proof of the uniqueness of the fixed point index*, Pacific J. Math. **35** (1970), 549–558.

[CaMaZa] A. Capietto, J. Mawhin and F. Zanolin, *Continuation theorems for periodic perturbations of autonomous systems*, Trans. Amer. Math. Soc. **329** (1992), 41–72.

[CoLe] E. A. Coddington and N. Levinson, *Theory of Ordinary Differential Equation*, McGraw –Hill Book Company Inc., New York, 1955.

[Cop] W. A. Coppel, *Stability and asymptotic behavior of differential equations*, Heath Math. Monograph, Boston, 1965.

[DuGr] J. Dugundji and A. Granas, *Fixed Point Theory*, Springer–Verlag, New York, 2003.

[Fur] M. Furi, *Second order differential equations on manifolds and forced oscillations*, Topological Methods in Differential Equations and Inclusions (A. Granas and M. Frigon, eds.), vol. 472, Kluwer Acad. Publ. Ser. C, 1995.

[FuPe1] M. Furi and M. P. Pera, *Global branches of harmonic solutions to periodic* ODEs *on manifolds*, Boll. Un. Mat. Ital. A (7) **11** (1997), 709–722.

[FuPe2] ———, *Global branches of periodic solutions for forced differential equations on nonzero Euler characteristic manifolds*, Boll. Un. Mat. Ital. C (6); Anal. Funz. Appl. **3** (1984), 157–170.

[FuPe3] ———, *A continuation principle for forced oscillations on differentiable manifolds*, Pacific J. Math. **121** (1986), 321–338.

[FuPe4] ———, *Global bifurcation of fixed points and the Poincaré translation operator on manifolds*, Ann. Mat. Pura Appl. **173** (1997), 313–331.

[FuPe5] ———, *Carathéodory periodic perturbations of the zero vector field on manifolds*, Topol. Methods Nonlinear Anal. **10** (1997), 79–92.

[FuPe6] ———, *A continuation principle for periodic solutions of forced motion equations on manifolds and application to bifurcation theory*, Pacific J. Math. **160** (1993), 219–244.

[FuPeSp1] M. Furi, M. P. Pera and M. Spadini, *Forced oscillations on manifolds and multiplicity results for periodically perturbed autonomous systems*, J. Comput. Appl. Math. **113** (2000), 241–254.

[FuPeSp2] ———, *Multiplicity of forced oscillations on manifolds and applications to motion problems with one-dimensional constraints*, Set-Valued Anal. **9** (2001), 67–73.

[FuSp1] M. Furi and M. Spadini, *On the fixed point index of the flow and applications to periodic solutions of differential equations on manifolds*, Boll. Un. Mat. Ital. A (7) **10** (1996), 333–346.

[FuSp2] ———, *On the set of harmonic solutions to periodically perturbed autonomous differential equations on manifolds*, Nonlinear Anal. **29** (1997), 963–470.

[FuSp3] ———, *Branches of forced oscillations for periodically perturbed second order autonomous* ODEs *on manifolds*, J. Differential Equations **154** (1999), 96–106.

[Gra] A. Granas, *The Leray–Schauder index and the fixed point theory for arbitrary* ANRs, Bull. Soc. Math. France **100** (1972), 209–228.

[GuPo] V. Guillemin and A. Pollack, *Differential Topology*, Prentice–Hall Inc., Englewood Cliffs, New Jersey, 1974.

[Hir] M. W. Hirsch, Graduate Texts in Math. **33** (1976), Springer–Verlag Berlin.

[Kra] M. A. Krasnosel'skiĭ, *The Operator of Translation along the Trajectories of Differential Equations*, Transl. Math. Monographs, vol. 19, Amer. Math. Soc., Providence, Rode Island, 1968.

[Ler] J. Leray, *Thèorie des points fixes: indice total et nombre de Lefschetz*, Bull. Soc. Mat. France **87** (1959), 221–233.

[Llo] N. G. Lloyd, *Degree Theory*, vol. 73, Cambridge Univ. Press, 1978.

[Maw] J. Mawhin, *Topological degree and boundary value problems for nonlinear differential equations*, C.I.M.E. Course on Topological Methods for Ordinary Differential Equations, Lecture Notes in Math. (M. Furi and P. Zecca, eds.), vol. 1537, Springer–Verlag, 1991, pp. 74–142.

[Mil] J. W. Milnor, *Topology from the Differentiable Viewpoint*, Univ. Press of Virginia, 1965.

[Nir] L. Nirenberg, *Topics in Nonlinear Functional Analysis*, Courant Institute of Mathematical Sciences, New York, 1974.

[Nus] R. D. Nussbaum, *The fixed point index and fixed points theorems*, C.I.M.E. Course on Topological Methods for Ordinary Differential Equations, Lecture Notes in Math. (M. Furi and P. Zecca, eds.), vol. 1537, Springer–Verlag, 1991, pp. 143–205.

[Pei] M. M. Peixoto, *On an approximation theorem of Kupka and Smale*, J. Differential Equations **3** (1966), 214–227.

[Pug] C. C. Pugh, *A generalized Poincaré index formula*, Topology **7** (1968), 217–226.

[Sev] F. Severi, *Conferenze di Geometria Algebrica*, Zanichelli, Bologna, 1927.

[Spd1] M. Spadini, *Perturbazioni periodiche di equazioni differenziali ordinarie su varietà differenziabili*, Doctoral dissertation (1997), University of Florence, Florence.

[Spd2] ———, *Harmonic solutions of periodic Carathéodory perturbations of autonomous ODEs on manifolds*, Nonlinear Anal. **41** (2000), 477–487.

[Spd3] ———, *Harmonic solutions to perturbations of periodic separated variables ODEs on manifolds*, Electron. J. Differential Equations (2003).

[Spa] E. H. Spanier, *Algebraic Topology*, McGraw–Hill Series in Higher Math., McGraw–Hill, New York, 1966.

[Srz1] R. Srzednicki, *On rest points of dynamical systems*, Fund. Math. **126** (1985), 69–81.

[Srz2] ———, *Generalized Lefschetz theorem and a fixed point index formula*, Topology Appl. **81** (1997), 207–224.

[Srz3] ———, *On periodic solutions inside isolating chains*, J. Differential Equations **165** (2000), 42–60.

[Tro] A. J. Tromba, *The Euler characteristic of vector fields on Banach manifolds and a globalization of Leray–Schauder degree*, Adv. Math. **28** (1978), 148–173.

21. ON THE EXISTENCE OF EQUILIBRIA AND FIXED POINTS OF MAPS UNDER CONSTRAINTS

Wojciech Kryszewski

1. Introduction

Since 1941, when S. Kakutani [K] showed that every upper semicontinuous set-valued self map admitting closed convex values of a closed ball in $\mathbb{R}^n$ has a fixed point, a lot of attention has been paid to the different aspects of the fixed point theory for set-valued maps (see e.g. [Go2]); in one direction the development has led to substantial weakening in the assumption that the values of the mapping be subsets of its domain.

Let E, F be Banach spaces, $K \subset E$ be closed and $\varphi\colon K \multimap F$ (i.e. $\varphi\colon K \to 2^F$) be a set-valued map. In what follows we study the existence of equilibria of φ (i.e. points $x \in K$ such that $0 \in \varphi(x)$ (1)) or, assuming that $E = F$, fixed points of φ. The problem may be stated in a different, and in a sense equivalent, way: assuming that $\varphi\colon E \multimap F$, we look for equilibria (or fixed points) belonging to a certain set $K \subset E$ of constraints. Therefore we refer to the mentioned problem as to the *constrained equilibrium*, or *fixed point problem*. Existence of equilibria or fixed points, unconstrained as well as constrained, play an important role in many nonlinear problems.

We shall study several aspects of the constrained problem. On one hand in the above setting, if $E = F$, then fixed points and equilibria are strictly related: a fixed point of φ is an equilibrium of $I_K - \varphi$ or $\varphi - I_K$ (where $I_K\colon K \to E$ is the inclusion) and vice-versa; on the second hand fixed point problems may be put in a purely nonlinear setting. One may study their existence assuming that $\varphi\colon K \multimap X$ where K is a subset (usually of a special type) of a metric space X.

This work was supported in parts by the Faculty of Mathematics and Computer Sciences Research Statute Fund and the KBN Grant 2/P03A/024/16.

(1) The appellation "equilibrium" originates from the calculus of variations where φ is typically a set of subgradients, i.e. the generalized gradient of a convex or a locally Lipschitz real function; it is also motivated by control problems where an equilibrium state is the stationary solution to a differential inclusion $x'(t) \in \varphi(x(t))$.

Methods used in these two situations may be (and usually are) different. While in the first case there is a variety of results concerning different types of mappings and their domains, in the second case most of the results concern maps defined on sets K with nonempty interiors contained, for instance, in an absolute neighbourhood retract X, and eventually they involve various type of the fixed-point index theory. We leave this type of results aside and confine ourselves to results concerning arbitrary (but satisfying some additional geometric conditions) closed (or compact) sets rather. Therefore we shall mainly study a situation described at the beginning. Being restricted to a Banach space setting and widely using the Banach space geometry, we will provide results for mappings satisfying not only topological conditions but also assumptions of a more (geo)metric nature. This seems to be natural for topics belonging to the so-called infinite-dimensional topology. We shall see that even when dealing with mappings that are continuous (or merely upper semicontinuous), defined on sets being absolute neighbourhood retracts, and satisfying some additional topological assumptions (such as compactness etc.), they are also subject to certain "boundary" conditions having geometrical character rather; in particular these conditions are not *absolute*, i.e. for instance, they depend on a particular location of a domain in an ambient space and are sensitive with respect to homeomorphisms (save translations) and homotopies. This of course may be considered as a restriction or a disadvantage; on the other hand such a situation allows to study some less stringent regularity properties. Given this mixed topological and geometric nature of constrained fixed point or equilibrium problem, we shall present results lying on a borderline between topological and metric fixed point theory. In particular we shall study set-valued set-contractions, condensing or compact maps and others as well as those satisfying, apart form the usual upper semicontinuity, Lipschitz conditions (e.g. contractions, nonexpansive and accretive maps), paying attention, however, only to results which rely on simple assumptions and leaving aside some more specialistic considerations. We shall also try to show the intimate relation of the constrained equilibrium problems with the theory of differential equations and inclusions pertaining to the existence, viability theory and the solution sets characterization. In the course of the paper we shall try to stay in the infinite-dimensional setting, therefore we will not present results having purely finite-dimensional nature.

Of course, the presentation reflects a personal view-point of the author and can not be considered as the full image of the state of art in the theory. There are many interesting directions which are not covered in this work. For instance, we shall not give account of results relying on the various aspects of the Conley index theory or the Ważewski principle; on the other hand it seems that set-valued aspects of this theory (see e.g. [GQ], [CGQ]) are still not very well understood. In the paper we deal mainly with set-valued mappings; but it is clear that all results remain valid for single-valued transformations; moreover, the importance of the topic does not rely on the presented set-valued setting.

The chapter is organized as follows. Below the reader will find some remarks concerning notation and preliminaries. In Section 1 we shall deal with results having a topological nature rather while in Section 2 we study maps being subject to (geo)metric conditions. The last Section 3 deals with some constrained coincidence problems.

Let (X, d) be a metric space ([2]) and $K \subset X$. If $x \in X$, the *distance of x to K* is defined by

$$d_K(x) = d(x, K) := \inf_{y \in K} d(x, y),$$

it is clear that $d_K: X \to \mathbb{R}$ is a 1-Lipschitz function. For any $\varepsilon > 0$ and $A \subset X$, we put

$$B_A(K, \varepsilon) := \{x \in A : d_K(x) < \varepsilon\}, \quad D_A(K, \varepsilon) := \{x \in A : d_K(x) \le \varepsilon\}$$

(the subscript A is omitted if $A = X$, unless it leads to ambiguity). The interior, the closure and the boundary of $K \subset X$ will be denoted by $\operatorname{int} K$, $\operatorname{cl} K$ and $\operatorname{bd} K$, respectively. If $x \in \operatorname{cl} K$, then the notation $y \xrightarrow{K} x$ means that y converges to x remaining in K, i.e. $y \in K$ and $d(x, y) \to 0$. If $A, B \subset X$, then we define the *Hausdorff distance*:

$$d_H(A, B) := \max\{\sup_{a \in A} d(a, B), \sup_{b \in B} d(b, A)\} \le \infty.$$

It is easy to see that $d_H(A, B) = \inf\{r \ge 0 : A \subset B(B, r),\ B \subset B(A, r)\}$.

Throughout the paper $(E, \|\cdot\|)$ denotes a *real Banach space* ([3]). The convex and the closed convex envelope of $Z \subset E$ will be denoted by $\operatorname{conv} Z$ and $\operatorname{cl\,conv} Z$, respectively. The Banach space of all bounded linear operators from E into a Banach space F is denoted by $\mathcal{L}(E, F)$ and E^* (resp. E^*_w) stands for the topological dual of E with the norm topology (resp. with the weak* topology); by $\langle \cdot , \cdot \rangle$ we denote the duality pairing: $\langle p, x\rangle := p(x)$ for $p \in E^*$, $x \in E$. Given $U \in \mathcal{L}(E, F)$, $\|U\|$ is the norm of U. The notation $x_n \rightharpoonup x$ (resp. $p_n \overset{*}{\rightharpoonup} p$) means that $(x_n) \subset E$ converges weakly to $x \in E$ (resp. $(p_n) \subset E^*$ converges to $p \in E^*$ in E^*_w); w^*-cl refers to the *weak* closure* in E^*. Given $K \subset E$ (resp. $K \subset E^*$), the *support function* $\sigma_K: E^* \to \mathbb{R} \cup \{\infty\}$ (resp. $\sigma_K: E \to \mathbb{R} \cup \{\infty\}$) is

([2]) The distance in a metric space will always, unless it leads to an ambiguity, be denoted by d.

([3]) Most of the presented results stay true for a real or complex *normed* space E; usually completeness plays no role. However, for simplicity we always stay in the real Banach space setting.

given by $\sigma_K(p) := \sup_{x\in K}\langle p,x\rangle$, $p \in E^*$, (resp. $\sigma_K(x) := \sup_{p\in K}\langle p,x\rangle$, $x \in E$). The (negative) *polar cone* of $K \subset E$ (resp. $K \subset E^*$) is the closed convex cone $K^- := \{p \in E^* : \sigma_K(p) \le 0\}$ (resp. $\{x \in E : \sigma_K(u) \le 0\}$); by the Hahn–Banach theorem, $K^{--} := (K^-)^- = \operatorname{cl}\operatorname{conv}\bigcup_{\lambda\ge 0}\lambda K$ (resp. $K^{--} = w^*\text{-}\operatorname{cl}\operatorname{conv}\bigcup_{\lambda\ge 0}\lambda K$).

A real function γ defined on the family of bounded subsets of E is a *measure of noncompactness* if $\gamma(B) = \gamma(\operatorname{cl}\operatorname{conv} B)$ for any bounded subset $B \subset E$. Two examples of such measures are of importance: given a bounded $B \subset E$,

$$\alpha(B) := \inf\{\varepsilon > 0 : B \text{ admits a finite covering by sets of diameter } \le \varepsilon\},$$
$$\beta(B) := \inf\{\varepsilon > 0 : B \text{ admits a finite covering by } \varepsilon\text{-balls}\}.$$

are the *Kuratowski* and the *Hausdorff* measures of noncompactness, respectively. Recall that both of these measures are *regular*, i.e. $\gamma(B) = 0$ if and only if B is relatively compact; *monotone*, i.e. if $B \subset B'$ then $\gamma(B) \le \gamma(B')$ and *nonsingular*, i.e. $\gamma(\{a\} \cup B) = \gamma(B)$ for any $a \in E$ (for details see [AKPRS]). In what follows we shall always speak of regular, monotone and nonsingular measures of noncompactness.

For $a, b \in \mathbb{R}$ and $K \subset E$, $C([a,b],K)$ is the set of continuous maps $[a,b] \to K$; $C([a,b],E)$ is the Banach space of continuous maps $[a,b] \to E$ equipped with the maximum norm. By $L^1([a,b],E)$ we mean the Banach space of all (Bochner) *integrable* maps $f\colon [a,b] \to E$, i.e. $f \in L^1([a,b],E)$ if and only if f is strongly measurable and $\|f\|_{L^1} := \int_a^b \|f(s)\|\,ds < \infty$. Recall that strong measurability is equivalent to the usual measurability in case E is separable. The Lebesgue measure of a measurable set $\Delta \subset \mathbb{R}$ is denoted by $|\Delta|$ (see e.g. [Yo]).

Denote by $\check{H}^*(\,\cdot\,)$ (resp. $H^*(\,\cdot\,)$ or $H_*(\,\cdot\,)$) the functor of the Čech cohomology (resp. singular cohomology or homology) with rational coefficients (4). Assume that X is a metric space such that the graded vector space $\{\check{H}^q(X)\}_{q\ge 0}$ is of a *finite type*, i.e. $\check{H}^q(X) = 0$ for almost all $q \ge 0$ and the *Betti numbers* $\beta^q(X) := \dim \check{H}^q(X) < \infty$ for all $q \ge 0$. Then the *Euler characteristic* $\chi(X) := \sum_{q\ge 0}(-1)^q\beta^q(X)$ is defined (see e.g. [Br]). The Euler characteristic is a topological invariant (it is invariant under homeomorphisms) and homotopy equivalent spaces have the same Euler characteristic. For instance, any compact polyhedron (or polytope) P has the Euler characteristic because $\{\check{H}^q(P)\}_{q\ge 0}$ is of finite type. In particular, if X is a compact neighbourhood retract in a metric space Y (i.e. there is a so-called *neighbourhood retraction* $r\colon U \to X$, i.e. a continuous map defined on a neighbourhood U of X in Y such that $r(x) = x$ for $x \in X$), then $\chi(X)$ is well-defined since X is homotopy equivalent to a compact polyhedron. If X is contractible (in particular, if $X \subset E$ is convex), then $\chi(X) = 1$; but,

(4) The reader not familiar with cohomology theory may not worry: its applications here are synthetic; one may accept necessary facts without much deeper insight.

for example, the Euler characteristic $\chi(S^n)$, of the n-dimensional unit sphere S^n, is 2 when n is even and 0 when n is odd.

Given a continuous mapping $f: X \to X$ of a space X of a finite type, one defines the (classical) Lefschetz number $\lambda(f)$ of f (see [Br]) (5). The number $\lambda(\,\cdot\,)$ is a homotopy invariant; moreover, $\chi(X)$ is nothing else but the Lefschetz number $\lambda(I_X)$ of the identity $I_X: X \to X$, $\chi(X) = \lambda(I_X)$.

The celebrated *Lefschetz–Hopf theorem* states that if X is a compact absolute neighbourhood retract (see e.g. [Bor]), then a continuous $f: X \to X$ has a fixed point provided $\lambda(f) \neq 0$. In particular, if f is homotopic to the identity and $\chi(X) \neq 0$, then f has a fixed point.

An interesting approach to the Euler characteristic of compact epi-Lipschitz sets (see Subsection 2.3) is given by Cornet in [Co2].

Given metric (or topological) spaces X and Y, a *set-valued map* $\varphi: X \multimap Y$ assigns to any $x \in X$ a *nonempty* subset $\varphi(x) \subset Y$. The map φ is *upper semicontinuous* (resp. *lower semicontinuous*), if the *inverse image* $\varphi^{-1}(A) := \{x \in X : \varphi(x) \cap A \neq \emptyset\}$ is closed (resp. open) in X whenever A is closed (resp. open) in Y. If φ is both lower and upper semicontinuous, then we say that it is *continuous*. A map $\varphi: X \multimap Y$ is *H-upper* (resp. *H-lower*) *semicontinuous* if, for each $x_0 \in X$ and $\varepsilon > 0$, there is δ such that $\varphi(x) \subset B(\varphi(x_0), \varepsilon)$ (resp. $\varphi(x_0) \in B(\varphi(x), \varepsilon)$) for all $x \in B(x_0, \delta)$. Upper semicontinuity (resp. H-lower semicontinuity) implies H-upper semicontinuity (resp. lower semicontinuity) and these notions coincide for maps with compact values. A map φ with bounded values is *contractive* (resp. *non-expansive*) if there is $k \in [0,1)$ such that, for all $x, y \in X$, $d_H(\varphi(x), \varphi(y)) \leq k d(x,y)$ (resp. $d_H(\varphi(x), \varphi(y)) \leq d(x,y)$). Such maps are H-continuous, i.e. both H-upper and H-lower semicontinuous.

A map $\varphi: X \multimap Y$ with closed values is *compact* if it is upper semicontinuous and the *image* $\varphi(X) = \bigcup_{x \in X} \varphi(x)$ is relatively compact. A map $\varphi: K \multimap E$ with compact values, where $K \subset E$, is a *k-set contraction* ($k \geq 0$) (resp. *condensing*) with respect to a measure of noncompactness γ, provided φ is upper semicontinuous and, for every bounded $B \subset K$, the set $\varphi(B)$ is bounded and $\gamma(\phi(B)) \leq \nu\gamma(B)$ (resp. $\gamma(\varphi(B)) < \gamma(B)$ provided B is not compact) (6). Any contraction with compact values is a set-contraction with respect to the Hausdorff or Kuratowski

(5) More generally, given a homomorphism $\varphi = \{\varphi^q\}: \check{H}^*(X) \to \check{H}^*(X)$, one defines $\lambda(\varphi) = \sum_{q \geq 0} (-1)^q \operatorname{trace}(\varphi^q)$; then $\lambda(f) := \lambda(f^*)$. In case X is a compact neighbourhood retract in a Banach space E and $r: U \to X$, where U is an open neighbourhood of X, is the retraction onto X, then $\lambda(f)$ may be defined by means of the topological degree theory; namely $\lambda(f)$ is the Leray–Schauder degree of the map $I - f \circ r$ on U with respect to 0. This definition is correct for it does not depend on the choice of r. Another, less intrinsic, definition of the Euler characteristic of a smooth manifold may be found in [Hi].

(6) If φ is a k-set contraction with $0 \leq k < 1$ with respect to γ, then we say that it is a set-contraction.

measure of noncompactness γ. For further results on set-valued maps, see [Go2], [AF] etc.

Any set-valued map $\varphi: X \multimap Y$, between metric spaces, may be identified with its *graph*

$$\mathrm{Gr}(\varphi) := \{(x, y) \in X \times Y : y \in \varphi(x)\}.$$

Let $p_\varphi: \mathrm{Gr}(\varphi) \to X$ and $q_\varphi: \mathrm{Gr}(\varphi) \to Y$ be the projections of $\mathrm{Gr}(\varphi)$ onto X and into Y, respectively, i.e. $p_\varphi := \pi_X|_{\mathrm{Gr}(\varphi)}$, $q_\varphi = \pi_Y|_{\mathrm{Gr}(\varphi)}$ where $\pi_X: X \times Y \to X$ and $\pi_Y: X \times Y \to Y$ are the projections. It is clear that both p_φ and q_φ are continuous; moreover, for each $x \in X$, $\varphi(x) = q_\varphi(p_\varphi^{-1}(x))$. This observation allows to study a bit more general situation. Namely we shall consider diagrams of the form

$$X \xleftarrow{p} Z \xrightarrow{q} Y \tag{1.1}$$

where Z is a topological space, $p: Z \to X$ is a continuous *surjection* and $q: Z \to Y$ is continuous. Observe that a diagram (1.1) determines a set-valued map $\varphi_{pq}: X \multimap Y$ given by $\varphi_{pq}(x) := q(p^{-1}(x))$. It is clear that there are different diagrams of the form (1.1) determining the same map. In case no ambiguity as concerns the domain Z may arise, such diagrams will be denoted by (p, q). Observe that a diagram of the form (1.1) may be identified with a continuous map $f := (p, q): Z \to X \times Y$; in this case $p = \pi_X \circ f$, $q = \pi_Y \circ f$.

The topological regularity of a map φ_{pq} corresponds well to the properties of a diagram $X \xleftarrow{p} Z \xrightarrow{q} Y$ or the map $f := (p, q): Z \to X \times Y$, determining it. For instance: the graph $\mathrm{Gr}(\varphi_{pq})$ is closed (in $X \times Y$) if and only if the following property is satisfied: given a sequence (z_n) in Z, if $(x, y) = \lim_{n\to\infty} f(z_n)$, then there is $z \in Z$ such that $f(z) = (x, y)$. The latter property holds if, for any compact $K \subset X \times Y$, the preimage $f^{-1}(K)$ is compact, i.e. f is a *proper* map ([7]). The upper (resp. lower) semicontinuity of φ_{pq} is reflected by the closeness (resp. openness) properties of p. Precisely, it is easy to see that φ_{pq} is upper (resp. lower) semicontinuous if and only if p is closed (open) with respect to the weakest topology on Z under which q is continuous. Hence, if $p: Z \to X$ is a closed (resp. open) surjection with respect to the original topology in Z, then the map φ_{pq} determined by $f = (p, q)$ (with an arbitrary but continuous $q: Z \to Y$) is upper (resp. lower) semicontinuous. Note that if $X \subset Y$, then φ_{pq} has a *fixed point* (i.e. $x \in X$ such that $x \in \varphi_{pq}(x)$) if and only if p and q have a *coincidence* (i.e. a point $z \in Z$ such that $p(z) = q(z)$).

Taking the above remarks into account, it seems that diagrams of the form (1.1) provide a richer structure than set-valued mappings only. In particular, such a setting is very useful with regard to the so-called set-valued maps *admissible*

([7]) Observe that f is proper if so is p or q.

in the sense of Górniewicz, see [Go1] and [Kr1], [Kr2]. A map $\varphi\colon X \multimap Y$ is *admissible* if it is represented $\varphi = \varphi_{pq}$ by a pair $X \xleftarrow{p} Z \xrightarrow{q} Y$ where Z is paracompact, q is continuous and $p\colon Z \to X$ is a *Vietoris map*, i.e. it is surjective, *perfect* (closed and, for each $x \in X$, the fiber $p^{-1}(x)$ is compact) and, for each $x \in X$, $\check{H}^*(p^{-1}(x)) = \check{H}^*(pt)$ where pt is a one-point space. Admissible maps are upper semicontinuous with compact values which are continuous images of *acyclic sets* (i.e. having cohomology of a point). The class of admissible maps is quite large. It contains single-valued continuous maps and numerous classes of convex as well as nonconvex maps. Moreover, it satisfies basic stability properties with respect to some elementary operations: the composition of admissible maps is admissible; the Cartesian product, the sum of two admissible maps and the product by a continuous map (provided their values lie in a linear metric space) are admissible.

A Vietoris–Begle theorem (see [Sp]), stating that a Vietoris map $p\colon Z \to X$ induces an isomorphism $p^*\colon \check{H}^*(X) \to \check{H}^*(Z)$, allows to define the *Lefschetz number* $\lambda(\varphi) = \lambda(q^* \circ p^{*-1})$ of an admissible map $\varphi\colon X \multimap X$ determined by a pair $X \xleftarrow{p} Z \xrightarrow{q} Y$ provided X is a space of a finite type (e.g. X is a compact absolute neighbourhood retract). Following Leray and Granas, Górniewicz in [Go1] defines also the *generalized Lefschetz number* $\Lambda(\varphi)$ of a compact admissible map $\varphi\colon X \multimap X$ defined on an absolute neighbourhood retract X. The nontriviality of these homotopy invariants (for the suitable notion of homotopy see [Go1] or [Kr2]) often implies the existence of fixed points. The variety of such conditions is provided in [Go1]. We shall need the following result.

(1.2) THEOREM. *If φ is an admissible self-map of an absolute retract X, then φ has a fixed point provided X is compact and $\lambda(\varphi) \neq 0$ or φ is compact and $\Lambda(\varphi) \neq 0$. If $\varphi\colon X \multimap X$ is homotopic to the identity, the absolute neighbourhood retract X is compact and $\chi(X) \neq 0$ (resp. X is acyclic* (8) *and φ is compact), then $\lambda(\varphi) \neq 0$ (resp. $\Lambda(\varphi) \neq 0$).*

(1.3) EXAMPLE. Suppose that $K \subset E$ is a compact neighbourhood retract, with $\chi(K) \neq 0$, and $\varphi\colon K \multimap K$ is admissible with a representing pair $K \xleftarrow{p} Z \xrightarrow{q} K$. There is $\delta > 0$ such that if, for each $z \in Z$, $\|p(z) - q(z)\| < \delta$, then φ has a fixed point.

To see this let $r\colon U \to K$ be a neighbourhood retraction. Since K is compact, there is $\delta > 0$ such that $B(K,\delta) \subset U$. Consider the pair $K \times [0,1] \xleftarrow{P} Z \times [0,1] \xrightarrow{Q} E$ where $P(z,t) = (p(z),t)$ and $Q(z,t) = r((1-t)q(z) + tp(z))$ for $z \in Z$, $t \in [0,1]$. Then P is a Vietoris map and Q is well-defined since, for all $t \in [0,1]$, $(1-t)q(z) + tp(z) \in U$ provided $\|p(z) - q(z)\| < \delta$ on Z. Thus

(8) For instance X is convex or contractible.

$\lambda(\varphi) = \lambda(I_K) = \chi(K) \neq 0$ since (P, Q) provides a homotopy joining φ to the identity I_K on K.

Let X, Y be metric spaces, $A \subset X$, let $\psi\colon A \multimap Y$ be a set-valued map and suppose that $x_0 \in \operatorname{cl} A$. Then, the (topological) *lower limit*

$$\operatorname*{Lim\,inf}_{x\to x_0} \psi(x) := \{y \in Y : \inf_{\eta>0} \sup_{x\in B_A(x_0,\eta)} d(y, \psi(x)) = 0\}$$

and the *upper limit*

$$\operatorname*{Lim\,sup}_{x\to x_0} \psi(x) := \{y \in Y : \sup_{\eta>0} \inf_{x\in B_A(x_0,\eta)} d(y, \psi(x)) = 0\}.$$

Therefore we see that

$$y \in \operatorname*{Lim\,inf}_{x\to x_0} \psi(x) \Leftrightarrow \forall\,(x_n)_{n=1}^{\infty},\ x_n \xrightarrow{A} x_0\ \exists\, y_n \in \psi(x_n),\ y_n \to y$$

and similarly

$$y \in \operatorname*{Lim\,sup}_{x\to x_0} \psi(x) \Leftrightarrow \exists\,(x_n)_{n=1}^{\infty},\ x_n \xrightarrow{A} x_0\ \exists\, y_n \in \psi(x_n),\ y_n \to y.$$

These limits admit also the following description

$$\begin{aligned}
\operatorname*{Lim\,inf}_{x\to x_0} \psi(x) &= \bigcap_{\varepsilon>0} \bigcup_{\eta>0} \bigcap_{x\in B_A(x_0,\varepsilon)} B(\psi(x), \varepsilon),\\
\operatorname*{Lim\,sup}_{x\to x_0} \psi(x) &= \bigcap_{\varepsilon>0} \bigcap_{\eta>0} \bigcup_{x\in B_A(x_0,\eta)} B(\psi(x), \varepsilon) = \bigcap_{\eta>0} \operatorname{cl}\Big(\bigcup_{x\in B_A(x_0,\eta)} \psi(x)\Big).
\end{aligned}$$

Let K be a closed subset in a Banach space E and $x \in K$. The *contingent* (or *Bouligand*) cone $T_K(x)$ to K at x is defined by

$$T_K(x) := \operatorname*{Lim\,sup}_{h\to 0+} \frac{K - x}{h}$$

and the *Clarke* tangent cone $C_K(x)$ by

$$C_K(x) := \operatorname*{Lim\,inf}_{h\to 0^+,\, y \xrightarrow{K} x} \frac{K - y}{h}.$$

It is easy to see that

$$\begin{aligned}
T_K(x) &= \Big\{v \in E : \liminf_{h\to 0+} \frac{d_K(x + hv)}{h} = 0\Big\},\\
C_K(x) &= \Big\{v \in E : \lim_{h\to 0^+,\, y \xrightarrow{K} x} \frac{d_K(y + hv)}{h} = 0\Big\}.
\end{aligned}$$

The sets $T_K(x)$ and $C_K(x)$ are closed *cones*, i.e. given $v \in T_K(x)$ (or $v \in C_K(x)$) and $\lambda \geq 0$, $\lambda v \in T_K(x)$ (or $\lambda v \in C_K(x)$). Additionally $C_K(x)$ is convex. In general $C_K(x) \subset T_K(x)$; if K is convex, then

$$T_K(x) = C_K(x) = S_K(x) \tag{1.4}$$

where

$$S_K(x) := \operatorname{cl}\left(\bigcup_{h>0} \frac{K-x}{h}\right).$$

If $T_K(x) = C_K(x)$ for each $x \in K$, then we say that K is *tangentially regular*; K is *sleek* if the cone map $K \ni x \mapsto T_K(x)$ is lower semicontinuous; for instance convex closed sets are sleek. In view of the following inclusion (which holds for any closed $K \subset E$)

$$\operatorname*{Lim\,inf}_{y \xrightarrow{K} x} T_K(y) \subset C_K(x), \quad x \in K,$$

(see [AF, Theorem 4.1.9]), we see that sleek sets are tangentially regular.

If $K \subset E$ is closed, $x \in K$, then by the *normal cone* $N_K(x)$ we understand the (negative) polar cone to $C_K(x)$, i.e. $N_K(x) = C_K(x)^-$. In case K is convex, it can be easily characterized by

$$N_K(x) = \{p \in E^* : \text{for all } y \in K\ \langle p, y - x\rangle \leq 0\}. \tag{1.6}$$

The map π_K assigning to each $y \in E$ the (possibly empty) set $\{z \in K : \|y - z\| = d_K(y)\}$ is called the *metric projection.* Observe that, for each $x \in K$, $\pi_K^{-1}(x) = \{y \in E : x \in \pi_K(y)\}$ is always nonempty. We define the *Mordukhovic normal cone* or *proximally normal cone* $N(x, K)$ by

$$N(x,K) := \operatorname*{Lim\,sup}_{y \xrightarrow{K} x} \left(\bigcup_{\lambda>0} \lambda(\pi_K^{-1}(y) - y)\right)$$

and its elements are called *proximal normals.*

In view of the Hahn–Banach theorem, the (*normalized*) *duality map*

$$J = J_E \colon E \multimap E^*, \quad J(x) := \{p \in E^* : \langle p, x\rangle = \|p\|^2 = \|x\|^2\}, \quad x \in E,$$

is well-defined. It is not difficult to show that the function J is *upper demicontinuous* (i.e. upper semicontinuous as a map $J \colon E \multimap E^*_w$) and has convex and weakly*-compact values. Let $G(x) = \|x\|^2/2$, $x \in E$; G is clearly convex. One of the main properties of J states that, for each $x \in E$,

$$J(x) = \partial G(x), \tag{1.7}$$

i.e. the *subdifferential* of the convex function G (see e.g. [AE]). Hence, for all $x \in E$, $p \in J(x)$ if and only if, for all $y \in E$, $\langle p, y - x\rangle \leq G(y) - G(x)$ (for other properties of J, see e.g. [M1]).

The following generalizes [AF, Proposition 4.1.2] and shows the relation between normal cones.

(1.8) PROPOSITION. *Let $K \subset E$ be closed and $x \in K$. If $v \in \lambda(\pi^{-1}(x) - x)$ for some $\lambda > 0$ (resp. $v \in N(x, K)$), then there is $p_0 \in J(v)$ such that, for all $w \in \operatorname{cl}\operatorname{conv} T_K(x)$ (resp. $w \in C_K(x)$), $\langle p_0, w\rangle \le 0$. In particular, $p_0 \in N_K(x)$.*

PROOF. We prove the second part: the first one is easier. Take $v \in N(x, K)$ and $w \in C_K(x)$. There are sequences $x_n \xrightarrow{K} x$, $\lambda_n > 0$ and $y_n \in E$ such that $x_n \in \pi_K(y_n)$ and $v = \lim_{n\to\infty} \lambda_n(y_n - x_n)$. Take a sequence $h_n \to 0^+$ such that $\lambda_n h_n \to 0$. Since $w \in C_K(x)$, there is a sequence $w_n \to w$ such that $z_n := x_n + h_n w_n \in K$. For any $n \ge 1$, take $p_n \in J(\lambda_n(y_n - z_n))$. Then

$$\begin{aligned}\langle p_n, w_n\rangle &= \frac{1}{\lambda_n h_n}\langle p_n, \lambda_n(y_n - x_n) - \lambda_n(y_n - z_n)\rangle \\ &\le \frac{1}{\lambda_n h_n}(G(\lambda_n(y_n - x_n)) - G(\lambda_n(y_n - z_n))) \\ &= \frac{\lambda_n}{2h_n}(\|y_n - x_n\|^2 - \|y_n - z_n\|^2) \le 0.\end{aligned}$$

Since $\lambda_n(y_n - z_n) = \lambda_n(y_n - x_n) - \lambda_n h_n w_n \to v$ and J is upper demicontinuous with weak* compact values, passing to subsequences if necessary, we may suppose that $p_n \overset{*}{\rightharpoonup} p \in J(v)$. At the same time $w_n \to w$; hence $\langle p, w\rangle \le 0$.

We have shown that

$$\sup_{w\in C_K(x)} \inf_{p\in J(v)} \langle p, w\rangle \le 0.$$

In view of the Sion min-max theorem ([9]),

$$0 \ge \sup_{w\in C_K(x)} \inf_{p\in J(v)} \langle p, w\rangle = \inf_{p\in J(v)} \sup_{w\in C_K(x)} \langle p, w\rangle = \sup_{w\in C_K(x)} \langle p_0, w\rangle$$

for some $p_0 \in J(v)$, since the function $E^* \ni p \mapsto \sup_{w\in C_K(x)}\langle p, w\rangle$ is weak* lower semicontinuous and $J(v)$ is weakly* compact. □

(1.9) COROLLARY. *If E^* strictly convex, then, for each $x \in K$, $J(N(x, K)) \subset N_K(x)$. If E is a Hilbert space ([10]), then $N(x, K)^- = C_K(x)$.*

The first part follows since in case E^* is strictly convex, the J is single-valued. The second part is left to the reader. Some indications as to the proof are given e.g. in [Sm]; moreover, the proof involves the so-called *Walkup–Wets formula*, see [WW] and [AF, Theorem 1.1.8].

([9]) See [Si]: Given a convex subset X of a topological vector space, a convex compact subset Y of a topological vector space, a function $F: X \times Y \to \mathbb{R}$ such that $F(\cdot, y): X \to \mathbb{R}$ is concave and upper semicontinuous for any $y \in Y$ and $F(x, \cdot): Y \to \mathbb{R}$ is convex and lower semicontinuous for any $x \in X$, the min-max equality $\sup_{x\in X}\inf_{y\in Y} F(x, y) = \inf_{y\in Y}\sup_{x\in X} F(x, y)$ holds true.

([10]) In this case J is an isomorphism $E \to E^*$.

Given a locally Lipschitz continuous function $f\colon E \to \mathbb{R}$, by $f^\circ(x;u)$ we denote the Clarke *generalized directional derivative* of f at $x \in E$ in the direction $u \in E$

$$f^\circ(x;u) := \limsup_{y\to x,\, h\to 0+} \frac{f(y+hu)-f(y)}{h}.$$

It is well-known that, for each $x \in E$, the function $E \ni u \mapsto f^\circ(x;u)$ is Lipschitz, subadditive and positively homogeneous. The *generalized gradient* of f at x is defined by

$$\partial f(x) := \{p \in E^* : \langle p, u\rangle \le f^\circ(x;u) \text{ for all } u \in E\}.$$

Hence $f^\circ(x;\,\cdot\,)$ is the support function $\partial f(x)$:

$$f^\circ(x;u) = \sigma_{\partial f(x)}(u), \quad u \in E,$$

and, the (negative) polar cone

$$\partial f(x)^- = \{u \in E : f^\circ(x;u) \le 0\}.$$

Hence, for all $x \in E$, the set $\partial f(x)$ is convex and weak*-compact. The function $E \times E \ni (x,u) \mapsto f^\circ(x;u)$ is upper semicontinuous; in other words the set-valued map $E \ni x \mapsto \partial f(x) \subset E^*$ is upper hemicontinuous and upper demicontinuous (see Subsection 2.1).

If $K \subset E$ is closed, $x \in K$, then $C_K(x) = \partial d_K(x)^-$ and $N_K(x) = \partial d_K(x)^{--}$. Proofs for all facts mentioned above and other details can be found in e.g. [AE] or [AF].

2. Constrained equilibria: a topological perspective

When studying the constrained equilibrium or fixed problem, i.e. the existence of zeros of a map $\varphi\colon K \multimap F$, where $K \subset E$ (or, if $E = F$, zeros of $\varphi - I_K$ or $I_K - \varphi$), there is a need for some special assumptions. Namely one has to impose conditions ruling out phenomena that appear, for instance, in the case of a translation $\varphi(x) := x + w$, $x \in K$, where $w \notin -K$. Such conditions, known as *tangency* or *inwardness* will be introduced throughout the rest of this paper. Various assumptions of this type were present in different branches of analysis and geometry: for instance one studies tangent fields on manifolds, the so-called *Nagumo* conditions implying the existence of viable solutions to constrained differential equations and others. In what follows we shall try to put these assumptions into the general context of a widely understood convex and non-smooth analysis and topology.

2.1. Convex compact results. The best known equilibrium result is the following pioneering result of Browder in [Bro] (with some modification due to Halpern, see [HB2], [H2]).

(2.1) THEOREM. *Assume that K is a compact convex subset of a Banach space E and $\varphi: K \multimap E$ is upper semicontinuous with closed convex values. If φ satisfies the weak tangency condition with respect to K* (11), *i.e.*

$$\varphi(x) \cap T_K(x) \neq \emptyset \quad \textit{for all } x \in K, \tag{2.1.1}$$

then φ has an equilibrium: there is $\overline{x} \in K$ such that $0 \in \varphi(\overline{x})$.

In particular, we get the following corollary yielding a generalization of the Kakutani, Bohnenblust and Karlin theorems.

(2.2) COROLLARY. *Given a convex compact set $K \subset E$ and an upper semicontinuous map $vp: K \multimap E$ with closed convex values, if φ is weakly inward, i.e.*

$$(\varphi(x) - x) \cap T_K(x) \neq \emptyset \quad \textit{for all } x \in K, \tag{2.2.1}$$

or weakly outward

$$(x - \varphi(x)) \cap T_K(x) \neq \emptyset \quad \textit{for all } x \in K, \tag{2.2.2}$$

then φ has a fixed point.

Observe that (2.1.1), (2.2.1) and (2.2.2) are in fact boundary conditions: if $x \in \operatorname{int} K$, then $T_K(x) = E$ and they hold automatically. It is also clear that if, for $x \in K$, $\varphi(x) \subset K$ (i.e. $\varphi: K \multimap K$), then (2.2.1) is satisfied.

Theorem (2.1) and Corollary (2.2) were generalized many times: Ky Fan [F2], [F3] proved that Browder's result remains true under weaker assumptions concerning regularity of φ. Finally Cornet [Co1] shown that the weak tangency condition (2.1.1) may be substantially relaxed. Below we present the result of Cornet with a different proof using the following well-known Browder–Fan fixed point theorem, see [Bro].

(2.3) THEOREM. *Let K be a convex compact subset of a Banach space E* (12) *and let $S: K \multimap K$ have convex values and open fibres (i.e. for each $y \in K$, $S^{-1}(y) := \{x \in K : y \in S(x)\}$ is open). Then S has a fixed point.*

PROOF. Since K is compact, there are $y_1, \ldots, y_n \in K$ such that

$$K = \bigcup_{i=1}^{n} S^{-1}(y_i).$$

Let a partition of unity $\{\lambda_i: K \to [0,1]\}_{i=1}^n$ be subordinated to the open covering $\{S^{-1}(y_i)\}_{i=1}^n$, i.e. for each $i = 1, \ldots, n$, $\operatorname{supp} \lambda_i := \operatorname{cl}\{x \in K : \lambda_i(x) \neq 0\} \subset$

(11) The terminology given here and below may differ from the one used elsewhere.
(12) It suffices to assume that E is a Hausdorff topological vector space.

$S^{-1}(y_i)$ and, for each $x \in K$, $\sum_{i=1}^n \lambda_i(x) = 1$. Let $s: K \to K$ be given by the formula $s(x) = \sum_{i=1}^n \lambda_i(x) y_i$, $x \in K$. It is easy to see that s is a continuous selection of S, i.e. for each $x \in K$, $s(x) \in S(x)$. Since $s : \operatorname{conv}\{y_1, \dots, y_n\} \to \operatorname{conv}\{y_1, \dots, y_n\}$ has a fixed point in view of the Brouwer fixed point theorem, we conclude the proof. □

(2.4) REMARK. Let us note that Theorem (2.3) provides means to prove the Schauder fixed point principle. In fact, given a continuous $f: K \to K$ and $\varepsilon > 0$, one defines $S_\varepsilon: K \multimap K$ by $S_\varepsilon(x) = B(f(x), \varepsilon) \cap K$. Then S_ε has convex values and, for each $y \in K$, $S^{-1}(y) = f^{-1}(B(y, \varepsilon))$. By Theorem (2.3), f has an ε-fixed point; passing with $\varepsilon \to 0$, the compactness of K and the continuity of f shows that f has a fixed point.

Let X be a topological space. Recall ([AF, Definition 2.6.2]) that a set-valued $\varphi: X \multimap E$ (resp. $\varphi: X \multimap E^*$) is *upper hemicontinuous* if, for each $p \in E^*$ (resp. $u \in E$), the real function $X \ni x \mapsto \sigma_{\varphi(x)}(p) \in \mathbb{R} \cup \{\infty\}$ (resp. $x \mapsto \sigma_{\varphi(x)}(u)$) is upper semicontinuous. For instance it is not difficult to show that the duality map $J: E \multimap E^*$ is upper hemicontinuous. It is also clear that, if $\varphi: X \multimap E$ (resp. $\varphi: X \multimap E^*$), where E is endowed with weak (resp. weak*) topology, is upper semicontinuous ([13]), then it is upper hemicontinuous. The converse is not valid in general; however, in view of the Castaing theorem (see [AE, Theorem 3.2.10]), it holds for maps having convex and weakly (resp. weakly*) compact values. Observe that if X is compact, φ is upper hemicontinuous and has bounded values, then $\varphi(X)$ is bounded in E (resp. in E^*). If $\varphi: X \multimap E$ (resp. $\varphi: X \multimap E^*$) is locally bounded (i.e. each point $x \in X$ has a neighbourhood U such that $\varphi(U)$ is bounded) and upper hemicontinuous, then the function $X \times E^* \ni (x, p) \mapsto \sigma_{u \in \varphi(x)} \langle p, u \rangle$ (resp. $X \times E \ni (x, u) \mapsto \sigma_{p \in \varphi(x)} \langle p, u \rangle$) is upper semicontinuous. The graph of an upper hemicontinuous map with closed convex values is closed in $X \times E$ (resp. $X \times E^*$) when E (resp. E^*) is endowed with weak (resp. weak*) topology. For other details on upper hemicontinuity, see [AE].

Let us now show the next result essentially due to Browder.

(2.5) THEOREM. *Let $K \subset E$ be compact convex. If an upper hemicontinuous map $\varphi: K \multimap E^*$ has weak*-compact convex values, then it admits a generalized equilibrium, i.e. there is $x_0 \in K$ such that $\varphi(x_0) \cap N_K(x_0) \neq \emptyset$.*

PROOF. Observe that $x_0 \in K$ is a generalized equilibrium of φ if and only if

$$\sup_{y \in K} \inf_{p \in \varphi(x_0)} \langle p, y - x_0 \rangle \leq 0. \tag{2.6}$$

([13]) As already mentioned, in this situation, one speaks of the *upper demicontinuity*; clearly any upper semicontinuous map $\varphi: X \multimap E$ (or $\varphi: X \multimap E^*$) is upper demicontinuous.

Indeed, the necessity of (2.6) is clear; if (2.6) holds, then by the Sion min-max theorem,

$$0 \geq \inf_{p \in \varphi(x_0)} \sup_{y \in K} \langle p, y - x_0 \rangle = \sup_{y \in K} \langle p_0, y - x_0 \rangle$$

for some $p_0 \in \varphi(x_0)$, since the function $E^* \ni p \mapsto \sup_{y \in K} \langle p, y - x_0 \rangle$ is weak* lower semicontinuous and $\varphi(x_0)$ is weak*-compact. By (1.6), $p_0 \in N_K(x_0)$.

Suppose to the contrary that φ has no generalized equilibria, i.e. by (2.6), for any $x \in K$, there is $y \in K$ such that $\inf_{p \in \varphi(x)} \langle p, y - x \rangle > 0$, i.e.

$$S(x) := \{y \in K : \sigma_{\varphi(x)}(x - y) = \sup_{p \in \varphi(x)} \langle p, x - y \rangle < 0\} \neq \emptyset.$$

It is clear that, for each $x \in K$, $S(x)$ is convex and, for any $y \in K$, $S^{-1}(y)$ is open in view of the upper hemicontinuity of φ. Hence, by Theorem (2.3), there is $\overline{x} \in K$ such that $\overline{x} \in S(\overline{x})$, a contradiction. □

A similar result holds for proximally normal cones.

(2.7) PROPOSITION. *Let $K \subset E$ be compact convex and let $\varphi: K \multimap E$ be upper semicontinuous with compact convex values. Then there is $x_0 \in K$ such that $\varphi(x_0) \cap (\pi_K^{-1}(x_0) - x_0) \neq \emptyset$; in particular $\varphi(x_0) \cap N(x_0, K) \neq \emptyset$.*

PROOF. It is clear that, for each $y \in E$, $\pi_K(y)$ is nonempty compact convex and $\pi_K: E \multimap K$ is upper semicontinuous. The map $\psi: K \times E \multimap K \times E$, given by $\psi(x, y) = \pi_K(y) \times (\varphi(x) + x)$ for $x \in K$ and $y \in E$, is upper semicontinuous, compact and has compact convex values. Hence, by the Bohnenblust–Karlin (or Fan–Glicksberg) fixed point theorem (see [BoK] or [Gl] and [F1]), there is $(x_0, y_0) \in K \times E$ such that $(x_0, y_0) \in \psi(x_0, y_0)$. Hence $x_0 \in \pi_K(y_0)$ and $y_0 \in \varphi(x_0) + x_0$, i.e. $\varphi(x_0) \cap (\pi_K^{-1}(x_0) - x_0) \neq \emptyset$. □

Application of Proposition (2.7) to the map $\varphi - I_K$ yields immediately the next result of Ky Fan (see [F2]).

(2.8) COROLLARY. *For any upper semicontinuous map $\varphi: K \multimap E$ with compact convex values, where $K \subset E$ is convex compact, there are points $x_0 \in K$ and $y_0 \in \varphi(x_0)$ such that $\|y_0 - x_0\| = d_K(y_0)$.*

(2.9) REMARK. Observe that if φ in Proposition (2.7) is admissible, then the assertion still hold true: instead of Bohnenblust–Karlin theorem one may use Theorem (1.2). Similar observation concerns Corollary (2.8).

(2.10) REMARK. The author is not sure whether Proposition (2.7) and Corollary (2.8) are true for upper hemicontinuous map φ with closed convex values. The second assertion of Proposition (2.7) certainly stays true in case E is a Hilbert space and if, additionally, φ has bounded values. In this case, for each $x \in K$, one may identify $N(x, K)$ with $N_K(x)$ and φ has weakly compact values.

(2.11) REMARK. In Theorem (2.5) if, for each $x \in K$, $0 \notin \varphi(x)$, then the existing generalized equilibrium $x_0 \in \operatorname{bd} K$. This information is significant if E is finite-dimensional for it carries information as to the localization of generalized equilibria (if $\dim E = \infty$, then K being compact has empty interior and $K = \operatorname{bd} K$). Similar observation is valid with regard to Proposition (2.7).

Theorem (2.5) and Proposition (2.7) imply some results concerning equilibria. We start with some trivial ones.

(2.12) COROLLARY. *An upper semicontinuous $\varphi\colon K \multimap \mathbb{R}^n$ with compact convex values defined an a compact convex $K \subset \mathbb{R}^n$ has an equilibrium in* $\operatorname{int} K$ *provided, or each $x \in \operatorname{bd} K$,*

$$\varphi(x) \cap N_K(x) = \emptyset. \tag{2.12.1}$$

Note that condition (2.12.1), contrary to the weak tangency condition (2.1.1) which prescribes directions for ϕ, forbids certain directions to ϕ.

(2.13) COROLLARY. *If $K \subset E$ is compact and convex, $\varphi\colon K \multimap E$ is upper semicontinuous, has closed convex values or is admissible and tangent, i.e.*

$$\varphi(x) \subset T_K(x) \quad \textit{for all } x \in K, \tag{2.13.1}$$

then φ has an equilibrium.

PROOF. By Proposition (2.7) (or Remark (2.9)), there is $x_0 \in K$ and $y_0 \in \varphi(x_0) \cap N(x_0, K)$. By Proposition (1.8), there is $p_0 \in J(y_0) \cap N_K(x_0)$; thus $\|y_0\|^2 = \langle p_0, y_0 \rangle \leq 0$ since $y_0 \in T_K(x_0)$. □

The next result, being a modification of a result essentially due to Cornet (see [Co1]), provides the above mentioned direct generalization of the Browder–Fan theorem (2.1) and Corollary (2.13).

(2.14) THEOREM. *Suppose that K is a convex compact subset of a Banach space E, F is a Banach space, $A \in \mathcal{L}(E, F)$ (i.e. $A\colon E \to F$ is a bounded linear operator) and let $M := A(K)$. If an upper hemicontinuous map $\varphi\colon K \multimap E$ with closed convex values satisfies the normality condition, i.e.*

$$\sup_{p \in N_M(A(x))} \inf_{y \in \varphi(x)} \langle p, y \rangle \leq 0 \quad \textit{for all } x \in K, \tag{2.14.1}$$

then there is $\overline{x} \in K$ *such that* $0 \in \varphi(\overline{x})$.

Before we give a proof let us discuss the normality condition (2.14.1) and its relation to the usual (weak) tangency condition (2.1.1). First (see [AF, p. 141 and Proposition 4.2.9]) observe that, for all $x \in K$,

$$A^{*-1}(N_K(x)) = N_{A(K)}(A(x)) \quad \text{and} \quad \operatorname{cl} A(T_K(x)) = T_{A(K)}(A(x)). \tag{2.15}$$

(2.16) LEMMA (comp. [AE, Lemma 6.4.7]). *Let* K, A *and* $M = A(K)$ *be as above. Consider the following conditions:*

$$\varphi(x) \cap T_M(A(x)) \neq \emptyset \quad \text{for all } x \in K, \tag{2.16.1}$$

$$\varphi(x) \cap \operatorname{cl} A(T_K(x)) \neq \emptyset \quad \text{for all } x \in K, \tag{2.16.2}$$

$$\sup_{p \in A^{*-1}(N_K(x))} \inf_{y \in \varphi(x)} \langle p, y\rangle \leq 0 \quad \text{for all } x \in K, \tag{2.16.3}$$

$$\sup_{p \in A^{*-1}(\partial d_K(x))} \inf_{y \in \varphi(x)} \langle p, y\rangle \leq 0 \quad \text{for all } x \in K, \tag{2.16.4}$$

$$\sup_{p \in \partial d_M(Ax)} \inf_{y \in \varphi(x)} \langle p, y\rangle \leq 0 \quad \text{for all } x \in K, \tag{2.16.5}$$

$$\inf_{y \in \varphi(x)} d_M^\circ(A(x); y) \leq 0 \quad \text{for all } x \in K. \tag{2.16.6}$$

Then (2.16.1) $\Leftrightarrow$ (2.16.2) $\Rightarrow$ (2.14.1) $\Leftrightarrow$ (2.16.3) $\Rightarrow$ (2.16.4), (2.14.1) $\Rightarrow$ (2.16.5) $\Leftrightarrow$ (2.16.6). *If values of* φ *are bounded, then* (2.16.4) $\Rightarrow$ (2.16.3) *and* (2.16.5) $\Rightarrow$ (2.14.1). *All these conditions are equivalent if* φ *has weakly compact values.*

PROOF. Equivalencies (2.16.1) $\Leftrightarrow$ (2.16.2) and (2.14.1) $\Leftrightarrow$ (2.16.3) follow in view of (2.15). To show (2.16.5) $\Leftrightarrow$ (2.16.6) observe that, by the Sion min-max equality,

$$\sup_{p \in \partial d_M(A(x))} \inf_{y \in \varphi(x)} \langle p, y\rangle = \inf_{y \in \varphi(x)} \sup_{p \in \partial d_M(A(x))} \langle p, y\rangle = \inf_{y \in \varphi(x)} d_M^\circ(A(x); y).$$

Implications (2.14.1) $\Rightarrow$ (2.16.5) and (2.16.3) $\Rightarrow$ (2.16.4) are obvious. If values of φ are bounded, then for each $x \in K$, the function $F^* \ni p \mapsto \inf_{y \in \varphi(x)} \langle p, y\rangle$ is weak* continuous; moreover, $N_M(A(x)) = \partial d_M(A(x))^{--} = w^*\text{-cl} \bigcup_{\lambda \geq 0} \lambda \partial d_M(A(x))$. This shows the implication (2.16.5) $\Rightarrow$ (2.14.1). Using the continuity of $A^* \colon F_w^* \to E_w^*$, the implication (2.16.4) $\Rightarrow$ (2.16.3) follows in a similar manner.

The implication (2.16.1) $\Rightarrow$ (2.14.1) is obvious; conversely suppose that φ has weakly compact values, (2.14.1) holds but (2.16.1) does not. Then there is $x_0 \in K$ such that $\varphi(x_0) \cap T_M(A(x_0)) = \emptyset$. Since $\varphi(x_0)$ is weakly compact, by the separation theorem, there is $p_0 \in F^*$ such that

$$\inf_{y \in \varphi(x_0)} \langle p_0, y\rangle > \sup_{v \in T_M(A(x_0))} \langle p_0, v\rangle.$$

Since $T_M(A(x_0))$ is a cone, we see that $\sup_{v\in T_M(A(x_0))}\langle p_0, v\rangle = 0$; hence $p_0 \in N_M(A(x_0))$ and $\inf_{y\in\varphi(x_0)}\langle p_0, y\rangle > 0$, a contradiction. □

In particular if $E = F$ and $A = I$ is the identity, then (2.14.1) is weaker than the weak tangency (2.1.1). Hence Theorem (2.14) may indeed be considered as an extension of Theorem (2.1).

(2.17) REMARK. Suppose that $E = \mathbb{R}^n$, $F = \mathbb{R}^m$. Condition (2.16.1) is then implied by the existence of solutions to the Cauchy problem $A(u'(t)) \in \varphi(u(t))$, $u(0) = x_0$ for all $x_0 \in K$. Indeed suppose that $u\colon [0,\varepsilon) \to K$, $\varepsilon > 0$, is a solution, i.e. there is an integrable function $w\colon [0,\varepsilon) \to \mathbb{R}^n$ such that $u(t) = x_0 + \int_0^t w(s)\,ds$ and $A(w(s)) \in \varphi(u(t))$ for almost all $t \in [0,\varepsilon)$. There clearly exists a sequence $t_k \to 0^+$ such that $u(t_k) \in K$. Fix $p \in \mathbb{R}^m$; the function $x \mapsto \sigma_{\varphi(x)}(p)$ is upper semicontinuous, then for all $\eta > 0$, there is $0 < \delta < \varepsilon$ such that

$$\langle A^*(p), w(s)\rangle = \langle p, A(w(s))\rangle \le \sigma_{\varphi(u(s))}(p) < \sigma_{\varphi(x_0)}(p) + \eta\|p\|$$

provided $s \in [0,\delta)$. Hence, for all sufficiently large k,

$$\frac{1}{t_k}\int_0^{t_k} \langle A^*(p), w(s)\rangle\,ds \le \sigma_{\varphi(x_0)}(p) + \eta\|p\|.$$

We conclude that $\langle p, A(v_k)\rangle = \langle A^*(p), v_k\rangle \le \sigma_{\varphi(x_0)}(p)\varepsilon\|p\|$ where $v_k := t_k^{-1}(u(t_k) - x_0)$. The sequence $(A(v_k))$, being bounded, converges to some $y_0 \in \operatorname{cl}(A(T_K(x_0)) = T_{A(K)}(A(x_0))$ satisfying the inequality $\langle p, y_0\rangle \le \sigma_{\varphi(x_0)}(p)$. Since η and p are arbitrary, we gather that $y \in \varphi(x_0)$.

PROOF OF THEOREM (2.14). Suppose to the contrary that φ has no equilibria, i.e. for each $x \in K$, $0 \notin \varphi(x)$. The separation theorem implies the existence of $p_x \in F^*$ such that $\inf_{y\in\varphi(x)}\langle p_x, y\rangle > 0$. The upper hemicontinuity (or rather weak upper semicontinuity) of φ implies that, for each $x \in K$, the set $U(x) := \{z \in K : \inf_{y\in\varphi(z)}\langle p_x, y\rangle > 0\}$ is an open neighbourhood of x (for $U(x) = \{z \in K : \sigma_{\varphi(z)}(-p_x) < 0\}$). Since $\mathcal{U} = \{U(x)\}_{x\in K}$ is an open covering of K, there is a (locally finite) partition of unity $\{\lambda_x\colon K \to [0,1]\}_{x\in K}$ subordinated to $\mathcal{U}$. In particular, for each $x \in K$, $\operatorname{supp}\lambda_x \subset U(x)$. Let

$$p(z) = \sum_{x\in K}\lambda_x(z)p_x, \quad z \in K.$$

Then $p\colon K \to F^*$ is continuous and, for each $x \in K$,

$$\inf_{y\in\varphi(x)}\langle p(x), y\rangle > 0. \tag{2.18}$$

By Theorem (2.5), there is $x_0 \in K$ such that $A^*p(x_0) \in N_K(x_0)$. Hence, by (2.16.3), $\inf_{y\in\varphi(x_0)}\langle p(x_0), y\rangle \le 0$, a contradiction with (2.18). □

(2.19) REMARK. It would be interesting to prove Theorem (2.14) under assumption (2.16.6) replacing (2.14.1). In view of Lemma (2.16), it holds when values of φ are additionally bounded. If $E = F$ and $A = I$, then condition (2.16.6) reads

$$\inf_{y\in\varphi(x)} d_K^\circ(x;y) \le 0 \quad \text{for all } x \in K \tag{2.19.1}$$

and is weaker than (2.14.1) and the usual weak tangency (2.1.1). Below (see Theorem (2.57)) we shall see that (2.19.1) is sufficient for the existence of equilibria.

(2.20) REMARK. Note that in the course of the above proof we have used the upper hemicontinuity of φ only in a very restrictive manner. It is sufficient that given $p \in F^*$, the set $\{x \in K : \inf_{y\in\varphi(x)}\langle p, y\rangle > 0\}$ is open; the "full" upper hemicontinuity is not necessary.

In general we have the following result.

(2.21) PROPOSITION. *Suppose that $K \subset E$ is closed convex and, $A\colon E \to F$ is a bounded linear operator such that*

(2.21.1) *$M = A(K)$ is compact, and*
(2.21.2) *$A|_K\colon K \to M$ is an open map.*

If $\varphi\colon K \multimap F$ is upper hemicontinuous with closed convex values and, for each $x \in K$, condition (2.14.1) is satisfied, then φ admits an equilibrium.

PROOF. Consider an auxiliary set-valued map $T\colon M \multimap K$ given by $T(x) = A^{-1}(x) \cap K$. According to assumption (2.21.2), T is lower semicontinuous and, obviously, has closed convex values. Hence, in view of the Michael theorem, there is a continuous $f\colon M \to K$ such that $f(x) \in T(x)$ for $x \in M$. It is easy to see that $A \circ f(x) = x$ on M. Let $\psi\colon M \multimap F$ be given by $\psi(x) = \varphi(f(x))$. Then ψ is upper hemicontinuous and has closed convex values. Observe that ψ satisfies (2.19.1) (with ψ and M replacing φ and K), i.e. for each $x \in M$, $\inf_{y\in\psi(x)} d_M^\circ(x;y) \le 0$. Hence, there is $x_0 \in M$ such that $0 \in \psi(x_0) = \varphi(f(x_0))$. □

Assumption (2.21.2) above is actually equivalent to the lower semicontinuity of T. This is, by all means, a very strong assumption. To see that in general the map T may behave badly let $K = \{(x,y,z) \in \mathbb{R}^3 : y^2 \le x \le 1,\ xz \ge y^2\}$ and let $A\colon \mathbb{R}^3 \to \mathbb{R}^2$ be given by $A(x,y,z) = (x,y)$. Then $M = A(K) = \{(x,y) \in \mathbb{R}^2 : y^2 \le x \le 1\}$, but $T(x,y) = A^{-1}(x,y) \cap K$ is not lower semicontinuous. Indeed, $(0,0,0) \in T(0,0)$; if $z_n \in T(n^{-2}, n^{-1})$, then $z_n \ge 1$. Hence there is no sequence in

$T(n^{-2}, n^{-1})$ converging to $(0,0,0)$. This shows that Proposition (2.19) is rather weak.

Theorem (2.14) yields some consequences in the theory of the constrained coincidences (comp. [AE, Theorem 6.4.10]). Replacing φ by $\varphi - A$ we get

(2.22) COROLLARY. *If K, M, A and φ are as in Theorem (2.14), for all $x \in K$ and $p \in N_M(A(x))$,*

$$\inf_{y \in \varphi(x)} \langle p, y \rangle \le \langle p, A(x) \rangle,$$

then there is $\overline{x} \in K$ such that $A(\overline{x}) \in \varphi(\overline{x})$.

(2.23) COROLLARY. *Under the assumptions of Theorem (2.14) suppose that $y_0 \in A(K)$. Then there is $\overline{x} \in K$ such that $A(\overline{x}) - y_0 \in \varphi(\overline{x})$.*

PROOF. Suppose that $y_0 = A(x_0)$, $x_0 \in K$. For $x \in K$, let $\psi(x) := \varphi(x) - A(x) + y_0 = \varphi(x) - A(x - x_0)$. Then ψ is upper hemicontinuous. Let $p \in F^*$, $A^*(p) \in N_K(x)$ and $\varepsilon > 0$. Take $y \in \varphi(x)$ such that $\langle p, y \rangle < \varepsilon$. For $z = y - A(x - x_0)$, $\langle p, y - A(x - x_0) \rangle = \langle p, y \rangle - \langle A^*(p), x - x_0 \rangle < \varepsilon$. Hence, by Theorem (2.14), $0 \in \psi(\overline{x})$ for some $\overline{x} \in K$. □

Later on (see Section 3) we shall return to similar constrained coincidence problems.

Note that results stated in this subsection admit the so-called *analytical formulations* which might be useful. To illustrate the setting we prove the following *Ky Fan inequality* being the counterpart of Theorem (2.14).

(2.24) PROPOSITION. *Let $K \subset E$ be compact convex and let $A: E \to F$ be a bounded linear operator. Suppose that a function $f: K \times F^* \to \mathbb{R} \cup \{\infty\}$ satisfies the following conditions:*

(2.24.1) *for all $p \in F^*$, $f(\cdot, p): K \to \mathbb{R} \cup \{\infty\}$ is upper semicontinuous,*
(2.24.2) *for any $x \in K$, $f(x, \cdot): F^* \to \mathbb{R} \cup \{\infty\}$ is quasi-convex,*
(2.24.3) *for each $x \in K$ and $p \in A^{*-1}(N_K(x))$, $f(x, p) \ge 0$.*

Then there is $x_0 \in K$ such that $f(x_0, p) \ge 0$ for all $p \in F^*$.

PROOF. Assume that the conclusion fails and define $\varphi: K \multimap E^*$ by the formula: $\varphi(x) = A^*(\psi(x))$ where $\psi(x) := \{p \in F^* : f(x,p) < 0\}$ for $x \in K$. It is clear that ψ has convex values and open fibres $\psi^{-1}(p)$, $p \in F^*$. The same properties are met by φ. Therefore, by the arguments used in the proof to Theorem (2.3), φ admits a continuous selection $s: K \to E^*$. Hence, by Theorem (2.5), there is $x_0 \in K$ such that $s(x_0) \in N_K(x_0)$, i.e. there is $p_0 \in \psi(x_0)$ such that $A^*(p_0) \in N_K(x_0)$. Hence, by (2.24.3), $0 > f(x_0, p_0) \ge 0$, a contradiction. □

(2.25) REMARK. Observe that Theorem (2.14) and Proposition (2.24) are, in fact, equivalent. For given an upper hemicontinuous $\varphi: K \multimap F$ with closed convex values, consider $f(x,p) = \sigma_{\varphi(x)}(-p) = -\inf_{y\in\varphi(x)}\langle p, y\rangle$, $x \in K$, $p \in F^*$. Then, for each $p \in F^*$, $f(\cdot\,, p)$ is upper semicontinuous, for each $x \in K$, $f(x, \cdot\,)$ is convex. Assumption (2.14.1) implies that, for all $x \in K$ and $p \in A^{*-1}(N_K(x))$, $f(x,p) \geq 0$. Hence all assumptions of Proposition (2.24) are satisfied; its assertion says that there is $x_0 \in K$ such that $f(x_0, -p) = \sigma_{\varphi(x)}(p) \geq 0$: this means that $0 \in \varphi(x_0)$.

2.2. Relaxing compactness. Now we shall try to discuss the possibilities to relax the assumption of compactness in Theorems (2.1) and (2.14). As we shall see this requires a more delicate treatment. Perhaps the first result in this direction is that of Aubin.

(2.26) PROPOSITION (comp. [AE, Theorem 6.4.17]). *Suppose that $K \subset \mathbb{R}^n$ is closed and convex, an upper hemicontinuous $\varphi\colon K \multimap \mathbb{R}^n$ has closed convex values, satisfies the weak tangency condition and is coercive in the following sense:*

$$\limsup_{x\in K,\ \|x\|\to\infty} \sigma_{\varphi(x)}(x) < 0. \tag{2.26.1}$$

The φ has an equilibrium in K.

PROOF. By (2.26.1), there is $r > 0$ such that $B(0, r) \cap K \neq \emptyset$ and

$$\sup_{x\in K,\ \|x\|\geq r} \sigma_{\varphi(x)}(x) < 0.$$

Hence, for all $x \in K$, $\|x\| = r$ and all $y \in \varphi(x)$, $\langle x, y\rangle \leq 0$, i.e. $\varphi(x) \subset T_D(x)$ where $D := D(0, r)$ is the closed ball of radius r around 0. Observe that, for all $x \in K \cap D$, $\varphi(x) \cap T_{K\cap D}(x) \neq \emptyset$ since, as it is easy to see,

$$T_{K\cap D}(x) = T_D(x) \cap T_K(x). \tag{2.27}$$

Thus, by Theorem (2.1), there is $x_0 \in K \cap D$ such that $0 \in \varphi(x_0)$. □

A general principle implying (2.27) is the following:

(2.28) LEMMA (see [AF, p. 141]). *Given a continuous operator $A\colon E \to F$, closed convex sets $K \subset E$, $C \subset F$, if the transversality condition holds, i.e. $0 \in \operatorname{int}(A(K) - C)$, then, for each $x \in K \cap A^{-1}(C)$,*

$$T_{K\cap A^{-1}(C)}(x) = T_K(x) \cap A^{-1}(T_C(A(x))).$$

It immediately gives the next result.

(2.29) PROPOSITION. *Suppose that* $K \subset E$ *is closed convex and* $\operatorname{int} K \neq \emptyset$. *If an upper hemicontinuous* $\varphi: K \multimap E$ *with compact convex values is compact (i.e.* $\operatorname{cl} \varphi(K)$ *is compact) weakly inward* (2.2.1), *i.e.* $\varphi(x) \cap (x + T_K(x)) \neq \emptyset$, *for each* $x \in K$, *then* φ *has a fixed point.*

PROOF. Suppose, without loss of generality, that $0 \in \operatorname{int} K$. There is a compact convex set $C \subset E$ such that $0 \in C$ and $\varphi(K) \subset C$. Hence, for any $x \in K \cap C$, $\varphi(x) \subset x + T_C(x)$. Since $C \cap \operatorname{int} K \neq \emptyset$, we see by Lemma (2.28) (with $A = I$) that, for each $x \in K \cap C$, $T_K(x) \cap T_C(x) = T_{K \cap C}(x)$ and $(\varphi(x) - x) \cap T_{K \cap C}(x) \neq \emptyset$. By Theorem (2.1) we conclude the proof. □

The above results motivate the following definition: we say that a bounded linear operator $A: E \to F$ and a closed convex set $C \subset F$ *control directions admitted by* $\varphi: K \multimap F$ if, for each $x \in K \cap A^{-1}(C)$, we have $\varphi(x) \cap T_C(A(x)) \neq \emptyset$.

For instance, if $K \subset E$, $C \subset F$ are closed convex, $\varphi: K \multimap F$ and, for each $x \in K$, $\varphi(x) \cap C \neq \emptyset$, then C and A control directions admitted by $\Phi = \varphi - A$.

This allows the following *non-compact* generalization of the Cornet and Browder–Fan Theorems (2.1), (2.14).

(2.30) THEOREM. *Suppose that* $K \subset E$ *is closed convex,* F *is a Banach space,* $A: E \to F$ *is a linear bounded operator which, together with a compact set* $C \subset F$, *controls directions admitted by an upper hemicontinuous* $\varphi: K \multimap F$ *having closed convex values. If the restriction* $A|_K$ *is proper* ([14]), $0 \in \operatorname{int}(A(K) - C)$ *and, for all* $x \in K \cap A^{-1}(C)$ *and all* $p \in A^{*-1}(N_K(x))$,

$$\inf_{y \in \varphi(x) \cap T_C(A(x))} \langle p, y \rangle \leq 0,$$

then φ *has an equilibrium.*

PROOF. By Lemma (2.28), for any $x \in K \cap A^{-1}(C)$,

$$\begin{aligned} N_{K \cap A^{-1}(C)}(x) &= T_{K \cap A^{-1}(C)}(x)^{-} \\ &= (T_K(x) \cap A^{-1} T_C(A(x)))^{-} = N_K(x) + A^*(N_C(A(x)). \end{aligned}$$

Let $x \in K \cap A^{-1}(C)$ and $A^*(p) \in N_{K \cap A^{-1}(C)}(x)$. Then, there are $p_1 \in N_K(x)$ and $q \in N_C(A(x))$ such that $A^*(p) = p_1 + A^*(q)$. Hence $p - q \in A^{*-1}(N_K(x))$ and

$$\inf_{y \in \varphi(x)} \langle p, y \rangle \leq \inf_{y \in \varphi(x) \cap T_C(A(x))} \langle p, y \rangle \leq \inf_{y \in \varphi(x) \cap T_C(A(x))} \langle p - q, y \rangle \leq 0.$$

The properness of A implies that $A^{-1}(C)$ is compact; this, by Theorem (2.14), ends the proof. □

([14]) This holds for instance if A is a semi-Fredholm operator and K is bounded. Recall that $A \in \mathcal{L}(E, F)$ is *semi-Fredholm* if the range $\operatorname{Im}(A)$ is closed and $\dim \operatorname{Ker}(A) < \infty$.

The result stated above, although sufficient on many occasions, in practice requires to know that $\operatorname{int} K \neq \emptyset$. From that reason stems the necessity to establish a result which relaxes this assumption, too. To discuss and explain an approach due to Deimling (see also [R3]), let us make the following observations.

Let, as usual, $K \subset E$ be a closed set of a Banach space E and let $x \in K$. If $\varepsilon > 0$ and $u\colon [0,\varepsilon] \to K$ is a continuous function such that $u(0) = x$ and the right derivative $v = u'_+(0)$ exists, then it is easy to see that $v \in T_K(x)$. Similarly as in Remark (2.17), given an upper semicontinuous map $\varphi\colon K \multimap E$ with compact convex values such that there exists a solution $u\colon [0,\varepsilon] \to K$ of the Cauchy problem:

$$\begin{cases} u'(t) \in \varphi(u(t)), \\ u(0) = x, \end{cases} \tag{2.31}$$

i.e. there is an (Bochner) integrable function $w\colon [0,\varepsilon] \to K$ such that $w(t) \in \varphi(u(t))$ and $u(t) = x + \int_0^t w(s)\,ds$ on $[0,\varepsilon]$, then $\varphi(x) \cap T_K(x) \neq \emptyset$. Indeed, for an arbitrary sequence $h_n \to 0^+$ and any $n \in \mathbb{N}$, $u(h_n) = x + h_n v_n \in K$ where

$$v_n = \frac{1}{h_n} \int_0^{h_n} w(s)\,ds.$$

It is clear that $v_n \in \operatorname{cl\,conv}\{w(s) : s \in [0,h_n]\} \subset \operatorname{cl\,conv} \varphi(\{u(s) : s \in [0,h_n]\})$. The upper semicontinuity of φ and the compactness of its values implies that, passing to subsequences if necessary, $v_n \to v \in \varphi(x)$. Hence $v \in \varphi(x) \cap T_K(x)$.

This means that in order to verify the weak tangency condition for φ it is sufficient to show that, for each $x \in K$, problem (2.31) admits a solution. The converse implication does hold under some additional assumptions. In particular one has the following result.

(2.32) PROPOSITION (see e.g. [D2, Theorem 9.1]). *Suppose that K is closed, bounded in E, an upper semicontinuous $\varphi\colon K \multimap E$ with convex compact values is k-set contractive ($k \geq 0$) with respect to the Kuratowski or Hausdorff measure of noncompactness γ. If φ is weakly tangent to K, then problem (2.31) admits a solution.*

(2.33) THEOREM (Deimling, [D2, Theorem 11.5], [D1]). *Let K be a closed bounded convex subset of a Banach space E and let an upper semicontinuous map $\varphi\colon K \multimap E$ with compact convex values be condensing with respect to the Kuratowski or Hausdorff measure of noncompactness γ. If φ is weakly inward, i.e. for each $x \in K$, $\varphi(x) \cap x + T_K(x) \neq \emptyset$, then φ has a fixed point.*

PROOF. We may assume that $0 \in K$ since a translation does not destroy any of the assumptions. Moreover, we may suppose that φ is a set-contraction. For if we take $k \in (0,1)$, then it easy to see that $k\varphi$ is a set-contraction which also satisfies the weak inwardness condition. Assuming that the result holds for set-contractions and taking a sequence $k_n \to 1^-$, for each $n \in \mathbb{N}$ we get $x_n \in K$ such that $x_n \in k_n\varphi(x_n)$. For any $\varepsilon > 0$ there is $N \in \mathbb{N}$ such that $k_n^{-1} - 1 < \varepsilon$ for $n \geq N$. For such n, $x_n \in B(k_n^{-1}x_n, \varepsilon) \subset B(\varphi(x_n), \varepsilon)$. Hence

$$\gamma(\{x_n\}_{n=1}^{\infty}) = \gamma(\{x_n : n \geq N\}) \leq \varepsilon + \gamma(\varphi(\{x_n : n \geq N\})) \leq \varepsilon + \gamma(\varphi(\{x_n\}_{n=1}^{\infty}).$$

This show that the set $\operatorname{cl}\{x_n\}_{n=1}^{\infty}$ is compact. Passing to a subsequence if necessary, we may assume that $x_n \to x_0$ and $x_0 \in \varphi(x_0)$.

Suppose that, for a bounded set $B \subset K$, $\gamma(\varphi(B)) \leq k\gamma(B)$, where $0 \leq k < 1$. Let $\varepsilon_n = 2^{-n}$ and

$$K_0 = K, \quad K_n = C_n \cap K_{n-1} \quad \text{where } C_n := \operatorname{cl}\operatorname{conv}\left[B(\varphi(K_{n-1}), \varepsilon_n) \cup B(0, \varepsilon_n)\right]$$

for $n \in \mathbb{N}$. It is clear that, for all $n \geq 0$, $0 \in K_n$, $K_{n+1} \subset K_n \subset K$ and

$$\gamma(K_n) \leq \gamma(C_n) \leq \gamma(\varphi(K_{n-1})) + 2\varepsilon_n \leq k\gamma(K_{n-1}) + 2\varepsilon_n.$$

Thus, by induction

$$\gamma(K_n) \leq k^n\gamma(K_0) + 2\sum_{i=1}^{n} k^{n-i}\varepsilon_i.$$

This shows that $\gamma(K_n) \to 0$; hence, by the Kuratowski theorem, the set

$$C := \bigcap_{n=0}^{\infty} K_n$$

is nonempty and compact convex. The map φ is weakly inward to $K = K_0$; suppose that so it does with respect to K_{n-1} $(n \geq 1)$. We shall show that φ is weakly inward to K_n as well. Let $x \in K_n$ and take $y \in \varphi(x)$ such that $y - x \in T_{K_{n-1}}(x)$. Since $0 \in \operatorname{int} C_n \cap K_{n-1}$ and $\varphi(x) \subset C_n$, we have that $y - x \in T_{C_n}(x)$ and, by Lemma (2.28),

$$y - x \in T_{C_n}(x) \cap T_{K_{n-1}}(x) = T_{K_n}(x).$$

Having this we shall prove that φ is weakly inward to C. To see this take $x \in C$ and observe that the Cauchy problem

$$\begin{cases} u'(t) \in \varphi(u(t)) - u(t), \\ u(0) = x, \end{cases} \tag{2.34}$$

has a solution $u_n\colon [0,1] \to K_n$ in view of Proposition (2.32). It is easy to see that the family $\{u_n\}_{n=1}^{\infty}$ is equicontinuous and, for each $t \in [0,1]$, the orbit $\{u_n(t)\}_{n=1}^{\infty}$

is relatively compact. Hence, by the Ascoli–Arzela theorem, we may assume that $u_n \to u \in C([0,1], C)$ uniformly on $[0,1]$. Obviously $u(0) = x$. It is also standard (see e.g. [BaK2, Appendix] for an argument in a more general situation) to see that, for almost all $t \in [0,1]$, the orbit $\{u_n'(t)\}_{n=1}^{\infty}$ is relatively compact. By a result due to Diestel [DRS, Corollary 2.6], passing to a subsequence if necessary, we infer that $u_n \rightharpoonup w \in L^1([0,1], E)$ weakly in L^1. Hence $u(t) = x + \int_0^t w(s)\,ds$, i.e. $u' = w$ almost everywhere on $[0,1]$. The application of the so-called convergence theorem [AE, Theorem 3.2.6] (or [AF, Theorem 7.2.2]) shows that $w(s) \in \varphi(u(s)) - u(s)$ almost everywhere on $[0,1]$, i.e. u is a solution to (2.34). By the remarks preceding Proposition (2.32), this implies that $(\varphi(x) - x) \cap T_C(x) \neq \emptyset$. In virtue of Theorem (2.1), φ has a fixed point. □

(2.35) Remark. Usually a separable setting is sufficient for the study of compact (or condensing) maps; for instance if K is closed convex in E and $\varphi\colon K \multimap K$ is compact, then $\operatorname{cl}\varphi(K) \subset E'$, where E' is a closed separable subspace of E, and $\varphi\colon K' \multimap K'$ where $K' = K \cap E'$. Hence, at most instances, it is sufficient to consider only *separable* spaces. It is not the case for the Deimling theorem: given a weakly tangent and compact map $\varphi\colon K \multimap E$, we still have that $\operatorname{cl}\varphi(K) \in E'$, but, in general, φ would be not weakly tangent to K'.

In the sequel we shall see that, in the case of a separable space, Theorem (2.33) follows as a special case from a much more general result (see Section 3).

(2.36) Remark. In case φ is compact, then the boundedness of K is not necessary.

Finally let us state the following result (comp. [L]).

(2.37) Theorem. *Let E be reflexive, $K \subset E$ closed and convex and suppose that a compact map $\varphi\colon K \multimap E$ is admissible and φ is (strongly) inward to K, i.e.*

$$\varphi(x) \subset x + T_K(x) \quad \text{for all } x \in K. \tag{2.37.1}$$

Then φ has a fixed point.

Proof. First recall that due to the Kadec and Troyanski theorem, there is an equivalent norm in E such that (with this new norm) E and E^* are locally uniformly convex and strongly convex. In particular the metric projection $\pi_K\colon E \to K$ is continuous and single-valued. Let $C := \operatorname{cl}\operatorname{conv}\varphi(K)$; then C is convex compact and consider map $\psi\colon K \times C \multimap K \times C$ by the formula $\psi(x,y) = \pi_K(y) \times \varphi(x)$, $x \in K$, $y \in C$. The same argument as in the proof of Proposition (2.7) (see also Remark (2.9)) shows that there exists $(x_0, y_0) \in K \times C$ such that $y_0 \in \varphi(x_0)$ and $x_0 = \pi_K(y_0)$. By Corollary (1.9), $J(y_0 - x_0) \in N_K(x_0)$. On the other hand $y_0 - x_0 \in T_K(x_0)$. Hence $y_0 = x_0$. □

The author is not sure whether Theorem (2.37) holds for non-reflexive spaces.

Theorem (2.33)gives sufficient conditions for the so-called *essentiality* of set-valued maps (comp. [DG]) which generalizes the so-called Leray–Schauder nonlinear alternative of Granas and a result due to Aubin [AF, Theorem 3.2.8].

(2.38) PROPOSITION. *Suppose that* $K \subset E$ *is closed convex and* $\operatorname{int} K \neq \emptyset$. *If a compact map* $\varphi\colon K \multimap E$ *has compact convex values and is weakly inward to* K, *then it is essential with respect to the boundary* $\operatorname{bd} K$, *i.e. any compact map* $\psi\colon K \multimap E$ *with compact convex values such that* $\psi|_{\operatorname{bd} K} = \varphi|_{\operatorname{bd} K}$ *has a fixed point. In particular, if a compact map* $\Phi\colon K \times [0,1] \multimap E$ *with compact convex values is such that* $\Phi(\,\cdot\,,0) = \varphi$, *for all* $x \in \operatorname{bd} K$ *and* $t \in [0,1]$, $x \notin \Phi(x,t)$, *then* $\Phi(\,\cdot\,,1)$ *has a fixed point.*

PROOF. The first part is obvious. For each $x \in K$, $\psi(x) \cap (x + T_K(x)) \neq \emptyset$: for $x \in \operatorname{bd} K$ it follows by assumption; if $x \in \operatorname{int} K$, then $T_K(x) = E$.

As concerns the second assertion, the proof uses the method of Borsuk. Let $B = \{x \in K : x \in \Phi(x,t) \text{ for some } t \in [0,1]\}$. The upper semicontinuity of Φ implies that B is closed; moreover, $\operatorname{bd} K \cap B = \emptyset$. Take an Urysohn function $t\colon K \to [0,1]$ separating $\operatorname{bd} K$ from A, i.e. t is continuous and $t|_{\operatorname{bd} K} \equiv 1$, $t|_B \equiv 1$. It is easy to see that $\psi(x) = \Phi(x, t(x))$ defines a compact map $\psi\colon K \multimap E$ with compact convex values and $\psi_{\operatorname{bd} K} = \varphi|_{\operatorname{bd} K}$. Therefore there is $x_0 \in K$ such that $x_0 \in \psi(x_0) = \Phi(x_0, t(x_0))$. Hence $x_0 \in B$, $t(x_0) = 1$ and $x_0 \in \Phi(x_0, 1)$. □

The next result follows in the same spirit.

(2.39) THEOREM. *Suppose* K *is convex closed,* $U \subset K$ *is open, a set-valued* $\varphi\colon K \to E$ *is compact, for all* $x \in K$, $\varphi(x) \cap (x + T_K(x)) \neq \emptyset$ *and the fixed point set* $\{x \in K : x \in \varphi(x)\} \subset U$. *If* $\psi\colon \operatorname{cl}_K U \multimap E$ *is compact* ([15]), *for each* $x \in U$, $\psi(x) \cap (x + T_K(x)) \neq \emptyset$, *then at least one of the following properties is satisfied:*

(2.39.1) *there is* $x \in \operatorname{bd}_K U$ *and* $t \in (0,1)$ *such that* $x \in (1-t)\varphi(x) + t\psi(x)$,
(2.39.2) *there is* $x_0 \in \operatorname{cl}_K U$ *such that* $x_0 \in \psi(x_0)$.

PROOF. Suppose that (2.39.1) does not hold and $x \notin \psi(x)$ for $x \in \operatorname{bd}_K U$. The homotopy $\Phi\colon \operatorname{cl}_K U \times [0,1] \multimap E$ given by $\Phi(x,t) = (1-t)\varphi(x) + t\psi(x)$ is compact and has compact convex values. The set $B = \{x \in \operatorname{cl}_K U : x \in \Phi(x,t) \text{for some} t \in [0,1]\}$ is clearly closed and $B \cap (K \setminus U) = \emptyset$. Take an Urysohn function $t\colon K \to [0,1]$ such that $t|_{K \setminus U} \equiv 0$ and $t|_B \equiv 1$. The map $\Phi(\,\cdot\,, t(\,\cdot\,))$ is actually defined on K, has compact convex values, is compact and satisfies the weak tangential inwardness condition, i.e. for each $x \in K$, $\Phi(x, t(x)) \cap (x + T_K(x)) \neq \emptyset$. Indeed, for $x \in K \setminus U$, $\Phi(x, t(x)) = \varphi(x)$; for $x \in U$, there are $y_1 \in \varphi(x)$ and $y_2 \in \psi(x)$ such that

([15]) In what follows $\operatorname{cl}_K U$, $\operatorname{bd}_K U$ denotes the closure and the boundary of U relative to K.

$y_i \in x + T_K(x)$, hence $y = (1 - t(x))y_1 + t(x)y_2 \in x + T_K(x)$ and $y \in \Phi(x, t(x))$. By Theorem (2.33), there is $x_0 \in K$ such that $x_0 \in \Phi(x_0, t(x_0))$. Thus $x_0 \in B$, $t(x_0) = 1$ and $x_0 \in \psi(x_0)$. □

(2.40) REMARK. Proposition (2.38) and Theorem (2.39) hold true if in addition to hypotheses stated there:

(2.40.1) K is compact convex and the involved maps are merely upper hemicontinuous with closed convex values;

(2.40.2) E is reflexive, K is closed convex and the involved maps are compact admissible and (strongly) inward, i.e. have values at $x \in K$ contained in $x + T_K(x)$.

(2.41) REMARK. The above results may be put into the context of the *fixed point index theory* defined for compact maps $\varphi\colon \mathrm{cl}_K U \multimap E$ (being admissible or having compact convex values) satisfying the above considered (strong or weak) inwardness condition and having no fixed point on the boundary $\mathrm{bd}_K U$. A variant of such a theory valid for a special class of convex sets $K \subset E$ was provided in [HS] (also cf. [O1], [O2]).

Another direction in the study of constrained equilibrium problem without compactness relies on a certain generalization of the fixed point theorem (2.3) due to Ben-El-Mechaiekh, Deguire and Granas (see [BDG1]).

(2.42) PROPOSITION. *Let $K \subset E$ be convex and $S\colon K \multimap K$ be such that:*

(2.42.1) *for all $y \in K$, $S^{-1}(y)$ is open (in K);*

(2.42.2) *for all $x \in K$, $S(x)$ is convex;*

(2.42.3) *there is a compact subset $L \subset K$ with the following property: given a finite subset $N \subset K$, there exists a compact convex $C_N \subset K$, $N \subset C_N$ such that $S(x) \cap C_N \neq \emptyset$ for $x \in C_N \setminus L$.*

Then S has a fixed point.

PROOF. Similarly as in the proof to Theorem (2.3) one constructs (via possibly infinite but locally finite partitions of unity) a continuous selection $s\colon K \to K$ of S. Since L is compact, one may assume that the restriction $s|_L$ has values in the convex polyhedron $\mathrm{conv}\, N$ spanned by a finite set $N \subset K$. Let $C_N \subset K$ be the compact convex set provided by condition (2.42.3) and such that $\mathrm{conv}\, N \subset C_N$. Let $S_N\colon C_N \multimap C_N$ be given by $S_N(x) = S(x) \cap C_N$, $x \in C_N$. Observe that S_N is well-defined (has nonempty values). Indeed if $x \in C_N \cap L$, then $s(x) \in S(x) \cap \mathrm{conv}\, N \subset S(x) \cap C_N$; if $x \in C_N \setminus L$, then $S_N(x) \neq \emptyset$ by (2.42.3). Now the existence of a fixed point follows from Theorem (2.3). □

We may now state the corresponding existence result (see [BK2]).

(2.43) THEOREM. *Let $K \subset E$ be convex, F is a Banach space, $\varphi\colon K \multimap F$ be upper hemicontinuous with closed convex values and $A\colon E \to F$ be a bounded linear operator. Assume that there exists a compact $L \subset K$ and, for each finite $N \subset K$, a compact convex $C_N \subset K$, $N \subset C_N$ such that φ satisfies the following tangency conditions: for all $x \in C_N \setminus L$, $\varphi(x) \cap T_{A(C_N)}(A(x)) \neq \emptyset$ and, for all $x \in L$, $\varphi(x) \cap T_{A(L)}(A(x)) \neq \emptyset$. Then φ has an equilibrium and, for any $x_0 \in K$, the map $\varphi - A(\cdot - x_0)$ has an equilibrium.*

It is obvious that the tangency conditions considered above may be relaxed as the usual weak tangency was weakened (see Lemma (2.16)). If K is compact, then putting $L = \emptyset$ and $C_N = K$ for any N we obtain Theorem (2.14).

The proof of the first assertion of Theorem (2.43) uses Proposition (2.42) in a similar way as the proof of Theorem (2.14) relies on Theorem (2.3) (via Theorem (2.5)). We leave the details to the interested reader (comp. [BK2]). We shall only provide the proof of the second assertion (comp. Corollary (2.23)).

PROOF. Given $x_0 \in K$, define $\psi\colon K \multimap F$ by $\psi(x) = \varphi(x) - A(x - x_0)$ for $x \in K$. It is evident that, for each $x \in K$, $\psi(x)$ is closed convex and ψ is upper hemicontinuous. We shall show that ψ satisfies the same conditions as φ does. Let $N \subset K$ be finite and let $M = N \cup \{x_0\}$. Let $D_N := C_M \supset N$. For any $x \in D_N \setminus L$, $\varphi(x) \cap T_{A(D_N)}(A(x)) \neq \emptyset$ and $A(x_0) - A(x) \in T_{A(D_N)}(A(x))$; hence $\psi(x) \cap T_{A(D_N)}(A(x)) \neq \emptyset$ because the latter is a convex cone. The proof that $\psi(x) \cap T_{A(L)}(A(x)) \neq \emptyset$ goes similarly. □

Let us mention that most of the above results have their counterparts in Hausdorff locally convex spaces and even more general spaces, e.g. spaces having sufficiently many linear functionals, or even vector spaces equipped with any topology that induces the Euclidean topology on finite polytopes. Some results in this direction may be found in [L], [R1] and see a good survey in [P]. Moreover, results stated above also admit analytical description analogous to the one presented in Proposition (2.24) and Remark (2.25).

2.3. Beyond convexity. As already observed the existence of an equilibrium for the map $\varphi\colon K \multimap E$ is equivalent to the existence of a fixed point for the map $I \pm \varphi$; hence any equilibrium theorem has a fixed point counterpart and vice versa. Taking this observation into account, when generalizing criteria for the existence of equilibria for inward or tangent mappings, one can turn to the various extensions (of homotopical or homological nature) of the known fixed point principles of self maps to nonconvex sets. To make it, one has to clarify the following issues:

(2.44.1) Which properties of sets are suitable substitutes for convexity?
(2.44.2) What would then be the counterpart of tangency conditions?

One of the most natural as well as convenient classes to be considered in extending results to nonconvex sets is the class of absolute retracts, and more generally that of absolute neighbourhood retracts. Recall (see e.g. [Bor]) that a closed set $K \subset E$ is a (neighbourhood) retract of E if there exists a continuous map $r\colon E \to K$ ($r\colon U \to K$, where U is an open neighbourhood of K) such that $r(x) = x$ on K (the retraction r is by no means unique).

As mentioned in Introduction it is well known that every continuous $f\colon K \to K$, where K is a compact retract, has a fixed point; while if K is a compact neighbourhood retract, then f has a fixed point provided the Lefschetz number $\lambda(f)$ of f is nontrivial. In particular, assuming that the Euler characteristic $\chi(K) \neq 0$ (what always holds true if K is a retract), it follows that a map f being homotopic to the identity has a fixed point. Observe that the Euler characteristic $\chi(S^1)$ of the unit one-dimensional sphere is 0 and it is easy to find a tangent equilibrium-free map on S^1!

Below we shall study possible ways to address question (2.44.2) above. If the set $K \subset E$ is not convex, then neither weak (2.1.1) nor strong tangency (2.13.1) for a map $\varphi\colon K \multimap E$ guarantees the existence of equilibria.

(2.45) Example. Let $K := \{x = (x_1, x_2, x_3) \in \mathbb{R}^3 : |x| \leq \sqrt{2}$ and $\sqrt{x_1^2 + x_2^2} \geq x_3\}$, $S := \{x \in K : x_1^2 + x_2^2 = 1$ and $x_3 = 1\}$ and $Z = \{x \in \mathbb{R}^3 : x_1^2 + x_2^2 \leq 1$ and $x_3 = 1\}$. It is easy to see that K is a compact retract. Next, for $x \in K$, put

$$\varphi(x) = \begin{cases} Z & \text{for } x \in D \setminus S, \\ \operatorname{conv}\{Z \cup \{(-x_2, x_1, 0\}\} & \text{for } x \in S. \end{cases}$$

Clearly $\varphi\colon K \multimap E$ is upper semicontinuous with compact convex values and $\varphi(x) \cap T_K(x) \neq \emptyset$ on D. But is easy to see that φ has no equilibria. Observe also that, for $x = 0$, the set $S(x)$ of all solutions to the Cauchy problem (2.31) is homeomorphic to the unit sphere S^1; hence it is not an R_δ-set (this will be explained later on). Notice that, for all $x \in K$, $x \neq 0$, the Bouligand and the Clarke tangent cones $T_K(x)$ and $C_K(x)$ coincide; however $T_K(0) \neq C_K(0)$ and $\varphi(0) \cap C_K(0) = \emptyset$.

Below we shall see that if, in the above example, should we take another map φ that satisfies the tangency condition (2.1.1) with the Bouligand cones replaced by the Clarke cones, then φ would possess equilibria. However, it is not true that such a procedure would be a general remedy.

(2.46) Example. Let $K := S_1 \cup S_{-1}$ where $S_i := \{z = (x, y) \in \mathbb{R}^2 : (x - i)^2 + y^2 = 1\}$. It is readily seen that K is a neighbourhood retract in $\mathbb{R}^2$ and $\chi(K) \neq 0$. Let

$$f(x, y) = \begin{cases} (y, 1 - x) & \text{for } (x, y) \in S_1, \\ (-y, 1 + x) & \text{for } (x, y) \in S_{-1}. \end{cases}$$

For all $z \in K$, $f(z) \in T_K(z) = C_K(z)$ but f has no zeros. At the same time the set of all solutions to (2.31) (with φ replaced by f and $x = (0,0)$) is even not connected.

It seems therefore that in order to express a correct substitute for the "boundary" conditions implying the existence of equilibria or fixed points in the convex setting, which would be sufficient to study equilibria in the nonconvex case and allowing also to characterize the set of solutions to the Cauchy problem (2.31), one should replace the Bouligand cones (which were present in the convex case) by Clarke cones and take care of the geometry of the involved (neighbourhood) retracts.

If one agrees to use less orthodox conditions than the usual tangency (expressed in terms of either Bouligand or Clarke cones), then perhaps the following simple result may be interesting. For simplicity we consider the Hilbert space setting (although if E together with E^* are locally uniformly convex, then the below result still holds true).

(2.47) PROPOSITION. *Let K be a compact neighbourhood retract in a Hilbert space $(H, \langle\cdot,\cdot\rangle)$ with a given neighbourhood retraction $r: U \to K$. If $\chi(K) \neq 0$ and an upper hemicontinuous $\varphi: K \multimap H$ with closed convex values is such that, for all $x \in K$ and $u \in r^{-1}(x)$,*

$$\inf_{y \in \varphi(x)} \langle u - x, y \rangle \leq 0, \tag{2.47.1}$$

then φ has an equilibrium in K.

In view of the relationship between the normal and proximally normal cones, condition (2.47.1) is somehow similar to the normality condition (2.14.1). If K is convex, then the most natural choice for r is the orthogonal projection $\pi_K: H \to K$ and then (2.47.1) is *identical* with (2.14.1), see Corollary (1.9).

PROOF. Assume that φ has no zeros. As in the proof of Theorem (2.14), we construct a continuous map $f: K \to H$ such that $\inf_{y \in \varphi(x)} \langle f(x), y \rangle > 0$ for all $x \in K$. Since K is compact, there is $\delta > 0$ such that $B(K, \delta) \subset U$. Take $\mu > 0$ such that $\mu \|f(x)\| < \delta$ for $x \in K$. The map $g: K \to K$ given by $g(x) := r(x + \mu f(x))$ ($x \in K$) is well-defined and homotopic to the identity on K (through the homotopy $h(x,t) = r(x + t\mu f(x))$, $x \in K$, $t \in [0,1]$). Hence there is $x_0 \in K$ such that $x_0 = r(x_0 + \mu f(x_0))$, i.e. $x_0 + \mu f(x_0) \in r^{-1}(x_0)$ and $\inf_{y \in \varphi(x_0)} \langle f(x_0), y \rangle \leq 0$, a contradiction. □

(2.48) REMARK. Condition (2.47.1) depends strongly on the choice of a retraction r. In order to get rid of the special structure of the ambient space in the above Proposition, in [BK1] the following approach was presented. Given a compact neighbourhood retract $K \subset E$ and a retraction $r: U \to K$, the authors define, for any $x \in K$, the so-called *retraction normal cone* $N_K^r(x) \subset E^*$ (the construction reminds to some extent that of the proximally normal cone) and show that an upper hemicontinuous $\varphi: K \multimap E$ with closed convex values satisfying the normality condition analogous to (2.14.1) (namely (2.14.1) with $N_K^r(x)$ replacing $N_K(x)$) has equilibria. The proof differs from that of Theorem (2.5) since $N_K^r(x)$ admits no description similar to (1.6). The retraction cones as well as retraction tangent cones (defined as negative polar cones) depend on the chosen retraction r. However, [BK1] shows that in case K is compact convex ([16]), then $N_K(x) = N_K^r(x)$ for *any* retraction $r: E \to K$. This proves the consistency of the approach of [BK1] with the one used in the convex setting.

As was said above, in order to deal with the usual tangency conditions one has to worry about the geometry of the retract K. We shall present two different (although related) approaches to this problem.

Let (X, d) be a metric space. We say that a set $K \subset X$ is an *$\mathcal{L}$-retract* (of X) if there is a neighbourhood retraction $r: U \to K$ and a constant $L \geq 1$ such that, for all $x \in U$,

$$d(r(x), x) \leq L d_K(x). \tag{2.49}$$

Clearly any $\mathcal{L}$-retract is a neighbourhood retract and is closed. The class of $\mathcal{L}$-retracts has been introduced and studied in [BK1].

Before we study the existence of equilibria on $\mathcal{L}$-retracts, let us provide some examples of such sets.

(2.50) EXAMPLE.

(2.50.1) Suppose that $K \subset X$ is closed and bi-Lipschitz homeomorphic with a closed convex set $A \subset E$ (i.e. there is a Lipschitz homeomorphism $h: K \to A$ with Lipschitz inverse $g = h^{-1}: A \to K$). Then K is an $\mathcal{L}$-retract.

To see this let $f: X \to A$ be an extension of h given by the Arens–Dugundji formula (see [BP, Proposition II.3.1]) and put $r(x) = g \circ f(x)$ for $x \in X$. In [BK1] it is shown that (2.49) holds for all $x \in X$ and $L = 3L_g L_h + 1$ where L_h and L_g are the Lipschitz constants of h and g, respectively.

(2.50.2) If $K \subset E$ is closed and convex, then for each $\varepsilon > 0$, there is $r: E \to K$ such that $\|r(x) - x\| \leq (1 + \varepsilon) d_K(x)$.

To see this, for $x \in E$, let $\psi(x) := \{y \in K : \|y - x\| \leq (1 + \varepsilon) d_K(x)\}$. It is easy to see that ψ has closed convex values and is lower semicontinuous and, for

([16]) In view of the Dugundji extension theorem, each closed and convex set $K \subset E$ is a retract of E.

$x \in K$, $\psi(x) = \{x\}$. In view of the Michael selection theorem, ψ has a continuous selection $r: E \to K$.

(2.50.3) Following [Pl] (where the finite-dimensional case was presented) we say that $K \subset E$ is a *proximate retract* if there are a neighbourhood U of K and a retraction $r: U \to K$ such that $\|r(x) - x\| = d_K(x)$ (we say that r is a *metric retraction* or *projection*). Proximate retracts in a Hilbert space (under the name *φ-convex sets*) have been studied in detail in [CG] (see also the extensive bibliography therein) and some equivalent conditions were formulated. In particular, proximate retracts are tangentially regular. For instance, sets with $C^{1,1}$-boundary are proximate retracts. Obviously each proximate retract is an $\mathcal{L}$-retract.

(2.50.4) If $K \subset E$ is a neighbourhood retract with Lipschitz continuous neighbourhood retraction $r: U \to K$. Then (2.49) holds with $L = \ell + 1$ where ℓ is the Lipschitz constant of r. In particular, if K is a compact neighbourhood with a locally Lipschitz retraction $r: U \to K$, then K is an $\mathcal{L}$-retract.

(2.50.5) Below we shall study the class of *epi-Lipschitz* sets introduced by Rockafellar [Ro]. Each epi-Lipschitz set is an $\mathcal{L}$-retract.

As we see the class of $\mathcal{L}$-retracts is pretty large and, as it appears, it behaves well as concerns the constrained equilibrium theory.

The first result in this direction is due to Plaskacz [Pl] who proved that *if $K \subset \mathbb{R}^n$ is a proximate retract with nontrivial Euler characteristic, $\varphi: K \multimap \mathbb{R}^n$ is upper semicontinuous with compact convex values satisfying the weak tangency condition (involving Bouligand or Clarke cones: it is equivalent in view of the tangential regularity of K), then φ has an equilibrium* (this result was repeated by Stern in [St]).

The next step was done by Clarke, Ledyaev and Stern in [CLS] who proved that *given $K \subset E$ such that either*

(2.51.1) *K is bi-Lipschitz homeomorphic with a compact convex set, or*

(2.51.2) *$E = \mathbb{R}^n$ and K is epi-Lipschitz and homeomorphic with a compact convex set, an upper semicontinuous $\varphi: K \multimap E$ with closed convex values satisfying the weak tangency condition (i.e. $\varphi(x) \cap C_K(x) \neq \emptyset$ for each $x \in K$),*

then φ has an equilibrium ([17]).

Observe that in [Pl] and in both cases considered in [CLS], the set K is a compact $\mathcal{L}$-retract and $\chi(K) = 1$. The decisive contribution to the problem was done by Ben-El-Mechaiekh and the present author in [BK1] and the following result was obtained.

([17]) Epi-Lipschitz in $\mathbb{R}^n$ having nontrivial Euler characteristic have been studied in [CC] in the context of generalized equilibria.

(2.52) THEOREM. *Let $K \subset E$ be a compact $\mathcal{L}$-retract with $\chi(K) \neq 0$. If $\varphi\colon K \multimap E$ is upper semicontinuous with closed convex values and weakly tangent to K, i.e.*

$$\varphi(x) \cap C_K(x) \neq \emptyset \quad \textit{for all } x \in K, \tag{2.52.1}$$

then φ has an equilibrium.

This result constitutes a direct generalization of Theorem (2.1) and the results of Plaskacz, Clarke, Ledyaev and Stern and also those by Czarnecki and Cornet from [CC]).

(2.53) REMARK. Example (2.45) above shows that condition (2.52.1) cannot be replaced by (2.13.1) or (2.1.1): note that the set K from Example (2.45) is a compact $\mathcal{L}$-retract and $\chi(K) = 1$. However, if in Theorem (2.52) $\varphi = f$ is single-valued and continuous, then (2.1.1) (i.e. $f(x) \in T_K(x)$ on K) is sufficient. More generally: if a lower semicontinuous map $\varphi\colon K \multimap E$ is strongly tangent, i.e. $\varphi(x) \subset T_K(x)$ on K, then by lower semicontinuity

$$\varphi(x) \subset \operatorname*{Lim\,inf}_{y \xrightarrow{K} x} T_K(y) \subset C_K(x)$$

in view of (1.5).

We shall not give the proof of Theorem (2.52) since we may still do better. After [CK1] we shall present a result generalizing both Theorems (2.52) and (2.14). To this end we first show the existence of generalized equilibria on $\mathcal{L}$-retracts. We shall deal with upper hemicontinuous maps only.

(2.54) THEOREM. *Suppose $K \subset E$ is a compact $\mathcal{L}$-retract with the nontrivial Euler characteristic. Any upper hemicontinuous set-valued map $\varphi\colon K \multimap E^*$ with convex weak* compact values admits a generalized equilibrium i.e. a point $x_0 \in K$ such that $\varphi(x_0) \cap N_K(x_0) \neq \emptyset$.*

PROOF. First observe that, in analogy to (2.6), $x \in K$ is a generalized equilibrium of φ if and only if

$$\sup_{y \in C_K(x)} \inf_{p \in \varphi(x)} \langle p, y\rangle \leq 0. \tag{2.55}$$

We shall actually prove that:

(GE) there is $x_0 \in K$ and $\delta > 0$ such that, for all $y \in E$, if $d_K^\circ(x_0; y) < \delta$, then $\inf_{p \in \varphi(x_0)} \langle p, y\rangle \leq 1$.

Note that (GE) implies (2.55): for if (GE) is satisfied, $y \in C_K(x_0)$ (i.e. $d_K^\circ(x_0; y) \le 0 < \delta$) and $\inf_{p\in\varphi(x_0)}\langle p, y\rangle > 0$, then, for some $\lambda > 0$, $\inf_{p\in\varphi(x_0)}\langle p, \lambda y\rangle > 1$ but still $d_K^\circ(x_0; \lambda y) \le 0 < \delta$.

Suppose to the contrary that (GE) does not hold true, i.e. for all $x \in K$ and $\delta > 0$, there is $y \in E$ such that $d_K^\circ(x; y) < \delta$ but $\inf_{p\in\varphi(x)}\langle p, y\rangle > 1$. Let $r: U \to K$, where U is a neighbourhood of K, be a retraction verifying condition (2.49). Since K is compact and φ has bounded values, there is $M > 0$ such that $\sup_{p\in\varphi(x)} \|p\| \le M$ for all $x \in K$. Take $0 < \delta < (ML)^{-1}$ and consider an auxiliary map

$$S(x) = \{y \in E : d_K^\circ(x; y) < \delta \text{ and } \inf_{p\in\varphi(x)}\langle p, y\rangle > 1\}, \quad x \in K.$$

For any $x \in K$, $S(x)$ is nonempty and convex; moreover, for each $y \in E$, the set $S^{-1}(y) = \{x \in K : y \in S(x)\}$ is open since the function $d_K^\circ(\,\cdot\,; y)$ is upper semicontinuous and, in view of the upper hemicontinuity, the function

$$K \ni x \mapsto \inf_{p\in\varphi(x)}\langle p, y\rangle$$

is lower semicontinuous. Arguing as in the proof of Theorem (2.3), there is a continuous $g: K \to E$ such that $g(x) \in S(x)$ and $g(x) \neq 0$ for all $x \in K$.

The boundedness of g on K implies that there is $\mu > 0$ such that $x + n^{-1}g(x) \in U$ for all $x \in K$ and $n \ge N$ (where $N^{-1} \le \mu$). For such n, consider the map $g_n: K \to K$ given by

$$g_n(x) = r(x + n^{-1}g(x))$$

for $x \in K$. This map is homotopic to the identity via the homotopy

$$K \times [0,1] \ni (x,t) \mapsto r(x + tn^{-1}g(x)).$$

The Lefschetz number $\lambda(g_n) = \chi(K) \neq 0$, i.e. there is $x_n \in K$ such that $x_n = g_n(x_n)$.

Since $g(x_n) \in S(x_n)$,

$$\begin{aligned} n^{-1} &< \inf_{p\in\varphi(x_n)}\langle p, n^{-1}g(x_n)\rangle \le M\|x_n + n^{-1}g(x_n) - g_n(x_n)\| \\ &= M\|x_n + n^{-1}g(x_n) - r(x_n + n^{-1}g(x_n))\| \le ML d_K(x_n + n^{-1}g(x_n)). \end{aligned}$$

Passing to subsequences if necessary, we may assume that $x_n \to x_0 \in K$ and

$$n^{-1} < ML(n^{-1}\|g(x_n) - g(x_0)\| + d_K(x_n + n^{-1}g(x_0))).$$

Hence

$$1 \le ML \limsup_{n\to\infty} n d_K(x_n + n^{-1}g(x_0)) \le ML d_K^\circ(x_0; g(x_0)) < 1,$$

a contradiction. □

It is easy to see that an analogue to Proposition (2.7) is true as well.

(2.56) THEOREM. *Suppose that* $K \subset E$ *is a compact proximate retract such that* $\chi(K) \neq 0$. *If* $\varphi\colon K \multimap E$ *is admissible, then there is* $x_0 \in K$ *such that* $\varphi(x_0) \cap N(x_0, K) \neq \emptyset$. *If, additionally,* φ *is (strongly) tangent to* K, *i.e.*

$$\varphi(x) \subset C_K(x) \quad \text{for all } x \in K, \tag{2.56.1}$$

then φ *has an equilibrium.*

PROOF. Let $r\colon U \to K$ be a metric projection provided by the definition of a proximate retract. Take $\mu > 0$ such that, for $x \in K$, $x + \mu\varphi(x) \subset U$. By Theorem (1.2), there is $x_0 \in K$ such that $x_0 \in r(x_0 + \mu\varphi(x_0))$ (observe that the map $K \ni x \mapsto r(x + \mu\varphi(x))$ is admissible). Hence $\varphi(x_0) \cap \mu^{-1}(r^{-1}(x_0) - x_0) \subset N(x_0, K)$. □

A version of the second part of the above result valid for $\mathcal{L}$-retracts will be proved in Section 4 (see Theorem (4.4)).

Now we return to an extension of Theorem (2.52).

(2.57) THEOREM. *Suppose that* $K \subset$ *is a compact* $\mathcal{L}$*-retract with* $\chi(K) \neq 0$.

(2.57.1) *Let* F *be a Banach space,* $A \in \mathcal{L}(E, F)$ *and let* $\varphi\colon K \multimap F$ *be an upper hemicontinuous mapping with closed convex values satisfying the normality condition* (2.14.1) *or* (2.16.3), *i.e.* $\inf_{y\in\varphi(x)}\langle p, y\rangle \leq 0$ *for all* $x \in K$ *and* $p \in A^{*-1}(N_K(x))$.

(2.57.2) *Let an upper hemicontinuous map* $\varphi\colon K \multimap E$ *have closed convex values and satisfy condition* (2.19.1), *i.e.* $\inf_{y\in\varphi(x)} d_K^{\circ}(x; y) \leq 0$ *for all* $x \in K$.

Then, in both cases (2.57.1) *and* (2.57.2), φ *admits an equilibrium.*

PROOF. The proof of the first part is identical to that of Theorem (2.14) (instead of Theorem (2.5) one applies Theorem (2.54)). It would be interesting to know whether condition (2.16.5) is sufficient.

When (2.57.2) holds, then again suppose to the contrary that $0 \notin \varphi(x)$ on K. The separation theorem implies the existence of a bounded linear form (of a sufficiently large norm) $p_x \in E^*$ such that $\inf_{y\in\varphi(x)}\langle p_x, y\rangle > 1$. As in the proof of Theorem (2.14), one constructs a continuous $p\colon K \to E^*$ such that, for all $x \in K$,

$$\inf_{y\in\varphi(x)}\langle p(x), y\rangle > 1. \tag{2.58}$$

As in the proof of Theorem (2.54), condition (GE) (with p replacing φ) is satisfied for some $x_0 \in K$ and $\delta > 0$. By the assumption, there is $y \in \varphi(x_0)$ such that $d_K^{\circ}(x_0; y) < \delta$; hence $\langle p(x_0), y\rangle \leq 1$, a contradiction with (2.58). □

(2.59) REMARK. It is clear that Theorem (2.57) admits an analytical approach analogous to that presented in Proposition (2.24). Theorem (2.38) holds for a compact $\mathcal{L}$-retract $K \subset \mathbb{R}^n$. Theorem (2.39) is also true (see Remark (2.40.1)) for a compact $\mathcal{L}$-retract and an upper hemicontinuous map with closed convex values as well as admissible maps (satisfying the strong tangency condition (2.56.1), comp. Theorem (4.4)).

(2.60) REMARK. It is worth to mention here that it is possible to construct the topological degree theory generalizing that from [HS], i.e. an integer-valued homotopy invariant $\deg(\varphi, U)$ defined for an upper semicontinuous map $\varphi\colon \operatorname{cl}_K U \multimap E$ with closed convex values (resp. an admissible map) defined on the (relative) closure of an open subset U of a compact $\mathcal{L}$-retract $K \subset E$ satisfying the weak tangency condition: for all $x \in \operatorname{cl}_K U$, $\varphi(x) \cap C_K(x) \neq \emptyset$ (resp. the strong tangency). The construction provided in [CK3] is rather involved, requires several additional lemmas and yields an invariant satisfying all of the usual properties of the topological degree. The interesting normalization property states that $\deg(\varphi, K) = \chi(K)$. This formula may be treated as a generalization of the celebrated Poincaré–Hopf formula (see [Hi]) relating the (global) index of zeros of a tangent field on a compact manifold with its Euler characteristic. In [CK3], the authors show various applications of the constructed degree, e.g. concerning the continuation and bifurcation of equilibria as well as the existence of periodic orbits of constrained differential inclusions.

In both Theorems (2.54) and (2.57) the nontriviality of $\chi(K)$ is crucial. In [CC] the authors show that this topological assumption is, in a sense, necessary for the existence of equilibria. Namely, they prove that if K is a compact epi-Lipschitz subset of $\mathbb{R}^n$ (hence a compact $\mathcal{L}$-retract), then there exists a nonzero single-valued continuous map tangent to K. The corresponding result for general $\mathcal{L}$-retracts in E (or even in $\mathbb{R}^n$) is not known; however we conjecture it to be true.

Unfortunately there is still no way to proceed with the constrained fixed problem of set-valued maps defined on arbitrary (noncompact) $\mathcal{L}$-retracts. The method proposed above strongly relies on the compactness. That is why below we shall establish criteria for the existence of equilibria of maps defined on sets of a slightly less general nature than $\mathcal{L}$-retracts. In this case we will be able to relax the compactness assumption which was yet unavoidable in Theorem (2.54) and (2.57). In that we shall (to some extent) mimic the approach presented in Theorems (2.29) and (2.30).

2.4. Epi-Lipschitz and regular constraints. To explain our approach let us first observe that the very main assumption (save the nontriviality of the Euler characteristic) in Theorem (2.52) (which implicitly is contained also in Theorem (2.57)), namely that of weak tangency (2.52.1) can hardly be verified in practice, it is well-known that, in general, the Clarke cones $C_K(\cdot)$ are not easy to compute. On the other hand, in practical applications, the constrained equilibrium (or fixed point) problem has often the following (and natural) form: *given a function* $f\colon \mathrm{Dom}(f) \to \mathbb{R}$ *defined on an open domain* $\mathrm{Dom}(f)$ *and an upper hemicontinuous set-valued map* $\varphi\colon E \multimap E$*, find conditions (on* φ *and* f*) sufficient for the existence of solutions to the following problem*

$$\text{(P)} \qquad 0 \in \varphi(x), \quad f(x) \le 0.$$

It is clear that solutions to (P) are equilibria of φ constrained to the set

$$\text{(S)} \qquad K := \{x \in \mathrm{Dom}(f) : f(x) \le 0\}.$$

We say that K is *represented* by f.

The answer to this question by means of Theorem (2.52) or (2.57) must be preceded by the careful study of the set K. Firstly, one should check whether K is a compact $\mathcal{L}$-retract and $\chi(K) \ne 0$. It is rather obvious that sets of the form (S) are rarely compact (they may even not be closed) and it is not clear altogether whether they belong to the class of $\mathcal{L}$-retracts. Secondly, the verification of the (weak) tangency condition is necessary; this may be a difficult task. In this case, one would prefer to replace this condition by "tangency" expressed in terms of the controlled constraint f, for instance conditions involving the negative polar cone $\partial f(\cdot)^-$ of the generalized gradient of f. It is easy to see and to provide examples showing that in general nothing whatsoever can be said about relations between the cone fields $C_K(\cdot)$ and $\partial f(\cdot)^-$ on K, save the following (comp. [AE, Propositions 16, 7.3]).

(2.61) PROPOSITION. *Given a set* K *of the form* (S) *and* $x_0 \in \mathrm{bd}\, K$*, if* $0 \notin \partial f(x_0)$*, then* $\partial f(x_0)^- \subset C_K(x_0)$.

Observe that the problem outlined above is in some sense more general than studied before. Given an arbitrary $K \subset E$, $K = \{x \in E : f(x) \le 0\}$, where $f = d_K$, and, for each $x \in K$, $C_K(x) = \partial d_K(x)^- = \partial f(x)^-$. Hence, given K and considering $f = d_K$ as its representing function we may translate next results to the previous setting.

Let $f\colon E \to \mathbb{R}$ be a locally Lipschitz function and let $C \subset E$ be closed convex. For any $x \in C$, let

$$|\!|\!|\partial f(x)|\!|\!|_C := \sup_{u \in -T_C(x) \cap D} \inf_{p \in \partial f(x)} \langle p, u\rangle$$

where $D := \{u \in E : \|u\| \le 1\}$ is the closed unit ball in E. In case $C = E$ we write $|\!|\!|\partial f(x)|\!|\!| := |\!|\!|\partial f(x)|\!|\!|_E$.

After [CK1, Proposition 3.1, Corollary 3.2] we have:

(2.62) LEMMA. *For each* $x \in C$,

(2.62.1) $0 \leq \|\partial f(x)\|_C = \sup_{u \in -T_C(x)} \langle p_0, u \rangle \leq \|p_0\|$ *for some* $p_0 \in \partial f(x)$;
(2.62.2) $\|\partial f(x)\|_C = -\inf_{u \in T_C(x)} f^\circ(x; u)$;
(2.62.3) $\|\partial f(x)\|_C = 0$ *if and only if* $0 \in \partial f(x) + N_C(x)$; *moreover,*

$$\|\partial f(x)\|_C \leq \inf_{q \in \partial f(x) + N_C(x)} \|q\|;$$

(2.62.4) *if* F *is a Banach space,* $M \subset F$ *is closed convex,* $A\colon F \to E$ *is a linear continuous bijection such that* $C = A(M)$, *then, for all* $x \in M$,

$$\|\partial(f \circ A)(x)\|_M \leq \|A\| \|\partial f(x)\|_C;$$

(2.62.5) *the function* $C \ni x \mapsto \|\partial f(x)\|_C$ *is lower semicontinuous;*
(2.62.6) *for any* $p \in \partial f(x)$, $\|\partial f(x)\|_C \leq \|p\|$; *in particular, if* $\|\partial f(x)\|_C > 0$, *then*

$$\partial f(x)^{--} = \bigcup_{\lambda \geq 0} \lambda \partial f(x).$$

For any $a > 0$, let $f^a := \{x \in E : f(x) < a\}$ be the (open) *sublevel set* of f. Additionally let $f^a_C := f^a \cap C$.

We say that the set K, represented by f, i.e. $K = \{x \in E : f(x) \leq 0\}$, is *strictly regular with respect to* C if $K \cap C \neq \emptyset$ ([18]), there is $a = a(C) > 0$ and a constant $m_C > 0$ such that

$$\|\partial f(x)\| \geq m_C \quad \text{for all } x \in f^a_C \setminus K. \tag{2.63}$$

In case K is strictly regular with respect to E, then we say that it is *strictly regular*.

(2.64) REMARK. Above, for the simplicity of exposition we considered a function f defined on the whole space E. A more general definition (where f is defined only on an open subset of E) is given in [CK1]. Moreover, in [CK1] the authors study the concept of a *regular* set. We shall not dwell upon this issue since the equilibrium theory on regular sets lacks the elegance of that on strictly regular sets.

(2.65) REMARK. If K is strictly regular with respect to C, then $\{x \in f^a : 0 \in \partial f(x) + N_C(x)\} \subset K$; this means that f has no critical points with respect to C in $f^a \setminus K$.

([18]) Hence, for all $a > 0$, $f^a_C \neq \emptyset$.

(2.66) REMARK. For an *a priori* given closed subset K of E, there are many (locally) Lipschitz functions g representing it; the most obvious is the distance function d_K. In case $E = \mathbb{R}^n$, there is even a C^∞-function with this property. However, in general such functions would not satisfy property (2.63) with respect to a given closed convex C. That is why we consider this property to be decisive. Let us note in however that if $K \subset \mathbb{R}^n$ is strictly regular with f as its representing function, then it is possible (see [CK2]) to define a locally Lipschitz function $g\colon \mathbb{R}^n \to \mathbb{R}$ such that $K = \{x \in \mathbb{R}^n : g(x) \le 0\}$, g is C^∞ on $\mathbb{R}^n \setminus K$ and $\partial g(x) \subset \partial f(x)$ on $\mathbb{R}^n$. Hence a strictly regular set $K \subset \mathbb{R}^n$ may always be represented by a locally Lipschitz function $g\colon \mathbb{R}^n \to \mathbb{R}$ such that $g \in C^\infty(\mathbb{R}^n \setminus K)$ and such that $\|\nabla g(x)\| \ge m$ for $x \in g^a$ $(a > 0)$.

Let us collect some examples of strictly regular sets.

(2.67) EXAMPLE.

(2.67.1) The set K in Example (2.45) is strictly regular; the set K in Example (2.46) is not.

(2.67.2) Suppose $K \subset E$ is *proximinal*, i.e. there is $a > 0$ such that, for $y \in E$, $d_K(y) < a$ (i.e. $y \in d_K^a$), $\pi_K(y) = \{z \in K : \|y - z\| = d_K(y)\} \neq \emptyset$. If, for any $y \in d_K^a$,

$$\pi_K(y) \cap \operatorname*{Lim\,inf}_{z \to y} \pi_K(z) \neq \emptyset,$$

then K, represented by d_K, is strictly regular.

Indeed, take $y \in d_K^a \setminus K$, let $\overline{x} \in \pi_K(y) \cap \operatorname{Lim\,inf}_{z\to y} \pi_K(z)$, put $u := \overline{x} - y$ and take sequences $y_n \to y$, $h_n \to 0^+$. There is $\overline{x}_n \in \pi_K(y_n)$ such that $\overline{x}_n \to \overline{x}$. Hence

$$d_K(y_n + h_n u) - d_K(y_n) \le h_n\|(\overline{x}_n - y_n) - u\| + d_K(y_n + h_n(\overline{x}_n - y_n)) - d_K(y_n).$$

One checks easily that $d_K(y_n + h_n(\overline{x}_n - y_n)) = (1 - h_n)d_K(y_n)$; therefore $d_K^\circ(y; u) \le -\|u\|$ and $\|\partial d_K(y)\| \ge 1$.

(2.67.3) A proximate retract K (with the metric retraction defined on an ε-neighbourhood of K) is strictly regular; in this case π_K is a single-valued and continuous.

(2.67.4) Let $K \subset E$, represented by d_K, be closed and convex. Then:

(2.67.4.1) K is strictly regular;

(2.67.4.2) if $C \subset E$ is convex, $0 \in \operatorname{int}(K - C)$ and C or K is bounded, then K is strictly regular with respect to C.

To see (2.67.4.1) take $x \notin K$. It is well-known (see e.g. [AE]) that

$$d_K^\circ(x; u) \le d_K(x + u) - d_K(x), \quad u \in E. \tag{2.68}$$

For any $\varepsilon > 0$, let $x_\varepsilon \in K$ be such that $\|x - x_\varepsilon\| < d_K(x) + \varepsilon$ and put $u_\varepsilon := x_\varepsilon - x$. Then

$$d_K^\circ\left(x; \frac{u_\varepsilon}{\|u_\varepsilon\|}\right) \leq \frac{-d_K(x)}{\|u_\varepsilon\|} < \frac{-d_K(x)}{d_K(x) + \varepsilon}.$$

By (2.62.1), $\|\partial d_K(x)\| \geq 1$.

In case (2.67.4.2) suppose that the closed ball $D(0, \varepsilon) \subset K - C$ where $\varepsilon > 0$. Take $x \in C \setminus K$ and choose $y \in K$ such that $0 < \|x - y\| < 2d_K(x)$. There is $\mu > 0$ such that $\|\mu(y - x)\| = \varepsilon$. Hence $\mu(x - y) \in D(0, \varepsilon) \subset K - C$, i.e., there are $\overline{y} \in K$ and $\overline{x} \in C$ such that $\mu(x - y) = \overline{y} - \overline{x}$. It is easy to check that

$$z := \left(1 - \frac{\mu}{\mu+1}\right)\overline{x} + \frac{\mu}{\mu+1}x = \left(1 - \frac{\mu}{\mu+1}\right)\overline{y} + \frac{\mu}{\mu+1}y \in K \cap C.$$

Let $u := z - x$. Clearly $u \in T_C(x)$ and $\mu(z - x) = \mu(\mu+1)^{-1}(\overline{x} - x)$, hence $\mu\|u\| < \|\overline{x} - x\| \leq R$ where R is the diameter of C (in case K is bounded, then $\mu\|u\| \leq R' + \varepsilon$ where R' is the diameter of K). By (2.68), $d_K^\circ(x; u) \leq d_K(x+u) - d_K(x) = -d_K(x)$ and

$$d_K^\circ\left(x; \frac{u}{\|u\|}\right) \leq -\frac{d_K(x)}{\|u\|} < -\frac{\|x - y\|}{2\|u\|} = -\frac{\varepsilon}{2\mu\|u\|} \leq -\frac{\varepsilon}{2R}.$$

Hence $\|\partial d_K(x)\|_C \geq \varepsilon/2R$ (19).

An important role in optimization is played by the so-called epi-Lipschitz sets. This notion (in the finite-dimensional context) has been introduced by Rockafellar [Ro]. The corresponding notion for subsets of a Banach space has been studied in [BK1].

We say that a closed set $K \subset E$ is *epi-Lipschitz* if, for every $x_0 \in D$, there are a neighbourhood U of x_0 (in E), a Banach space Z, a topological isomorphism $A\colon Z \times \mathbb{R} \to E$ with $A(z_0, \lambda_0) = x_0$, a locally Lipschitz function $g\colon Z \to \mathbb{R}$ such that $D \cap U = U \cap A(\operatorname{Epi} g)$ where $\operatorname{Epi} g := \{(z, \lambda) : g(z) \leq \lambda\}$ is the *epigraph* of g.

(2.69) PROPOSITION ([Cl2], [Ro], [CC]). *If $K \subset E = \mathbb{R}^n$ is closed, then the following conditions are equivalent:*

(2.69.1) *K is epi-Lipschitz;*

(2.69.2) *for any $x \in K$, $\operatorname{int} C_K(x) \neq \emptyset$ (or, equivalently, the Clarke normal cone $N_K(x) := C_K(x)^-$ is pointed, i.e. $N_K(x) \cap (-N_K(x)) = \{0\}$);*

(2.69.3) *for any $x \in \partial K$, $C_K(x) = \partial\Delta_K(x)^-$, where, for $y \in \mathbb{R}$, $\Delta_K(y) := d_K(y) - d_{E\setminus \operatorname{int} K}(y)$ and $0 \notin \partial\Delta_K(x)$.*

Implication (2.69.1) $\Rightarrow$ (2.69.2) is obvious. Implication (2.69.3) $\Rightarrow$ (2.69.2) follows since if $0 \notin \partial\Delta_K(x)$ for $x \in \partial K$, then $\partial\Delta_K(x)^- \subset C_K(x)$ (see [Cl2,

(19) Is it possible to avoid the boundedness assumption?

Theorem 2.4.7]); hence, $N_K(x) \subset (\partial\Delta_K(x)^-)^- = \bigcup_{\lambda\geq 0}\lambda\partial\Delta_K(x)$ and $N_K(x)$ is pointed. Both these facts hold if $\dim E = \infty$. But in order to prove implications (2.69.2) $\Rightarrow$ (2.69.1) and (2.69.2) $\Rightarrow$ (2.69.3) one needs finite dimensional arguments. The author does not know whether they hold when $\dim E = \infty$. The partial answer is given by the following result obtained in [BaK2].

(2.70) PROPOSITION. *Suppose that* $\dim E \leq \infty$ *and let* $K \subset E$ *be closed. Then* (2.69.1) $\Rightarrow$ (2.69.3). *Moreover, if* K *is epi-Lipschitz and compact (i.e.* $K \subset \mathbb{R}^n$ *since evidently* $\operatorname{int} K \neq \emptyset$*), then it is strictly regular.*

More generally (see [CK1]) we have the following example.

(2.71) EXAMPLE. Suppose that $K \subset E$ is epi-Lipschitz and let $C \subset E$ be compact convex and such that $M \cap C \neq \emptyset$. One checks easily that if $0 \notin A^*N_C(A(z_0,\lambda_0)) + \partial g(z_0) \times \{-1\}$ for any $x_0 = A(z_0,\lambda_0) \in \operatorname{bd} K \cap C$ (where A comes from the definition), then K is strictly regular w.r.t. C. It follows that also infinite-dimensional epi-Lipschitz sets may be studied in the framework of the present paper.

The basic tool for the further study of strictly regular sets is a version of the so-called *deformation lemma*. Deformation results for C^1-functionals form a very well-known topic in variational calculus, see [Wi] for example. A result in this direction for locally Lipschitz functions on reflexive Banach spaces was established for the first time by Chang [C1] and, independently, in a finite-dimensional context, by Bonniseau–Cornet [BC]. Below, after [CK1] (see also [CK2]), we provide a result generalizing both [C1] and [BC].

(2.72) THEOREM. *If a closed set* $K \subset E$*, represented by a locally Lipschitz function* $f: E \to \mathbb{R}$*, is strictly regular with respect to a closed convex set* $C \subset E$*, then* $K \cap C$ *is a strong deformation of* f_C^b *for any* $b \in [0,a]$ *where* $a = a(C)$.

Recall that a closed subset A of a (topological) space X is a *strong deformation retract* of X if there is a retraction $r: X \to A$ such that $i \circ r: X \to X$, where $i: A \hookrightarrow X$ is the inclusion, is homotopic relative to A with the identity I_X on X, see e.g. [Sp, Chapter 1.4]. If A is a strong deformation retract of X, then i induces an isomorphism $\check{H}^*(X) \cong \check{H}^*(A)$ (obviously this holds for any (co)homology theory).

PROOF. We shall only give a sketch of a rather long and tedious proof. Firstly one constructs (see [CK1, Theorem 4.1]) a locally Lipschitz vector field $V: E\setminus K \to E$ such that:

(2.73.1) for all $x \in E \setminus K$, $\|V(x)\| \leq 2$;
(2.73.2) for all $x \in C \setminus K$, $V(x) \in -T_C(x)$ and

(2.73.3) for all $x \in f_C^a \setminus C$,

$$\inf_{p \in \partial f(x)} \langle p, V(x) \rangle > \frac{1}{2} m_C.$$

It is easily seen that V plays a role of a *pseudo-gradient field* studied in the context of deformation results is the calculus of variations (see e.g. [Wi]).

For any $x \in E \setminus K$, the Cauchy problem

$$\begin{cases} \frac{d}{dt}\sigma(t,x) = -V(\sigma(t,x)), \\ \sigma(0,x) = x, \end{cases}$$

has a unique solution $\sigma(\,\cdot\,, x)$ in $E \setminus K$ defined on the (right) maximal interval of existence $[0, T(x))$; moreover σ is continuous on the set $U := \{(t,x) \in [0,\infty) \times (E \setminus K) : 0 \le t < T(x)\}$. It is well-known that the function $E \setminus K \ni x \mapsto T(x) \in [0, \infty]$ is lower semicontinuous; hence U is open (in $[0,\infty) \times (E \setminus K)$).

For each $x \in C \setminus K$, $-V(x) \in T_C(x)$; hence in virtue of the Nagumo type viability theorem (comp. e.g. [Bre] or [M2]),

$$\sigma(t,x) \in C \quad \text{for } x \in C \setminus K,\ t \in [0, T(x)). \tag{2.74}$$

For all $x \in f_C^a \setminus K$ and $0 \le t < T(x)$, $\sigma(t,x) \in f_C^a$. Indeed, by (2.74), it is enough to show that $f(\sigma(t,x)) < a$. The function $g = f \circ \sigma(\,\cdot\,, x)$, defined on $[0, T(x))$, is positive, differentiable almost everywhere and

$$g'(t) \le \sup_{p \in \partial f(\sigma(t,x))} \left\langle p, \frac{d}{dt}\sigma(t,x) \right\rangle$$

for almost all $t \in [0, T(x))$. Suppose to the contrary that, for some $t_0 \in [0, T(x))$, $g(t_0) = a$ and $f(\sigma(t,x)) < a$ on $[0, t_0)$. Hence $g'(t) \le -m_C/2$ for almost all $t \in [0, t_0)$. Thus, for all such t,

$$-a < -f(x) = -g(0) < g(t) - g(0) = \int_0^t g'(s)ds \le -\frac{1}{2} m_C t. \tag{2.75}$$

If $t \to t_0^-$, then by continuity, $f(\sigma(t_0,x)) = g(t_0) \le g(0) - m_C t_0/2 < a$, a contradiction.

It also follows that (2.75) holds for all $t \in [0, T(x))$. Passing in (2.75) with $t \to T(x)$, we obtain

$$T(x) \le \frac{2f(x)}{m_C} < \infty. \tag{2.76}$$

If $x \in f_C^a \setminus K$, then the inequality

$$\|\sigma(t_2, x) - \sigma(t_1, x)\| \leq \int_{t_1}^{t_2} \left\| \frac{d}{dt}\sigma(t, x) \right\| dt \leq 2(t_2 - t_1),$$

satisfied for all $t_1 < t_2$ from $[0, T(x))$, shows that $\sigma^*(x) := \lim_{t \to T(x)^-} \sigma(t, x) \in f_C^a$ is well-defined. Clearly $f(\sigma^*(x)) = 0$; for if $f(\sigma^*(x)) > 0$, then $\sigma(\,\cdot\,, x)$ can be extended beyond $T(x)$.

Using (2.75), one shows that the function $f_C^a \ni x \mapsto T(x) < \infty$ is continuous and defines a map $\eta\colon f_C^a \times [0,1] \to f_C^a$ by

$$\eta(x, t) = \begin{cases} x & \text{for } (x, t) \in (K \cap C) \times [0, 1], \\ \sigma(tT(x), x) & \text{for } t \in [0, 1),\ x \in f_C^a \setminus K, \\ \sigma^*(x) & \text{for } t = 1,\ x \in f_C^a \setminus K. \end{cases}$$

After verifying the continuity of η it is clear that η provides a homotopy (relative to $K \cap C$) joining the identity (on f_C^a) to $i \circ r$, where the retraction $r\colon f_C^a \to K \cap C$ is given on f_C^a by

$$r(x) = \begin{cases} x & \text{for } x \in K \cap C, \\ \sigma^*(x) & \text{for } x \in f_C^a \setminus K, \end{cases} \tag{2.77}$$

and $i\colon K \cap C \hookrightarrow f_C^a$ is the inclusion. This concludes the proof. □

Observe that, by (2.76), for $x \in f_C^a$,

$$\|r(x) - x\| \leq \frac{4f(x)}{m_C}. \tag{2.78}$$

Compact strictly regular sets belong to the class of $\mathcal{L}$-retracts. More generally the following holds.

(2.79) PROPOSITION. *Suppose K is strictly regular with respect to C. Then*

(2.79.1) *$K \cap C$ is a neighbourhood retract in E,*

(2.79.2) *for each $x \in K \cap C$ and $u \in T_C(x)$,*

$$d^\circ_{K\cap C}(x; u) \leq \frac{4}{m_C} \limsup_{y \xrightarrow{K\cap C} x,\, h \to 0^+} \frac{f(y + hu)}{h}, \tag{2.79.2.1}$$

(2.79.3) *if f is Lipschitz continuous on a neighbourhood of $K \cap C$ ([20]), then $K \cap C$ is an $\mathcal{L}$-retract.*

PROOF. (2.79.1) Since C is convex, there is a retraction $\rho\colon E \to C$ such that

$$\|\rho(x) - x\| \leq 2d_C(x), \quad x \in E,$$

([20]) This is always the case if $K \cap C$ is compact.

(see e.g. [BP, Corollary II.3.4] or [BK1]). Put $U := \rho^{-1}(f_C^a)$ and let

$$R := r \circ (\rho|U): U \to K \cap C$$

where $r: f_C^a \to K \cap C$ is defined by (2.77). It is clear that U is an open neighbourhood (in E) of $K \cap C$ and R is a well-defined retraction of U onto $K \cap C$. Moreover, $U \cap C = f_C^a$. In view of (2.78), for $x \in f_C^a$,

$$\|R(x) - x\| = \|r(\rho(x)) - x\| = \|r(x) - x\| \leq \frac{4f(x)}{m_C}. \tag{2.80}$$

(2.79.2) A bit involved proof here uses inequality (2.80); for the details see [CK1, Proposition 4.3].

(2.79.3) Suppose that f is Lipschitz (with the constant L) on a neighbourhood V of $K \cap C$. For any $x \in U_R \cap V$,

$$\begin{aligned}\|R(x) - x\| &\leq \|r(\rho(x)) - \rho(x)\| + \|\rho(x) - x\| \\ &\leq \frac{4}{m_C} f(\rho(x)) + 2d_C(x) \leq \frac{4}{m_C}[f(\rho(x)) - f(z)] + 2d_C(x) \\ &\leq \frac{4L}{m_C}\|\rho(x) - z\| + 2d_C(x) \leq \left(\frac{8L}{m_C} + 2\right) d_C(x) + \frac{4L}{m_C}\|x - z\|,\end{aligned}$$

where z is any point from $K \cap C$. Since $d_C(x) \leq d_{K\cap C}(x)$ and $z \in K \cap C$ is arbitrary,

$$\|R(x) - x\| \leq \left(\frac{12L}{m_C} + 2\right) d_{K\cap C}(x). \qquad \square$$

(2.81) REMARK. If K is strictly regular w.r.t. C, then:

(2.81.1) since $K \cap C$ is a neighbourhood retract of E, its Euler characteristic $\chi(K \cap C)$ is well-defined provided $K \cap C$ is compact;

(2.81.2) in view of (2.79.2.1), for any $x \in K \cap C$,

$$T_C(x) \cap \partial f(x)^- \subset C_{K\cap C}(x).$$

As said above, we shall now consider the existence of solutions to problem (P) where $f: E \to \mathbb{R}$ is a locally Lipschitz, the set $K := \{x \in E : f(x) \leq 0\}$ is nonempty and $\varphi: K \multimap E$ is an upper hemicontinuous set-valued map with closed convex values. If K is strictly regular and compact, then in view of (2.79.3), K is an $\mathcal{L}$-retract and, by Theorem (2.57), φ has an equilibrium in K (being a solution to (P)) provided $\chi(K) \neq 0$ and φ satisfies the weak tangency condition.

Unfortunately, sets of the form (S) are rarely compact. To relax compactness we shall assume that there exists a closed convex set C which controls directions admitted by φ, i.e. for any $x \in K \cap C$, $\varphi(x) \cap T_C(x) \neq \emptyset$. In what follows, given a closed convex set $C \subset E$ and a set $A \subset C$, denote by $\operatorname{cl}_C A$, $\operatorname{int}_C A$ and $\operatorname{bd}_C A$ the *closure*, the *interior* and the *boundary* of A *relative to* C, respectively. Clearly $x \in \operatorname{bd}_C A$ if and only if $x \in \operatorname{cl}_C A$ and there is a sequence (x_n) in $C \setminus A$ converging to x.

We have (see [CK1, Theorem 5.2]) the main result of this subsection.

(2.82) THEOREM. *Let $K \subset E$, represented by a locally Lipschitz function $f: E \to \mathbb{R}$, be nonempty and strictly regular with respect to a closed convex set $C \subset E$ which controls directions admitted by an upper hemicontinuous set-valued map $\varphi: K \multimap E$ with closed convex values. If $K \cap C$ is compact, the Euler characteristic $\chi(K \cap C) \neq 0$ and*

$$(2.82.1) \qquad \inf_{u \in \varphi(x) \cap T_C(x)} \limsup_{y \xrightarrow{K \cap C} x,\, h \to 0^+} \frac{f(y+hu)}{h} \le 0 \quad \text{for all } x \in \operatorname{bd}_C(K \cap C),$$

then φ has an equilibrium in K.

Before an easy proof let us discuss the hypotheses and the meaning of assumption (2.82.1).

(2.83) REMARK. Remember that the entire space controls directions admitted by any map. Moreover, given a map $\varphi: K \multimap E$ such that, for each $x \in K$, $\varphi(x) \cap C \neq \emptyset$ and $K \cap C \neq \emptyset$, then $(\varphi(x) - x) \cap T_C(x) \neq \emptyset$ on $K \cap C$. Hence in case C is compact and $\varphi(x) \cap C \neq \emptyset$ on K, then Theorem (2.82) corresponds nicely to Theorem (2.30) (especially in view of Example (2.67.4)).

(2.84) REMARK. First observe that $\operatorname{bd}_C(K \cap C) \subset \operatorname{bd} K \cap C \subset K \cap C$. Hence, replacing $\operatorname{bd}_C(K \cap C)$ by any of these supersets, we would obtain conditions implying (2.82.1). Next, consider the following, more familiar and friendly looking, conditions: for all $x \in \operatorname{bd}_C(K \cap C)$,

$$(2.84.1) \qquad \varphi(x) \cap T_C(x) \cap \partial f(x)^- \neq \emptyset,$$

$$(2.84.2) \qquad \sup_{p \in \partial f(x)^{--}} \inf_{u \in \varphi(x) \cap T_C(x)} \langle p, u \rangle \le 0,$$

$$(2.84.3) \qquad \sup_{p \in \partial f(x)} \inf_{u \in \varphi(x) \cap T_C(x)} \langle p, u \rangle \le 0,$$

$$(2.84.4) \qquad \inf_{u \in \varphi(x) \cap T_C(x)} f^\circ(x; u) \le 0.$$

Arguing as in the proof of Lemma (2.16) we see that (2.84.1) $\Rightarrow$ (2.84.2) $\Rightarrow$ (2.84.3) $\Leftrightarrow$ (2.84.4). If $\varphi(x)$ is bounded, then (2.84.3) $\Rightarrow$ (2.84.2) and, if $\varphi(x)$ is weakly compact and convex, then (2.84.4) $\Rightarrow$ (2.84.1). Since, in general, one has

$$\limsup_{y \xrightarrow{K \cap C} x,\, h \to 0^+} \frac{f(y+hu)}{h} \le f^\circ(x; u)$$

for any $x \in K\cap C$ and $u \in E$, condition (2.84.4) is stronger than (2.82.1). Observe also that condition (2.84.1), in view of Remark (2.81.2) implies that φ is weakly tangent to $K \cap C$ (this corresponds to Theorem (2.52)).

All above conditions are stated in terms of problem's (P) data as required and are sufficient for the existence of equilibria. But condition (2.82.1) is, in general, the weakest among all others.

(2.85) REMARK. The relevance of the assumption concerning the nontriviality of the Euler characteristic has been already discussed in the context of equilibria on $\mathcal{L}$-retracts. In case of a general $\mathcal{L}$-retract this assumption is hard to verify. But in the present setting we are in a better situation. To see this let K and C be as in Theorem (2.82).

If, for example $0 \in K\cap C$ and the representing function f is *convex in rays*, i.e. $f(\lambda x) \le \lambda f(x)$ for any x and $\lambda \in [0,1]$, then $K\cap C$, being star-shaped around 0, is contractible and $\chi(K\cap C) = 1$. In view of Theorem (2.72), For each $b \in [0,a]$, $\chi(f_C^a) = \chi(K\cap C)$. In particular, if $C \subset f^a$, then $f_C^a = C$ and, thus, $\chi(K\cap C) = 1$.

In general, in order to provide means to compute $\chi(K\cap C)$, assume that the representing function f is C^1-smooth, let $b \in (0,a)$ and suppose that each point $z \in f_C^b$ such that $-f'(z) \in N_C(z)$ is isolated. Since all such points belong to $K\cap C$, we infer that there are finitely many of them: $\{z_1,\ldots,z_k\}$. For any $j = 1,\ldots,k$, the so-called *critical group* $c^*(f,z_j)$ is defined (see [C2, Sections 4 and 6.2]) being a graded vector space over rationals. For $q \ge 0$, let $m^q := \sum_{j=1}^k \dim c^q(f,z_j)$ be the *Morse type number*. One of the simple consequences of the so-called *Morse inequalities* (again see [C2, Section 6.2]), applied to our situation, asserts that $\chi(f_C^a) = \sum_{q\ge 0}(-1)^q m^q$.

If $E = \mathbb{R}^n$, then, using a suitable approximation procedure [CK2] and the Sard theorem, the similar computation may be provided for arbitrary (locally Lipschitz) function f representing K.

PROOF OF THEOREM (2.82). Observe that instead of (2.82.1), we may assume that

$$\inf_{u\in\varphi(x)\cap T_C(x)} \limsup_{y \xrightarrow{K\cap C} x,\, h\to 0^+} \frac{f(y+hu)}{h} \le 0 \quad \text{for all } x \in K\cap C. \tag{2.86}$$

To this end, note that $K\cap C = \operatorname{bd}_C(K\cap C)\cup \operatorname{int}_C(K\cap C)$ and if $x \in \operatorname{int}_C(K\cap C)$, then

$$\limsup_{y \xrightarrow{K\cap C} x,\, h\to 0^+} \frac{f(y+hu)}{h} \le 0$$

for any $u \in T_C(x)$. Indeed, if $y_n \xrightarrow{K\cap C} x$, $h_n \to 0^+$, then there is a sequence $u_n \to u$ such that $y_n + h_n u_n \in C$. For large n, $f(y_n + h_n u) \le f(y_n + h_n u_n) + h_n\|u_n - u\| \le h_n\|u_n - u\|$ since $y_n + h_n u_n \in K\cap C$ and the desired inequality follows immediately.

In view of Proposition (2.79.3), $K \cap C$ is a compact $\mathcal{L}$-retract; by (2.79.2.1), (2.86) and since C controls directions admitted by φ, for any $x \in K \cap C$,

$$\inf_{u \in \varphi(x)} d^{\circ}_{K\cap C}(x)(u) \leq \frac{4}{m_C} \inf_{u \in \varphi(x) \cap T_C(x)} \limsup_{y \xrightarrow{K\cap C} x,\, h \to 0+} \frac{f(y+hu)}{h} \leq 0.$$

All assumptions of Theorem (2.57) are satisfied and φ has an equilibrium (in $K \cap C$). □

(2.87) REMARK. In view of Proposition (2.79) (see (2.79.2.1)) it is clear that if K is compact and strictly regular condition (2.86) is stronger than (2.19.1); hence Theorem (2.82) (for strictly regular and compact K) provides yet another special version of Theorem (2.57) with assumptions stated in terms of the constraint f. The advantage of Theorem (2.82) is that it enables us to relax the compactness; it was unfortunately not possible in the context of $\mathcal{L}$-retracts.

The facts stated in this subsection generalize most of the above results: each compact convex set $K \subset E$ is strictly regular (for $f = d_K$) (see Example (2.67)), $\chi(K) = 1$, conditions imposed in [Co1] are equivalent to (2.84.2) and imply (2.82.1). It also extends to the infinite-dimensional context results from [CLS], [CC] (except for the necessity aspects), [Co2] and those of [BC] being valid for compact epi-Lipschitz subsets of $\mathbb{R}^n$.

In the setting of Theorem (2.82) it is also possible to obtain the existence of equilibria (or fixed points) of admissible maps.

(2.88) THEOREM. *Let K and C be as in Theorem* (2.82). *Suppose $\varphi \colon K \multimap E$ is an admissible map such that $\varphi(x) \subset T_C(x)$ for all $x \in K \cap C$. If, for any $x \in \operatorname{bd}_C(K \cap C)$, $\varphi(x) \subset \partial f(x)^-$, then φ has an equilibrium.*

The proof is similar. In view of Remark (2.81.2), for all $x \in K \cap C$, $\varphi(x) \subset C_{K\cap C}(x)$. By Remarks (2.59), (2.60), the conclusion follows.

Now we pass to the generalized equilibrium problem.

(2.89) REMARK. Suppose K, represented by f, is strictly regular with respect to a closed convex C. Let $\Phi \colon K \multimap E^*$ be an upper hemicontinuous map with weak*-compact convex values. Theorem (2.54) together with Proposition (2.79.3) imply that there is $x_0 \in K \cap C$ such that $\Phi(x_0) \cap N_{K\cap C}(x_0) \neq \emptyset$ provided $K \cap C$ is compact and $\chi(K \cap C) \neq 0$. Reasoning as at the beginning of the proof of Theorem (2.82) and using (2.79.2.1), we see that if $x_0 \in \operatorname{int}_C(K \cap C)$ (clearly $\operatorname{int} K \cap C \subset \operatorname{int}_C(K \cap C)$), then $N_{K\cap C}(x_0) = N_C(x_0)$.

Under stronger assumptions we get the following result.

(2.90) THEOREM. *Assume that K, C and Φ are as above. If $0 \notin \partial f(x)+N_C(x)$ whenever $x \in \operatorname{bd}_C(K\cap C)$, then at least one of the following conditions is satisfied*

(2.90.1) $$\exists x_0 \in \operatorname{int}_C(K\cap C)\ \Phi(x_0)\cap N_C(x_0) \neq \emptyset;$$

(2.90.2) $$\exists x_0 \in \operatorname{bd}_C(K\cap C)\ \Phi(x_0)\cap [N_C(x_0)+\partial f(x_0)^{--}] \neq \emptyset.$$

PROOF. By Remark (2.89), there is $x_0 \in K\cap C$ and $p_0 \in \Phi(x_0)$ such that $p_0 \in N_{K\cap C}(x_0)$. If $x_0 \in \operatorname{int}_C(K\cap C)$, then $N_{K\cap C}(x_0) = N_C(x_0)$, hence $p_0 \in N_C(x_0)$. If $x_0 \in \operatorname{bd}_C(K\cap C)$, in view of Remark (2.81.2)

$$\begin{aligned} p_0 \in N_{K\cap C}(x_0) &\subset [T_C(x_0)\cap \partial f(x_0)^-]^- \\ &= w^*\text{-cl}\,[N_C(x_0)+\partial f(x_0)^{--}] = N_C(x_0)+\partial f(x_0)^{--}. \end{aligned}$$

The first equality is standard and the second one follows from the following abstract lemma (for the proof see [CK]). □

(2.91) LEMMA. *Let $A \subset E^*$ be a weak*-closed cone, let $C \subset E^*$ be weak*-compact, $B = C^{--}$, and suppose that $0 \notin A+C$. Then $w^*\text{-cl}\,(A+B) = A+B$.*

Let us conclude with the following application. Suppose that $V\colon \mathbb{R}^n \to \mathbb{R}$ is a continuously differentiable *coercive* functional, i.e. such that

$$\lim_{|x|\to\infty} V(x) = \infty.$$

Hence, for any $\alpha \in \mathbb{R}$, the (closed) sublevel set $V_\alpha := \{x \in \mathbb{R}^n : V(x) \le \alpha\}$ is compact. Further suppose that there is $r_0 > 0$ such that the gradient $\nabla V(x) \neq 0$ for $x \in \mathbb{R}^N$ with $|x| \ge r_0$. Therefore, for any $r \ge r_0$, the Brouwer degree $\deg(\nabla V, B(0,r), 0)$ of ∇V on the ball $B(0,r)$ with respect to 0 is defined and, in fact, does not depend on r. Let us put

$$\operatorname{ind} V := \deg(\nabla V, B(0,r), 0), \quad r \ge r_0.$$

The coercivity of V implies that $\operatorname{ind} V = 1$ (see [KZ, Theorem 12.9]). For an arbitrary $M > \sup_{|x|\le r_0} V(x)$, we have $D(0,r_0) \subset V_M$ and, for $x \in \operatorname{bd} V_M = \{x \in \mathbb{R}^n : V(x) = M\}$, $\nabla V(x) \neq 0$. Hence V_M is a C^1-smooth manifold with boundary $\operatorname{bd} V_M$ and, for $x \in \operatorname{bd} V_M$, $T_{V_M}(x) = \{y \in \mathbb{R}^n : \langle y, \nabla V(x)\rangle \le 0\} = \{\nabla V(x)\}^-$.

The celebrated Poincaré–Hopf formula (see e.g. [C2, Theorem 3.1]) asserts that the Euler characteristic $\chi(V_M) = \operatorname{ind} V = 1$ since one may (in view of the Sard theorem) assume that critical points of V (i.e. zeros of ∇V) are isolated.

(2.92) THEOREM. *Suppose that $\varphi\colon \mathbb{R}^n \multimap \mathbb{R}^n$ is upper hemicontinuous and has convex closed values. If*

(2.92.1) $$\inf_{y\in\psi(x)} \langle \nabla V(x), y\rangle \le 0 \quad \text{for } |x| \ge r_1,$$

then φ has an equilibrium ([21]).

([21]) If condition (2.92.1) is satisfied, then we sometimes say that V is the guiding potential for φ.

PROOF. Taking large M (such that $M > \sup_{|x|\le r} V(x)$ where $r = \max\{r_0, r_1\}$), we see that the set $K := V_M$ is a compact set of the form (S) (it is represented by $f(x) = V(x) - M$ such that $0 \notin \partial f(x)$ for $x \in \operatorname{bd} K$); in particular K is strictly regular set with $\chi(K) \neq 0$. Our assumption implies that condition (2.84.1) is satisfied. Hence the assertion follows from Theorem (2.82). □

One may easily replace the differentiability of V by the assumption that V is locally Lipschitz and consider its generalized gradient instead of ∇V in order to obtain direct generalizations of Theorem (2.92) (the presented proofs remain unchanged). In this case the only (serious) difficulty is to show that $\chi(V^M) = \operatorname{ind} V$ (a version of the Poincaré–Hopf formula). It may be done by means of the smoothing technique due to Ćwiszewski and the present author in [CK2].

3. Metric aspects of the theory

When discussing the existence of constrained equilibria or fixed points of a map $\varphi\colon K \multimap E$ satisfying the Lipschitz condition

$$d_H(\varphi(x), \varphi(y)) \le k\|x - y\|, \quad x, y \in K$$

one has to distinguish two cases: $0 \le k < 1$ and $k \ge 1$. The first one may be studied in a fairly general setting of metric spaces. The second one requires more attention and seems to be interesting only for a noncompact domain. If K is compact, then we have the following result.

(3.2) THEOREM. *Let $K \subset E$ be a compact neighbourhood retract and $\varphi\colon E \multimap E$ with convex compact values satisfy condition* (3.1) *with $k \ge 1$ and the* (*strong*) *tangency* (2.13.1). *Then φ has an equilibrium.*

The proof (of a slightly more general result) is postponed to Section 4 (see Theorem 4.26).

3.1. Set-valued contractions. We shall start our investigations by the study of fixed points of set-valued contractions $\varphi\colon K \multimap X$ where K is a closed subset of a metric space X. We shall discuss only contractions understood in the sense recalled in Introduction leaving the so-called *generalized contractions* (in the sense of Browder and others) aside.

First we shall improve the notion of the (weak) inwardness a bit. Given a closed set $K \subset E$, the *inward set* $I_K(x)$ of K at x is defined by

$$I_K(x) := \operatorname{cl}\Big(\bigcup_{h\in(0,1)} \frac{K - x}{h}\Big).$$

It is easy to see that $I_K(x) = \mathrm{cl}\,\{v = \lambda(y - x) : y \in K,\ \lambda > 1\}$. Moreover, it is standard to see that

$$I_K(x) = \left\{ v \in E : \inf_{h\in(0,1)} \frac{d_K(x + hv)}{h} = 0 \right\} \tag{3.3}$$

and, if K is convex, then $I_K(x) = T_K(x)$ for all $x \in K$.

In general however $I_K(x)$ is not a cone and $T_K(x) \subsetneq I_K(x)$, $x \in K$. For instance if $K = S^1 \subset \mathbb{R}^2$ is the unit sphere, then $T_K(x) = \{v \in \mathbb{R}^2 : \langle v, x\rangle = 0\}$ while $I_K(x) = \{v \in \mathbb{R}^2 : \|v + x\| \geq 1 \text{ and } \langle v, x\rangle \leq 0\}$, for any $x \in K$.

The first result definitely worth mentioning is the following one (essentially due to Deimling [D2, Theorem 11.4]; see also [X2, Theorem 1.3.2] and [MY]).

(3.4) THEOREM. *Let K be a closed subset of a Banach E space and $\varphi\colon K \multimap E$ be a contraction with closed values such that each $x \in K$ admits a nearest point in $\varphi(x)$* (22). *If φ is tangentially inward, i.e. for all $x \in K$,*

$$\varphi(x) \subset x + T_K(x), \tag{3.4.1}$$

then φ has a fixed point. In case φ is single-valued a fixed point is unique.

As mentioned above, if K is not convex, then $T_K(x) \subsetneq I_K(x)$. The question of Deimling [D2] in condition (3.4.1) the cone $T_K(x)$ may be replaced $I_K(x)$ was answered in the positive by Xu [X2, Theorem 1.3.4].

(3.5) THEOREM. *If K, E and φ satisfy the same assumptions as above and φ is inward, i.e. for each $x \in K$,*

$$\varphi(x) \subset x + I_K(x), \tag{3.5.1}$$

then φ has a fixed point.

Quite recently Lim (in [L2]) answered another question of Deimling [D2] (see also [K2]) by removing the assumption that each $x \in K$ admits a nearest point in $\varphi(x)$. He has shown

(3.6) THEOREM. *Let K, E satisfy the above assumptions and let $\varphi\colon K \multimap E$ be a contraction with closed values satisfying* (3.5.1). *Then φ has a fixed point.*

(3.7) REMARK. The strong inwardness condition (3.5.1) in Theorems (3.5) and (3.6) can not be replaced by the weak inwardness:

$$\varphi(x) \cap (x + I_K(x)) \neq \emptyset \quad \text{for all } x \in K$$

(22) I.e. there is $y \in \varphi(x)$ such that $\|x - y\| = d(x, \varphi(x))$; it holds for instance if φ has compact values.

even if K is convex. To see this consider the following examples.

(3.7.1) (Deimling, [D2, Example 11.1]) Let $\varphi(x) = \{y = (y_1, y_2) \in \mathbb{R}^2 : \|y\| = 2,\ y_1 \leq 1\}$ for $x \in K := \{y \in \mathbb{R}^2 : \|y\| \leq 1\}$. Then φ is a contraction (in fact constant map) with compact contractible values such that $\varphi(x) - x \cap T_K(x) \neq \emptyset$, but φ has no fixed points.

(3.7.2) (Xu, [X2, p. 26]) Let $K = [0,1] \subset \mathbb{R}$ and $\varphi(x) = \{-1, 2\}$. Then φ is a constant map without fixed points but, for all $x \in K$, $\varphi(x) - x \cap T_K(x) \neq \emptyset$.

Nevertheless there is a result of Downing and Kirk [DK] (see also [R4]) which explains what sort of conditions are violated in the above examples.

(3.8) THEOREM. *If $K \subset E$ is closed convex $\varphi\colon K \multimap E$ is a contraction with compact values such that, for each $x \in K$, $\varphi_1(x) \cap (x + T_K(x)) \neq \emptyset$ where $\varphi_1(x) := \{y \in \varphi(x) : \|x - y\| = d(x, \varphi(x))\}$, then φ has a fixed point.*

In what follows we shall give a result which subsumes all mentioned above in that it holds for closed subset in a complete metric spaces, it allows boundary conditions relaxing weak inwardness and pertains even deeper into the nature of the above examples.

Our starting point was the following generalization of the Banach principle due to Clark [Cl1].

(3.9) THEOREM. *Let (X, d) be a complete metric space, $k \in [0,1)$ and $f\colon X \to X$ be a continuous mapping satisfying the following property: there is a number $k \in [0,1)$ and, for each $x \in X$, if $x \neq f(x)$, there is $y \in X$, $y \neq x$ such that*

$$d(x,y) + d(y, f(x)) = d(x, f(x)), \tag{3.9.1}$$

$$d(f(x), f(y)) \leq k d(x,y). \tag{3.9.2}$$

Then f admits a fixed point.

A map f satisfying above conditions (3.9.1), (3.9.2) is called (by Clarke) a *directional contraction* and condition (3.9.1) means that y belongs to the half open metric segment joining x to $f(x)$. This segment may be essentially larger than the usual linear segment joining x with $f(x)$.

This nice result generalizes the Banach principle for metric spaces being *metrically convex* (see e.g. [GK, p. 23]) i.e. for instance Banach spaces. Clarke provides examples of directional contractions which are not contractions even in the local sense.

It is to be noted here that the Lim's proof of Theorem (3.6) involves an interesting argument based on the transfinite induction, while Xu as well as Deimling and others make use of the following well-known and famous Caristi theorem which was stated implicitly in [C] (see the survey article by Kirk [K1] discussing various aspects of this result and the bibliography annotations; see also a certain generalization given in Section 4, Proposition (4.2)).

(3.10) THEOREM. *Let* (X, d) *be a complete metric space, a function* $e: X \to \mathbb{R} \cup \{\infty\}$ *be bounded from below, lower semicontinuous and not identically equal to* ∞. *If* $\psi: X \multimap X$ *is such that, for any* $x \in X$, *there is* $y \in \psi(x)$, *with*

$$e(y) + d(x, y) \leq e(x), \tag{3.10.1}$$

then ψ *has a fixed point. Additionally, if for each* $x \in X$ *and all* $y \in \psi(x)$, *inequality* (3.10.1) *holds, then there is* $x_0 \in X$ *such that* $\psi(x_0) = \{x_0\}$.

Clarke's theorem relies on the Ekeland principle (being in fact equivalent to the Caristi theorem) and the results itself was extended to the constrained set-valued maps with compact values by Song in [So]. We shall discuss his method in detail giving certain generalizations.

The Clarke approach suggests the following simple observation saying that the weak inwardness (3.7.1) implies a condition similar to the one from the Clarke theorem.

(3.11) PROPOSITION. *Let* X *be a normed space,* $K \subset X$ *and* $\varphi: K \multimap X$. *Suppose that* $x \in K$ *and* $x \notin \varphi(x)$. *If* $z \in \varphi_1(x) \cap (x + I_K(x))$ (23), *then, for each* $\alpha \in (0, 1)$, *there is* $y \in K$, $y \neq x$, *such that*

$$\alpha\|x - y\| + d(y, \varphi(x)) \leq d(x, \varphi(x)). \tag{3.11.1}$$

PROOF. Fix $\alpha > 0$. Since $z - x \in I_K(x)$, by (3.3),

$$\inf_{h \in (0,1)} \frac{d_K(x + h(z - x))}{h} = 0.$$

Next, since $x \notin \varphi(x)$ and $z \in \varphi(x)$, we see that $\|x - z\| > 0$ and there is $h \in (0, 1)$ such that

$$d_K(x + h(z - x)) < \frac{1 - \alpha}{1 + \alpha} h\|x - z\|.$$

Hence there is $y \in K$ such that

$$\|x + h(z - x) - y\| \leq \frac{1 - \alpha}{1 + \alpha} h\|x - z\|.$$

(23) φ_1 has the same meaning as in Theorem (3.8).

It is clear that $y \neq x$ and

$$\begin{aligned}\alpha\|x-y\| + d(y,\varphi(x)) &\le \alpha\|x-y\| + \|y-z\| \\ &\le \alpha(h\|x-z\| + \|x+h(z-x)-y\|) \\ &\quad + (\|y-x-h(z-x)\| + (1-h)\|x-z\|) \\ &\le (1+\alpha)\|y-x-h(z-x)\| + (1-h+\alpha h)\|x-z\| \\ &\le ((1-\alpha)h + 1 - h + \alpha h)\|x-z\| = \|x-z\| = d(x,\varphi(x)).\end{aligned}$$

This ends the proof. □

(3.12) REMARK. Suppose X, $K \subset X$, φ and x are as above. If $\varphi(x) \subset x + I_K(x)$, then for each $\alpha \in (0,1)$, there is $y \in K$, $y \neq x$ such that (3.11.1) is satisfied.

The proof of this fact is much more involved (it uses the transfinite induction in the spirit of the mentioned paper by Lim [L2] or [K2]) and will not be presented here.

(3.13) LEMMA. *If (Y,d) is a metric space, $X \subset Y$, $\varphi: X \multimap Y$ is H-upper semicontinuous (resp. lower semicontinuous), then the real function $X \ni x \mapsto f(x) := d(x,\varphi(x))$ is lower semicontinuous (resp. upper semicontinuous).*

PROOF. Let $r \in \mathbb{R}$, $A = \{x \in X : f(x) > r\}$ and $B = \{x \in X : f(x) \geq r\}$. To prove the lower (resp. upper) semicontinuity of f one has to show that A is open (resp. B is closed). Let $S(x) = \{x\} \times \varphi(x)$. Then $S: X \multimap X \times Y$ is lower semicontinuous provided so is φ. Let $V = \{(x,y) \in X \times Y : d(x,y) \geq r\}$. It is clear that V is closed and $x \in B$ if and only if $S(x) \subset V$. Hence if φ is lower semicontinuous, then B is closed. This ends the proof when φ is lower semicontinuous.

In case of H-upper semicontinuity in order to show that A is open take $x_0 \in A$. Then $\rho := f(x_0) > r$. Let $\varepsilon = (\rho - r)/3$. There is $0 < \delta < \varepsilon$ such that $\varphi(x) \subset B(\varphi(x_0),\varepsilon)$ if $x \in B(x_0,\delta)$. For such x, we easily see that $f(x) > r$, i.e. $B(x_0,\delta) \subset A$. □

Now we are ready to state and prove the main results of this section. The following theorem is a slight extension of a result essentially due to Song (see [So]).

(3.14) THEOREM. *Suppose that K is a closed subset of a complete metric space X, an H-upper semicontinuous map $\varphi: K \multimap X$ has closed values and there are numbers $0 \leq k < \alpha \leq 1$ satisfying the following property: for each $x \in K$, if $x \notin \varphi(x)$, then there is $y \in K$, $y \neq x$ such that*

(3.14.1) $$\alpha d(x,y) + d(y,\varphi(x) \leq d(x,\varphi(x)),$$

(3.14.2) $$d_H(\varphi(x),\varphi(y)) \leq k d(x,y).$$

Then φ has a fixed point.

The reader will easily see the similarities of conditions (3.14.1), (3.14.2) to Clarke's conditions of directional contractions.

PROOF. Let, for $x \in K$,

$$e(x) = \frac{1}{\alpha - k} d(x, \varphi(x)).$$

By Lemma (3.13), e is lower semicontinuous and, obviously $e \geq 0$. Suppose to the contrary that φ has no fixed points. Define $\psi: K \to K$ assigning to each $x \in K$ a point $y \in K$, $y \neq x$, satisfying condition (3.14.1). Then, for $y = \psi(x)$, by (3.14.2) and (3.14.1),

$$\begin{aligned} d(y, \varphi(y)) &\leq d(y, \varphi(x)) + d_H(\varphi(x), \varphi(y)) \\ &\leq d(y, \varphi(x)) + kd(x, y) \leq d(x, \varphi(x) - (\alpha - k)d(x, y). \end{aligned}$$

Hence $d(x, y) \leq e(x) - e(y)$. By the Caristi Theorem (3.10), ψ has a fixed point, a contradiction. □

In view of Proposition (3.11) and Remark (3.12) we get the extension of the Xu, Downing–Kirk and Lim theorems. Observe that even if $X = E$ is a Banach space, then the "boundary" condition (3.14.1) is much weaker than the weak inwardness from the Downing–Kirk and the strong inwardness (3.5.1) from the theorems by Xu and Lim.

To see this better, let E be a Banach space and, for each $\alpha \in (0, 1]$, and $x \in K$, put $J_K^\alpha := \{z \in E :$ there exists $y \in K,\ y \neq x,\ \alpha\|x - y\| + \|y - z\| \leq \|x - z\|\}$. Proposition (3.11) shows in fact that if $z \in x + I_K(x)$ and $z \neq x$, then $z \in J_K^\alpha(x)$. In general the set $\{x\} \cup J_K^\alpha(x)$ is much larger that $I_K(x)$.

Using this notation we have the following:

(3.15) COROLLARY. *If K is a closed subset of a Banach space E, $\varphi: K \to E$ is a k-contraction ($k \in [0, 1)$) such that, for each $x \in K$, either $x \in \varphi(x)$ or $\varphi_1(x) \cap J_K^\alpha(x) \neq \emptyset$ (where φ_1 has the same meaning as in Theorem (3.8)) for some $k < \alpha \leq 1$, then φ has a fixed point.*

In the next parts of this section we shall turn our attention to the existence of equilibria (and fixed points) of set-valued maps satisfying conditions of a different (geo)metric nature. In particular, we shall discuss nonexpansive mappings.

3.2. One sided Lipschitz estimates. One of the principal applications of the metric fixed point theory in a functional context has been to study monotone and accretive operators (which arise as a natural generalization of monotone maps).

In order to proceed we need to recall some auxiliary notions. Let $J\colon E \multimap E^*$ be the (normalized) duality mapping and consider the *semi-inner products* $\langle\,\cdot\,,\,\cdot\,\rangle_\pm\colon E\times E\to\mathbb{R}$ given by the formulae: for $x,y\in E$,

$$(3.16)\qquad \langle x,y\rangle_+ := \sup_{p\in J(x)}\langle p,y\rangle,\quad \langle x,y\rangle_- = \inf_{p\in J(x)}\langle p,y\rangle.$$

Let us collect some properties of these functions. For each $x\in E$, $\langle x,\,\cdot\,\rangle_\pm\colon E\to\mathbb{R}$ is Lipschitz continuous, subadditive and positively homogenous; for each $y\in E$, $\langle\,\cdot\,,y\rangle_+$ (resp. $\langle\,\cdot\,,y\rangle_-$) is upper (resp. lower) semicontinuous on E and positively homogenous; for any $x,y\in E$ and $\omega\in\mathbb{R}$, $|\langle x,y\rangle_\pm|\le\|x\|\|y\|$, $-\langle x,y\rangle_\pm=\langle x,-y\rangle_\mp$ and $\langle x,y+\omega x\rangle_\pm=\langle x,y\rangle_\pm+\omega\|x\|^2$.

It is also convenient to put

$$(3.17)\qquad [x,y]_\pm := \lim_{h\to 0^\pm}\frac{\|x+hy\|-\|x\|}{h},\qquad x,y\in E.$$

Then, for all $x,y\in E$,

$$\langle x,y\rangle_\pm=\|x\|[x,y]_\pm.$$

Finally (see [Ka]), observe that given a function $u\colon[a,b]\to E$, if u is right- (left-) differentiable at $t\in[a,b]$, then so is $\|u\|$ and

$$\frac{d^\pm}{dt}\|u(t)\|=[u(t),u'_\pm(t)]_\pm.$$

Let $D(A)\subset E$. We say that $A\colon D(A)\multimap E$ is *accretive* if, for all $x,y\in D(A)$, $u\in A(x)$, $v\in A(y)$ and $\lambda\ge 0$,

$$\|x-y\|\le\|x-y+\lambda(u-v)\|.$$

If, additionally, the range $R(I+A)=E$, then we say that A is *m-accretive*.

A fundamental result of Kato in [Ka] states that $A\colon D(A)\multimap E$ is accretive if and only if $\langle x-y,u-v\rangle_+\ge 0$ for all $(x,u),(y,v)\in\mathrm{Gr}(A)$. In this setting, if A is m-accretive and, for some $(x,u)\in E\times E$,

$$\langle x-y,u-v\rangle_+\ge 0\quad\text{for all } (y,v)\in\mathrm{Gr}(A),$$

then $u\in A(x)$.

It motivates us to say that $A\colon D(A)\multimap E$ is *strongly accretive* if, for all $x,y\in K$ and $u\in A(x)$, there exists $v\in A(y)$ such that $\langle x-y,u-v\rangle_-\ge 0$.

If A is single valued, then strong accretivity implies accretivity (since $\langle\,\cdot\,,\,\cdot\,\rangle_-\le\langle\,\cdot\,,\,\cdot\,\rangle_+$); if additionally A is continuous, then accretivity and strong accretivity coincide, see [Ba, Theorem 1]. If E^* is strictly convex, then accretivity entails strong accretivity (for then the duality map is single-valued and semi-inner products $\langle\,\cdot\,,\,\cdot\,\rangle_\pm$ coincide). In general strong accretivity seems to be independent of accretivity.

Our interest in strongly accretive maps is motivated by part (3.19.2) of the following simple lemma.

(3.19) LEMMA.

(3.19.1) *Given* $\omega \in \mathbb{R}$, $A - \omega I\colon D(A) \multimap E$ *is strongly accretive if and only if, for all* $x, y \in D(A)$ *and* $u \in A(x)$,

(3.19.1.1)
$$\langle x - y, u - v\rangle_- \geq \omega \|x - y\|^2$$

for some $v \in A(y)$ (24).

(3.19.2) *If* $K \subset E$, *a set-valued map* $\varphi\colon K \multimap E$ *is nonexpansive and has compact values, then* $A = I - \varphi\colon K \multimap E$ *is strongly accretive* (*and not necessarily accretive unless it is single-valued*).

PROOF. (3.19.1) For all $x, y \in D(A)$, $u \in A(x)$ and $v \in A(y)$,

$$\langle x-y, u-\omega x-v+\omega y\rangle_- = \langle x-y, u-v+\omega(y-x)\rangle_- = \langle x-y, u-v\rangle_- - \omega\|x-y\|^2.$$

Hence (3.19.1.1) holds if and only if $A - \omega I$ is strongly accretive.

(3.19.2) Let $x, y \in K$ and $u \in x - \varphi(x)$, i.e. $u = x - a$ where $a \in \varphi(x)$. Since φ is nonexpansive,

$$d(a, \varphi(y)) \leq \sup_{z \in \varphi(x)} d(z, \varphi(y)) \leq d_H(\varphi(x), \varphi(y)) \leq \|x - y\|,$$

i.e. there is $b \in \varphi(y)$ such that $\|a-b\| = d(a, \varphi(y)) \leq \|x-y\|$. Hence $\langle x-y, a-b\rangle_+ \leq \|x - y\|^2$ and, putting $v := y - b$,

$$\langle x - y, u - v\rangle_- = \|x - y\|^2 - \langle x - y, a - b\rangle_+ \geq 0. \qquad \square$$

Let us first state and prove the following result (the first part of which is due to Martin in [M1] (see also [M2, Lemma 6.4] and [MP, Theorem 4.2])).

(3.20) THEOREM. *Let* $K \subset E$ *be closed,* $\omega > 0$ *and let* $A\colon D(A) \multimap E$. *Assume that* $K \subset \operatorname{cl} D(A)$. *If one of the following conditions is satisfied:*

(3.20.1) *A is single-valued continuous and such that* $-A - \omega I$ *is accretive and, for each* $x \in K$,

(3.20.1.1)
$$A(x) \in T_K(x);$$

(24) Equivalently: for all $x, y \in D(A)$ and $u \in A(x)$,

(3.19.1.2)
$$[x - y, u - v]_+ \leq -\omega\|x - y\|.$$

(3.20.2) *$-A-\omega I$ is m-accretive and, for each $x \in K$,*

$$\liminf_{h\to 0^+} \frac{d(x, R((I+hA)|_K))}{h} = 0, \tag{3.20.2.1}$$

then there is $x_0 \in K$ such that $0 \in A(x_0)$.

PROOF. As shown by Martin, in case (3.20.1), for every $x \in K$, the Cauchy problem $u'(t) = A(u(t))$, $x(0) = x$ has a unique (a.e. differentiable) solution $u(\,\cdot\,; x)\colon [0, \infty) \to K$. Moreover, for any $x, y \in K$,

$$\|u(t;x) - u(t;y)\| \le e^{-\omega t}\|x-y\|.$$

Let, for $t \ge 0$ and $x \in K$, $S(t)x = u(t;x)$. Hence the family $\{S(t)\colon K \to K\}_{t\ge 0}$ is defined; $S(0) = I$ and $S(t+s) = S(t)\circ S(s)$ for any $t, s \ge 0$. This means that the family $\{S(t)\}_{t>0}$ of contractions is commutative and, therefore, there exists $x_0 \in K$ such that $S(t)x_0 = x_0$ for all $t \ge 0$. Thus $u(\,\cdot\,;x_0) \equiv x_0$ is constant and $0 \in A(x_0)$.

In the second case, first observe that $-A$ is m-accretive. Crandall and Liggett in [CL] construct a semigroup $\{S(t) : \operatorname{cl} D(A) \to E\}_{t\ge 0}$ (by the so-called Crandall-Liggett formula) such that, for each $x, y \in \operatorname{cl} D(A)$ and $t \ge 0$, $\|S(t)x - S(t)y\| \le e^{-\omega t}\|x-y\|$ and, for any $x \in \operatorname{cl} D(A)$, $u := S(\,\cdot\,)x\colon [0,\infty) \to E$ is a so-called *integral solution* to the Cauchy problem $u'(t) \in A(u(t))$, $u(0) = x$, i.e. u is continuous, $u(0) = x$, $u(t) \in \operatorname{cl} D(A)$ for a.a. $t \ge 0$ and, for any $0 \le s < t$ and $(y, v) \in \operatorname{Gr}(-A)$,

$$\|u(t) - y\|^2 \le \|u(s) - y\|^2 + \int_s^t \langle u(\tau) - y, -v\rangle_+ \, d\tau.$$

As shown by Bothe in [Bo2, Theorem 1], condition (3.20.2.1) implies that, for each $x \in K$ and $t \ge 0$, $S(t)x \in K$. Hence, again as above, there is $x_0 \in K$ such that $S(t)x_0 = x_0$ for all $t \ge 0$, i.e. the constant function $u(t) \equiv x_0$ is an integral solution; therefore, for all $(y, v) \in \operatorname{Gr}(-A)$, $\langle x_0 - y, -v\rangle + \ge 0$. By the m-accretivity of $-A$, we gather that $0 \in A(x_0)$. □

The main question here is whether in the second case of Theorem (3.20) a natural from our view-point tangency condition: for all $x \in K$, $A(x) \cap T_K(x) \ne \emptyset$, implies (3.20.2.1) and thus yields the existence of equilibria of A. This is known in some special cases but in general seems to be unknown and deserves investigation.

The presented above method of differential equations (or inclusions) seems to a powerful tool for the study of equilibria. Studying the fixed points of the solution map of the corresponding differential equation, we are frequently able to derive the existence of equilibria. Note that in what was said above (see Remark (2.17), the arguments in front of Proposition (2.32) and the proof to Theorem (2.33)) the existence of solutions to such a differential equation (or inclusion) was sufficient for a suitable tangency condition. Here the tangency condition was used to establish the existence of solutions what, in turn, together with the properties of the solution map, implied the existence of equilibria. These methods will also be used below and in the sequel.

Our main purpose now is to prove the following theorem which corresponds to the above one and complements the result of Deimling, see [D2, Theorem 10.4].

(3.21) THEOREM. *Let $A\colon E \multimap E$ be such that, for some $\omega > 0$, $-A - \omega I$ is strongly accretive* ([25]) *and let $K \subset E$ be closed. If E is reflexive, A is continuous, bounded on bounded sets, has compact convex values and, for any $x \in K$, $A(x) \subset T_K(x)$, then there is $\overline{x} \in K$ such that $0 \in A(\overline{x})$.* ([26])

PROOF. For an arbitrary $x_0 \in K$, let us consider an auxiliary Cauchy problem

$$\begin{cases} x'(t) \in A(x(t)) & \text{for } t \in J := [0,1], \\ x(0) = x_0. \end{cases} \tag{3.22}$$

By a *solution* (resp. *ε-solution* where $\varepsilon > 0$) to (3.22) we understand a function $x\colon J \to E$, for which there is $w \in L^1(J,E)$ such that $x(t) = x_0 + \int_0^t w(s)\,ds$ for all $t \in J$ ([27]) and such that, for a.a. $t \in J$, $w(t) \in A(x(t))$ (resp. $d(w(t), A(x(t))) \le \varepsilon$). We say that an ε-solution is *polygonal* if there is a countable partition $[0,1) = \bigcup_{i=0}^{\infty} J_i$ of the interval J, where J_i are disjoint intervals (of the form $[a_i, b_i)$), such that x' is constant on J_i.

Claim 1. For each $\varepsilon > 0$, the problem (3.22) has a polygonal ε-solution x such that $d(x(t), K) < \varepsilon$ on J and $x(a_i) \in K$ ($i \ge 0$).

Choose $v_0 \in A(x_0)$. By the same methods as in the proof of Lemma (3.13), we see that the map $d(\cdot\,, A(\cdot))$ is continuous; hence there is $0 < \delta_1 < \varepsilon$ such that $d(v, A(x)) < \varepsilon$ if $\|v - v_0\|, \|x - x_0\| < \delta_1$. Since $v_0 \in T_K(x_0)$, we find $\overline{v}_0 \in E$ and $t_1 > 0$ such that $\|\overline{v}_0 - v_0\| < \delta_1$, $x_1 := x_0 + t_1\overline{v}_0 \in K$ and $t_1\|\overline{v}_0\| < \delta_1$. For $t \in [0, t_1]$, define $x(t) := x_0 + t\overline{v}_0$. Then $x'(t) = \overline{v}_0$; hence $d(x'(t), A(x(t))) < \varepsilon$ and $d(x(t), K) \le t\|\overline{v}_0\| < \delta_1 < \varepsilon$ on $[0, t_1]$.

Taking $v_1 \in A(x_1) \subset T_K(x_1)$ and $\delta_2 \in (0, \varepsilon)$ such that $d(v, A(x)) < \varepsilon$ if $\|v - v_1\|, \|x - x_1\| < \delta_2$, we choose $\overline{v}_1 \in E$ and $t_2 > 0$ such that $\|\overline{v}_1 - v_1\| < \delta$, $x_2 := x_1 + t_2\overline{v}_1 \in K$, $t_2\|\overline{v}_1\| < \delta$ and put $x(t) := x_1 + (t - t_1)\overline{v}_1$ for $t \in [t_1, t_1 + t_2]$. Then again $d(x'(t), A(x(t))) < \varepsilon$ and $d(x(t), K) < \varepsilon$ on $[t_1, t_1 + t_2]$.

([25]) Recall condition (3.19.1.2).

([26]) Added in proof: a similar result in a slightly different terminology has been obtained by T. Donchev, see [Do].

([27]) It is clear that x is differentiable a.e. and $x'(t) = w(t)$ for a.a. $t \in J$; integrals are understood in the sense of Bochner.

One may continue the procedure in a similar way in order to obtain a polygonal ε-solution x defined on an interval $[0, \bar{t})$ where $\bar{t} \le 1$. Clearly x is Lipschitz continuous (since A is bounded); hence $x(\bar{t}) = \lim_{t\to\bar{t}} x(t) \in K$ exists. If $\bar{t} < 1$, then, as above, x may be extended beyond $\bar{t}$. Using the Kuratowski-Zorn lemma argument, this means that the maximal polygonal ε-solutions exists on the whole interval J and ends the proof of Claim 1.

Consider now a decreasing sequence $(\varepsilon_j)_{j=1}^{\infty}$ of positive numbers such that $\varepsilon_j \to 0$ and, for any $j \ge 1$, if $\rho_j \colon J \to \mathbb{R}$ is the solution of the problem $\rho'(t) = -\omega\rho(t) + 4\varepsilon_j$, $\rho(0) = 0$, then $\sup_{t\in J} |\rho_j(t)| < 2^{-j}$.

Claim 2. For each $j \ge 1$, there is a polygonal ε_j-solution $x_j \colon J \to E$ to (3.22) such that, for all $t \in J$, $d(x_j(t), K) < \varepsilon_j$ and $\|x_j(t) - x_{j+1}(t)\| \le \rho_j(t)$.

By Claim 1, there is an ε_1-solution x_1 to (3.22) and a countable partition $\{J_i\}_{i=0}^{\infty}$ of J such that, for all $n \in \mathbb{N}$, x_1' is constant on J_i, $d(x_1'(t), A(x_1(t))) < \varepsilon_1$ and $d(x_1(t), K) < \varepsilon_1$ on J. Without loss of generality we suppose that $0 \in J_0$. For $v_0 \in A(x_0)$, we choose $0 < \delta < \varepsilon_2$ such that, if $\|v - v_0\| < \delta$, $\|x - x_0\| < \delta$, then $d(v, A(x)) < \varepsilon_2$; next we choose $\widetilde{v}_0 \in E$ and $s_1 > 0$, $s_1 \in J_0$ such that $\|\widetilde{v}_0 - v_0\| < \delta$, $y_1 = x_0 + s_1\widetilde{v}_0 \in K$, $s_1\|x_1'(0)\| < \varepsilon_1$, $s_1\|\widetilde{v}_0\| < \delta$. Moreover, s_1 is such that the following condition holds: for $t \in [0, s_1]$,

$$\omega t \le 2(1 - e^{-\omega t}). \tag{3.23}$$

For $t \in [0, s_1]$, put $y(t) := x_0 + t\widetilde{v}_0$. Now we see that $d(y'(t), A(y(t))) < \varepsilon_2$, $d(y(t), K) < \varepsilon_2$ for $t \in [0, s_1]$ and, by (3.23),

$$\|x_1(t) - y(t)\| = t\|x_1'(0) - \widetilde{v}_0\| < t(\varepsilon_1 + \varepsilon_2) < 2t\varepsilon_1 \le \rho_1(t)$$

(recall that, for $t \in [0, s_1]$, $x_1(t) = x_0 + tx_1'(0)$). Suppose that a polygonal ε_2-solution y satisfying the above estimate exists on some interval $[0, \bar{t})$. Again y is Lipschitz; thus $y(\bar{t}) := \lim_{t\to\bar{t}} y(t) \in K$ exists. In order to extend y beyond $\bar{t}$, take i such that $\bar{t} \in J_i = [a, b)$ (on which x_1' is constant) and take $v \in A(y(\bar{t})) \subset T_K(y(\bar{t}))$ such that

$$\langle x_1(\bar{t}) - y(\bar{t}), \xi - v\rangle_+ \le -\omega\|x_1(\bar{t}) - y(\bar{t})\|^2$$

where $\xi \in A(x_1(\bar{t}))$ is such that $\|x_1'(\bar{t}) - \xi\| < \varepsilon_1$. Thus

$$[x_1(\bar{t}) - y(\bar{t}), x_1'(\bar{t}) - v]_+ \le -\omega\|x_1(\bar{t}) - y(\bar{t})\| + \varepsilon_1 < -\omega\|x_1(\bar{t}) - y(\bar{t})\| + \varepsilon_1 + \varepsilon_2.$$

The function $(x, y) \mapsto [x, y]_+ + \omega\|x - y\|$ is upper semicontinuous (as a real function); thus there is $0 < \delta < \min\{b - \bar{t}, \varepsilon_2\}$ such that

$$[x_1(t) - x, x_1'(t) - w]_+ + \omega\|x_1(t) - x\| < \varepsilon_1 + \varepsilon_2$$

provided $\|x - y(\bar{t})\|, \|v - w\| < \delta$ and $|t - \bar{t}| < \delta$. Since $v \in T_K(y(\bar{t}))$, we may take $s > 0$ and $\widetilde{v} \in E$ such that, $y(\bar{t}) + s\widetilde{v} \in K$, $s < \delta$, $\|v - \widetilde{v}\| < \delta$ and $s\|\widetilde{v}\| < \delta$. Then we extend y by putting $y(t) = y(\bar{t}) + (t - \bar{t})\widetilde{v}$ for $t \in [\bar{t}, \bar{t} + s]$. Clearly, for $t \in [\bar{t}, \bar{t} + s]$ we have

$$[x_1(t) - y(t), x_1'(t) - y'(t)]_+ \le -\omega\|x_1(t) - y(t)\| + \varepsilon_1 + \varepsilon_2.$$

The function $\|x_1 - y\|$ is Lipschitz continuous, i.e. absolutely continuous and almost everywhere differentiable. Hence, for a.a. $t \in [0, \bar{t} + s]$,

$$\frac{d}{dt}\|x_1(t) - y(t)\| = [x_1(t) - y(t), x_1'(t) - y'(t)]_+ < \varepsilon_1 + \varepsilon_2 - \omega\|x_1(t) - y(t)\|.$$

By the elementary properties of differential inequalities (see e.g. [Ha, Theorem 4.1]) we get that, for all $t \in [0, \bar{t} + s]$,

$$\|x_1(t) - y(t)\| \le \rho_1(t).$$

Again by the maximality (Zorn lemma) arguments we get the existence of a polygonal ε_2-solution $x_2 := y$ defined on J and satisfying the required estimate.

Inductively: having an ε_i-solution x_i, one construct as above a polygonal ε_{i+1}-solution x_{i+1} such that $\|x_i(t) - x_{i+1}(t)\| \le \rho_i(t)$ on J.

Given $x_0 \in K$, let $S(x_0)$ denote the set of all solutions $x\colon J \to E$ to (3.22) which survive in K, i.e. $x(t) \in K$ on J.

Claim 3. For each $x_0 \in K$, $S(x_0)$ is nonempty and compact in $C(J, E)$.

Take a sequence (x_i) of polygonal ε_i-solutions obtained in Claim 2. Therefore, for $n < m$ and $t \in J$,

$$\|x_n(t) - x_m(t)\| \le \sum_{j=n}^{m-1} \rho_j(t) \le 2^{n-1}.$$

This shows that (x_i) is a Cauchy sequence; thus $x_j \to x \in C(J, E)$ (uniformly); in particular $x(0) = x_0$ and $x(t) \in K$ on J.

Since E is reflexive and there is a bound for $\|x_j'(t)\|$ we gather that (for a subsequence) $x_j' \to w \in L_1(J, E)$ (weakly in $L_1(J, E)$) and $x(t) = x_0 + \int_0^t w(s)\, ds$. By the convergence theorem (see [AF, Theorem 7.2.2]), we gather that $x\colon J \to K$ is a solution, i.e. $x \in S(x_0)$. In a similar way we show that $S(x_0)$ is closed.

Now we shall prove that, for each $\mu > 0$ there is a compact μ-net for $S(x_0)$. To this end fix $\varepsilon > 0$. It is easy to see that there are points $0 = t_0 < \ldots < t_n = 1$ such that if $z\colon J \to E$ is a an *approximate* solution to (3.22) *corresponding to the partition* $\{t_0, \ldots, t_n\}$, i.e. z is a continuous a.e. differentiable function such that

$z(0) = x_0$ and, for all $0 \le i \le n$, $z'(t) \in A(z(t_i))$ for a.a. $t \in [t_i, t_{i+1})$, then $d_H(A(z(t)), A(z(t_i))) \le \varepsilon$ for $t \in [t_i, t_{i+1})$.

The set Z of all approximate solution corresponding to the partition $\{t_0, \dots, t_n\}$ is compact: to see this let $X_0 := \{x_0\}$, $X_{k+1} := (t_{k+1} - t_k)\operatorname{cl}\operatorname{conv} A(X_k)$ $(k = 1, \dots, n-1)$. Sets X_k are compact. It is clear that Z is equicontinuous and $Z(t) := \{x(t) : x \in Z\} \subset X_{k+1}$ if $t \in [t_k, t_{k+1})$. Hence $Z(x_0)$ is compact in view of the Ascoli–Arzela theorem.

We are now going to show that, for any $x \in S(x_0)$ there is $z \in Z$ such that $\|x(t) - z(t)\| \le \rho(t)$ on J where $\rho'(t) = -\omega\rho(t) + \varepsilon$ and $\rho(0) = 0$.

Suppose that z is already defined on $[0, t_i]$ where $0 \le i < n$. For $m \in \mathbb{N}$, $u \in E$ and $t \in [t_i, t_{i+1}]$, consider

$$G_m(t,u) := \operatorname{cl}\{y \in A(z(t_i)) : \\ [x(t) - u, x'(t) - y]_+ < -\omega\|x(t) - u\| + d_H(A(z(t_i)), A(u)) + m^{-1}\}.$$

Using the assumptions, it is easy to see that $G_m : [t_i, t_{i+1}] \times E \multimap E$ has nonempty compact and convex values. The Lusin property of x' shows that G_m is almost lower semicontinuous (in the sense of [D2, Definition 3.3]). Hence, in view of [D2, Theorem 9.3], a problem

$$\begin{cases} u' \in G_m(t,u), \\ u(0) = z(t_i), \end{cases}$$

has a solution $z_m \colon [t_i, t_{i+1}] \to E$. Hence

$$[x(t) - z_m(t), x'(t) - z_m'(t)]_+ \le -\omega\|x(t) - z_m(t)\| + \varepsilon + m^{-1}$$

for $t \in [t_i, t_{i+1}]$. Consequently, for $t \in [t_i, t_{i+1}]$,

$$\|z_m(t) - x(t)\| \le \rho_m(t)$$

where $\rho_m'(t) = -\omega\rho_m(t) + \varepsilon + n^{-1}$ and $\rho_m(t_i) = \|z(t_i) - x(t_i)\|$. In this way we define z_m on J and $z_m \in Z$. The compactness of Z implies that (passing to a subsequence if necessary), $z_m \to z \in Z$ and $\|z(t) - x(t)\| \le \rho(t)$ on J. It is clear that if ε is sufficiently small, then $\|x(t) - z(t)\| < \mu$.

Claim 4. Given a solution $x \in S(x_0)$, a point $y_0 \in K$ and $\varepsilon > 0$, there is a solution $y \in S(y_0)$ such that $\|x(t) - y(t)\| \le \rho(t) + \varepsilon$ for $t \in J$, where $\rho'(t) = -\omega\rho(t) + \varepsilon$ and $\rho(0) = \|x_0 - y_0\|$.

Fix $\mu = \varepsilon/2$. First we show that there exists a polygonal μ-solution $z \colon J \to E$ such that $\|x(t) - z(t)\| < \mu$ on J. Using the compactness of the trajectory $x(J)$, there is $0 < \lambda < \mu$ such that $d_H(A(u), A(y)) < \mu$ if $u \in x(J)$ and $\|u - y\| < \lambda$.

Next take $n \in \mathbb{N}$ such that $\|x(t)-x(t')\| < \lambda/2$ provided $t, t' \in J$ and $|t-t'| \leq n^{-1}$. Let $t_i := in^{-1}$ $(i = 0, 1, \ldots, n)$ and $x_i := x(t_i)$. For $t \in [t_i, t_{i+1}]$, we define

$$z(t) := x_i + \frac{t - t_i}{t_{i+1} - t_i}(x_{i+1} - x_i).$$

Then, for all $i = 0, \ldots, n-1$ and $t, t' \in [t_i, t_{i+1}]$, $\|z(t) - x(t')\| < \lambda < \mu$. In particular $\|z(t) - x(t)\| < \mu$. Obviously, on $J_i = [t_i, t_{i+1})$ $(i = 0, \ldots, n)$, $z'(t)$ is constant. Moreover, for all $t \in [t_i, t_{i+1}]$,

$$z'(t) = n(x_{i+1}-x_i) = n\int_{t_i}^{t_{i+1}} x'(s)\, ds \in \operatorname{cl}\operatorname{conv} A(x([t_i, t_{i+1}])) \subset A(z(t))+B(0,\mu).$$

There is $v_0 \in A(y_0) \subset T_K(y_0)$ such that $[x_0 - y_0, \xi - v_0]_+ \leq -\omega\|x_0 - x_1\|$, where $\xi \in A(x_0)$ is such that $\|\xi - z'(0)\| < \mu$. Hence (recall that $z(0) = x_0$)

$$[z(0) - y_0, z'(0) - v_0]_+ < 2\mu - \omega\|z(0) - x_1\|.$$

The continuity implies, that there is $0 < \delta < t_1$ such that

$$[z(t) - x, z'(t) - w]_+ + \omega\|z(t) - x\| < 2\mu$$

provided $\|x - x_1\|, \|v_0 - w\| < \delta$ and $0 \leq t < \delta$. Since $v_0 \in T_K(y_0)$, we may take $s > 0$ and $\widetilde{v}_0 \in E$ such that, $y_0 + s\widetilde{v}_0 \in K$, $s < \delta$, $\|v - \widetilde{v}\| < \delta$ and $s\|\widetilde{v}\| < \delta$. Then we define $\overline{y}(t) = x_1 + t\widetilde{v}_0$ for $t \in [0, s]$. Clearly, for $t \in [0, s]$ we have

$$[z(t) - \overline{y}(t), z'(t) - \overline{y}'(t)]_+ \leq -\omega\|z(t) - \overline{y}(t)\| + 2\mu.$$

Arguing as before we show that there is a polygonal δ-solution $\overline{y}$ defined on J and such that

$$\|z(t) - \overline{y}(t)\| \leq \rho(t).$$

Passing with $\delta \to 0$ and proceeding as before we get a sequence of finer approximate solutions converging to the solution y of (3.22) ($y(0) = y_0$) such that $y(t) \in K$ and $\|z(t) - y(t)\| < \rho(t)$ on J. Hence $\|x(t) - y(t)\| \leq \rho(t) + \varepsilon$.

Let, for $x_0 \in K$, $P(x_0, t) := \{x(t) : x \in S(x_0)\} \subset K$.

From Claim 4, passing with $\varepsilon \to 0$, we gather that, for all $x_0, y_0 \in K$, any solution $x \in S(x_0)$ and every $t \in J$,

$$d(x(t), P(y_0, t)) \leq e^{-\omega t}\|x_0 - y_0\|.$$

In other words, we have proved that, for each $t \in J$, the set-valued map $P(\,\cdot\,, t)\colon K \multimap K$ has compact values and, if $x, y \in K$, then

$$d_H(P(x,t), P(y,t)) \leq e^{-\omega t}\|x - y\|.$$

Therefore, by the Nadler theorem (see [N]), for each $t \in J$, $t > 0$, $P(\cdot, t)$ has a fixed point $x_t \in K$, i.e. $P(x_t, t) = x_t$. In particular there is a sequence $(x_n)_{n=0}^{\infty}$ in K such that, for all $n \geq 0$, $P(x_n, 2^{-n}) = x_n$. It is clear that, for all $n \geq 0$, $x_n = P(x_n, 1)$. In other words, points x_n belong to the fixed point set of $P(\cdot, 1)$. The latter map is a k-contraction $k = e^{-\omega}$ with respect to the Kuratowski measure of noncompactness. Hence (passing to a subsequence if necessary) we may assume that $\overline{x} = \lim_{n\to\infty} x_n \in K$ exists. It is now easy to see that $0 = A(\overline{x})$. □

If in Theorem (3.21), A is upper semicontinuous and accretive, then (suitably modifying the proof of [D2, Theorem 10.4] along lines proposed above), (3.22) has a unique solution $u(\cdot\,; x_0)$ and $\|u(t; x_0) - u(t; y_0)\| \leq e^{-\omega t}\|x_0 - y_0\|$. Thus applying the method form the proof of Theorem (3.20) or given in the above proof, we obtain the existence of $\overline{x} \in K$ such that $0 \in A(\overline{x})$. In other words one gets the following improvement of (3.20.1).

(3.24) THEOREM. *Let K be a closed subset of a reflexive Banach space E and let an upper semicontinuous map $A\colon K \multimap E$ have closed convex bounded values. If, for each $x \in K$, $A(x) \cap T_K(x) \neq \emptyset$ and, for some $\omega > 0$, $-A - \omega I$ is accretive, then there is $\overline{x} \in K$ such that $0 \in A(\overline{x})$.*

Now we shall see how Theorems (3.20), (3.24) and (3.21) help to obtain further results concerning equilibria. We follow some arguments of Ray (presented in [GK, p. 142]). The second part of this result may be considered as the constrained Minty–Browder in [Mi].

(3.25) THEOREM. *Let $K \subset E$ be a closed convex having the fixed point property for nonexpansive (single-valued) maps. Let $A\colon D(A) \multimap E$ and $K \subset D(A)$. If one of the following conditions is satisfied:*

(3.25.1) *A is single-valued continuous (or A is upper semicontinuous with closed convex bounded values and E is reflexive), $-A$ is accretive and,*

$$A(x) \in T_K(x) \quad (\textit{resp. } A(x) \cap T_K(x) \neq \emptyset) \quad \textit{for each } x \in K,$$

(3.25.2) *$-A$ is m-accretive and, for each $x \in K$,*

$$\liminf_{h\to 0+} \frac{d(x, R((I + hA)|_K))}{h} = 0,$$

then there is $x_0 \in K$ such that $0 \in A(x_0)$.

PROOF. Take $t \in (0, 1)$ and set $\lambda = (1-t)^{-1}t$. It is easy to check that (in both cases) the operator $J_\lambda\colon (I - \lambda A)(K) \to K$ given by $J_\lambda(y) = (I - \lambda A)^{-1} = \{x \in D(A) : y \in x - \lambda A(x)\}$ is well-defined and, in fact, single-valued; moreover it is nonexpansive: for all $y, z \in R(I - \lambda A)$, $\|J_\lambda(y) - J_\lambda(z)\| \leq \|y - z\|$. Suppose for a while that

$$K \subset (I - \lambda A)(K). \tag{3.26}$$

Then $J_\lambda \colon K \to K$ and has a fixed point $x_0 \in A$; therefore $0 \in A(x_0)$. Hence it remains to show (3.26).

To this aim take $y \in K$ and define $B(x) := tA - (1-t)(x-y)$. We see easily that $-B(x) - (1-t)x = -tA(x) - (1-t)y$ is accretive and m-accretive if so is $-A$. In the first case we easily check that, for each $x \in K$, $B(x) \in T_K(x)$ (or $B(x) \cap T_K(x) \neq \emptyset$). In the second case take $x \in K$ and an arbitrary $u \in K$, then, for $h > 0$ and $v = tw - (1-t)(u-y) \in B(u)$ (where $w \in A(u)$),

$$\frac{1}{h}\|x - (u - hv)\| = (1+\alpha)\frac{1}{h}\left\|\frac{x+\alpha y}{1+\alpha} - \left(u - \frac{ht}{1+\alpha}w\right)\right\|$$

where $\alpha = (1-t)h$. Hence

$$\liminf_{h \to 0+} \frac{d(x, R((I-hB)|_K))}{h} = \liminf_{h \to 0+} \frac{d(\overline{x}, R((I-hA)|_K))}{h} = 0$$

where $\overline{x} = (1+\alpha)^{-1}(x+\alpha y) \in K$. By Theorem (3.20) or (3.24), in both cases $0 \in B(z)$ for some $z \in K$. Thus $y \in z - \lambda A(z)$. □

In order to derive an analogous result for strongly accretive maps, recall that a map $\varphi \colon D \multimap E$, where $D \subset E$, is *demiclosed* if, for any sequence (x_n, y_n) in $\mathrm{Gr}(\varphi)$, the following condition holds: if $x_n \rightharpoonup x$ (weakly), $y_n \to y$, then $x \in D$ and $y \in \varphi(x)$.

One says that a Banach space E satisfies the *Opial condition* if whenever (x_n) in E converges weakly to x_0, them for $x \in E$, $x \neq x_0$,

$$\liminf_{n \to \infty} \|x_n - x_0\| < \liminf_{n \to \infty} \|x_n - x\|.$$

It is well-known that, for instance, any uniformly convex space with weakly sequentially continuous duality mapping satisfies this condition, but also ℓ^p, for any $1 < p < \infty$ do as well (hence the Opial condition is independent of uniform convexity).

We have:

(3.27) EXAMPLE.

(3.27.1) If $\varphi \colon D \multimap E$ is *strongly upper semicontinuous*, i.e. for a sequence $(x_n, y_n) \in \mathrm{Gr}(\varphi)$, if $x_n \rightharpoonup x$, then there is a subsequence (y_{n_k}) converging strongly to some $y \in \varphi(x)$, then φ is demiclosed. In particular, if E' is a Banach space,

$j: D \to E'$ is completely continuous (i.e. maps bounded sets into compact ones) and $\Phi: E' \multimap E$ is upper semicontinuous, then $\varphi = \Phi \circ j$ is strongly upper semicontinuous.

(3.27.2) If each bounded closed convex subset of E has the fixed point property for nonexpansive maps, $D \subset E$ and $A: D \multimap E$ is m-accretive, then A is demiclosed (this is a simple consequence of [R2]).

(3.27.3) In particular, if E is a Hilbert space and φ is upper semicontinuous and monotone (i.e. accretive), then φ is demiclosed.

(3.27.4) If $K \subset E$ is closed and convex , $\varphi: K \multimap E$ is nonexpansive with compact values, then $f = I - \varphi$ is demiclosed provided

(3.27.4.1) E satisfies the Opial condition or

(3.27.4.2) E is uniformly convex.

Indeed (comp. a single-valued case from [GK, Theorem 10.3] and [LD]) let $(x_n, y_n) \in \mathrm{Gr}(f)$, $x_n \rightharpoonup x$ and $\lim_{n\to\infty} \|y_n - y\| = 0$. Since these limits are preserved under the translation $x \mapsto x - y$, we may suppose that $y = 0$. There is $u_n \in \varphi(x_n)$ such that $y_n = x_n - u_n$ and $\|x_n - u_n\| \to 0$. Since $\varphi(x)$ is compact, for each $n \in \mathbb{N}$, there is $z_n \in \varphi(x)$ such that $\|x_n - z_n\| = d(x_n, \varphi(x))$. Passing to a subsequence if necessary, we may suppose that $z_n \to z \in \varphi(x)$. Then we have

$$\|x_n - z_n\| \le d(x_n, \varphi(x)) \le d(x_n, \varphi(x_n)) + d_H(\varphi(x), \varphi(x_n)) \le \|x_n - u_n\| + \|x - x_n\|.$$

Hence $\|x_n - z\| \le \|x_n - z_n\| + \|z_n - z\| \le \|x_n - u_n\| + \|x - x_n\| + \|z_n - z\|$ and

$$\liminf_{n\to\infty} \|x_n - z\| \le \liminf_{n\to\infty} \|x - x_n\|.$$

By Opial's condition $z = x$, i.e. $x \in \varphi(x)$.

In the second case the mentioned single-valued argument in [GK] does not work. Without loss of generality we may assume that K is bounded (for (x_n) is bounded). Since $\inf_{x\in K} d(x, \varphi(x)) = 0$, arguing as in Lim [L1] (see also the footnote ([29])) we gather that φ has a fixed point (see also [GK, pp. 167–168]).

We say that a closed $K \subset E$ is *tangentially simple* if there is $x_0 \in K$ such that, for each $x \in K$ and $v \in T_K(x)$, $v - (x - x_0) \in T_K(x)$.

Observe that if K is tangentially simple, $x \in K$ and $v \in T_K$, then, for any $\lambda > 0$, $v - \lambda(x - x_0) \in T_K(x)$. Indeed, since $T_K(x)$ is a cone, $v/\lambda \in T_K(x)$; hence $v - \lambda(x - x_0) = \lambda(v/\lambda - (x - x_0)) \in T_K(x)$.

(3.28) EXAMPLE. Suppose that K is *starshaped around* $x_0 \in K$, i.e. $x_0 + t(x - x_0) \in K$ for $x \in K$ and $t \in [0, 1]$. Then K is tangentially simple. To this end suppose that $x \in K$ and $v \in T_K(x)$. There are sequences $h_n \to 0^+$ and $v_n \to v$ such that $y_n := x + h_n v_n \in K$. Hence $x_0 + (1 - h_n)(y_n - x_0) = x + h_n(v_n - (y_n - x_0)) \in K$. This means that $v - (x - x_0) \in T_K(x)$ since $w_n = v_n - (y_n - x_0) \to v - (x - x_0)$.

In particular any closed convex K is tangentially simple.

(3.29) COROLLARY. *Let $A: E \multimap E$, where E is reflexive, have convex compact values, be continuous, bounded on bounded sets and such that $-A$ is strongly accretive. If K is closed, bounded and tangentially simple, then there is $z \in E$ belonging to the weak closure of K such that $0 \in A(z)$ provided A is demiclosed and $A(x) \subset T_K(x)$ for each $x \in K$.*

PROOF. For any $n \in \mathbb{N}$, let

$$A_n(x) = A(x) - \frac{1}{n}(x - x_0), \quad x \in E.$$

Then $-A_n(x) - n^{-1}x = -A(x) - n^{-1}x_0$ is strongly accretive and, for any $x \in K$, $A_n(x) \subset T_K(x)$. By the Theorem (3.21), for each $n \in \mathbb{N}$, there is $x_n \in K$ such that $0 \in A_n(x_n) = A(x_n) - n^{-1}(x_n - x_0)$. It is clear that (passing to a subsequence) $x_n \rightharpoonup z \in E$ and $y_n := n^{-1}(x_n - x_0) \to 0$. Since $(x_n, y_n) \in \mathrm{Gr}(A)$ we gather that $0 \in A(z)$. □

3.3. Nonexpansive maps. It is interesting to note that the results of the previous section allow to obtain fixed point results for constrained nonexpansive set-valued maps (see [GK, p. 143] for the single-valued results). Our first result is a direct consequence of Corollary (3.29) and constitutes an extension of [Ya, Theorem 1].

(3.30) THEOREM. *If $\varphi: E \multimap E$ is nonexpansive with compact convex values, K is closed bounded and tangentially simple, E is reflexive and satisfies the Opial condition (or E is uniformly convex and K is additionally convex), then φ has a fixed point (belonging to the weak closure of K, i.e. to K is K is convex) provided, for each $x \in K$, $\varphi(x) \subset x + T_K(x)$.*

PROOF. Let $A = \varphi - I$. Then $-A$ is strongly accretive and $A(x) \subset T_K(x)$ for all $x \in K$. Moreover, A is demiclosed. Therefore all assumptions of the above Corollary (3.29) are satisfied and the assertion follows. □

In fact, if K is convex, then one can do better. First let us mention a result due to Xu (see [X2, Theorem 2.3.1]) which extends the result of Kirk and Massa [KM] involving assumptions concerning the compactness of asymptotic centers of bounded sequences in E. Next let us state another result of Xu.

(3.31) THEOREM (comp. [X2, Theorem 2.3.3]). *If E is uniformly convex, $K \subset E$ is closed convex and bounded, $\varphi: K \multimap E$ is nonexpansive with compact values and such that $\varphi(x) \subset x + T_K(x)$ for all $x \in K$, then φ has a fixed point.*

We provide an independent proof under an additional assumption: the metric projection $\pi := \pi_K: E \to K$ is nonexpansive (observe that, since E is uniformly convex, π is well-defined and single valued continuous ([28])).

([28]) It is known that π is *uniformly continuous* on bounded neighbourhoods of K.

PROOF. The map $\psi := \pi \circ F: K \multimap K$ is nonexpansive, has compact values and, in view of the already mentioned result of [L1] ([29]), there is $x \in K$ such that $x \in \psi(x)$. Hence there is $y \in \varphi(x)$ such that $x = \pi(y)$, i.e. $y - x \in \pi^{-1}(x) - x$. By Proposition (1.8), there is $p \in J(y - x) \cap N_K(x)$. Hence, as in the proof of Corollary (2.13), we gather that $y = x$. □

(3.32) REMARK. In the course of the proof we have additionally proved that if E and K satisfy the assumptions of the Theorem, $\varphi: K \multimap E$ has compact values and is nonexpansive, then there is $x \in K$ such that $J(\varphi(x) - x) \cap N_K(x) \neq \emptyset$. This means that $A = \varphi - I$ has a "generalized equilibrium". It seems that the question of the existence of generalized equilibria for maps of the form $\varphi - I$, where φ is nonexpansive or strictly contractive, has not been studied.

In this place we end the study of the metric aspects of the set-valued equilibria theory; some other results may be found in [X2], [Pe] and others.

4. Coincidences under constraints

4.1. Internal coincidences. As stated in Introduction set-valued maps $\varphi: X \multimap Y$ correspond well to diagrams of the form $X \xleftarrow{p} Z \xrightarrow{q} Y$ where Z is a metric space, $p: Z \to X$, $q: Z \to Y$ are continuous (single-valued) maps and p is surjective. If $X \subset Y$ and a pair (p, q) determines φ (i.e. $\varphi(x) = q(p^{-1}(x))$ for all $x \in X$), then a fixed point x_0 of φ corresponds to a coincidence of p and q: there is $z_0 \in Z$ such that $p(z_0) = x_0 = q(z_0)$. Conversely, given $z_0 \in Z$ such that $p(z_0) = q(z_0)$, the point $x_0 = p(z_0)$ is a fixed point of φ.

The presented theory of constrained fixed points suggests to study the *constrained coincidence* problem. Namely suppose that $K \subset E$ is closed, maps $p: Z \to K$ and $q: Z \to E$ are given: we are to find conditions ensuring the existence of $z \in Z$ such that $p(z) = q(z)$. Unless p is a surjection this problem cannot be translated to the previous setting since the pair (p, q) *does not* determine a set-valued map. Therefore we shall present some results in this direction. In the next subsection we shall outline some results concerning a different coincidence problem (the so-called *external problem*): given a set-valued map $\varphi: K \multimap E$ and an operator $A: E \to E$ we look for points $x \in K$ such that $A(x) \in \varphi(x)$.

We shall start with a result that constitutes a counterpart of Theorem (3.14).

(4.1) THEOREM. *Let K be a closed subset of a complete metric space (X, d), Z be a topological space and let $p: Z \to K$, $q: Z \to X$ be continuous; moreover assume that p is perfect* ([30]). *Suppose that there are numbers $0 \leq k < \alpha \leq 1$*

([29]) Lim asserts that if E and K are as above, $\psi: K \multimap K$ is nonexpansive and has compact values, then ψ has a fixed point.

([30]) Recall that a continuous $p: Z \to X$ is *perfect* if p is a closed map and, for each $x \in X$, the fiber $p^{-1}(x)$ is compact.

with the following property: if $z \in Z$, $p(z) \neq q(z)$, then there is $w \in Z$ such that $p(w) \neq p(z)$ and

$$\alpha d(p(z), p(w) + d(p(w), q(z)) \leq d(p(z), q(z)), \tag{4.1.1}$$

$$d(q(z), q(w)) \leq kd(p(z), p(w)). \tag{4.1.2}$$

Then there is $z_0 \in Z$ such that $p(z_0) = q(z_0)$.

In order to present a proof we shall need a slight generalization of the Caristi theorem.

(4.2) PROPOSITION. *Let (X, d) be a complete metric space, Z a topological space and let $p: Z \to X$ be perfect map. Let a function $e: Z \to \mathbb{R} \cup \{\infty\}$ be bounded from below, lower semicontinuous and not identically equal to ∞. If $\varphi: Z \multimap Z$ is such that, for any $z \in Z$, there is $w \in \varphi(z)$ such that*

$$e(w) + d(p(z), p(w)) \leq e(z), \tag{4.2.1}$$

then there is $z_0 \in Z$ such that $p(z_0) \in p(\varphi(z_0))$. If, for each $z \in Z$ and all $w \in \varphi(z)$, inequality (4.2.1) holds, then there is $z_0 \in Z$ such that $p(\varphi(z_0)) = \{p(z_0)\}$ ([31]).

PROOF. The result follows easily by the use of arguments adapted from the proof of the Ekeland principle. For each $z \in Z$, let

$$W(z) := \{w \in Z : e(w) + d(p(z), p(w)) \leq e(z)\}.$$

It is clear that, for $z \in Z$: $z \in W(z)$, $W(z)$ is closed since p is continuous and e is lower semicontinuous and if $w \in W(z)$, then $W(w) \subset W(z)$. Take an arbitrary $z_1 \in Z$ such that $e(z_1) < \infty$ and assume that, for some integer $n \geq 2$, a point $z_{n-1} \in Z$ is defined. We then choose $x_n \in W(z_{n-1})$ such that

$$e(z_n) < \inf_{u \in W(z_{n-1})} e(u) + n^{-1}.$$

Then $W(z_n) \subset W(z_{n-1})$ and, for any $w \in W(z_n)$,

$$e(w) \geq \inf_{u \in W(z_{n-1})} e(u) > e(z_n) - n^{-1}$$

$$d(p(z_n), p(w)) \leq e(z_n) - e(w) < n^{-1}.$$

Hence the diameter $\operatorname{diam} p(W(z_n)) < 2^{-1}$. The family $\{p(W(z_n))\}_{n=1}^{\infty}$ is decreasing and consists of closed sets whose diameters converge to 0. By the completeness and the Cantor theorem, there is $x_0 \in X$ such that

$$\{x_0\} = \bigcap_{n=1}^{\infty} p(W(z_n)).$$

([31]) It is clear that if $Z = X$ and p is the identity, then we get the Caristi theorem.

This implies that, for each $n \geq 1$, $Z_n := p^{-1}(x_0) \cap W(z_n) \neq \emptyset$. The family $\{Z_n\}$ is decreasing and consists of compact sets (since $p^{-1}(x_0)$ is compact). Hence there is $z_0 \in Z$ such that $z_0 \in Z_n$ for all $n \geq 1$. This means that $p(z_0) = x_0$, $z_0 \in W(z_n)$ and $W(z_0) \subset W(z_n)$ for all $n \geq 1$. There is $w_0 \in \varphi(z_0) \cap W(z_0)$. Thus, for all $n \geq 1$, $p(w_0) \in p(W(z_n))$, i.e. $p(w_0) = x_0 = p(z_0)$. □

Let us now show in passing that in certain cases the lower semicontinuity of e is not necessary.

(4.3) PROPOSITION. *Let X, Z and p be as above, $e: Z \to \mathbb{R} \cup \{\infty\}$ be bounded from below and not identically ∞. If $\varphi: Z \multimap Z$ has the closed graph and, for each $z \in Z$, there is $w \in \varphi(z)$ such that*

$$e(w) + d(p(z), p(w)) \leq e(z),$$

then there is $z_0 \in Z$ such that $p(z_0) \in p(\varphi(z_0))$.

PROOF. The sequence $(p(z_n))$ (where, for $n \geq 1$, z_n is constructed as above) satisfies the Cauchy property. Hence $x_0 = \lim_{n\to\infty} p(z_n)$ exists. For each $n \geq 1$, there is $w_n \in \varphi(z_n) \cap W(z_n)$; hence $p(w_n) \to x_0$. The map p, being perfect is also proper, i.e. the preimages of compact sets are compact. In particular the sets $\{z_n\}$ and $\{w_n\}$ are precompact, i.e. they have accumulation points z_0 and w_0, respectively. The closeness of the graph imply that $w_0 \in \varphi(z_0)$. However, at the same time $p(z_0) = p(w_0)$. □

PROOF OF THEOREM (4.1). Suppose to the contrary that there are no coincidences; hence for all $z \in Z$, $p(z) \neq q(z)$. Let $\varphi(z) = \{w \in Z : \text{(4.1.1) is satisfied}\}$. Hence, for any $z \in Z$, $p(z) \notin p(\varphi(z))$.

For any $z \in Z$, let $e(z) = d(p(z), q(z))$ ($e: Z \to \mathbb{R}$ is obviously continuous and $e \geq 0$). Let $z \in Z$; for $w \in \varphi(z)$, we have by (4.1.1) and (4.1.2)

$$\begin{aligned} e(w) &= d(p(w), q(w)) \leq d(p(w), q(z)) + d(q(z), q(w)) \\ &\leq d(p(z), q(z)) - \alpha d(p(z), p(w)) + k d(p(z), p(w)) \\ &= e(z) - (\alpha - k) d(p(z), q(w)). \end{aligned}$$

This implies that

$$d(p(z), p(w)) \leq \frac{1}{\alpha - k}(e(z) - e(w)).$$

By Proposition (4.2), there is $z_0 \in Z$ such that $p(z_0) \in p(\varphi(z_0))$, a contradiction. □

Now we shall consider a result corresponding to Theorem (2.52). Let $K \subset E$ be a compact $\mathcal{L}$-retract. Since K is compact, there is $\delta > 0$ and a retraction $r: U = B(K, \delta) \to K$ such that, for each $x \in U$, $\|r(x) - x\| \leq L d_K(x)$ where $L \geq 1$.

Let Z be a topological space and consider a continuous map $p: Z \to K$ satisfying the following property:

(H) there is $\varepsilon > 0$ such that, for any continuous map $f: Z \to K$, if $\|p(z) - f(z)\| < \varepsilon$, then there is $z \in Z$ such that $p(z) = f(z)$.

It follows from Example (1.3) that if $\chi(K) \neq 0$ and $p: Z \to K$ is a Vietoris map, then it satisfies condition (H) (with $\varepsilon = \delta$). When looking for other conditions ensuring (H) one encounters an interesting paper of Halpern [H3]. In this paper, given topological spaces X, Y and continuous maps $p, f: X \to Y$, the author defines a *coincidence Lefschetz number* $L(p, f)$ as the Lefschetz number $\lambda(\psi)$ of the homomorphism $\psi: H_*(Y) \to H_*(Y)$ defined via the formula:

$$\psi(c) = p_*((f^*(b/c) \frown a), \quad c \in H_*(Y),$$

where $p_*: H_*(X) \to H_*(Y)$ and $f^*: H^*(Y) \to H^*(X)$ are the homomorphism induced on (singular) homology and cohomology by p and f, respectively, $a \in H_n(X)$ ($n \geq 0$ is fixed), $b \in H^n(Y \times Y)$ is such that $\ell^*(b) = 0$ (where $\ell: Y \times Y \setminus \Delta \to Y \times Y$ is the inclusion and Δ is the diagonal in $Y \times Y$) and $/, \frown$ stand for the *slash* and *cap* products, respectively (see [Sp]).

Halpern shows that if $L(p, f) \neq 0$, then p and f have a coincidence. Observe that in our setting, if $\|p(z) - f(z)\| < \delta$, then p and f are homotopic to each other (the homotopy is provided by $Z \times [0, 1] \ni (z, t) \mapsto r((1 - t)p(z) + tf(z)) \in K$). Hence $L(p, f) = \lambda(\psi)$ where $\psi(c) = b/c \frown p_*(a)$. If $a \in H_n(X)$ and $b \in H^n(Y \times Y)$ are chosen in a way ensuring that $\lambda(\psi) \neq 0$, then (H) (again with $\varepsilon = \delta$) follows and we are done.

An approach similar to that of Halpern, but subsuming the case of Vietoris maps was presented by Saveliev [S]. However, in his setting the spaces involved are subject to some more restrictive assumptions. Nevertheless, changing suitably our situation one may easily produce conditions implying (H).

(4.4) THEOREM. *Suppose that $K \subset E$ is a compact $\mathcal{L}$-retract and let Z be compact. If $p: Z \to K$ satisfies condition* (H), *$q: Z \to E$ is continuous and such that, for each $z \in Z$, $q(z) \in p(z) + C_K(p(z))$, then there is $z_0 \in Z$ such that $p(z_0) = q(z_0)$.*

Observe that if p is a surjection (as in the case of a Vietoris map), then (p, q) determines a set-valued map $\varphi := q(p^{-1}(\,\cdot\,))$ and the above assumption means that, for all $x \in K$, $\varphi(x) \subset C_K(x)$. Therefore Theorem (4.4) constitutes a generalization of Theorem (2.56).

PROOF. Without loss of generality we may assume that $\delta \leq \varepsilon$. There is $M > 0$ such that, for all $z \in Z$, $\|q(z) - p(z)\| \leq M$. Take $\mu \leq \delta/(M(L + 1))$. Then

$p(z) + \mu(q(z) - p(z)) \in B(K, \delta)$ on Z. For $n \in \mathbb{N}$ such that $n^{-1} < \mu$, consider a map $f_n: Z \to K$ given by

$$f_n(z) = r(p(z) + n^{-1}(q(z) - p(z))), \quad z \in Z.$$

Then clearly $\|p(z) - f_n(z)\| < \delta$. Therefore by hypothesis (H), there is $z_n \in Z$ such that $p(z_n) = f_n(z_n)$. Passing to subsequences if necessary, we may assume that $z_n \to z_0$. Observe that

$$\begin{aligned} n^{-1}\|q(z_n) - p(z_n)\| &= \|p(z_n) + n^{-1}(q(z_n) - p(z_n)) - f_n(z_n)\| \\ &\leq L d_K(p(z_n) + n^{-1}(q(z_n) - p(z_n))). \end{aligned}$$

Next we see that

$$\begin{aligned} &d_K(p(z_n) + n^{-1}(q(z_n) - p(z_n))) \\ &\quad \leq n^{-1}\|q(z_n) - q(z_0) + p(z_0) - p(z_n)\| + d_K(p(z_n) + n^{-1}(q(z_0) - p(z_0))). \end{aligned}$$

Therefore, for sufficiently large $n \in \mathbb{N}$,

$$\|q(z_n) - p(z_n)\| \leq L[n d_K(p(z_n) + n^{-1}(q(z_0) - p(z_0))) + \|(q-p)(z_n) - (q-p)(z_0)\|].$$

Letting $n \to \infty$, we obtain

$$\|q(z_0) - p(z_0)\| \leq L \limsup_{y \xrightarrow{K} p(z_0),\, h \to 0+} \frac{d_K(y + h(q(z_0) - p(z_0)))}{h} = 0.$$

This concludes the proof. □

It is easy to see that in a similar manner one may formulate results concerning the existence of coincidences of maps $p: Z \to K$, $q: Z \to E$ where K is convex compact or strictly regular.

4.2. External coincidences and differential inclusions. In what was said above we saw that there is a relationship of the constrained fixed point or equilibrium problem with the theory of differential inclusions. Methods involving the techniques of differential equations or inclusions were used in Theorem (2.33) in order to establish the suitable tangency condition, while in Theorem (3.21), for instance, they were used to get the existence of fixed points or equilibria. The methodology of this approach may be explained as follows. Let $K \subset E$ be closed and a set-valued map $\varphi: K \multimap E$ be given. For $x \in K$, we consider the problem

$$\begin{cases} u'(t) \in \varphi(u(t)), \\ u(0) = x. \end{cases}$$

We know already that the existence of a (local) solution $u: [0,\varepsilon] \to K$ to (4.5) implies that $\varphi(x) \cap T_K(x) \neq \emptyset$. Conversely, quite often tangency implies existence of solutions. Anyway assume that, for all $x \in K$, the set $S(x)$ of all solutions to (4.5) is nonempty; i.e. $S: K \multimap C([0,1], K)$ is a well-defined set-valued map. For $t \in [0,1]$, let $e_t: C([0,1], K) \to K$ be the *evaluation operator* $e_t(u) = u(t)$, $u \in C([0,1], K)$ and define the *Poincaré operator* $P_t: K \to K$ by the following formula

$$P_t(x) = e_t \circ S(x).$$

It is easy to see that if, for some $t \in [0,1]$, there is $x_t \in K$ such that $x_t \in P_t(x_t)$, then there is a solution $u_t \in S(x_t)$ such that $u_t(0) = x_t = u_t(t)$. Consequently, if this is the case for $t = t_n = 2^{-n}$, then letting $n \to \infty$ and assuming that $x_n := x_{t_n}$ has an accumulation point x_0, one may expect that $0 \in \varphi(x_0)$. This procedure works on many occasions. However let us enumerate several restrictions. First, to establish the existence of $x_t \in K$ one needs a suitable fixed point theory for a class of maps embracing the Poincaré operator P_t. Generally speaking the structure of this map may be fairly bad and hardly anything can be done with regard to this problem. Secondly, even if we can control the structure of P_t, then one has to establish the necessary topological (or metric) conditions of P_t such as compactness or contractivity. As simple examples (see Examples (2.45)or (2.46)) show, both problems are serious.

A classical result of Aronszajn [Ar] states the solution set for the Cauchy problem in $\mathbb{R}^N$ (involving a bounded continuous right hand side) is a compact R_δ-set; it is also true for differential inclusions, see e.g. [HV1], [HV2], the surveys [DMNZ], [Go3] and [D2].

Recall that a compact metric space X is an R_δ*-set* if there is an absolute neighbourhood retract Y containing X as a closed subspace such that X is contractible in each of its open neighbourhoods (i.e. given an open neighbourhood V of X in Y, there is a continuous map $h: X \times [0,1] \to V$ with $h(x,0) = x$ and $h(x,1) = x_0 \in V$, for all $x \in X$) ([32]). The R_δ-property is a homotopy invariant: if compacta X_1, X_2 are homotopy equivalent and $X_1 \in R_\delta$, then so is X_2. Similarly, given an absolute neighbourhood retract Z containing X as a closed subspace, if $X \in R_\delta$, then X is contractible in each of its neighbourhoods in Z. It is clear that if X is a subset of a metric space T and there is a decreasing family $\{X_n\}_{n\geq 1}$ of closed contractible sets such that $\gamma(X_n) \to 0$, where γ is a regular, monotone and nonsingular measure of noncompactness on T, then $X \in R_\delta$. The celebrated result of Hyman [H] states that if $X \in R_\delta$, then there is a decreasing sequence of contractible compacta $\{X_n\}$ containing X as a closed subspace such that $X = \bigcap_{n\geq 1} X_n$. R_δ-sets are connected,

---

([32]) R_δ-sets are sometimes called *cell-like sets*.

have trivial shape and are acyclic with respect to any continuous (co)homology theory (for instance $\check{H}^*$), i.e. have the same (co)homology as a one-point space.

Throughout the rest of this subsection we assume that:

(4.6.1) $\varphi\colon K \multimap E$, where K is a closed subset of a Banach space E, is an upper semicontinuous set-valued map with compact convex values having the linear growth (i.e. there is $c \geq 0$ such that, $\sup_{y\in\varphi(x)} \|y\| \leq C(1 + \|x\|)$) for $x \in K$.

This implies that, given a continuous $u\colon [0,1] \to K$, the map $[0,1] \ni t \multimap \varphi(u(\cdot))$ possesses an integrable selection, i.e. there is $w \in L^1(J,E)$ such that $w(s) \in \varphi(u(s))$ on J. The set of such w is denoted by $N_\varphi(u)$.

(4.6.2) A closed densely defined linear operator $A\colon E \supset D(A) \to E$ is the *infinitesimal generator* of a C_0-semigroup $\mathcal{U} = \{U(t)\}_{t\geq 0}$ of bounded linear operators on E ($A \equiv 0$ and/or $E = \mathbb{R}^N$ is not excluded) such that $\|U(t)\| \leq \exp(\omega t)$, where $\omega \in \mathbb{R}$ for $t \geq 0$ ([33]).

(4.6.3) The set K is *invariant* with respect to $\mathcal{U}$, i.e. $U(t)K \subset K$ for $t \geq 0$.

This condition may be stated in terms of A only (see [M2, Proposition VII.5.3, Remark VII.5.2] and (see [MP]) holds if and only if $\liminf_{t\to 0^+} d_K(U(t)x)/t = 0$ for all $x \in K$.

We shall see that in such a fairly general setting it is possible to overcome the difficulties enumerated above and we shall study solutions to the Cauchy problem for the following semilinear differential inclusion

$$\text{(CP)} \qquad \begin{cases} u'(t) \in Au(t) + \varphi(u(t)), & t \in J := [0,1],\ u \in K, \\ u(t_0) = x \in K. \end{cases}$$

Problem (CP) with $D = E$ was studied by many authors (see the monographs [KOZ], [MP], [HP] and the rich bibliography therein) and a diversity of results has been obtained. Here, making some additional assumptions on K, we shall address the question of the R_δ-characterization of the set of all solutions (understood in an appropriate sense) to (CP).

Given $x \in E$ and $f \in L^1(J,E)$, the function

$$M(x_0; f)(t) = U(t)x_0 + \int_0^t U(t-s)f(s)\,ds, \quad t \in [0,1]$$

is, by definition, the *mild solution* to the initial value problem

$$\text{(4.7)} \qquad \begin{cases} u'(t) = Au(t) + f(t), \\ u(0) = x. \end{cases}$$

([33]) It is clear (using an appropriate renorming procedure) that this does not restrict generality (for details cf. [M2, Chapter VII], [Yo], [Pa]).

Note that even the continuity of f does not imply that (4.7) has a *strong* solution, i.e. an almost everywhere (a.e.) differentiable function $u: J \to E$ such that $u' \in L^1(J, E)$, $u(t_0) = x$ and $u'(t) = Au(t) + f(t)$ a.e. on J; however if $\mathcal{U}$ is *uniformly continuous* (i.e. $\|I - U(t)\| \to 0$ when $t \to 0^+$), or the function

$$v: t \mapsto \int_{t_0}^{t} U(t-s) f(s)\, ds$$

is differentiable a.e. with $v' \in L^1(J, E)$ and $x_0 \in D(A)$, then the mild solution is a (unique) strong solution (see [Pa]).

A continuous function $u: J \to K$ is a *mild solution* to (CP) if there is $w \in N_\varphi(u)$ such that $u = M(x; w)$; hence the set $S(x)$ of all mild solutions to (CP) coincides with the set of fixed points of the set-valued operator $M(x;\,\cdot\,) \circ N_\varphi$ defined on $C(J, K)$.

The following existence result due to Bothe [Bo1, Theorem 7.2, Corollary 7.1] seems to be the most general.

(4.8) THEOREM. *Assume that φ satisfies the weak tangency condition, i.e. for all $x \in K$,*

$$\varphi(x) \cap T_K(x) \neq \emptyset. \tag{4.8.1}$$

Then the set-valued map $S: K \multimap C(J, K)$ assigning to $x \in K$ the set $S(x)$ of all mild solutions of (CP), *is upper semicontinuous with nonempty, compact values provided one of the following conditions holds:*

(4.8.2) *there is $k \geq 0$ such that, for any bounded $\Omega \subset K$,*

$$\beta(\varphi(\Omega)) \leq k\beta(\Omega); \tag{4.8.2.1}$$

(4.8.3) *the semigroup $\mathcal{U}$ is compact* ([34]).

(4.9) REMARK. If $f: K \to E$ is single-valued and locally Lipschitz and satisfies (4.8.1), then (CP) has a unique mild solution which depends continuously on $x \in K$. Indeed under these assumption f satisfies all hypotheses of Theorem (4.8) locally; hence a local unique mild solution exits. To establish the result one applies the usual continuation method.

(4.10) REMARK. If φ has compact values and is Lipschitz with constant $k \geq 0$, i.e. $d_H(\varphi(x), \varphi(y)) \leq k\|x - y\|$ for $x, y \in K$, then (4.8.2.1) is satisfied and φ has the linear growth.

([34]) $\mathcal{U}$ is *compact* if $U(t)$ is compact for all $t > 0$.

(4.11) REMARK. Observe that, for any $x \in K$, $y \in E$ and $h > 0$,

$$d_K(U(h)x + hy) \le h\|U(h)y - y\| + d_K(U(h)(x + hy)).$$

This, together with the invariance of K with respect to $\mathcal{U}$, implies that $T_K(x) \subset T_K^{\mathcal{U}}(x)$ and condition (4.8.1) implies that, for all $x \in K$,

$$\varphi(x) \cap T_K^{\mathcal{U}}(x) \neq \emptyset \tag{4.11.1}$$

where

$$T_K^{\mathcal{U}}(x) := \left\{ y \in E : \liminf_{h \to 0^+} \frac{d_K(U(h)x + hy)}{h} = 0 \right\}$$

(in general $T_K^{\mathcal{U}}(x)$ is not a cone). Theorem (4.8) has been actually proved under assumption (4.11.1) instead of (4.8.1). Condition (4.11.1) is strictly weaker than our (4.8.1). To see this consider $E = \mathbb{R}$, $U(t) = e^{-t}$ for $t \ge 0$ and let $K = [-1, 1]$. Then K is invariant with respect to $U(t)$ (i.e. $U(t)K \subset K$) but $T_K(1) = (-\infty, 0] \subset (-\infty, 1] = T_K^{\mathcal{U}}(1)$. It is clear that if $A \equiv 0$, then $\mathcal{U}$ consists of identities and $T_K^{\mathcal{U}}(x) = T_K(x)$.

Both conditions (4.8.1) and (4.11.1) are natural (comp. Remark (2.17)) since we have

(4.12) PROPOSITION. *Condition* (4.11.1) *is necessary for the existence of solutions to* (CP). *Precisely, if φ is upper semicontinuous and, for every $x \in K$,* (CP) *has a mild solution, then* (4.11.1) *is satisfied.*

Next, in view of the inequality

$$|d_K(x + h(Ax + y)) - d_K(U(h)x + hy)| \le \|x + hAx - U(h)x\|$$

valid for all $x \in D(A)$, we see that (4.11.1) implies that, for all $x \in K \cap D(A)$,

$$[Ax + \varphi(x)] \cap T_K(x) \neq \emptyset. \tag{4.13}$$

In case $E = \mathbb{R}^N$, (then A is defined everywhere and bounded), condition (4.13) is sufficient and necessary for the existence. The simple, constructive and based on the technique of the so-called proximal aiming, proof of this fact is given in [BaK1]. Moreover, (4.13) may be relaxed (instead of $T_K(\cdot)$ one can consider the convex envelope $\operatorname{conv} T_K(\cdot)$).

As indicated above, when dealing with (CP) from the view-point of the R_δ-characterization of $S(x)$, $x \in K$, we shall impose additional conditions on K: we shall assume that K is convex, epi-Lipschitz or strictly regular.

The following result holds:

(4.14) THEOREM. *If K is convex and $\varphi\colon K \to E$ is weakly tangent to K, i.e. condition (4.8.1) holds, then for each $x \in K$ the set $S(x)$ is an R_δ-set in $C(J,E)$ provided that (4.8.2.1) is satisfied or $\mathcal{U}$ is compact.*

For the proof we shall need the following result (see [BaK2]) which might be of interest on its own.

(4.15) LEMMA. *If $\varphi\colon K \multimap E$ is upper semicontinuous with closed convex values and tangent to K, then, for any $\varepsilon > 0$, there is a locally Lipschitz map $f\colon K \to E$ such that, for all $x \in K$, $f(x) \in \varphi(B(x,\varepsilon)\cap K) + B(0,\varepsilon)$ (35) and $f(x) \in T_K(x)$.*

PROOF OF THEOREM (4.14). Take $x \in K$; by Lemma (4.15), there is a locally Lipschitz $f_n\colon K \to E$ such that $f_n(x) \in T_K(x)$ and $f_n(x) \in \varphi_n(x) := \operatorname{cl}\operatorname{conv}\varphi(B_K(x,n^{-1})) + B(0,n^{-1})$ on K $(n \geq 1)$. Clearly, for $n \geq 1$, $\varphi(x) \subset \varphi_n(x) \subset \varphi_{n+1}(x)$ on K; thus

$$\emptyset \neq S_n(x) \supset S_{n+1} \quad \text{and} \quad S(x) \subset \bigcap_{n\geq 1} S_n(x),$$

where $S_n(x)$ stands for the set of all mild solutions to (CP) with φ replaced by φ_n.

Let $u_n \in S_n(x)$ for all $n \geq 1$. Using (4.8.2.1) (or the compactness of the semigroup $\mathcal{U}$) one shows then (see [BaK2, Appendix]) that there exists a subsequence $(u_{n_m})_{m\geq 1}$ such that $u_{n_m} \to u_0 \in S(x)$ in $C([0,1],E)$ as $m \to \infty$. It follows that

$$S^k(x_0,t_0) = \bigcap_{n=1}^{\infty} \operatorname{cl} S_n(x), \quad \beta_0(S_n(x)) \to 0 \quad \text{as } n \to \infty,$$

where β_0 stands for the Hausdorff measure of noncompactness in $C(J,E)$.

For each $n \geq 1$, $z \in [0,1]$ and $y \in K$, the problem

$$\begin{cases} v' \in Av + f_n(t,v), \\ v(z) = y, \end{cases} \tag{4.16}$$

admits a unique solution $v_n(\,\cdot\,; z,y)\colon J \to K$ which depends continuously on (z,y) (see Remarks (4.9)–(4.11)). Define the homotopy $h\colon [0,1]\times \operatorname{cl} S_n(x) \to C(J,E)$ by the formula

$$h(\lambda,u)(s) := \begin{cases} u(s) & \text{if } s \in [0,\lambda], \\ v(s;\lambda,u(\lambda)) & \text{if } s \in [\lambda,1], \end{cases}$$

for $u \in \operatorname{cl} S_n(x_0)$ and $\lambda \in [0,1]$. It is easy to see that h is continuous. Moreover, that $h([0,1]\times S_n(x)) \subset S_n(x)$. The continuity of h implies that $h([0,1]\times \operatorname{cl} S_n(x_0)) \subset \operatorname{cl} S_n(x)$. Finally observe that $h(0,u) = v(\,\cdot\,;0,x)$ and $h(1,u) = u$ for every $u \in \operatorname{cl} S_n(x_0)$, i.e. $\operatorname{cl} S_n(x_0)$ is contractible. Hence $S(x) \in R_\delta$. □

In a similar manner we obtain the following results (see [BaK2]).

(35) This means that f is an ε-graph-approximation of φ.

(4.17) THEOREM. *If K is an epi-Lipschitz set φ is weakly tangent to K, i.e.* (2.52.1) *holds:* $\varphi(x) \cap C_K(x) \neq \emptyset$ *on K, then for each $x \in K$, the set $S(x)$ of all mild solutions to the initial value problem* (CP) *is an R_δ-set provided the compactness condition* (4.8.2.1) *is satisfied or the semigroup $\mathcal{U}$ is compact.*

PROOF. Again one constructs, for each $n \geq 1$, a locally Lipschitz map $f_n\colon K \to E$ such that each point $x \in K$ has a neighbourhood W with $f_n(W)$ lying in a compact subset of E, $f_n(x) \in C_D(x)$ and

$$f_n(x) \in \varphi_n(x) := \operatorname{cl} \operatorname{conv} \varphi(B_K(x, n^{-1})) + B(0, n^{-1}) \tag{4.18}$$

for all $x \in K$. The construction reminds that from Lemma (4.15) but it also makes a strong use of the fact that $C_K(x) = \partial\Delta_K(x)^-$ (see the discussion of epi-Lipschitz sets) and $0 \notin \partial\Delta_K(x)$ on $\operatorname{bd} K$. Namely

$$f_n(x) = \sum_{s \in S} \lambda_s(x) w_s(t)$$

where $\{\lambda_s\}_{s \in S}$ is an appropriate locally Lipschitz partition of unity and, for $s \in S$, $w_s\colon J \to E$ is a measurable finite-valued function such that $\Delta_D^\circ(y; w_s(t)) < 0$ for all $t \in [0,1]$ and $y \in \operatorname{supp} \lambda_s$. Having this the proof concludes similarly as above.□

(4.19) THEOREM. *Assume that K is compact strictly regular and φ satisfies* (2.52.1). *Then, for any $x \in K$, the set $S(x)$ is an R_δ-set provided condition* (4.8.2.1) *is satisfied or $\mathcal{U}$ is compact.*

The proof of Theorem (4.19) is technically involved and long. It resembles the above sketched arguments with important modifications: one shows that φ may be appropriately extended to a map φ_n defined on a neighbourhood $K_n := \{x \in E : d_K(x) \leq \eta_n\}$ (where $\eta_n \to 0$ as $n \to \infty$) possessing a locally Lipschitz selection $f_n\colon K_n \to E$ such that $d^\circ_{K_n}(x; f_n(x)) \leq 0$ for $x \in K_n$ for any $n \geq 1$. Finally one shows that $S(x)$ is the intersection of the decreasing family of the contractible sets $S_n(x)$ of all mild solution of (CP) with K replaced by K_n and φ by φ_n.

The result in [BaK2] concerns in fact strictly regular sets which are actually non-compact, but in this case the weak tangency condition (2.52.1) must be replaced by the so-called *uniform tangency condition* which requires much more attention.

Being well-prepared we may turn our attention to the existence of coincidences, i.e. points $x \in K \cap D(A)$ such that $-A(x) \in \varphi(x)$. Under the assumptions of Theorem (4.14), (4.17) and (4.19), for all $t \in [0,1]$, the operator $P_t\colon K \multimap K$ is well-defined and admissible (being the composition of an upper semicontinuous map $S(\cdot)$ having acyclic compact values with single-valued and the continuous e_t).

We shall need an auxiliary result depending on the next standing assumption:

$$\text{there is } \theta \in \mathbb{R} \text{ such that, for any } t > 0,\ \|U(t)\|^\beta := \beta(U(t)B(0,1)) \leq e^{\theta t}. \tag{4.20}$$

Observe that since $\|U(t)\|^\beta \le \|U(t)\|$, the number θ always exists and $\theta \le \omega$. Recall that for a given bounded set Ω in E the estimate $\beta(U(t)\Omega) \le \|U(t)\|^\beta \beta(\Omega)$ holds.

We shall work with the measure of noncompactness $\widetilde{\beta}_0$ given by

$$\widetilde{\beta}_0(\Omega) := \sup\{\beta(C) : C \subset \Omega \text{ countable}\},$$

for each bounded subset Ω of E. It can be shown that $\widetilde{\beta}_0$ is regular, monotone and nonsingular (see [AKPRS, Section 1.4]).

Using these notions we have (see [BaK2], comp. [B]) the following theorem.

(4.21) THEOREM. *Assume that P_t is well-defined and let Ω be a bounded subset of K.*

(4.21.1) *If the semigroup $\mathcal{U}$ is compact, then $P_t(\Omega)$ is relatively compact. Thus P_t is a compact operator.*

(4.21.2) *Suppose that* (4.20) *and* (4.8.2.1) *hold* (*but we do not assume that the semigroup $\mathcal{U}$ is compact*). *Then*
$\beta(P_t(\Omega)) \le \exp(t(\theta + k))\beta(\Omega)$ *if E is separable;*
$\widetilde{\beta}_0(P(\Omega)) \le \exp(t(\theta + 2k))\widetilde{\beta}_0(\Omega)$ *if E is a weakly compactly generated;*
$\widetilde{\beta}_0(P(\Omega)) \le \exp(t(\theta + 4k))\widetilde{\beta}_0(\Omega)$ *if E is arbitrary.*

This result gives us means to establish conditions necessary for P_t to be a ν-set contraction. For instance if E is an arbitrary Banach space and $\theta + 4k < 0$, then P_t (if defined) is a ν-set contraction with $\nu := \exp(t(\theta + 4k))$.

Observe now that (in the situation of Theorems (4.14), (4.17) and (4.19)) P_t is an admissible map which being homotopic to the identity on K; the homotopy is provided by an admissible mapping $K \times [0,1] \ni (x, \lambda) \mapsto e_{\lambda t} \circ S(x)$. Hence we are in a position to use all results from Theorem (1.2). We therefore obtain the final result (see [BaK2]).

(4.22) THEOREM. *There is a point $x_0 \in K \cap D(A)$ such that $0 \in A(x_0) + \varphi(x_0)$ provided:*

(4.22.1) *K is compact, strictly regular, $\chi(K) \neq 0$ and* (2.52.1) *holds, i.e. for all $x \in K$, $\varphi(x) \cap C_K(x) \neq \emptyset$; or*

(4.22.2) *K is compact convex and φ is weakly tangent, i.e. satisfies condition* (4.8.1)*; or*

(4.22.3) *K is convex, the semigroup $\mathcal{U}$ is compact and φ is weakly tangent; or*

(4.22.4) *K is epi-Lipschitz bounded and acyclic, $\mathcal{U}$ is compact and φ satisfies condition* (2.52.1)*; or*

(4.22.5) *K is convex, φ is a weakly tangent k-contraction such that $\theta + k < 0$ if E is separable, $\theta + 2k < 0$ in case E is weakly compactly generated and $\theta + 4k < 0$ for a general Banach space E; or*

(4.22.6) *K is convex and weakly compact, φ is weakly tangent, strongly upper semicontinuous and $\mathcal{U}$ is nonexpansive and uniformly continuous.*

PROOF. We shall prove only the compact cases (4.22.1), (4.22.2). By Theorem (1.2), for each $n \in \mathbb{N}$, there is a 2^{-n}-periodic solution u_n of (CP). The compactness of K shows that, passing to a subsequence if necessary, $u_n \to u$ in $C([0,1], E)$ and u is a mild solution to (CP). Clearly $u(t) \equiv x_0 \in K$. Thus

$$x_0 = U(t)x_0 + \int_0^t U(t-s)w(s)\,ds, \quad t \in [0,T], \tag{4.23}$$

where $w(s) \in \varphi(x_0)$ on $[0,1]$. The function $v(t) = \int_0^t w(s)\,ds$ is differentiable a.e. Suppose that for $z \in (0,1)$ $v'(z) = w(z) =: y_0 \in \varphi(x_0)$ exists. By (4.23), for $h > 0$,

$$x_0 = U(h)x_0 + \int_z^{z+h} U(z+h-s)w(s)\,ds.$$

Hence

$$\frac{U(h)x_0 - x_0}{h} = \frac{1}{h}\int_z^{z+h} (w(s) - U(z+h-s)w(s))\,ds - \frac{v(z+h)-v(z)}{h}.$$

The first term in the right-hand side tends to 0 as $h \to 0^+$ and the second one converges to $-y_0$. Hence $x_0 \in D(A)$ and $A(x_0) = -y_0$. □

(4.24) REMARK. Observe that if $A = -I$, then $U(t)x = e^{-t}x$; hence $\theta = -1$. In case E is separable, (4.22.5) implies that φ has a fixed point provided it is a weakly tangent set-contraction. This is nothing else but the Theorem (2.33)of Deimling. However, in order to establish the existence of fixed points of a k-set contraction in the case of an arbitrary Banach space E by means of Theorem (4.22) we need $k < 1/4$.

(4.25) REMARK. Theorems (4.8), (4.14), (4.17), (4.19) and (4.22) have been already generalized. In his doctoral dissertation, Ćwiszewski studies the existence and the structure of solutions to the following problem

$$\begin{cases} u'(t) = -A(u(t)) + \varphi(u(t)), \\ u(t) \in K, \\ u(0) = x \in K, \end{cases} \tag{4.25.1}$$

where $K \subset E$ is closed, $A\colon E \supset D(A) \multimap E$ is an m-accretive operator and $\varphi\colon K \multimap E$ is an upper semicontinuous map with compact convex values. Moreover, he assumes that, for all $t \geq 0$, $S_A(t)(K) \subset K$ (where $S_A(\cdot)$ is the semigroup given by the Crandall–Liggett formula [CL]) and, for all $x \in K$, $\varphi(x) \cap C_K(x) \neq \emptyset$. The solutions to (4.25.1) are considered in the *integral* sense (comp. Proof of (3.20.2)). After establishing the existence, he shows analogues of Theorems (4.14), (4.17), (4.19) and (4.22) under appropriate additional hypotheses (for instance, in case φ is not single-valued, then it is necessary to assume that E^* is uniformly convex). The existence of equilibria (i.e. points $x \in D(A) \cap K$ such that $0 \in -A(x) + \varphi(x)$) is, in fact, put into the context of a suitable topological degree theory. Results of [Cw] will be published elsewhere.

We end the paper with the following observation complementing Theorem (4.22) and generalizing Theorem (3.2).

(4.26) THEOREM. *Let $K \subset E$ be a compact neighbourhood retract with $\chi(K) \neq 0$. If $\varphi: U \multimap E$, where U is an open neighbourhood of K, has compact convex values and is Lipschitzian ((3.1) holds for all $x, y \in U$) and satisfies the strong tangency condition (2.13.1) on K, then there is $x_0 \in K \cap D(A)$ such that $0 \in A(x_0) + \varphi(x_0)$.*

PROOF. Without loss of generality we may assume that φ is defined on E, upper semicontinuous with compact convex values, satisfies the linear growth, the strong tangency on K and the Lipschitz condition on a neighbourhood of K. To see this consider an open neighbourhood V of K such that $\operatorname{cl} V \subset U$ and an Urysohn function $\eta: E \to [0,1]$ such that $\eta|_{\operatorname{cl} V} \equiv 1$, $\eta|_{E \setminus U} \equiv 0$. The map $\widetilde{\varphi}(x) = \eta(x)\varphi(x)$ satisfies the enumerated properties.

For each $x \in K$, the set $\widetilde{S}(x)$ of mild solutions $u: [0,1] \to E$ to (CP) (in E) is an R_δ-set and the map $K \ni x \mapsto \widetilde{S}(x)$ is upper semicontinuous. We shall see that $\widetilde{S}(x) = S(x)$ on K where $S(x)$ has the same meaning as in Theorem (4.8). To this end take $x \in K$, let $u \in \widetilde{S}(x)$ and suppose that, for $t \in [0,T]$, $u(t) \in U$. It is clear that

$$u(t) = U(t)x + \int_0^t U(t-s)w(s)\,ds,$$

where $w(s) \in \varphi(u(s))$ on $[0,T]$. The function $U(\,\cdot\,)x$ is continuously differentiable and $[0,T] \ni t \mapsto \int_0^t U(t-s)w(s)\,ds$ is absolutely continuous; hence $f: [0,T] \to \mathbb{R}$ given by $f(t) = d_K(u(t))$ is absolutely continuous and differentiable almost everywhere. Fix $t \in [0,T)$ for which $f'(t)$ exists, let $z \in \pi_K(u(t))$ and pick $v \in T_K(z) \subset T_K^{\mathcal{U}}(z)$. For each (sufficiently small) $h > 0$,

$$u(t+h) = U(h)u(t) + \int_t^{t+h} U(t+h-s)w(s)\,ds$$

and

$$\begin{aligned} f(t+h) - f(t) &= d_K(u(t+h)) - d_K(u(t)) \\ &\le d_K(U(h)z + hv_h) + \|U(h)z - U(h)u(t)\| - \|z - u(t)\| \\ &\le d_K(U(h)z + hv_h) + (e^{\omega h} - 1)f(t) \end{aligned}$$

where

$$v_h := \frac{1}{h}\int_t^{t+h} U(t+h-s)w(s)\,ds.$$

Without loss of generality, by the upper semicontinuity, $v_h \to w \in \varphi(u(t))$ as $h \to 0^+$. Hence,

$$\begin{aligned} f'(t) &= \liminf_{h\to 0^+} \frac{f(t+h)-f(t)}{h} \\ &\leq \liminf_{h\to 0^+} \frac{d_K(U(h)z+hv)+(e^{\omega h}-1)f(t)}{h} + \|v-w\| \leq \omega f(t) + \|v-w\|. \end{aligned}$$

Taking inf over $T_K(z)$ and since $\varphi(z) \subset T_K(z)$ we get

$$\begin{aligned} f'(t) &\leq \omega f(t) + d(w, T_K(z)) \\ &\leq \omega f(t) + d_H(\varphi(u(t)), \varphi(z)) \leq (\omega+k)\|u(t)-z\| = (\omega+k)f(t). \end{aligned}$$

Since $f(0) = 0$, the Gronwall inequality implies that $f(t) \equiv 0$ and, therefore $u(t) \in K$. The existence of equilibria follows now exactly as in the proof of Theorem (4.22). □

At this point we end this survey. As indicated at the beginning the presentation is probably far from being complete. It seems the discussed topic is developing rapidly and so is the literature. The author expresses his conviction that there is still a lot of open questions which deserve a deep study.

References

[AKPRS] R. R. Akhmerov, M. I. Kamenskiĭ, A. S. Potapov, A. E. Rodkina and B. N. Sadovskiĭ, *Measures of Noncompactness and Condensing Operators*, Birkhäuser, 1992.

[Ar] N. Aronszajn, *Le correspondant topologique de l'unicite dans la théorie des équations differentielles*, Ann. of Math. **43** (1942), 730–738.

[AE] J.-P. Aubin and I. Ekeland, *Applied Nonlinear Analysis*, Wiley–Interscience, New York, 1987.

[AF] J.-P. Aubin and H. Frankowska, *Set-Valued Analysis*, Birkhäuser, Boston–Basel–Berlin, 1990.

[B] R. Bader, *The periodic problem for semilinear differential inclusions in Banach spaces*, Comment. Math. Univ. Carolin. **39** (1998), 671–684.

[BaK1] R. Bader and W. Kryszewski, *On the solution set of constrained differential inclusions with applications*, Set-Valued Anal. **9** (2001), 289–313.

[BaK2] ______, *On the solution sets of differential inclusions and the periodic problem in Banach spaces*, Nonlinear Anal. **54** (2003), 707–754.

[Ba] V. Barbu, *Continuous perturbations of nonlinear m-accretive operators in Banach spaces*, Boll. Un. Mat. Ital. (4) **6** (1972), 270–278.

[BDG1] H. Ben-El-Mechaiekh, P. Deguire and A. Granas, *Une alternative non linéaire en analyse convexe et applications*, C. R. Acad. Sci. Paris Sér. I **295** (1982), 257–259.

[BDG2] ______, *Points fixex et coincidences pour les applications multivoque II*, C. R. Acad. Sci. Paris Sér. I **295** (1982), 381–384.

[BK1] H. Ben-El-Mechaiekh and W. Kryszewski, *Equilibria of set-valued maps on nonconvex domains*, Trans. Amer. Math. Soc. **349** (1997), 4159–4179.

[BK2] ———, *Equilibrium for perturbations of multifunctions by convex processes*, Georegian Math. J. **3** (1996), 201–215.

[BP] C. Bessaga and A. Pe-czyński, *Selected topics in infinite-dimensional topology*, Monografie Mat., vol. 58, PWN, Warszawa, 1975.

[BC] J.-M. Bonnisseau and B. Cornet, *Fixed point theorems and Morse's lemma for lipschitzian functions*, J. Math. Anal. Appl. **146** (1990), 318–332.

[Bor] K. Borsuk, *Theory of Retracts*, Monografie Mat., vol. 44, PWN, Warszawa, 1967.

[Bo1] D. Bothe, *Multivalued differential equations on graphs and applications*, Ph. D. dissertation, Universität Paderborn (1992).

[Bo2] ———, *Flow invariance for perturbe nonlinear evolution equations*, Abstr. Appl. Anal. **1** (1996), 379–395.

[BoK] H. F. Bohnenblust and S. Karlin, *On a theorem of Ville*, Contributions to the Game Theory, vol. 1, Princeton, 1950.

[Bre] H. Brezis, *On a characterization of flow invariant sets*, Comm. Pure Appl. Math. **23** (1970), 261–263.

[Br] R. Brown, *The Lefschetz Fixed Point Theorem*, Scott, Foresman and Comp., Glenview Ill., London, 1971.

[Bro] F. E. Browder, *The fixed point theory for multivalued mappings in topological vector spaces*, Math. Ann. **177** (1968), 283–301.

[CGQ] P. Cardaliaguet, G. Gabor and M. Quincampoix, *Equilibria and strict equilibria of multivalued maos on noninvariant sets*, Ann. Polon. Math. **82** (2003), 19–37.

[C] K. Caristi, *Fixed point theorems for mappings satisfying inwardness conditions*, Trans. Amer. Math. Soc. **215** (1976), 241–251.

[C1] K. C. Chang, *Variational methods for nondifferentiable functionals and its applications to partial differential equations*, J. Math. Anal. Appl. **80** (1981), 102–129.

[C2] ———, *Infinite Dimensional Morse Theory and Multiple Solution Problems*, Birkhäuser, Boston, 1993.

[Cl1] F. H. Clarke, *Pointwise contractional criteria for the existence of fixed points*, Bull. Canad. Math. Soc. **21** (1978), 7–11.

[Cl2] ———, *Optimization and Nonsmooth Analysis*, Wiley-Interscience, New York, 1983.

[CLS] F. H. Clarke, Yu. S. Ledyaev and R. J. Stern, *Fixed points and equilibria in nonconvex sets*, Nonlinear Anal. **25** (1995), 145–161.

[CG] G. Colombo and V. V. Goncharov, *Variational inequalities and regularity properties of closed sets in Hilbert spaces*, J. Convex Anal. **8** (2001), 197–221.

[Co1] B. Cornet, *Paris avec handicaps et théoremes de surjectivité de correspondances*, C. R. Acad. Sci. Paris Sér. A **281** (1975, 479–482).

[Co2] ———, *Euler characteristic and fixed point theorem*, Positivity **6** (2002), 243–260.

[CC] B. Cornet and M. O. Czarnecki, *Necessary and sufficient conditions for the existence of (generalized) equilibria on compact epi-Lipschitzian domain*, Comm. Appl. Nonlinear Anal. **7** (2000), 21–53.

[CL] M. G. Crandall and T. M. Liggett, *Generation of semi-groups of nonlinear transformations on general Banach spaces*, Amer. J. Math. **93** (1971), 265–298.

[Cw] A. Ćwiszewski, *Zagadnienia ró.zniczkowe z ograniczeniami na stan; stopień topologiczny zaburzeń operatorŵ akretywnych*, Ph. D. Thesis (2003), Nicolaus Copernicus University, Toruń.

[CK1] A. Ćwiszewski and W. Kryszewski, *Equilibria of set-valued maps: a variational approach*, Nonlinear Anal. **48** (2002), 707–746.

[CK2] ———, *Approximate smoothing of locally Lipschitz functionals*, Boll. Un. Mat. Ital. B **5** (2002), 289–320.

[CK3] ———, *Degree theory for tangent set-valued vector fields on closed sets*, submitted.

[D1] K. Deimling, *Fixed points of weakly inward multis*, Nonlinear Anal. **10** (1986), 1261–1262.

[D2] ______, *Multivalued differential equations*, Walter de Gruyter, Berlin, New York, 1992.

[DHP] K. Deimling, S. C. Hu and J. Prüss, *Fixed points of weakly inward multivalued maps*, Nonlinear Anal. **10** (1986), 465–469.

[DRS] J. Diestel, W. M. Ruess and W. Schachermayer, *Weak compactness in $L^1(\mu, X)$*, Proc. Amer. Math. Soc. **118** (1993), 447–453.

[Do] T. Donchev, *Properties of one-sided Lipschitz multivalued maps*, Nonlinear Anal. **49** (2002), 13-20.

[DK] D. Downing and W. A. Kirk, *Fixed point theorems for set-valued mappings in metric and Banach spaces*, Math. Japonica **22** (1977), 99–112.

[DMNZ] R. Dragoni, J. W. Macki, P. Nistri and P. Zecca, *Solution Sets of Differential Operators in Abstract Spaces*, Addison Wesley Longman Ltd., Harlow, 1996.

[DG] J. Dugundji and A. Granas, *Fixed Point Theory*, Monografie Mat., vol. 61, PWN, Warszawa, 1982.

[F1] K. Fan, *Fixed points and minimax theorems in locally convex topological spaces*, Proc. Nat. Acad. Sci. U.S.A. **38** (1952), 121–126.

[F2] ______, *Extensions of two fixed point theorem of F. E. Browder*, Math. Z. **112** (1969), 234–240.

[F3] ______, *Some properties of convex sets related to fixed point theorems*, Math. Ann. **266** (1984), 519–537.

[GQ] G. Gabor and M. Quincampoix, *On existence of equilibria of set-valued maps*, Boll. Un. Mat. Ital. B (8) **6** (2003), 309–321.

[Gl] I. L. Glicksberg, *A further generalization of the Kakutani fixed point theorem, with application to Nash equilibrium points*, Proc. Amer. Math. Soc. **3** (1952), 170–174.

[GK] K. Goebel and W. A. Kirk, *Topics in Metric Fixed Point Theory* (1990), Cambridge Univ. Press, Cambridge.

[Go1] L. Górniewicz, *Homological methods in fixed point theory of multi-valued maps*, Dissertationes Math. **129** (1976), 1–66.

[Go2] ______, *Topological Fixed Point Theory*, Kluwer, 1999.

[Go3] ______, *Topological structure of solution sets: Current results*, Arch. Math. (Brno) **36** (2000), 343–382.

[H1] B. Halpern, *Fixed point theorem for outward maps*, Ph. D. Thesis (1965), UCLA.

[H2] ______, *Fixed point theorems for set-valued maps in infinite dimensional spaces*, Math. Ann. **189** (1970), 87–98.

[H3] ______, *A general coincidence theory*, Pacific J. Math. **77** (1978), 451–471.

[HB1] B. R. Halpern and G. M. Berman, *A fixed point theorem for inward and outward maps*, Trans. Amer. Mat. Soc. **62** (1968), 353–358.

[HB2] ______, *A fixed point theorem for inward and outward maps*, Trans. Amer. Math. Soc. **130** (1968), 353–358.

[Ha] P. Hartman, *Ordinary differential eqiuations* (1982), Birkhaüser, Boston.

[HV1] C. Himmelberg and F. Van Vleck, *On the topological triviality of solution sets*, Rocky Mountain J. Math. **10** (1980), 247–252.

[HV2] ______, *A note on the solution sets of differential inclusions*, Rocky Mountain J. Math. **12** (1982), 621–625.

[Hi] M. Hirsch, Springer-Verlag, New York, Heidelberg, Berlin (1976).

[HP] S. Hu and N. S. Papageorgiou, *Handbook of Set-Valued Analysis*, vol. I, Kluwer Academic Publishers, 1999; vol. II, 2001.

[HS] S. Hu and Y. Sun, *Fixed point index for weakly inward maps*, J. Math. Anal. Appl. **172** (1993), 266–273.

[H] D. M. Hyman, *On decreasing sequences of compact absolute retracts*, Fund. Math. **64** (1969), 91–97.

[KOZ] M. Kamenskiĭ, V. Obukhovskiĭ and P. Zecca, *Condensing multivalued maps and semilinear differential inclusions in Banach spaces*, Walter de Gruyter (to appear).

[K] S. Kakutani, *A generalization of Brouwer's fixed point theorem*, Duke J. Math. **8** (1941), 457–459.

[Ka] T. Kato, *Nonlinear semigroups and evolution equations*, J. Math. Soc. Japan, **19** (1967), 508–520.

[K1] W. A. Kirk, *History and methods of metric fixed point theory*, Antipodal Points and Fixed Points, Lecture Notes, vol. 28, Seul National University, Research Institute of Mathematics, 1995.

[K2] ———, *Transfinite methods in metric fixed-point theory*, Abstr. Appl. Anal. **5** (2003), 311–324.

[KM] W. A. Kirk and S. Massa, *Remarks on asymptotic and Chebyshev centers*, Houston J. Math. **16** (1990), 357–364.

[Ko] H. N. Ko, *Fixed points theorems for point-to-set mappinga and the set of fixed points*, Pacific J. Math. **42** (1972), 369–379.

[KZ] M. A. Krasnosel'skiĭ and P. P. Zabreĭko, *Geometrical Methods in Nonlinear Analysis*, Nauka, Moscow, 1975. (Russian)

[Kr1] W. Kryszewski, *Topological and approximation methods in the degree theory of set-valued maps*, Diss. Math. **336** (1994), 1–102.

[Kr2] ———, *Homotopy properties of set-valued mappings*, Habilitation Thesis (1997), Nicolaus Copernicus University, Torunń.

[LD] E. Lami Dozo, *Multivalued nonexpansive mappings and Opial's condition*, Proc. Amer. Math. Soc. **38** (1973), 286–292.

[L] M. Lassonde, *On the use of* KKM *multifunctions in fixed point theory and related topics*, J. Math. Anal. Appl. **97** (1983), 151–201.

[L1] T. C. Lim, *A fixed point theorem for multivalued nonexpansive mappings in a uniformly convex Banach space*, Bull. Amer. Math. Soc. **124** (1996), 3345–3349.

[L2] ———, *A fixed point theorem for weakly inward multivalued contractions*, J. Math. Anal. Appl. **247** (2000), 323–327.

[M1] R. H. Martin, *Differential equations on closed sets in Banach spaces*, Trans. Amer. Math. Soc. **179** (1973), 399–414.

[M2] R. H. Martin, *Nonlinear Operators and Differential Equations in Banach Spaces*, Wiley–Interscience, New York, 1976..

[MY] C. Martinez-Yanez, *A remark on weakly inward multivalued mappings*, Nonlinear Anal. **16** (1991), 847–848.

[Mi] G. Minty, *A theorem on monotone sets in Hibert spaces*, J. Math. Anal. Appl. **11** (1967), 434–439.

[MT] N. Mizoguchi and W. Takahashi, *Fixed point for multivalued mappings on complete metric spaces*, J. Math. Anal. Appl. **141** (1989), 177–188.

[MP] D. Motreanu and N. H. Pavel, *Tangency, flow invariance for differential equations and optimization problems* (1999), Marcel Dekker Inc., New York.

[N] S. B. Nadler Jr., *Multi-valued contraction mappings*, Pacific J. Math. **30** (1969), 475–488.

[O1] D. O'Regan, *A continuation theory for weakly inward maps*, Glasgow Math. J. **40** (1998), 311–321.

[O2] ———, *Fixed points and random fixed points for weakly inward approximable maps*, Proc. Amer. Math. Soc. **38** (1999), 89–100.

[P] S. Park, *Eighty years of the Brouwer fixed point theorem*, Lecture Notes, vol. 28, Seul National University, Research Institute of Mathematics, 1995.

[Pa] A. Pazy, Springer–Verlag 1983.

[Pe] J.-P. Penot, *Fixed point theorems without convexity*, Bull. Soc. Math. France Mém. **60** (1979), 129–152.

[Pl] S. Plaskacz, *On the solution sets of differential inclusions*, Boll. Un. Mat. Ital. A (7) **6** (1992), 387–394.

[Pr] R. Precup, *On some fixed point theorems of Deimling*, Nonlinear Anal. **23** (1994), 1315–1320.

[R1] S. Reich, *Fixed points in locally convex spaces*, Math. Z. **125** (1972), 17–31.

[R2] ———, *The range and sums of accretive and monotone operators*, J. Math. Anal. Appl. **68** (1979), 310–317.

[R3] ———, *Fixed points of condensing functions*, J. Math. Anal. Appl. **41** (1973), 460–467.

[R4] ———, *Approximate selections, best approximations, fixed point and invariant sets*, J. Math. Anal. Appl. **62** (1978), 104–113.

[Ro] R. T. Rockafellar, *Clarke's tangent cones and boundaries of closed sets in* $\mathbb{R}^n$, Nonlinear Analysis **3** (1070), 145–154.

[S] P. Saveliev, *A Lefschetz type coincidence theorem.*

[Si] M. Sion, *On general minimax theorem*, Pacific J. Math. (1958), 171–176.

[Sm] G. V. Smirnov, *Introduction to the theory of differential inclusions*, Grad. Texts in Maths. **41** (2002), Amer. Math. Soc. Providence.

[So] W. Song, *A generalization of Clarke's fixed point theorem*, Appl. Math. J. Chinese Univ. Ser. **10 B** (1995), 13–23.

[Sp] E. Spanier, *Algebraic Topology*, McGraw-Hill, New York, 1966; Glenview, Illinois, 1972.

[St] R. J. Stern, *On zeros of multifunctions in non-lipschitzian sets*, manuscript (1996).

[WW] D. W. Walkup and R. Wets, *Continuity of some convex-cone-valued mapping*, Proc. Amer. Math. Soc. **18**, 229–235.

[Wi] M. Willem, *Minimax Theorems*, Birkhäuser, Boston, 1996.

[X1] H.-K. Xu, *Metric fixed point theory for multivalued mappings*, Diss. Math. **389** (2000), 1–39.

[X2] ———, *Metric fixed point theory for multivalued mappings*, Diss. Math. **389** (2000), 1–39.

[X3] ———, *Multivalued nonexpansive mappings in Banach spaces*, Nonlinear Anal. **43** (2001), 693–706.

[Ya] K. Yanagi, *On some fixed point theorems for multivalued mappings*, Pacific. J. Math. **87** (1980), 233–240.

[Yo] K. Yosida, *Functinal Analysis*, Springer–Verlag, Berlin, 1966.

22. TOPOLOGICAL FIXED POINT THEORY AND NONLINEAR DIFFERENTIAL EQUATIONS

JEAN MAWHIN

1. Introduction

The development of fixed point theory is very closely linked with the study of various problems in ordinary differential equations. On one side, in his famous memoir of 1890 on the three-body problem crowned by King Oscar Prize in [Po3], Poincaré reduced the study of the T-periodic solutions of a differential system in $\mathbb{R}^n$

$$x' = f(t, x) \tag{1.1}$$

to the study of the fixed points of the operator P_T defined on $\mathbb{R}^n$ by

$$P_T(y) = p(T; 0, y),$$

where $p(t; s, y)$ denotes the solution of (1.1) verifying the initial condition

$$x(s) = y.$$

This is a period where fixed point theory was in its infancy, to say the least, and, as observed by P. S. Aleksandrov in [Al]:

- As regards theorems on the existence of fixed points under this or that continuous mapping, Poincaré already understood the significance of these theorems as a means of proving existence theorems in analysis.

As early as 1883, Poincaré stated in [Po1] a theorem shown much later equivalent to a fixed point theorem for continuous mappings of a closed ball into itself (see [Br]), published by Brouwer the very same year 1912 as Poincaré's fixed point theorem for area-preserving mappings of an annulus (see [Po4]).

Ordinary and partial differential equations are also at the source of fixed point theory in infinite-dimensional spaces. The method of successive approximations or iterations for finding solutions requires the equation to be put in a fixed point form. Picard's systematic application of this method to various differential equations problems from 1890 (see [Pi]) led to Banach fixed point theorem for contractions ([B]), the same year as boundary value problems for nonlinear ordinary differential equations were motivating Birkhoff–Kellogg's extension of Brouwer fixed point theorem to some function spaces (see [BiK]). In 1930, nonlinear elliptic problems were at the origin of Schauder's extension of Brouwer fixed point theorem to arbitrary Banach spaces (see [Sch]).

Those close connections did not stop, and it would be hopeless to try to estimate the number of papers on differential equations based upon fixed point theory. We shall highlight, in this paper, some recent ones, related to topological fixed point theory, without any hope of being exhaustive.

2. Poincaré's operator and fixed point theorems

2.1. Poincaré's operator. Let $f\colon \mathbb{R}\times\mathbb{R}^n \to \mathbb{R}^n, (t,x)\mapsto f(t,x)$ be locally Lipschitzian in x and continuous. For each $(s,y)\in\mathbb{R}\times\mathbb{R}^n$, there exists a unique solution $x(t)=p(t;s,y)$ of the *Cauchy or initial value problem*

$$x' = f(t,x), \quad x(s)=y, \tag{2.1}$$

defined on a maximal interval $]\tau_-(s,y),\tau_+(s,y)[$, for some

$$-\infty \le \tau_-(s,y) < s < \tau_+(s,y) \le \infty.$$

Moreover, p is continuous on its open set of definition

$$G = \{(t,s,y)\in\mathbb{R}\times\mathbb{R}\times\mathbb{R}^n : \tau_-(s,y) < t < \tau_+(s,y)\}$$

(see e.g. [CL], [Ku]).

Assume in addition that f is T*-periodic* with respect to t for some $T>0$, i.e. that $f(t,x)=f(t+T,x)$ for all $t\in\mathbb{R}$.

(2.2) Definition. A T*-periodic solution* of

$$x' = f(t,x), \tag{2.2.1}$$

is a solution of (2.2.1) defined over $\mathbb{R}$ and such that $x(t)=x(t+T)$ for all $t\in\mathbb{R}$.

As already observed by Poincaré, $x(t)=p(t;0,y)$ *is a* T*-periodic solution of* (2.2.1) *if and only if* $y\in\mathbb{R}^n$ *is such that* $\tau_+(0,y)>T$ *and* $y=p(T;0,y)$, i.e. if and only if y is a fixed point of the *Poincaré's operator* P_T defined on $\{y\in\mathbb{R}^n : \tau_+(0,y)>T\}$ by

$$P_T(y) = p(T;0,y).$$

The difficulty in applying Poincaré method lies in the fact that the maximal interval of existence of $p(t;0,y)$ may me difficult to estimate and that P_T is rarely explicitely known. The first difficulty can be overcome by finding conditions under which all solutions exist for $t \in [0,\infty[$, for example by exhibiting a *positively invariant* bounded set, i.e. a bounded set G such that $p(t;0,y) \in G$ for $t > 0$ whenever $y \in G$. If this set has additional properties, Brouwer fixed point theorem (see [Br]), that we recall now, can be used.

2.2. Applications of Brouwer fixed point theorem.

(2.3) LEMMA. *Let $C \subset \mathbb{R}^n$ be homeomorphic to a closed ball. Then each continuous mapping $F\colon C \to C$ has at least one fixed point.*

Lemma (2.3) was first applied in 1943, to some forced Liénard equations, by Lefchetz ([L]) and Levinson ([Le1]), as well as by Chevalley in an unpublished work, and by Nagumo ([N]) in 1944.

(2.4) THEOREM. *Assume that there exists a bounded open positively invariant set $\Delta \subset \mathbb{R}^n$, with closure $\overline{\Delta}$ homeomorphic to a ball, such that each solution $p(t;0,y)$ with $y \in \overline{\Delta}$ belongs to $\overline{\Delta}$ for all $t \geq 0$. Then the system* (2.2.1) *has at least one T-periodic solution.*

PROOF. By assumption, P_T maps continuously $\overline{\Delta}$ into itself, and has a fixed point by Lemma (2.3). □

In applying this idea, Lefschetz and Nagumo construct the domain Δ, for the system

$$x_1' = x_2 - g(x_1), \quad x_2' = -f(x_1) + e(t), \tag{2.5}$$

equivalent to the Liénard equation

$$y'' + f'(y)y' + g(y) = e(t),$$

through a function $V(x_1, x_2)$ such that, for some $R > 0$,

$$\overline{\Delta} = \{(x_1, x_2) : V(x_1, x_2) \leq R\},$$

is homeomorphic to a closed disc and

$$\partial_{x_1} V(x_1, x_2)[x_2 - g(x_1)] + \partial_{x_2} V(x_1, x_2)[-f(x_1) + e(t)] < 0$$

whenever $V(x_1, x_2) \geq R$. The same idea is applied by Lefschetz to a system of the form

$$x' = g(x) + h(t, x), \tag{2.6}$$

where the g_i are homogeneous polynomials of the same degree p and the h_i are polynomials with continuous T-periodic coefficients.

(2.7) THEOREM. *Assume that* $\deg h_i < p$, $(1 \leq i \leq n)$ *and that there exists a positive definite quadratic form*

$$V(x) = \frac{1}{2} \sum_{i,j=1}^{n} a_{ij} x_i x_j$$

such that

$$\langle V'(x), g(x) \rangle < 0 \tag{2.7.1}$$

whenever $|x| = 1$. *Then the system* (2.6) *has at least one* T*-periodic solution.*

PROOF. Notice that $\langle V'(x), g(x) \rangle < 0$ is an homogeneous polynomial of degree $p+1$ and that, by (2.7.1), there exists $\alpha > 0$ such that

$$\langle V'(x), g(x) \rangle \leq -\alpha |x|^{p+1}$$

for all $x \in \mathbb{R}^n$. On the other hand, there exists $\rho > 0$ such that

$$|\langle V'(x), h(t,x) \rangle| \leq \frac{\alpha}{2} |x|^{p+1}$$

whenever $t \in [0, T]$ and $|x| \geq \rho$. Consequently,

$$\langle V'(x), g(x) + h(t,x) \rangle \leq -\frac{\alpha}{2} |x|^{p+1} < 0$$

whenever $t \in [0, T]$ and $|x| \geq \rho$. The closure of the ellipsoid $\Delta = \{x \in \mathbb{R}^n : V(x) < \rho\}$ is mapped into itself by the Poincaré's map P_T. □

2.3. Dissipative systems. We now consider the class of dissipative systems or systems of class D as defined by Levinson in 1944 (see [Le2]).

(2.8) DEFINITION. We say that system (2.2.1) is *dissipative* or *of class* D if there exists $R > 0$ such each solution $p(t; s, y)$ of (2.2.1) satisfies

$$\limsup_{t \to \infty} |p(t; s, y)| < R. \tag{2.8.1}$$

Then, each solution $p(t; 0, y)$ exists for $t \in [0, \infty[$, and Poincaré's operator is defined for each $y \in \mathbb{R}^n$.

Clearly, condition (2.2.1) implies that P_T^n maps the closed ball $B[R]$ into itself for n sufficiently large, but this only implies the existence of a nT-periodic solution of (2.2.1). To prove the existence of a T-periodic solution, one can rely, following [KZ], upon a more refined fixed point theorem than Brouwer's one, obtained by F. Browder in 1959 (see [Bro]). Levinson's original proof, for $n = 2$, required a more restrictive concept of dissipativeness. See [Pl] for other properties and for examples of dissipative systems.

(2.9) LEMMA. *Assume that* $F: \mathbb{R}^n \to \mathbb{R}^n$ *is continuous and that there exists a bounded, convex open set* Ω *and a positive integer* n_0 *such that, for all* $n \geq n_0$, $F^n(\overline{\Omega}) \subset \overline{\Omega}$ *and* $F^n(x) \neq x$ *whenever* $x \in \partial\Omega$. *Then* F *has at least one fixed point in* Ω.

(2.10) THEOREM. *Each dissipative system* (2.2.1) *has at least one* T*-periodic solution.*

PROOF. By the dissipativeness condition (2.8.1), for each $z \in P_T(B[R])$, there exists an integer $m(z)$ such that $P_T^{m(z)}(z) \in B(R)$. Hence there is an open neighbourhood $V(z)$ of z such that $P_T^{m(z)}(V(z)) \subset B(R)$. By compactness of $P_T(B[R])$, we extract a finite subcovering $\{V(z_j)\}_{1 \leq j \leq s}$ of $P_T(B[R])$, and set $m_0 = \max_{1 \leq j \leq s} m(z_j)$. Choose $R_1 > 0$ such that $\bigcup_{j=0}^{m_0} P_T^j(B[R]) \subset B[R_1]$, and let $\Omega = B(R_1 + 1)$. By a similar reasoning, for each $y \in \overline{P_T(\Omega)}$, there is a positive integer $k(y)$ and an open neighbourhood $W(y)$ such that $P_T^{k(y)}(W(y)) \subset B(R)$. Then $P_T^n(W(y)) \subset B[R_1]$ if $n \geq m_0 + k(y)$. Again, extract a finite covering $\{W(y_j)\}_{1 \leq j \leq r}$ of $\overline{P_T(\Omega)}$. If $n_0 = m_0 + \max_{1 \leq j \leq r} k(y_j)$, then $P_T^n(\overline{\Omega}) \subset B[R_1]$ whenever $n \geq n_0$, and P_T has a fixed point by Lemma (2.9). □

3. Poincaré's operator and Brouwer degree

3.1. Brouwer degree. Lemma (2.3) is easily proved using Brouwer degree [Br], an integer which counts algebraically the number of zeros of F in $\overline{\Omega}$ in a way which invariant for small perturbations of F. Let $\Omega \subset \mathbb{R}^n$ be open and bounded, and let $V_0(\Omega, \mathbb{R}^n)$ be the class of continuous mappings $F: \overline{\Omega} \to \mathbb{R}^n$ such that $0 \notin F(\partial\Omega)$. To each $F \in V_0(\Omega, \mathbb{R}^n)$, Brouwer [Br] has associated in 1912 an integer $d_B[F, \Omega]$, the *degree of* F *in* Ω, satisfying the following basic properties, in which, for any set $\Delta \subset \mathbb{R}^n \times [0,1]$, and any $\lambda \in [0,1]$, we write

$$\Delta_\lambda = \{x \in \mathbb{R}^n : (x, \lambda) \in \Delta\}.$$

(3.1) PROPOSITION (Addition-excision). *If* Ω_1 *and* Ω_2 *are disjoint open subsets in* Ω *such that* $0 \notin F(\overline{\Omega} \setminus (\Omega_1 \cup \Omega_2))$, *then* $F \in V_0(\Omega_i, \mathbb{R}^n)$, $(i = 1, 2)$ *and*

$$d_B[F, \Omega] = d_B[F, \Omega_1] + d_B[F, \Omega_2].$$

(3.2) PROPOSITION (Homotopy invariance). *If* Γ *is open and bounded in* $\mathbb{R}^n \times [0,1]$, $\mathcal{H}: \overline{\Gamma} \to \mathbb{R}^n$ *is continuous and such that* $\mathcal{H}(x, \lambda) \neq 0$ *for each* $x \in (\partial\Gamma)_\lambda$ *and each* $\lambda \in [0,1]$, *then* $d_B[\mathcal{H}(\cdot, \lambda), \Gamma_\lambda]$ *is independent of* λ *on* $[0,1]$.

(3.3) PROPOSITION (Normalization). *If* $0 \in \Omega$ *then* $d_B[I, \Omega] = 1$.

The following propositions are simple consequences of the basic properties.

(3.4) Proposition (Excision). *If $F \in V_0(\Omega, \mathbb{R}^n)$ and if $\Omega_1 \subset \Omega$ is an open set such that $0 \notin F((\overline{\Omega} \setminus \Omega_1))$, then $F \in V_0(\Omega_1, \mathbb{R}^n)$ and $d_B[F, \Omega] = d_B[F, \Omega_1]$.*

(3.5) Proposition (Existence). *If $F \in V_0(\Omega, \mathbb{R}^n)$ and $d_B[F, \Omega] \neq 0$, then F has at least one zero in Ω.*

(3.6) Proposition (Boundary dependence). *If $F \in V_0(\Omega, \mathbb{R}^n)$, $G \in V_0(\Omega, \mathbb{R}^n)$ are such that $F(x) = G(x)$ for each $x \in \partial\Omega$, then $d_B[F, \Omega] = d_B[G, \Omega]$.*

The definition of Brouwer degree also implies that

$$d_B[SF, \Omega] = (\operatorname{sign} \det S)\, d_B[F, \Omega].$$

if $f \in V_0(\Omega, \mathbb{R}^n)$ and $S: \mathbb{R}^n \to \mathbb{R}^n$ is any linear isomorphism.

Finally, we mention an useful result for odd mapping.

(3.7) Proposition (Borsuk–Ulam Theorem). *If Ω is a bounded open symmetric neighbourhood of 0 and $F \in V_0(\Omega, \mathbb{R}^n)$ is odd, then $d_B[F, \Omega] = 1 \mod 2$.*

Notice finally that if X and Z are oriented vector spaces of the same finite dimension n, $\Omega \subset X$ is an open bounded set and $V_0(\Omega, Z)$ denotes the class of continuous mappings $F: \overline{\Omega} \to Z$ such that $0 \notin F(\partial\Omega)$, a corresponding Brouwer degree, with the same properties, can be defined by

$$d_B[F, \Omega] := d_B[WFU, U^{-1}(\Omega)],$$

where $U: \mathbb{R}^n \to X$ and $W: Z \to \mathbb{R}^n$ are orientation-preserving isomorphisms. See e.g. [D], [GD], [Ll], [Ze] for various definitions and more properties of Brouwer degree.

3.2. Krasnosel'skiĭ–Perov theorem. Brouwer degree can be applied to $I - P_T$ to obtain information about the existence and multiplicity of the fixed points of P_T. One way of doing this leads to an interesting existence theorem whose special cases can be traced to Berstein and Halanay in [BeH] and due, in its general form, to Krasnosel'skiĭ and Perov [KP1], [KP2] (see also [K], [KZ]).

(3.8) Theorem. *Assume that there exists an open bounded set $G \subset \mathbb{R}^n$ such that the following conditions hold:*

(3.8.1) *For each $y \in \overline{G}$, $p(t; 0, y)$ exists over $[0, T]$.*
(3.8.2) *For each $\lambda \in]0, 1]$ and each $y \in \partial G$, $p(\lambda T; 0, y) \neq y$.*
(3.8.3) *For each $y \in \partial G$, $f(0, y) \neq 0$.*

Then

$$d_B[I - P_T, G] = (-1)^n d_B[f(0, \cdot), G].$$

If we assume moreover that

(3.8.4) $d_B[f(0,\,\cdot\,), G] \neq 0$,

then equation (2.2.1) *has at least one T-periodic solution x such that $x(0) \in \overline{G}$.*

PROOF. Define the continuous mapping $h\colon \overline{G} \times [0,1] \to \mathbb{R}^n$, $(y,\lambda) \mapsto h(y,\lambda)$ by

$$h(y,\lambda) = \frac{y - p(\lambda T; 0, y)}{\lambda} \quad (\lambda \in \,]0,1]), \quad h(y,0) = -Tf(0,y).$$

Clearly,

$$h(y,1) = y - P_T(y), \quad \lim_{\lambda\to 0+} h(y,\lambda) = \lim_{\lambda \to 0+} \frac{-Tp'(\lambda T; 0, y)}{\lambda} = -Tf(0,y).$$

Moreover, by Assumptions (3.8.2) and (3.8.3), $h(y,\lambda) \neq 0$ on $\partial G \times [0,1]$. The homotopy invariance implies that

$$d_B[I - P_T, G] = d_B[-Tf(0,\,\cdot\,), G] = (-1)^n d_B[f(0,\,\cdot\,), G].$$

If Assumption (3.8.4) also holds, the existence property implies that $I - P_T$ has as least one zero in G. □

3.3. Brouwer degree of gradient mappings. To show how techniques of differential equations can help in computing the Brouwer degree of some mappings we use Theorem (3.8) to prove Amann's extension [Am] of a result of Krasnosel'skiĭ in [K] on the degree of gradient mappings. If $V \in C^1(\Omega, \mathbb{R})$ and $c \in \mathbb{R}$, let

$$V^c := \{x \in \Omega : V(x) < c\}, \quad \widetilde{V^c} := \{x \in \Omega : V(x) \leq c\}.$$

Notice that

$$\partial V^c \subset V^{-1}(c) := \{x \in \Omega : V(x) = c\} \quad \text{and} \quad \overline{V^c} \subset \widetilde{V}_c,$$

and the inclusions may be strict. However, if we assume that c is not a critical value of V, i.e. that $V'(x) \neq 0$ whenever $x \in V^{-1}(c)$, then, by the open mapping theorem, we have $\partial V^c = V^{-1}(c)$, $\overline{V^c} = \widetilde{V^c} = V^c \cup V^{-1}(c)$.

(3.9) THEOREM. *Let $\Omega \in \mathbb{R}^n$ be open and $V \in C^1(\Omega, \mathbb{R})$, with locally Lipschitzian gradient V'. Assume that there exists $\beta \in \mathbb{R}$ such that $\widetilde{V^\beta}$ is compact, and $\alpha < \beta$, $r > 0$, $y_0 \in \Omega$ such that $\widetilde{V^\alpha} \subset B[y_0; r] \subset V^\beta$, and $V'(x) \neq 0$ for every $x \in \widetilde{V^\beta} \setminus V^\alpha$. Then*

$$d_B[V', V^\beta] = 1.$$

PROOF. For each $c \leq \beta$, $\widetilde{V^c} \subset \widetilde{V^\beta}$ is bounded and positively invariant for the associated gradient system

(3.10) $$x' = -V'(x).$$

Indeed, if $p(t;0,y)$ is the solution of (3.10) with $p(0;0,y) = y$, then,

$$\frac{d}{dt}V(p(t;0,y)) = \langle V'(p(t;0,y)), p'(t;0,y)\rangle = -|V'(p(t;0,y))|^2 \leq 0, \tag{3.11}$$

so that if $y \in \widetilde{V^c}$, then $V(p(t;0,y)) \leq V(y) \leq c$ for $t \in [0, \tau_+(0, x(0))[$. As $\widetilde{V^c}$ is bounded, $\tau_+(0, x(0)) = \infty$. Notice also that if $y \in \partial V^\beta$, $V(y) = \beta$, so that, by (3.11),

$$\frac{d}{dt}V(p(t;0,y))|_{t=0} = -|V'(y)|^2 < 0,$$

$V(p(t;0,y)) < V(y)$ for all $t > 0$, and hence $p(t;0,y) \neq y$ for all $t > 0$. It follows from Theorem (3.8) with $G = V^\beta$ that, for all $t > 0$, one has

$$d_B[I - p(t;0,\cdot), V^\beta] = (-1)^n d_B[-V', V^\beta] = d_B[V', V^\beta]. \tag{3.12}$$

We now show that $d_B[I - p(t;0,\cdot), V^\beta] = 1$ for t sufficiently large. Let $\gamma := \min_{\widetilde{V^\beta}\setminus V^\alpha} |V'(x)|^2 > 0$. If $y \in \partial V^\beta$, $p(t;0,y) \in \widetilde{V^\beta}$ for all $t \geq 0$, and if $p(t;0,y) \in \widetilde{V^\beta} \setminus V^\alpha$ for $t \in [0,\tau]$, with $V(p(\tau;0,y)) = \alpha$, then

$$\beta - \alpha \geq V(y) - V(p(\tau;0,y)) = \int_\tau^0 \frac{d}{dt}V(p(t;0,y))\,dt = \int_0^\tau |V'(p(t;0,y))|^2\,dt \geq \gamma\tau.$$

Consequently, for $\tau > (\beta - \alpha)/\gamma$, $p(\tau;0,y) \in V^\alpha \subset B[y_0;r]$. Thus $p(\tau;0,\cdot)$ maps ∂V^β into $B[y_0,r]$, and, for each $\lambda \in [0,1]$ and each $y \in \partial V^\beta$, we have

$$|(1-\lambda)(y - y_0) + \lambda(y - p(\tau;0,y))| \geq |y - y_0| - |p(\tau;0,y) - y_0| > 0,$$

as $|y - y_0| > r$ for $y \in \partial V^\beta$. The homotopy invariance implies that

$$\deg_B[I - p(\tau;0,\cdot), V^\beta] = \deg_B[I - y_0, V^\beta] = 1. \qquad \square$$

The following consequence of Theorem (3.9) is a result of Krasnosel'skiĭ in [K] for the Brouwer degree of the gradient of a coercive real function.

(3.13) Definition. We say that $V\colon \mathbb{R}^n \to \mathbb{R}$ is *coercive* if

$$V(x) \to \infty \quad \text{as } |x| \to \infty.$$

V is called *anticoercive* if $-V$ is coercive.

(3.14) Corollary. *Let $V \in C^1(\mathbb{R}^n, \mathbb{R})$, with locally Lipschitzian gradient V', be such that $V'(x) \neq 0$ for some $r_0 > 0$ and all $x \in \mathbb{R}^n$ with $|x| \geq r_0$. If V is coercive, then $d_B[V', B(r)] = 1$ for all $r \geq r_0$.*

Proof. Let $\alpha := \max_{|x|\leq r_0} V(x)$ and $r > 0$ such that $\widetilde{V^\alpha} \subset B[r]$. If $\beta > \max_{|x|\leq r} V(x,)$ the assumptions of Theorem (3.9) with $y_0 = 0$ are satisfied. Thus, $d_B[V', V^\beta] = 1$, and the result follows by excision. $\square$

Another consequence of Theorem (3.9) was first derived by Rothe in [Ro].

(3.15) COROLLARY. *Let U be an open neighbourhood of $y_0 \in \mathbb{R}^n$ and $V \in C^1(U, \mathbb{R})$, with V' locally Lipschitzian. If y_0 is an isolated critical point of V at which V has a local minimum, then for all sufficiently small $\rho > 0$ one has*

$$d_B[V', B(\rho)] = 1.$$

PROOF. Without loss of generality, take $U = B(\rho)$, $y_0 = 0$, $V(0) = 0$, $\rho > 0$ such that $V(x) > 0$ for all $x \in U$.

Fix $0 < r_1 < r_2 < \rho$ and let $\beta = \min_{r_1 \le |x| \le r_2} V(x) > 0$. There exists $r > 0$ such that $B[r] \subset V^\beta$. Take $\alpha = (1/2) \min_{r \le |x| \le r_2} V(x)$, so that $\alpha < \beta$ and all conditions of Theorem (3.9) are satisfied. The result follows by excision. □

(3.16) REMARK. All the above results hold when V' is only assumed to be continuous. This can be shown using a partition of unity argument in the proofs above. We use freely this fact in the sequel.

3.4. Brouwer degree and stability. Consider now the autonomous differential system

$$x' = f(x), \tag{3.17}$$

where $f: \mathbb{R}^n \to \mathbb{R}^n$ is locally Lipschitzian and $f(0) = 0$.

(3.18) DEFINITION. The solution 0 is said to be *stable* for (3.17) if, for each $\varepsilon > 0$ we can find $\delta > 0$ such that $|p(t; 0, z)| \le \varepsilon$ whenever $t \ge 0$ and $|z| \le \delta$. 0 is said (uniformly) *asymptotically stable* if it is stable and *attractive*, i.e. there exists $\beta > 0$ such that for each $\eta > 0$ one can find $T > 0$ such that $|p(t; 0, z)| \le \eta$ whenever $t \ge T$ and $|z| \le \beta$.

See [RM] for details.

(3.19) THEOREM. *If 0 is an isolated zero of f and is asymptotically stable for (3.17), then there exists $\rho > 0$ such that*

$$d_B[-f, B(\rho)] = 1.$$

PROOF. We claim that there exists some $\rho_0 \in]0, \beta]$ such that, for each $\rho \in]0, \rho_0]$, each z with $|z| = \rho$ and each $t > 0$ we have $p(t; 0, z) \ne z$. If it is not the case, we can find, for each integer $k \ge 1$ such that $1/k \le \beta$, some $\rho_k \in]0, 1/k]$, some z_k with $|z_k| = \rho_k$ and some $t_k > 0$ such that $p(t_k; 0, z_k) = z_k$, i.e. such that $p(t; 0, z_k)$ is t_k-periodic. But, this implies that $p(t; 0, z_k) \not\to 0$ for $t \to \infty$, contradicting the attractivity of 0. Consequently $p(\lambda T; 0, z) \ne z$ whenever $z \in B[\rho_0] \setminus \{0\}$, $\lambda \in]0, 1]$, and $T > 0$. Hence, by Theorem (3.9),

$$d_B[-f, B(\rho)] = d_B[I - p(T; 0, \cdot), B(\rho)]$$

for all $T > 0$ and all $\rho \in \,]0, \rho_0]$. For $T > 0$ associated to $\eta = \rho/2$ in the definition of attractivity, we have $|p(T; 0, z)| \le \rho/2$ for all $z \in B[\rho]$, and hence $|z - \lambda p(T; 0, z)| \ge \rho/2$ whenever $|z| = \rho$ and $\lambda \in [0, 1]$, so that

$$d_B[I - p(T; 0, \cdot\,), B(\rho)] = d_B[I, B(\rho)] = 1. \qquad \square$$

More delicate results on the degree when 0 is only assumed to be stable can be found in [Th1], [Th2] and [O].

For further applications of algebraic topological tools to Poincaré's operator, see [Fi1], [Fi2].

4. Degree for some nonlinearly perturbed Fredholm mappings

4.1. Boundary value problems and fixed points in function spaces. For $J = [0, T]$, $f: J \times \mathbb{R}^n \to \mathbb{R}^n$ continuous, $M, N: \mathbb{R}^n \to \mathbb{R}^n$ constant matrices, consider the two-point nonlinear boundary value problem

$$x' = f(t, x), \quad Mx(0) + Nx(T) = 0. \tag{4.1}$$

By Poincaré's observation, the study of T-periodic solutions of (2.2.1) is a special case of (4.1) with $M = -N = I$, and his approach to the periodic problem can be extended to (4.1) when the initial value problem (2.1) has a unique solution. Indeed, $p(t; 0, y)$ is a solution of (4.1) if $y \in \mathbb{R}^n$ is such that $p(t; 0, y)$ exists over $[0, T]$ and

$$My + Np(T; 0, y) = 0. \tag{4.2}$$

But one can use another approach, which does not rely upon the initial value problem (2.1). Assume that we can find $a \in C(J, \mathcal{L}(\mathbb{R}^n, \mathbb{R}^n))$ such that the linear boundary value problem

$$x' + a(t)x = h(t), \quad Mx(0) + Nx(T) = 0$$

has a unique solution x_h for each $h \in C([0, T], \mathbb{R}^n)$. By a classical argument based upon Ascoli–Arzelá's theorem, the application $R: h \to x_h$ is linear and completely continuous on $C(J, \mathbb{R}^n)$. Then, the problem (4.1), written in the equivalent form

$$x' + a(t)x = f(t, x) + a(t)x, \quad Mx(0) + Nx(T) = 0,$$

is equivalent to the fixed point problem in $C(J, \mathbb{R}^n)$

$$x = R[F + A](x),$$

where $A: C(J, \mathbb{R}^n) \to C(J, \mathbb{R}^n)$ and $F: C(J, \mathbb{R}^n) \to C(J, \mathbb{R}^n)$ are respectively defined by

$$Ax(t) = a(t)x(t), \quad Fx(t) = f(t, x(t)), \quad (t \in J).$$

Instead of reducing (4.1) to a *finite-dimensional implicit* nonlinear system, we reduce it to an *infinite-dimensional explicit* fixed point problem, having a structure for which Brouwer's topological fixed point and degree techniques have been extended by Schauder in 1930 [Sch] and Leray and Schauder in 1934 (see [LS]). Now, the natural abstract formulation of boundary value problems for ordinary differential equations like (4.1) usually leads to an operator which is the sum of a Fredholm linear mapping of index zero and a nonlinear mapping having some compactness properties. It is therefore appropriate to develope once for all a degree theory for this class of mappings.

4.2. L-compact perturbations of Fredholm mappings of index zero. Let X and Z be real normed vector spaces.

(4.3) DEFINITION. A linear mapping $L: D(L) \subset X \to Z$ is called *Fredholm* if its kernel $N(L) := L^{-1}(0)$ has finite dimension and if its range $R(L) := L(D(L))$ is closed and has finite codimension. The *index* of L is $\dim N(L) - \operatorname{codim} R(L)$.

If L is Fredholm of index zero, there exist continuous projectors $P: X \to X$, $Q: Z \to Z$ such that $R(P) = \ker L$, $N(Q) = R(L)$, and a bijection $J: N(L) \to R(Q)$. Then $L + JP: D(L) \to Z$ is a bijection. Denote by $\mathcal{F}(L)$ the set of linear continuous mapppings of finite rank $A: X \to Z$ such that $L + A: D(L) \to Z$ is a bijection. For each $A \in \mathcal{F}(L)$, and $N: \Delta \subset X \to Z$, the equation

$$Lx + Nx = 0 \tag{4.4}$$

is equivalent to the fixed point problem

$$x = (L + A)^{-1}(A - N)x. \tag{4.5}$$

Notice that $(L + A)^{-1}A$ is always a completely continuous operator. Let E be a metric space and let $G: E \to Z$.

(4.6) LEMMA. *If there exists $A \in \mathcal{F}(L)$ such that $(L+A)^{-1}G$ is compact on E, then the same is true for any $B \in \mathcal{F}(L)$.*

PROOF. Let $B \in \mathcal{F}(L)$. Then

$$\begin{aligned}(L + B)^{-1}G &= (L + B)^{-1}(L + A)(L + A)^{-1}G \\ &= (L + B)^{-1}(L + B + A - B)(L + A)^{-1}G \\ &= [I + (L + B)^{-1}(A - B)](L + A)^{-1}G.\end{aligned}$$

As $A - B$ is continuous and has finite rank, $(L + B)^{-1}(A - B)$ is continuous and has finite rank, and hence $(L + B)^{-1}(A - B)(L + A)^{-1}G$ is compact on E. □

Lemma (4.6) justifies the following definition, introduced in [Vo1], and equivalent to an earlier one given in [Ma1].

(4.7) DEFINITION. We say that $G: E \to Z$ is *L-compact* on E if there exist $A \in \mathcal{F}(L)$ such that $(L + A)^{-1}G: E \to X$ is compact on E. If $G: X \to Z$ is L-compact on each bounded set $B \subset X$, we say that G is *L-completely continuous on X*.

For $E \subset X$, $X = Z$ and $L = I$, this concept reduces to the classical one of compact mapping introduced by Schauder [Sch]. If $K_{PQ} := (L|_{D(L)\cap N(P)})^{-1}(I - Q)$ is the right inverse of L associated to projectors P and Q, then $G: E \to Z$ is L-compact on E if and only if $QG: E \to Z$ is continuous, $QG(E)$ is bounded and $K_{PQ}G: E \to X$ is compact. If $L: D(L) \to Z$ is invertible, one can take $A = 0$ in Definition (4.7) and the L-compactness of G on E is equivalent to the compactness of $L^{-1}G$ on E. The following useful property of linear L-completely continuous mappings, first proved in a different way in [LM], is a consequence of Riesz theory of linear compact operators.

(4.8) PROPOSITION. *If $C: X \to Z$ is linear, L-completely continuous on X and if $N(L + C) = \{0\}$, then $L + C: D(L) \to Z$ is bijective and, for each L-compact mapping $G: E \to Z$, the mapping $(L + C)^{-1}G: E \to X$ is compact on E.*

4.3. Leray–Schauder and coincidence degrees. Leray and Schauder have extended in [LS] the Brouwer degree to the class $V_I(\Omega, X)$ of compact perturbations $F = I + N$ of the identity in a Banach space X, such that $0 \notin F(\partial\Omega)$, for some open bounded subset $\Omega \subset X$. This *Leray–Schauder degree $d_{LS}[F, \Omega]$ of F in Ω* reduces to the Brouwer degree when X is finite-dimensional, and satisfies the three basic properties of Brouwer degree if, in the homotopy invariance, one restricts $\mathcal{H}$ to mappings of the form $I + \mathcal{N}$, with $\mathcal{N}: \overline{\Gamma} \to X$ compact and $\Gamma \subset X \times [0, 1]$ open and bounded. See [D], [GD], [Ll], [Ze] for recent expositions.

The following extension of Leray–Schauder degree is convenient to study nonlinear boundary value problems of the type (4.1). Denote by $V_L(\Omega, Z)$ the set of mappings $F = L + N$, where $L: D(L) \subset X \to Z$ is linear Fredholm of index zero, $\Omega \subset X$ is open and bounded, $N: \overline{\Omega} \to Z$ is L-compact on $\overline{\Omega}$, $0 \notin F(D(L) \cap \partial\Omega)$, and by $\mathcal{C}(L)$ the set of linear completely continuous mappings $C: X \to Z$ such that $N(L + C) = \{0\}$. By Proposition (4.8), $L + C: D(L) \to Z$ is bijective and $(L + C)^{-1}G$ is compact over $E \subset X$ whenever $G: E \to Z$ is L-compact. Furthermore, $\mathcal{F}(L) \subset \mathcal{C}(L)$.

(4.9) LEMMA. *If $C \in \mathcal{C}(L)$ and $B \in \mathcal{C}(L)$, $\Delta_{B,C} := (L+B)^{-1}(C-B)$ is completely continuous on X and*

$$I + (L+B)^{-1}(N-B) = (I + \Delta_{B,C})[I + (L+C)^{-1}(N-C)].$$

PROOF. We have

$$\begin{aligned} I+ &(L+B)^{-1}(N-B) \\ &= I + (L+B)^{-1}(L+C)(L+C)^{-1}(N-C+C-B) \\ &= I + [I + (L+B)^{-1}(C-B)](L+C)^{-1}(N-C) + (L+B)^{-1}(C-B) \\ &= (I + \Delta_{B,C})[I + (L+C)^{-1}(N-C)]. \end{aligned}$$

The complete continuity of $\Delta_{B,C}$ on X follows from Proposition (4.6). □

As $N(I + \Delta_{B,C}) = N(L+C) = \{0\}$, we obtain

$$d_{LS}[I+(L+B)^{-1}(N-B), \Omega] = d_{LS}[I+\Delta_{B,C}, B(r)] \cdot d_{LS}[I+(L+C)^{-1}(N-C), \Omega],$$

where $r > 0$ is arbitrary. The relation in $\mathcal{C}(L)$ defined by $B \sim C$ if and only if $d_{LS}[I + \Delta_{B,C}, B(r)] = +1$ is an equivalence relation over $\mathcal{C}(L)$. If we fix an orientation on $N(L)$ and on $\operatorname{coker} L = Z/R(L)$, we can for example define $\mathcal{C}_+(L)$ as the class containing the application $C = \pi_Q^{-1}\Lambda P$, where $\Lambda\colon N(L) \to \operatorname{coker} L$ is an orientation preserving isomorphism and π_Q is the restriction to $R(Q)$ of the canonical projection $\pi\colon Z \to \operatorname{coker} L$. Setting

$$J = \pi_Q^{-1}\Lambda\colon N(L) \to R(Q),$$

it is easy to compute that

$$I + (L+JP)^{-1}(N-JP) = I - P + J^{-1}QN + K_{P,Q}N.$$

This justifies the following definition.

(4.10) DEFINITION. If $F \in V_L(\Omega, Z)$, the *degree of F in Ω with respect to L* is defined by

$$d_L[F, \Omega] = d_{LS}[I + (L+C)^{-1}(N-C), \Omega]$$

for any $C \in \mathcal{C}_+(L)$.

This degree, introduced in 1972 in [Ma1], is also refered as *coincidence degree of L and $-N$ in Ω*, and systematic expositions are given in [GM1], [Ma3], [Ma4]. The present definition [Ma4], a minor modification of the one in [Vo1], allows a unification of the approaches in [Vo1] and in [PV]. The coincidence degree d_L reduces to Leray–Schauder degree d_{LS} if $X = Z$ and $L = I$, and satisfies the first two basic properties of Brouwer or Leray–Schauder degrees in the following form.

(4.11) PROPOSITION (Addition–excision). *If $F \in V_L(\Omega, Z)$ and Ω_1 and Ω_2 are disjoint open subsets in Ω such that $0 \notin F[D(L) \cap (\overline{\Omega} \setminus (\Omega_1 \cup \Omega_2))]$, then $F \in V_L(\Omega_1, Z)$, $F \in V_L(\Omega_2, Z)$ and*

$$d_L[F, \Omega] = d_L[F, \Omega_1] + d_L[F, \Omega_2].$$

(4.12) PROPOSITION (Homotopy invariance). *If Γ is open and bounded in $X \times [0,1]$, $\mathcal{H}: (D(L) \times [0,1]) \cap \overline{\Gamma} \to Z$ has the form $\mathcal{H}(x, \lambda) = Lx + \mathcal{N}(x, \lambda)$, where $\mathcal{N}: \overline{\Gamma} \to Z$ is L-compact on $\overline{\Gamma}$, and if $\mathcal{H}(x, \lambda) \neq 0$ for each $x \in D(L) \cap (\partial\Gamma)_\lambda$ and each $\lambda \in [0,1]$, then $d_L[\mathcal{H}(\cdot, \lambda), \Gamma_\lambda]$ is independent of λ on $[0,1]$.*

The excision and existence properties, and the Borsuk–Ulam theorem, are similar to the corresponding results for Brouwer degree. The normalization property is replaced by the following one.

(4.13) PROPOSITION (Normalization). *If $F \in V_L(\Omega, Z)$ is the restriction to $\overline{\Omega}$ of a linear one-to-one mapping from $D(L)$ into Z, then $|d_L[F - b, \Omega]| = 1$ if $b \in F(D(L) \cap \Omega)$.*

The computation of d_L is reduced to that of some Brouwer degree for some classes of nonlinear perturbations with finite-dimensional range.

(4.14) PROPOSITION (Reduction). *If $F \in V_L(\Omega, Z)$ and $F = L + G$ with $G(\overline{\Omega}) \subset Y$, Y a finite-dimensional direct summand of $R(L)$, then $G|_{N(L)} \in V_0(\Omega \cap N(L), Y)$ and*

$$|d_L[F, \Omega]| = |d_B[G|_{N(L)}, \Omega \cap N(L)]|.$$

PROOF. Choose the projector $Q: Z \to Z$ such that $N(Q) = R(L)$, $R(Q) = Y$, and take $A = JP$, where $J: N(L) \to Y$ is an isomorphism such that $A \in \mathcal{C}_+(L)$. Then, as $(L + A)^{-1} = J^{-1}Q + K_{P,Q}$, we have, using a reduction property of Leray–Schauder degree (see [LS]),

$$\begin{aligned} d_L[F, \Omega] &= d_{LS}[I + (L + A)^{-1}(G - A), \Omega] = d_{LS}[I + J^{-1}G - P, \Omega] \\ &= d_B[J^{-1}G|_{N(L)}, \Omega \cap N(L)] = (\operatorname{sign} \det J^{-1})\, d_B[G|_{N(L)}, \Omega \cap N(L)]. \quad \square \end{aligned}$$

4.4. Continuation and existence theorems. Let Γ be open and bounded in $X \times [0,1]$, with $\Gamma_0 := \{x \in X : (x, 0) \in \Gamma\}$ nonempty. For $\mathcal{N}: \overline{\Gamma} \to Z$ L-compact on Γ, let

$$\mathcal{S} = \{(x, \lambda) \in \overline{\Gamma} : x \in D(L) \text{ and } Lx + \mathcal{N}(x, \lambda) = 0\},$$

and, for $\lambda \in [0,1]$, let $\mathcal{S}_\lambda = \{x \in X : (x, \lambda) \in \mathcal{S}\}$. The existence, homotopy invariance, normalization and reduction properties lead to existence theorems, which result from the following *Leray–Schauder type alternative theorem*.

(4.15) LEMMA. *If* $L+\mathcal{N}(\cdot,0) \in V_L(\Gamma_0, Z)$ *and* $d_L[L+\mathcal{N}(\cdot,0),\Gamma_0] \neq 0$, *then there exists a closed connected subset* Σ *of* $\mathcal{S}$ *joining* $\Gamma_0 \times \{0\}$ *to either* $\Gamma_1 \times \{1\}$ *or* $\{(x,\lambda) \in \partial\Gamma : \lambda \in]0,1[\}$.

Using a topological result of Kuratowski–Whyburn, one gets a version of Lemma (4.15) for a possibly unbounded set Γ.

(4.16) COROLLARY. *If* Γ *is possibly unbounded, but* Γ_0 *is bounded and the conditions of Lemma* (4.15) *hold, then there exists a closed connected subset* Σ *of* $\mathcal{S}$ *intersecting* $\Gamma_0 \times \{0\}$ *and such that either*

(4.16.1) Σ *joins* $\Gamma_0 \times \{0\}$ *to* $\Gamma_1 \times \{1\}$,
(4.16.2) *or* Σ *joins* $\Gamma_0 \times \{0\}$ *to* $\{(x,\lambda) \in \partial\Gamma : \lambda \in]0,1[\}$,
(4.16.3) *or* Σ *is unbounded.*

Returning to Γ bounded, we have the following *Leray–Schauder type continuation theorem.*

(4.17) COROLLARY. *If* $L+\mathcal{N}(\cdot,0) \in V_L(\Gamma_0, Z)$ *and*

(4.17.1) $d_L[L+\mathcal{N}(\cdot,0),\Gamma_0] \neq 0$,
(4.17.2) $Lx+\mathcal{N}(x,\lambda) \neq 0$ *for each* $x \in (\partial\Gamma)_\lambda$ *and each* $\lambda \in]0,1]$,

then a closed connected subset Σ *of* $\mathcal{S}$ *joins* $\Gamma_0 \times \{0\}$ *to* $\Gamma_1 \times \{1\}$.

A first consequence of Corollary (4.17) is a *Poincaré–Bohl type existence theorem.*theorem!Poincaré–Bohl type existence

(4.18) THEOREM. *Let* Ω *open and bounded in* X, $H \in V_L(\Omega, Z)$ *and* $F = L+N$ *with* $N:\overline{\Omega} \to Z$ *L-compact. Assume that the following conditions are satisfied.*

(4.18.1) $d_L[H,\Omega] \neq 0$.
(4.18.2) $\lambda Fx + (1-\lambda)Hx \neq 0$ *for each* $(x,\lambda) \in (D(L) \cap \partial\Omega) \times]0,1[$.

Then equation $Lx+Nx = 0$ *has at least one solution in* $D(L) \cap \overline{\Omega}$.

PROOF. For each $x \in D(L) \cap \overline{\Omega}$ and each $\lambda \in [0,1]$, one has, if $H = L+K$ with K L-compact on $\overline{\Omega}$,

$$\lambda Fx + (1-\lambda)Hx = Lx + \lambda Nx + (1-\lambda)Kx = Lx + \mathcal{N}(x,\lambda),$$

with $\mathcal{N}:\overline{\Omega} \times [0,1] \to Z$ L-compact. If we take $\Gamma = \Omega \times [0,1]$ in Lemma (4.15), Γ is bounded as well as $\mathcal{S}$. Thus, either equation $Lx+Nx = 0$ has a solution in $D(L) \cap \partial\Omega$, or the set Γ satisfies all the conditions of Corollary (4.17). □

Taking for H an invertible linear mapping in Theorem (4.18)gives a result which can be found in [Ma3] in the present form, but many special cases or variants of it have been widely applied to ordinary and partial differential equations since the pioneering work of Leray and Schauder (see [RM], [GM1], [Ma3]–[Ma5] for references). It can be refered as a *Leray–Schauder–Schaefer type existence theorem* (see [S]).

(4.19) THEOREM. *Let $F = L + N$, with $N:\overline{\Omega} \to Z$ L-compact, $A: X \to Z$ linear L-completely continuous mapping and $z \in (L + A)(D(L) \cap \Omega)$ be such that*

(4.19.1) $N(L + A) = \{0\}$,
(4.19.2) $Lx + (1 - \lambda)(Ax - z) + \lambda Nx \neq 0$ *for each* $x \in D(L) \cap \partial\Omega$ *and each* $\lambda \in]0, 1[$.

Then equation $Lx + Nx = 0$ has at least one solution in $D(L) \cap \overline{\Omega}$.

PROOF. Take $H = L + A - z$ in Theorem (4.18) and use Proposition (4.13).□

Taking $H = L + G$ with $G(\overline{\Omega})$ transversal to $R(L)$ in Theorem (4.18) provides a result, first given in [Ma3], covering situations with degree different from ± 1.

(4.20) THEOREM. *Let $F = L+N$, with $N:\overline{\Omega} \to Z$ L-compact and let $G:\overline{\Omega} \to Y$ be L-compact on $\overline{\Omega}$, with Y a direct summand of $R(L)$. Assume that the following conditions hold.*

(4.20.1) $Lx + (1 - \lambda)Gx + \lambda Nx \neq 0$ *for each* $x \in D(L) \cap \partial\Omega$ *and each* $\lambda \in]0, 1[$.
(4.20.2) $Gx \neq 0$ *for each* $x \in N(L) \cap \partial\Omega$.
(4.20.3) $d_B[G|_{N(L)}, \Omega \cap N(L)] \neq 0$.

Then equation $Lx + Nx = 0$ has at least one solution in $D(L) \cap \overline{\Omega}$.

PROOF. Let $H = L + G$ and let $Q: Z \to Z$ be the continuous projector such that $R(Q) = Y$ and $N(Q) = R(L)$. Then $QG = G$ and $Hx = 0$ if and only if

$$QHx = 0, \quad (I - Q)Hx = 0,$$
$$\text{i.e.} \quad Gx = 0,\ Lx = 0, \quad \text{or} \quad Gx = 0,\ x \in N(L).$$

Hence, by assumption (4.20.2), $H \in V_L(\Omega, Z)$ and Proposition (4.14) gives

$$|d_L[H, \Omega]| = |d_B[G|_{N(L)}, \Omega \cap N(L)]|.$$

The result follows then from assumptions (4.20.1), (4.20.3) and Theorem (4.17).□

A useful consequence of Theorem (4.20) is the following *continuation theorem*, first proved in [Ma1]. Other proofs or variants can be found in [FMV], [FP1], [FP2], [GGKK], [M], [Vo2], and applications to various problems are given in [GM1], [Ma3], [Ma4], [Ma5] and their references. See also [Ma8] for a bibliography of the many recent applications to population dynamics.

(4.21) THEOREM. *Let $F = L + N$, with $N: \overline{\Omega} \to Z$ L-compact. Assume that the following conditions are satisfied.*

(4.21.1) $Lx + \lambda Nx \neq 0$ *for each* $(x, \lambda) \in [(D(L) \setminus N(L)) \cap \partial\Omega] \times]0, 1[$.
(4.21.2) $Nx \notin R(L)$ *for each* $x \in N(L) \cap \partial\Omega$.
(4.21.3) $d_B[QN|_{N(L)}, \Omega \cap N(L)] \neq 0$.

Then equation $Lx + Nx = 0$ *has at least one solution in* $D(L) \cap \overline{\Omega}$.

PROOF. Take, in Theorem (4.20), $Y = R(Q)$ and $G = QN$, which is L-compact on $\overline{\Omega}$. Assumption (4.21.2) immediately implies that $QNx \neq 0$ for each $x \in N(L) \cap \partial\Omega$, and assumptions (4.20.2) and (4.20.3) are fulfilled. On the other hand, if $Lx + (1 - \lambda)QNx + \lambda Nx = 0$ for some $x \in D(L) \cap \partial\Omega$ and some $\lambda \in]0, 1[$, then, applying Q and $I - Q$ to both members of this equation, we obtain

$$QNx = 0, \quad Lx + \lambda Nx = 0.$$

The first of those equations and assumption (4.21.2) imply that $x \in (D(L) \setminus N(L)) \cap \partial\Omega$ and the second one then contradicts assumption (4.21.1). The existence of a solution follows from Theorem (4.20). □

5. Bound sets and Floquet boundary conditions

5.1. Bound sets for Floquet boundary value problems. In Section 1, the applications of Brouwer fixed point theorem correspond to systems (2.2.1) for which the vector field $f(t, x)$ on the boundary of some set $G \subset \mathbb{R}^n$ points inward G. In his seminal paper [W] of 1947 about a new topological principle to study the asymptotic behavior of the solutions of ordinary differential equations, Ważewski has introduced the concept of *regular polyfacial set with respect to the differential system* (2.2.1). This concept is defined as follows. Let p and q be nonnegative integers with $p + q > 1$, let $L_k \in C^1(J \times \mathbb{R}^n, \mathbb{R})$ and denote $L'_k = (\partial_t L_k, \partial_{x_1} L_k, \dots, \partial_{x_n} L_k)$, $1 \leq k \leq p + q$. Let

$$\begin{aligned} \Gamma &= \{(t, x) \in J \times \mathbb{R}^n : L_k(t, x) < 0 \ (1 \leq k \leq p + q)\}, \\ \pi_j &= \{(t, x) \in J \times \mathbb{R}^n : L_k(t, x) \leq 0 \ (1 \leq k \leq p + q), \ L_j(t, x) = 0\}. \end{aligned}$$

Γ is called a *regular polyfacial set with respect to* (2.5) if the following conditions hold

$$\begin{aligned} &\langle L'_k(t, x), (1, f(t, x)) \rangle > 0 \quad \text{for } t \in J \text{ and } x \in \pi_j \quad (1 \leq j \leq p), \\ &\langle L'_k(t, x), (1, f(t, x)) \rangle < 0 \quad \text{for } t \in J \text{ and } x \in \pi_j \quad (p + 1 \leq j \leq p + q). \end{aligned}$$

The sets π_j $(1 \leq j \leq p)$ are called the *positive faces* and the sets π_j $(p + 1 \leq j \leq p + q)$ are called the *negative faces*. Some variants, with supplementary conditions to make them manifolds with corners, have been recently defined and used by Mrozek in [Mr1], [Mr2] and Srzednicki in [Srz1], [Srz2], under the name of *blocks with corners of type* (p, q), in their topological study of rest points and periodic solutions. The related concept of *bound set* was introduced independently, thirty years after Ważewski's paper, by Gaines and the author [GM1], [GM2], to generalize and unify a number of existence results for nonlinear ordinary differential equations with various boundary conditions.

(5.1) DEFINITION. Let $\Gamma \subset J \times \mathbb{R}^n$ be open in the relative topology on $J \times \mathbb{R}^n$ and bounded. We say that Γ is a *bound set* for (2.2.1) if, for any $(\tau, u) \in \partial\Gamma$ there exist $V_{\tau,u} \in C^1(J \times \mathbb{R}^n, \mathbb{R})$ such that:

(5.1.1) $\Gamma \subset \{(t, x) \in J \times \mathbb{R}^n : V_{\tau,u}(t, x) < 0\}$,
(5.1.2) $V_{\tau,u}(\tau, u) = 0$,
(5.1.3) $\langle V'_{\tau,u}(\tau, u), (1, f(\tau, u))\rangle \neq 0$,
where $V'_{\tau,u}(t, x) = (\partial_t V_{\tau,u}(t, x), \partial_{x_1} V_{\tau,u}(t, x), \dots, \partial_{x_n} V_{\tau,u}(t, x))$.

We refer to [Ry] for some comparaison with Ważewski's concept, and to [FZ1]–[FZ3], [Ma2], [Z1], [Z2] for the relations with flow invariance and various extensions.

Following [Ma7], we study an extension of the concept of bound set suitable for the study of nonlinear differential equations with *Floquet boundary conditions*

$$x(T) = Cx(0), \tag{5.2}$$

where $C \in GL_n(\mathbb{R})$ generates a finite group $\langle C\rangle < GL_n(\mathbb{R})$. For simplicity, we restrict ourself to the case of an autonomous bound set (the V_u do not depend upon t).

(5.3) DEFINITION. G is an *autonomous bound set* for the Floquet boundary value problem

$$x' = f(t, x), \quad x(T) = Cx(0), \tag{5.3.1}$$

if ∂G is invariant under the action of $\langle C\rangle$ and if, for each $u \in \partial G$, there exists $V_u \in C^1(\mathbb{R}^n, \mathbb{R})$ such that the following conditions hold:

(5.3.2) $G \subset \{x \in \mathbb{R}^n : V_u(x) < 0\}$,
(5.3.3) $V_u(u) = 0$,
(5.3.4) $\langle V'_u(u), f(t, u)\rangle \neq 0$ for each $t \in J$,
(5.3.5) $0 \notin [\langle V'_u(u), f(0, u)\rangle, \langle V'_{Cu}(Cu), f(T, Cu)\rangle]$.

(5.4) REMARK. When the $(n \times n)$-matrix N is invertible, linear boundary conditions of the form (4.1) are equivalent to the Floquet boundary conditions (5.2) with $C = -N^{-1}M$.

(5.5) EXAMPLE. For $C = I$, $\langle C \rangle = \{I\}$, (5.3.1) is the *periodic* boundary value problem, the invariance condition is satisfied for each G and condition (5.3.5) is a consequence of (5.3.4). So Definition (5.3) reduces to the concept of bound set for a periodic problem given in [GM1], [GM2]. Furthermore $N(C - I) = \mathbb{R}^n$.

(5.6) EXAMPLE. For $C = -I$, $\langle C \rangle = \{I, -I\}$, (5.3.1) is the *antiperiodic* boundary value problem, the invariance condition means that ∂G is symmetric with respect to 0, and condition (5.3.5) becomes

$$0 \notin [\langle V'_u(u), f(0,u)\rangle, \langle V'_{-u}(-u), f(T,-u)\rangle].$$

Furthermore $N(C - I) = \{0\}$.

(5.7) EXAMPLE. If $n = 2$, $p \in \mathbb{N}_0$, $0 \leq k \leq p$, $\mathbb{R}^2 \simeq \mathbb{C}$, $C = \alpha = e^{2k\pi i/(p+1)}$, $\langle C \rangle = \{1, \alpha, \alpha^2, \ldots, \alpha^p\}$ is the cyclic group of order $p+1$, a special case of (5.3.1) is the Floquet boundary value problem

$$z' = f(t,z), \quad z(T) = \alpha z(0)$$

introduced in [Ma6]. The invariance condition is satisfied if $\overline{G}$ is a regular polygon with $m(p+1)$ sides two of which are orthogonal to the real line ($m \geq 1$), and conditions (5.3.4) and (5.3.5) hold if $m = 2$ and if the signs of the inner product of $f(t,z)$ with the outer normal to the side containing z alternate.

5.2. Existence theorems for Floquet boundary value problems. For $h \in C(J, \mathbb{R}^n)$, consider the linear Floquet boundary value problem

$$x' = h(t), \quad x(T) = Cx(0). \tag{5.8}$$

Elementary considerations show that if $N(C - I) = \{0\}$, then (5.8) has a unique solution for each $h \in C(J, \mathbb{R}^n)$, and, if $N(C - I) \neq \{0\}$, then (5.8) has at least one solution if and only if

$$\overline{h} := \frac{1}{T}\int_0^T h(t)\,dt \in R(C - I).$$

Consequently, the linear operator L defined in $X = C(J, \mathbb{R}^n)$ by

$$D(L) = \{x \in C^1(J, \mathbb{R}^n) : x(T) = Cx(0)\}, \quad Lx = x'$$

is Fredholm of index zero. Let S be a projector in $\mathbb{R}^n$ with range $R(C - I)$. Define $\overline{f}\colon \mathbb{R}^n \to \mathbb{R}^n$ by

$$\overline{f}(u) = \frac{1}{T}\int_0^T f(t,u)\,dt.$$

(5.9) THEOREM. *Assume that G is a bound set for* (5.3.1) *and that either*

(5.9.1) $N(C-I)=\{0\}$ *and* $0\in G$, *or*

(5.9.2) $\overline{f}(u)\notin R(C-I)$ *for* $u\in\partial G\cap N(C-I)$ *and*

$$d_B[(I-S)\overline{f}|_{N(C-I)}, G\cap N(C-I)]\neq 0.$$

Then (5.3.1) *has at least one solution x such that $x(t)\in G$ for all $t\in J$.*

PROOF. We apply Theorem (4.19) with $A=0$ or Theorem (4.20), according to $N(C-I)=\{0\}$ or $N(C-I)\neq\{0\}$, to the abstract formulation of (5.3.1) as an equation $Lx=Nx$ in $X=C(J,\mathbb{R}^n)$ with L defined above and $Nx=f(\cdot,x(\cdot))$ L-completely continuous on X. Define the open bounded set

$$\Gamma(G)=\{x\in C(J,\mathbb{R}^n): x(t)\in G \text{ for all } t\in[0,T]\},$$

and consider the homotopy

$$x'=\lambda f(t,x),\quad x(T)=Cx(0),\quad \lambda\in\,]0,1]. \tag{5.10}$$

Let $x\in\overline{\Gamma(G)}$ be a possible solution of (5.10) for some $\lambda\in\,]0,1]$, so that $x(t)\in\overline{G}$ for all $t\in J$. If $x\in\partial\Gamma(G)$, then $x(\tau)\in\partial G$ for some $\tau\in J$, and $V_{x(\tau)}(x(\cdot))$ reaches its maximum at τ. If $\tau\in\,]0,T[$, this implies that

$$0=\frac{d}{dt}V_{x(\tau)}(x(t))\Big|_{t=\tau}=\lambda\langle V'_{x(\tau)}(x(\tau)),f(\tau,x(\tau))\rangle,$$

a contradiction with (5.3.4). If $\tau=0$, then

$$0\geq\frac{d}{dt}V_{x(0)}(x(t))\Big|_{t=0}=\lambda\langle V'_{x(0)}(x(0)),f(0,x(0))\rangle. \tag{5.11}$$

Now, $x(T)=Cx(0)\in\partial G$, and hence we have $V_{x(T)}(x(T))=0$, which shows that $V_{x(T)}(x(\cdot))$ reaches its maximum at $t=T$. Consequently

$$0\leq\frac{d}{dt}V_{x(T)}(x(t))\Big|_{t=T}=\lambda\langle V'_{Cx(0)}(Cx(0)),f(T,Cx(0))\rangle. \tag{5.12}$$

From (5.11) and (5.12) we get a contradiction to (5.3.5). If $\tau=T$, then

$$0\leq\frac{d}{dt}V_{x(T)}(x(t))\Big|_{t=T}=\lambda\langle V'_{Cx(0)}(Cx(0)),f(T,Cx(0))\rangle. \tag{5.13}$$

Now, $x(0)=C^{-1}x(T)\in\partial G$, and hence $V_{x(0)}(x(0))=0$, which shows that $V_{x(0)}(x(\cdot))$ reaches its maximum at $t=0$. Consequently

$$0\geq\frac{d}{dt}V_{x(0)}(x(t))\Big|_{t=0}=\lambda\langle V'_{x(0)}(x(0)),f(0,x(0))\rangle. \tag{5.14}$$

From (5.13) and (5.14) we get again a contradiction to (5.3.5). Thus (5.10) has no solution on $\partial\Omega$. If $N(C-I)=\{0\}$, then $N(L)=\{0\}$ and the conclusion follows from Theorem (4.19) with $A=0$. If $N(C-I)\neq\{0\}$, and if $u\in\partial G\cap N(C-I)$, then condition (5.9.2) imply that all assumptions of Theorem (4.20) are satisfied.□

(5.15) COROLLARY. *Assume that G is an open bounded convex neighbourhood of 0 satisfying the following conditions:*

(5.15.1) *∂G is invariant under the action of $\langle C\rangle$.*

(5.15.2) *For each $u \in \partial G$, there is an outer normal $n(u)$ such that $\langle n(u), f(t,u)\rangle \neq 0$ for each $t \in J$, and such that*

$$0 \notin [\langle n(u), f(0,u)\rangle, \langle n(Cu), f(T,Cu)\rangle].$$

(5.15.3) *Either $N(C-I) = \{0\}$ or $\overline{f}(u) \notin R(C-I)$ for $u \in \partial G \cap N(C-I)$ and*

$$d_B[(I-S)\overline{f}|_{N(C-I)}, G \cap N(C-I)] \neq 0.$$

Then (5.3.1) *has at least one solution x such that $x(t) \in G$ for all $t \in J$.*

PROOF. For any such convex set G and any $u \in \partial G$, there exists at least one outer normal $n(u)$, i.e. some $n(u) \in \mathbb{R}^n \setminus \{0\}$ such that $\langle n(u), u\rangle > 0$ and $G \subset \{x \in \mathbb{R}^n : \langle x-u, n(u)\rangle < 0\}$, and we can take $V_u(x) = \langle x-u, n(u)\rangle$. □

For the periodic case, Corollary (5.15) generalizes a result of Gustafson and Schmitt (see [GS]).

(5.16) COROLLARY. *Assume that there exist $V \in C^1(\mathbb{R}^n, \mathbb{R})$ such that the following conditions hold:*

(5.16.1) *V^0 is non-empty and bounded and $V^{-1}(0)$ is invariant under the action of $\langle C\rangle$.*

(5.16.2) *$\langle V'(u), f(t,u)\rangle < 0$ for each $u \in V^{-1}(0)$ and each $t \in J$.*

(5.16.3) *Either $N(C-I) = \{0\}$ and $V(0) < 0$ or $V'(u) \notin R(C-I)$ for $u \in \partial G \cap N(C-I)$ and*

$$d_B[(I-S)V'|_{N(C-I)}, V^0 \cap N(C-I)] \neq 0.$$

Then (5.3.1) *has at least one solution x such that $V(x(t)) < 0$ for each $t \in J$.*

PROOF. Take $G = V^0$ and observe that, as $V'(u) \neq 0$ for each $u \in V^{-1}(0)$, one has $\partial G = V^{-1}(0)$. Choosing $V_u(x) = V(x)$ for each $u \in \partial G$, we easily check that G is an autonomous bound set for (5.3.1). □

(5.17) REMARK. Taddei [T1] has extended Theorem (5.9) to cases where the functions V_u need not to be differentiable, and to more general boundary value problems of type (4.1).

5.3. Floquet boundary value problems for complex-valued differential equations. Let $\beta\colon J \to \mathbb{R}$ be continuous and positive, $q \in [0, \infty[$, $p \geq 1$ be an integer, and $h\colon J \times \mathbb{C} \to \mathbb{C}$ be continuous. Let $\alpha \in \mathbb{C}$ be such that $\alpha^{p+1} = 1$ ($(p+1)$th root of unity). Consider the Floquet boundary value problem

(5.18) $$z' = \beta(t)|z|^q\overline{z}^p + h(t,z), \quad z(T) = \alpha z(0).$$

(5.19) LEMMA. *Assume that* $\beta_0 := \min_{t\in[a,b]} \beta(t) > 0$, *and that*

$$|h(t,z)| \le \gamma |z|^{p+q} + \delta, \tag{5.19.1}$$

for some

$$0 \le \gamma < \beta_0 \cos^p \frac{\pi}{2(p+1)} \sin \frac{\pi}{2(p+1)},$$

some $\delta \ge 0$ *and all* $(t,z) \in J \times \mathbb{C}$. *Then there exists an open bounded set* G *such that* (5.18) *has at least one solution* x *with* $x(t) \in \overline{G}$ *for all* $t \in J$.

PROOF. For each $R > 0$, let G_R be the open bounded convex neighbourhood of the origin whose closure is the regular polygon with $2(p+1)$ sides, two of which are orthogonal to the x-axis, inscribed in the circle of radius R. Elementary computations show that, if $n(z)$ denotes the outer unit normal at un point $z = \rho\exp(i\theta)$ belonging to the side of $\overline{G_R}$ for which

$$\frac{(2k-1)p\pi}{2(p+1)} \le \theta \le \frac{(2k+1)p\pi}{2(p+1)},$$

then, for each $t \in J$,

$$\langle n(z), \beta(t)|z|^q\overline{z}^p\rangle \ge \beta_0 R^{p+q} \cos^p \frac{\pi}{2(p+1)} \cos \frac{p\pi}{2(p+1)} \qquad (k \text{ even}),$$
$$\langle n(z), \beta(t)|z|^q\overline{z}^p\rangle \le -\beta_0 R^{p+q} \cos^p \frac{\pi}{2(p+1)} \cos \frac{p\pi}{2(p+1)} \qquad (k \text{ odd}).$$

As $\cos p\pi/(2(p+1)) = \sin \pi/(2(p+1))$, we easily see that, using (5.19.1) and letting

$$f(t,z) = \beta(t)|z|^q\overline{z}^p + h(t,z),$$

we have, for all $t \in J$,

$$\langle n(z), f(t,z)\rangle \ge R^{p+q}\left[\cos^p \frac{\pi}{2(p+1)} \sin \frac{\pi}{2(p+1)} - \gamma\right] - \delta \qquad (k \text{ even}),$$
$$\langle n(z), f(t,z)\rangle \le -R^{p+q}\left[\cos^p \frac{\pi}{2(p+1)} \sin \frac{\pi}{2(p+1)} - \gamma\right] + \delta \qquad (k \text{ odd}).$$

Consequently, there exists $R_0 > 0$ depending only upon p, q, γ, δ such that for all $R \ge R_0$ one has, on the side of $\overline{G_R}$ characterized by

$$\frac{(2k-1)p\pi}{2(p+1)} \le \theta \le \frac{(2k+1)p\pi}{2(p+1)},$$
$$\langle n(z), f(t,z)\rangle > 0 \quad (t \in J,\ k \text{ even}), \qquad \langle n(z), f(t,z)\rangle < 0 \quad (t \in J,\ k \text{ odd}).$$

As the signs of the inequalities alternate along the sides of $\overline{G_R}$, they are the same on two sides related by $z' = \alpha z$. Finally, $N(I-\alpha) = \{0\}$ if $\alpha \ne 1$ and

$N(I-\alpha) = \mathbb{C} \simeq \mathbb{R}^2$ if $\alpha = 1$. In this last case, it is easy to see that, for all sufficiently large R, one has

$$d_B[\overline{f}, G_R] = -p,$$

and the result follows from Corollary (5.15). □

(5.20) REMARK. Condition (5.19.1) holds in particular if

$$\lim_{|z|\to\infty} \frac{|h(t,z)|}{|z|^{p+q}} = 0 \tag{5.20.1}$$

uniformly in $t \in J$.

We now extend Lemma (5.19) to a wider class of functions β, by using a technique introduced in [Ma6]. Let $\eta \in C^1([a,b], \mathbb{C}\setminus\{0\})$ with $\eta(0) = \eta(T)$, $q \in [0,\infty[$, $p \geq 1$ be an integer, and $h: J \times \mathbb{C} \to \mathbb{C}$ be continuous. Let $\alpha \in \mathbb{C}$ be such that $\alpha^{p+1} = 1$. We consider the Floquet boundary value problem

$$z' = \eta(t)|z|^q\overline{z}^p + h(t,z), \quad z(T) = \alpha z(0). \tag{5.21}$$

(5.22) THEOREM. *Assume that* $\eta_0 := \min_{t\in J}|\eta(t)| > 0$, $q+p > 1$, *and that* (5.19.1) *holds for some*

$$0 \leq \gamma < \eta_0 \cos^p \frac{\pi}{2(p+1)} \sin \frac{\pi}{2(p+1)}, \tag{5.22.1}$$

some $\delta \geq 0$ *and all* $(t,z) \in J \times \mathbb{C}$. *Then (5.21) has at least one solution* z *such that* $\theta^{-1/(p+1)}(t)z(t) \in G_R$ *where* G_R *is given by Theorem* (5.19) *and* $\theta = \eta/|\eta|$.

PROOF. Let $m = \operatorname{ind}\eta$, the winding number of $\eta(t)$, and $\theta := \eta/|\eta| \in C^1(S^1, S^1)$, so that $\operatorname{ind}\theta = \operatorname{ind}\eta = m$. Notice that

$$\arg\frac{\eta(0)}{\eta(T)} = \arg\eta(0) - \arg\eta(T) = -m, \qquad \frac{|\eta(0)|}{|\eta(T)|} = 1,$$
$$\left(\frac{\eta(0)}{\eta(T)}\right)^{1/(p+1)} = e^{-2im\pi/(p+1)}, \qquad \overline{\theta} = \frac{1}{\theta}.$$

The change of variable $z(t) = \theta^{1/(p+1)}(t)w(t)$, where $\theta^{1/(p+1)}(t)$ is a fixed determination of the $(p+1)$th root of $\theta(t)$, leads to the equivalent boundary value problem

$$\begin{aligned} w' &= |\eta(t)||w|^q\overline{w}^p - \frac{\theta'(t)}{(p+1)\theta(t)}w + \theta^{-1/(p+1)}(t)h(t,\theta^{1/(p+1)}(t)w), \\ w(T) &= \alpha e^{-2im\pi/(p+1)}w(0). \end{aligned}$$

to which Lemma (5.19) can be applied as $p+q > 1$. □

Under assumption (5.20.1), Theorem (5.19) with $\eta(t) = \exp(2\pi imt/T)$ was first proved by Srzednicki [Srz3]–[Srz4] using generalized Ważewski–Conley blocks, Poincaré's operator and Lefschetz fixed point theorem. Through degree theory and a priori estimates in terms of integral norms, it was shown in [MMZ], [Ma6] that the existence result for (5.21) is valid when (5.19.1) holds and γ satisfies the more general condition $0 \leq \gamma < \eta_0$. In particular, Theorem (5.22) implies that the equation

$$z' = \overline{z}^2 + h(t)$$

has a T-periodic solutions for each continuous T-periodic $h: \mathbb{R} \to \mathbb{C}$. On the other hand, it has been shown that equation

$$z' = z^2 + h(t)$$

has no T-periodic solution for some h (see [C], [CO], [GF], [Mi], [Zo1], [Zo2]) and further recent studies of this class of equations are given in [BoM], [Srz6], [Srz7], [T2]. For similar results concerning T-periodic of quaternionic-valued differential equations, see [CM].

6. Guiding functions and periodic solutions

6.1. Generalized guiding functions. We now concentrate on the periodic problem

$$x' = f(t,x), \quad x(0) = x(T), \tag{6.1}$$

with $f: J \times \mathbb{R}^n \to \mathbb{R}^n$ continuous, and give several extensions of Corollary (5.16) in this setting. We set

$$C^{\#}_J := \{x \in C(J, \mathbb{R}^n) : x(0) = x(T)\},$$

and, for $G \subset \mathbb{R}^n$,

$$\Gamma(G) := \{x \in C^{\#}_J : x(t) \in G \text{ for all } t \in J\}.$$

$\Gamma(G)$ is open (resp. closed) if G is open (resp. closed). Problem (6.1) can be written as an abstract equation $Lx = Nx$ in $C^{\#}_J$, if we define $D(L) = C^{\#}_J \cap C^1(J, \mathbb{R}^n)$, $L: D(L) \subset C^{\#}_J \to C(J, \mathbb{R}^n)$, $x \mapsto x'$, $N: C^{\#}_J \to C(J, \mathbb{R}^n)$, $x \mapsto f(\cdot, x(\cdot))$. L is a linear Fredholm mapping of index zero, with kernel $N(L)$ the subspace of constant functions in $C^{\#}_J$, and range $R(L)$ the subspace of functions y with zero mean value, so that the corresponding operators P and Q can be chosen as

$$Px = Qx = \overline{x} := \frac{1}{T} \int_0^T x(t)\, dt.$$

(6.2) DEFINITION. $V \in C^1(G,\mathbb{R})$ is a *generalized guiding function on* G for (6.1) if the following conditions hold:

(6.2.1) $V'(x) \neq 0$ for each $x \in G$.
(6.2.2) $\langle V'(x), f(t,x)\rangle \leq 0$, for each $t \in J$ and each $x \in G$.

When $G = \mathbb{R}^n \setminus B(\rho)$ for some $\rho > 0$, the concept can be found in [Ma3], and is a slight extension of Krasnosel'skiĭ's classical *guiding function* (see [K]), which requires the strict inequality in condition (6.2.2).

We need a result on the computation of the degree associated to an autonomous gradient system [Ma3].

(6.3) LEMMA. *Let $V \in C^1(\mathbb{R}^n,\mathbb{R})$ be such that V^0 is non-empty and bounded, V is a generalized guiding function on $V^{-1}(0)$, and let $G\colon \Gamma(V^0) \to C(J,\mathbb{R}^n)$, $x \mapsto V'(x)$. Then*

$$|d_L[L+G,\Gamma(V^0)]| = |d_B[V',V^0]|.$$

PROOF. We consider the homotopy

$$\mathcal{H}\colon \overline{\Gamma}(V^0) \times [0,1] \to C(J,\mathbb{R}^n), \quad (x,\lambda) \mapsto Lx + \lambda Gx + (1-\lambda)QGx.$$

If $\mathcal{H}(x,\lambda) = 0$, then x is T-periodic and

$$|x'(t)|^2 = -\lambda\langle V'(x(t)), x'(t)\rangle - (1-\lambda)\left\langle \frac{1}{T}\int_0^T V'(x(s))\,ds, x'(t)\right\rangle,$$

so that

$$\int_0^T |x'(t)|^2\,dt = 0.$$

Consequently, x is constant and satisfies $V'(x) = 0$. Hence $x \notin V^{-1}(0)$, and, using the homotopy invariance of d_L and Proposition (4.14), we get

$$\begin{aligned}|d_L[L+G,\Gamma(V^0)]| &= |d_L[L+QG,\Gamma(V^0)]| \\ &= |d_B[QG|_{N(L)}, V^0 \cap N(L)]| = |d_B[V',V^0]|. \qquad \square\end{aligned}$$

Notice that Lemma (6.3) is a special case of the following deeper result of Capietto, Zanolin and the author [CMZ] on the computation of d_L for the arbitrary autonomous system (3.17).

(6.4) PROPOSITION. *Assume that there exists an open bounded set $\Omega \subset C_J^\#$ such that (3.17) has no solution on $\partial\Omega$. Let $N\colon C_J^\# \to C_J^\#$, $x \mapsto f(x(\cdot))$. Then*

$$d_L[L-N,\Omega] = (-1)^n d_B[f,\Omega \cap \mathbb{R}^n],$$

where $\mathbb{R}^n$ is identified with the subspace of constant functions in $C_J^\#$.

The proof is this result given in [BM] uses the invariance of the degree with respect to the S^1-symmetry coming from the fact that each time translation of a solution is a solution.

(6.5) THEOREM. *Assume that there exist $V \in C^1(\mathbb{R}^n, \mathbb{R})$ such that the following conditions hold:*

(6.5.1) *V^0 is non-empty and bounded.*
(6.5.2) *V is a generalized guiding function for (6.1) on $V^{-1}(0)$.*
(6.5.3) *$d_B[V', V^0] \neq 0$.*

Then problem (6.1) has at least one solution $x \in \Gamma(V^0 \cup V^{-1}(0))$.

PROOF. Following Theorem (4.18), we introduce the homotopy

$$x' = -(1-\lambda)V'(x) + \lambda f(t,x), \quad x(0) = x(T), \quad \lambda \in [0,1[. \tag{6.6}$$

Let $x \in \overline{\Gamma(V^0)}$ be a possible solution of (6.6) with some $\lambda \in [0,1[$. We show that $x \in \Gamma(V^0)$, i.e. $V(x(t)) < 0$ for all $t \in J$. If it not the case, then $V(x(t)) \leq 0$ for all $t \in J$ and $V(x(\tau)) = 0$ for some $\tau \in J$. Thus $x(\tau) \in V^{-1}(0)$ and, by (6.2.1), $V'(x(\tau)) \neq 0$. Consequently,

$$\begin{aligned} 0 = \frac{d}{dt}V(x(t))\bigg|_{t=\tau} &= \langle V'(x(\tau)), x'(\tau)\rangle \\ &= -(1-\lambda)|V'(x(\tau))|^2 + \lambda\langle V'(x(\tau)), f(\tau, x(\tau))\rangle < 0, \end{aligned}$$

a contradiction. Thus, either (6.1) has a solution on $\partial\Gamma(V^0)$, in which case Theorem (6.5) is proved, or, for each $\lambda \in [0,1]$, problem (6.6) has no solution on $\partial\Gamma(V^0)$. By the homotopy invariance, Theorem (4.14), Lemma (6.3), and assumptions (6.5.1) and (6.5.3) we find, with $G(x) := V'(x(\cdot))$,

$$|d_L[L - N, \Gamma(V^0)]| = |d_L[L + G, \Gamma(V^0)]| = |d_B[V', V^0]| \neq 0,$$

and the result follows from the existence property. □

(6.7) REMARK. By assumptions (6.5.1) and (6.5.2) and condition (6.2.1), ∂V^0 is a compact hypersurface (i.e. submanifold of dimension $n-1$) of $\mathbb{R}^n$, and $V'(x)$ is the normal to ∂V^0 at $x \in \partial V^0$. Thus $d_B[V', V^0]$ is the degree of the *Gauss map* $g: \partial V^0 \to S^{n-1}$, $x \mapsto V'(x)/|V'(x)|$. Now, $d_B[V', V^0] = \chi(\overline{V^0})$ and, when n is odd, $d_B[V', V^0] = \chi(\partial V^0)/2$, where χ denotes the *Euler–Poincaré characteristic* [GP], [Mil]. Thus, assumption (6.5.3) can be replaced by $\chi(\overline{V^0}) \neq 0$, and, for n odd, by $\chi(\partial V^0) \neq 0$.

Assumption (6.5.3) may be difficult to check in general. We give some special cases in which it is automatically satisfied or easy to verify.

(6.8) COROLLARY. *Assume that there exist an even generalized guiding function V for (6.1) on $V^{-1}(0)$, such that V^0 is a bounded neighbourhood of 0. Then problem (6.1) has at least one solution $x \in \Gamma(V^0 \cup V^{-1}(0))$.*

PROOF. A consequence of Borsuk–Ulam theorem. □

(6.9) COROLLARY. *Assume that there exists $p \in \mathbb{R}^n$ and $V \in C^1(\mathbb{R}^n, \mathbb{R})$ such that the following conditions hold:*

(6.9.1) V^0 *is bounded and* $p \in V^0$.
(6.9.2) V *is a generalized guiding function on* $V^{-1}(0)$ *for* (6.1).
(6.9.3) $\langle V'(x), x - p\rangle > 0$ *for each* $x \in \partial V^0$.

Then problem (6.1) *has at least one solution* $x \in \Gamma(V^0 \cup V^{-1}(0))$.

PROOF. By Poincaré–Bohl's theorem, $d_B[V', V^0] = d_B[I - p, V^0] = 1$. □

Corollary (6.9) is due to Zanolin (see [Z3]).

(6.10) COROLLARY. *Assume that there exist $\rho > 0$ and $V \in C^1(\mathbb{R}^n, \mathbb{R})$ such that the following conditions hold:*

(6.10.1) V^0 *is bounded and* $B[\rho] \subset V^0$.
(6.10.2) V *is a generalized guiding function on* $(V^0 \cup V^{-1}(0)) \setminus B(\rho)$ *for* (6.1).
(6.10.3) $d_B[V', B(\rho)] \neq 0$.

Then problem (6.1) *has at least one solution* $x \in \Gamma(V^0 \cup V^{-1}(0))$.

PROOF. By the excision property of degree, $d_B[V', V^0] = d_B[V', B(\rho)]$. □

Lemma (3.14) allows the following slight extension [Ma3] of the fundamental theorem on coercive guiding functions of Krasnosel'skiĭ–Perov [K], [KP2], [KP1], [KS].

(6.11) COROLLARY. *Assume that there exists $\rho > 0$ and a coercive (resp. anticoercive) generalized guiding function V on $\mathbb{R}^n \setminus B(\rho)$ for* (6.1). *Then problem* (6.1) *has at least one solution x such that $V(x(t)) \leq \max_{B[\rho]} V$ for all $t \in J$.*

PROOF. If V coercive, take $r > \max_{B[\rho]} V$ and $W = V - r$. $W^0 = V^r$ is bounded, and $B[\rho] \subset W^0$. W is a generalized guiding function on $(W^0 \cup W^{-1}(0)) \setminus B(\rho)$ for (6.1). Lemma (3.14) implies $d_B[W', B(\rho)] = 1$. If V anticoercive, x is a solution of (6.1) if and only if $y(t) = x(T - t)$ solves

$$y' = -f(T - t, y), \quad y(0) = y(T). \tag{6.12}$$

Condition (6.2.2) can be written, letting $t = T - s$,

$$\langle -V'(x), -f(T - s, x)\rangle \leq 0,$$

for all $x \in \mathbb{R}^n$ such that $|x| \geq \rho$, so that $-V$ is a coercive generalized guiding function for (6.12). □

6.2. Averaged guiding functions. If we consider the periodic problem for the scalar linear differential equation

$$(6.13)\qquad x' = c(t)x + h(t), \quad x(0) = x(T),$$

with $c, h \in C_J^{\#}$, then $V(x) = x^2/2$ is a coercive generalized guiding function for (6.13) outside of a sufficiently large interval $]-\rho, \rho[$ if $c(t) < 0$ for all $t \in J$, and an anticoercive one if $c(t) > 0$ for all $t \in J$. Hence, by Corollary (6.11), (6.13) has at least one solution for each h if either inequality for c is satisfied. But elementary direct calculations show that the same conclusion is true if (and only if) either $\int_0^T c(t)\,dt < 0$ or $\int_0^T c(t)\,dt > 0$. To be able to obtain existence theorems for nonlinear systems implying this sharp result for (6.13), we introduce the class of averaged guiding functions [MaWa].

(6.14) DEFINITION. Let $G \subset \mathbb{R}^n$ be non-empty. $V \in C^1(G, \mathbb{R})$ is an *averaged guiding function on* G for (6.1) if the the following conditions hold:

(6.14.1) $V'(x) \neq 0$ for each $x \in G$.
(6.14.2) $\int_0^T \langle V'(x(t)), f(t, x(t))\rangle\,dt \le 0$ for all $x \in \Gamma(G)$.

Any generalized guiding function on G for (6.1) is an averaged guiding function on G for (6.1), and the converse is true when (6.1) is autonomous. The following result is a partial extension of Theorem (6.5) (see [MaWa]).

(6.15) THEOREM. *Assume that there exists* $V \in C^1(\mathbb{R}^n, \mathbb{R})$ *such that the following conditions hold:*

(6.15.1) V^0 *is non-empty and* V^r *is bounded, where*

$$r = \max\left\{T \max_{u\in V^{-1}(0)} |V'(u)|^2, \max_{u\in V^{-1}(0)} \int_0^T |\langle V'(u), f(t,u)\rangle|\,dt\right\}.$$

(6.15.2) V *is an averaged guiding function on* $\mathbb{R}^n \setminus V^0$ *for* (6.1).
(6.15.3) $d_B[V', V^0] \neq 0$.

Then problem (6.1) *has at least one solution* $x \in \Gamma(V^r \cup V^{-1}(r))$.

PROOF. We again introduce the homotopy (6.6). Let $\lambda \in [0, 1[$ and x be a possible solution of (6.6). If $V(x(t)) \ge 0$ for all $t \in J$, integrating the identity

$$(6.16)\qquad \frac{d}{dt}V(x(t)) = -(1-\lambda)|V'(x(t))|^2 + \lambda\langle V'(x(t)), f(t, x(t))\rangle$$

on J and using the periodicity and assumption (6.15.2) gives

$$0 = -(1-\lambda)\int_0^T |V'(x(t))|^2\,dt + \lambda\int_0^T \langle V'(x(t), f(t, x(t))\rangle\,dt < 0,$$

a contradiction. Thus, there exists $t_0 \in J$ such that

$$V(x(t_0)) < 0, \tag{6.17}$$

i.e. such that $x(t_0) \in V^0$. If

$$V(x(t)) \leq 0 \quad \text{for all } t \in J, \tag{6.18}$$

we have an a priori bound for x by assumption (6.15.1). If it is not the case, then $V(x(t_1)) > 0$ for some $t_1 \in J$, and hence

$$\max_{t \in J} V(x(t)) = V(x(\tau)) > 0. \tag{6.19}$$

From (6.17) and (6.19) follows the existence of $\sigma \in J$ such that either

$$\sigma < \tau, \quad V(x(\sigma)) = 0, \quad 0 < V(x(t)) \quad \text{for } t \in \,]\sigma, \tau], \quad \text{or} \tag{6.20.1}$$

$$\tau < \sigma, \quad V(x(\sigma)) = 0, \quad 0 < V(x(t)) \quad \text{for } t \in [\tau, \sigma[. \tag{6.20.2}$$

For definiteness, we consider the case where (6.20.1) holds, the other one being similar. Assume first that $\tau \neq 0$ (hence $\tau \neq T$), and, for each positive integer n such that $\tau + 1/n < T$, define the continuous function $x_n \colon J \to \mathbb{R}^n$ as follows:

$$x_n(t) = \begin{cases} x(\sigma) & \text{if } t \in [0, \sigma], \\ x(t) & \text{if } t \in \,]\sigma, \tau], \\ x(\tau + n(\sigma - \tau)(t - \tau)) & \text{if } t \in \,]\tau, \tau + 1/n], \\ x(\sigma) & \text{if } t \in \,]\tau + 1/n, T]. \end{cases} \tag{6.21}$$

In the case where $\tau = 0$, so that $V(x(\cdot))$ reaches its maximum at 0 and T, we define x_n, for n so large that $1/n < \sigma$, as follows:

$$x_n(t) = \begin{cases} x(\tau + n(\sigma - \tau)t) & \text{if } t \in [0, 1/n], \\ x(\sigma) & \text{if } t \in \,]1/n, \sigma], \\ x(t) & \text{if } t \in \,]\sigma, T]. \end{cases} \tag{6.22}$$

In any case, (x_n) is a sequence in $C_J^{\#}$, such that $0 \leq V(x_n(t)) \leq V(x(\tau))$, converging pointwise on J to ξ defined by

$$\xi(t) = \begin{cases} x(t) & \text{if } t \in [\sigma, \tau], \\ x(\sigma) & \text{if } t \in J \setminus [\sigma, \tau], \end{cases}$$

which is such that $0 \leq V(\xi(t)) \leq V(x(\tau))$. From assumption (6.15.2), we have

$$\int_0^T \langle V'(x_n(t)), f(t, x_n(t)) \rangle \, dt \leq 0, \tag{6.23}$$

for all sufficiently large n, and Lebesgue dominated convergence implies that

$$\int_0^T \langle V'(\xi(t)), f(t,\xi(t))\rangle\, dt \le 0. \tag{6.24}$$

Consequently, we get, using (6.16),

$$\begin{aligned} 0 &> \left(\int_\sigma^\tau + \int_{J\setminus[\sigma,\tau]}\right)[-(1-\lambda)|V'(\xi(t))|^2 + \lambda\langle V'(\xi(t)), f(t,\xi(t))\rangle]\, dt \\ &= \int_\sigma^\tau \frac{d}{dt} V(x(t))\, dt \\ &\quad + \int_{J\setminus[\sigma,\tau]} [-(1-\lambda)|V'(\xi(t))|^2 + \lambda\langle V'(\xi(t)), f(t,\xi(t))\rangle]\, dt \\ &= V(x(\tau)) + \int_{J\setminus[\sigma,\tau]} [-(1-\lambda)|V'(x(\sigma))|^2 + \lambda\langle V'(x(\sigma)), f(t,x(\sigma))\rangle]\, dt. \end{aligned}$$

Therefore, using assumption (6.15.1),

$$\begin{aligned} V(x(\tau)) &< \int_{J\setminus[\sigma,\tau]} [(1-\lambda)|V'(x(\sigma))|^2 - \lambda\langle V'(x(\sigma)), f(t,x(\sigma))\rangle]\, dt \\ &\le \max\left\{T \max_{u\in V^{-1}(0)} |V'(u)|^2, \max_{u\in V^{-1}(0)} \int_0^T |\langle V'(u), f(t,u)\rangle|\, dt\right\} = r. \end{aligned} \tag{6.25}$$

Hence, $V(x(t)) < r$ for all $t \in J$, which, combined with (6.18), holds for all possible solutions of (6.6). If (6.1) has a solution on $\partial\Gamma(V^r)$, Theorem (6.15) is proved. If not, by homotopy invariance, Lemma (6.3), Theorem (4.18) and excision, one finds

$$|d_L[L-N, \Gamma(V^r)]| = |d_L[L+G, \Gamma(V^r)]| = |d_B[V', V^r]| = |d_B[V', V^0]| \ne 0,$$

and the existence of a solution for (6.1) in $\Gamma(V^r)$ follows. □

The following result extends earlier ones of [Ma3] and [KKM2].

(6.26) Corollary. *Problem* (6.1) *has at least one solution if there exist* $\rho > 0$ *and a coercive* (*resp. anticoercive*) *averaged guiding function* V *on* $\mathbb{R}^n \setminus B(\rho)$ *for* (6.1).

Proof. If V is coercive, let $R > \max_{|u|\le\rho} V(u)$, so that $B[\rho] \subset V^R$. Let $W = V(\cdot) - R$. Then W is a coercive averaged guiding function on $\mathbb{R}^n \setminus W^0$, and W^r is bounded for any $r > 0$. From Lemma (3.14) and excision, we get

$$d_B[W', W^0] = d_B[W', B(\rho)] = 1.$$

Thus W satisfies the conditions of Theorem (6.15) and the result follows. Assume now V is anticoercive. x is a solution of (6.1) if and only if $y(t) = x(T-t)$ is a solution of (6.12). Condition (6.14.2) becomes, after letting $t = T - s$,

$$\int_0^T \langle -V'(x(s)), -f(T-s, x(s)) \rangle \, ds \leq 0,$$

for all $x \in \Sigma_{\mathbb{R}^n \setminus B(\rho)}$ (notice that $\Sigma_{\mathbb{R}^n \setminus B(\rho)}$ is invariant under the substitution $s \to T - s$), so that $-V$ is a coercive averaged guiding function on $\mathbb{R}^n \setminus B(\rho)$ for (6.12), and the result follows from the first part of the proof. □

It is easy to state the special cases of Corollary (6.26) obtained using $V(x) = \pm|x|^a/a$ for some $a > 0$ or $V(x) = \pm \ln|x|$ (modified if necessary in a neighbourhood of 0 to make it of class C^1 on $\mathbb{R}^n$). We give an example when $n = 1$.

(6.27) COROLLARY. *Problem (6.1) with $n = 1$ has at least one solution if, for some $\rho > 0$,*

$$\int_0^T \operatorname{sign} x(t) f(t, x(t)) \, dt \tag{6.27.1}$$

does not change sign for all $x \in C_J^\#$ such that $\min_{t\in J} |x(t)| \geq \rho$.

PROOF. When (6.27.1) is non negative, let $V \in C^1(\mathbb{R}, \mathbb{R})$ be a C^1 extension of the function defined for $|x| \geq \rho/2$ by $V(x) = |x|$. Then V is coercive, $V'(x) = \operatorname{sign} x$ for $|x| \geq \rho/2$, and, as $\min_{t\in J} |x(t)| \geq \rho$ means either $x(t) \geq \rho$ for all $t \in J$, or $x(t) \leq -\rho$ for all $t \in J$, V is an averaged guiding function on $\mathbb{R} \setminus]-\rho, \rho[$ for (6.1). When (6.27.1) is non positive, use the anticoercive function $-V$. □

Conditions of type (6.27.1) were first introduced by Villari in [Vi]. We mention two special cases of Corollary (6.27). The first one slightly extends a result of Reissig (see [Re]), and the second one is a variant of a result of Gossez (see [G1], [G2]) given in [Ma3].

(6.28) COROLLARY. *If $g: \mathbb{R} \to \mathbb{R}$ is continuous and such that, for some $\rho > 0$, $g(x)\operatorname{sign} x$ does not change sign when $|x| \geq \rho$, then problem*

$$x' + g(x) = p(t), \quad x(0) = x(T), \tag{6.28.1}$$

has at least one solution for each continuous p having mean value zero.

(6.29) COROLLARY. *Assume that $f(t, \cdot)$ is monotone for each $t \in J$. Then (6.1) has at least one solution if and only if there exists $c \in \mathbb{R}$ such that*

$$\int_0^T f(t, c) \, dt = 0. \tag{6.29.1}$$

PROOF. We assume for definiteness $f(t, \cdot)$ nonincreasing. Let u be a solution of (6.1). Then

$$\int_0^T f(t, u(t))\, dt = 0.$$

If $m = \min_{t\in J} u(t)$, $M = \max_{t\in J} u(t)$, we have $f(t, M) \leq f(t, u(t)) \leq f(t, m)$ for all $t \in J$, and hence,

$$\int_0^T f(t, M)\, dt \leq 0 \leq \int_0^T f(t, m)\, dt.$$

The intermediate value theorem gives c verifying ((6.27)). If c satisfies ((6.27)), take $\rho > |c|$. For $x \in C_J^{\#}$ such that $x(t) \geq \rho$ for all $t \in J$, one has $x(t) > c$ and

$$\operatorname{sign} x(t) f(t, x(t)) \leq \operatorname{sign} x(t) f(t, c), \tag{6.30}$$

for all $t \in J$. When $x(t) \leq -\rho$ for all $t \in J$, one has $x(t) < c$ and (6.30) holds as well. Integrating (6.30) on J, we can apply Corollary (6.27). □

(6.31) COROLLARY. *Problem* (6.28.1) *with g monotone has a solution if and only if the mean value of p belongs to the range of g.*

6.3. Asymptotic averaged guiding functions. The following concept of asymptotic averaged guiding function is motivated by the version given in Proposition VI.7 of [Ma3] of a result of [MaW].

(6.32) DEFINITION. $V \in C^1(\mathbb{R}^n, \mathbb{R})$ is an *asymptotic averaged guiding function* for (6.1) if there exists $\alpha \in L^1(J, \mathbb{R}_+)$ such that the following conditions hold:

(6.32.1) $\langle V'(x), f(t, x)\rangle \leq \alpha(t)$ for $t \in J$ and every $x \in \mathbb{R}^n$.
(6.32.2) $\int_0^T \limsup_{|x|\to\infty} \langle V(x), f(t, x)\rangle\, dt < 0$.

Any C^1 extension to $\mathbb{R}$ of the function V defined for $|x| \geq 1$ by $V(x) = \ln|x|$ is a guiding (resp. generalized guiding, asymptotic averaging guiding) function on $\mathbb{R}$ for the scalar linear differential equation

$$x' = c(t)x,$$

with $c \in C(J, \mathbb{R})$, if and only if $c(t) < 0$ for all. $t \in J$, (resp. $c(t) \leq 0$ for $t \in J$, $\int_0^T c(t)\, dt < 0$). On the other hand, the function $V(x) = x^2/2$ is a guiding function on $\mathbb{R}$, but not an asymptotic averaged guiding function, for the scalar nonlinear differential equation

$$x' = -\frac{x}{1 + x^4}.$$

Thus there is no general relation between generalized guiding functions and asymptotic averaged guiding functions. However, asymptotic averaged guiding functions are averaged guiding functions outside of a large ball.

(6.33) THEOREM. *For all sufficiently large* ρ, *any asymptotic averaged guiding function for* (6.1) *is an averaged guiding function on* $\mathbb{R}^n \setminus B(\rho)$ *for* (6.1).

PROOF. Let V be an asymptotic averaged guiding function for (6.1). Using Definition (6.32) and Fatou's lemma, we have

$$\limsup_{|x|\to\infty} \int_0^T \langle V'(x), f(t,x)\rangle\, dt < 0,$$

and, for some $\rho_0 > 0$, $V'(x) \neq 0$ whenever $|x| \geq \rho_0$. We show that there exists $\rho \geq \rho_0$ such that

$$\int_0^T \langle V'(x(t)), f(t, x(t))\rangle\, dt \leq 0$$

for all $x \in \Sigma_{\mathbb{R}^n \setminus B(\rho)}$. If not, a sequence $(x_n)_{n \geq \rho_0}$ in $C_J^{\#}$ exists, such that

$$\min_{t\in J} |x_n(t)| \geq n$$

and

$$\int_0^T \langle V'(x_n(t)), f(t, x_n(t))\rangle\, dt > 0,$$

whenever $n \geq \rho_0$. By Fatou's lemma,

$$\int_0^T \limsup_{n\to\infty} \langle V'(x_n(t)), f(t, x_n(t))\rangle\, dt \geq 0,$$

and hence, as $|x_n(t)| \to \infty$ for each $t \in J$,

$$\int_0^T \limsup_{|x|\to\infty} \langle V'(x), f(t, x)\rangle\, dt \geq 0,$$

a contradiction with condition (6.32.2). □

(6.34) COROLLARY. *Problem* (equ) *has at least one solution if there exists a coercive (resp. anticoercive) asymptotic averaged guiding function for* (6.1).

(6.35) EXAMPLE. Consider the periodic problem

(6.35.1) $$x' = A(t)((\ln|x|)x) + b(t,x), \quad x(0) = x(T),$$

where $A \in C(J, \mathcal{L}(\mathbb{R}^n, \mathbb{R}^n))$, $b\colon J \times \mathbb{R}^n \to \mathbb{R}^n$ is continuous and such that for some $\alpha, \beta \geq 0$, all $t \in J$ and all $x \in \mathbb{R}^n$,

$$\langle x, b(t,x)\rangle \leq \alpha|x|^2 + \beta|x|,$$

and the function $(\ln|x|)x$ is continuously extended by 0 at $x = 0$. Any C^1 extension V to $\mathbb{R}^n$ of the function defined for $|x| \geq 1$ by $V(x) = \ln(\ln|x|)$, is coercive and, for all $t \in J$ and all $x \in \mathbb{R}^n$ with $|x| \geq 1$, we have

$$\langle V'(x), A(t)((\ln|x|)x) + b(t,x)\rangle = \frac{\langle x, A(t)x\rangle}{|x|^2} + \frac{1}{|x|^2 \ln|x|}\langle x, b(t,x)\rangle,$$

so that (6.32.1) easily follows and

$$\int_0^T \limsup_{|x|\to\infty} \langle V'(x), A(t)(\ln|x|x) + b(t)\rangle\,dt = \int_0^T \sup_{|y|=1} \langle A(t)y, y\rangle\,dt = \int_0^T \mu[A(t)]\,dt,$$

with $\mu[A(t)]$ the *logarithmic norm* of $A(t)$ (see e.g. [Ma3]). Thus (6.35.1) has a solution if $\int_0^T \mu[A(t)]\,dt < 0$ or, by taking $-V$ instead of V, if $\int_0^T \mu[A(t)]\,dt > 0$.

(6.36) REMARK. The results of this section hold if the continuity of f is remplaced by Carathéodory conditions [MaT], [MaWa].

For other extensions of the method of guiding functions, see [BKFFM], [Fi3], [KKM1], [KKMP], [Ra].

References

[Al] P. S. Aleksandrov, *Poincaré and topology*, Russian Math. Surveys **27** (1972), 157–168.

[Am] H. Amann, *A note on degree theory for gradient mappings*, Proc. Amer. Math. Soc. **85** (1982), 591–595.

[B] S. Banach, *Sur les opérations dans les ensembles abstraits et leur application aux équations intégrales*, Fund. Math. **3** (1922), 133–181.

[BKFFM] A. A. Barabanova, M. A. Krasnosel'skiĭ, I. V. Fomenko, H. I. Freedman and J. Mawhin, *Vector Lyapunov functions in problems of nonlinear oscillations*, Automat. Remote Control **57** (1996), 322–328.

[BM] Th. Bartsch and J. Mawhin, *The Leray–Schauder degree of S^1-equivariant operators associated to autonomous neutral equations in spaces of periodic functions*, J. Differential Equations **92** (1991), 90–99.

[BeH] I. Berstein and A. Halanay, *Index of a singular point and the existence of periodic solutions of systems with small parameter*, Dokl. Akad. Nauk SSSR **111** (1956), 923–925. (Russian)

[BiK] G. D. Birkhoff and O. D. Kellogg, *Invariant points in function space*, Trans. Amer. Math. Soc. **23** (1922), 96–115.

[BoM] A. Borisovich and W. Marzantowicz, *Multiplicity of periodic solutions for the planar polynomial equation*, Nonlinear Anal. **43** (2001), 217–231.

[Br] L. E. J. Brouwer, *Ueber Abbildungen von Mannigfaltigkeiten*, Math. Ann. **71** (1912), 97–115.

[Bro] F. E. Browder, *On a generalization of the Schauder fixed point theorem*, Duke Math. J. **26** (1959), 291–303.

[C] J. Campos, *Möbius transformation and periodic solutions of complex Riccati equations*, Bull. London Math. Soc. **9** (1997), 205–213.

[CM] J. Campos and J. Mawhin, *Periodic solutions of quaternionic-valued ordinary differential equations*, Ann. Mat. Pura Appl. (to appear).

[CO] J. Campos and R. Ortega, *Nonexistence of periodic solutions of a complex Riccati equation*, Differential Integral Equations **9** (1996), 247–250.

[CMZ] A. Capietto, J. Mawhin and F. Zanolin, *Continuation theorems for periodic perturbations of autonomous systems*, Trans. Amer. Math. Soc. **329** (1992), 41–72.

[CL] E. A. Coddington and N. Levinson, *Theory of Ordinary Differential Equations*, McGraw–Hill, New York, 1955.

[D] K. Deimling, *Nonlinear Functional Analysis*, Springer, Berlin, 1985.

[FZ1] M. L. Fernandes and F. Zanolin, *Remarks on strongly flow-invariant sets*, J. Math. Anal. Appl. **128** (1987), 176–188.

[FZ2] ______, *Repelling conditions for boundary sets using Liapunov-like functions,* I. *Flow-invariance, terminal value problems and weak persistence*, Rend. Sem. Mat. Univ. Padova **80** (1988), 95–116; II. *Persistence and periodic solutions*, J. Differential Equations **86** (1990), 33–58.

[FZ3] ______, *On periodic solutions, in a given set, for differential systems*, Riv. Mat. Pura Appl. **8** (1991), 107–130.

[Fi1] V. V. Filippov, *A mixture of the Leray–Schauder and Poincaré–Andronov methods in problem on periodic solutions of ordinary differential equations*, Differential Equations **35** (1999), 1736–1739.

[Fi2] ______, *Remarks on periodic solutions of ordinary differential equations*, J. Dynam. Control Systems **6** (2000), 431–451.

[Fi3] ______, *Remarks on guiding functions and periodic solutions*, Differential Equations **37** (2001), 485–497.

[FMV] M. Furi, M. Martelli and A. Vignoli, *On the solvability of nonlinear operator equations in normed spaces*, Ann. Mat. Pura Appl. (4) **124** (1980), 321–343.

[FP1] M. Furi and M. P. Pera, *An elementary approach to boundary value problems at resonance*, Nonlinear Anal. **4** (1980), 1081–1089.

[FP2] ______, *Co-bifurcating branches of solutions for nonlinear eigenvalue problems in Banach spaces*, Ann. Mat. Pura Appl. **135** (1983), 119–132.

[GF] S. R. Gabdrakhmanov and V. V. Filippov, *On the absence of periodic solutions to the equation* $z'(t) = z^2 + h(t)$, Differential Equations **33** (1997), 741–744.

[GM1] R. E. Gaines and J. Mawhin, *Coincidence Degree and Nonlinear Differential Equations,* Lecture Notes Math., vol. 586, Springer, Berlin, 1977.

[GM2] *Ordinary differential equations with nonlinear boundary conditions*, J. Differential Equations **26** (1977), 200–222.

[GGKK] K. Gęba, A. Granas, T. Kaczyński and W. Krawcewicz, *Homotopie et équations non linéaires dans les espaces de Banach*, C. R. Acad. Sci. Paris Sér. I **300** (1985), 303–306.

[G1] J. P. Gossez, *Existence of periodic solutions for some first order ordinary differential equations*, Equadiff 78, Firenze 1978, (R. Conti, G. Sestini and G. Villari, eds.), Univ. di Firenze, Firenze, 1978, pp. 361–379.

[G2] ______, *Periodic solutions for some first order ordinary differential equations with monotone nonlinearities*, Bull. Math. Soc. Sci. Math. R. S. Roumanie **24** (1980), 25–33.

[GD] *Fixed Point Theory, Springer*, New York, 2003.

[GP] V. Guillemin and A. Pollack, *Differential Topology*, Prentice-Hall, Englewoods Cliffs, N.J., 1974.

[GS] G. Gustafson and K. Schmitt, *A note on periodic solutions for delay-differential systems*, Proc. Amer. Math. Soc. **42** (1974), 161–166.

[K] M. A. Krasnosel'skiĭ, *The Operator of Translation along the Trajectories of Differential Equations*, Amer. Math. Soc., Providence, 1968.

[KKM1] A. M. Krasnosel'skiĭ, M. A. Krasnosel'skiĭ and J. Mawhin, *On some conditions of existence of forced periodic oscillations*, Differential Integral Equations **5** (1992), 1267–1273.

[KKM2] ______, *Differential inequalities in problems of forced nonlinear oscillations*, Nonlinear Anal. **25** (1995), 1029–1036.

[KKMP] A. M. Krasnosel'skiĭ, M. A. Krasnosel'skiĭ, J. Mawhin and A. Pokrowskiĭ, *Generalizing guiding functions in a problem of high frequency forced oscillations*, Nonlinear Anal. **22** (1994), 1357–1371.

[KP1] M. A. Krasnosel'skiĭ and A. I. Perov, *On a certain principle for the existence of bounded, periodic and almost periodic solutions of systems of ordinary differential equations*, Dokl. Akad. Nauk SSSR **123** (1958), 235–238. (Russian)

[KP2] ______Some criteria for existence of periodic solutions of a system of ordinary differential equations, Proc. Intern. Sympos. Nonlinear Vibrations, Kiev, 1961, vol. II, Izdat. Akad. Nauk Ukrain. SSR, Kiev, 1961, pp. 202–211. (Russian)

[KS] M. A. Krasnosel'skiĭ and V. V. Strygin, *Some criteria for the existence of periodic solutions of ordinary differential equations*, Soviet Math. Dokl. **5** (1964), 763–766.

[KZ] M. A. Krasnosel'skiĭ and P. P. Zabreĭko, *Geometrical Methods of Nonlinear Analysis*, Springer, Berlin, 1984.

[Ku] J. Kurzweil, *Ordinary Differential Equations*, Elsevier, Amsterdam, 1986.

[LM] B. Laloux and J. Mawhin, *Coincidence index and multiplicity*, Trans. Amer. Math. Soc. **217** (1976), 143–162.

[L] S. Lefschetz, *Existence of periodic solutions for certain differential equations*, Proc. Nat. Acad. Sci. **29** (1943), 1–4.

[LS] J. Leray and J. Schauder, *Topologie et équations fonctionnelles*, Ann. Sci. Ecole Norm. Sup. (3) **51** (1934), 45–78.

[Le1] N. Levinson, *On the existence of periodic solutions for second order differential equations with a forcing term*, J. Math. Phys. **22** (1943), 41–48.

[Le2] ______, *Transformation theory of nonlinear differential equations of the second order*, Ann. Math. (2) **45** (1944), 723–737; **49** (1948), 738.

[Ll] N. G. Lloyd, *Degree Theory*, Cambridge Univ. Press, Cambridge, 1978.

[MMZ] R. Manásevich, J. Mawhin, and F. Zanolin, *Periodic solutions of complex-valued differential equations and systems with periodic coefficients*, J. Differential Equations **126** (1996), 355–373.

[M] M. Martelli, *Continuation principles and boundary value problems*, Topological Methods for Ordinary Differential Equations, Lect. Notes Math., vol. 1537, Springer, Berlin, 1993, pp. 32–73.

[Ma1] J. Mawhin, *Equivalence theorems for nonlinear operator equations and coincidence degree theory for some mappings in locally convex topological vector spaces*, J. Differential Equations **12** (1972), 610–636.

[Ma2] ______Functional analysis and boundary value problems, Studies in Ordinary Differential Equations (J. K. Hale, ed.), Math. Assoc. of America, Washington, 1977, pp. 128–168.

[Ma3] ______, *Topological Degree Methods in Nonlinear Boundary Value Problems*, CBMS, vol. 40, Amer. Math. Soc., Providence, R.I., 1979.

[Ma4] ______, *Topological degree and boundary value problems for nonlinear differential equations*, Topological Methods for Ordinary Differential Equations, Lecture Notes in Math., vol. 1537, Springer, Berlin, 1993, pp. 74–142.

[Ma5] ______, *Continuation theorems and periodic solutions of ordinary differential equations*, Topological Methods in Differential Equations and Inclusions (A. Granas and M. Frigon, eds.), Kluwer, Dordrecht, 1995, pp. 291–376.

[Ma6] ______, *Nonlinear complex-valued differential equations with periodic, Floquet or nonlinear boundary conditions*, Equadiff 95, International Conference on Differential Equations (Lisbon 1995) (L. Magalhaes, C. Rocha, L. Sanchez, eds.), World Scientific, Singapore, 1997, pp. 154–164.

[Ma7] ______, *Bound sets and Floquet boundary value problems for nonlinear differential equations*, Univ. Iagel. Acta Math. **36** (1998), 41–53.

[Ma8] ______, *Les héritiers de Pierre-François Verhulst: une population dynamique*, Acad. Roy. Belg. Bull. Cl. Sci. (6) **13** (2002), 349–378.

[MaT] J. Mawhin and H. B. Thompson, *Periodic or bounded solutions of Carathéodory systems of ordinary differential equations*, J. Dynam. Differential Equations **15** (2003), 327–333.

[MaW] J. Mawhin and W. Walter, *Periodic solutions of ordinary differential equations with one-sided growth restrictions*, Proc. Roy. Soc. Edinburgh Sect. A **82** (1978), 95–106.

[MaWa] J. Mawhin and J. R. Ward, *Guiding-like functions for periodic or bounded solutions of ordinary differential equations*, Discrete Contin. Dynam. Systems **8** (2002), 39–54.

[Mi] D. Miklaszewski, *An equation $z' = z^2 + p(t)$ with no 2π-periodic solution*, Bull. Belg. Math. Soc. Simon Stevin (2) **3** (1996), 239–242.

[Mil] J. Milnor, *Topology from the Differentiable Viewpoint*, The University Press of Virginia, Charlottesville, 1965.

[Mr1] M. Mrozek, *The fixed point index of a translation operator of a semiflow*, Univ. Iagel. Acta Math. **27** (1988), 13–22.

[Mr2] ______, *Periodic and stationary trajectories of flows and ordinary differential equations*, Univ. Iagel. Acta Math. **27** (1988), 29–37.

[N] M. Nagumo, *On the periodic solution of an ordinary differential equation of second order* (1944), Zenkoku Shijou Suugaku Danwakai, 54–61 (Japanese); English transl. in Mitio Nagumo Collected Papers, Springer, 1993, pp. 279–283.

[O] R. Ortega, *Some applications of the topological degree to stability theory*, Topological Methods in Differential Equations and Inclusions (A. Granas and M. Frigon, eds.), Kluwer, Dordrecht, 1995, pp. 377-409.

[PV] J. Pejsachowicz and A. Vignoli, *On the topological coincidence degree for perturbations of Fredholm operators*, Boll. Un. Mat. Ital. (5) **17-B** (1980), 1457–1466.

[Pi] E. Picard, *Mémoire sur la théorie des équations aux dérivés partielles et la méthode des approximations successives*, J. Math. Pures Appl. **6** (1890), 145–210.

[Pl] V. A. Pliss, *Nonlocal Problems of the Theory of Oscillations*, Academic Press, New York, 1966.

[Po1] H. Poincaré, Sur certaines solutions particulières du problème des trois corps, C.R. Acad. Sci. Paris Sér. I **97** (1883), 251–252.

[Po2] ______, *Sur les courbes définies par les équations différentielles*, J. Math. Pures Appl. (4) **2** (1886), 151–217.

[Po3] ______, *Sur le problème des trois corps et les équations de la dynamique*, Acta Math. **13** (1890), 1–270.

[Po4] ______, *Sur un théorème de géométrie*, Rend. Circ. Mat. Palermo **33** (1912), 375–407.

[Ra] D. Rachinskiĭ, *Multivalent guiding functions in forced oscillation problems*, Nonlinear Anal. **26** (1996), 631–639.

[Re] R. Reissig, *On the existence of periodic solutions of a certain non-autonomous differential equations*, Ann. Mat. Pura Appl. (4) **85** (1970), 235–240.

[Ro] E. Rothe, *A relation between the type number of a critical point and the index of the corresponding field of gradient vectors*, Math. Nachr. **4** (1950-51), 12–27.

[RM] N. Rouche and J. Mawhin, *Ordinary Differential Equations. Stability and Periodic Solutions*, Pitman, Boston, 1980.

[Ry] K. P. Rybakowski, *Some remarks on periodic solutions of Carathéodory RFDES via Ważewski principle*, Riv. Mat. Univ. Parma (4) **8** (1982), 377–385.

[S] H. Schaefer, *Ueber die Methode der a-priori Schranken*, Math. Ann. **129** (1955), 415–416.

[Sch] J. Schauder, *Der Fixpunktsatz in Funktionalraümen*, Studia Math. **2** (1930), 171–180.

[Srz1] R. Srzednicki, *On rest points of dynamical systems*, Fund. Math. **126** (1985), 69–81.

[Srz2] ______, *Periodic and constant solutions via topological principle of Ważewski*, Univ. Iagel. Acta Math. **26** (1987), 183–190.

[Srz3] ______, *A Geometric Method for the Periodic Problem in Ordinary Differential Equations, Sém. Anal. Moderne*, vol. 22, Université de Sherbrooke, 1992.

[Srz4] ______, *Periodic and bounded solutions in blocks for time-periodic non-autonomous ordinary differential equations*, Nonlinear Anal. **22** (1994), 707–737.

[Srz5] ______, *On periodic solutions of planar polynomial differential equations with periodic coefficients*, J. Differential Equations **114** (1994), 77–100.

[Srz6] ______, *On solutions of two-point boundary value problems inside isolating segments*, Topol. Methods Nonlinear Anal. **13** (1999), 73–89.

[Srz7] ______, *On periodic solutions inside isolating chains*, J. Differential Equations **165** (2000), 42–60.

[T1] V. Taddei, *Bound sets for first order differential equations with general linear two-point boundary conditions*, Dynam. Contin. Discrete Impuls. Systems Ser. A **9** (2002), 133–150.

[T2] ______, *Periodic solutions for certain systems of planar complex polynomial equations*, Bull. Belg. Math. Soc. Simon Stevin **8** (2001), 623–637.

[Th1] K. Thews, *Der Abbildungsgrad von Vektorfelden zu stabilen Ruhelagen*, Arch. Math. (Basel) **52** (1989), 71–74.

[Th2] ______, *On a topological obstruction to regular forms of stability*, Nonlinear Anal. **22** (1994), 347–351.

[Vi] G. Villari, *Soluzioni periodiche di una classe di equazioni differenziali*, Ann. Mat. Pura Appl. (4) **73** (1966), 103–110.

[Vo1] P. Volkmann, *Zur Definition des Koinzidenzgrades*, preprint (1981).

[Vo2] ______, *Démonstration d'un théorème de coincidence par la méthode de Granas*, Bull. Soc. Math. Belgique B **36** (1984), 235–242.

[W] T. Ważewski, *Sur un principe topologique pour l'examen de l'allure asymptotique des intégrales des équations différentielles*, Ann. Soc. Polon. Math. **20** (1947), 279–313.

[Z1] F. Zanolin, *Bound sets, periodic solutions and flow-invariance for ordinary differential equations in $\mathbb{R}^n$: some remarks*, Rend. Istit. Mat. Univ. Trieste **19** (1987), 76–92.

[Z2] ______, *Some remarks about persistence for differential systems and processes*, Advanced Topics in the Theory of Dynamical Systems, (Trento, 1987), Academic Press, San Diego, 1989, pp. 253–266.

[Z3] ______, *Continuation theorems for the periodic problem via the translation operator*, Rend. Sem. Mat. Univ. Polit. Torino **54** (1996), 1–23.

[Ze] E. Zeidler, *Nonlinear Functional Analysis and its Applications*, vol. I, Springer, New York, 1986.

[Zo1] H. Żołądek, *The method of holomorphic foliations in planar periodic systems. The case of Riccati equation*, J. Differential Equations **165** (2000), 143–173.

[Zo2] ______, *Periodic planar systems without periodic solutions*, Qualitative Theory Dynamical Systems **2** (2001), 45–60.

23. FIXED POINT RESULTS BASED ON THE WAŻEWSKI METHOD

Roman Srzednicki — Klaudiusz Wójcik — Piotr Zgliczyński

1. Introduction

Topological methods are frequently used in proving results on qualitative properties of differential equations, especially in problems of the existence of solutions satisfying some boundary value data. They are usually based on fixed point theorems or on properties of the Brouwer and Leray–Schauder degrees. The most common approach applies those tools to integral operators corresponding to the considered equations in infinite-dimensional spaces of functions satisfying the prescribed boundary data. Another approach, which is restricted to equations representing some evolution in time, applies them to translation operators along solutions. In the case of ordinary differential equations those operators are finite-dimensional; they are infinite-dimensional if one considers parabolic partial differential equations or delay differential equations. Frequently, the existence of a required solution is a consequence of the fact that the translation operator preserves some compact convex subset of the phase space of the equation, hence the Brouwer or Schauder fixed point theorem applies. In particular, such an approach applies to dissipative equations. Using the Lefschetz fixed point theorem, the same idea can be applied to compact subsets being absolute neighbourhood retracts. However, for non-dissipative equations usually there are no reasonable compact subsets which are invariant with respect to the translation operator. The aim of this note is to describe a method which sometimes can be applied in that context. It is based on the concept of isolating segment and applies the Lefschetz fixed point theorem, the fixed point index, and the retract method of Ważewski. It provides results on the existence of periodic solutions, and, after suitable arrangements, also on the existence of solutions of some other two-point boundary value problems, homoclinic and multibump solutions, and chaotic dynamics of various kinds. Our aim is to present those results and illustrate them using some concrete equations.

The first author is supported by KBN grant 2/P03A/041/24.

Finally, we make some comments on a numerical algorithm leading to construction of isolating segments.

We use a standard notation concerning fixed points; in particular, Fix(f) denotes the set of fixed points of a map $f\colon U \to X$, where $U \subset X$. If U is open and Fix(f) is compact then ind(f) denotes the fixed point index (compare [Do]). The singular homology functor with coefficients in the field of rational numbers $\mathbb{Q}$ is denoted by H. The nth iterate $f \circ \ldots \circ$ of f is denoted by f^n.

2. Local semi-flows and the retract method of Ważewski

In 1947, Tadeusz Ważewski published the paper [Wa] in which he presented a new topological method of proving the existence of solutions remaining in a given set for positive values of time. Below we briefly describe some basic notions concerning the method. Actually, our presentation will use a contemporary approach to the method which slightly differs from the original one.

Let X be a topological space and let D be an open subset of $X \times [0, \infty)$. A continuous map $\phi\colon D \to X$ is called a *local semi-flow* on X if for every $x \in X$ the set $\{t \in [0,\infty) : (x,t) \in D\}$ is equal to an interval $[0, \omega_x)$ for some $\omega_x > 0$ or $\omega_x = \infty$,

$$\phi(x, 0) = x, \tag{2.1}$$

and if $(x,t) \in D$, $(\phi(x,t), s) \in D$ then $(x, t+s) \in D$ and

$$\phi(x, s+t) = \phi(\phi(x,t), s). \tag{2.2}$$

We write also $\phi_t(x)$ instead of $\phi(x,t)$, hence $\phi_0 = \mathrm{id}$ and $\phi_{s+t} = \phi_t \circ \phi_s$. Obviously, ϕ is called a *semi-flow* if $D = X \times [0, \infty)$. In a natural way a more restrictive notion notions of *local flow* and *flow* are defined — one should extend the above definition symmetrically to negative values of t. In this case one assumes that the set $\{t \in \mathbb{R} : (x,t) \in D\}$ is equal to an open interval (α_x, ω_x) with some α_x and ω_x,

$$-\infty \le \alpha_x < 0 < \omega_x \le \infty,$$

and the equations (2.1) and (2.2) hold. Ordinary differential equations deliver the most natural examples of local flows: a smooth vector-field $v\colon M \to TM$ on a manifold M determines a local flow ϕ such that the orbit $t \mapsto \phi_t(x_0)$ of $x_0 \in M$ is the unique solution of the initial value problem

$$\dot{x} = v(x), \quad x(0) = x_0.$$

Let ϕ be a local semi-flow on X; in this case X is called the *phase space* of ϕ. Usually, the t parameter is interpreted as the time.

Let $x \in X$. The map $t \mapsto \phi_t(x)$ is called an *orbit* of x and its image

$$\phi^+(x) := \{\phi_t(x) : t \in [0, \omega_x)\}$$

is called the *positive semi-trajectory* of x. In the case of a local flow, one defines the *trajectory* as

$$\phi(x) := \{\phi_t(x) : t \in (\alpha_x, \omega_x)\}.$$

If the orbit of x constant then it is called *stationary* and x is called a *stationary point*; if it is a periodic map then it is called a *periodic orbit* and x is a *periodic point*. In the case of a local flow generated by a vector-field v a point x_0 is stationary if and only if $v(x_0) = 0$.

It is a natural question to ask for the existence of periodic or stationary points of a given local semi-flow. That question is closely connected to problems in the fixed point theory: a periodic point of period $T > 0$ is a fixed point of the map ϕ_T and a stationary point is a fixed point of ϕ_t with each $t \geq 0$. In order to apply results of the topological fixed point theory one usually assumes that the whole phase space is compact or the points of interest are located in some its compact subset. We distinguish a class of subsets (not invariant, in general) which are particularly convenient to deal with the problem. To this aim we recall some facts related to the Ważewski retract method.

Let $W \subset X$. Define the *exit set* of W as

$$W^- := \{x \in W : \phi(x, [0,t]) \not\subset W \text{ for all } t \in (0, \omega_x)\}.$$

We call W a *Ważewski set* for ϕ if it is closed and its exit set W^- is closed as well. (That notion was introduced by Charles Conley, compare [C]. Actually, the original Conley's definition is more general.) A compact Ważewski set is called here a *block*; an example of a block for a local flow generated by some planar vector-field is drawn in Figure 1. In the case ϕ is a local flow we say that a block W is *isolating* if the boundary of W is equal to the union of W^- and the *entry set*

W^+ defined as the exit set of W with respect to the local flow obtained from ϕ by the reversal of time $t \to -t$. Obviously, the block in Figure 1 is isolating.

Let another subset of W (called the *asymptotic part* of W) be defined as

$$W^* := \{x \in W : \text{there exists } t \in (0, \omega_x) : \phi_t(x) \notin W\},$$

The main property of the notion of Ważewski set is given in the following lemma.

(2.3) LEMMA. *If W is a Ważewski set then the mapping*

$$\sigma : W^* \ni x \to \sup\{t \in [0, \omega_x) : \phi(x, [0,t]) \subset W\} \in [0, \infty)$$

is continuous.

The mapping σ in the lemma is called the *escape-time map.* As a consequence of the continuity of σ one instantly gets the following corollary.

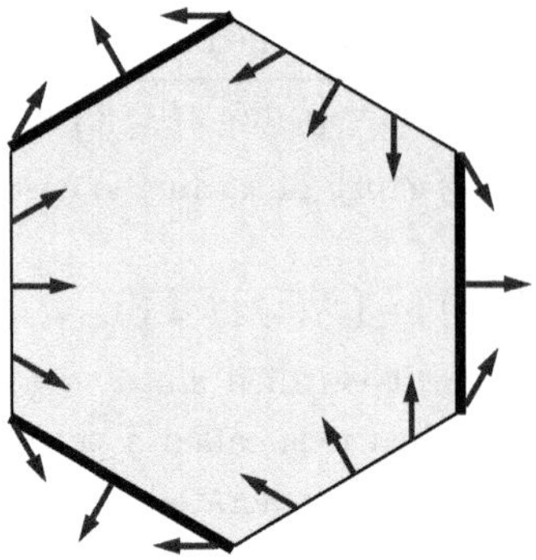

FIGURE 1. A block for a planar vector-field. The exit set consists of three thickened sides of the hexagon.

(2.4) COROLLARY. *If W is a Ważewski set then W^- is a strong deformation retract of W^*.*

That corollary can be reformulated into a version of the Ważewski retract theorem:

(2.5) THEOREM. *If W^- is not a strong deformation retract of a Ważewski set W then there exists an $x \in W$ such that $\phi^+(x) \subset W$.*

Actually, if ϕ is a local flow and W is a block then x can be chosen such that the whole trajectory $\phi(x)$ is contained in W. As we see below, in some cases one can get a stationary trajectory x; the block in Figure 1 represents such a case.

The Ważewski retract method consists in applications of Lemma (2.3) and its consequences to problems in differential equations. In particular, simple results on asymptotic behavior of solutions can be derived from Theorem (2.5). Other applications, like results on the existence of solutions of two-point boundary value problems, require more advanced theorems.

3. Stationary points in blocks

Results on the existence of periodic or stationary points in Ważewski sets which we present here are based on the Lefschetz fixed point theorem. We consider compact Ważewski sets (i.e. blocks) only.

(3.1) THEOREM ([S2]). *Let W be a block and let $T > 0$. Then the set*

$$U := \{x \in W : \phi_t(x) \in W \setminus W^- \text{ for all } t \in [0, T]\}$$

is an open subset of W and the set of fixed points $\mathrm{Fix}(\phi_T|_U)$ *of the restriction*

$$\phi_T|_U : U \to W$$

is compact. Moreover, if W and W^- are ANR*s then*

$$\operatorname{ind}(\phi_T|_U) = \chi(W) - \chi(W^-). \tag{3.1.1}$$

In particular, if

$$\chi(W) - \chi(W^-) \neq 0 \tag{3.1.2}$$

then ϕ_T has a fixed point in W.

We do not provide a proof of the above result here since it is a corollary of a more general Theorem (5.1). It follows by Theorem (3.1) that the fixed point index does not distinguish the essential periodic orbits in W from the stationary ones:

(3.2) Corollary. *If W is a block, W and W^- are* ANR*s, and* (3.1.2) *holds then there exists a stationary point in W.*

Let a local flow on an n-dimensional manifold M be generated by a smooth vector-field v. In this case Corollary (3.2) states that v has a zero in W provided (3.1.2) holds. Moreover, using Theorem (3.1) one can prove that that if $M = \mathbb{R}^n$ and v has no zeros on the boundary of W then the Brouwer degree of v in the interior of W is given by

$$\deg(0, v, \operatorname{int} W) = (-1)^n(\chi(W) - \chi(W^-)) \tag{3.3}$$

(compare [S1]). If the block W is a smooth n-dimensional submanifold of M with boundary and W^- is an $n-1$ dimensional submanifold of ∂W with boundary, then the formulas (3.1.1) and (3.3) are particular cases of the generalized Poincaré-Bendixson formula (see [G] for the history and references related it). For extension of that formula to the non-smooth case we refer to [F].

(3.4) Example. Let a planar local flow ϕ has a block W represented in Figure 1. By above results, there exists a stationary point of ϕ inside the block, since

$$\chi(W) = 1, \quad \chi(W^-) = 3.$$

An example of equation which generates such a local flow is given by

$$\dot{z} = \overline{z}^2 + f(z)$$

(written in the complex-number notation) where $f\colon \mathbb{C} \to \mathbb{C}$ is a smooth such that $f(z)/|z|^2 \to 0$ if $|z| \to \infty$ (compare [S2]; see also Example (5.10) below).

4. Local processes and segments

By a *local semi-process* on a topological space X we mean a continuous map $\Phi: D \to X$, where D is an open subset of $\mathbb{R} \times X \times [0, \infty)$, such that the map

$$\phi: D \ni ((\sigma, x), t) \to (\sigma + t, \Phi(\sigma, x, t)) \in \mathbb{R} \times X$$

is a local semi-flow on $\mathbb{R} \times X$. (In that case ϕ is called a local semi-flow *generated* by Φ.) In a similar way we define a *local process* Φ if the corresponding map ϕ is a local flow. In particular, for a local semi-process (or a local process) Φ one has

$$\Phi(\sigma, x, 0) = x, \quad \Phi(\sigma, x, s + t) = \Phi(\sigma + s, \Phi(\sigma, x, s), t),$$

whenever it is defined (compare [H]). In the sequel we write $\Phi_{(\sigma,t)}(x)$ instead of $\Phi(\sigma, x, t)$; in that notation

$$\Phi_{(\sigma,0)} = \mathrm{id}, \quad \Phi_{(\sigma,s+t)} = \Phi_{(\sigma+s,t)} \circ \Phi_{(\sigma,s)}.$$

The space $\mathbb{R}\times X$ is called the *extended phase space* of Φ. The notion of local process is motivated by properties of solutions of non-autonomous differential equations: if $v: \mathbb{R} \times M \to TM$ is a smooth time-dependent vector-field on a manifold M then the system of equations

$$\dot{x} = v(t, x), \quad \dot{t} = 1$$

generates a local flow on $\mathbb{R}\times M$, hence a local process Φ on M such that for $t_0 \in \mathbb{R}$ and $x_0 \in M$ the map $\tau \mapsto \Phi_{(t_0,\tau-t_0)}(x_0)$ is the solution of the initial value problem

$$\dot{x} = v(t, x), \quad x(t_0) = x_0.$$

Let $T > 0$. A local semi-process Φ is called *T-periodic* if $\Phi_{(\sigma,t)} = \Phi_{(\sigma+T,t)}$ for each σ and t. In that case the map $\Phi_{(0,T)}$, called the *Poincaré map*, satisfies

$$\Phi_{(0,nT)} = \Phi^n_{(0,T)}.$$

Observe that if v is a smooth time-dependent vector-field which is T-periodic with respect to t then the local process Φ is generated by v is T-periodic. Moreover, in this case fixed points of the Poincaré map correspond to initial points of T-periodic solutions of the equation $\dot{x} = v(x, t)$.

In order to establish results on fixed points which refer to local semi-processes (hence also to non-autonomous equations) we introduce a special class of blocks, called *segments* in the extended phase space. At first we introduce the following notation: we denote by π_1 and π_2 the projections of $\mathbb{R}\times X$ onto $\mathbb{R}$ and, respectively, X, and if Z is a subset of $\mathbb{R} \times X$ and $t \in \mathbb{R}$, $[t_1, t_2] \subset \mathbb{R}$ then we put

$$Z_t := \{z \in X : (t, z) \in Z\}, \quad Z_{[t_1,t_2]} = Z \cap ([t_1, t_2] \times X).$$

Let Φ be a local semi-process on X and let ϕ be the corresponding local semi-flow on $\mathbb{R} \times X$. Assume that $a < b$. A set $W \subset [a,b] \times X$ is called a *segment* over $[a,b]$ if it is a block with respect ϕ such that the following conditions hold:

(4.1.1) there exists a compact subset W^{--} of W^- (called the *essential exit set*) such that

$$W^- = W^{--} \cup (\{b\} \times W_b), \quad W^- \cap ([a,b) \times X) \subset W^{--},$$

(4.1.2) there exists a homeomorphism $h\colon [a,b] \times W_a \to W$ such that $\pi_1 \circ h = \pi_1$ and

$$h([a,b] \times W_a^{--}) = W^{--}.$$

If Φ is a local process then a segment is called *isolating* if it is an isolating block for ϕ.

The notion of segment is explained in a simple case in Figure 2.

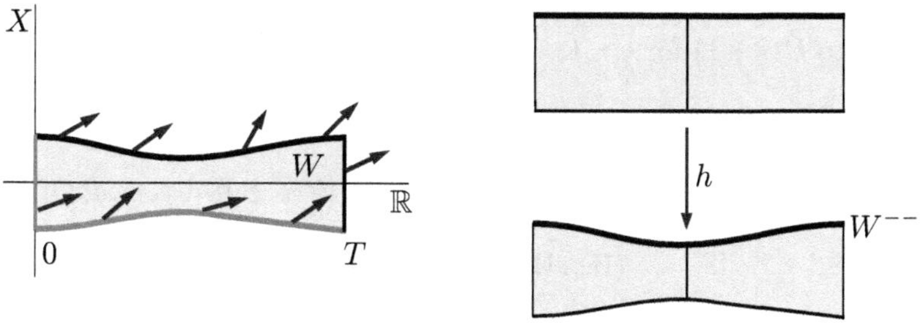

FIGURE 2. A segment W over $[0,T]$ and a monodromy homeomorphism h.

Intuitively, W consists of the left-hand side $\{a\} \times W_a$, the right-hand side $\{b\} \times W_b$, and the main part located over the open interval (a,b). The condition (4.1.2) means that (W, W^{--}) is a pair of trivial bundles over $[a,b]$ with the fibre (W_a, W_a^{--}). Because of the specific behavior of ϕ (it moves along the time-axis with speed 1), it is clear that that the right-hand side must belong to the exit set.

Let a homeomorphism h satisfies (4.1.2). We define the corresponding *monodromy map*

$$m\colon (W_a, W_a^{--}) \to (W_b, W_b^{--}), \quad m(x) = \pi_2 h(b, \pi_2 h^{-1}(a,x)).$$

The monodromy map is actually a homoeomorphism. It can be proved that a different choice of the homeomorphism satisfying (4.1.2) provides the monodromy map homotopic to m. It follows, in particular, that the isomorphism in homologies

$$\mu_W := H(m)\colon H(W_a, W_a^{--}) \to H(W_b, W_b^{--})$$

is an invariant of the segment W.

Segments can be glued in a natural way: if W is a segment over $[a, b]$, Z is a segment over $[b, c]$ and

$$(W_b, W_b^{--}) = (Z_b, Z_b^{--}) \tag{4.2}$$

then their union $W \cup Z$ is a segment over $[a, c]$ and its monodromy map is a composition of monodromy maps of W and Z.

(4.3) Remark. Even if the condition (4.2) is not satisfied, the union of W and Z is a useful object (with respect to problems considered in this note) provided $W \cup Z$ is a block with respect to the local semi-process generated by Φ. In this case W and Z are called *contiguous* and $W \cup Z$ is called a *chain*. More generally, a chain is a union of a finite number of segments $U(1), \dots, U(r)$ such that $U(i)$ is contiguous to $U(i + 1)$ for $i = 1, \dots, r - 1$ (see [S6]).

5. Detection of periodic solutions

Let W be a segment over $[a, b]$. The segment W is called *periodic* if

$$(W_a, W_a^{--}) = (W_b, W_b^{--}).$$

In that case, if $H(W_a, W_a^{--})$ is of finite type then the Lefschetz number $\Lambda(\mu_W)$ is correctly defined. It is called the *Lefschetz number of the periodic segment* W. Our results on the existence of periodic solutions of non-autonomous differential equations are based on the following theorem:

(5.1) Theorem ([S2]). *Let W be a periodic segment over $[a, b]$. Then the set*

$$U =: U_W := \{x \in W_a : \Phi_{(a,t-a)}(x) \in W_t \setminus W_t^{--} \text{ for all } t \in [a, b]\}$$

is open in W_a and the set of fixed points of the restriction $\Phi_{(a,b-a)}|_U\colon U \to W_a$ is compact. Moreover, if W and W^{--} are ANR*s then*

$$\operatorname{ind}(\Phi_{(a,b-a)}|_U) = \Lambda(\mu_W).$$

In particular, if

$$\Lambda(\mu_W) \neq 0 \tag{5.1.1}$$

then $\Phi_{(a,b-a)}$ has a fixed point in W_a.

If Φ is the local process generated by a time-dependent vector-field v then x_0 is a fixed point of $\Phi_{(a,b-a)}$ if and only if $\tau \mapsto \Phi_{(a,\tau-a)}(x_0)$ is a solution of the periodic problem

$$\dot{x} = v(t,x), \quad x(a) = x(b), \tag{5.2}$$

hence Theorem (5.1) immediately implies:

(5.3) COROLLARY. *Let W be a periodic segment and let W and $W^{-}-$ be* ANR*s. If* (5.1.1) *holds then the periodic problem* (5.2) *has a solution which passes through W_a at time a. In particular, if W is a segment over $[0,T]$,* (5.1.1) *holds, and the vector-field v is T-periodic in t then the equation $\dot{x} = v(t,x)$ has a T-periodic solution passing through W_0 at time* 0.

(5.4) REMARK. Using the notion of chain (see Remark (4.3)), a direct generalization of Theorem (5.1) is presented in [S6].

PROOF OF THEOREM (5.1). By the definition of segment, W is an ANR if and only if W_a is an ANR and the same holds for W^{--} and W_a^{--}. We define maps $m_s\colon W_s \to W_a$ ($s \in [a,b]$), by

$$m_s(x) = \pi_2 h(b, \pi_2 h^{-1}(s,x)).$$

In particular $m_a = m$ and $m_b = \mathrm{id}$. Let σ be the escape-time map for W (see Lemma (2.3); here obviously $W = W^*$). Consider a homotopy $H\colon W_a \times [0,1] \to W_a$, $H_t := H(\,\cdot\,,t)$, given by

$$H_t(x) := \begin{cases} m_{a+\sigma(a,x)}(\Phi_{(a,\sigma(a,x))}(x)) & \text{if } \sigma(a,x) \le (1-t)(b-a), \\ m_{a+(1-t)(b-a)}(\Phi_{(a,(1-t)(b-a))}(x)) & \text{if } \sigma(a,x) \ge (1-t)(b-a). \end{cases}$$

In particular, $H_1 = m$. Moreover, it is easy to check that $H_t(x) = m(x)$ if $t \in [0,1]$ and $x \in W_a^{--}$, hence $H_t(W_a^{--}) = W_a^{--}$ if $t \in [0,1]$. By the homotopy property of the Lefschetz number we get

$$\Lambda(m) = \Lambda(H_1) = \Lambda(H_0). \tag{5.5}$$

Since

$$H_0(x) = m_{a+\sigma(a,x)}(\Phi_{(a,\sigma(a,x))}(x)), \tag{5.6}$$

for every $x \in W_a$, one has $H_0(x) = \Phi_{(a,b-a)}(x)$ if $\sigma(a,x) = b-a$, hence

$$H_0|_U = \Phi_{(a,b-a)}|_U. \tag{5.7}$$

Let us observe that

$$U = \{x \in W_a : \sigma(a,x) = b-a,\ \Phi_{(a,b-a)}(x) \in W_a \setminus W_a^{--}\}, \tag{5.8}$$

so by (5.6), $U = (H_0)^{-1}(W_a \setminus W_a^{--})$ and consequently U is open in W_a. If $x \in W_a \setminus W_a^{--}$ and $H_0(x) = x$ then necessarily $\sigma(a,x) = b-a$ (since in the other case $H_0(x) \in W_a^{--}$), hence $x \in U$ by (5.8), and thus

$$\mathrm{Fix}(\Phi_{(a,b-a)}|_U) = \mathrm{Fix}(H_0) \cap \{x \in W_a : \sigma(a,x) = b-a\}.$$

In particular $\mathrm{Fix}(\Phi_{(a,b-a)}|_U)$ is compact. Put $V := \{x \in W_a : \sigma(a,x) < b-a\}$. It follows that V is open in W_a, $W_a^{--} \subset V$, and $H_0(V) = W_a^{--}$. One can easy check that

$$\mathrm{Fix}(H_0) = \mathrm{Fix}(H_0|_U) \cup \mathrm{Fix}(H_0|_V) = \mathrm{Fix}(\Phi_{(a,b-a)}|_U) \cup \mathrm{Fix}(m|_{W_a^{--}}).$$

Since both the sets $\mathrm{Fix}(\Phi_{(a,b-a)}|_U)$ and $\mathrm{Fix}(m|_{W_a^{--}})$ are compact and disjoint, by the Lefschetz fixed point theorem and the additivity of the fixed point index we get

$$\Lambda(H_0) = \mathrm{ind}(H_0|_U) + \mathrm{ind}(H_0|_V), \tag{5.9}$$

By the commutativity of the fixed point index and the Lefschetz fixed point theorem we obtain

$$\mathrm{ind}(H_0|_V) = \mathrm{ind}(H_0|_{W_a^{--}}) = \Lambda(H_0|_{W_a^{--}}) = \Lambda(m|_{W_a^{--}}).$$

Combining (5.5), (5.7) and (5.9) we get

$$\mathrm{ind}(\Phi_{(a,b-a)}|_U) = \Lambda(m) - \Lambda(m|_{W_a^{--}}) = \Lambda(\mu_W),$$

so the proof is complete. □

As it was already pointed out, Theorem (3.1) is a particular case of Theorem (5.1). Indeed, a local flow ϕ on X generates a local process Φ given by

$$\Phi_{(a,t)} := \phi_t$$

for each $a \in \mathbb{R}$. If B is a block for ϕ then $[a,b] \times B$ is a segment for Φ and its proper exit set is equal to $[a,b] \times B^-$; since the identity is a monodromy, one has

$$\Lambda(\mu_W) = \chi(B) - \chi(B^-).$$

In the following examples (taken from [S2]) we provide some natural applications of the obtained results.

(5.10) EXAMPLE. Consider a planar non-autonomous equation

$$\dot{z} = \overline{z}^n + f(t,z), \tag{5.10.1}$$

where n is an integer, $n \geq 1$, $z \in \mathbb{C}$ and $f: \mathbb{R} \times \mathbb{C} \to \mathbb{C}$ is a smooth function T-periodic with respect to t for some $T > 0$. Assume that

$$\frac{f(t,z)}{|z|^n} \to 0, \quad \text{as } |z| \to \infty \text{ uniformly in } t. \tag{5.10.2}$$

Then the equation (5.10.1) has a T-periodic solution.

Indeed, by (5.10.2) the term $\overline{z}^n$ becomes dominating as $|z| \to \infty$, hence the behavior of solutions of (5.10.1) near infinity resembles the phase portrait of the autonomous equation

$$\dot{z} = \overline{z}^n. \tag{5.10.3}$$

For the local flow generated by the latter equation there exists a family of isolating blocks $\{B_r\}_{r>0}$, where B_r is an equilateral $2(n+1)$-gon centered at zero with the diameter equal to $2r$ and the exit set B_r^- consists of $n+1$ disjoint sides of B_r, one of which intersects perpendicularly the positive real semi-axis (compare Figure 1 in the case $n = 2$). It follows that for r sufficiently large the prism $[0,T] \times B_r$ is an isolating segment for (5.10.1). It is depicted in Figure 3 for $n = 2$.

FIGURE 3. An isolating segment over $[0,T]$ for the equation (5.10.1) with $n = 2$.

Its essential exit set $[0,T] \times B_r^-$ consists of $n+1$ faces of the prism (in the picture they are marked in gray). Since the identity is a monodromy map for the segment, one concludes that its Lefschetz number is equal $-n$, hence there exists a T-periodic solution of the equation (5.10.1) by Corollary (5.3).

(5.11) EXAMPLE. Let us modify the previous example by multiplying the leading term of the right-hand side of the equation by e^{it}:

$$\dot{z} = e^{it}\overline{z}^n + f(t,z). \tag{5.11.1}$$

Here we assume that $n \geq 2$ and $f: \mathbb{R} \times \mathbb{C} \to \mathbb{C}$ is smooth and 2π-periodic in t. As before we assume that (5.10.2) holds. It follows by results in [S2] that for r sufficiently large the set

$$W := \{(t, z) \in [0, 2\pi] \times \mathbb{C} : e^{-it/(n+1)} z \in B_r\}$$

is an isolating segment over $[0, 2\pi]$ for (5.11.1) with the essential exit set

$$W^{--} = \{(t, z) \in [0, 2\pi] \times \mathbb{C} : e^{-it/(n+1)} z \in B_r^-\},$$

where B_r and B_r^- are defined in Example (5.10). It means that W is a twisted prism with a $2(n+1)$-gon base centered at the origin and its time sections W_t are obtained by rotating the base with the angular velocity $1/(n+1)$ over the time interval $[0, 2\pi]$. The set W^{--} consists of $n+1$ disjoint ribbons winding around the prism, as is shown in Figure 4 in the case $n = 2$.

FIGURE 4. An isolating segment over $[0, 2\pi]$ for the equation (5.11.1) with $n = 2$.

One can choose the rotation by the angle $2\pi/3$ as a monodromy map of the segment, hence the Lefschetz number of the segment W is equal to 1. It follows by Corollary (5.3) that (5.11.1) has a 2π-periodic solution.

(5.12) EXAMPLE. Now we consider a special case of (5.11.1) (recall that $n \geq 2$):

$$\dot{z} = e^{it}\bar{z}^n + \bar{z}. \tag{5.12.1}$$

The zero solution is 2π-periodic, hence one should look for a nontrivial one. By the previous example, there is a large segment W for the equation such that

$$\Lambda(\mu_W) = 1.$$

Since the term $\bar{z}$ on the right-hand side of (5.12.1) dominates as $|z| \to 0$, it can be proved that there is another segment Z for that equation: it is a prism having

FIGURE 5. The isolating segment Z for the equation (5.12.1).

a sufficiently small square centered at the origin as a base (see Figure 5). Moreover, $Z \subset W$ and $\Lambda(\mu_Z) = -1$.

If there is no 2π-periodic solution of (5.12.1) then 0 is the only fixed point of the Poincaré map $\Phi_{(0,2\pi)}$ for the equation, hence by Theorem (5.1),

$$\Lambda(\mu_W) = \mathrm{ind}(\Phi_{(0,2\pi)}|_{\mathrm{int}\, W_0}) = \mathrm{ind}(\Phi_{(0,2\pi)}|_{\mathrm{int}\, Z_0}) = \Lambda(\mu_Z)$$

which is a contradiction. Thus (5.12.1) has a nonzero 2π-periodic solution.

(5.13) EXAMPLE. Finally, we consider another special case of (5.11.1):

$$\overline{z} = e^{it}\overline{z}^n + z. \tag{5.13.1}$$

As before, we have a large segment W over $[0, 2\pi]$ which is a twisted prism having an equilateral $(2n+1)$-polygon as a base. There exist also a smaller segment $Y \subset W$ being a cylinder over a disc centered at 0 such that Y^{--} is equal to the whole boundary, hence its Lefschetz number is also equal to 1 and the argument used in Example (5.12) fails here. However, if we glue three copies of W along the interval $[0, 6\pi]$ we obtain a new segment $\widetilde{W}$ such that its monodromy map comes from the full 2π-rotation, hence it is equal to the identity and thus $\Lambda(\mu_{\widetilde{W}}) = -n$.

A similar gluing of three copies of Y provides the segment $\widetilde{Y}$ for which again $\Lambda(\mu_{\widetilde{Y}}) = 1$. Thus, by the argument in the previous example we conclude that there exists a nonzero 6π-periodic solution of the equation (5.13.1).

Results on the existence of periodic solutions of planar non-autonomous equations extending those presented in the above examples can be found in [S3].

6. Detection of chaotic dynamics

In order to formulate results on chaotic dynamics we use the notion of *shift on r symbols*, where r is some positive integer. It is a pair (Σ_r, σ), where Σ_r, called the *shift space*, is defined as

$$\Sigma_r := \{0, \dots, r-1\}^{\mathbb{Z}},$$

i.e. the set of bi-infinite sequences of r symbols, and the *shift map* σ is given by

$$\sigma\colon \Sigma_r \ni (\dots s_{-1}.s_0 s_1 \dots) \to (\dots s_0.s_1 s_2 \dots) \in \Sigma_r.$$

In the above notation the dot . marks the 0th term of a sequence, hence σ moves the sequence by one position to the left. The shift on r symbols is a model example of complicated dynamics; in particular it satisfies all three conditions from the classic definition of chaos: sensitive dependence on initial conditions, topological transitivity, and the density of periodic orbits (compare [De]). The term *chaotic dynamics* for a non-autonomous T-periodic equation (and, more generally, for a T-periodic local process Φ) is used by us if the following two conditions are satisfied. The first condition is the existence of a semi-conjugacy between the Poincaré map $\Phi_{(0,T)}$ restricted to some compact subset I of the phase space of the equation and the shift map, i.e. there exists a continuous surjective map $g\colon I \to \Sigma_r$ such that

$$\sigma \circ g = g \circ \Phi_{(0,T)} \tag{6.1}$$

holds (that condition is often called *symbolic dynamics*). The second condition asserts that for infinitely many of periodic sequences $c \in \Sigma_r$ the counter-image $g^{-1}(c)$ contains an initial point of a periodic solution of the equation.

Below we present results on the existence of chaotic dynamics based on a proper configuration of segments. We consider a local process Φ on a topological space X and we assume that it is T-periodic for some $T > 0$. Our first result is a simple consequence of Theorem (5.1).

(6.2) THEOREM ([S4]). *Let $W(0), \dots, W(r)$ be periodic segments over $[0,T]$. Assume that $W(0)_0$ and $W(0)_0^{--}$ are* ANR*s and*

(6.2.1) $(W(0)_0, W(0)_0^{--}) = \dots = (W(r)_0, W(r)_0^{--})$,

(6.2.2) *there exists $s \in (0,T)$ such that $W(i)_s \cap W(j)_s = \emptyset$ for every $i,j = 0,\dots,r$,*

(6.2.3) *there exists $n \in \mathbb{N}$ such that*

$$H_n(W(0)_0, W(0)_0^{--}) = \mathbb{Q}, \quad H_k(W(0)_0, W(0)_0^{--}) = 0 \quad \textit{for all } k \neq n.$$

Then there are a compact set $I \subset X$, invariant for the Poincaré map $\Phi_{(0,T)}$, and a continuous surjective map $g\colon I \to \Sigma_{r+1}$ such that (6.1) holds and for every k-periodic sequence $c \in \Sigma_{r+1}$ there exists $x \in g^{-1}(c)$ such that

$$\Phi^k_{(0,T)}(x) = x.$$

An example of segments satisfying the assumptions of the above theorem for $r = 1$ and $n = 1$ is shown in Figure 6.

Before we give a proof the above theorem we define an operation of gluing of periodic segments. If W and Z are periodic segments over $[0,T]$ having the same cross-sections at 0, i.e.

$$(W_0, W_0^{--}) = (Z_0, Z_0^{--})$$

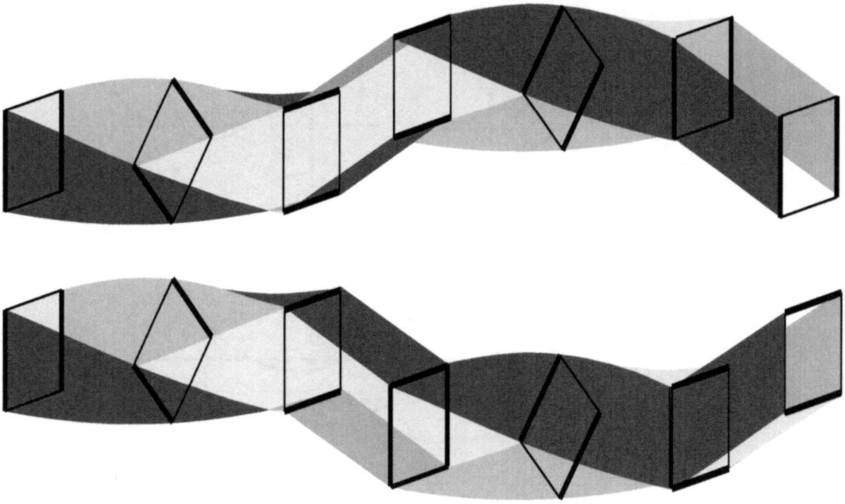

FIGURE 6. Two periodic segments satisfying (6.2.1)–(6.2.3).

holds. Put

$$WZ := \{(t,x) \in [0,2T] \times X : x \in W_t \text{ if } t \in [0,T],\ x \in Z_{t-T} \text{ if } t \in [T,2T]\}.$$

(see Figure 7). It is a periodic segment over $[0, 2T]$.

If $Z(1), \dots, Z(r)$ are periodic segments over $[0,T]$ having the same cross-sections at 0 then we define recurrently another periodic segment

$$Z(1)\dots Z(r) := (Z(1)\dots Z(r-1))Z(r).$$

If $Z(i) = W$ for each $i = 1, \dots, r$ then w put $W^r := Z(1)\dots Z(r)$.

PROOF OF THEOREM (6.2). The required set I is defined as

$$I := \{x \in W(0)_0 : \text{for all } k \in \mathbb{Z} \text{ there exists } i = 0, \dots, r : \\ \Phi_{(0,kT+t)} \in W(i)_t \text{ for all } t \in [0,T]\}.$$

By (6.2.2), the map g given by

$$g(x) = c \text{ if and only if } \Phi_{(0,kT+s)} \in W(c_k)_s \text{ for all } k \in \mathbb{Z}$$

is continuous and provides the required semi-conjugacy (6.1) because the considered local process is T-periodic. Since the set of periodic sequences in Σ_{r+1} is dense, in order to prove the surjectivity of g (and also the remaining claim of the theorem) it suffices to prove that for each periodic sequence $c \in \Sigma_{r+1}$

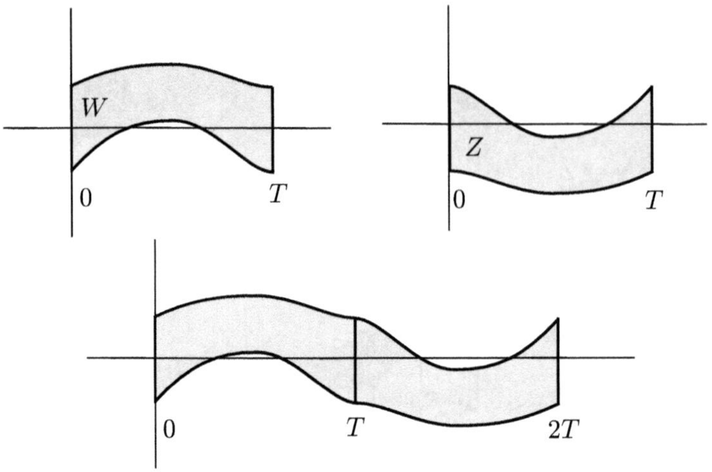

FIGURE 7. Periodic segments W and Z satisfying (6.2.2) and the segment WZ.

there exists a corresponding periodic point of the Poincaré map. Let c be such a k-periodic sequence; it is uniquely determined by a sequence $(c_0, \dots, c_{k-1})$ in the set $\{0, \dots, r\}^{\{0,\dots,k-1\}}$. Define a periodic segment $W := W(c_0)\dots W(c_{k-1})$ over $[0, kT]$. Since the homologies of (W_0, W_0^-) are one-dimensional by (6.2.3), and μ_W is an automorphism (since each monodromy map is a homeomorphism), $\Lambda(\mu_W) \neq 0$. Thus, by Theorem (5.1), there exists an $x \in W_0$ such that

$$\Phi^k_{(0,T)}(x) = \Phi_{(0,kT)}(x) = x.$$

It follows by the T-periodicity of Φ that $x \in I$ and $g(x) = c$, hence the result follows. □

(6.3) EXAMPLE. One can verify the existence of two isolating segments satisfying the assumptions of Theorem (6.2) for the planar equation

$$\dot{z} = \frac{1}{2}e^{-i\kappa t}z\left(\frac{1}{2}i\kappa(z+1) + e^{i\kappa t}(\overline{z}+1)\right)\left(\frac{1}{2}i\kappa(z-1) + e^{i\kappa t}(\overline{z}-1)\right)$$

provided $\kappa > 0$ is small enough. They are similar to those in Figure 6; for an explanation we refer to [S4].

The other results stated in this note assert the existence of chaotic dynamics in presence of two periodic segments, one of which contains the other.

(6.4) THEOREM ([SW], [W1]). *Let Z and W be periodic segments over $[0, T]$ which satisfy*

(6.4.1) $$Z \subset W, \quad (Z_0, Z_0^{--}) = (W_0, W_0^{--}).$$

Assume that Z_0 *and* Z_0^{--} *are* ANR*s. Assume moreover, that there exists an* $n_0 \in \mathbb{N} \setminus \{1\}$ *such that*

(6.4.2) $\mu_Z = \mu_W^{n_0} = \mathrm{id}_{H(Z_0, Z_0^{--})}$,

(6.4.3) $\Lambda(\mu_W) = \Lambda(\mu_W^i)$ *for* $i \in \{1, \dots, n_0 - 1\}$,

(6.4.4) $\Lambda(\mu_W) \neq \chi(W_0, W_0^{--})$ *and* $\chi(W_0, W_0^{--}) \neq 0$,

Then there are a compact set $I \subset X$*, invariant for the Poincaré map* $\Phi_{(0,T)}$*, and a continuous surjective map* $g: I \to \Sigma_2$ *such that the equation* (6.1) *holds and*

(6.4.5) *if* n_0 *is even then for each* n*-periodic sequence* $c \in \Sigma_2$ *there exists* $x \in g^{-1}(c)$ *such that* $\Phi^n_{(0,T)}(x) = x$,

(6.4.6) *if* n_0 *is odd then for each* n*-periodic sequence* $c \in \Sigma_2$ *such that the symbol* 1 *appears* k *times in* $(c_0, \dots, c_{n-1})$ *and* k *is not an odd multiplicity of* n_0*, there exists* $x \in g^{-1}(c)$ *such that* $\Phi^n_{(0,T)}(x) = x$.

The theorem appeared first in [SW] in the case $n_0 = 2$ and then in [W1] in a full generality. In order to present a sketch of its proof we introduce the following convenient notation for the segments W and Z. For a finite sequence

$$c = (c_0, \dots, c_{n-1}) \in \{0, 1\}^{\{0, \dots, n-1\}}$$

we write $W^n(c)$ for the segment $W(0) \dots W(n-1)$, where $W(i) = W$ if $c_i = 1$ and $W(i) = Z$ if $c_i = 0$ (see Figure 8). In particular, if $c_i = 1$ for all $i = 0, \dots, n-1$ then $W^n(c) = W^n$.

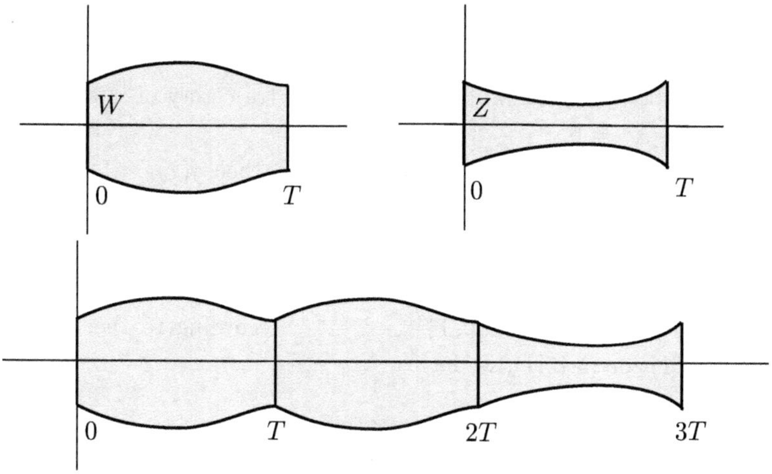

FIGURE 8. Periodic segments W and Z satisfying (6.4.1), and the segment $W^3((1,1,0))$.

SKETCH OF A PROOF OF THEOREM (6.4). The following idea of the proof comes from [W1] (compare also [SW]). Let

$$I := \bigcap_{n=-\infty}^{\infty} \{x \in W_0 : \Phi_{(0,t+nT)}(x) \in W_t \text{ for all } t \in [0,T]\}$$

be the set of all points in W_0 whose full trajectories are contained in the bigger segment W. It follows that I is compact. Let σ_Z be the escape-time map for the smaller segment Z (see Lemma (2.3)). It follows by (6.4.1) that $\sigma(0,x)$ is defined for every $x \in W_0$ and if $x \in I$ then either

(6.5.1) $$\sigma_Z(0,x) < T,$$

or

(6.5.2) $$\sigma_Z(0,x) = T \quad \text{and} \quad \Phi_{(0,T)}(x) \in W_0 \setminus W_0^{--}.$$

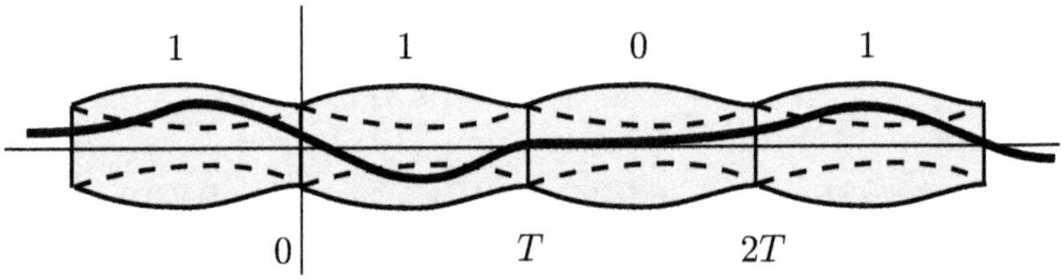

FIGURE 9. Coding of the trajectory of an $x \in I$.

For $x \in I$ we define $g(x) \in \Sigma_2$ by the following rule:

(6.6.1) if on the time interval $[iT,(i+1)T]$ the trajectory of x is contained in Z, then $g(x)_i = 0$,

(6.6.2) if $\Phi^i_{(0,T)}(x)$ leaves Z in time less then T, then $g(x)_i = 1$.

It follows by (6.5.1) and (6.5.2) that the map $g\colon I \to \Sigma_2$ is continuous and satisfies (6.1). By compactness of I and density of the set of periodic sequences in the shift space Σ_2 it is sufficient to show that (6.4.5) and (6.4.6) hold.

Let $c = (c_0,\dots,c_{n-1}) \in \{0,1\}^{\{0,\dots,n-1\}}$. According to the notation in the statement of Theorem (5.1), we have

$$U_{W^n(c)} := \{x \in W_0 : \Phi_{(0,t)}(x) \in W^n(c)_t \setminus W^n(c)_t^{--} \text{ for all } t \in [0,nT]\}.$$

We define $U_{W^n(c),c} \subset U_{W^n(c)}$ as follows: $x \in U_{W^n(c)}$ belongs to the set $U_{W^n(c),c}$ if and only if for each $i \in \{0,\dots,n-1\}$ such that $c_i = 1$ there exists $t \in (0,T)$ such that $\Phi_{(0,t_iT+t)}(x) \in W_t \setminus Z_t$, i.e. $\Phi_{(0,t_iT)}(x) \in Z_0$ leaves Z in less time then T. It is easy to check that $U_{W^n(c),c}$ is open in W_0 and the sets $U_{W^n(c),c}$ over all n-element sequences c from $\{0,1\}^{\{0,\dots,n-1\}}$ form open and disjoint covering of U_{W^n}. We define

$$F_c := \{x \in g^{-1}(c) : \Phi^n_{(0,T)}(x) = x\} \subset I.$$

The set F_c consists of all fixed points of the nth iterate of the Poincaré map $\Phi^n_{(0,T)}$ whose trajectories are coded by the sequence c. It is easy to check that F_c is compact and it is equal to $\operatorname{Fix}(\Phi^n_{(0,T)}|_{U_{W^n(c),c}})$, so $\operatorname{ind}(\Phi^n_{(0,T)}|_{U_{W^n(c),c}})$ is defined. It can be proved that

(6.7) $\operatorname{ind}(\Phi^n_{(0,T)}|_{U_{W^n(c),c}})$

$$= \begin{cases} \Big(\sum\limits_{s:\, n_0|s} (-1)^{k-s}\binom{k}{s}\Big)(\chi(Z_0, Z_0^{--}) - \Lambda(\mu_W)) & \text{if } k \geq 1, \\ \chi(Z_0, Z_0^{--}) & \text{if } k = 0, \end{cases}$$

where the symbol 1 appears exactly k-times in the sequence c. The equation (6.7) is a consequence of Theorem (5.1), elementary properties of the fixed point index, and some combinatorial calculations. We skip its proof here referring the reader to [W1]. One can check that

$$\sum_{s:n_0|s} (-1)^{k-s}\binom{k}{s} = 0$$

if and only if n_0 is odd and k is an odd multiplicity of n_0, hence the proof of Theorem (6.4) is finished. □

(6.8) Example. Consider the following planar non-autonomous equation

$$\dot{z} = (1 + e^{i\kappa t}|z|^2)\bar{z}^n, \tag{6.8.1}$$

where $\kappa > 0$ is a real parameter and $n \geq 1$ is an integer. The right-hand side of the equation (6.8.1) is $2\pi/\kappa$ periodic. It was proved in [W1] (and in [SW] in the case $n = 1$; see also [WZ3]) that for sufficiently small κ there are two periodic segments $Z(n)$ and $W(n)$ over $[0, 2\pi/\kappa]$ which satisfy all assumptions of Theorem (6.4) with $n_0 = n + 1$. We describe briefly how the segments look like. For a small $|z|$ the dynamics generated by (6.8.1) is close to the one of the autonomous equation (5.10.3), hence by a similar argument then the one in Example (5.10) (but now we are near the origin, not infinity) we conclude the existence of a periodic isolating segment

$$Y(n) := [0, 2\pi/\kappa] \times B_r.$$

for (6.8.1), independent of the choice of $\kappa > 0$, with $r > 0$ sufficiently small. Recall that B_r is an equilateral $2(n+1)$-gon centered at the origin with the diameter equal to $2r$. The essential exit set is given by

$$Y(n)^{--} = [0, 2\pi/\kappa] \times B_r^-,$$

where B_r^- consists of $n+1$ disjoint sides of B_r. If $|z|$ is large then the term $e^{i\kappa t}|z|^2\bar{z}^n$ dominates in (6.8.1) and it follows by results in [S2] that for R sufficiently large and each $\kappa > 0$ the set

$$W(n) := \{(t, z) \in [0, 2\pi/\kappa] \times \mathbb{C} : e^{-it\kappa/(n+1)}z \in B_R\}$$

is an isolating segment over $[0, 2\pi]$ for (6.8.1) with the essential exit set

$$W(n)^{--} = \{(t, z) \in [0, 2\pi/\kappa] \times \mathbb{C} : e^{-it\kappa/(n+1)}z \in B_R^-\}.$$

It is obvious that if $R > \sqrt{2}r$ then $Y(n) \subset W(n)$ for every $n \in \mathbb{N}$ and $\kappa > 0$, but in this case the zero-sections $Y(n)_0$ and $W(n)_0$ are not equal each to the other, hence the condition (6.4.1) is not satisfied. In order to get (6.4.1) we modify the smaller segment $Y(n)$. It can be done for $0 < \kappa < \kappa_0$, where κ_0 is sufficiently small and as result we obtain a new segment, which we denote by $Z(n)$. Its construction can be described as follows. Like for $Y(n)$, the time t-section $Z(n)_t$ is a regular $2(n+1)$-gon based prism centered at the origin and the essential exit set $Z(n)^{--}$ consists of $n+1$ disjoint parts. However, contrary to $Y(n)$, the diameter of $Z(n)_t$ decreases linearly from $2R$ to $2r$ as t passes through the interval $[0, \Delta]$ to some $\Delta < \pi/\kappa$, then stays constant in $[\Delta, 2\pi/\kappa - \Delta]$, and then increases linearly from $2r$ to $2R$ in $[2\pi/\kappa - \Delta, 2\pi/\kappa]$.

In particular, the segments $Z(2)$ and $W(2)$ are similar to the ones shown in Figure 10.

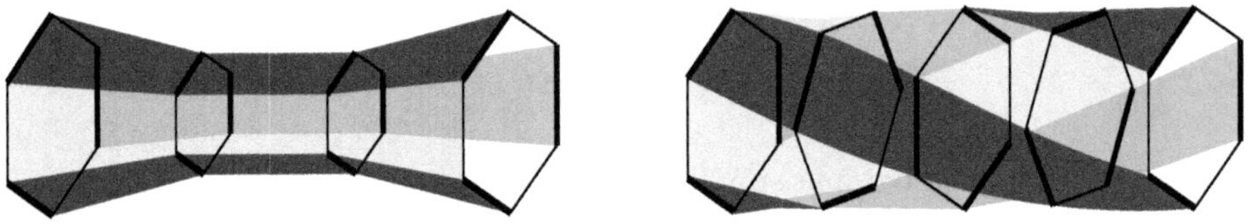

FIGURE 10. Isolating segments $Z(2) \subset W(2)$ for the equation (6.8.1) with $n = 2$. The shaded faces are the exit sets Z^{--} and W^{--}.

It follows that

$$\Lambda(\mu_{W(2)}) = \Lambda(\mu_{W(2)}^2) = 1, \quad \chi(Z(2)_0, Z(2)_0^{--}) = -2, \quad \mu_{W(2)}^3 = \mathrm{id}_{H(Z(2)_0, Z(2)_0^{--})}.$$

By generalizing those equations to the case of arbitrary n one can get the following conclusion:

(6.9) THEOREM ([W1]). *For every $n \in \mathbb{N}$ there exists $\kappa_0 > 0$ such that for each $0 < \kappa < \kappa_0$ the local process generated by the equation* (6.8.1) *satisfies the assumptions of Theorem* (6.4) *with $n_0 = n + 1$.*

Assume that Φ is a T-periodic local process on $\mathbb{R}^n$ generated by a time-dependent vector-field f such that $f(t,-x) = -f(t,x)$. In this case we present a modified version of the previous theorem.

(6.10) THEOREM ([W2]). *Let W and Z be two periodic segments over $[0,T]$ such that the condition* (6.4.1) *holds. Assume that Z_0 and Z_0^{--} are* ANR*s and*

(6.10.1) $(Z_0, Z_0^{--}) = (-Z_0, -Z_0^{--})$ *i.e. the pair (Z_0, Z_0^{--}) is symmetric with respect to the origin,*

(6.10.2) $\mu_W = H(-\mathrm{id}_{(Z_0,Z_0^{--})})\colon H(Z_0,Z_0^{--}) \to H(Z_0,Z_0^{--})$,

(6.10.3) $\mu_Z = \mathrm{id}_{H(Z_0,Z_0^{--})}$,

(6.10.4) $\Lambda(\mu_W) \neq \chi(Z_0,Z_0^{--}), \quad \chi(Z_0,Z_0^{--}) \neq 0$.

Then there are a compact set I invariant with respect to the Poincaré map $\Phi_{(0,T)}$ and a continuous surjective map $g\colon I \to \Sigma_2$ such that the equation (6.1) *holds and*

(6.10.5) *for each k-periodic sequence $c \in \Sigma_2$ there exists a fixed point $x \in g^{-1}(c)$ of $\Phi^k_{(0,T)}$,*

(6.10.6) *exists a fixed point $x \in g^{-1}(c)$ of $\Phi^{2k}_{(0,T)}$ such that $\Phi^k_{(0,T)}(x) = -x$.*

AN IDEA OF A PROOF. By (6.10.2) and (6.10.3), we get

$$\mu_W \circ \mu_W = \mu_Z = \mathrm{id}_{H(Z_0,Z_0^{--})},$$

hence the existence of g and (6.10.5) follow by Theorem (6.4) with $n_0 = 2$. The proof of (6.10.6) is based on the version of Theorem (5.1) concerning the existence of antiperiodic solutions inside of a periodic segment given in [S5]. We skip it here referring to [W2]. □

(6.11) EXAMPLE. As an example of applications of Theorem (6.10) we consider the equation (6.8.1) with $n = 1$, i.e. the equation

$$\dot{z} = (1 + e^{i\kappa t}|z|^2)\bar{z}. \tag{6.11.1}$$

In this case we present a more detailed description of the periodic isolating segments Z and W over $[0, 2\pi/\kappa]$ which appear Example (6.8). They are shown in Figure 11.

The larger segment and its essential exit set are of the form

$$W = \{(t,z) \in [0,2\pi/\kappa] \times \mathbb{C} : |\Re(e^{-it\kappa/2}z)| \le R,\ |\Im(e^{-it\kappa/2}z)| \le R\},$$
$$W^{--} = \{(t,z) \in W : |\Re(e^{-it\kappa/2}z)| = R\}.$$

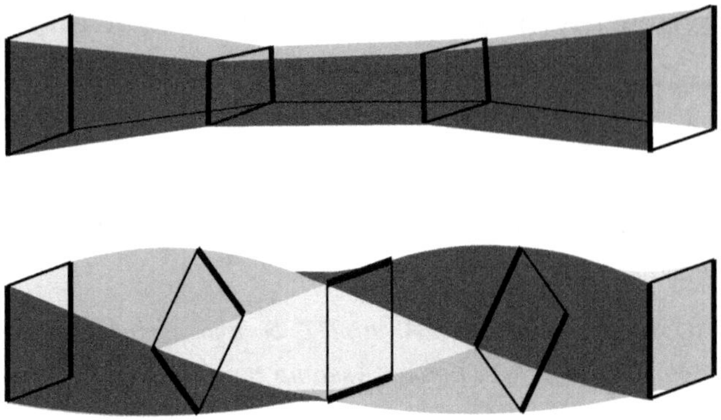

FIGURE 11. Isolating segments for (6.11.1): Z at the top and W at the bottom.

In order to construct the smaller segment, we set

$$\omega = \frac{R-r}{\Delta} \tag{6.11.2}$$

Let $s\colon \mathbb{R} \to \mathbb{R}$ be a $2\pi/\kappa$-periodic function such that

$$s(t) := \begin{cases} R - \omega t & \text{if } t \in [0, \Delta], \\ r & \text{if } t \in [\Delta, 2\pi/\kappa - \Delta], \\ R - \omega(2\pi/\kappa - t) & \text{if } t \in [2\pi/\kappa - \Delta, 2\pi/\kappa]. \end{cases}$$

Then smaller segment and its essential exit set are given by

$$Z = \{(t,z) \in [0, 2\pi/\kappa] \times \mathbb{C} : |\Re z| \le s(t),\ |\Im z| \le s(t)\},$$
$$Z^{--} = \{(t,z) \in Z : |\Re z| = s(t)\}.$$

For the proof of the following result we refer the reader to Lemma 19 in [WZ1] (some more restrictive estimates were given earlier in [SW]).

(6.12) LEMMA ([WZ1]). *Assume* $\kappa \in (0, 0.495]$, *then for* $R = 1.15$, $r = 0.5946$, *and* $\Delta = 0.935$, *the above sets* W *and* Z *are periodic isolating segments over* $[0, 2\pi/\kappa]$ *for* (6.11.1) *which satisfy the assumptions of Theorem* (6.10).

As an conclusion we get the following precise information on the range of values of the parameter κ for which a chaotic dynamics occur:

(6.13) COROLLARY ([SW], [WZ1]). *The local process generated by the equation* (6.11.1) *satisfies the conclusion of Theorem* (6.10) *if* $0 < \kappa \le 0.495$.

(6.14) REMARK. Results on the existence of chaotic dynamics using isolating chains (see Remark (4.3)), similar to Theorem (6.4), are presented in [P]. Examples in that paper are based on equations different from the ones considered above and to which theorems given here cannot be directly applied.

7. Detection of homoclinic and multibump solutions

Results on isolating segments can be applied in proofs of other properties of time-periodic non-autonomous equations, like the existence of solutions asymptotic to zero at $\pm\infty$ or approaching zero in some intervals, as we indicate using the previously considered equation (6.11.1). In the following result we gather several properties of that equation, including the ones stated in Corollary (6.13).

(7.1) THEOREM ([W3]). *Let Φ be the local process generated by* (6.11.1). *Put*

$$T := \frac{2\pi}{\kappa}. \tag{7.1.1}$$

If $0 < \kappa < 0.495$ then there exists a compact set I such that $\Phi_{(0,T)}(I) = I$ and a continuous map $g\colon I \to \Sigma_2$ with the following properties:

(7.1.2) $\sigma \circ g = g \circ \Phi_{(0,T)}$,

(7.1.3) $g(I) = \Sigma_2$,

(7.1.4) *if $c \in \Sigma_2$ is n-periodic sequence, then $g^{-1}(c)$ contains a point x such that $\Phi^n_{(0,T)}(x) = x$,*

(7.1.5) *if $c \in \Sigma_2$ is n-periodic sequence, then $g^{-1}(c)$ contains a point x such that $\Phi^n_{(0,T)}(x) = -x$ and $\Phi^{2n}_{(0,T)}(x) = x$,*

(7.1.6) *for each $c \in \Sigma_2$ such that $c_i = 0$ for $i \geq i_0$, $g^{-1}(c)$ contains a point x such that $\lim_{t\to\infty} \Phi_{(0,t)}(x) = 0$,*

(7.1.7) *for each $c \in \Sigma_2$ such that $c = 0$ for $i \leq i_0$, $g^{-1}(c)$ contains a point x such that $\lim_{t\to-\infty} \Phi_{(0,t)}(x) = 0$,*

(7.1.8) *for each $c \in \Sigma_2$ such that $c_i = 0$ for $|i| \geq i_0$, $g^{-1}(c)$ contains a point x such that $\lim_{t\to\pm\infty} \Phi_{(0,t)}(x) = 0$,*

(7.1.9) *for each $t_1 < t_2$ and $\varepsilon > 0$ there is infinitely many geometrically distinct subharmonic (i.e. kT-periodic for some $k \in \mathbb{N}$) solutions z of* (6.11.1) *such that $|z(t)| < \varepsilon$ for $t \in [t_1, t_2]$.*

Solutions satisfying (7.1.8) are called *homoclinic* to the zero solution, while solutions satisfying (7.1.9) belong to the class of *multibump* solutions.

As we pointed out above, the properties (7.1.2)–(7.1.5) are already proved. For a proof of the other properties we extend a notation used in the proof of Theorem (6.4) in Section 6. For a moment we consider T arbitrary, i.e. (7.1.1) is not necessarily satisfied. Assume that W and Z are periodic segments over $[0,T]$ satisfying (6.4.1). Let V be a periodic isolating segment over $[0,lT]$ (where $l \in \mathbb{N}$) for which there are integers $0 \leq k_0 < \ldots < k_{n-1} \leq l-1$ such that

$$V_{t+k_iT} = W_t, \quad \text{for } i \in \{0,\ldots,n-1\},\ t \in [0,T],$$

hence $V_{[k_iT,(k_i+1)T]}$ is equal to the segment W translated to $[k_iT,(k_i+1)T]$. For a finite sequence $c = (c_0,\ldots,c_{n-1}) \in \{0,1\}^{\{0,\ldots,n-1\}}$ we define $V(c)$ as the periodic segment over $[0, lT]$ obtained from V by replacing $V_{[k_iT,(k_i+1)T]}$ by the translated copy of Z for each $i \in \{0, \ldots, n-1\}$ such that $c_i = 0$, i.e.

$$V(c)_t := \begin{cases} Z_{t \bmod T} & \text{if } t \in [k_iT, (k_i+1)T] \text{ and } c_i = 0, \\ V_t & \text{otherwise,} \end{cases}$$

(see Figure 12).

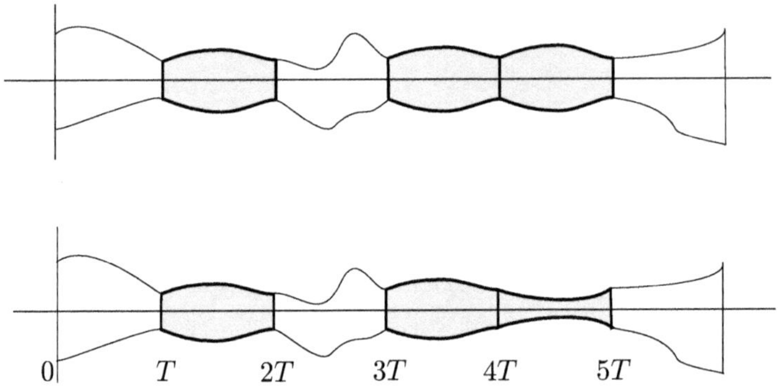

FIGURE 12. A segment V (top) and $V((1,1,0))$ (bottom) with $k_0 = 1$, $k_1 = 3$, and $k_2 = 4$.

In particular, if $V = W^n$ then that notation coincides with the one given in Section 6. Using the notation in the statement of Theorem (5.1), we put

$$U_{V(c)} := \{x \in V_0 : \Phi_{(0,t)}(x) \in V(c)_t \setminus V(c)_t^{--} \text{ for all } t \in [0, lT]\}.$$

Similarly like in the proof of Theorem (6.4), w define $U_{V(c),c} \subset U_{V(c)}$ by the following rule: $x \in U_{V(c)}$ belongs to the set $U_{V(c),c}$ if and only if for each $i \in \{0, \ldots, n-1\}$ such that $c_i = 1$ there exists $t \in (0, T)$ such that

$$\Phi_{(0,k_iT+t)}(x) \in W_t \setminus Z_t.$$

(7.2) LEMMA. *Let V be the segment given above. Assume that Z_0 and Z_0^{--} are* ANRs, *and*

(7.2.1) $\mu_W \circ \mu_W = \mu_Z = \mathrm{id}_{H(Z_0, Z_0^{--})}$,

(7.2.2) $\Lambda(\mu_W) \neq \chi(Z_0, Z_0^{--})$ *and* $\chi(Z_0, Z_0^{--}) \neq 0$.

(7.2.3) *for each* $c = (c_0, \ldots, c_{n-1}) \in \{0,1\}^{\{0,\ldots,n-1\}}$ *such that* 1 *appears exactly* k *times in* c,

$$\Lambda(\mu_{V(c)}) = \Lambda(\mu_W^k).$$

Then the set $U_{V(c),c}$ *is open in* V_0, *the set of fixed points of the restriction*

$$\Phi_{(0,lT)}|_{U_{V(c),c}}: U_{V(c),c} \mapsto V_0$$

is compact, and

$$\operatorname{ind}(\Phi_{(0,lT)}|_{U_{V(c),c}}) = \begin{cases} (-2)^{k-1}(\Lambda(\mu_W) - \chi(Z_0, Z_0^{--})) & \text{if } k \geq 1, \\ \chi(Z_0, Z_0^{--}) & \text{if } k = 0. \end{cases}$$

An Idea of a Proof. One can use a similar argument as the one in the proof of (6.7). For details we refer the reader to the proof of Lemma 1 in [SW]. □

Now we return to the local process generated by (6.11.1) and from now we assume that κ, R, r, and Δ are the numbers given in Lemma (6.12), and W and Z are the corresponding isolating segments. We built some other segments related to the equation. Let $r_1 > 0$. Put

$$P(r_1) := \{(t, z) \in \mathbb{R} \times \mathbb{C} : |\Re z| \leq r_1,\ |\Im z| \leq r_1\}$$

By Lemma 3 in [SW] we have

(7.3) Lemma. *Let* $\kappa > 0$ *be arbitrary and let* $r_1 \leq 1/3$. *For all* $a < b$ *the set* $P(r_1)_{[a,b]}$ *is a periodic isolating segment for* (6.11.1) *and its essential exit set is given by*

$$P(r_1)^{--}_{[a,b]} = \{(t, z) \in P(r_1)_{[a,b]} : |\Re z| = r_1\}.$$

Through reminder of this section we assume (7.1.1), i.e. $T = 2\pi/\kappa$. For ω defined by (6.11.2), $\gamma > 0$, and $t \geq 0$ set

$$s_V(t) := \begin{cases} R - \omega t & \text{if } t \in [0, \Delta], \\ r & \text{if } t \in [\Delta, T], \\ re^{-\gamma(t-T)} & \text{if } t \geq T \end{cases}$$

and extend the definition to the whole real line by $s_V(t) := s_V(-t)$ for $t < 0$. Using s_V we define a set $V := \{(t, z) \in \mathbb{R} \times \mathbb{C} : |\Re z| \leq s_V(t),\ |\Im z| \leq s_V(t)\}$. In particular, $V_{[0,T]} \subset Z$. The following result essentially appeared in [WZ2] as Lemma 9:

(7.4) Lemma. *Let* $\gamma = 0.25$. *Then for all* $a < b$ *the set* $V_{[a,b]}$ *is an isolating segment for* (6.11.1) *and*

$$V^{--}_{[a,b]} = \{(t, z) \in V_{[a,b]} : |\Re z| = s_V(t)\}.$$

The next lemma is helpful in a proof of the property (7.1.9).

(7.5) LEMMA. *Let $t_1 < t_2$ and $\varepsilon > 0$ be fixed. Then there exists a periodic segment N for (6.11.1) over the interval $[\mu T, \nu T]$ for some integers $\mu < \nu$ such that*

(7.5.1) $\mu T < t_1 < t_2 < \nu T$,
(7.5.2) $N_{\mu T} = Z_0$, $N_{\mu T}^{--} = Z_0^{--}$,
(7.5.3) $|z| < \varepsilon$ *for each* $t \in [t_1, t_2]$ *and* $z \in N_t$.

The required segment is schematically shown in Figure 13.

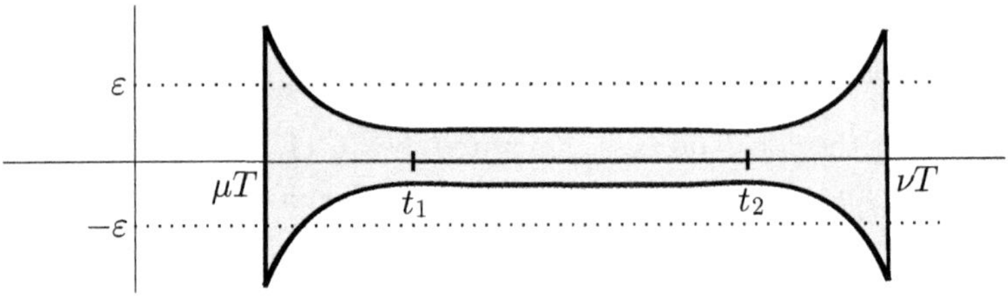

FIGURE 13. Segment N.

PROOF. Without loss of generality we can assume that $t_1 = i_1 T$ and $t_2 = i_2 T$ for some integers $i_1 < i_2$, and $\varepsilon < 1/3$. Let an integer $p \geq 2$ be such that

$$r_1 := re^{-\gamma(p-1)T} < \varepsilon/4,$$

where γ is given in Lemma (7.4). For a real number s define the time-s translation along the time axis in $\mathbb{R} \times \mathbb{C}$ as

$$\tau_s(t, z) := (t + s, z). \tag{7.6}$$

Then we can put $\mu := i_1 - p$, $\nu := i_2 + p$, and

$$N := \tau_{\mu T}(V_{[0,pT]}) \cup P(r_1)_{[i_1 T, i_2 T]} \cup \tau_{\nu T}(V_{[-pT,0]}),$$

where $P(r_1)$ is taken from Lemma (7.3).

PROOF OF THEOREM (7.1). Recall that by Corollary (6.13) it suffices to prove (7.1.6)–(7.1.9). In order to deal with solutions that are asymptotic to the trivial solution we will use the set V defined above. We describe only the main idea of the proof of assertion (7.1.8). Proofs of (7.1.6) and (7.1.7) are similar. Let i_0 be a positive integer and let c be a sequence of symbols 0 and 1 such that $c_i = 0$ if $|i| \geq i_0$. Denote by $d = (d_0, \dots, d_{2i-1})$ the shifted fragment of c given by $d_i = c_{i-i_0}$. For $k \in \mathbb{N}$ define a periodic segment $Y(k)$ over $[0, 2(i_0 + k)T]$ as

$$Y(k) := \tau_{kT}(V_{[-kT,0]}) \cup \tau_{kT}(W^{2i_0}(d)) \cup \tau_{(2i_0+k)T}(V_{[0,kT]}),$$

where the translations τ are given by (7.6) and V satisfies Lemma (7.4). By an application of Lemma (7.2) to the segment $Y(k)$, for every k we get a point $y_k \in Y(k)_0$ such that

$$\Phi^{2(k+i_0)}_{(0,T)}(y_k) = y_k$$

and $\Phi_{(0,t)}(y_k) \in Y(k)_t$ for $t \in [0, 2(i_0 + k)T]$, and if $k \leq s \leq k + 2i_0 - 1$ and $Y(k)_{[sT,(s+1)T]} = W$ then $\Phi_{(0,t)}(y_n) \notin Z_t$ for some time $t \in (sT, (s+1)T)$. Put

$$x_k := \Phi_{(0,(k+i_0)T)}(y_k).$$

An accumulation point x of the sequence $\{x_k\}$ has the required property (see Figure 14).

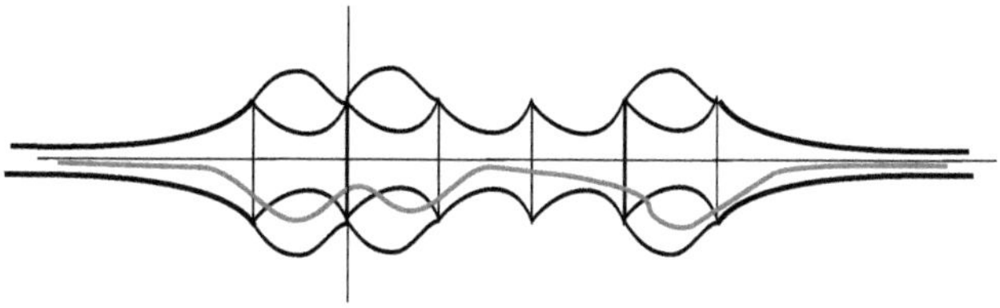

FIGURE 14. A trajectory homoclinic to the trivial solution coded by the sequence $\ldots 0001.1001000 \ldots$

The proof of (7.1.9) is based on the same idea as the proof of (7.1.6)–(7.1.8). In order to deal with multibump solutions described in (7.1.9), for fixed $t_1 < t_2$ and $\varepsilon > 0$ we consider the auxiliary segment N like in Lemma (7.5). The result follows by application of Lemma (7.2)to the segments being the union of N followed by the translated copy of $W^n(c)$, where $c \in \{0,1\}^{\{0,\ldots,n-1\}}$ is a finite sequence. □

8. Continuation theorem

We assume that X is a metric space with a distance function ρ. By the same letter ρ we denote also a corresponding distance on $\mathbb{R} \times X$. By $B(D, \delta)$ we denote an open ball of the radius δ around the set D contained either in X or in $\mathbb{R} \times X$.

Let Φ be a local semi-process on X generating the local semi-flow ϕ on $\mathbb{R} \times X$. Let $T > 0$ and W and Z be two subsets of $\mathbb{R} \times X$. We consider the following conditions:

(8.1.1) W and Z are T-periodic segments for Φ which satisfy (6.4.1), and Z_0 and Z_0^{--} are ANRs,

(8.1.2) there exists $\eta > 0$ such that for every $w \in W^{--}$ and $z \in Z^{--}$) there exists $t > 0$ such that for $0 < \tau < t$ holds $\phi_\tau(w) \notin W$, $\rho(\phi_t(w), W) > \eta$, $\phi_\tau(z) \notin Z$, and $\rho(\phi_t(z), Z) > \eta$).

Let K be a positive integer and let $E(1),\dots,E(K)$ be disjoint closed subsets of the essential exit set Z^{--} which are T-periodic, i.e. $E(l)_0 = E(l)_T$, and such that

$$Z^{--} = \bigcup_{l=1}^{K} E(l).$$

(In applications we will use the decomposition of Z^{--} into connected components). For $n \in \mathbb{N}$ and every finite sequence $c = (c_0,\dots,c_{n-1}) \in \{0,\dots,K\}^{\{0,\dots,n-1\}}$ and $D \subset W_0$, we define D_c as a set of points x satisfying the following conditions:

(8.2.1) $\Phi_{(0,lT)}(x) \in D$ for $l \in \{0,\dots,n\}$,

(8.2.2) $\Phi_{(0,t+lT)}(x) \in W_t \setminus W_t^{--}$ for $t \in [0,T]$ and $l \in \{0,\dots,n-1\}$,

(8.2.3) for each $l = 0,\dots,n-1$, if $c_l = 0$, then $\Phi_{(0,lT+t)}(x) \in Z_t \setminus Z_t^{--}$ for $t \in (0,T)$,

(8.2.4) for each $l = 0,\dots,n-1$, if $c_l > 0$, then $\Phi_{(0,lT)}(x)$ leaves Z in time less than T through $E(c_l)$.

Now let

$$[0,1] \times \mathbb{R} \times X \times [0,\infty) \ni (\lambda,\sigma,x,t) \to \Phi^{\lambda}_{(\sigma,t)}(x) \in X$$

be a continuous family of T-periodic semi-processes on X. Let ϕ^λ denotes the local semi-flow on $\mathbb{R}\times X$ generated by the semi-process Φ^λ. We say that the conditions (8.1.1) and (8.1.2) are satisfied *uniformly* (with respect to λ) if they are satisfied with Φ replaced by Φ^λ and the same η in (8.1.2) is valid for all $\lambda \in [0,1]$.

We write D_c^λ for the set defined by the conditions (8.2.1)–(8.2.4) for the semi-process Φ^λ.

(8.3) LEMMA. *If D is open in W_0, then D_c^λ is also open in W_0.* □

The main result of this section is the following:

(8.4) THEOREM (Continuation theorem, comp. [WZ1]). *Let Φ^λ be a continuous family of T-periodic semi-processes such that* (8.1.1) *and* (8.1.2) *hold uniformly. Then for every $n > 0$ and every finite sequence $c = (c_0,\dots,c_{n-1}) \in \{0,\dots,K\}^{\{0,\dots,n-1\}}$ the fixed point indices* $\operatorname{ind}(\Phi^\lambda_{(0,nT)}|_{(W_0\setminus W_0^{--})_c^\lambda})$ *are correctly defined and equal each to the other (i.e. do not depend on $\lambda \in [0,1]$).*

SKETCH OF A PROOF. We follow the argument in [WZ1]. Let $0 < \beta < \eta$ be such that

$$\beta < \rho(E(l),E(j)) \quad \text{for } l \neq j.$$

One can check that there exists a $\delta > 0$ such that for each $\lambda \in [0,1]$ the following condition hold: every point x from $\overline{B(W_0^{--},\delta)} \cap W_0$ leaves W in time $\tau < T$, i.e.

$$\sigma_W^\lambda(0,x) < T \quad \text{for } x \in \overline{B(W_0^{--},\delta)} \cap W_0 \tag{8.5}$$

(where σ_W^λ is the escape-time function for ϕ^λ). By decreasing δ we can assume that $\beta > 2\delta$. We define two open subsets of W_0 by

$$D := W_0 \setminus \overline{B(W_0^{--}, \delta)}, \quad C := W_0 \setminus \overline{B(W_0^{--}, \delta/2)}.$$

It follows that

$$W_0 \cap B(D, \delta/2) \subset C, \quad W_0 \cap B(C, \delta/2) \subset W_0 \setminus W_0^{--}.$$

Let us fix $\lambda_0 \in [0,1]$. There exists a set Λ open in $[0,1]$, $\lambda_0 \in \Lambda$, such that for every λ_1, $\lambda_2 \in \Lambda$,

$$\rho(\Phi^{\lambda_1}_{(0,t)}(x), \Phi^{\lambda_2}_{(0,t)}(x)) \leq \frac{\delta}{2}$$

for $0 \leq t \leq nT$ and $x \in W_0$. One can check using (8.1.2) that

(8.6) $$D_c^\lambda \subset C_c^{\lambda_0} \subset (W_0 \setminus W_0^{--})_c^\lambda, \quad \text{for } \lambda \in \Lambda.$$

From the choice of δ, the definition of the set D, and (8.5) one can conclude that

(8.7) $$\Phi^\lambda_{(0,nT)}(x) \neq x, \quad \text{for} \quad x \in (W_0 \setminus W_0^{--})_c^\lambda \setminus D_c^\lambda.$$

From (8.6) we deduce that for λ, $\lambda_0 \in \Lambda$ all sets ∂D_c^λ, ∂C_c^λ, and $\partial C_c^{\lambda_0}$ are contained in $(W_0 \setminus W_0^{--})_c^\lambda \setminus D_c^\lambda$, and, in consequence, by (8.7) the fixed point index for the map $\Phi^\lambda_{(0,nT)}$ relative to those sets is correctly defined. By (8.6), (8.7), and the excision property of the fixed point index we obtain that for $\lambda \in \Lambda$,

(8.8) $$\operatorname{ind}(\Phi^\lambda_{(0,nT)}|_{D_c^\lambda}) = \operatorname{ind}(\Phi^\lambda_{(0,nT)}|_{C_c^{\lambda_0}}) = \operatorname{ind}(\Phi^\lambda_{(0,nT)}|_{(W_0 \setminus W_0^{--})_c^\lambda}).$$

In particular, for $\lambda = \lambda_0$ we can assert that

(8.9) $$\operatorname{ind}(\Phi^{\lambda_0}_{(0,nT)}|_{D_c^{\lambda_0}}) = \operatorname{ind}(\Phi^{\lambda_0}_{(0,nT)}|_{C_c^{\lambda_0}}).$$

Combining (8.6) with (8.7) we see that for $\lambda \in \Lambda$ and $x \in \partial C_c^{\lambda_0}$,

$$\Phi^\lambda_{(0,nT)}(x) \neq x,$$

hence by the homotopy property of the fixed point index we get

(8.10) $$\operatorname{ind}(\Phi^\lambda_{(0,nT)}|_{C_c^{\lambda_0}}) = \operatorname{ind}(\Phi^{\lambda_0}_{(0,nT)}|_{C_c^{\lambda_0}}), \quad \lambda \in \Lambda$$

and finally from (8.8)–(8.10) we conclude

(8.11) $$\operatorname{ind}(\Phi^\lambda_{(0,nT)}|_{D_c^\lambda}) = \operatorname{ind}(\Phi^{\lambda_0}_{(0,nT)}|_{D_c^{\lambda_0}}), \quad \lambda \in \Lambda.$$

By (8.11), the fixed point index $\operatorname{ind}(\Phi^\lambda_{(0,nT)}|_{D_c^\lambda})$ is locally constant with respect to λ, and consequently

$$\operatorname{ind}(\Phi^0_{(0,nT)}|_{D_c^0}) = \operatorname{ind}(\Phi^1_{(0,nT)}|_{D_c^1}).$$

Therefore

$$\operatorname{ind}(\Phi^0_{(0,nT)}|_{(W_0 \setminus W_0^{--})_c^0}) = \operatorname{ind}(\Phi^1_{(0,nT)}|_{(W_0 \setminus W_0^{--})_c^1})$$

by (8.8), and the proof is complete. □

9. Model semi-processes and applications of Continuation Theorem

In this section we show that a more complete description of the dynamics generated by the equations of the form (6.8.1) (where $n = 1, 2$) can be obtained by adapting the continuation method to the context of previous sections. The use of model semi-processes as the terminal objects of continuation for those equations enables us to dig deeper into the structure of the set of periodic solutions than does the method based on the Lefschetz Fixed Point Theorem alone. As in Section 7, in the sequel we put $T := 2\pi/\kappa$.

Let Φ be the local process generated by (6.11.1) in $\mathbb{R}^2$. Let W and Z be the T-periodic isolating segments for Φ described in Lemma (6.12). In particular, for $R = 1.15$, $Z_0 = W_0 = [-R, R] \times [-R, R]$.

For $0 < c < a < b < R$ we put $J_{-1} = [-b, -a]$, $J_0 = [-c, c]$ and $J_1 = [a, b]$. Consider a function $f\colon J_{-1} \cup J_0 \cup J_1 \to [-R, R]$ having the graph shown in Figure 15. We stress that $f(-x) = -f(x)$ and $R = f(c) = f(a) = f(-b)$.

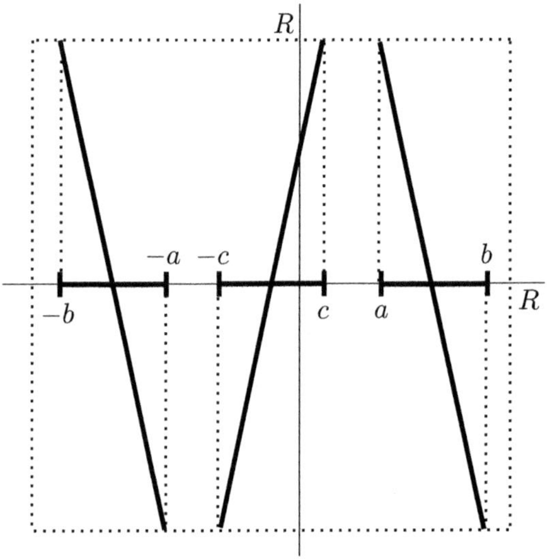

FIGURE 15. Function f in the model for (6.11.1).

Let Z^{+1}, Z^{-1} be two connected components of Z^{--}, the right one ($x > 0$) and the left one ($x < 0$), respectively. Let us observe that after a suitable modifications outside of some large ball we can assume that Φ is a (global) process.

(9.1) LEMMA. *There exists a semi-process Φ^M on $\mathbb{R}^2$ such that*

(9.1.1) $\{J_{-1} \cup J_0 \cup J_1\} \times [-R, R] = \{z \in W_0 : \Phi^M_{(0,t)}(z) \in W_t$ *for all* $t \in [0, T]\}$,

(9.1.2) $J_0 \times [-R, R] = \{z \in W_0 : \Phi^M_{(0,t)}(z) \in Z_t$ *for all* $t \in [0, T]\}$,

(9.1.3) *for* $l = +1, -1$,

$$J_l \times [-R, R] = \{z \in W_0 : z \text{ leaves } Z \text{ through } Z^l \text{ in time} \leq T\},$$

(9.1.4) *for* $z = (x, y) \in \{J_{-1} \cup J_0 \cup J_1\} \times [-R, R]$ *the Poincaré map is given by*

$$\Phi^M_{(0,T)}(x, y) = (f(x), 0),$$

(9.1.5) Z *and* W *are periodic isolating segments over* $[0, T]$ *for a family of* T*-periodic semi-processes* Φ^λ *such that* (8.1.1) *and* (8.1.2) *hold uniformly, and*

$$\Phi^0 = \Phi, \quad \Phi^1 = \Phi^M,$$

(9.1.6) $\operatorname{ind}(\Phi^M_{(0,T)}|_{(W_0 \setminus W_0^{--})^1_c}) \neq 0$, *for every* $c \in \{-1, 0, 1\}^{\{0,\dots,n-1\}}$.

Φ^M is called a *model semi-process.* We do not provide its construction (actually, intuitive but complicated a little) here, referring the reader to the proof of Theorem 20 in [WZ1]. As a corollary we get the following improvement of a part of Theorem (7.1), in which the shift on two symbols is replaced by the one on three symbols:

(9.2) THEOREM (com. [WZ1]). *Let* Φ *be a local process generated by the equation* (6.11.1) *with* $0 < \kappa \leq 0.495$. *Then there are a compact set* $I \subset \mathbb{C}$, *invariant with respect to* $\Phi_{(0,T)}$, *and a continuous surjective map* $g: I \to \Sigma_3$ *such that*

(9.2.1) $\sigma \circ g = g \circ \Phi_{(0,T)}$,
(9.2.2) *if* $c \in \Sigma_3$ *is* n*-periodic then* $g^{-1}(c)$ *contains an* n*-periodic point of* $\Phi_{(0,T)}$.

PROOF. For $I := \{x \in W_0 : \Phi_{(0,t+kT)}(x) \in W_t \text{ for } t \in [0, T],\ k \in \mathbb{Z}\}$, we define a semi-conjugacy $g: I \to \Sigma_3$ by

$$g(x)_l := \begin{cases} 0 & \text{if } \Phi_{(0,t+lT)}(x) \in Z_t \text{ for all } t \in (0, T), \\ 1 & \text{if } \Phi_{(0,lT)}(x) \text{ leaves } Z \text{ in time less than } T \text{ through } Z^{-1}, \\ 2 & \text{if } \Phi_{(0,lT)}(x) \text{ leaves } Z \text{ in time less than } T \text{ through } Z^{+1}. \end{cases}$$

By Theorem (8.4) and Lemma (9.1), all fixed point indices for periodic sequences of symbols are nontrivial, hence $g(I)$ contains all periodic sequences from Σ_3. But the set of all periodic sequences is dense in Σ_3, so $g(I) = \overline{g(I)} = \Sigma_3$. □

Now we study the dynamics generated by the equation (6.8.1) with $n = 2$, i.e.

$$\dot{z} = (1 + e^{i\kappa t}|z|^2)\bar{z}^2. \tag{9.3}$$

The segments $Z := Z(2)$ i $W := W(2)$ for (9.3) are shown in Figure 10. It follows that $Z_0 = W_0$ is a hexagon centered at the origin and the exit set Z^{--} has three

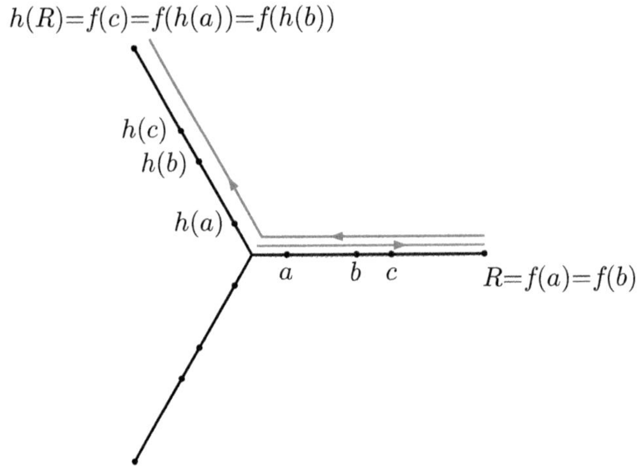

FIGURE 16. Map f in the model for (9.3). The gray lines above the segment $[0,R]$ represents the image of $[0,a]$ (the lower line) and of $[b,c]$ (the upper line). Actually, the latter image is contained in $[0,R] \cup [0,h(R)]$. The arrows indicate the orientation of the image of f when going from 0 to a on the lower line and from b to c on the upper line.

components E_1, E_2 i E_3. Let R be equal to the radius of the inscribed circle of Z_0 and let $h\colon \mathbb{C} \ni z \to ze^{i2\pi/3} \in \mathbb{C}$ be the rotation.

We define $S := [0,R] \cup h([0,R]) \cup h^2([0,R])$. For $0 < a < b < c < R$ we put

$$J := [0,a] \cup h([0,a]) \cup h^2([0,a]), \quad J_k := h^{k-1}([b,c]), \quad k \in \{1,2,3\}.$$

(9.4) Let $f\colon J \cup J_1 \cup J_2 \cup J_3 \to S$ be a continuous function, symmetric with respect to h, with the graph shown in Figure 16.

Let us observe that

(9.4.1) $f(0) = 0$,
(9.4.2) $f(h^k(a)) = f(h^k(b)) = h^k(R)$ and $f(h^k(c)) = h^{k+1}(R)$ for $k \in \{0,1,2\}$,
(9.4.3) the restrictions of f

$$f\colon [0,h^{k-1}(a)] \to [0,h^{k-1}(R)], \quad f\colon J_k \to [h^{k-1}(R),0] \cup [0,h^k(R)]$$

are homeomorphisms.

Let $r\colon W_0 \to S$ be the retraction schematically shown in Figure 17. We put

$$K := r^{-1}(J \cup J_1 \cup J_2 \cup J_3).$$

The following result was proved in [WZ3].

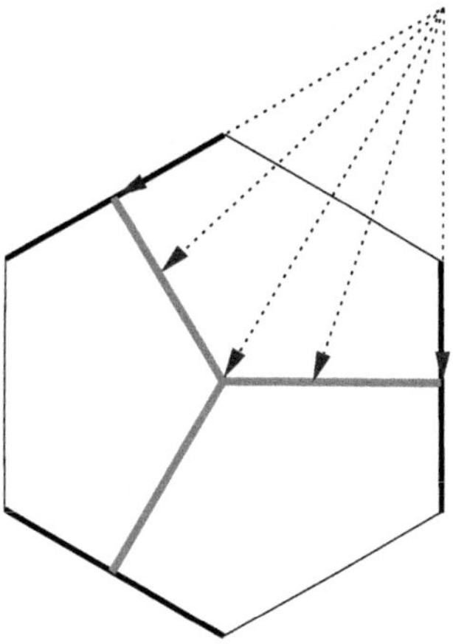

FIGURE 17. The retraction $r\colon W_0 \to S$.

(9.5) LEMMA. *There exists a model semi-process* Φ^M *for* (9.3) *such that*

(9.5.1) *Z and W are periodic isolating segments for a family of T-periodic semi-processes Φ^λ such that* (8.1.1) *and* (8.1.2) *hold uniformly and*

$$\Phi^0 = \Phi, \quad \Phi^1 = \Phi^M,$$

(9.5.2) $\{z \in W_0 : \Phi^M_{(0,t)}(z) \in W_t \text{ for all } t \in [0,T]\} = K$,

(9.5.3) $\{z \in W_0 : \Phi^M_{(0,t)}(z) \in Z_t \text{ for all } t \in [0,T]\} = r^{-1}(J)$,

(9.5.4) $\{z \in K : z \text{ leaves } Z \text{ through } E_k \text{ at time} \leq T\} = r^{-1}(J_k)$,

(9.5.5) *the Poincaré map for Φ^M, denoted by P_M, is of the form*

$$P_M(z) := \Phi^M_{(0,T)}(z) = f(r(z)), \quad z \in K.$$

We want now to investigate the symbolic dynamics on sets J and J_k for the model map. The symbol 0 will correspond to J and the symbols k for $k = 1, 2, 3$ will corresponds to J_k. To simplify the notation we set $J_4 := J_1$. Observe that for $k = 1, 2, 3$, $P_M(J_k)$ covers J_k, J_{k+1} and part of J. The image of $P_M(J)$ covers J and all J_k's. But if we want to see where we can go from J_k and J under P_M^2 we need to consider the parts of J, what they cover and which part is covered by J_k.

We define a set $\Pi \subset \Sigma_4 = \{0, 1, 2, 3\}^{\mathbb{Z}}$ as follows: a sequence c belongs to Π if the following conditions hold:

(9.6.1) if $c_i = k$ for some $k \in \{1, 2, 3\}$, then $c_{i+1} = 0$ or $c_{i+1} = k$ or $c_{i+1} = k \pmod 3 + 1$,

(9.6.2) if $c_p = 0$ for $p \leq i$, then $c_{i+1} \in \{0, 1, 2, 3\}$,

(9.6.3) if $c_i = 0$ and $p < i$ is such that $c_p = k \neq 0$ and for $p < s \leq i$ $c_s = 0$, then $c_{i+1} = 0$ or $c_{i+1} = k$ or $c_{i+1} = (k \bmod 3) + 1$.

The condition (9.6.1) says, for example, that the symbol 1 in the sequence c can be followed by the symbols 0, 1, 2. It is a consequence of the fact that $f(J_1)$ covers J, J_1 i J_2 in a proper way. Hence, if the trajectory coded by the sequence $c \in \Pi$ leaves the segment Z in the time interval $[kT,(k+1)T]$ through the component Z_1^{--}, then, in the time interval $[(k+1)T,(k+2)T]$, it can stay in Z or leaves Z through Z_1^{--} or Z_2^{--}. By the continuation theorem we get:

(9.7) THEOREM ([WZ3]). *Let Φ be a local process generated by* (9.3). *There exists a $\kappa_0 > 0$ such that for $0 < \kappa \leq \kappa_0$ there are a compact set $I \subset \mathbb{C}$ invariant under the Poincaré map $\Phi_{(0,T)}$ and a continuous map $g\colon I \to \Sigma_4$ such that*

(9.7.1) $\sigma \circ g = g \circ \Phi_{(0,T)}$,

(9.7.2) $\Pi \subset g(I)$,

(9.7.3) *if $c \in \Pi$ is n-periodic, then $g^{-1}(c)$ contains an n-periodic point for $\Phi_{(0,T)}$.*

PROOF. The set I is defined like in the proof of Theorem (9.2). We define a continuous map $g\colon I \to \Sigma_4$ by

$$g(x)_l := \begin{cases} 0 & \text{if } \Phi_{(0,t+lT)}(x) \in Z_t \text{ for all } t \in (0,T), \\ k & \text{if } \Phi_{(0,lT)}(x) \text{ leaves } Z \text{ in time less than } T \text{ through } E_k. \end{cases}$$

It is easy to check that the condition (9.7.1) is satisfied and (9.7.2) follows by (9.7.3), hence it is enough to prove (9.7.3). Let Π_l denotes the projection of Π onto the coordinates $0,\dots,l-1$. This means that if $\alpha = (\alpha_0,\dots,\alpha_{l-1}) \in \{0,1,2,3\}^l$ then $\alpha \in \Pi_l$ if and only if there exists $c \in \Pi$ such that $\alpha_i = c_i$ for $i = 0,\dots,l-1$. For $\alpha \in \Pi_l$ such that $\alpha_0 \neq 0$ we define $s(\alpha) := \max\{j : \alpha_j \neq 0\}$.

Let $c \in \Pi$ be a l-periodic sequence. If $c_i = 0$ for all i, then $g(0) = c$. Let us observe that it is enough to consider $c \in \Pi$, such that $c_0 \neq 0$.

Let l be the principal period of c and let $\alpha = (c_0,\dots,c_{n-1}) \in \Pi_l$. One can easily check that there exists a closed interval $A \subset J_{\alpha_0}$, such that for the semi-process Φ^M we get

$$(W_0 \setminus W_0^{--})_\alpha = r^{-1}(\operatorname{int} A), \quad f^l(A) = [0, h^{s(\alpha)-1}(R)] \cup [0, h^{s(\alpha) \bmod 3}(R)],$$

and $f^l|_A$ is a homeomorphism. Observe that

$$\text{either } c_0 = s(\alpha) \text{ or } c_0 = (s(\alpha) \bmod 3) + 1.$$

It follows the set $r^{-1}(A)$ is topologically a product of a segment A and another interval B, the map $f^l \circ r$ sends $A \times B$ onto $[0, h^{s(\alpha)-1}(R)] \cup [0, h^{s(\alpha) \bmod 3}(R)]$ containing A in its interior, hence it is easy to see that

$$\operatorname{ind}(P_M^l|_{(W_0 \setminus W_0^{--})_\alpha}) = \operatorname{ind}(f^l|_{\operatorname{int} A}) = \pm 1 \neq 0.$$

By Theorem (8.4) we have $\mathrm{ind}(\Phi^l_{(0,T)}|_{(W_0 \setminus W_0^{--})_\alpha}) \neq 0$, hence there exists an $x \in (W_0 \setminus W_0))_\alpha$, such that $\Phi^l_{(0,T)}(x) = x$. Observe that $g(x) = c$ and the proof is finished. □

We will finish this section with the following example.

(9.8) EXAMPLE. Consider a time-dependent Hamiltonian system

$$\dot{x} = -\frac{\partial H}{\partial y}, \quad \dot{y} = \frac{\partial H}{\partial x} \tag{9.8.1}$$

where

$$\begin{aligned} H(x,y,t) &= x^3 y + x y^3 + H_1(x,y,t), \\ H_1(x,y,t) &= -\frac{1}{2} y^2 \sin(\kappa t) - xy\cos(\kappa t) + \frac{1}{2} x^2 \sin(\kappa t). \end{aligned}$$

One can prove that for $0 < \kappa < \kappa_0$ there are two periodic segments Z and W for (9.8.1) that look like the ones shown in Figure 18.

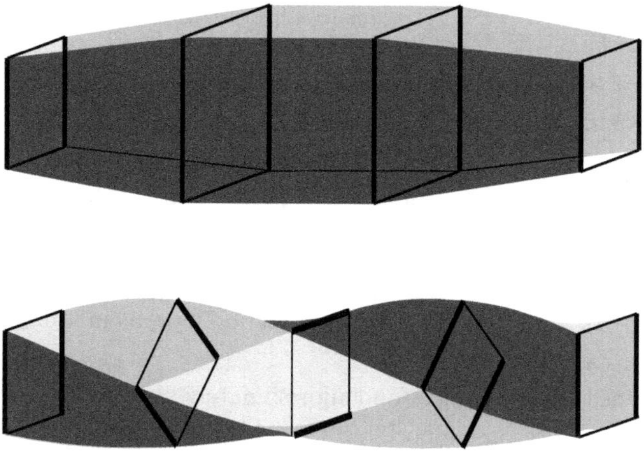

FIGURE 18. Isolating segments for (9.8.1): Z at the top and W at the bottom.

We see that when compared to the previous examples we have a slightly different geometry here. Previously the bigger segment was rotating, while here the smaller one rotates. Both these situation are manifestly homeomorphic. An application of the continuation theorem give us, like in the case of the equation (6.11.1), the following result:

(9.9) THEOREM ([WZ3]). *Let Φ be a local process generated by (9.8.1). Then there exists a $\kappa > 0$ such that for $0 < \kappa < \kappa_0$ there are a compact set $I \subset \mathbb{R}^2$, invariant with respect to the Poincaré map $\Phi_{(0,T)}$, and a surjective continuous map $g: I \to \Sigma_3$ such that $\Phi_{(0,T)}$ is semi-conjugated to the shift map on three symbols by the map g. Moreover, for each n-periodic sequence $c \in \Sigma_3$, $g^{-1}(c)$ contains an n-periodic point of the Poincaré map.*

10. Rigorous numerical shadowing using isolating segments

The aim of this section is to describe a potential application of isolating segments to rigorous numerical shadowing for non-autonomous equations. A minor modifications of the presented argument can provide similar application in the autonomous case. The question of reliability of numerical simulations of chaotic dynamical systems is usually addressed by shadowing algorithms, see for example [GHY,SY]. When applied to ordinary differential equations, these algorithms require rigorous estimates for Poincaré maps. Here we outline a possible shadowing algorithm based on the concept of isolating segment, without such requirement.

Let us consider an non-autonomous equation

$$\dot{x} = f(t,x) \tag{10.1}$$

in $\mathbb{R}^d$, where f is Lipschitz with respect to x. Our point of departure is a pseudoorbit (an approximate numerical trajectory) $\omega = \{\omega_0, \dots, \omega_N\}$ for (10.1) obtained for $t_0 < \dots < t_N$ (if our aim is to obtain the periodic orbit then we have $\omega_0 = \omega_N$, $t_0 = 0$, $t_N = kT$, where $k \in \mathbb{N}$, and f is T-periodic in t). Our goal is to show that there is true orbit nearby ω. We construct the following auxiliary objects:

(10.2.1) local sections $\Pi_0, \dots, \Pi_N$ for (10.1) defined by $\Pi_i := \{t = t_i\}$, such that $\omega_i \in \Pi_i$. For periodic orbit for T-periodic problem (10.1) we require that $t_N = kT$ for some $k \in \mathbb{N}$.

(10.2.2) d linearly independent and approximately invariant vector fields $X_1, \dots,$ X_d along the pseudoorbit ω. By this we understand that the following conditions are satisfied for $l = 1, \dots, d$

$$X_{l,i} \in \mathbb{R}^d, \quad dP_{i \to i+1} X_{l,i} \approx \lambda_{l, i \to i+1} X_{l,i+1}$$

where $P_{i \to j}: \Pi_i \to \Pi_j$ denote the Poincaré map between sections Π_i and Π_j for $i < j$. To obtain the vectors $X_{l,i}$ for a periodic trajectory we perform the diagonalization of the return map and then we evolve forward the unstable eigenvectors and backward the stable ones.

(An effective procedure of construction of $X_{l,i}$ for a very long (not periodic) pseudoorbit is described in [SY] for the case of planar periodically forced ODE.)

For any $i < j$ we set $\lambda_{l,i\to j} = \lambda_{l,i\to i+1} \cdot \lambda_{l,i+1\to i+2} \cdot \dots \cdot \lambda_{l,j-1\to j}$. The l-th direction is called *unstable* if $|\lambda_{l,0\to N}| > 1$ and is called *stable* if $|\lambda_{l,0\to N}| < 1$. We assume that each direction is either stable or unstable. Let U be the set of unstable directions. We would like to construct segment $W = \bigcup W_{[t_i,t_{i+1}]}$ isolating a trajectory close to ω, so that W_{t_i} is a parallelogram

$$\begin{aligned}
W_{t_i} &:= \Big\{x \in \Pi_i : x = \omega_i + \sum_{l=1}^{d} [-1,1] s_{l,i} X_{l,i}\Big\},\\
W_{[t_i,t_{i+1}]} &:= \Big\{(t,x) : t = (1-\alpha)t_i + \alpha t_{i+1},\\
&\qquad x = (1-\alpha)\Big(\omega_i + \sum_{l=1} c_l s_{l,i} X_{l,i}\Big) + \alpha\Big(\omega_{i+1} + \sum_{l=1} c_l s_{l,i+1} X_{l,i+1}\Big),\\
&\qquad c_l \in [-1,1], \alpha \in [0,1]\Big\},
\end{aligned}$$

where $s_{l,i}$ are some positive numbers. We introduce the candidate for the monodromy homeomorphism $h_i\colon [0,1]\times[-1,1]^d \to W_{[t_i,t_{i+1}]} \subset [t_i,t_{i+1}]\times\mathbb{R}^d$ for $W_{[t_i,t_{i+1}]}$ by

$$\begin{aligned}
&h_i(\alpha, c_1, \dots, c_d)\\
&:= \Big((1-\alpha)t_i + \alpha t_{i+1}, (1-\alpha)\Big(\omega_i + \sum_{l=1} c_l s_{l,i} X_{l,i}\Big) + \alpha\Big(\omega_{i+1} + \sum_{l=1} c_l s_{l,i+1} X_{l,i+1}\Big)\Big).
\end{aligned}$$

We expect h_i to be a homeomorphism. This happens when a small time step used in the construction of the pseudorbit ω. If this is the case, then also $X_{l,i}$ and $X_{l,i+1}$ are almost collinear and the set of vectors spanning $h_i([0,1]\times[-1,1]^d)$, given approximately by $(1,0), (0, X_{1,i}), \dots, (0, X_{d,i})$, is linearly independent.

Before we describe $W^{--}_{[t_i,t_{i+1}]}$, we define the faces of $W_{[t_i,t_{i+1}]}$ and W_{t_i} for $l = 1,\dots,d$ by

$$\begin{aligned}
W^{l+}_{t_i} &:= \{x = h(0, c_1, \dots, c_l) : c_l = 1,\ c_i \in [-1,1] \text{ for } i \neq l\},\\
W^{l-}_{t_i} &:= \{x = h(0, c_1, \dots, c_l) : c_l = -1,\ c_i \in [-1,1] \text{ for } i \neq l\},\\
W^{l+}_{[t_i,t_{i+1}]} &:= \{x = h(\alpha, c_1, \dots, c_l) : c_l = 1,\ c_i \in [-1,1] \text{ for } i \neq l, \alpha \in [0,1]\},\\
W^{l-}_{[t_i,t_{i+1}]} &:= \{x = h(\alpha, c_1, \dots, c_l) : c_l = -1,\ c_i \in [-1,1] \text{ for } i \neq l, \alpha \in [0,1]\}.
\end{aligned}$$

Now we put

$$\begin{aligned}
W^{--}_{t_i} &:= \bigcup_{l\in U} (W^{l+}_{t_i} \cup W^{l-}_{t_i}), &\quad W^{--}_{[t_i,t_{i+1}]} &:= \bigcup_{l\in U} (W^{l+}_{[t_i,t_{i+1}]} \cup W^{l-}_{[t_i,t_{i+1}]}),\\
W^{++}_{t_i} &:= \bigcup_{l\notin U} (W^{l+}_{t_i} \cup W^{l-}_{t_i}), &\quad W^{++}_{[t_i,t_{i+1}]} &:= \bigcup_{l\notin U} (W^{l+}_{[t_i,t_{i+1}]} \cup W^{l-}_{[t_i,t_{i+1}]}).
\end{aligned}$$

An algorithm which successively builds the segment can be described as follows:

0. Input values $s_{l,i}$, $l = 1, \ldots, d$; Output values $s_{l,i+1}$.
1. For each l we choose $s_{l,i+1}$,
   1.1. $s_{l,i+1} < |\lambda_{l,i\to i+1}|$, for unstable directions,
   1.2. $s_{l,i+1} > |\lambda_{l,i\to i+1}|$ for stable directions.
   Intuitively, this means that in the face unstable face is tilted towards the pseudoorbit and the stable face is tilted away from the pseudoorbit. This should give is exit and entry points on W^{--} and W^{++}, respectively.) To assure that the size of W_{t_i} not change to much and to make sure that in the periodic case $W_{t_N} = W_{t_0}$ it is desirable to choose $s_{l,i+1}$, such that
   $$s_{l,1} \cdot s_{l,2} \cdot \ldots \cdot s_{l,i+1} \approx 1.$$
2. Verification of the algorithm — the following conditions should be checked:
   2.1. the map h_i is a homeomorphism. For this purpose it is enough to check that a certain set of $d+1$ vectors is linearly independent,
   2.2. for $l = 1, \ldots, d$ we check that for each $x \in W^{--}_{[t_i,t_{i+1}]}$ the vector $(1, f(x))$ is pointing inside $W_{[t_i,t_{i+1}]}$ and for each $x \in W^{++}_{[t_i,t_{i+1}]}$ the vector $(1, f(x))$ is pointing inside $W_{[t_i,t_{i+1}]}$. For this purpose for each face in $W_{[t_i,t_{i+1}]}$ we have to evaluate the product of $(1, f(x))$ and the normal vector to the face. In interval arithmetic this might be doable in one step for the whole face at once.

We consider the construction successful if we are able to perform N steps (in the case of periodic orbit we also want to have $W_{t_N} = W_{t_0}$). In view of the theory developed in previous sections it is easy that if N steps of the algorithm has been successfully completed, then there exists $x_0 \in W_0$ and a time $T > 0$, such that $\phi([0,T], (0, x_0)) \in \bigcup_i W_{[t_i,t_{i+1}]}$ and $\phi(T, (0, x_0)) \in W_{t_N}$, where ϕ is the local flow induced by (10.1). Moreover, if the starting pseudoorbit was periodic and $W_{t_0} = W_{t_N}$, then $\phi(T, (0, x_0)) = (0, x_0)$, which means that x_0 is periodic.

The following converse statement is quite obvious: *if the periodic pseudorbit ω is in fact a periodic hyperbolic orbit (with all real eigenvalues) for* (10.1) *and if the time step is small enough then the algorithm described above yields an isolating segment.*

Of course to make the above statement into the theorem one needs to specify how to choose $s_{l,i}$, but it is quite obvious that it is enough to take $s_{l,i+1} = (1 \pm \varepsilon_l)\lambda_{l,i\to i+1}$ for some small $\varepsilon_l > 0$, with the plus sign for unstable directions and minus sign for stable ones.

References

[C] C. Conley, *Isolated Invarinat Sets and the Morse Index*, CBMS Regional Conference Series in Mathematics, vol. 38, Amer. Math. Soc., Providence, R.I., 1978.

[De] R. L. Devaney, *An Introduction to Chaotic Dynamical Systems*, Addison-Wesley, Menlo Park, 1985.

[Do] A. Dold, *Lectures on Algebraic Topology*, Springer-Verlag, Berlin, 1980.

[F] R. Fessler, *A generalized Poincaré index formula*, J. Differential Equations **115** (1995), 304–324.

[G] D. H. Gottlieb, *A De Moivre like formula for fixed point theory*, Contemp. Math. **72** (1988), 99–105.

[GHY] C. Grebogi, S. Hammel and J. Yorke, *Do numerical orbits of chaotic dynamical processes represent true orbits?*, J. Complexity **3** (1987), 136–145.

[H] J. Hale, *Theory of Functional Differential Equations*, Springer-Verlag, New York, 1977.

[P] L. Pieniążek, *Isolating chains and chaotic dynamics in planar nonautonomous* ODE*s*, Nonlinear Anal. **54** (2003), 187–204.

[SY] T. Sauer and J. Yorke, *Rigorous verification of trajectories for the computer simulation of dynamical systems*, Nonlinearity **4** (1991), 961–979.

[S1] R. Srzednicki, *On rest points of dynamical systems*, Fund. Math. **126** (1985), 69–81.

[S2] ———, *Periodic and bounded solutions in blocks for time-periodic nonautonomuous ordinary differential equations*, Nonlinear Anal. **22** (1994), 707–737.

[S3] ———, *On periodic solutions of planar polynomial differential equations with periodic coefficients*, J. Differential Equations **114** (1994), 77–100.

[S4] ———, *On geometric detection of periodic solutions and chaos*, ,,Nonlinear Analysis and Boundary Value Problems for Ordinary Differential Equations", Proceedings of the Conference held in CISM Udine, October 2–6, 1995, CISM Lecture Notes (F. Zanolin, ed.), vol. 371, Springer–Verlag, Wien, New York, 1996, pp. 197–209.

[S5] ———, *On solutions of two-point boundary value problems inside isoltaing segments*, Topol. Methods Nonlinear Anal. **13** (1999), 73–89.

[S6] ———, *On periodic solutions inside isolating chains*, J. Differential Equations **165** (2000), 42–60.

[SW] R. Srzednicki and K. Wójcik, *A geometric method for detecting chaotic dynamics*, J. Differential Equations **135** (1997), 66–82.

[Wa] T. Ważewski, *Sur un principe topologique de l'examen de l'allure asymptotique des intégrales des équations différentielles ordinaires*, Ann. Soc. Polon. Math. **20** (1947), 279–313.

[W1] K. Wójcik, *On detecting periodic solutions and chaos in the time periodically forced* ODE*'s*, Nonlinear Anal. **45** (2001), 19–27.

[W2] ——— Isolating segments and anti-periodic solutions, Monatsh. Math. **135** (2002), 245–252.

[W3] ———, *Remark on complicated dynamics of some planar system*, J. Math. Anal. Appl. **271** (2002), 257–266.

[WZ1] K. Wójcik and P. Zgliczyński, *Isolating segments, fixed point index and symbolic dynamics*, J. Differential Equations **161** (2000), 245–288.

[WZ2] ———, *Isolating segments, fixed point index and symbolic dynamics* II. *Homoclinic solutions*, J. Differential Equations **172** (2001), 189–211.

[WZ3] ———, *Isolating segments, fixed point index and symbolic dynamics* III. *Applications*, J. Differenital Equations **183** (2002), 262–278.

AUTHORS

Jan Andres
Department of Mathematical Analysis
Faculty of Science, Palacký University
Tomkova 40
779 00 Olomouc-Hejčín, Czech Republic
andres@inf.upol.cz

Robin Brooks
Department of Mathematics
Bates College
Lewiston, ME 04240, USA
rbrooks.javanet@rcn.com

Robert F. Brown
Mathematics Department
University of California, Los Angeles
Los Angeles, CA 90095-1555, USA
rfb@math.ucla.edu

Davide L. Ferrario
Dipartimento di Matematica e Applicazioni
University of Milano–Bicocca
via Cozzi, 53
20125 Milano, Italy
ferrario@matapp.unimib.it

Massimo Furi
Dipartimento di Matematica
Applicata "G. Sansone"
Via S. Marta
3, I-50139 Florence, Italy
furi@dma.unifi.it

Daciberg L. Gonçalves
Departamento de Matemática
IME-USP, Caixa Postal 66281
Agência Cidade de São Paulo
05311-970 São Paulo SP, Brasil
dlgoncal@ime.usp.br

Lech Górniewicz
Julisz P. Schauder Center
for Nonlinear Studies
Nicolaus Copernicus University
Chopina 12/18
87-100 Toruń, Poland
gorn@mat.uni.torun.pl

Evelyn L. Hart
Department of Mathematics
Colgate University
13 Oak Drive
Hamilton NY 13346-1398, USA
ehart@mail.colgate.edu

Philip R. Heath
Department of Mathematics
and Statistics
Memorial University of Newfoundland
St. John's
Newwfoundland, A1C 5S7, Canada
pheath@math.mun.ca

Jorge Ize
IIMAS-FENOMEC
Universidad Nacional Autónoma
de México
jil@mym.iimas.unam.mx

Jerzy Jezierski
Institute of Applied Mathematics
Univeristy of Agriculture
Nowoursynowska 159
02-776 Warszawa, Poland
jezierski@acn.waw.pl

Boju Jiang
Department of Mathematics
Peking University
Beijing 100871, China
jiangbj@math.pku.edu.cn

Michael R. Kelly
Department of Mathematics
and Computer Science
Loyola University
6363 St Charles Avenue
New Orleans, LA 70118, USA
kelly@loyno.edu

Edward C. Keppelmann
Department of Mathematics
and Statistics
University of Nevada
Reno, NV 89509, USA
keppelma@unr.edu

Wojciech Kryszewski
Faculty of Mathematics
and Computer Science
Nicolaus Copernicus University
Chopina 12/18
87-100 Toruń, Poland
Wojciech.Kryszewski@mat.uni.torun.pl

Wacław Marzantowicz
Faculty of Mathematics
and Computer Science
Adam Mickiewicz University
Umultowska 87
61-614 Poznań, Poland
marzan@amu.edu.pl

Takashi Matsuoka
Department of Mathematics
Naruto University of Education
Naruto 772-8502, Japan
matsuoka@naruto-u.ac.jp

Jean Mawhin
Département de Mathématique
Université de Louvain
B-1348 Louvain-la-Neuve, Belgium
Mawhin@math.ucl.ac.be

Jacobo Pejsachowicz
Dipartimento di Matematica
Politecnico di Torino
Torino, To, Italy
jacobo@polito.it

Maria Patrizia Pera
Dipartimento di Matematica
Applicata "G. Sansone"
Via S. Marta
3, I-50139 Florence, Italy
pera@dma.unifi.it

Sławomir Rybicki
Faculty of Mathematics
and Computer Science
Nicolaus Copernicus University
Chopina 12/18
87-100 Toruń, Poland
Slawomir.Rybicki@mat.uni.torun.pl

Robert Skiba
Faculty of Mathematics
and Computer Science
Nicolaus Copernicus University
Chopina 12/18
87-100 Toruń, Poland
robo@mat.uni.torun.pl

Marco Spadini
Dipartimento di Matematica
Applicata "G. Sansone"
Via S. Marta
3, I-50139 Florence, Italy
spadini@dma.unifi.it

Roman Srzednicki
Mathematics Department
Jagiellonian University
Reymonta 4
30-059 Kraków, Poland
Roman.Srzednicki@im.uj.edu.pl

Peter Wong
Department of Mathematics
Bates College
Lewiston, ME 04240, USA
pwong@bates.edu

Klaudiusz Wójcik
Mathematics Department
Jagiellonian University
Reymonta 4
30-059 Kraków, Poland
wojcik@im.uj.edu.pl

Piotr Zgliczyński
Mathematics Department
Jagiellonian University
Reymonta 4
30-059 Kraków, Poland
zgliczyn@im.uj.edu.pl

Xuezhi Zhao
Department of Mathematics
Capital Normal University
Beijing 100037, China
zhaoxve@mail.cnu.edu.cn

INDEX

GPSR Compliance

The European Union's (EU) General Product Safety Regulation (GPSR) is a set of rules that requires consumer products to be safe and our obligations to ensure this.

If you have any concerns about our products, you can contact us on ProductSafety@springernature.com

In case Publisher is established outside the EU, the EU authorized representative is:

Springer Nature Customer Service Center GmbH
Europaplatz 3
69115 Heidelberg, Germany

Batch number: 10584428

Printed by Printforce, the Netherlands